INTERACTING ELECTRONS

Theory and Computational Approaches

Recent progress in the theory and computation of electronic structure is bringing an unprecedented level of capability for research. Many-body methods are becoming essential tools vital for quantitative calculations and understanding materials phenomena in physics, chemistry, materials science, and other fields. This book provides a unified exposition of the most-used tools: many-body perturbation theory, dynamical mean-field theory, and quantum Monte Carlo simulations.

Each topic is introduced with a less technical overview for a broad readership, followed by in-depth descriptions and mathematical formulation. Practical guidelines, illustrations, and exercises are chosen to enable readers to appreciate the complementary approaches, their relationships, and the advantages and disadvantages of each method. This book is designed for graduate students and researchers who want to use and understand these advanced computational tools, get a broad overview, and acquire a basis for participating in new developments.

RICHARD M. MARTIN is Emeritus Professor at the University of Illinois at Urbana-Champaign and Consulting Professor at Stanford University. He has made extensive contributions to the field of modern electronic structure methods and the theory of interacting electron systems and he is the author of the companion book, *Electronic Structure: Basic Theory and Methods*.

LUCIA REINING is CNRS Senior Researcher at École Polytechnique Palaiseau and founding member of the European Theoretical Spectroscopy Facility. Her work covers many-body perturbation theory and time-dependent density functional theory. She is a recipient of the CNRS silver medal and fellow of the American Physical Society.

DAVID M. CEPERLEY is a Founders and Blue Waters Professor at the University of Illinois at Urbana-Champaign where he has pioneered the quantum Monte Carlo method, including the development of variational, diffusion, and path-integral Monte Carlo. He is a member of the US National Academy of Sciences and recipient of the Rahman Prize for Computational Physics of the APS and the Feenberg Medal for Many-Body Physics.

Interacting Electrons
Theory and Computational Approaches

Richard M. Martin
University of Illinois Urbana-Champaign

Lucia Reining
École Polytechnique Palaiseau

David M. Ceperley
University of Illinois Urbana-Champaign

CAMBRIDGE
UNIVERSITY PRESS

University Printing House, Cambridge CB2 8BS, United Kingdom

One Liberty Plaza, 20th Floor, New York, NY 10006, USA

477 Williamstown Road, Port Melbourne, VIC 3207, Australia

314-321, 3rd Floor, Plot 3, Splendor Forum, Jasola District Centre, New Delhi - 110025, India

79 Anson Road, #06-04/06, Singapore 079906

Cambridge University Press is part of the University of Cambridge.

It furthers the University's mission by disseminating knowledge in the pursuit of
education, learning and research at the highest international levels of excellence.

www.cambridge.org
Information on this title: www.cambridge.org/9780521871501

© Cambridge University Press 2016

First published 2016

A catalogue record for this publication is available from the British Library

Library of Congress Cataloging in Publication data
Names: Martin, Richard M., 1942– author. | Reining, Lucia, author. | Ceperley, David, author.
Title: Interacting electrons : theory and computational approaches / Richard M. Martin
(University of Illinois, Urbana-Champaign), Lucia Reining (Ecole Polytechnique, Palaiseau),
David M. Ceperley (University of Illinois, Urbana-Champaign).
Description: New York, NY : Cambridge University Press, [2016] | ?2016
Identifiers: LCCN 2015041121 | ISBN 9780521871501 (hardback) |
ISBN 0521871506 (hardback)
Subjects: LCSH: Electronic structure. | Electrons. | Many-body problem. |
Perturbation (Quantum dynamics) | Quantum theory. | Monte Carlo method.
Classification: LCC QC176.8.E4 M368 2016 | DDC 539.7/2112–dc23
LC record available at http://lccn.loc.gov/2015041121

ISBN 978-0-521-87150-1 Hardback

To our families

Contents

Part II Foundations of theory for many-body systems

Part IV Stochastic methods

Part V Appendices

Preface

Recent progress in the theory and computation of electronic structure is bringing an unprecedented level of capability for research. It is now possible to make quantitative calculations and provide novel understanding of natural and man-made materials and phenomena vital to physics, chemistry, materials science, as well as many other fields. Electronic structure is indeed an active, growing field with enormous impact, as illustrated by the more than 10,000 papers per year.

Much of our understanding is based on mean-field models of independent electrons, such as Hartree–Fock and other approximations, or density functional theory. The latter is designed to treat ground-state properties of the interacting-electron system, but it is often also used to describe excited states in an independent-electron interpretation. Such approaches can only go so far; many of the most interesting properties of materials are a result of interaction between electrons that cannot be explained by independent-electron descriptions. Calculations for interacting electrons are much more challenging than those of independent electrons. However, thanks to developments in theory and methods based on fundamental equations, and thanks to improved computational hardware, many-body methods are increasingly essential tools for a broad range of applications. With the present book, we aim to explain the many-body concepts and computational methods that are needed for the reader to enter the field, understand the methods, and gain a broad perspective that will enable him or her to participate in new developments.

What sets this book apart from others in the field? Which criteria determine the topics included? We want the description to be broad and general, in order to reflect the richness of the field, the generality of the underlying theories, and the wide range of potential applications. The aim is to describe matter all the way from isolated molecules to extended systems. The methods must be capable of computing a wide range of properties of diverse materials, and have promise for exciting future applications. Finally, practical computational methods are an important focus for this book.

Choices have to be made since the number of different approaches, their variations, and applications is immense, and the book is meant to be more than an overview. We therefore cannot focus on such important areas as quantum chemistry methods, e.g. coupled cluster theory and configuration interaction methods, nor do we cover all of the developments in lattice models, or explore the vast field of superconductivity. Rather, we concentrate on three methods: many-body perturbation theory, dynamical mean-field theory, and stochastic quantum Monte Carlo methods. Our goals are to present each of these widely used methods with its strengths and weaknesses, to bring out their relationships, and to illuminate paths for combining these and other methods in order to create more powerful approaches. We believe that it is a unique feature of this book to have all three methods addressed in a coherent way: the conceptual structure, the actual methods, and selected applications.

The book is written so that a reader interested in many-body theory and methods should not have to go through the details of density functional theory and methods for independent-particle calculations. Summaries of pertinent material are given in this book, and extensive exposition can be found in the companion book by one of us (R. M. Martin, *Electronic Structure: Basic theory and methods*, Cambridge University Press, Cambridge, 2004, reprinted 2005 and 2008). The two books are complementary and can be read independently.

The organization of the book follows naturally: Part I (Chapters 1–4) contains background material including history, experimental motivations, and some classic concepts and models referred to throughout the book. Part II (Chapters 5–8) discusses theoretical foundations useful in all many-body treatments: mean-field theories, correlation functions and Green's functions, many-body wavefunctions, the concepts of quasi-particles, and functionals used in Green's function methods. Part III contains two distinct but related ways to use Green's functions, namely many-body perturbation theory including GW and the Bethe–Salpeter equation (Chapters 9–15), and dynamical mean-field theory (Chapters 16–21). Finally, Part IV (Chapters 22–26) discusses stochastic (quantum Monte Carlo) methods. The descriptions of these three approaches are written so that each can be read independently. Altogether, we try to clarify concepts, differences, and points of contact between the various methods, and possibilities for combining the methods as a basis for future progress in understanding and predicting properties of interacting electron systems.

Acknowledgments

This book has benefitted from the support of many colleagues and friends, and from the working environment provided by institutions and initiatives. The book reflects thoughts, discussions, and struggles for understanding that we have shared with our students and teachers, post docs, and peers. It is not possible to name everyone; people who have contributed directly are: Alexei Abrikosov, James W. Allen, Ole Krough Andersen, Ferdi Aryasetiawan, Friedhelm Bechstedt, Arjan Berger, Silke Biermann, Nils Blümmer, Fabien Bruneval, Michele Casula, Bryan Clark, Rodolfo Del Sole, Thomas P. Devereaux, Stefano Di Sabatino, Jacques Friedel, Christoph Friedrich, Matteo Gatti, Ted Geballe, Antoine Georges, Rex Godby, Emmanuel Gull, Matteo Guzzo, Walter Harrison, Kristjan Haule, Karsten Held, Markus Holzmann, Mark Jarrell, Borje Johansson, Erik Koch, Gabriel Kotliar, Wei Ku, Jan Kuneš, Giovanna Lani, Alexander I. Lichtenstein, Andrew McMahan, Andrea Marini, Andrew Millis, Elisa Molinari, Giovanni Onida, Eva Pavarini, Warren E. Pickett, John Rehr, Claudia Rödl, Pina Romaniello, Mariana Rossi, Alice Ruini, Sashi Satpathy, Sergej Savrasov, Richard Scalettar, Fausto Sirotti, Francesco Sottile, Adrian Stan, Zhi-Xun Shen, Mark van Schilfgaarde, Liu Hao Tjeng, Dieter Vollhardt, Jess Wellendorff, Philipp Werner, Jonas Weinen, Weitao Yang, Amy Young, Haijun Zhang, Shiwei Zhang, Jianqiang Sky Zhou.

Of great benefit was the continuous and stimulating scientific exchange in the framework of a series of European collaborative projects, culminating in the Nanoquanta network of excellence and the European Theoretical Spectroscopy Facility (ETSF).

Our institutions have supported our work through this long project: the Department of Physics, the Frederick Seitz Materials Research Laboratory at the University of Illinois Urbana-Champaign, and the Materials Computation Center, supported by the National Science Foundation; the Laboratoire des Solides Irradiés in Palaiseau and its responsible institutions – the Centre National de la Recherche Scientifique (CNRS), the École Polytechnique, and the Commissariat à l'énergie atomique et aux énergies alternatives (CEA); the Department of Applied Physics, Standford University.

Additional support was provided by the Aspen Center of Physics, Agence Nationale pour la Recherche Scientifique (ANR) in France, ICMAB – Universitat Autonoma Barcelona, SIESTA Foundation.

Last not least, we are deeply thankful to our families, friends, and colleagues for their great patience and their continuous encouragements.

Notation

Abbreviations

w.r.t.	with respect to
+c.c.	add complex conjugate of preceding quantity
Re, Im	real and imaginary parts
BZ	first Brillouin zone
1D	one dimension or one-dimensional (similar for 2D and 3D)

Acronyms most used

BSE	Bethe–Salpeter equation
DFT	density functional theory
DMFA	various approximations in DMFT
DMFT	dynamical mean-field theory
EXX	exact exchange in DFT
GGA	generalized gradient approximation
GWA	GW approximation
HF	Hartree–Fock
HFA	Hartree–Fock approximation
KS	Kohn–Sham
LDA	local density approximation
MBPT	many-body perturbation theory
QMC	quantum Monte Carlo
RHF	restricted Hartree–Fock
RPA	random phase approximation
TDDFT	time-dependent density functional theory
UHF	unrestricted Hartree–Fock

General physical quantities

E	energy
μ	chemical potential
E_F	Fermi energy (chemical potential at $T = 0$)
T	temperature (also computer time)
β	inverse temperature, $1/k_B T$
Z	partition function; same notation for canonical and grand canonical (also renormalization factor for spectral weight)
S or A	action

S	entropy
F	Helmholz free energy
G	Gibbs free energy (when identified), otherwise G denotes a Green's function
Ω	grand potential

Time and frequency

t	time
τ	imaginary time
$\Delta\tau$	imaginary time step
ω	frequency (real or complex as specified or as clear from context)
z	frequency defined in complex plane
ω_n	Matsubara frequencies
z_n	position of ω_n in complex plane, $z_n = \mu + i\omega_n$

Coordinates and operators

$\mathbf{r}$	electron position
σ	electron spin
x	combined space/spin coordinates $(\mathbf{r}, \sigma)$
1	combined space/spin/time coordinates (x_1, t_1)
$\bar{1}$	variable to be integrated
R	N-electron coordinates $R = \{\mathbf{r}_1 \ldots \mathbf{r}_N\}$
N	number of electrons
$\mathbf{R}_I$	position of ion I
Z_I	charge of ion I
$\hat{O}$	general operator in Schrödinger picture
$\langle \hat{O} \rangle$	expectation value
$\hat{O}_H(t)$	operator in Heisenberg picture, for static $\hat{H}$, $\hat{O}_H = e^{i\hat{H}t}\hat{O}e^{-i\hat{H}t}$
$c^\dagger$ and c	creation and annihilation operators for fermions
$\hat{\psi}^\dagger, \hat{\psi}$ and $\hat{\psi}^\dagger(t), \hat{\psi}(t)$	field operators for fermions; time argument denotes Heisenberg picture, for static $\hat{H}$, $\hat{\psi}^\dagger(t) = e^{i\hat{H}t}\hat{\psi}^\dagger e^{-i\hat{H}t}$, $\hat{\psi}(t) = e^{i\hat{H}t}\hat{\psi}e^{-i\hat{H}t}$
$b^\dagger$ and b	creation and annihilation operators for bosons
$\hat{n}$	particle number operator
$\hat{\mathbf{S}}, \mathbf{m}$	atomic moments or spin, $\mathbf{m} = \langle \hat{\mathbf{S}} \rangle$
T	time-ordering operator
T_τ	time ordering in imaginary time τ
$\hat{\mathcal{A}}$	antisymmetrization operator
Tr	trace
$\mathfrak{Tr}$	$\mathfrak{Tr}F = -\frac{i}{2\pi}\lim_{\eta \to 0^+} \mathrm{Tr} \int_{-\infty}^{\infty} d\omega\, e^{i\eta\omega} F(\omega)$ for $T = 0$ $\mathfrak{Tr}F = \mathrm{Tr}\,\beta^{-1}\sum_n F(i\omega_n)$ for $T \neq 0$ for a general function F

Hamiltonian and eigenstates

$\hat{H}$	many-body hamiltonian
$\hat{V}_{ee}$	electron–electron interaction
$v_c(\|\mathbf{r}_i - \mathbf{r}_j\|)$	Coulomb interaction
E_α	energy of many-body state labeled α
E_0	ground-state energy
$\Psi_\alpha(R)$ or $\Phi_\alpha(R)$	many-body wavefunction (R denotes $\mathbf{r}_i$, $i = 1$); often subscript α is omitted
$V(R)$	many-body potential energy function
$\Psi_T(R)$ or $\Psi(R)$	trial wavefunction
$U(R)$	N-electron Jastrow factor
$u(r_{ij})$ or $u(\mathbf{r}_i, \mathbf{r}_j)$	Jastrow factor for a pair of electrons
$\Theta(R)$	N-electron phase of the trial wavefunction
$E_L(R\|\Psi)$	local energy of a trial function
E_V	variational energy of a trial function
σ^2	variance of the local energy of a trial function
$\mathcal{O}$	overlap of trial function with exact wavefunction
$\hat{H}_0$	independent-particle hamiltonian
$\chi_m(\mathbf{r})$	single-particle basis function, $m = 1, \ldots, N_{\text{basis}}$
$h_{m,m'}$ or $h^0_{m,m'}$	matrix element of independent-particle hamiltonian
$v(\mathbf{r})$	single-particle potential
$v_H(\mathbf{r})$	Hartree potential
$\psi_i(\mathbf{r})$	single-particle wavefunction, $i = 1, \ldots, N_{\text{states}}$, also orbitals in a Slater determinant
ε_i	independent-particle eigenvalue, $i = 1, \ldots, N_{\text{states}}$
$f(\varepsilon)$	Fermi function

Parameters of model hamiltonians

t, t'	hopping matrix elements
U	on-site interaction in Hubbard-type models
J	intra-atomic exchange interaction, also interatomic Heisenberg exchange constant

Correlation functions, Green's functions, response functions, self-energy, etc.

$C_{AB}(t)$	correlation function for quantities A and B
$n(\mathbf{r})$	density (also spin-resolved $n(\mathbf{r}, \sigma)$)
$n_0(\mathbf{r})$	ground-state density (also $n_0(\mathbf{r}, \sigma)$)
$\rho(\mathbf{k})$	momentum distribution (also $\rho(\mathbf{k}, \sigma)$)
$\rho(\mathbf{r}, \mathbf{r}')$	one-electron density matrix (also $\rho(\mathbf{r}, \sigma; \mathbf{r}', \sigma')$)

$n(\mathbf{r}, \mathbf{r}')$	pair distribution function (also $n(\mathbf{r}, \sigma; \mathbf{r}', \sigma')$)
$g(\mathbf{r}, \mathbf{r}')$	normalized pair distribution (also $g(\mathbf{r}, \sigma; \mathbf{r}', \sigma')$)
$S(\mathbf{q}, \mathbf{q}'), S(\mathbf{k})$	structure factor (also $S(\mathbf{q}, \sigma; \mathbf{q}', \sigma')$ or $S_{\mathbf{kk}'}^{\sigma\sigma'}$)
$S(\omega)$	dynamic structure factor with similar notation for momenta and spin

G	one-body Green's function expressed as a function of arguments, e.g., $\mathbf{r}, \mathbf{r}'$, σ, t, τ, ω, z or as a matrix in a basis, $G_{m,m'}$
G^0 or G_0	independent-particle Green's function
A	spectral function
Σ	self-energy (irreducible)
Γ	vertex function
Z	renormalization factor of spectral weight (also partition function)

$\mathcal{G}$	Green's function for embedded site or cell
$\mathcal{G}_0$	embedding Green's function for embedded site or cell
Δ	hybridization function for impurity, embedded site or cell

G_2	two-particle Green's function
L	two-particle correlation function, $L = -G_2 + GG$
χ	general linear response function, susceptibility
χ_0	general linear response function for independent particles
ϵ	dielectric function
W	screened interaction
P	polarizability (irreducible)

Functionals

$F[f]$	general functional of the function f
$\mathcal{D}$	domain of a functional
$E_{\mathrm{xc}}[n]$	exchange–correlation energy in Kohn–Sham theory
$\Phi[G]$	interaction functional of Green's function G
$F[\Sigma]$	Legendre transform of $\Phi[G]$
$\Psi[G, W]$	interaction functional of G and screened interaction W
$\mathcal{J}(1, 1')$	external probe field – non-local in space and time
$\Omega[\mathcal{J}]$	grand-potential functional of $\mathcal{J}$
$\Gamma[Q]$	effective action functional, Legendre transform of $\Omega[\mathcal{J}]$

Notation for crystals

Ω	volume of cell (primitive cell or supercell)
$\mathbf{T}$	lattice translations

$\mathbf{G}$	reciprocal lattice vectors
$\mathbf{k}$ or $\mathbf{q}$	wavevector in first Brillouin zone
	$\mathbf{k}$ for electrons; $\mathbf{q}$ for interactions, susceptibilities
$\psi_{\mathbf{k}}(\mathbf{r}) = e^{i\mathbf{k}\cdot\mathbf{r}}u_{\mathbf{k}}(\mathbf{r})$	Bloch function in crystal, with $u_{\mathbf{k}}(\mathbf{r})$ periodic
$\varepsilon_{\mathbf{k}}$	eigenvalues that define bands as a function of $\mathbf{k}$
$G_{\mathbf{k}}, \Sigma_{\mathbf{k}}$	Green's function and self-energy as functions of $\mathbf{k}$
$\chi_{\mathbf{q}}$ or $\chi(\mathbf{q})$	general susceptibility as a function of $\mathbf{q}$
G_{ii} or G_{00}	on-site Green's function at site i
$\mathbf{L}_i$	primitive vector of supercell $i = \{1, 2, 3\}$
$\mathbf{K}$	total wavevector for many-body wavefunction
$\hat{H}(\mathbf{K})$	$\mathbf{K}$-dependent hamiltonian with periodic eigenvectors
	$u_{\mathbf{K}}(R), R = \{\mathbf{r}_1 \ldots \mathbf{r}_N\}$
Δ_{cf}	crystal field splitting
e_g and t_{2g}	crystal field states in cubic symmetry

Notation for stochastic methods

$\mathcal{S}, s, R$	state vector, i.e., variables describing the instantaneous state of a random walk
$R^{(n)}, s^{(n)}$	the nth many-body configuration in a random walk
$\Psi^{(n)}$ or Ψ_τ	projected trial wavefunction after n iterations or imaginary time τ
$\Pi(s)$	many-body distribution function to be sampled
$\langle\tilde{O}\rangle_\Pi$	average of $\tilde{O}$ over distribution Π
$P(s \to s'), P_{s's}$	transition probability random walk
$T(s \to s')$	trial transition probability
$A(s \to s')$	acceptance probability
$\tilde{O}$	estimator for property O
$\bar{O}$	estimated mean value of property O
ν_O	variance of property O
σ_O	standard error of $\bar{O}$
κ_O	autocorrelation time for property O in a random walk
$\Delta\tau$	time step in VMC, DMC, or PIMC
$\mathcal{P}$	electron permutation
$G(R, R'), \hat{G}$	many-body projector
$\mathcal{G}(R \leftarrow R')$	importance-sampled projector
$\Psi_G(R)$	guiding function for importance sampling
$\rho(R, R'; \beta)$	many-body density matrix

PART I

INTERACTING ELECTRONS: BEYOND THE INDEPENDENT-PARTICLE PICTURE

1

The many-electron problem: introduction

The calculation of a wavefunction took about two afternoons, and five wavefunctions were
calculated in the whole . . . .

Wigner and Seitz, 1933

Summary

In order to explain many important properties of materials and phenomena, it is
necessary to go beyond independent-particle approximations and directly account for
many-body effects that result from electronic interaction. The many-body problem is
a major scientific challenge, but there has been great progress resulting from theoretical developments and advances in computation. This chapter is a short introduction
to the interacting-electron problem, with some of the history that has led up to the
concepts and methods described in this book.

The many-body interacting-electron problem ranks among the most fascinating and fruitful
areas of research in physics and chemistry. It has a rich history, starting from the early days
of quantum mechanics and continuing with new intellectual challenges and opportunities.
The vitality of electronic structure theory arises in large part from the close connection
with experiment and applications. It is spurred on by new discoveries and advances in
techniques that probe the behavior of electrons in depth. In turn, theoretical concepts and
calculations can now make predictions that suggest new experiments, as well as provide
quantitative information that is difficult or not yet possible to measure experimentally.

This book is concerned with the effects of interactions between electrons beyond
independent-particle approximations. Some phenomena cannot be explained by any
independent-electron method, such as broadening and lifetime of excited states and two-
particle bound states (excitons) that are crucial for optical properties of materials. There are
many other examples, such as the van der Waals interaction between neutral molecules that
arises from the dipole–induced dipole interaction. This force, which is entirely due to correlation between electrons, is an essential mechanism determining the functions of biological

systems. Other properties, such as thermodynamically stable magnetic phases, would not exist if there were no interactions between electrons; even though mean-field approximations can describe average effects, they do not account for fluctuations around the average. Ground-state properties, such as the equilibrium structures of molecules and solids, can be described by density functional theory (DFT) and the Kohn–Sham independent-particle equations. However, present approximations are often not sufficient, and for many properties the equations, when used in a straightforward way, do not give a proper description, even in principle. A satisfactory theory ultimately requires us to confront the problem of interacting, correlated electrons.

It is challenging to develop robust, quantitative theoretical methods for interacting, correlated electrons, but there has been great progress. The exponential growth of computational capability and the development of algorithms have turned theoretical ideas into practical methods. The field has gone a long way, from calculations for single atoms or the homogeneous electron gas, to applications that impact not only the traditional fields of solid-state physics, materials science and chemistry, but also diverse areas ranging from engineering to biology, from archeology to astrophysics. Progress in all these disciplines can be greatly enhanced by quantitative calculations based on the fundamental laws of quantum mechanics. In this chapter we will see how the bridge is made between theoretical foundations and computational realizations.

1.1 The electronic structure problem

What we want to describe is the behavior of atoms, molecules, and condensed matter, which is governed by the quantum statistical mechanics for electrons and nuclei interacting via the Coulomb potential. The essential ingredients[1] are contained in the hamiltonian

$$\hat{H} = -\frac{\hbar^2}{2m_e} \sum_i \nabla_i^2 - \sum_{i,I} \frac{Z_I e^2}{|\mathbf{r}_i - \mathbf{R}_I|} + \frac{1}{2} \sum_{i \neq j} \frac{e^2}{|\mathbf{r}_i - \mathbf{r}_j|}$$
$$- \sum_I \frac{\hbar^2}{2M_I} \nabla_I^2 + \frac{1}{2} \sum_{I \neq J} \frac{Z_I Z_J e^2}{|\mathbf{R}_I - \mathbf{R}_J|}, \tag{1.1}$$

where electrons are denoted by lowercase subscripts and coordinates $\mathbf{r}_i$ and nuclei are denoted by uppercase subscripts, coordinates $\mathbf{R}_I$, charge $Z_I e$, and mass M_I. Throughout much of the book, we will use *atomic units* defined such that $\hbar = m_e = e = 4\pi/\epsilon_0 = 1$.

---

[1] The hamiltonian in Eq. (1.1) displays the *dominant* terms for the problems that are addressed throughout this book. Other terms are included when needed: externally applied electric and magnetic fields can be treated as an added scalar potential, a vector potential in the kinetic energy operator as $(i\nabla_j + \frac{e}{c}\mathbf{A})^2$, and Zeeman terms for the spin in a magnetic field. Time-dependent electromagnetic fields are included later in perturbation theory. Spin–orbit coupling and other relativistic effects are important in heavy atoms; however, they are one-particle terms that originate deep in the core and can be included in an ion core potential (or pseudopotential) as explained in [1, Chs. 10 and 11]. Corrections to the Coulomb potential due to quantum electrodynamics are small on the scale of interest for the chosen applications and are neglected.

Thus the unit of length is a Bohr ≈ 0.0529 nm and energy is a Hartree = 2 Rydberg $\approx$ 27.211 eV (or the equivalent in temperature of 315 775 K).

In Eq. (1.1) there is only one term that can be regarded as small: the nuclear kinetic energy, proportional to the inverse mass of the nuclei $1/M_I$. For the most part we ignore this term to concentrate on the electronic problem with fixed nuclei, using the Born–Oppenheimer (adiabatic) approximation [2]. Once we have neglected the kinetic energy of the nuclei and fixed their positions,[2] the final term, the interaction of nuclei with each other, is a constant that can be added to the zero of energy. Hence, the hamiltonian essential for the theory of interacting electrons consists of the first three terms, the kinetic energy of the electrons, the electron–nucleus interaction, and the electron–electron interaction,[3] which we can write as

$$\hat{H} = \hat{T}_e + \hat{V}_{en} + \hat{V}_{ee}. \tag{1.2}$$

This equation encapsulates the electronic structure problem: the first and last terms are universal for all problems, and the information specific to any system is contained in the middle term, a potential that acts equally on all electrons. There is a vast collection of approximations and techniques to deal with this hamiltonian because there are so many phenomena and materials, and because the two-body electron–electron interaction term $\hat{V}_{ee}$ makes the problem so difficult. However, in recent years it has become possible to compute many properties with methods derived directly from Eqs. (1.1) and (1.2). These methods and the resulting properties are the topic of this book.

1.2 Why is this problem hard?

Let us first address the question of why the straightforward solution of the many-body Schrödinger equation is so difficult. The underlying reason is the dimensionality of the problem, "the curse of dimensionality." A many-body wavefunction does not factorize because of the electron–electron interaction. In three dimensions and for N electrons it is a complex-valued function of $3N$ variables. Suppose we try to write the complete many-electron wavefunction for N electrons in a basis. With this aim, we introduce a single-particle basis set; there must be many basis functions to describe the wavefunction accurately. Let the total number of single-particle basis functions be M, with $M/N > 1$. Since electrons are indistinguishable fermions, the total basis for the many-body wavefunction consists of all Slater determinants of matrices with size $N \times N$ that can be constructed from the single-particle basis set. The number of these determinants is

---

[2] Quantum effects in the nuclear motion are important in some cases, e.g., hydrogen, helium, and other light nuclei. Those quantum effects can be treated directly, for example, by using path integral methods as described in Ch. 25 or approximately, for example, by assuming harmonic phonons. For heavier elements, the dynamics of the nuclei can be handled with the classical molecular dynamics or Monte Carlo methods.

[3] For bulk systems, there is a difficulty in dividing up the various terms that involve the Coulomb interaction because each term is individually infinite. This can be handled by adding a uniformly charged background to each such term, so each one is charge neutral.

$$\binom{M}{N} = \frac{M!}{N!(M-N)!} \approx e^{CN}, \qquad (1.3)$$

where $C > 0$. This is typically a huge number; for example, consider two carbon atoms, with their $N = 12$ electrons. Suppose we represent the orbitals with only 36 basis functions. This small representation yields more than 10^9 determinants (see Ex. 6.1).

The wavefunction for each eigenstate is a linear combination of these determinants. Though there are often other symmetries, such as inversion or lattice translation, they do not help enough to reduce the size of this linear combination to a reasonable amount. It follows that to store a complete wavefunction and to manipulate it becomes exponentially difficult as the number of electrons becomes large. The straightforward application of methods such as the configuration interaction method [3], used in quantum chemistry to compute properties of small molecules by exact diagonalization of the hamiltonian matrix, is therefore limited to small numbers of electrons.

The difficulty in computing and storing one wavefunction (typically the ground state) is only the tip of the iceberg. For systems at a non-zero temperature, all states that are thermally occupied need to be computed and one often wants the response to external perturbations including spectra, thermodynamic functions, and more. To solve these problems requires computation not only of the ground state, but also of many excited states. The number of states required will scale exponentially with the system size. Finally, if we want to do a molecular dynamics simulation, the calculations must be made many millions of times with the forces on all of the atoms calculated at each step.

The path to deal with these issues is to realize that it is not essential to tabulate the complete many-body wavefunction. What is usually needed are single numbers, such as the energy or its derivatives, one-particle properties such as the electron density, and spectra derived from one- or two-particle, or higher-order, correlation functions. All these are expectation values. In Ch. 5 we introduce some of their most important properties. Much of the rest of this book is then devoted to alternative ways to approach the problem: methods to compute measurable properties without precise knowledge of the many-body wavefunctions.

One way this can be done is in terms of Green's functions, introduced in Ch. 5, which represent directly the experimentally measurable spectra. These are developed in Part III, along with examples of applications. Another approach is through techniques that directly simulate the many-electron system. Instead of trying to represent explicitly the complete many-body wavefunction, it is sampled instead. These methods, called quantum Monte Carlo (QMC) methods, are discussed in Part IV.

There is no universal approach to find feasible, well-founded methods to deal with the many-body problem. There are many different materials, many phenomena, and many different criteria to obtain the accuracy needed for a useful result. For some problems, such as the relative stability of dissimilar structures that may be separated by very small differences in energy, we might require the accuracy in the energy difference to be much less than the ambient temperature, 300 K ≈ 0.026 eV $\approx 9.5 \times 10^{-4}$ a.u., which is often called "chemical accuracy." In other cases, one may find it satisfactory to calculate a bandgap to say 0.1 eV, or to detect the presence or absence of a satellite in an energy window of 1 eV. The nature of the problem and the required precision, together with the available computer power, will finally determine the appropriate methods to use.

1.3 Why is the independent-electron picture so successful?

Underlying all independent-particle approaches is the idea that each electron interacts with an effective potential that mimics to some extent the effects of the other electrons. A single-particle wavefunction can be expressed in a basis of dimension M, the same size as assumed in the many body problem in the previous section. However, there is a great simplification since an independent-electron state needs only one determinant. One can work with wave-function methods that require only $M \times M$ matrix operations that scale as M^3 or better, as opposed to the exponential scaling of the interacting many-body problem.

The independent-particle picture is extremely successful: it is used to classify solids into metals, semiconductors, and insulators, and to establish much of the present-day understanding of solids.[4] Quantitative calculations were put into practice more than 80 years ago, for example, numerical mean-field calculations for atoms in 1928 by the Hartrees [5].[5] Today, essentially all practical, quantitative many-body methods depend on information from independent-particle calculations as input, and the analysis is often in terms of single-body concepts. It is important to recognize the reasons for the success of independent-particle concepts and techniques. These provide invaluable lessons for the much more difficult problems encountered in many-body theory, and they reveal the cases where it is essential to go beyond independent-particle pictures.

Independent fermions

Any reasonable independent-particle approach recognizes the fact that electrons "see" each other, since an orbital can only be occupied by one electron of a given spin; by "independent" we mean "independent except for the requirement of the exclusion principle." Indeed, the very origin of the Fermi–Dirac statistics and the term "fermion" is an example of the power of independent-particle arguments. In his 1926 paper [7],[6] Fermi arrived at the conclusion "it is required to admit that an electronic orbit is already 'occupied' when it contains only one electron" by considering electrons in heavy atoms, where he argued that the nuclear potential dominates over electron–electron interactions. From the observation that

---

[4] Further background and references can be found in [1] and an exposition of the history of the theory of solids can be found in the book [4].

[5] D. R. Hartree was aided by his father W. R. Hartree, a businessman with an interest in mathematics, who carried out calculations on a desk calculator. Together they published numerous calculations on atoms. D. R. went on to become one of the pioneers of computer science and the use of electronic computers, and he published a book on the calculation of the electronic structure of atoms [6].

[6] The paper by Fermi (see [8] for a translation) refers to earlier work by Stoner [9] on "The distribution of electrons among atomic levels" and Pauli [10] on the interpretation of atomic spectra. Apparently the development of the determinant formulation by Dirac [11] in 1926 was independent, though he refers to Fermi.

the lowest s state holds only two electrons, Fermi drew sweeping conclusions on the statistics of particles in an ideal monatomic gas. This was recognized by Dirac in his 1930 book on quantum mechanics [12], where he says "....a special statistics, which was first studied by Fermi, so we shall call particles for which only antisymmetric states occur in nature *fermions*." In the independent-particle approximations, the antisymmetry of the full many-body wavefunction is hidden, but it is implicitly present through the use of Fermi–Dirac statistics.

Symmetries and conservation laws

The antisymmetry of the wavefunctions is a general feature of systems of fermions, as is time reversal in the absence of external magnetic fields. In specific systems the hamiltonian may have additional symmetries, such as inversion and rotational invariance, and continuous or discrete translational symmetry in a homogeneous gas or a crystal, respectively. Coming together with a symmetry there is a conservation law, e.g., conservation of momentum for translation symmetry. It may be detrimental for a theory to violate these conservation laws, and, on the contrary, a key for success is to take them into account. In independent-particle methods each state of a many-particle system can be specified by the occupation numbers of the independent-particle states, and it is not difficult to take the symmetries into account. In a many-body theory that treats the interaction between particles, these states are mixed and the problem is not so simple.

Two examples show the issues that one may face in many-body calculations. In cases where there is a discrete quantum number, such as plus or minus parity, the many-body states can be readily constructed to have the correct symmetry. So long as the energy levels are discrete, the wavefunction and energy for each state can be found by exact diagonalization, at least in principle. Nevertheless, it is very useful to identify independent-particle states that are as close as possible to the actual states to facilitate the calculation and to interpret the results. However, in condensed matter the conserved quantity can take a continuous range of values, namely momentum that can take any value in a homogeneous system or "crystal momentum" that varies continuously in the Brillouin zone. For a given energy there can be an infinite number of ways the momentum can be shared among the electrons. In an independent-particle picture each of the possible states with energy E and momentum $\mathbf{K}$ consists of independent excitations with momenta $\mathbf{k}_1, \mathbf{k}_2, \ldots$ where $\mathbf{k}_1 + \mathbf{k}_2 + \cdots = \mathbf{K}$ and $\varepsilon_1 + \varepsilon_2 + \cdots = E$. In a many-body theory, however, these states are mixed by the interaction and it is not always easy to develop approximations that are guaranteed to obey the symmetry. In fact, a fundamental paper for many-body Green's function methods by Baym [13] is devoted to the conditions for "conserving approximations" that are described in Sec. 8.5. Again, independent-particle pictures can provide guidance. For example, the Bloch theorem applied to independent particles leads to the band structure for a crystal $\varepsilon_{\mathbf{k}}$; as brought out in Sec. 2.4 and other places in this book, the spectral function including interactions can often be understood as a broadened version of the band structure.

Fermi liquid theory and continuity

Perhaps the most direct use of independent-particle concepts to characterize interacting systems is Fermi liquid theory, proposed by Landau [14–16] and summarized in Sec. 3.4. This theory, originally proposed for liquid ^{3}He, assumes the low-temperature properties are determined entirely by free-particle-like low-energy excitations near the Fermi surface, called quasi-particles, with renormalized masses and weak effective interactions. It is not obvious that excitations in liquid ^{3}He should behave like a weakly interacting gas; the same reasoning applies to electrons in solids, where the magnitude of the interaction ($\approx e^2/R \approx$ several eV, with R a typical atomic size) is not small.

According to Alexey Abrikosov,[7] Landau arrived at his proposal by asking himself, "What is conserved?" The fact that there are fundamental features such as conservation laws unchanged by the interaction suggests continuity. Suppose that we could turn on the electron–electron interaction continuously.[8] Then there should be a one-to-one correspondence between the excitations of the non-interacting system and the real excitations. Although the momentum is shared among electrons, the total current is the same as for non-interacting particles since the total momentum is conserved. The reasoning should apply so long as the system evolves continuously, i.e., there is no phase transition as the interaction is turned on. Continuity is a critical part of the reasoning regarding the Fermi surface and bandgaps in interacting systems (e.g., Chs. 13 and 16–21).

The Fermi surface is determined by a step in the momentum distribution at zero temperature. The existence of such a step, as well as the shape of the surface in **k**-space where the step occurs, are straightforward to understand for independent electrons, but less obvious in the interacting case, because of the electron–electron scattering. Both features are *assumed* in Fermi liquid theory, but in order to determine them, one must resort to other methods, which is not so easy in a many-body theory. Fortunately, the "Luttinger theorem" (Sec. 3.6) states that, so long as the actual interacting system can be considered to evolve continuously from a non-interacting system, the volume enclosed by the Fermi surface does not change. This is sufficient to determine the surface in liquid ^{3}He and the electron gas, since the surface is spherical and is determined by the volume. The problem is not as simple in a crystal, but the conservation of the volume is a welcome guidepost for many-body calculations.

The situation changes if the connection is broken. For bulk systems, discontinuities in properties can only occur at phase transitions. Phases can be classified according to symmetry and described by order parameters, as formulated by Landau [18, 19]. More recent developments on topological order are providing qualitatively new classifications and phenomena (see, e.g., Sec. 6.3 and [20, 21]). Although the continuous connection to one and the same independent-particle theory may be lost, nevertheless the continuity arguments can still be applied so long as one is careful to connect to an independent-particle theory with an effective potential with the same symmetry as the actual phase. To determine the

---

[7] Private discussions in 2005.
[8] This can now be done in cold atom experiments using the Feshbach resonance [17].

stable phase and the transition points is therefore one of the main challenges for many-body calculations.

1.4 Development of theoretical approaches to the many-body problem

When is the independent-particle picture not sufficient?

There are many examples of cases where the independent-particle picture clearly meets its limits. Some phase transitions and ordered states would never occur if the electrons did not interact. One is the Wigner crystal transition [22] (Sec. 3.1), where the electrons break translation symmetry at low density. Other examples are ordered magnetic states: there is no reason for spins to order antiferromagnetically without an electron–electron interaction. In Ch. 2 we will see several examples of failures of the independent-particle picture. In each case an independent-particle or to be precise, *one single* independent-particle description does not succeed in capturing the essential effects.

A different type of example is the density of states at the Fermi energy. In Bardeen's thesis as a graduate student with Wigner, he showed that the Hartree–Fock approximation introduces a singularity at the Fermi surface [23]. This conclusion applies to all metals since it depends only on analysis of an integral where the integrand has a singularity due to the long-range Coulomb interaction (see also Sec. 11.3). The observed smooth dispersion of excitations can be explained only if the interaction is screened, i.e., correlation of electrons that reduces the effective interactions. Screening is an essential feature of the methods in this book. The Green's function methods are developed in terms of a screened Coulomb interaction called W in the GW method (Chs. 11–13) and the screened effective interaction U in applications of dynamical mean-field theory (Chs. 19–21). Screening is also a key ingredient in the variational Monte Carlo method (VMC, Ch. 23), where the correlation (Jastrow) factor is often derived from the random phase approximation (RPA): to lowest order in the electron–electron interaction it corrects the Hartree–Fock wavefunction for effects of the interaction. All these developments go beyond a static mean-field description.

Genesis of theoretical methods for many-body calculations

Many methods in use today can be traced back to their roots more than 50 years ago. By 1930 there were accurate calculations by Hylleraas [24] of the ground-state energies of two-electron atoms and ions. In condensed matter physics, quantitative calculations on sodium were reported by Wigner and Seitz in 1933 and 1934 [25, 26] (see also [22]). The first paper [25] provided the picture of correlation in which one electron in a Wigner–Seitz cell prevents another electron from occupying the same cell. In their 1934 paper [26] they explained the effects in terms of the "Fock picture," called "one-electron," and the correction that lowers the energy. As phrased by Wigner [22], "This energy will be called the 'correlation energy'." Of course, this is hard to calculate and they proposed an interpolation between the low- and high-density limits that compares very well with the

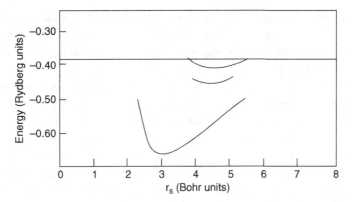

Figure 1.1. Total energy of sodium as a function of the Wigner–Seitz radius: reproduction of Fig. 4 from the 1934 paper by Wigner and Seitz [26]. The bottom curve is the energy of the lowest state in the band, and the top curve is the total energy in the Hartree–Fock approximation. The middle curve (from Fig. 6 of that paper) is the total energy including the estimated correlation energy. The horizontal line is the calculated atomic energy [26]. We can see that correlation increases the binding energy by more than a factor of 2; in fact, the result is within 15% of the measured binding energy.

best-known calculations of the correlation energy using diffusion Monte Carlo (Ch. 24) (see, e.g., [1, Ch. 5]). The results for Na are reproduced in Fig. 1.1, showing that the binding energy is greatly increased by correlation. Remarkably, the final result for the binding energy is close to the measured value.

Many of the most important developments in theoretical methods occurred in the late 1940s and 1950s. Two notable developments were the invention of path-integral methods [27] and Feynman diagrams [28]. With the advent of modern computers, the former has become an important non-perturbative method for studying correlated particles. The latter provides the graphical interpretation that has brought to life the summation of diagrams that are the basis for an untold number of quantum many-body techniques. In the same period was the advent of the Dyson equation and the functional formulation of the quantum many-body problem by Schwinger, Tomanaga, and others.[9]

Essential features of screening in a metal were captured by the RPA of Bohm and Pines in 1951–53 [30], which leads to the Lindhard dielectric function [31] for the electron gas. In 1958, Hubbard [32] showed that the RPA follows from the diagrammatic bubble expansion for the dielectric function and in 1959, Ehrenreich and Cohen [33] showed that it was tightly linked to the time-dependent Hartree approximation.

The late 1950s and early 1960s witnessed developments in diagrammatic techniques, including the work of Gell-Mann and Brueckner in 1957 [34], Galitskii and Migdal in 1958 [35], and Luttinger and Ward [36, 37], among many others. The book by Abrikosov, Gorkov, and Dzyaloshinski [38] in 1963 is remarkable for elucidating the physics and

---

[9] The 1949 paper by Dyson [29] gives references to Feynman, Schwinger, and Tomanaga.

in-depth coverage of the latest major development at the time. Baym and Kadanoff [13, 39] and earlier work of Martin and Schwinger [40] provided a general framework for characterizing functionals of the Green's functions and a concise statement of conditions for conserving approximations.

The years 1964–65 were especially significant in the development of the most widely used methods today. Density functional theory (DFT) by Kohn, Hohenberg, and Sham [41, 42] came after the more general functionals of Luttinger and others, but it introduced new ideas that led to practical methods. It was a major turning point that, with the advent of powerful computers, revolutionized computational materials modeling. A further landmark is Hedin's 1965 paper [43]. The abstract says "there is not much new in principle in this paper," but that work identified aspects of the many-body problem that are especially relevant for excitations in solids, and it established the "GW" approximation. McMillan (while a student of Bardeen) carried out variational QMC simulations of ^{4}He [44], the prototype for modern calculations, by realizing the relation between quantum many-body theory and the classical many-body computer simulation techniques that had been developed in the previous decade. The period 1963–65 witnessed the papers by Gutzwiller and Hubbard (see Sec. 3.2) that defined the models for interacting electrons on lattices and provided much of the conceptual structure for methods such as dynamical mean-field theory (DMFT). Together, the acronyms DFT, GW, QMC, and DMFT encompass most of the first-principles computations for materials today.

1.5 The many-body problem and computation

Moore's law (that the number of transistors per chip doubles roughly every year) has changed science in a profound way. Theory is no longer only a "pencil and paper" activity and we are no longer limited to the calculation of a wavefunction that requires two afternoons, as described in the quote from Wigner and Seitz at the beginning of this chapter. For electronic structure theory the advent of computers has been a game-changer, with the capability for quantitative calculations of properties of materials and the development of new approaches not envisioned in the early days of quantum mechanics before the invention of computing machines.

The combination of theory and numerical calculations allows us to get closer and closer to reality, with fewer approximations and more understanding. Quantitative calculations within independent-particle approximations using computers started in the 1950s, but it was only with the computers of the 1970s that calculations could be done with a sufficiently large basis that different methods could give the same answer – thus moving from debates about numerical approximations to addressing the deeper physics: how to deal with the interacting-electron problem. Several aspects highlight this important evolution:

- The great advances in the theory of interacting-electron systems in the 1950s and 1960s were applied to model systems such as the homogeneous electron gas. But it was only in the 1970s that computers and algorithms made possible QMC (Chs. 22–25) calculations of its correlation energy that are the benchmark to which theories are compared. Now QMC calculations provide access to detailed materials properties of the interiors of the giant planets and exoplanets, and they provide pressure standards for experiments under extreme conditions to mention just a few applications.

- Excitations are essential for many important phenomena in science, such as the electronic states of semiconductors, the optical properties of materials, the defect energy levels that give rise to solid-state lasers, and much more, including fundamental studies of materials that are today responsible for Moore's law. Thanks to the power of computers, and the algorithms developed to take advantage of this power, starting in the 1980s it became possible to calculate excitations in semiconductors and other materials using many-body perturbation methods such as GW (Chs. 9–15). Now such calculations can be done for problems requiring hundreds of atoms, such as surfaces and impurities.
- Many phenomena manifest the presence of interacting electrons, such as magnetic or metal–insulator phase transitions, and renormalization of low-energy excitations. As an example, it was known in the 1960s that the behavior of magnetic impurities in metals, the Kondo effect, requires non-perturbative methods because expansions in the interaction diverge as the temperature goes to zero. Wilson's numerical renormalization group approach in the 1970s paved the way for understanding this and related problems in many fields of physics. These are the forerunners of methods such as DMFT (Chs. 16–21) that are now being developed to provide quantitative calculations for materials with strong interactions as a function of temperature.
- The structures and other ground-state properties of materials can be calculated using DFT, however, it was only in the 1970s that it was feasible to carry out calculations to determine how well the theory would work in practice. The success of computations has led to an explosion of work so that the field is active and growing, with enormous impact, as illustrated by the more than 10,000 papers per year. By using efficient algorithms (classical molecular dynamics, fast Fourier transforms, and clever iterative methods), the field moved from crystals with simple structures to complex systems, thermal simulations of liquids, nanostructures, and large molecules in solution, to mention just a few applications.

The ambition to treat the many-body interacting-electron problem for a broad range of systems, including relaxation to find the minimum energy structures, molecular dynamics, etc., raises the challenge: to create efficient methods for large systems with many atoms, which is especially difficult for approaches beyond DFT. Some examples are proteins, a crystal with defects, impurities, or any kind of disorder, or a liquid where there is no periodicity. This poses the question of how the computer time for a calculation of a certain property, e.g., an optical absorption spectrum, using a given algorithm, will depend on the number of electrons N in the system or the unit cell. Indeed, one of the key methods of analysis in understanding the future of electronic structure methods is *computational complexity*.[10] We ask the question: "how does the computer time T to

---

[10] Do not be confused with "complex systems," though of course there is a relation.

perform this calculation depend on the number of electrons, N, i.e., what is $T(N)$?" To precisely define $T(N)$ for a given property, one must specify many things: for example, what types of atoms and what accuracy is required, what computer and program. However, the asymptotic form for scaling as a function of N is often independent of these details, and gives a simple criterion to compare algorithms.

Straightforward exact methods (i.e., with a bounded error) scale as $T \sim \exp(CN)$, as given in Eq. (1.3).[11] Moore's law would then not be sufficient to give access to a broad range of materials within a reasonable time: approximate approaches such as those described in this book are therefore crucial. We can consider several representative present-day methods. One has to be careful in the comparison: they are designed to calculate different properties, and with a different target accuracy, so the following are just examples.[12] Quantum chemistry methods such as the coupled cluster methods [45] can be applied to the calculation of excitation energies; coupled cluster calculations scale as a power law, e.g. N^7, when single and double excitation operators are included (CCSD). Many-body perturbation theory methods, such as GW for quasi-particle energies or Bethe–Salpeter for optical spectra, scale between N^2 and N^5, according to the implementations (Chs. 12 and 14). Ground-state energies, static correlation functions, etc. can be computed by QMC methods using the fixed-node approximation with computational effort proportional to N^3 in general, but with a lower power of N if electrons are localized. In DMFT the effects of interactions are approximated by calculations on parts of the system i each with N_i electrons, and the scaling for the calculation of the local spectral function can vary from N_i^3 to exponential in N_i depending on the method (Ch. 18). Independent particle methods such as the Kohn–Sham equations scale as N^2 or N^3, but in practical iterative methods the dominant computational cost scales as $N \ln(N)$.

There are tradeoffs between different methods: approximations, speed, and memory needed; how accurate they are for different properties and problems. It is important to realize that there is no best method; each method has its strengths and weaknesses in terms of accuracy, speed, and which properties are accessible. We hope to provide some guidelines later in this book. In practical calculations there are many factors involved in choosing a computational method, not only the size of the system, e.g., how robust the method is, and very practical questions such as the availability of a reliable, applicable code. There is also a question of the prefactor and crossover of methods: the asymptotic analysis is not necessarily the most relevant factor. In addition, parallelization, memory usage, and other criteria may be important. For quantitative calculations usually the most relevant question is whether the calculation can be done sufficiently accurately with the available resources in a timely fashion.

---

[11] There are certain quantum problems (e.g., 1D electron systems) that have a much better scaling.

[12] In the limit of large size, many properties are determined by a finite region around each atom, i.e., different parts of the system act independently, and for these properties it should in principle be possible to cast each method in a form that is "order N"; this refers to the idea of computing properties in computer time linear in the number of electrons.

1.6 The scope of this book

What is the role of this book, and which choices set it apart from others in the field?

There are many fruitful approaches to the many-body theory of electrons in physics or chemistry, far beyond the reach of any one book to cover. Since our main focus is on condensed matter, we only consider theoretical approaches that are applicable to extended systems. The methods must be able to describe wide ranges of materials and properties, applications to realistic problems must be feasible in practice, and there must be promising avenues for future developments.

We want to explain the methods in enough detail that a reader can appreciate the inner workings of actual calculations. Most of all, we want the reader to appreciate the physics behind the methods, the capabilities, and the limitations, and the relations among the methods. Our goal is for the reader to develop a deep understanding that will lead to further questions and ideas, a basis for the development of methods for the future.

With these criteria we must limit the topics to a very small number, each focused on the major methods within a class. We have chosen three approaches: two that use Green's functions, namely many-body perturbation theory and dynamical mean-field theory, and one that uses simulation methods, quantum Monte Carlo. The parts describing the three principal methods can be read independently of each other, but there are many pointers to help the reader understand their relationships.

Part I of the book is devoted to background that introduces the problem, summarizes selected experimental observations that are the motivation for all this work, and a few of the classic models and concepts that are the basis for understanding the role of interactions and correlation in many-electron systems.

Part II provides background for mean-field methods (Ch. 4) and the mathematical tools that are common to all the approaches, namely correlation functions (Ch. 5). These chapters contain the crucial physical information in a compact way. They are followed by Chs. 6–8, which bring out general properties of wavefunctions and salient concepts and developments common to the Green's function methods.

Part III is devoted to the Green's function methods, along with examples of applications. These approaches are formulated in terms of the one- and two-particle Green's functions, correlation functions that are directly linked to experimentally measurable quantities. For example, photoemission measures the spectra for addition and removal of electrons and optical experiments probe properties that involve excitation of the electronic system. In general, it is not possible to solve the corresponding equations exactly, and an advantage of using Green's function methods is that they provide ways to choose approximations based on physical reasoning. They are especially relevant for condensed matter, where the spectra are continuous and cannot be represented by a set of discrete eigenfunctions.

The Green's function methods are grouped into two categories. Chapters 9–15 cover many-body perturbation theory, where certain quantities are expanded in powers of the electron–electron interaction. This is a well-developed field that includes such methods as RPA for screening of the interactions, the GW approximation for one-particle spectra, and the Bethe–Salpeter methods for calculation of two-particle response functions, in particular, optical spectra.

The other approach is complementary; in a broad sense it starts from localized atomic-like systems where interaction effects are strong, and builds the solid by perturbation theory in terms of the hopping matrix elements between the atoms. The methods and approximations in the framework of DMFT in Chs. 16–20 are designed to treat strong interactions by self-consistent methods. They are applied to elements and compounds involving transition metals, lanthanides, actinides, and others.

Chapter 21 is devoted to combining the Green's function methods, as a way to make a first-principles theory applicable to the strongly interacting problems.

Part IV is devoted to quantum Monte Carlo simulation methods that work with the complete many-body system by sampling the wavefunction instead of trying to represent it explicitly as in certain quantum chemistry approaches. It is clearly not possible for the simulation to visit every many-body state or configuration of electrons, but unbiased results can be obtained if it is theoretically possible for the simulation to reach each state. For non-zero temperatures, one does not need to calculate all of the excited states, but instead one samples the many-body density matrix. However, because of the fermion sign problem, no way has been found to perform such a calculation for general quantum many-body simulations without introducing an approximation.

Altogether the chapters summarize the strengths and weaknesses of the methods. With an eye toward future developments, the concluding remarks in Ch. 26 contain an invitation to consider the power of combining approaches, be it different theories, or theory and experiment.

SELECT FURTHER READING

Anderson, P. W. *Concepts in Solids: Lectures on the Theory of Solids* (W. A. Benjamin, New York, 1963). Reprinted (Westview Press, Boulder, CO, 2003).

Anderson, P. W. *Basic Notions of Condensed Matter Physics* (Addison-Wesley, Reading, MA., 1984). Reprinted in the Advanced Book Classics series (Addison-Wesley, Reading, MA., 1997). Expansion of the second half of "Concepts in Solids" with reprints of many original papers.

Kohn, W., "Nobel Lecture: Electronic structure of matter – wave functions and density functionals," *Rev. Mod. Phys.* **71**, 1253, 1999. Introduction to the many-electron problem.

Martin, R. M. *Electronic Structure: Basic theory and methods* (Cambridge University Press, Cambridge, 2004). Theory of electrons focused on density functional theory methods.

Mott, N. F. and Jones, H. *The Theory of the Properties of Metals and Alloys* (Clarendon Press, Oxford, 1936). Reprinted in paperback (Dover Press, New York, 1955).

Peierls, R. E. *Quantum Theory of Solids* (Oxford University Press, Oxford, 1955).

Seitz, F. *The Modern Theory of Solids* (McGraw-Hill, New York, 1940). Reprinted in paperback (Dover Press, New York, 1987).

2

Signatures of electron correlation

Real knowledge is to know the extent of one's ignorance.

Confucius, 500 BCE

Summary

The topic of this chapter is a small selection of the vast array of experimentally observed phenomena chosen to exemplify crucial roles played by the electron–electron interaction. Examples in the present chapter bring out the effects of correlation in ground and excited states as well as in thermal equilibrium. These raise challenges for theory and quantitative many-body methods in treating interacting electrons, the topics of the following chapters.

The title of this book is *Interacting Electrons*. Of course, there are no non-interacting electrons: in any system with more than one electron, the electron–electron interaction affects the energy and leads to correlation between the electrons. All first-principles theories deal with the electron–electron interaction in some way, but often they treat the electrons as independent fermions in a static mean-field potential that contains interaction effects approximately. As described in Ch. 4, the Hartree–Fock method is a variational approximation with a wavefunction for fermions that are uncorrelated, except for the requirement of antisymmetry. The Kohn–Sham approach to DFT defines an auxiliary system of independent fermions that is chosen to reproduce the ground-state density. It is exact in principle and remarkably successful in practice. However, many properties such as excitation energies are not supposed to be taken directly from the Kohn–Sham equations, even in principle. Various other methods attempt to incorporate some effect of correlation in the choice of the potential.

This chapter is designed to highlight a few examples of experimentally observed phenomena that demonstrate qualitative consequences of electron–electron interactions *beyond* independent-particle approximations. Some examples illustrate effects that cannot be accounted for in any theory where electrons are considered as independent particles. Others are direct experimental measurements of correlation functions that would vanish if the electrons were independent. In yet other cases, a phenomenon can be explained in terms of independent particles in some effective potential, but it is deeply unsatisfying if one has to invent a different potential for every case, even for different properties in the same material. A satisfactory theory ultimately requires us to confront the problem of interacting, correlated electrons.

This already shows the difficulty of defining what is meant by correlation. We will clarify this in the first section of this chapter. We will then see different signatures of correlation in different circumstances: in the ground state and thermodynamic equilibrium (Sec. 2.2), magnetism as intrinsically associated with correlation (Sec. 2.3), when electrons are removed from or added to a system (Secs. 2.4 and 2.5), or when the system is excited (Sec. 2.6). The Kondo effect and heavy fermions in Sec. 2.7 are striking manifestations of correlation, as are metal–insulator transitions in Sec. 2.8. We conclude the chapter with a glance at what happens in lower dimensions in Sec. 2.9.

2.1 What is meant by correlation?

The literature contains countless uses of the term "correlation." When a new phenomenon in condensed matter is observed, there is a tendency to call it a "correlation effect," or even "strong correlation." Before identifying signatures of correlation, let us say what we mean by this term. We do not aim to propose a universal definition, but simply to clarify what is intended when we employ this expression, the important thing being to arrive at an understanding of real many-electron systems.

For a precise, quantitative definition we adopt the historical convention that the correlation energy is the difference between the expectation values of the Hartree–Fock (HF) and the exact wavefunctions. The HF wavefunction is a single determinant of one-electron orbitals and the ground state is determined by varying the orbitals to minimize the energy. The determinant enforces the principle that two electrons with the same spin cannot be in the same single-particle state (the Pauli principle), which is also a type of correlation. Thus our definition of correlation refers to only that beyond what is already contained in Hartree–Fock. While the single determinant wavefunction is a theoretical construction that is not experimentally accessible, it can be calculated with high precision and comparisons with experiments provide direct confirmation of effects beyond Hartree–Fock. For example, we call the difference between the exact ground-state energy and the HF expectation value of the hamiltonian the *correlation energy*.

It is possible that the best (in the sense of the minimum ground-state energy) single determinant breaks the symmetries of the hamiltonian; to allow such solutions is called unrestricted Hartree–Fock (UHF). An example is a magnetically ordered ground state such as in Sec. 2.3, where time-reversal symmetry is broken even though all terms in the hamiltonian are symmetric under time reversal. Restricted HF (RHF) means that the single determinant is required to have the symmetries of the hamiltonian even if there exists a lower-energy solution. In this book we will define correlation to be relative to the restricted solution, which provides a unique definition independent of the various ways that a broken symmetry could occur. With this definition, an unrestricted solution with a lower energy captures some part of the correlation energy, but, of course, there are additional contributions missing in a mean-field UHF calculation. We will attempt to dispel confusion by clarifying what is meant by the term "correlation" whenever it is used.

Excitations reveal different aspects of correlation. Eigenvalues of the Hartree–Fock equations correspond to the energies for removing or adding orbitals to the determinant, keeping all other orbitals fixed; this relation is known as Koopmans' theorem [46, 47]. As discussed in Secs. 2.4 and 2.7 and throughout the book, the HF eigenvalues are often a very poor approximation to the actual spectra for adding or removing electrons, and there can be major improvements by including correlation among the electrons.

In addition, there are a number of phenomena that cannot be explained by *any* independent-particle theory with a static potential. For example, independent particles occupy discrete eigenstates;[1] a finite lifetime broadening and additional satellites in spectra cannot be explained in an independent-particle theory and must be due to interactions between the particles. Often the most compelling signatures of correlation are experimental demonstrations that many-particle systems exhibit behavior different from the sum of one-particle properties. An example is the comparison of different excitation spectra. The energies to remove an electron (create a hole) can be measured in photoemission, and to add an electron in inverse photoemission. Optical absorption creates an electron–hole pair, and the electron–hole interaction can cause large shifts in the energies with respect to the difference of electron addition and removal (i.e., excitonic effects as illustrated in Sec. 2.6). Such observations do not provide a way to take the correlation into account, but they show that a satisfactory explanation can be found only by a theory that goes beyond an independent-particle approach.

The term "correlation" is also used in a more general sense, and there are various correlation functions for quantities such as the density and magnetic moments associated with sites in a lattice. We will attempt to be clear in the meaning of the functions and the general descriptions of such correlations whenever they are used.

2.2 Ground-state and thermodynamic properties

We start with a short section pointing to signatures of correlation in the ground state. These are often quantitative rather than qualitative and more difficult to discern than correlation effects in excited states, but still, they can be very important. Properties of the electronic ground state and thermodynamic equilibrium include the structure, energy, entropy, spin density in a magnetic system, and many other properties, as well as derivatives such as bulk modulus, vibrational frequencies (in the adiabatic approximation), static dielectric and magnetic susceptibility, etc. The correlation energy is crucial in any quantitative description of molecules and solids. For example, the comparison of Hartree–Fock and experiment for the hydrogen molecule shows that correlation is responsible for ≈ 1 eV out of a total binding energy of ≈ 5 eV [1]. The early work of Wigner and Seitz in Fig. 1.1 has already

---

[1] In a crystal there is a continuum of energies, but there is a discrete set of eigenstates for a given momentum.

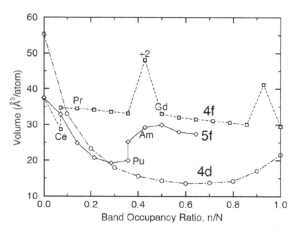

Figure 2.1. Volume per atom of the lanthanides (4*f*) and actinides (5*f*) compared with 4*d* transition metals as a function of fractional occupation of the *d* or *f* states. The smooth parabolic curve for the 4*d* series indicates a gradual filling of the *d* bands with maximum bonding at half-filling. In contrast, the 4*f* series lanthanides retain atomic-like character with little effect on the volume as the *f* shell is filled. (The jumps for Eu and Yb also support the atomic-like picture with non-monotonic changes in 4*f* occupation, denoted by the label "+2," due to added stability of the half-filled and the filled shells.) The anomalous elements are Ce and Pu; both have complex phase diagrams (see Fig. 20.1 for Ce), spectra (e.g., Fig. 2.9), and magnetic properties that are prototypes of strong interactions and signatures of correlation. (Figure provided by A. K. McMahan.)

established that the binding energy of sodium is off by a factor of 2 if correlation is not included. The question is not *whether* to include interactions, but *how* to include them.

Consider the trends in Fig. 2.1, which show the volume per atom for 4*d*, 4*f*, and 5*f* transition metal elements. For the 4*d* series, the smooth variation as a function of atomic number indicates the filling of delocalized states with a gradual evolution of the density and a maximum binding energy when the band is half-filled. Already in the 1970s it was established that this is well described by DFT in the local density approximation (LDA) [48]. But the figure also shows the very different behavior of the lanthanide elements where the volume hardly changes, indicating that the 4*f* states are localized and do not participate in the bonding. However, two elements, Ce and Pu, are anomalous; here the actinides change from a delocalized to a localized behavior. A more complete picture emerges from the phase diagram as a function of temperature and pressure for Ce, shown in Fig. 20.1. There is a first-order transition that terminates in a critical point where the two types of behavior must merge into one. Taken together, the strong temperature dependence, magnetic behavior, excitation spectra (e.g., for Ce shown in Fig. 2.9), and many other properties point to strong effects of correlation in the lanthanides and actinides.

Van der Waals dispersion interaction

A compelling example of correlation is the weak attraction between atoms and molecules, even when there is no covalent bonding or average electrostatic interaction [49]. The

Table 2.1. Cohesive energy of rare gas
atoms at $T = 0$ in eV [51, 52]

He	Ne	Ar	Kr	Xe
0.001	0.02	0.08	0.11	0.17

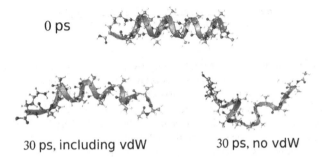

0 ps

30 ps, including vdW   30 ps, no vdW

Figure 2.2. Alanine helix structures in alanine polypeptides: the α-helix structure is stable in experiment up to 725 K. This compact structure is favored by the van der Waals dispersion interaction. This has been illustrated [53] using density functional molecular dynamics with a functional including or excluding the van der Waals contribution. The figure shows snapshots from the simulations at the start (top) and after 30 ps (bottom). At 700 K inclusion of the van der Waals contribution preserves the α-helix, whereas its neglect leads to an opening of the structure. (Figure provided by Mariana Rossi.)

attractive force, often called the "London" (named after Fritz London, who gave its mathematical formulation [50]), or "van der Waals dispersion," or simply "dispersion" force, is due to quantum fluctuations of the electric dipole on one molecule that induces dipoles on another molecule, oriented such that the molecules will attract each other. The force is purely due to correlation of the electrons and would vanish if the electrons were independent. Since the interaction is induced dipole–dipole, the non-relativistic asymptotic form is proportional to R^{-6} with the inter-molecular distance R, and its magnitude is very weak as indicated in Tab. 2.1.

An example where these weak interactions have a large impact is biopolymer folding, where a system with a huge number of atoms prefers a unique state that is crucial for its function – when a protein takes a wrong geometry, this can have disastrous consequences, an example is mad cow disease. Because it is always attractive, the van der Waals contribution favors compact structures. As an example, Fig. 2.2 displays the evolution of a polyalanine (Ac–Ala$_n$n–LysH$^+$) structure during a DFT molecular dynamics simulation including or excluding the van der Waals contribution to the exchange–correlation functional [53]. The figure shows typical structures found during the simulation. The van der Waals contribution greatly enhances the stability of the α-helix, consistent with experiment. Without the van der Waals contribution there is a tendency to realize more open structures.

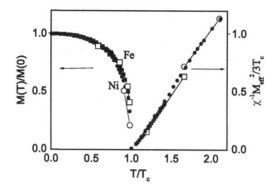

Figure 2.3. Magnetization (left curve) and magnetic susceptibility (right curve) of the $3d$ transition metals Fe (squares) and Ni (circles). Closed symbols are experiment and open symbols are from calculations [55], as discussed in Ch. 19. The curves have been normalized to the zero-temperature magnetization $M(0)$ and transition temperature $T_C = 1043$ K and 631 K for Fe and Ni, respectively. At low temperature, these are metals with delocalized states near the Fermi energy. Above the Curie temperature, however, the magnetic susceptibility indicates that these materials act as a collection of thermally disordered atomic-like moments. The fact that these two types of behavior occur in the same material is a signature of correlation among the electrons. To capture such effects is a goal of dynamical mean-field theory, Chs. 16–21. (From [55].)

Many-body perturbation theory yields the correct R^{-6} form for the dispersion interaction (Sec. 13.5), but in order to describe the strength correctly also, one has to go beyond the simplest (random-phase) approximation for the calculation of the polarizability that determines the charge fluctuations. The dipole–dipole correlation functions that give rise to the dispersion forces can be calculated directly using Quantum Monte Carlo (QMC) methods. An outstanding goal for the future is to develop many-body methods capable of treating large systems like the polypeptide discussed above, including weak van der Waals bonds as well as strong bonds and reactions, quantum motion of light nuclei as well as molecular dynamics, electronic excitation and energy transfer, and other properties directly relevant to the functions of such systems.

2.3 Magnetism and local moments

Magnetism is closely linked to electron–electron correlation. If there were no interactions, just the Pauli principle, the ground state of a crystal would always have bands with electrons in doubly occupied delocalized states. However, interactions between the electrons can lead to magnetic order and fluctuations that indicate "local moment" behavior. This can be illustrated by the ferromagnetic transition metals Fe and Ni, as shown in Fig. 2.3. At low temperature there is an average ordered magnetic moment and the materials are metals, indicating that the electrons are in delocalized states. However, the magnetization decreases with increasing temperature until it vanishes at the transition temperature T_c. For $T > T_c$ there is no magnetic order but a magnetic susceptibility that diverges at $T = T_c$ and is approximated by the mean-field Curie–Weiss form

$\chi(T) = (M_{\text{eff}}^2/3)(T - T_c)^{-1}$, where M_{eff} is a magnetic moment associated with each transition metal atom, called a local moment. The experimental data for Fe and Ni obey this relation well, as shown in the figure, with effective local moments $M_{\text{eff}} = 3.13 \ \mu_B$ for Fe and $1.62 \ \mu_B$ for Ni [54], not far from the values expected for the atomic d shells.

Theoretical pictures to explain the phenomena take many forms, ranging from mean-field theories of electrons in delocalized bands (termed "itinerant") to models in which the electrons are not mobile but act as local moments. The former is the purview of approaches such as the mean-field Stoner picture [56], sufficient for an approximate description of the ordered state, even though there are no fluctuations. The latter is the domain of statistical mechanics of spin systems that can describe the magnetic order, fluctuations, and behavior at the phase transition, even though there are no delocalized states and no metallic behavior.

The challenge is to bring these two pictures together to treat the entire range of temperatures, bridging the gap between band-like and atomic-like behaviors. Calculations for Fe and Ni reported in Sec. 20.4 illustrate steps toward this goal. However, this is but the tip of the iceberg: in the lanthanides the transition temperatures are much lower. Anomalous cases such as Ce bring out the issues more clearly. For example, in "heavy fermion" materials (see Secs. 2.7 and 20.2) there is local moment behavior with no magnetic order that persists down to much lower temperatures, of order 10 K and less. Only at lower temperatures do the electrons cooperate to create a metallic Fermi liquid, which may then give rise to a state with magnetic or some other order. The metal–insulator transitions in Sec. 2.8 are intimately tied to local moments, either ordering or forming disordered states with conductivities decreased by orders of magnitude. Thus the challenge is not to merely describe Fe and Ni, but rather to understand a vast range of phenomena in a coherent picture. Steps toward understanding and quantitative theories are primary topics of Chs. 19–21.

2.4 Electron addition and removal: the bandgap problem and more

Because the ground state is an integral of many possible electronic processes, observations of it reveal only "average" electronic correlation. Interesting effects are detected more easily when one looks directly at these events. For example, one can excite a system by adding or removing an electron. Such a process is called a *one-particle excitation* – but to what extent can we link the observation to a *single* particle state in the sample?

Figure 2.4 illustrates the process of electron removal. One can imagine that the missing electron acts as a perturbation due to a positive charge that appears at a given time. Because of the interaction, this perturbation can induce oscillations of the electron gas, depicted schematically by outgoing spherical waves. Alternatively, in a particle picture, electron–hole pairs will be created. In any case, the induced excitations will change the kinetic energy of the outgoing electron.

Photoemission and related spectroscopies: a qualitative overview

Probably the most relevant experiment that measures the spectrum for removal of an electron is photoemission spectroscopy (PES) [57], in which an incident photon is absorbed

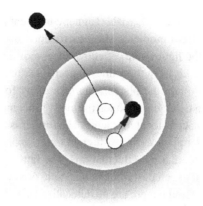

Figure 2.4. Schematic illustration of electron removal and the resulting reaction of the remaining electrons, seen alternatively as wave-like oscillations or displacement of individual particles. This is a screening effect that "dresses" the hole.

and the kinetic energy of an emitted electron is measured. The advent of synchrotron radiation has revolutionized the field and made possible meV resolution. The most complete information is angle-resolved photoemission spectroscopy (ARPES) [58, 59], illustrated in Fig. 2.5. In particular, the kinetic energy and angles of emission are often used to deduce energy–momentum relations of electrons inside the material. The complementary experiment, inverse photoemission (IPES) [58], involves an incident electron that fills an empty state with emission of a photon to conserve energy. It is most often performed with a detector that is limited to a fixed photon energy range, and is then called Bremsstrahlung isochromat spectroscopy (BIS).

One can consider PES and IPES to be the most direct measurements available for determining one-electron excitations.[2] There are caveats: a real experiment involves effects due to the surface, secondary electron background, and other factors.[3] However, these are often quantitative contributions that do not change the qualitative picture in which we are interested here. Thus, we will examine how a direct or inverse photoemission spectrum (from the ideal theoretical point of view, a *spectral function*, see Chs. 5 and 7) may display correlation effects. The middle panel of Fig. 2.5 is a schematic view of the independent-particle picture, where electrons occupy the valence bands and the conduction bands are empty. For fixed momentum, energy and momentum conservation tell us that the possible measured kinetic energies of the photoelectrons are sharp δ-function peaks.

---

[2] Other experiments, like scanning tunneling spectroscopy (see Sec. 2.7) or pump–probe experiments, also give direct information on the one-particle spectra.

[3] At very high energies the technique becomes more bulk sensitive, but there are still other factors. The very description of photoemission as a one-particle excitation is an approximation, concentrating on the hole and its effect on the material, whereas the full problem is at least a two-particle one, including the excited electron, its propagation to the detector, and its interaction with the hole. See also footnote 6 in this chapter, [57, 59] and references cited therein.

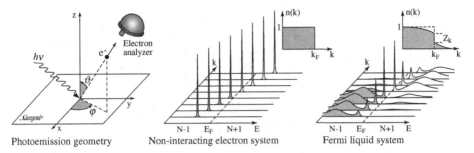

Figure 2.5. Schematic illustration of ARPES. The middle figure shows an independent-particle spectrum of δ-functions with occupation 1 or 0. The right figure exemplifies a spectrum for interacting electrons with "quasi-particle" peaks (i.e., peaks due to dressed one-particle excitations) having fractional weight and "satellites" or "sidebands" due to additional excitations that are induced in the system. Examples for satellites in photoemission are shown in Figs. 2.8 and 2.9 for plasmons in Na and atomic-like excitations in Ce, and throughout the following chapters. The insets depicting occupation numbers $n(k)$ are discussed in more detail in Secs. 5.2 and 7.5. (Provided by A. Damascelli, similar to figure in [59].)

However, the remaining electrons react to the hole that is left behind, as illustrated in Fig. 2.4. There are an infinite number of possible excitations, leading to a continuous energy spectrum with broadening and extra structures, as shown in the right panel of Fig. 2.5. All these features can be related to the one-particle Green's function and the *self-energy*, which are central to much of our understanding of interacting electrons, brought out especially in Chs. 5, 7 and throughout Part III.

Band structures

The concept of a sharp "band structure" is directly linked to an independent-particle picture, and it reflects the idea of "quasi-particles" that act like electrons with energies that are affected by interaction with other electrons. Indeed, experimental studies using different techniques have shown that for numerous materials there are quasi-particle bands with little broadening (long lifetimes), i.e., well-defined peaks similar to the schematic diagram in the middle panel of Fig. 2.5, or the peaks close to the Fermi level in the right panel. An example is the experimental results for Ge, where the dots shown in the middle panel of Fig. 2.6 are the maxima of peaks measured by ARPES and inverse ARPES. Even far from the Fermi energy the peaks are sufficiently well-defined to identify the bands.

Also shown in Fig. 2.6 are the calculated results of three approaches. On the right are independent-particle bands in the Hartree–Fock approximation; these show a gap that is much too large and a valence band that is too broad. By definition, the discrepancy from experiment is due to correlation effects that renormalize the energies and lead to a narrowing of the gap and bandwidth. On the left are the eigenvalues of the Kohn–Sham equations in the LDA. Even though there is no rigorous theoretical justification, they are remarkably close to the actual bands, except that the gap is too small, as discussed in the following

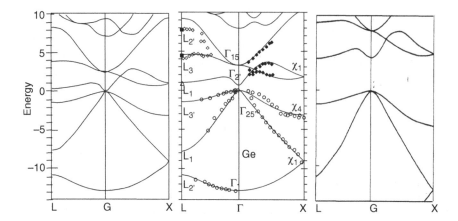

Figure 2.6. Bands of Ge from photoemission [60] and inverse photoemission [61] measurements (circle symbols in the middle panel) and from calculations in the local density (left), GW [62] (middle), and Hartree–Fock [63] (right) approximations. The LDA results, based on the assumption that the Kohn–Sham eigenvalues can be interpreted as addition and removal energies, are dramatically wrong: they give Ge as a metal. In contrast, the Hartree–Fock bands [63] have gaps that are much too large. This sets the theoretical challenge to go beyond Hartree–Fock and include screening; more recent GWA calculations (Ch. 13) provide remarkably accurate results for a large range of materials, as shown in Fig. 13.5.

paragraphs. In the center panel are results for the quasi-particle energies calculated in the "GW" approximation, where correlation is included in the form of dynamical screening with respect to Hartree–Fock. The results for Ge and many other materials show a much better agreement with experiment. Chapters 11 and 12 explain the theory, methods, and selected applications of the GW approach.[4]

Quantitative or qualitative? The "bandgap problem"

The fundamental gap (the lowest energy gap) in an insulator or semiconductor is *not* an independent-particle concept. The bottom of the conduction band is the lowest energy possible for adding an electron $(E(N + 1) - E(N)$ for a system of N electrons) and the top of the valence band is the lowest energy for removing an electron $E(N) - E(N - 1)$. The states involved are all ground states with N, $N - 1$, and $N + 1$ electrons; there is no possibility of decay and the many-body states have infinite lifetimes, even including interactions. Thus, without recourse to theoretical models, the gap can be defined by the difference in ground-state energies to add and remove electrons:

$$E_{gap} = [E(N + 1) - E(N)] - [E(N) - E(N - 1)]. \qquad (2.1)$$

---

[4] Accurate band structures can also be calculated using QMC. Results for Si are shown in Fig. 13.3.

Since there can be a gap with or without interactions, the precise value of a gap is quantitative; it is not possible to detect the effects of interaction by just looking at the measured gap. We need a calculation to establish the role of correlation.

For example, consider the gap in Ge shown in Fig. 2.6. If the Kohn–Sham eigenvalues are interpreted as electron addition and removal energies, the resulting bandgap comes out too small; within DFT-LDA Ge is found to be a metal! Because Kohn–Sham calculations of materials have become so ubiquitous, this quantitative bias has acquired a name: "the bandgap problem." However, in general it is a misuse of the Kohn–Sham equations to interpret the bands as electron addition and removal energies, and the "problem" is not, by itself, a measure of the effects of correlation.[5]

On the contrrary, the Hartree–Fock bands have a meaning: by Koopmans' theorem these are one-electron addition or removal energies if all the other electrons are not allowed to adjust. But the HF gap (see Fig. 2.6) is much too large, since it is missing the correlation. As discussed above, screening put in with a GW calculation gives accurate gaps, hence, screening is often the dominant effect of correlation on the gap.

Are there other cases where a bandgap is a *qualitative* feature that cannot be given by any independent-particle method, or not even by a many-body perturbation method such as the GW approximation? This is the issue posed by Mott (see Sec. 3.2) and the subject of innumerable papers. We return to this in Sec. 2.8 on Mott insulators and metal–insulator transitions, Sec. 3.6 related to the Luttinger theorem, Secs. 13.3 and 20.7 where we discuss the nature of a gap in paramagnetic NiO, and other places in the book.

Interaction effects in one-particle excitations: lifetimes

Interaction modifies the band structure, i.e., the *position* of quasi-particle peaks in the photoemission spectrum, in a quantitative manner. The very fact that measured peaks are broad, independent of the actual value of the broadening, demonstrates that electrons are not independent. Electron dynamics and the finite lifetime of excitations can also be measured in real time, for example, by two-photon photoemission using femtosecond lasers [65, 66]. A first pulse puts the system into an intermediate state; a second, time-delayed pulse is then used for the photoemission experiment. From the resulting time-dependent spectra the lifetime of the excited electron can be deduced. This can be seen in the example of several metals in Fig. 2.7: experiments [64] show that close to the Fermi energy the lifetime is long, whereas at an energy of 2 eV above the Fermi level the quasi-particle excitation decays within 10 fs. A big difference is found between Ag and the other transition metals, where the strong localization of the d electrons leads to narrow bands and a high density of states at the Fermi level. This increases the possibilities for scattering due to electron–electron interaction, decreasing the lifetime compared to Ag. The narrow bands with large interactions are the ingredients that lead

---

[5] There are subtleties in this problem that are discussed in Sec. 4.4.

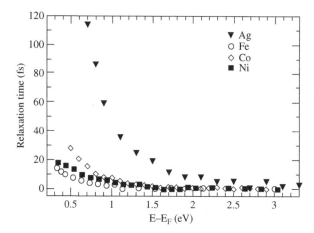

Figure 2.7. Electron lifetimes in transition metals measured with two-photon photoemission. The fact that the lifetime away from the Fermi level is finite can only be explained with interactions. Note that the lifetimes are much shorter in the transition metals that have narrow d bands at the Fermi level and display many other phenomena linked to correlation. (From [64].)

to local moments (see Sec. 2.3) and other phenomena in transition metal elements and compounds.

2.5 Satellites and sidebands

A full spectrum has even more striking features that cannot be explained in an independent electron picture. As illustrated in the right panel of Fig. 2.5, in some cases the coupling of the electrons to other excitations leads to distinct additional peaks in photoemission. The peaks are called "satellites" or "sidebands." They can also occur in inverse photoemission or other electronic spectra. Different kinds of satellites exist, according to the dominant process that is involved; we will look at two examples and there are others in Chs. 15 and 20.

Plasmons

Figure 2.8 shows the measured photoemission spectrum of bulk sodium [67]. This "simple" metal has one partially occupied band due to the $3s$ states. The other states are tightly bound and, concerning valence electron spectroscopy, can be considered to be part of an inert core with a binding energy beyond 30 eV. Therefore, one would expect just one broadened peak in the energy window of Fig. 2.8, close to the Fermi level. However, the figure clearly shows two additional peaks. Their distance from the quasi-particle is about one and two times the plasmon energy in sodium, respectively. They cannot be explained in any static band picture, but are due to the dynamic response of the system

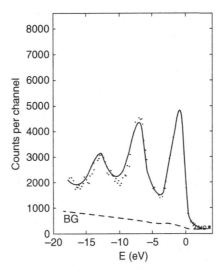

Figure 2.8. Experimental photoemission spectrum of bulk sodium (from [67]). The spectrum exhibits a quasi-particle band close to the Fermi level, and two prominent satellites between −5 and −15 eV that stem from plasmon excitations and cannot be explained in any independent-particle picture.

to the photoemission hole depicted in Fig. 2.4. This perturbation due to the hole can excite electron–hole pairs and plasmons: the photoelectron leaves the system with less energy, leading to the additional peaks, plasmon satellites.[6] Their presence is a clear qualitative signature of strong interaction effects, in a material as "simple" as metallic sodium.

Plasmon satellites are a general interaction feature that has been seen in many metals, semiconductors, and insulators. Examples are for silicon, discussed in Secs. 13.1 and 15.7, and vanadium dioxide, in Ch. 13. This transition metal oxide clearly differs from a homogeneous electron gas; the notion of a "plasmon excitation" is less well defined. This is, however, not a fundamental difficulty, since the above discussion holds in principle for any kind of excitation that can be created by the photoemission hole, be it plasmons, interband transitions, local atomic excitations, or even when other excitations like magnons or phonons come into play (see also Sec. 13.4).

Strong atomic-like interactions

Materials with partially occupied f states, the lanthanide and actinide transition elements, demonstrate dramatic effects of the large interactions between electrons in the f states.

---

[6] Their intensity should be interpreted with caution, because the outgoing photoelectron undergoes scattering processes that can lead to additional plasmon excitations called *extrinsic* effects, which means that they are not contained in the intrinsic hole spectral function. Moreover, there is *interference* between extrinsic and intrinsic contributions.

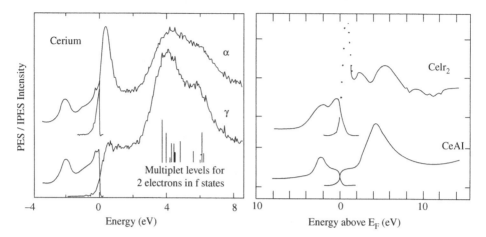

Figure 2.9. Electron removal (photoemission) and addition (inverse photoemission) spectra for $4f$ electrons in Ce [68] and two Ce compounds [69]. In the high-volume γ phase of Ce and the antiferromagnet CeAl, the strong interaction is evidenced by the peaks below the Fermi energy E_F for the removal of one $4f$ electron and above E_F for the addition of a second electron, and the multiplet structure like that in a Ce atom. Similar peaks are observed in the "volume-collapsed" α phase (see Sec. 20.2) and the compound CeIr$_2$ that are paramagnetic, and in addition there is a peak at the Fermi energy. In the absence of interactions there would only be the single peak at the Fermi energy; the division of the spectrum into three pieces is a many-body effect that cannot be described by an independent-particle approach.

Similar effects occur in other materials, but are not so pronounced. Figure 2.9 shows the direct and inverse photoemission spectra of cerium in its α and γ phases (see Figs. 2.1 20.1) and two cerium compounds. The s, p, d states form wide bands that constitute the broad background in the figures. If there were no interactions, the $4f$ states would form 14 (counting spin) bands that are partially filled. This would lead to a peak at the Fermi energy narrower than the s, p, d bands since the f states are more localized and have smaller overlap with their neighbors.

However, none of the spectra looks like that.[7] There are two peaks due to the f states, one ≈ 2 eV below the Fermi energy and a broad peak ≈ 4 eV above the Fermi energy. Such a spectrum can be understood if f states are considered to be atomic-like with strong interactions. If the energy for one f electron $\varepsilon_f \approx 2$ eV is below the Fermi energy, the state will be occupied. The energy to add a second electron will then be $\varepsilon_f + U$, where U is the interaction between electrons in an f state. The peak at ≈ 4 eV corresponds to $U \approx 6$ eV, similar to values found in many materials and justified by theory in Ch. 19. The peaks in the spectra above and below the Fermi energy are often called sidebands or upper and lower "Hubbard bands" because the interpretation as atomic-like states corresponds to the bands found by Hubbard and illustrated in Fig. 3.3 for large interaction U. The peak above the Fermi energy is broad because there is a range of energies for different configurations

---

[7] The spectra must be interpreted with caution, especially due to surface effects and possible structure due to other states in the same energy range.

of two electrons in the f shell of a Ce atom, called multiplets, that are shown by the bars in the figure on the left. The spread in energy is due to interactions and it cannot be explained by any non-interacting electron calculation.[8]

The lessons to be learned from the spectra in Fig. 2.9 are described in more detail in Sec. 20.2. However, we can already see the most important points. The spectra of CeAl and Ce in the high-volume γ phase are as expected for a magnetically ordered state having one of the 14 f states full, the other 13 empty at an energy $\varepsilon_f + U$, and nothing at the Fermi level. The other compound CeIr$_2$ and the "volume-collapsed" α phase of Ce are not magnetically ordered even at $T = 0$, and the spectra show an additional peak at the Fermi level. This illustrates the difficulty one has to face: the localized atomic-like behavior explains the two sidebands above and below the Fermi energy, but a peak at the Fermi energy requires delocalized band-like behavior. Indeed, the "three-peak" structure is a hallmark of correlation beyond any single independent-particle picture; the paradigm is the Anderson impurity model (Fig. 3.4) and the Kondo effect (Sec. 2.7).

2.6 Particle–hole and collective excitations

The previous section was devoted to electron addition and removal. However, many important spectroscopic measurements directly excite the electron system without adding or subtracting electrons from the sample: a good example is optical absorption. Optical experiments [72] have the advantage of exquisite energy resolution, sensitivity, and polarization-dependent selection rules that can be used to determine the symmetries of states. In the independent-particle picture an incoming photon can be absorbed by promoting an electron into an empty conduction state, with a spectrum governed by the joint density of states of the valence and conduction bands. However, observations are often different. Figure 2.10 presents the optical spectrum of the ionic material LiF. Dots are experimental data from [70]. The arrow indicates the fundamental electron addition–removal quasi-particle gap, Eq. (2.1). The joint density of states starts at this energy and there would be no absorption below the bandgap if there were no interactions between quasi-particles. Instead, experiment shows a clear and strong peak well within the bandgap.

Indeed, as depicted in Fig. 2.11, the electron that is promoted interacts with the hole that is created. This changes the spectral line shape and can lead to bound states (excitons). One can understand that the excitation energy will be lowered due to the electron–hole attraction, and that localization of the electron–hole pair will tend to increase the oscillator strength of the corresponding transitions. The bound electron–hole pair is a hydrogen-like two-particle problem, which explains the Rydberg series, for example in rare gas solids (see Fig. 14.10 for solid argon), although the interaction is screened.

Also, dipole-forbidden excitonic transitions can be observed experimentally. Powerful experimental tools include inelastic X-ray scattering (IXS) spectroscopy [73–75] and

---

[8] Multiplets occur in all atoms with open shells and more than one electron. Some aspects of the multiplet spectra can be approximated using an unrestricted Hartree–Fock method, where there are different possible solutions, with the lowest energy state obeying Hund's rule of maximum S, maximum L.

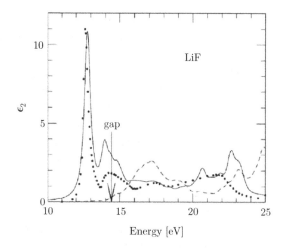

Figure 2.10. Optical spectrum of the wide-bandgap insulator LiF. The dots are experimental data from [70], the arrow indicates the quasi-particle bandgap that can be obtained from electron addition and removal spectroscopy, Eq. (2.1). The peaks at energies below the quasi-particle gap cannot be explained by a model with non-interacting quasi-particles. Calculations including the electron–hole interaction via the solution of the Bethe–Salpeter equation [71] (continuous line, see Ch. 14) reproduce the experimental peak inside the gap. Neglecting (dashed line) this interaction, the peak is absent.

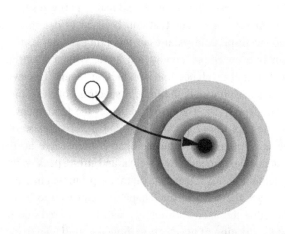

Figure 2.11. Schematic view of the absorption process in which an electron–hole pair is created and both particles are screened by the other electrons. With respect to a situation where hole and electron are added *separately* to the many-body ground state, there is a correction that can be formulated as an electron–hole interaction. This is then a correlated two-body problem.

electron energy loss spectroscopy (EELS) [76, 77]. IXS is in essence Raman scattering with photons that have high energy and momentum, allowing the study of the dispersion of excitations over several Brillouin zones [73]. It can measure the dynamical structure factor $S(\mathbf{q}, \omega)$ (see Sec. 5.8), with plasmon structure, longitudinal optical phonons, interband and

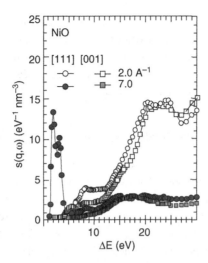

Figure 2.12. IXS spectrum as a function of energy for various momenta q for NiO [78]. For all values and directions of momentum transfer one observes the continuum with an onset at ≈ 4 eV, corresponding to the quasi-particle gap. However, for large momentum transfer in the [111] direction, a strong double peak appears far below the bandgap. It stems from dipole-forbidden d–d excitations and can be understood alternatively in an atomic picture or in terms of the interaction between electrons and holes (see text).

excitonic transitions, etc. An illustration is given in Fig. 2.12, showing the spectra of NiO for various values and directions of momentum transfer [78].

There is no contradiction between the two pictures, but only a question of which is more convenient to use. As expected, the continuum spectra start at ≈ 4 eV, the quasi-particle gap of NiO shown in Fig. 20.8. However, for large momentum transfer in the [111] direction, a strong double peak is observed within the bandgap, at 1.7 and 2.9 eV [79]. As in the case of LiF, one can explain these features starting from a quasi-particle band picture and adding a large electron–hole interaction. Alternatively, one may understand the exciton as an excitation in localized atomic-like states with no change in the number of electrons. This avoids the need to first invoke electron addition and removal bands and then add a strong electron–hole attraction (see Secs. 14.9, 14.11 and Ex. 14.1).

Finally, it is interesting to compare absorption and loss spectra over a wider energy range, where more delocalized electrons are also seen. The basic understanding of absorption is a sum of transitions, which justifies the independent particle picture as a first rough description. In loss spectra (e.g., the dynamic structure factor), however, *collective* excitations of the system dominate the measured spectrum. For atomic-like transitions this is not an important point. However, the long-range collective oscillations of an electron gas are very different from individual particle–hole excitations, because of the Coulomb interaction. They are similar to the collective oscillations of a system of coupled oscillators. Therefore, for example, the difference between the left (absorption) and right (loss) spectra of silicon in Fig. 5.6 in Ch. 5 is a pure interaction effect. It is already captured by the RPA (see Secs. 11.2 and 14.3). Successful current *ab initio* approaches to electronic spectra, including excitonic effects, start from a good quasi-particle band structure, mostly obtained in the GW approximation (see Chs. 10–13) and in addition, take the electron–hole correlation into account. This is done by going beyond the RPA, most often by solving the electron–hole Bethe–Salpeter equation introduced in Ch. 14.[9]

2.7 The Kondo effect and heavy fermions

The Kondo effect is one of the clearest examples of a low-energy phenomenon that emerges in a system of interacting electrons. The experimental observation is a resistance minimum in a metal at low temperature, first pointed out in 1934 [82]. The normal behavior of a metal is that the resistance decreases monotonically as T decreases; however, in some materials there is an upturn at low T so that there is a minimum in the resistance. The upturn puzzled theorists for decades,[10] until Kondo [83] showed that the scattering of conduction electrons by impurity spins increases as the temperature decreases, i.e., the impurity spins and metal electrons become more and more correlated as $T \rightarrow 0$. This is a strong-coupling problem that was finally solved in 1975 by Wilson using the numerical renormalization group [84] and later by exact solutions [85]. From this work, we now understand that there emerges a new energy scale, called the Kondo temperature, T_K. The right side of Fig. 2.13 shows measured values of T_K for the series of $3d$ transition metal impurities in Cu and Au hosts.[11] The extremely large variation of T_K, by 5 orders of magnitude, is explained by the exponential dependence of the Kondo temperature on the ratio of the interaction to the coupling to neighbors (see Sec. 3.5).

A direct way to observe the Kondo effect is by using a scanning tunneling microscope to measure the low-energy resonance in the spectrum of dI/dV near an atom with a magnetic moment on the surface (or in subsurface layers) of a metal. The left side of Fig. 2.13 shows the spectra for a single Co atom on a Cu surface [80], and the decrease in the magnitude with distance from the Co atom. Note that the Kondo temperature observed for Co on the Cu surface (≈ 50 K) is an order of magnitude less than for Co in bulk Cu (≈ 500 K), which is consistent with the expected trend that the states on the surface are more localized. It is also found that T_K is intermediate for Co in layers near the surface [86].

The Kondo effect is now one of the classic models described in Sec. 3.5; together with the Anderson impurity model (Sec. 3.5), it has become part of the conceptual fabric for understanding condensed matter. This is the starting point for approaching other problems, for example, lanthanide or actinide compounds with very narrow bands called "heavy

---

[9] As in the case of electron addition and removal, dynamical screening can lead to satellite structures. The description of satellites in two-particle excitations is even more complicated than in the case of one-particle excitations (see Sec. 14.10).

[10] As recalled by Ted Geballe, he asked John Bardeen in 1955 to name the most important unsolved problems in condensed matter physics. Bardeen replied "superconductivity" and, after a characteristic pause, "the resistance minimum" now called the Kondo effect.

[11] One can find a range of values for T_K in a material because the Kondo effect is a crossover, not a sharp transition. Nevertheless, the trends are well-established.

Table 2.2. Specific heat coefficient $\gamma = C_v/T$ of some heavy fermion materials compared with Na and Pd [87]

Crystal	Na	Pd	CeCu$_2$Si$_2$	UBe$_{13}$	CeAl$_3$
γ (mJ mole^{-1} K^{-2})	1.5	10	1000	1100	1620

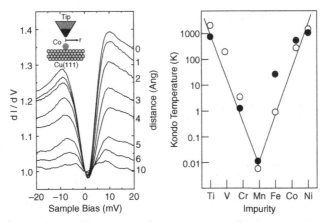

Figure 2.13. Left: spectrum of dI/dV vs. voltage V measured by STM at 4 K near a single Co atom on a Cu(111) surface. The form is that expected for the Kondo resonance and the effect is maximum if the tip is directly over the Co atom (top curve) and decreases with distance from the Co atom shown on the right vertical axis (in Angstrom). Fits to a theoretical form for a large sampling of Co atoms yields a Kondo temperature of $T_K \approx 53$ K. (From [80].) Right: Kondo temperatures for the series of $3d$ transition metals in Cu (filled circles) and Au (open circles) hosts. Note the extremely large variation interpreted as the exponential dependence of the Kondo temperature on the material parameters (see text). (From [81].)

fermion" materials [87–89] with specific heat coefficient γ much greater than ordinary materials, as illustrated in Tab. 2.2. The narrowing of the bands stems from interaction effects similar to the impurity problem; in compounds Ce and U the f states are so localized that the impurity effects persist in a crystal with a Ce or U atom in every cell. As an example, $T_K \approx 2$ K for a Ce impurity replacing a La atom in LaAl$_3$, which evolves into a narrow band when La is fully replaced by Ce in the heavy fermion material CeAl$_3$. Similarly, the peak at the Fermi energy in the photoemission spectrum for Ce in Fig. 2.9 is an analog of the resonance in the tunneling spectra for Co on Cu shown in Fig. 2.13.

2.8 Mott insulators and metal–insulator transitions

One of the most dramatic phenomena observed in condensed matter is a metal–insulator transition (see, e.g., the review [91]). This can occur in many ways: it may be associated

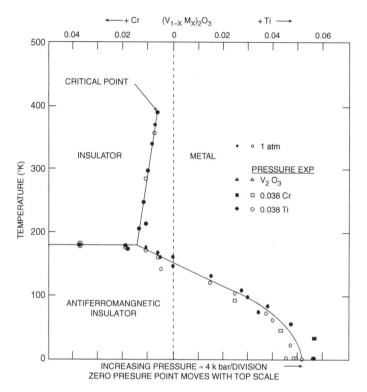

Figure 2.14. Phase diagram of $(V_{1-x} M_x)_2O_3$ as a function of pressure or doping with M denoting Cr or Ti. With the addition of $\approx 1\%$ Cr (argued to act like a negative pressure) there is a first-order transition that ends in a critical point. This has the characteristics expected for a Mott transition; however, there are many possible mechanisms as described in the text and Sec. 20.6. (From [90].)

with a transition to a magnetically ordered state, for example, the paramagnetic-to-antiferromagnetic transition observed in many transition metal oxides [91]. It may occur when there is a lattice distortion, for example, dimerization that occurs in VO and VO_2 (see Sec. 13.4), a "Peierls transition." In each of these cases there is a broken symmetry and a doubling of the unit cell. Perhaps the most intriguing case is a transition without change of symmetry, often called a "Mott transition."[12]

The material often considered to be the classic example of metal–insulator transitions is V_2O_3, which has the phase diagram as a function of temperature and pressure or alloying shown in Fig. 2.14. There has been controversy for decades over the origin of the effects, and some are recounted in Sec. 20.6. Here the purpose is to identify the experimentally observed phenomena. At low temperature V_2O_3 is an antiferromagnetic insulator (AFI) that transforms to a paramagnetic metal (PM) with increasing temperature or pressure (the

---

[12] The terms "Mott insulator" and "Mott transition" are used in many ways in the literature, from strict definitions that signify a state of matter qualitatively different from normal metals and insulators to a general categorization as any case in which the dominant localization mechanism is argued to be the electron–electron interaction. The theoretical concepts are discussed in more detail in Sec. 3.2 and various places in the book.

right side of the figure). The transition also occurs in $(V_{1-x}Ti_x)_2O_3$, where the reduced lattice constant is argued to act like an effective pressure. At the transition, the conductivity changes by many orders of magnitude. This is readily understood since it is accompanied by a broken symmetry and an increase in the number of atoms per unit cell in the antiferromagnetic phase.

If V is substituted by Cr in $(V_{1-x}Cr_x)_2O_3$, the lattice constant increases (like an effective negative pressure). In a narrow range around $x \approx 0.01$ there emerges a first-order transition in the paramagnetic phase above ≈ 180 K that terminates in a critical point at ≈ 400 K. There is no change in symmetry since the phases are connected continuously above the critical point, where there is no distinction.[13] This is called a metal–insulator transition because there is a change in conductivity that can be very large. For example, if pressure is applied to a sample with $x \approx 0.02$ there is an abrupt jump in the conductivity by over a factor of 100 at room temperature [92]. Note that this transition occurs only in a range of temperature that does not extend to $T = 0$. One should take care in interpreting the observations using theoretical concepts that apply rigorously only at $T = 0$.

The opening and closing of a gap can be observed by photoemission, as demonstrated in Fig. 2.15 [93]. The figure shows the remarkable sequence from the ordered insulator (AFI) at low temperature with a well-resolved gap, to a paramagnetic metal with no gap at the Fermi energy, to the high-temperature paramagnetic insulator (PI) with a gap, albeit smaller and less well-defined than in the ordered phase. The spectra for the paramagnetic metal show clearly two features: a quasi-particle peak at or near the Fermi energy and a satellite at ≈ 1.5 eV. A study [94] with photons from 20 to 6000 eV supports the interpretation that the two features in the spectra are indeed two parts of the same bulk V $3d$ spectrum, i.e., a band-like feature and a satellite.

Other experiments find effects that bear upon the phenomena at the atomic scale. X-ray absorption measurements [95] and the local moment inferred from high-temperature susceptibility [96] have shown that there is a transfer of occupation between crystal field states at the transition and identify spin-1 local moments that can only be accounted for by correlation (see Sec. 2.3). As discussed further in Sec. 20.6, experiments and quantitative theory are unraveling details that clarify the mechanisms and suggest intriguing possibilities.

Doping a Mott insulator

Some of the most interesting phenomena in physics occur for materials where the parent compound is a Mott insulator (usually an antiferromagnetic insulator) and the electron concentration is changed by some form of doping. In an ordinary insulator or semiconductor, removing an electron causes the Fermi level to shift into the valence band, with only minor changes in the bands. However, if the gap itself is due to interactions, then

---

[13] This is analogous to the water–steam transition, where there is a large change in the mean free path of molecules at ordinary pressure caused by the change in density, but there is no change of symmetry. One cannot distinguish the two phases above the critical point.

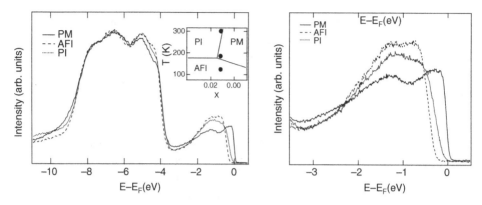

Figure 2.15. Photoemission spectra [93] for $(V_{1-x}Cr_x)_2O_3$ with $x = 0.012$ chosen so that all three phases are accessed by varying temperature as shown in the inset. The spectra were taken with 500 eV photons to be bulk sensitive, as illustrated in Fig. 2.17. The left side shows the bands filled with 18 electrons, primarily oxygen $2p$, well separated from the $3d$ bands occupied by two electrons per V atom. The spectra on the right, the $3d$ bands plotted on an expanded scale, show the sequence of insulator–metal–insulator phases as the temperature is increased. (Figure provided by S.-K. Mo. Modified from [93].)

removing an electron also removes an interaction which shifts a state (which can accommodate two electrons) from the conduction to the valence band. An example is NiO doped with Li (replacing Ni^{2+} with Li^{1+} to create holes), where the experimental results shown in Fig. 20.9 indicate that weight is transferred from the conduction to the valence band and effectively the gap collapses.

One of the major paradigms that has emerged is "Mott–Hubbard" vs. "charge transfer" in the transition metal oxides. As depicted schematically in Fig. 19.6, the former denotes materials where the oxygen p states are strongly bound and the phenomena are governed by only the d bands that form narrow bands. In the latter, the states at the Fermi level have a large admixture of oxygen character. A consequence is that doping with holes leads to states at the Fermi energy that are mobile due to the large band width caused by the coupling of the d and oxygen states. The foremost example is materials with CuO_2 planes that are antiferromagnetic insulators in the stoichiometric form and have a large admixture of Cu d and oxygen p states. With hole doping these materials exhibit high-temperature superconductivity, certainly among the most fascinating phenomena discovered in recent decades. The mechanisms are not known and cannot be covered here, but an example of the striking effects in the electronic spectra is illustrated in Sec. 2.9.

2.9 Lower dimensions: stronger interaction effects

It is intuitive to think that stronger localization of electrons leads to stronger interaction effects. In addition, in lower dimensions there can be no order such as antiferromagnetism at non-zero temperature, opening possibilities for other phenomena to be manifest.

Thus, materials with chain- or plane-like structures are fertile places for the formation of unusual states. The prime example is high-temperature superconductivity in materials with square planes formed from CuO_2 and intermediate layers that control the density of carriers. The parent compounds are antiferromagnetic insulators with strong coupling between neighbors within a plane but low transition temperatures due to weak coupling between planes. With doping the order disappears and there emerges a superconducting state with nodes in the gap and transition temperatures T_c far greater than any known example of phonon-mediated superconductivity. This is much too large a field that is much too unsettled to be treated in any detail. Nevertheless, the theoretical models for these materials are based on characteristics of the electronic structure that are the topics of this book.

Electronic spectra in planar systems

Two-dimensional systems are ideal for ARPES, since the momentum **k** in two dimensions is determined directly by the measured energies and momenta of the emitted electrons. Photoemission is sensitive to surface effects and there are fewer problems in interpretation of the data if the crystals cleave cleanly, exposing the 2D planes. The high-temperature superconductors formed by materials with CuO_2 planes have been the subject of many studies (see, e.g., [59, 97]). For large doping (far from half-filling) the materials exhibit a well-defined Fermi surface and Fermi liquid behavior for temperatures $T > T_c$. As the doping is decreased, the transition temperature T_c increases and the electronic states do not act like a normal Fermi liquid for $T > T_c$. At yet lower doping (closer to half-filling, called "underdoped"), T_c decreases and remarkable effects are observed in the electronic properties near the Fermi energy. The first indications were from magnetic resonance and resistivity measurements, and later the electronic spectra were observed directly by ARPES. Figure 2.16 shows an example of results[14] for an underdoped sample of $Bi_2Sr_2CaCu_2O_{8+x}$ for temperature above the superconducting transition. There is a well-defined Fermi surface (the peak at the Fermi energy μ) around the diagonal direction in the BZ, as expected for a metal. However, the Fermi surface signature vanishes in the region near the zone boundary at $(0, \pi)$ where the spectra have a pronounced dip at E_F, which has been termed a "pseudogap." The Fermi surface appears to terminate as shown on the right side of the figure: an anomalous, weak intensity for part of the Fermi surface, or a "Fermi arc," instead of a connected surface that occurs in a normal metal (see, e.g., [99]). The effect is certainly due to electron interactions, but the mechanism is not agreed upon and it is paramount to understand the extent to which such striking behavior can be captured by a Hubbard-type model, as discussed in Sec. 17.6.

---

[14] The ARPES data is only for emission of electrons with intensity that decreases like the Fermi function, $f(x) = 1/(1 + \exp x)$, with $x = (\varepsilon - \mu)/k_B T$, for energies near and above the Fermi energy. The curves shown in Fig. 2.16 have been symmetrized about the Fermi energy to remove the Fermi factor (since $f(x) + f(-x) = 1$). In addition, the spectrum for a superconductor is symmetric about the Fermi energy so that the analysis corresponds to that done for the superconducting state. See [98] for further motivations and details.

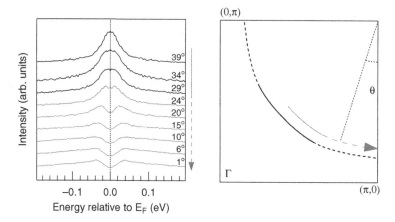

Figure 2.16. Spectral function measured by ARPES for an "underdoped" sample of $Bi_2Sr_2CaCu_2O_{8+x}$ (Bi-2212) at temperature $T = 102$ K, above the superconducting $T_c = 92$ K. The left panel shows the variation in the spectra as a function of the position $\mathbf{k}$ in the Brillouin zone shown on the right. As explained in the text, this is the measured intensity $I(\mathbf{k}, \omega)$ symmetrized about the Fermi energy μ. For angle $\theta > \approx 24°$ there is a peak at E_F, the expected signature of a Fermi surface. But for smaller θ the peak at μ disappears and there is a depression of intensity, called a "pseudogap," that occurs in a region of the BZ around the $(0, \pi)$ point. (Provided by M. Hashimoto and Z.-X. Shen, similar to Fig. 3 in [97].)

Quantitative effects of increased interaction and reduced coordination

Another way to understand the role of dimensionality is to identify trends that can be observed when one studies a given material in different dimensions. The effective Coulomb interaction is stronger in lower dimensions and, at the same time, screening is weaker when the average density of the system is smaller on the scale of typical distances between the objects that are to be screened. In low dimensions there can even be antiscreening, i.e., an increase instead of a decrease in the interaction at certain distances (see Sec. 13.3 and Fig. 13.9). Globally, lower coordination leads to narrower bands and increased effects of interaction.

Surfaces provide many examples of the effects of reduced dimensions, and the effects upon electronic spectra can be studied by photoemission which allows one to look at the same system in the bulk and at the surface. The mean free path of a photoelectron depends on its kinetic energy; therefore, the photon energy determines the effective probing depth of an experiment [102–104]. Of course, one has to be careful before drawing conclusions, since the surface may show additional differences from the bulk, like defects or steps, that can also depend on sample preparation.

Figure 2.17 illustrates the photoemission spectra of S_rVO_3 for the $3d$ states as a function of the photon energy [100]. There are two structures in the energy range near the Fermi energy, both due to V $3d$ states as established by V $2p$–V $3d$ resonant photoemission. The peak nearest the Fermi energy is the quasi-particle and the second peak below is an associated sideband that can only be explained in an interacting picture, as discussed in Secs. 2.5

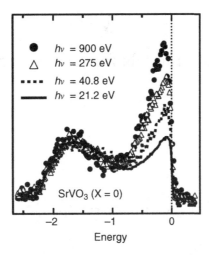

Figure 2.17. Photoemission spectra for the $3d$ states in $SrVO_3$ as a function of the photon energy [100]. The spectra are normalized by the incoherent spectral weight. With increasing photon energy the experiment becomes more bulk sensitive and the quasi-particle/satellite ratio increases, suggesting weaker correlation effects. Similar studies were done for other vanadates [100] and for V_2O_3 [93, 101] to establish that the data in Fig. 2.15 are for the bulk. (From [100].)

and 2.8. The quantity of interest is the ratio between the quasi-particle and sideband intensities, as a measure of the importance of interactions. A strong photon energy dependence can be observed. The data for the highest photon energy is most representative of the bulk. Since the sideband-over-quasi-particle ratio is highest at the lowest photon energy (most surface sensitive), this suggests that, as intuitively expected, interaction effects are stronger at the surface than in the bulk.

2.10 Wrap-up

The examples in this chapter have been chosen to illustrate the essential role of correlation, defined as the effect of interaction beyond that contained in Hartree–Fock, in the properties of materials. In every case there are features that cannot be explained by an independent-particle approach, not even in principle. The purpose of this overview is to provide motivation for theoretical methods that can provide understanding and quantitative results in a unified way. These are but a handful of the phenomena due to correlation. Much more could be included, and much more is yet to be discovered.

3

Concepts and models for interacting electrons

The art of being wise is the art of knowing what to overlook.

William James

Summary

This chapter is devoted to idealized models and theoretical concepts that underlie the topics in the rest of this book. Among the most dramatic effects are the Wigner and Mott transitions, exemplified by electrons in a homogeneous background of positive charge and by the Hubbard model of a crystal. Fermi liquid theory is the paradigm for understanding quasi-particles and collective excitations in metals, building on a continuous link between a non-interacting and an interacting system. The Luttinger theorem and Friedel sum rule are conservation laws for quantities that do not vary at all with the interaction. The Heisenberg and Ising models exemplify the properties of localized electronic states that act as spins. The Anderson impurity model is the paradigm for understanding local moment behavior and is used directly in dynamical mean-field theory.

The previous chapters discuss examples of experimental observations where effects of interactions can be appreciated with only basic knowledge of physics and chemistry. The purpose of this chapter is to give a concise discussion of models that illustrate major characteristics of interacting electrons. These are prototypes that bring out features that occur in real problems, such as the examples in the previous chapter. They are also pedagogical examples for the theoretical methods developed later, with references to specific sections.

3.1 The Wigner transition and the homogeneous electron system

The simplest model of interacting electrons in condensed matter is the homogeneous electron system, also called homogeneous electron gas (HEG),[1] an infinite system of electrons

---

[1] The system is often referred to as the one-component quantum plasma or as jellium, emphasizing that it can be considered as element zero in the periodic table.

with a uniform compensating positive charge background. It was originally introduced as a model for alkali metals. Now the HEG is a standard model system for the development of density functionals and a widely used test system for the many-body perturbation methods in Chs. 10–15. It is also an important model for quantum Monte Carlo calculations, described in Chs. 23–25.

To define the model, we take the hamiltonian in Eq. (1.1) and replace the nuclei by a rigid uniform positive charge with density equal to the electron charge density n. The non-relativistic hamiltonian in atomic units is:

$$\hat{H} = -\frac{1}{2} \sum_i \nabla_i^2 + \frac{1}{2} \sum_{i \neq j} \frac{1}{|\mathbf{r}_i - \mathbf{r}_j|} + E_0, \tag{3.1}$$

where E_0 is the energy resulting from the neutralizing background.[2] The electron density n is usually given by the average mean distance between electrons r_s, defined as $4\pi r_s^3/3 = 1/n$ with n in atomic units.[3] Besides the r_s parameter, for completeness, we have to specify the temperature, and we may want to add a magnetic field to the hamiltonian, or to fix the spin polarization.

The HEG system is an important model not only in 3D, but also in 1D and 2D (e.g., electrons on a 2D surface, as well as the systems that can be formed at semiconductor surfaces and interfaces or on a 1D wire[4]). Chapter 5 of [1] has an extensive discussion of the solutions in the Hartree and Hartree–Fock approximations, as well as results for the energy and correlation functions of the HEG. There is a detailed volume devoted to various topics of the electron gas [106] and extensive expositions such as [107].

Wigner [22] noted that at low density (large r_s) the potential energy dominates the kinetic energy, since the kinetic energy scales with density as r_s^{-2} and the potential energy as r_s^{-1} (see [1] or Ex. 3.1). Hence, as $r_s \to \infty$, the stable phase will minimize the potential energy by forming a crystal structure, the so-called "Wigner crystal." In 3D the most stable crystal structure (lowest Madelung energy) for a pure $(1/r)$ interaction is b.c.c.; in 2D it is a hexagonal lattice. In contrast, at high density (i.e., at small r_s), the kinetic energy will dominate and as $r_s \to 0$ the system becomes a non-interacting Fermi gas. At the density when the crystalline order is lost, there must be a phase transition. Of course, other phases besides the free Fermi gas and the Wigner crystal are possible at intermediate density; conjectured phases include polarized fluids [110], superconducting phases, striped phases [111] and with magnetic fields, the topological quantum Hall and fractional quantum Hall phases [106].

Note that the scaling argument involving density does not depend on the statistics of the particles, the dimensionality of space, or indeed on quantum mechanics: there is also a

---

[2] Note that special care is needed to have a well-defined thermodynamic limit. See footnote 3 in Ch. 1 and App. F in [1].

[3] The model is also applicable to systems of positrons, muons, protons, electrons in a semiconductor, etc. One can scale the mass or interaction between the particles and the distance scale such that the hamiltonian is only multiplied by a constant. One is still left with the same model, but with a different r_s. See Ex. 3.1.

[4] Note that in 1D the $1/r$ interaction must be cut off at short distances so it will be well behaved. Physically this cutoff arises from the finite thickness of a wire [105].

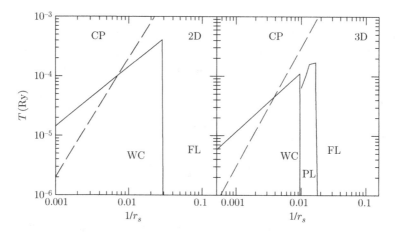

Figure 3.1. Phase diagram of the 2D (left panel) and 3D (right panel) HEG as estimated with quantum Monte Carlo [108, 109], showing the stability of the Wigner crystal (WC), where r_s is the mean inter-electron distance in atomic units, and T is the temperature in Rydbergs (a.u./2). The dashed lines represent the Fermi energy for non-interacting electrons given as a temperature T. They separate the regions of quantum degeneracy (lower right) from classical electrons (upper left). CP shows the classical plasma regime. FL is the degenerate Fermi liquid regime, and PL is a partially spin-polarized Fermi liquid stable in 3D.

Wigner crystal transition for bosons or "Boltzmannons" (particles without exchange statistics), and at temperatures where we can neglect quantum mechanics. In the HEG system, there are no band or lattice effects to drive the transition. This justifies the statement that the formation of the Wigner crystal is the purest example of a "Mott transition," i.e., a transition where the only essential ingredient is the electron–electron interaction. In addition to density, the Wigner crystal state can be stabilized by using very strong magnetic fields [106].

The phase diagram of the HEG in 2D and 3D is shown in Fig. 3.1.[5] The most accurate estimates of the transition density at zero temperature come from diffusion quantum Monte Carlo methods, as discussed in Ch. 24, namely $r_s = 106$ in 3D [109] and $r_s = 35$ in 2D [108], and for $T > 0$ from classical and path integral Monte Carlo calculations, Ch. 25. Only the Wigner crystal in the classical (upper left) portion of the 2D phase diagram has been observed experimentally, because at densities appropriate to the quantum Wigner crystal the effect of disorder dominates.

Figure 5.1 shows a picture of the density of a Wigner crystal in 2D generated by a PIMC calculation (this method is described in Ch. 25). The momentum distribution is shown in Fig. 5.2, the pair correlation functions and structure factor in Fig. 5.4, and tests of the wavefunctions in Fig. 6.6.

---

[5] In this figure the electron–electron interaction is still proportional to $1/r$ in 2D. We define the density parameter in 2D as $\pi r_s^2 = 1/n_2$, where n_2 is the number of electrons per unit area.

In the quantum Wigner crystal phase at very low temperatures, the electron spins can become ordered because of the exchange of electrons on different lattice sites. We discuss the physics of this in Sec. 3.3. Path integral methods (see Ch. 25) have been used to determine the exchange frequencies and, from that, determine the low-temperature magnetic order in the Wigner crystal. A conventional antiferromagnetic ordering is found in 3D [112], since the b.c.c. lattice is bipartite[6] and pair exchange dominates. However, the 2D hexagonal lattice is not bipartite. Also, exchange of three or more spins becomes important [113] because of strong electron correlation in the Wigner crystal. Both effects frustrate Néel ordering. It is quite possible that the Wigner crystal in 2D does not have spin order even at very low temperatures near melting; this is called a spin liquid [114].

3.2 The Mott transition and the Hubbard model

One of the most compelling arguments for the crucial role of interactions was expressed succinctly by Mott in a 1949 paper [115].[7] The essence is brought out by a Gedanken-experiment with hydrogen atoms brought together in a crystalline array and not allowed to form molecules. In the atom, the first electron is bound by 0.5 a.u., whereas a second electron is barely bound, because there is an intra-atomic interaction energy $U \approx 1/2$ a.u. (see also Ex. 3.2). For a sufficiently large lattice constant, the crystal will certainly be an insulator because an energy U is required for an electron to move from one H atom to another that already has an electron. In contrast, if the atoms are so close together that they overlap strongly, then the electrons can move from atom to atom and the crystal will be a metal with a characteristic bandwidth W, much like in the non-interacting case.

The proposal by Mott is that at a critical lattice constant between the two limits there must be a metal–insulator transition, often called a Mott transition. The Mott argument is so simple that it should apply to all types of atoms with partially filled shells, and it should depend in some way on the ratio U/W of interaction and bandwidth. Although there are features in common with the Wigner transition described in the previous section, there are essential differences. In a crystal the nuclei provide the attractive potential that stabilizes the structure and tend to localize the electrons near each nucleus, so that a metal–insulator transition can occur at a much higher density and temperature. There are many possible mechanisms for transitions due to the different crystal structures and multiple bands.

However, there has been controversy and many proposals concerning the issue of whether or not a Mott transition necessarily involves some broken symmetry in the insulating state. In particular, the picture painted above ignores the fact that the electrons have spin. If the spins order, for example, forming an antiferromagnetic structure with a larger unit cell, then the insulating state can be explained by an independent-particle model with a

---

[6] A bipartite lattice is a lattice that can be divided into two sublattices A and B, such that the nearest neighbors of all sites of A belong to B, and vice versa. An example is the square lattice in 2D; the black and white squares of a checkerboard illustrate that it is bipartite.

[7] See, for example, the discussions in *Rev. Mod. Phys.* devoted to conferences on magnetism [116] and metal–insulator transitions [117], Mott's papers [118–120], the book by Herring [121], and an extensive review by Imada *et al.* [91].

potential that has the symmetry of the ordered state. This approach is associated with Slater, following his 1951 paper [122] and other works such as in [117], and is used in practice in spin-ordered unrestricted Hartree–Fock and spin-density functional calculations. However, a Mott insulator without ordering of the spins, often called a spin liquid [114], is more difficult to understand; the existence and possible properties of such a state that does not order even at zero temperature is among the most fundamental problems in condensed matter physics. The issue is the same as for the Wigner crystal in the previous section, but in a solid, there are many more possibilities due to the various structures and states for the electrons.

The terms "Mott insulator" and "Mott transition" are used in many ways in the literature. One definition of a Mott insulator is a general categorization including any crystal that is argued to be an insulator only because of electron–electron interaction. We will use the more narrow definition that includes only cases where order is not essential for the insulating gap, e.g., ones that act as insulating spin liquids at some temperature even if they order at low temperature. The prototypical example is NiO in Secs. 13.4 and 20.7. It is also important to recognize the difference between $T \neq 0$ and $T = 0$. It is only at $T = 0$ that there is a rigorous distinction between metals and insulators. For example, the arguments invoked in the Luttinger theorem on the volume enclosed by the Fermi surface (Sec. 3.6 and App. J) apply at $T = 0$, whereas Fermi surfaces and gaps are smeared out at $T \neq 0$. It is therefore important to always define what is meant by terms such as "metal," "insulator," "Mott transition," etc., and to distinguish between zero and non-zero temperature. We will try to be clear in the use of the terms and describe the actual phenomena that are observed in representative examples.

The Hubbard model

The Hubbard model[8] is one of the simplest models for interacting electrons on a lattice: the Hilbert space for the electrons is restricted to one state per site with on-site energy ε_0 and on-site repulsive interaction[9] $U \geq 0$. The hamiltonian can be written as

$$\hat{H} = \sum_{i,\sigma} \varepsilon_0 \hat{n}_{i\sigma} + \frac{1}{2} U \sum_{i,\sigma} \hat{n}_{i\sigma} \hat{n}_{i-\sigma} - \sum_{i \neq j, \sigma} t_{ij} c_{i\sigma}^\dagger c_{j\sigma}. \tag{3.2}$$

This is customarily called the Hubbard model if the only hopping matrix elements are nearest-neighbor, all with the same value t. The physics is determined by three dimensionless parameters: the chemical potential, the interaction U, and the temperature, all in units of the hopping t. Models with longer-range hopping, interactions, and/or multiple bands can be considered as generalized Hubbard models.

---

[8] This is the common terminology in recognition of the extensive series of papers by J. Hubbard on "Electron correlations in narrow energy bands" in 1963–67 [123–128]. Similar models have been proposed by other authors, including Gutzwiller [129–131], Anderson [132], and Kanamori [133].

[9] In the Hubbard model this effective interaction is often referred to as the Coulomb term, however, in contrast to the HEG and to real materials, it is short-ranged (actually on site), and, thus, it does not lead to certain phenomena observed with a long-range interaction, like plasmons.

The Hubbard model is a testing ground for interacting electrons in condensed matter, including the GW approximation (see in particular Sec. 11.7), DMFT in Chs. 16–18, and quantum Monte Carlo methods (Sec. 25.5). The one-dimensional Hubbard model has been solved exactly by Lieb and Wu [134] using the Bethe ansatz [135]. This is an important benchmark showing that for all U/t, the solution at $T = 0$ is a metal with Fermi surface the same as for $U = 0$ (agreeing with the Luttinger theorem), except for half-filling $n_\uparrow = n_\downarrow = 1/2$, where it is an insulator with a gap for any $U > 0$. Thus there is no Mott transition as a function of U/t. It is not known how to extend the solution to higher dimensions, nevertheless, it provides quantitative tests for approximate methods, e.g., for DMFT in Fig. 17.4.

In a bipartite lattice with only nearest-neighbor hopping, there is electron–hole symmetry (dispersion the same but opposite sign in upper and lower parts of the band) and the model can be solved numerically using QMC methods at half-filling (one electron per site) where there is no sign problem; see Sec. 25.5. For the 3D cubic lattice the ground state at half-filling is antiferromagnetic, as illustrated in Fig. 17.8. In 2D there can be long-range order only at $T = 0$; there are QMC results for the Heisenberg spin model [136] (see Sec. 3.3) which should be the large-U limit. These and cases with partially filled bands are examples of the application of DMFT methods in Chs. 16, 17, and 21.

The two-site Hubbard dimer

The simplest non-trivial Hubbard system is the two-site model, called the "Hubbard dimer." The hamiltonian is given by Eq. (3.2) with sites i and $j = 1, 2$, and we set $\varepsilon_0 = 0$ for simplicity; it is written out explicitly in Ex. 3.5. This is a simplified version of the Heitler–London [137] picture for the hydrogen molecule in Sec. 6.4, and it provides insights for understanding interacting-electron problems. Here we describe the solutions which provide instructive tests for approximate methods: the broken-symmetry Hartree–Fock approximation in Ch. 4, the GW and DMFT methods in Chs. 11 and 16, and the Hubbard approximations in App. K and later in this section.

The exact solutions can be found by following the steps in Ex. 3.5, which involve nothing more than 2×2 determinants. For one or three electrons, the states are simply the bonding and antibonding spin-1/2 states for a single particle or a single hole, respectively, with energies $\pm t$ and $U \pm t$. Two electrons can form singlet and triplet states. The triplet states are non-bonding and independent of the interaction, since it is only between electrons with opposite spins. There are four singlet states with energies that are shown in Fig. 3.2: two with odd parity and energies 0 and U, and two with even parity and energies $\frac{1}{2}(U \pm \sqrt{(16t^2 + U^2)})$. The ground state for two electrons is always a singlet with energy that varies from $-2t$ for $U = 0$ to $-4t^2/U$ in the large-interaction regime where $U/t \gg 1$. From the energy differences and eigenvectors for different particle numbers, one can determine the spectra for adding and subtracting electrons and thus the single-particle Green's function.

Temperature dependence of the electronic properties is often a crucial feature that signifies strong correlation. This is illustrated by the magnetic susceptibility $\chi(T)$, worked out in Ex. 3.6 for two electrons; for large U/t, $\chi(T)$ varies from a constant for $T \ll 4t^2/U$

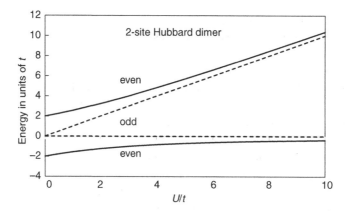

Figure 3.2. Energies for the four singlet states of the Hubbard dimer with two electrons that can be classified as even or odd about the center of the dimer bond. The analytic expressions for the energies are given in the text and derived in Ex. 3.5.

to a Curie law $\propto 1/T$ for decoupled spin-1/2 electrons for $T \gg 4t^2/U$. The model also illustrates the way that correlation of spin and occupation (see Ex. 3.7) can be used to quantify the crossover from nearly independent-particle behavior to strongly correlated behavior.

Two scenarios for the Mott transition

Two approaches have provided the conceptual framework for much of the work attempting to understand a Mott transition without a broken symmetry. Here we give the main results. Details are given in App. K.

One approach due to Gutzwiller is in terms of the ground-state wavefunction. As described in Sec. K.1, it consists of two separate approximations: the Gutzwiller wavefunction [129, 130] for the ground state (see Eq. (K.2)),

$$|\Psi_G\rangle = \prod_i^N [1 - (1 - g)\hat{n}_{i\uparrow}\hat{n}_{i\downarrow}]|\Psi_{HF}\rangle \equiv \hat{P}_G|\Psi_{HF}\rangle, \tag{3.3}$$

and the Gutzwiller approximation [131] for the energy for this wavefunction. Equation (3.3) has the form of a variational wavefunction, where $(1 - g)\hat{n}_{i\uparrow}\hat{n}_{i\downarrow}$ is a projection operator that reduces the number of doubly occupied sites for $0 < g \le 1$. This is widely used for interacting particles on a lattice and is analogous to the Slater–Jastrow wavefunction for continuum problems described in Sec. 6.6. The ground-state (variational) energy for this type of wavefunction can be found using variational Monte Carlo (Ch. 23). The Gutzwiller approximation stems from the assumption that the occupations on neighboring sites are uncorrelated. This leads to an analytic form for the energy given in Eqs. (K.4) and (K.5), governed by the fraction of doubly occupied sites. For a fractionally filled band this model leads to a metal with reduced kinetic energy that is interpreted as an increased

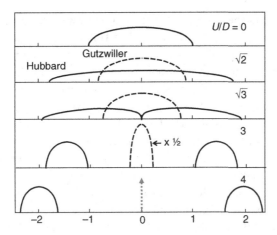

Figure 3.3. Spectral function for a half-filled band with semicircular density of states (Eq. (16.22)) for a non-interacting band with bandwidth $2D$ and different values of U/D. The scale is defined with $\mu = 0$. The "Hubbard III" alloy approximation (see Sec. K.2) is shown by the solid lines that broaden as U/D increases until there is a metal–insulator transition with the opening of a gap for $U/D > U_c^H/D = \sqrt{3}$. The dashed lines depict the Gutzwiller approximation (see Sec. K.1) in which the density of states narrows as U/D increases until the width vanishes at $U_c^G/D = 3.395$, above which the solution does not change. (The height for $U/D = 3$ is scaled by a factor of 1/2.) As discussed in the text, this can be interpreted as a metal–insulator transition with mass diverging as $U \to U_c^G$ from below or as the occupation of the band $n_\uparrow = n_\downarrow \to 1/2$ for $U > U_c^G$. Note the remarkable similarity to Fig. 16.4, where we see that DMFT brings together these two pictures with the three-peak structure that consists of two Hubbard-like sidebands and the narrow central feature like the "Kondo" peak in Fig. 3.4.

mass (see dashed curves in Fig. 3.3). For a half-filled band there is a critical value of the repulsion U_c^G where the mass diverges; for $U > U_c^G$ the solution of the equations has zero kinetic energy and no double occupancy. There is, however, a solution for $U > U_c^G$ for any other occupation with the property that the effective mass diverges as $n_\uparrow = n_\downarrow \to 1/2$, which is depicted by the dashed vertical arrow in the lower panel of Fig. 3.3. This was interpreted by Brinkman and Rice [138] as a way to understand the metal–insulator Mott transition. For the semicircular non-interacting density of states (defined in Sec. 16.9 and used in several examples), the critical value of the interaction $U_c^G = 32D/3\pi \approx 3.395D$ where D is the half-width of the band.

A different approach to the interacting-electron problem, devised by Hubbard in a series of four seminal papers [123–126], is formulated in terms of the spectrum of excitations, which is naturally a Green's function method. This corresponds to the proposal by Mott that the metal–insulator transition is governed by the energies for addition and removal of electrons from the atoms compared with the hopping matrix elements between atoms. As described in more detail in Secs. 16.4 and K.2, the guiding principle in Hubbard's methods is to first treat the interacting-electron problem for the system of decoupled atoms including interactions and then add the coupling between the atoms as a perturbation. For the Hubbard model in the decoupled limit ($t = 0$ in Eq. (3.2)), the spectrum consists of peaks at ε_0 and $\varepsilon_0 + U$ that have come to be known as the lower and upper "Hubbard bands." In order to include the hopping t, one proposal of Hubbard is to consider the interacting system to be analogous to an alloy, with each electron propagating in a disordered array of sites with energy increased by U if a site is occupied by an opposite-spin electron. The analysis following Hubbard's original paper is summarized in Sec. K.2, where the equations are the same as the coherent potential approximation (CPA) widely used for actual alloys [139–142], except that Hubbard added a "resonance-broadening correction" to attempt to take into account the fact that the electrons are not fixed like an alloy. In this picture the band is not narrowed but rather the spectrum of energies broadens as U increases until a gap opens, signalling a metal–insulator transition. For the semicircular density of states in Eq. (16.21), Hubbard found the result $U_c^H = \sqrt{3}D$, derived in Sec. K.3.

The spectra for these two different approaches are shown in Fig. 3.3: narrowing of the band until it disappears (Gutzwiller, dashed curves) and widening of the band until a gap opens (Hubbard, continuous curves). The two types of behavior can be compared with DMFT, which can be considered to be an exact solution in this case since the density of states corresponds to a Bethe lattice in the limit of an infinite number of neighbors, as discussed in Sec. 16.9. As shown in Figs. 16.4 and 16.5 and the related text, both types of behavior are found, and there is a first-order phase transition with a coexistence region between two critical values of U_c, remarkably close to the values found by Gutzwiller and Hubbard. Thus these two scenarios provide useful semi-quantitative results as well as alternative descriptions that illuminate what is happening at a Mott metal–insulator transition.

3.3 Magnetism and spin models

Spin models have a long history in the theory of magnetic behavior and phase transitions in condensed matter (see, e.g., [143, 144]). Since spins really represent electrons in the localized limit, this is relevant for the topics of this book, in particular, the local moment behavior in many examples of d and f states in Chs. 2 and 20. In the regime of large interaction between electrons in these localized states, any change in electron number is an excitation with high energy, but a reorientation of the moment requires little energy (zero energy for degenerate states in isolated atoms). Throwing away all high-energy excitations, the only degrees of freedom are the orientations of the spins that can be described by a spin model. Of course, there is no conductivity since the spins are strictly localized.

The resulting spin systems can be represented by Heisenberg models with continuous rotations or Ising models with only discrete spin directions along axes fixed by the crystal structure. Such models can be defined by a hamiltonian of the form

$$\hat{H} = -\sum_{i<j} J_{ij} \hat{\mathbf{S}}_i \cdot \hat{\mathbf{S}}_j, \tag{3.4}$$

where the J's are exchange constants that couple spins $\hat{\mathbf{S}}$ on sites i and j. There is an extensive literature for various lattices and forms of J_{ij}. For example, a spin-1/2 model can be derived by considering the low-energy excitations of the Hubbard model in the limit of large U compared to the bandwidth. The exchange is due to virtual hopping of an electron to a neighboring site occupied by an electron with opposite spin, so that $J \propto t^2/U$ to leading order in the hopping matrix element t. In Ex. 3.5 the exact expression is derived explicitly for the two-site model. The exchange constants for the Wigner crystal and ^{3}He have been calculated using path-integral Monte Carlo as discussed in Sec. 25.2.

One example illustrates a benchmark relevant for our purposes. The numerically exact solution for spin 1/2 on a square lattice can be found using Monte Carlo methods [136]: at $T = 0$ it is an antiferromagnet (like the Hubbard model) with average moment per site 0.31 ± 0.02. This is lower than the classical value 0.5. Indeed, the classical state with alternating spin-up and spin-down states minimizes the $S_i^z S_j^z$ contribution in Eq. (3.4), but it is not an eigenstate of the hamiltonian because of the missing contributions in the x and y directions. The resulting superposition of states is interpreted as "quantum fluctuations." This serves as a reminder of effects not included in a mean-field calculation. In a static mean-field approximation, such as Hartree–Fock, the symmetry may have the correct antiferromagnetic order but the spin on each site has $\sigma_z = \pm 1/2$ along a chosen axis z; there are not any quantum fluctuations.

3.4 Normal metals and Fermi liquid theory

The term "normal state" has come to denote states of matter that evolve continuously from a system of non-interacting particles as the electron–electron interaction is turned on. This is embodied in Fermi liquid theory (FLT), a wonderfully successful theory of normal metals developed in 1956–57 by Landau [14–16] and others.[10] The first application was to ^{3}He, in which interactions are of the same order of magnitude as the independent-particle kinetic energy, and yet the liquid behaves in some aspects like a gas of weakly interacting particles with a renormalized mass. With this formulation, relations between different quantities can be explained and many properties of actual systems can be described with a few parameters. In this book we do not make specific use of the parameters, but we do lean heavily on the idea of *continuity* from a non-interacting to an interacting system. Of particular importance is the concept of "*quasi-particles*": these are weakly interacting and long-lived particles in one-to-one correspondence with the states of the non-interacting system. See Sec. 7.5 for a more detailed view on how quasi-particles emerge from the many-body system. It is important to define properties such as the effective mass, used in many places to characterize the states near the Fermi surface (e.g., in the "heavy fermion" materials; see Sec. 2.7 and Ch. 20). The quasi-particles conserve charge, spin, and momentum, but their effective magnetic moment and mass are renormalized with respect to the values in the non-interacting system.

FLT is a theory of excitations in a metal with energy ε near the Fermi level and momentum $\mathbf{k}$ near the Fermi surface, where $\varepsilon_{\mathbf{k}} \approx \mu$. If we add a particle to the non-interacting system in state $\mathbf{k}$ with energy $\varepsilon_{\mathbf{k}}^0$, and then turn on the interactions, the resulting state can be

---

[10] Reprints of Landau's papers [14, 16] can be found in [145]. Pedagogical development of Fermi liquid theory can be found in many texts, such as [38, 106, 107, 146–149] and a review on He [150].

considered to describe a quasi-particle with energy ε_k added to the ground state of the real system. In the non-interacting system, the occupation of the states is given by the Fermi function

$$f(\varepsilon) = \frac{1}{e^{(\varepsilon - \mu)/k_B T} + 1}. \tag{3.5}$$

For low-energy states near the Fermi energy, a quasi-particle has a long lifetime and a well-defined energy (the arguments are given in Sec. 7.5), so that the occupations are still given by Eq. (3.5) with $\varepsilon = \varepsilon_k$. The quasi-particle velocity at the Fermi energy is given by $v_F = d\varepsilon_k/dk$ at $k = k_F$ and the momentum per unit velocity defines an effective mass $m^* = k_F/v_F$. (For simplicity we give only the expressions for an isotropic system, but the theory can be generalized to a crystal with an anisotropic Fermi surface.) The low-temperature specific heat is determined by the number of states per unit energy. It is the same as for a non-interacting system [51, 151] except with $\varepsilon_k^0 \to \varepsilon_k$,

$$C = \frac{\pi^2 k_B^2 T}{3} N(0) = \frac{\pi^2 k_B^2 T}{3} N^0(0) \frac{m^*}{m_e}, \tag{3.6}$$

where m_e is the electron mass and $N(0)$ and $N^0(0)$ are the densities of states for interacting and non-interacting particles at the Fermi energy (indicated by the argument (0)). This provides the relation of the effective mass in Fermi liquid theory to the experimental specific heat, e.g., as reported in Tab. 2.2. Calculation of m^* for the HEG using quantum Monte Carlo is discussed in Sec. 23.7. The increase in specific heat due to mass renormalization is but one example of a possible interpretation of experimental findings in terms of a few, physically meaningful parameters.

In the absence of interaction between quasi-particles, the energy of the system would simply be the sum of the energies of the isolated quasi-particles. The difference in the actual energy from this sum is therefore the consequence of the remaining interactions. Suppose that the density of quasi-particles is low. One can then expand the energy in terms of a change in the quasi-particle distribution $\delta n_{k,\sigma}$. This yields

$$E - E_0 = \sum_{k,\sigma} \frac{\delta E}{\delta n_{k,\sigma}} \delta n_{k,\sigma} + \frac{1}{2} \sum_{k,\sigma,k',\sigma'} f^{\sigma,\sigma'}(k,k') \delta n_{k,\sigma} \delta n_{k',\sigma'} + \cdots \tag{3.7}$$

The first term sums the energies of isolated quasi-particles. The second term contains the second derivatives[11] $f^{\sigma,\sigma'}(k,k')$ that represent the effective interaction between quasiparticles. The momentum k can be restricted to the Fermi surface. Therefore, in an isotropic, spin-symmetric system such as the unpolarized HEG, $|k| = |k'| = k_F$, and the interactions can be expanded in Legendre polynomials $f^{\sigma,\sigma'}(k,k') = \sum_\ell f_\ell^{\sigma,\sigma'} P_\ell(\cos\theta)$ and separated into components that are symmetric and antisymmetric in the spins $f_\ell^s = \frac{1}{2}(f_\ell^{\uparrow\uparrow} + f_\ell^{\uparrow\downarrow})$ and $f_\ell^a = \frac{1}{2}(f_\ell^{\uparrow\uparrow} - f_\ell^{\uparrow\downarrow})$. From these, the dimensionless parameters $F_\ell^{s(a)} = \frac{m^* k_F}{\pi^2} f_\ell^{s(a)}$ can be defined. These are sufficient to relate mass, compressibility, magnetic susceptibility, and other properties to a few fundamental parameters (see Ex. 3.3). For example, the speed

---

[11] Note that the derivatives are calculated by keeping all other quasi-particle occupation numbers fixed.

of sound in an isotropic medium is $v_s^2 = \frac{k_F^2}{3m_e m^*}[1 + F_0^s]$. For a repulsive interaction F_0^s is positive, and hence the interaction increases the speed of sound. If the effective mass is known from a measurement of the specific heat, then measurement of the speed of sound determines the interaction parameter F_0^s. The magnetic susceptibility is given by $\chi_S = 2\mu_B^2 N(0)/[1 + F_0^a]$ (see Ex. 4.4 for the relation to the Stoner mean-field theory). There are similar relations for other physical quantities.

The long-range Coulomb interaction

The Fermi liquid theory of Landau does not apply directly to electrons because it is based on the assumption that the interactions are short ranged. The extension to charged particles was done by Silin [152].

The long-range Coulomb interaction is a notorious source of problems in many-body physics. For an infinite system representing condensed matter, a direct expansion in the bare interaction diverges as shown by Gell-Mann and Brueckner [34]. However, there it was also shown that divergent terms can be summed up to all orders, leading to a regularization. The physics contained in these sums is the screening of the Coulomb interaction, leading in the electron gas to an effective short-range interaction already within the RPA of Bohm and Pines [30]. It is therefore natural to work with the screened instead of the bare interaction; this idea is, for example, the basis of the GW approximation in Chs. 9–13. It is still not trivial to move from the Landau FLT of neutral objects to a theory for charged species, which can be seen by looking at Eq. (3.7). The variations in that equation are defined by keeping the system frozen, so that screening at large distances does not occur. The problem can be overcome by splitting the potential of the charged particles into a classical long-range potential created by the additional charge and induced short-range contributions. The latter can be treated within the standard FLT formalism. The former potential is linear in the change in particle number because of Gauss' law, so that the corresponding contribution to the energy appears in the second-order term. This adds an explicit long-range component to the effective quasi-particle interaction.

There are many experimental confirmations of FLT finds, and it is firmly linked to rigorous many-body theory [153, 154]. FLT parameters have been calculated to first and second order in the screened Coulomb interaction W, and there are comparisons of different approximations in the GW framework [43].

3.5 The Kondo effect and the Anderson impurity model

The Kondo effect and the Anderson impurity model [155] occupy special places in the understanding of electronic structure.[12] As discussed in Sec. 2.7, the Kondo effect is a

---

[12] The Kondo effect and the Anderson impurity model are described in detail in the book by Hewson [146] and in many texts, e.g., Mahan [149, pp. 57–59, 977 ff], and Doniach and Sondheimer [156, p. 186 ff]. Since the low-energy excitations in the Anderson model correspond to spin-flips of the impurity, the low-energy properties are equivalent to the Kondo model of a spin coupled to the electrons. The exact relation is given by the Schrieffer–Wolfe transformation [157].

minimum in the resistance as a function of temperature in a metal with magnetic impurities. Above a characteristic temperature T_K the resistance decreases as T decreases, like in an ordinary metal, but below T_K it increases to a maximum as $T \to 0$. The increase was a major puzzle for decades and the solution is among the milestones of condensed matter physics. As shown by Kondo [83], the essential difficulty is that an expansion in the interaction of the electrons and spins diverges for $T \to 0$, no matter how small the interaction. The solution requires non-perturbative methods first provided by Wilson's numerical renormalization group [84] and later an exact solution [85] using the Bethe ansatz [135]. Numerically exact solutions for realistic problems with multiple orbitals and the full atomic hamiltonian can be found with continuous-time Monte Carlo methods in Ch. 18, for example by calculations for Co on Cu surfaces and in bulk Cu as described in Sec. 20.1. Understanding these problems is a key to further developments, including DMFT in Chs. 16–21.

The impurity spins are actually electrons in states localized around the impurity, and the Anderson impurity model (AIM) is the simplest model that captures the essential features:[13] a single state with spin degeneracy ($\uparrow$ and $\downarrow$) and interaction $Un_\uparrow n_\downarrow$, coupled to the host metal. The hamiltonian can be written as

$$\hat{H}_{AIM} = \hat{H}_{site} + \hat{H}_{hyb} + \hat{H}_{host}$$
$$= \sum_\sigma \varepsilon_0 c_{0\sigma}^\dagger c_{0\sigma} + U\hat{n}_{0\uparrow}\hat{n}_{0\downarrow} + \sum_{\ell\sigma} t_{0\ell}(c_{\ell\sigma}^\dagger c_{0\sigma} + c_{0\sigma}^\dagger c_{\ell\sigma}) + \sum_{\ell\sigma} \varepsilon_\ell c_{\ell\sigma}^\dagger c_{\ell\sigma}, \quad (3.8)$$

where the impurity state with energy ε_0 is indicated by the subscript 0, the host is represented by a continuum of states labeled ℓ with energies ε_ℓ, and the matrix elements $t_{0\ell}$ denote the hybridization of the impurity and host states.[14] The interaction U is only in the localized states; the delocalized host states are treated as non-interacting.

To simplify the discussion, the matrix elements and the host density of states are taken to be constants, $t_{0\ell} = t$ and $\bar{\rho}$, and we consider the half-filled case where the spectrum is symmetric about the Fermi energy. If there were no interactions ($U = 0$), the resulting spectrum of energies (see Sec. 7.3) would be a lorentzian resonance with width $\Delta = \pi\bar{\rho}t^2$ for both $\uparrow$ and $\downarrow$ spin, as shown at the bottom of Fig. 3.4. If we now include interactions $U \neq 0$, there is competition between the interaction U and the independent-particle electron terms with characteristic energy Δ. Perturbation theory can describe many aspects but one must go to ever higher order as U/Δ increases, and there is always an instability above some value of U/Δ [160].

In order to consider the strong-interaction regime, it is instructive to first consider the solution in the Hartree–Fock approximation. For small U there is no effect except a shift in

---

[13] The generalization to N-fold states was proposed by Coqblin and Schrieffer [158] for degenerate $3d$ or $4f$ states at a high-symmetry site. In the limit of large N, the problem can be solved analytically [159], which provides an illuminating way to understand both the resistance minimum and the spectrum with a peak at the Fermi energy and high-energy satellites as in Fig. 3.4.

[14] In many papers the energy is explicitly referred to the Fermi energy ($\varepsilon_0 \to \varepsilon_0 - \mu$) and often ε_0 is set to 0, so that the on-site term becomes $-\mu c_{0\nu}^\dagger c_{0\nu}$. The original paper of Anderson and much of the subsequent literature uses the notation V instead of t. In this book we reserve V to denote potentials.

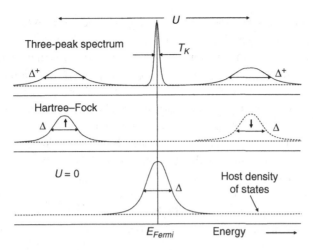

Figure 3.4. Schematic illustration of the spectrum (the sum of ↑ or ↓ spins) for the symmetric spin-1/2 AIM. (Bottom) The non-interacting case with $U = 0$ with a half-filled lorentzian resonant level with width Δ at the Fermi energy. (Middle) The unrestricted Hartree–Fock spectrum for U large enough to give a broken-symmetry magnetic solution with ↑ occupied and ↓ empty. (Top) The full spectrum with the characteristic three-peak structure, a narrow Kondo peak of width T_K at the Fermi energy, and two atomic-like peaks at energies split by $\approx U$. The width of the peaks denoted Δ^+ is greater than Δ, as discussed in the text.

the energy, which has no effect since the energy is chosen so the impurity state is always half-filled. However, as worked out in Ex. 4.8 and shown in Fig. 4.1, above a critical U/Δ there is a broken-symmetry unrestricted (UHF) solution that has lower energy and a magnetic moment $\langle \mathbf{m} \rangle$ with fixed orientation. The spectrum is depicted in the middle panel of Fig. 3.4, where the ↑ spin is occupied with energy below the Fermi energy and ↓ is empty with energy increased by $\approx U$. The UHF solution is an artifact; the true solution has zero average moment in any direction $\langle \mathbf{m} \rangle = 0$, but a non-zero mean square value $\langle |\mathbf{m}|^2 \rangle \neq 0$. Nevertheless, the UHF solution is useful because it indicates the regime where many-body interaction effects are dominant, there is a "local moment," and the spectrum has peaks above and below the Fermi energy.

In the local moment (large U) regime, the actual spectrum for low temperature has the characteristic "three-peak" form sketched in the top panel of Fig. 3.4. One way to understand the spectrum is to first consider the decoupled impurity states. The high-energy peaks appear naturally as the energies to remove or add electrons separated by U. The energies are similar to those in the UHF approximation; however, there is no broken symmetry and ↑ and ↓ spins are equivalent. Coupling to the host broadens the peaks more than in the UHF approximation: in the broken symmetry solution ↑ and ↓ spin states are broadened independently by the hybridization $t_{0\ell}$, but in the singlet state an electron with spin σ can also scatter to $-\sigma$ state with a spin-flip of another electron. The spin-flip scattering leads to the narrow peak at the Fermi energy with width $\sim T_K$, often referred to as the Abrikosov–Suhl resonance. This is a manifestation of the Kondo effect; it occurs in the interacting many-body system only for temperature $T < \approx T_K$ because it is a collective phenomenon of the impurity and host states absent in a non-interacting system.

From the exact or numerical solutions and the analytic solution in the large degeneracy limit [159], the characteristic energy scale is $T_K \propto \exp(-\pi U/2\Delta)$ in the large U local moment regime. The argument of the exponent $\propto -1/\Delta$ shows that an expansion in Δ is not possible. This is the reason a non-perturbative method is required, and it explains the variation in T_K over orders of magnitude as shown in Tab. 2.2.

The non-perturbative solution is the forerunner of DMFT in Chs. 16–21, and a three-peak structure in the spectrum like that in Fig. 3.4 can be considered as a signature of strong interactions and the development of a low-energy scale similar to the Kondo effect. These effects can carry over to lattices with interactions on every site, where there are no exact solutions, and many challenging problems. The remarkable consequences include metal–insulator transitions, heavy fermions, and other phenomena that are a principal topic of Chs. 16–21.

3.6 The Luttinger theorem and the Friedel sum rule

In this book we will attempt to clarify basic guiding principles at critical points where the questions arise. Is a given theoretical approach firmly rooted in fundamental principles? Does a result of a computation agree with those principles even though it involves approximations? This section is devoted to one such principle. It is an example of the continuity between a non-interacting and an interacting system that is emphasized in Secs. 1.3 and 3.4, and is used explicitly or implicitly in much of what is presented in this book. In the present case, we examine quantities that remain completely constant when the interaction is turned on.

The Luttinger theorem can be stated as a sum rule requiring that the volume in **k**-space enclosed by the Fermi surface is conserved independent of interactions. The volume depends only on the number of electrons per cell, and is the same for a non-interacting system and an interacting system. In order to avoid misunderstandings, it is critical to realize that the very statement of the theorem assumes zero temperature. At $T = 0$ there is a sharp Fermi surface where the states have infinite lifetime. In the non-interacting system we define the Fermi surface as the boundary between occupied and empty states. In the interacting system the occupation is spread in **k**-space outside the Fermi surface, but there is still a step in the momentum distribution that defines the Fermi surface, as shown in Figs. 5.2, 7.3, and 11.7. However, at $T > 0$, although $\mu(T)$ is still well defined, the Fermi surface is blurred and its volume is no longer precisely defined; it is a quantitative question whether or not a surface is sufficiently well defined to confidently measure its volume. One has to be careful in interpreting results for metal–insulator transitions and other phenomena when performing experiments or calculations at $T > 0$, since gaps may be smeared out.

When applying the theorem, all electrons are counted: even the spins in a Heisenberg or Ising model are really electrons that must be included. Each filled band contributes one electron per cell (for each spin) and a volume equal to the Brillouin zone volume. Missing electrons in a band (holes) can be counted as negative volume. So long as the theorem is obeyed, a crystal with a fractional number of electrons of either spin per cell must be a metal with a Fermi surface. An integer number of electrons of each spin can be an insulator or a semi-metal with positive and negative volumes, with the total volume an integer multiple of the volume of the Brillouin zone.

The derivation by Luttinger and Ward (LW) [36, 37] is based on a diagrammatic expansion. The steps in the arguments are summarized in App. J. Rather than trying to prove that a particular way to sum diagrams converges, a different approach is to argue that it should apply to any state that evolves continuously from a system of non-interacting particles as the electron–electron interaction is turned on.[15] This is the condition for the "normal state" of matter in Sec. 3.4. In any case, the arguments may break down if there is a non-analytic point as a function of the interaction strength, for example, if there is a phase transition. Nevertheless, modifications to the Luttinger theorem can be made if each phase can be continuously connected to some non-interacting state. An example is a transition from a metal with fractionally filled bands (continuously connected to a non-interacting metal) to an antiferromagnetic insulator with a larger unit cell and integer number of electrons per cell (connected to non-interacting electrons in a spin-dependent mean-field potential with the same antiferromagnetic order).

There is great interest in the possibility of a Mott insulator or spin liquid with no broken symmetry (see Secs. 3.1 and 3.2), which would violate the original statement of the theorem since there is no Fermi surface. As discussed in App. J, there is a proposal to extend the theorem to include a surface of zeros of the Green's function within the gap [161].

Friedel sum rule and conductivity

The effect of an impurity in a metal is to modify the electronic states locally; at large distance the states are unchanged except for a phase shift η. The consequences can be expressed most simply for high-symmetry crystals where the states can be classified by angular momentum l. As worked out in [1], App. J, and texts such as [149], the scattering cross-section for states at the Fermi energy is $\propto \sum_\ell (2\ell + 1) \sin^2(\eta_\ell)$. The maximum possible scattering for each ℓ is for $\eta_\ell = \pi/2$, which is called the "unitary limit" [162].

The Friedel sum rule states that the sum of phase shifts of scattering waves at the Fermi energy around an impurity in a metal equals π times the number of electrons bound by the impurity, which applies for each spin separately. Since charges are screened in the metal, the number of excess (or missing) electrons $\Delta n = \Delta n_\uparrow + \Delta n_\downarrow$ around the impurity is just the number required to neutralize the nuclear charge, i.e., $\Delta n = \Delta Z$, where ΔZ is the difference between the number of protons in the impurity and host atoms. Thus, the nuclear charge difference ΔZ alone is sufficient to place constraints on the phase shifts independent of interactions and the details of the specific problem. The sum rule was originally derived by Friedel in 1952 [163] for electrons that interact only via the

---

[15] See Sec. 8.2 for more in-depth discussion of issues that may arise as the interaction is turned on.

average Hartree Coulomb potential. However, it can be extended [164–166] to interacting electrons based on the same arguments as for the Luttinger theorem, except that here the conserved quantity is the sum of phase shifts instead of the volume enclosed by the Fermi surface. This holds only at $T = 0$, where the phase shifts are well defined for the same reasons as the requirement that the Fermi surface be well defined for the Luttinger theorem.

For a single band with s symmetry one can draw remarkable conclusions. There is only one phase shift which is uniquely determined: an impurity with $\Delta Z = Z_{imp} - Z_{host} = 1$ causes a $\pi/2$ phase shift for each spin, i.e., the maximum possible scattering. This happens whether it is Mg in Na or the spin-$1/2$ Anderson impurity model with a small Kondo temperature T_K. In the former case, this leads to scattering that is only weakly temperature-dependent. But if T_K is small, the scattering is weak for $T > T_K$, and increases to a maximum at $T = 0$. Thus, from the sum rule alone one can understand qualitatively the signature of the Kondo effect that the resistance increases at low temperature. Furthermore, the phase shifts at the Fermi energy can occur only if there is a resonance in the spectrum, so that we also understand that there must be a peak near the Fermi energy, as illustrated in the top panel of Fig. 3.4.

SELECT FURTHER READING

Baym, G. and Pethick, C. *Landau Fermi-Liquid Theory* (John Wiley, Chichester, 2004). A good pedagogical exposition.

Fradkin, E. *Field Theories of Condensed Matter Physics* (Cambridge University Press, Cambridge, 2013). Chapter 2 starts with "All theories of strongly correlated systems begin with the Hubbard model . . . "

Giuliani, G. and Vignale, G. *Quantum Theory of the Electron Liquid* (Cambridge University Press, Cambridge, 2005). Extensive analysis of the homogeneous electron system.

Hewson, A. C. *The Kondo Problem to Heavy Fermions* (Cambridge University Press, Cambridge, 1993). Paperback edition with corrections, 1997.

Imada, M., Fujimori, A., and Tokura, Y., "Metal–insulator transitions," *Rev. Mod. Phys.* **70**, 1039–1263, 1998. A comprehensive review of metal–insulator transitions.

Mott, N. F. *Metal–Insulator Transitions* (Taylor and Francis, London, 1990).

Pines, D. *The Many Body Problem* (Advanced Book Classics, originally published in 1961) (Addison-Wesley, Reading, MA, 1997). Classic book with presentation of ideas, physical reasoning, and reprints of many important papers.

Pines, D. and Nozières, P. *The Theory of Quantum Liquids, Vol. 1* (W. A. Benjamin, New York, 1966). Vol. 1 and 2 reprinted in the Advanced Book Classics series (Westview Press, Boulder, CO, 1999). A pedagogical exposition of Fermi liquid theory and charged quantum liquids.

Exercises

3.1 Consider a bulk system composed of a single species of particles with charge Q, mass M, dielectric constant ϵ, and density n with a rigid homogeneous background charge so overall the system is neutral. What units of length and energy can be used to reduce the hamiltonian to that of Eq. (3.1)? Find the effective r_s parameter in terms of Q, M, ϵ, and density. Show that the kinetic energy dominates as $r_s \to 0$ and the potential energy dominates as $r_s \to \infty$.

3.2 Show that for the hydrogen atom with the wavefunction $\psi(r) = (\pi a_0^3)^{-1/2} e^{-r/a_0}$, the inter-atomic "Hubbard U," defined as the intra-orbital Coulomb interaction energy, is given by $U = \frac{5}{8} e^2 / a_0$.

3.3 Derive the expressions in Sec. 3.4 for the sound velocity (also relate this to the compressibility) and magnetic susceptibility in terms of the effective mass and Fermi liquid parameters.

3.4 This exercise is to show the divergence of individual terms in expansion of correlation energy in powers of e^2 for the HEG. Consider the direct term in the expansion of the correlation energy to the lowest order beyond Hartree–Fock in the HEG. The Hartree–Fock wavefunction is the filled Fermi sea and the energy is first order in the interaction. In the next order one has sums over all excited states generated by the interaction (see Eq. (4.10)):

$$H_{\text{int}} = \frac{1}{2} \sum_{k,\sigma,k',\sigma',q} v_c(q) \left(c_{k+q,\sigma}^{\dagger} c_{k'-q,\sigma'}^{\dagger} c_{k',\sigma'} c_{k,\sigma} \right), \tag{3.9}$$

where $v_c(q)$ is the Coulomb interaction in Fourier space. Give the standard perturbation expression for the energy to second order in $v_c(q) \approx e^2$, and identify the direct and the exchange terms. Consider the direct term and show that it diverges. This is sufficient to show that the perturbation series formally does not converge; the series must be resummed to get a finite expression.

3.5 Derive the solutions for the two-site Hubbard dimer that are discussed in Sec. 3.2. The results are used in Exs. 3.6 and 3.7 and are the tests for other methods and approximations in Exs. 4.7, 11.4, and 16.4.

The hamiltonian is a finite-size version of Eq. (3.2), with a hopping term t between sites and an interaction term on each site (for convenience we set the on-site energy $\varepsilon_0 = 0$):

$$H = -t \sum_{\sigma} (c_{1\sigma}^{\dagger} c_{2\sigma} + c_{2\sigma}^{\dagger} c_{1\sigma}) + U(n_{1\uparrow} n_{1\downarrow} + n_{2\uparrow} n_{2\downarrow}). \tag{3.10}$$

(a) Derive the exact energies and eigenstates for all possible cases with 0, 1, 2, 3, and 4 electrons that are given in Sec. 3.2. The solutions for 0 and 1 electrons are straightforward. Show that the solutions for 4 and 3 electrons are equivalent to 0 and 1 but shifted in energy (an example of particle–hole duality).

(b) For two electrons, find the energies for all singlet and triplet states. First give reasons that for two electrons the lowest state is always a singlet, which can be done without quantitative calculations. Show that the energies for the odd-parity states can be found by inspection and for the even-parity states by diagonalizing a 2×2 matrix, and derive the expressions given in Sec. 3.2.

(c) From the results of (a) and (b) find the energies for electron removal and addition for a molecule with one electron in the ground state. This is used in Ex. 11.4 and selected results are given in Fig. 11.10.

(d) Show that for two electrons and small U ($U \ll t$) one has the usual bonding and antibonding states, as is appropriate for a hydrogen molecule at the equilibrium distance.

(e) For two electrons and large U ($U \gg t$) show that the low-energy states have energy 0 for the three triplet states and $-4t^2/U$ for the singlet ground state, and that the energies are the same as for a two-site spin-1/2 Heisenberg model with antiferromagnetic coupling of magnitude $|J| = 4t^2/U$.

3.6 Using the results of Ex. 3.5, calculate the magnetic susceptibility $\chi(T)$ for the Hubbard dimer with two electrons as a function of temperature. As pointed out in the text, this is greatly simplified by the fact that one has only to deal with the triplet that is occupied thermally.

   (a) Before doing the calculations, explain how to set up the problem so that it applies for any values of t and U and for a two-site Heisenberg model for any J.

   (b) Derive an expression $\chi(T)$ valid for all t, U, and T.

   (c) Show that for large T, compared with the singlet–triplet splitting, the susceptibility approaches that for two decoupled spin-1/2 particles.

   (d) For strong interactions ($U/t \gg 1$) show that the excitation energies are the same as for a two-site spin-1/2 Heisenberg model with antiferromagnetic coupling of magnitude $|J| = 4t^2/U$.

   (e) Discuss the senses in which this illustrates that strong interactions lead to low-energy scales.

3.7 Use the exact solution in Ex. 3.5 to calculate the density and spin correlation functions: $\langle n^2 \rangle$ for a site, and $\langle n_1 n_2 \rangle$ and $\langle \sigma_1 \sigma_2 \rangle$ for intersite correlation, for the case with two electrons in a singlet state. Note that this is a model for the Heitler–London picture in Sec. 6.4.

   (a) Give the analytic expressions and discuss the small U/t and large U/t regimes.

   (b) Describe how such a calculation can be used to quantify the nature of the bonding that varies from weak to strong interactions.

PART II

FOUNDATIONS OF THEORY FOR
MANY-BODY SYSTEMS

4

Mean fields and auxiliary systems

It is a characteristic of wisdom not to do desperate things.

Henry David Thoreau

Summary

The essence of a mean-field method is to replace an interacting many-body problem
with a set of independent-particle problems having an effective potential. It can be
chosen either as an approximation for effects of interactions in an average sense or as
an auxiliary system that can reproduce selected properties of an interacting system.
The effective potential can have an explicit dependence on an order parameter for
a system with a broken symmetry such as a ferro- or antiferromagnet. Mean-field
techniques are vital parts of many-body theory: the starting points for practical
many-body calculations and often the basis for interpreting the results. This chapter provides a summary of the Hartree–Fock approximation, the Weiss mean field,
and density functional theory that have significant roles in the methods described in
this book.

Mean-field methods denote approaches in which the interacting many-body system is
treated as a set of non-interacting particles in a self-consistent field that takes into account
some effects of interactions in some way. In the literature such methods are often called
"one-electron"; however, in this book we use "non-interacting" or "independent-particle"
to refer to mean-field concepts and approaches. We reserve the terms "one-electron"
or "one-body" to denote quantities that involve quantum mechanical operators acting
independently on each body in a many-body system.[1] Mean-field approaches are relevant
for the study of interacting, correlated electrons because they lead to approximate

---

[1] For example, the term "one-body Green's function" G, defined in Eqs. (5.76)–(5.78) and used throughout the
book, denotes the propagator for adding and removing an electron in the full interacting many-body system.
This is different from the independent-particle Green's function for a non-interacting system, specified by a
subscript or superscript, G_0 or G^0.

formulations that can be solved more easily than more sophisticated approaches; when judiciously chosen, mean-field solutions can yield useful, physically meaningful results, and they can provide the basis and conceptual structure for investigating the effects of correlation. The particles that are the "bodies" in a many-body theory can be the original particles with their bare masses and interactions, or, most often, they may be the solutions of a set of mean-field equations chosen to facilitate the solution of the many-body problem. A large part of many-body theory in condensed matter involves the choice of the most appropriate independent particles. Hence, it is essential to define clearly the particles that are created and annihilated by the operators $c_i^\dagger$ and c_j in which the many-body theory is formulated.

It is important to recognize the various roles that mean fields play in correlated many-body systems. A mean field can incorporate effects that must be treated accurately if there is to be any hope of quantitative predictions. The average potential acting on one particle, obtained by replacing operators with their expectation values (see, e.g., Eq. (4.9)), is a mean field that often is very large and must be treated by a self-consistent method. For example, the static average Coulomb potential (the Hartree potential) is determined by the electron density $n(\mathbf{r}) = \langle \hat{n}(\mathbf{r}) \rangle = \langle \sum_i \delta(\mathbf{r} - \mathbf{r}_i) \rangle$, which can be included in a self-consistent mean-field calculation. A broken-symmetry phase can be described by an order parameter, for example, the spin density in a ferromagnetic or antiferromagnetic phase. A mean field with broken symmetry can be a trick to capture part of the effects of correlation; even when the results are unphysical they can be used to construct better approximations for the solution with the correct symmetry. Several examples are given in Sec. 4.7.

The primary topics of this chapter are the mean-field methods most relevant for the interacting-electron approaches treated in this book: the Hartree–Fock approximation (HFA), the Weiss mean field, and density functional theory (DFT). As summarized in Sec. 4.1, the HFA restricts the many-body wavefunction to a single determinant with no correlation, but makes no approximation to the full N-body hamiltonian. This is equivalent to replacing pairs of creation and annihilation operators with expectation values. The Weiss mean-field approximation for interacting spins in Sec. 4.2 replaces correlated spin variables with averages. It is the classic example of mean-field theory and it is relevant for this book because spin systems are models for the magnetic behavior in solids that ultimately is a consequence of electron interactions, and because it is an important precedent of dynamical mean-field theory.

In contrast, DFT leads to a different type of mean-field method; it is included here in Sec. 4.3 because the Kohn–Sham auxiliary system consists of non-interacting particles in a potential that is analogous to a mean field. In principle, it can be utilized to determine the density $n(\mathbf{r})$ and the total energy of the interacting system exactly. Practical calculations involve approximations such as the LDA (local density), LSDA (local number and spin densities), and so forth. The problems with the use of the Kohn–Sham equations to interpret electronic band structure, etc., are the topic of Sec. 4.4. Extensions of the Kohn–Sham approach and time-dependent DFT are described briefly in Secs. 4.5 and 4.6.

Currently these methods provide the starting point for most practical many-body calculations for electrons in condensed matter. Moreover, Part III can be considered as building on the HFA, for example, by adding dynamical screening of the bare Coulomb interaction in the Fock exchange term due to excitations of the many-electron system. In Part IV, practical variational and projector Monte Carlo methods depend on nodes of the many-body wavefunction that are determined by trial functions which are often taken from mean-field calculations. In addition to practical uses, DFT provides an intellectual stimulus for ideas that are now used in many areas of physics, for example, DMFT in Ch. 16.

4.1 The Hartree and Hartree–Fock approximations

There are two basic independent-particle approaches: the Hartree–Fock approximation to the interacting many-body system and the non-interacting electron methods in which there is a *local* effective potential. The latter are often referred to as "Hartree-like," after D. R. Hartree [5] who approximated the Coulomb interaction between electrons by an average local mean-field potential. The approaches are similar in that each assumes the electrons are uncorrelated except that they must obey the exclusion principle; however, they are different in spirit and in their interpretation. Hartree-like theories modify the original inter-acting many-body problem and treat a system of non-interacting electrons in an effective potential. This approach was placed on a firm footing by Kohn and Sham who showed that one can define an auxiliary system of non-interacting electrons that, in principle, leads to the exact ground-state density and energy.

The HFA[2] treats directly the system of interacting fermions, with the approximation that the many-body wavefunction is restricted to be an antisymmetrized uncorrelated product function that can be written as a single determinant[3] which explicitly respects the exclusion principle,

$$\Psi_{HF} = \frac{1}{(N!)^{1/2}} \begin{vmatrix} \psi_1(\mathbf{r}_1,\sigma_1) & \psi_1(\mathbf{r}_2,\sigma_2) & \psi_1(\mathbf{r}_3,\sigma_3) & \cdots \\ \psi_2(\mathbf{r}_1,\sigma_1) & \psi_2(\mathbf{r}_2,\sigma_2) & \psi_2(\mathbf{r}_3,\sigma_3) & \cdots \\ \psi_3(\mathbf{r}_1,\sigma_1) & \psi_3(\mathbf{r}_2,\sigma_2) & \psi_3(\mathbf{r}_3,\sigma_3) & \cdots \\ \cdot & \cdot & \cdot & \cdots \\ \cdot & \cdot & \cdot & \cdots \end{vmatrix}, \tag{4.1}$$

where ψ denotes a single-particle "spin orbital." Symmetries in space and spin are given in references such as [169] and [170]. For our purposes it is sufficient to consider orbitals that are eigenfunctions of $\hat{S}_z$ with spin $\sigma = \pm 1/2$ quantized along an axis. Then $\psi_i(\mathbf{r},\sigma)$ is a product of a space orbital $\psi_i^\sigma(\mathbf{r})$ and a spin function. The interaction energy is calculated exactly for this wavefunction so that the resulting total energy

---

[2] The HFA was developed as a practical self-consistent method by the two Hartrees [167], who combined their previous methods [5] with the uncorrelated antisymmetric wavefunction and exchange energy due to Fock [168]. (They also said, "the equations were suggested about the same time by Slater.") The original Hartree equation can also be obtained with a product ansatz for the wavefunction in a way similar to Hartree–Fock, but without taking into account the antisymmetry of the wavefunction.

[3] Multireference Hartree–Fock is a generalization to a sum of a few determinants (or configurations) that are especially important, for example, in bond breaking.

is variational, i.e., it is an upper bound to the true ground-state energy E_0. Although the HFA fails qualitatively for metals (see Sec. 11.3 and Ex. 4.2), and may have large quantitative errors in other cases, it is widely used as the starting point for correlated many-body methods. The total energy of a system of electrons in an external potential v_{ext} is

$$\langle \Psi_{HF} | \hat{H} | \Psi_{HF} \rangle = -\int d\mathbf{r} \sum_{i\sigma_i} \psi_i^{\sigma_i *}(\mathbf{r}) \frac{\nabla^2}{2} \psi_i^{\sigma_i}(\mathbf{r}) + \int d\mathbf{r} v_{ext}(\mathbf{r}) n(\mathbf{r}) + E_H + E_x, \quad (4.2)$$

where the first term on the right-hand side is the kinetic energy for independent particles,

$$E_H = \frac{1}{2} \int d\mathbf{r} d\mathbf{r}' \frac{n(\mathbf{r}) n(\mathbf{r}')}{|\mathbf{r} - \mathbf{r}'|} \quad (4.3)$$

is the Hartree contribution to the total energy, and

$$E_x = -\frac{1}{2} \sum_{\sigma} \sum_{i,j}^{occ} \int d\mathbf{r} d\mathbf{r}' \psi_j^{\sigma *}(\mathbf{r}') \psi_i^{\sigma}(\mathbf{r}') \frac{1}{|\mathbf{r} - \mathbf{r}'|} \psi_j^{\sigma}(\mathbf{r}) \psi_i^{\sigma *}(\mathbf{r}) \quad (4.4)$$

is the corresponding Fock term. Minimization of the total energy with respect to the orthonormal single-body orbitals ψ_i^{σ} leads to the Hartree–Fock equations

$$\left[-\frac{1}{2} \nabla^2 + v_{ext}(\mathbf{r}) + \sum_{j,\sigma_j}^{occ} \int d\mathbf{r}' \psi_j^{\sigma_j *}(\mathbf{r}') \psi_j^{\sigma_j}(\mathbf{r}') \frac{1}{|\mathbf{r} - \mathbf{r}'|} \right] \psi_i^{\sigma}(\mathbf{r})$$

$$- \sum_j^{occ} \int d\mathbf{r}' \psi_j^{\sigma *}(\mathbf{r}') \psi_i^{\sigma}(\mathbf{r}') \frac{1}{|\mathbf{r} - \mathbf{r}'|} \psi_j^{\sigma}(\mathbf{r}) = \varepsilon_i^{\sigma} \psi_i^{\sigma}(\mathbf{r}). \quad (4.5)$$

The potential in the first line,

$$v_H(\mathbf{r}) \equiv \sum_{j,\sigma_j}^{occ} \int d\mathbf{r}' \psi_j^{\sigma_j *}(\mathbf{r}') \psi_j^{\sigma_j}(\mathbf{r}') \frac{1}{|\mathbf{r} - \mathbf{r}'|} = \int d\mathbf{r}' \frac{n(\mathbf{r}')}{|\mathbf{r} - \mathbf{r}'|}, \quad (4.6)$$

has the Hartree form of a local potential that acts on each function equally at each point $\mathbf{r}$. It arises from the charge of all the electrons, including each electron acting on itself. However, this is merely a convenient division since the second line cancels the unphysical self-interaction.[4] It involves the non-local Fock operator

$$\Sigma_{x\sigma}(\mathbf{r}, \mathbf{r}') = -\sum_j^{occ} \psi_j^{\sigma *}(\mathbf{r}') \frac{1}{|\mathbf{r} - \mathbf{r}'|} \psi_j^{\sigma}(\mathbf{r}) = -\frac{\rho(\mathbf{r}, \sigma; \mathbf{r}', \sigma)}{|\mathbf{r} - \mathbf{r}'|}, \quad (4.7)$$

an integral operator acting on ψ_i^{σ}, and it contains only like spins. Equation (4.5) is a coupled set of integro-differential equations that can be solved exactly only in special

---

[4] It is now customary to define the effective "Hartree potential" with an unphysical self-interaction term so that the potential is orbital independent. However, as noted in Ch. 1, Hartree [5] actually defined a different potential for each state to remove the self-term for each electron that occupies the orbital, apparently the first self-interaction correction (SIC), now a widely used concept (see Secs. 4.5, 11.4, and 19.6).

cases such as spherically symmetric atoms and the homogeneous electron gas. In general, one introduces a basis; then the equations can be written in terms of the expansion coefficients of the orbitals and integrals involving the basis functions. The HF ground-state wavefunction is the determinant built from the N lowest-energy single-particle states.

Excited states can be represented by choosing other combinations of single-particle spin orbitals $\psi_i^{\sigma_i}$ to build the determinant. The eigenvalues of the HF equations (Eq. (4.5)) correspond to total energy differences, namely, to the energies to add or subtract electrons that would result from increasing the size of the matrix by adding an empty orbital or decreasing the size by removing an orbital, if all other orbitals are frozen. This is called Koopmans' theorem [46]. For a fixed number of electrons, excited states can be formed by substituting the ground-state orbitals with empty eigenstates of Eq. (4.5) forming single, double, triple, etc. excitations. The difference between the total energy calculated in such a state and in the ground state may be considered as a first approximation to the excitation energy.

The absence of relaxation often leads to large overestimations of the addition, removal, and excitation energies. In contrast, "ΔSCF" (self-consistent field) calculations determine total energy differences by performing two separate *self-consistent* calculations for each couple of N, $N + 1$, or $N - 1$-particle states. Such calculations allow all the orbitals to relax, and thereby include an aspect of correlation. In finite systems the results can be much more accurate than when eigenvalues from one calculation are used. For infinite systems, however, the self-consistent mean field includes only the average relaxation due to a single particle in extended orbitals, which becomes negligible so that ΔSCF energy differences would merely reproduce the eigenvalues (see also the discussion of the static COHSEX approximation to GW in Sec. 11.4).

In second quantization the origin of the HF "mean field" becomes most evident. One starts from the full expression for a two-body interaction operator (Eq. (A.11)),

$$\hat{V}_{ee} = \frac{1}{2} \sum_{mm'n'n,\sigma\sigma'} v_{mm'n'n} c_{m\sigma}^\dagger c_{m'\sigma'}^\dagger c_{n'\sigma'} c_{n\sigma}, \qquad (4.8)$$

where $m, m', n, n' = 1, \ldots, N_{states}$ refers to the set of independent-particle basis functions, and $v_{mm'n'n}$ denotes the 4-center matrix element of the interaction. For example, the latter can be a matrix element of the bare Coulomb interaction, or an effective interaction U as given in Eq. (19.3) for localized orbitals. The HFA corresponds to neglect of correlation. Then the four operators in Eq. (4.8) factor into products of single-body terms,

$$c_{m\sigma}^\dagger c_{m'\sigma'}^\dagger c_{n'\sigma'} c_{n\sigma} \; \rightarrow \; \langle c_{m'\sigma'}^\dagger c_{n'\sigma'} \rangle c_{m\sigma}^\dagger c_{n\sigma} - \delta_{\sigma\sigma'} \langle c_{m\sigma}^\dagger c_{n'\sigma} \rangle c_{m'\sigma}^\dagger c_{n\sigma}, \qquad (4.9)$$

where the first term yields the Hartree potential that involves the density operator and the second term yields the exchange term that couples like spins only. Equation (4.9) displays clearly how the unphysical self-interaction for $n' = n, m' = m, \sigma = \sigma'$ cancels exactly in the HFA.

In a plane-wave basis the Coulomb matrix elements are determined by the Fourier components $v_c(q) = 4\pi/q^2$. Then the interaction can be expressed as (see Ex. 4.1)

$$\hat{V}_{ee} = \frac{1}{2} \sum_q \sum_{k\sigma, k'\sigma'} v_c(q) \, c^\dagger_{(k+q)\sigma} c^\dagger_{(k'-q)\sigma'} c_{k'\sigma'} c_{k\sigma}, \tag{4.10}$$

where q is the momentum transfer in the interaction. In Ex. 4.1 you are asked to indicate the factorized approximation of this expression that corresponds to the HFA and to discuss the reciprocal space expressions for the homogeneous electron gas and for a periodic system. Exercise 4.2 is to derive the dispersion for the electron gas in the Hartree–Fock approximation and reproduce the result found by Bardeen (see Sec. 1.4), which shows the unphysical behavior of the Hartree–Fock approximation.

For a lattice model, the interaction is often most conveniently written in terms of site operators; for example, in the one-band Hubbard model defined in Eq. (3.2) the Hartree–Fock approximation can be written as a sum over sites i,

$$U \sum_i \hat{n}_{i\uparrow}\hat{n}_{i\downarrow} \to U \sum_i \left[\langle \hat{n}_{i\uparrow} \rangle \hat{n}_{i\downarrow} + \langle \hat{n}_{i\downarrow} \rangle \hat{n}_{i\uparrow} \right], \tag{4.11}$$

where U is the on-site interaction and $\hat{n}_{i\sigma} = c^\dagger_{i\sigma} c_{i\sigma}$.

4.2 Weiss mean field and the Curie–Weiss approximation

The classic, textbook example of mean-field theory is the Weiss field[5] introduced by P. Weiss in 1907 [171] to describe a classical system of interacting spins with thermal fluctuations. The mean field is an approximation for the effects of interactions of a spin with its neighbors, which leads to the Curie–Weiss approximation for the susceptibility (see Ex. 4.4) and to phase transitions to an ordered state as the temperature is lowered. It is relevant here because the spins are really electrons in disguise, and the magnetic properties are examples of what we wish to derive from the fundamental theory of interacting electrons. In many ways DMFT can be viewed as the extension of the classical Weiss field to a dynamical quantum field, and the relationships are brought out in Ch. 16, especially in Tab. 16.1.

The Heisenberg model for a system of spins S_i is defined in Sec. 3.3; the hamiltonian is given by $\hat{H} = -\sum_{i<j} J_{ij}\hat{S}_i \cdot \hat{S}_j - \sum_i \mathbf{h}_i^{\text{appl}} \cdot \hat{S}_i$, where the J_{ij} are exchange interactions and $\mathbf{h}_i^{\text{appl}}$ is an external magnetic field applied at site i. The expressions for the partition function and free energies are given in Sec. H.4, along with expressions for the average moments $\mathbf{m}_i = \langle \hat{S}_i \rangle$, where $\langle \ldots \rangle$ denotes a thermal average. The simplest approximation is the Weiss mean field, $\hat{H} \to -\sum_i \mathbf{h}_i^{\text{eff}} \cdot \hat{S}_i$, with

---

[5] Perhaps it would be more correct to call this the Néel field in the present context. Weiss considered the average field to be the sum of infinitesimal contributions from all other spins, which can only lead to a ferromagnetic solution. It was Néel who showed that local interactions could have either sign, and the average of neighbors considered here can lead to antiferromagnetic ordering. (This was pointed out in the obituary for Louis Néel by J. Friedel; *Physics Today*, 54(10), 88, 2001.)

$$\mathbf{h}_i^{\text{eff}} = \mathbf{h}_i^{\text{appl}} + \sum_j J_{ij}\mathbf{m}_j, \tag{4.12}$$

which is the average field due to the neighbors, assuming the fluctuations are uncorrelated. For a lattice with all sites equivalent and n_z nearest neighbors with interaction $J_{ij} = J > 0$, this leads to the Curie–Weiss self-consistent equation for the magnetization $m_i = m$,

$$m = B(\beta[h^{\text{appl}} + n_z J m]), \tag{4.13}$$

where B is the Brillouin function and $\beta = 1/k_B T$. In the absence of an applied field ($h^{\text{appl}} = 0$), the solution always has a transition from a high-temperature paramagnetic phase with $m = 0$ to a low-temperature ferromagnetic phase with $m \neq 0$. For $J < 0$ the analogous equation depends on the lattice structure and for many cases the solution is an antiferromagnet.

There are various proposals for going beyond the mean-field approximation. Bethe and Peierls [172, 173] proposed a finite-sized cluster with a central site and n_z neighbors embedded in a self-consistent mean field. For the Ising model with small n_z, this leads to remarkably accurate predictions [174] for the transition temperature. There are analogous extensions for clusters and extended forms in DMFT in Chs. 17 and 21.

4.3 Density functional theory and the Kohn–Sham auxiliary system

Density functional theory is presented in detail in [1] and many other references, e.g., [175–177], where it is emphasized that it is a theory of the *interacting many-electron* system.[6] In this book we focus on some aspects of DFT that are motivated by the conceptual relation of DFT to the methods presented in later chapters, and by its use in practice as a starting point for Green's function and QMC calculations. The theory reflects the many-body nature of the problem; indeed, DFT is explored up to the strong-interaction limit where electrons form a Wigner crystal (see, e.g., [180] and earlier work cited there). DFT is included in this "mean-field" chapter because the Kohn–Sham construction of an auxiliary system provides a framework in which the solution can be found using mean-field methods. The theory is reviewed here in some detail because of its importance for quantitative many-body calculations and its role in subsequent developments of many-body methods.

The Hohenberg–Kohn formulation and Levy–Lieb constrained search

The total energy of an interacting many-electron system can be expressed as a sum of terms involving only the electrons plus the effects[7] of the external potential $v_{\text{ext}}(\mathbf{r})$ due to the nuclei and other sources,

$$E = \langle \Psi | \hat{H} | \Psi \rangle = \langle \Psi | \hat{T} + \hat{V}_{ee} | \Psi \rangle + \int d\mathbf{r}\, v_{\text{ext}}(\mathbf{r}) n(\mathbf{r}). \tag{4.14}$$

---

[6] Density functional theory for classical liquids was developed before the work of Hohenberg and Kohn. Perhaps the first density functional theory was by van der Waals in 1894 [178] to describe capillary action in liquids. This was revived by Cahn and Hilliard in 1958 [179].

[7] It is assumed that there is no magnetic field, but Zeeman terms can be added in spin-density functional theory to describe magnetic systems. Time dependence and other extensions are considered in Secs. 4.5 and 4.6.

This expression is defined for a range of many-body wavefunctions $\Psi(\mathbf{r}_1, \mathbf{r}_2, \ldots)$ and density $n(\mathbf{r})$; the minimum $\Psi = \Psi_0$ for all possible Ψ that are normalized and obey the antisymmetry requirement is the ground state, with the ground-state density $n = n_0$ and the total ground-state energy $E = E_0$. Since Ψ_0 is determined by the external potential $v_{\text{ext}}(\mathbf{r})$, all terms in Eq. (4.14) for $\Psi = \Psi_0$ are determined by $v_{\text{ext}}(\mathbf{r})$ and each can be considered to be a *functional* of v_{ext}, i.e., it depends on the entire function. This is expressed by the notation $E[v_{\text{ext}}]$ with square brackets (see Ch. 8 and App. H). Equation (4.14) with $\Psi = \Psi_0$ is a good example of a functional, since the state of a quantum system at any point $\mathbf{r}$ depends on the potential everywhere, not only at the point $\mathbf{r}$, and the ground-state energy can be defined for a range of potentials, not only the potential for a particular system.

The accomplishment of Hohenberg and Kohn (HK) [181] was to show that the energy can also be considered to be a functional of the density,

$$E_{HK}[n] = F_{HK}[n] + \int d\mathbf{r}\, v_{\text{ext}}(\mathbf{r}) n(\mathbf{r}), \tag{4.15}$$

where $F_{HK}[n] = \langle \hat{T} \rangle + \langle \hat{V}_{ee} \rangle$ is a universal functional of the density, the same for all electron systems. This is an example of a Legendre transformation (see App. H), where the left-hand side is a functional of n if v_{ext} is determined by n, i.e., the transformation is invertible so that there is a one-to-one relation between n and v_{ext}. HK provided an existence proof for the functional $E_{HK}[n]$ of n restricted to densities that can be generated as ground state of some local potential called "v-representable." The minimum for all such densities is for $n = n_0$ and $E = E_0$, the ground-state density and total energy. Since all properties of the system are determined by v_{ext}, which is determined by n, it follows that each property is a functional of the density. The energy functional is special in the sense that the ground state is determined by minimization of the energy.

The restriction to v-representable densities is overcome in an alternative approach, the two-step constrained search formulation of Levy [182] and Lieb [183] (LL), who divided the minimization argument into two steps. The first step is to define a functional of density $n(\mathbf{r})$ as the minimum in the space of all allowed Ψ with density $n(\mathbf{r})$,[8]

$$E_{LL}[n] = \min_{\Psi \to n(\mathbf{r})} \langle \Psi | \hat{T} + \hat{V}_{ee} | \Psi \rangle + \int d\mathbf{r}\, v_{\text{ext}}(\mathbf{r}) n(\mathbf{r}). \tag{4.16}$$

Thus the LL functional can be understood simply as the energy for the best possible many-body wavefunction for a given density $n(\mathbf{r})$. Note that $n(\mathbf{r})$ and Ψ need not be the density and eigenstate of any potential; the only requirement is that the density is never negative and that it is normalized to N electrons. Hence the problem of v-representability of the HK formulation is replaced by the much less difficult question of "n-representability." The second step is to minimize the total energy $E_{LL}[n]$ with respect to the density $n(\mathbf{r})$.

The HK and LL constructions are examples of the useful fact that different functionals can be used to find the same solution at the minimum.

---

[8] It can be easy to construct various many-body wavefunctions that have the same density, for example, any plane wave $e^{i\mathbf{k}\cdot\mathbf{r}}$ has the same uniform density. It is not hard to generalize this to many-body systems that need not be homogeneous.

The Kohn–Sham auxiliary system

The choice of the density as the basic quantity allows one to express explicitly two important contributions to the total energy: the energy in the external potential and the average Coulomb Hartree term. The rest of the functional is unknown and very difficult to approximate in general. The success of DFT in practice is due to the ingenious reformulation by Kohn and Sham (KS) [42]: the construction of an "auxiliary system" that is soluble and is chosen to reproduce some properties of the full interacting many-body system *but not all properties*. The big step forward stems from the fact that a further important contribution to the total energy, the non-interacting kinetic energy, can be formulated explicitly. Then only the remainder has to be approximated. The price to pay is to move away from an explicit functional of the density, by introducing orbitals, since the non-interacting kinetic energy is known as a functional of the non-interacting density *matrix*.

With this in mind, we can write the expression for the energy in Eq. (4.15) as

$$E_{KS}[n] = T_{ip}[n] + \int d\mathbf{r}\, v_{ext}(\mathbf{r})n(\mathbf{r}) + E_H[n] + E_{xc}[n], \tag{4.17}$$

where T_{ip} is the kinetic energy for independent particles with density n, E_H the Hartree potential energy Eq. (4.3), and the exchange–correlation energy is defined as

$$E_{xc}[n] = \langle \hat{T} \rangle - T_{ip}[n] + \langle \hat{V}_{ee} \rangle - E_H[n]. \tag{4.18}$$

This expresses the meaning of E_{xc} explicitly as the difference between the exact kinetic energy $\langle \hat{T} \rangle$ and T_{ip} plus the difference between the exact interaction energy $\langle \hat{V}_{ee} \rangle$ and the Hartree approximation. It is a functional of the density n following the Hohenberg–Kohn theorem. Thus the original HK functional has been redefined in Eq. (4.17) as a sum of the first three terms that can be calculated explicitly plus $E_{xc}[n]$, which includes all contributions of exchange and correlation to the ground-state energy.

Now we set up the KS auxiliary system, defined as an independent-particle system with the same density as the original, interacting one. The properties of this auxiliary system of independent "electrons" can be derived from the solution of independent-particle equations with some as-yet-undetermined potential $v_{eff}(\mathbf{r})$,

$$\left(-\frac{1}{2}\nabla^2 + v_{eff}(\mathbf{r})\right)\psi_i(\mathbf{r}) = \varepsilon_i\psi_i(\mathbf{r}), \tag{4.19}$$

with auxiliary eigenvalues ε_i and eigenvectors ψ_i and density given by

$$n(\mathbf{r}) = \sum_i f_i|\psi_i(\mathbf{r})|^2, \tag{4.20}$$

where f_i denotes the occupation numbers of the states.

The condition of minimum energy for fixed particle number yields the effective potential of Eq. (4.19),

$$v_{eff}(\mathbf{r}) = v_{ext}(\mathbf{r}) + \frac{\delta E_H[n]}{\delta n(\mathbf{r})} + \frac{\delta E_{xc}[n]}{\delta n(\mathbf{r})}$$
$$\equiv v_{ext}(\mathbf{r}) + v_H([n], \mathbf{r}) + v_{xc}([n], \mathbf{r}). \tag{4.21}$$

The set of Eqs. (4.19)–(4.21) are the famous Kohn–Sham equations that must be solved self-consistently since the potential is a functional of the density. At the solution, $v_{\text{eff}}(\mathbf{r}) = v_{KS}(\mathbf{r})$ is the Kohn–Sham potential that (except for an irrelevant additive constant) is in one-to-one correspondence with the density n.

This is an in-principle-exact method for the calculation of the ground-state density and energy if it is possible to find a density that is the same as in the original interacting electron system, i.e., the actual density is "non-interacting v-representable." This has only been proven for lattice problems [184] and special cases, but one may expect that the transition from a dense discrete lattice to a continuum does not spoil the results [185]. Experience indicates that the assumption is valid for many systems.

Approximate exchange–correlation functionals

Once the principles of the approach are settled, it still remains to find good approximations for the unknown exchange–correlation contribution. The reason KS-DFT is so successful is that even very simple approximate functionals can be remarkably accurate for ground-state properties of many systems. Here we mention two main categories that are widely used, including most of the examples shown in this book.

- **The local density approximation (LDA).** This approximation was introduced in the original work of Kohn and Sham. The exchange–correlation energy density is supposed to depend locally on the charge density,[9] and it is taken from results for the HEG at every value of the density:

$$E_{\text{xc}}^{\text{LDA}}[n] = \int d\mathbf{r}\, n(\mathbf{r}) \epsilon_{\text{xc}}^{\text{HEG}}(n(\mathbf{r})). \qquad (4.22)$$

The exchange energy density of the HEG is $\epsilon_{\text{x}}^{\text{HEG}}(n) = -\frac{3}{4}(\frac{3}{\pi}n)^{(1/3)}$. The correlation energy density has been calculated numerically for a set of densities using quantum Monte Carlo [109] and interpolated in various forms. Although originally expected to be valid only for slowly varying densities, the LDA has also been very successful for inhomogeneous systems, not least because it satisfies several exact constraints, which

---

[9] The terms "local" and "non-local" are used in several ways that should be clear from the context. In the present chapter, it is important to distinguish two uses. One is the functional dependence upon the density of a potential $v_{\text{eff}}(\mathbf{r}, [n])$ at a point $\mathbf{r}$ in the Kohn–Sham equations, Eqs. (4.19)–(4.21). In the local density approximation, the value of the potential at $\mathbf{r}$ depends only on the density at that same point; however, in general there is a non-local dependence of the value at point $\mathbf{r}$ upon the density at other points $\mathbf{r}'$. A different use of the terms is to distinguish a "local potential" and a "non-local potential." The former acts only at one point $\mathbf{r}$ and is denoted $v(\mathbf{r})$ or similar. An example is the local effective potential $v_{\text{eff}}(\mathbf{r})$ that multiplies the wavefunction at the same point $\mathbf{r}$ in the Kohn–Sham equations. A hallmark of the Kohn–Sham approach is that $v_{\text{eff}}(\mathbf{r})$ is always a local potential whether or not it depends on the density at other points. In contrast, a non-local potential is really an operator that connects the wavefunction at two points at the same time. An example is the Hartree–Fock equation, Eq. (4.5), where the equation for $\psi_i^{\sigma}(\mathbf{r})$ involves (the first term on the second line) an integral over the same function $\psi_i^{\sigma}(\mathbf{r}')$ at other points $\mathbf{r}'$. Such an equation acts differently on different orbitals $\psi_i^{\sigma}(\mathbf{r})$. Note the difference from *explicitly* state-dependent *local* potentials, such as self-interaction corrected ones (Sec. 4.5), that lead to a different hamiltonian for each state and therefore to non-orthogonal states.

leads to physically meaningful results. It is, however, not self-interaction free, which requires corrections especially for systems with localized electrons.

- **Generalized gradient approximations (GGAs)**. In the spirit of an expansion around homogeneous density, a dependence on gradients and higher derivatives of the density has been included in the functional. Whereas a straightforward expansion has not solved the problem, several generalized forms have been proposed. They usually contain parameters that are fixed by requirements to fulfill exact constraints and/or to match other calculations or experiments on a set of prototype systems, for exchange (see, e.g., [186]) and correlation (see, e.g., [187]). Often structural and bonding properties are improved with respect to the LDA. Also GGAs suffer in general from a self-interaction error.
- **Further developments**. There are other types of functionals that have been classified according to "Jacob's ladder" [188], a term used to represent the notion of improving upon the Hartree approximation to approach the "heaven" of chemical accuracy: starting from the LDA and GGAs further ingredients are added, for example, second-order gradients in meta-GGAs, the exact exchange energy (expressed in occupied orbitals, see Sec. 4.5) in hybrid functionals (Secs. 4.5, 11.4, 12.2), and approximations to correlation like the RPA (Secs. 4.5, 11.5). The price to pay for improved accuracy seems to be a tendency toward more complex ingredients, away from explicit density functionals.

Functionals for the total energy in the Kohn–Sham auxiliary system

Just as in the original Hohenberg–Kohn formulation, the total energy in the Kohn–Sham methods can be written as a functional of the density, the potential, or a combination of the two. These functionals take the same value at the stationary point, the physical solution. In practice, rewriting the expression for the energy may suggest different approximations. Moreover, the different functionals behave differently away from the stationary point, which is important when the equations are evaluated only approximately; the Harris–Weinert–Foulkes functional explained in Sec. H.3 is an example where the use of approximate input densities often yields satisfactory energies.

In the Kohn–Sham auxiliary system the external potential is fixed, but the effective potential $v_{\text{eff}}(\mathbf{r})$ is a variable function that is the Kohn–Sham potential only at the solution. The expression for the energy in Eq. (4.17) can be written in a form that is useful for comparison with the many-body functionals in Ch. 8,

$$E_{KS}[\ldots] = E_{\text{ip}}[v_{\text{eff}}] - \int d\mathbf{r}\{v_H(\mathbf{r}) + v_{\text{xc}}(\mathbf{r})\}n(\mathbf{r}) + \{E_H[n] + E_{\text{xc}}[n]\}, \qquad (4.23)$$

where the notation $E_{KS}[\ldots]$ indicates that this expression can be used to define a functional of the potential v_{eff}, density n, or both considered as independent variables, which are defined in Sec. H.3. The first term is the total energy of an independent-particle system with the potential $v_{\text{eff}} = v_{\text{ext}}(\mathbf{r}) + v_H(\mathbf{r}) + v_{\text{xc}}(\mathbf{r})$, which is the sum of eigenvalues $E_{\text{ip}} = \sum_i f_i \varepsilon_i$ weighted by the occupation f_i; thus E_{ip} is determined by v_{eff}. The middle term subtracts the contribution to the eigenvalues of the terms due to electron–electron interactions, and the last term in curly brackets includes the electron–electron contribution correctly in the total energy. The last term is written as a functional of n since the Hartree energy depends only on the density $n(\mathbf{r})$ and $E_{xc}[n]$ is taken to be a functional of n. The corresponding expression for the grand potential Ω_{KSM} in the Mermin approach [189] is given in Eq. (8.3).

Equation (4.23) has the generic form of a Legendre transformation (Eq. (H.6)). One can include the external potential energy in the second and third terms, which provides a straightforward way to construct different functionals of $v_{\mathrm{eff}}(\mathbf{r})$, $n(\mathbf{r})$, or both. All the functionals must have the same minimum energy solution but they have different variations away from the minimum. They are explained in detail in [1, Ch. 9], and their properties are summarized in Sec. H.3. They are included in this book because analogous forms have a prominent role in the functionals of the Green's function in Sec. 8.2.

4.4 The Kohn–Sham electronic structure

KS-DFT is an in-principle-exact and, with the above approximations, an in-practice-useful method to determine the ground-state density and total energy. Within the Kohn–Sham approach, good approximations to the density functional expression of other observables are in general not known. It is therefore tempting to proceed in analogy with the density Eq. (4.20), and use the KS orbitals and eigenvalues from Eq. (4.19) in independent-particle expressions to determine important quantities such as bandgaps or response functions. However, whereas Eq. (4.20) yields by definition the exact interacting density, there is no such theorem for other quantities. Nevertheless, it is sometimes used and may even be helpful.

In particular, one finds that KS eigenvalues are frequently interpreted as total energy differences, namely electron addition or removal energies. However, this is correct only for the eigenvalue of the highest occupied state in a finite system, that corresponds indeed to the ionization energy when the exact KS potential is used [190, 191]. This is sometimes called the "DFT Koopmans' theorem." To avoid confusion, let us summarize some important facts.

- The disagreement between KS band structures and measured ones, in particular concerning the bandgap, should *not* be called a correlation effect: KS eigenvalues are *not* defined to be electron addition or removal energies (see also Sec. 2.4), even when the exact functional is used.[10]
- Additional errors may be due to approximate functionals, in particular self-interaction errors for localized states. An illustration for the relative weight of the two errors in silicon can be found in [192].

---

[10] To be precise, the situation is more subtle for many of the most widely used approximate functionals, like the LDA. It has been shown that the exact functional is a non-analytic functional of the density, but for all analytic functionals the eigenvalues do correspond to total energy differences; see the discussion at the end of this section.

- The fact that the KS potential is by definition static and local makes it too inflexible to describe more than just one (i.e., the highest occupied) state correctly. When, as is frequently the case, the whole valence band manifold is reasonably well described, this can be understood because states are relatively similar; moreover, they are constrained by the requirement that the density is correct (see also [193]).
- The success in describing bandgaps of certain functionals such as DFT+U, SIC, or hybrid functionals can be ascribed to the fact that eigenvalues are calculated from non-local or state-dependent potentials, beyond standard KS (see also Sec. 4.5). This is *per se* not an indication for a better description of correlation.
- In the one-electron limit, exact unoccupied KS eigenvalues are far from electron addition energies. However, KS eigenvalue differences equal the exact excitation energies of the system (see Ex. 4.6), which eases the description, of, e.g., optical excitation of small molecules [194].

Still, one may use a KS or generalized KS electronic structure as starting point for a more adequate calculation of observables, such as the band structure including the bandgap. This is indeed often done with success, including many illustrations in this book. Hence, the Kohn–Sham eigenvalues merit a closer look. An attempt to give these eigenvalues a meaning is made by Janak's theorem [195]. Introducing the concept of fractional particle number[11] with occupations f_i, the theorem states

$$\frac{\partial E}{\partial f_i} = \varepsilon_i. \tag{4.24}$$

From this one can find the meaning of the highest occupied eigenvalue ε_N, since (with all states of lower energy fully occupied)

$$E_N - E_{N-1} = \int_0^1 df_N \frac{\partial E}{\partial f_N} = \int_{N-1}^N df_N \varepsilon_N. \tag{4.25}$$

Thus the ionization energy $E_{N-1} - E_N$ equals minus the highest occupied eigenvalue ε_N ("DFT Koopmans' theorem") if the latter does not depend on f_N. This is indeed true on the interval of integration, since for the exact functional $E(N)$ is a series of straight lines, with slope discontinuities at integer N [196].

One may use the same reasoning to integrate between E_N and E_{N+1}. This yields the electron affinity as

$$E_{N+1} - E_N = \varepsilon_{N+1}. \tag{4.26}$$

One could think to continue the series for all N and obtain a Koopmans' theorem for all eigenvalues. There is, however, a flaw in this reasoning: in order for the derivation to be true, ε_N is the highest occupied eigenvalue of the N-electron system, but ε_{N+1} belongs to the $N+1$-electron system. They are *not* eigenvalues of the same Kohn–Sham hamiltonian!

---

[11] The following discussion is based on a generalization of DFT to fractional particle number, justified in the framework of ensemble DFT where states with different particle number are included in the same ensemble [196].

This has an immediate consequence for the calculation of a bandgap. Note that DFT should yield the exact result for the bandgap calculated as a difference of total ground-state energies,

$$E_g = E_{N+1} - E_N - (E_N - E_{N-1}). \tag{4.27}$$

If we express this in terms of Kohn–Sham eigenvalues,

$$E_g = \varepsilon_{N+1}(N+1) - \varepsilon_N(N), \tag{4.28}$$

where we have indicated the particle number of the system in parentheses. Hence, with respect to the Kohn–Sham gap $\varepsilon_g^{KS} = \varepsilon_{N+1}(N) - \varepsilon_N(N)$, that is calculated with the N-electron Kohn–Sham hamiltonian, there is a correction

$$E_g = \varepsilon_g^{KS} + \varepsilon_{N+1}(N+1) - \varepsilon_{N+1}(N). \tag{4.29}$$

This correction $\Delta = \varepsilon_{N+1}(N+1) - \varepsilon_{N+1}(N)$ is called the *derivative discontinuity*, since it stems from the fact that $E(N)$ changes slope at integer particle number [190].

In a solid, the addition of one particle changes the density only by an infinitesimal $(1/N)$ amount. The discontinuity is then simply a rigid jump in the Kohn–Sham exchange–correlation potential, independent of position. If Δ were small, with a good functional at hand, one could use the Kohn–Sham band structure to describe real materials. However, in general Δ should rather be of the order of an eV [192]: this can be considered to be the origin of the famous "bandgap problem."

Turned around, this also indicates a problem of many approximations: one might think to use Eq. (4.27) directly to determine bandgaps, since then DFT is, in principle, correct. However, this presupposes that one uses a functional that contains the derivative discontinuity. Otherwise the total energy difference will equal the poor result obtained when Kohn–Sham eigenvalues are used. Standard functionals such as the LDA have no derivative discontinuity, contrary to orbital functionals such as EXX (see Sec. 4.5). More generally [197–199], it has been shown that the exact exchange–correlation functional cannot be an explicit and differentiable functional of the electron density. All explicit analytic functionals of the density suffer from the absence of the discontinuity, contrary to analytic functionals of the density *matrix*, such as EXX. In this sense, one may also blame the approximate functional for the "bandgap problem."

4.5 Extensions of the Kohn–Sham approach

Exchange–correlation energy as a functional of the wavefunctions

The success of the Kohn–Sham approach is due in large part to the fact that it gives up the idea of "density only": because the non-interacting kinetic energy is expressed in terms of independent-particle wavefunctions, it can be calculated exactly. It is one of the most important contributions to the total energy, and its full description allows one to capture much of the actual properties of the many-electron system. This does not violate the deeper sense of DFT: the ψ_i are determined by the potential $v_{KS}(\mathbf{r})$, which is in one-to-one correspondence with the density. Thus it gets around the fact that it is very difficult to find good approximations for the kinetic energy expressed explicitly in terms of the density. For example, the Thomas–Fermi approximation is an expression for the kinetic energy in terms of the density at each point, i.e., a local kinetic energy functional. However, the shell structure in atoms is completely missing and molecules are not even bound in this approximation [200].

In the KS scheme the exchange–correlation energy is still expressed as an explicit functional of the density $E_{xc}[n]$, such as the LDA and GGAs. Although this has been very successful, there is room for improvement. It is reasonable to expect that the next step is to express the exchange–correlation energy, or a part of it, as an explicit functional of the density matrix or the wavefunctions, $E_{xc}[\psi_i]$, instead of the density. This is called a generalized Kohn–Sham scheme [201], and functionals in terms of orbitals are called "orbital functionals." Of course, the orbitals are functionals of the density, so there is no contradiction with the original concept. Several approaches that are related to the many-body methods are useful as starting points for many-body calculations, and are given below in this section.

A typical feature of these methods is that they improve the treatment of localized d and f states, that suffer in many cases from large errors due to the local or semi-local approximation for the density dependence of the exchange potential [202]. For example, the cancellation of the self-interaction term between Hartree and Hartree–Fock is lost in the functionals such as LDA and the GGAs. As stated in footnote 4 on p. 62, corrections to this particular error are called "self-interaction correction" [203].

Local effective orbital-dependent potentials

One approach is to keep the Kohn–Sham requirement of a local potential even though the resulting wavefunctions are used in an orbital-dependent energy functional. This is the sense of the "optimized effective potential" (OEP) that predates DFT by many years [204] (see, e.g., [205] and [1, Sec. 8.7]). Originally, it was an approximate mean field, but it can be combined with DFT to construct the potential as a derivative of the energy with respect to the density. An example is the "exact exchange" (EXX) method that includes the Hartree–Fock expression for the exchange energy in terms of the wavefunctions (Eq. (4.4)). However, the wavefunctions $\psi_i^{\sigma_i}$ are determined by a Kohn–Sham equation (4.19) with a *local* exchange–correlation potential $V_{EXX}(\mathbf{r}) = \delta E_{EXX}/\delta n(\mathbf{r})$. The resulting Kohn–Sham gap is smaller than the one obtained with the corresponding non-local Fock exchange (see Ex. 4.6), and in some cases significantly closer to experiment compared with those from explicit density functionals like LDA. This is a result of error canceling, due to the neglect of the derivative discontinuity on one side,[12] and correlation contributions on the other side.

---

[12] Note that, as mentioned at the end of the previous section, functionals of the density matrix do have a derivative discontinuity.

Nevertheless, it can be useful as an estimate, or as a starting point for further calculations. More about OEPs can be found in Sec. I.1.

Non-local effective potentials

A different approach goes outside the standard Kohn–Sham framework. A large improvement can be obtained for the excited states, provided that one gives more flexibility to the effective potential. In particular, one can work with an effective non-local or state-dependent potential, instead of the local Kohn–Sham potential. This leads to eigenvalues that yield a better description of excited-state energies, such as the opening of a bandgap in many antiferromagnetic materials.[13] The non-local or state-dependent effective potential can be regarded alternatively as a generalization of the Kohn–Sham idea, or as an approximation to the self-energy in the many-body Green's function methods in Part III (see Sec. 11.4). Here we summarize some aspects of widely used functionals; more extended descriptions and many references can be found in [1, Ch. 8], and other references such as [206]; selected examples are in Secs. 13.3 and 13.4, and in Chs. 19 and 20.

Self-interaction corrections are formulated to explicitly remove the unphysical self-interaction for localized states. This is physically motivated and was actually used by Hartree [5] in calculations for atoms in the 1920s. The resulting potentials are different for every state, which explains why eigenvalues give a better description of addition and removal energies than the eigenvalues of the original Kohn–Sham equation, similar to what happens with the explicitly non-local potentials.

Hybrid functionals are mixtures of explicitly density-dependent functionals and non-local Fock exchange. Unlike EXX they require an actual Hartree–Fock-like calculation, involving a non-local potential. Like the SIC, they yield eigenvalues closer to addition and removal energies than when a local Kohn–Sham potential is used. Hybrid functionals are widely used for molecules where they constitute a relatively simple addition to the Hartree–Fock methods already in use, and their application in extended systems is expanding, especially as a starting point for Green's function calculations (see Sec. 13.4).

DFT+U methods (also called LDA+U) include a Hubbard-like interaction for localized states in solids, e.g., d and f states. These can be considered as semi-empirical approaches with U chosen to fit spectral energies, or as approximations to the many-body methods in Part III (static approximation to GW in Sec. 11.4 and DMFT in Chs. 19 and 20).

The consequences of the non-local potentials can be illustrated by antiferromagnetic NiO, where LDA and GGA calculations lead to a very small bandgap with states at the top of the valence band composed primarily of Ni $3d$ orbitals (see Sec. 20.7). In contrast, methods with an effective non-local exchange potential shift the filled $3d$ states to lower energy so that the highest occupied states are mainly O $2p$ in character, and the lowest unoccupied states are mainly $3d$ with the bandgap greatly increased to values close to experiment. This has important consequences for quantitative many-body calculations for NiO and related

---

[13] As an example, consider the exchange term. The corresponding local potential in EXX leads to a much smaller gap than its non-local counterpart, Hartree–Fock.

materials, where the results often depend on having a good starting point from the simpler calculations using independent-particle methods, as brought out in Chs. 12, 13, and 20.

A further step: RPA correlation

The first systematic correlation contribution in the framework of orbital functionals comes from the RPA. Contrary to exact exchange, it involves both occupied and empty orbitals as well as eigenvalues. The DFT-RPA correlation energy can be obtained from the adiabatic connection fluctuation dissipation (ACDF) theorem [207, 208]

$$E_c[n_0] = \frac{1}{2\pi} \int_0^1 d\lambda \int d\mathbf{r}\, d\mathbf{r}'\, v_c(\mathbf{r} - \mathbf{r}') \int_0^\infty d\omega \left[\chi_0(\mathbf{r}', \mathbf{r}; i\omega) - \chi_\lambda(\mathbf{r}', \mathbf{r}; i\omega) \right]. \quad (4.30)$$

Here v_c is the Coulomb interaction, χ_0 is the causal version of the density–density response function (Eq. (5.149)) of the Kohn–Sham auxiliary system, and χ_λ the response of a system with scaled interaction λv_c, and with ground-state density equal to that of the real system. When χ is approximated by the RPA (see Sec. 11.2), $\chi_\lambda = (\chi_0^{-1} - \lambda v_c)^{-1}$, the λ-integration can be done analytically, leading to a logarithm. This leads to the RPA total energy including the kinetic and exchange contributions Eq. (11.37). One of the big advantages of the RPA is its ability to describe van der Waals interactions (see Sec. 13.5). In Sec. 11.5 the RPA expression is derived from the Klein functional of Green's function theory. The RPA total energy is often evaluated using wavefunctions and eigenvalues from other functionals, but in principle a local or non-local effective potential can be derived and the calculations done self-consistently.

Spin functionals

A very important addition to DFT is the generalization to more than one type of density. The most prominent example is spin densities, i.e., one density for up and one for down spins. In principle, the total density alone is sufficient to describe the system, but the use of two densities makes it easier to find a good approximation in many cases. The Kohn–Sham approach with an independent-particle auxiliary system is constructed to reproduce not only the total density, but also both components separately, by using different potentials for up and down spins. Examples of spin-dependent problems include finite systems with a net spin, for example a hydrogen atom, and magnetic materials (see Ex. 4.4 for the susceptibility). The spin density is incorporated in essentially all functionals, e.g., LDA often designates both spin-independent and spin-dependent LSDA functionals. Some results for the ferromagnets Fe and Ni and antiferromagnetic transition metal oxides and lanthanides are given in Chs. 13, 15, and 20. Often these stem from spin density functional calculations that are used as a starting point for further many-body calculations.

Anomalous pair density for superconductivity

Another example of a generalization to more than one type of density is the introduction in [209] of the "anomalous density matrix" $\chi(\mathbf{r}, \mathbf{r}')$ that characterizes a superconducting phase in addition to the density $n(\mathbf{r})$. In later work [210, 211] n and χ were combined with the probability distribution function of the nuclei $\Gamma(\mathbf{R}_1, \mathbf{R}_2, \ldots)^{14}$ to formulate a DFT for phonon-mediated superconductors. The auxiliary system is composed of electrons, pairs, and nuclei with a local Kohn–Sham potential for the electrons $v_s^e[n, \chi, \Gamma](\mathbf{r})$, a non-local anomalous potential $\Delta_s[n, \chi, \Gamma](\mathbf{r}, \mathbf{r}')$, and nuclear potential $v_s^n[n, \chi, \Gamma](\mathbf{R}_1, \mathbf{R}_2, \ldots)$.

This approach can be compared with the Green's function-based Eliashberg theory [212], where an anomalous Green's function is introduced as the order parameter in the broken-symmetry superconducting phase. The introduction of the anomalous Green's function allows one to find relatively simple approximations. Eliashberg theory neglects vertex corrections, i.e., the self-energy is built to lowest order in the dressed phonon and Coulomb interaction, in the same spirit as the (purely electronic) GW approximation described in Ch. 11. Hence, phonon and electron spectra are intrinsic ingredients in the theory. In contrast, density functional theory for superconductivity addresses the equilibrium densities and energy, and excitation spectra are not given directly by the generalized Kohn–Sham solution.

A quantity of major interest is the transition temperature to the superconducting phase T_c. This can in principle be predicted by both the Green's function and the DFT approach; for the latter this is illustrated in practice by calculations like in [210, 211, 213].

4.6 Time-dependent density and current density functional theory

The Hohenberg–Kohn theorems, as originally formulated, apply only to static systems that are time-reversal invariant. Time-dependent density functional theory (TDDFT) and current density functional theory extend the theorems of density functional theory to time-dependent potentials and to include magnetic fields [106, 214–216]. This opens the way to calculations of spectra for excitations that conserve the number of electrons, such as optical spectra including excitonic effects, scattering of charged particles, and (with generalization to spin density) magnetic excitations such as spin waves.

A theorem for TDDFT has been derived in analogy with the original Hohenberg–Kohn and Kohn–Sham arguments [217, 218]. For a finite system, the conclusion is that, given the initial wavefunction at one time, the evolution at all later times is governed by a potential that is a unique functional of the time-dependent density. Since the potential at time t depends on the density at all previous times, the evolution has memory. One can then write time-dependent Kohn–Sham equations that can be solved in real time with time propagation methods or in frequency space in the linear and low-order non-linear response regimes [214].

TDDFT has become a widely used approach with two approximations: (i) the exchange–correlation functional is adiabatic, i.e., $v_{KS}(\mathbf{r}, t)$ is determined by the density at the same

---

[14] The nuclei require special consideration because they are localized around lattice sites that define the reference frame for the electrons. The additional information needed involves the vibrations of the nuclei around the lattice sites, which are strongly correlated. This was formulated in [210, 211] in terms of the probability distribution function of the nuclei $\Gamma(\mathbf{R}_1, \mathbf{R}_2, \ldots)$.

time t; and (ii) the potential is derived from a ground-state functional.[15] With these approximations, the method leads to a straightforward generalization of the Kohn–Sham equations, and to a Dyson-like equation for the linear density–density response function with a static kernel f_{Hxc}, the density derivative of the ground-state Hartree and exchange–correlation potential (see Eq. (14.58) in Sec. 14.11). Many calculations have been carried out in the adiabatic LDA, denoted TDLDA or ALDA, where $f_{\text{Hxc}}^{\text{ALDA}}(\mathbf{r}, \mathbf{r}') = v_c(\mathbf{r} - \mathbf{r}') + \delta(\mathbf{r} - \mathbf{r}')dv_{\text{xc}}^{\text{LDA}}(\mathbf{r})/dn(\mathbf{r})$ is frequency independent and has a local exchange–correlation component. ALDA calculations were carried out [220] even before TDDFT was formalized with theorems, and there are numerous results for small systems and plasmon spectra in solids, in spite of its limitations (e.g., absence of excitonic effects in solids). In general, it is a challenging task to go beyond the local and adiabatic approximation.

There are parallels between TDDFT and the time-dependent Hartree–Fock (TDHF) approximation, which has a long history (e.g., in calculations for nuclei [221] and atoms [220]). One can make similar comparisons between the two as between static DFT and Hartree–Fock: TDDFT is in principle exact for certain excitations, whereas TDHF is always an approximation. The former includes the time evolution of a density and a local potential, the latter of a density matrix and a non-local potential.

The theory has many interesting features and difficult issues. The original theory is a *current density functional* theory. The original references [217, 218] dealt with the current and constructed a density functional theory using the continuity equation, $\nabla \cdot \mathbf{j}(\mathbf{r}, t) = -\partial n(\mathbf{r}, t)/\partial t$, to relate the density to the current density. This is valid for finite systems and for longitudinal currents (density waves) in extended systems. However, it is not sufficient in general. A simple counter-example is a uniform ring of charge that can flow around the ring. Since the density is always uniform, the state of the system is determined only if the current is specified in addition to the density.[16]

A theory that deals with currents and magnetic fields can be constructed using both time-dependent external potentials and vector potentials that add a term $\mathbf{p} \cdot \mathbf{A}_{\text{ext}}$ in the hamiltonian. Following the same reasoning as in the original Hohenberg–Kohn proofs, the properties of the system are a functional of both the density n and the current density $\mathbf{j} = -\frac{e}{m}\mathbf{p}$ [222–224]. The Kohn–Sham auxiliary system can be constructed in a way analogous to the density functional case, except that now there is an effective vector potential $\mathbf{A}_{\text{eff}}(\mathbf{r}, t)$ as well as an effective potential $v_{\text{eff}}(\mathbf{r}, t)$. Similarly, the exchange–correlation energy will be replaced by a dynamic functional of density and current density.

Many developments of the static Kohn–Sham approach can be generalized to the time-dependent case. This includes much of the extensions outlined in the previous section, as well as possible combinations with the Green's function method (see, e.g., [225] and Sec. 14.11).

---

[15] In the adiabatic case one can work with the total energy. In the general time-dependent case the quantum mechanical action is the appropriate quantity, in analogy to classical mechanics. For issues linked to the question of symmetry and causality, see [219].

[16] Often such uncertainties arise in infinite systems. For example, in periodic crystals the charge density is not sufficient to define the electric polarization; see the discussion about the Berry phase in Sec. 6.3 and extensive discussion in [1, Ch. 22].

4.7 Symmetry breaking in mean-field approximations and beyond

In Sec. 2.1 the correlation energy has been defined to be the difference between the true energy and the *restricted* Hartree–Fock energy, where the HF solution is constrained to have the same symmetry as the hamiltonian. We make this distinction because there may be *unrestricted* solutions with lower symmetry and lower energy than the restricted solution. Because we define the correlation energy in terms of the restricted energy, these solutions capture some fraction of the correlation energy. However, the energy of any Hartree–Fock solution is above the true energy, even if it is an unrestricted broken-symmetry solution.

There are other aspects of broken-symmetry calculations that apply not only to Hartree–Fock but also to other mean-field approximations and many-body calculations, including exact solutions in some cases. The issue is whether a broken-symmetry solution is physically meaningful and, even if it is fundamentally incorrect, whether it can be employed to derive useful information or understanding. It is useful to consider three classes of problems.

Broken symmetry in condensed matter may correspond to an actual state. For example, a crystal has broken rotation and translation symmetry even though the hamiltonian for nuclei and electrons is homogeneous in space if we neglect physical boundaries. The Wigner crystal in Ch. 3 is an example of such a broken-symmetry state. In 1962, Overhauser proved that the RHF solution is unstable with respect to spin fluctuations at any density [226]. The global minimum energy state within HF is a spontaneously broken-symmetry state. Symmetry breaking is thought to occur in the true many-body state only at very low density, $r_s \sim 100$ (see Sec. 3.1).

In every theoretical calculation for a crystal the first item of business is to determine the crystal structure, either theoretically (often using density functional theory) or using the experimentally determined structure. Given the crystal structure, there may be other broken symmetries, for example a magnetically ordered state such as a ferromagnet or antiferromagnet. The Curie–Weiss mean-field approximation for spins in Sec. 4.2 can give a transition to such a state. Similarly, unrestricted Hartree–Fock or spin density functional calculations may find such a state. Examples are the mean-field density functional calculations for Ni in Fig. 13.11, showing the results for ↑ (majority) and ↓ (minority) spins; the DFT+U calculations for NiO shown in Fig. 13.11 are for an antiferromagnetic state. This figure also shows the results for GW calculations that have the same broken symmetry, but are many-body calculations of the spectra.

Broken symmetry in a finite system with a degenerate ground state may merely correspond to choosing one of the degenerate states. Perhaps the simplest example is a hydrogen atom where the exact energy is 1 Ryd for either ↑ or ↓. Since for an open-shell atom with many electrons the ground state is normally degenerate, a calculation can be carried out for any one of the degenerate states. In this case the UHF solution satisfies Hund's rule of maximum spin, maximum angular momentum; it already contains part of the correlation energy and is an appropriate starting point for a many-body method to determine a correlated wavefunction, such as discussed in Ch. 6. The other states have the same energy and can be derived by rotation operations in real and spin space.

Broken symmetry can be erroneous but useful. There are also cases where an unrestricted, broken-symmetry state has lower energy than a restricted state (whether in Hartree–Fock or in approximations to density functional or many-body Green's function approaches), and yet it is fundamentally incorrect. Perhaps the simplest example is the two-site "Hubbard dimer" that is used as an example in Ch. 3; one can readily show that the exact ground state (Ex. 3.5) is a singlet with no broken symmetry. However, as worked out in Ex. 4.7, for large interaction U a UHF calculation finds the lowest-energy energy state to have broken symmetry with mainly ↑ on one site and ↓ on the other. Nevertheless, this can still be useful. The joint probability of finding an ↑ electron on one site and a ↓ electron on the same or another site is a well-defined and important property; even though the values from UHF are approximate, they are useful. A second way the broken-symmetry states can be used is as a basis to determine the correct state; each is a single determinant and the exact ground state is a symmetric linear combination of determinants. In the large-U regime, the solution in Ex. 3.5 is primarily a linear combination of the two determinants.

An example where a broken-symmetry solution is wrong, but very useful, is the Anderson impurity/Kondo model in Sec. 3.5, defined in Eq. (3.8). The Hartree–Fock approximation is obtained by replacing the interaction with a shift in the site energy $\varepsilon_{0\sigma}$ for spin σ by the average interaction, $\varepsilon_{0\sigma}^{HF} = \varepsilon_{0\sigma} + U\langle\hat{n}_{0-\sigma}\rangle$. If the matrix elements t and the density of states ρ of the host are constant, the resulting spectral function is a lorentzian for each spin σ centered on the energy $\varepsilon_{0\sigma}^{HF}$ and width $\Delta = \pi\bar{\rho}t^2$ (see Sec. 7.3 and Ex. 7.1). There are two types of solution: a non-magnetic one having equal occupation of the two spin states as in the non-interacting system, or a broken-symmetry magnetic state with unequal occupations of the two spin states. The latter case is illustrated in the middle panel of Fig. 3.4, where a large interaction U leads to two lorentzians separated by $\approx U$, respectively almost full and almost empty. The condition for a broken-symmetry solution is worked out in Ex. 4.8 and shown in Fig. 4.1, which depicts the transition from non-magnetic to broken-symmetry magnetic solutions as the width Δ decreases. The most

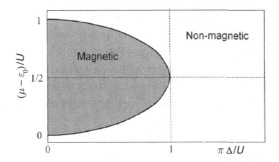

Figure 4.1. Solution for the Anderson impurity model in the unrestricted Hartree–Fock approximation derived in Ex. 4.8. The shaded area labeled "magnetic" indicates the local moment regime as discussed in the text.

favorable case for magnetism is half-filling; in contrast, there is no magnetic solution if the level is far below the Fermi energy μ (full) or far above (empty). The broken-symmetry UHF solution is unphysical since there can be no phase transition for a single impurity; nevertheless, the UHF solution identifies, at least qualitatively, the regime where local moments are formed, i.e., the regime where the Anderson impurity model leads to a Kondo effect. Local moment behavior is observed in many examples in Chs. 2 and 20.

An example where the broken symmetry may not be quite so apparent is the static disorder approximations to a disordered paramagnetic state (the "Hubbard III" approximation in Sec. 3.2 and App. K) or the combination with DFT, e.g., in the disordered local moment (DLM) method in Sec. 19.6. In these methods the system is approximated as an average over states that have static disorder, i.e., with broken translational symmetry; the symmetry is restored by an average that does not treat the dynamics of the system.

4.8 Wrap-up

Mean-field approaches replace the interacting many-body problem with a system of independent particles, with each particle subject to an effective static potential that takes into account in some fashion the effects of all other particles. This may be either an approximation, like Hartree–Fock, or an auxiliary system of non-interacting particles chosen to reproduce only certain properties, like the ground-state density in Kohn–Sham DFT.

These methods are, nevertheless, an essential part of quantitative many-body calculations. In Green's function perturbation methods it is very advantageous to start from a mean-field independent-particle G^0 that is as close as possible to the final Green's function G. In many-body perturbation methods like GW, the quality of the final result depends directly on the accuracy of the expansion in the difference $G - G^0$. Non-perturbative methods like DMFT are most often applied to a subset of orbitals only. This requires a different description of the rest of the system, and determination of the effective interaction in the subsystem. These tasks are often carried out using mean-field approaches, and the DMFT results depend crucially on this input. In a fixed-node projector quantum Monte Carlo calculation, the outcome of the calculation is determined by the choice of nodes, which are often found using mean-field methods. This is a two-way street: improved many-body calculations in turn are the means by which the quality and usefulness of a mean-field calculation can be assessed, and sometimes also improved.

It is important to note that in this chapter we have considered *static* mean fields. A full dynamical Green's function for an interacting many-body system can be cast in terms of a self-energy that is a type of dynamical field. This is brought out in general terms in Ch. 7 and used throughout Part III of this book.

SELECT FURTHER READING
Good reviews of Hartree–Fock theory can be found in:

Many-Particle Theory by E. K. U. Gross, E. Runge, and O. Heinonen (Adam Hilger, Bristol, 1991).

Modern Quantum Chemistry: Introduction to Advanced Electronic Structure Theory by A. Szabo and
N. S. Ostlund, 1996 (Dover, Mineola, NY, unabridged reprinting of 1989 version).

Extensive discussions of DFT can be found in:

Density-Functional Theory of Atoms and Molecules by R. G. Parr and W. Yang (Oxford University
Press, Oxford, 1989).
Density Functional Theory: An Approach to the Quantum Many-Body Problem by R. M. Dreizler
and E. K. U. Gross (Springer-Verlag, Berlin, 1990).
Density Functional Theory by E. K. U. Gross and R. M. Dreizler (Plenum, New York, 1995).
The companion book *Electronic Structure: Basic Theory and Methods* by R. M. Martin, (Cambridge
University Press, Cambridge, 2004).
A very readable paper with many references is "Perspective on density functional theory," by K.
Burke, *J. Chem. Phys.* **136**, 150901, 2012.

An overview of TDDFT can be found in

Time-Dependent Density Functional Theory, Lecture Notes in Physics, Vol. 706, by M. A. L. Marques, C. A. Ullrich, F. Nogueira, A. Rubio, K. Burke, and E. K. U. Gross (Springer-Verlag,
Berlin, 2006).
Ullrich, C. *Time-Dependent Density-Functional Theory: Concepts and Applications*, (Oxford Graduate Texts, 2012). A graduate-level text on the concepts and applications of TDDFT, including
many examples and exercises, and an introduction to static DFT.
For broken symmetry, see "More is different: Broken symmetry and the nature of the hierachical
structure of science," by P. W. Anderson, *Science* **177**, 393–396, 1972 [228] and references
therein, and texts such as *Principles of Condensed Matter Physics* by P. M. Chaikin and
T. C. Lubensky (Cambridge University Press, Cambridge, 1995).

Exercises

4.1 Derive the expression Eq. (4.10) for the interaction in a plane wave basis. Approximate
Eq. (4.10) by factorizing the four operators pairwise, which leads to the HFA. Discuss the
two resulting terms, in particular the q-dependence of the Hartee and Fock contributions,
respectively, in the case of the homogeneous electron gas and of a periodic system.

4.2 For the 3D homogeneous electron gas derive the expression for the dispersion ε_k in the Hartree–
Fock approximation. Show that: (1) for the Coulomb interaction (use the results of Ex. 4.1)
the slope $d\varepsilon_k/dk$ diverges at the Fermi momentum; (2) for a short-range potential there is no
divergence.

Discuss the implications of these results for the effective mass of electrons at the Fermi
surface and the importance of screening in the theory of metals (see also Sec. 11.3).

4.3 Show that the Hubbard interaction form Eq. (4.11) follows from Eq. (4.8) with the approximation of non-overlapping orbitals and an on-site interaction, and using the Pauli principle.

4.4 Derive the expressions below for the magnetic susceptibility χ in the two mean-field
approximations:
A. The Curie–Weiss expression for spins in an effective mean field due to neighbors, $\chi(T) =$
$(M^2/3)/(T - T_c)$ for $T > T_c$.

B. The Stoner itinerant model for which $\chi = \mu_B^2 N(0)/(1 - I\,N(0))$, where $N(0)$ is the density of states at the Fermi energy and I is the Stoner parameter [56, 121] given by $I = dv_m/dm$, with $m = n_\uparrow - n_\downarrow$ and $v_m = v_\uparrow - v_\downarrow$ the potential difference for the two spins. (*Hint*: Find the expression for dm/dv_m in terms of $N(0)$.) This applies for mean-field band theories such as the Kohn–Sham method. Show the relation to Fermi liquid parameters in Sec. 3.4.

4.5 Show that $E[v, n]$ is stationary as a functional of the density, and (a) $\delta E[v, n]/\delta n(\mathbf{r}) = 0$ leads to the condition for the potential $v(\mathbf{r}) = v_{KS}(\mathbf{r})$ in Eq. (4.21). (b) Stationarity with respect to the potential, $\delta E[v, n]/\delta v(\mathbf{r}) = 0$, leads to the requirement that the density be given by $n(\mathbf{r}) = \sum_i f_i |\psi_i(\mathbf{r})|^2$ in Eq. (4.20). (c) The functional $E_{HWF}[n]$ defined in Sec. H.3 is a saddle point at the solution and there is no lower bound.

4.6 Determine the exact Kohn–Sham potential for one electron. Show that the "DFT Koopmans' theorem" is valid. What is the meaning of the unoccupied KS eigenvalues? Consider the cases of electron addition and of an excitation of the electron. Compare to Hartree–Fock.

4.7 The two-site Hubbard model (Hubbard dimer) can be solved exactly (see Ex. 3.5) and used to illustrate the good and bad of a broken-symmetry solution.
  (a) Carry out the unrestricted HF calculation for the two-site Hubbard model with two electrons. Show that as U increases there is a transition where the lowest-energy state has broken symmetry.
  (b) For large U/t show that the lowest energy for a broken-symmetry state is $-2t^2/U$, which can be compared to the exact result $-4t^2/U$ for the singlet (see Ex. 3.5, part (d)). Discuss why there is a factor of 2 error. *Hint:* The broken-symmetry state has spins frozen as $\uparrow$ and $\downarrow$, which is not a proper state of total angular momentum.
  (c) Explain how to form a state with proper symmetry using the broken-symmetry solutions and a sum of determinants. Show that it has the correct energy as given by Ex. 3.5.
  (d) Calculate the magnetic field strength at which the broken-symmetry solution is in fact the correct ground state. Do this by adding a staggered field h, opposite at the two sites 1 and 2 with a term in the hamiltonian $h(\sigma_z(1) - \sigma_z(2))$. Finally, find the actual magnitude of the magnetic field in Tesla using the values $t = U = 1\,eV$.
  (e) Give arguments (analogous to the reasoning for the Anderson impurity model in Ex. 4.8) why the unrestricted broken-symmetry solution is a useful indicator of a local moment regime, even though it is not the correct solution. The results of part (d) may be useful.

4.8 The object of this problem is to carry out the UHF solution of the spin-1/2 Anderson model for an impurity in a metal, and to show that within this approximation there is a transition from a singlet state with no magnetic moment to the "local moment regime" as shown in Fig. 4.1. The hamiltonian is given by Eq. (3.8); for the host use the "flat" density of states where $\Delta = \pi \bar{\rho} t^2$, in which case the added density of states for each spin $\delta \rho_\sigma$ due to the impurity is a lorentzian. It is helpful to define the dimensionless parameters $x = (\mu - \varepsilon_0)/U$ and $y = U/\Delta$. In the UHF approximation, the Green's function for each spin is given by the same expressions as for $U = 0$, Eq. (7.14), except that the energy is shifted as discussed in the text.
  (a) Show that in the HF approximation the number of localized electrons of each spin type is given by

$$n_\sigma = \int_{-\infty}^{\mu} d\omega \, \delta\rho_\sigma(\omega) = \frac{1}{\pi} \cot^{-1}\left(\frac{\varepsilon_0 + Un_{-\sigma} - \mu}{\Delta}\right). \tag{4.31}$$

(b) In terms of the variables x and y, show that the equations that determine $n_\uparrow$ and $n_\downarrow$ can be written:

$$\cot(\pi n_\uparrow) - y(n_\downarrow - x) = 0 \quad \text{and} \quad \cot(\pi n_\downarrow) - y(n_\uparrow - x) = 0. \tag{4.32}$$

(c) Show that for some values of x and y there is only one non-magnetic solution with $n_\uparrow = n_\downarrow$, whereas for other values there are three solutions. This is readily done by graphing $n_\downarrow$ as a function of $n_\uparrow$ from the first equation and $n_\uparrow$ as a function of $n_\downarrow$ from the second equation. Plot the solutions for $n_\downarrow$ vs. $n_\uparrow$ and show examples of a case with only one solution and a case with three solutions. In the case of three solutions one must show that a magnetic solution with $n_\uparrow \neq n_\downarrow$ is lowest in energy.

(d) Show that the boundary between the non-magnetic and magnetic regimes is given by the condition that the two curves have the same slope at the point $n_\downarrow = n_\uparrow \equiv n_c$, and that this leads to the relation of critical values of n_c and y_c given by $y_c = \sin^2(\pi n_c)/\pi$.

(e) Finally, show that the boundary of the magnetic regime has the form found by Anderson [155] and shown in Fig. 4.1.

5

Correlation functions

Preparation, I have often said, is rightly two-thirds of any venture.

Amelia Earhart

Summary

Correlation functions provide a direct way to characterize and analyze many-body systems, both theoretically and experimentally. In this chapter we review the properties of one- and two-body correlation functions in quantum systems, with emphasis on several key quantities: static density correlations that determine the energy and thermodynamic potentials, dynamic correlation functions such as response functions that describe excitations of the system, and Green's functions that are basic tools in the theory of interacting many-body systems.

Correlation functions are central quantities in the description of interacting many-body systems, both in the theoretical formulation and in the analysis of experiments. In contrast to single numbers like the total energy, correlation functions reveal far more information about the electrons, how they arrange themselves, and the spectra of their excitations. In contrast to the many-body wavefunctions that contain all the information on the system, correlation functions extract the information most directly relevant to experimentally measurable properties. Dynamic current–current correlation functions are sufficient to determine the electrical and optical properties: one-body Green's functions describe the spectra of excitations when one electron is added to or removed from the system, static and dynamic correlations are measured using scattering techniques, and so forth. In this chapter we present the basic definitions and properties of correlation functions and Green's functions that are the basis for much of the developments in the following chapters.

In general, a correlation function quantifies the correlation between two or more quantities at different points in space $\mathbf{r}$, time t, or spin σ. Very often the correlation function can be specified as a function of the Fourier-transformed variables, momentum (wavevector) $\mathbf{k}$, and frequency ω. It is useful to distinguish between a dynamic correlation function which describes the correlation between events at different times and a static or equal-time correlation function, by which we mean that of a property measured or computed with "snapshots" of the system. Also, the different correlation functions can be classified by the number of particles and/or fields involved.

The electron density $n(\mathbf{r}, t)$, a one-body quantity, can be thought of as a correlation function of the position of the electrons at a given time with respect to a fixed reference frame, typically determined by the nuclei, external fields such as due to the walls of the container, or the center of mass of a molecule. The one-body density matrix expresses the change in wavefunction as one electron goes from one place to another; its Fourier transform is the momentum distribution of the electrons. Two-body functions include the pair correlation function and its spatial Fourier transform, the static structure factor. Different thermodynamic phases can be characterized by order parameters; these can be defined as the limiting behavior of a correlation function at large distances in time or space, for example, of the density, the magnetization, or the relative phase of the many-body wavefunction. There are also three- and higher-body correlation functions; these arise in the non-linear response of a system, but for reasons of space we will not discuss them here.

Dynamic correlation functions reveal other information; the dynamic generalization of the density matrix is the one-body Green's function, which adds or removes particles and relates to experiments such as photoemission. Two-particle Green's functions are needed to describe dielectric functions and optical properties: in an interacting system, an equilibrium density fluctuation at position $\mathbf{r}$ and time t will be related to a density fluctuation at $(\mathbf{r}', t')$. Through the fluctuation–dissipation theorem, this is proportional to measurable quantities: one can perturb with a potential that couples to the density at $(\mathbf{r}, t)$ and measure the response to the density at $(\mathbf{r}', t')$. This can be formulated in terms of electron–hole pairs, hence the two-particle Green's function. The linear response to external perturbations is most relevant because many probes (e.g., scattering of particles and interaction with light) are very weak on the scale of microscopic forces.

We first define correlation functions, and then delve into equal-time correlation functions followed by dynamic correlation functions, response functions, and Green's functions.

5.1 Expectation values and correlation functions

In quantum statistical physics, expectation values are calculated by integrating over a distribution:

$$\langle \hat{O} \rangle = \sum_\alpha w_\alpha \langle \alpha | \hat{O} | \alpha \rangle, \tag{5.1}$$

where $\hat{O}$ is an operator, α is a many-body state, and w_α is the probability of state α in the distribution. Often, the theoretical development is explicitly formulated for zero temperature, where $\langle \hat{O} \rangle = \langle 0 | \hat{O} | 0 \rangle$ is the expectation value in the ground state ($\alpha = 0$), i.e., Eq. (5.1) with $w_\alpha = \delta_{\alpha,0}$. For systems in thermodynamic equilibrium in the grand-canonical ensemble at non-zero temperature, we have:

$$\langle \hat{O} \rangle = \frac{1}{Z} \mathrm{Tr} \left\{ e^{-\beta(\hat{H} - \mu \hat{N})} \hat{O} \right\} \quad \text{with} \quad Z = \mathrm{Tr} \left\{ e^{-\beta(\hat{H} - \mu \hat{N})} \right\} \equiv e^{-\beta\Omega}, \tag{5.2}$$

where $\mathrm{Tr}\{\ldots\}$ denotes the trace over all states, $\beta = 1/[k_B T]$ is the inverse temperature divided by Boltzmann's constant, μ is the chemical potential, Z is the grand-canonical partition function, and Ω is the grand potential. In terms of the eigenstates of the hamiltonian $\hat{H}$ with energy $E_\alpha = \langle \alpha | \hat{H} | \alpha \rangle$ and number of particles $N_\alpha = \langle \alpha | \hat{N} | \alpha \rangle$, the expectation values can be written as the sum over states in Eq. (5.1), with

$$w_\alpha = e^{-\beta(E_\alpha - \mu N_\alpha)}/Z \quad \text{and} \quad Z = \sum_\alpha e^{-\beta(E_\alpha - \mu N_\alpha)}. \tag{5.3}$$

By definition, a correlation function links at least two quantities; let us represent these by operators $\hat{A}(1), \hat{B}(2), \ldots$, where 1 is a compact notation denoting position and spin, $x_1 = (\mathbf{r}_1, \sigma_1)$, and time t_1. The quantities are said to be correlated[1] if the expectation value of the product is not equal to the product of individual expectation values, $\langle \hat{A}\ \hat{B} \ldots \rangle \neq \langle \hat{A} \rangle \langle \hat{B} \rangle \ldots$ As discussed in Sec. 5.4, dynamic correlation functions are defined by Heisenberg operators $\hat{O}_H(1) = \exp(i\hat{H}t_1)\hat{O}(x_1)\exp(-i\hat{H}t_1)$ that carry the time dependence, with $\hat{O}$ and $|\alpha\rangle$ denoting time-independent operators and eigenstates. We define a *static* correlation function at equal times. Unequal-time correlation functions are called *dynamic*. Both dynamic and static correlation functions can be written

$$C_{AB}(1,2) = \sum_\alpha w_\alpha \langle \alpha | \hat{A}_H(1) \hat{B}_H(2) | \alpha \rangle \equiv \langle \hat{A}_H(1) \hat{B}_H(2) \rangle. \tag{5.4}$$

Very often the uncorrelated product $\langle \hat{A}_H \rangle \langle \hat{B}_H \rangle$ is subtracted from the correlation function so it just represents the fluctuations (e.g., Eqs. (5.23) and (5.134)). The subscript H will be omitted for static correlation functions and also for some dynamic operators when it simplifies the notation and it is clear. In the following sections we first consider static correlation functions.

5.2 Static one-electron properties

The density

One of the most important examples of a one-body expectation value is the electron (spin-resolved) density $n(x)$, where $x = (\mathbf{r}, \sigma)$ denotes position $\mathbf{r}$ and spin σ. In first quantized notation,[2] the density operator can be expressed as $\hat{n}(x) = \sum_i \delta(x - x_i)$ and the density is given by

$$n(x) = \langle \hat{n}(x) \rangle = N \sum_\alpha w_\alpha \int dx_2 \cdots dx_N |\Psi_\alpha(x, x_2, x_3, \ldots, x_N)|^2, \tag{5.5}$$

---

[1] This is the usage of the word correlated in statistics. Note that with this definition, non-interacting particles can appear to be correlated because they are identical bosons or fermions. For example, the two-particle correlation function is non-zero even for independent particles in the Hartree–Fock approximation, see Eq. (10.7).

[2] Expressions in terms of creation and annihilation operators, and of Green's functions, are given in later sections.

where N is the number of electrons. The positions and spins of the other electrons are integrated out. When the ground-state many-body wavefunction Ψ_0 is given by a single Slater determinant of orthonormal orbitals $\psi_i(x, t)$, at zero temperature the expression reduces to the independent-particle form $n(x) = \sum_i |\psi_i(x)|^2$.

The spin-resolved density is central to the theory of interacting electrons. Density functional theory rests upon the fact that the potential acting on the electrons due to the nuclei and other electrostatic fields involves only the density (in the absence of magnetic fields). One of the primary goals of a quantitative many-body theory must be to describe the density accurately while also calculating many other properties of the system.

One of the purest examples of electron correlation is given by the Wigner crystal, see Sec. 3.1. For sufficiently low density and temperature it is energetically favorable for a homogeneous system of charged particles to form a crystal, whether the particles are classical, or are fermions or bosons, and whether they are in two dimensions or three dimensions. In a translationally invariant system, the density averaged over degenerate solutions will be uniform. However, any disorder, always present in nature, will "pin" the crystal and result in a non-uniform density. Figure 5.1 shows the density computed with path-integral Monte Carlo for a 2D Wigner crystal; it forms a hexagonal lattice. The periodic boundary conditions preserve the translational symmetry, so the exact density would be uniform. However, the system *spontaneously breaks symmetry*, and has a

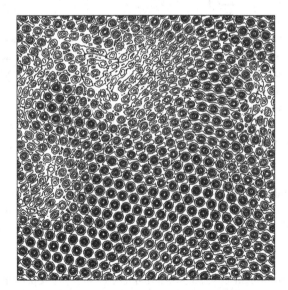

Figure 5.1. A contour plot of the spatial density in a realization of a 2D Wigner crystal as computed with path-integral Monte Carlo at a surface density of 3.2×10^{12} cm^{-2} ($r_s = 60$) and temperature 16 K (10^{-4} Ry). For the definition of the model, see Sec. 3.1. The disorder results from the initial conditions of the Monte Carlo random walk. The spatial boundary conditions frustrate the formation of a perfect crystal.

non-uniform density because of initial conditions, the finite length of the simulation, and the random number generator, just as an electron system in nature would but for different reasons. Because of the symmetry breaking, the figure shown is only one possible realization; another simulation would have the lattice displaced or rotated and the defects in different arrangements. Two-body correlation functions, discussed in the next section, give properties insensitive to the realization.

The one-body density matrix

The one-body spin-resolved density matrix $\rho(x, x')$ is the correlation of the wavefunction of one particle at two different points in space and spin at the same time:[3]

$$\rho(x, x') = N \sum_\alpha w_\alpha \int dx_2 \ldots dx_N \Psi_\alpha^*(x, x_2, x_3, \ldots, x_N) \Psi_\alpha(x', x_2, x_3, \ldots, x_N). \quad (5.6)$$

Evaluating it on the diagonal ($x = x'$) gives the spin-resolved density $\rho(x, x) = n(x)$; moreover $\rho(x, x')$ contains information about the phase of the wavefunction, a purely quantum dependence, and, as we will see, the momentum distribution.

One can represent the density matrix in terms of its eigenfunctions, $\varphi_i(x)$, known as the "natural orbitals" and eigenvalues n_i, the "occupation numbers," defined as

$$\sum_{\sigma'} \int d^3 r' \rho(x, x') \varphi_i(x') = n_i \varphi_i(x). \quad (5.7)$$

The orbitals and occupation numbers are the many-body generalizations of the orbitals and occupation numbers in mean-field or Kohn–Sham theory. For a single determinant wavefunction in an independent-particle description at zero temperature, the density matrix is "idempotent," i.e., the eigenvalues are either 1 or 0, corresponding to states that are occupied or empty. In contrast, for a generic wavefunction, the eigenvalues are only restricted to lie in the range $0 \le n_i \le 1$; the interactions mix the independent-particle states and lead to non-integer occupation numbers. That is, the density matrix is, in general, no longer idempotent.

For a bulk fermion system, the density matrix is localized as a function of $\delta r = |\mathbf{r} - \mathbf{r}'|$, but metals and insulators differ in the amount of localization: the density matrix decays exponentially with distance for an insulator and as δr^{-3} in a 3D metal. In contrast, bosons in a superfluid are characterized by off-diagonal long-range order [229], which means that as $\delta r \to \infty$ the density matrix goes to a non-zero constant, the condensate fraction.

Higher-order density matrices can be defined. For example, a superconductor is characterized by infinite-range order in the off-diagonal part of the two-body density matrix.

---

[3] The definition is given in Eq. (5.120) in terms of the Green's function.

The momentum distribution

The momentum distribution[4] is the probability density of observing a single particle with momentum $\mathbf{k}$. It is obtained by integrating out all but one of the momenta from the $3N$-dimensional momentum distribution:

$$\rho(\mathbf{k}_1) = (2\pi)^{-3N} \sum_\alpha w_\alpha \int d\mathbf{k}_2 \ldots d\mathbf{k}_N \left| \int dx_1 \ldots dx_N \Psi_\alpha(x_1, \cdots x_N) e^{-i\sum_j \mathbf{k}_j \cdot \mathbf{r}_j} \right|^2. \quad (5.8)$$

Performing the integrals $d\mathbf{k}_2 \ldots d\mathbf{k}_N$ gives

$$\rho(\mathbf{k}) = \frac{1}{(2\pi)^3 N} \int d\mathbf{r} \int d\mathbf{r}'' e^{-i\mathbf{k}\cdot\mathbf{r}''} \rho(\mathbf{r}, \mathbf{r} - \mathbf{r}''), \quad (5.9)$$

i.e., it is the Fourier transform of the (spin-integrated) one-body density matrix given in Eq. (5.6). Since $\rho(\mathbf{k})$ is a probability distribution function it is normalized as $\int d\mathbf{k}\, \rho(\mathbf{k}) = 1$. The total electronic kinetic energy is given by

$$\langle \hat{T} \rangle = \frac{N}{2} \int d\mathbf{k}\, k^2\, \rho(\mathbf{k}). \quad (5.10)$$

The momentum distribution provides a clear example of the effect of interactions to change properties radically. As discussed in the previous subsection, the states of non-interacting electrons are either occupied or empty: in a homogeneous system, the eigenstates are plane waves[5] with momentum $\mathbf{k}$ and occupation 1 for $k \leq k_F$, and 0 for $k > k_F$. Interactions cause correlations among the electrons to lower the total energy, always increasing the kinetic energy by mixing in higher-momentum states so the occupation number is no longer either 0 or 1 but varies between these limits (for fermions). For a Fermi liquid, there remains a discontinuity defining the Fermi surface[6] and preserving the volume within it, as discussed in Sec. 3.6. However, the magnitude of the discontinuity at k_F is reduced; it is a measure of the quasi-particle strength Z, related to the self-energy $\Sigma(\omega)$ in Ch. 7. Figure 5.2 shows the momentum distribution of the unpolarized 3D homogeneous electron gas (see Sec. 3.1) as a function of r_s defined as $4\pi r_s^3/3 = 1/n$. The jump Z at the Fermi surface is a measure of correlation; it is close to 1 in a situation of weak correlation ($r_s = 1$) and decreases for decreasing density (larger r_s, i.e., stronger correlation).

Compton scattering measures the spectra of scattered X-rays and was one of the first experiments to verify the fermion character of electrons. When the energy and momen-

---

[4] Note that the momentum distribution is often denoted n_k in the literature. We use $\rho(\mathbf{k})$ to avoid confusion with occupation numbers denoted $n_{\mathbf{k}}$. In the homogeneous electron gas, for example in Fig. 5.2, the two quantities coincide.

[5] The occupation of a state equals the momentum distribution times $(2\pi)^3 n$, where n is the electron density.

[6] In one dimension there is a power law singularity. There is no discontinuity in a superconductor since it is in a different state than normal metals.

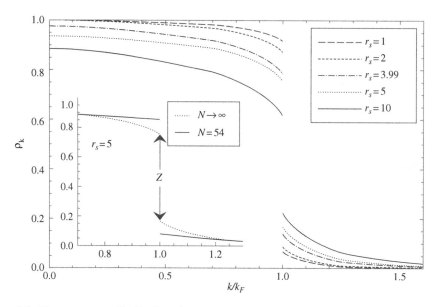

Figure 5.2. The momentum distribution of the unpolarized 3D homogeneous electron gas (Sec. 3.1) calculated in QMC for various densities r_s. The inset shows the extrapolation at $r_s = 5$ from a system of $N = 54$ electrons to the thermodynamic limit [231]. The jump at the Fermi surface Z, the quasi-particle strength, is the renormalization factor, see Sec. 7.5.

tum transferred to the electron are large, ignoring the directional dependence of $\rho(\mathbf{k})$ and applying the "impulse approximation,"[7] the measured spectrum is related to $\rho(\mathbf{k})$ as

$$J(q) = \frac{3}{8\pi k_F^3} \int_0^q d\mathbf{k}\, k\, \rho(\mathbf{k}). \qquad (5.11)$$

One can invert this transform to recover the momentum distribution, as has been done in Fig. 11.7 for sodium, and compare with DFT, GW, and QMC calculations [230].

5.3 Static two-particle correlations: density correlations and the structure factor

The static (equal-time) probability $n(x, x')$ of finding a pair of particles of spin σ at point $\mathbf{r}$ and of spin σ' at point $\mathbf{r}'$ is given by

$$n(x, x') = \sum_{i \neq j} \langle \delta(x - x_i)\delta(x' - x_j) \rangle = N(N-1) \int dx_3 \cdots dx_N |\Psi_0(x, x', x_3, \ldots, x_N)|^2, \qquad (5.12)$$

where the second equality holds for zero temperature. It is useful to define the normalized pair distribution,

$$g(x, x') = \frac{n(x, x')}{n(x)n(x')}. \qquad (5.13)$$

---

[7] In this approximation the scattering process is so fast that the electron system does not react and the excited electron can be approximated as a free electron.

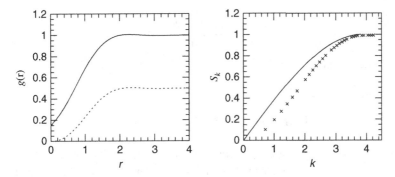

Figure 5.3. The pair correlation function (left panel) of the 3D homogeneous electron gas at $r_s = 2$. The upper curve is the total, the lower curve for parallel spins. The derivative at $r = 0$ is related to the cusp condition on the many-body wavefunction, see Sec. 6.1. The right panel shows the total spin-independent structure factor. The solid curve is for non-interacting electrons. Note that its value at $k = 0$ is $N/2$, as in Eq. (5.18).

This is unity for uncorrelated particles[8] and generally $g(x, x') \to 1$ at large $|\mathbf{r} - \mathbf{r}'|$. The pair correlation function is proportional to the probability of finding another particle a displacement $\mathbf{r}$ away,

$$g_{\sigma',\sigma''}(\mathbf{r}) = \frac{\Omega}{N(N-1)} \int_\Omega d\mathbf{r}' d\mathbf{r}'' n(x', x'') \delta(\mathbf{r}' - \mathbf{r}'' - \mathbf{r}), \tag{5.14}$$

and generally depends on the spins of the electrons. We define the total $g(\mathbf{r}) = \sum_{\sigma',\sigma''} g_{\sigma',\sigma''}(\mathbf{r})$. The left panel of Fig. 5.3 shows $g(\mathbf{r})$ for total and like-spins for the 3D HEG.

We can calculate the electron–electron interaction energy per volume as:

$$\langle \hat{V}_{ee} \rangle = \frac{N(N-1)}{2\Omega} \int_\Omega d\mathbf{r}(g(\mathbf{r}) - 1)/r. \tag{5.15}$$

Using Green's functions, this expression is also derived in Eq. (5.128). For bulk systems, there is a difficulty in dividing up the various terms that involve the Coulomb interaction because the electron–electron and electron–ion interaction energy are individually infinite. This can be handled by adding a uniformly charged background to each such term, so each one is charge neutral as described in [1]; this accounts for the second term, i.e., the -1 in parentheses in the above equation.[9]

The structure factor

The static structure factor is related to the Fourier transform of the pair correlation function and can be measured by X-ray or neutron scattering. First, define an operator whose average value is the Fourier transform of the instantaneous spin-resolved density as

---

[8] Note that even non-interacting electrons are correlated by the Pauli principle; see Ex. 5.3.
[9] Note that Eq. (5.15) is the same result as Eq. (5.128), except for the fact that there the $-1/r$ term is not added.

$$\hat{\varrho}_{\mathbf{k}}^{\sigma} = \frac{1}{\sqrt{N}} \sum_{i=1}^{N} e^{i\mathbf{k}\mathbf{r}_i} \delta_{\sigma,\sigma_i}. \qquad (5.16)$$

Then the static structure factor is defined as

$$
\begin{aligned}
S_{\mathbf{k},\mathbf{k}'}^{\sigma,\sigma'} &= \langle \hat{\varrho}_{-\mathbf{k}}^{\sigma} \hat{\varrho}_{\mathbf{k}'}^{\sigma'} \rangle \\
&= \frac{1}{N} \int_{\Omega} d\mathbf{r} d\mathbf{r}' \, n(x,x') e^{i(\mathbf{k}\cdot\mathbf{r}-\mathbf{k}'\cdot\mathbf{r}')} + \frac{\delta_{\sigma,\sigma'}}{N} \int_{\Omega} d\mathbf{r} n(x) e^{i(\mathbf{k}-\mathbf{k}')\cdot\mathbf{r}}. \qquad (5.17)
\end{aligned}
$$

As shown, it is related to the Fourier transform of Eq. (5.12) and the density. Note that unless $\mathbf{k} = \mathbf{k}'$ the structure factor will vanish in a system with translation invariance, or in a perfect crystal for $\mathbf{k} - \mathbf{k}'$ away from the reciprocal lattice vectors. In general, $S_{\mathbf{k},\mathbf{k}'}$ has the periodicity and symmetry of the crystal, but certain long-wavelength properties are independent of an underlying lattice. In the rest of this section we only consider the diagonal part of $S_{\mathbf{k},\mathbf{k}'}$, that is with $\mathbf{k} = \mathbf{k}'$, and write $S_{\mathbf{k}} = \sum_{\sigma\sigma'} S_{\mathbf{k},\mathbf{k}}^{\sigma,\sigma'}$. The normalization has been chosen so that $\lim_{k\to\infty} S_{\mathbf{k}} = 1$. Also note that electrons randomly arranged will have $S_{\mathbf{k}} = 1$ for all values of $\mathbf{k}$.

For non-interacting spin (1/2) fermions at zero temperature and in the absence of an external potential, the structure factor has the form

$$
S_{\mathbf{k}} = \begin{cases}
N/2 & k = 0 \\
\frac{k}{4k_F}\left(3 - \left(\frac{k}{2k_F}\right)^2\right) & 0 < k \leq 2k_F \\
1 & 2k_F < k
\end{cases} \qquad (5.18)
$$

where k_F is the Fermi momentum. This function is plotted in Fig. 5.3. It is linear in k at small k, an incorrect behavior for any charged system as shown by the following inequality. For a homogeneous system of electrons at zero temperature, the structure factor is bounded by the dielectric function:

$$S_{\mathbf{k}} \leq \frac{k^2}{2\omega_p}\left(1 - \frac{1}{\epsilon_k}\right)^{\frac{1}{2}}. \qquad (5.19)$$

Here $\omega_p^2 = 4\pi n$ is the plasma frequency and ϵ_k is the static dielectric susceptibility (see Ex. 5.1). Hence, the structure factor of an interacting charged system must vanish quadratically at small k [106, 232].

The structure factor essentially measures fluctuations in the density over a region of size $2\pi/k$. The Pauli principle for electrons reduces the fluctuations of charge to avoid electrons being in the same location, so that $S_{\mathbf{k}} \propto k$ at long wavelength. In any metal the fluctuations are further reduced by the Coulomb interaction, so that $S_{\mathbf{k}} \to k^2/(2\omega_p)$. Finally, in an insulator the k^2 coefficient is further reduced by the factor $(1 - \epsilon_k^{-1})^{\frac{1}{2}}$ because electrons are localized and cannot move as in a metal.

A crystalline solid is characterized by a periodic density, i.e., long-range order in the density–density correlation function. This is readily described in terms of the structure factor where the periodicity leads to Bragg peaks; see Fig. 5.4. The Bragg peak is the "order parameter" which characterizes the crystalline phase. Wigner crystallization, while

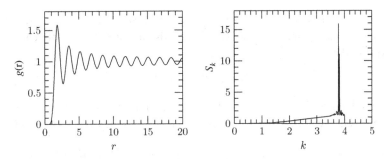

Figure 5.4. The pair correlation function (left panel) of a Wigner crystal in 2D at $r_s = 60$. The right panel shows the corresponding structure factor; the peak shows ordering at the first reciprocal lattice vector.

clearly seen in the density plot, becomes even clearer in the structure factor since it remains invariant if the crystal is translated.

5.4 Dynamic correlation functions

In the rest of this chapter we discuss dynamic correlation functions which characterize various physical properties such as dielectric response and optical spectral features. First, we consider general properties of the dynamic correlations arising from quantum and thermal fluctuations and their frequency spectrum. Then we look at response functions, which are important for analysis of experiment and related to intrinsic correlations by the fluctuation–dissipation theorem. In the subsequent sections we introduce Green's functions to describe interacting systems. Green's functions are a type of correlation functions and are closely related to the intrinsic fluctuations. We will look specifically at one-body Green's functions in Secs. 5.6 and 5.7, whereas two-particle correlation will be discussed in Sec. 5.8. More details about Green's functions can be found in Apps. C and D; more material can be found in the books [156, 233].

In the following we will use the compact notation introduced in Sec. 5.1: $1 \rightarrow (x_1, t_1)$, with $x_1 \rightarrow \mathbf{r}_1, \sigma_1$.

Dynamic correlation in time

A dynamic correlation function is given by Eq. (5.4), $C_{AB}(1,2) = \langle \hat{A}_H(1)\hat{B}_H(2) \rangle$, with $\hat{A}_H(1)$ and $\hat{B}_H(2)$ in the Heisenberg picture where the wavefunctions are stationary and the time dependence is carried by the operators.[10] As in the static case, the expectation values are defined by Eq. (5.1), in particular, Eqs. (5.2) and (5.3) for the grand-canonical

---

[10] The general definition of Heisenberg operators, denoted $\hat{A}_H(t)$, and Schrödinger operators, denoted $\hat{A}$ with no subscript, is given in App. B.

ensemble. In the following we focus on the case of a static hamiltonian $\hat{H}$.[11] Then the Heisenberg representation of any operator $\hat{O}$ depends only on the time difference $t - t_0$ and can be expressed as

$$\hat{O}_H(t) = e^{i\hat{H}(t-t_0)} \hat{O} e^{-i\hat{H}(t-t_0)}. \tag{5.20}$$

Here t_0 is a reference time that drops out in the final formula and can be set to zero. The correlation function can be expanded in terms of the eigenstates $|\alpha\rangle$ of $\hat{H}$ using Eq. (5.4),

$$C_{AB}(x_1, x_2; t_1 - t_2) = \sum_{\alpha} w_\alpha e^{iE_\alpha(t_1-t_2)} \langle\alpha|\hat{A}(x_1)e^{-i\hat{H}(t_1-t_2)}\hat{B}(x_2)|\alpha\rangle, \tag{5.21}$$

where $\hat{A}(x_1)$ and $\hat{B}(x_2)$ are time-independent operators in the Schrödinger picture and C_{AB} is a function only of the time difference $t_1 - t_2$. For a system in the grand-canonical ensemble at equilibrium the weights are $w_\alpha = e^{-\beta(E_\alpha - \mu N_\alpha)}/Z$. Inserting the identity expanded in terms of a complete set $\sum_\lambda |\lambda\rangle\langle\lambda|$ of eigenstates of $\hat{H}$ between the operators in Eq. (5.21) gives

$$C_{AB}(x_1, x_2, t_1 - t_2) = \sum_{\alpha,\lambda} w_\alpha A_{\alpha\lambda}(x_1) B_{\lambda\alpha}(x_2) e^{i(E_\alpha - E_\lambda)(t_1 - t_2)}, \tag{5.22}$$

with $A_{\alpha\lambda}(x_1) = \langle\alpha|\hat{A}(x_1)|\lambda\rangle$, and similarly for B. A major difference from the static correlation functions is the appearance of the complex phase factor; this makes quantum Monte Carlo calculations of dynamic correlation functions very difficult, as discussed in Sec. 25.6.

One often subtracts from the full expression the uncorrelated term

$$C_{AB}^0(x_1, x_2) \equiv \langle\hat{A}_H(1)\rangle\langle\hat{B}_H(2)\rangle = \sum_{\alpha} w_\alpha A_{\alpha\alpha}(x_1) \sum_{\lambda} w_\lambda B_{\lambda\lambda}(x_2). \tag{5.23}$$

Note that this term does not depend on the time difference. Also, note that for states with fixed particle number this term is non-zero only if $\hat{A}$ and $\hat{B}$ conserve particle number. Hence, as we see in the next section, there is no such term in the one-electron Green's function.

It is also useful to define the correlation function of the commuted operators

$$\tilde{C}_{AB}(1, 2) = \pm C_{BA}(2, 1), \tag{5.24}$$

and in the following we drop the subscript AB. The sign choice $\pm$ refers to operators that commute in the time ordering like bosons (upper sign) or fermions (lower sign), i.e., without or with a change of sign, respectively.

Three combinations of C and $\tilde{C}$ are important.

The **retarded** correlation function with the property $C^R(1, 2) = 0$ for $t_1 < t_2$ is

$$C^R(1, 2) = \Theta(t_1 - t_2)\langle[\hat{A}(1), \hat{B}(2)]_{\mp}\rangle = \Theta(t_1 - t_2)\left(C(1, 2) - \tilde{C}(1, 2)\right). \tag{5.25}$$

---

[11] With the appropriate modification of the time-evolution operators the definitions remain valid in the non-equilibrium case, but the expressions depend on two times instead of time differences, which does not allow one later to work with a single frequency.

The **advanced** form is

$$C^A(1,2) = -\Theta(t_2 - t_1)\langle[\hat{A}(1), \hat{B}(2)]_{\mp}\rangle = \Theta(t_2 - t_1)\left(\tilde{C}(1,2) - C(1,2)\right). \quad (5.26)$$

The **time-ordered** form with both orderings in time is

$$C^T(1,2) = \langle T\left[\hat{A}(1)\hat{B}(2)\right]\rangle = \Theta(t_1 - t_2)C(1,2) + \Theta(t_2 - t_1)\tilde{C}(1,2)$$
$$= C^R(1,2) + \tilde{C}(1,2) = C^A(1,2) + C(1,2). \quad (5.27)$$

Here, T denotes time ordering of operators as defined in Eq. (B.28).

Dynamic correlation in frequency space

The fluctuation spectrum is given by the Fourier transform defined in Eq. (C.7). Starting from Eq. (5.22) this yields

$$C(x_1, x_2, \omega) = \int_{-\infty}^{\infty} dt \sum_{\alpha,\lambda} w_\alpha A_{\alpha\lambda}(x_1) B_{\lambda\alpha}(x_2) e^{i(E_\alpha - E_\lambda)t} e^{i\omega t}$$
$$= 2\pi \sum_{\alpha,\lambda} w_\alpha A_{\alpha\lambda}(x_1) B_{\lambda\alpha}(x_2) \delta(E_\alpha - E_\lambda + \omega). \quad (5.28)$$

There is a similar expression for $\tilde{C}(x_1, x_2, \omega)$,

$$\tilde{C}(x_1, x_2, \omega) = \pm 2\pi \sum_{\alpha,\lambda} w_\alpha B_{\alpha\lambda}(x_2) A_{\lambda\alpha}(x_1) \delta(E_\lambda - E_\alpha + \omega). \quad (5.29)$$

Hence, for a finite system, the fluctuation spectrum is a set of δ-functions. In the thermodynamic limit, the peaks merge to form a continuum, a weighted density of transitions.

An important case is when the operators are adjoint ($\hat{B} = \hat{A}^\dagger$) and evaluated at the same point $x_2 = x_1$, so that

$$C_{AA^\dagger}(x, x, \omega) = 2\pi \sum_{\alpha,\lambda} w_\alpha |A_{\alpha\lambda}(x)|^2 \delta(E_\alpha - E_\lambda + \omega) \geq 0, \quad (5.30)$$

as required for a spectral density. The same is true for the diagonal elements $C_{A^\dagger A}^{kk}(\omega)$ of the matrix $C_{A^\dagger A}(x_1, x_2, \omega)$ in any basis, and it also holds in many other cases for symmetry reasons. This property is used in the analytic continuation of imaginary-time functions into real frequencies in Sec. 25.6. Note that $E_\lambda - E_\alpha$ can be either positive or negative; the expressions are not separated into positive and negative energies, unlike in the case of zero temperature that we will examine below.

In thermal equilibrium with Eq. (5.3), exchange of the indices α, λ in the sum in Eq. (5.29) and comparison with Eq. (5.28) yields

$$\tilde{C}(x_1, x_2; \omega) = \pm e^{-\beta(\omega - \mu)} C(x_1, x_2; \omega), \quad (5.31)$$

where μ is the chemical potential corresponding to the change in particle number caused by $\hat{B}$ (which for states with fixed particle number is minus the change caused by $\hat{A}$). In particular, for the one-body Green's function, μ has the usual definition of a chemical potential, the change in free energy of the system when the number of electrons is changed by one. In the two-particle correlation function L (Sec. 5.8), μ is zero when one looks at the creation of electron–hole pairs.

Spectral functions

We now construct a spectral function A that is defined as

$$
2\pi A(x_1, x_2, \omega) = \left[C(x_1, x_2, \omega) - \tilde{C}(x_1, x_2, \omega) \right]
$$
$$
= \pm \tilde{C}(x_1, x_2, \omega) \left[\mp 1 + e^{\beta(\omega - \mu)} \right]
$$
$$
= C(x_1, x_2, \omega) \left[1 \mp e^{-\beta(\omega - \mu)} \right], \tag{5.32}
$$

where the last two lines hold in thermal equilibrium. If the commutator of the operators $\hat{A}$ and $\hat{B}$ is a scalar function

$$
[\hat{A}(x_1), \hat{B}(x_2)]_{\mp} = f(x_1, x_2), \tag{5.33}
$$

the spectral function obeys the sum rule

$$
\int d\omega A(x_1, x_2, \omega) = f(x_1, x_2). \tag{5.34}
$$

In the following we will no longer display the spatial and spin arguments (x_1, x_2) so as to focus on the dynamic behavior.

The spectral function contains all the information of the various correlation functions. To show this, we have to move into the complex frequency plane, where the Fourier transform of the Heavyside Θ-function is defined. Using Eqs. (C.9) and (5.32), we can express the Fourier transform of the retarded and advanced expressions in Eqs. (5.25) and (5.26) as

$$
C^R(\omega) = i \lim_{\eta \to 0^+} \int d\omega' \frac{A(\omega')}{\omega - \omega' + i\eta} \quad \text{and} \quad C^A(\omega) = i \lim_{\eta \to 0^+} \int d\omega' \frac{A(\omega')}{\omega - \omega' - i\eta}. \tag{5.35}
$$

We can see that $C^R(\omega)$ has poles slightly below the real axis, whereas the poles of C^A are situated slightly above. The analytic structure of $C^R(\omega)$ is depicted in the left panel of Fig. 5.5. There are no other poles, and Eq. (5.35) can be analytically continued into the whole complex plane, $\omega \to z$. We denote the continued function by $\mathcal{C}$:

$$
\mathcal{C}(z) = i \int_{-\infty}^{\infty} d\omega' \frac{A(\omega')}{z - \omega'}. \tag{5.36}
$$

We use the symbol z for the argument to stress the fact that we have moved into the complex plane. $\mathcal{C}$ is analytic everywhere, except along the real axis for frequencies where there are isolated singularities and branch cuts, a continuum of poles (see App. C.5). In terms of $\mathcal{C}(z)$, the retarded and advanced correlation functions are given by

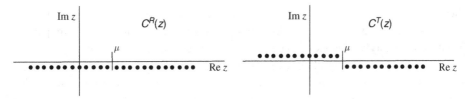

Figure 5.5. Analytic structure of correlation functions as a function of complex frequency z. As explained after Eq. (5.31), μ is the chemical potential for adding or removing one or more particles, and zero for an excitation with no change in particle number. Left: Poles for a retarded correlation function, such as a causal response function discussed in Sec. 5.5. The poles at frequencies z, corresponding for example to excitation energies of the N-electron system, are displaced into the lower half-plane by the infinitesimal $\mathrm{Im}\, z = -\eta$. This form follows from the requirement of causality and it leads to the Kramers–Kronig relations, Eq. (5.53). Right: The structure of the time-ordered correlation function $C^T(z)$ at zero temperature in the complex plane, with two branch cuts (see Sec. C.5) on the real axis. For frequencies $\mathrm{Re}\,(z) < \mu$, the poles are displaced into the upper plane by $+i\eta$. Hence, for example, when C is a one-particle Green's function, $\mathrm{Im}\, z = \eta > 0$ denotes energies for which particles can be removed. For addition of particles at real frequency $\mathrm{Re}\, z > \mu$, the imaginary part $\mathrm{Im}\, z = -\eta$.

$$C^R(\omega) = \lim_{\eta \to 0^+} C(\omega + i\eta),$$

$$C^A(\omega) = \lim_{\eta \to 0^+} C(\omega - i\eta). \tag{5.37}$$

Moreover, with Eq. (5.32) the spectral function is related to the correlation functions by

$$2\pi A(\omega) = \lim_{\eta \to 0^+} [C(\omega + i\eta) - C(\omega - i\eta)] = C^R(\omega) - C^A(\omega)$$

$$= 2\mathrm{Im}\,[iC^R(\omega)] \quad \text{when} \quad A(\omega) \text{ is real.} \tag{5.38}$$

The last equality applies with the conditions specified after Eq. (5.30). It is here written in terms of $\mathrm{Im}\,[iC^R(\omega)]$ to be consistent with the expressions given later in terms of Green's functions, where $A(\omega) = \frac{1}{\pi}|\mathrm{Im}\, G(\omega)|$.

With Eqs. (5.27), (C.12), and (5.32) one also obtains the frequency-dependent time-ordered correlation function. In thermal equilibrium it reads

$$C^T(\omega) = i\left[\mathcal{P}\int d\omega' \frac{A(\omega')}{\omega - \omega'} - i\pi \left\{\begin{array}{c} \cotanh \\ \tanh \end{array}\right\} \left[\frac{\beta(\omega - \mu)}{2}\right] A(\omega)\right], \tag{5.39}$$

where $\mathcal{P}$ indicates a principal value integral, cotanh is for the case where $\hat{A}$ and $\hat{B}$ commute as bosons, and tanh if they commute like fermions. The corresponding expression for the retarded correlation function Eq. (5.35) reads

$$C^R(\omega) = i\left[\mathcal{P}\int d\omega' \frac{A(\omega')}{\omega - \omega'} - i\pi A(\omega)\right]. \tag{5.40}$$

There is a substantial difference between Eqs. (5.39) and (5.40): the relation between the time-ordered correlation function and the spectral function depends on a thermal factor, whereas Eq. (5.40) has no such factor. The thermal factor prevents one from casting C^T in a form like Eq. (5.35) that would allow an analytic continuation to complex frequencies. This different analytic behavior is a potential source of error.

A spectral representation like Eqs. (5.28) or (5.29) displays the physical content of the correlation functions as sums over fluctuations between eigenstates of the system. However, an explicit calculation of the spectral sum is not convenient except for very small numbers of particles. An advantage of the so-called *Lehmann representation*, Eq. (5.36), is that it facilitates the passage from a set of discrete excitation energies to the thermodynamic limit where the δ-peaks of the spectral function merge to form continuous spectra. In addition, the spectral function is especially useful for non-zero temperature (Sec. D.2). It allows one to perform calculations on the imaginary frequency axis, far from the poles, where the functions are smooth, using for example the Matsubara technique [234] (see App. D). This is a powerful approach to calculate integrals of correlation functions. The correlation functions themselves determined at the Matsubara frequencies can then in principle be continued to the real axis to get the spectral function and thus the correlation functions for real frequencies; this is however a delicate procedure, as discussed in Sec. 25.6.

Correlation at zero temperature and fixed particle number

As the temperature goes to zero with a fixed number N of electrons, the thermal equilibrium averages are replaced by expectation values in the ground state. Here we summarize the most important relations for this situation. The spectral function Eq. (5.32) using Eqs. (5.28) and (5.29) becomes

$$A(\omega) = \sum_\lambda \delta(\omega - \varepsilon_\lambda) \begin{cases} A_{0\lambda} B_{\lambda 0} & \text{for } \omega > \mu \\ \mp B_{0\lambda} A_{\lambda 0} & \text{for } \omega < \mu \end{cases}, \tag{5.41}$$

where $\varepsilon_\lambda = \pm(E_\lambda - E_0)$ for $\omega \gtrless \mu$, with E_0 the ground-state energy. Now one finds

$$C^T(\omega) = i \left[\mathcal{P} \int d\omega' \frac{A(\omega')}{\omega - \omega'} - i\pi \operatorname{sgn}(\omega - \mu) A(\omega) \right]$$

$$= i \lim_{\eta \to 0^+} \int d\omega' \frac{A(\omega')}{\omega - \omega' + i\eta \operatorname{sgn}(\omega' - \mu)}. \tag{5.42}$$

The right panel of Fig. 5.5 shows the structure of the time-ordered function at zero temperature in the complex plane with poles slightly above the real axis for $\omega < \mu$ and slightly below for $\omega > \mu$. Hence at zero temperature one can perform the analytic continuation of C^T into the complex plane; it is identical to $\mathcal{C}$. The inverse relation for real $A(\omega)$ is

$$A(\omega) = \frac{1}{\pi} \begin{cases} +\operatorname{Im}\left[iC^T(\omega)\right] & \text{for } \omega > \mu \\ -\operatorname{Im}\left[iC^T(\omega)\right] & \text{for } \omega < \mu \end{cases}. \tag{5.43}$$

5.5 Response functions

A response function describes the response of a system to an external perturbation. In a quantum system, the response can be expressed as the change in the expectation value of an operator $\langle \hat{A}_H(1) \rangle$ at a point x_1 and time t_1 caused by a perturbation at an earlier time t_2 at a point x_2. Suppose that the perturbation has the form $\int dx_2\, \phi(2)\hat{B}(x_2)$ for a general time-dependent external field $\phi(2)$ that couples to an operator $\hat{B}$ acting on the system.[12] Then the response is determined by the evolution of the system in the presence of the perturbed hamiltonian $\hat{H} + \int dx_2\, \phi(x_2, t)\hat{B}(x_2)$, where $\hat{H}$ is the unperturbed hamiltonian. We suppose that $\hat{H}$ is independent of time and $\phi(x, t) = 0$ for times $t < t_0$, so that the expectation value of $\hat{A}$ at time t_1 can be expressed as

$$\langle \hat{A}_H(1) \rangle^+ = \langle U_S^\dagger(t_1, t_0)\hat{A}(x_1)U_S(t_1, t_0) \rangle, \tag{5.44}$$

where the time-evolution operator U_S is given in Eq. (B.5)

$$U_S(t_1, t_0) = e^{-i\int_{t_0}^{t_1} dt_2 [\hat{H} + \int dx_2\, \phi(2)\hat{B}(x_2)]}, \tag{5.45}$$

and the superscript $\langle \cdots \rangle^+$ of the expectation value denotes that the time dependence includes the effects of the perturbation.

Expansion of Eq. (5.44) in powers of ϕ, using Eq. (5.45), leads to[13]

$$\left\langle \hat{A}_H(1) \right\rangle^+ =$$

$$\left\langle \left[1 + i\int_{t_0}^{t_1} dt_2 \int dx_2 \phi(2)\hat{B}_H(2) + \cdots \right] \hat{A}_H(1) \left[1 - i\int_{t_0}^{t_1} dt_2 \int dx_2 \phi(2)\hat{B}_H(2) + \cdots \right] \right\rangle$$

$$= \left\langle \hat{A}_H(1) \right\rangle + i\int_{t_0}^{t_1} dt_2 \int dx_2 \phi(2) \left\langle \left[\hat{B}_H(2), A_H(1) \right]_- \right\rangle + \cdots , \tag{5.46}$$

where each operator on the right is a Heisenberg operator involving only the unperturbed hamiltonian $\hat{H}$, i.e., $\hat{O}_H(t) = e^{-i\hat{H}(t-t_0)}\hat{O}e^{i\hat{H}(t-t_0)}$. Hence the expectation value of $\hat{A}_H$ is obtained as an expansion in powers of ϕ. The coefficients are called *response functions*. Most spectroscopic experiments measure the dominant first-order contribution, the linear response, but important insight can also be obtained from experiments probing higher orders, such as second harmonic generation that is linked to the second-order response function. In this book we focus on linear response, with a brief excursion to second order in Sec. 14.11.

Linear response

To lowest order in the external potential ϕ, Eq. (5.46) reads

$$\delta\langle \hat{A}_H(1) \rangle = \langle \hat{A}_H(1) \rangle^+ - \langle \hat{A}_H(1) \rangle = \int_{-\infty}^{\infty} dt_2 \int dx_2\, \chi_{AB}(1, 2)\phi(2) \tag{5.47}$$

with the linear response function

$$\chi_{AB}(1, 2) = -i\Theta(t_1 - t_2)\langle [A_H(1), \hat{B}_H(2)]_- \rangle. \tag{5.48}$$

---

[12] More general forms of the perturbation can be used, but this is not needed here for our purpose.

[13] This is not a simple expansion since the operators $\int dx_2 \phi(2)\hat{B}(x_2)$ and $\hat{H}$ may not commute. The full derivation is given in Sec. B.3, where the same issues arise in the separation of single-body and interaction terms in the interaction representation.

The function $\chi_{AB}(1,2)$ exemplifies the general form of a response function: a correlation function of a commutator of Heisenberg operators that is a retarded function as defined in Eq. (5.25). More precisely, we can use the relation

$$- iC_{AB}^R(1,2) \rightarrow \chi_{AB}(1,2) \tag{5.49}$$

in order to derive all properties of the response functions from Sec. 5.4. Note that the commutator is the one with the upper sign in Sec. 5.4. From causality, χ is non-vanishing only for measurement at time t_1 after the perturbation time t_2, and χ only depends on $t_1 - t_2$, since the unperturbed hamiltonian is assumed to be time-independent.

With Eqs. (5.28), (5.32), and (5.35), the Fourier transform of χ_{AB} is

$$\chi_{A,B}(\omega) = \sum_{\alpha,\lambda} w_\alpha \left[\frac{A_{\alpha\lambda}B_{\lambda\alpha}}{\omega - (E_\lambda - E_\alpha) + i\eta} - \frac{B_{\alpha\lambda}A_{\lambda\alpha}}{\omega + (E_\lambda - E_\alpha) + i\eta} \right]. \tag{5.50}$$

Its analytic continuation $\chi(z)$ is a complex function that has poles at $z = E_\lambda - E_\alpha - i\eta$, which are in the lower half of the plane; this is the characteristic structure of a causal response function as shown in the left panel of Fig. 5.5. Using Eq. (5.38) we find:

$$A(\omega) = \sum_{\alpha,\lambda} [w_\alpha - w_\lambda] A_{\alpha\lambda}B_{\lambda\alpha}\delta(\omega - (E_\lambda - E_\alpha)). \tag{5.51}$$

When $A(\omega)$ is real as specified after Eq. (5.30), the spectral function is proportional to the imaginary part of χ,

$$A(\omega) = -\frac{1}{\pi}\text{Im}\,\chi(\omega). \tag{5.52}$$

Kramers–Kronig relations

The analytic properties of a response function lead to the Kramers–Kronig (KK) relations (see also [1, App. D]) relating the real and imaginary parts of diagonal elements of $\chi(\omega)$:

$$\text{Re}\chi(\omega) = -\frac{1}{\pi}\mathcal{P}\int_{-\infty}^{\infty} d\omega' \frac{\text{Im}\chi(\omega')}{\omega - \omega'},$$

$$\text{Im}\chi(\omega) = \frac{1}{\pi}\mathcal{P}\int_{-\infty}^{\infty} d\omega' \frac{\text{Re}\chi(\omega')}{\omega - \omega'}. \tag{5.53}$$

The relations can be derived [235, 236] (Ex. 5.2) by closing the contour in the upper plane where $\chi(z)$ is analytic. This assumes that $\chi(z)$ vanishes fast enough as $|z| \rightarrow \infty$ so that the integral on the half circle vanishes.

The KK relations are very useful in relating the energy loss (dissipation or absorption function, i.e., the spectral function that with Eq. (5.52) in general stems from the imaginary part) to the reaction function (real part) in experiments and calculations. The spectral function is non-zero only at energies where transitions can occur. As shown by the first of Eq. (5.53) and since with Eq. (5.52) Im χ < 0, the real part must be negative for frequencies below the onset of absorption and decay as $1/\omega$ for $\omega \rightarrow \infty$.

Fluctuation–dissipation theorem

The fluctuation–dissipation theorem [237, 238] links the response of a system to a time-varying perturbation with the fluctuations that naturally occur in statistical equilibrium. One can either observe the intrinsic fluctuations in the system (equilibrium correlation functions) or apply a perturbation and observe the consequences.

When a time-dependent external perturbation is applied to the system, power is dissipated because the system is excited. The sum of possible excitations is given by the spectral function Eq. (5.51), hence with Eq. (5.52), by the imaginary part of the response function. In thermal equilibrium and for $B = A^\dagger$ we obtain

$$\text{Im}\,\chi(\omega) = \pi(1 - e^{\beta\omega}) \sum_{\alpha\lambda} w_\alpha |A_{\lambda\alpha}|^2 \delta(\omega + E_\lambda - E_\alpha)$$

$$= \frac{1 - e^{\beta\omega}}{2} \int dt\, \langle \hat{A}^\dagger(0)\hat{A}(t)\rangle e^{i\omega t}. \tag{5.54}$$

This is a general formula relating dissipation to fluctuations in time. These may be fluctuations of density, current, spin density, or other properties of the system. Important consequences of the theorem include the Kubo formulas [238, 239] that are indispensable tools in the theory of transport and response functions. For example, the longitudinal conductivity in the x-direction is related to the current–current correlation function χ_{jj} as

$$\text{Re}\,\sigma_{xx}(q_x, \omega) = \frac{\text{Im}\,\chi_{j_x j_x}(q_x, \omega)}{\omega}. \tag{5.55}$$

Density–density response

We will now concentrate on correlation and response functions for charge and spin densities. They are related to many experiments, and moreover, the charge density–density correlation is one of the main ingredients of the Green's function description of extended systems in Chs. 10–21. As we can see from Eq. (5.54), the equal-time correlation functions in Sec. 5.3 are integrals over frequency of dynamic response functions.

The response to an external potential that couples to the charge or spin densities is given by the linear density–density response function χ. It is obtained from Eq. (5.48) with $\phi(2)\hat{B}(2) \to v_{\text{ext}}(2)\hat{n}(2)$ and $\hat{A}(1) \to \hat{n}(1)$, where the operators are defined in the Heisenberg picture. The induced change in charge or spin density $\delta n(1) = n([v_{\text{ext}}], \mathbf{r}_1, \sigma_1, t_1) - n([v_{\text{ext}} = 0], \mathbf{r}_1, \sigma_1)$ to first order in the external potential $v_{\text{ext}}(2)$ is given by

$$\delta n(1) = \int d2\, \chi(1,2) v_{\text{ext}}(2) \quad \text{with} \quad \chi(1,2) = -i\Theta(t_1 - t_2)\langle[\hat{n}(1), \hat{n}(2)]_-\rangle. \tag{5.56}$$

The general expression for the change in total potential $\delta v_{\text{tot}}(1)$ is the sum of external and induced potentials. Suppose that the induced potential is created by the induced density through an interaction $v(1, 3)$ that in general is non-local in space, spin, and time.[14] Then the relation to linear order can be written as

$$\delta v_{\text{tot}}(1) = v_{\text{ext}}(1) + \int d3\, v(1,3)\delta n(3) = v_{\text{ext}}(1) + \int d3 \int d2\, v(1,3)\chi(3,2)v_{\text{ext}}(2). \tag{5.57}$$

The most important function for our purposes is the response of the total charge density $n(\mathbf{r}, t) = \sum_\sigma n_\sigma(\mathbf{r}, t)$ to a spin-independent external potential. Since the Coulomb interaction v_c is spin-independent and instantaneous (neglecting relativistic effects),[15] the change in the total potential to linear order in the external potential is expressed by

$$\delta v_{\text{tot}}(\mathbf{r}_1, t_1) = \int d\mathbf{r}_2 \int dt_2\, \epsilon^{-1}(\mathbf{r}_1, \mathbf{r}_2, t_1 - t_2)v_{\text{ext}}(\mathbf{r}_2, t_2), \tag{5.58}$$

where the inverse dielectric function is defined as

$$\epsilon^{-1}(\mathbf{r}_1, \mathbf{r}_2, t_1 - t_2) = \delta(\mathbf{r}_1 - \mathbf{r}_2)\delta(t_1 - t_2) + \int d\mathbf{r}_3\, v_c(|\mathbf{r}_1 - \mathbf{r}_3|)\chi(\mathbf{r}_3, \mathbf{r}_2, t_1 - t_2) \tag{5.59}$$

and $\chi(\mathbf{r}_1, \mathbf{r}_2, t_1 - t_2) = \sum_{\sigma_1, \sigma_2} \chi(1, 2)$. Written in frequency space one then has

$$\delta v_{\text{tot}}(\mathbf{r}_1; \omega) = \int d\mathbf{r}_2\, \epsilon^{-1}(\mathbf{r}_1, \mathbf{r}_2; \omega)v_{\text{ext}}(\mathbf{r}_2; \omega) \tag{5.60}$$

with

$$\epsilon^{-1}(\mathbf{r}_1, \mathbf{r}_2; \omega) = \delta(\mathbf{r}_1 - \mathbf{r}_2) + \int d\mathbf{r}_3\, v_c(|\mathbf{r}_1 - \mathbf{r}_3|)\chi(\mathbf{r}_3, \mathbf{r}_2; \omega). \tag{5.61}$$

In a periodic system Eq. (5.61) reads

$$\epsilon^{-1}_{\mathbf{GG'}}(\mathbf{q}; \omega) = \delta_{\mathbf{GG'}} + v_c(\mathbf{q} + \mathbf{G})\chi_{\mathbf{GG'}}(\mathbf{q}; \omega), \tag{5.62}$$

where the $\mathbf{G}$ are reciprocal lattice vectors, $\mathbf{q}$ is a vector in the first Brillouin zone, and the Fourier transform of the Coulomb interaction $v_c(\mathbf{q} + \mathbf{G}) = \frac{4\pi}{|\mathbf{q} + \mathbf{G}|^2}$. With the general sum rule Eq. (5.34), the diagonal of the inverse dielectric function obeys the f-sum rule

$$\int_0^\infty d\omega\, \omega \text{Im}\, \epsilon^{-1}_{\mathbf{GG}}(\mathbf{q}; \omega) = -2\pi^2 n, \tag{5.63}$$

where n is the electron number density.

From the density–density response or inverse dielectric function one obtains the screened Coulomb interaction

$$W_{\mathbf{GG'}}(\mathbf{q}; \omega) = \epsilon^{-1}_{\mathbf{GG'}}(\mathbf{q}; \omega)v_c(\mathbf{q} + \mathbf{G'}) = v_c(\mathbf{q} + \mathbf{G}) + v_c(\mathbf{q} + \mathbf{G})\chi_{\mathbf{GG'}}(\mathbf{q}; \omega)v_c(\mathbf{q} + \mathbf{G'})$$
$$\equiv v_c(\mathbf{q} + \mathbf{G}) + W^p_{\mathbf{GG'}}(\mathbf{q}; \omega). \tag{5.64}$$

---

[14] This is still not the most general form of a response equation; for example, one may be interested in the induced density matrix due to a potential created by a density matrix perturbation. Such response functions appear in the framework of many-body perturbation theory; see, e.g., Secs. 11.1 and 11.7.

[15] For simplicity we write the expressions without spin polarization, but they are readily generalized to a ferromagnetic system where spin ↑ and ↓ are not equivalent.

Here we have defined the *polarization contribution* W^p. It differs from χ only through static factors. Hence, its analytic properties are the same. In particular, one can introduce a spectral function and perform analytic continuations; see Sec. 11.3.

The relations between n, χ, v, and ϵ are not limited to the causal response but also hold for the corresponding time-ordered or contour-ordered functions (see App. E), with the necessary caution at non-zero temperature where one cannot perform the analytic continuation of the time-ordered ones, as explained in Sec 5.4.

Dynamic structure factor

An important application for the density–density response is scattering theory, for example in the interpretation of X-ray and neutron scattering. Starting from time-dependent perturbation theory, the Born approximation neglects the contribution of the scattered field to the perturbation. Then one can use the Fermi golden rule to get the probability of inelastic transitions, which leads to the *dynamic structure factor*

$$S_{\mathbf{kk'}}^{\sigma\sigma'}(\omega) = \sum_{\alpha\neq\lambda} w_\alpha \varrho_{\alpha\lambda}^\sigma(-\mathbf{k})\varrho_{\lambda\alpha}^{\sigma'}(\mathbf{k}')\delta(\omega - (E_\lambda - E_\alpha)), \tag{5.65}$$

where $\varrho_{\alpha\lambda}(\mathbf{k}) = \langle\alpha|c_\mathbf{k}^\dagger c_{\mathbf{k'}}|\lambda\rangle$ for each spin. The density response is the sum over σ and σ', which is equivalent to Eq. (25.46) that is written in a form appropriate for a finite cell of N particles. The static structure factor in Eq. (5.17), $\langle\varrho^\sigma(-\mathbf{k})\varrho^{\sigma'}(\mathbf{k}')\rangle$, is the integral over frequencies of the dynamic expression in Eq. (5.65). Note, however, that the static structure factor also includes the $\alpha = \lambda$ contribution that is excluded in Eq. (5.65). This contribution would give rise to the elastic peak at $\omega = 0$, and it is responsible for the non-zero value of the static structure factor at $k = 0$ in Eq. (5.18). In a crystal, momentum conservation requires that $\mathbf{k} - \mathbf{k'}$ is a reciprocal lattice vector, so that the dynamic structure factor can be written as $S_{\mathbf{G,G'}}(\mathbf{q},\omega)$.

Link to experiment

The susceptibilities χ, the inverse dielectric function ϵ^{-1}, and the dynamic structure factor are directly linked to experimental spectra. Some important points are as follows.

- **The loss function.** As explained above, the imaginary part of χ is linked to dissipation. More precisely, $-\mathrm{Im}\,\epsilon_{\mathbf{GG'}}^{-1}(\mathbf{q};\omega) = -v_c(\mathbf{q}+\mathbf{G})\mathrm{Im}\,\chi_{\mathbf{GG}}(\mathbf{q};\omega)$ is the *loss function*. This is what one measures, e.g., in momentum-resolved electron energy-loss experiments, with a momentum transfer $\mathbf{Q} = \mathbf{q}+\mathbf{G}$.[16]
- **The dynamic structure factor.** The spectrum of equilibrium fluctuations can be measured by IXS [240]. More precisely, IXS measures the diagonal in momentum space $S(\mathbf{k},\omega) = \sum_{\sigma\sigma'} S_{\mathbf{kk}}^{\sigma\sigma'}(\omega)$ of the dynamic structure factor Eq. (5.65) summed over

---

[16] Off-diagonal elements of $\epsilon_{\mathbf{GG'}}^{-1}(\mathbf{q};\omega)$ are usually not accessed, although they contribute indirectly to experimental results such as spatially resolved electron energy loss.

spins. In a crystal and with $\mathbf{Q} = \mathbf{q} + \mathbf{G}$, we denote $S(\mathbf{Q}, \omega) = S_{\mathbf{G,G}}(\mathbf{q}, \omega)$. Using the fluctuation–dissipation theorem Eq. (5.54), this is

$$S(\mathbf{Q}, \omega) = -\frac{1}{\pi(1 - e^{-\beta\omega})} \text{Im} \chi_{\mathbf{G,G}}(\mathbf{q}, \omega) \qquad (5.66)$$

for a momentum transfer $\mathbf{Q} = \mathbf{q} + \mathbf{G}$. The spin-resolved version can be measured by neutron scattering. Coherent IXS (CIXS) [241] also allows one to study some off-diagonal $(\mathbf{G} \neq \mathbf{G}')$ elements of $S(\omega)$, but most experiments concentrate on the diagonal elements. Therefore the term "dynamic structure factor" often just refers to the diagonal.

- **Optical absorbtion spectrum Im $\epsilon_M(\mathbf{q}, \omega)$.** The *macroscopic dielectric function* $\epsilon_M(\mathbf{q}, \omega)$ is the inverse of the long-wavelength part of ϵ^{-1}, i.e., $\epsilon_M(\mathbf{q}, \omega) = [\epsilon_{00}^{-1}(\mathbf{q}; \omega)]^{-1}$. For $\mathbf{q} \to 0$, its imaginary part is measured in absorption spectra of light,[17] as illustrated for bulk silicon in the upper-left panel of Fig. 5.6. In order to relate to measurements for large momentum $\mathbf{Q} = \mathbf{q} + \mathbf{G}$, it is useful to define

$$\epsilon_M(\mathbf{Q}, \omega) = \frac{1}{\epsilon_{\mathbf{G=G}'}^{-1}(\mathbf{q}, \omega)} = \frac{1}{1 + v_c(\mathbf{Q})\chi_{\mathbf{G=G}'}(\mathbf{q}, \omega)}. \qquad (5.67)$$

The loss function expressed in terms of ϵ_M reads

$$-\text{Im} \, \epsilon^{-1}(\mathbf{Q}, \omega) = \frac{\text{Im} \, \epsilon_M(\mathbf{Q}, \omega)}{[\text{Re} \, \epsilon_M(\mathbf{Q}, \omega)]^2 + [\text{Im} \, \epsilon_M(\mathbf{Q}, \omega)]^2}. \qquad (5.68)$$

There is a crucial difference from the absorption given by $\text{Im} \, \epsilon_M$ in Eq. (5.67). In the expression (5.68), ϵ_M is screened and zeros of the real part of $\epsilon_M(\mathbf{Q}, \omega)$ in the denominator create plasmons, collective oscillations of the electrons. The effect is important, as one can appreciate from the example of bulk silicon in Fig. 5.6. The top-left panel shows the absorption spectrum ($\text{Im} \, \epsilon_M(\mathbf{Q} \to 0, \omega)$), the top-right panel displays the loss function for the same vanishing momentum transfer. Note that the plasmon peak in $-\text{Im} \, \epsilon^{-1}(\mathbf{Q}, \omega)$ is above 17 eV, far from the main absorption structures around 4 eV (note the change in energy scale between the panels). The two bottom panels show the same comparison for a larger momentum transfer $\mathbf{Q}$. Now the difference is smaller, because screening decreases with increasing momentum transfer (corresponding to smaller distances). The difference between the two spectra can also be understood by rewriting Eq. (5.67) as

$$\epsilon_M(\mathbf{Q}, \omega) = 1 - \frac{v_c(\mathbf{Q})\chi_{\mathbf{G=G}'}(\mathbf{q}, \omega)}{1 + v_c(\mathbf{Q})\chi_{\mathbf{G=G}'}(\mathbf{q}, \omega)} = 1 - v_c(\mathbf{Q})\bar{\chi}_{\mathbf{G=G}'}(\mathbf{q}, \omega) \qquad (5.69)$$

with $\bar{\chi}$ determined by the scalar relation

$$\bar{\chi}_{\mathbf{GG}}(\mathbf{q}, \omega) = \chi_{\mathbf{GG}}(\mathbf{q}, \omega) - \chi_{\mathbf{GG}}(\mathbf{q}, \omega)v_c(\mathbf{Q})\bar{\chi}_{\mathbf{GG}}(\mathbf{q}, \omega). \qquad (5.70)$$

---

[17] In general, ϵ is a tensor relating the electric field and polarization vectors. However, it can be reduced to scalar relations for the principal axes when a transformation to principal axes can be performed for the real and imaginary parts. This allows one to determine optical spectra by calculating the longitudinal response to a potential instead of the transverse response to a field. See [1, App. E].

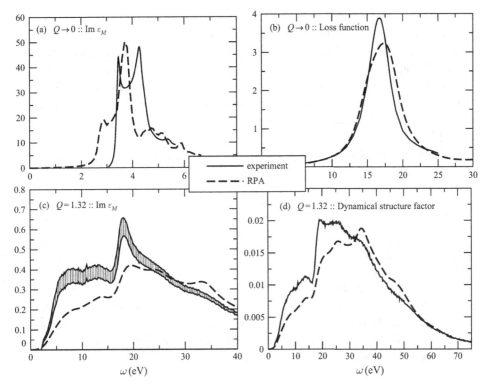

Figure 5.6. Electronic spectra of bulk silicon. Figure by Francesco Sottile, with data of the upper-right and the two lower panels taken from [242–244]. Left panels: Im $\epsilon_M(\omega)$ (corresponding to absorption when $\mathbf{Q} \to 0$, where the experiment is from [245]). Upper right: Loss function in 1/eV. Lower right: Dynamic structure factor in 1/eV. Top panels are for vanishing momentum transfer, bottom panels for $|\mathbf{Q}| = 1.32$ a.u. The result for $\epsilon_M(\mathbf{Q})$ at large $\mathbf{Q}$ has been obtained from the measured Im $\frac{1}{\epsilon_M}$ using the KK relation Eq. (5.53), which explains the error bar. Im ϵ_M and $-$Im $\frac{1}{\epsilon_M}$ are similar for large momentum transfer, but very different for small momentum transfer, where the long-range part of the Coulomb interaction plays an important role.

For small momentum transfer $\mathbf{Q}$, the difference between χ and $\bar{\chi}$ is large because of the long-range nature of the Coulomb interaction. This is the difference between the density–density response χ that contains the plasmons and is measured in loss spectroscopies and the macroscopic dielectric function measured for example in absorption. For larger $\mathbf{Q}$ the Coulomb component $v_c(\mathbf{Q})$ decreases and the two spectra become more and more similar. Note that this is independent of the level of approximation that is used to calculate the response function χ, and holds even when χ is based on an independent-particle approximation.

5.6 The one-particle Green's function

We will now look at the most widely used correlation function of this book: the one-body Green's function G. As outlined in the next section, most of the observables we are

interested in can be derived directly from G. The remaining ones are almost always linked
to the response functions in the previous section or the two-body Green's functions that
are discussed in Sec. 5.8. Some general properties of Green's functions are summarized
in Apps. C and D, along with examples and derivations of relevant aspects for our purposes. In particular, the conventions used here for Fourier transforms are given in Sec. C.2.
Examples of many aspects of Green's functions are brought out in Ch. 7 on particles
and quasi-particles. Detailed derivations can be found in texts such as [47] and [156],
or [233].[18]

Everything we need is outlined in the previous section; one simply has to make the
substitution

$$- iC \rightarrow G$$
$$\hat{A}_H(x_1, t_1) \rightarrow \hat{\psi}(x_1, t_1)$$
$$\hat{B}_H(x_2, t_2) \rightarrow \hat{\psi}^\dagger(x_2, t_2), \tag{5.71}$$

where $\hat{\psi}$ stands for the Heisenberg operator defined in Eq. (B.13). Since $\hat{\psi}^\dagger$ adds one particle, μ is the usual chemical potential. We will look specifically at the case of *fermions* here.
Hence the fermion commutation rules apply, corresponding to the lower sign in the expressions of Sec. 5.4. Moreover, we assume *thermal equilibrium*. The relations for bosons
and/or statistical factors other than thermal equilibrium are readily derived using the results
of the previous sections.

Propagating a particle

The building blocks of the Green's functions are the *propagators*, analogous to Eqs. (5.21)
and (5.24):

$$G^>(x_1, t_1; x_2, t_2) = -i\langle \hat{\psi}(x_1, t_1)\hat{\psi}^\dagger(x_2, t_2)\rangle \tag{5.72}$$

and

$$G^<(x_1, t_1; x_2, t_2) = i\langle \hat{\psi}^\dagger(x_2, t_2)\hat{\psi}(x_1, t_1)\rangle. \tag{5.73}$$

These are dynamic correlation functions; for example, with respect to the previous section
we have

$$G^>(x_1, t_1; x_2, t_2) = -iC_{\hat{\psi}\hat{\psi}^\dagger}(x_1, t_1; x_2, t_2), \tag{5.74}$$

and the equivalent for all other one-particle Green's functions.

The propagators express correlation between creation (annihilation) of a particle at time
t_2 and annihilation (creation) of a particle at time t_1. To see this more clearly, look for
example at $G^>$ for zero temperature, expressed in the Schrödinger picture

$$G^>(x_1, t_1; x_2, t_2)_{T \rightarrow 0} = -i\langle \Psi_0(t_1)|\hat{\psi}_S(x_1)\hat{\psi}_S^\dagger(x_2)|\Psi_0(t_2)\rangle. \tag{5.75}$$

---

[18] Note however the differences in notation, in particular a factor $1/(2\pi)$ in the spectral function.

At time t_2 the system is in state $|\Psi_0(t_2)\rangle$. At that time an electron is added in point x_2, and $G^>$, being the scalar product of the two states $\hat{\psi}_S^\dagger(x_2)|\Psi_0(t_2)\rangle$ and $\hat{\psi}_S^\dagger(x_1)|\Psi_0(t_1)\rangle$, gives the probability amplitude to find an extra particle in x_1 at time t_1.

From these propagators we can now build retarded, advanced, and time-ordered Green's functions corresponding to Eqs. (5.25), (5.26), and (5.27):

$$G^R(x_1,t_1;x_2,t_2) = \Theta(t_1 - t_2)\left(G^>(x_1,t_1;x_2,t_2) - G^<(x_1,t_1;x_2,t_2)\right), \qquad (5.76)$$

$$G^A(x_1,t_1;x_2,t_2) = \Theta(t_2 - t_1)\left(G^<(x_1,t_1;x_2,t_2) - G^>(x_1,t_1;x_2,t_2)\right), \qquad (5.77)$$

and the time-ordered

$$\begin{aligned} G^T(x_1,t_1;x_2,t_2) &= -i\left\langle \mathrm{T}\left[\hat{\psi}(x_1,t_1)\hat{\psi}^\dagger(x_2,t_2)\right]\right\rangle \\ &= G^R(x_1,t_1;x_2,t_2) + G^<(x_1,t_1;x_2,t_2) \\ &= G^A(x_1,t_1;x_2,t_2) + G^>(x_1,t_1;x_2,t_2). \end{aligned} \qquad (5.78)$$

For *independent particles* the Green's functions are diagonal in the basis of the single-particle orbitals; their matrix elements read

$$G_{\ell\ell}^{0>} = -ie^{-i\varepsilon_\ell(t_1-t_2)}\left[1 - \frac{1}{1 + e^{\beta(\varepsilon_\ell-\mu)}}\right], \qquad (5.79)$$

$$G_{\ell\ell}^{0<} = ie^{-i\varepsilon_\ell(t_1-t_2)}\frac{1}{1 + e^{\beta(\varepsilon_\ell-\mu)}}, \qquad (5.80)$$

$$G_{\ell\ell}^{0R} = -i\Theta(t_1 - t_2)e^{-i\varepsilon_\ell(t_1-t_2)} \qquad G_{\ell\ell}^{0A} = i\Theta(t_2 - t_1)e^{-i\varepsilon_\ell(t_1-t_2)}, \qquad (5.81)$$

$$G_{\ell\ell}^{0T} = -ie^{-i\varepsilon_\ell(t_1-t_2)}\left[\Theta(t_1 - t_2)\frac{e^{\beta(\varepsilon_\ell-\mu)}}{1 + e^{\beta(\varepsilon_\ell-\mu)}} - \frac{\Theta(t_2 - t_1)}{1 + e^{\beta(\varepsilon_\ell-\mu)}}\right], \qquad (5.82)$$

where the ε_ℓ are eigenvalues of the single-particle hamiltonian $\hat{h}$. At zero temperature, the last expression becomes

$$G_{\ell\ell\,T=0}^{0T} = -ie^{-i\varepsilon_\ell(t_1-t_2)}\left[\Theta(t_1 - t_2)\Theta(\varepsilon_\ell - \mu) - \Theta(t_2 - t_1)\Theta(\mu - \varepsilon_\ell)\right]. \qquad (5.83)$$

By taking the derivative with respect to t_1 as suggested in Ex. 5.4, one finds that G^{0R}, G^{0A}, and G^{0T} obey the same *equation of motion*,

$$\left[i\frac{\partial}{\partial t_1} - h_{\ell\ell}\right]G_{\ell\ell}^0 = \delta(t_1 - t_2). \qquad (5.84)$$

Hence, in operator form

$$i\frac{\partial}{\partial t_1} - \hat{h} = [G^0]^{-1}. \qquad (5.85)$$

Equation (5.84) illustrates why G^0 is called a Green's function.[19]

---

[19] Note that such an equation does not exist for $G^>$ and $G^<$; these are rather building blocks that have to be combined with the Θ-functions in order to lead to a Green's function equation like Eq. (5.84).

For *interacting particles* with Green's functions given in Eqs. (5.76)–(5.78), the time derivative also leads to a δ-function in time due to the Heavyside Θ-functions in the Green's functions. The anticommutation relation of the creation and annihilation operators moreover leads to a term $\delta(x_1 - x_2)\delta(t_1 - t_2)$, as one expects in general from a Green's function (see Ex. 5.4). The resulting equation of motion is given in Eq. (10.3) and used in Ch. 10.

Since the various Green's functions, G^R, G^A, or G^T, are interrelated, one can in principle choose which one to work with. Note that G^{0R} and G^{0A} only contain information about a single particle, whereas the thermal occupation factor and chemical potential, that refer to the whole system, appear in $G^{0\gtrless}$ and G^{0T}. As we will see in Ch. 9, the time-ordered form G^T is suitable for perturbation expansions at zero temperature, so we will mostly refer to this choice in the applications in Chs. 11–14. Moreover, many formulae remain valid even at non-zero temperature and/or out of equilibrium when one replaces ordering in real time by ordering on a contour as explained in App. E. Therefore, in Chs. 10–15 the superscript on G^T is omitted and the symbol G is used, whenever this does not cause confusion.

The Green's functions in frequency space

The Fourier transforms with respect to time of the propagators Eqs. (5.72), (5.73) in analogy to Eqs. (5.28), (5.29) are

$$G^>(x_1, x_2, \omega) = -2\pi i \sum_{\alpha\lambda} w_\alpha f_{\alpha\lambda}(x_1) f^*_{\alpha\lambda}(x_2)\delta(\omega + E_\alpha - E_\lambda) \tag{5.86}$$

and

$$G^<(x_1, x_2, \omega) = +2\pi i \sum_{\alpha\lambda} w_\lambda f_{\alpha\lambda}(x_1) f^*_{\alpha\lambda}(x_2)\delta(\omega + E_\alpha - E_\lambda), \tag{5.87}$$

where E_α are eigenvalues of the many-body hamiltonian $\hat{H}$ and f are the Dyson amplitudes

$$f_{\alpha\lambda}(x_1) = \langle\alpha|\hat{\psi}(x_1)|\lambda\rangle. \tag{5.88}$$

Here $|\alpha\rangle$ and $|\lambda\rangle$ are eigenstates of $\hat{H}$ and $|\lambda\rangle$ contains one particle more than $|\alpha\rangle$. The Dyson amplitudes measure the overlap between an eigenstate of $\hat{H}$ and a state where a particle (electron or hole) has been added to another eigenstate. In the case of independent particles, this is just the wavefunction of the additional particle. The equivalent of Eq. (5.31) links the propagators through

$$G^<(x_1, x_2, \omega) = -e^{-\beta(\omega - \mu)}G^>(x_1, x_2, \omega). \tag{5.89}$$

The retarded and advanced Green's functions are obtained by analytic continuation following Eq. (5.36),

$$G(z) = \int_{-\infty}^{\infty} d\omega' \frac{A(\omega')}{z - \omega'} \tag{5.90}$$

and

$$G^R(\omega) = \lim_{\eta\to 0^+} G(\omega + i\eta) \qquad G^A(\omega) = \lim_{\eta\to 0^+} G(\omega - i\eta). \tag{5.91}$$

Note that we do not introduce a new symbol for the Green's function in the complex plane; the analytic continuation is understood from the argument. The analytic continuation is particularly important for the Green's functions; for example, it is used to define quasi-particles (see Sec. 7.7) and to set the initial conditions in the equation of motion for the Green's function (see Sec. G.1).

The *spectral function* of G corresponding to Eq. (5.32) is given by

$$
A(x_1, x_2, \omega) = \frac{i}{2\pi} \left[G^>(x_1, x_2, \omega) - G^<(x_1, x_2, \omega) \right] = -\frac{i}{2\pi} G^<(x_1, x_2, \omega) \left[1 + e^{\beta(\omega - \mu)} \right]
$$

$$
= \frac{i}{2\pi} \left[G^R(x_1, x_2, \omega) - G^A(x_1, x_2, \omega) \right]. \tag{5.92}
$$

In particular, we have that

$$
A_{\ell\ell}(\omega) = -\frac{1}{\pi} \operatorname{Im} G^R_{\ell\ell}(\omega), \tag{5.93}
$$

where $A_{\ell\ell}$ is a matrix element of the spectral function in an orthonormal basis.

Using Eqs. (5.86) and (5.87), the spectral function can be expressed in the form

$$
A(x_1, x_2, \omega) = \sum_{\alpha\lambda} f_{\alpha\lambda}(x_1) f^*_{\alpha\lambda}(x_2) \delta(\omega + E_\alpha - E_\lambda) [w_\alpha + w_\lambda]. \tag{5.94}
$$

This expression illustrates the physics of the spectral function most clearly: it is the spectrum of the poles of the Green's function corresponding to electron addition and removal energies, weighted by the statistical factor and the Dyson amplitudes $f_{\alpha\lambda}$. From Eq. (5.34) and the commutation rules for field operators follows the sum rule

$$
\int d\omega \, A_{\ell\ell}(\omega) = 1. \tag{5.95}
$$

Using Eqs. (5.71) or (5.74), one can apply all findings of Sec. 5.4 to the Green's functions. In particular, Eq. (5.39) becomes

$$
G^T(\omega) = \left[\mathcal{P} \int d\omega' \frac{A(\omega')}{\omega - \omega'} - i\pi \tanh \frac{\beta(\omega - \mu)}{2} A(\omega) \right], \tag{5.96}
$$

whereas

$$
G^R(\omega) = \left[\mathcal{P} \int d\omega' \frac{A(\omega')}{\omega - \omega'} - i\pi A(\omega) \right], \tag{5.97}
$$

without the thermal factor that appears in G^T.

To obtain the expressions for *non-interacting particles*, one can expand the field operators as defined in Eq. (A.12): $\hat{\psi}(x) = \sum_\ell \psi_\ell(x) c_\ell$ in terms of the single-particle eigenfunctions $\psi_\ell(x)$ and annihilation operators c_ℓ, where ℓ labels the single-particle states; similarly for $\hat{\psi}^\dagger(x)$. This shows that for a Slater determinant, the Dyson amplitudes are zero unless $|\alpha\rangle$ and $|\lambda\rangle$ differ only by one orbital. In the latter case, the $f_{\alpha\lambda}(x_1)$ equals that single-particle orbital, and each total energy is replaced by a sum over the eigenvalues ε_ℓ of the occupied single-particle states.

The limit of zero temperature and fixed particle number

For zero temperature and fixed number of particles N, in analogy with Eq. (5.28), the Fourier transform of the propagator Eqs. (5.86) and (5.87) becomes

$$G^{\gtrless}(x_1, x_2, \omega) = \mp 2\pi i \sum_\lambda f_\lambda(x_1) f_\lambda^*(x_2) \delta(\omega - \varepsilon_\lambda), \tag{5.98}$$

where $G^>$ is for *electron addition*,

$$\varepsilon_\lambda = E(N+1, \lambda) - E_0 > \mu,$$
$$f_\lambda(x_1) = f_{0\lambda}(x_1), \tag{5.99}$$

and $G^<$ for *electron removal*,

$$\varepsilon_\lambda = E_0 - E(N-1, \lambda) < \mu,$$
$$f_\lambda(x_1) = f_{\lambda 0}(x_1). \tag{5.100}$$

Here E_0 is the N-electron ground-state energy, and the amplitudes $f_{\alpha\lambda}$ are defined as in Eq. (5.88).

The spectral function becomes

$$A(x_1, x_2, \omega) = \sum_\alpha f_\lambda(x_1) f_\lambda^*(x_2) \delta(\omega - \varepsilon_\lambda), \tag{5.101}$$

which leads to

$$G^{\gtrless}(x_1, x_2, \omega) = \mp i 2\pi A(x_1, x_2, \omega) \Theta(\pm[\omega - \mu]). \tag{5.102}$$

As for the general correlation function, at vanishing temperature the time-ordered $G^T(z)$ can be analytically continued to $G(z)$, Eq. (5.90). Therefore for zero temperature Eq. (5.93) can also be written as

$$A_{\ell\ell}(\omega) = \frac{1}{\pi} |\text{Im } G_{\ell\ell}^T(\omega)|. \tag{5.103}$$

The spectral representation in terms of excitations of G^T at zero temperature is

$$G^T(x_1, x_2, \omega) = \lim_{\eta \to 0^+} \sum_\lambda \frac{f_\lambda(x_1) f_\lambda^*(x_2)}{\omega - \varepsilon_\lambda + i\eta \, \text{sgn}(\varepsilon_\lambda - \mu)}, \tag{5.104}$$

and its analytic structure is given in the right panel of Fig. 5.5.

Independent particles at zero temperature and fixed particle number

The Green's function for independent particles at zero temperature is obtained as before by expanding the field operators in single-particle orbitals ψ_ℓ; this yields for the time-ordered Green's function G_0^T,

$$G_0^T(x_1, x_2, \omega) = \lim_{\eta \to 0^+} \sum_\ell \frac{\psi_\ell(x_1) \psi_\ell^*(x_2)}{\omega - \varepsilon_\ell + i\eta \, \text{sgn}(\varepsilon_\ell - \mu)}. \tag{5.105}$$

G_0^T is diagonal in the single-particle basis, and we have

$$G_{0\,\ell\ell}^T(\omega) = \frac{1}{\omega - \varepsilon_\ell + i\eta\,\mathrm{sgn}(\varepsilon_\ell - \mu)}. \tag{5.106}$$

The spectral function is

$$A(x_1, x_2, \omega) = \sum_\ell \psi_\ell(x_1)\psi_\ell^*(x_2)\delta(\omega - \varepsilon_\ell), \tag{5.107}$$

or

$$A_{\ell\ell}(\omega) = \delta(\omega - \varepsilon_\ell), \tag{5.108}$$

and the trace of the spectral function is simply the spectrum of independent-particle eigenvalues, the density of states.

The effects of interaction become clear from a comparison of Eq. (5.108) and matrix elements of the fully interacting spectral function Eq. (5.101) in the same basis,

$$A_{\ell\ell}(\omega) = \sum_\lambda |\langle\ell|f_\lambda\rangle|^2 \delta(\omega - \varepsilon_\lambda), \quad \text{where} \quad \langle\ell|f_\lambda\rangle = \int dx_1\,\psi_\ell^*(x_1)f_\lambda(x_1). \tag{5.109}$$

The independent-particle spectral function has a single δ-peak since $\langle\ell|\ell'\rangle = \delta_{\ell\ell'}$, whereas in general $\langle\ell|f_\lambda\rangle$ has non-vanishing contributions from many λ. For example, in a homogeneous system there are several ways to construct a many-body wavefunction with total momentum $\mathbf{k}$. Therefore the interacting $A_{\ell\ell}(\omega)$ consists of many peaks, that merge to form a continuum in the case of an infinite system. This explains the spectral functions shown in Figs. 2.5, 2.8, or 2.9 and discussed in Sec. 7.4, where interaction transforms the non-interacting δ-peak into a broadened structure accompanied by satellites; the latter stem from collective excitations contained in $|\lambda\rangle$ that still conserve the total momentum of the state.[20]

5.7 Useful quantities derived from the one-particle Green's function

The knowledge of the one-particle Green's function allows one to access much useful information. Most importantly, the expectation value of any one-body operator can be written in terms of the Green's function. This follows from the definition of the expectation value Eq. (5.2), the expression of a one-body operator in terms of field operators (see App. A)

$$\hat{O} = \int dx_1\,\hat{\psi}^\dagger(x_1)O(x_1)\hat{\psi}(x_1), \tag{5.110}$$

and the definition of the propagator Eq. (5.73), yielding

$$\langle\hat{O}\rangle = -i\int dx_1\,O(x_1)G^<(x_1, t; x_1, t) = -i\int dx_1\,O(x_1)\lim_{t'\to t^+}G^T(x_1, t; x_1, t'), \tag{5.111}$$

---

[20] Note that this consideration can be extended to include phonons; the total electron–phonon system can be excited with conserved momentum, leading also to phonon sidebands.

where

$$t^+ = \lim_{\eta \to 0} t + \eta, \qquad \eta > 0. \tag{5.112}$$

Equation (5.111) can also be extended to non-local operators as

$$\langle \hat{O} \rangle = -i \int dx_1 \, dx_2 \, O(x_1, x_2) G^<(x_1, t; x_2, t). \tag{5.113}$$

In the following, with the Fourier transforms, Eq. (C.7), we make use of

$$G^<(t;t) = \lim_{t' \to t^+} G^T(t;t') = \frac{1}{2\pi} \int d\omega G^<(\omega) = \frac{1}{2\pi} \lim_{\eta \to 0^+} \int d\omega e^{i\eta\omega} G^T(\omega), \tag{5.114}$$

where we have omitted the space and spin coordinates. The expectation value becomes

$$\langle \hat{O} \rangle = \frac{1}{2\pi} \lim_{n \to 0^+} \int dx_1 \, dx_2 \, O(x_1, x_2) \int d\omega e^{i\eta\omega} G^T(x_1, x_2, \omega). \tag{5.115}$$

Finally, with Eq. (5.92) the expectation value Eq. (5.113) can be expressed as

$$\langle \hat{O} \rangle = \int dx_1 \, dx_2 \, O(x_1, x_2) \int d\omega \, \frac{A(x_1, x_2, \omega)}{1 + e^{\beta(\omega - \mu)}} \tag{5.116}$$

$$= \frac{1}{\beta} \sum_\nu \int dx_1 \, dx_2 \, O(x_1, x_2) G(x_1, x_2, z_\nu), \tag{5.117}$$

where the spectral function is weighted by the Fermi distribution $1/(1 + e^{\beta(\omega - \mu)})$ in Eq. (5.116), and in Eq. (5.117) it is expressed as a sum over Matsubara frequencies (see App. D). The various expressions in time and frequency space are equivalent, and can be chosen according to convenience. In the following we will use the general formula of Eq. (5.116). All expressions are readily converted to the Matsubara form following the example of Eqs. (5.116) and (5.117).

Some important expectation values are as follows.

- **Density.** With the local operator $\hat{n}(x) = \hat{\psi}^\dagger(x)\hat{\psi}(x)$, the (spin-resolved) density is given by

$$n(x) = -iG^T(x, t, ; x, t^+) = \int d\omega \frac{A(x, x, \omega)}{1 + e^{\beta(\omega - \mu)}} \tag{5.118}$$

$$= \sum_\lambda \Theta(\mu - \varepsilon_\lambda)|f_\lambda(x)|^2 \qquad \text{for } T \to 0. \tag{5.119}$$

  It is a sum over squared amplitudes for annihilation of a particle in the ground state, i.e., the probability that the ground state contains a particle at point x.

- **Density matrix.** The density matrix, given by Eq. (5.6), is proportional to the equal time limit of the Green's function,

$$\rho(x_1, x_2) = -iG^T(x_1, t, ; x_2, t^+) \tag{5.120}$$

$$= \sum_\lambda \Theta(\mu - \varepsilon_\lambda)f_\lambda(x_1)f_\lambda^*(x_2) \qquad \text{for } T \to 0. \tag{5.121}$$

With the correspondence $f_\lambda \leftrightarrow \psi_\ell$, and $\varepsilon_\lambda \leftrightarrow \varepsilon_\ell$, one can easily see the connection to the non-interacting case. A diagonal element in a basis reads, following Eq. (5.117),

$$\rho_{\ell\ell} = \int d\omega \frac{A_{\ell\ell}(\omega)}{1 + e^{\beta(\omega-\mu)}}. \tag{5.122}$$

From this equation one can in particular deduce the momentum distribution, Eq. (5.8). For zero temperature the latter yields

$$\rho(\mathbf{k}) = \sum_\lambda \Theta(\mu - \varepsilon_\lambda) |\langle \mathbf{k} | f_\lambda \rangle|^2, \tag{5.123}$$

where the f_λ are projected on plane waves. This shows immediately that for independent electrons in the homogeneous electron gas, where the plane waves are eigenstates, $\rho(\mathbf{k}) = 1$ for $|\mathbf{k}|^2/2 < \mu$ and $\rho(\mathbf{k}) = 0$ for $|\mathbf{k}|^2/2 > \mu$. In a non-interacting but inhomogeneous system the f_λ are single-particle states $|\lambda\rangle$ different from $|\mathbf{k}\rangle$. Hence $\langle \mathbf{k} | f_\lambda \rangle$ is no longer a δ-function and therefore $\rho(\mathbf{k})$ is modified, even without interaction. However, the density matrix is still diagonal in the single-particle eigenstates with eigenvalues, i.e., occupation numbers, that are 0 or 1. In the interacting system one cannot find a basis where this is possible, and the distribution is modified due to scattering between electrons, as shown in Figs. 5.2 and 11.7, and discussed in Sec. 7.5.

- **Kinetic energy.** Equations (5.113) to (5.117) yield for the kinetic energy

$$\langle \hat{T} \rangle = i \int dx \lim_{x' \to x} \left[\frac{\nabla_\mathbf{r}^2}{2} G(x, t, x', t^+) \right] \tag{5.124}$$

$$= -\int dx \lim_{x' \to x} \left[\frac{\nabla_\mathbf{r}^2}{2} \int d\omega \frac{A(x, x', \omega)}{1 + e^{\beta(\omega-\mu)}} \right] \tag{5.125}$$

$$= -\sum_\lambda \Theta(\mu - \varepsilon_\lambda) \int dx f_\lambda^*(x) \frac{\nabla^2}{2} f_\lambda(x) \qquad \text{for } T \to 0. \tag{5.126}$$

Combining Eqs. (5.122) and (5.125) yields Eq. (5.10).

- **Coulomb potential energy.** The potential energy due to interactions between particles[21] $\langle \hat{V}_{ee} \rangle = \langle \int dx dx' \hat{\psi}^\dagger(x) \hat{\psi}^\dagger(x') v_c(\mathbf{r}, \mathbf{r}') \hat{\psi}(x') \hat{\psi}(x) \rangle$ depends on the two-body interaction v_c; for this reason it can be expressed in terms of the pair probability $n(x, x')$, Eq. (5.12). It might seem that the one-body Green's function alone should not contain the needed information. Indeed, G is not sufficient to specify a two-body correlation function, nevertheless, it is sufficient to determine the energy. The reason lies in the fact that the time derivative of a field operator in the Heisenberg representation Eq. (B.14),

$$i \frac{\partial}{\partial t} \hat{\psi}(x, t) = \hat{h}(x) \hat{\psi}(x, t)$$

$$+ \int dx' v_c(\mathbf{r}, \mathbf{r}') \hat{\psi}^\dagger(x', t) \hat{\psi}(x', t) \hat{\psi}(x, t), \tag{5.127}$$

---

[21] Note that the Coulomb potential energy is divergent and must be compensated by a neutralizing background, as done in Eq. (5.15) and used throughout the book. We do not specify this contribution here.

creates a linear term with the independent-particle part $h(x) = -\frac{\nabla_r^2}{2} + v_{ext}(x)$ of the hamiltonian, plus a term involving the Coulomb interaction v_c and three field operators. Hence one can express the expectation value of the interaction contribution using the time derivative of the one-body Green's function and Eq. (A.11):

$$
\begin{aligned}
\langle \hat{V}_{ee} \rangle &= \frac{1}{2} \int dx \lim_{\substack{t' \to t^+ \\ x' \to x}} \left[\frac{\partial}{\partial t} + ih(x) \right] G(x, t, x', t') \\
&= \frac{1}{2} \int dx \lim_{x' \to x} \int d\omega \, [\omega - h(x)] \frac{A(x, x', \omega)}{1 + e^{\beta(\omega - \mu)}} \\
&= \frac{1}{2\beta} \sum_\nu \int dx \lim_{x' \to x} [z_\nu - h(x)] G(x, x', z_\nu),
\end{aligned}
\tag{5.128}
$$

where again $h(x) = -\frac{\nabla_r^2}{2} + v_{ext}(x)$.

- **Total energy.** With $E = \langle \hat{H}_0 + \hat{V}_{ee} \rangle$ one obtains for the total energy the Galitskii–Migdal formula [35]

$$
E = \frac{1}{2} \int dx \lim_{\substack{t' \to t^+ \\ x' \to x}} \left[\frac{\partial}{\partial t} - ih(x) \right] G(x, t, x', t').
\tag{5.129}
$$

- **Grand partition function and grand potential.** In general there are no simple expressions for Z and $\Omega = -\beta^{-1} \ln Z$ (see Eq. (5.2)) in terms of the Green's function, contrary to the energy Eq. (5.129). The grand potential Ω implicitly contains the contribution of entropy $S = -k_B \text{Tr}\,[\hat{\rho} \ln \hat{\rho}]$, which measures disorder as encoded in the density matrix ρ [246]. When the density matrix has eigenvalues 0 or 1, the entropy is zero, whereas it is maximum when all eigenvalues are equal. The former is the case for zero temperature and vanishing interaction. An explicit expression for Ω can be found using the adiabatic connection approach where one starts with the non-interacting hamiltonian $\hat{H}_0$ and introduces a coupling constant λ that scales the interaction

$$
\hat{H}_\lambda = \hat{H}_0 + \lambda \hat{V}_{ee}.
\tag{5.130}
$$

Then

$$
\frac{\partial}{\partial \lambda} \ln Z = \frac{1}{Z} \text{Tr} \left[\frac{\partial}{\partial \lambda} e^{-\beta(\hat{H}_\lambda - \mu \hat{N})} \right] = -\frac{\beta}{\lambda} \langle \lambda \hat{V}_{ee} \rangle_\lambda,
\tag{5.131}
$$

where the subscript λ at the expectation value indicates that the weight factors of the thermal average contain $\lambda \hat{V}_{ee}$.[22] Finally one can integrate from the non-interacting $Z^0 = Z(\lambda = 0)$ to the fully interacting $Z = Z(\lambda = 1)$, and with Eq. (5.128) and $\Omega = -\frac{1}{\beta} \ln Z$,

$$
\Omega = \Omega^0 + \frac{1}{2} \int_0^1 \frac{d\lambda}{\lambda} \int dx \lim_{x' \to x} \int d\omega \, [\omega - h(x)] \frac{A_\lambda(x, x', \omega)}{1 + e^{\beta(\omega - \mu)}}.
\tag{5.132}
$$

---

[22] This is the reason why we have to keep λ explicitly as prefactor of $\hat{V}$; otherwise, there would be inconsistency between the potential and the weights, and the result could not be identified with that of a system with scaled interaction. Also note that the fact that $\hat{V}_{ee}$ does not commute with the rest of the hamiltonian is not a problem because the trace is cyclic.

Luttinger and Ward have written Ω as the sum of a functional Φ that can be expressed as a series of diagrams in terms of the Green's function G, and further terms involving G and the self-energy (Eq. (8.8)). Their derivation, done in terms of Matsubara frequencies, can be found in App. D.

An alternative approach is to use thermodynamic relations to calculate the entropy S and thus the grand potential $\Omega = E - k_B TS - \mu N$. The entropy can be determined by an integration of specific heat divided by the temperature, $\Delta S(T) = \int_{T_0}^{T} C(T')/T' dT'$ starting from a state with known entropy. Useful expressions in terms only of the energy are given in Ex. 17.8, and an example of a calculation starting from $T \to \infty$ where the entropy is known is given in Sec. 20.2.

Link to experiment

Electron addition and removal and the associated spectral function are closely related to experiments such as tunneling, photoemission, and inverse photoemission. In each of these experiments an electron is either added to, or removed from, the material. Of course, there is no electron appearing or disappearing, and one should strictly speaking always look at an electron–hole, i.e., a two-particle Green's function that is introduced in the next section. However in photoemission, for example, an electron goes from the sample to the detector. In the *sudden approximation* [247], the state after photon absorption is described as a product of the outgoing free photoelectron and the remaining $(N - 1)$-electron system, in an excited state. The latter is described by the one-body spectral function of the hole left by the electron. Similar considerations hold for the closely related inverse photoemission and tunneling spectroscopies. The spectra are then given by $\int dx_1 \, dx_2 \, A(x_1, x_2; \omega) p(x_1, x_2)$, where $\hat{p}$ is a perturbation operator that depends on the specific measurement. For example, if one excites the sample with light or soft X-rays, $\hat{p}$ is the dipole operator. According to Eq. (5.94) the spectrum is then a weighted sum of dipole transitions between many-body states.

The sudden approximation is a severe approximation. It neglects, for example, the fact that the photoelectron, before arriving at the detector, must travel in the material and pass through a surface or interface region which can affect the measured results substantially, since the electron interacts strongly with the material and can lose energy, e.g., through plasmon emission. This is called the *extrinsic* effects, as opposed to the *intrinsic* one-body spectral function. Interaction between electron and hole also leads to interference effects. These factors are instead contained in the two-body spectral function corresponding to a one-step picture of photoemission, where one does not describe the creation of the photo-hole separately from the rest. They are discussed in detail in reviews such as [59, 247]. Modifications to the spectra can be important, but changes are mostly quantitative rather than qualitative, for example, the weight of satellites (see Fig. 2.5) may be changed rather than new structures be created. In the comparisons made in this book we will always, unless stated differently, assume that measured addition and removal spectra are essentially given by the intrinsic one-body spectral function. Many calculations make even further approximations, in particular, they approximate the transition matrix elements that stem from the spatial and spin dependence of the spectral function $A(x_1, x_2; \omega)$, often by replacing them with a constant that depends on symmetry.

5.8 Two-particle Green's functions

In the previous sections we have examined the one-body Green's function as a particular case of a correlation function. The topic of the present section is the two-body Green's function. Again, the expressions in Sec. 5.4 can be applied directly.

Definitions

The time-ordered two-particle Green's function is

$$G_2(1, 2, 1', 2') = (-i)^2 \langle T[\hat{\psi}(1)\hat{\psi}(2)\hat{\psi}^\dagger(2')\hat{\psi}^\dagger(1')] \rangle. \tag{5.133}$$

The two-particle correlation function L is then obtained from G_2 by subtracting an uncorrelated contribution as defined in Eq. (5.23),[23]

$$L(1, 2, 1', 2') = -G_2(1, 2, 1', 2') + G(1, 1')G(2, 2'). \tag{5.134}$$

Note that L contains fluctuations with respect to the uncorrelated propagation of two particles.

The one-particle Green's function contains information about electrons or holes, according to the time ordering of the two field operators. In the case of the two-particle correlation function many combinations of time orderings are possible. Some are of particular importance in practice, and we will focus on them in the following.

Propagation of electron–hole pairs

Let us first look at the case where creation and annihilation field operators alternate such that one can observe the propagation of electron–hole pairs. This is given by the equal time limits $t'_1 = t_1^+$ and $t'_2 = t_2^+$, where the meaning of t^+ is defined in Eq. (5.112). We can now define the electron–hole propagators

$$G_{eh}^>(1, 2; 1'2') = (-i)^2 \langle \hat{\psi}^\dagger(1')\hat{\psi}(1)\hat{\psi}^\dagger(2')\hat{\psi}(2) \rangle|_{t'_1 = t_1^+, t'_2 = t_2^+} \tag{5.135}$$

and

$$G_{eh}^<(1, 2; 1'2') = (-i)^2 \langle \hat{\psi}^\dagger(2')\hat{\psi}(2)\hat{\psi}^\dagger(1')\hat{\psi}(1) \rangle|_{t'_1 = t_1^+, t'_2 = t_2^+}. \tag{5.136}$$

---

[23] Different conventions for the sign and order of indices of L can be found. Here we adopt the choice of [248, 249] which is different from [13, 300]. With this convention, there is a factor $(-i)^2$ between Green's function and correlation function, in line with the factor $(-i)$ associated with a pair of field operators in the one-body Green's function.

Now all the results of Sec. 5.4 can be used, with the prescription

$$(-i)^2 C(1, 2) \rightarrow G_{eh}(1, 2, 1', 2')$$
$$\hat{A}_H(x_1, t_1) \rightarrow \hat{\psi}^\dagger(x_{1'}, t_1^+)\hat{\psi}(x_1, t_1)$$
$$\hat{B}_H(x_2, t_2) \rightarrow \hat{\psi}^\dagger(x_{2'}, t_2^+)\hat{\psi}(x_2, t_2). \tag{5.137}$$

Note that here $\hat{A}$ and $\hat{B}$ commute like bosons in the time-ordered product, and the chemical potential for the creation of an electron–hole pair is $\mu = 0$.

Just as for the one-body propagators $G^\gtrless$ in Sec. 5.6, the electron–hole propagators are probability amplitudes. $G^>_{eh}$ is the probability amplitude to find an electron–hole pair in (x_1', x_1) at time t_1 when such a pair is created with coordinates (x_2', x_2) at time t_2, and similarly for $G^<_{eh}$. For $t_2 = t_1$ and $x_i' = x_i$, $G^<_{eh}$ is the pair probability Eq. (5.12).

With Eqs. (5.134) and (5.120) we can also define

$$L^\gtrless_{eh}(1, 2; 1'2') = -G^\gtrless_{eh}(1, 2; 1'2') + \rho(x_1, x_1')\rho(x_2, x_2'), \tag{5.138}$$

where the $L^\gtrless_{eh}$ differs from $G^\gtrless$ only by a static contribution.[24] Note that

$$L^<_{eh}(1, 2; 1', 2') = L^>_{eh}(1', 2'; 1, 2). \tag{5.139}$$

The time-ordered electron–hole correlation function is

$$L^T_{eh}(1, 2; 1', 2') = \Theta(t_1 - t_2)L^>_{eh}(1, 2; 1', 2') + \Theta(t_2 - t_1)L^<_{eh}(1, 2; 1', 2')$$
$$= -\Theta(t_1 - t_2)G^>_{eh} - \Theta(t_2 - t_1)G^<_{eh} + \rho(x_1, x_1')\rho(x_2, x_2'), \tag{5.140}$$

and its retarded version is given by

$$L^R_{eh}(1, 2; 1', 2') = \Theta(t_1 - t_2)\left[L^>_{eh}(1, 2; 1', 2') - L^<_{eh}(1, 2; 1', 2')\right]. \tag{5.141}$$

Electron–hole correlation function in frequency space

When the hamiltonian does not depend on time, $L^\gtrless_{eh}$ depends only on the time difference $t = t_1 - t_2$. We can then perform the Fourier transform to frequency space

$$L^\gtrless_{eh}(x_1, x_1'; x_2, x_2'; \omega) = \int dt\, e^{i\omega t} L^\gtrless_{eh}(x_1, x_1'; x_2, x_2'; t). \tag{5.142}$$

With Eq. (5.139) it follows that

$$L^<_{eh}(x_1, x_1'; x_2, x_2'; \omega) = L^>_{eh}(x_2, x_2'; x_1, x_1'; -\omega). \tag{5.143}$$

The electron–hole spectral function obtained from Eqs. (5.32) and (5.28) reads

$$A_{eh}(x_1, x_1'; x_2, x_2'; \omega) = (1 - e^{-\beta\omega}) \sum_{\alpha, \lambda} w_\alpha f^{eh}_{\alpha\lambda}(x_1, x_1') f^{eh*}_{\alpha\lambda}(x_2', x_2)\delta(E_\alpha - E_\lambda + \omega), \tag{5.144}$$

---

[24] This is true when the external potential does not depend on time, otherwise the density n itself is time-dependent. Note that there is no such contribution for the one-body Green's function because in that case the operators that are correlated ($\hat{\psi}$ and $\hat{\psi}^\dagger$) do not conserve particle number, whereas here the pairs ($\hat{\psi}^\dagger\hat{\psi}$) do.

where E_α and E_λ are total energies of the system in the ground or excited states with the same number of electrons and

$$f_{\alpha\lambda}^{eh}(x_1, x_1') = \langle \alpha | \hat{\psi}^\dagger(x_1') \hat{\psi}(x_1) | \lambda \rangle. \tag{5.145}$$

Peaks in the electron–hole spectral function correspond to excitations that do not change particle number. We will often use the terminology "neutral excitations" when we refer to these processes. With Eq. (5.35) the retarded electron–hole correlation function is

$$L_{eh}^R(x_1, x_1'; x_2, x_2'; \omega) = i \sum_{\alpha,\lambda} \left(1 - e^{-\beta(E_\lambda - E_\alpha)}\right) w_\alpha \frac{f_{\alpha\lambda}^{eh}(x_1, x_1') f_{\alpha\lambda}^{eh*}(x_2', x_2)}{\omega - E_\lambda + E_\alpha + i\eta}. \tag{5.146}$$

Note that because of the bosonic prefactor, the contribution $\alpha = \lambda$ drops out. Instead, for the time-ordered correlation function, this is achieved by the extra static contribution in Eq. (5.140).

Link to measurable quantities

For the elements $x_1' = x_1$ and $x_2' = x_2$ of the retarded electron–hole correlation function L_{eh}^R one obtains from Eq. (5.141) the density–density response function

$$\chi(1,2) \equiv -iL^R(1,2;1^+,2^+) = -i\Theta(t_1 - t_2)\langle[\hat{n}(1), \hat{n}(2)]_-\rangle, \tag{5.147}$$

the same as Eq. (5.56). This highlights the link between the induced density as response to an external potential, the creation of electron–hole pairs, and the correlation of density fluctuations in the system.[25] Similarly, for example, the current–current response can be obtained from

$$\chi_{j_\alpha j_\beta}(1,2) = \delta_{\alpha,\beta}\delta(1,2)n(1) - \frac{1}{4}\left[(\nabla_{1\alpha} - \nabla_{1'\alpha})(\nabla_{2\beta} - \nabla_{2'\beta})L^R(1,2,1',2')\right]\Big|_{1'=1^+,2'=2^+}. \tag{5.148}$$

From Eq. (5.140) the time-ordered density–density correlation function, or reducible polarizability, is obtained as

$$\chi^T(1,2) \equiv -iL^T(1,2;1^+,2^+) = -i\langle T[\hat{n}(1)\hat{n}(2)]\rangle - in(1)n(2). \tag{5.149}$$

The time-ordered χ^T appears as a fundamental quantity in the many-body expansions in Ch. 11, whereas here χ is the retarded, measurable quantity. The two are closely linked, as outlined in Sec. 5.4. There is no problem in working in a time-ordered framework to derive the corresponding causal results.

The measurable quantities are functions of two arguments only, diagonal elements of the full four-argument, two-particle correlation function. However, it is not easy to calculate the observable functions directly when many-body effects are important, because the explicit expressions are not known. Very often it is therefore convenient to calculate the

---

[25] Note that Eq. (5.147) is the same as $-i\Theta(t_1 - t_2)\langle[(\hat{n}(1) - \langle\hat{n}(1)\rangle), (\hat{n}(2) - \langle\hat{n}(2)\rangle)]_-\rangle$.

four-point correlation function L, and then to take the elements of interest. This is worked out in Ch. 14.

Two electrons or two holes

Another important piece of information contained in the two-particle Green's function is the simultaneous propagation of two electrons or two holes. In order to access this, one can work with the propagators

$$G_{ee}(1,2;1'2') = (-i)^2 \langle \hat{\psi}(1)\hat{\psi}(2)\hat{\psi}^\dagger(1')\hat{\psi}^\dagger(2') \rangle_{|t_1'=t_2'^+, t_1=t_2^+} \qquad (5.150)$$

and

$$G_{hh}(1,2;1'2') = (-i)^2 \langle \hat{\psi}(1')^\dagger \hat{\psi}(2')^\dagger \hat{\psi}(1)\hat{\psi}(2) \rangle_{|t_1'=t_2'^+, t_1=t_2^+}. \qquad (5.151)$$

G_{ee} is the probability amplitude to find an electron pair in (x_1, x_2) at time t_1 when such a pair is created with coordinates (x_1', x_2') at time t_1'; similarly for G_{hh}. As in the case of the one-body Green's function there is no uncorrelated contribution when particle number is conserved, and combinations of G_{ee} and G_{hh} yield directly the correlation functions, besides a sign.

Insertion of a complete set of many-body states yields

$$G_{ee}(x_1, x_1'; x_2, x_2'; \omega) = -2\pi \sum_{\alpha\lambda} w_\alpha f_{\alpha\lambda}^{ee}(x_1, x_2) f_{\alpha\lambda}^{ee*}(x_2', x_1') \delta(\omega - E_\lambda + E_\alpha), \qquad (5.152)$$

where E_α, E_λ are eigenvalues of the many-body hamiltonian $\hat{H}$ with a difference in particle number $N_\lambda - N_\alpha = 2$, and

$$f_{\alpha\lambda}^{ee}(x_1, x_2) = \langle \alpha | \hat{\psi}(x_1)\hat{\psi}(x_2) | \lambda \rangle. \qquad (5.153)$$

The Fourier transform of this contribution to G_2 will therefore exhibit poles at the addition energies of two particles. The propagation of two holes is described in an analogous way. Direct experimental evidence of such energies is obtained, e.g., in Auger electron spectroscopy (AES) [251, 252] or in double charge transfer (DCT) spectroscopy [253–255] experiments.

Other equal time limits

A particular equal time limit of L, the Fourier transform of which has poles at electron addition and removal energies, appears in the equation of motion for the one-particle Green's function G, namely, $t_2 = t_1^+$, $t_{2'} = t_1^{++}$, Eq. (10.6). Special equal time limits also display discontinuities of G_2 in terms of G that stem from the commutation relations of the field operators; for example for $t_1 \to t_{1'} > t_2, t_{2'}$ one gets

$$\lim_{t_1 \to t_{1'}^+} G_2^T(1,2,1',2') - \lim_{t_1^+ \to t_{1'}} G_2^T(1,2,1',2') = -iG^T(2,2')\delta(x_1 - x_{1'}). \qquad (5.154)$$

When all times are equal one has the static correlation functions. In particular, the two-body density matrix is

$$\rho(x_1, x_2; x_{1'}, x_{2'}) = (-i)^2 G_2(x_1 t^+, x_2 t; x_{1'} t^{++}, x_{2'} t^{+++}).\tag{5.155}$$

This is the non-local generalization of the pair probability Eq. (5.12) in analogy to the one-body density matrix from the density. It allows one to express the total energy as

$$E = -i \int dx h(x) G(xt, x_1 t^+)\bigg|_{|x_1| \to x} + \frac{1}{2} \int dx dx_1 v_c(\mathbf{r}, \mathbf{r}_1) \rho(x, x_1; x, x_1),\tag{5.156}$$

where $\rho(x, x_1; x, x_1) = n(x, x_1)$, the pair correlation function in Eq. (5.12). This is the same result as that obtained combining Eqs. (5.125) or (5.10), (5.14), and (5.15) for infinite volume, and without the background correction, i.e., the last $1/r$ term in Eq. (5.15). A more complete overview of different time orderings can be found in [248, 256].

SELECT FURTHER READING

Abrikosov, A. A., Gorkov, L. P., and Dzyaloshinski, I. E. *Methods of Quantum Field Theory in Statistical Physics* (Prentice-Hall, Englewood Cliffs, NJ, 1963).

Doniach, S. and Sondheimer, E. H. *Green's Functions for Solid State Physicists* (reprinted in Frontiers in Physics Series, No. 44) (W. A. Benjamin, Reading, MA, 1974).

Economou, E. N. *Green's Functions in Quantum Physics*, 2nd edn. (Springer-Verlag, Berlin, 1992).

Fetter, A. L. and Walecka, J. D. *Quantum Theory of Many-Particle Systems* (McGraw-Hill, New York, 1971).

Gross, E. K. U., Runge, E., and Heinonen, O. *Many-Particle Theory* (Adam Hilger, Bristol, 1991). See especially Part II on general properties of Green's functions, which emphasizes the one-particle Green's function for fermions.

Mahan, G. D. *Many-Particle Physics*, 3rd edn. (Kluwer Academic/Plenum Publishers, New York, 2000).

Martin, P. C. *Measurements and Correlation Functions* (Gordon and Breach Science Publishers, New York, 1968). A wonderful introduction to correlation functions and the relation to measurable quantities.

Exercises

5.1 Prove the inequality in Eq. (5.19) starting with the inequality $\int_{-\infty}^{\infty} d\omega S(k, \omega)[\omega^{-\frac{1}{2}} - b\omega^{\frac{1}{2}}]^2 \geq 0$. Find the value of b which minimizes the right-hand side. Then use the f-sum rule, Eq. (5.63), and the relation between the dynamic structure factor and the dielectric susceptibility Eq. (5.66).

5.2 Derive the KK relations, Eq. (5.53), from the analytic properties of the response functions. *Hint:* An integral along the real axis can be closed in the upper plane with a contour that is at $|z| \to \infty$. Since the contour encloses no poles, the line integral vanishes; also, the integral at infinity vanishes. The integral along the axis can be broken down into the principal value parts and the residue parts, leading to Eq. (5.53). See [235, 236].

5.3 Evaluate Eq. (5.13) using a Hartree–Fock two-particle wavefunction, and show that the pair distribution is different from unity because of the Pauli principle. Also show that the pair correlation function for a completely uncorrelated situation would be different from the result for the non-interacting HEG shown in Fig. 5.3.

5.4 Calculate the time derivative of G^R, G^A, and G^T for independent particles to derive Eq. (5.84). Then take the time derivative of Eqs. (5.76)–(5.78), and, by using the anticommutation rules of the field operators, show that a term $\delta(t_1 - t_2)\delta(x_1 - x_2)$ appears. The full expression for this equation of motion is given in Ch. 10.

6

Many-body wavefunctions

In no wave function of the type (1) [product of single determinants for each spin] is there a statistical correlation between the positions of the electrons with antiparallel spin. The purpose of the aforementioned generalization of (1) is to allow for such correlations. This will lead to an improvement of the wave function and, therefore, to a lowering of the energy value.

E. Wigner, *Phys. Rev.* 46, 1002 (1934)

Summary

Although the exact many-body wavefunction cannot be written down in general, we do know of some of its properties. For example, there are differences between the wavefunctions of insulators and metals and the cusp condition gives the behavior as any two charges approach each other. In this chapter we also discuss approximate wavefunctions, ways to judge their accuracy and how to include electronic correlation. Examples of many-body wavefunctions are the Slater–Jastrow (pair product) wavefunction and its generalization to pairing and backflow wavefunctions.

In other places in this book, we argue that it is not necessary to look at the many-body wavefunctions explicitly because they are unwieldy; the one- and two-body correlation functions discussed in Ch. 5 are sufficient to determine the energy and give information on the excitation spectra. However, these correlation functions do not always contain all information of interest. In principle, the ground-state energy of a many-electron system is a functional of the density, but the very derivation of the theorem invokes the many-body wavefunction, as expressed explicitly in Eq. (4.16). The effects of antisymmetry are manifest in the correlation functions, but antisymmetry is most simply viewed as a property of the wavefunction; electronic correlation is fundamentally a result of properties of the many-body wavefunction.

Studying many-body wavefunctions provides a very useful, different point of view of many-body physics from the approaches based on correlation functions. Many of the most important discoveries in physics have come about by understanding the nature of wavefunctions, such as the Laughlin wavefunction for the fractional quantum Hall effect, the

BCS wavefunction for superconductors, p-wave pairing in superfluid ^{3}He, and the Heitler–London approach for molecular binding. The role of Berry's phases has brought out the importance of the phase of the wavefunction in determining properties of quantum systems. This has led to new understanding of the classification of insulators, metals, superconductors, vortices, and other states of condensed matter. Last but not least, a good many-body wavefunction is needed for efficient zero-temperature quantum Monte Carlo simulations.

When we say that we want a many-body wavefunction, we do not mean that we want to tabulate its values,[1] but only to find an accurate analytic approximation or alternatively an algorithm to calculate the wavefunction as needed. Examples of analytic approaches are given later in this chapter. Since we cannot, in general, perform the many-dimensional integrals to estimate observables analytically, simulation techniques (see Ch. 23) or methods based on perturbation techniques are needed.

In this chapter, we discuss what we know about the many-body wavefunction, written as $\Phi(R)$. This is not only of considerable practical importance, but also a challenging intellectual goal that can lead to an understanding of quantum many-body systems. We primarily discuss the ground-state wavefunction; much of what is discussed applies to excited states as well. In this chapter we will use a first quantized picture, assuming that there are a fixed number of electrons, N, and we discuss properties of the exact wavefunction, $\Phi_0(X)$, with energy E_0. An approximate "trial wavefunction" is denoted $\Psi(X)$. We write the coordinate and spin of electron i as $x_i = (\mathbf{r}_i, \sigma_i)$ and the coordinates and spins of all the electrons as $X = \{x_1, \ldots, x_N\}$. R refers to the $3N$ set of particle coordinates $\{\mathbf{r}_1, \mathbf{r}_2, \ldots\}$. We start by reviewing some exact properties.

6.1 Properties of the many-body wavefunction

What properties are possessed by an exact wavefunction, $\Phi(X)$? We mention a few.

- **Antisymmetry.** From the Pauli principle, all proper wavefunctions are antisymmetric under exchange of electrons, e.g.,

$$\Phi(x_1, x_2, \ldots) = -\Phi(x_2, x_1, \ldots). \tag{6.1}$$

For a general exchange that maps electrons $(1, 2, 3, \ldots)$ into the permuted set $(\mathcal{P}_1, \mathcal{P}_2, \mathcal{P}_3, \ldots)$, the wavefunction picks up the sign of the permutation, written as $(-1)^{\mathcal{P}}$. If there is no magnetic field or spin–orbit coupling present, the energy of the electronic system can be found by considering $N_\uparrow$ up spins and $N_\downarrow$ down spins, so that the z-component of total spin is $S_z = \frac{1}{2}\hbar(N_\uparrow - N_\downarrow)$. We can treat the electrons as if they

---

[1] Tabulation of a many-body wavefunction quickly becomes difficult. For N electrons, the wavefunction is a $3N$-dimensional function, neglecting spin. If each dimension requires a basis of 100 functions, specifying the wavefunction will require 100^{3N} complex numbers; an exponentially large amount of information. Symmetry drastically reduces this estimate, but the exponential dependence remains, as discussed in Sec. 1.2. Complete knowledge, in the sense of physically recording the values of the wavefunction for all arrangements of the electrons, is hopeless. This argument also implies that experiments will not be able to measure the complete many-body wavefunction.

are two different types of particles, and exchange can be limited to spatial permutations between electrons with the same spin.

- **Other symmetries.** For any operator $\hat{A}$ with $[\hat{H}, \hat{A}] = 0$, the wavefunction can be chosen to be an eigenfunction of both $\hat{A}$ and $\hat{H}$. For example, since an atom has spherical symmetry, total angular momentum is conserved and every state can be chosen to be an eigenfunction of the total angular momentum operator. For a homogeneous extended system, translational symmetry leads to momentum being a good quantum number, and for a periodic solid, to total crystal momentum (modulo a reciprocal lattice vector) being a good quantum number.[2]

- **Continuity.** Wavefunctions and all of their derivatives are continuous except at places in configuration space where the potential is non-analytic. In the electronic hamiltonian, Eq. (1.1), the non-analytic points are the coincidence points: two electrons at the same position $\mathbf{r}_i = \mathbf{r}_j$ or an electron at an ionic position $\mathbf{r}_i = \mathbf{R}_j$. One can also have non-analytic points at boundaries, for example if the wavefunctions are required to vanish on a surface. The requirement of continuity does not apply to hamiltonians of lattice models, such as the Hubbard model.

- **The cusp condition.** For charged particles in the continuum, the derivative of the wavefunctions as any two particles approach each other is given by the cusp condition [257, 258]:

$$\lim_{r_{ij} \to 0} \frac{\partial}{\partial r_{ij}} \int \frac{d\Omega}{4\pi} \Phi(R) = \frac{e_i e_j \mu_{ij}}{2} \Phi(R). \tag{6.2}$$

In this expression, particles i and j are being brought together, and it is assumed that all other particles are fixed. The integral is over all directions Ω separating the two particles, e_i is the charge, and μ_{ij} is the effective mass of the pair with $\mu_{ij} = \frac{1}{2}$ for two electrons in atomic units. If the two electrons have the same spin, then both sides of the equation vanish.

- **Real wavefunctions.** If there is no magnetic field, and we have periodic, antiperiodic, or open boundary conditions, the wavefunction can be made real. If there is a magnetic field or twisted boundary conditions, the wavefunction will be complex, in general. We now discuss boundary conditions.

6.2 Boundary conditions

For typical condensed-matter problems, the physical system we want to model is macroscopic, but many correlated methods such as quantum Monte Carlo are set up to calculate a system with only a relatively small number of electrons, because the computational burden grows rapidly with the number of electrons. This makes the choice of boundary conditions (BC) crucial, since we want to reach the thermodynamic limit as quickly as possible. In addition, boundary conditions can affect the solution in qualitative, often subtle, ways. An

---

[2] We can generalize to multiple operators $\hat{A}_i$ assuming that $[\hat{H}, \hat{A}_i] = [\hat{A}_i, \hat{A}_j] = 0$ for all $\{i, j\}$.

example from one-particle theory is the electric polarization as discussed in Sec. 6.3 and in [1, Ch. 22]. We mention three types of boundary condition: open, periodic, and twisted.

Open boundary conditions imply that $\Phi \to 0$ if any electron i has $|\mathbf{r}_i| \to \infty$.[3] This is the appropriate boundary condition for a bound state of an isolated system such as a molecule or cluster. A related example is an electron scattering from a molecule with N electrons. At large distances the wavefunction must reduce to a product of the scattering wavefunction $f(\mathbf{r})$ of a single electron and the wavefunction of the N-electron system.

Approximating an extended system as an isolated finite cluster with open BC typically causes significant surface effects, so open BC are not usually very efficient for modeling an extended system. But for insulators, the effect of surfaces decreases exponentially away from the surface, so that calculations of a finite cluster using open BC are often used.

Periodic boundary conditions (PBC) are the common choice to represent a macroscopic system by a small system. We first define the "supercell," the volume defined by three cell vectors: $(\mathbf{L}_1, \mathbf{L}_2, \mathbf{L}_3)$. Particle coordinates are restricted to be inside the supercell, and if they move outside, they are returned to the cell by shifting by a cell vector. We then construct a potential energy function, $V(R)$, that is periodic in the supercell so that, e.g.,

$$V(\mathbf{r}_1 + \mathbf{L}_m, \mathbf{r}_2, \ldots, \mathbf{r}_N) = V(\mathbf{r}_1, \mathbf{r}_2, \ldots, \mathbf{r}_N), \tag{6.3}$$

where $\mathbf{L}_m$ denotes any one of the cell vectors.[4] The modulus of the wavefunction, $|\Phi(R)|$, must have the same periodicity. Since the cell vectors $\mathbf{L}_m$ correspond to operators that commute with the hamiltonian, the solutions can be classified by a total supercell "momentum" $\mathbf{K}$:

$$\Phi_{\mathbf{K}}(\mathbf{r}_1 + \mathbf{L}_m, \mathbf{r}_2, \ldots, \mathbf{r}_N) = e^{i\mathbf{K} \cdot \mathbf{L}_m} \Phi_{\mathbf{K}}(\mathbf{r}_1, \mathbf{r}_2, \ldots, \mathbf{r}_N). \tag{6.4}$$

This is the many-body generalization of Bloch's theorem, see Sec. 6.3 and [1, Ch. 4]. The BC "twist" $\theta_m = \mathbf{K} \cdot \mathbf{L}_m$ is the change in phase of the wavefunction as a single electron follows a path around the periodic boundaries and back to its starting place. Without loss of generality, we can assume that $|\theta_m| \leq \pi$. Twists with $\theta_m = 0$ are PBC and denoted as the "Γ point," while $\theta_m \neq 0$ are twisted boundary conditions.

Consider non-interacting electrons in a cube of side L with no external potential. The single-particle spatial orbitals are plane-wave states: $\phi(\mathbf{r}) = L^{-3/2} \exp(i\mathbf{kr})$. To satisfy the boundary conditions we must have $\mathbf{k}_n = (2\pi \mathbf{n} + \boldsymbol{\theta})/L$. In the ground state, the N states with the lowest energies are then occupied. Figure 6.1 shows the occupied states for 13 spin-up electrons in 2D for PBC ($\boldsymbol{\theta} = 0$) and for a non-zero twist. The occupied states lie within a circle centered at the origin with radius k_F. One can see a "finite size" effect in PBC: the filled states do not represent a Fermi disk very well!

---

[3] The wavefunction for a bound state with a non-zero ionization energy must go exponentially to zero as any particle is removed from the cluster.

[4] The Coulomb potential is made periodic typically by using the Ewald summation methods, as discussed in [1, App. F].

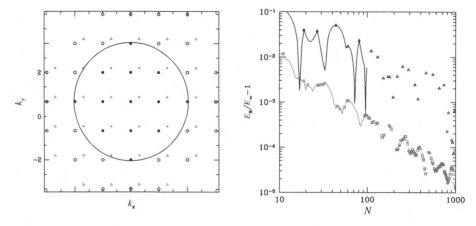

Figure 6.1. Occupied states for 13 fermions in a 2D square in the ground state (left panel). The occupied states (closed symbols) and empty states (open symbols) with zero twist (circles, PBC) and a twist equal to $2\pi(0.3, 0.15)$ (triangles). The circle shows the infinite system's Fermi surface. The right panel shows the error in the non-interacting energy using PBC (upper symbols) and TABC (lower symbols) in 3D versus the number of electrons.

Twist averaging is used to approach the thermodynamic limit quicker. With twist averaged boundary conditions (TABC), one averages over all possible boundary conditions. The expectation of an operator $\hat{O}$ is then calculated as:

$$\langle \mathcal{O} \rangle_{TABC} \equiv (L/(2\pi))^3 \int_{-\pi/L}^{\pi/L} d^3\mathbf{K} \int dR \; \Phi_\mathbf{K}^*(R) \hat{O} \Phi_\mathbf{K}(R). \tag{6.5}$$

TABC is equivalent to integrating over a Brillouin zone in single-particle calculations, see [1, p. 92], thus removing finite size effects for independent particles. Only finite size effects due to correlation remain and need to be corrected for; see Sec. 23.5. TABC are necessary for a system with a Fermi surface, see Fig. 6.1, but less so for an insulator. However, the variation of the wavefunction with the twist provides quantitative information and qualitative characteristics for insulators.

6.3 The ground-state wavefunction of insulators

We now discuss wavefunctions of insulators; how the insulating properties are encoded in the wavefunction.[5] The essential difference between a metal and an insulator is that a metal can carry current and the $T \rightarrow 0$ conductivity is non-zero. Also, it is associated with the nature of the excitations at the Fermi energy. In recent years it has been found that the distinction can be formulated in terms of the variation of the ground-state wavefunction with boundary conditions, i.e., the variation as a function of the twist, whereas the previous

---

[5] For a review with many earlier references, see [21, 259].

section dealt with the twist average. Together with other work, this has led to a new under-standing of insulating states of matter, e.g., electric polarization, quantum Hall insulators, and topological insulators. The theory of insulators in terms of the wavefunction is usually described in terms of independent-particle electron concepts; here we emphasize that they apply as well to interacting-electron systems and novel states of matter. Other examples, such as the fractional quantum Hall effect, are due to interactions and would not exist in independent-particle systems.

A way to distinguish metals and insulators was proposed by Kohn [260] in terms of the *energy* of the ground state as a function of the twist in the BC as defined in Eq. (6.4). In an ideal metal, if the system of N electrons has a total momentum $N\mathbf{K}$, the center of mass $\mathbf{R}_{cm}$ has a net velocity and the energy increases as $2D\Omega K^2$, where D is the "Drude weight" and Ω is the volume.[6] In an insulator, however, $\mathbf{R}_{cm}$ is localized, and, hence, the energy is independent of the BC in the thermodynamic limit. Of course, all the electrons can be translated by a lattice vector; localization means that the tunneling of all the electrons to a state translated by a lattice constant is exponentially small. In this limit the mean square fluctuation $\langle|\Delta\mathbf{R}_{cm}|^2\rangle$ diverges in a metal, whereas it is finite in an insulator [262]. All insulators share this property, and there is no distinction between different insulating materials or types of insulator.

A more powerful way to characterize insulators is to use the properties of the wavefunc-tion instead of the energy. Let us define a generalized Bloch function $\Phi_\mathbf{K}(\mathbf{r}_1, \mathbf{r}_2, \ldots, \mathbf{r}_N) = e^{i\mathbf{K}\cdot\mathbf{R}}u_\mathbf{K}(\mathbf{r}_1, \mathbf{r}_2, \ldots)$, where $\mathbf{R} = \sum_i \mathbf{r}_i$ and $u_\mathbf{K}$ is a periodic function under the translation $\mathbf{r}_i \rightarrow \mathbf{r}_i + \mathbf{L}$. Then $\Phi_\mathbf{K}$ satisfies the twisted BC in Eq. (6.4). If $\Phi_\mathbf{K}$ is an eigenstate of $\hat{H} = \hat{H}_{kinetic} + \hat{V}$, then $u_\mathbf{K}$ is an eigenstate of the hamiltonian with the modified kinetic energy operator $\hat{H}_{kinetic}(\mathbf{K}) = \frac{1}{2}(i\nabla + \mathbf{K})^2$ (Ex. 6.2). Note that in a system with PBC, the position operator $\hat{\mathbf{r}}$ and the center of mass $\mathbf{R}_{cm}$ need to be defined; only displacements of the center of mass $\Delta\mathbf{R}_{cm}$ are meaningful, as discussed in [1, Ch. 22]. The displacement can be related to a change in polarization $\Delta\mathbf{P} = e(N/\Omega)\Delta\mathbf{R}_{cm}$, an experimentally measurable quantity; however, in a system with PBC, the displacement $\Delta\mathbf{R}_{cm}$ *cannot be determined from the particle density alone*. This quandary was resolved by King-Smith and Vander-bilt (KV) [263], who showed that it is related to the contour integral in Fig. 6.2 using the ground-state wavefunction, i.e., a Berry phase. If the hamiltonian is parameterized by $0 \leq \lambda \leq 1$ that specifies some quantity, e.g., a displacement of atoms from position 0 to position 1, and we fix the wavefunction phase by $u_{-\pi/L} = u_{\pi/L}$, then the integrals over λ cancel, resulting in an expression that depends only on the wavefunctions at the end points: $\lambda = 1$ and $\lambda = 0$,

$$\Delta R_{cm,i} = \frac{2\Omega}{(2\pi)^3}\int d\mathbf{K}\,\mathrm{Im}\left[\langle u_\mathbf{K}^1|\partial_{K_i}u_\mathbf{K}^1\rangle - \langle u_\mathbf{K}^0|\partial_{K_i}u_\mathbf{K}^0\rangle\right], \tag{6.6}$$

plus a multiple of L corresponding to a multiple of 2π in the phase. Here $i = 1, 2, 3$ denotes the vector components. This is the only non-perturbative way known to calculate a

---

[6] A superconductor can be distinguished from a metal by the dependence of the current–current correlation function on the order of limits for $\omega \rightarrow 0$ and $L \rightarrow \infty$ [261].

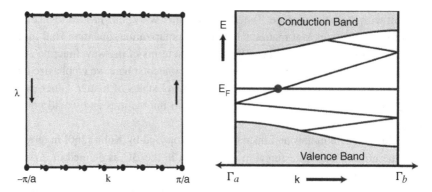

Figure 6.2. Left: Two-dimensional contour integral for calculation of $\Delta R_{cm,i}$ in one direction i, where λ represents a change in the hamiltonian, e.g., positions of atoms in the unit cell. The notation is for the original derivation for independent particles where k denotes the crystal momentum in the BZ in the direction i [259, 263]. Right: The interface between a topological and an ordinary insulator has an odd number of edge states that extend from valence to conduction bands, as shown for k between two boundary Kramers degenerate points $\Gamma_a = 0$ and $\Gamma_b = \pi/a$. They are determined by the nature of the bulk band structure and exist independent of the details of the interface. (From [21].)

polarization and it is now standard in density functional calculations. The original derivation was for independent particles where $L = a$, the lattice constant, but the same form holds for a many-body system with L denoting a supercell [264].

In addition, the mean-square fluctuation of the center of mass position $\Delta \mathbf{R}_{cm}$ can be expressed in terms of the ground-state wavefunction as [262]

$$\langle \Delta R_{cm,i}^2 \rangle = \frac{\Omega}{(2\pi)^3} \int d\mathbf{K} \left[\mathrm{Re} \langle \partial_{K_i} u_\mathbf{K} | \partial_{K_i} u_\mathbf{K} \rangle - \langle \partial_{K_i} u_\mathbf{K} | u_\mathbf{K} \rangle \langle u_\mathbf{K} | \partial_{K_i} u_\mathbf{K} \rangle \right]. \tag{6.7}$$

This has been shown to be equal to an integral over the frequency-dependent conductivity, with the consequence that the fluctuation diverges in a metal but is finite in an insulator. This is consistent with the conclusions from the analysis above in terms of the energy; however, in this approach there is no distinction among different metals, but the relation provides a quantitative measure of localization in insulators.

The derivations given above assume that the ground state is non-degenerate; however, for PBC the path around the contour shown in the figure can connect one degenerate state with another. If there is a q-fold degeneracy, the states can be chosen so that each line integral acquires a phase $2\pi/q$. This has been termed "quantum order" [20] because the wavefunctions differ only by a phase. This has no counterpart in classical systems and cannot be classified by the Landau theory of phase transitions. It is termed "topological" because the state depends on the topology, e.g., on a torus the order is specified by the winding around the torus. This occurs in the fractional quantum Hall effect [265] due to interactions where it leads to fractional quantum numbers, and it is an on-going effort to find other systems that may exhibit such properties (see, e.g., [266]).

Topological insulators [21] have the remarkable property that there exist edge currents at a boundary with an ordinary insulator (or the vacuum at a surface of a topological insulator). This is depicted on the right side of Fig. 6.2 as a state that extends between the valence and conduction bands crossing the Fermi level. An odd number of such crossings leads to boundary states that have dissipation-less currents. The existence of these states depends only on qualitative features of the bulk bands, which has been worked out for independent particles including spin, an essential ingredient. The topological character can be determined by the wavefunction at a set of points in the BZ where $\mathbf{k}$ and $-\mathbf{k}$ coincide, which has been used to develop a classification of topological insulators. The conclusions carry over to interacting electrons, which is an example of the rule of continuity (Sec. 1.3) that the non-interacting state evolves continuously as interactions are turned on, so long as there is no phase transition. For example, the gap is not only an independent-particle concept. The lowest conduction band state is defined in the many-body system as the ground state of the system with one extra electron. Similarly, the top of the valence band is defined by removing an electron. The special points in the BZ are determined by symmetry and survive in the interacting system. The edge states are also precisely defined because there is no mechanism for decay so long as the bulk is insulating.

6.4 Correlation in two-electron systems

To start the discussion of correlated wavefunctions, we look at the simplest example, namely those with two electrons, e.g., the hydrogen molecule or the helium atom with fixed nuclear positions. The ground-state wavefunction of two electrons in a local external potential must be a non-degenerate singlet of the form $\Phi = \Phi(\mathbf{r}_1, \mathbf{r}_2)(|\uparrow\rangle|\downarrow\rangle - |\downarrow\rangle|\uparrow\rangle)$ where $\Phi(\mathbf{r}_1, \mathbf{r}_2)$ is symmetric ("bosonic") under exchange of the two electrons. By choosing the phase of the wavefunction, $\Phi(\mathbf{r}_1, \mathbf{r}_2)$ can be made real and positive and have all the symmetries of the potential.

The Hartree–Fock (uncorrelated) wavefunction has the spatial form

$$\Psi_{HF}(\mathbf{r}_1, \mathbf{r}_2) = \psi_1(\mathbf{r}_1)\psi_2(\mathbf{r}_2). \tag{6.8}$$

It is a "restricted" wavefunction (RHF) if we assume that the orbitals $\psi_k(\mathbf{r})$ have the spatial symmetries of the nuclear potential. This implies $\psi_1(\mathbf{r}) = \psi_2(\mathbf{r})$. There may exist a lower-energy UHF solution that breaks these symmetries, as discussed in Ch. 4. This leads to some difficulty in defining correlation: do we use the RHF or the UHF energy as a reference? As discussed in Sec. 2.1, we define correlation with respect to the RHF solution.

We use the hydrogen molecule to illustrate the properties of the exact wavefunction and the difference between the RHF and the exact wavefunction. Fix two protons on the x-axis with a separation equal to the distance that minimizes the energy $a = 1.4$ a.u. In the left panel of Fig. 6.3 we show the wavefunction versus the position of electron 2 on the x-axis as electron 1 is held fixed: the solid line corresponds to $x_1 = -1$, the dotted line $x_1 = 1$. The maxima are when electron 2 is at the position of one of the protons (i.e., $x_2 = \pm 0.7$). We see a strong left–right correlation: when electron 1 is to the left of the origin, electron 2 prefers to be on the right and vice versa.

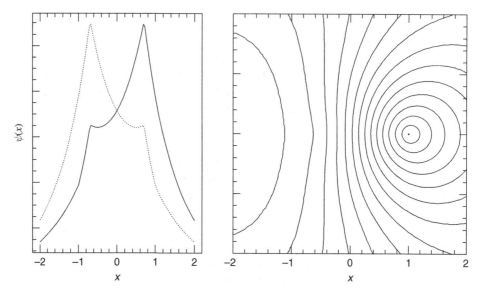

Figure 6.3. The H_2 correlated wavefunction: the protons are fixed on the x-axis a distance 1.4 apart at the coordinates $(\pm 0.7, 0, 0)$. The left panel is $\Phi(\mathbf{r}_1, \mathbf{r}_2)$ with both electrons on the x-axis: $x_1 = -1$ (solid line) or $x_1 = 1$ (dotted line). The second electron varies along the x-axis. The right panel is a contour plot of the log of the ratio of the RHF wavefunction to the exact wavefunction with one electron at $(1, 0, 0)$ and the second electron moves in the $x-y$ plane. The spacing of the contour lines is 0.05.

In order to see the correlation more clearly, we divide by the RHF wavefunction and take the logarithm. The right panel shows a contour plot of the correlated wavefunction, $-\ln(\Phi/\Psi_{RHF})$, as electron 2 varies in the $x-y$ plane with electron 1 held fixed at $x_1 = 1$. Note that the contour spacing is only 0.05: the effects of correlation are much weaker than the variations of the HF wavefunction. Immediately apparent in the right panel is the "cusp"; the repulsive peak centered at $x_2 = 1$. It has a spherical shape around the cusp. One makes a major improvement over Ψ_{HF} by multiplying the HF function by a function of r_{12}: $\Psi_2 = \Psi_{HF} \exp(-u(|\mathbf{r}_1 - \mathbf{r}_2|))$. In fact, the correlation energy of H_2 is 1.11 eV; putting in this "spherical correlation" gains 95% of the correlation energy. The energy is accurate to about 0.05 eV. Note that to achieve this improvement, one must reoptimize the orbitals $\psi(\mathbf{r})$; because $u(r)$ is repulsive, it pushes electrons out of the bond region. To minimize the energy, one needs to pull them back in by increasing the magnitude of $\psi(\mathbf{r})$.

The RHF trial function works well for the equilibrium molecular bond length. However, another approach is needed as the bond is broken; i.e., for $a \gg 1.4$. Heitler and London (HL) [137] assumed that the spatial part of the molecular wavefunction has the form

$$\Psi_{HL}(R) = \psi_1(\mathbf{r}_1)\psi_2(\mathbf{r}_2) + \psi_2(\mathbf{r}_1)\psi_1(\mathbf{r}_2), \tag{6.9}$$

where they took the orbitals to be the $1s$ state of the hydrogen atom located on proton 1 (ψ_1) or proton 2 (ψ_2). These are two *configurations* (i.e., different choices of orbitals in the determinant) in the quantum chemistry terminology. Clearly, the left–right correlation is added by the presence of these two configurations. One can get additional correlation by multiplying Ψ_{HL} by an explicit function of r_{12}. This approach is discussed, e.g., in Sec. 3.2, as a simple solvable model for the Mott transition. We discuss later in this chapter a many-body generalization.

The HL approach can also be generalized by expanding in the complete set of single Slater determinants of one-body orbitals:

$$\Psi_{CI}(R) = \sum_{k=1}^{M} c_k \det{}_k(\psi_i(\mathbf{r}_j)), \tag{6.10}$$

where $\det_k(R)$ is specified by choosing a specific set of orbitals (a configuration) and the coefficients, c_k, determined by minimizing the energy, leading to a numerical eigenvalue problem. The determinants can be ordered by their symmetry and by their non-interacting energy. In quantum chemistry this approach is known as configuration interaction (CI). In the limit that all configurations are included in the sum, the expansion is exact. It provides invaluable direct information on the states of the quantum system, and, in principle, all information. Furthermore, CI can provide very accurate quantitative results for cases where the first few determinants dominate the ground state. Correlations are included by mixing low-energy excitations. It is possible to perform the exact diagonalization for systems whose Hilbert space has dimension up to $\approx 10^9$ configurations if one is only interested in a small subset of the eigenfunctions. However, the expansion into determinants scales poorly with the number of electrons. Returning to the argument given in Sec. 1.2, suppose we have a system of N atoms and suppose we need to include p excitations on each atom. Then without considering symmetry, one needs $M = p^N$ determinants. When we expand in determinants, the number of determinants scales exponentially with the number of atoms; even 30 fully correlated electrons are difficult to treat using today's computers. If the basis set is truncated, then one has the "size consistency" problem. By that we mean that calculations assuming a fixed number of determinants of a system with more electrons will be less accurate than those for a smaller system. Consider the description of an H_2 molecule as the two ions are pulled apart: two determinants provide an exact description at large separations, but many determinants are needed for small separations, thus the error in the molecular binding depends on the separation, unless the expansion is fully converged at all separations. One approach to ameliorate this problem is to consider forms of the wavefunction for which the correlations are much more succinctly described, e.g., Ψ_{HL} as opposed to Ψ_{CI}.

6.5 Trial function local energy, Feynman–Kac formula, and wavefunction quality

We now discuss "trial" many-body wavefunctions (written as $\Psi(R)$), a method for systematically improving trial functions, and how to measure the quality of an approximate trial function. We can define a local error in a trial wavefunction at a given point in many-body phase space, $R = \{\mathbf{r}_1, \mathbf{r}_2, \ldots, \mathbf{r}_N\}$, by its "local"[7] or "residual" energy:

---

[7] This is not to be confused with other uses of the term "local energy," e.g., see [1, p. 419].

$$E_L(R|\Psi) = \frac{\hat{H}\Psi(R)}{\Psi(R)}. \qquad (6.11)$$

Note that the local energy is a function of the hamiltonian, the trial function, and the positions of all the electrons. It is "local" in the full phase space of N electrons (not in 3D).

The local energy gives a quantitative measure of how well the trial wavefunction solves the Schrödinger equation at a given point in phase space. Its fluctuations measure the error of the trial function since if the trial function were exact, the real part of the local energy would equal the exact energy for any arrangement of the electrons and its imaginary part would vanish. The local energy provides a way to derive conditions on a trial wavefunction, an analytical or a numerical method to improve wavefunctions, and plays a key role in the variational (Ch. 23) and diffusion quantum Monte Carlo methods (Ch. 24).

For real non-negative trial functions the Feynman–Kac (FK) formula[8] connects the local energy of a trial function to the exact wavefunction:

$$\Phi(R_0) = \frac{\Psi(R_0)}{\langle \Psi|\Phi\rangle} \langle \exp(-\int_0^\infty dt(E_L(R_t|\Psi) - E_0)))\rangle_{\Psi^2}. \qquad (6.12)$$

The brackets mean to average the exponential of the integrated local energy over an infinite number of drifting random walks, all starting at the point R_0, the argument of the wavefunction on the left-hand side. A given random walk, $\{R_t\}$, specifies the positions of the electrons as a function of t, where t is the elapsed time since the beginning of the walk. At large time, the local energy on average will approach a constant, E_0, the ground-state energy, the integral will converge, and we can stop the random walk. During the random walk, the trial function biases the walk toward increasing values of Ψ with a "force" given by $F = 2\nabla \ln \Psi$. For a further explanation of the drifting random walks and the origin of this remarkable formula, see Sec. 24.5 and Eq. (24.34).

To use the FK formula we need either to make approximations or do random walks on the computer. However, two qualitative features emerge from a simple argument. First, in regions of phase space where the local energy is lower (higher) than the average energy, the trial function needs to be increased (decreased). Second, because of the diffusive nature of random walks, the correction to the trial function is smoother than the local energy. As mentioned above, the local energy as a function of time decays to the ground-state energy so that we can cut off the upper limit of the time integral at some value $\tau(R_0)$. For a good trial function, the exponent will be small and we can take the averaging into the exponent. This is known as the cumulant approximation. Then, taking logarithms, the correction to the trial function can be written as:

$$\ln[\Phi(R_0)/\Psi(R_0)] \sim -\tau(R_0)(\overline{E}_L(R_0|\Psi) - E_0)), \qquad (6.13)$$

where $\overline{E}_L(R_0|\Psi)$ is the average of the local energy over a spatial region of size $\tau(R_0)^{\frac{1}{2}}$, the distance a random walk can cover in time $\tau(R_0)$. The size of this region is proportional to how rapidly in 3N-dimensional coordinate space the local energy returns to its average

---

[8] Here we discuss the generalized FK formula; in the usual FK formula, $\Psi(R) = constant$.

value. Notice that we have derived an explicit relation between an individual term in the local energy and a needed correction to $\ln(\Psi)$. We discuss examples of the use of this formula later in the chapter: motivating the Slater–Jastrow and backflow wavefunctions and the van der Waals interaction. In another example of using the local energy, the Lanczos method [1, Ch. 23] uses $E_L(R|\Psi)\Psi(R)$ to generate an improved wavefunction for a lattice model.

Now let us try to measure the overall quality of a trial function with a few numbers. Three such "moments" are commonly used. We define them here: they can be computed either with exact diagonalization or with quantum Monte Carlo techniques. The different measures are related to each other but useful for different purposes. In the rest of this section, we assume that Ψ is normalized: $\int dR|\Psi(R)|^2 = 1$.

- **The variational energy**:

$$E_V[\Psi] = \int dR \Psi^*(R)\hat{H}\Psi(R) = \langle E_L(R|\Psi)\rangle_{|\Psi|^2} \geq E_0. \qquad (6.14)$$

The brackets $\langle \cdots \rangle$ mean the average of the local energy over the distribution $|\Psi(R)|^2$. The variational energy, E_V, is an upper bound to the ground-state energy, E_0. Note that for the upper-bound property, the trial function needs to have the same symmetries as the ground state; e. g., if the trial function is antisymmetric, then E_0 is the exact fermion ground-state energy. Also, in order to prove the inequality, we must have that $\Psi(R)$ be continuous with a continuous derivative. If the objective of a calculation is to find the best energy (i.e., the least upper bound), one needs to vary the parameters in $\Psi(R)$ to minimize E_V, thus the name variational energy. The variational theorem applies not only to the global ground-state energy, but also to all states that are ground states of a given symmetry. Suppose we look at a non-degenerate ground state and make a single particle–hole excitation. Often an excitation will have a different value of total linear or angular momentum or spin and, thus, the variational theorem will also apply to the excitation. For the generalization of the variational theorem to all excitations, see Sec. 23.7. Note that there is not a rigorous inequality constraining energy differences between two states.

- **The variance of the local energy**:

$$\sigma^2[\Psi] = \int dR|(\hat{H} - E_V)\Psi(R)|^2 = \langle|E_L(R|\Psi) - E_V|^2\rangle_{|\Psi|^2} \geq 0. \qquad (6.15)$$

If the trial function were an exact eigenfunction, its variance would vanish. Hence, the variance is an absolute measure of the quality of the trial function, not requiring knowledge of the exact ground-state energy or wavefunction, and is not limited to ground states. The variance controls how quickly quantum Monte Carlo estimates converge, so variance minimization can lead to more efficient calculations. Also, note that the variance can be infinite, even though the variational energy and other physical properties are well-defined. This is a potential problem in quantum Monte Carlo evaluations. Knowledge of E_V and σ^2 can be used to find both an upper and a lower bound [267] to the exact energy:

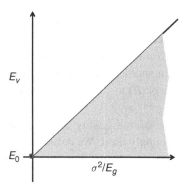

Figure 6.4. The gray area shows feasible values of the variational energy and variance, see Eq. (6.16). Exact wavefunction values are the lower-left vertex. In practical problems one does not know either the exact energy E_0 or the gap E_g, but this plot constructed with different approximate trial functions can be useful as in the right panel of Fig. 6.6.

$$E_V - \frac{\sigma^2}{E_g} \le E_0 \le E_V. \tag{6.16}$$

Here, E_g is the gap to the next excited state (with energy E_1) of the same symmetry. For the inequality to be rigorous, the gap must be underestimated: $E_g \le E_1 - E_0$. Figure 6.4 shows the allowable values of energy and variance.

- **The overlap with the exact wavefunction**:

$$\mathcal{O}[\Psi] = \text{Re} \int dR \Psi^*(R) \Phi(R) = 1 - \frac{1}{2} \int dR |\Psi(R) - \Phi(R)|^2. \tag{6.17}$$

Maximizing the overlap is equivalent to finding the trial function that minimizes the average square difference with the exact wavefunction. The overlap is the preferred quantity to optimize if you want to use the trial wavefunction to calculate a property other than the ground-state energy.

As shown in Exs. 6.4 and 6.5, the three measures emphasize different parts of the energy spectrum with the overlap giving the most weight to low-energy states, and the variance to high-energy states. The three properties $(\ln(\mathcal{O}), (E_V - E_0), \sigma^2)$ are intimately related so that if one of them is extensive, proportional to N as $N \to \infty$, that implies the other two are extensive.

6.6 The pair product or Slater–Jastrow wavefunction

In this section we discuss the most common form for a correlated wavefunction. Although the use of this wavefunction was originated by Hylleraas [268], Bijl [269], and Dingle [270], it is usually referred to as the Slater–Jastrow (SJ) form [271]. The Gutzwiller wavefunction [129] for lattice models can be considered a special form of a Slater–Jastrow wavefunction, as can the Laughlin [272] wavefunction for quantum Hall liquids.

One of the motivations [269] for the SJ form comes from the attempt to make a microscopic model of bulk helium: the many-fermion system ^{3}He or the many-boson system ^{4}He. To see most clearly the need for this wavefunction, let us assume that the interatomic potential is a hard-sphere interaction (diameter d): the potential energy is infinitely repulsive if any two spheres overlap, $r_{ij} \leq d$. One needs to have a wavefunction that vanishes upon overlap of any pair. To describe many-body ^{3}He, we can use the Slater determinant to make an antisymmetric function, and then multiply by a symmetric function $\prod_{i<j} f(r_{ij})$. To keep the spheres from overlapping we set[9] $f(r) = 0$ for $r \leq d$, thereby exactly satisfying the necessary boundary conditions in the complex many-body phase space.

For mathematical convenience we use the logarithm of $u(r) = -\ln(f(r))$ and allow it to become a general 6D function of (x_i, x_j), possibly depending on the spins of electrons (i, j):

$$\Psi_{SJ}(R) = \det(\psi_k(x_i)) \exp(-\sum_{i<j} u(x_i, x_j)). \tag{6.18}$$

Here $u(x_i, x_j)$ is the *(Jastrow) correlation factor*, a real symmetric function of its arguments, and ψ_k are the *orbitals*. Note that $u(x_i, x_j)$ is defined with a minus sign[10] so that it will have the same sign as the potential. This is more than an accident: we show in Ch. 25 that the many-body density matrix (the finite-temperature generalization of Φ) interpolates smoothly between the classical Boltzmann distribution and the SJ wavefunction. In the high-temperature limit $u(r_{ij}) \rightarrow v_{ee}(r_{ij})/(2k_B T)$.

To justify the Jastrow form for electrons, we start with a Hartree form $\prod_k \psi_k(x_k)$ and use the Feynman–Kac formula in Eqs. (6.12) and (6.13) to find how interactions will change it. Let us suppose that the orbitals are an exact solution to the external potential due to the nuclei, then the local energy needed in the Feynman–Kac formula equals the interaction potential: $\sum_{i<j} r_{ij}^{-1}$. Then, from Eq. (6.13), an approximate trial function, which improves over the Hartree–Fock function, is:[11]

$$\Psi(R_0) = \det(\psi_k(x_i)) \exp(-\tau \sum_{i<j} \langle r_{ij}^{-1} \rangle), \tag{6.19}$$

where $\langle \ldots \rangle$ means to average over random walks starting at the point R_0 in configuration space and τ is the time it takes for the local energy of the random walk (the interaction potential) to reach its average value. The averaging by the random walk makes $u(r)$ smoother than r^{-1}. Usually it is difficult to carry out the averaging process, either analytically or computationally, so both the orbitals and the correlation factor u and τ are determined in another way, such as by minimizing one of the measures defined in Sec. 6.5: the variational energy, variance, or overlap.[12]

---

[9] The alternative form $1 + \sum_{ij} f(r_{ij})$ can be used for the two particle problem but it does not satisfy the conditions for $N > 2$.

[10] In this book, we will always define the correlation factor, u, as above, but in the literature the minus sign is often absent.

[11] We also antisymmetrize the result. The pair term is symmetric, so it is unchanged, but the Hartree form changes to a Slater determinant.

[12] How this minimization can be done is discussed in Ch. 23.

Note that the calculation of the properties of the SJ wavefunction is much more difficult than that of a determinant. For example, to normalize the trial function is equivalent to finding the partition function of a classical liquid.[13] Early work focused on doing the integrals approximately, see, e.g., Boys and Handy [273]. Since McMillan [44] introduced the variational Monte Carlo method, stochastic methods have been used[14] to evaluate the integrals; this separates the problem of determining a good approximation to the exact wavefunction from the evaluation of its properties.

We now use "local energy" arguments to establish properties of good Jastrow correlation factors. Consider bringing electrons 1 and 2 together, keeping the other electrons fixed. Assume electrons 1 and 2 are in different spin states. Clearly, one term of the local energy, namely the potential energy between the two electrons, diverges; it equals r_{12}^{-1}, but since the local energy for a good trial function should be constant, we need a term in the kinetic energy that equals $-r_{12}^{-1}$ to cancel the potential energy term. Using the SJ trial function, the terms in the local energy involving only the distance separating electrons 1 and 2 are:

$$E_L(R|\Psi) = 1/r_{12} + \nabla^2 u(r_{12}) - (\nabla u(r_{12}))^2 = 1/r_{12} - f(r_{12})^{-1}\nabla^2 f(r_{12}), \qquad (6.20)$$

where $f(r) = \exp(-u(r))$. An intuitive result emerges: for small r_{12}, $f(r)$ will equal the solution to a two-body Schrödinger equation with the potential $1/r$. In the limit $r \to 0$ this gives the "cusp condition" written in Eq. (6.2): the value of the derivative of the Jastrow correlation factor at the origin is

$$\left.\frac{du(r)}{dr}\right|_0 = -\frac{1}{2}. \qquad (6.21)$$

Exercise 6.6 generalizes this to two dimensions and for electrons in a spin-triplet state.

Now consider the large-distance behavior of the optimal Jastrow correlation factor. This is important for the low-energy response properties and the van der Waals interaction. For a translationally invariant system, such as the electron gas (see Ch. 3), the RHF orbitals are plane waves and, in the thermodynamic limit, the correlation factor depends only on the distance between two electrons, $u(\mathbf{r}_i, \mathbf{r}_j) = u(r_{ij})$. Analysis (see Ex. 6.7), either by using the local energy or the RPA method, shows that for any metal in three dimensions: $\lim_{r\to\infty} u(r) = 4\pi n^{-\frac{1}{2}}r^{-1}$, where n is the (itinerant) electron density. We note that the e–e correlation extends over long distances; it decays as r^{-1} in 3D. For an insulator, the r^{-1} form remains, but the coefficient is reduced.

A remarkable formula for the Jastrow correlation factor of a homogeneous system was obtained by Gaskell [274]:

$$2nu_k = -\frac{1}{S_{0k}} + \left[\frac{1}{S_{0k}^2} + \frac{4nv_k}{k^2}\right]^{1/2}, \qquad (6.22)$$

---

[13] However, electrons are easier to treat than a classical liquid since the integrals exist and can often be treated by perturbation theory.

[14] However, deterministic methods such as correlated basis function theory, the hypernetted chain approach, and "r_{ij}" methods have also been used.

where $v_k = 4\pi k^{-2}$ and u_k are the Fourier transforms of the interaction and the Jastrow factor $u(r)$ and S_{0k} is the structure factor for uncorrelated fermions, see Eq. (5.17). This form combines together both the correct short-range and long-range behavior for systems with a variety of interactions or dimensionality. For the 2D and 3D HEG, this zero-parameter wavefunction can hardly be improved upon [275] by optimization, even in the strongly correlated regime, though the RPA assumption on which it is based is not accurate there. It is found to obtain about 95% of the correlation energy in the HEG. Other properties such as correlation functions are also good. For other hamiltonians, even though it has the correct limiting behaviors, it can often be improved at intermediate distances. Although it is desirable to have a simple analytic form for $u(r)$, current methods parameterize an arbitrary radial function using a spline (piecewise polynomials with continuity conditions imposed at the "knots"). The simplest general basis functions are piecewise cubic polynomials, with continuity of $u(r)$ and $du(r)/dr$ imposed [276]. One imposes the cusp condition at $r = 0$ and continuity of $u(r)$ and its derivative at an upper cutoff, typically chosen to be at $L/2$ in PBC.

For a molecule or extended inhomogeneous system the SJ wavefunction of Eq. (6.18) does not have translation invariance of the electron gas; the correlation factor, $u(\mathbf{r}_i, \mathbf{r}_j)$ is a six-dimensional function, and the practical difficulty in representing it becomes important. The basis set problem (what functions are best for a given problem, which are most compact, fastest to compute, etc.) becomes significant. It is only recently that robust QMC optimization methods have been introduced (see Ch. 23) to determine parameters in such functions with many parameters. We can expand the Jastrow part of the trial function as:

$$U(R) = \sum_{i,I} u_{eI}(\mathbf{r}_{iI}) + \sum_{i<j} u_{ee}(\mathbf{r}_{ij}) + \sum_{i<j,I} u_{eeI}(\mathbf{r}_{iI}, \mathbf{r}_{jI}, \mathbf{r}_{ij}). \tag{6.23}$$

It contains a 1-electron term, a 2-electron term, etc. Lowercase indices refer to electrons, uppercase indices to ions. Each function is then expanded in terms of a basis, and parameters optimized. Note that there is arbitrariness in the SJ wavefunction: one can multiply the orbitals by a function of position and then subtract a corresponding factor from u_{eI} in such a way that the trial function is unchanged. Conventionally, the orbitals are often fixed by a mean-field calculation. Since the mean-field density is usually quite accurate, once a repulsive e–e term u_{ee} is turned on, one needs to compensate with an attractive one-body term u_{eI}. The most common form of two-body term for a molecule is $u_{ee}(r) = (a_1 + a_2 r)^{-1}$. The cusp condition (Eq. (6.2)) then allows one to determine the relation between a_1 and a_2.

Figure 6.5 shows how much correlation energy is recovered by a SJ wavefunction for atoms up to Ar. As the lower panel shows, each electron has a correlation energy of between 0.5 and 1 eV. The solid line in the upper panel shows that the SJ wavefunction obtains between 75% and 90% of the correlation energy, except for the H, He, and Li atoms. H and He, are exactly described by a general SJ wavefunction, and for Li the SJ wavefunction gets 97.5% of the correlation energy. The upper dotted line shows a fixed-node result (see Sec. 24.4). This shows the most energy that can be obtained from a HF Slater determinant times a completely general Jastrow factor, $U(R)$. This energy error mirrors the SJ result,

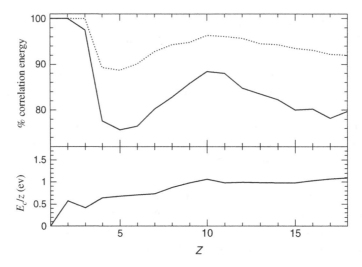

Figure 6.5. The bottom panel shows the correlation energy per electron (E_c/Z) of the atoms from H to Ar (plotted as the atomic number, Z). The top panel shows the percentage of correlation energy [277] recovered by a SJ wavefunction (solid line) and as determined by the QMC fixed-node procedure (dotted line) using a HF determinant. For an atom this should be very close to the optimal SJ energy since the trial function nodes are highly constrained by the spherical symmetry of an atom.

but is much more accurate. Because the trial function is restricted to a single determinant, even a perfectly optimized Jastrow factor does not equal the exact wavefunction.

There is a connection between the long-range van der Waals interaction between two separated molecules (see Sec. 2.2) and the long-range ($1/r$) tail of the Jastrow correlation factor. Consider two molecules, interacting but so far apart that we can neglect their electron density overlap. Let Φ_∞ be the exact wavefunction when the molecules are separated and $\delta V(R)$ be the electrostatic interactions of the charges on one molecule with the charges on the other molecule. Subtract any average interaction in case the molecules have net charges or dipole moments, so that $\langle \delta V(R) \rangle = 0$, where the expectation value is with respect to the wavefunction Φ_∞. The Feynman–Kac formula Eq. (6.13) gives the change in Φ_∞ caused by the perturbation $\delta V(R)$:

$$\Psi(R) \approx \Phi_\infty(R)\exp(-\tau\langle\delta V(R)\rangle). \tag{6.24}$$

Here τ is a parameter with units of inverse energy.[15] For the two molecules well separated, $\langle\delta V(R)\rangle = \delta V(R)$. Then we must choose the factors in Eq. (6.23) (the electron–electron u_{ee} and the electron–ion u_{el}) to decay as $\tau e_i e_j/r_{ij}$. If and only if this holds will the SJ wavefunction have van der Waals interactions correctly built in at large r. The parameter τ is a function[16] of the polarizabilities of the two molecules.

---

[15] It is the imaginary time needed for the interaction energy to relax to its average value; see Ch. 24.

[16] For the interaction of two atoms we find that $\tau^{-1} = \tau_1^{-1} + \tau_2^{-1}$, where τ_i is the relaxation time of atom i.

Wavefunctions for insulators and the Wigner crystal

The description of the trial function in the previous sections did not discuss in detail the spin-orbitals that enter into the Slater determinant. For a perfect ionic lattice, there is the well-known Wannier transformation, which changes the Bloch orbitals to site functions as discussed in [1, Ch. 21]. As shown by Kohn and others, Wannier functions are exponentially localized about lattice sites for an insulator rather than being delocalized.[17] The localization destroys the Fermi surface and causes a gap in the low-energy spectrum.

Returning again to the HEG, as discussed in Sec. 3.1, Wigner showed that it must crystallize at a sufficiently low temperature and density, now known from QMC to be for $r_s > 106$, a very low density for real materials. The best trial function for the Wigner crystal has orbitals that break the translational symmetry. The simplest form is:

$$\psi_i(\mathbf{r}) = \exp(-C(\mathbf{r} - \mathbf{Z}_i)^2), \tag{6.25}$$

where $\mathbf{Z}_i$ is a set of lattice sites and C is a parameter that controls the localization.[18] Note that the orbitals on different sites are not orthogonal to each other. One might wonder whether making the Jastrow factor sufficiently repulsive could cause localization and a metal–insulator transition. But it has been found that a plane-wave determinant, even with a strong correlation factor, does not provide a good description of the Wigner crystal. In our earlier discussion of the H_2 molecule, we defined the difference between a UHF and a RHF determinant as whether the Slater determinant has the symmetries of the hamiltonian. It appears that this distinction applies also to a Slater–Jastrow wavefunction: for the Wigner crystal one can get a much better trial wavefunction by using an unrestricted Slater–Jastrow function that breaks the translation invariance of the hamiltonian.

6.7 Beyond Slater determinants

Some properties are completely wrong in the SJ wavefunction, such as the spin-dependent Fermi liquid parameters [279], ordered phases such as a superconductor, and ground states that need to be described with several determinants. One can see in Fig. 6.5 that the SJ ansatz for open-shell atoms is considerably less accurate than for closed-shell ones. For quantum Monte Carlo methods, having better trial functions, in particular the antisymmetric part is crucial for achieving reliable results; see Ch. 24. We discuss two other approaches.

---

[17] A practical point is that computation using localized functions can be much more efficient for large systems than using Bloch (delocalized) orbitals. One only needs to represent the orbital in a finite region around a given ion. If there is lattice symmetry, this can further reduce the memory requirements.

[18] It has been found that spherical gaussians work as well as more complicated functions [278]. A 3D b.c.c. lattice (the lattice that minimizes the potential energy) is bipartite (i.e., separates into two sublattices); hence, one can place the two species of spin on the two sublattices.

Paired electrons

Suppose we replace the spin-orbitals in the Slater determinant with functions that pair electrons. Assuming that there are equal numbers of spin up and down electrons,[19] we can use the generalized valence bond (GVB) form:

$$\Psi_{GVB} = e^{-U(R)} \det[v(\mathbf{r}_i \uparrow, \mathbf{r}_j \downarrow)], \tag{6.26}$$

where $v(\mathbf{r}, \mathbf{r}')$ is the pairing function or "geminal" [280] and $U(R)$ is a Jastrow function. The pairing function is a six-dimensional symmetrical function that can be expanded in terms of single-particle orbitals as:

$$v(\mathbf{r} \uparrow, \mathbf{r} \downarrow) = \sum_{k=1}^{M} \psi_k^*(\mathbf{r} \uparrow)\psi_k(\mathbf{r} \downarrow). \tag{6.27}$$

If we expand the determinant in terms of the orbitals $\psi_k(\mathbf{r})$, we obtain the expansion of $\Psi_{GVB}(R)$ in Slater determinants:

$$\Psi_{GVB} = e^{-U(R)} \sum_{\{I_i\}} \det(\psi_{I_i}(\mathbf{r}_{j\uparrow})) \det(\psi_{I_i}(\mathbf{r}_{j\downarrow})), \tag{6.28}$$

where the sum is over all selections of $N/2$ integers from the set of positive integers less than or equal to M. Clearly we must have $M \geq N/2$. For $M = N/2$ there is only one possible selection, so the GVB form is equivalent to a single Slater determinant. As $M > N/2$, the number of determinants grows rapidly.

To see how this changes the trial function, consider the case of a single Be atom with four electrons. The ground state is $(1s2s)^2$ but there are three other configurations $\{(1s2p_x)^2,$ $(1s2p_y)^2, (1s2p_z)^2\}$ very close in energy. Because of the near degeneracy, the single Slater determinant (using just the $1s, 2s$ orbitals) is inaccurate, as seen in Fig. 6.5: the VMC energy only recovers 77.6% of the correlation energy. However, if we let v include the sum over the five orbitals $(1s, 2s, 2p_x, 2p_y, 2p_z)$ then we can recover 97.5% of the correlation energy. Even more impressive, consider the system Be_N and suppose the atoms are well separated from each other. To recover this amount of correlation energy would require 4^N Slater determinants but only a single geminal determinant. Of course, each orbital requires a sum over N states but the complexity of the calculation is the same as for a single Slater determinant. For other applications of geminals, see [281]. This trial function can also describe superconductivity (Cooper pairs): if we neglect the Jastrow factor, it is the BCS wavefunction for fixed number of electrons (i.e., in the canonical ensemble) [282]. One can also pair electrons in a spin-triplet state:

$$\Psi_P(R) = e^{-U(R)} \hat{A} \prod_{i=1}^{N/2} v_p(r_{2i}, r_{2i-1}). \tag{6.29}$$

---

[19] Unpaired electrons can also be accommodated.

The antisymmetry operator is:

$$\hat{A}\Psi(X) = (N!)^{-1} \sum_{\mathcal{P}} (-1)^{P} \Psi(\hat{\mathcal{P}}X), \tag{6.30}$$

where $\hat{\mathcal{P}}$ is an operator that permutes spatial and spin coordinates. This function Ψ_P is a pfaffian and can be calculated in $\mathcal{O}(N^3)$ time [283]. Note that the square of a pfaffian is a determinant, but the pfaffian is a more general form, encompassing various forms of determinants.

Backflow and three-body trial functions

Let us now consider how to go beyond the Slater–Jastrow form using the FK formula, Eq. (6.12). So that we have a tractable function at the end, we start the procedure from the Hartree–Jastrow function $\Psi_{HJ} = e^{-U^{(2)}} \prod_i \psi_i(\mathbf{r}_i)$, where $U^{(2)}$ is the sum of one- and two-body Jastrow factors. We then use Eq. (6.13) to suggest corrections:

$$\Psi_{BF-3B}(R) = \hat{A}\Psi_{HJ}(R) \exp[-\tau \bar{E}_L(R|\Psi)]. \tag{6.31}$$

Note that $\bar{E}_L(R|\Psi)$ is a smoothed local energy of Ψ_{HJ}. The FK approximation introduces two additional terms in the SJ trial function: "backflow" and "three-body" functions.

In a backflow determinant, the particle coordinates in the Slater determinants are mapped into "quasi-particle" coordinates:

$$\Psi_{BF-3B}(R) = e^{-U^{(2)}-U^{(3)}} \det[\psi_k(\mathbf{x}_i, \sigma_i)], \tag{6.32}$$

where the quasi-particle coordinates are defined by $\mathbf{x}_i = \mathbf{r}_i + \sum_j \eta(r_{ij})(\mathbf{r}_i - \mathbf{r}_j)$. The backflow function $\eta(r)$ is similar to the Jastrow correlation factor, however, it decays more quickly [284], as r^{-3} in 3D. Backflow has the important property of changing the many-body nodes of the trial function, an important aspect for quantum Monte Carlo.

The three-body term[20] has the functional form of a squared force:

$$U^{(3)}(R) = -\sum_i \left[\sum_j \xi(r_{ij})(\mathbf{r}_i - \mathbf{r}_j)\right]^2. \tag{6.33}$$

This term can be important at strong correlation, for example in the electron gas at large r_s [279, 285]. Figure 6.6 shows how the energy and variance of the electron gas improve as these terms in the trial function are added in.

Exercises

6.1 Use Eq. (1.3) to find the total number of states for two carbon atoms with $N_s = 3$, the number of basis functions per electron (the result is greater than 10^9). How much does this increase if N_s is doubled? Using Stirling's formula $n! \approx \sqrt{2\pi n}(n/e)^n$ for large n, find an expression for the coefficient C in Eq. (1.3) in terms of the number of electrons per atom, z, and N_s.

---

[20] This is not the only three-body term, but the most important one for the HEG.

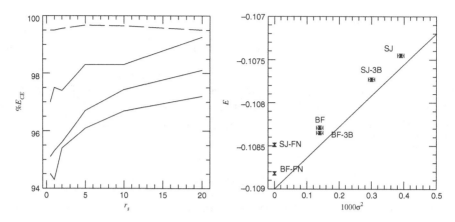

Figure 6.6. Accuracy of various trial functions in the 3D HEG. Left panel: The percentage correlation energy [286] for (from the bottom) SJ function, SJ-3B, fixed-node SJ, and (dashed line) BF. The BF fixed-node energy is considered to be the exact answer, in reality roughly 0.1% of the energy is missing. Right panel: Energy (Ryd/electron) versus the variance of the local energy at $r_s = 10$. Each symbol represents one variational calculation: from higher to lower energies: SJ, 3B, BF, and BF-3B [285]. The two points at $\sigma^2 = 0$ are the fixed-node results using a Slater determinant (SJ-FN) and a backflow determinant (BF-FN). The line is a guide to the eye.

6.2 As stated in Sec. 6.3, a many-body eigenstate for a crystal can be written as a generalized Bloch function $\Phi_\mathbf{K}(\mathbf{r}_1, \mathbf{r}_2, \ldots) = e^{i\mathbf{K}\cdot\mathbf{R}} u_\mathbf{K}(\mathbf{r}_1, \mathbf{r}_2, \ldots)$, where u is a periodic function of each argument. Show that $u_\mathbf{K}$ is an eigenstate of the modified hamiltonian with $i\nabla \to (i\nabla + K)$ with the same energy.

6.3 Given a closed-shell state of the 2D electron gas, with N-spinless (only up spins) non-interacting electrons in a square with periodic boundary conditions, for what values of $N < 30$ is the ground state non-degenerate (i.e., a closed shell)? Taking $N = 9$, what is the lowest-energy excitation that couples to the ground state? What states are occupied?

6.4 The error in a trial function can be expanded to a density of states as:

$$c(E) = \sum_\alpha \delta(E - E_i)|\langle \Psi | i \rangle|^2. \tag{6.34}$$

Here $(E_i, |i\rangle)$ are the exact eigenvalues and states. (a) How is $c(E)$ normalized? (b) Express the variational energy and variance of Ψ in terms of $c(E)$. (c) Express the overlap in terms of $c(E)$, by first choosing the phase of Ψ to get maximum overlap. (d) For a pair product trial function with the wrong cusp condition, what is the smallest value of k for which $\int dE E^k c(E)$ diverges?

6.5 Consider the wavefunction defined as $\phi(\tau) = \exp(-\tau\hat{H})\Psi$ with $\hat{H}$ a hamiltonian. Let $E(\tau)$ be its variational energy. (a) Show that $E(0) = E_V$ and $E(\infty) = E_0$. (b) Show that $dE(\tau)/d\tau|_0 = -\sigma^2$, the trial function variance. (c) Show that $2\ln(\mathcal{O}) = \int_0^\infty d\tau(E(\tau) - E_0)$, where $\mathcal{O}$ is the trial function overlap defined in Eq. (6.17). This evolution in imaginary time is what projector Monte Carlo does (see Ch. 24). This proves that if $(E(\tau) - E_V)$ is extensive (proportional to the number of atoms), then σ^2 and $\ln(\mathcal{O})$ are as well.

6.6 Consider a homogeneous system of particles in D dimensions interacting with a potential r^{-n}. For what values of n and D can you derive a cusp condition on the wavefunction as in Eq. (6.2)? How does the cusp condition change if the particles are in a relative p or d state?

6.7 Derive the long-distance behavior of the Jastrow factor for the 3D HEG.

6.8 In Eq. (6.25), what is the r.m.s. deviation of an electron from a lattice site? Ignore antisymmetry and the Jastrow factor.

6.9 What is the local energy function for the helium atom wavefunction of the form $\psi(r_1)\psi(r_2)\exp(-u(r_{12}))$? Write the expression for the local energy as a function of ψ and u. What are the conditions on $\psi(r)$ and $u(r)$ needed to keep the local energy bounded everywhere?

7

Particles and quasi-particles

Art is the elimination of the unnecessary.

Pablo Picasso

Summary

Condensed matter is constituted by a huge number of electrons and nuclei interacting with Coulomb potentials. The topic of this chapter is a way to deal with the full, coupled problem by separating it into a part that is tractable and the rest that is approximated. This naturally leads to the appearance of dynamical fields, and to the concept of "quasi-particles" that have the same quantum numbers as non-interacting electrons. The quasi-particles obey equations where the potential and bare interactions in the hamiltonian are replaced by dynamical self-energies and screened interactions that can describe many of the effects of correlation. In this chapter we discuss the intuitive concepts behind this approach in general terms to motivate the more rigorous formulations and approximations used in the Green's function methods of the following chapter and in Part III.

The first equations of this book, Eqs. (1.1) and (1.2), express the fundamental theory of matter in terms of electrons and nuclei that interact with Coulomb potentials. For systems with only a few electrons, a solution of the problem can be obtained by exact diagonalization, using configuration interaction methods. For many-electron systems one has to resort to other methods. For example, QMC stochastic simulations (Chs. 22–25) are among the most accurate methods known to calculate certain expectation values, such as equilibrium thermodynamic properties, total energies, the density, and various static correlation functions like those in Ch. 5. Other properties such as excitation spectra are more difficult to access with QMC. Straightforward perturbation theory is, in general, not appropriate since interactions among the electrons are of the same order of magnitude as the independent-particle terms. Hence, we must develop other approaches.

One strategy that gives access to equilibrium thermodynamic as well as dynamical excitation properties is to separate the interacting many-body problem into a part that is tractable by some means and the rest of the problem that is dealt with more approximately. The present chapter is devoted to this idea. It is a prelude to many-body Green's function methods; the aim is to provide a unified framework for the developments in Chs. 8–21, along with a qualitative description of the most relevant quantities. We will call the part that is tractable a "small" system (not necessarily small in real space, but in the sense that it can be described by a number of degrees of freedom that is small enough to handle, or that is simple enough to be tractable with a feasible computational effort). The central concept is that the effect of coupling to the larger (more difficult to treat) system can be expressed through an effective dynamical field that modifies how the small system evolves; i.e., the larger system "dresses" the smaller system. This idea is very general and relevant in many circumstances, well beyond electronic structure. Throughout Chs. 9–21 it will appear numerous times, though in different forms, ranging from screened Coulomb interactions (Ch. 11), to an impurity in a bath (Ch. 16), to the self-energy of one-particle excitations, and many others.

The self-energy of interacting electrons is indeed such a dynamical field, created by the coupling of a particle to other degrees of freedom (other excitations). This and the closely linked idea of *quasi-particles* introduced by Landau (see Secs. 1.3 and 3.4) are among the most powerful concepts in many-body theory. It can be very useful to construct objects that behave like dressed, weakly interacting particles, with effective properties that are due to the correlations in the original system of interacting particles. In this way, one obtains a description of excitations of the interacting many-body problem as objects with the same quantum numbers as non-interacting "bare" particles, but with modified energies and lifetimes that take into account the effects of interactions with other particles. Whereas electrons and nuclei are the fundamental particles with fixed charges and masses, quasi-particles are a construction chosen to represent approximately the particle-like excitations of an actual interacting system. There is a close link to Fermi liquid theory, summarized in Sec. 3.4. To work with quasi-particles, and to represent the bare interactions by effective quantities, is an example of "elimination of the unnecessary."

The first section of this chapter introduces the formulation of the general approach; in particular, it explains the appearance of a dynamic (i.e., frequency-dependent) self-energy that contains all the coupling effects that one does not want to treat explicitly. Section 7.2 is devoted to the Dyson equation, one of the fundamental equations of this book. This is followed in Sec. 7.3 by an example meant to illustrate the main points. All considerations are very general, not limited to electrons. The subsequent Sec. 7.4 describes characteristic features of the spectral function and self-energy. It is still general, but it also contains some points that are specific to interacting electrons. Sections 7.5 and 7.6 focus on the intuitive quasi-particle picture, its mathematical link to the Green's function, and how to find quasi-particles in practice. Section 7.7 contains more specific links to later chapters; it may be considered material for more advanced readers. Finally, the last section gives a wrap-up of this chapter.

7.1 Dynamical equations and Green's functions for coupled systems

In many cases we want to work with equations that treat explicitly only one or a few selected degrees of freedom, while taking into account their coupling to a much larger system. Thus we work in a reduced Hilbert space with equations that contain effective potentials and interactions, different from the original bare ones. This can be visualized as a subsystem embedded in a larger system. A straightforward analysis is sufficient to provide a basis for theoretical methods that are used extensively in this book. It shows how a frequency-dependent self-energy can appear in a static problem, and it gives hints for strategies for approximations.

Origin of a frequency-dependent self-energy

Suppose we have an $m \times m$ linear eigenvalue problem[1]

$$\begin{bmatrix} S & C_1 \\ C_2 & R \end{bmatrix} \times \begin{bmatrix} \phi_1 \\ \phi_2 \end{bmatrix} = \omega \begin{bmatrix} \phi_1 \\ \phi_2 \end{bmatrix}, \tag{7.1}$$

where S is a matrix in the "small" space of dimension $n \times n$ and R is defined in the remainder of the space of size $(m - n) \times (m - n)$. From Eq. (7.1) it follows that the vector ϕ_2 of dimension $m - n$ is

$$\phi_2 = -(R - \omega)^{-1} C_2 \phi_1. \tag{7.2}$$

Substituting in the equation for ϕ_1, one gets

$$\left[S - C_1 (R - \omega)^{-1} C_2 \right] \phi_1 \equiv \tilde{S}(\omega) \phi_1 = \omega \phi_1, \tag{7.3}$$

provided $(R - \omega)^{-1}$ exists. This is an equation for ϕ_1 determined by a frequency-dependent $n \times n$ matrix $\tilde{S}(\omega) \equiv S - C_1 (R - \omega)^{-1} C_2$. Note that this is merely a transformation of the problem, since Eqs. (7.2) and (7.3) together determine both ϕ_1 and ϕ_2. No information is lost if the equations are solved completely.

However, if only the solution for the small space of dimension $n \times n$ is retained, part of the information concerning the eigenvectors of the original problem is lost. The only aspect of the larger space that is needed is the *projection* on the smaller space $C_1 (R - \omega)^{-1} C_2$. The price to pay is that one has to work with the frequency-dependent operator $\tilde{S}(\omega)$ and solve the non-linear equation $\tilde{S}(\omega) \phi_1 = \omega \phi_1$. The advantage of this may not be obvious; however, it is at the core of all the practical methods in the following Chs. 8–21. In each case, the spaces are chosen by physical reasoning and the larger space is either approximated or specified by other information.

This idea can also be formulated in terms of Green's function techniques. First, we write down the formal expressions and follow with examples later in the chapter. The equation for the Green's function for the entire system, G^{tot}, can be written as (see App. C)

$$[\omega - H] G^{tot} = I, \tag{7.4}$$

---

[1] These equations are the basis for many techniques in numerical analysis and adapted for methods in electronic structure theory; examples are discussed in [1, Apps. L and M]. Often they are applied to hermitian problems but that is not assumed here.

where each term is a matrix, I is the identity matrix:

$$H = \begin{bmatrix} H_S & H_{SR} \\ H_{RS} & H_R \end{bmatrix}, \quad G^{tot} = \begin{bmatrix} G_S & G_{SR} \\ G_{RS} & G_R \end{bmatrix}, \quad I = \begin{bmatrix} I & 0 \\ 0 & I \end{bmatrix}, \quad (7.5)$$

and ω is the frequency that can be complex.[2] As above, H_S is a submatrix of size $n \times n$, etc. Thus, G^{tot} is the inverse[3] of the matrix $\omega - H$ and algebraic manipulations like those above lead to the result that G_S (the part of G^{tot} in the smaller $n \times n$ space) can be written

$$G_S(\omega) = \left(\omega - H_S - H_{SR} [\omega - H_R]^{-1} H_{RS} \right)^{-1}$$
$$= \left([G_S^0(\omega)]^{-1} - H_{SR} G_R^0(\omega) H_{RS} \right)^{-1}, \quad (7.6)$$

where $G_S^0 = (\omega - H_S)^{-1}$ and $G_R^0 = (\omega - H_R)^{-1}$ are the Green's functions for the decoupled small system and the rest of the system, respectively.

Finally, the expressions can be written in the elegant form of a Dyson equation,

$$G_S(\omega) = \left([G_S^0(\omega)]^{-1} - \Sigma_S(\omega) \right)^{-1}, \quad (7.7)$$

where $\Sigma_S(\omega)$ is defined by

$$\Sigma_S(\omega) = [G_S^0(\omega)]^{-1} - [G_S(\omega)]^{-1} = H_{SR} G_R^0(\omega) H_{RS}. \quad (7.8)$$

This is a very general formulation and $\Sigma_S(\omega)$ can denote the effect on system S due to many factors: hybridization $\Delta(\omega)$ due to single-body hopping, a self-energy $\Sigma(\omega)$ due to interactions, screening of interactions described by polarizability $P(\omega)$, and other examples.

One of the main purposes of this chapter is to bring out the close relationships among all the methods in Chs. 8–21 which utilize the philosophy and strategy embodied in these equations. The GW approximation (Chs. 10–13) can be understood as single electrons coupled to all others through a dynamically screened mean field. It contains the screened

---

[2] Throughout the book we follow the notation of Ch. 5, where ω denotes frequency (or energy since we set $\hbar = 1$) that can be real or complex. When it is important to specify the behavior in the complex plane, we will use the symbols z or $\omega \pm i\eta$, where ω and η are real and the sign of the small imaginary part $i\eta$ determines the retarded, advanced, and time-ordered Green's functions (see Sec. 5.6). Matsubara frequencies in the finite-temperature formalism (App. D) are denoted $i\omega_n$. We will point out specifically cases where the expressions require that ω is real.

[3] To be precise, the Green's function is more than just the inverse of the matrix: it contains information about the boundary conditions. This is familiar in many problems, e.g., in electrostatics, where the hamiltonian is always the same Coulomb interaction but the solution depends on the boundary conditions. This is further discussed in Sec. G.1. Here it is sufficient to write the expressions for G and its inverse with the understanding that the way it is used in any specific case must be specified.

interaction $W(\omega)$, which is itself determined from the bare interaction by an equation analogous to Eq. (7.7): as expressed in Eq. (7.11) below, $W(\omega)$ is the analog of $G(\omega)$, and the polarizability of the system $P(\omega)$ plays the role of $\Sigma(\omega)$. In dynamical mean-field theory in Chs. 16–21 the goal is to determine the self-energy $\Sigma(\omega)$ using an auxiliary system of an atom (a small interacting system) embedded in a bath (a dynamical mean field described by $\Delta(\omega)$). Alternatively, the auxiliary system can be viewed as a single non-interacting system (atom + bath) with interactions added only for the states on the atom. We can build upon the foundations in this chapter to develop methods based on different approximations but adhering to the same principles so they can supplement one another and finally arrive at steps toward unified methods in Ch. 21.

Strategies for approximations and auxiliary systems

Since one cannot find the exact solution of the full problem, it is useful to find suitable approximations to the difference between the actual Green's function $G_S(\omega)$ and the decoupled $G_S^0(\omega)$. Concerning strategies to accomplish this, there are several important considerations.

- Finding an acceptable approximation is easier when the correction to G_S^0 is small. Therefore, it is best to work in a basis where the full matrix is nearly block diagonal, i.e., where the coupling terms in the hamiltonian H_{SR} are small.[4] Finding a good starting point is an important step in practice. For example, one usually does not start from bare electrons and include all Coulomb interactions in Σ, but rather from particles in a static Hartree–Fock or Kohn–Sham potential.
- The dynamical self-energy can often be approximated efficiently by replacing the static bare coupling with a dynamically screened one, while restricting the space to a small system. Thus the renormalized coupling represents the effects of the "larger system." Eq. (7.43) illustrates this point; the GW approximation with its dynamically screened Coulomb interaction is a practical example.
- Note that the full information on the large system R is not needed in Eq. (7.8), but only the projection $H_{SR} G_R^0(\omega) H_{RS}$ of R onto the subspace S of dimension $n \times n$. A general approach is to identify a different system R', called an auxiliary system, with the same projection so that it can reproduce exactly some chosen property of system S. Even if it is not feasible to find the exact results, this can facilitate the search for useful approximations, and it is a widely used concept (see Ch. 4).

7.2 The self-energy and the Dyson equation

Let us take a closer look at the relation between the Green's function G_S and the generalized self-energy Σ_S; this is one of the fundamental links throughout this book. It is very

---

[4] In some cases there are also canonical transformations that allow one to minimize the coupling, such as in a system of harmonic oscillators where this can be done exactly. See also the Lang–Firsov transformation in Eq. (21.15).

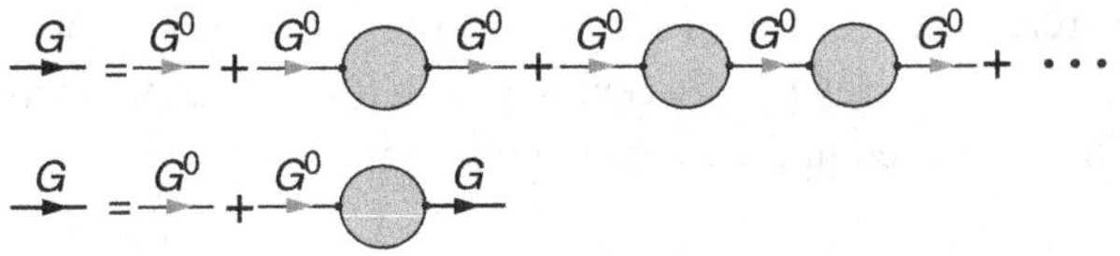

Figure 7.1. Schematic depiction of the Dyson equation. In the top line the Green's function is expressed as a sum of repeated processes (Eq. (7.9)), which is an expansion of the form in the bottom line (Eq. (7.10)). The circles represent a self-energy. The relations are not restricted to one-particle G and Σ for an electron. The form is general and applies to any Green's function and derived quantities: the screened interaction W and polarization P, response functions, etc.

important to note that the quantity we are interested in is in general G_S, not G^{tot}: our aim is to calculate *certain properties* of a system, not to solve the complete problem contained in G^{tot}. Therefore, in the following we drop the subscript S for simplicity with the understanding that in the Dyson equation all quantities G, G^0, and Σ are defined in the "small" space S; all effects of the rest of the space are hidden in the self-energy Σ.

It is instructive to rewrite Eq. (7.7) as

$$G(\omega) = \left[1 - G^0(\omega)\Sigma(\omega)\right]^{-1} G^0(\omega)$$
$$= G^0(\omega) + G^0(\omega)\Sigma(\omega)G^0(\omega) + G^0(\omega)\Sigma(\omega)G^0(\omega)\Sigma(\omega)G^0(\omega) + \cdots \quad (7.9)$$

This expansion is illustrated in Fig. 7.1 and has the interpretation of a bare propagation G^0 and an effective coupling Σ. The self-energy Σ includes all processes that cannot themselves be expressed in terms of subprocesses connected by a bare propagation. This defines the "proper," or "irreducible," self-energy [47], as opposed to the reducible $\Sigma^{red} = \Sigma + \Sigma G^0 \Sigma + \Sigma G^0 \Sigma G^0 \Sigma + \cdots$ that leads to $G = G^0\Sigma^{red}G^0$. The infinite series of effective couplings generates all possible terms in the interaction (see also Sec. 9.5). Eq. (7.7) or (7.9) can also be written as the compact Dyson equation

$$G(\omega) = G^0(\omega) + G^0(\omega)\Sigma(\omega)G(\omega), \quad (7.10)$$

where the dressed G appears on both sides, which implicitly includes the infinite sum in Eq. (7.9).

The Dyson equation in the form of Eq. (7.9) or (7.10) is very general. It is an integral equation that links a bare quantity (here G^0) to the dressed solution (here G) through a *kernel* (here the self-energy), which contains the effects of coupling (for example, due to some additional potential or interaction). It is used frequently throughout the book as a compact way to include all the effects of coupling in principle, or as a convenient way to express the Green's function in terms of an approximation to the proper self-energy. The self-energy is the central quantity in much of the theory of interacting electrons, in which case the coupling is due to the Coulomb interaction. The "GW" method (Chs. 10–13) is named for the approximation $\Sigma_{xc} = iGW$ to the exchange–correlation part of the one-particle self-energy, and DMFT (Chs. 16–21) is formulated in terms of approximations where Σ is assumed to be short range in real space. Explicit rules for the construction of Σ

are given in many references for Green's functions [38, 47, 149, 227] and are summarized in Chs. 9 and 10. The self-energy plays a central role in functionals for many-body systems, as discussed in Ch. 8 and used in the following chapters.

Dyson equation for screening

As stressed before, the formulation of Green's functions, self-energies, subspaces, and Dyson equations is completely general. In particular, analogous Dyson equations also hold for two-particle Green's functions and derived quantities that determine the "dressing" of an electron or hole due to screening by the other electrons, including the screened interaction $W(\omega)$. Physically, one can imagine that a particle inserted in a system interacts with another particle via the bare Coulomb interaction v_c. However, through v_c, it also induces a polarization of the system that is then felt by the other particle as an additional dynamic potential. If we define $P(\omega)$ as the proper, or irreducible, polarizability (as opposed to the reducible polarizability $\chi(\omega)$ in Sec. 5.8 that contains processes connected by bare Coulomb lines), then the interaction becomes $W(\omega) = v_c + v_c \chi(\omega) v_c = v_c + v_c P(\omega) v_c + \cdots$ This is an infinite series, equivalent to Eq. (7.9), and hence has a form analogous to Eq. (7.10),

$$W(\omega) = v_c + v_c P(\omega) W(\omega) = \left[v_c^{-1} - P(\omega) \right]^{-1}. \tag{7.11}$$

Here v_c is the bare quantity, $W(\omega)$ is the dressed solution, and $P(\omega)$ constitutes the kernel that includes the effects of coupling. Equation (7.11) is one of the fundamental equations used, e.g., in the GW formalism, Eq. (11.12), and depicted in the second line of Fig. 11.1. It is also represented by the schematic diagram in Fig. 7.1, with the bare G^0 replaced by the bare interaction v_c and the circles representing the polarizability P. Like $G(\omega)$, the imaginary part of ω determines the retarded, advanced, and time-ordered forms of $P(\omega)$ and $W(\omega)$.

The screened interaction $W(\omega)$ is central throughout Part III of this book:

- The dynamical screening is the only difference between the GWA, where the exchange–correlation part of the self-energy is $\Sigma_{xc} = iGW$, and the independent-particle Hartree–Fock approximation that has only an exchange self-energy $\Sigma_x = iGv_c$, determined by the unscreened interaction v_c.
- Electron–boson coupling models and the related *ab initio* cumulant Green's function presented in Ch. 15 involve electronic screening when the boson is made of electron–hole excitations.
- The application of DMFT to actual materials relies on the effective interactions between electrons in localized states (usually d or f states) that are screened by electrons in the rest of the system (typically s and p states). The frequency dependence of the screening is discussed in Ch. 21.

Finally, note that the many-electron problem can be formulated in terms of the dressed Green's function and the screened Coulomb interaction W, see Sec. 8.3.

7.3 Illustration: a single state coupled to a continuum

In this section we work out a simple, soluble example of a single state coupled to a continuum of non-interacting states. The qualitative features are quite general and apply, e.g., to the problem of interacting electrons with a single discrete independent-particle state interacting with a continuum of many-body excitations.

Consider the independent-particle problem of a state labeled 0 with energy ε_0 and matrix elements $t_{0\ell}$ that couple it to a set of states with eigenvalues ε_ℓ. The hamiltonian can be written

$$\hat{H} = \hat{H}_0 + \hat{H}_{0,rest} + \hat{H}_{rest}$$
$$= \varepsilon_0 c_0^\dagger c_0 + \sum_{\ell=1} t_{0\ell}(c_\ell^\dagger c_0 + c.c.) + \sum_\ell \varepsilon_\ell c_\ell^\dagger c_\ell, \tag{7.12}$$

where the operators $c^\dagger$ and c create and annihilate particles in the respective spaces (called S and R in Sec. 7.1). Thus they automatically generate the matrix operations in Eqs. (7.1)–(7.6).

The Green's function for the entire system is defined by the matrix equation $[\omega - \hat{H}]G(\omega) = I$, where ω can be a real or a complex frequency. The formal expressions of Sec. 7.1 can be written explicitly as the coupled equations for the components of the Green's function,

$$(\omega - \varepsilon_0)G_{00}(\omega) - \sum_{\ell=1} t_{0\ell}G_{\ell 0}(\omega) = 1,$$
$$(\omega - \varepsilon_\ell)G_{\ell 0}(\omega) - t_{\ell 0}G_{00}(\omega) = 0, \tag{7.13}$$

plus an equation for $G_{\ell\ell'}$ that is not needed here. These equations can be solved for $G_{00}(\omega)$ with the result (see Ex. 7.1)

$$G_{00}(\omega) = \frac{1}{\omega - \varepsilon_0 - \Delta(\omega)}, \tag{7.14}$$

where $\Delta(\omega)$ is an example of the generalized definition of a self-energy in Eq. (7.8), called a hybridization function[5]

$$\Delta(\omega) = \sum_\ell \frac{t_{0\ell}^2}{\omega - \varepsilon_\ell} \rightarrow \int d\varepsilon \rho(\varepsilon) \frac{t(\varepsilon)^2}{\omega - \varepsilon}, \tag{7.15}$$

and $\rho(\varepsilon)$ is the unperturbed density of states of the rest of the system. If the sum is over N states, $t(\varepsilon) = t_{0\ell}\sqrt{N}$ for $\varepsilon = \varepsilon_\ell$; the factor of $\sqrt{N}$ arises because the matrix element to each state must decrease as the number of states per unit energy increases to the continuum limit (Ex. 7.2). In this limit the poles merge to form a branch cut on the real axis (see

---

[5] The notation Δ is used to highlight the fact that here the generalized self-energy stems from a coupling to a non-interacting system. Note that in the following chapters of this book we will reserve Σ to denote effects due to interactions.

Sec. C.5). Using Eqs. (5.91) and (5.92), the spectral function $A(\omega)$ of the state 0 coupled to the continuum is

$$A(\omega) = \frac{1}{\pi} \frac{|\mathrm{Im}\,\Delta(\omega)|}{[\omega - \varepsilon_0 - \mathrm{Re}\,\Delta(\omega)]^2 + [\mathrm{Im}\,\Delta(\omega)]^2}. \tag{7.16}$$

Without the coupling to the continuum, $A(\omega) = \delta(\omega - \varepsilon_0)$; with the coupling, the energy of the peak is shifted by $\mathrm{Re}\,\Delta(\omega)$ and there is broadening due to $\mathrm{Im}\,\Delta(\omega)$. To get $\mathrm{Im}\,\Delta(\omega)$, we have to take the hybridization function slightly off the real axis and use Eq. (C.12), which yields

$$\lim_{\eta \to 0^+} \Delta(\omega \pm i\eta) = \mathcal{P} \int d\varepsilon \rho(\varepsilon) \frac{t^2(\varepsilon)}{\omega - \varepsilon} \mp i\pi \rho(\omega) t^2(\omega). \tag{7.17}$$

If the coupling and density of states of the continuum are approximated as constants, $t(\varepsilon) = \bar{t}$ and $\rho(\varepsilon) = \bar{\rho}$, then $\Delta(\omega \pm i\eta)_{|\eta \to 0^+}$ is a purely imaginary constant $\Delta = \mp i\pi \bar{\rho} \bar{t}^2$. With these assumptions the spectral function in Eq. (7.16) is a lorentzian:

$$A(\omega) = \frac{1}{\pi} \frac{|\Delta|}{(\omega - \varepsilon_0)^2 + |\Delta|^2}. \tag{7.18}$$

This can be interpreted as a resonance with a lifetime for an initial state to decay into the continuum. Moreover, structure in $\Delta(\omega)$ can lead to additional structure in the spectral function. Note the close analogy to the spectral function of Eq. (7.22), where a more detailed discussion can be found.

The analysis above maps directly onto the problem of an impurity embedded in a crystal, where ε_0 is the energy of a state on the decoupled impurity and the "rest" is the continuum of states of the host crystal. An interpretation of $|\Delta|$ in Eq. (7.18) is the rate for a particle initially in the impurity state to escape into the continuum. It is not hard to imagine that this analysis can be generalized to the Anderson impurity model in Sec. 3.5, where particles interact on the impurity site. This can be formulated with the state 0 replaced by a local interacting system with several eigenstates, for example, a single site of a Hubbard model with four eigenstates $(0, \uparrow, \downarrow, \text{and } \uparrow\downarrow)$ as in Sec. 3.5. Embedding the interacting system in a solid with a continuum of states is the basis for DMFT in Chs. 16–21.

The example may also illustrate the decay of an excitation due to coupling to other excitations. This coupling of excitations is the basis of the very existence of a self-energy in an interacting many-body problem, as will be outlined in the following section.

7.4 Interacting systems: the self-energy and spectral function

Let us now consider interacting systems using the general approach of coupled systems and the Dyson equation in Secs. 7.1 and 7.2. We will see ways in which the problem is analogous to the coupling between non-interacting subspaces in the previous section, even though it is much more difficult (in general, not feasible) to find the exact self-energy. In this section the equations are written in the notation for the one-particle Green's function G and the self-energy for electrons, since that is the primary use in this book; however, everything is also valid for interactions other than the Coulomb interaction, and for particles or fields other than electrons. The formal properties of Green's function are worked out in detail in Ch. 5, and practical approaches are the topic of Part III of this book. The purpose of this section is to bring out ways that many important properties, such as the schematic spectral functions in Figs. 2.5 and 7.2, can be appreciated without complicated mathematical derivations.

What is hidden in the self-energy of interacting electrons?

In order to focus on the effects of interactions, we consider a system with hamiltonian $\hat{H} = \hat{H}_0 + \hat{V}$, where $\hat{H}_0 = \sum_{ij} h_{ij} c_i^\dagger c_j$ is an independent-particle hamiltonian and $\hat{V}$ is an interaction that can be written in the form $\hat{V} = \frac{1}{2} \sum_{ijk\ell} v_{ijk\ell} c_i^\dagger c_j^\dagger c_k c_\ell$, as given, e.g., in Eq. (4.8) and App. A. Consider the one-electron Green's function for $T = 0$. The general expressions are given in Sec. 5.6, where any of the Green's functions can be derived from the propagators $G^<$ and $G^>$. For a system of N electrons, the Green's function for electron addition can be written in a basis of states labeled $i, j, \ldots$ as

$$G_{ij}^{\text{el}}(\omega) = \langle N, 0| c_i [\omega - \hat{H} + E_N]^{-1} c_j^\dagger |N, 0\rangle, \tag{7.19}$$

where 0 denotes the ground state with energy E_N, and a similar expression for holes with c interchanged with $c^\dagger$. These are the propagators we are interested in: they correspond to G_S in Sec. 7.1 and they are given by the matrix elements of $[\omega - \hat{H} + E_N]^{-1}$ between states where a particle or hole has been added to the many-body ground state.

In the absence of interaction, $c_j^\dagger |N, 0\rangle$ and $c_j |N, 0\rangle$ can be chosen to be eigenstates of the independent-particle hamiltonian $\hat{H}_0$, so that Eq. (7.19) is simply $1/(\omega - (E_{N+1,i} - E_N)) = 1/(\omega - \varepsilon_i^0)$, where ε_i^0 is a single-particle eigenvalue, and there is a corresponding expression for a hole in $G^<$. There is no coupling to states other than $c_i^\dagger |N, 0\rangle$ (or $c_i |N, 0\rangle$): in the language of Eqs. (7.4)–(7.8), this is G_S^0. In the presence of interaction there is coupling between the state with an added particle or hole and other, excited, states of the $(N \pm 1)$-electron system. The effects of coupling must be contained in the self-energy.

The scheme of Eqs. (7.4)–(7.8) is based on a separation of some fixed basis set into a "small" part and the rest; it does not consider a change in the hamiltonian, e.g., switching on and off the interaction part. In the present case of interacting-electron Green's functions, however, we are interested in the difference between interacting and non-interacting G and G^0. The analysis still applies but now the decoupled systems are interacting and the challenge is to take this into account. Nevertheless, it is clear that the essence of the self-energy is still the coupling of excitations caused by the presence of the interaction $\hat{V}$.

The correspondence can be seen by examining the perturbation expansion of $[\omega - \hat{H} + E_N]^{-1}$ and the ground state in Eq. (7.19) in powers of $\hat{V}[\omega - \hat{H}_0 + E_N]^{-1}$. The zero-order term corresponds to $G^0(\omega)$ in Eq. (7.9) and Fig. 7.1. In first order the Hartree and Fock contributions appear. They are sometimes excluded from the definition of the self-energy and included as a mean-field contribution in G^0. In the higher-order terms, the interaction $\hat{V}$ provides coupling to excited states corresponding to the coupling terms H_{SR} in Eq. (7.6).

This is the physics contained in what is called the *correlation contribution* to the self-energy. The series in this form can be used directly for finite systems; for infinite systems $(N \to \infty)$ the development of useful, feasible perturbation expansions is a major topic of Part III.

Despite all this complexity, the equations can be cast in a form that has the structure of Eq. (7.9) and Fig. 7.1, where Σ_{ij} is the sum of all processes that start from a state $c_j^\dagger |N, 0\rangle$ and return the system to a state $c_i^\dagger |N, 0\rangle$, passing through excited states with $N+1$ particles (or the states with $N-1$ particles and $c^\dagger$ replaced by c), and which cannot be factored into parts linked only by G^0. This defines the proper self-energy, which can be concatenated to form a series as in Eq. (7.9), and the Dyson Eq. (7.10). This agrees with the general expression for the self-energy in Eq. (7.8) that is related to the Green's functions by

$$\Sigma(\omega) = [G^0(\omega)]^{-1} - G^{-1}(\omega), \tag{7.20}$$

where each term is a matrix, G^0 is a non-interacting or mean-field Green's function, and a time-ordered G is built from the propagators as described in Sec. 5.6.

Some widely used approximations such as GW explicitly show the coupling (albeit approximate) of excitations: in GW the screened Coulomb interaction W contains neutral (electron–hole) excitations that cause Σ to be frequency dependent. Coupling to excitations is also the content of certain model hamiltonians such as the electron–boson model, Eq. (11.46), where an isolated fermion couples to excitations of the electron system, such as plasmons, or other boson excitations as in the model in Sec. 21.6.

Consequences of the frequency-dependent self-energy

The consequences of $\Sigma(\omega)$ can be illustrated by writing the expressions in a single-particle basis. Here we consider a crystal where the independent-particle Green's function can be written[6] $G^0_\mathbf{k}(\omega) = \left[\omega - \varepsilon^0_\mathbf{k}\right]^{-1}$ (see, e.g., Eq. (5.106)), with $\varepsilon^0_\mathbf{k}$ the single-particle eigenenergies. The interacting Green's function becomes

$$G_\mathbf{k}(\omega) = \left[\omega - \varepsilon^0_\mathbf{k} - \Sigma_\mathbf{k}(\omega)\right]^{-1}. \tag{7.21}$$

Thus $\Sigma_\mathbf{k}(\omega)$ modifies the bare energy $\varepsilon^0_\mathbf{k}$, adding a non-trivial shift (through the real part of the frequency-dependent Σ) and an imaginary part (through the imaginary part of Σ). As in the case of the hybridization function $\Delta(\omega)$ in Eq. (7.14), there can be a rich structure in $G_\mathbf{k}(\omega)$ due to the frequency dependence of the self-energy $\Sigma_\mathbf{k}(\omega)$.

The theory for Green's functions of interacting systems is not limited to weakly interacting particles. Dyson equations provide a concise formulation for self-energies that can modify the spectrum to an arbitrary extent. The only requirement is that the excitations have the same symmetry as those of a system with Green's function G^0, including the crystal momentum $\mathbf{k}$, spin in cases where hamiltonian is spin-independent, etc. This is an

---

[6] We do not indicate band indices here. When the self-energy is not diagonal in the independent-particle bands, one has to interpret Eq. (7.20) and the subsequent equations as matrix equations.

example of the continuity principle emphasized in Sec. 1.3. For non-interacting particles the energy is given by $\varepsilon_\mathbf{k}^0$ and the Green's function has a pole for $\omega = \varepsilon_\mathbf{k}^0$; however, there is a continuum of energies for excitations formed by multiparticle excitations with total momentum $\mathbf{k}$, and interactions couple all these excitations. In the thermodynamic limit, the δ-function peaks of the spectral function merge to form broad structures.

These conclusions also apply to other problems, for example, a self-energy of an oscillator, where anharmonic coupling to a bath with a continuum of excitations leads to damping (see Ex. 7.13). It is instructive to look more carefully at the expressions for phonons and the differences from electrons that are the subject of Ex. 7.5.

General features of the spectral function

Let us now look closer at the spectrum of an interacting system and the example shown in Fig. 7.2. For simplicity we consider temperature $T = 0$. Because the spectrum is that of a sharp state that scatters into a continuum of excitations, a diagonal element of the spectral function is analogous to Eq. (7.16), with the hybridization function Δ replaced by the self-energy Σ,

$$A_\mathbf{k}(\omega) = \frac{1}{\pi}|\mathrm{Im}\, G_\mathbf{k}(\omega)| = \frac{1}{\pi}\frac{|\mathrm{Im}\, \Sigma_\mathbf{k}(\omega)|}{[\omega - \varepsilon_\mathbf{k}^0 - \mathrm{Re}\, \Sigma_\mathbf{k}(\omega)]^2 + [\mathrm{Im}\, \Sigma_\mathbf{k}(\omega)]^2}, \qquad (7.22)$$

which is defined for real ω and is the same as Eq. (5.103). For vanishing self-energy, $A_\mathbf{k}(\omega)$ is a δ-function peak centered at $\varepsilon_\mathbf{k}^0$, which is the energy spectrum for a single non-interacting particle and is depicted by the vertical bar in Fig. 7.2. The figure is simplified; it is obtained from the model of Exs. 7.6 and 7.7 but shows some general trends, and is quite similar to the right panel in Fig. 2.5.

The real part of Σ leads to a shift of the peak, as can be seen from the denominator of $A_\mathbf{k}$. The peak remains sharp if the imaginary part of Σ around that energy is small. In that case the zero of the real part,

$$\varepsilon_\mathbf{k} = \varepsilon_\mathbf{k}^0 + \mathrm{Re}\, \Sigma_\mathbf{k}(\omega = \varepsilon_\mathbf{k}), \qquad (7.23)$$

is the new peak position $\varepsilon_\mathbf{k}$. Because of the one-to-one correspondence with the non-interacting particle peak, this structure is called a *quasi-particle* peak. Since in practice $\varepsilon_\mathbf{k}^0$ may contain static mean-field contributions of the interaction, the division between the independent-particle and the real part of the self-energy is not unique; Fig. 7.2 is typical for a situation where the independent-particle energy corresponds to the Hartree–Fock solution, and the self-energy consists only of the correlation part. In that case $\varepsilon_\mathbf{k}$ tends to be closer to the Fermi energy than $\varepsilon_\mathbf{k}^0$ (see Ex. 7.7). If $\mathrm{Im}\, \Sigma_\mathbf{k}(\omega)$ is approximately constant in that energy region, the spectrum has a lorentzian form as in Eq. (7.18). This can be a useful approximation over some range of energies, as depicted in Fig. 7.2. If the broadening is very large, the quasi-particle peak may be completely washed out and a single particle-like picture is no longer appropriate to interpret observations.

The excitations that may occur when a particle is added or removed from the system show up as a broad background and/or as distinct structures in $\mathrm{Im}\, \Sigma_\mathbf{k}(\omega)$. These structures

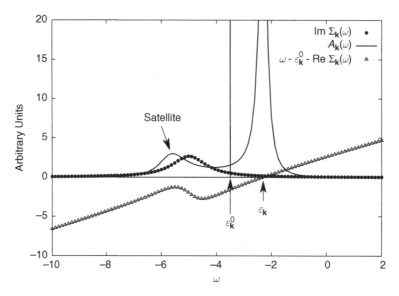

Figure 7.2. Schematic illustration of an element of the spectral function for removal of electrons below the Fermi energy ($\omega = 0$). In the independent-particle case, the spectral function is diagonal in the appropriate basis, and each diagonal element **k** is a δ-function, situated at $\varepsilon_{\mathbf{k}}^0$. Through the interaction, the real part of the self-energy (empty triangles represent the function $\omega - \varepsilon_{\mathbf{k}}^0 - \mathrm{Re}\,\Sigma_{\mathbf{k}}(\omega)$) shifts the main peak to the quasi-particle energy $\varepsilon_{\mathbf{k}}$. The imaginary part (full circles) is small but non-vanishing around $\varepsilon_{\mathbf{k}}$, which leads to broadening. Still, in this example one can relate the resulting structure to the non-interacting one and interpret it as a dressed particle, i.e., a *quasi-particle*. The self-energy contains in principle all possible excitations of the system in the presence of an additional particle, which leads to structure in Im Σ. This structure shows up as satellites in the spectral function; there is one satellite in this figure, indicated by an arrow. The total weight is conserved, which means that the quasi-particle has lost weight to the satellite. Note the similarity to the schematic spectrum for photoemission in Fig. 2.5.

lead to additional peaks in $A(\omega)$ (see Ex. 7.6), termed "satellites" or "sidebands." Since the total weight of the spectral function is conserved, as required by the sum rule of Eq. (5.95), the satellites take weight from the quasi-particle peak. When the quasi-particle is no longer the dominant structure in the spectrum, this is one signature of "strong correlation." The effect of mixing of excitations from the point of view of mixing of Slater determinants is explored in Ex. 7.3.

The schematic spectral function Fig. 7.2 contains a well-defined quasi-particle and one satellite. Overall, this figure reflects much of what is observed in advanced calculations and in experiment. For illustration, one may look at the middle panel of Fig. 11.5 for the homogeneous electron gas, or Fig. 13.1 for the spectrum of bulk silicon. In those examples, the satellites are due to plasmons. In other systems, atomic-like excitations dominate, an example being Ce with its strongly localized $4f$ electrons in Fig. 2.9. In still other cases, the many-body nature of the problem makes a simple interpretation in terms of electron–hole excitations difficult (see, e.g., the example of Ni in Secs. 13.4, 15.4, and 20.4).

7.5 Quasi-particles

Many calculations concentrate only on the quasi-particle part of the spectrum, because the quasi-particle energies yield the band structure of a solid, a fundamental ingredient for the understanding of materials. Indeed, as discussed in this section, the quasi-particle spectra should be well-defined near the Fermi energy in a metal or the band edges in a semiconductor or insulator. In this section we will see how quasi-particles emerge from the interacting spectral function and we will investigate quasi-particle properties in more detail, whereas Sec. 7.6 discusses how one may find the quasi-particles in a practical calculation.

The idea of quasi-particles appears in Fermi liquid theory (FLT), Sec. 3.4, as particles that are dressed. The dressing modifies their effective mass, but keeps many characteristic features of the bare particles, such as their symmetries. Quasi-particles can be understood to be directly born from the non-interacting δ-function peaks. In the following we will concentrate on those parts of the spectral function that correspond to such modified particle-like peaks. However, here the definition of a quasi-particle is slightly different than in FLT. What we call "quasi-particle peaks" are broadened structures because the excitations have a finite lifetime, and part of the quasi-particle weight is transferred to a satellite part of the spectrum that has no analogy in independent-particle systems. When these effects are neglected, the formalism yields the spectrum of stable, independent quasi-particles like Landau's quasi-particles (Sec. 3.4) that have the entire weight, but modified energy, effective mass, etc.

Quasi-particles near the Fermi energy

Let us concentrate on a region close to the Fermi energy E_F in a metal at zero temperature, where sharp quasi-particle peaks exist, in order to discuss the modifications brought about by the self-energy. The lifetime increases when the quasi-particle energy approaches E_F, i.e., Im Σ decreases as

$$\text{Im } \Sigma_{\mathbf{k}}(\omega) \sim (\omega - E_F)^2. \tag{7.24}$$

This reflects a decrease of the scattering rate, since the number of excitations to which an electron at energy ω can decay is proportional to $(\omega - E_F)^2$ due to the Fermi exclusion principle (see Ex. 7.4). This is the signature of a "normal Fermi liquid" described in Sec. 3.4. Thus the quasi-particle states have well-defined energies $\varepsilon_{\mathbf{k}}$ near E_F and the Fermi surface can be defined as the surface in $\mathbf{k}$-space where $\varepsilon_{\mathbf{k}} = E_F$, as depicted in Figs. 7.3 and 2.5. For non-interacting particles at $T = 0$, the occupation numbers $n_{\mathbf{k}}$ (defined in Eq. (5.7)) as a function of $\mathbf{k}$ jump from one to zero at the Fermi surface.[7] In the interacting system, there is a distribution with fractional $n_{\mathbf{k}}$ as shown schematically by the shaded regions on

---

[7] Here we refer to occupation numbers of single-particle states. This coincides with the momentum distribution in a homogeneous electron gas, but not in a crystal where the single-particle eigenstates are not plane waves. Therefore calculations and measurements of a momentum distribution in a crystal may find a discontinuity that is different from one. This does not reflect a correlation effect then, but scattering due to the crystal potential. For details, see the discussion after Eq. (5.123).

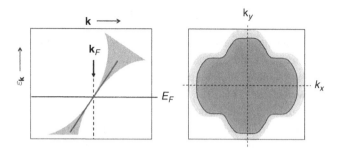

Figure 7.3. Schematic illustration of the width and dispersion of the quasi-particle structure in the spectral function Eq. (7.22) for addition and removal of electrons close to the Fermi level. In the left panel, the solid line indicates the quasi-particle energy $\varepsilon_{\mathbf{k}}$ as a function of $\mathbf{k}$ near the Fermi momentum $\mathbf{k}_F$ for a single direction in $\mathbf{k}$-space. The shaded area represents the width of $A_{\mathbf{k}}(\omega)$ around $\omega = \varepsilon_{\mathbf{k}}$, given by Im $\Sigma_{\mathbf{k}}(\varepsilon_{\mathbf{k}}) \propto (\varepsilon_{\mathbf{k}} - E_F)^2$ in Eq. (7.24); so long as the width is small on the energy scale of interest, there is a peak in the spectral function that can be identified as a quasi-particle. Because the lifetime must go to infinity for states at the Fermi energy, the energies are well-defined and their variation with $\mathbf{k}$ defines the Fermi surface, depicted on the right. The shaded regions indicate the spread of the occupation numbers $n_{\mathbf{k}}$ in an interacting system (see, e.g., Fig. 5.2 for the homogeneous electron gas). Still, there is a discontinuity and hence a sharp surface, indicted by the solid line. The shape may be affected by interactions but the volume does not change (the Luttinger theorem in Sec. 3.6).

the right side of Fig. 7.3; nevertheless, there is still a discontinuity that defines the Fermi surface, shown as the solid line in the right side of Fig. 7.3.

Equation (7.23) is an implicit equation for $\varepsilon_{\mathbf{k}}$, because the real part of Σ is taken at the quasi-particle energy. However, as shown in the third part of Ex. 7.6, the real part of the self-energy is approximately linear close to the Fermi energy. Therefore one can expand to linear order:[8]

$$\text{Re } \Sigma_{\mathbf{k}}(\omega) \approx \text{Re } \Sigma_{\mathbf{k}}(\varepsilon_{\mathbf{k}}^0) + (\omega - \varepsilon_{\mathbf{k}}^0)\frac{\partial \text{Re } \Sigma_{\mathbf{k}}(\omega)}{\partial \omega}\bigg|_{\omega = \varepsilon_{\mathbf{k}}^0}. \qquad (7.25)$$

With the renormalization factor

$$Z_{\mathbf{k}} = \frac{1}{1 - \frac{\partial \text{Re } \Sigma_{\mathbf{k}}(\omega)}{\partial \omega}\big|_{\omega = \varepsilon_{\mathbf{k}}^0}}, \qquad (7.26)$$

the quasi-particle condition Eq. (7.23) reads[9]

$$\varepsilon_{\mathbf{k}} = \varepsilon_{\mathbf{k}}^0 + Z_{\mathbf{k}}\text{Re } \Sigma_{\mathbf{k}}(\omega = \varepsilon_{\mathbf{k}}^0). \qquad (7.27)$$

[8] The expansion can alternatively be performed around $\omega = \varepsilon_{\mathbf{k}}$ or $\varepsilon_{\mathbf{k}}^0$. The former is the logical choice for an expansion around the quasi-particle like in the following section; the latter, used here, allows one to avoid the solution of an implicit equation for the quasi-particle energies in practical calculations, see Sec. 12.2. We will not explicitly distinguish the two corresponding $Z_{\mathbf{k}}$. When even the linear term is neglected, one has the "on-shell" approximation.

[9] Note that, in principle, there can be more than one solution to this condition. We will come back to this problem in Sec. 15.7.

Of course, Eqs. (7.25) and (7.27) should be used only over a limited energy range according to the precision that is required. Often this is a useful approximation for band structure calculations near the Fermi energy; however, over a wider energy range the self-energy may have more structure, as illustrated in Fig. 7.2 and many examples in this book.

When the linear expansion is used around the quasi-particle energy ε_k in the spectral function of Eq. (7.22), and when the imaginary part is taken at ε_k, one finds that around the quasi-particle energy

$$A_k(\omega) \approx Z_k \frac{Z_k \operatorname{Im} \Sigma_k(\varepsilon_k)}{(\omega - \varepsilon_k)^2 + [Z_k \operatorname{Im} \Sigma_k(\varepsilon_k)]^2}, \tag{7.28}$$

which is a lorentzian[10] like Eq. (7.18) with weight Z_k, positioned at energy ε_k and broadened by $[Z_k \operatorname{Im} \Sigma_k(\varepsilon_k)]$. The renormalization factor Z_k is the "quasi-particle weight" defined to be the spectral weight for adding or removing a bare electron, which is the relevant quantity for the amplitude in photoemission, tunneling, and other experiments in which an electron (not a quasi-electron!) is added or removed. By the same reasoning Z_k scales the step in n_k at the Fermi energy. The weight that is missing in the quasi-particle is transferred to the satellite part of the spectrum. Note that Z_k is exclusively due to the frequency dependence of Σ and must hence be equal to one in any independent-particle approximation.

The renormalization factor also changes the effective mass,[11] which is inversely proportional to the velocity $d\varepsilon_k/dk|_{\varepsilon_k=E_F}$ at the Fermi energy. With Eq. (7.23), the derivative of the quasi-particle energy with respect to momentum becomes

$$\frac{d\varepsilon_k}{dk} = \frac{d\varepsilon_k^0}{dk} + \frac{\partial \operatorname{Re} \Sigma_k(\omega)}{\partial k} + \left. \frac{\partial \operatorname{Re} \Sigma_k(\omega)}{\partial \omega} \right|_{\omega=\varepsilon_k} \frac{d\varepsilon_k}{dk}. \tag{7.29}$$

This can be solved for $d\varepsilon_k/dk$:

$$\frac{d\varepsilon_k}{dk} = Z_k \left[\frac{d\varepsilon_k^0}{dk} + \left. \frac{\partial \operatorname{Re} \Sigma_k(\omega)}{\partial k} \right|_{\omega=\varepsilon_k} \right] = Z_k \frac{\partial \varepsilon_k}{\partial k}. \tag{7.30}$$

As one can see from the appearance of Z_k in this equation, the dispersion is changed by the *frequency dependence* of $\Sigma_k(\omega)$, as well as by its k dependence. For localized states the self-energy is sometimes only weakly k-dependent, and the major effect may come from the frequency dependence.

This explains why the deviation of the measured dispersion from predictions of an independent-particle calculation is often taken to be a measure of correlation. However, it is risky to draw conclusions about correlation strength from the effective mass alone. There are many different independent-particle calculations (LDA, Hartree–Fock, etc.), each with

---

[10] When the linear variation of the imaginary part of the self-energy is also taken into account, one obtains an additional contribution leading to an asymmetry, like in a Fano profile [287].

[11] As discussed in Sec. 3.4, the effective mass is $m^* = k_F/(d\varepsilon/dk)$ for an isotropic system. In general, for an anisotropic crystal it is a tensor and the specific heat (see Eq. (3.6)) is determined by an average over the Fermi surface.

a different dispersion, and, therefore, each has a different self-energy correction. Other observations, such as the weight transfer from the quasi-particle to sidebands, are more conclusive.

Quasi-particles in time

The spectral function $A(\omega)$ is defined and used for real frequencies ω, see, e.g., Eq. (5.36). Equation (7.28) suggests that the continuation $\mathcal{A}(z)$ of $A(\omega)$ into the complex plane has a pole at a finite distance $iZ_\mathbf{k} \mathrm{Im}\, \Sigma_\mathbf{k}(\varepsilon_\mathbf{k})$ from the real axis. One has to be careful here: this does not imply that the Green's function itself has a complex pole as one might think, since for complex frequency it is no longer true that $\mathcal{A}_\mathbf{k}(z)$ is the imaginary part of $G_\mathbf{k}(z)$ (see Ex. 7.10). The pole of $\mathcal{A}_\mathbf{k}(z)$ is *not* a pole of the Green's function $G_\mathbf{k}(z)$; remember that the poles of the Green's function in the physical complex plane are *real* total energy differences (see Sec. 5.6), with just an infinitesimal imaginary part, which form a continuum, a branch cut in the thermodynamic limit. One can however define a continuation G^{cont} of the Green's function that has the complex poles of the continued $\mathcal{A}(z)$,

$$G_\mathbf{k}^{cont}(z) = G_\mathbf{k}(z) \pm 2\pi i \, \mathcal{A}_\mathbf{k}(z); \quad \mathrm{Im}\,[z] \gtrless 0. \tag{7.31}$$

It corresponds to an analytic continuation of $G_\mathbf{k}(z)$ *across* the branch cut for z on the real axis, where G is discontinuous (see Eqs. (5.91) and (5.92)). Indeed, $A(\omega)$ for real ω is the discontinuity in $G_\mathbf{k}(z)$; a full cycle in the complex plane does not bring the Green's function back to its starting value, but adds the discontinuity given by $A(\omega)$. One can imagine this as moving on a spiral staircase, bringing one to a higher Riemann sheet; see Sec. C.5, [148, 288] for details, and Ex. 7.11 for an exercise illustrating the spiral staircase.

One of the uses of the continuation of the Green's function in Eq. (7.31) lies in its potential for interpreting many-body excitations in terms of the approximate concept of quasi-particles. For example, one may look at the evolution of a quasi-particle in time. This can be done using a Fourier transform defined on an appropriate contour with consideration of several contributions. (See, for example, [233].) One finds that for a time difference t after the insertion of a particle, the analytic continuation of the Green's function is dominated by the contribution from the complex pole $\varepsilon_\mathbf{k} - i|Z_\mathbf{k} \mathrm{Im}\, \Sigma_\mathbf{k}(\varepsilon_\mathbf{k})|$ and can be approximated as

$$G_\mathbf{k}(t) \approx -iZ_\mathbf{k} e^{-i\varepsilon_\mathbf{k} t - |Z_\mathbf{k} \mathrm{Im}\, \Sigma_\mathbf{k}(\varepsilon_\mathbf{k})| t}. \tag{7.32}$$

The finite quasi-particle width $|Z_\mathbf{k} \mathrm{Im}\, \Sigma_\mathbf{k}(\varepsilon_\mathbf{k})|$ leads to a damping, and, hence, a limited lifetime. This is consistent with the observation that the finite width of the peak in $A_\mathbf{k}(\omega)$ stems from a series of infinitely close-lying excitations, into which the quasi-particle decays. Supposing that no other pole contributes, this result is valid in a time range

$$\frac{1}{|\varepsilon_\mathbf{k} - \mu|} \ll t \ll \frac{1}{|Z_\mathbf{k} \mathrm{Im}\, \Sigma_\mathbf{k}(\varepsilon_\mathbf{k})|}. \tag{7.33}$$

The lower limit expresses the time necessary to dress the bare particle that was inserted, and hence to form the quasi-particle. The upper limit is given by the quasi-particle decay

[233]. Both diverge for $\varepsilon_{\mathbf{k}} \to \mu$, but for a normal Fermi liquid their difference is large enough to allow the quasi-particles to be well-defined.

7.6 Quasi-particle equations

The idea of describing the strongly interacting system in terms of weakly interacting quasi-particles is appealing. How far can this idea be pushed, i.e., how close can we get to something that resembles the independent-particle framework? In this section the so-called quasi-particle equations are derived from the Dyson equation. A rigorous derivation is somewhat tedious, but the overall principles allow one to understand some otherwise confusing points, such as the difference between the poles of G, that are *real* total energy differences, and the *complex* quasi-particle energies. To make the arguments precise, we denote frequencies in the complex plane by z as in Eq. (7.31).

Ideally, one would be able to find an effective one-particle Schrödinger-like equation with eigenvalues $\tilde{\varepsilon}_{\mathbf{k}}$ and wavefunctions $\tilde{\psi}_{\mathbf{k}}$ that allow one to express useful quantities in a simple way. In particular, one would like to have an effective hamiltonian with eigenvalues that approximate the spectrum for quasi-particles near the Fermi energy. Moreover, there should be a way to include aspects such as the renormalization $Z_{\mathbf{k}}$ in Eq. (7.26). Of course, this will always be an approximation, because only the quasi-particle part of the spectrum is described.

Let us start from the observation that in the Dyson Eq. (7.10), the self-energy plays the role of a potential, but with a frequency dependence. The one-particle Schrödinger-like equation involving this potential should hence have eigenvalues $\tilde{\varepsilon}(z)$ and eigenfunctions $\tilde{\psi}(z)$ that also depend parametrically on the complex frequency z. We can expand the eigenfunctions in the single-particle basis $\psi_i(\mathbf{r})$; for a crystal, this reads

$$\tilde{\psi}_{n\mathbf{k}}(\mathbf{r}, z) = \sum_m b_{n\mathbf{k}}^m(z)\psi_{m\mathbf{k}}(\mathbf{r}). \tag{7.34}$$

In this basis, the Schrödinger-like equation is[12]

$$[\varepsilon_{\mathbf{k}}^0 + \Sigma_{\mathbf{k}}^{cont}(z)]b_{\mathbf{k}}(z) = \tilde{\varepsilon}_{\mathbf{k}}(z)b_{\mathbf{k}}(z), \tag{7.35}$$

where $\Sigma_{\mathbf{k}}^{cont}(z)$ is the analytic continuation of the self-energy that, inserted in the Dyson equation, yields G^{cont} of Eq. (7.31). Note that $\Sigma_{\mathbf{k}}^{cont}(z)$ is in general non-hermitian; hence there exists an adjoint equation to Eq. (7.35) yielding the left eigenfunctions $\bar{b}_{\mathbf{k}}(z)$. The resulting Dyson equation is consistent with a Green's function constructed as [289][13]

$$G_{\mathbf{k}}^{cont}(z) = \frac{b_{\mathbf{k}}(z)\bar{b}_{\mathbf{k}}(z)}{z - \tilde{\varepsilon}_{\mathbf{k}}(z)}. \tag{7.36}$$

---

[12] To simplify the notation, the band indices m and n in Eq. (7.34) are omitted; in general, Σ^{cont} and b are matrices in the band indices.

[13] This is called a "bilinear representation."

We are interested in the quasi-particle solution where z is close to the complex pole at energy $\tilde{\varepsilon}_{\mathbf{k}} = \tilde{\varepsilon}_{\mathbf{k}}(z = \tilde{\varepsilon}_{\mathbf{k}})$. Expansion to first order in frequency around such a pole leads to[14]

$$G_{\mathbf{k}}^{QP}(z) = Z_{\mathbf{k}} \frac{b_{\mathbf{k}}(\tilde{\varepsilon}_{\mathbf{k}}) \overline{b}_{\mathbf{k}}(\tilde{\varepsilon}_{\mathbf{k}})}{z - \tilde{\varepsilon}_{\mathbf{k}}}, \tag{7.37}$$

where $\tilde{\varepsilon}_{\mathbf{k}}$ is obtained from Eq. (7.35) at $z = \tilde{\varepsilon}_{\mathbf{k}}$,

$$[\varepsilon_{\mathbf{k}}^0 + \Sigma_{\mathbf{k}}^{cont}(\tilde{\varepsilon}_{\mathbf{k}})] b_{\mathbf{k}}(\tilde{\varepsilon}_{\mathbf{k}}) = \tilde{\varepsilon}_{\mathbf{k}} b_{\mathbf{k}}(\tilde{\varepsilon}_{\mathbf{k}}). \tag{7.38}$$

This is the desired *quasi-particle equation*.

As can be seen from Eq. (7.37), the poles of $G_{\mathbf{k}}^{QP}(z)$ have fractional weight since the renormalization factor Z is less than one in an interacting system. Thus the total charge is not conserved and these objects cannot be considered as quasi-particles in the sense of Landau. This difficulty is most often avoided in practice since the overall renormalization factor Z is usually set to one when the resulting G^{QP} is used for calculations. Moreover, often one is interested only in the real part of the quasi-particle energies and the imaginary part of $\tilde{\varepsilon}_{\mathbf{k}}$ is neglected. One obtains then a spectrum of ideal quasi-particles, with sharp peaks and a one-to-one correspondence with the non-interacting spectrum. This makes Eq. (7.38) hermitian, and it finally reads

$$\left[\varepsilon_{\mathbf{k}}^0 + \mathrm{Re}\, \Sigma_{\mathbf{k}}(\varepsilon_{\mathbf{k}})\right] b_{\mathbf{k}}(\varepsilon_{\mathbf{k}}) = \varepsilon_{\mathbf{k}} b_{\mathbf{k}}(\varepsilon_{\mathbf{k}}), \tag{7.39}$$

which is equivalent to Eq. (7.23). Solving the simplified Eq. (7.39) is the most straightforward and most common way to determine the quasi-particle energies and hence a band structure. In Ch. 12 more details and approximations are given. A more direct way is to calculate the spectral function including the imaginary part of Σ (e.g., from Eq. (7.22)), and detect the maxima of the quasi-particle peaks, similar to what is done when a band structure is extracted from an experimental spectrum.

The two approaches should lead to the same result for the lowest-energy excitations in an insulator, those closest to the Fermi energy, because the corresponding peaks are sharp (i.e., the states at the top of the occupied bands and the bottom of the empty bands have infinite lifetime), as emphasized in Sec. 2.4. As explained in Sec. 7.5, broadening of the peaks is due to decay to other states with conservation of energy and momentum; that is not possible close to the gap. At a distance from the band edges approximately equal to the gap, there is an onset of broadening and the two approaches may differ. Quasi-particle energies calculated from Eq. (7.39) have a tendency to be too far from the Fermi level with respect to peak positions in $A(\omega)$ [290]. This difference stems from variations in the imaginary part of the self-energy, which are larger for energies further from the Fermi energy. As an example, for bulk silicon the difference is of the order of 0.02 eV around 10 eV above the Fermi level and 1 eV for states around 70 eV [290]. Hence, band structure calculations limited to a restricted region around the Fermi level are, in general, not much influenced.

---

[14] Note that here the real and imaginary part of the self-energy are taken into account in Z, and the derivative is taken at the quasi-particle energy.

Indeed, in a region of a few eV around the Fermi level and for materials that are considered to be only moderately correlated, the approximation of Eq. (7.39) is commonly used, e.g., for the calculation of the bands in Fig. 13.3.

A still different result may be obtained with methods such as quantum Monte Carlo that can calculate very precisely differences in total energy of the ground state with N particles and a state with $N \pm 1$ particles and momentum $\mathbf{k}$. However, the spectrum away from the Fermi energy is not simply a series of sharp poles with positions given by total energy differences. As discussed in Secs. 13.1 and 24.8, this might explain observed overestimations of bandwidths from QMC calculations.

7.7 Separating different contributions to a Dyson equation

The topic of this section is a general approach for breaking a Dyson equation into coupled equations in ways that may be advantageous for calculations. It is important to distinguish this from the decoupling of equations due to symmetry. For example, translation symmetry in a crystal leads to conservation of crystal momentum $\mathbf{k}$ so that there are separate Dyson equations for each $\mathbf{k}$. This is already used in Eq. (7.21), where we have separate equations for each $\mathbf{k}$ given by $G_{\mathbf{k}}(\omega) = \left[\omega - \varepsilon_{\mathbf{k}}^0 - \Sigma_{\mathbf{k}}(\omega) \right]^{-1}$, and in the following equations where all quantities are labeled $\mathbf{k}$. A pictorial representation is given in Fig. 7.1, which applies for each $G_{\mathbf{k}}$ and $G_{\mathbf{k}}^0$, where the shaded circles denote the self-energy $\Sigma_{\mathbf{k}}$. Analogous forms apply for the screened interaction $W(\mathbf{q}, \omega) = v_c(\mathbf{q}) + v_c(\mathbf{q})P(\mathbf{q}, \omega)W(\mathbf{q}, \omega)$.[15] In this section we discuss in general terms when and how it may be useful to break any one of the Dyson equations into two or more coupled equations.

A set of coupled Dyson equations

There is a simple mathematical property of a Dyson equation that can be used to create two or more coupled equations, with the first one being used as input for the following ones. This turns out to be very useful for efficient calculations in various contexts. We omit the label $\mathbf{k}$ or $\mathbf{q}$ with the understanding that the arguments apply to each momentum separately.

Suppose that in the Dyson Eq. (7.10) the self-energy is composed of more than one part; for simplicity we write the equations for two parts $\Sigma = \Sigma^1 + \Sigma^2$. Introducing an intermediate G^1 allows one to break up the equation as (see Ex. 7.9)

$$G^1(\omega) = G^0(\omega) + G^0(\omega)\Sigma^1(\omega)G^1(\omega), \tag{7.40}$$

$$G(\omega) = G^1(\omega) + G^1(\omega)\Sigma^2(\omega)G(\omega), \tag{7.41}$$

as depicted in Fig. 7.4. Thereby one can determine the fully dressed Green's function G in terms of a modified Green's function G^1, which means that one only has to consider G^1 and the part Σ^2 of the self-energy in the second equation. Thus, G^1 may already contain contributions of the interaction.

---

[15] Here $\mathbf{q}$ is also a momentum in the Brillouin zone; we use the notation $\mathbf{q}$ to distinguish the momentum transfer in an interaction from the momentum $\mathbf{k}$ in the one-particle Green's function.

Figure 7.4. Dyson equation expressed in terms of two contributions to Σ as defined in Eqs. (7.40) and (7.41). The formulation is exact if the sum $\Sigma^1 + \Sigma^2$ equals the exact self-energy. In principle any separation is possible, but in order to be useful, this should lead to a decoupling of some kind, and hence to two *smaller* problems. This is the case if Σ^1 and Σ^2 arise from independent mechanisms, in other words, if they belong to completely separate subspaces, but in general such a separation is an approximation. Note that here only two equations are shown, but the concept holds for any number of subspaces.

The same considerations hold concerning any Dyson equation, including the screening equation. The screened Coulomb interaction W contains the polarizabilities of all parts of a system, but if one expresses the proper polarizability as a sum $P = P^1 + P^2$, the screening can be calculated in a two-step form like Eq. (7.41),

$$W^1(\omega) = v_c + v_c P^1(\omega) W^1(\omega), \tag{7.42}$$

$$W(\omega) = W^1(\omega) + W^1(\omega) P^2(\omega) W(\omega). \tag{7.43}$$

This is illustrated in Fig. 7.4, with G replaced by W, G_0 replaced by v_c, and Σ replaced by P. The Coulomb interaction v_c is independent of frequency; however, the screened interactions $W(\omega)$ and $W^1(\omega)$ are functions of ω. This is an example of the consequences pointed out after Eq. (7.3): when one includes contributions from a larger space, in general the quantities become frequency dependent.

Examples for the use of coupled Dyson equations

In principle one can use any separation in Eqs. (7.40)–(7.43). However, in order to be useful this should lead to a decoupling of some kind, and, hence, to two or more problems that are *smaller in some sense*. Exercise 7.12 gives an illustration of what this could mean. There are different situations where such a separation is of practical interest; most often this involves an approximation. The following are some examples illustrating the break-up of Dyson equations as presented in the previous subsection that will be worked out in later chapters.

- The separation of the self-energy into a static part Σ^1 and the rest, a dynamical contribution Σ^2 that will usually be evaluated approximately, for example by neglecting self-consistency. One example in this book is Σ^1 chosen to be a Kohn–Sham potential, and Σ^2 the difference from the self-energy in the GW approximation, evaluated perturbatively using the wavefunctions from the Kohn–Sham equations. This is an example of what is called "G_0W_0" (see Secs. 11.3 and 12.2). Another choice for Σ^1 along the same lines is a static, but self-consistently evaluated approximation to GW, such as COHSEX or quasi-particle self-consistent GW [291–293] (see Sec. 12.4).
- The kernel Eq. (10.22) of the Bethe–Salpeter equation (10.23) for the two-particle correlation function L is composed of two terms: the variation v_c of the Hartree potential with respect to the density, which also appears in the RPA, (see Sec. 11.2), and the variation

of the exchange–correlation self-energy with respect to the one-particle Green's function. Because of their different space–spin structure, it can be convenient to treat the two terms separately.

- Application of DMFT in Chs. 16–21 to real materials requires that we treat a subset of localized orbitals (usually d and f orbitals of transition metal and actinide materials) on a different level of description than the other, delocalized ones (often the states deriving from s and p atomic orbitals), which implies a division of the self-energy into parts. This is done most completely in Ch. 21, where Figs. 21.1 and 21.2 are the application of Fig. 7.4 and the equivalent diagram for the interaction.
- A way to use the separation Eqs. (7.42) and (7.43) for the screened interaction is to build P^2 from transitions between certain chosen orbitals, and to include all other transitions in P^1. This separation is especially clear in the RPA, where P is a simple sum over electron–hole pairs, as can be seen for example in Eq. (12.7). The first equation yields the effective screened interaction W^1 that acts in the subspace of the chosen orbitals. Such a separation is done for the determination of the local interaction in a subspace of d or f orbitals, screened by the s and p orbitals: see Sec. 21.5.

Self-consistency relations

The pair of Eqs. (7.40) and (7.41), or Eqs. (7.42) and (7.43), suggests that one defines Σ^1 and solves the first equation, and then obtains the final result from the second equation involving Σ^2. However, in many cases the solution of the second equation affects the first. Indeed, the methods presented in Part III of this book rely on the fact that quantities like the self-energy or polarizability are expressed in terms of the actual, measurable functions themselves. For example, Σ^1 may be given in terms of G, which then requires a self-consistent solution of Eqs. (7.40) and (7.41). This is important in DMFT, where Eq. (7.41) corresponds to the solution of a single site in a bath G^1 that has to be determined self-consistently, because Σ^1 depends on the final Green's function itself. In Ch. 8 we take a closer look at the formulation in terms of functionals of the full Green's function or other measurable properties.

7.8 Wrap-up

In this chapter we have seen that frequency-dependent effective fields appear whenever one formulates the problem in terms of a few degrees of freedom of interest embedded in the rest. This is true even for particles in a static external potential and with instantaneous, frequency-independent, interactions. This situation is generic for the many-body problem: for example, the one-electron self-energy is a frequency-dependent field containing all possible allowed excitations of the system, crucial for describing correctly one-particle excitations from the ground state. The presence of such a frequency-dependent field modifies the spectral function: instead of δ-function peaks in the absence of coupling, one finds broadened quasi-particle peaks and often additional structure. This is a general finding, not limited to the Coulomb interaction, but valid for any kind of coupling. A crucial ingredient of the many-body methods in the following chapters is the screened interaction W; it is a dynamic quantity obeying analogous equations, and with similar effects of coupling. The GW, DMFT, and other methods are examples of this approach using approximations to create practical methods.

SELECT FURTHER READING

Abrikosov, A. A., Gorkov, L. P., and Dzyaloshinski, I. E. *Methods of Quantum Field Theory in Statistical Physics* (Prentice-Hall, Englewood Cliffs, NJ, 1963). Chapter 1 includes the Landau–Fermi liquid theory.

Farid, B., "Ground and low-lying excited states of interacting electron systems; a survey and some critical analysis," p. 103 in *Electron Correlation in the Solid State*, edited by N. H. March (Imperial College Press, London, 1999).

Gross, E. K. U., Runge, E., and Heinonen, O. *Many-Particle Theory* (Adam Hilger, Bristol, 1991). See especially Part II on general properties of Green's functions, which emphasizes the one-particle Green's function for fermions.

Pines, D. and Nozières, P. *The Theory of Quantum Liquids*, Vol. 1 (W. A. Benjamin, New York, 1966). Vols. 1 and 2 reprinted in the Advanced Book Classics series (Westview Press, Boulder, CO, 1999). A pedagogical exposition of Fermi liquid theory and charged quantum liquids.

Exercises

7.1 Show that the expressions for the Green's functions in Eq. (7.13) follow from the definition of $[\omega - \hat{H}]G(\omega) = I$, and that the solutions for G_{00} and Δ in Eqs. (7.14) and (7.15) follow from solving the two simultaneous equations. Also, derive these equations by using directly the general result of Eqs. (7.7) and (7.8).

7.2 Explain the fact that $t_{0\ell} \propto 1/\sqrt{N}$ in the continuum limit, so that $t(\varepsilon)$ and the integral in Eq. (7.15) are well-behaved. You may think for example of the matrix element for a single state coupled to normalized plane waves.

7.3 Calculate the spectral function for non-interacting electrons, where eigenstates of the hamiltonian are single Slater determinants, and show that its diagonal elements consist of one single peak if the appropriate basis is chosen. Suppose then that additional Slater determinants are mixed in because the interaction is turned on, and show that the spectral function must have more peaks. Alternatively, suppose that the system is indeed non-interacting, but that one has chosen a different single-particle basis. Show that in this case diagonal elements of the true spectral function are no longer single δ-function peaks, but that the total spectrum (trace of $A(\omega)$) is basis invariant.

7.4 At zero temperature the scattering rate for an electron with momentum $\mathbf{k}$ and energy $\omega = \varepsilon_{\mathbf{k}}$ is proportional to the density of states for an electron plus an electron–hole pair with conservation of energy and momentum.

(a) For a spherical Fermi surface show that the density of such states increases as $|\omega - E_F|^2$, and find the expression in terms of the Fermi velocity $v_F = \nabla \varepsilon_{\mathbf{k}}$ for k near the Fermi surface. Without working out explicit formulas, show that in any metal in any dimension the density of such states is $\propto |\omega - E_F|^2$.

(b) For an insulator with a gap in the one-particle spectrum, show that there is a gap ω_{min} in the spectrum of multiparticle excitations. Explain why the onset is a discrete exciton energy and the continuum starts at the lowest gap ω_{gap}. Show that in the continuum the density of states increases as $\sqrt{\omega - \omega_{gap}}$ in three dimensions.

(c) From this show that there is an energy range for electrons in the conduction band (holes in the valence band) where the states are well-defined with infinite lifetime even in the presence of electron–electron interactions (neglecting phonons and any other scattering mechanism).

(d) Estimate the range of energies where states cannot decay in silicon based on the band structure in Fig. 13.2. Be sure to take into account **k**-conservation in the scattering process.

(e) Look up the definition of the Auger effect and show the relation to this problem.

7.5 It is illuminating to see the difference between electrons and phonons for the scattering rate as a function of frequency. Consider long-wavelength acoustic phonons in a three-dimensional solid. For simplicity assume an isotropic model with only one sound mode $\omega(k) = vk + bk^2$, where b is a small constant that determines the dispersion of the sound velocity, and consider a small anharmonic interaction γ whereby one phonon can decay into the two-phonon continuum.

(a) Without working out any formulas, argue that the decay rate of a phonon depends on the sign of b with infinite lifetime for $b > 0$.

(b) Suppose that $b < 0$. Find the power law for the lifetime ($\propto \omega^m \approx k^m$): give the value of m.

(c) Summarize the reasons for the differences between the energy dependence of the lifetimes for low-energy electrons and phonons.

7.6 The self-energy for interacting electrons contains excitations of the system with respect to addition or removal of a particle. Starting from Eq. (11.29), the correlation part Σ_c, which is the correction with respect to Hartree–Fock, can be modeled as $\Sigma_c(\omega) = \lambda/[\omega - \varepsilon + \omega_p - i\gamma] + \lambda/[\omega + \varepsilon - \omega_p + i\gamma]$. This describes a situation where a collective excitation with energy ω_p (for example, a plasmon) may occur when a particle is removed or added to the system. λ is the coupling strength for this process, γ a broadening due to other decay processes and/or dispersion, and ε a typical one-particle excitation energy.

(a) Calculate and plot the spectral function Eq. (7.22) for this self-energy, for various choices of parameters including $\varepsilon = 0$. Compare with Fig. 7.2.

(b) Suppose that γ is small such that results can be evaluated at first order. Determine the position of the quasi-particle peak and its width with respect to the Hartree–Fock energy, and the position and nature of the satellite. Relate the results to the discussion about plasmarons in Sec. 15.7 and show that the effect can be produced with a suitable choice of parameters.

(c) Show that the real part of this self-energy is approximately linear far from the resonances. This fact is used in several places, including practical calculations (see Sec. 12.2). What is the behavior of Re Σ far above the resonances? Show that at high energy the system behaves like independent particles in the Hartree–Fock mean field.

7.7 Explain roughly the position $\varepsilon_{\mathbf{k}}^0$ of the independent-particle peak in Fig. 7.2 defined as the Hartree–Fock energy, by using the sum rules for the spectral function, Eqs. (5.95) and (11.30). To do so, approximate both the quasi-particle peak and the satellite by δ-functions, and suppose that about 30% of the weight is transferred to the satellite.

7.8 Take the model of a single state coupled to a continuum in Sec. 7.3 and use in Eq. (7.15) a density of states with a peak at $-\varepsilon_0$ and at $+\varepsilon_0$. Evaluate the hybridization function and the spectral function. Compare with the model self-energy in Ex. 7.6 and explain the relation to dynamical screening.

7.9 Starting from the Dyson equation, derive Eqs. (7.40) and (7.41) by working with the inverse of the Green's functions.

7.10 Take a spectral function of the form of Eq. (7.28) and calculate the corresponding Green's function using Eq. (5.36). Show that the continuation of the spectral function has a pole in the complex plane, but not the Green's function.

7.11 Suppose that the spectral function has the simple form $A(\omega) = \Theta(a-\omega)\Theta(\omega+a)$, i.e., $A(\omega) = 1$ for $-a < \omega < a$ and zero otherwise. Show that the corresponding Green's function from Eq. (5.36) has a logarithmic form. What happens when the argument of the logarithm circles the origin in the complex plane?

7.12 Extend Eq. (7.5) to a 3×3 matrix representing three spaces S, T, and R, and calculate the self-energy as before. Show that when the coupling between T and R is neglected, the self-energy is the sum of two contributions that represent coupling of S to T and of S to R, respectively.

7.13 Take a system of coupled harmonic oscillators with a small anharmonicity. Explain why you can describe this by a hamiltonian consisting of a sum of independent bosons and an interaction term. Formulate the Green's function that you would like to calculate in order to obtain the spectrum of the system, and discuss the self-energy.

8

Functionals in many-particle physics

> To avoid confusion, it is necessary to be quite explicit about what is assumed
> and what is to be proved.
>
> A. Klein, *Phys. Rev.* 121, 950 (1961)

Summary

The topic of this chapter is functionals that provide a concise formulation for thermo-dynamic quantities and Green's function methods in interacting many-body systems. Building upon the instructive examples of density functional theory and the Hartree–Fock approximation, functionals of the Green's function are developed to provide a framework for practical methods used extensively in many-body perturbation theory (Chs. 9–15), dynamical mean-field theory (Chs. 16–20), and their combination in Ch. 21. Functional concepts give a firm foundation for those methods, for example, they give the conditions for approximations to obey conservation laws. Additional background and specific aspects are in App. H.

Many physical quantities can be expressed as *functionals*, i.e., they depend on an entire function.[1] For example, in the quantum variational method to determine a wavefunction, the total energy of a system of many electrons is a functional of a many-body trial wave-function $\Psi(\mathbf{r}_1, \mathbf{r}_2, \ldots)$: it is the expectation value of the hamiltonian. At the variational minimum, assuming the trial wavefunction is completely flexible, one finds the many-body Schrödinger equation for the ground state. Many of the properties we are interested in are, formally, simple functionals of the ground- or excited-state wavefunctions.

In the search for feasible ways to calculate properties such as excitation spectra, we are led to different types of functional. The idea is to deal only with a physical, measurable quantity Q of interest and a variational functional $E[Q]$ of this quantity. The variational character of E allows one to derive equations that determine Q_0, the value of Q for the system in a desired state, typically the ground state or equilibrium state at $T \neq 0$. Other

---

[1] See App. H for definitions and a summary of salient properties of functionals.

Figure 8.1. Motivation for the use of functionals. The existence of a variational functional $E[Q]$ allows one to determine a given quantity of interest Q_0, for example, the ground-state density $n_0(\mathbf{r})$ in density functional theory. Other properties may also be functionals of that quantity and hence can be computed. In practice, all functionals must be approximated and the choice of Q is crucial in order to find good approximations.

properties should be different functionals $F_m[Q]$ that can then be evaluated at $Q = Q_0$. This strategy is outlined in Fig. 8.1.

To appreciate the power of such approaches, one can consider the impact of density functional theory, which is formulated as a functional of the density $n(\mathbf{r})$ of electrons; it is defined for a range of densities and the variational minimum for the energy functional $E[n]$ leads to equations to determine the actual ground-state density $n_0(\mathbf{r})$ and total energy E_0. The Kohn–Sham method provides an in-principle exact solution for $n_0(\mathbf{r})$ and E_0 in terms of a set of equations that can actually be solved in practice. In principle, all other properties are functionals of the density, as shown by Hohenberg and Kohn. Although neither the functional for the exchange and correlation energy nor the density functionals for other properties are known, the approach has led to approximations that have proven to be very useful.

The subject of this chapter is functionals of the Green's function G and self-energy Σ that provide powerful formulations of many-body theories. They are similar to density functionals, but with G playing the role of the quantity Q. However, they have a crucial advantage with respect to DFT: the expectation value of any one-body operator is given by a simple expression in terms of the one-body Green's function, whereas little is known concerning the calculation of most physical quantities from the density. For example, as explained in Sec. 5.7, an intrinsic electron addition or removal spectral function is simply the imaginary part of G. Whereas we do not know the explicit form of the exchange–correlation energy as a functional of the density, the energy can be expressed as a series of diagrams in terms of the Green's functions. This suggests useful approximations and provides paths for improvement of any given approximation.

Chapter 7 is devoted to the general properties of G and Σ independent of any particular problem. The present chapter brings us closer to a framework that can be used in practice to determine G and Σ, in a general form that applies to many problems independent of the details. The functional formulation provides much more than just a starting point from which equations are derived. It suggests compact approximations and provides criteria that must be satisfied for an approximation in order not to violate essential conservation laws, i.e., conserving approximations [13, 39]. This chapter lays the groundwork for the following chapters on Green's function methods that delve deeply into the details required for actual solutions for particular systems.

The topic of Sec. 8.1 is density functional theory and the Hartree–Fock approximation, with the expressions in Ch. 4 rewritten in terms of their respective independent-particle Green's functions. This provides background for functionals and motivation for the explicit expressions in the following sections. In Sec. 8.2 we describe the functionals of the many-body Green's function and self-energy that are central to the Green's function methods in this book. The form derived by Luttinger and Ward is explained in general terms, that provide insights into the use in later chapters. Section 8.3 is the logical extension to functionals of G and the screened interaction W. Of course, there are usually approximations in actual calculations. A general approach for constructing approximate functionals is the subject of Sec. 8.4, which has important consequences such as the conditions given in Sec. 8.5 that must be satisfied by an approximation in order to ensure that it obeys conservation laws. Finally, Sec. 8.6 is a wrap-up.

8.1 Density functional theory and the Hartree–Fock approximation

In this section we review some aspects of DFT to bring out the salient characteristics of functionals and to cast the Kohn–Sham approach in terms of the Kohn–Sham Green's function G_{KS}. At the solution, G_{KS} is constructed with the Kohn–Sham eigenvalues and eigenfunctions following Eq. (5.105) for $T = 0$ or Eq. (D.25) in the Matsubara formalism. The Hartree plus exchange–correction potential v_{Hxc} plays the role of a static and local self-energy Σ, and the Kohn–Sham Green's function obeys the Dyson equation

$$G_{KS} = G_0 + G_0 v_{Hxc} G_{KS} \quad \text{or} \quad G_{KS}^{-1} = G_0^{-1} - v_{Hxc}, \tag{8.1}$$

where G_0 is the Green's function in absence of the Coulomb interaction.

The Hartree–Fock approximation exemplifies many of the same features of functionals, although, contrary to DFT, it is an approximation to the many-body functionals of G and Σ. Together, DFT and Hartree–Fock provide stepping stones between Ch. 7, where G and Σ are defined, and the many-body functionals of G and Σ that are the topic of the following sections. Relations among the functionals are brought out in Tab. 8.1 later.

As summarized in Sec. 4.3, the variational total energy functional of the interacting many-body system with trial many-body wavefunction Ψ can be written in the form

$$E = \langle \Psi | \hat{H} | \Psi \rangle = \langle \Psi | \hat{T} + \hat{V}_{ee} | \Psi \rangle + \int d\mathbf{r} v_{ext}(\mathbf{r}) n(\mathbf{r}), \tag{8.2}$$

where $\hat{H}$ is the hamiltonian of Eq. (1.1), which can be expressed as $\hat{H} = \hat{T} + \hat{V}_{ee} + \sum_i v_{ext}(\mathbf{r}_i)$ with $\hat{T}$ the kinetic energy operator, $\hat{V}_{ee}$ the electron–electron interaction, $v_{ext}(\mathbf{r})$ the external potential acting on the electrons, e.g., the potential due to the nuclei. The solution is determined by the external potential $v_{ext}(\mathbf{r})$ since all other operators in the hamiltonian are the same in any electronic system. Since the density and potential v_{ext} are directly coupled only through the last bilinear term, the energy can be expressed as a functional of the density $n(\mathbf{r})$ as the independent variable by a Legendre transform, as described in Sec. 4.3 and App. H. Such a transform is valid only if it is invertible, i.e., v_{ext} is determined by n. This was demonstrated by Hohenberg and Kohn by showing that there is a one-to-one relation of v_{ext} and n, and an alternative proof is in Sec. H.5. Since v_{ext} is determined by n, it follows that in principle all properties of the many-body system are determined by the density, i.e., each is a functional of $n(\mathbf{r})$. However, this provides no way to construct the functional for the energy or any other property (except to solve the original many-body problem!). Such a functional must be exceedingly complex and it is unknown, perhaps unknowable.

The Kohn–Sham auxiliary system and the Mermin generalization

The Kohn–Sham formulation of DFT provides insights valuable for the formulation of Green's function functionals. The choice of the density as the fundamental variable, denoted as Q in Fig. 8.1, is analogous to the Green's function in the many-body functionals. This, as well as correspondences of other quantities, is noted in Tab. 8.1 later. To take advantage of the analogy and make the parallels apparent, we will rewrite the Kohn–Sham approach in terms of the Green's function G_{KS}. To be general, we will write the expressions for the finite-temperature Mermin functional; they reduce to the Kohn–Sham forms in the $T = 0$ limit.

As described in Sec. 4.3, the Kohn–Sham auxiliary system of non-interacting particles is constructed to determine the density, total energy, and properties that can be derived from the energy, such as forces. To achieve this, Kohn and Sham rewrote the Hohenberg–Kohn functional as the energy of non-interacting particles with density $n(\mathbf{r}) = \sum_i f_i |\psi_i(\mathbf{r})|^2$ (Eq. (4.20)) plus the Hartree energy and a functional that includes the effects of exchange and correlation. The Mermin functional[2] for the grand-potential in the grand-canonical ensemble (see Eq. (5.2)) is

$$\Omega_{KSM}[\ldots] = \Omega_{xc}[n] + E_H[n] - \int d\mathbf{r}\, v_{Hxc}(\mathbf{r})n(\mathbf{r}) + \Omega_{ip}[v_{eff}], \qquad (8.3)$$

where Ω_{xc} corresponds to E_{xc} and Ω_{ip} is the grand potential for independent Kohn–Sham particles with potential $v_{eff}(\mathbf{r}) = v_{ext}(\mathbf{r}) + v_{xc}(\mathbf{r}) + v_H(\mathbf{r})$. In order to make the expressions compact, we have defined $v_{Hxc}(\mathbf{r}) = v_{xc}(\mathbf{r}) + v_H(\mathbf{r})$. The Kohn–Sham equations have the same form as the $T = 0$ expressions in Eq. (4.19), even though the potential depends on temperature. The notation $\Omega_{KSM}[\ldots]$ indicates that the right-hand side of the equation can be considered as a functional of the potential v_{eff}, density n, or both considered as independent variables (see Sec. H.3). It can also be considered a functional of v_{xc} instead of v_{eff}, since $v_{ext}(\mathbf{r})$ is fixed for a given system and $v_H(\mathbf{r})$ is an explicit functional of $n(\mathbf{r})$.

The grand potential for Ω_{ip} for independent particles at temperature T, or inverse temperature $\beta = 1/(k_B T)$, and the chemical potential μ can be written out explicitly as (see, e.g., [149])

---

[2] The Mermin functional [189] in Eq. (8.3) for the grand potential $\Omega = E - TS - \mu N$ is the same as the Kohn–Sham expression with only the change of symbols from E to Ω. However, there is an enormous difference. The total energy E is an integral over all the electronic states, but the entropy S is governed by the low-energy excitations, which are often more difficult to describe with a functional of the density. Green's functions provide a systematic way to calculate Ω, for example, using the expressions for energy and entropy given in Sec. 5.7.

$$\Omega_{\mathrm{ip}} = -\beta^{-1} \ln Z_{\mathrm{ip}} = -\beta^{-1} \sum_i \ln(1 + e^{-\beta(\varepsilon_i - \mu)}), \qquad (8.4)$$

where ε_i is an eigenvalue of the Kohn–Sham equations. This expression can be derived by using the fact that the partition function for independent fermions Z_{ip} is a product of states i that are filled or empty with probabilities given by the Fermi function (see Ex. 8.1).

In order to make the connection to functionals of the Green's function in the following sections, it is useful to go through the steps to express the Kohn–Sham equations in terms of the independent-particle Kohn–Sham Green's function G_{KS}. Each of the terms in Eq. (8.3) can be expressed in terms of G_{KS}. The spin-resolved density is $n(x) = -iG_{KS}(x, t; x, t^+)$, as given by Eq. (5.118) or $n(x) = -G_{KS}(x, \tau; x; \tau^+)$ in Eq. (D.7) for the imaginary time expression. The equal time limit corresponds to an integral over frequency. It is convenient to define the notation $\mathfrak{T}\mathfrak{r}$: for $T \neq 0$, $\mathfrak{T}\mathfrak{r} = \mathrm{Tr}\,\beta^{-1} \sum_n$ and for $T = 0$, $\mathfrak{T}\mathfrak{r} = -\frac{i}{2\pi} \lim_{\eta \to 0^+} \mathrm{Tr} \int_{-\infty}^{\infty} d\omega\, e^{i\eta\omega}$. The $T = 0$ integral can also be performed on the imaginary frequency axis using contour deformation, as explained in Sec. C.3 (see also Sec. 5.7). Here, Tr is the trace over states in a basis including the sum over spin. (Note that $\mathfrak{T}\mathfrak{r}$ has units of energy.) For example, the number of particles is $N = \int d\mathbf{r}\, n(\mathbf{r}) = \mathfrak{T}\mathfrak{r}\, G = \mathfrak{T}\mathfrak{r}\, G_{KS}$, and the integral over $\mathbf{r}$ in Eq. (8.3) can be expressed as

$$\int d\mathbf{r} v_{Hxc}(\mathbf{r}) n(\mathbf{r}) = \mathfrak{T}\mathfrak{r}(v_{Hxc} G_{KS}). \qquad (8.5)$$

Finally, we want to write the term Ω_{ip} in a way that makes clear that it represents a system that has a non-interacting Green's function G_0 and a potential $v_{Hxc}(\mathbf{r})$ due to interactions. In Sec. D.4 we derive the expressions for independent particles. Applying the transformations for both G_{KS} and G_0, and defining Ω^0 to be the grand potential in the absence of Coulomb interactions, it follows that the difference $\Omega_{\mathrm{ip}} - \Omega^0$ can be evaluated as $-\mathfrak{T}\mathfrak{r} \ln(-G_{KS}^{-1}) + \mathfrak{T}\mathfrak{r} \ln(-G_0^{-1}) = -\mathfrak{T}\mathfrak{r} \ln(G_{KS}^{-1} G_0)$. Together with the Dyson equation for G_{KS} in Eq. (8.1), this leads to the final expression[3,4]

$$\Omega_{KSM} = \Omega_{xc}[n] + E_H[n] - \mathfrak{T}\mathfrak{r}(v_{Hxc} G_{KS}) - \mathfrak{T}\mathfrak{r} \ln(1 - G_0 v_{Hxc}) + \Omega^0. \qquad (8.6)$$

This is exactly the same form as the Luttinger–Ward functional in Eq. (8.8) given in the next section, except that the dynamic self-energy $\Sigma(\omega)$ is replaced by the static v_{xc}; see the comparison in Tab. 8.1. Of course, the equations in an independent-particle method are much easier to deal with, since $v_{xc}(\mathbf{r})$ is a static potential and the Green's function can be found in terms of wavefunctions and independent-particle energies derived by matrix diagonalization. However, the functional $E_{xc}[n]$ is unknown; there are very useful approximations but no systematic method to improve them. The thermodynamic functional $\Omega_{xc}[n]$ is yet more difficult. In contrast, the Luttinger–Ward functionals in the following section are more difficult to work with, but there are definite procedures to find the corresponding functional $\Phi[G]$. That is the topic of Part III of this book.

---

[3] See footnote 8 on page 176 for the definition of the logarithm of an operator and other issues.

[4] The derivation in Sec. D.4 involves the transformation of the expression for Ω^0 into a sum over Matsubara frequencies using contour integration in the complex plane. The expression in terms of a logarithm can also be derived by coupling constant integration as done in Sec. 8.4 and Ex. 8.4.

The Hartree–Fock functional

The Hartree–Fock approximation (see Sec. 4.1) can also be formulated as a functional in which the Hartree and exchange energies can be written out explicitly. Correlation between electrons is ignored by definition. If we follow the same steps as in Eqs. (8.3)–(8.6), we find a functional of the same form as Eq. (8.6) except that Ω_{xc} is replaced by Ω_x, the Green's function is G_{HF} constructed with Hartree–Fock eigenvalues and orbitals, and the local potential v_{xc} is replaced by the non-local exchange potential Σ_x, which can be viewed as a static self-energy. Thus, $\mathfrak{Tr}(v_{Hxc}G_{KS})$ becomes $\mathfrak{Tr}((v_H + \Sigma_x)G_{HF})$. The analog to the exchange contribution in Eq. (8.5) is now

$$\sum_\sigma \int d\mathbf{r} d\mathbf{r}' \Sigma_x^\sigma(\mathbf{r}, \mathbf{r}') \rho_{HF}^\sigma(\mathbf{r}', \mathbf{r}) = \mathfrak{Tr}(\Sigma_x G_{HF}), \tag{8.7}$$

where $\rho_{HF}^\sigma(\mathbf{r}, \mathbf{r}') = -iG_{HF}^\sigma(\mathbf{r}, t; \mathbf{r}', t^+)$ is the independent-particle density matrix. The HFA is a precedent for the Luttinger–Ward functionals in the next section; it is the lowest-order approximation to the sum of diagrams. The correspondence is indicated in Tab. 8.1.

8.2 Functionals of the Green's function G and self-energy Σ

In this section we move to functionals of the full Green's function or self-energy. The expressions for the Kohn–Sham and Hartree–Fock methods provide useful ways to understand the form of the resulting equations. However, the present section is *not* a simple extension of independent-particle methods; the construction of the functionals is a major achievement and the calculations of G or Σ are much more involved than in the independent-particle Kohn–Sham and Hartree–Fock methods.

In many-body perturbation theory, expressions for physical quantities can be derived by diagrammatic expansions in terms of the bare one-particle Green's functions G_0 and the bare interactions v_c, constructed using well-defined rules. The methods are described in detail in texts such as [38, 47, 149, 227], and aspects most relevant for our purposes are summarized in Chs. 9 and 10. For example, expressions can be derived for the proper self-energy Σ as an infinite sum of diagrams involving G_0, and the Green's function for the interacting system G is then given by the Dyson equation $G = \left[G_0^{-1} - \Sigma \right]^{-1}$.

It is less obvious that one can construct a theory in which the properties are expressed in terms of the one-particle Green's function G or the self-energy Σ. Three bodies of work have been especially important in the development of the structure for such functional methods in many-body theory, all predating density functional theory by several years. Here we follow especially the developments by Luttinger and Ward [36] and Baym and Kadanoff [13, 39], noting places where clarification is provided by Klein's formulation [294].[5] In the following, when it comes to specific formulas, the finite-temperature

---

[5] Luttinger and Ward (LW) [36] derived a functional of the Green's function in terms of a diagrammatic expansion. The functional was attributed to LW by Baym [13] and by Abrikosov, Gorkov, and Dzyaloshinski [38], who present the theory with physical arguments. An alternative formulation is due to Klein [294], who built

Table 8.1. Comparison of the Kohn–Sham density functional, Green's function, and Hartree–Fock methods. The Kohn–Sham method involves an auxiliary system of particles designed to determine only $n(\mathbf{r})$ and energy or grand potential. The $G - \Sigma$ functionals can also be viewed as an auxiliary system of equations designed to determine only $G(\mathbf{r}, \mathbf{r}', \omega)$ and energy or grand potential. Note the analogous roles of v_{xc} and Σ that determine these quantities in the respective methods. In all cases there is a local static external potential $v_{ext}(\mathbf{r})$ and Hartree potential $v_H(\mathbf{r})$; here we indicate only terms due to interactions beyond the Hartree potential. The Hartree–Fock method is the lowest-order approximation to the $G - \Sigma$ functionals with a static Σ_x. The Kohn–Sham–Mermin functional $\Omega_{xc}[n]$ is unknown but the Luttinger–Ward $\Phi[G]$ is given explicitly as a series (see Chs. 9 and 10)

Theory	Kohn–Sham DFT	$G - \Sigma$ functionals	Hartree–Fock approximation
Quantity calculated	Static density $n(\mathbf{r})$	Dynamic Green's function $G(\mathbf{r}, \mathbf{r}', \omega)$	Static density matrix $\rho_{HF}(\mathbf{r}, \mathbf{r}')$
Approach	Auxiliary system of independent particles with static local potential $v_{xc}(\mathbf{r})$	System described by dynamic non-local self-energy $\Sigma(\mathbf{r}, \mathbf{r}', \omega)$	Approximation of independent particles with static non-local potential $\Sigma_x(\mathbf{r}, \mathbf{r}')$
Exact formulation	Universal $\Omega_{xc}[n]$ $v_{xc}(\mathbf{r}) = \frac{\delta \Omega_{xc}}{\delta n(\mathbf{r})}$	Luttinger–Ward $\Phi[G]$ $\Sigma(\mathbf{r}, \mathbf{r}', \omega) = \frac{\delta \Phi}{\delta G(\mathbf{r}, \mathbf{r}', \omega)}$	None
Canonical approximation	LDA Sec. 4.3	$\Sigma = iGW$ Ch. 10	Hartree–Fock equations Sec. 4.1
Other approximation examples	GGAs Sec. 4.3	Vertex corrections T-matrix Ch. 15	Multireference HF

formulation is adopted. This gives compact expressions that encompass $T = 0$ and are useful for computations as a function of temperature; finite temperature is sometimes also used as a trick to evaluate quantities like the total energy that require integration over otherwise sharp poles.

The Luttinger–Ward interaction functionals $\Phi[G]$ and $F[\Sigma]$

Here we discuss the functional constructed by Luttinger and Ward [36] in general terms, which is sufficient to bring out important properties. The actual diagrammatic procedures

upon the earlier work of Klein and Prange [295], emphasizing the importance of the stationarity conditions in deriving results using approximate Green's functionals and self-energies. The work of Baym and Kadanoff [13, 39] provided a general framework for characterizing functionals of the Green's functions and a concise statement of conditions needed for conservation laws. The 1959 paper of Martin and Schwinger [40] presents many of the ideas and lists earlier references.

for constructing the functionals are presented in Chs. 9 and 10. The approach of LW, presented in more detail in Sec. 9.8, was to show that the grand potential Ω can be expressed in a form involving the full G and proper self-energy Σ,[6,7,8]

$$\Omega[\ldots] = \Phi[G] - \mathfrak{Tr}(\Sigma G) - \mathfrak{Tr}\ln(1 - G_0\Sigma) + \Omega^0, \tag{8.8}$$

where $\Phi[G]$ can be expressed as a sum of diagrams involving the Green's function G. In the last two terms G_0 and Ω^0 represent the non-interacting system and Σ is the self-energy due to interactions. However, the interactions are over-counted, as is already clear in the Hartree and Fock contributions, for example, in the Kohn–Sham expression Eq. (8.6). The second term $-\mathfrak{Tr}(\Sigma G)$ in Eq. (8.8) cancels the interaction terms in $-\mathfrak{Tr}\ln(1 - G_0\Sigma)$, and $\Phi[G]$ includes interactions with the correct counting.

As explained in Secs. 9.8 and 10.4, LW expressed $\Phi[G]$ as the sum of diagrams. Each diagram represents an integral over a number of space–spin–time points, called "interaction vertices," where a Coulomb interaction v_c is connected to two full G. LW have shown that only "skeleton" diagrams contribute; these are the set of closed-link diagrams that cannot be divided into two disjoint parts by removing one Green's function line. The fact that $\Phi[G]$ can be written in this way means that it is a functional of G.[9] Furthermore, the construction leads directly to the result that the proper self-energy is given by the functional derivative

$$\Sigma = \frac{\delta\Phi}{\delta G}. \tag{8.9}$$

Graphically, this corresponds to opening any of the G-lines in each of the diagrams for Φ, which leads to a sum of closed-link diagrams that define the proper self-energy. For example, Fig. 10.3 is the self-energy diagram corresponding to the second-order terms in the Φ-functional given in Fig. 10.2. Note that Eq. (8.9) should also contain a constant factor that depends on the usage, namely β in Matsubara sums, or i in real frequency methods. For compactness we do not display the factor here and in the corresponding functional derivatives in Secs. 8.3 and 8.5 and Chs. 9 and 10.

The first two terms on the right-hand side of Eq. (8.8) have the form of a Legendre transform (see Sec. H.2) of $\Phi[G]$, which can be written as a functional of Σ:

$$F[\Sigma] = \Phi[G] - \mathfrak{Tr}(\Sigma G). \tag{8.10}$$

---

[6] Here and in following expressions we do not specify the Hartree term to simplify the notation. To retrieve the Hartree contribution explicitly, one has to interpret Σ as the sum of the Hartree potential and the exchange–correlation self-energy Σ_{xc}, and Φ as the exchange–correlation functional plus E_H.

[7] For many purposes the last two terms can be combined so that Eq. (8.8) becomes

$$\Omega[\ldots] = \Phi[G] - \mathfrak{Tr}(\Sigma G) - \mathfrak{Tr}\ln(\Sigma - G_0^{-1}) \quad \text{or} \quad \Omega[\ldots] = \Phi[G] - \mathfrak{Tr}(\Sigma G) - \mathfrak{Tr}\ln(-G^{-1}),$$

where the first form is given by LW and the second by Baym, Klein, and other references. However, the sum over Matsubara frequencies in $\mathfrak{Tr}\ln(-G^{-1})$ is only conditionally convergent; it is defined only up to a constant (see Sec. D.4) but it is sufficient to generate the equations for the Green's functions and self-energy that require only derivatives of Ω where the constant drops out. See Ex. 8.5.

[8] Functions of operators or matrices, for example $\ln(-G_{KS}^{-1})$, have to be interpreted by their Taylor expansion, e.g., $\ln(O) = -\sum_n \frac{(1-O)^n}{n}$. As pointed out by LW, there is a branch cut in the definition of the logarithm and expressions depend on the choice; here and in LW the cut is on the negative axis. See also Sec. D.4.

[9] If the Coulomb interaction is considered as variable, Φ is also a functional of v_c; this will be used in Sec. 8.3.

It follows immediately that $G[\Sigma]$ is given by

$$G = -\frac{\delta F}{\delta \Sigma}. \tag{8.11}$$

There are many ways to build upon the LW formulation of the interacting many-body problem, and it is important to identify some key features before launching into specific cases. The LW construction consists of two important steps. First, in the original perturbation series in terms of v_c and G^0, the contributions are grouped such that the full G appears. In this way the self-energy is given as a functional of G instead of the bare G^0. Second, a coupling constant integration is performed, starting from the non-interacting system, in order to obtain the functional for the grand potential. This is shown in more detail in Sec. 9.8. Both steps contain hypotheses that should be kept in mind.

Let us start by considering the coupling constant integration. It supposes continuity, thus the LW $\Phi[G]$ is strictly defined only for G's that are continuously connected to some non-interacting system. This embodies the continuity principle stated in Sec. 1.3 and is the criterion in Sec. 3.4 for a normal state of matter that can be considered to evolve continuously from a system of non-interacting particles as the electron–electron interaction is turned on. This defines the *domain* of the functional: it is important to keep in mind that a functional is defined only if it is specified, i.e., the range of functions that are allowed (see Sec. H.1). Since the diagrammatic expansion for $\Phi[G]$ is the same for all two-body interactions, $\Phi[G]$ is a *universal* functional of G for all functions G in the domain. It may be possible to evaluate the sum of diagrams (or an approximation to the infinite sum) for G's outside the domain of the LW functional, but it must be established what physical problem corresponds to the new choice of domain.

Even if we are satisfied to stay within the LW domain, there are still issues to be faced. These are linked to the first step of the LW construction. It is of course plausible to reorder the terms in the perturbation series to make the physically meaningful G appear, but this does not correspond to a rigorous mathematical proof that the result can be used for calculations. One should keep in mind that reordering terms in an infinite series can change the result if the series is not absolutely convergent.

The problem can also be highlighted from a slightly different point of view. If we consider only the $\Phi[G]$ and equations that involve G, the problem is to constrain G to those functions that can be generated from some G_0, i.e., to invert $G[G_0]$ to find $G_0[G]$. If the inversion $G_0[G]$ exists and if it is unique, then the unique functional $\Sigma[G^0]$ can be written as $\Sigma[G^0[G]] \equiv \Sigma[G]$, where we keep the same symbol Σ – although of course the functional expression is not the same. The inversion of a functional is not an easy task, and is a critical issue in the theory of functionals. The same question applies for density functionals, and has been answered in several ways: the original Hohenberg–Kohn existence proof, the Levy–Lieb construction in Sec. 4.3, and the invertibility proof in Sec. H.5. To our knowledge there is no such demonstration for functionals of G and/or Σ in general, and this is a topic of ongoing research; see, for example, the work on Hubbard-type models in [296].

Other issues relate to the convergence. To our knowledge it has not been established that the skeleton series converges; furthermore, there is the possibility that it converges to the incorrect result, since it corresponds to a reordering of the original diagrammatic expansion in terms of G^0 as outlined above.

Finally, in the absence of rigorous proofs, the practical approach is to construct approximate functionals and apply them within some domain, always being alert to the possibility of unphysical or incorrect solution of the equations. In actual calculations with approximate functionals, warning flags appear immediately since even the simplest approximations lead to non-linear equations. For example, as worked out in Sec. G.2, the Dyson equation in the Hartree–Fock approximation is a quadratic equation, with multiple solutions. In some cases an instability is physically meaningful and corresponds to a phase transition. Other cases may not be so clear, and one must exercise great caution and care to draw meaningful conclusions.

Functionals for the grand potential Ω

The expression in Eq. (8.8) provides a way to calculate Ω in terms of the exact Φ or F; however, this does not uniquely specify a functional that may be defined on different domains of functions G and/or Σ that include the solution (see App. H). The different possible functionals may have different behaviors away from the stationary points; the only requirement is that the functionals have the same value at the stationary points that describe the physical system.[10] A way to understand the relation of different choices is to note that Eq. (8.8) has the classic form given in Eq. (H.6), which encompasses both a Legendre transform and its inverse: the first term on the right-hand side depends only on G, the last term depends only on Σ, and the middle term is the simple bilinear form of a Legendre transform involving the product ΣG. Analogies with the case of DFT can be found by comparing with Sec. H.3.

The generalized G–Σ-functional $\Omega[G, \Sigma]$ is the expression given in Eq. (8.8) with the interpretation that G and Σ are considered as independent variables. It follows immediately from Eq. (H.7) that stationarity with respect to G leads to the relation

$$\frac{\delta\Omega[G, \Sigma]}{\delta G} = \frac{\delta\Phi[G]}{\delta G} - \Sigma = 0 \;\Rightarrow\; \Sigma = \frac{\delta\Phi[G]}{\delta G}, \tag{8.12}$$

as given by Eq. (8.9). Furthermore, stationarity with respect to Σ leads to the relation in Eq. (8.11), $G = -\delta F/\delta\Sigma$, and to the Dyson equation

$$\frac{\delta\Omega[G, \Sigma]}{\delta\Sigma} = -G - (\Sigma - G_0^{-1})^{-1} = 0 \;\Rightarrow\; G = (G_0^{-1} - \Sigma)^{-1}. \tag{8.13}$$

Thus, Eq. (8.8) defines a functional $\Omega[G, \Sigma]$ for all allowed functions G and Σ and the two stationarity conditions lead to the correct solution that satisfies the relations for G and

---

[10] At the stationary point, all functionals have the same energy as in the Galitskii–Migdal formula of Eq. (5.129), which however is not stationary.

Σ, Eqs. (8.12) and (8.13). This is analogous to the generalized $E[v, n]$ functional for DFT, where the solution is a saddle point with v and n considered as independent variables.

The Luttinger–Ward $\Omega_{LW}[\Sigma]$ and Klein $\Omega_{Klein}[G]$ functionals are constructed with the requirement that the Dyson Eq. (8.13) be satisfied for all G and Σ in the domain of definition. The difference is that $\Omega_{LW}[\Sigma]$ was constructed in the original work of LW regarding G as a functional of Σ,

$$\Omega_{LW}[\Sigma] = \Phi[G[\Sigma]] - \mathfrak{Tr}(\Sigma G[\Sigma]) - \mathfrak{Tr}\ln(1 - G_0\Sigma) + \Omega^0, \tag{8.14}$$

whereas Klein did the converse with $\Sigma[G]$ regarded as a functional of G, which leads to the form

$$\Omega_{Klein}[G] = \Phi[G] - \mathfrak{Tr}(G_0^{-1}G - 1) - \mathfrak{Tr}\ln(G_0 G^{-1}) + \Omega^0. \tag{8.15}$$

The former definition is analogous to the Kohn–Sham functional $\Omega_{KS}[v]$, where the density is constrained to be consistent with the potential; the latter is analogous to $\Omega_{KS}[n]$. However, there is a very important difference: the Kohn–Sham solutions are minima, whereas for both $\Omega_{LW}[\Sigma]$ and $\Omega_{Klein}[G]$ it is only known that the solutions are stationary points.

Functionals $\Omega_F[\Sigma]$ and $\Omega_\Phi[G]$ can be constructed in terms of $F[\Sigma]$ and $\Phi[G]$, respectively, with the requirement that Σ and G obey the functional relationships of Eqs. (8.9) and (8.11) over the entire domain. Since the relationships are satisfied in the definition of $F[\Sigma]$, the functional of Σ is the same as Eq. (8.8) with Eq. (8.10),

$$\Omega_F[\Sigma] = F[\Sigma] - \mathfrak{Tr}\ln(1 - G_0\Sigma) + \Omega^0. \tag{8.16}$$

This functional is the basis for the variational analysis in Secs. 16.6 and I.2. The functional of G follows from Eq. (8.8) with Eq. (8.9),

$$\Omega_\Phi[G] = \Phi[G] - \mathfrak{Tr}G\frac{\delta\Phi}{\delta G} - \mathfrak{Tr}\ln(1 - G_0\frac{\delta\Phi}{\delta G}) + \Omega^0. \tag{8.17}$$

Further analysis of these functionals can be found in [296, 297]. Note that the Dyson equation is not satisfied in the entire domain of the functional, but only at the stationary solution. These are analogous to the Harris–Weinert–Foulkes functional in DFT defined in Sec. H.3.

8.3 Functionals of the screened interaction W

The construction of functionals of observable quantities or other dressed objects is a powerful concept that is not limited to the one-body Green's function. More on the theoretical foundations can be found in Sec. 8.4, and the concept is made more concrete in Sec. 9.7; here we will briefly show that the idea allows several extensions. In particular, one may note that the bare system is described not only by G^0, but also by the interaction v_c, which means that for a complete description one should consider G to be a functional of both G^0 and v_c, $G = G[G^0, v_c]$. More generally, the Green's function and the interaction play a parallel role throughout the equations, with $G^0 \leftrightarrow v_c$, $G \leftrightarrow W$, and $\Sigma \leftrightarrow P$, as one can see, e.g., by comparing Eqs. (7.10) and (7.11). This suggests that one can also work with the screened, non-local, and frequency-dependent interaction $W(\mathbf{r}, \mathbf{r}', \omega)$ instead of the bare

Coulomb interaction $v_c(\mathbf{r} - \mathbf{r}')$ [43, 298]. The result is a new functional that depends both on G and W:

$$\Psi[G, W] = \Phi[G, \tilde{v}[G, W]] - \frac{1}{2}\mathfrak{Tr}'[PW - \ln(1 + PW)], \tag{8.18}$$

where $\mathfrak{Tr}' = \mathfrak{Tr}$ without the sum over spin, and for which

$$\left.\frac{\delta\Psi}{\delta W}\right|_G = -\frac{1}{2}P \text{ and } \left.\frac{\delta\Psi}{\delta G}\right|_W = \Sigma, \tag{8.19}$$

where the derivatives are taken at fixed G and W, respectively. Hence, $\Psi[G, W]$ is a generalized functional and its variations yield the irreducible polarizability P and the irreducible self-energy Σ. Like the self-energy, P is given as a sum of skeleton diagrams in terms of the dressed Green's function and the screened Coulomb interaction, without self-energy or polarization insertions. The grand potential can be written in a form that shows the parallel with the functionals of G and Σ,

$$\Omega[G, \Sigma, W, P] = \Psi[G, W] + \mathfrak{Tr}'\frac{1}{2}[PW + \ln(1 - v_cP)] - \mathfrak{Tr}[\Sigma G - \ln(1 - G_0\Sigma)] + \Omega^0. \tag{8.20}$$

The solution is stationary with respect to G, Σ, W, and P, and thus errors in Ω are second order in the errors in all the quantities.

To get an idea why the functional has the form of Eq. (8.18) and for a check, one can consider Eq. (8.18) to be given, and verify Eq. (8.19). To this end one has to calculate

$$\frac{\delta\Psi}{\delta W} = \frac{\delta\Phi}{\delta W} - \frac{1}{2}P + \frac{1}{2}\frac{P}{1 + PW}. \tag{8.21}$$

The Φ-functional is a functional of the bare Coulomb interaction, but in analogy to the G–Σ case, here we consider the bare interaction to be a functional of the screened W in some domain around the actual screened interaction. Since we allow variations in this domain, the bare interaction is in general not the Coulomb interaction; for clarity, we call it $\tilde{v}$. The existence of the functional $\tilde{v}[W]$ supposes that the relation $\tilde{v} \to W$ can be inverted. With the Dyson Eq. (7.11) for the screened Coulomb interaction,

$$W = v_c + v_c\chi v_c = v_c + v_cPW \text{ or } v_c = W/(1 + PW) \tag{8.22}$$

(in schematic form omitting the arguments), $\tilde{v}[W]$ is given by $\tilde{v} = W/(1 + PW)$. We can now use $\tilde{v}$ in a chain rule when differentiating with respect to W [298], which yields

$$\frac{\delta\Phi}{\delta W} = \frac{\delta\Phi}{\delta\tilde{v}}\frac{\delta\tilde{v}}{\delta W} = -\frac{1}{2}\chi\frac{1}{(1 + PW)^2}, \tag{8.23}$$

where we have used

$$\frac{\delta\Phi}{\delta\tilde{v}} = \frac{\delta\Omega}{\delta\tilde{v}} = -\frac{1}{2}\chi. \tag{8.24}$$

The last equality can be shown by using the definition of the grand potential and of expectation values, Eq. (5.2). Derivation of the grand potential with respect to the Coulomb interaction creates four field operators in the expectation value, which leads to the density–density correlation function χ. The result of Eq. (8.23) yields the desired first expression in Eq. (8.19). The rest can be derived along similar lines.

In analogy to the functions for Ω in the previous section defined in terms of G and/or Σ, one can use the Ψ-functional to define various functionals of G, Σ, W, and/or P. The Ψ-functional exemplifies the essence of the functionals approach. It is the contribution of interactions to the grand-potential functional $\Omega[G, \Sigma, W, P]$ that is defined for any Green's function G and interaction W. The actual system appears only in the construction of Ω in Eq. (8.20), and it is only at the stationary point of Ω that the calculations enforce the condition that $\tilde{v} = v_c$ and that G, Σ, W, P, and all other quantities are derived for the actual system with its bare interaction v_c and Green's function G_0.

The $\Psi[G, W]$ provides a natural framework for the GW approximation, which is just the leading diagram of the series (see Fig. 10.5). It is in this sense analogous to Hartree–Fock in the Φ-functional, with the difference that the screened instead of the bare interaction appears. Therefore, it contains much more physics and is much more powerful. Extended DMFT methods in Sec. 17.5 are an example of the use of a Ψ-functional. First-principles DMFT methods described in Ch. 21 also require a screened interaction, which can be formulated in terms of the $\Psi[G, W]$-functional [297]. Indeed, all tricks that can be applied to the Green's function functionals also apply to the functionals of the screened interaction. For more details, refer to [298].

Extensions

The $\Psi[G, W]$-functional is in principle exact, and good approximations are obtained retaining only a small number of terms, provided that W represents the dominant effects. This is often true in solids with significant screening, but not necessarily in the low-density regime where short-range correlations due to particle–particle scattering are important. It may then be smarter to work with another kind of effective interaction, given by the so-called four-point vertex $^4\tilde{\Gamma}$ defined as

$$L(1, 2, 1', 2') = G(1, 2')G(2, 1') + G(1, \bar{3})G(\bar{4}, 1')^4\tilde{\Gamma}(\bar{3}, \bar{5}; \bar{4}, \bar{6})G(\bar{6}, 2')G(2, \bar{5}), \quad (8.25)$$

where $\bar{n}$ stands for $\int d\mathbf{r}_n dt_n \Sigma_{\sigma_n}$. This equation for the two-particle correlation function L defined in Eq. (5.134) highlights the meaning of $^4\tilde{\Gamma}$ as a scattering matrix. Note that we use a superscript 4 to distinguish $^4\tilde{\Gamma}$ from the three-point vertex $\tilde{\Gamma}$ in Hedin's equations in Ch. 11 and in following chapters. In some other works, $^4\tilde{\Gamma}$ is called a T-matrix [248, 299], but we have to distinguish this from the T-matrix as defined in Sec. 10.5.

The four-point vertex obeys the Dyson equation (14.11). Functionals of G and $^4\tilde{\Gamma}$ can be derived in analogy with the $\Psi[G, W]$-functional.[11] The lowest-order self-consistent approximation that would be the analog of the GWA in the case of W is the T-matrix or Bethe–Goldstone approximation, where $^4\tilde{\Gamma} \approx T$ in Eq. (10.45). A DMFT-like approach

[11] For the derivation and more details, see in particular [300].

is also possible; this leads to the dynamical vertex approximation DΓA explained in Sec. 21.7.

8.4 Generating functionals

In this section we go back to the definition of the grand potential and introduce a compact formalism that encompasses many aspects that one encounters in many-body physics. These include the functional relations seen earlier in this chapter, tricks for the derivation of diagrammatic series, the description of non-equilibrium phenomena, linear response theory, and a generalized functional formalism for the construction of auxiliary systems. The resulting formulas are used as starting point for the derivation of the perturbative approaches in Chs. 10–15, and they are the basis for the actual calculations for dynamical mean-field theory in Ch. 18.

Generation of expectation values

Let us start with the general definition of an expectation value, Eq. (5.2), of an operator $\hat{O}$,

$$\langle \hat{O} \rangle = \frac{1}{Z} \text{Tr} \left\{ e^{-\beta(\hat{H}-\mu\hat{N})} \hat{O} \right\}, \quad \text{with} \quad Z = \text{Tr} \left\{ e^{-\beta(\hat{H}-\mu\hat{N})} \right\} = e^{-\beta\Omega}. \quad (8.26)$$

This expectation value follows from the *generating function*

$$\Omega(u) = -\frac{1}{\beta} \ln \text{Tr} \left\{ e^{-\beta(\hat{H}-\mu\hat{N})} e^{-u\hat{O}} \right\}, \quad (8.27)$$

where u couples $\hat{O}$ to the system. This defines a u-dependent grand potential $\Omega(u)$ that becomes the unperturbed grand potential Ω at $u = 0$. Taking the derivative of $\Omega(u)$ with respect to u, we find

$$\beta \frac{\partial \Omega}{\partial u} \bigg|_{u=0} = \frac{1}{Z} \text{Tr} \left\{ e^{-\beta(\hat{H}-\mu\hat{N})} \hat{O} \right\} = \langle \hat{O} \rangle. \quad (8.28)$$

An example is the expression for the total number of particles $-\partial\Omega/\partial\mu = \langle \hat{N} \rangle$. The simple relations of Eqs. (8.27) and (8.28) are the basis of much of what follows here and in Chs. 10–21; we only have to generalize the relations to more general operators and couplings: u becomes a function that we denote $\mathcal{J}$, and $\Omega[\mathcal{J}]$ a functional.

The operators we are interested in usually depend on some space, spin, and time coordinates. For example, the expectation value of an operator that is a time-ordered product of two Heisenberg operators[12] can be expressed as

$$\langle \text{T}_C \hat{Q}(1, 1') \rangle = \frac{1}{Z} \text{Tr} \left\{ \text{T}_C \, e^{-\beta(\hat{H}-\mu\hat{N})} \hat{Q}(1, 1') \right\}, \quad (8.29)$$

---

[12] Note that the Heisenberg field operators are defined in absence of $\mathcal{J}$.

where T_C denotes time ordering on some contour (see App. E).[13] Recall from Sec. 5.1 that "1" is a compact notation for position and spin, $x_1 = (\mathbf{r}_1, \sigma_1)$ and time t_1; thus, $(1, 1')$ denotes $(x_1, t_1; x_{1'}, t_{1'})$, or equivalently for imaginary time. The one-particle Green's function is an example of such an expectation value as defined in Eq. (5.78). For a general time ordering the expression reads

$$G(1, 1') = -(i)\frac{1}{Z}\text{Tr}\left\{T_C\left[e^{-\beta(\hat{H}-\mu\hat{N})}\hat{\psi}(1)\hat{\psi}^\dagger(1')\right]\right\}, \tag{8.30}$$

where the $\hat{\psi}$ and $\hat{\psi}^\dagger$ are field operators in the Heisenberg picture. The prefactor (i) is needed for the real-time Green's function, and it is to be omitted for the imaginary-time Green's function following Eq. (D.7).

We now generalize Eq. (8.27) to[14]

$$\Omega[\mathcal{J}] \equiv -\frac{1}{\beta}\ln\text{Tr}\left\{e^{-\beta(\hat{H}-\mu\hat{N})}T_C\,e^{-\int\int d1\,d1'\,\mathcal{J}(1,1')\hat{Q}(1,1')}\right\}, \tag{8.31}$$

where the time-ordered exponential of operators is defined as

$$T_C\,e^{-\mathcal{J}(\bar{1},\bar{1}')\hat{Q}(\bar{1},\bar{1}')} = 1 + \sum_{n=1}^{\infty}\frac{(-1)^n}{n!}\mathcal{J}(\bar{1},\bar{1}')\ldots\mathcal{J}(\bar{n},\bar{n}')T_C\left[\hat{Q}(\bar{1},\bar{1}')\ldots\hat{Q}(\bar{n},\bar{n}')\right], \tag{8.32}$$

with all arguments integrated over, as indicated by the symbol $\bar{n}$. We then have

$$Q(1, 1') = \langle T_C\,\hat{Q}(1, 1')\rangle = \beta\frac{\delta\Omega[\mathcal{J}]}{\delta\mathcal{J}(1, 1')}\bigg|_{\mathcal{J}=0}. \tag{8.33}$$

In particular, one can get the Green's function by choosing[15]

$$\hat{Q}(1, 1') \to (i)\hat{\psi}^\dagger(1')\hat{\psi}(1'), \quad \text{hence} \quad G(1, 1') = \beta\frac{\delta\Omega[\mathcal{J}]}{\delta\mathcal{J}(1', 1)}\bigg|_{\mathcal{J}=0}. \tag{8.34}$$

Coupling to other sources

Any product operator $\hat{Q}$ with the corresponding $\mathcal{J}$ is eligible to be treated in this framework. For example, one can create an anomalous Green's function using $\hat{Q} = \hat{\psi}\hat{\psi}$ or

---

[13] This formula allows any kind of time ordering, in real or imaginary time, or on a contour like the Keldysh contour. The expressions derived in this section are general, although more natural for $T \neq 0$.

[14] Now one has to be careful to distinguish the forwards and backwards time evolution. If $\mathcal{J}$ is local in time and space it can be a physical time-dependent potential. This allows one in principle to follow the evolution of the system in time, which is a true non-equilibrium problem. If this is the goal, one must evolve forwards and backwards in real time, like on the Keldysh contour (see [300] and App. E). If instead, as is the case here, $\mathcal{J}$ is introduced just as a trick and set to zero at the end of the derivation, then there are several possible choices. For example, in [248] the interaction picture is used, with $\mathcal{J}$ a fictitious potential term and a time evolution that is always forwards; in the original work of Hedin [43], there is forwards and backwards time evolution, but with different potentials.

[15] The order of indices is a choice that we make, consistent with the literature. In this way, if $\mathcal{J}$ is local the additional term in Eq. (8.31) reduces to the usual contribution of a static potential to the hamiltonian. The results for observables do not depend on this choice when all expressions are evaluated consistently.

$\hat{Q} = \hat{\psi}^{\dagger}\hat{\psi}^{\dagger}$, leading to non-vanishing expectation values in the superconducting state. More generally, one can introduce more than one source term $\mathcal{J}_1, \mathcal{J}_2, \ldots$, each coupling to a different operator $\hat{Q}_1, \hat{Q}_2, \ldots$, which allows one to calculate different expectation values from the same grand potential. We do not follow that route in this book.

Higher-order variations

Derivatives of $\Omega[\mathcal{J}]$ can be defined to any order. For example, if the one-particle Green's function is evaluated from the equations corresponding to Eqs. (8.31) and (8.33), the two-particle correlation function L that is linked to the two-particle Green's function as $L = -G_2 + GG$ (Eq. (5.134)) is obtained as a second derivative (see Ex. 8.7),

$$L(1, 2, 1', 2') = \left[\frac{\delta^2 \beta \Omega[\mathcal{J}]}{\delta \mathcal{J}(1', 1)\delta \mathcal{J}(2', 2)} \right]_{\mathcal{J}=0} = \left[\frac{\delta G(1, 1')[\mathcal{J}]}{\delta \mathcal{J}(2', 2)} \right]_{\mathcal{J}=0}. \tag{8.35}$$

As explained in Sec. 10.4, conservation laws can be expressed in terms of the symmetry of L with respect to interchange of the arguments $(1, 1')$ and $(2, 2')$ that can be seen if we change the order of the derivatives in Eq. (8.35). The last form in Eq. (8.35), the variation of $G(1, 1')$ with respect to $\mathcal{J}(2, 2')$, is a generalized time- or contour-ordered response function: the propagation of a particle $(1' \to 1)$ changes due to a disturbance where a particle is inserted at $2'$ and removed at 2. With Eq. (5.149) the density–density response $\chi(1, 2) = -iL(1, 2, 1^+, 2^+)$ is a diagonal element of $L(1, 2, 1', 2')$.

Note that one could also have created L directly by introducing a four-variable source term; this is, however, not useful for our purpose. Alternatively, one could have introduced two sources for the two anomalous Green's functions $\hat{Q}_1 = \hat{\psi}^{\dagger}\hat{\psi}^{\dagger}$ and $\hat{Q}_2 = \hat{\psi}\hat{\psi}$, and taken the derivative with respect to $\mathcal{J}_1$, and $\mathcal{J}_2$; because of time ordering, this would also yield L.

Building a functional of measurable quantities

One of the key points of the Green's function approaches in this book is the fact that one can formulate the theory in terms of functionals of the measurable quantities themselves. Here we address the question of how to construct a functional of the expectation value Q. To do so, let us first introduce the functional

$$\tilde{\Omega}[\mathcal{J}, Q] \equiv -\frac{1}{\beta} \ln \text{Tr} \left\{ e^{-\beta(\hat{H}-\mu\hat{N})} T_C \, e^{-\mathcal{J}(\bar{1},\bar{1}')(\hat{Q}(\bar{1},\bar{1}')-Q(\bar{1},\bar{1}'))} \right\}, \tag{8.36}$$

where a bar indicates integration, e.g., $f(\bar{1})g(\bar{1}, 2) \equiv \int d1 f(1)g(1, 2)$. With this functional, the expectation value of $\hat{Q}$ takes a given value $\langle \hat{Q} \rangle = Q$ for a source term $\mathcal{J} = \mathcal{J}_Q$ chosen such that

$$\frac{\delta \tilde{\Omega}[\mathcal{J}, Q]}{\delta \mathcal{J}} \bigg|_{\mathcal{J}=\mathcal{J}_Q} = 0. \tag{8.37}$$

The functional

$$\Gamma[Q] \equiv \beta\tilde{\Omega}[\mathcal{J}_Q[Q], Q] = \beta\Omega[\mathcal{J}_Q] - \mathcal{J}_Q(\bar{1}, \bar{1}')Q(\bar{1}, \bar{1}') \qquad (8.38)$$

is the *effective action functional* of the measurable quantity Q itself. As can be seen on the right side of Eq. (8.38), it is a Legendre transform (see Sec. H.2), where Q is considered to be the independent variable and $\mathcal{J}_Q$ is a functional of Q. The transformation of the independent variables in Eq. (8.38) is well-defined only if the relation $Q[\mathcal{J}_Q]$ can be inverted to give $\mathcal{J}_Q$ as a functional of Q (see Sec. H.2) over the domain of definition. For example, this can be demonstrated for density functional theory where $\mathcal{J}$ is a static, local potential $v(x)$ and $\hat{Q}(1, 1') = \hat{n}(1)\delta(1 - 1')$ is a local operator in space and time, as worked out in Ex. 8.3 and Sec. H.5 (see also Ex. H.2). To our knowledge there are no rigorous proofs for functionals of the Green's function; see also the discussion at the end of Sec. 8.2.

The source $\mathcal{J}_Q$ can be derived from $\Gamma[Q]$ by

$$\mathcal{J}_Q(1, 1') = -\frac{\delta\Gamma[Q]}{\delta Q(1, 1')}. \qquad (8.39)$$

It follows from Eq. (8.39) that the functional $\Gamma[Q]$ is extremal,

$$\frac{\delta\Gamma[Q]}{\delta Q(1, 1')} = 0, \qquad (8.40)$$

when $\mathcal{J}_Q = 0$. Then $Q = Q_0$ is the equilibrium expectation value of $\hat{Q}$ given by Eq. (8.33). Furthermore, at this point Γ yields the grand potential Ω,

$$\frac{1}{\beta}\Gamma[Q_0] = \Omega[\mathcal{J} = 0] = \Omega. \qquad (8.41)$$

For zero temperature, Eq. (8.40) expresses the requirement that $Q = Q_0$, where Q_0 is the ground-state expectation value of $\hat{Q}$, and $\frac{1}{\beta}\Gamma[Q_0] = E_0$ the ground-state energy.

Building an auxiliary system

In order to put these formal relations into practice, approximations are needed. One idea is to introduce an auxiliary system. The motivation comes from Eq. (8.37): this equation tells us that one may obtain a desired expectation value Q of an operator $\hat{Q}$ by carefully choosing a source term $\mathcal{J}_Q$ that is added to the original hamiltonian. In particular, if Q is the expectation value that would result from the full hamiltonian $\hat{H}$ without the external source, one may hope to obtain Q from a *simpler* hamiltonian $\hat{H}_s$, but with $\mathcal{J}_Q \neq 0$. The source $\mathcal{J}_Q$ is an auxiliary field that is designed to reproduce Q, but not necessarily other expectation values.

We now split the effective action of Eq. (8.38) as

$$\Gamma[Q] = \Gamma_s[Q] + \Delta\Gamma[Q]. \qquad (8.42)$$

Here $\Gamma_s[Q]$ is the effective action functional calculated with a different hamiltonian $\hat{H} \rightarrow \hat{H}_s$, for which the solution can be obtained. The last term $\Delta\Gamma$ is the correction term that is needed to retrieve the full result. It follows with Eq. (8.39) that

$$\frac{\delta\Gamma}{\delta Q} = 0, \quad \text{where} \quad \mathcal{J} = -\frac{\delta\Gamma_s[Q]}{\delta Q} = \frac{\delta\Delta\Gamma[Q]}{\delta Q}. \tag{8.43}$$

This is the value of the auxiliary field that must be chosen to obtain the desired expectation value Q. Similarly, the condition $\delta\tilde{\Omega}[\mathcal{J}, Q]/\delta\mathcal{J} = 0$ leads to

$$\frac{\delta F[\mathcal{J}]}{\delta\mathcal{J}} = -Q, \tag{8.44}$$

where F is the Legendre transform $F \equiv \Delta\Gamma[Q] - \mathcal{J}Q$. Comparison with the equations in Sec. 8.2 shows that when Q is the Green's function and $\hat{H}_s$ is chosen to be the non-interacting hamiltonian, $\mathcal{J}$ corresponds to the self-energy that plays the role of an auxiliary field that restores the interacting Green's function. In that case the correction term $\Delta\Gamma$ is the functional $\Phi[G]$. In other words, the formulation in terms of a Green's function and a self-energy can be interpreted as a picture where a non-interacting particle evolves in a dynamical non-local potential.

Other choices of $\hat{H}_s$ lead to other auxiliary fields; for example, a site embedded in a medium that is the source for particles, where Fig. 16.1 depicts a particle hopping from the medium to the site or vice versa, in general at different times. This can be formulated as a dynamical hybridization function $\Delta(\omega)$ or the inverse Green's function for the medium $\mathcal{G}_0^{-1}(\omega)$ in Eqs. (16.8) and (18.3). As explained in Sec. 18.3, the partition function for the embedded system in Eq. (18.6) has the form corresponding to Eq. (8.31) with $\mathcal{J}(1, 1')\hat{Q}(1, 1')$ replaced by $\psi_\sigma^\dagger(\tau)\mathcal{G}_0^{-1}(\tau - \tau')\psi_\sigma(\tau')$.

A way to construct a functional in practice: the coupling constant integration

At this point we still have the question: how to construct the functional $\Delta\Gamma[Q]$? An intuitive strategy that one finds over and over again in all fields of physics is to start with a part that can be calculated straightforwardly, and to treat the rest as well as one can. One possibility to do so is the so-called *coupling constant integration*, used in Secs. 5.7 and 9.8. Let us again call the first part Γ_s, and let us introduce a parameter λ that scales the difficult part. For $\lambda = 0$ one has the easy part, and for $\lambda = 1$ the full problem is obtained. The scaling is done such that the expectation value Q is constant as λ is varied. Hence we have

$$\Gamma[Q] = \Gamma_s[Q] + \int_0^1 d\lambda \, \frac{d\Gamma^\lambda[Q]}{d\lambda}. \tag{8.45}$$

With $\frac{d\Gamma^\lambda}{d\lambda} = \beta\frac{d\Omega^\lambda}{d\lambda}$, one is then left with the task to calculate or approximate certain expectation values. For example, if Γ_s corresponds to a non-interacting hamiltonian and Q to the one-body Green's function, λ scales the interaction and $\frac{d\Omega^\lambda}{d\lambda}$ leads to the appearance of a two-particle Green's function. LW have used the coupling constant approach to derive the in-principle-exact result Eq. (8.14) (see Sec. 9.8); most often one has to approximate the λ-dependent part. In DFT, where Q is the density, this is the λ-dependent density–density response function (Eq. (4.30)) that is then approximated, for example in the RPA (see Sec. 11.5).

8.5 Conservation laws and conserving approximations

There exist many possibilities to invent approximations, and it is useful to have guidelines that tell us what a good approximation might be. In particular, one will wish to construct theories that obey the fundamental laws of physics. One of the cornerstones, or maybe *the* cornerstone, is symmetry and the associated conservation laws. It is difficult to imagine how to follow the evolution of a system if particle number, momentum, etc. are not conserved. Of course, the exact solution is conserving, but is there a way to enforce the laws in practical calculations where one must make approximations? An elegant way to do this is the topic of this section.

In independent-particle methods the question is relatively simple: the laws are obeyed so long as the hamiltonian obeys all the symmetries of the problem, e.g., in Kohn–Sham DFT or the Hartree–Fock approximation.[16] However, if one is given an approximate self-energy and Green's function, it is not at all apparent how to verify that the resulting dynamics obeys all conservation laws. Think of the spectral function: beyond the independent-particle picture, weight is distributed from the quasi-particle peak to the incoherent part of the spectrum (see Sec. 7.4), and if this is done in an approximate way one may violate the sum rule of Eq. (5.95) on the spectral function, which means that one changes the number of particles. We also want the equations to be conserving at every point in space during the evolution of a system. For example, charge is conserved with respect to a current flow $\mathbf{j}$ through the continuity equation $\partial n / \partial t = -\nabla \cdot \mathbf{j}$. However, if there is screening of the charge by the many-body system of electrons, it is not so easy to show that an approximate form of the screening will affect the current and the density in such a way that the particle number is conserved.

A conceptual advance was made by Baym and Kadanoff [13, 39]: in two papers in 1961 and 1962, BK derived the conditions for "conserving approximations" based solely on the symmetry of the Green's functions, independent of the details of a particular approximation. Section 10.4 in Ch. 10 formulates conservation laws by closely following their reasoning in terms of the two-particle Green's function. However, we can already get the essence by simply looking at the Dyson Eq. (8.13), which can be rewritten alternatively as

$$G_0^{-1} G = 1 + \Sigma G, \qquad \text{or} \qquad (8.46)$$

$$G G_0^{-1} = 1 + G \Sigma. \qquad (8.47)$$

Let us now derive one of the equations that express conservation laws, for example the continuity equation. For this we need the time derivative of the charge density, which can be expressed in terms of the Green's function as

---

[16] This is not true for unrestricted Hartree–Fock or DFT (such as the unrestricted form of "LDA+U"), which may break the symmetry.

$$\frac{\partial n(x_1,t)}{\partial t} = -i \left\{ \frac{\partial}{\partial t_1} + \frac{\partial}{\partial t_{1'}} \right\} G(x_1,t_1,x_{1'},t_{1'}) \Bigg|_{t_1 \to t, t_{1'} \to t^+} . \tag{8.48}$$

Since $G_0^{-1} = -i\frac{\partial}{\partial t} - \hat{h}$ (Eq. (5.85)), subtraction of Eq. (8.47) from Eq. (8.46) yields the continuity equation, if and only if

$$\int d2 \Sigma(1,2) G(2,1^+) = \int d2 G(1,2) \Sigma(2,1^+). \tag{8.49}$$

A relation similar to the continuity equation links momentum and forces, as worked out in Sec. 10.4, angular momentum and torque, etc. All these equations start with a time derivative of the quantity of interest, that reflects possible changes, and the associated symmetry conditions can be derived from the Dyson equation in a way similar to the example of the continuity equation. The momentum–force relation, for example, gives rise to

$$\int dx_1 \int d2\, \Sigma(1,2)\nabla_{\mathbf{r}_1} G(2,1^+) = \int dx_1 \int d2\, G(1,2)\nabla_{\mathbf{r}_1} \Sigma(2,1^+). \tag{8.50}$$

More details can be found in [301]. These relations are fulfilled if the self-energy $\Sigma(1,2)$ is proportional to a closed set of diagrams where the Green's function $G(2,1)$ has been removed, such that by putting it back, for example on the left side of Eq. (8.49), the original expression is retrieved (although not integrated over all variables, and maybe with a prefactor). This is like cutting a part of a chocolate cake; the closed expression is the cake, $G(2,1)$ is the part removed, and the self-energy is the rest. The left and the right-hand side of Eq. (8.49) cut two different parts of the cake, but since the piece is put back, the result is the same. Mathematically, the "chocolate cake" can be thought to be an integrated quantity $\Phi' \equiv \int d1 \int d2 \phi(2,1)$ that is a functional of G. If an approximate self-energy is $\Sigma'(1,2)[G] = \delta\Phi'[G]/\delta G(2,1^+)$ and if its holds that $[\delta\Phi'[G]/\delta G(2,1^+)]G(2,1^+) \propto \phi(2,1)$, the conservation laws are fulfilled.

This can be stated as follows:

In order for a theory with approximate Σ', and G' obtained from the Dyson equation, to be conserving, there must exist a functional $\Phi'[G]$ such that $\Sigma'(1,2)[G] = \delta\Phi'[G]/\delta G(2,1^+)$.

This is the same requirement as Eq. (8.9), which holds for the exact self-energy functional. The picture of the chocolate cake illustrates that the functional Φ is not completely arbitrary, but must obey certain construction rules. These constraints can best be understood using an example; in Ex. 8.8 the lowest order, the Hartree–Fock approximation, is examined. The construction rules for the exact functional have been given by Luttinger and Ward; this is the topic of Sec. 9.8. A physically meaningful approximate functional is then a subset of the contributions that are contained in the exact formalism. The translation of this statement into practice will become clearer with the help of Chs. 9 and 10.

The fact that the self-energy is the derivative of a functional implies that

$$\frac{\delta\Sigma(1,1')}{\delta G(2',2^+)} = \frac{\delta\Sigma(2,2')}{\delta G(1',1^+)}, \tag{8.51}$$

which must then also be fulfilled by an approximate $\Sigma'[G]$ if it is to obey conservation laws. As pointed out by Baym [13], this relation can be understood as a vanishing of a curl, from which it follows that there must exist a scalar functional $\Phi'[G]$ from which the self-energy can be derived. This is analogous to the existence of a central potential in classical mechanics that follows only from rotational invariance.

A good example to see how the relation of Σ to Φ impacts fundamental laws is number conservation; see also the relation to the Luttinger theorem in App. J. In equilibrium the number of particles is given by the derivative of the grand potential with respect to the chemical potential μ, as specified after Eq. (8.28). Using Eq. (8.8) and the relation $G^{-1}(\omega) = \omega + \mu - \hat{h} - \Sigma$ in the grand-canonical ensemble, one obtains

$$-N = \frac{d\Omega}{d\mu} = \mathfrak{Tr}\left\{\frac{\delta\Phi}{\delta G}\frac{dG}{d\mu} - \frac{d(\Sigma G)}{d\mu} + G\left(-1 + \frac{d\Sigma}{d\mu}\right)\right\}. \tag{8.52}$$

The chain rule has been used in the first term, where $d\Phi/d\mu = (\delta\Phi/\delta G)(dG/d\mu)$, and in the last term. Combining terms in Eq. (8.52), we find the correct expression (see Eq. (5.118)),

$$N = \mathfrak{Tr}(G) = \beta^{-1}\sum_i\sum_n G_{ii}(\omega_n) \quad \text{or} \quad -\sum_i\frac{i}{2\pi}\int d\omega e^{i\eta\omega}G_{ii}(\omega) \text{ for } T = 0, \tag{8.53}$$

if and only if there is the needed cancellation of terms in Eq. (8.52). This happens if Σ and G are consistent and related by $\Sigma = \delta\Phi/\delta G$, even for approximate G and Σ.

For a Green's function that does not obey this relation, such as those from calculations that are not carried out self-consistently, one can define a density corresponding to Eq. (8.53); however, the energy and other thermodynamic variables will not be correct, and the number of particles may not be conserved as G evolves in the presence of external potentials. For example, take the GW approximation of Ch. 11 and Sec. 12.2, where $\Sigma(1,1') = iG(1,1')W(1,1')$ with W a screened Coulomb interaction. This approximation fulfills the condition $\Sigma G = G\Sigma$ at $(x' = x, t' - t = 0^+)$ that is necessary for the continuity equation, if W is symmetric in $1 \leftrightarrow 1'$. However, if G in Σ is replaced by some G_0, the condition is violated. Indeed, $\Sigma = iG_0W$ cannot be obtained from any universal $\Phi[G]$, i.e., any functional that does not depend on G_0. The violation is usually of the order of one percent or less in the HEG at metallic densities or in typical semiconductors (see Sec. 11.5) when one uses a reasonable guess for G^0, such as a KS Green's function. However, total energies are sensitive to sum rules (see Sec. 11.5), and have indeed been found to improve when one goes from $\Sigma = iG_0W$ to the self-consistent $\Sigma = iGW$, for example in the HEG Sec. 11.6.

8.6 Wrap-up

This chapter is dedicated to the properties of functionals and the formulation of ones most relevant for the theory of interacting electrons. Density functional theory provides motivation and many useful guidelines, and it leads up to the functionals of the Green's function and self-energy that are a foundation for much of the work in Chs. 9–21. Most important are the functionals that take into account the interaction: $\Phi[G]$ and its Legendre transform $F[\Sigma]$ defined in Sec. 8.2, and the functional of G and the screened interaction $\Psi[G, W]$ in Sec. 8.3. As shown by Luttinger and Ward, these can be expressed as an expansion in diagrams that is explained in more detail in Ch. 9. These form a natural basis for the GW approximation and other methods that are the workhorses for much of present-day calculations for electrons in condensed matter.

A general feature is the transformation of functionals of the bare Green's function and interactions into functionals of dressed quantities, e.g., the actual Green's function G instead of the bare G_0. Working with dressed quantities instead of bare ones allows us to find more compact expressions, since many terms are resummed. To work with such renormalized quantities is also more physical and, in some cases, avoids divergencies; this is the case, for example, when a screened Coulomb interaction is used in the self-energy, leading to the GW approximation that cures many shortcomings of Hartree–Fock (see Sec. 11.3). The functional approach also provides a way to construct approximations that automatically fulfill conservation laws (Sec. 8.5).

The grand potential Ω can be expressed in different ways, as discussed in Sec. 8.2. Using the fact that $\Phi[G]$, $F[\Sigma]$, and $\Psi[G, W]$ are defined for ranges of their arguments (the domains of definition), one can construct various functionals for Ω that all have the same value at the stationary solution, but different variation as functionals of G, Σ, and/or W. Much insight can be gained by starting from the grand potential, divided into a universal term and a term that depends on the particular system, e.g., as used in the variational derivation of DMFT in Sec. 16.6. The theory of generating functionals in Sec. 8.4 is a compact formalism that encompasses these functionals in an approach that applies to many fields of physics.

SELECT FURTHER READING

Abrikosov, A. A., Gorkov, L. P., and Dzyaloshinski, I. E. *Methods of Quantum Field Theory in Statistical Physics* (Prentice-Hall, Englewood Cliffs, NJ, 1963). A concise presentation of the work of Luttinger and Ward.

Baym, G., "Self-consistent approximations in many-body systems," *Phys. Rev.* **127**, 1391–1401, 1962. The original paper on functionals and conserving approximations that is very readable.

Negele, J. W. and Orland, H. *Quantum Many-Particle Systems* (Advanced Book Classics, originally published in 1988 by Westview Press, Boulder, CO; Addison-Wesley, Reading, MA). A basic reference for functionals in quantum statistical physics.

Stefanucci, G. and van Leeuwen, R. *Non-equilibrium Many-Body Theory of Quantum Systems* (Cambridge University Press, Cambridge, 2013). Very precise and contains many details about functionals and the derivations.

The review articles by Georges *et al.* [302] and Kotliar *et al.* [303] listed at the end of Ch. 16 have extensive discussion of functionals of the Green's function.

Exercises

8.1 Derive the expression in Eq. (8.4) using the hints given after that equation. Argue why it is easier to express the free energy for the grand-canonical ensemble at fixed μ than for the microcanonical ensemble with fixed number of particles.

8.2 Give the explicit form for the Hartree–Fock total energy functional formed from Eq. (8.7) plus the contributions of the kinetic energy and potential terms. Show that the Hartree–Fock equations, Eq. (4.5), follow from the stationary point of the functional. As part of this exercise, show explicitly that at the solution the total energy is the same as $\langle \Phi_{HF}|\hat{H}|\Phi_{HF}\rangle$ given in Eqs. (4.2) to (4.4).

8.3 Show that the Hohenberg–Kohn derivation for DFT is an example of the use of effective action functionals by choosing $\hat{Q} = \hat{n}$, the density operator, and $\mathcal{J}$ a local potential. The proof that the stationary point is a variational minimum is given in Sec. H.5 and Ex. H.2. Discuss how this demonstration relates to the statement after Eq. (4.15) that the HK theorem is restricted to "v-representable" densities.

8.4 Derive the last two terms in Eq. (8.6), $-\mathfrak{Tr}\ln(1-G_0 v_{Hxc})+\Omega^0$, by coupling-constant integration where the potential v_{Hxc} is scaled be a factor λ.

8.5 Show that the expressions for Ω in Eq. (8.16) and the equation in footnote 7 on page 176 lead to the same equations for Σ and G. The derivations of Eqs. (16.16) and (I.9) are examples where the more compact form given in the footnote is used.

8.6 Look at the diagrammatic expansions in Ch. 9, and try to find effective quantities other than G or W in terms of which it could be interesting to build a functional. In particular, try to understand the idea behind building a functional of $^4\tilde{\Gamma}$ defined in Eq. (8.25).

8.7 Derive Eq. (8.35).

8.8 Use the Hartree–Fock self-energy to check the symmetry conditions of Eqs. (8.49) and (8.50) that guarantee the continuity equation and the momentum–force relation. This is helpful to understand that the functional Φ is not completely arbitrary. Would the relations be fulfilled for any kind of interaction?

PART III

MANY-BODY GREEN'S FUNCTION METHODS

9

Many-body perturbation theory: expansion in the interaction

We especially need imagination in science.

Maria Mitchell

Summary

The many-body problem consists of two parts: the first is the non-interacting system in a materials-specific external potential; the second is the Coulomb interaction that makes the problem so hard to solve. The most straightforward idea is to use perturbation theory, with the Coulomb interaction as perturbation. This is conceptually simple, but it turns out to be difficult in practice, since the Coulomb interaction is often not small compared with typical energy differences, it is long-ranged and in the thermodynamic limit there is an infinite number of particles, contributing with an infinite number of mutual interaction processes. The present chapter outlines how one can deal with this problem. It contains an overview of facts that one can also find in many standard textbooks on the many-body problem, but that are useful to keep in mind in order to look at later chapters from a sound and well-established perspective.

The many-body problem is a tough one, and it has many facets. Sorting it out is like putting together a huge puzzle. The eight introductory chapters of this book provide pieces of the puzzle, and ideas on what one might do about it. In the present chapter we choose to go in one of the possible directions, in order to arrive at something tangible. The chapter gives the general framework and the main ideas; specific approximations are the topic of Chs. 10–15.

The idea is to start from an independent-particle problem and add the Coulomb interaction as a perturbation. This is not easy: first, the interaction is responsible for a rich variety of phenomena that are completely absent otherwise, such as the finite lifetime of quasi-particles, or additional structures in spectra due to the fact that a quasi-particle excitation may transfer its energy to other elementary excitations, for example plasmons. Second, because of the two-body Coulomb interaction, the problem scales badly with the number of electrons, and straightforward perturbation theory for the many-body hamiltonian with the Coulomb interaction as perturbation rapidly becomes intractable or even useless, especially in large systems.

To get started, Sec. 9.1 recalls why things are not so easy. The following sections try to solve one problem after the other, starting from Sec. 9.2 where the Green's function is reformulated in a way that is appropriate for a perturbation expansion. This leads to first results in Sec. 9.3. Since the formulas that one obtains are rather complex, it is also useful to introduce a compact way to represent and interpret them: this is provided by Feynman diagrams. Still one can do better, and Sec. 9.4 gives two useful theorems: these are Wick's theorem and the linked cluster theorem. The expressions are considerably simplified; in particular, they can be reformulated in terms of a Dyson equation with a well-defined self-energy as shown in Sec. 9.5. The developments are made at zero temperature; the analogous approach for non-vanishing temperature is formulated in Sec. 9.6. Section 9.7 contains a resummation of diagrams leading to an expression in terms of the dressed rather than the bare Green's function, as derived by Luttinger and Ward. The same idea is also applied to the interaction. The expression in terms of the dressed Green's function is then used to obtain the Luttinger–Ward functional for the grand potential in Sec. 9.8. Finally, a wrap-up Sec. 9.9 concludes the chapter.

9.1 The Coulomb interaction and perturbation theory

The aim of this and the following chapters of Part III is to develop methods for calculating materials properties from Green's functions. It is clear that we will not be able to solve the many-body problem exactly. Instead, the hope is to be able to use some kind of perturbation theory, most logically with the interaction as perturbation, since one can in general solve the non-interacting problem. Green's functions are used all over physics, and the first reflex could be to apply ideas used in other problems. Let us recall some of them.

Green's functions from scattering theory

Green's functions are used in scattering theory, where particles scatter from a localized perturbation. Suppose the system without the perturbation is described by $\hat{H}_0$, and $\hat{V}$ indicates the perturbation. For energies ω in the continuum, only the wavefunctions of the scattered states have to be determined. The relation between an unperturbed state $|\phi^0\rangle$ and an eigenstate of the full hamiltonian $|\phi\rangle$ at the same energy ω is

$$|\phi\rangle = |\phi^0\rangle + (\omega - \hat{H}_0)^{-1}\hat{V}|\phi\rangle. \tag{9.1}$$

This is the Lippmann–Schwinger equation [304]. It can be reformulated as

$$|\phi\rangle = (1 - G_0\hat{V})^{-1}|\phi^0\rangle \quad \text{with} \quad G_0 \equiv (\omega + i\eta - \hat{H}_0)^{-1}\big|_{\eta \to 0^+}. \tag{9.2}$$

The Green's function G_0 depends only on the unperturbed system. It incorporates the boundary condition that the scattering contribution $|\phi\rangle - |\phi^0\rangle$ only contains outgoing contributions; this is achieved by adding the infinitesimal η, which makes the solution causal (see the retarded Green's function in Eq. (5.91)).

Equation (9.2) is equivalent to $|\phi\rangle = GG_0^{-1}|\phi^0\rangle$ with the definition of the full Green's function $G \equiv (1 - G_0\hat{V})^{-1}G_0$, which fulfills the Dyson equation $G = G_0 + G_0\hat{V}G$. Its perturbation expansion is the Born series $G = G_0 + G_0\hat{V}G_0 + G_0\hat{V}G_0\hat{V}G_0 + \cdots$

One could think to treat the electron–electron scattering due to the Coulomb interaction in an analogous way. In that case, using the notation of Sec. 7.4 the full retarded Green's function is $G^{tot} = (\omega + i\eta - \hat{H})^{-1}$, where $\hat{H}$ is the full many-body hamiltonian. If we split the hamiltonian $\hat{H} = \hat{H}_0 + \hat{V}$, where now $\hat{V}$ is the interaction contribution, in strict analogy to the scattering problem the Green's function obeys

$$G^{tot}(\omega) = G_0^{tot}(\omega) + G_0^{tot}(\omega)\hat{V}G^{tot}(\omega), \tag{9.3}$$

and its perturbation expansion is

$$G^{tot}(\omega) = G_0^{tot}(\omega) + G_0^{tot}(\omega)\hat{V}G_0^{tot}(\omega) + G_0^{tot}(\omega)\hat{V}G_0^{tot}(\omega)\hat{V}G_0^{tot}(\omega) + \cdots \tag{9.4}$$

However, G^{tot} is not what we want. As one can see from the definitions in Sec. 5.6 and the discussion in Sec. 7.4, the aim is to calculate expectation values of G^{tot} in states $c_i^{\dagger}|\Phi_0\rangle$ or $c_i|\Phi_0\rangle$, where $|\Phi_0\rangle$ is the N-particle ground state. So, even if we knew G^{tot}, we would still have a problem, because the *interacting* ground state appears.

Particles embedded in an interacting system

What hinders us from introducing a Dyson equation such as Eq. (7.7) and constructing the self-energy according to Eq. (7.8)? Again, the problem is that the Green's functions we are interested in are *specific elements* of the full many-body Green's function G^{tot} of the system defined with the *interacting* eigenstates. To use directly the construction rule for the self-energy, Eq. (7.8), we should be able to calculate matrix elements of $\hat{H}$, and this not only between states $c_i^{\dagger}|\Phi_0\rangle$ or $c_i|\Phi_0\rangle$ as above: Eq. (7.8) also requires knowledge of the rest $\hat{H}_R$ and $\hat{H}_{RS}$, which would in the present case mean matrix elements involving also excited many-body states. Hence, the explicit expression for the self-energy is very useful for understanding, but it doesn't solve the problem, unless one can use it as a starting point for approximations. The first thing that comes to mind is usually perturbation theory: can one use a simple perturbative scheme to approximate the many-body wavefunctions and expectation values? This discussion can be found in many places. Here we follow [227], and give the main arguments in the following.

Perturbation theory for energy and wavefunctions

As for Eq. (9.3), we decompose the many-body hamiltonian into the non-interacting part $\hat{H}_0$ and the interaction $\hat{V}$. There are different techniques to obtain a perturbation expansion [246, 305]. In the Brillouin–Wigner approach the perturbed Schrödinger equation is written as a spectral representation in terms of the eigenvalues E_{ip} and eigenfunctions Φ_{ip} of the unperturbed equation where $\hat{V} = 0$. The expansion of the secular equation leads to

$$E = E_{ip} + \sum_{n=1}^{\infty} \left\langle \Phi_{ip} \left| \hat{V} \left(\frac{1 - \hat{P}}{E - \hat{H}_0} \hat{V} \right)^{(n-1)} \right| \Phi_{ip} \right\rangle \tag{9.5}$$

and

$$|\Phi\rangle = \sum_{n=0}^{\infty} \left(\frac{1 - \hat{P}}{E - \hat{H}_0} \hat{V} \right)^{n} |\Phi_{ip}\rangle \tag{9.6}$$

for the energy E and wavefunction Φ in terms of the non-interacting solutions E_{ip} and Φ_{ip}. Here $\hat{P} = |\Phi_{ip}\rangle\langle\Phi_{ip}|$ is the projector on $|\Phi_{ip}\rangle$; higher-order contributions to the wavefunction are orthogonal to $|\Phi_{ip}\rangle$.

In the Rayleigh–Schrödinger approach the eigenvalues and eigenfunctions of the perturbed Schrödinger equation are expanded in orders of the perturbation, which leads to slightly different expressions:

$$E = E_{ip} + \sum_{n=1}^{\infty} \left\langle \Phi_{ip} \left| \hat{V} \left(\frac{1 - \hat{P}}{E_{ip} - \hat{H}_0} (E_{ip} - E + \hat{V}) \right)^{(n-1)} \right| \Phi_{ip} \right\rangle \tag{9.7}$$

and

$$|\Phi\rangle = \sum_{n=0}^{\infty} \left(\frac{1 - \hat{P}}{E_{ip} - \hat{H}_0} (E_{ip} - E + \hat{V}) \right)^{n} |\Phi_{ip}\rangle. \tag{9.8}$$

In both approaches the energy E appears self-consistently, and one has to sort out terms order by order in the interaction. For higher orders this is increasingly complicated. However, this is not the main problem, since one would of course hope to get away with a low-order approximation.[1] Instead, there is a specific obstacle in the many-body problem.

In the Brillouin–Wigner result, the first-order ($n = 1$) correction to the energy is proportional to N^2/Ω, where the square of the particle number N^2 stems from the sum over all possible interacting pairs of particles, and the volume Ω from the normalization of the wavefunction. In the thermodynamic limit with $N \to \infty$ and $\Omega \to \infty$ with $N/\Omega = const.$, this term is hence proportional to the number of particles, as is the energy of the non-interacting system, which is what one would require: the approach is *size extensive*.[2] Instead, terms of higher order are either independent of N or vanish with increasing N. To understand this, it is convenient to insert complete sets of non-interacting states between the operators in Eqs. (9.5)–(9.8), as suggested in Ex. 9.1, and to examine the single contributions. Each energy denominator in Eq. (9.5) contains a first-order energy correction

---

[1] Usual textbook expressions give most often the Rayleigh–Schrödinger formula to first and second-order. The second-order expression for the wavefunction has a contribution from the first-order correction to the energy E that appears in $(E_{ip} - E + \hat{V})$ in the numerator of Eq. (9.8); see Ex. 9.1.

[2] Note that the total energy diverges with $N \to \infty$, but the energy per particle is finite.

and introduces therefore a factor $1/N$, which yields $1/N^{n-1}$. Each matrix element yields a factor $1/\Omega$, which gives $1/\Omega^n$, and the summations over all pairs of particles due to the matrix elements a factor N^{2n}. However, since the Coulomb interaction only depends on the difference of electronic positions, momentum is conserved, and not all combinations should be counted, as one can see for example by looking at Eq. (4.10). This fact reduces the scaling of higher-order terms by at least a factor of N. In other words, the second- and higher-order corrections taken individually are vanishingly small compared with the first order. When the first-order correction is not sufficient, one cannot expect the second or third-order terms to give significant improvements, but one would need to sum a huge number of terms. Also, the expansion of the eigenstates is not promising, since the perturbed wavefunction consists of a superposition of independent-particle states where a finite number of particles are excited (see details in the next subsection). This will be a big deviation from the true eigenstate for many electrons, which will contain a huge number of excitations. Therefore, straightforward Brillouin–Wigner perturbation theory with the full Coulomb interaction as perturbation does not seem to be a generally suitable approach to the many-body problem.

In the Rayleigh–Schrödinger approach, the energy denominators contain differences of *unperturbed* energies. Since the corresponding matrix elements are non-vanishing only when the excitation involves a pair of particles, the energy differences are independent of N. For this reason, the Rayleigh–Schrödinger energy expansion does not exhibit the problem of absence of terms proportional to N at higher orders. Still, the perturbed eigenstate is merely a superposition of states containing a limited number of excited particles, and will hence for a large number of particles be far from the true wavefunction at any finite order, as in the Brillouin–Wigner case.

One might argue that we are not interested in eigenstates, but in expectation values. In spite of the fact that the Hartree–Fock wavefunction has vanishing overlap with the true many-body ground state in the thermodynamic limit (see Ex. 6.5), expectation values like the density or energy differences can be reasonable. This suggests following the lines of Ch. 8, and formulating the problem in terms of the quantities one is interested in, instead of passing explicitly through many-body wavefunctions. Before we come back to this, let us add a last idea.

Expansion around a mean-field hamiltonian: Møller–Plesset perturbation theory

Maybe the separation into a non-interacting part and the interaction is too simplistic. Since one knows how to include part of the interaction in a mean-field theory, why not start perturbation theory from there?

In Møller–Plesset (MP) perturbation theory [47, 306] the unperturbed part of the hamiltonian is the non-interacting $\hat{H}_0$ plus the Hartree–Fock mean field, which includes part of the Coulomb interaction. The remainder $\Delta\hat{V}$ is treated in Rayleigh–Schrödinger perturbation theory. The approach remains size-extensive. The unperturbed solution for the energy is the sum of occupied Hartree–Fock eigenvalues ε_i. The unperturbed wavefunction is a Slater determinant of Hartree–Fock one-electron orbitals. The first-order correction to the energy from Eq. (9.7) is $E^{(1)} = \langle \Phi_{ip} | \Delta \hat{V} | \Phi_{ip} \rangle$. This is a difference of 4-center matrix elements of the Coulomb interaction in the single-particle states (see Eqs. (4.8) and (4.9)),

$$E^{(1)} = - \sum_{i<j} \left[v_{ijji} - v_{ijij} \right],$$

(9.9)

so that $E_{ip} + E^{(1)}$ is just the Hartree–Fock total energy Eq. (4.2).

One has to go to second-order perturbation theory in order to find correlation: this is usually called MP2. If one takes the second-order contribution to Eq. (9.7),

$$E^{(2)} = \left\langle \Phi_{ip} \left| \Delta \hat{V} \frac{(1 - \hat{P})}{E_{ip} - \hat{H}_0} (E_{ip} - E + \Delta \hat{V}) \right| \Phi_{ip} \right\rangle,$$

(9.10)

and inserting complete sets of independent-particle eigenstates into the product of operators it becomes clear that only doubly excited determinants contribute: single excitations yield vanishing contributions following the Brillouin theorem, which states that the Hartree–Fock wavefunction doesn't mix with singly excited determinants.[3] Higher excitations cannot contribute to this order because $\Delta \hat{V}$ stems from a two-body interaction. This leads to

$$E^{(2)} = \sum_{i<j}^{occ} \sum_{a<b}^{empty} \frac{|v_{ijab} - v_{ijba}|^2}{\varepsilon_i + \varepsilon_j - \varepsilon_a - \varepsilon_b}.$$

(9.11)

The computer time for a straightforward evaluation of this expression scales as N^5 with the number of electrons. Recent work suggests that Monte Carlo evaluation of the integrals can make this more efficient [308]. MP2 contains important features, for example, van der Waals dispersion forces. When it is not sufficient, one can in principle go to higher orders; indeed, MP3 and MP4 are methods of use in quantum chemistry, sometimes with additional approximations like the neglect of triple excitations that occur in the fourth order. However, scaling gets worse, the computer programs become hard to write, and the results do not necessarily improve. In the framework of quantum chemistry other methods like coupled cluster [309, 310], that sum contributions to infinite order (see also footnote 7 on p. 229), become competitive. Those methods are not the topic of the present book. MP2 and higher orders can instead be considered as approximations in the diagrammatic perturbation framework, on which we concentrate in this and the following six chapters. Diagrams will be explained in Sec. 9.3; the diagrams in Fig. 10.2 represent the MP2 expression if the Green's function lines are interpreted as Hartree–Fock ones.

Lessons to be learnt

Do we really have to expand in the Coulomb interaction? This is the case if the only thing one is able to solve is the non-interacting problem. However, with the advent of bigger

---

[3] Sometimes the expressions are evaluated using other mean-field solutions, e.g., from a Kohn–Sham calculation. Then the Brillouin theorem no longer applies, leading to *singles corrections*, as used e.g. in [307].

computers there is more freedom. For example, it may be possible to solve the interacting problem in a small portion of some space, and to use perturbation theory with a different expansion parameter. In a Hubbard model this could be the hopping to neighboring or further distant sites. Also in this case a perturbation series can be derived and indeed, many findings of the present chapter apply. In particular, one may consider starting from a set of isolated atoms as an unperturbed problem and dealing with hybridization as perturbation. This connects to Chs. 16–20, and more specifically to Sec. 18.6. Still, the following chapters demonstrate that it is worthwhile to explore what one can do with perturbation theory in the Coulomb interaction. Indeed, the difficulties exposed in this section should not discourage exploration of this approach, since they give rise to some interesting hints.

- From the scattering theory expansion, we can conclude that it is most natural to head for an expression in terms of G_0 rather than, e.g., in terms of the non-interacting ground-state wavefunction.
- The problems of perturbation theory for energies and wavefunctions in the thermodynamic limit suggest concentrating directly on the observables themselves, e.g., the one-body Green's function, instead of the wavefunctions with which to calculate expectation values.
- As often in perturbation theory, the explicit calculation of more than first or second order becomes rapidly intractable. This can be seen when one tries to go beyond MP2.
- One can expect that an infinite number of terms has to be resummed in order to obtain meaningful corrections in the thermodynamic limit. This can be seen from the discussion of Eqs. (9.5)–(9.8).
- In the Green's function the interaction appears in two ways: in the many-body ground-state wavefunction for the zero-temperature case or the statistical weight at non-vanishing temperature, and in the time evolution of the field operators. This is a major complication, which we have to deal with in a clever way.

Let us start with the last item point, and try to get rid of the uncomfortable expectation value over an interacting wavefunction at zero temperature: this is the topic of the next section.

9.2 Connecting the interacting and non-interacting systems

The aim of the present section is to overcome a problem that is specific for zero temperature, $T = 0$: the explicit appearance of the interacting ground state in the expression of the Green's function. One might argue that this is just a pathology of zero temperature, since otherwise expectation values are given as a trace, Eq. (5.2), which allows one to use any basis of states, including the non-interacting one. However, then the statistical weight spoils the formulation in terms of a time-ordered expectation value. The topic of Sec. 9.6 is a way to include the statistical weight in a perturbation expression. To get the ideas sorted out, here we will instead start with the $T = 0$ case, and try to find expressions that do not contain the *interacting* many-body ground state.

Adiabatic switching on

One successful idea is to start with the non-interacting system in its ground state and examine what happens if the interaction is switched on. This has to happen very slowly, such that the system always remains in its ground state: we are not interested in non-equilibrium properties here. This adiabatic switching on of the interaction [311] is mathematically expressed using a time-dependent interaction

$$v_\eta(t) = \lim_{\eta \to 0^+} v_c e^{-\eta|t|}, \tag{9.12}$$

that vanishes at times $t = \pm\infty$ and is equal to the true Coulomb interaction v_c at $t = 0$. Here the expression also contains an adiabatic switching off; this symmetry in time will be useful later in the actual expansion of the Green's function.

If this is possible, then one can simply follow the time evolution of the system from $t_0 = -\infty$ in the past, where it was non-interacting, to the present, at $t = 0$, where it is the real interacting system. This is most conveniently done in the interaction picture described in App. B. With this aim we split the hamiltonian into a non-interacting part $\hat{H}_0$ and the interaction contribution $\hat{V}(t)$. The non-interacting $\hat{H}_0$ is static, but the interaction part $\hat{V}(t)$ is now time-dependent through Eq. (9.12). In the interaction picture, operators $\hat{O}$ and states $|\alpha >$ evolve according to Eq. (B.19),

$$\hat{O}_I(t) = e^{i\hat{H}_0(t-t_0)}\hat{O}_S(t)e^{-i\hat{H}_0(t-t_0)}, \tag{9.13}$$

with $\hat{O}_S$ the operator in the Schrödinger picture, and

$$|\alpha_I(t)\rangle = \hat{U}_I(t, t_0)|\alpha_I(t_0)\rangle. \tag{9.14}$$

The time-evolution operator $\hat{U}$ in the interaction picture is given by Eq. (B.30):

$$\hat{U}_I(t, t_0) = \mathrm{T}\left[e^{-i\int_{t_0}^t dt' \hat{V}_I(t')}\right]$$

$$= 1 + \sum_{n=1}^{\infty} \frac{1}{n!}(-i)^n \int_{t_0}^t dt_1 \dots \int_{t_0}^t dt_n \mathrm{T}\left[\hat{V}_I(t_1)\dots\hat{V}_I(t_n)\right], \tag{9.15}$$

where T denotes time ordering as defined in Eq. (B.28). This is a convenient expression, since it is sorted in orders of $\hat{V}$. This does of course not imply that one will always obtain a well-behaved converging series in $\hat{V}$, but as we will see, Eq. (9.15) is very useful to derive practical expressions.

Applied to our problem, for the moment Eq. (9.14) tells us just how to connect the non-interacting ground state $|\Phi_{ip}\rangle$ at $t_0 = -\infty$ to the state at time t that we call $|\Phi^{\eta-}(t)\rangle$. We also introduce the state $|\Phi^{\eta+}(t)\rangle$ obtained by propagating backwards from $t_0' = +\infty$ where the system is also supposed to be in the non-interacting ground state $|\Phi_{ip}\rangle$. The time-evolution operator in the presence of the interaction is denoted $\hat{U}^\eta$. We then have

$$\left\langle \Phi_H^{\eta+}\left|\hat{O}_H(t)\right|\Phi_H^{\eta-}\right\rangle = \left\langle \Phi_I^{\eta+}\left|\hat{O}_I(t)\right|\Phi_I^{\eta-}\right\rangle = \left\langle \Phi_{ip}\left|\hat{U}_I^\eta(\infty, t)\hat{O}_I(t)\hat{U}_I^\eta(t, -\infty)\right|\Phi_{ip}\right\rangle. \tag{9.16}$$

Note that we have chosen to use the state that connects to $+\infty$ for the bra, and the one that connects to $-\infty$ for the ket. This allows for the use of time ordering.

The limit of vanishing η should correspond to the original Coulomb interaction. This limit requires attention however. Take the case where the interaction is a simple real or complex number c_v that does not have to be treated as an operator. The eigenstates of the full hamiltonian are then equal to the non-interacting ones. Since $\hat{H}_0$ and c_v commute, $\hat{V}_I(t) = e^{i\hat{H}_0(t-t_0)}c_v e^{-\eta|t|}e^{-i\hat{H}_0(t-t_0)} = c_v e^{-\eta|t|}$, and each time integral in Eq. (9.15) yields a factor $\frac{1}{\eta}$. Hence, the time-evolution operator to time $t = 0$ becomes

$$U_I^\eta(0, -\infty) = 1 + \sum_{n=1}^{\infty} \frac{1}{n!} (-i)^n \left(\frac{-c_v}{\eta} \right)^n = e^{ic_v/\eta} : \tag{9.17}$$

in the limit $\eta \to 0$, the time evolution yields a divergent phase. The same contribution arises from $\hat{U}_I(\infty, 0)$. The two contributions do not cancel because in Eq. (9.16) the time always goes forward. To maintain the time ordering but get rid of the divergence, one has to introduce an appropriate normalization, namely

$$\frac{\left\langle \Phi_I^{\eta+} \middle| \hat{O}_I(t) \middle| \Phi_I^{\eta-} \right\rangle}{\left\langle \Phi_I^{\eta+} \middle| \Phi_I^{\eta-} \right\rangle} = \frac{\left\langle \Phi_{ip} \middle| \hat{U}_I^\eta(\infty, t)\hat{O}_I(t)\hat{U}_I^\eta(t, -\infty) \middle| \Phi_{ip} \right\rangle}{\left\langle \Phi_{ip} \middle| \hat{U}^\eta(\infty, -\infty) \middle| \Phi_{ip} \right\rangle}. \tag{9.18}$$

This does the job: the time-evolution operators in the denominator cancel the divergent phase of the numerator. It does not introduce new problems, since the denominator is also time-ordered, and if the forward and backward evolution yield the same state, it does not alter the result. In other words, whereas the limit $\eta \to 0$ taken separately for the numerator and denominator leads to divergencies, these problems cancel in the ratio. For non-degenerate states this holds also with the true interaction.

We can now use the expansion of the time-evolution operator Eq. (9.15). By regrouping the integrals in a suitable way, and with the definition

$$\hat{S}_\eta \equiv \hat{U}_I^\eta(\infty, -\infty), \tag{9.19}$$

this leads to

$$\lim_{\eta \to 0} \left\langle \Phi_H^{\eta+} \middle| \hat{O}_H(t) \middle| \Phi_H^{\eta-} \right\rangle = \lim_{\eta \to 0} \sum_{p=0}^{\infty} (-i)^p \frac{1}{p!} \int_{-\infty}^{\infty} dt_1 \cdots \int_{-\infty}^{\infty} dt_p$$

$$\times e^{-\eta(|t_1|+\cdots+|t_p|)} \frac{\left\langle \Phi_{ip} \middle| T\left[\hat{V}_I(t_1) \cdots \hat{V}_I(t_p)\hat{O}_I(t) \right] \middle| \Phi_{ip} \right\rangle}{\left\langle \Phi_{ip} \middle| \hat{S}_\eta \middle| \Phi_{ip} \right\rangle}. \tag{9.20}$$

Note here that $\hat{V}_I$ is the original Coulomb interaction operator, and the adiabatic switching is shown explicitly by the prefactors $e^{-\eta|t|}$. The fundamental hypothesis of the adiabatic switching on is that in the limit of vanishing η the states $|\Phi^{\eta\pm}\rangle$ become the interacting ground state $|\Phi_0\rangle$. Then one has with $\hat{S} \equiv \lim_{\eta \to 0} \hat{S}_\eta$

$$\left\langle \Phi_0 \middle| \hat{O}_H(t) \middle| \Phi_0 \right\rangle = \frac{\left\langle \Phi_{ip} \middle| T \left[\hat{O}_I(t)\hat{S} \right] \middle| \Phi_{ip} \right\rangle}{\left\langle \Phi_{ip} \middle| \hat{S} \middle| \Phi_{ip} \right\rangle}. \tag{9.21}$$

This also holds when $\hat{O}_I$ is a time-ordered product of operators. Therefore, the Green's function Eq. (5.78) at vanishing temperature becomes[4]

$$G(x, t; x', t') = -i \frac{\left\langle \Phi_{ip} \middle| T \left[\psi(x, t)\psi^\dagger(x', t')\hat{S} \right] \middle| \Phi_{ip} \right\rangle}{\left\langle \Phi_{ip} \middle| \hat{S} \middle| \Phi_{ip} \right\rangle}. \tag{9.22}$$

This formal expression is the starting point for the diagrammatic expansion of the Green's function.

How safe is this procedure? The underlying hypothesis is *continuity* between the non-interacting and the interacting system. Equation (9.18) is useful only if the fictitious time evolution really connects the non-interacting ground state to the ground state of the interacting system. Such a connection is known as the Gell-Mann–Low theorem [311], but with an important limitation: the Gell-Mann–Low theorem only states that the non-interacting ground state evolves into *some eigenstate* of the interacting system, which is not necessarily the ground state – there can be level crossing. However, in many cases we do in fact obtain the ground state, so we will use Eq. (9.22) in the following. This is also a good starting point to understand things in a simple way before extending the approach to non-vanishing temperature where the problem disappears.[5]

9.3 Telling the story of particles: diagrams

The expression in Eq. (9.22) is what we wanted. Moreover, Eq. (9.19) with Eq. (9.15) describes the expansion as a series in powers of the interaction.

Expansion of the Green's function

With the Coulomb interaction v_c as perturbation, the corresponding contribution in the interaction picture reads

$$\hat{V}_I(t) = \frac{1}{2} \int \int dx \, dx' \, \hat{\psi}_I^\dagger(x, t)\hat{\psi}_I^\dagger(x', t)v_c(\mathbf{r}, \mathbf{r}')\hat{\psi}_I(x', t)\hat{\psi}_I(x, t), \tag{9.23}$$

where the field operators in the interaction picture are $\hat{\psi}_I(x, t) = e^{i\hat{H}_0(t-t_0)}\hat{\psi}(x)e^{-i\hat{H}_0(t-t_0)}$. With the compact notation $(x_1, t_1) \rightarrow 1$ and the Coulomb interaction[6] $v_c(1, 1') =$

---

[4] In the following, we omit the hat on the field operators to make the notation less clumsy.
[5] One can work at non-vanishing temperature and take the limit $T \rightarrow 0$. When the result is different from that for $T = 0$, the former is the correct solution; see Sec. 9.6.
[6] Throughout the book, we use the definition $t_1^+ = \lim_{\eta \to 0^+} t_1 + \eta$.

$\delta(t_{1'} - t_1^+)v_c(\mathbf{r}_1 - \mathbf{r}_{1'})$, and using the definition of Eq. (9.19) and the expansion of Eq. (9.15), the numerator of Eq. (9.22) becomes

$$-i\left\langle \Phi_{ip}\left|T\left[\hat{\psi}(x,t)\hat{\psi}^\dagger(x',t')\hat{S}\right]\right|\Phi_{ip}\right\rangle = G^0(x,t,x',t')$$

$$+\sum_{p=1}^{\infty}\left(\frac{i}{2}\right)^p\frac{1}{p!}\int\int d1d1'\ldots\int\int dpdp'\ v_c(1,1')\ldots v_c(p,p')$$

$$\times G^0_{2p+1}(x,t,1,1',\ldots,p,p';x',t',1,1',\ldots,p,p'), \tag{9.24}$$

with the non-interacting one-body Green's function

$$G^0(x,t,x',t') = \left\langle\Phi_{ip}\left|T\left[\hat{\psi}_I(x,t)\hat{\psi}_I^\dagger(x',t')\right]\right|\Phi_{ip}\right\rangle \tag{9.25}$$

and the higher-order non-interacting Green's functions defined as

$$G^0_s(1,\ldots,s;1',\ldots,s') = (-i)^s\left\langle\Phi_{ip}\left|T\left[\hat{\psi}_I(1)\hat{\psi}_I(2)\ldots\hat{\psi}_I(s)\hat{\psi}_I^\dagger(s')\ldots\hat{\psi}_I^\dagger(2')\hat{\psi}_I^\dagger(1')\right]\right|\Phi_{ip}\right\rangle, \tag{9.26}$$

which is consistent with the definition of the interacting counterpart Eq. (10.8) and the special case Eq. (5.133) for $s = 2$.

The zeroth order of Eq. (9.24) in v_c is the non-interacting Green's function. To first order, there is a term G^0_3 containing six (three creation and three annihilation) field operators: it describes the propagation of three particles. Going to higher orders, each v_c adds another two particles. What are these higher-order non-interacting Green's functions? In principle, one can work them out one by one, using the fact that $|\Phi_{ip}\rangle$ is a single Slater determinant,[7] and expanding the field operators in the corresponding single-particle basis. For example, as shown in Ex. 9.2, the result for G^0_2 is

$$G^0_2(1,2,1',2') = G^0(1,1')G^0(2,2') - G^0(1,2')G^0(2,1') = \begin{vmatrix} G^0(1,1') & G^0(1,2') \\ G^0(2,1') & G^0(2,2') \end{vmatrix}. \tag{9.27}$$

This is a determinant, the last expression in Eq. (9.27). It corresponds to the RPA for G_2, which leads for example to the RPA form of the irreducible polarizability $P = -iGG$ in Eq. (11.16).[8] With the compact notation $x,t \to c$ and $x',t' \to c'$, the first contribution that is useful here, namely G^0_3, reads[9]

$$G^0_3(c,1,1';c',1^+,1'^+)$$
$$= G^0(c,c')G^0(1,1^+)G^0(1',1'^+) - G^0(c,c')G^0(1,1')G^0(1',1)$$
$$- G^0(c,1)G^0(1,c')G^0(1',1'^+) + G^0(c,1')G^0(1,c')G^0(1',1)$$
$$- G^0(c,1')G^0(1,1^+)G^0(1',c') + G^0(c,1)G^0(1,1')G^0(1',c'). \tag{9.28}$$

---

[7] Note that the requirement of antisymmetry of the wavefunction is taken into account even for the non-interacting Green's function.

[8] In Ch. 10 it is shown that when G^0_2 from Eq. (9.27) is used to build the self-energy in Eq. (10.5), the Hartree–Fock approximation is obtained.

[9] Note that the infinitesimal time differences are needed only to define the Green's functions when equal arguments appear, in order to ensure the correct order of field operators.

This can again be written as a determinant,

$$G_3^0(c, 1, 1'; c', 1^+, 1'^+) = \begin{vmatrix} G^0(c, c') & G^0(c, 1) & G^0(c, 1') \\ G^0(1, c') & G^0(1, 1^+) & G^0(1, 1') \\ G^0(1', c') & G^0(1', 1) & G^0(1', 1'^+) \end{vmatrix}. \tag{9.29}$$

It is extremely tedious to derive this result just from scratch, and you would certainly not want to go for higher orders in the same way. In the next section Wick's theorem helps us to do this more efficiently. For now, let us see what we have, and how we can work with it.

Every non-interacting m-body Green's function is composed of all possible combinations of propagating single particles. Even without knowing the final result, one can guess that with increasing order m these contributions will be more and more numerous. How can we keep track of this, and at the same time give the result some physical meaning? One possible answer is to use *diagrams*, as will be explained in the next subsection.

Drawing the story: diagrams

Diagrams may be considered to be the comic strip of many-body perturbation theory. They are a compact, graphical notation for the numerous terms and integrations. There exist different possibilities, for example a representation in real or in momentum space, and the way time is dealt with. Here we use a representation where we treat space, spin, and time on the same footing, consistent with the compact notation for the arguments of the Green's functions adopted above. The link between the explicit formulas and the graphical notation is defined by what are often called "Feynman rules." These rules, presented in the following and summarized in the last subsection of Sec. 9.4 and in Tab. 9.1, allow one to switch quickly between the integrals and their graphical representation. A more detailed introduction can be found in [312].

The zeroth-order contribution to Eq. (9.24) is the non-interacting one-body Green's function $G^0(x, t; x', t')$. The corresponding propagation between space–time–spin points $c' = (x', t')$ and $c = (x, t)$ is represented by an arrow from c' to c, as shown in Fig. 9.1. Here we adopt the convention that the arrow points from the second to the first argument of the Green's function, consistent with the insertion of a particle and subsequent propagation. It is important to remain consistent with this convention. Since space, spin, and time are combined in one coordinate, in this book we do not orient the diagrams in a particular way with respect to time.

To first order in the interaction, Eq. (9.24) contains the six contributions of Eq. (9.28). Now, besides $G^0(x, t; x', t')$, the Coulomb interaction v_c has to be depicted. We represent it by a dashed line. The six $p = 1$ contributions to Eq. (9.24) are shown in Fig. 9.2, in the same order as in Eq. (9.28). The equation contains integrations: the *vertices* where this happens are indicated by dots. Some Green's functions that appear in the first column have two arguments that are equal. Consistently with Eq. (9.28), it is to be understood that $G^0(c, c)$ corresponds to $G^0(x, t, x, t^+)$. This is proportional to the density $n(x, t)$; note that the direction of the arrow is meaningless in that case. In the first diagram of the second

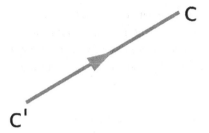

Figure 9.1. Graphical representation of $G^0(x, t, x', t')$: a Feynman diagram. Space, spin, and time are combined in one coordinate, $x, t \to c$.

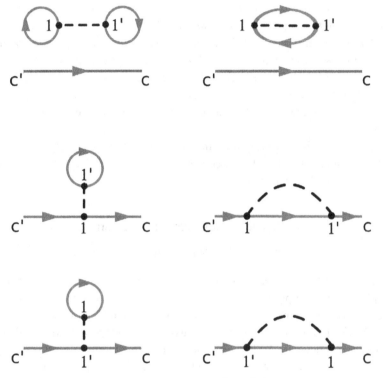

Figure 9.2. Propagation of an interacting electron from $c' = (x', t')$ to $c = (x, t)$: first-order processes contributing to the numerator of Eq. (9.22). The diagrams in the first row are *disconnected*, the others are *connected* diagrams. Continuous line: non-interacting Green's function. Closed circle: non-interacting density. Dashed line: bare Coulomb interaction v_c. Dots: internal points called *vertices*; these are integrated over. To obtain the final results, one has to use the Feynman rules specified in the text and summarized in the last subsection of Sec. 9.4.

column, two Green's functions form a closed loop, with an insertion v_c. Since the Coulomb interaction is instantaneous, an equal time limit has to be taken. This is why above the Coulomb interaction v_c is defined with a $\delta(t'_1 - t_1^+)$: it guarantees that the two Green's functions have opposite time ordering and form an electron–hole pair, consistent with the expansion. Finally, we have to take care of the prefactors of Eq. (9.28). To first order there is a factor i that we assign to the Coulomb interaction, since in the general case we have i^p for the pth order in v_c. Three of the terms have a prefactor -1; in Fig. 9.2 these are the first term in the second column and the second and third terms in the first column. The common feature is that they contain one closed loop composed of one or two Green's functions.

The second and third contributions in the first column are equivalent because they differ only in the order of indices that are integrated over. The same holds for the second and third contributions in the second column. In other words, of the six contributions, only four are *topologically distinct* and have to be calculated. The two contributions in the first line are *disconnected*, because they can be separated into pieces without cutting a line. For the others this is not possible: they are *connected*. These diagrams illustrate the physical meaning that is hidden in the formula. In the connected diagrams, a particle is traveling and at some point scattering when the total charge density (the loop) occurs, or there is an exchange event (the graphs on the right side of the figure). In the disconnected diagrams, a particle is traveling independently from fluctuations that occur in the system: these are called *vacuum fluctuations*. As will be shown in the next section, these drop out in the final expression.

Now we have in principle everything in hand to go on and calculate higher orders as well as the denominator of Eq. (9.22). However, as anticipated, let us first introduce two theorems that make life much easier.

9.4 Making the story easier: two theorems

Wick's theorem

Wick's theorem [313] is a generalization to fermions and time-ordered products of an earlier idea [314] to handle products of creation and annihilation operators by carrying all annihilation operators to the right. A product where all annihilation operators are to the right is called in *normal order*; its vacuum expectation value is zero. We denote normal order with the operator $\mathcal{N}$. The vacuum $|0\rangle$ is defined with respect to the operators $\hat{b}_i$ that are considered, i.e., $\hat{b}_i|0\rangle = 0$ for all i. In the process of ordering the operators one has to keep track of the commutations, hence for fermions, e.g., $\mathcal{N}(\hat{c}_i^\dagger \hat{c}_j) = \hat{c}_i^\dagger \hat{c}_j$ and $\mathcal{N}(\hat{c}_j \hat{c}_i^\dagger) = -\hat{c}_i^\dagger \hat{c}_j$. The second concept that is used is *contraction* of two operators $\hat{A}$ and $\hat{B}$, defined as

$$\hat{A}\hat{B} = \hat{A}\hat{B} - \mathcal{N}(\hat{A}\hat{B}). \tag{9.30}$$

This is either zero or a commutator. If the commutator is simply a number instead of an operator, we have

$$\hat{A}\hat{B} = \langle 0|\hat{A}\hat{B}|0\rangle = \langle 0|\hat{A}\hat{B} - \mathcal{N}(\hat{A}\hat{B})|0\rangle = \langle 0|\hat{A}\hat{B}|0\rangle, \tag{9.31}$$

since the expectation value of the normal order vanishes by definition. For products of more operators, one finds

$$\hat{A}\hat{B}\hat{C}\ldots\hat{X}\hat{Y}\hat{Z} = \mathcal{N}(\hat{A}\hat{B}\hat{C}\ldots\hat{X}\hat{Y}\hat{Z})$$

$$+ \mathcal{N}(\dot{A}\dot{B}\hat{C}\ldots\hat{X}\hat{Y}\hat{Z}) + N(\dot{A}\hat{B}\dot{C}\ldots\hat{X}\hat{Y}\hat{Z}) + \cdots \text{ (one pair, contracted)}$$

$$+ \mathcal{N}(\dot{A}\ddot{B}\hat{C}\dot{D}\ldots\hat{X}\hat{Y}\hat{Z}) + \cdots \text{ (two pairs, all combinations)}$$

$$+ \cdots \text{(three pairs, all combinations)}$$

$$\cdots, \tag{9.32}$$

where two operators with the same number of dots are always contracted, and for fermions the permutations to bring them together yield a sign. Finally, Eq. (9.31) can be used to replace all the contractions by an expectation value.

In Wick's theorem this is extended to time-ordered products, which can be used here to get the exact expression for G^0_{2n+1} written as a determinant,[10]

$$G^0_{2n+1}(c, 1, 1', 2, 2', \ldots, n, n'; c', 1, 1', 2, 2', \ldots, n, n')$$
$$= \sum_{\mathcal{P}} (-1)^P G^0(c, \tilde{c}') \ldots G^0(n, \tilde{n}) G^0(n', \tilde{n}'). \tag{9.33}$$

The set $\left\{\tilde{c}', \tilde{1}, \tilde{1}', \ldots, \tilde{n}'\right\}$ is a permutation $\mathcal{P}$ of $\{c', 1, 1', \ldots, n'\}$, and P is the number of permutations of two indices between the two sets. Odd permutations lead to a minus sign. When a given permutation keeps an index $\tilde{m} = m$, this leads to a term $G^0(m, m)$. As in the previous section, in the general case considered here such a diagonal element has to be interpreted as $G^0(m, m^+)$, which is a factor i times the non-interacting density. The number of permutations P is odd when an odd number of closed loops appear, where a closed loop is either a Green's function closed in itself, i.e., a density, or a loop of the type $G^0(m, m')G^0(m', m)$. This loop represents an electron and a hole. It looks like a bubble when one draws the diagrams, as one can see in the upper-right diagram of Fig. 9.2. One therefore speaks about *bubble diagrams*.

Equation (9.33) confirms the special case of Eq. (9.28), and shows the general result is a determinant. The result is intuitive: one would indeed expect that non-interacting particles propagate independently, as expressed by the products. One would also expect that for fermions a factor $(-1)^P$ appears due to exchange. This expression is obviously extremely helpful in evaluating higher orders of Eq. (9.24). We can also use it to write down the denominator of Eq. (9.22), which becomes

$$\left\langle \Phi_{\text{ip}} \left| \hat{S} \right| \Phi_{\text{ip}} \right\rangle = 1$$

$$+ \sum_{p=1}^{\infty} \left(\frac{i}{2}\right)^P \frac{1}{p!} \int\int d1 d1' \ldots \int\int dp dp' \, v_c(1, 1') \ldots v_c(p, p')$$

$$\times G^0_{2p}(1, 1', \ldots, p, p'; 1, 1', \ldots, p, p') \tag{9.34}$$

---

[10] The G^0 are expectation values calculated with single Slater determinants. This is the vacuum for $\hat{c}_\ell$ where the state ℓ is not occupied, and for $\hat{c}^{hole}_\ell \equiv \hat{c}^\dagger_\ell$ for the states ℓ that are contained in the Slater determinants.

with

$$G_{2p}^0(1,1',\ldots,p,p';1,1',\ldots,p,p') = \sum_P (-1)^P G^0(1,\tilde{1})G^0(1',\tilde{1'})\ldots G^0(p,\tilde{p})G^0(p',\tilde{p'}).$$

(9.35)

Interestingly, the first-order contribution is just the two vacuum diagrams in the first-order contribution to the numerator, the first row of Fig. 9.2.

Linked cluster theorem

Wick's theorem helps to decompose the expressions in terms of processes involving only the one-body non-interacting G^0, and the interaction. This leads to a plethora of diagrams, and one might ask whether all of them are important. In particular, we have just noticed, thanks to Eq. (9.35), that to first order the denominator is a sum of vacuum fluctuations that also appear in the numerator. In the following we will see that this leads to cancellations in all orders, which is a realization of the *linked cluster theorem* [47, 315] that is important in many problems concerning statistical physics.

To understand how these cancellations come about, it is useful to make the simple Fig. 9.3 on which one may draw lines with a pencil to fix the ideas. The figure symbolizes the structure of a contribution to Eq. (9.22) decomposed using Wick's theorem Eq. (9.33). On the left side are the points that appear in a third-order term of the numerator, with the connections by the Coulomb interaction shown, but without specifying the Green's function lines. Seven Green's function lines have to be placed such that one line starts at each of the points $c', 1, 1', 2, 2', 3, 3'$ and one line ends at each of $c, 1, 1', 2, 2', 3, 3'$. The points c and c' are different in that they are not reached by a Coulomb line. On the right side one has the analogous figure for the denominator. It has the same points, except for the fact that c and c' are absent. One can imagine extending the series to higher orders by having more pairs of points n, n' connected by interactions on both sides.

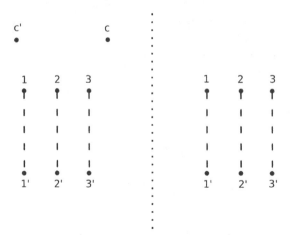

Figure 9.3. Graphical support to understand the linked cluster theorem.

If, on the numerator side, one connects points c and c' by a Green's function line, this contribution is necessarily disconnected from the remaining diagrams. These remaining diagrams cancel with the denominator, since both have the same structure. This yields the zero-order contribution to $G(c, c')$, namely $G^0(c, c')$. Suppose we now draw lines such that c and c' connect to some other point. One can then identify a connected complex containing c and c', and there can be a part that is disconnected from it. For example, one may have c connected to 1 and c' connected to $1'$, a connection between 1 and $1'$, with no connection between $1, 1'$ and the rest. That rest constitutes a subset of points and Coulomb lines that is increasingly large with increasing order, if we fix the connected part. Since all these points are integrated over, and since orders are summed to infinity, for each connected complex that we can identify in the numerator, the remaining disconnected part is just equal to the denominator, so that it cancels.

The combinatorial prefactors are taken care of in the following way: at a given order p, there are $p!/q!(p - q)!$ possibilities to have a connected complex containing c, c', and $(p - q)$ Coulomb lines, hence a disconnected rest of order q. The factor $p!$ cancels with the prefactor $1/p!$ in Eq. (9.24). The two vertices of each Coulomb line are integrated over. Therefore, by exchanging their label one obtains a topologically equivalent diagram. For $(p-q)$ Coulomb lines this yields $(p-q)!$ possibilities to have a topologically equivalent diagram for the connected complex, hence this factor also cancels if we limit the diagrams to the topologically distinct ones. These are for example the ones in the last row of Eq. (9.28) and Fig. 9.2. The powers of $1/2$ in Eq. (9.24) disappear if we consider that n and n' are equivalent in the integration, since the Coulomb interaction is symmetric. Finally only $1/q!$ remains, which is what we need to cancel the corresponding order in the denominator.

We can conclude that in Eq. (9.22) *the denominator cancels, and only the topologically distinct connected terms of the numerator have to be retained.*

Application to the calculation of the Green's function

We now have everything in hand to write down the rules for constructing the perturbation series of the Green's function and to draw the corresponding Feynman diagrams. There is a contribution to each order, the zeroth-order one being simply G^0. To order p, we have to:

- Construct diagrams by adding one Green's function line to each end of one interaction line (dashed line). We need p such building blocks.
- Each end point of an interaction line is a vertex; we give it a dot.
- Combine the resulting building blocks consisting of two Green's functions and one interaction in such a way that every vertex is saturated by two Green's function lines and the whole diagram is closed, except for one Green's function line that constitutes a single leg, and one interaction vertex that connects only to one Green's function. Here a Green's function line closed on itself, i.e., the density, counts as two lines. The direction of two joining Green's function lines must be continuous.
- Add a G^0 to saturate the last vertex; this creates a second single leg. In this way, one obtains a diagram with two external points c and c'.

- Draw all connected diagrams that are topologically distinct, which have p interaction lines and two external points. With the rule above, they always have $2p + 1$ lines that represent a G^0.
- Translate the diagrams into an equation.
  - Label each vertex as a space–spin–time point $m = (\mathbf{r}_m, \sigma_m, t_m)$.
  - An arrow from m' to m is $G^0(m, m')$.
  - A dashed line is $iv_c(m, m') = iv_c(\mathbf{r}_m - \mathbf{r}_{m'})\delta(t_{m'} - t_m)$.
  - Vertices are integrated over.
  - Every closed loop gives an extra factor -1.

To first order, we obtain the two contributions given by the last row of Fig. 9.2. The first one is a modification of the propagation due to the Hartree potential, $-iG^0 v_c G^0 G^0 = G^0 v_H G^0$, the second one is the corresponding exchange term, with opposite sign.

To second or higher order, we can have combinations of several of those events; an example is given by the two subsequent events on the left side of Fig. 9.4. We also have a new kind of diagram appearing, shown on the right side of the same figure. It contains the creation of an electron–hole pair (the bubble), in other words, the propagating particle polarizes the system. This is a diagram that plays a central role in Chs. 10–15.

All connected and topologically distinct diagrams contributing to the Green's function up to second order in the interaction are shown in Fig. 9.5. Table 9.1 summarizes the rules on how to convert diagrams into formulas. They are applied in Ex. 9.3. The final compact complete formula for the Green's function reads

$$
G(c, c') = \sum_{p=0}^{\infty} i^p \int v_c(1, 1') \ldots v_c(p, p')
\begin{vmatrix}
G^0(c, c') & G^0(c, 1) & \cdots & G^0(c, p') \\
G^0(1, c') & G^0(1, 1^+) & \cdots & G^0(1, p') \\
\vdots & \vdots & \ddots & \vdots \\
G^0(p', c') & G^0(p', 1) & \cdots & G^0(p', p'^+)
\end{vmatrix}_{cd}.
\tag{9.36}
$$

As indicated by the subscript cd, only connected and topologically distinct diagrams have to be retained.

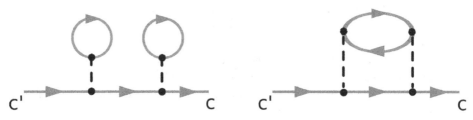

Figure 9.4. Two of the second-order contributions to G: two subsequent interactions with the density, and a polarization diagram.

Table 9.1. Conversion of diagrams into equations. Each closed loop carries a factor (-1). In the last column the expressions are given in the space of momentum and frequency, which are carried by the Green's functions and internally integrated as $\frac{1}{\beta} \sum_{\omega_n \sigma} \int d\mathbf{k}$. It is understood that $G_\mathbf{k}^0$ is scalar in a homogeneous system, and a matrix in an inhomogeneous system.

	$T = 0$	$T \neq 0$ (space, imaginary time)	$T \neq 0$ (reciprocal space, Matsubara frequency)
n	$\mathbf{r}_n, \sigma_n, t_n$	$\mathbf{r}_n, \sigma_n, \tau_n$	
$n \bullet$	$\int d x_n \int_{-\infty}^\infty d t_n$	$\int d x_n \int_0^\beta d \tau_n$	
$n \quad \overset{\longleftarrow}{} \quad m$	$G^0(n, m)$	$G^0(n, m)$	$G_{\mathbf{k}\sigma}^0(\omega_n)$
$n \quad \overset{\blacktriangleleft\!\!-}{} \quad m$	$G(n, m)$	$G(n, m)$	$G_{\mathbf{k}\sigma}(\omega_n)$
$\underset{m}{\bigcirc} \quad \underset{m}{\bigcirc}$	$n^0(\mathbf{r}_m)$ and $n(\mathbf{r}_m)$	$n^0(\mathbf{r}_m)$ and $n(\mathbf{r}_m)$	n^0 and n
$n \;\text{-----}\; m$	$i v_c(\mathbf{r}_1, \mathbf{r}_2)\delta(t_1^+ - t_2)$	$-v_c(\mathbf{r}_1, \mathbf{r}_2)\delta(\tau_1^+ - \tau_2)$	$-v_c(\mathbf{k})$

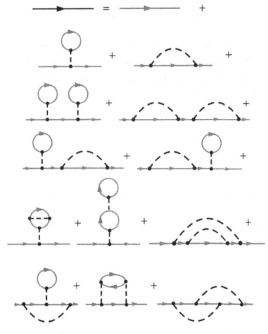

Figure 9.5. All diagrams contributing to the interacting Green's function up to second order in the Coulomb interaction.

9.5 Dyson equation for the one-particle Green's function, and the self-energy

We are now ready to cast our results in a form similar to Eqs. (9.3) and (9.4), but for the one-body Green's function. Starting from first order, each of the diagrams shown above has an ingoing and an outgoing non-interacting Green's function line. Therefore, as depicted schematically in Fig. 9.6, the final expression for the interacting Green's function can be written as the matrix equation

$$G = G^0 + G^0 \Sigma_{\text{red}} G^0, \tag{9.37}$$

where we do not show the matrix indices. Σ_{red} is the *reducible*, or improper, self-energy. From its definition, it is the sum of all Feynman diagrams with no external lines. It is called reducible because it contains lower-order patterns that are simply linked by one Green's function line. For example, the first diagram in Fig. 9.4 can be cut in the middle, and one is left with two, in this example identical, first-order diagrams. We now introduce the *proper*, or irreducible, self-energy Σ, also defined in Sec. 7.2. It is the sum of only those self-energy diagrams that cannot be separated into pieces by cutting a single Green's function line. Turned around, the improper self-energy is obtained by concatenating the proper contributions through G^0-lines. This reads

$$\Sigma_{\text{red}} = \Sigma + \Sigma G^0 \Sigma + \Sigma G^0 \Sigma G^0 \Sigma + \cdots = \Sigma + \Sigma G^0 \Sigma_{\text{red}}. \tag{9.38}$$

Combining Eqs. (9.37) and (9.38), one obtains the Dyson Eq. (7.10):

$$G = G^0 + G^0 \Sigma [G^0] G, \tag{9.39}$$

which is graphically represented in Fig. 9.7. Note that at this point the self-energy is given as a functional of G^0. The numerical result of the diagrammatic series depends of course on the system through G^0, but the functional form, i.e., the dependence on G^0 and v_c, is universal. In Ch. 8 the self-energy is instead presented as a functional of the fully interacting G; we will discuss this in Sec. 9.8.

Figure 9.6. One-particle Green's function and the reducible self-energy.

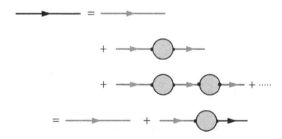

Figure 9.7. Dyson equation for the one-particle Green's function.

Figure 9.8. First-order contributions to the proper self-energy in terms of the bare Green's function G^0 and the bare interaction v_c.

The diagrammatic representation nicely illustrates how a low-order approximation to the proper self-energy creates contributions to G of all orders. For example, Fig. 9.8 shows the first-order contributions to the proper self-energy. These are just the Hartree potential, on the left, and the Fock exchange, on the right (also shown in Fig 10.1), evaluated with the non-interacting G^0. Inserted into the Dyson equation, the resulting Green's function contains an infinite number of interaction lines.

9.6 Diagrammatic expansion at non-vanishing temperature

What happens to this derivation at non-vanishing temperature? It is important to make sure that the main results are general, and not limited to $T = 0$; in a physical system in equilibrium in the thermodynamic limit there is always a temperature different from zero, and actually, as we will discuss briefly at the end of this section, one has to be careful when working at zero temperature.[11] In the following, we give a brief answer to the above question; details can be found in many textbooks such as [233], and much of what we have seen for zero temperature also applies here. There is, however, one seeming complication: the weight factor $e^{-\beta(\hat{H}-\mu\hat{N})}$ that appears in the definition Eq. (5.2) of correlation functions at non-vanishing temperature prevents one from following the same derivation as at $T = 0$. Otherwise, things are in a way much simpler, because with Eq. (5.2) the Green's function is defined with a trace, which means that one can perform the calculations over a set of *non-interacting* states: there is no need anymore for the adiabatic switching on. In contrast, the problem of the weight factor can be overcome with a simple trick, which is to interpret β as an imaginary-time. In order to use the idea of time ordering while still having the original order of operators, one can imagine being on a contour where one end is given by the piece on the imaginary-time axis. This is shown in Fig. E.3. For equilibrium problems, the ones that we consider in this book, a further possible step is to move completely to the imaginary-time axis. As explained in App. D, this is obtained by defining a new Green's function on the imaginary-time axis, replacing the time arguments t in the definition of the real-time Green's functions with $-i\tau$ and considering ordering in τ. Moreover, energies are defined with respect to the chemical potential μ. With this trick, the exponentials involving time and the exponentials involving temperature can be treated on the same footing, and a strategy similar to that for $T = 0$ can be applied. The imaginary-time Green's function is convenient to calculate integrals such as the total energy, since

---

[11] This can be considered to be a general rule: whenever one has several limits, like the thermodynamic limit and the limit $T \to 0$ in the present case, one should check whether the order of limits could cause problems.

it avoids the poles on the real frequency axis. To get real spectra is more difficult (see Sec. D.5).

The task is to expand Eq. (D.7), the imaginary-time temperature Green's function

$$G(x, x'; \tau - \tau') = \frac{\text{Tr}\{e^{-\beta(\hat{H} - \mu\hat{N})} \text{T}_\tau \left[\hat{\psi}(x, \tau)\hat{\psi}^\dagger(x', \tau') \right]\}}{\text{Tr}\{e^{-\beta(\hat{H} - \mu\hat{N})}\}}. \tag{9.40}$$

For a static external potential, $G(x, x'; \tau - \tau')$ is a periodic function, with a period equal to 2β. As given in Eqs. (D.8) and (D.9), we have for $0 < \tau - \tau' \le \beta$,

$$G(x, x'; \tau - \tau') = -G(x, x'; \tau - \tau' - \beta) \tag{9.41}$$

and for $-\beta \le \tau - \tau' < 0$,

$$G(x, x'; \tau - \tau') = -G(x, x'; \tau - \tau' + \beta). \tag{9.42}$$

As in the case of $T = 0$, we now separate the hamiltonian into a non-interacting part $\hat{H}_0$ and the interaction part that we treat approximately. The strict analogy of working with real time at $T = 0$ or imaginary time and temperature is shown in Sec. B.2. An important relation is

$$e^{-\beta(\hat{H} - \mu\hat{N})} = e^{-\beta(\hat{H}_0 - \mu\hat{N})} \hat{U}_I(\beta, 0), \tag{9.43}$$

where the imaginary-time time-evolution operator in the interaction picture is

$$\hat{U}_I(\tau, \tau') = e^{(\hat{H}_0 - \mu\hat{N})\tau} e^{-(\hat{H} - \mu\hat{N})(\tau - \tau')} e^{-(\hat{H}_0 - \mu\hat{N})\tau'}. \tag{9.44}$$

Proceeding as in the case of real time and zero temperature in App. B, one can show that $\hat{U}_I$ satisfies

$$\frac{\partial}{\partial \tau} \hat{U}_I(\tau, \tau') = -\hat{V}_I(\tau)\hat{U}_I(\tau, \tau'), \tag{9.45}$$

with the solution

$$\hat{U}_I(\tau, \tau') = 1 + \sum_{n=1}^{\infty} (-1)^n \frac{1}{n!} \int_{\tau'}^{\tau} d\tau_1 \dots \int_{\tau'}^{\tau} d\tau_n \text{T}_\tau \left[\hat{V}_I(\tau_1) \dots \hat{V}_I(\tau_n) \right], \tag{9.46}$$

in close analogy to Eq. (9.15). This leads to the final expression for the Green's function,

$$G(x, x'; \tau - \tau') = -\frac{\left\langle \text{T}_\tau \left[\hat{\psi}(x, \tau)\hat{\psi}^\dagger(x', \tau')\hat{U}_I(\beta, 0) \right] \right\rangle_0}{\left\langle \hat{U}_I(\beta, 0) \right\rangle_0}. \tag{9.47}$$

Here $\langle \dots \rangle_0$ denotes the zero-order ensemble average obtained using the *unperturbed* density matrix $\hat{\rho}^0 = e^{-\beta(\hat{H}_0 - \mu\hat{N})}$. This is similar to the zero-temperature expression of Eq. (9.22). Note that the denominator is naturally introduced through the partition function $\text{Tr}\{e^{-\beta(\hat{H} - \mu\hat{N})}\}$ that appears in $T \ne 0$ averages such as the Green's function Eq. (9.40). An equivalent to Wick's theorem can be obtained by working with zero-order ensemble averages instead of ground-state expectation values. Again this leads to the diagrammatic perturbation expansion; for the nth-order contribution, for which one has to draw all topo-logically distinct connected diagrams with two external points, n interaction lines, and $(2n+1)$ directed lines, one simply has to assign different labels, integrations, and prefactors. These rules are summarized in Tab. 9.1.

There is one fundamental difference between $T = 0$ and $T \neq 0$: at non-vanishing tem-perature, independent-particle states in G^0 are occupied with a certain probability, so that one cannot assign states as *particle* or *hole* states. Therefore, one can have diagrams that would otherwise be absent. Take the example of a homogeneous system, where we evaluate the diagrams in reciprocal space, following the last column of Tab. 9.1. The independent-particle state corresponding to a given momentum $\mathbf{k}$ is either above or below the Fermi momentum $\mathbf{k}_F$, and the independent-particle Green's function $G^0_{\mathbf{k}}$ at $T = 0$ is therefore either going forward or backward in time, as one can see from the time-ordered expres-sion of Eq. (5.83). This means that a bubble diagram must be composed of two different momenta, one above and one below $\mathbf{k}_F$. At non-vanishing temperature $G^0_{\mathbf{k}}$ has in general both contributions, and one can form a bubble that has no momentum transfer. Diagrams that only exist at non-vanishing temperature are called *anomalous diagrams* [316]. One would expect that for $T \to 0$ these diagrams vanish and the zero-temperature result is recovered; however, this is not always true for a system in the thermodynamic limit. In the case of discrepancy, the *correct* result is the one obtained by performing the limit of vanishing temperature, and not the one that results from the zero-temperature expansion, since the Gell-Mann–Low theorem does not guarantee that the correct state is obtained. The problem can be linked to the order of limits: one has to make the number of particles go to infinity before taking the limit of vanishing temperature. A more detailed explanation and a simple example can be found in [316].

In the following, we do not specify temperature but use the results that are valid over the whole range. Note that the equations for the Green's function are presented here in real space, but they can be expressed in any basis. In particular, in a translationally invariant system both the independent-particle and the interacting Green's function are diagonal in momentum space. Moreover, in a static system at $T \neq 0$ one can work with Matsubara frequencies ω_n (see App. D). The last column of Tab. 9.1 contains the corresponding pre-scriptions. Each vertex redistributes momentum and energy, but the sum is conserved, and all internal variables are integrated.

9.7 Self-consistent perturbation theory: from bare to dressed building blocks

In the previous sections the Green's function G and proper self-energy Σ are expressed in terms of diagrams that involve interaction vertices and bare propagators G^0. The resulting expressions are universal. However, these are not the functionals discussed in Ch. 8: in Eq. (9.39) the self-energy is introduced as a functional[12] of the bare G^0, whereas the topic

---

[12] Note that we use the generic symbol Σ for $\Sigma[G]$ and $\Sigma[G^0]$, although it is understood that the functional expressions in terms of G and G^0 are not the same.

of Ch. 8 is functionals of the dressed Green's function, and in Sec. 8.3 also functionals of the screened instead of the bare interaction.

There are several motivations to work with dressed instead of bare quantities. First, the dressed quantities are the physical ones: the bare G^0 does not have an immediate meaning, whereas the dressed G gives direct access to the density, spectral function, and other measurable quantities. Similarly, the screened Coulomb interaction W is the effective interaction experienced by classical charges in a medium.

Second, the dressed quantities are solutions of Dyson equations that contain terms to all orders. In this way parts of the perturbation series are implicitly summed to infinity, which may be important in case of convergence issues; see Sec. G.3 for more discussion concerning this point. The fact that the screened Coulomb interaction is generally weaker than the bare one is an additional point in favor of dealing with W instead of v_c.

Finally, working with dressed quantities is supported by the pragmatic argument that it reduces the number of diagrams that have to be explicitly calculated, if one stops at a given order. For these reasons, many-body perturbation theory in condensed matter is often formulated in terms of the dressed Green's function G and screened Coulomb interaction W, similarly to density functional theory, that is formulated in terms of the *interacting* density.

As always when there are convergence issues, one still has to be careful, as explained in Sec. G.3. However, the formulation in terms of functionals of the dressed Green's function and interaction is a success story, and it lays the ground for much of the work presented in Chs. 10–21.

Expansion in the dressed Green's function

It was a great accomplishment of Luttinger and Ward [36] to show that the expressions for Σ can be written in terms of diagrams involving the fully interacting Green's function G itself.[13] Like the expansion in G^0, this expression is universal. LW noted that higher-order diagrams for the self-energy contain insertions of lower-order self-energy contributions. For example, several of the second-order diagrams in Fig. 9.5 contain a Green's function line that is "dressed" with an interaction line, such that one can identify either the Hartree or the exchange first-order self-energy contribution. LW called diagrams where there are no such self-energy insertions *skeleton diagrams*. All diagrams for Σ may be obtained by drawing all skeleton diagrams and then inserting all possible proper self-energy parts into each line. This is equivalent to saying:

$$\Sigma = \text{all possible skeleton diagrams with } G^0 \text{ replaced by } G. \qquad (9.48)$$

The skeleton diagrams are composed of building blocks that are similar to the bare ones: one interaction combined with two Green's function lines. The only difference

---

[13] Note that the original papers by LW use the notation G for self-energy and S for Green's functions, which can lead to confusion.

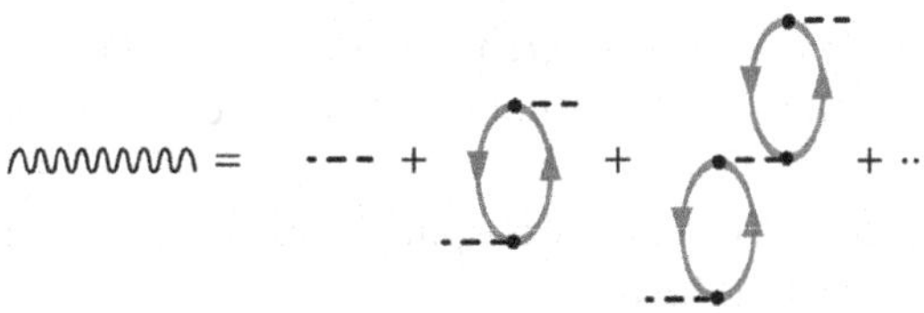

Figure 9.9. RPA expansion of the irreducible polarizability: bubble diagram contributions to the screened Coulomb interaction.

is that now the Green's function lines represent the interacting G, and that fewer combinations contribute. The sum of skeleton diagrams can be interpreted as a sum of symmetry-conserving scattering processes; in particular, momentum is conserved at the vertices.

At this point, we have the explicit expression for the self-energy as a functional of the dressed Green's function as required in Ch. 8.[14] This is very important: Eq. (9.48) is explicitly or implicitly at the basis of much of the following chapters.

Expansion in the dressed interaction

In Sec. 8.3 the self-energy $\Sigma[G, W]$ is also formulated as a functional of the screened Coulomb interaction W introduced in Sec. 5.5, instead of the bare v_c. The screened interaction fulfills a Dyson Eq. (7.11), where v_c plays the role of G^0 and the irreducible polarizability P replaces the self-energy Σ. The next-to-last diagram in Fig. 9.5 shows how this comes about: this diagram is the same as the first-order exchange one, but with an effective interaction that is given by the two bare interaction lines dressed with a bubble insertion. To higher order, more and more insertions appear. If one only considers the bubbles, the resulting screened interaction is the wiggly line in Fig. 9.9: in this case the irreducible, or proper, polarization P is simply given by a bubble. This approximation for P, that is of order zero in the interaction, gives rise to an infinite series of bubbles and interaction lines through the Dyson equation for W. This is the random phase approximation (RPA) that is looked at more deeply in Chs. 11 and 14. At higher order in v_c, additional diagrams contribute to P, but the Dyson screening equation (7.11) for W is still valid. The polarizability beyond the RPA can be obtained from the Bethe–Salpeter equation that is introduced in Ch. 14.

9.8 The Luttinger–Ward functional

Luttinger and Ward also showed how to formulate the grand potential Ω, the free energy in the grand-canonical ensemble, and the energy at zero temperature, in terms of Σ and

---

[14] Note that this is an implicit equation for Σ, since the self-energy is also contained in the Green's function G by the Dyson equation. Turned around, the Dyson equation $G = G^0 + G^0 \Sigma[G]G$ with the self-energy functional of G instead of G^0 is a non-linear equation, which may lead to a problem of multiple solutions, as mentioned in Sec. 10.3 and discussed in App. G.

G. This leads to the *LW functional*, whereby Ω and other properties can be considered as functionals of G and Σ, as introduced in Ch. 8. and used extensively in the following chapters.

To derive the LW result, one can use the coupling-constant integration Eq. (5.132). Using the Dyson equation, Eq. (5.132) becomes

$$\Omega = \Omega^0 + \frac{1}{2}\int_0^1 \frac{d\lambda}{\lambda}\mathfrak{Tr}(\Sigma_\lambda G_\lambda), \tag{9.49}$$

with Ω^0 the non-interacting grand potential. The trace $\mathfrak{Tr}$ is defined in Ch. 8 as $\mathfrak{Tr} = \mathrm{Tr}\,\beta^{-1}\sum_n$ for $T \neq 0$, and for $T = 0$, $\mathfrak{Tr} = \mathrm{Tr}\,\frac{-i}{2\pi}\int_{-\infty}^{\mu}d\omega$. The symbol Tr denotes the trace over states in a basis, including the sum over spin. We now decompose $\Sigma_\lambda = \Sigma[G_\lambda, \lambda v_c]$ in orders of λ, as $\Sigma_\lambda = \sum_{n=1}^\infty \lambda^n \Sigma^{(n)}[G_\lambda, v_c]$. Partial integration yields

$$\Omega = \Omega^0 + \Phi[G] - \frac{1}{2}\sum_{n=1}^\infty \int_0^1 d\lambda\,\frac{\lambda^n}{n}\frac{d}{d\lambda}\mathfrak{Tr}\left\{\Sigma^{(n)}[G_\lambda, v_c]G_\lambda\right\}, \tag{9.50}$$

where we have defined

$$\Phi[G] = \sum_n \frac{1}{2n}\mathfrak{Tr}\left[G\Sigma^{(n)}[G]\right]. \tag{9.51}$$

Note that $\Phi[G]$ is a *universal* functional, independent of G^0. The last term in Eq. (9.50) is

$$T_2 = -\sum_{n=1}^\infty \int_0^1 d\lambda\,\frac{\lambda^n}{2n}\mathfrak{Tr}\left\{\frac{\delta\Sigma_n}{\delta G_\lambda}\frac{dG_\lambda}{d\lambda}G_\lambda + \Sigma_n\frac{dG_\lambda}{d\lambda}\right\}. \tag{9.52}$$

The derivative of Σ with respect to G eliminates one Green's function line. Since the building blocks of the self-energy consist of a Coulomb interaction and two Green's functions, except one block that has only one Green's function, there are $2n-1$ possibilities per order to eliminate a Green's function line. Therefore, we have

$$\Sigma_n = \frac{1}{2n-1}\frac{\delta\Sigma^{(n)}}{\delta G}G \quad \text{or} \quad \Sigma_n = \frac{1}{2n}\left[\frac{\delta\Sigma^{(n)}}{\delta G}G + \Sigma^{(n)}\right]. \tag{9.53}$$

With this, T_2 defined in Eq. (9.52) becomes

$$T_2 = -\int_0^1 d\lambda\,\mathfrak{Tr}\left[\Sigma_\lambda \frac{dG_\lambda}{d\lambda}\right] = -\mathfrak{Tr}\,[\Sigma_\lambda G_\lambda] + \int_0^1 d\lambda\,\mathfrak{Tr}\left[\frac{d\Sigma_\lambda}{d\lambda}G_\lambda\right]. \tag{9.54}$$

By expanding the Dyson equation for G_λ, one can show that the second integrand is the derivative with respect to λ of the expansion of $-\mathfrak{Tr}\ln(1 - G^0\Sigma_\lambda)$. The final result is

$$\Omega = \Omega^0 + \Phi[G] - \mathfrak{Tr}\,[\Sigma G] - \mathfrak{Tr}\,[\ln(1 - G^0\Sigma)]. \tag{9.55}$$

This is the same as Eq. (8.8), with Φ given by summation of diagrams as expressed in Eq. (9.51). With the help of Eq. (9.53), the relation of Eq. (9.51) leads to

$$\Sigma_{xc} = \frac{\delta \Phi[G]}{\delta G}, \tag{9.56}$$

as in Eq. (8.9).

9.9 Wrap-up

This chapter has shown that it is possible to formulate a perturbation theory for the many-body problem in terms of the Coulomb interaction. This is not straightforward. First, the Green's function of the interacting system has to be linked to the independent-particle one. This has been achieved by using an adiabatic connection based on the Gell-Mann–Low theorem at zero temperature, or by treating the inverse non-vanishing temperature as an imaginary time. Second, the numerous terms resulting from the expansion of the interacting Green's function are grouped such that redundancies and cancellations can be used to simplify the expressions. The result is an infinite sum involving many integrals over space and time points, that can be written in compact graphical notation using Feynman diagrams. If the expansion converges, this is the explicit solution of the many-body problem. However, the result will be useful only if we can get by with a small number of terms. Unfortunately, an accurate approximation needs many terms. In the thermodynamic limit, any *finite* number of terms would cause problems. Therefore, one has to find ways to resum some of the diagrams. This leads to the use of Dyson equations, where the interacting Green's function is dressed by a self-energy, or the screened Coulomb interaction determined by the irreducible polarizability. Finally, the perturbation series shows explicitly how one can express the self-energy and the grand potential in terms of the dressed Green's function and screened interaction, making the Luttinger–Ward functional and the Ψ-functional of Sec. 8.2 explicit.

The following chapters do not explicitly develop perturbation theory in the Coulomb interaction in the way it has been done here, but they make use of the insight that emerges from these results to derive the expressions in an alternative way, gaining inspiration for approximations to be applied in practice.

SELECT FURTHER READING

Economou, E. N. *Green's Functions in Quantum Physics*, 2nd edn. (Springer-Verlag, Berlin, 1992). A compact presentation of perturbation theory – just what is needed.

Fetter, A. L. and Walecka, J. D. *Quantum Theory of Many-Particle Systems* (McGraw-Hill, New York, 1971). Overview and precise discussion of many-body perturbation theory.

Pines, D. *The Many-Body Problem* (Frontiers in Physics – a lecture notes series, W. A. Benjamin, New York, 1961). A short summary of history and main tools, followed by a collection of historical publications.

Stefanucci, G. and van Leeuwen, R. *Non-equilibrium Many-Body Theory of Quantum Systems* (Cambridge University Press, Cambridge, 2013). Presents in detail and very precisely the general case of arbitrary temperature and (time-dependent) external potential.

Zagoskin, A. M. *Quantum Theory of Many-Body Systems* (Springer-Verlag, Berlin, 1998). A nice and broad summary.

Exercises

9.1   Work out the first two orders of the Brillouin–Wigner and Rayleigh–Schrödinger perturbation series, Eqs. (9.5)–(9.8), by inserting complete sets of non-interacting states between the operators. Using the fact that the Coulomb interaction is a two-body interaction, specify the results using a single Slater determinant for Φ_{ip}.

9.2   Without knowing Wick's theorem, calculate G_2^0. You will then love Wick's theorem later! Use that in the presence of a single Slater determinant the four field operators must create and destroy particles pairwise. Try to also derive the result for G_3^0, Eq. (9.28).

9.3   (i) Convert the diagram of Fig. 9.10 into a formula. (ii) Use the points and labels of Fig. 9.11 to show that

$$E = G^0(1,3)v_c(3,3')G^0(1,3')v_c(1,2)G^0(2,2^+)G^0(3,4)G^0(3',4)$$
$$\times v_c(4,5)G^0(5,6)G^0(5,6')G^0(6,6')v_c(6,6') \tag{9.57}$$

is the energy of a fish. It is understood that all arguments are integrated over.

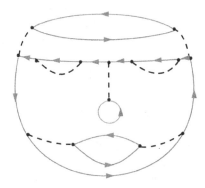

Figure 9.10. A friendly diagram, to be converted into a formula.

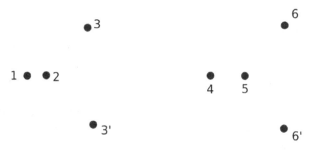

Figure 9.11. Graphical support for the energy of a fish.

9.4 Derive the rules in $\mathbf{k}$- and frequency space in the last column of Tab. 9.1.

9.5 Use Fig. 9.3 to show that the final result for the Green's function consists only of linked diagrams.

9.6 Draw the third-order diagrams for Σ in terms of v_c and G^0 and, from this, find self-energy insertions and skeleton diagrams. Identify pieces of W.

9.7 Derive Eq. (9.55) from Eq. (9.54).

10

Many-body perturbation theory via functional derivatives

There is no known technique to solve functional differential equations [like Eq. (10.12)] in an efficient way.

From the book of Kadanoff and Baym [317]

Summary

In this chapter the equation of motion for the one-body Green's function G is examined. In that equation a two-particle correlation function appears, since the interaction causes the propagation of particles to be correlated instead of independent. A closed scheme is derived by expressing the two-particle correlation function as a functional derivative of G with respect to an external perturbation. By defining a self-energy and reformulating the problem in the form of a Dyson equation as introduced in Ch. 7, a convenient starting point for many physically meaningful approximations is obtained.

The present chapter is a bridge between the general introduction to various aspects and ingredients of the many-body methods in Part I, more specific points of perturbation theory in Ch. 9, and the actual realization of the ideas in terms of concrete approximations with tangible physical content and results in Chs. 11–15. To start with, it is useful to recall some facts.

Any problem defined by a hamiltonian can be equivalently formulated in terms of a Green's function. The full many-body Green's function is defined as $(\omega - \hat{H})^{-1}$, with $\hat{H}$ the many-body hamiltonian. However, as pointed out in several places in this book, starting from the very first chapter, the aim is not to solve the complete many-body problem but merely to get an answer to some of the many questions that one might ask. Therefore our target is not the full many-body Green's function, but only certain matrix elements. The one-body Green's function G introduced in Sec. 5.6, and the two-body correlation function L given in Sec. 5.8, are such matrix elements. As outlined in Ch. 5, the knowledge of G is sufficient to calculate and understand many properties of materials and measurable quantities such as the total energy and the electron addition and removal spectra. Many of the examples presented in Ch. 2 can be explained in terms of G or L. Instead of dealing with the full many-body Schrödinger equation and hamiltonian $\hat{H}$, one can then work with a Dyson equation and a self-energy to obtain G. As discussed in Ch. 7, this is a completely general framework, not limited to the one-body Green's function.

However, this framework alone does not tell us how to calculate the self-energy and hence the Green's functions. Chapters 8 and 9 lay the ground to find ways, with a formulation in terms of functionals (Ch. 8) and an introduction to many-body perturbation theory and diagrams (Ch. 9). Here we adopt a strategy that is based on the functional formalism. There is no contradiction, since one finds the same physics, expressed through the same diagrams as those shown in Ch. 9. The functional approach simply gives a different point of view and reasoning, and one may have a personal preference for one or the other choice. We have opted for the functional derivation as it displays the physics in a direct and very compact way, it suggests useful resummations of diagrams, and it facilitates combinations with other approaches. Moreover, it is a way to have conservation laws built into the approximations. Using this formalism one can obtain, from one central equation in a straightforward manner, all the most widely used approximations and on the same footing, be it Hartree–Fock, different flavors of GW, or T-matrix approximations, as demonstrated in Sec. 10.5. This allows one to better understand and compare the physics of the various approximations, that will be worked out in subsequent chapters. Ultimately, it also lays the ground for new ways to go beyond the standard schemes.

The chapter can be read in two ways. First, it is meant for those who wish to develop the theory further and therefore get insight into the physics, but also to learn about the technical side of the approach. Second, one may read it without going into the details of every single equation, in order to learn which approximations and ideas underlie current methods, and the questions that motivate what we are doing, such as *why* do we write an equation of motion? *why* is it convenient to introduce a self-energy? or *why* is there a functional derivative appearing?

The chapter starts with the equation of motion for the one-body Green's function and the infinite chain of higher-order Green's functions that is generated by the Coulomb interaction. Section 10.2 shows that this chain can be replaced by a functional derivative of the one-body Green's function with respect to an external perturbation. The subsequent section is dedicated to the Dyson equation and an explicit expression for the self-energy. Section 10.4 discusses conservation laws. This topic is also touched upon in Sec. 8.5, but it can be deepened and illustrated further thanks to the relation between the self-energy and the two-body Green's function. Some important approximations for the self-energy are derived in Sec. 10.5, giving at the same time specific insight and a general strategy to include more physics. Finally, a wrap-up concludes the chapter.

10.1 The equation of motion

The many-body Schrödinger equation is a differential equation that can be solved in principle using Green's function techniques. In practice it is neither feasible nor desirable to calculate the full Green's function for a system with many electrons, and we are interested only in certain matrix elements, mainly the one- and two-body Green's functions that contain much of the physics, as outlined in particular in Secs. 5.7 and 5.8. The problem we are then facing is introduced in Sec. 7.1: the full Green's function G^{tot} is determined by the many-body hamiltonian $\hat{H}$ through Eq. (7.4) – that is $(\omega - \hat{H})G^{tot}(\omega) = I$, where I is the identity matrix. The corresponding equations governing the few-body Green's functions are more complex. Expressions such as Eq. (7.6) tell us what is in principle contained in such an effective approach, but they do not constitute a practical alternative to a full solution. *We first have to find the equations* that determine the one- or two-body Green's function we are interested in; this can be at least as challenging as solving them later.

The aim is to find a closed equation or set of equations where the various terms are known or can be approximated reasonably well. In this section we will determine the equations governing the one-body and higher-order Green's functions. To do so, we first derive their equations of motion [40]; these are simply the first-order differential equations in time, in the spirit of Eq. (5.84), that correspond in frequency space to the desired form $\left[\omega - \hat{h}_{\text{eff}}(\omega)\right] G(\omega) = I$ with a frequency-dependent effective hamiltonian $\hat{h}_{\text{eff}}(\omega)$.

To do so, one has to calculate the time derivative of the Green's function $\partial G/\partial t$, starting from Eq. (5.78) for the time-ordered[1] G. This, in turn, requires the equation of motion for the field operators in the Heisenberg representation[2] $\hat{\psi}(\mathbf{x}, t)$, which one obtains from the general equation of motion for operators, Eq. (B.12). It implies the calculation of the commutator $[\hat{\psi}, \hat{H}]$ with the hamiltonian of Eq. (1.2). In second quantization, for a static external potential $v_{\text{ext}}(x_1)$, where $x_1 = (\mathbf{r}_1, \sigma_1)$ and neglecting relativistic effects, the hamiltonian (see Eq. (A.15)) reads

$$\hat{H} = \int dx_1 \hat{\psi}^\dagger(x_1) h(x_1) \hat{\psi}(x_1) + \frac{1}{2} \int \int dx_1 dx_2 \hat{\psi}^\dagger(x_1) \hat{\psi}^\dagger(x_2) v_c(x_1, x_2) \hat{\psi}(x_2) \hat{\psi}(x_1),$$

(10.1)

in terms of the field operators in the Schrödinger picture and with the independent-particle contribution $h(x_1) = -\nabla_{\mathbf{r}_1}^2/2 + v_{\text{ext}}(x_1)$ and the Coulomb interaction v_c.[3]

With Eq. (B.12) the equation of motion for the field operators is

$$i\frac{\partial}{\partial t} \hat{\psi}_H(x, t) = h(x) \hat{\psi}_H(x, t) + \int dx' v_c(x, x') \hat{\psi}_H^\dagger(x', t) \hat{\psi}_H(x', t) \hat{\psi}_H(x, t). \qquad (10.2)$$

Together with $\partial \Theta(t - t_0)/\partial t = \delta(t - t_0)$ this yields the *Heisenberg equation of motion* for the one-body Green's function, the first-order inhomogeneous differential equation

$$\left[i\frac{\partial}{\partial t_1} - h(x_1)\right] G(1, 1') + i \int dx_2 v_c(x_1, x_2) G_2(1, 2, 1', 2^+)\bigg|_{t_2 = t_1^+} = \delta(1, 1'), \qquad (10.3)$$

where the compact notation $(1) \equiv (x_1, t_1)$ has been adopted. The two-particle Green's function G_2 containing four field operators is defined in Eq. (5.133). An adjoint equation holds

---

[1] Here it is specified that we are using the time-ordered framework, which is appropriate at $T = 0$. However, the derivations are valid also when other contours are used, see App. E.

[2] We do not put a subscript H to indicate the Heisenberg picture, since it should be clear from the context.

[3] The Coulomb interaction only depends on $(\mathbf{r}_1 - \mathbf{r}_2)$. However, all equations in this chapter are also valid for more complicated interactions, as long as we are dealing with an instantaneous two-body interaction; see, e.g., [318] for the case of a spin-dependent interaction.

for the derivative with respect to $t_{1'}$ (Ex. 10.1). Note the infinitesimal time differences,[4] e.g., $G_2 = G_2(t_1, t_1 + \eta, t_1', t_1 + 2\eta)$, $\eta \to 0^+$ in Eq. (10.3). They must be specified since the order of field operators is important: the Green's functions are discontinuous in time, see, e.g., Eq. (5.154).

The two-particle Green's function, appears because of the two-body interaction v_c in Eq. (10.1). It contains the information that an electron is not propagating alone in an external potential, but interacting with other electrons. Instead of the full G_2, only some components with particular equal space–spin and time arguments are needed in Eq. (10.3), since $v_c(1, 2) \equiv v_c(x_1, x_2)\delta(t_1 - t_2)$ is an instantaneous two-body interaction. Unfortunately, even for these selected components of G_2 no simple form is known, and therefore the equation of motion for G is not of closed form. Hence, much of the full complexity of the many-body problem is hidden in the G_2 term. Still, Eq. (10.3) is a useful starting point for further discussions and developments, as we will see in the following.

If one neglects the interaction, Eq. (10.3) is the equation of motion Eq. (5.84) for a one-particle Green's function G^0 of independent electrons in the external potential v_{ext}:

$$\left[i\frac{\partial}{\partial t_1} - h(1)\right] G^0(1, 1') = \delta(1, 1').$$

(10.4)

We then have Eq. (5.85), $[G^0]^{-1} = i\partial/\partial t - \hat{h}$. The solution G^0 of this differential equation is well-defined if initial conditions are specified as discussed in App. G (see also Ex. G.1); then one can write Eq. (10.3) as

$$G(1, 1') = G^0(1, 1') - iG^0(1, \bar{2})v_c(\bar{2}, \bar{3})G_2(\bar{2}, \bar{3}^+, 1', \bar{3}^{++}).$$

(10.5)

Note that here and from now on, for the sake of compactness, we will represent integrals by placing a bar, like $\bar{2}$, over integrated variables. For example, $f(1, \bar{2})g(\bar{2})$ stands for $\int d2 f(1, 2)g(2)$.

One can make a part of the interaction term explicit by introducing the two-particle correlation function $L(1, 2, 1', 2') = -G_2(1, 2, 1', 2') + G(1, 1')G(2, 2')$, Eq. (5.134). Because of the special equal arguments of G_2 in Eq. (10.5) the uncorrelated product term GG contains the one-particle density $n(\bar{3})$ that gives rise to the Hartree potential $v_H(2) = -iv_c(2^+, \bar{3})G(\bar{3}, \bar{3}^+)$ (see Eq. (4.6)). Altogether the equation of motion becomes

$$G(1, 1') = G_0(1, 1') + G_0(1, \bar{2})v_H(\bar{2})G(\bar{2}, 1') + iG_0(1, \bar{2})v_c(\bar{2}, \bar{3})L(\bar{2}, \bar{3}^+, 1', \bar{3}^{++}).$$

(10.6)

The simplest approximation to L is another product of one-body Green's functions, namely

$$L_0(1, 2, 1', 2') = G(1, 2')G(2, 1'),$$

(10.7)

which is the first diagram on the right side of Fig. 14.1. It yields the Hartree–Fock approximation;[5] see Ex. 10.2.

---

[4] For the sake of clarity, here we display the infinitesimals explicitly, instead of including some of them in the definition of v_c as in Sec. 9.3.

[5] One may wonder why Hartree–Fock does not correspond to a vanishing correlation function. One has a non-zero correlation because electrons are indistinguishable fermions; therefore, G_2^0 is given by a linear combination of all possible products of two one-body G^0.

Beyond Hartree–Fock, there are correlation contributions that can no longer be expressed as a product of two one-body Green's functions. One could imagine finding an improved description of G_2 or, equivalently, of L, by writing down the equation of motion for G_2. However, by the same mechanism that the time derivative of G has created additional field operators and led to the appearance of G_2, the equation of motion for G_2 yields a term involving the three-particle Green's function G_3, and so on. For order s and with the definition

$$G_s(1,\ldots,s;1',\ldots,s') = (-i)^s \left\langle T\left[\psi(1)\psi(2)\ldots\psi(s)\psi^\dagger(s')\ldots\psi^\dagger(2')\psi^\dagger(1')\right]\right\rangle \quad (10.8)$$

one finds a set of equations called "Martin–Schwinger hierarchy" [40], as derived in Ex. 10.1,

$$[(G^0)^{-1}G_s](1,2,\ldots,s;1',\ldots,s') = \sum_{j=1}^{s}(-1)^{(j-s)}\delta(1,j')$$

$$G_{s-1}(2,\ldots,s;1',\ldots,j'-1,j'+1,\ldots,s') - iv_c(1,\bar{p})G_{s+1}(1,2,\ldots,s,\bar{p}^+;1',\ldots,s',\bar{p}^{++}).$$

$$(10.9)$$

Note that for vanishing interaction, Wick's theorem, Eqs. (9.33) and (9.35), for the independent-particle Green's function of order s is recovered by iterating Eq. (10.9). Including interaction, Eq. (10.9) yields an infinite chain of equations that has somehow to be truncated. For example, starting from some highest order $s > s_{max}$ one might replace all Green's functions by uncorrelated products of lower-order ones. All lower-order Green's functions could then be calculated from Eq. (10.9). The simplest example is the Hartree–Fock approximation obtained above using Eq. (10.7) or Eq. (10.9) for $s = 2$ and $v_c = 0$. The next section will introduce a general strategy that allows one to formally eliminate higher-order Green's functions without such a drastic approximation, and obtain an expression that is a useful starting point for systematic improvements.

10.2 The functional derivative approach

Beyond the Hartree and Hartree–Fock approximations, the equation of motion gives an active role to the electrons. For example, it contains the reaction of the electrons to the propagation of another electron, contained in G_2. This links to Sec. 5.8, where the relation between the two-particle correlation function and response functions is pointed out. It suggests a way to eliminate the undesired higher-order Green's functions from the equation of motion: L can be expressed as a variation of the one-particle Green's function with respect to a fictitious perturbation that vanishes at the end of the calculation, as given by Eq. (8.35). Probing the system in this way leads to a set of first-order functional–differential integral equations for G, a starting point for numerous successful approximations [40, 320, 321]. These equations are sometimes called *Schwinger–Dyson equations*.

Let us first concentrate on Eq. (10.6) and use the expression in Eq. (8.35) for L in terms of a functional derivative with respect to a general non-local perturbation $\mathcal{J}(3',3)$. In

Eq. (10.6), only some selected components of L, and therefore only the derivative of G with respect to a particular component of $\mathcal{J}$, appear, namely

$$L(2, 3, 1', 3^+) = \left[\frac{\delta G(2, 1')[\mathcal{J}]}{\delta \mathcal{J}(3^+, 3)} \right]_{\mathcal{J}=0}. \tag{10.10}$$

Hence only a local potential $u(3) \equiv \mathcal{J}(3^+, 3)$ is needed to represent the equations. It plays the role of an ordinary time-dependent external potential added to the system.[6] When it is not set to zero, the equations describe the corresponding non-equilibrium dynamics, provided that contour Green's functions and integrations are used as explained in App. E. We do not make use of this extension in this book, but for the derivations and understanding it should be kept in mind.

For the Green's function G_u in the presence of u we have

$$\frac{\delta G_u(2, 1')}{\delta u(3)} = L_u(2, 3, 1', 3^+), \tag{10.11}$$

where L_u is the two-particle correlation function written as a functional of u. This allows us to formulate Eq. (10.6) in the presence of u in the compact form

$$G_u(1, 1') = G^0(1, 1') + G^0(1, \bar{2}) \left\{ [u(\bar{2}) + v_H(\bar{2})]G_u(\bar{2}, 1') + iv_c(\bar{2}, \bar{3}) \frac{\delta G_u(\bar{2}, 1')}{\delta u(\bar{3}^+)} \right\}. \tag{10.12}$$

Note that the non-interacting Green's function G_u^0 in the presence of u fulfills the Dyson equation $G_u^0 = G^0 + G^0 u G_u^0$, so that one can make use of Eqs. (7.40) and (7.41) to derive Eq. (10.12).

With this trick a set of coupled non-linear first-order functional–differential integral equations in terms of one unknown, the one-particle Green's function, has been obtained. It replaces the prescription for the diagrammatic construction of G in terms of G^0 of Sec. 9.4. For equilibrium properties one only needs the solution at vanishing fictitious potential u.

Which solution?

As has been pointed out in [317] and quoted at the start of this chapter, "there is no known technique to solve functional differential equations like Eq. (10.12) in an efficient way." In general the equation is not solved directly, but explicitly or implicitly iterated. Once the starting point of an iteration and the iteration procedure are fixed, the result at each step is well-defined. In Ex. 10.5 it is shown that the diagrammatic expansion of Sec. 9.4 can be obtained by iterating Eq. (10.12) starting from $G_u = G^0$ and setting $u = 0$ at the end.

The iteration implicitly defines the boundary condition of the differential equation, that has otherwise in general multiple solutions. The additional requirement to specify the solution arises because in going from Eq. (10.6) to Eq. (10.12), the two-particle correlation

---

[6] We use the symbol u instead of v_{ext} as used for an external potential in other places like Ch. 5, in order to stress the fact that the applied potential u is fictitious and set to zero at the end of the calculation.

function L, which is well-defined in Eqs. (5.134) and (5.133), is replaced by a functional derivative of G, which is itself the unknown quantity of the equations. Different functionals $\tilde{G}_u$ could satisfy Eq. (10.12), and there is no reason to suppose that they will take the same value at $u \to 0$: by transforming Eq. (10.6) into a functional differential equation, information is lost. Iterating the equations starting from G^0 is a way to use the continuity principle put forward in Sec. 1.3; it connects the interacting solution to the non-interacting one.

Of course, this does not tell us whether and how such a series will converge; see related discussions in [34], and in Sec. 8.2 and App. G. It may diverge, or have conditional or asymptotic convergence. In general one must go beyond straightforward perturbation theory and try to group terms or find partial resummations in order to find meaningful results; the formulation in terms of a compact differential equation should be used in a more intelligent way than just iterating starting from the non-interacting solution. Moreover, the limit of vanishing potential may be non-analytic when there is an instability to a broken symmetry solution, so that the result may depend on the way the limit is performed. For example, this could determine which magnetic structure is found. Therefore, although in Eq. (10.12) the problem appears to be formulated in a closed and compact form, physical insight will still be needed in order to obtain meaningful and definitive answers.

As the numerous prototypical examples in Chs. 11–15 illustrate, the combination of intuition and straightforward mathematical developments has indeed led to many successful applications.

10.3 Dyson equations

One of the problems of iterating Eq. (10.12) can be understood by neglecting the last term, which yields the Dyson equation in the Hartree approximation, $G_H = G^0 + G^0 v_H G_H$. The first thing that one expects from the interaction is that it modifies quasi-particle energies with respect to the poles of G^0. In the Hartree approximation, this is the only modification of the spectrum, since peaks remain sharp and no additional structures appear in the spectral function. Suppose for simplicity that the Hartree solution G_H and G^0 are diagonal in the same basis; the solution for a given matrix element is then the scalar expression

$$G_H = \frac{1}{\omega - \varepsilon^0 - v_H}, \tag{10.13}$$

which shows the expected shift of the pole. Now imagine iterating the Dyson equation by starting from $G_H = G^0$. This yields

$$G_H = \frac{1}{\omega - \varepsilon^0} + \frac{v_H}{(\omega - \varepsilon^0)^2} + \frac{v_H^2}{(\omega - \varepsilon^0)^3} + \cdots, \tag{10.14}$$

which is a series of higher-order poles, instead of the desired shift that appears in the solution Eq. (10.13). This is a good reason to solve Dyson equations instead of calculating finite series. Another motivation is given in the next subsection.

The self-energy

In order to highlight the physics that governs the propagation of an electron, it is most useful to look at the inverse of the Green's function. From Eq. (10.12) one has

$$G_u^{-1}(1, 1') = G_0^{-1}(1, 1') - [u(1) + v_H(1)] \, \delta(1, 1') - i v_c(1, \bar{3}) \frac{\delta G_u(1, \bar{2})}{\delta u(\bar{3}^+)} G_u^{-1}(\bar{2}, 1'). \quad (10.15)$$

Now all contributions that modify the non-interacting propagation are simply additive: the additional external potential u, the Hartree potential, and the remainder that by definition contains the exchange–correlation effects. This suggests that we can define the exchange–correlation *self-energy* Σ_{xc}

$$\Sigma_{xc}(1, 1') = i v_c(1, \bar{3}) \left[\frac{\delta G(1, \bar{2})}{\delta u(\bar{3}^+)} \right] G^{-1}(\bar{2}, 1') = -i v_c(1, \bar{3}) G(1, \bar{2}) \left[\frac{\delta G^{-1}(\bar{2}, 1')}{\delta u(\bar{3}^+)} \right], \quad (10.16)$$

where here and in the following we omit subscripts u, noting that the expressions are valid also for $u \neq 0$, when they are interpreted on an appropriate contour (see App. G). The second part of this equation is derived in Ex. 10.8. The definition of the self-energy Eq. (10.16) allows one to formulate Eq. (10.12) in the form of a Dyson equation, Eq. (7.10) or Eq. (8.13). This has an obvious advantage: if one is able to approximate Σ_{xc} even to low order in the interaction, solving the Dyson equation for the Green's function creates contributions to all orders, as one can see from the expansion in Eq. (7.9). For example, the first-order approximation to Σ_{xc} in Eq. (10.16) is the Fock self-energy

$$\Sigma_x(1, 1') = i v_c(1^+, 1') G(1, 1'), \quad (10.17)$$

consistent with Eq. (4.7), since the density matrix is related to the Green's function by Eq. (5.120). The solution of the Dyson equation leads then to a Green's function that has interaction contributions up to infinite order. This is an essential idea; it applies to all Dyson-like equations.[7] It is a generalization of the argument in the previous subsection, where it was limited to the Hartree approximation.

Let us now examine the time structure of the G_2 that appears in Eq. (10.5). Since the Coulomb interaction is instantaneous,[8] G_2 contains two equal-time limits, and at $u = 0$ only one time difference appears; similarly for L in Eq. (10.6). In frequency space Eq. (10.6) reads

$$G(\omega) = G_0(\omega) + G_0(\omega) v_H G(\omega) + i G_0(\omega) v_c L_p(\omega), \quad (10.18)$$

---

[7] Note that introducing a self-energy is not the only possibility to follow such a strategy. Schematically the self-energy in Eq. (10.16) is proportional to $\frac{1}{G} \frac{dG}{du} = \frac{d \ln G}{du}$, which suggests an exponential solution for G. Thus one can try to approximate the exponent and then take the exponential. Since $e^x = 1 + x + x^2/2 + \cdots$, a low-order approximation to the exponent creates contributions to all orders. Approaches like the coupled cluster expansion in quantum chemistry [309, 310], the Jastrow factor in variational Monte Carlo (see Sec. 6.6), or the cumulant expansion (see Sec. 15.7) make use of the effective creation of many diagrams from few terms in an exponent.

[8] Remember that this is true because retardation effects are neglected for the purpose of this book.

where we have denoted the frequency Fourier transform of the equal-time limit of L that appears in Eq. (10.6) with a subscript p. As worked out in Ex. 10.6, $L_p(\omega)$ has poles situated at the electron addition and removal energies, $\varepsilon_s = E_{N+1,s} - E_N$ and $\varepsilon_s = E_N - E_{N-1,s}$, like the one-body Green's function $G(\omega)$. Let us now assume that G is diagonal in the same basis as G^0. Then we have, for a given matrix element,

$$G^{-1} = G_0^{-1} - v_H - iv_c L_p G^{-1}. \tag{10.19}$$

If L_p and G are dominated by one pole, this pole cancels, and the correction with respect to the Hartree case is just a shift of energies. This suggests that when one is interested in a quasi-particle peak, a self-energy defined as $iv_c L_p G^{-1}$ in Eq. (10.16) may be approximated in a relatively simple way, simpler than, e.g., L_p itself, and that it is therefore convenient to use a Dyson equation.

Beyond the quasi-particle approximation there are more poles, which create additional structures in the spectral function, called satellites (see Sec. 7.4), and the argument does not hold. Interestingly, alternatives to the Dyson equation, such as the cumulant form mentioned in footnote 7 on p. 229 and worked out in Sec. 15.7, often lead to better results for satellites.

Dyson equation for the two-particle correlation function

One can also write a Dyson equation for the two-particle Green's function or correlation function. If one does not concentrate on some selected components, but wishes to describe the most general situation, a fully non-local source $\mathcal{J}$ is needed. Starting from Eq. (8.35) one can transform the functional derivative as

$$L(1,2,1',2') = \frac{\delta G(1,1')[\mathcal{J}]}{\delta \mathcal{J}(2',2)} = -G(1,\bar{3})\frac{\delta G^{-1}(\bar{3},\bar{3}')[\mathcal{J}]}{\delta \mathcal{J}(2',2)}G(\bar{3}',1'). \tag{10.20}$$

We can now use Eqs. (10.15) and (10.16) to evaluate the functional derivative of the inverse Green's function, which yields

$$L(1,2,1',2') = G(1,2')G(2,1') + G(1,\bar{3})\frac{\delta \Sigma(\bar{3},\bar{3}')[\mathcal{J}]}{\delta \mathcal{J}(2',2)}G(\bar{3}',1'), \tag{10.21}$$

with $\Sigma = v_H + \Sigma_{xc}$. As above, the desired equilibrium solution is obtained for $\mathcal{J} \to 0$, on which we will concentrate in the following.

The first term on the right-hand side of Eq. (10.21) is L_0, defined in Eq. (10.7). Since Σ_{xc} can be expressed as a functional of G (see Ch. 8 and Sec. 9.8), and G as a functional of $\mathcal{J}$, one can expand the derivative with respect to $\mathcal{J}$ as a chain rule (see end of Sec. H.1). We now define the effective two-particle interaction $i\Xi$ as

$$\Xi(3,2,3',2') \equiv -i\delta(3,3')\delta(2'^+,2)v_c(3^+,2) + \frac{\delta \Sigma_{xc}(3,3')}{\delta G(2',2)}. \tag{10.22}$$

Then, Eq. (10.21) becomes the *Bethe–Salpeter equation*

$$L(1,2,1',2') = L_0(1,2,1',2') + L_0(1,\bar{3}',1',\bar{3})\Xi(\bar{3},\bar{4},\bar{3}',\bar{4}')L(\bar{4}',2,\bar{4},2'). \tag{10.23}$$

Equation (10.23) is a Dyson equation as defined in Sec. 7.2. It is depicted diagrammatically in Fig. 14.1. The interaction *kernel* of the equation, Ξ, is given in Eq. (10.22) as a sum of two terms. One can therefore use Eq. (7.41), and break this expression into its irreducible contribution $\tilde{L}$ that doesn't contain the derivative of the Hartree potential

$$\tilde{L}(1,2,1',2') = L_0(1,2,1',2') + L_0(1,\bar{3}',1',\bar{3}) \frac{\delta \Sigma_{xc}(\bar{3},\bar{3}')}{\delta G(\bar{4}',\bar{4})} \tilde{L}(\bar{4}',2,\bar{4},2') \qquad (10.24)$$

and

$$L(1,2,1',2') = \tilde{L}(1,2,1',2') - i\tilde{L}(1,\bar{3},1',\bar{3})v_c(\bar{3}^+,\bar{4})L(\bar{4},2,\bar{4}^+,2'). \qquad (10.25)$$

With Eq. (10.20), L is the variation of G with respect to a non-local external potential $\mathcal{J}$. With $v_H(2',2) \equiv \delta(2',2)v_H(2)$ can also express $\tilde{L}$ as

$$\tilde{L}(1,2,1',2') = \frac{\delta G(1,1')}{\delta[\mathcal{J}(2',2) + v_H(2',2)]}. \qquad (10.26)$$

This equation is derived in Ex. 10.4. It shows that $\tilde{L}$ is the functional derivative with respect to $v_H + \mathcal{J}$, the sum of $\mathcal{J}$ and the Hartree potential. In the case of a local external potential u instead of the general non-local $\mathcal{J}$, this sum corresponds to the total classical potential v_{cl} that is used in Sec. 11.1.

The separation of the Hartree and exchange–correlation kernels can be advantageous in certain situations because of the simpler structure of Eq. (10.25), as one can see for instance in Secs. 11.1 and 11.2.

Self-consistent Dyson equations

Using Eqs. (10.15) and (10.16), the self-energy can be expressed as a functional of the bare G^0, or as a functional of the self-consistent dressed G.[9] In Ex. 10.7 the self-energy is expanded as a series in G, consistent with the discussions in Sec. 9.7. When Σ is a functional of the dressed G, the Dyson equation is a non-linear equation. Such an equation can have multiple solutions. One has to be careful to find the good solution, as explained in Secs. 12.4 and G.2.

10.4 Conservation laws

In the previous section the problem of finding an approximation to G or G_2 has been moved to finding an approximation to Σ or Ξ. This is a difficult task, and one needs some further guidance to select approximations that are physically sound. As discussed in Sec. 8.5, an important requirement should be that approximations are *conserving*, i.e., observables calculated from the Green's functions should obey the same conservation laws

---

[9] The equivalent is true for other Dyson equations, for example for the two-particle correlation function.

as the many-body Hamiltonian:[10] conservation of particle number, momentum, angular momentum, and energy. There may be additional conserved quantities for systems with additional symmetries.

In Sec. 8.5 criteria for such conserving approximations are introduced in terms of the self-energy. Baym and Kadanoff [13, 299] have formulated these criteria in terms of the two-particle Green's function. Suppose one chooses an approximate G_2 in order to determine G from Eq. (10.3) or Eq. (10.5). The one-particle Green's function G is fully conserving when:

[A] For one and the same G_2 the *same* one-particle Green's function fulfills the equation of motion Eq. (10.3) and its adjoint (see Ex. 10.1), or, equivalently,[11]

$$G(1, \bar{2})v_c(\bar{2}, \bar{3})G_2(\bar{2}, \bar{3}^+, 1', \bar{3}^{++}) = G_2(1, \bar{3}^{--}, \bar{2}, \bar{3}^-)v_c(\bar{3}, \bar{2})G(\bar{2}, 1'). \qquad (10.27)$$

This condition on G_2 is sufficient to yield number conservation in the resulting G.[12]

[B] The approximate G_2 must have the symmetry $G_2(1, 2, 1^+, 2^+) = G_2(2, 1, 2^+, 1^+)$ that one can deduce from Eq. (5.133) for the exact G_2. This condition, together with [A], guarantees the conservation of total momentum, total angular momentum, and total energy, as well as number conservation.

As an example, following [299] let us look at the total momentum, the integral of the current density, in the presence of a local external potential $u(1)$,

$$\mathbf{P}(t_1) = -\frac{1}{2} \int dx_1 \left\{ (\nabla_1 - \nabla_{1'})G_u(1, 1') \right\} \Big|_{1'=1^+}. \qquad (10.28)$$

Its time derivative should be equal to the total force applied to the system:

$$\mathbf{F}(t_1) = -\int dx_1 n(1)\nabla u(1). \qquad (10.29)$$

For the time derivative of $\mathbf{P}(t_1)$, the time derivative of the two field operators in $G(1, 1^+)$ is needed. This can be achieved by subtracting its adjoint from the equation of motion Eq. (10.3), generalized for the presence of the external potential u. Subsequently, one has to apply $(\nabla_1 - \nabla_{1'})$, take $1' \to 1^+$, and integrate over x_1. This yields

$$\frac{d}{dt_1}\mathbf{P}(t_1) = \frac{i}{2} \int dx_1 \left\{ (\nabla_1 - \nabla_{1'}) \left[u(1)G_u(1, 1') - G_u(1, 1')u(1') \right] \right\} \Big|_{1'=1^+}$$
$$+ \frac{1}{2} \int dx_1 \{ (\nabla_1 - \nabla_{1'})(v_c(1, \bar{3})G_{2,u}(1, \bar{3}^+, 1', \bar{3}^{++})$$
$$- v_c(1', \bar{3})G_{2,u}(1, \bar{3}^{--}, 1', \bar{3}^-)) \} \Big|_{1'=1^{+++}}. \qquad (10.30)$$

---

[10] In practical applications further approximations may be needed that violate conservation laws, e.g., one may be interested in the low-energy part of a spectrum only and therefore adopt an approximation that alters only the high-energy part such that a sum rule is not fulfilled. See, for example, [322] for photoemission spectra, or the discussion of plasmon pole models in Sec. 12.2.

[11] In analogy to the abbreviation 3^+ the expression 3^- stands for $(x_3, t_3 - \eta)$ with $\eta \to 0^+$.

[12] As shown in Sec. 8.5, if one subtracts from the equation of motion Eq. (10.3) its adjoint equation one obtains the continuity equation, if the above symmetry is present. Otherwise there is a source term.

The use of multiple time infinitesimals is necessary in order to keep the additional field operators stemming from the time derivative in the correct positions. Since $v_c(1', 3)$ is instantaneous, the two G_2 are equal. Therefore, the contributions of ∇ acting on G_2 cancel. The remaining G_2 term is $\int dx_1 \left\{ \left[\nabla_1 v_c(1, \bar{3}) \right] G_{2,u}(1, \bar{3}^+; 1^{+++}, \bar{3}^{++}) \right\}$. If symmetry condition [B] holds, interchange of the integration variables x_1 and $\bar{x}_3$ leads to a minus sign that stems from the derivative of the Coulomb interaction. Hence in this case the term vanishes, the right-hand side of Eq. (10.30) reduces to Eq. (10.29), and the force is correctly due to the external potential terms only. Otherwise G_2 would create an additional force, entirely due to internal interactions: its invariance under interchange of arguments guarantees an "action–reaction" principle, so that the system cannot build up total momentum without an external perturbation.

When L is obtained from Eq. (10.20), conservation laws for G automatically imply conservation laws for L. In practice, however, one may derive L from a given approximation to the Bethe–Salpeter Eq. (10.23) rather than from the functional derivative Eq. (10.20). This sets a constraint on the kernel:

[C] The explicit condition for the effective two-particle interaction is $\Xi(1, 2, 1', 2') = \Xi(2, 1, 2', 1')$.

The symmetry in both $(11')$ and $(22')$ guarantees gauge invariance in electrical transport [323].[13]

Condition [C] translates into a symmetry condition Eq. (8.51) for the functional derivative of the self-energy that we repeat here:

$$\frac{\delta \Sigma_{\text{xc}}(1, 1')}{\delta G(2', 2^+)} = \frac{\delta \Sigma_{\text{xc}}(2, 2')}{\delta G(1', 1^+)}. \tag{10.31}$$

Such a symmetry is satisfied by a self-energy that can be written as

$$\Sigma_{\text{xc}}(1, 1') = \frac{\delta \Phi[G]}{\delta G(1', 1^+)}, \tag{10.32}$$

where Φ is a functional of G that is gauge-invariant and constant with respect to translations in space and time, and rotations: the prescription by Luttinger and Ward [36] in Secs. 9.7 and 9.8 for the diagrammatic construction of Φ in terms of scattering processes that conserve symmetry fulfills this requirement. The crucial point to note here is:

If one determines G from a Dyson equation, with a self-energy that is derived from a functional $\Phi[G]$ that is constructed as a sum of symmetry-conserving scattering processes, and if L is the second derivative of Φ with respect to an external potential, G and L automatically fulfill the same conservation laws as the many-body hamiltonian.

---

[13] It is understood that G appearing in L_0 and the self-energy used to calculate Ξ are linked by the Dyson equation for G.

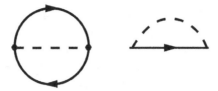

Figure 10.1. Left side: lowest-order contribution to the Φ-functional Eq. (10.33), the Fock term. Right side: the Fock self-energy, obtained by summing all possibilities to remove a line corresponding to the Green's function G, which cancels the prefactor $\frac{1}{2}$ in Eq. (10.33). This is the same diagram as the second contribution to Fig. 9.8, but here the lines are dressed Green's functions, whereas the bare G^0 appears in Fig. 9.8.

A general discussion about functionals can be found in Ch. 8, in particular in Sec. 8.2 it is pointed out that Φ is the part of the grand potential that is due to interactions. More details can be found in [301]. For completeness, we summarize here the important result of Sec. 9.7 that the contribution to Φ of a given order n in v_c is given by the sum of all skeleton[14] self-energy diagrams $\Sigma^{(n)}$ of order n, each closed with a Green's function G and integrated over all variables. The sum over all orders, weighted by a factor $\frac{1}{2n}$, is given by Eq. (9.51), namely

$$\Phi[G] = \sum_n \frac{1}{2n} \mathfrak{Tr}\left[G\Sigma^{(n)}[G] \right], \tag{10.33}$$

where $\mathfrak{Tr}$ is defined as in Ch. 8 and after Eq. (9.51). To lowest order, Eq. (10.33) yields the Fock contribution depicted in Fig. 10.1. Also the following orders can easily be constructed, but there is no known closed expression that would represent the sum of the whole series. Therefore the problem remains – one has to decide which diagrams to choose for a given problem. In other words, one still has to work hard to possibly find a functional that is not only conserving, but also leads to results close to reality.

10.5 A starting point for approximations

The aim of this section is to introduce a general expression for the self-energy that is at the same time compact and displays the content in an intuitive way. This allows one to derive approximations combining mathematics and physical insight, as the examples will show.

Interpretation of the self-energy as generalized potential

In order to make an appropriate choice of higher-order diagrams, it is helpful to rewrite the full self-energy $\Sigma = v_H + \Sigma_{xc}$, including the Hartree potential, in several equivalent ways, starting from Eq. (10.16):

---

[14] A skeleton diagram does not contain any self-energy insertion other than itself.

$$\Sigma(1,2) = v_H(1,2) - iv_c(1,\bar{4})G(1,\bar{3})\left[\frac{\delta G^{-1}(\bar{3},2)}{\delta u(\bar{4}^+)}\right] \tag{10.34}$$

$$= v_H(1,2) + iv_c(1^+,\bar{4})G(1,\bar{3})\left[\delta(\bar{3},2)\delta(\bar{3},\bar{4}) + \frac{\delta\Sigma(\bar{3},2)}{\delta u(\bar{4})}\right] \tag{10.35}$$

$$= v_H + \Sigma_x + iv_c(1^+,\bar{4})G(1,\bar{3})\left[\frac{\delta(v_H + \Sigma_{xc})(\bar{3},2)}{\delta G(\bar{6},\bar{5})}\frac{\delta G(\bar{6},\bar{5})}{\delta u(\bar{4})}\right] \tag{10.36}$$

$$= v_H + \Sigma_x + iG(1\bar{3})\Xi(\bar{3},\bar{5};2,\bar{6})v_c(1^+,\bar{4})L(\bar{6},\bar{4};\bar{5},\bar{4}^+), \tag{10.37}$$

where $G^{-1} = G_0^{-1} - u - \Sigma$ has been used and arguments of v_H and Σ_x are omitted. Let us pause to consider the implications of the different expressions.

- Writing Eq. (10.34) in terms of the inverse Green's function leads to a differential Eq. (10.35) for the self-energy, involving only the bare Coulomb interaction v_c, the external perturbation u, and the *dressed* G.
- The derivative of the bare u gives rise to the exchange Σ_x in Eq. (10.36). The rest of the right side is, besides the Hartree potential, by definition the correlation self-energy Σ_c. It is obtained through a chain rule, using the fact that Σ is a functional of G.
- The derivative of G in Eq. (10.36) is the two-particle correlation function L, Eq. (10.10). Since L fulfills the Bethe–Salpeter Eq. (10.23), Eqs. (10.36), (10.22), and (10.23) form a set that can be iterated, yielding Σ as a functional of G (see also Ex. 10.7).
- Equation (10.37) gives a physical picture of the self-energy as a generalized potential: in $v_H(1) = -iv_c(1,\bar{3})G(\bar{3},\bar{3}^+)$, the density, which is the diagonal of the Green's function, creates a classical Coulomb potential. In $\Sigma_x(1,2) = iG(1,2)v_c(1^+,2)$, the Green's function gives rise to non-local exchange. The remaining correlation contribution can be interpreted as a generalized *induced* non-local potential, caused by the propagation of a particle. This can be understood as follows: the induced potential v_{ind} due to an external charge n_{ext} and the classical Coulomb interaction v_c reads $v_{ind} = v_c\chi v_c n_{ext}$, where χ is the density–density correlation function Eq. (5.149). The correlation self-energy Σ_c in Eq. (10.37) is $\Sigma_c = v_c L(i\Xi)G$. Here the Green's function corresponds to n_{ext}, and the correlation function L to χ. The term $v_c L$ is hence the generalization of the screening contribution $v_c\chi$. The role of a generalized interaction is played by $i\Xi$, given as $i\Xi = i\delta\Sigma/\delta G$ in Eq. (10.22). It consists of the classical Coulomb interaction $v_c = i\delta v_H/\delta G$ and the effective "exchange–correlation interaction" $i\delta\Sigma_{xc}/\delta G$.

The interpretation of Σ_c in terms of an effective induced potential suggests various approximations to the self-energy; these are combinations of an approximation to L and an approximation to Ξ. In situations where screening is important one needs a good approximation for L, whereas in situations where it is important that the effective

Table 10.1. Self-energies obtained by approximating the effective interaction $i\Xi$ and the two-particle correlation function L that yields the screening in the exact expression Eq. (10.37). To go beyond Hartree–Fock, both Ξ and L must be taken into account. Different flavors of the GW approximation are obtained when the effective interaction is the classical Coulomb interaction v_c, and some screening is included. In standard applications of GW one uses GW^{RPA}. More advanced effective interactions, for example including scattering contributions in a T-matrix approximation, can be used to include physics not contained in the GW approximation. The screened Coulomb interaction W, the scattering matrix T, the vertex function $\tilde{\Gamma}$, and all acronyms are defined in the text of this section and/or in Ch. 15, where approximations beyond GW are studied. The variation $\delta W / \delta G$ is neglected

$L \rightarrow$ $i\Xi \downarrow$	0	L_0	L^{RPA}	L beyond RPA
0	HF	HF	HF	HF
v_c	HF	GW^{L_0}	GW^{RPA}	GW^{TC-TC}
$v_c + i\delta \Sigma_{xc}^{GW}/\delta G$	HF	GW+SOSEX	$GW\tilde{\Gamma}^{(1)}$	
$-T$	HF	TMA	screened TMA	

interaction is not simply the classical Coulomb interaction[15] one would concentrate on improving Ξ.

In the following we use Eq. (10.37) in order to introduce some widely used approximations. They are summarized in Tab. 10.1.

Hartree–Fock approximation

When $L = 0$ or when variations of the Hartree potential and of the exchange-correlation self-energy are neglected ($\Xi = 0$), there is no correlation and one finds the HF approximation $\Sigma_{xc} \approx \Sigma_x$ (first row and column of Tab. 10.1; right side of Fig. 10.1). The HF self-energy can be derived from the Φ-functional on the left side of Fig. 10.1 (see also Tab. 8.1) by summing all possibilities of removing one Green's-function line. This is a quick check to see that self-consistent Hartree–Fock is a conserving approximation.

First glance at GW: the screened interaction approximation

Let us now add the correlation contribution. First we retain only the Hartree potential in the calculation of the effective two-particle interaction Ξ, Eq. (10.22); this means that the interaction is purely classical. Then Eq. (10.37) reads

$$\Sigma_{xc}(1,2) = \Sigma_x(1,2) + v_c(1^+, \bar{3})G(1,2)L(\bar{4}, \bar{3}; 4^+, \bar{3}^+)v_c(2, \bar{4}^+). \tag{10.38}$$

---

[15] The atomic limit of the Hubbard molecule is an example where the correlation part of the interaction is crucial; see Sec. 11.7.

With Eq. (10.11) we have

$$- iL(3, 2; 3^+, 2^+) = \frac{\delta n(3)}{\delta u(2)} = \chi(3, 2), \tag{10.39}$$

which is the polarizability introduced in Sec. 5.8, consistent with Eqs. (5.147) and (5.149). Note that here we are in a time- or contour-ordered framework; we will not specify this by superscripts in order to keep the notation simple. The time- or contour-ordered polarizability is symmetric, $\chi(4, 2) = \chi(2, 4)$, because of the symmetry of L. Its causal version is the density–density response to an external potential (see Eq. (5.50)). Hence $-iv_cLv_c$ is the classical induced potential created by a classical charge: the *test charge–test charge* contribution to the screened Coulomb interaction W defined as (see Eq. (5.6))

$$W(1, 2) = v_c(1, 2) + v_c(1, \bar{3})\chi(\bar{3}, \bar{4})v_c(\bar{4}, 2), \tag{10.40}$$

where we have used the symmetry of v_c and χ.[16]

Altogether this yields the entries labeled "GW" in the second row of Tab. 10.1, where

$$\Sigma_{xc}(1, 2) = iG(1, 2)W(1^+, 2). \tag{10.41}$$

Now the interaction is screened, contrary to Hartree–Fock where with Eq. (10.17) $\Sigma_x = iGv_c$. For the calculation of W the polarizability χ, or equivalently L, is in general approximated. The simplest approximation would be $L \approx L_0$, which yields the entry GW^{L_0} in Tab. 10.1. This may be meaningful in finite systems, but would be a very bad choice in extended systems, as discussed in Secs. 5.5 and 14.3. The simplest way to go beyond this is to approximate the *irreducible* $\tilde{L}$ in Eq. (10.25) by neglecting the kernel $\delta\Sigma_{xc}/\delta G$ of Eq. (10.24), hence $\tilde{L} \approx L_0$. This yields the RPA for χ and W, discussed in Secs. 11.2 and 14.3, and leads to the GW^{RPA} in Tab. 10.1. Approximations that go beyond the RPA W are called GW^{TC-TC} in the table, for example one may obtain $\tilde{L}$ from the Bethe–Salpeter equation or TDDFT, as explained in Ch. 14. The case GW^{RPA} has been discussed in [299]. There it is called *shielded interaction approximation*. It has been obtained as the first term in a series in W by Hedin [43] (see Ch. 11). Today $\Sigma = iGW^{RPA}$ is known as *the* GW approximation. It is the topic of Chs. 11–13.

It is intuitive to expect that screening the Hartree–Fock improves the performance of the approximation, especially in extended systems. The effects of screening are discussed in Sec. 11.3. The GWA is conserving, as can be seen from the diagrams: for the GW^{L_0} version the Φ-functional that creates the self-energy is depicted on the left-hand side of Fig. 10.2. By removing one of the four equivalent Green's-function lines one obtains the corresponding self-energy, Fig. 10.3. There are four possibilities, but the multiplicity is canceled by the prefactor $1/4$ of the second-order diagram in Eq. (10.33). The corresponding second-order exchange diagram on the right-hand side of Figs. 10.2 and 10.3 is not included in the GW approximation; it is discussed in Sec. 15.4.

---

[16] For the same reason also W is symmetric, $W(1, 2) = W(2, 1)$.

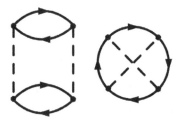

Figure 10.2. Second-order contribution to the Φ-functional. Left: direct term, contained in the GW approximation. The interaction is screened with one bubble. Right: exchange term. This term is discussed in Sec. 15.4.

Figure 10.3. Second-order contribution to the self-energy. Left: direct term, contained in the GW approximation. Right: exchange term. This term is discussed in Sec. 15.4.

Figure 10.4. Left: one of the higher-order contributions to the Φ-functional in the GW approximation; more bubbles are added. Right: corresponding self-energy.

When L is evaluated in the RPA, $\delta\Sigma_{xc}/\delta G \approx 0$ in Eq. (10.22) and one can set $1' = 1^+, 2' = 2^+$ in the Bethe–Salpeter Eq. (10.23) for L to obtain directly a Dyson equation for χ, similar to the TDDFT linear response Eq. (14.58) but without an exchange–correlation contribution. The first term is given by the two Green's functions in L_0 that are now attached at their end points: this is the *bubble* of Sec. 9.4. Iterating the RPA Dyson equation for χ yields an infinite series of these bubbles, connected by interaction lines. For example, the first second-order contribution to the Φ-functional in Fig. 10.2 or to the self-energy in Fig. 10.3 contains one bubble and two interaction lines. The equivalent third-order contribution is given in Fig. 10.4: now we have two bubbles that are connected by v_c. Adding bubbles doesn't destroy the conserving property of the approximation to any order, as one can quickly check by removing Green's-function lines. In extended systems it is important to sum *all* bubble diagrams, i.e., to solve the Dyson equation for L.

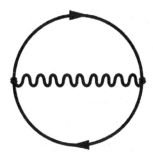

Figure 10.5. $\Psi[G, W]$-functional in the GW approximation: the wiggly line is a screened Coulomb interaction W. The corresponding self-energy is shown in Fig. 11.4.

In Sec. 8.3 an alternative functional [298] is introduced: $\Psi[G, W]$ is the universal exchange–correlation functional in terms of the screened instead of the bare Coulomb interaction. The GW approximation reads

$$\Psi^{GW}[G, W] = -\frac{1}{2}\mathfrak{Tr}\,(GWG).\qquad(10.42)$$

The GWA Ψ-functional is depicted in Fig. 10.5. The figure shows how compact the diagrammatic representation becomes when one moves from a functional of v_c to a functional of W: the infinite series of diagrams containing bubbles is replaced by just one diagram in terms of W.

The T-matrix approximation

One can in principle iterate Eq. (10.37) by inserting a given approximation to Σ_{xc} in the calculation of Ξ (Eq. (10.22)) to get a higher-order expression. This is sometimes done starting from the GW self-energy, as discussed in Sec. 15.2. It is indicated by an entry $GW\tilde{\Gamma}^{(1)}$ in the third row of Tab. 10.1. Then one gets a new effective interaction Ξ that goes beyond the classical Coulomb interaction between charge densities.

Alternatively, one can derive another Dyson-like equation for an improved effective interaction, in what is often called a *T-matrix approximation* [321, 324]. Let us start from an ansatz that generalizes the idea of the GW self-energy, and write Σ as an integral of the Green's function and a generalized effective interaction T,[17]

$$\Sigma(1, 2) = iG(\bar{4}, \bar{3})T(1, \bar{3}; 2, \bar{4}).\qquad(10.43)$$

---

[17] Note that there are different definitions and uses of a T-matrix. Sometimes the four-point vertex $^4\tilde{\Gamma}$ Eq. (8.25) that appears in Eq. (10.50) is called T, for example in the reformulation of the Bethe–Salpeter equation in [248]. Moreover, conventions may differ by a prefactor $\pm i$. One has to distinguish between these various expressions, although they are of course closely linked. Here the aim is to define a T-matrix that is the appropriate effective interaction in the self-energy and therefore yields a simple expression for Σ.

T is linked to Ξ through the functional derivative of the self-energy as

$$\Xi(1,3;2,4) = iT(1,3;2,4) + iG(\bar{5},\bar{6})\frac{\delta T(1,\bar{6};2,\bar{5})}{\delta G(4,3)}. \tag{10.44}$$

To derive the TMA, we insert Eq. (10.43) into the left-hand side of Eq. (10.37) and Eq. (10.44) into the right-hand side, neglecting the variation $\delta T/\delta G$.[18] This yields an integral equation for GT. The simplest non-trivial approximation $L \approx L_0$ leads to

$$iG(\bar{4},3)T(1,\bar{3};2,\bar{4}) = [v_H + \Sigma_x](1,2) - G(1,\bar{3})T(\bar{3},\bar{5};2,\bar{6})G(\bar{6},\bar{2}^+)G(\bar{2},\bar{5})v_c(1^+,\bar{2}). \tag{10.45}$$

This is the T-matrix approximation, entry TMA in the last row of Tab. 10.1. The lowest-order contributions to T are the classical Hartree interaction

$$T_H(1,3;2,4) = -v_c(1^+,3)\delta(1,2)\delta(4^+,3) \tag{10.46}$$

and the effective exchange interaction

$$T_x(1,3;2,4)) = v_c(1^+,2)\delta(1,4)\delta(2,3). \tag{10.47}$$

They are, besides a factor i, equal to the first-order terms in Ξ.

Since T is integrated, beyond first order Eq. (10.45) has several solutions, depending on which G is extracted on the right side. The physics of this choice can be understood from the discussion in Sec. 15.1. Here we take as example one possible class of solutions given by

$$T(1,4;2,6) = T_H(1,4;2,6) + T_x(1,4;2,6)$$
$$+ iG(1\bar{3})T(\bar{3},\bar{5};2,6)G(4,\bar{5})v_c(1^+,4); \tag{10.48}$$

for other solutions see Sec. 15.4. One can decouple the Hartree and exchange-like contributions by splitting T into $T_1 + T_2$, with

$$T_1 = T_H + iGT_1Gv_c \quad \text{and} \quad T_2 = T_x + iGT_2Gv_c. \tag{10.49}$$

Note that the relation between T_H (Eq. (10.46)) and T_x (Eq. (10.47)) implies $T_2(1,4;2,6) = -T_1(1,4;6,2)$.

Equations (10.49) have a four-point Dyson-like form, like the Bethe–Salpeter equation (10.23). To see their meaning, one can iterate them, as suggested by Ex. 10.9. This leads, for example, to the diagrams in Fig. 10.6. Visibly, there are no bubbles, hence no screening. Instead, the diagrams look like *ladders*: these indicate scattering between particles (in Fig. 10.6, these are two holes). Here the two particles have the same direction in time: contrary to the GWA, which contains correlated electron–hole pairs in W, the TMA Eqs. (10.49) correlate two electrons or two holes. Note, however, that the difference appears

---

[18] This approximation is similar in spirit to most applications of the Bethe–Salpeter equation, where one starts from the GW self-energy with the effective interaction W and neglects $\delta W/\delta G$ in the calculation of the kernel, as discussed in Sec. 14.5.

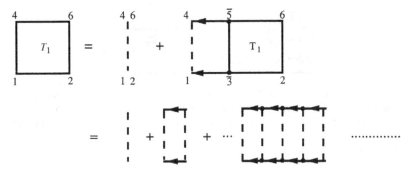

Figure 10.6. T-matrix approximation to an effective four-point interaction: a series of two-particle scattering events appears. The T_1 contribution on the left side of Eq. (10.49) for the case of particle–particle scattering is shown. The second line of the graph represents an iteration of the equation starting from $T_1 = T_H$. Note the similarity with the electron–hole scattering diagrams of the electron–hole Bethe–Salpeter equation in Ch. 14.

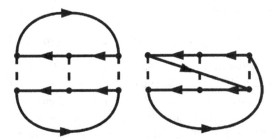

Figure 10.7. A third-order contribution to the TMA for the Φ-functional: one can see the ladder structure representing hole–hole scattering. Left: direct term T_1. Right: exchange term T_2. The corresponding self-energy can be depicted, for example, by omitting the Green's-function line at the bottom of the graph. Note that if the Coulomb interaction in the center of the left figure is omitted, one has a second-order contribution that is indistinguishable from the second-order bubbles contribution Fig. 10.2.

only starting from the third order. You can see this by eliminating the ladder leg in the center of the left diagram in Fig. 10.7: this yields just the same as the left diagram[19] in Fig. 10.2.

The TMA is conserving, as can be seen from the corresponding Φ-functional in Fig. 10.7. It can also be derived [300] as the lowest-order approximation to the functional of the four-point vertex $^4\tilde{\Gamma}$ mentioned in Sec. 8.3. This is then analogous to the GW approximation being the lowest-order approximation to the $\Psi[G, W]$-functional. The T-matrix approximation is discussed further in Secs. 15.4 and 15.6.

[19] Note that the exchange diagram on the right side is missing in the second-order contribution to the GW approximation.

Figure 10.8. The self-energy expressed in terms of an effective interaction. Left panel: the derivative of the potential-like terms and the two-particle correlation function are highlighted, as in Eq. (10.37). Right panel: the four-point vertex function is highlighted, as in Eq. (10.50); with respect to the left panel, besides $^4\tilde{\Gamma}$, only one-particle Green's functions and the bare Coulomb interaction appear.

Alternative formulation

Finally, we give an alternative expression for the self-energy in terms of the four-point vertex $^4\tilde{\Gamma}$ (Eq. (8.25)), which is used in Sec. 21.7. This alternative formulation is obtained by combining Eqs. (10.23) and (8.25), which yields the relation $\Xi = {}^4\tilde{\Gamma}L_0L^{-1}$ and, with Eq. (10.37), the self-energy as

$$\Sigma(1,2) = v_H + \Sigma_x + iv_c(1^+,\bar{4})G(1\bar{3})^4\tilde{\Gamma}(\bar{3},\bar{5};2,\bar{6})L_0(\bar{6},\bar{4};\bar{5},\bar{4}^+). \qquad (10.50)$$

Note that the relation between the Bethe–Salpeter Eq. (10.23) and Eq. (8.25), and hence between the kernels Ξ and $^4\tilde{\Gamma}$ of those equations, is analogous to the relation between Eqs. (9.39) and (9.37), and hence their kernels, the proper self-energy Σ and its counterpart $\tilde{\Sigma}$. The two Eqs. (10.37) and (10.50) are depicted in the diagrams of Fig. 10.8. They differ by the appearance of L in Eq. (10.37), instead of L_0 in Eq. (10.50): in Eq. (10.50) all two-particle correlation terms are included in the four-point interaction vertex.

10.6 Wrap-up

In the quest for the one-body Green's function G, it is often convenient to work with a Dyson equation, an integral equation with a kernel that is the self-energy Σ. In such a way, one can do better than the straightforward perturbation expansion of G in the Coulomb interaction, which is known to have convergence problems [34]. The reason is that even a low-order approximation for Σ creates contributions to all orders in the interaction in G. These contributions are conveniently resummed to infinity in the Dyson equation.

The equation of motion for G leads to an explicit expression for the self-energy, unfortunately in terms of the still unknown two-particle Green's function. In order to obtain a closed framework, the two-particle Green's function is then expressed in terms of a variation of the one-body G with respect to an external potential, inspired by the relations of linear response. In this way one obtains a set of coupled functional–differential equations for G or equivalently for Σ, from which all diagrammatic contributions can be obtained, for example by iteration.

The formulation suggests the concepts of screening and effective interactions, which allows one to design approximations using straightforward mathematics, based on physical arguments and guided by conservation laws. In this chapter some widely used approximations have been elucidated, starting from the central equation (10.37) that gives Σ in terms of the two-particle correlation function L and an effective two-particle interaction $i\Xi$; they are summarized in Tab. 10.1.

In particular, the Hartree–Fock approximation is obtained by neglecting the two-particle correlation function L, consistent with the definition that correlation is everything beyond Hartree–Fock, as stated in Ch. 1. Different flavors of the GW approximation result when the effective two-particle interaction is simply the classical Coulomb interaction, but there is screening through some approximate L. This highlights the physics contained in the GWA: an additional electron or hole has a Hartree and an exchange interaction with the system charges. On top of this, the additional particle polarizes the system, but to do this it just "sees" the classical charge density of the system, inducing a sort of mean-field response. There is no explicit correlation between the additional charge and the system electrons. The original and most widely used version, which may be called *the* GWA, is obtained when W is evaluated in the RPA. The GW approximation is the topic of Chs. 11–13. When exchange–correlation contributions are included in the effective interaction $i\Xi$, one obtains self-energies that go beyond the GW approximation. The corrections to GW are called vertex corrections in Ch. 11, and discussed in Ch. 15. One example is the T-matrix approximation, which highlights two-particle correlation.

It is now time to make a choice and see how the expressions work out in practice. We start with the widely used GW approximation, which is the topic of the next three chapters.

SELECT FURTHER READING

Baym, G. and Kadanoff, L. P., "Conservation laws and correlation functions," *Phys. Rev.* **124**, 287, 1961 and Baym, G., "Self-consistent approximations in many-body systems," *Phys. Rev.* **127**, 1391, 1962. Fundamental papers for this chapter, with an accent on conservation laws.

Csanak, C., Yaris, R., and Taylor, H. S., "Green's function technique in atomic and molecular physics," *Adv. At. Mol. Phys.* **7**, 287, 1971. Compact presentation of the functional derivative approach and various approximations.

Kadanoff, L. P. and Baym, G. *Quantum Statistical Mechanics* (W. A. Benjamin, New York, 1964). A book giving a compact overview of the formalism, many approximations, and link to thermodynamics.

Martin, P. C. and Schwinger, J., "Theory of many-particle systems. I," *Phys. Rev.* **115**, 1342, 1959. This article introduces much of what is needed: many-particle Green's functions, their properties and equation of motion, the idea to use functional derivatives, and approximations. Moreover, the physics of the equations is discussed in depth.

Strinati, G., "Application of the Green's function method to the study of the optical properties of semiconductors," *Rivista del Nuovo Cimento* **11**, 1, 1988. Derivation of the main equations and link to spectroscopy.

Exercises

10.1 Calculate the time derivative $\partial G / \partial t$ of the one-particle Green's function using the equation of motion for the field operators in the Heisenberg representation Eq. (10.2). Pay attention to the time ordering: it can be expressed by a Θ-function that also has to be derived. Show that

Eq. (10.3) is obtained, and calculate its adjoint by differentiating with respect to t'. Calculate also the equation of motion for G_2, and derive the general equation of motion for the order-s Green's function, Eq. (10.9).

10.2 Show that when one approximates $G_2(1,2;1',2') \approx G(1,1')G(2,2') - G(1,2')G(2,1')$, a single-particle ansatz Eq. (5.105) for G is a solution of the equation of motion Eq. (10.3) at $T = 0$, and the Hartree–Fock equations are obtained.

10.3 By subtracting its adjoint from the equation of motion Eq. (10.3), show the continuity equation that relates the density $n(1) = -iG(11^+)$ and the current $j(1) = -i/2\left[(\nabla_1 - \nabla_{1'})G(1,1')\right]\big|_{1'=1+}$; see also Sec. 8.5.

10.4 Using chain rules, show that $\tilde{L}$ (Eq. (10.24)) is the variation of G with respect to the sum of external and Hartree potentials, Eq. (10.26).

10.5 Iterate the functional differential Eq. (10.12) by starting with $G = G^0$. Compare the result taken at external potential $u = 0$ that you obtain for the lowest orders in v_c with the perturbation result of Eq. (9.36). Try other starting points, for example the non-interacting Green's function G_u in the presence of the external potential u, solution of $G_u = G^0 + G^0 u G_u$, or the Hartree Green's function G_H, solution of $G_H = G_u + G_u v_H G_H$, where v_H is the Hartree potential. What do you observe comparing order by order in v_c?

10.6 By using the definitions of the Green's functions given in Sec. 5.8, show that the time arguments of the two-particle Green's function and two-particle correlation functions appearing in Eqs. (10.5) and (10.6), respectively, are such that the Fourier transform of these functions has poles at the electron addition and removal energies.

10.7 Iterate Eqs. (10.22), (10.23), and (10.37). Show that the first steps yield the first contributions to the self-energy derived as $\Sigma = \frac{\delta\Phi}{\delta G}$ from the universal functional Φ, Eq. (10.33).

10.8 Derive the last expression in Eq. (10.16) by using the fact that $GG^{-1} = 1$.

10.9 Iterate Eqs. (10.49) starting from the first order in v_c. This is most easily done by drawing the equations in the form of Feynman diagrams. Show that Fig. 10.6 is obtained from Eqs. (10.49).

11

The RPA and the GW approximation for the self-energy

Besides the proof of a modified Luttinger–Ward–Klein variational principle and a related self-consistency idea, there is not much new in principle in this paper.

L. Hedin, *Phys. Rev.* 139, A796–823 (1965)

Summary

In this chapter a set of equations is formulated that determine the self-energy and the one-body Green's function in terms of the screened Coulomb interaction between classical charges. The equations contain a correction to the classical picture in terms of a vertex function. The physical meaning of the various contributions is discussed. The simplest approximation for the vertex yields the random phase approximation for the polarizability and the GW approximation for the self-energy. Various aspects of the GWA are analyzed, with a focus on the physics that is added beyond Hartree–Fock. A brief summary of model cases illustrates the domain of validity and the limits of the GWA.

In this chapter we elaborate in more detail on the question of how to calculate the one-body Green's function from a Dyson equation with a self-energy kernel. In the previous chapter a scheme was introduced to design approximations to the self-energy. However, the question of where to stop, which pieces of physics to include and which to neglect, is not yet settled. Of course, there is no unique answer, besides the exact solution, but different strategies can be more or less advantageous in practice. In a system with a few electrons, for example, different aspects will be important than in a system with many electrons.

Here we are mostly interested in solids, or more generally in extended systems. In such systems, *screening* plays an essential role: the interaction between two charges is strongly modified, in general reduced, by the rearrangement of all the other charges. It is therefore most convenient to reformulate the equations such that screening appears explicitly. Some steps in this direction can be found in earlier chapters, in particular in Sec. 8.3, the formulation of the $\Psi[G, W]$-functional of the screened interaction W instead of the $\Phi[G, v_c]$-functional of the bare v_c. In Sec. 10.5 the screened interaction approximation for the self-energy is derived, with the self-energy as a product of the one-body Green's function and the screened interaction W.

This approximation can be understood as a logical step in the iteration of a system of equations formulated and applied to the homogeneous electron gas by Hedin in 1965; today it is called the GW approximation. Hedin recognized the continuity with earlier work; the quote at the start of this chapter cites the abstract of his fundamental publication [43], where he states that "besides the proof of a modified Luttinger–Ward–Klein variational principle and a related self-consistency idea, there is not much new in principle in this paper." Much work since then has been inspired by his contributions and way of looking at the problem. The GW approximation is widely used to compute electron addition and removal spectra of materials, in particular quasi-particle band structures and bandgaps. Chapters 12 and 13 are dedicated to practical aspects and applications of the approach.

Before moving on to those applications, it is useful to dig a little deeper into the GW approximation. This should allow us to do more with our calculations than just number crunching: one has to understand the physical content of an approximation, be able to analyze results within its framework, estimate the limits of the approximation and go beyond. After an introduction to Hedin's equation in Sec. 11.1 and to the RPA in Sec. 11.2, the GW approximation is derived in Sec. 11.3, where we concentrate on different aspects of the physics contained in this approximation. We describe links of the GWA to some widely used static mean-field approaches in Sec. 11.4, and touch upon the use of GW to determine ground-state properties, and the link to DFT, in Sec. 11.5. In Secs. 11.6 and 11.7 results for the homogeneous electron gas and small model systems, respectively, illustrate the main points and give a first idea about strengths and weaknesses of the approximation.

The chapter does not contain complicated mathematics, but rather extensive discussions of physical content. It can be read by anyone interested in many-body perturbation theory, independently of whether the aim is to be a knowledgeable user of GW calculations or to understand and develop the theory further.

11.1 Hedin's equations

In Sec. 10.5 several equivalent expressions for the self-energy, Eqs. (10.34)–(10.37), are derived. Here we formulate a set of self-consistent equations in a way proposed by Hedin that highlights important physical contributions, in particular the screened interaction W that appears explicitly as a fundamental quantity at any level of approximation. This is often crucial in extended systems, where screening plays an important role.

In the following discussion about the physics of the approximations, we will often mention "response" properties, although one may work in a time- or contour-ordered framework where this "response" is not causal. Nevertheless, since time-ordered and causal quantities are readily connected, as explained in Ch. 5, this point will not be an obstacle for the discussion. To keep the notation simple we will not use extra symbols, such as superscript T, but it should be clear from the context. Moreover, we do not specify the functional dependence on the fictitious external potential u that is set to zero at the end of the derivations.

Equation (10.37) gives the self-energy as the sum of the Hartree potential, the Fock term, and a correlation correction. Alternatively, one can start from Eq. (10.34), where the self-energy is expressed in terms of a variation with respect to the potential u introduced in Sec. 10.2. We now define the *reducible vertex function*

$$\Gamma(4, 2; 3) = -\frac{\delta G^{-1}(4, 2)}{\delta u(3)} = \delta(4, 3)\delta(2, 3) + \Xi(4, \bar{5}; 2, \bar{4})G(\bar{4}, \bar{6})G(\bar{7}, \bar{5})\Gamma(\bar{6}, \bar{7}; 3), \quad (11.1)$$

where the Dyson equation for G has been used to obtain the last equality. With Eq. (10.10) $L = \delta G/\delta \mathcal{J} = -G(\delta G^{-1}/\delta \mathcal{J})G$, Eq. (11.1) is equivalent to[1] the Bethe–Salpeter equation (10.23). For the self-energy (Eq. (10.34)), this yields $\Sigma = v_H + \Sigma_{\mathrm{xc}}$ with

$$\Sigma_{\mathrm{xc}}(1, 2) = iG(1, \bar{4})v_c(1^+, \bar{3},)\Gamma(\bar{4}, 2; \bar{3}). \quad (11.2)$$

The first, local, contribution to the vertex Γ in Eq. (11.1) leads to the Hartree–Fock self-energy; the remainder is called a *vertex correction*. In the following we mostly concentrate on the exchange–correlation self-energy Σ_{xc} and treat the Hartree term separately.

In order to correct the shortcomings of Hartree–Fock, one might try to iterate Eqs. (11.2) and (11.1) with the definition $\Xi = \delta\Sigma/\delta G$ given in Eq. (10.22). This would lead to an expansion in the bare Coulomb interaction v_c (see Ex. 11.1). However, in an extended system such a development in the bare interaction is in general not successful. To overcome the problem, a better strategy is to work as much as possible with *total* potentials, that are screened, instead of *bare* potentials. To this end one defines a new vertex function $\tilde{\Gamma}$ for which the vertex *corrections* are less important, i.e., where $\tilde{\Gamma}$ is closer to the local term in Eq. (11.1) given by the δ-functions. In this way, the vertex can be more efficiently approximated. The total potential including the self-energy is, however, a complicated object. The idea is therefore to add just the Hartree potential to the bare potential u. One can expect that including the induced Hartree contribution to the variation of the potential makes a significant difference, since it covers the strong classical electrostatic effects. At the same time, the total potential remains local, which keeps the structure of the equations on the same level of complexity as when one works with the bare u (to think more about this choice or possible alternatives, see Ex. 11.2 and App. F). We call the sum of the external potential u and the Hartree potential v_H the *total classical potential* $v_{\mathrm{cl}} \equiv u + v_H$. This leads to the definition of the *irreducible* vertex

$$\tilde{\Gamma}(4, 2; 3) = -\frac{\delta G^{-1}(4, 2)}{\delta v_{\mathrm{cl}}(3)}. \quad (11.3)$$

The variation of v_{cl} with respect to the external potential is the inverse dielectric function introduced in Sec. 5.5,

$$\epsilon^{-1}(1, 2) = \frac{\delta v_{\mathrm{cl}}(1)}{\delta u(2)} = \frac{\delta[u(1) + v_H(1)]}{\delta u(2)} = \delta(1, 2) + v_c(1, \bar{3})\chi(\bar{3}, 2), \quad (11.4)$$

where the polarizability χ is the variation of the density with respect to the external potential Eq. (10.39), or equivalently Eq. (5.56). The screened Coulomb interaction W given in Eq. (10.40) in terms of χ is then

---

[1] To be precise, it is equivalent to the Bethe–Salpeter equation for $L(1, 2, 1', 2' = 2^+)$.

$$W(1, 3) = \epsilon^{-1}(1, \bar{5})v_c(\bar{5}, 3). \tag{11.5}$$

We can now insert v_{cl} in a chain rule in the derivative in Eq. (10.16) and, using the definition Eq. (11.3), replace Eq. (11.2) by the desired expression

$$\Sigma_{xc}(1, 2) = iG(1, \bar{4})W(1^+, \bar{3})\tilde{\Gamma}(\bar{4}, 2; \bar{3}). \tag{11.6}$$

The appearance of the weaker, screened interaction in Eq. (11.6) instead of the bare v_c in Eq. (11.2) is a crucial step forward in the construction of valid approximations, in particular in extended systems, as will be discussed in Sec. 11.3.

The inverse of Eq. (11.4) is

$$\epsilon(1, 2) = \frac{\delta u(1)}{\delta v_{cl}(2)} = \frac{\delta[v_{cl}(1) - v_H(1)]}{\delta v_{cl}(2)} = \delta(1, 2) - v_c(1, \bar{3})P(\bar{3}, 2), \tag{11.7}$$

with the *irreducible* polarizability[2]

$$P(1, 2) = -i\frac{\delta G(1, 1^+)}{\delta v_{cl}(2)} = \frac{\delta n(1)}{\delta v_{cl}(2)}. \tag{11.8}$$

This is a diagonal element of the irreducible two-particle correlation function given in Eq. (10.26),

$$P(1, 2) = -i\tilde{L}(1, 2; 1^+, 2^+). \tag{11.9}$$

The full, or reducible, χ and the irreducible P are related by

$$\chi(1, 2) = \frac{\delta n(1)}{\delta u(2)} = \frac{\delta n(1)}{\delta v_{cl}(\bar{3})}\frac{\delta v_{cl}(\bar{3})}{\delta u(2)} = P(1, \bar{3})\epsilon^{-1}(\bar{3}, 2) = P(1, 2) + P(1, \bar{3})v_c(\bar{3}, \bar{4})\chi(\bar{4}, 2). \tag{11.10}$$

The last equality is the Dyson equation for χ. It is the two-coordinate diagonal of the four-coordinate relation Eq. (10.25) between the full reducible and irreducible two-particle correlation functions. The relations can be used in Eq. (10.40) to obtain a Dyson equation for W. These manipulations result in a closed set of equations, usually referred to as "*Hedin's equations*":

$$\Sigma_{xc}(1, 2) = iG(1, \bar{4})W(1^+, \bar{3})\tilde{\Gamma}(\bar{4}, 2; \bar{3}) \tag{11.11}$$

$$W(1, 2) = v_c(1, 2) + v_c(1, \bar{3})P(\bar{3}, \bar{4})W(\bar{4}, 2) \tag{11.12}$$

$$P(1, 2) = -iG(1, \bar{3})G(\bar{4}, 1)\tilde{\Gamma}(\bar{3}, \bar{4}; 2) \tag{11.13}$$

$$\tilde{\Gamma}(1, 2; 3) = \delta(1, 2)\delta(1, 3) + \frac{\delta \Sigma_{xc}(1, 2)}{\delta G(\bar{4}, \bar{5})}G(\bar{4}, \bar{6})G(\bar{7}, \bar{5})\tilde{\Gamma}(\bar{6}, \bar{7}; 3) \tag{11.14}$$

$$G(1, 2) = G^0(1, 2) + G^0(1, \bar{3})\Sigma(\bar{3}, \bar{4})G(\bar{4}, 2) \tag{11.15}$$

---

[2] To be consistent with the main GW literature, we adopt the symbol P for the irreducible polarizability. The *reducible* polarizability is the density–density correlation function χ, Eq. (5.149).

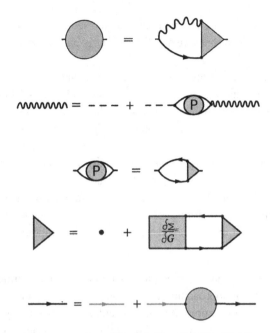

Figure 11.1. Diagrammatic representation of Hedin's Eqs. (11.11)–(11.15).

The diagrammatic form of these equations is depicted in Fig. 11.1. Note that G^0 is the non-interacting Green's function, since in Eq. (11.15) the Hartree potential v_H is included in the self-energy.

This way of formulating the problem is appealing, since it highlights the important physical ingredients:

- The *irreducible polarizability P*, Eq. (11.13), which contains the response of the system to the total classical perturbation according to Eq. (11.8), results from the creation of pairs, the two Green's functions. In particular, when the vertex is such that $\bar{t}_3 = \bar{t}_4$, as for example in the random phase approximation in the next section, these are electron–hole pairs, which correspond to a simple picture of polarization.[3] The *vertex function* $\tilde{\Gamma}$ in P contains the information that the two particles interact.

- The *effective classical interaction* between two charges in the system is not the bare interaction v_c, but its screened counterpart W. As explained before Eq. (7.11), the Dyson equation for W, Eq. (11.12), reflects the self-consistent response of the system to the external and induced classical potentials. Remember that P is the response to the total *classical* potential as expressed by Eq. (11.8), and W is screened by the induced Hartree

---

[3] The appearance of $\tilde{\Gamma}$ can introduce physics beyond the screening based on the creation of electron–hole pairs. The mathematical reason is that $\bar{t}_3$ and $\bar{t}_4$ in Eq. (11.13) can lie on opposite sides of t_1, which means that the two Green's functions can have the same time ordering, and hence both express the propagation of an electron (or both of a hole). The physical reason is that polarizing the many-body system always corresponds to multi-particle processes, not just to the creation of independent pairs, and one has to consider all possible interactions. In particular, one can have particle–particle scattering events. Such processes will be neglected in the present chapter. They are instead taken into account, e.g., in the T-matrix approximation, see Sec. 15.4.

potential. Thus, W is the potential that is created and seen by external classical charges,[4] and the dielectric function that screens $W = \epsilon^{-1} v_c$ is called the *test-charge test-charge* dielectric function.

- An electron in the system is more than an external test charge: it is a fermion that belongs to the system, indistinguishable from the other system electrons. Thus an electron must experience a total potential that contains an exchange–correlation component, which cannot be given by W. This correction appears in the *vertex* $\tilde{\Gamma}$ that changes the effective interaction in Eq. (11.11) from W to $W\tilde{\Gamma}$. The vertex correction in Eq. (11.14) stems from the functional derivative $\delta\Sigma_{xc}/\delta G$, right as the classical bare interaction is related to the Hartree potential by $v_c = \delta v_H/\delta n$. One may call the total effective interaction $W\tilde{\Gamma}$ a generalized *test-charge test-electron* screened interaction. This explains why the vertex has to be non-local with respect to its first two arguments: an induced Hartree potential is local, but an induced exchange–correlation component must be non-local, like the Fock term.

Equations (11.11)–(11.15) are in principle exact, but formidably difficult to solve. Consequently, one can think of approaching the solution by iterating the equations. As a first task, one has to find explicit expressions, in particular, the self-energy as a functional of G. As a second task, one has to evaluate the resulting expressions. Since the functional dependence of Σ_{xc} on G is universal (see Ch. 8), the Dyson equation that contains the link to a specific system through the non-interacting G^0 is not needed in the first task. One can circulate the remaining four equations as indicated on the left side of Fig. 11.2. Most conveniently one starts with an approximate self-energy, e.g., $\Sigma_{xc} = 0$ or some mean-field potential, or with an approximate vertex, e.g., $\tilde{\Gamma} = \delta$. Typically only few steps will be carried out, otherwise the resulting functional expressions become untractable.

Once the functional expressions have been obtained, one has to define the functions with which the functionals will be evaluated. One may start with a good guess for G, such as a Kohn–Sham or Hartree–Fock Green's function. This allows one to do a one-shot calculation of all expressions. Then one can circle around the pentagon, including now also the Dyson Eq. (11.15), in order to obtain a new G and evaluate all expressions again, up to self-consistency. This is schematized in the right panel of Fig. 11.2, for the case of the GWA functional. In that case, the vertex $\tilde{\Gamma}$ does not depend on G and the pentagon reduces to a quadrangle. If one stops after the one-shot calculation, this is called G_0W_0. The various levels of self-consistency in the GWA are discussed in Sec. 11.3.

Before concluding this section let us make a consideration that is not indispensable for the following, but may be interesting if one wishes to go beyond. Hedin's equations are in principle exact, but it is clear that in practice only one, or maybe two, iterations of the pentagon will be performed in order to obtain an approximate expression for the self-energy. This is the reason why it is so vital to formulate the equations in terms of the quantities that dominate the physics of the problem: here we put the emphasis on the screening of the

---

[4] The system itself instead contains internally all exchange–correlation effects, unless P is approximated in the RPA, Sec. 11.2.

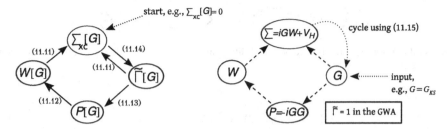

Figure 11.2. Hedin's coupled equations. Four of them determine the universal functional dependence of the self-energy. The fifth one is the Dyson equation that contains the information about the specific system. Approximate expressions for the functional form of the self-energy can be derived by circulating the first four equations, starting typically from an approximation for Σ_{xc}, or for $\tilde{\Gamma}$. This is shown in the left panel. Once the desired level of approximation of the functional expressions is reached, one can evaluate them for a given system. This is done by starting with some guess for G, that is used to build P and Σ. Keeping the form of the approximate functionals fixed, one can then circulate numerically the pentagon, including the Dyson equation. This is shown in the right panel for the case of the GWA, where $\tilde{\Gamma} = 1$ does not depend on G. If only one step is performed, one has the $G_0 W_0$ approach.

Coulomb interaction. What happens if one tries to use the approach for cases where other aspects are dominant? In principle, one could imagine cycling through the pentagon until these features have been picked up by the formula. In practice, this is neither feasible nor is there any guarantee of convergence. Stated roughly, *perturbation theory is only useful when one can stay at low order, and/or resum certain classes of diagrams to infinity*. This means that this formulation of Hedin's equations already points to a certain domain of applications, whereas for other cases one will have to change the paradigm. For example, one can reformulate the set of equations in terms of the two-particle vertex $^4\tilde{\Gamma}$ defined in Eq. (8.25) [300]. This is strictly equivalent, but suggests different low-order approximations, in particular the T-matrix approximation introduced in Sec. 10.5 and discussed further in Sec. 15.4. In that case, scattering events are privileged, which is important in the low-density regime, as opposed to screening in the high-density regime that is well-described by the GWA.

 With this remark in mind, let us move on to explore the GWA: as will become clear in this and subsequent chapters, this approximation often contains the essential physics of ground- and excited-state properties for a wide range of materials.

11.2 Neglecting vertex corrections in the polarizability: the RPA

The simplest approximation to $\tilde{\Gamma}(1, 2; 3)$ assumes this function to be diagonal in all coordinates and unity, $\tilde{\Gamma}(1, 2; 3) = \delta(1, 2)\delta(1, 3)$: *vertex corrections*[5] are neglected. The irreducible polarizability is then

---

[5] From now on, vertex corrections refer to the *irreducible* vertex Eq. (11.3).

Figure 11.3. Dyson equation for the polarizability χ in the bubble, or random phase, approximation, where $P \approx P_0$.

$$P(1, 2) \approx P_0(1, 2) \equiv -iG(1, 2^+)G(2, 1^+); \qquad (11.16)$$

it consists of non-interacting *electron–hole* pairs, as one can see from the opposite time ordering of the two Green's functions. The full reducible polarizability χ is obtained directly from Eq. (11.10). With the approximation Eq. (11.16), it reads

$$\chi(1, 2) = P_0(1, 2) + P_0(1, \bar{3})v_c(\bar{3}, \bar{4})\chi(\bar{4}, 2) = P_0(1, \bar{3})[1 - v_c P_0]^{-1}(\bar{3}, 2). \qquad (11.17)$$

The first relation in Eq. (11.17) is depicted in Fig. 11.3.[6] The approximation is called *bubble approximation* because Fig. 11.3 expanded order by order in v_c yields a series of closed electron–hole pair diagrams that look like bubbles. This is equivalent to the diagrammatic expansion of the screened interaction (Fig. 9.9). One also finds the name *random phase approximation*. Strictly speaking, RPA was introduced by Pines and Bohm [30] for the HEG discussed in Sec. 3.1, where it leads to the Lindhard dielectric function [325]. Pines and Bohm called it *random phase* approximation since the single-particle wavefunctions in the HEG are plane waves, so one finds sums over exponentials in the expression for the response function. The RPA of Bohm and Pines consisted of retaining only the dominant terms, where "random phases" were supposed to average to zero. Hubbard showed [326] that the bubble approximation gave results similar to the original RPA [30] for the HEG. The bubble expansion is, however, more general, not limited to the HEG. It is equivalent to the time-dependent Hartree approximation [325], when Eq. (11.16) is evaluated with Hartree Green's functions. Often one uses the term RPA whenever Eq. (11.16) is evaluated with independent-particle Green's functions. $P \approx P_0$ is then the response to the total independent-particle potential. We adopt this terminology in this book.

When G is an *interacting* one-particle Green's function, as it appears, e.g., in a self-consistent GWA calculation, P_0 in Eq. (11.16) is not a derivative of the density with respect to a potential. Hence it is, strictly speaking, not an RPA response function. We will then rather talk about an *RPA form*.[7] The problem of self-consistency in the RPA is touched upon in Sec. 12.4.

The impact of the RPA appears in many places in this book, in particular in total energy calculations, Sec. 13.5 (for example, as an approximation to Eq. (4.30)), when looking

---

[6] One can see this graphically starting from the Bethe–Salpeter equation (Fig. 14.1) when the kernel (Fig. 14.3) contains only the Coulomb part, by attaching the end points of the Green's-function lines.

[7] Note that the term RPA is still used in other ways in the literature; in the chemistry community the RPA is often extended to include the variation of the bare exchange. This is equivalent to having a bare interaction in the second contribution to the kernel of the Bethe–Salpeter equation (see Fig. 14.3). In that case one cannot reduce the equation directly to a two-argument equation, but one has first to solve it with four arguments. The approximation is called random phase approximation with exchange (RPAE), but sometimes simply RPA.

at spectra, Ch. 14, and when estimating the Jastrow factor in a correlated wavefunction, Sec. 6.6.

11.3 Neglecting vertex corrections in the self-energy: the GW approximation

The GW self-energy

With the approximation $\tilde{\Gamma}(1,2;3) = \delta(1,2)\delta(1,3)$, the self-energy becomes

$$\Sigma_{xc}^{GW}(12) = iG(12)W(1^+2), \tag{11.18}$$

which is equal to the screened interaction approximation Eq. (10.41). This is known as Hedin's GW approximation (GWA) [43]. Figure 11.4 shows the diagrammatic representation of this expression.

The GWA self-energy looks similar to the Fock operator $\Sigma_x = iGv_c$ (Eq. (10.17)), but it replaces the bare Coulomb potential v_c with the *dynamically screened potential W*. Screening is indeed the essential point of the GWA. This is consistent with the fact that the roots of the GWA can be traced back to developments made for the homogeneous electron gas, like the work of Quinn and Ferrel [327] who derived an expression for the inelastic lifetime, limited by electron–electron scattering, in the high-density limit. Further developments, such as extensions by Ritchie [328] and Quinn [329] to energies away from the Fermi surface, or by Adler [330] and Quinn [331] to periodic lattices, have appeared in the literature before the general formulations [43, 299] that essentially sketch the path followed in this book. Note that the GW self-energy is of first order in W; the higher-order contributions are contained in the vertex corrections. A way to understand the improvement of the GWA with respect to Hartree–Fock is to see it as the first term in an expansion with respect to the screened, and hence weaker, interaction.

The Coulomb interaction v_c, and therefore also W, is spin-independent. Therefore, Σ_{xc} carries the spin-dependence of G, which is spin-diagonal for a spin-independent hamiltonian. For this reason spin indices will appear as a single subscript for the rest of this chapter, or simply be suppressed when appropriate.

Non-local Fock exchange: a reminder

The GWA is often used to correct bandgaps after a calculation of Kohn–Sham eigenvalues, as illustrated for example in Fig. 13.5. It is therefore important to remind oneself of a fundamental difference: there is no general Koopmans' theorem (see Secs. 4.1 and 4.4) for the

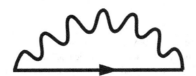

Figure 11.4. Self-energy diagram in the GWA.

local Kohn–Sham potential, but there is one for Hartree–Fock with its non-local exchange part. The fact that the bandgap opens with respect to the Kohn–Sham eigenvalue difference is entirely due to the non-local Fock exchange contained in the GWA. Correlation, in the form of screening, *reduces* bandgaps in the GWA framework as given by the example of the band structure of germanium in Fig. 2.6. The role of exchange is also to cancel the spurious self-interaction in the Hartree potential. This cancellation is often incomplete in approximate DFT, but restored by the Fock contribution $\Sigma_x = iGv_c$ in the GWA.[8]

Static screening of the exchange

It is useful to examine the various aspects that W brings into play. If we neglect for the moment its non-locality in time by replacing $W(\mathbf{r}, t; \mathbf{r}', t')$ with an instantaneous interaction $W(\mathbf{r}, \mathbf{r}'; \omega = 0)\delta(t' - t)$, the GWA Eq. (11.18) becomes

$$\Sigma_{xc\sigma}(\mathbf{r}, t; \mathbf{r}', t') = -\rho_\sigma(\mathbf{r}, \mathbf{r}')W(\mathbf{r}, \mathbf{r}'; \omega = 0)\delta(t' - t^+), \qquad (11.19)$$

where $\rho_\sigma(\mathbf{r}, \mathbf{r}')$ is the spin σ-component of the single-particle density matrix, Eqs. (5.6) and (5.121). The self-energy Eq. (11.19) is nothing else but the statically screened Fock operator.

Despite its simplicity, this screened exchange (SEX) approximation constitutes significant progress with respect to Hartree–Fock. Static screening contains the information that all electrons relax to minimize the energy of the system when an electron or hole is added. Screening reduces the bandgap in solids. This is due to the fact that there is usually a large difference between exchange contributions in the valence and conduction bands, due to a different localization of the electrons. These exchange contributions are reduced by the screening. The effect can be seen in the example of germanium in Fig. 2.6. Screening is quantitatively and qualitatively important in metals: for the homogeneous electron gas Eq. (11.19) reads

$$\Sigma_{xc\sigma}(\mathbf{k}) = -\frac{1}{(2\pi)^3} \int d\mathbf{q}\rho_\sigma(\mathbf{q})W(\mathbf{k} - \mathbf{q}; \omega = 0). \qquad (11.20)$$

In Hartree–Fock, the unscreened $v_c(\mathbf{k}-\mathbf{q}) = 4\pi/|\mathbf{k}-\mathbf{q}|^2$ favors contributions to the integral around $\mathbf{q} \simeq \mathbf{k}$. By picking $\rho_\sigma(\mathbf{q}) \simeq \rho_\sigma(\mathbf{k})$ this causes a singular variation of the self-energy and of the quasi-particle energy when $\mathbf{k}$ passes through the Fermi surface, and hence gives rise to an unrealistic excitation spectrum [332]. This unphysical behavior, which stems from the long-range nature of the Coulomb interaction, is a general problem of Hartree–Fock in condensed matter.[9] With screening, the interaction takes the form $W(\mathbf{k} - \mathbf{q}; \omega = 0) = 4\pi/(|\mathbf{k} - \mathbf{q}|^2 + \lambda^2)$, where λ corresponds to an inverse screening length, and the problem disappears [43, 287, 334].

---

[8] There is, however, self-screening in the GWA, as discussed in Sec. 11.7.

[9] In the next section hybrid functionals are mentioned, which add a portion of bare Fock exchange to Kohn–Sham correlation. In extended systems they are used together with the concept of range separation [333], i.e., the long-range part of the Coulomb interaction is treated differently, because of the above-mentioned problems.

The screened exchange approximation contains only contributions from occupied states in Σ_{xc}, because the $\delta(t'-t^+)$-function in Eq. (11.19) is placed asymmetrically with respect to the origin in time. In reality, however, the polarization contribution $W^P = W - v_c$ defined in Eq. (5.64) has a finite extension of the time range; the typical time to build up a screening cloud. $W^P(t^+, t')$ therefore has the same weight for $t > t'$ and $t < t'$, and if one wants to approximate this with a δ-function, one should place it symmetrically. With the definition of the time-ordered Green's function, Eq. (5.78), and $\Theta(t)\delta(t) = \delta(t)/2$, this yields

$$\Sigma_{xc\sigma}(\mathbf{r}, t; \mathbf{r}', t') = \delta(t'-t)\left[\Sigma_{x\sigma}(\mathbf{r}, \mathbf{r}') - \frac{1}{2}\{\rho_\sigma(\mathbf{r}, \mathbf{r}') - \rho_\sigma^{empty}(\mathbf{r}, \mathbf{r}')\}W^P(\mathbf{r}', \mathbf{r}; 0)\right],$$
(11.21)

where the Fock operator Σ_x is given in Eqs. (10.17) and (4.7), ρ^{empty} is a sum over empty states, or, with $\rho^{empty} = \delta(\mathbf{r}, \mathbf{r}') - \rho$,

$$\Sigma_{xc\sigma}(\omega) = -\rho_\sigma(\mathbf{r}, \mathbf{r}')W(\mathbf{r}', \mathbf{r}; \omega = 0) + \frac{1}{2}\delta(\mathbf{r}, \mathbf{r}')W^P(\mathbf{r}', \mathbf{r}; \omega = 0).$$
(11.22)

This static approximation, called "*Coulomb hole plus screened exchange*" (COHSEX), adds a "Coulomb hole" to the screened exchange term Eq. (11.19). The static Coulomb hole is the local induced potential that results from the adiabatically built up induced charge density around an external point charge [43, 335], due to the fact that the electrons repel each other. The static COHSEX approximation contains already much useful physics [287], such as the classical screening of a localized charge, the classical energy of a highly excited Rydberg state of a free atom in agreement with the result of Born and Heisenberg [336] (see Ex. 11.6), or the image potential of an electron outside a metal surface.

Dynamical contributions

It is now interesting to examine the frequency dependence of the self-energy, because this aspect distinguishes the GW approximation from any static mean-field theory, and in particular it makes it *qualitatively* different from Hartree–Fock. Although the considerations are general, let us for simplicity discuss this expression for vanishing temperature and fixed particle number. The frequency Fourier transform of Eq. (11.18) yields

$$\Sigma_{xc\sigma}(\mathbf{r}, \mathbf{r}'; \omega) = \lim_{\eta \to 0^+} \frac{i}{2\pi} \int d\omega' G_\sigma(\mathbf{r}, \mathbf{r}'; \omega + \omega')W(\mathbf{r}, \mathbf{r}'; \omega')e^{i\eta\omega'},$$
(11.23)

where the infinitesimal imaginary part in the exponent stems from the infinitesimal shift of the time argument of W in Eq. (11.18). The dynamical properties of W are determined by the dynamical properties of the polarizability, the density–density correlation function χ given by Eq. (5.149). Therefore, it has a spectral representation that can be obtained using the time-ordered counterpart of the retarded expression Eq. (5.146) for an electron–hole correlation function. This is equivalent to splitting $W = v_c + v_c\chi v_c \equiv v_c + W^P$ into the bare v_c and the dynamic polarization contribution W^P as defined in Eq. (5.64), and using the spectral representation Eq. (5.36) for W^P. The screened Coulomb interaction becomes

$$W(\mathbf{r}, \mathbf{r}'; \omega) = v_c(\mathbf{r}, \mathbf{r}') + \sum_{s \neq 0} \frac{2\omega_s W_s^P(\mathbf{r}, \mathbf{r}')}{\omega^2 - (\omega_s - i\eta)^2}, \tag{11.24}$$

where $s \neq 0$ labels an excited state of the many-body system. The amplitude $W_s^P(\mathbf{r}, \mathbf{r}') = V^s(\mathbf{r})V^{s*}(\mathbf{r}')$ is a product of the *fluctuation amplitudes* $V^s(\mathbf{r}) = \int d\mathbf{r}' v_c(\mathbf{r}, \mathbf{r}') f_{s0}^{eh}(\mathbf{r}', \mathbf{r})$. The V^s are potentials created by the spin-integrated local electron–hole amplitudes f_{s0}^{eh}, Eq. (5.145), for an excitation from the ground state. The corresponding energy difference is $\omega_s = E_s - E_0$, the energy of a neutral excitation.[10]

With the Lehmann representation of Eq. (5.104) for the Green's function, the exchange–correlation self-energy in the GWA becomes

$$\Sigma_{xc\sigma}(\mathbf{r}, \mathbf{r}'; \omega) = \Sigma_{x\sigma}(\mathbf{r}, \mathbf{r}') + \sum_{\lambda, s \neq 0} \frac{f_{\lambda\sigma}(\mathbf{r}) f_{\lambda\sigma}^*(\mathbf{r}') W_s^P(\mathbf{r}, \mathbf{r}')}{\omega + (\omega_s - i\eta)\,\mathrm{sgn}(\mu - \varepsilon_\lambda) - \varepsilon_\lambda}, \tag{11.25}$$

where $f_{\lambda\sigma}$ are the one-body amplitudes defined in Eqs. (5.99), (5.100), and (5.88), respectively, and ε_λ are electron addition and removal energies (see Eq. (5.100)). This expression can still be cast into a COHSEX form, with a dynamically screened exchange part

$$\Sigma_{SEX\sigma}(\mathbf{r}, \mathbf{r}'; \omega) = -\sum_\lambda \Theta(\mu - \varepsilon_\lambda) f_{\lambda\sigma}(\mathbf{r}) f_{\lambda\sigma}^*(\mathbf{r}') W(\mathbf{r}, \mathbf{r}'; \omega - \varepsilon_\lambda) \tag{11.26}$$

and a dynamical Coulomb hole

$$\Sigma_{COH\sigma}(\mathbf{r}, \mathbf{r}'; \omega) = \sum_\lambda f_{\lambda\sigma}(\mathbf{r}) f_{\lambda\sigma}^*(\mathbf{r}') \sum_{s \neq 0} \frac{W_s^P(\mathbf{r}, \mathbf{r}')}{\omega - (\omega_s - i\eta) - \varepsilon_\lambda}. \tag{11.27}$$

Note that the Coulomb hole is no longer given by a local operator, because the δ-function in Eq. (11.22) is replaced by a sum over all states with an energy-dependent weight.

Dynamical effects modify the real part of the self-energy and hence the quasi-particle spectrum obtained from Eq. (7.23). The imaginary part of W gives rise to an imaginary part of Σ. With Eq. (11.25) a diagonal element in a basis $\chi_m(\mathbf{r})$ reads

$$\mathrm{Im}\, \Sigma_{xc\sigma m}(\mathbf{r}, \mathbf{r}'; \omega) = \pi \sum_{\lambda, s \neq 0} \pm \int d\mathbf{r} d\mathbf{r}'\, \chi_m^*(\mathbf{r}) \chi_m(\mathbf{r}') f_{\lambda\sigma}(\mathbf{r}) f_{\lambda\sigma}^*(\mathbf{r}') W_s^P(\mathbf{r}, \mathbf{r}') \delta(\omega \pm \omega_s - \varepsilon_\lambda), \tag{11.28}$$

where the upper (lower) sign refers to $\varepsilon_\lambda < \mu$ ($\varepsilon_\lambda > \mu$). The GWA exhibits the main features that are discussed for a general self-energy in Ch. 7. In particular, $\mathrm{Im}\, \Sigma(\omega)$ passes through zero when ω equals the chemical potential. For a material with a gap, the imaginary part vanishes between $\varepsilon_R - \omega_{s0}$ and $\varepsilon_A + \omega_{s0}$, where ω_{s0} is the lowest excitation energy, ε_R is the highest removal, and ε_A the lowest addition energy: no quasi-particle decay is possible when energy conservation doesn't allow the additional electron or hole to excite

---

[10] This is an exact expression, however one can also write approximations for the screened Coulomb interaction as a sum over poles. The following discussion holds whether or not a vertex is included in the calculation of P, whereas the situation is more complex when the vertex $\tilde{\Gamma}$ in $\Sigma = iGW\tilde{\Gamma}$, Eq. (11.12), is taken into account. Note that here we display the spectral representation as a discrete sum for the sake of clarity, but the sum over excitations s is meant to include all discrete excitations, as well as the continuous part of the spectra.

a mode ω_s. In RPA ω_{s_0} is at least the bandgap $\varepsilon_A - \varepsilon_R$, so that the imaginary part of Σ_{xc} is zero within a range of at least three times the bandgap centered around the Fermi level. As can be seen from Eq. (11.28), Im $\Sigma_{xc}(\omega)$ is significant at frequencies ω equal to the sum or difference of an addition or removal energy and a neutral excitation of the system. These excitations are dominantly plasmons and often quite well described in RPA. The peaks in the imaginary part of the self-energy give rise to satellite structures in the spectral function. Consistent with the discussions in Sec. 7.4, in the GWA these satellites are due to peaks in the loss function, which is the essential ingredient of W (see Sec. 5.5).

An illustration of these points is given in Sec. 11.6 for the homogeneous electron gas. For further discussions, we suppose that the input G, with which the self-energy Eq. (11.23) is constructed, is calculated from an independent-particle approach. The $f_{\lambda\sigma}$ are then single-particle orbitals ψ_i^σ and the $\varepsilon_{\lambda\sigma}$ single-particle energies $\varepsilon_{i\sigma}$, and a diagonal matrix element of the exchange–correlation part of the self-energy reads

$$\int\int d\mathbf{r}\, d\mathbf{r}'\, \psi_k^{*\sigma}(\mathbf{r})\Sigma_{xc\sigma}(\mathbf{r},\mathbf{r}';\omega)\psi_k^\sigma(\mathbf{r}') = \Sigma_{xk\sigma} + \sum_{i,s\neq 0}\frac{W_{s,kk\sigma}^{p,ii}}{\omega + \omega_s\,\mathrm{sgn}(\mu - \varepsilon_i^\sigma) - \varepsilon_i^\sigma},$$

(11.29)

where ω_s stands for $\omega_s - i\eta$, $\Sigma_{xk\sigma}$ is a matrix element of the Fock operator and $W_{s,kk\sigma}^{p,ii} = \int\int d\mathbf{r}\, d\mathbf{r}'\, \psi_k^{*\sigma}(\mathbf{r})\psi_i^\sigma(\mathbf{r})W_s^p(\mathbf{r},\mathbf{r}')\psi_i^{*\sigma}(\mathbf{r}')\psi_k^\sigma(\mathbf{r}')$ are matrix elements of W_s^p between pairs of single-particle states as defined in Eq. (12.44). The dominant contributions to the sum come from states close to $i = k$. Since $\mathrm{sgn}(\mu - \varepsilon_i^\sigma)$ depends on whether ε_i lies above or below μ, for states k far from the Fermi level the satellites lie on one side. Instead, for k close to the Fermi level the spectral function becomes more symmetric. This can be observed, e.g., in Fig. 11.5 later. For $|\omega - \mu| < \omega_s + |\varepsilon_i - \mu|$ one can Taylor-expand Eq. (11.29) in ω around $\omega = \mu$, which means that close to the Fermi level the self-energy is approximately linear in frequency.

Equation (11.29) is useful for understanding some of the common approximations; for example, the plasmon pole models in Sec. 12.2 are obtained when the sum over excitations s is restricted to a limited number of poles of W, which are the characteristic excitation energies of the system usually dominated by plasmons, as discussed in Sec. 14.3. Other approximations make a link to static mean-field approaches; these will be discussed in the next section.

Spatial range of the self-energy

As brought out in Sec. 8.3, the self-energy can always be written as a functional of the dressed Green's function and the screened Coulomb interaction. The interplay of these two quantities leads to the important fact that the self-energy has a relatively limited spatial range. Looking at the GWA as a leading contribution, one can understand this property from general considerations.

Let us first concentrate on the static COHSEX approximation. Bare Fock exchange multiplies the $1/|\mathbf{r} - \mathbf{r}'|$ of the Coulomb interaction with the decay of the one-body density matrix. For insulators, the density matrix decays as $e^{-\gamma|\mathbf{r}-\mathbf{r}'|}$ [337, 338], ignoring a power-law prefactor [338, 339]. In a metal at zero temperature the density matrix decays with a power law [340]; this decay becomes exponential at finite temperature [337]. Hence, for example in the non-interacting homogeneous electron gas at zero temperature, exchange is an oscillatory function with an amplitude that falls off as $1/|\mathbf{r} - \mathbf{r}'|^3$, on a scale of $1/k_F$ where k_F is the Fermi momentum [341].

This qualitative difference between metals and insulators is compensated by the screening of the exchange term. The screened Coulomb interaction decays exponentially in the homogeneous electron gas. Adding screening cures the pathologies of Hartree–Fock that stem from the long-range Coulomb interaction, as discussed in Sec. 11.2. In insulators at large $|\mathbf{r} - \mathbf{r}'|$ screening modifies the Coulomb interaction from $1/|\mathbf{r} - \mathbf{r}'|$ to $1/(\epsilon|\mathbf{r} - \mathbf{r}'|)$ where ϵ is the dielectric constant, which does not alter the overall behavior. In other words, statically screened exchange is short-ranged in metals because of screening, and in insulators because of the decay of the density matrix. The static Coulomb hole in Eq. (11.22) is local and therefore does not change this result.

The inclusion of dynamical effects broadens the Coulomb hole, but does not alter this picture significantly, since at large distances the static contribution to screening dominates [342]. The full self-energy has of course contributions beyond the GWA; however, the GWA part is generally believed to be the most long-ranged contribution. This is reasonable since higher orders, including, e.g., screened ladder diagrams in T-matrix-like approximations discussed in Sec. 15.5, consist of higher-order chains of the Green's function and the screened Coulomb interaction.

The short-range character of the self-energy is important for conceptual and practical reasons; it supports the local approximation of single-site DMFT [343] explained in Chs. 16–20, and makes real space methods for GWA calculations [344, 345] efficient, as explained in Sec. 12.7. It has been discussed and illustrated frequently, e.g., [43, 341, 346].

Conservation laws, self-consistency, and sum rules

As can be seen in Sec. 10.5, the GWA is a conserving approximation, when it is evaluated self-consistently. However, there is a tradeoff between fundamental principles, realistic results, and computational limitations that does not necessarily always lead to the same conclusion. Therefore, three main levels of self-consistency are used in practice and discussed in this book. We briefly summarize the main points in the following; more about self-consistency can be found for the particular case of ground-state properties in Sec. 11.5, some indicative results are given in Sec. 11.6, and practical aspects in Sec. 12.4. See also, e.g., [347, 348] and references therein.

- Many *non-self-consistent* calculations are performed by using an independent-particle Green's function as input to calculate W and the self-energy. This corresponds to performing the calculations in the right panel of Fig. 11.2 only once; it is called the one-shot, or G_0W_0, approach. Here "G_0" refers to the input Green's function that is used to calculate the self-energy; it is most often the result of a mean-field calculation, and contains already some interaction effects. G_0W_0 destroys the conserving property of the

GWA. Indications concerning the magnitude of deviations from number conservation in practice are given in Sec. 11.5.

- When W is kept fixed, for example as obtained from an RPA calculation with some independent-particle mean-field input, but G is updated self-consistently, one speaks about GW_0. This scheme is number conserving when W_0 has the correct symmetries [349], as one can understand from the $\Psi[G, W]$-functional in Fig. 10.5. GW_0 is, however, not momentum and energy conserving [350], because for inhomogeneous and time-dependent systems the screened interaction is not invariant under space and time translations. This does not directly impact results for spectra or total energy in equilibrium.

- When W is calculated in the RPA form, but evaluated with a self-consistently calculated Green's function, one has a *fully self-consistent* GW calculation. It leads to a one-body G that fulfills all conservation laws and does not depend on the choice of a mean field. However, the self-consistent RPA W is a problematic quantity: as mentioned in the previous section, $P = -iGG$ is not a proper response function when built with the self-consistent Green's function, i.e., it is not the functional derivative of the density with respect to some potential. In particular, the resulting inverse dielectric function does not fulfill the f-sum rule of Eq. (5.63) [351]. This is an indication of the fact that full self-consistency should be accompanied by vertex corrections.

Besides conservation laws, sum rules can tell us much about the overall quality of approximations and calculations. The sum rule on the spectral function is given in Eq. (5.95). Moreover, in [352] the following equalities are proven for both G_0W_0 and GW_0 in the HEG.

- The first moment of the spectral function is the Hartree–Fock eigenvalue [148], where the HF calculation is done with the interacting density matrix:

$$\int_{-\infty}^{+\infty} d\omega\, \omega A_k(\omega) = \varepsilon_k^{HF}. \tag{11.30}$$

- The second moment yields the square of this result, plus a correction term given by the total spectral density of the self-energy:

$$\int_{-\infty}^{+\infty} d\omega\, \omega^2 A_k(\omega) = [\varepsilon_k^{HF}]^2 + \frac{1}{\pi}\int_{-\infty}^{+\infty} d\omega\, |\mathrm{Im}\, \Sigma_k(\omega)|. \tag{11.31}$$

- The total spectral density of the self-energy is independent of k and only determined by W_0, namely

$$\int_{-\infty}^{+\infty} d\omega\, |\mathrm{Im}\, \Sigma_k(\omega)| = \frac{1}{(2\pi)^3}\int_0^{+\infty} d\omega\, dq\, \mathrm{Im}\, W_0(q, \omega). \tag{11.32}$$

In particular, the last sum rule puts a constraint on the changes that can occur between G_0W_0 and GW_0 when keeping the same W_0: with respect to a G_0W_0 calculation, GW_0 can only *redistribute* spectral weight in the self-energy.

11.4 Link between the GWA and static mean-field approaches

The GWA is a prescriptive approach that naturally includes much of the physics of other, more approximate and in part empirical approaches. These simpler approaches can partially be understood as approximations to the GWA. The screening of the Coulomb interaction, a major ingredient of the GWA, reflects the response of the electron system to an additional charge. This relaxation of the system is also contained in certain static mean-field approaches, as will be worked out in the following.

Link to ΔSCF

Let us first look at a very localized state $\psi_\ell^\sigma(\mathbf{r})$, e.g., a core level, a localized level of a cluster that is spatially well separated from all other states, or an image state outside a crystal. If the overlap between the state $\ell\sigma$ and other states of the system is small, the dominant contribution to the correlation part $\Sigma_c = \Sigma_{xc} - \Sigma_x$ of the GWA self-energy, Eq. (11.29), evaluated at the energy ε_ℓ^σ of the state is

$$\Sigma_{c\ell\sigma}(\varepsilon_\ell) = \sum_{s\neq 0} \frac{W_{s,\ell\ell\sigma}^{p,\ell\ell}}{\omega_s \, \mathrm{sgn}(\mu - \varepsilon_\ell^\sigma)} = -\,\mathrm{sgn}(\mu - \epsilon_\ell^\sigma)\frac{1}{2} W_{\ell\ell\sigma}^{p,\ell\ell}(\omega = 0), \qquad (11.33)$$

with the polarization contribution $W^p = W - v_c$ that is defined in Eq. (5.64) and its spectral representation given in Eq. (11.24). The expression Eq. (11.33) is the classical self-energy contribution, namely the Coulomb interaction of an additional charge with the adiabatically built up Hartree potential that is induced by adding the charge. Since W^p is negative, this term shifts quasi-particle energies closer to the Fermi level, consistent with the fact that it is a relaxation contribution.

In the Δ self-consistent field (ΔSCF) approach introduced in Sec. 4.7, the total energy of a system with an additional electron or hole in a state $\ell\sigma$ is calculated by populating or depopulating the corresponding orbital, and the electronic system is allowed to relax. ΔSCF calculations are performed in some mean-field approximation, for example in the LDA or in Hartree–Fock. Because of the explicit relaxation, this approach contains the adiabatic response of the electron system to the additional charge, even in Hartree–Fock. The leading contribution to the total energy, the linear response term, is $\frac{1}{2} W_{\ell\ell\sigma}^{p,\ell\ell}(\omega = 0)$ for either an electron or a hole in a state $\ell\sigma$. This lowers the energy of an empty state, defined as $\varepsilon_\ell^\sigma = E_{N+1}^{\ell\sigma} - E_N$, and increases the energy of an occupied state, defined as $\varepsilon_\ell^\sigma = E_N - E_{N-1}^{\ell\sigma}$. The result is equivalent to Eq. (11.33). Therefore, one can understand that ΔSCF and GWA may yield similar results in finite systems; see, e.g., [353–355].[11]

Link to DFT+U

Another class of localized states are the d and f-electron levels, e.g., in transition metal oxides and rare earth compounds. When the approximate equation (11.33) is justified for both occupied (d_v) and empty (d_c) states separately, the relaxation closes the gap by

---

[11] To be precise, in ΔSCF all internal potentials are allowed to vary, including the exchange and correlation ones. This goes beyond the RPA. In order to simulate an RPA result, only the Hartree potential should be relaxed.

$\frac{1}{2}(W^{p,d_v d_v}_{d_v d_v} + W^{p,d_c d_c}_{d_c d_c})$. When the matrix elements are similar, their sum is the polarization contribution to the screened self-interaction of a d-orbital. Moreover, there is the Fock contribution that pushes down the valence state by $v^{dd}_{c,dd}$, which opens the gap. The conduction state is less affected by the Fock operator.[12] The total self-energy contribution to the gap is then $v^{dd}_{c,dd} + W^{p,dd}_{dd} = W^{dd}_{dd}$, the screened Coulomb interaction U of a localized orbital with itself. The DFT+U approach in Sec. 4.5 can hence be seen as a limiting case of the GWA [356]. U is often used as a parameter. However, a fixed U is not as flexible as the first principles screening in the GWA, which may explain for example why DFT+U for fixed U and with reasonable lattice distortions cannot reproduce the metal–insulator transition in VO_2 [357, 358], whereas the GWA is successful [359–362]. The DMFT in Chs. 16–20 starts from a partially screened U, and there is an effort towards an *ab initio* calculation of this parameter, for example in constrained RPA, as explained in Sec. 21.5.

Link to the self-interaction correction

The bare Fock contribution cancels the self-interaction contained in the Hartree term, as explained in Sec. 4.1. In the limit of localized orbitals this contribution lowers the occupied states with respect to the Hartree result by the bare Coulomb interaction energy of an orbital with itself. In many approximate mean-field calculations such as the LDA, this cancellation is incomplete. The *self-interaction correction* [363] introduced in Sec. 4.5 restores this cancellation. Moreover, it eliminates self-correlation contributions. The resulting potential is orbital-dependent, which places the method in the framework of generalized Kohn–Sham, beyond standard local potentials. This explains why bandgaps are larger than the traditional KS ones, and sometimes close to GWA results. In the GWA, the direct Coulomb self-interaction is canceled by the Fock term. However, the GWA contains a spurious self-correlation because an orbital contributes to the screen itself, as worked out in Sec. 11.7. It is difficult to make a more complete comparison, since approximate functionals like the LDA cannot be expressed in terms of diagrams comparable to the contributions that make up the GW self-energy.

Link to screened exchange generalized Kohn–Sham approaches

Generalized Kohn–Sham can also be based on screened exchange with an explicit screening function [201, 364], most often obtained from a model; this can be thought of as a modeling of the screened exchange approximation (Eq. (11.19)) to the GWA. The remaining static Coulomb hole and dynamical corrections are represented by an additional local exchange–correlation potential.

Link to hybrid functionals

Also, *hybrid functionals* (see Sec. 4.5) can include screening in an effective way [365]. These are generalized Kohn–Sham exchange–correlation potentials [201] that contain a

---

[12] One can suppose, for example, that valence and conduction states have opposite spins in the region where they overlap, as may be the case in an antiferromagnet.

part of non-local exchange. The prefactor of the exchange contribution and its limitation in solids to a short-range Coulomb contribution in range-separated hybrids such as HSE03 [365] can be interpreted as a remnant of the screening [366] of the exchange in GW.

A last word on COHSEX: correlation in the GWA

The COHSEX formula shows in a simplified way how the GWA treats correlation. The classical relaxation energy due to an external charge is an integral containing the square of its density, because the induced potential is proportional to the external charge. This is what is found for ΔSCF and DFT+U at the beginning of this section. The full COHSEX matrix element, instead, is linear in the density of the additional charge. To understand COHSEX one must *suppose* that the additional charge is added in *some* place $\mathbf{r}_0$ as a δ-function, its induced potential determined in the same place $\mathbf{r}_0$, and then the interaction energy calculated by weighting with the probability distribution $|\psi_\ell^\sigma(\mathbf{r}_0)|^2$ given by the single-particle wavefunction. Correlation is now more than the response to a known classical charge distribution: it contains the notion of *probability*. Contrary to ΔSCF, one does not have to know where to place the additional charge. The Coulomb hole follows the charge wheresoever it is, similar to the exchange–correlation hole in DFT.[13] It is a strong point of the Green's function methods that such a correlation effect can be expressed in a relatively simple way.[14] This is important for a solid, because in a small system one can often guess where the charge goes, which makes the problem more classical and explains the success of ΔSCF.[15]

11.5 Ground-state properties from the GWA

As shown in Sec. 5.7 the one-body Green's function, and hence the GWA, allow one to access ground-state properties. This is appealing since it makes the approach self-contained. Moreover it can be used, in principle, to cure specific problems of ground-state DFT, when approximate functionals fail.

Charge density

The charge density is often quite well described using relatively simple density functionals. Major shortcomings of the LDA and related functionals are charge delocalization and an

---

[13] Still, the medium responds on a mean-field level: there is no correlation between the additional charge and individual system particles. This is discussed further in Sec. 11.7 and Ch. 15.

[14] This is much more difficult in DFT: imagine a ΔSCF calculation in a solid, where the wavefunction of the added particle is extended. The derivative discontinuity of the exact Kohn–Sham potential, Eq. (4.29) would lead to the correct lowest addition and highest removal energies, but most approximate functionals fail; see the discussion in Sec. 4.4.

[15] One may see a relation to the fact that in simple small systems a single determinant can have a strong overlap with the many-body wavefunction, whereas this is not true, e.g., close to degeneracy or in a solid.

overestimate of hybridization, mainly because of spurious self-interaction. The GWA contains the full Fock operator and hence should lead to improvement when used to recalculate the charge density.

In the case of a fully dynamical GWA, self-consistency in G is mandatory to fulfill Eq. (10.32) and guarantee charge conservation [13, 299, 367]. When G is obtained from a $G_0 W_0$ calculation in which the highest occupied state is calculated from the quasi-particle condition of Eq. (7.23) as $\mu = \mu_0 + \text{Re} \, \Sigma_{k_F}(\mu)$, this conservation law is not fulfilled. Deviations in the relative particle number can be calculated using the expression for the density in terms of the spectral function, Eq. (5.118), which for a spin-unpolarized system at arbitrary temperature yields

$$\frac{\delta N}{N} = \frac{2}{N} \int d\mathbf{r} \int d\omega \left[\frac{A(\mathbf{r}, \mathbf{r}, \omega)}{1 + e^{\beta(\omega - \mu)}} - \frac{A_0(\mathbf{r}, \mathbf{r}, \omega)}{1 + e^{\beta(\omega - \mu_0)}} \right]. \tag{11.34}$$

Here A_0 is the spectral function of the independent-particle Green's function used to calculate the self-energy, therefore the integral involving A_0 and μ_0 yields the correct particle number N. The integral over A, and with the new chemical potential μ, instead, is guaranteed to yield the same number N only if G and hence also μ are calculated self-consistently. At vanishing temperature the deviations are of the order of 0.1% for the homogeneous electron gas at metallic densities but up to 6% in the dilute limit [368, 369], 0.05 to 0.3% in typical semiconductors [370], between 0.1 and 1% for H_2 at distances between 1 and 6 a.u. [350], or a few percent for a four-site Hubbard cluster with two electrons and moderate correlation strength, around $U/t = 4$ [367]. Whether or not such an error can be tolerated depends on the application and required precision; for example, often one may not care about a slight loss of intensity in quasi-particle spectra, but require charge conservation in transport properties.

One-body density matrix

Similarly, from the GWA Green's function the one-body density matrix ρ is obtained using Eq. (5.120). This is interesting since its eigenvalues are the occupation numbers, and their deviation from the independent-particle values 0 or 1 are a measure of correlation, as explained in Sec. 5.2. For any static potential or static non-local self-energy the independent-particle picture holds, but the GWA self-energy is frequency dependent because of the dynamical screening. Therefore, the eigenvalues of the GWA density matrix are fractional occupation numbers, that reflect correlation. This is illustrated in Fig. 11.7 later for the simple metal sodium.

Total energy

There are different ways to calculate the total energy. One possibility is to use the Galitskii–Migdal formula [371] given in Eq. (5.129). Equation (5.128) shows that the interaction energy is related to an integral over the spectral function. At vanishing temperature the integral covers frequencies up to the chemical potential. For spectral functions that are completely situated on one side of the chemical potential, the sum rule of Eq. (11.30) is then a strong constraint. The GWA obeys the sum rule, and this may well contribute to the success of the GWA in the calculation of total energies, independent of the quality of details of the spectral function; as can be seen, e.g., for bulk silicon in Fig. 13.4, the GWA spectral function can be rather bad. For matrix elements of the spectral function between independent-particle states close to the Fermi level, and the more the system is correlated, the more there are contributions on either side of the chemical potential, and deviations appear.

Alternatively, one can use the variational functionals [300, 372–374] introduced in Ch. 8. All expressions yield the same result and are equal to the Galitskii–Migdal total energy at $T = 0$, when Σ and the Green's function are calculated self-consistently; however, when an approximate Green's function is used, the results differ. The use of variational expressions has the advantage that small deviations in G from the self-consistent expression do not alter the result, and one can hope to omit the costly self-consistency procedure. Moreover, at $T \neq 0$ the grand potential also contains the entropy.

On the GWA level, contributions to the Φ-functional and the Ψ-functional are given in Figs. 10.2, 10.4, and 10.5. The GWA Luttinger–Ward Φ-functional reads

$$\Phi_{GW} = \frac{1}{2}\mathfrak{Tr}\,[Gv_H] + \frac{1}{2}\mathfrak{Tr}\,[G\Sigma_x] - \mathfrak{Tr}\left[\frac{1}{4}(v_cGG)^2 + \frac{1}{6}(v_cGG)^3 + \cdots\right]$$

$$= \frac{1}{2}\mathfrak{Tr}\,[Gv_H] + \frac{1}{2}\mathfrak{Tr}\,[G\Sigma_x] + \frac{1}{2}\mathfrak{Tr}\,[v_cGG + \ln(1 - v_cGG)], \qquad (11.35)$$

where the first term is the Hartree contribution, and the trace $\mathfrak{Tr}$ and logarithm are defined after Eq. (9.51) or before Eq. (8.5). From Φ one can obtain the total energy or grand potential as the stationary point of different functionals, e.g., Eq. (8.15) in terms of G, or Eq. (8.14) in terms of Σ and G.

GW total energies have interesting features. In particular, the terms v_cGG in Eq. (11.35) show that long-range polarization effects corresponding to $P_0 = -iGG$, Eq. (11.16), are included. These give rise to the van der Waals dispersion interaction, which is difficult to include in simple density functionals but important in many situations, for example the structure of proteins, as shown in Sec. 2.2.

Link to the RPA in a DFT framework

One of the strong points of the variational functionals is that they can yield reasonable results even with an approximate input. What happens when they are evaluated with an independent-particle Green's function G_{ip} that stems from some mean-field calculation? Although such a Green's function may contain some interaction contributions, it has a much simpler structure than a self-consistent G, and therefore makes calculations easier. For $G \to G_{ip}$ and within the GWA, the functional of the Green's function Eq. (8.15) reads [373, 375]

$$\Omega_{Klein}[G_{ip}] = \Phi[G_{ip}] - \mathfrak{Tr}(G_0^{-1}G_{ip} - 1) - \mathfrak{Tr}\ln(-G_{ip}^{-1}). \qquad (11.36)$$

As worked out in Sec. 8.1,

$$- \mathfrak{Tr} \ln(-G_{ip}^{-1}) = \Omega_{ip}.$$

In the case of Kohn–Sham, $G_{ip}^{-1} = G_{KS}^{-1} = G_0^{-1} - v_H - v_{xc}$. This leads to a grand potential

$$\Omega_{Klein}^{GW}[G_{KS}] = \frac{1}{2} \mathfrak{Tr}[G v_H] + \frac{1}{2} \mathfrak{Tr}[G_{KS} \Sigma_x] + \frac{1}{2} \mathfrak{Tr}[v_c G_{KS} G_{KS} + \ln(1 - v_c G_{KS} G_{KS})]$$
$$- \mathfrak{Tr}[(v_H + v_{xc}) G_{KS}] + \Omega_{ip}$$
$$= \Omega_x + \frac{1}{2} \mathfrak{Tr}[v_c G_{KS} G_{KS} + \ln(1 - v_c G_{KS} G_{KS})], \tag{11.37}$$

with Ω_x the Hartree–Fock grand potential, or at vanishing temperature and for fixed particle number to the total energy

$$E^{GW}[G_{KS}] = E_x[\psi^{KS}] + \frac{1}{2} \mathfrak{Tr}[v_c G_{KS} G_{KS} + \ln(1 - v_c G_{KS} G_{KS})]. \tag{11.38}$$

The first term in the last line is the Hartree–Fock energy functional evaluated with Kohn–Sham orbitals; the remainder is the correlation correction.[16] The total energy expression of Eq. (11.38) is called RPA in the framework of DFT [208, 376, 377], since it is obtained by using the RPA density–density response function χ^{RPA} in the expression for the correlation energy, Eq. (4.30), obtained from the adiabatic connection and fluctuation dissipation theorem.

DFT-RPA reflects good features of the GWA expression of Eq. (11.35), in particular it also describes the van der Waals dispersion. One can in principle derive the Kohn–Sham potential that corresponds to Eq. (11.38) (see Ex. 11.9), but the total energy is most often evaluated with Kohn–Sham Green's functions that stem from another functional, for example the exact exchange EXX introduced in Sec. 4.5. More discussion can be found, e.g., in [378–382]; selected applications are discussed in Sec. 13.5.

11.6 The GWA in the homogeneous electron gas

In Ch. 13, comparisons of GWA results with experiment are shown. Here and in the next section we will examine model systems. In the present section the HEG, representing simple metals, is used to illustrate characteristic features and trends. In the next section, two exactly solvable models, namely the Hubbard molecule (a sort of simplified hydrogen molecule) and a polaron model (simulating, e.g., core electrons) give insight into the range of validity of the approximation and the main shortcomings.

The homogeneous electron gas has been studied intensively in the GWA. Many results can be found, e.g., in [334, 383, 384]. For simplicity, the following discussions assume vanishing temperature.

---

[16] Note that this term contains the contribution from the interaction energy as well as the correlation contribution to the kinetic energy. The contribution from the interaction part of the self-energy in the GWA is $\Sigma_c G = (v_c GG)(v_c GG)/(1 - v_c GG)$, which, as opposed to the expansion in the first line of Eq. (11.35), has all prefactors equal to one [375].

In the non-spin-polarized HEG the GWA self-energy, Eq. (11.23), reads

$$\Sigma_{xc}(k, \omega) = \frac{i}{(2\pi)^4} \int d\mathbf{q}\, d\omega'\, G(|\mathbf{k} + \mathbf{q}|, \omega + \omega') W(q, \omega') e^{i\omega'\eta}\Big|_{\eta \to 0^+} \tag{11.39}$$

with the Green's function

$$G(k, \omega) = \frac{1}{\omega - \varepsilon_k - \Sigma_{xc}(k, \omega)}, \tag{11.40}$$

and the screened Coulomb interaction

$$W(q, \omega) = v_c(q) + v_c(q) P(q, \omega) W(q, \omega) \tag{11.41}$$

built with the RPA expression

$$P_0(q, \omega) = -\frac{2i}{(2\pi)^4} \int d\mathbf{k}\, d\omega'\, e^{i\omega'\eta} G(|\mathbf{k} + \mathbf{q}|, \omega + \omega') G(k, \omega')\Big|_{\eta \to 0^+}, \tag{11.42}$$

where the factor 2 is due to the summation over spin.

Starting from an independent-particle Hartree Green's function

$$G_0(k, \omega) = \frac{1}{\omega - \varepsilon_k + i\eta\, \text{sgn}(\varepsilon_k - \mu)}, \tag{11.43}$$

one obtains from Eq. (11.42) the Lindhard dielectric function $\epsilon_L(q, \omega)$ [325], and W becomes

$$W(q, \omega) = v_c(q) / \epsilon_L(q, \omega). \tag{11.44}$$

For small enough q, before the plasmon enters the electron–hole continuum and starts to broaden, the imaginary part of $W(q, \omega)$ is peaked at the dispersing plasmon energy $\omega_p(q)$; this small q-range dominates the integral in Eq. (11.39). For this reason a plasmon pole model, as explained in Sec. 12.2, was used in the original calculations of the HEG. For a density of $r_s = 5$ the resulting real and imaginary parts of the $G_0 W_0$ self-energy (upper panel), and the corresponding spectral functions (lower panel), are shown by the continuous lines in Fig. 11.5 [385]. In agreement with the discussions in Sec. 7.4 and as pointed out in Sec. 11.3, Eq. (11.28) shows that the imaginary part of $\Sigma_{xc}(k, \omega)$ must be zero at the Fermi energy and has structure dominated by the plasmons, mostly around $\varepsilon_k \pm \omega_p$. This is observed in Fig. 11.5. There is some broadening due to the integration over the dispersing plasmon and the sum over states in Eq. (11.28). The real part, linked through the Kramers–Kronig relation, is varying rapidly in correspondence with these structures; by contrast, it is smooth at lower energies and linear around the Fermi level, as worked out in Ex. 7.6.

In the following discussion we will focus on several aspects that may be related to points brought up in Chs. 2 and 7 and to results shown in Ch. 13.

Overall spectral function. The general formula for the spectral function in terms of the self-energy is given in Eq. (7.22). It has structure whenever the real part in the denominator vanishes, i.e., the quasi-particle condition Eq. (7.23) holds, and the imaginary part of the self-energy is small, or when the imaginary part of Σ is peaked and only mildly screened by the real part in the denominator. The quasi-particle condition is visualized by

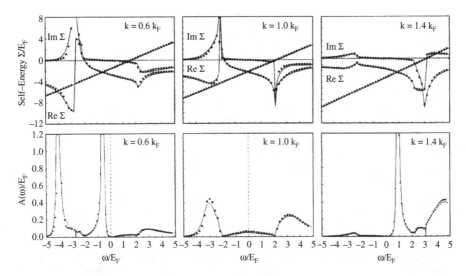

Figure 11.5. Real and imaginary parts of the self-energy of the homogeneous electron gas (upper panels) and spectral functions (lower panels) for three different momenta, from [385]. Solid lines, G_0W_0; dots, including a vertex correction discussed in Sec. 15.2. The diagonal line in the upper panel is the function $(\omega - \varepsilon_k)$ with ε_k the Hartree eigenvalue. The quasi-particle peaks in the spectral function appear where Re $\Sigma_{xc}(k,\omega)$ crosses this line. Here $r_s = 5$.

the intersection of the real part of the self-energy and the diagonal line $(\omega - \varepsilon_k)$ in the upper panel. The intersection close to the Fermi energy yields the quasi-particle peaks in the spectral function (lower panel), broadened by the imaginary part of Σ_{xc}. Since Im Σ is zero at the Fermi energy, the quasi-particle peak at $k = k_F$ is so sharp that it cannot be seen in the spectral function in this picture. Close to the Fermi level the spectral function is much more symmetric than at higher or lower energies. Qualitatively similar results are obtained for other values of r_s.

Bandwidth. By following the quasi-particle peaks as a function of k, the quasi-particle band dispersion is obtained. The result is shown in the left panel of Fig. 11.6. Note that this, as well as most results discussed in the following, are taken from [351] and obtained for $r_s = 4$. This is close to the density of sodium ($r_s = 3.99$) and allows one therefore to compare with experiments.[17] The measured bandwidth is 2.5–2.65 eV [388, 389], significantly narrower than the value of 3.23 eV that one finds for the free, i.e., non-interacting, HEG at the same density. This narrowing is attributed to correlation effects, since Hartree–Fock calculations yield an even larger bandwidth, more than 7 eV. The addition of dynamical screening in a G_0W_0 calculation leads to a strong band narrowing, both with respect to

---

[17] Note that the experimental bandwidth of simple metals is a controversial subject; see [386] and references therein. Definite theoretical benchmark data are not yet available, since QMC calculations, which yield a bandwidth about 1 eV larger than that extracted from experiment, may suffer from a poor description of the nodal surface, in particular for the bottom of the band [387], and because it is difficult to say which total energy difference is representative of the quasi-particle peak at the bottom of the band, as discussed in Sec. 13.1.

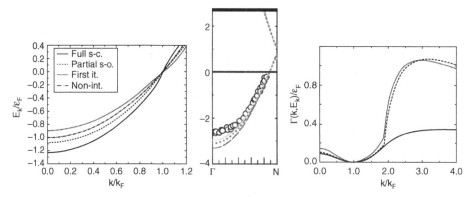

Figure 11.6. Effects of the real and imaginary parts of self-energy corrections for the HEG at $r_s = 4$. Left panel, from [351]: the dispersion of the quasi-particles is compared with a free-electron parabola (dash-dotted) for the cases G_0W_0 (dotted), GW_0 (dashed), and GW (solid). Quasi-particle energies are plotted in units of the free-electron bandwidth, and the Fermi level is set to zero. In these units the experimental bandwidth is about 0.8. Only the simplest G_0W_0 shows band narrowing; the fully self-consistent GW result has the largest bandwidth. The Hartree–Fock bandwidth is about 2, beyond the scale of the figure. Middle, from [291]: the quasi-particle band of sodium, where $r_s \approx 4$. Circles: experimental photoemission data from [388]. Dashed curve: LDA, which corresponds to the free-electron gas in the left panel. Dotted curve, G_0W_0 on top of LDA. Continuous curve, quasi-particle self-consistent (QSGW) discussed in Sec. 12.4. As in the HEG, G_0W_0 leads to band narrowing. Self-consistency with a *static* self-energy enhances this trend. Right panel, from [351]: the inverse lifetime of the quasi-particles $\frac{1}{\pi}|\text{Im} \Sigma(k, E_k)|$ at full self-consistency (GW, solid line) is compared with the corresponding quantities from GW_0 (dashed line) and G_0W_0 (dotted line) calculations. The last two results show a region with much shorter lifetimes caused by inelastic collision with plasmons. This physically intuitive feature is absent from the full GW result.

Hartree–Fock and with respect to the free-electron result. About half of the error in the free-electron result with respect to experiment is corrected by the G_0W_0 calculation; when one compares with Hartree–Fock and experiment, one can conclude that G_0W_0 recovers a large portion of the correlation effect. The middle panel of Fig. 11.6 contains experimental photoemission [388] and calculated results for sodium. Without the effect of the weak crystal potential, the potentials v_H and v_{xc} would be constant. Therefore, the LDA results correspond to the free-electron gas in the left panel. G_0W_0 on top of LDA leads to band narrowing, as observed for the HEG.

Quasi-particle broadening. The width of the quasi-particle peak corresponds to the inverse lifetime, displayed in the right panel of Fig. 11.6. One can see the expected parabolic increase with $(k - k_F)^2$ due to an enhanced probability of decay in electron–hole pairs as given in Eq. (7.24), and an abrupt increase of the inverse lifetime with the onset of energy loss through inelastic collisions with plasmons (see Ex. 11.7).

Satellites. For k around k_F, the peaks in $\text{Im} \Sigma_{xc}(k, \omega)$ give rise to plasmon satellites in the spectral function below and above the Fermi level. Intuitively one would expect that satellites are at a distance of about the plasmon frequency ω_p from the quasi-particle peak, since this is the energy loss that leads to quasi-particle decay. However, in the GWA they are further away, as one can understand from the denominator of the spectral function in

Eq. (7.22). For larger $|k - k_F|$ the quasi-particle equation $\omega = \varepsilon_k + \Sigma_{xc}(k, \omega)$ (Eq. (7.23)) has additional solutions that lead to strong peaks in the spectral function. In [383, 384] these were called *plasmarons* and analyzed as strongly coupled electron–plasmon excitations.[18] In the lower-left panel of Fig. 11.5 one can see that this structure completely masks the plasmon satellite. We will come back to this point in the discussion of the model hamiltonian in Eq. (11.46) and of the cumulant expansion in Sec. 15.7.

Quasi-particle renormalization. The satellites take weight from the quasi-particle peak. The quasi-particle renormalization factor Z (Eq. (7.26)) describes the remaining weight; for the HEG at $r_s = 3.99$, $G_0 W_0$ calculations yield a value of $Z = 0.65$ at the Fermi surface, in agreement with variational Monte Carlo calculations using backflow wavefunctions that are described in Sec. 6.7, and reasonably close to experiments on sodium that find $Z = 0.58$ [230, 390]. Since in any independent-particle calculation $Z = 1$, the transferred weight $(1 - Z)$ can be regarded as a dimensionless measure of correlation in a system. It appears that the HEG at $r_s = 4$ and the simple metals in the corresponding density range show significant correlation effects.

Momentum distribution. As discussed in Secs. 5.7 and 7.5, occupation numbers n_k are an indicator of correlation effects. The discontinuity of occupation numbers at the Fermi surface reflects the weight of the quasi-particles. In simple metals the momentum distribution $\rho(\mathbf{k})$ is close to the occupation numbers, as discussed in Secs. 5.7 and 7.5. Experimentally, one can deduce the momentum distribution from an X-ray Compton profile measurement [230]. QMC results for the momentum distribution in sodium are also available. Figure 11.7 shows a comparison of $\rho(\mathbf{k})$ calculated in LDA, $G_0 W_0$, QMC with Slater–Jastrow wavefunctions (Sec. 6.6), and determined experimentally [230]. The LDA result is very close to the step function. The small deviation reflects the effect of the lattice: since the system is not perfectly homogeneous, the momentum distribution $\rho(\mathbf{k})$ is slightly different from the occupation numbers $n_\mathbf{k}$, that are always 0 or 1 in an independent-particle

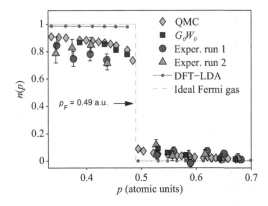

Figure 11.7. The momentum distribution function $\rho(\mathbf{k})$ of sodium (in the figure denoted $n(p)$). Shown are results of experiment, QMC with Slater–Jastrow wavefunctions, $G_0 W_0$, and LDA calculations. The step function is the free-electron gas result. (From [230].)

---

[18] Some other people call all plasmon-caused satellites plasmarons.

calculation. Experiment and all approaches beyond the independent-particle picture shown
in the figure have significant deviation from the step function and agree with each other;
similar agreement between G_0W_0 and QMC is also found at $r_s = 5$ [391].

Self-consistency. GW_0 quasi-particle energies and spectral functions have been calculated and compared with G_0W_0 in [352]. Self-consistency in G increases the bandwidth,
as can be seen in the left panel of Fig. 11.6. Figure 11.8 shows the GW_0 spectral function
for $r_s = 4$, at k_F (left) and $k = 0$ (right panel), taken from [351]. The overall features
of the GW_0 spectral functions are like those of G_0W_0 in Fig. 11.5. Self-consistency in
G decreases the distance between quasi-particle and satellite and makes the satellite less
sharp, which brings the spectrum closer to the photoemission spectra of sodium in Fig. 2.8.
Still, the satellite remains too far away; for comparison, the plasma frequency of sodium at
small wavevector is about $\omega_p = 6$ eV, which corresponds to about $1.8/\varepsilon_F$.

Also shown is the effect of self-consistency, including W, in the fully self-consistent
GW approach. As can be seen in Fig. 11.6, the fully self-consistent bandwidth is about
20% larger than the free-electron one. The satellite in the spectral function in Fig. 11.8
moves still closer to the Fermi level, but it is strongly damped and becomes practically
structureless, different from experimental observations.[19] This is caused by the continuous redistribution of weight to more and more satellites in the self-consistency procedure,
which washes out all plasmon features. This fact is also observed in the inverse lifetime in

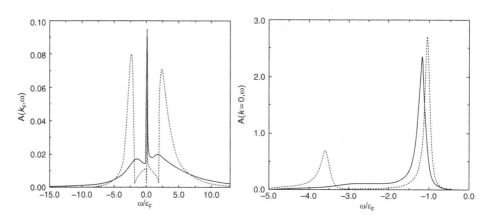

Figure 11.8. The one-electron spectral function $A_k(\omega)$ from the fully self-consistent GW calculation
(solid line) compared with that from the partially self-consistent GW_0 calculation (dashed line).
Left panel, the comparison is made at the Fermi surface $k = k_F$; right panel, same comparison at the
bottom of the band, $k = 0$. In the left panel the height of the quasi-particle peak cannot be
accommodated within the figure. Note in both figures the damping of plasmon satellites in the fully
self-consistent GW calculation. Note also in the right panel the markedly larger bandwidth in the
GW case. Here, $r_s = 4$. (From [351].)

[19] One has to be careful when comparing with experiment. Experimental satellite structure contains extrinsic and
interference effects that are beyond the intrinsic spectral function calculated here; see the subsection about
link to experiment in Sec. 5.7, and footnote 16 on p. 415.

the right panel of Fig. 11.6. At the same time, the quasi-particle peak becomes sharper and increases its weight, yielding $Z = 0.79$ in fully self-consistent GW [351] compared with $Z = 0.65$ in G_0W_0 [230] and $Z = 0.70$ in GW_0 [352]. This trend is not confirmed by experiment or QMC. Roughly speaking, with full self-consistency the system moves closer to Hartree–Fock because the average gap (defined as the distance between the center of mass of the density of states below and above the Fermi level, respectively) is increased due to the satellites. Full self-consistency in a calculation without any vertex correction hence seems to worsen the spectral properties. Alternatively, one can perform a self-consistent calculation with an effective static hamiltonian that does not transfer weight to satellites. This is done in the QSGW approach discussed in Sec. 12.4. By definition, the resulting spectral function has only the quasi-particle peaks. The resulting band [291] is given in the middle panel of Fig. 11.6, in comparison with the LDA and G_0W_0. Now, contrary to the full dynamical self-consistency, the bandwidth is further reduced with respect to G_0W_0. These results suggest that full dynamical self-consistency, including W, may be detrimental when vertex corrections are neglected, whereas QSGW avoids some of these problems.

Total energy. Concerning total energy calculations for the HEG, self-consistency has been shown to improve the results. Figure 11.9 shows a comparison of GWA calculations on various levels using the Galitskii–Migdal formulation, with QMC results [109] for both the correlation contribution to the kinetic energy (Eq. (5.126)) and the Coulomb potential energy (Eq. (5.128)) [392]. For both quantities self-consistent GW reproduces the QMC result with high precision, whereas G_0W_0 shows significant deviations; more recent results can be found, e.g., in [369]. However, it should be noted that a total energy is a

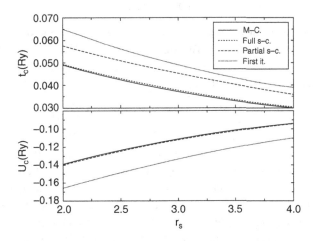

Figure 11.9. The correlation part of the kinetic (upper panel) and interaction (lower panel) energy per electron of the homogeneous electron gas as a function of r_s, obtained with different levels of self-consistency in the GWA compared with Monte Carlo results. The curve labeled "Partial s-c." is GW_0, "First it." corresponds to G_0W_0. (The figure is taken from [392], which contains the original GWA results and the QMC data from [109].)

single number that can hide many details and can benefit from error canceling. One would have to consider many more results to draw conclusions based on such numerical comparisons. Indeed, calculations on Hubbard clusters have shown [393] that self-consistency increases the total energy, but since the error in G_0W_0 depends on the filling, this increase of total energy constitutes an improvement for low filling (33%) but a worsening for larger filling (67%).

Lower dimensions. In lower dimensions interaction effects are enhanced because screening is weaker and there is confinement, so that the missing diagrams beyond the GWA should have a more dramatic effect. Total energies have been calculated for the two-dimensional gas [369], showing the same trends as in three dimensions but with increased error.

Calculations for the homogeneous electron gas in slabs and spheres of the HEG have also been used to illustrate other properties of the GWA, for example its capability to describe image states outside spheres [394] (see Sec. 13.3). Of particular interest is the fact that the GWA contains the van der Waals dispersion interaction, even when it is evaluated with independent-particle Green's functions, as discussed in Sec. 11.5. Concerning the HEG, this has been illustrated by calculating the interaction between slabs [395].

11.7 The GWA in small model systems

Complementary insight can be obtained by looking at systems for which the exact solution is known, such as the small model systems on which we concentrate in this section. This illustrates some key features of the approximations, and their drawbacks.

Hubbard dimer

We first consider a Hubbard molecule made up of two equivalent sites, each with one orbital and an on-site interaction U. This model does not reflect effects of the long-range Coulomb interaction, but allows one to probe possible problems of the GWA linked to short-range correlations. The hamiltonian is

$$H = -t \sum_{\sigma} \left[c_{1\sigma}^{\dagger} c_{2\sigma} + c_{2\sigma}^{\dagger} c_{1\sigma} \right] + U \sum_{i=1,2} \hat{n}_{i\downarrow} \hat{n}_{i\uparrow}, \tag{11.45}$$

the same as Eq. (3.10) in Ex. 3.5: it consists of a kinetic term proportional to the hopping parameter t, and the on-site interaction contribution that is proportional to U. At half-filling, with two electrons, this can be considered as a model for the hydrogen molecule H_2 and at quarter-filling, with one electron, for H_2^+. The exact solution for these two cases, as well as the G_0W_0 results, can be obtained in Exs. 3.5 and 11.4, respectively, or found, e.g., in [396]. The results only depend on U/t. This is a measure of the effective interaction strength, since the gap of the non-interacting dimer between bonding and antibonding states is $2t$, which means that the perturbation effect of the Coulomb interaction increases with decreasing t. One can expect that the larger U/t, the more it is problematic to treat U as a perturbation. Spectral functions for different sets of parameters are shown in Fig. 11.10,

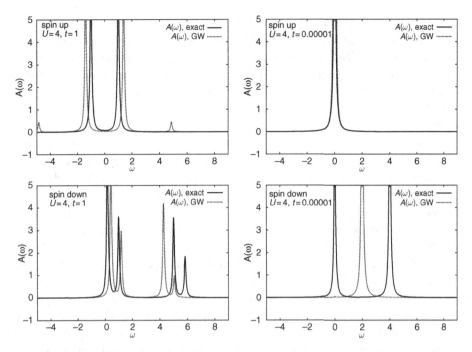

Figure 11.10. Spectral functions of the two-site Hubbard model with one spin-up electron (quarter-filling). Spin-up components are in the upper, spin-down components in the lower panels. On the left, $U/t = 4$; on the right, $U/t \to \infty$ with $U = 4$. The continuous curves are the exact result, G_0W_0 is given by the dashed lines. Note that in the atomic limit G_0W_0 finds only one average peak for the addition of a spin-down electron, because it cannot correlate the additional electron to the electron in the system, and hence it "sees" the average density on each site.

for the case where the system in its ground state contains one spin-up electron. In the left panel $U/t = 4$, whereas in the right panel $U/t \to \infty$ with $U = 4$. Solid lines are exact, dashed lines are G_0W_0 results. The upper panels are spin-up components of the spectral function, and the lower panels are spin-down ones. The exact spin-up spectral function corresponds to the non-interacting one, even for electron addition, since the additional spin-up electron has to go on an unoccupied state, and the Hubbard interaction is only on-site. The spin-down spectral function has quasi-particle peaks and satellites. The comparison between G_0W_0 and the exact solution illustrates several points.

The GWA is made for weak to moderate interaction strength. GW is by definition exact for $U = 0$. For $U/t = 4$ the GWA result shows the main features of the exact spectral function, but in the atomic limit $U/t \to \infty$, G_0W_0 and exact spectral functions are completely different.

The GWA suffers from a self-screening problem. The GWA result shows spurious satellites. Satellites are created by the dynamical W. In the GWA, screening is calculated from the *total* electron density (the one electron in the present case), which means that when the electron is removed, it screens itself and can leave the system in a spurious excited state. Therefore, the spectral function in the upper-left panel shows a spurious removal satellite. This self-screening, also called self-correlation, effect cannot be cured by moving from RPA to a more elaborate, even to the exact, irreducible polarizability and hence W: it is inherent in the GWA and the fact that the vertex $\tilde{\Gamma}$ is set to unity in $\Sigma_{xc} = iGW\tilde{\Gamma}$. Formally, it can be related to the fact that the GWA breaks the *crossing symmetry* (see Sec. 15.8) for the four-point vertex that is a consequence of the Pauli principle (see Ch. 13 in [397]).

More intuitively, as discussed in Sec. 11.1 the vertex should contain induced exchange–correlation components calculated from $\delta\Sigma_{xc}/\delta G$, on top of an induced Hartree potential. This vertex expresses the difference between the screening of a classical test charge and the screening of a fermion. In the extreme case of a single electron, the induced exchange component simply removes the self-screening, similar to the removal of the Hartree self-interaction by the Fock operator. Examples, analysis, and corrections can be found, e.g., in [398–400]; the problem is also discussed further in Sec. 15.2.

The GWA has a problem when the ground state is degenerate. In the atomic limit $t \to 0$, bonding and antibonding states become degenerate. For non-vanishing interaction $U/t \to \infty$, which is outside the regime where one would expect a perturbative expression such as the GWA to be valid. What does this mean in terms of physics? In the symmetric molecule there is equal probability of finding the electron on one of the two sites. In the absence of symmetry breaking, the electron density is half an electron on each site. However, with probability $1/2$ the electron *is* on either of the two sites, and one can imagine that the electron stays a long time on one site when the hopping t is small. Therefore, there are two exact electron addition energies: zero, when the additional electron goes to the unoccupied site and U, when it goes to the occupied one. Instead the GWA interprets the charge density as a classical static charge distribution, and will therefore yield only one electron addition energy, of $U/2$.[20] Here one would need a method that *correlates* the additional particle and the one in the system. In the present framework, this can only be obtained by a vertex in $\Sigma_{xc} = iGW\tilde{\Gamma}$. The particle–particle correlation function that is needed to describe correctly the physics of this situation is contained, e.g., in the T-matrix approximation discussed in Secs. 10.5 and 15.4, which is exact in the case considered here. DMFT in Chs. 16–20 is a different approach to describe the missing physics. The GWA total energy is unaffected by the problem of the atomic limit, because it is a one-electron property in the particular case of quarter-filling.

Also at half-filling the GWA works well for smaller U/t but fails in the atomic limit [396]. In this case, there is a problem of correlation between *spins*, since the ground state is a spin singlet. This means that a successful theory should tell an additional electron whether it meets a spin-up or a spin-down electron on a given site. Such a spin correlation is not contained in the GW diagrams. In this case even the T-matrix approximation is not sufficient to solve the problem. Correct spectral functions in the atomic limit can be

---

[20] This is the same result as Hartree–Fock. It has nothing to do with the self-screening problem, which disappears in the atomic limit, because electron–hole excitations are no longer possible in the model with one orbital per site.

obtained by breaking the symmetry, with a fixed position of the charge on one site for quarter-filling and a fixed position of the up and down spins in the case of half-filling. This is an example of the situation discussed in Sec. 4.7, where symmetry breaking can capture a part of the correlation effects in an unrestricted GWA calculation, in analogy to unrestricted Hartree–Fock.

Small Hubbard clusters are among the most extensively studied test systems for the GWA; more results pertinent for the questions raised here can be found, e.g., in [393, 401–403]. They are very simple systems, but instructive for the real case. For example, one can learn from the spin-singlet Hubbard dimer concerning the problem of paramagnetic insulators, like NiO discussed in Sec. 13.4.

The hydrogen molecule

The Hubbard dimer is an approximate version of the real H_2 molecule, with its long-range Coulomb interaction. This system can be solved numerically exactly using configuration interaction [3]. This allows one in particular to study the problem of the dissociation of molecules. Figure 11.11, taken from [404], shows the total energy of H_2 as a function of bond length. The configuration interaction (CI) result [405] is given for reference. The total energy has a minimum at the equilibrium distance and tends to the sum of the energy of two isolated hydrogen atoms at infinite distance. Also shown is the result of a restricted Hartree–Fock calculation using an exact exchange Kohn–Sham calculation as input. The total energy is too high, because the wavefunction is limited to a single Slater determinant: the error indicates the amount of correlation, which is absent in Hartree–Fock. Fully self-consistent GW does well around the equilibrium position. It corrects the large error of Hartree–Fock, and yields better results than DFT-RPA introduced in Sec. 11.5. However, it does not tend to the correct dissociation limit: the GWA total energy of two hydrogen atoms in the atomic limit is not the sum of the energies of two isolated atoms, which means that the restricted GWA is not size consistent, like restricted Hartree–Fock. Similar results have been obtained using the GWA Luttinger–Ward functional [350, 379].

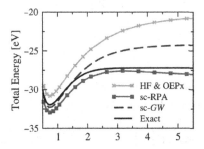

Figure 11.11. Total energy of the H_2 molecule as a function of bond length. From [404]. Fully self-consistent GW (dashed) and DFT-RPA (squares) are compared with accurate full configuration interaction calculations (continuous) taken from [405]. Restricted Hartree–Fock is shown by the crosses for comparison.

Polaron model for core electrons

One approach to describe properties of large systems is to concentrate on a small part and describe the rest as an effective medium. This is in the spirit of Ch. 7, or of DMFT in Chs. 16–20 when the effective medium is dynamic. The GWA can also be cast into this framework, since electrons are dressed by excitations of the many-body system. In this subsection we examine a model hamiltonian that reflects this idea.

In photoemission spectra of extended systems, important correlation effects are due to the excitation of plasmons. A model [406, 407] to simulate the intrinsic spectral function of an isolated level in a polarizable environment, such as a core state, is given by

$$\hat{H} = \varepsilon_0 c^\dagger c + c^\dagger c \sum_q g_q (b_q + b_q^\dagger) + \sum_q \omega_q b_q b_q^\dagger, \qquad (11.46)$$

where $c^\dagger$ creates a core electron of energy ε_0, $b_q^\dagger$ creates a plasmon of energy ω_q with a dispersion $\omega_q = \omega_p + q^2/2$, and the electron–plasmon coupling coefficient is $g_q = \sqrt{v_c(q)\omega_p^2/(2\omega_q)}$. This is a general model for electron–boson coupling and therefore also applies to the coupling between electrons and phonons.[21] It is similar to the Holstein–Hubbard model (Eq. (21.14)), but here we have no hopping and no Coulomb interaction, because the core electron is supposed to be isolated, and we are only interested in electron removal. The exact Green's function of the core electron can be calculated by a canonical transformation which removes the linear term in the boson operators [287, 406]. The core electron removal Green's function is

$$G(\tau \equiv t_1 - t_2) = i\Theta(-\tau)e^{-i\varepsilon_0\tau} e^{-i\sum_q \frac{g_q^2}{\omega_q^2}\tau} e^{\sum_q \frac{g_q^2}{\omega_q^2}(e^{i\omega_q\tau}-1)}. \qquad (11.47)$$

The first exponential gives the bare core Green's function. The second exponential leads to a quasi-particle shift, and the remaining term creates satellites. The simple exponential form allows one to identify the exponent that corrects the bare energy with a *cumulant*; this concept is explored in Sec. 15.7. Taylor expansion in g_q^2/ω_q^2 of the last term allows one to calculate the spectral function

$$A(\omega) = Z\delta(\omega - \varepsilon^{QP}) + Z\left[\frac{\beta(\varepsilon^{QP} - \omega)}{(\omega - \varepsilon^{QP})^2} + \cdots\right], \qquad (11.48)$$

with the quasi-particle energy $\varepsilon^{QP} = \varepsilon_0 + \sum_q \frac{g_q^2}{\omega_q^2}$ and the renormalization factor $Z = \exp(-\bar{n})$, with $\bar{n} = \sum_q g_q^2/\omega_q^2 = \int d\omega \beta(\omega)/\omega^2 = 0.201 r_s^{3/4}$ the mean number of plasmons that are excited. Typical values for $\bar{n}$ lie between 0.34 and 0.57, for r_s between 2 for aluminum and 4 for sodium. The function $\beta(\omega)$ is defined as

$$\beta(\omega) \equiv \frac{1}{\pi}\left(\frac{r_s^3}{12}\right)^{1/4}\frac{\omega_p^2}{\omega}\sqrt{\frac{\omega_p}{\omega - \omega_p}}\Theta(\omega - \omega_p). \qquad (11.49)$$

Full results can be found in [287].

---

[21] Indeed, the coupled electron–plasmon state has also been called an *electronic polaron*.

Equation (11.48) shows a sharp quasi-particle peak followed by a series of satellites that start at $\varepsilon^{QP} - \omega_p$, followed by $\varepsilon^{QP} - 2\omega_p$ and higher orders. The $G_0 W_0$ result is

$$A^{GW}(\omega) = Z^{GW}\delta(\omega - \varepsilon^{QP}) + A_s^{GW}(\omega). \qquad (11.50)$$

The GWA renormalization factor is $Z^{GW} = 1/(1 + \bar{n})$, which equals the exact result $\exp(-\bar{n})$ to first order in $\bar{n}$. It is comparable to the exact Z for the metallic densities. The quasi-particle peak is in the correct position, but the satellite spectrum, given by

$$A_s^{GW}(\omega) = Z^{GW}\left[\frac{\beta(\varepsilon^{QP} - \omega)}{(\omega - \varepsilon^0 - \mathrm{Re}[\Sigma_0(\omega - \Delta\varepsilon)])^2 + \pi^2\beta^2(\varepsilon^{QP} - \omega)}\right], \qquad (11.51)$$

has only one peak. Here Σ_0 is the $G_0 W_0$ self-energy of the model, with $\mathrm{Im}\,\Sigma_0(\omega) = \pi\beta(\varepsilon^0 - \omega)$. The parameter $\Delta\varepsilon$ is used to simulate self-consistency of the chemical potential, as described in Sec. 12.4. This improves the position and strength of the first satellite. If it is set to zero, one has $G_0 W_0$.

The biggest problem of $G_0 W_0$ and the exact result is the number of satellites. Since $G_0 W_0$ is of first order in W_0, this can create only one plasmon at a time, and no satellite series such as that of Fig 2.8. However, the average core electron energy $\int \omega A(\omega)d\omega$ equals the exact one, which is also equal to the non-interacting ε_0, consistent with the sum rule of Eq. (11.30). Full self-consistency in G, including its frequency dependence, can in principle create more satellites; this can be seen in the model results in Sec. 15.3. The effect is negligible in the HEG at metallic densities, as shown for example in Fig. 11.8. The problem of plasmon satellites in valence electron spectra is discussed further in Sec. 15.7, where the cumulant approach is introduced as a possible step forward.

11.8 Wrap-up

The GW approximation consists of a self-energy that is first order in the screened Coulomb interaction W. Dynamical screening is the only difference from Hartree–Fock, but it is a crucial difference. The effect of dynamical screening on the spectral function is a shift and broadening of peaks, and the appearance of satellites. These are due to the excitations contained in W, in other words, in the inverse dielectric function. Often the dominant excitations are plasmons. The GW approximation is well illustrated by Fig. 2.4: an additional hole or electron moves in an effective medium that is excited and acts back on the particle, like a boat on the water that creates waves. This can be described as a fermion coupled to bosons.

This picture anticipates the weak and the strong points of the approximation: on the downside, replacing the many-electron system by an effective medium and its dynamical response fails when explicit correlation between particles is crucial. This is the case, for example, for the Hubbard dimer in the dissociation limit, where explicit two-particle correlation is needed. Even when the picture of an effective medium, corresponding to an electron–boson coupling model, is appropriate, the $G_0 W_0$ approximation has problems. In particular, it cannot describe the simultaneous excitation of multiple bosons that lead to satellite series: it yields only one, average, satellite. Some improvement is found when G is calculated self-consistently.

On the upside, the GWA yields a good description of quasi-particles, which are very often the quantities of interest. This is illustrated by numerous results shown in Ch. 13. The reason is that the plasmons, which play the role of the bosons, are generally well described in the RPA. The fact that the GWA is conserving is also important. It can be related to the fact that errors in the spectral function average out in integrals such as the total energy, which is therefore often well described by the GWA. This also explains why self-consistency can be important for total energies, but detrimental for spectra.

These positive points are strong enough to motivate an intense use of the GWA. How this can be done, and the variety of results that can be obtained, are the topics of the next two chapters.

SELECT FURTHER READING

The main articles of L. Hedin contain much of the information given in this chapter. See, in particular, the original work in *Phys. Rev.* **139**, A796, 1965, as well as the review article *J. Phys. C* **11**, R489, 1999 and L. Hedin and S. Lundqvist, *Solid State Phys.* **23**, 1, 1969.

Two review articles with an introduction to the GWA, computational aspects and applications: Aryasetiawan, F. and Gunnarsson, O. E. "The GW method," *Rep. Prog. Phys.* **61**, 237, 1998; Aulbur, W. G., Jonsson, L., and Wilkins, J. W., "Quasiparticle calculations in solids," *Solid State Phys.* **54**, 1, 2000.

A recent book on the theory and computation of electronic excitations contains some extensions that are not included here, such as situations where the Green's functions are not spin diagonal: Bechstedt, F. *Many-Body Approach to Electronic Excitations; Concepts and Applications*, Springer Series in Solid-State Sciences No. 181 (Springer-Verlag, Berlin, 2015).

Exercises

11.1 Iterate Eqs. (11.2) and (11.1) with Eq. (10.22), starting from the Hartree–Fock self-energy. Which approximation for the self-energy is obtained at the next step? Can you find it in the diagrams of Ch. 10?

11.2 Section 11.1 discusses that the vertex correction can more easily be approximated when the total classical instead of the external potential is used in the variations. Work out what happens if one also includes exchange–correlation. Derive equations when one uses the Kohn–Sham v_{xc} and the full Σ_{xc}, respectively. Would it be an efficient strategy to work with Σ_{xc}? Why does v_{xc} not lead to the exact result, even when we take the exact Kohn–Sham potential? Some help can be found in App. F.

11.3 Start from the Hartree one-body Schrödinger equation and add a small time-dependent perturbation, in order to calculate the linear density–density response function χ to an external potential. Show that this response function obeys the Dyson equation $\chi = \chi_0 + \chi_0 v_c \chi$ (equivalent to Eq. (11.17)) where χ_0 is equal to P_0 (Eq. (11.16)), besides the fact that it is causal.

11.4 The two-site Hubbard model for one and two electrons is solved in Ex. 3.5. Use the energies and states to calculate the exact Green's function. Compare with the GWA solution (see, e.g., [396]).

11.5 In the half-filled two-site Hubbard model, examine the f-sum rule for $P = -iGG$ calculated with the fully interacting (exact) Green's function G and for $P_0 = -iG_0G_0$.

11.6 Consider a highly excited unoccupied Rydberg state localized in $\delta(\mathbf{r} - \mathbf{r}_0)$ outside a free atom situated in the origin. Making use of the fact that the overlap of its wavefunction with the charge density can be neglected because the charge density vanishes exponentially outside the atom, show from a multipole expansion that the classical result of Born and Heisenberg, $-\alpha/(2|\mathbf{r}_0|^4)$ for the polarization contribution to its energy, is contained in the COHSEX expression. (Here α is the dipole polarizability.) See [287].

11.7 Discuss the results for the inverse lifetime in the homogeneous electron gas shown in Fig. 11.6, using Eq. (11.28).

11.8 Derive COHSEX differently: suppose that at $\omega = \varepsilon_k$ the sum over states $\sum_i$ in Eq. (11.29) is dominated by contributions with $|\varepsilon_k - \varepsilon_i| < \omega_s$, where ω_s are the characteristic plasmon frequencies [287]. Then $|\varepsilon_k - \varepsilon_i|$ in the denominator can be neglected and the sum over states $\sum_i$ can be carried out using the closure relation. Show that this yields the matrix element $\langle k|\Sigma_c(\varepsilon_k)|k\rangle$ in the COHSEX approximation. Make suggestions for improvement.

11.9 Starting from the total energy Eq. (11.38), derive an equation that determines the corresponding local Kohn–Sham potential. This is an *optimized effective potential* (OEP), as introduced in Sec. 4.5. Show that this corresponds to a stationary point of the GWA functional with the constraint that the Green's function must stem from a local potential.

12

GWA calculations in practice

An idea that is developed and put into action is more important than an idea that exists only as an idea.

Buddha

Summary

In this chapter we sketch how GW calculations are performed in practice, touching upon approximations and numerical methods. Typical calculations are done in three steps: one has to determine the dynamically screened Coulomb interaction W, build the GW self-energy, and finally solve the quasi-particle or Dyson equation. All steps have their own difficulties. Choices have to be made, and the calculations are challenging for many materials. Computational approaches are constantly evolving, but many of the aspects contained in the chapter are expected to remain topical for quite some time.

GW calculations have become part of the standard toolbox in computational condensed-matter physics. Many details on foundations and putting into practice can be found in overviews and reviews, like [287, 334, 347, 408].

What does it mean to do a GWA calculation in practice? The formula for the GWA self-energy is as simple as its name, but GW calculations have a long history with continuous improvements. Modern GWA calculations are in the continuation of pioneering attempts to include correlations beyond Hartree–Fock using the concept of screening. Already in 1958 [409] correlation energies for the homogeneous electron gas were obtained from the study of the polarization of the gas due to an individual electron, and from the action of this polarization back on the electron, through the self-energy. These calculations, including several approximations, were limited to states close to the Fermi level. A GWA-like approach [410, 411] was applied to the electron gas in 1959, although these calculations didn't cover the range of densities $r_s \sim 2-5$ which is typical for simple metals. Hedin's work [43] is fundamental in that it presented the GWA as the first term of a series in terms of the screened Coulomb interaction, and it contained an extensive description of the homogeneous electron gas on the GWA level. Many more studies on the

HEG followed, including detailed investigation of the spectral functions [383, 384, 412], the importance of self-consistency [351, 352, 392], the electron gas in lower dimensions [369] (see also Ch. 11), and vertex corrections beyond the GWA (e.g., [413–415]), as discussed in Ch. 15. Calculations for realistic materials were first carried out in the tight-binding framework for diamond [416, 417],[1] and a few years later in an *ab initio* framework [192, 418–424]. Since then there has been much improvement of algorithms and huge increase of computer power, but these first calculations were a true challenge, and they opened the way to reliable calculations of band structure and spectra of many real materials.

The subject is still tricky, both theoretically and computationally. The success of the GW approximation, in particular in correcting the Kohn–Sham "bandgap problem," has motivated more and more ambitious applications to large and complex systems, many of significant technological interest, that bring new computational difficulties and make the field constantly face the limits of feasibility. At the same time, there is also more ambition concerning precision: once it has been demonstrated that the GWA increases bandgaps from Kohn–Sham eigenvalue differences towards realistic values, a smaller and smaller error bar is desired, in order to be predictive for materials properties, and to better assess the performance of the GW approximation itself. Finally, it should not be forgotten that "GW" does not tell us unambiguously which G and which W one has to use. This is a question that goes beyond numerics, but that is inherent in many-body perturbation theory, and for which guidelines are needed.

This chapter addresses a series of questions: What does it imply to do a GW calculation? What are the possible choices, both concerning ingredients and numerical realization? What are the successful approximations, and what is their consequence? Altogether, the chapter should help to go beyond a blind use of existing codes, and to develop an own-judgement of what is appropriate, what should be avoided, and how much one can trust results. With this aim Sec. 12.1 reviews the task to be carried out, and Sec. 12.2 summarizes the main approximations that can be found in the literature. Two questions merit particular attention: the separation into core and valence electrons, discussed in Sec. 12.3 and self-consistency, in Sec. 12.4. These require choices that can influence the results. The description of important technical points is divided into three sections: frequency integrations in Sec. 12.5, calculations in a basis in Sec. 12.6, and considerations concerning scaling and convergence in Sec. 12.7. A wrap-up concludes the chapter.

12.1 The task: a summary

The central target of a GWA calculation is the self-energy $\Sigma_{xc}(1, 2) = iG(1, 2)W(1^+, 2)$ of Eq. (11.18). This is simply a product of the one-body Green's function G and the screened Coulomb interaction W in space–spin–time coordinates. Most often, matrix elements of

---

[1] Note that the calculations in [416, 417] started from Hartree–Fock, and the corresponding vertex corrections beyond the GWA were included along the lines of Sec. 15.2.

the self-energy are calculated. In a basis of independent-particle states labeled $i, j, \ldots$, the matrix elements read

$$\Sigma_{\mathrm{xc},k\ell}(t_1, t_2) = i \sum_{ij} G_{ij}(t_1, t_2) W_{k\ell}^{ij}(t_1^+, t_2). \tag{12.1}$$

The screened interaction is a matrix element of *four* basis functions, see Apps. A, F, and Eq. (19.5).

The matrix elements are defined in Eqs. (12.40) and (12.42) with (12.44). In the following, we develop the specific expressions that have to be calculated. We give the results in two forms: using time-ordered Green's functions at vanishing temperature, and the Matsubara formalism for $T \neq 0$. In both cases, the general expressions of the previous chapters, in particular the Dyson equation, can be used to calculate systems in equilibrium.[2]

Let us first concentrate on the spin structure of Eq. (12.1). When the bare interaction is spin-independent, i.e., when no terms like spin–orbit interaction are added to the hamiltonian, and in the absence of non-collinear magnetic states, one can choose a reference frame in which $G_{\sigma\sigma'} = \delta_{\sigma\sigma'} G_\sigma$ is spin-diagonal, for example the z-axis along the magnetization direction in a ferromagnetic system. The bare Coulomb interaction v_c does not depend on spin coordinates. Therefore the screened interaction W is spin-independent also, since (omitting time arguments for simplicity) the Dyson Eq. (11.12) can be written as

$$W(\mathbf{r}_1, \mathbf{r}_2) = v_c(\mathbf{r}_1, \mathbf{r}_2) + \int d\mathbf{r}_3 d\mathbf{r}_4 v_c(\mathbf{r}_1, \mathbf{r}_3) P(\mathbf{r}_3, \mathbf{r}_4) W(\mathbf{r}_4, \mathbf{r}_2), \tag{12.2}$$

where the polarizability P is a sum over spins. This corresponds to Eq. (5.64), and is worked out in Ex. 12.6.

Usually the GWA is used together with the random phase form for P, introduced in Sec. 11.2. In this case $P \approx P_0(\mathbf{r}_3, \mathbf{r}_4) = -i \sum_\sigma G_\sigma(\mathbf{r}_3, \mathbf{r}_4) G_\sigma(\mathbf{r}_4, \mathbf{r}_3)$ from Eq. (11.16). The self-energy is spin-diagonal,

$$\Sigma_{\mathrm{xc},\sigma} = i G_\sigma W, \tag{12.3}$$

and its component $\Sigma_{\mathrm{xc},\sigma}$ is directly determined by the spin component of G_σ. Often one starts the calculation with a G^0 from a mean-field calculation that already includes some exchange–correlation contributions. They are included in an effective exchange–correlation potential v_{xc} that is also spin-diagonal, for example in a Kohn–Sham calculation. The Dyson Eq. (11.15) then holds separately for each spin component:

$$G_\sigma = G_\sigma^0 + G_\sigma^0 (\Sigma_{\mathrm{xc},\sigma} - v_{\mathrm{xc},\sigma}) G_\sigma. \tag{12.4}$$

Note that $v_{\mathrm{xc},\sigma}$ has to be subtracted, because it is already included in G^0.

The frequency Fourier transform of Eq. (11.18) is Eq. (11.23), which reads:

$$\Sigma_{\mathrm{xc},\sigma}(\mathbf{r}, \mathbf{r}'; \omega) = \lim_{\eta \to 0^+} \frac{i}{2\pi} \int d\omega' G_\sigma(\mathbf{r}, \mathbf{r}'; \omega + \omega') W(\mathbf{r}, \mathbf{r}'; \omega') e^{i\eta\omega'}. \tag{12.5}$$

---

[2] Using time-ordered Green's functions at non-zero temperature would require modifications [425, 426].

The frequency integration can be performed in various ways given in Sec. 12.5, for example as a sum over Matsubara frequencies (Eq. (12.31)) for calculations at non-vanishing temperature.

Calculations consist of several main steps: one has to determine a starting G, use it to calculate a polarizability P and then W from Eq. (12.2), and perform the product or convolution of G and W in Σ_{xc}. Finally, one has to solve the Dyson Eq. (12.4) to obtain a new Green's function, or the quasi-particle Eq. (7.38) or (7.39) if one is only interested in quasi-particle properties. If desired, the equations can be iterated to self-consistency. In spite of the apparent simplicity of the GWA, realistic calculations can be a formidable task. With respect to Kohn–Sham calculations, the main additional difficulties are the non-locality of the self-energy as in Hartree–Fock, and the determination and use of dynamical screening. The flowchart in Fig. 12.1 summarizes the main steps of a typical GWA calculation, together with choices that are either of a technical nature, or that concern approximations and therefore change the results (these are shown *in italic*).

12.2 Frequently used approximations

The GWA by itself is an approximation. Still, calculations may be too difficult for some systems, and additional approximations are made. Some frequently used ones are listed in the following.

Building on a single-particle Green's function G^0

To build $\Sigma_{xc} = iGW$ in Eq. (12.5), one needs a Green's function G. The simplest choice is $G = G^0$, where G^0 stems from a mean-field potential or some approximate *static* self-energy. In this case, G^0 has the spectral representation of an independent-particle Green's function,

$$G_\sigma^0(\mathbf{r}_1, \mathbf{r}_2; z) = \lim_{\eta \to 0^+} \sum_\ell \frac{\psi_{\ell\sigma}^0(\mathbf{r}_1)\psi_{\ell\sigma}^{0*}(\mathbf{r}_2)}{z - \varepsilon_{\ell\sigma}^0}, \tag{12.6}$$

where $\varepsilon_{\ell\sigma}^0$ and $\psi_{\ell\sigma}^0$ are single-particle eigenvalues and eigenfunctions. At $T = 0$ this represents the time-ordered Green's function on the real frequency axis $G^0(\omega)$, Eq. (5.105), and one has to replace $z \to \omega + i\eta \operatorname{sgn}(\varepsilon_{\ell\sigma}^0 - \mu)$ with $\eta \to 0^+$ in the denominator. In the grand-canonical ensemble at $T \neq 0$ and with $\varepsilon_{\ell\sigma}^0 \to \varepsilon_{\ell\sigma}^0 - \mu$, Eq. (12.6) is the Green's function at the Matsubara frequencies $z_m = i\omega_m$, Eq. (D.25).

The next step is the calculation of the irreducible polarizability $P_0 = -iG^0G^0$ (see Ex. 12.5),

$$P_0(\mathbf{r}, \mathbf{r}'; z) = \sum_{ij\sigma}(f_{i\sigma} - f_{j\sigma}) \frac{\psi_{i\sigma}^{0*}(\mathbf{r})\psi_{j\sigma}^0(\mathbf{r})\psi_{i\sigma}^0(\mathbf{r}')\psi_{j\sigma}^{0*}(\mathbf{r}')}{z - (\varepsilon_{j\sigma}^0 - \varepsilon_{i\sigma}^0)}. \tag{12.7}$$

At $T = 0$ this is the time-ordered function on the real frequency axis $P_0(\omega)$ with f_i the occupation numbers and $z \to \omega + i\eta \operatorname{sgn}(f_{i\sigma} - f_{j\sigma})$ in the denominator. At $T \neq 0$ the f_i stand for the thermal weight as in Eq. (5.146) and one gets the correlation function at the

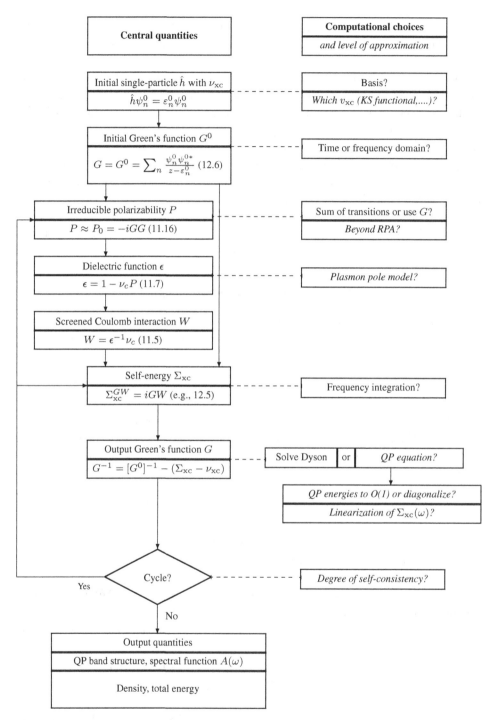

Figure 12.1. Flowchart for a typical GWA calculation.

Matsubara frequencies $z = z_m$. Subsequently, W_0 is obtained from Eq. (12.2) with $P = P_0$, and the self-energy $\Sigma_{xc} = iG^0 W_0$ is calculated from Eq. (12.5) or an equivalent expression. At $T = 0$ the result equals the sum of the dynamical screened exchange Eq. (11.26) and Coulomb hole Eq. (11.27), when the Dyson amplitudes f_λ and energies ε_λ are replaced by the single-particle $\psi^0_{\ell\sigma}$ and $\varepsilon^0_{\ell\sigma}$. This is called the $G_0 W_0$ approximation.

Which G^0? The results of a $G_0 W_0$ calculation depend on the wavefunctions ψ^0 and energies ε^0 that are used for the input G^0. It is important to choose a starting independent-particle hamiltonian $\hat{h}$ for which the band structure is as close as possible to the quasi-particle band structure that comes out of the $G_0 W_0$ calculation, or that yields the same density[3] [427]. This can be considered to be an approximate self-consistency condition. Possible starting hamiltonians $\hat{h}$ are the following.

- *Kohn–Sham hamiltonian.* The first *ab initio* GWA band structure calculations [418, 421] were performed starting from an LDA band structure. This, or other KS functionals like GGA (see Sec. 4.3 for some KS functionals), are frequently used starting points. DFT yields in principle a good density; moreover, from the pragmatic point of view, a GWA calculation is most often preceded by a Kohn–Sham ground-state calculation, and $G^0 = G_{KS}$ follows naturally. The biggest error comes from the eigenvalues; as pointed out in Sec. 4.4, the KS gap typically underestimates the quasi-particle gap by 50 to 100%.[4] Arguments have been given [428] to support the fact that the use of KS *wavefunctions* is a reasonable approximation, even in the LDA. For simple semiconductors the overlap between KS-LDA and quasi-particle wavefunctions, which are the solutions of the quasi-particle Eq. (7.39), is often better than 95%. However, the local density approximation has a tendency to delocalize density, overestimate hybridizations, and underestimate magnetic moments. Consequences of these errors are detected in $G_0 W_0$ results, especially in the presence of localized electrons; see the example of VO$_2$ in Sec. 13.4. Other KS functionals sometimes yield improvements, like GGA [429] or the EXX local Kohn–Sham potential [430] discussed in Sec. 4.5. The EXX is a particular case: it improves the density because it eliminates the self-interaction error,[5] and it also yields larger KS bandgaps, often close to quasi-particle ones, because of error canceling as explained in Sec. 4.5.
- *Hartree–Fock hamiltonian.* There is a conceptual advantage in starting from a Hartree–Fock calculation, since Koopmans' theorem assigns the correct physical meaning to the Hartree–Fock eigenvalues (namely, electron addition and removal energies, see Sec. 4.1). This route has been followed, for example, for large-gap semiconductors [416] and atoms [431],[6] where screening is weak to moderate. The fact that the Hartree–Fock

---

[3] The two options, similar band structure or similar density, are not strictly the same. Most often the first criterion is used.

[4] Note that the problem is evident in G^0, whereas in W_0 the lack of self-energy corrections is partially canceled by the neglect of the electron–hole attraction in the RPA. This is discussed in Secs. 14.7 and 15.3.

[5] This is particularly clear for the hydrogen atom, where exchange is the only contribution to the exact Kohn–Sham potential [398].

[6] In both works vertex corrections derived from the Fock operator were included.

gap is much too large does not suggest this choice for materials with smaller bandgap.

- *Generalized Kohn–Sham hamiltonian.* The Kohn–Sham approach has been generalized to non-local potentials [201]. This is discussed in Sec. 4.5. It allows one in particular to maintain the exact non-local exchange form of the Fock operator. Hybrid functionals, which combine some fraction of non-local exchange with conventional local Kohn–Sham exchange and correlation contributions [432], can well describe ground-state properties and at the same time yield band structures that are much closer to experiment than those of standard Kohn–Sham calculations. Especially in a range-separated form [433], they are a frequently used starting point for GWA calculations [434].

- *DFT+U.* The DFT+U approach [356, 435] opens a gap of magnitude U between localized states. With an empirical value for U, this is a pragmatic starting point, for example to calculate transition metal or lanthanide oxides [436–439].

Which W_0?

In the original version of the GWA presented in Ch. 11, screening is calculated in the RPA. Most equations remain unchanged if one uses instead a polarizability $P \neq P_0$ beyond the RPA. Whether or not this should be done is linked to the question of how much the GW approximation benefits from error canceling between the two missing vertex corrections, in P (Eq. (11.13)) and in Σ_{xc} (Eq. (11.11)). This is the topic of Sec. 15.3. Besides the potential mutual canceling of vertex corrections, it should be noted that the RPA calculated with KS ingredients, often yields a good approximation for the inverse dielectric function, close to experiment. This is a strong argument to use such a W_0. *Why* the DFT-RPA W_0 is often close to experiment is discussed in Secs. 14.11 and 15.3. To improve W, one possibility is to use TDDFT described in Sec. 14.11. This is an example of how one can combine methods.

Representing the screened Coulomb interaction W by few poles

One of the major difficulties of a GWA calculation is the frequency integration Eq. (12.5), even in the $G_0 W_0$ approximation, and even with the RPA. If W_0 is approximately represented as a sum over one or more discrete poles as in Eq. (11.24), the self-energy becomes a double sum over the poles of G^0 and the poles of W_0, as can be seen in Eq. (11.29). The Bethe–Salpeter equation in Ch. 14 is a way to obtain an explicit pole representation for the dielectric function as in Eqs. (14.42) or (14.44). The Bethe–Salpeter two-particle equation can in principle also be used for an RPA calculation, by setting the direct electron–hole attraction $\delta \Sigma_{xc}/\delta G$ to zero in Eq. (14.7). In practice, solving the BSE is in general computationally too demanding to produce all elements of $W(\mathbf{r}, \mathbf{r}')$ that are needed in the GWA, except for small systems (see, e.g., [431]). Therefore, the representation of W as a sum over poles is done approximately.

Plasmon pole models. Because of the Kramers–Kronig relations of Eq. (5.53) between the real and imaginary parts of a correlation function, it is sufficient to find a good approximation for the *imaginary part* of W_0. Early GWA calculations for the homogeneous electron gas [440] approximated Im W_0 by one single-plasmon branch with a dispersion $\omega^p(q)$ that has a quadratic $v_F^2 q^2/3$ and a $q^4/4$ contribution. The parameters of the model are determined by the density. Similarly, in crystals one can replace each element of the matrix in reciprocal space Im $W_{0G,G'}(\mathbf{q}; \omega)$ by just *one* pole that disperses with $\mathbf{q}$. This is motivated by the fact that the diagonal Im $W_{0G,G}(\mathbf{q}; \omega)$ is proportional to the loss function $-$Im $\epsilon_{G,G}^{-1}(\mathbf{q}; \omega)$. With $v_c(|\mathbf{q}|) = 4\pi/|\mathbf{q}|^2$, the screened Coulomb interaction $W = \epsilon^{-1}v_c$ is dominated by small values of $|\mathbf{q} + \mathbf{G}|$. In many materials, for small momentum transfer the loss spectrum is given approximately by one well-defined plasmon peak (see, e.g., the spectrum of bulk silicon in Fig. 5.6). This justifies the use of a *plasmon pole model*.

In the single-plasmon pole model, the time-ordered inverse dielectric function is

$$\text{Im}\, \epsilon_{GG'}^{-1}(\mathbf{q}; \omega) = -\mathcal{A}_{GG'}(\mathbf{q}) \left[\delta(\omega - \omega_{GG'}^p(\mathbf{q})) + \delta(\omega + \omega_{GG'}^p(\mathbf{q})) \right], \tag{12.8}$$

with the effective plasmon position ω^p and strength $\mathcal{A}$. The Kramers–Kronig relations of Eq. (5.53) yield

$$\text{Re}\, \epsilon_{GG'}^{-1}(\mathbf{q}; \omega) = \delta_{GG'} + \frac{2}{\pi} \frac{\mathcal{A}_{GG'}(\mathbf{q})\omega_{GG'}^p(\mathbf{q})}{\omega^2 - [\omega_{GG'}^p(\mathbf{q})]^2}. \tag{12.9}$$

The parameters $\mathcal{A}$ and ω^p can be determined in several ways. One can, for example [421], use the $\omega = 0$ limit and the f-sum rule of Eq. (5.63); another suggestion [441] is to calculate each element of $\epsilon^{-1}(i\omega)$ along the imaginary energy axis $i\omega$, where it is a smooth function, and to fit the two parameters of the model at two frequencies ω. These are typically $\omega = 0$ and an energy $i\tilde{\omega}$, with $\tilde{\omega}$ of the order of the plasmon energy of the system. The results of these two approaches differ often by a quite rigid shift of all quasi-particle energies. An explanation for this fact, and a hint when the difference should not be a rigid shift, is provided in Ex. 12.2. There is evidence [442] that the fitting method [441] yields results that are on average closer to those calculated without the plasmon pole approximation than the method of [421]. The fitting method has moreover the advantage that one can check easily the reliability of results by changing the fit frequency on the imaginary axis: if the model is appropriate, results should not change much.

One can extend the plasmon pole model to an arbitrary number of poles. A band Lanczos approach can, for example, be used to produce these parameters, yielding in principle any desired precision [443, 444]. If many frequencies are needed, this approach does not reduce much the computation of $W(\omega)$, but it is a way to approach the frequency integration. This integration is the topic of Sec. 12.6.

Restricting calculations to quasi-particle properties

When one is interested only in band structure, and maybe lifetime broadening, one can resort to the quasi-particle Eq. (7.38). This equation has in principle to be solved in the complex plane, where the self-energy is non-hermitian. Often, however, the imaginary part is neglected and only the real part of quasi-particle energies is calculated from

Eq. (7.39). It has the form of a one-body Schrödinger equation, which can be written in short notation as

$$\left[\hat{h} + \mathrm{Re}\,\hat{\Sigma}(\varepsilon_\ell)\right]|\psi_\ell(\varepsilon_\ell)\rangle = \varepsilon_\ell|\psi_\ell(\varepsilon_\ell)\rangle. \tag{12.10}$$

There is a different equation for each quasi-particle state ℓ, because Σ is energy-dependent. To solve Eq. (12.10) is sufficient for band structure calculations close to the Fermi level, which represents the goal of the vast majority of GWA calculations for real systems.

Solving the quasi-particle equation to first order

With an appropriate starting G^0, a solution of Eq. (12.10) to first order in the difference $\Sigma_{xc} - v_{xc}$ is justified. Here v_{xc} stands for a (local or non-local) exchange–correlation contribution to the independent-particle hamiltonian $\hat{h}$; it can, but doesn't have to be, the Kohn–Sham exchange–correlation potential. In the first-order approximation the quasi-particle energies are given by

$$\varepsilon_\ell = \varepsilon_\ell^0 + \langle\psi_\ell^0|\hat{\Sigma}_{xc}(\varepsilon_\ell) - \hat{v}_{xc}|\psi_\ell^0\rangle, \tag{12.11}$$

where $|\psi_\ell^0\rangle$ are the eigenstates of $\hat{h}$.

Linearizing the self-energy

Equation (12.11) is self-consistent in the quasi-particle (QP) energy ε_ℓ^{QP}. Close to the QP energy the self-energy is approximately linear, as shown in Ex. 7.6 and as can be observed for the example of silicon in Fig. 13.1. By expanding the self-energy $\Sigma_{xc}(\omega)$ in Eq. (12.11) to first order around $\omega = \varepsilon_\ell^0$, one obtains

$$\varepsilon_\ell = \varepsilon_\ell^0 + Z_\ell\langle\psi_\ell^0|\hat{\Sigma}_{xc}(\varepsilon_\ell^0) - \hat{v}_{xc}|\psi_\ell^0\rangle, \tag{12.12}$$

with

$$Z_\ell^{-1} = 1 - \langle\psi_\ell^0|\left.\frac{\partial\hat{\Sigma}_{xc}}{\partial\omega}\right|_{\varepsilon_\ell^0}|\psi_\ell^0\rangle, \tag{12.13}$$

which corresponds to Eq. (7.27) and to Eq. (7.26). Equation (12.12) shows that the quasi-particle renormalization factor Z that gives the weight of the quasi-particle in the full spectral function reduces the self-energy correction.

Model GW approaches

One cannot simplify much more on an *ab initio* level. Further approximations are made in model GWA calculations; these are beyond the scope of this book. Some widely used models can be found in [342, 445–448], the self-consistent approach of [449, 450], or even the very simple estimate for a self-energy scissor correction in [451].

12.3 Core and valence

Core–valence separation

The many-body problem is considerably simplified by separating particles in different regions of space, or with different characteristic energies. This is done in the Born–Oppenheimer approximation for electrons and nuclei, and it leads to Eq. (1.1), the first equation of this book. It is the topic of Ch. 7, where it is shown how one can obtain a description of a small system embedded in the rest. It is also at the basis of DMFT in Chs. 16–21, where an effective hamiltonian for strongly localized electrons in d or f states is used, determined by the s and p electrons of the system. Since we are mostly interested in the properties of *valence electrons*, in the present section we distinguish between tightly bound *core electrons* close to the nuclei and more loosely bound valence electrons.

The Green's function and the RPA polarizability can be written as a sum of the core and valence contributions[7]

$$G = G_c + G_v \qquad\qquad\qquad P = P_c + P_v. \qquad (12.14)$$

The Dyson equation for W can be split using Eqs. (7.42) and (7.43), as

$$W = W_v + W_v P_c W \qquad \text{with} \qquad W_v \equiv v + v P_v W_v. \qquad (12.15)$$

In order to avoid misunderstandings, here the bare Coulomb interaction is v without subscript c.

The core polarizability P_c is weak. To first order in P_c one obtains

$$W \approx W_v + W_v P_c W_v \qquad (12.16)$$

and the GW self-energy becomes

$$\Sigma_{xc} \approx iG_v W_v + iG_v W_v P_c W_v + iG_c(W_v + W_v P_c W_v). \qquad (12.17)$$

This is the valence-only self-energy, plus a term containing the effect of the core polarizability on the valence electrons, and the screened exchange of the core electrons. The screened core exchange is close to the bare exchange [452], since core electrons are localized and screening is inefficient at short distances. Then W_v can be replaced by v in the parentheses of Eq. (12.17). A simplified expression is obtained that allows one to reduce the number of electrons in a GWA calculation, as will be explained in the following.

Valence-only GW approximation

The simplest idea is to neglect all terms except $\Sigma_{xc} \approx iG_v W_v$. The argument is that P_c is small, and that valence and core orbitals have little overlap, which makes matrix elements

---

[7] This is straightforward for the Green's function and easy to see when P is calculated in the RPA Eq. (12.7), where P_c can be defined as the sum of all terms involving a core level. Beyond RPA one can still define a separation where the valence part contains only contributions that do not involve a core level, and P_c is by definition the rest. This is analogous to the discussion in Sec. 21.2.

of the last contribution to Eq. (12.17) between valence states negligible. However, it has been shown for example for atomic Na [334] and solid Al [453] that the neglected terms contribute of the order of 1 eV to quasi-particle energies. Therefore these terms must be taken into account at least approximately. Including them within LDA or GGA reduces the error to about 0.1–0.2 eV for gaps of elements with a stiff core that is well separated from the valence [454, 455]. In order to have a well-separated core and valence, in some cases a large number of electrons have to be treated as valence electrons. Examples are materials containing valence d electrons that overlap strongly with s and p electrons of the same shell; this is the topic of the next subsection.

Core–valence exchange

The form of exchange in Kohn–Sham functionals and in the GWA is completely different. The Kohn–Sham exchange potential is non-linear in the density and cannot be represented as a sum of core and valence contributions when these are not spatially distinct. The Fock exchange, instead, is a simple sum. A valence GWA calculation starting from Kohn–Sham replaces the valence-only Kohn–Sham exchange by the valence-only Fock exchange. Because of the different nature of the KS and Hartree–Fock exchanges, the neglect of core–valence exchange or its inclusion within an approximate Kohn–Sham potential leads to inconsistencies that can deteriorate results drastically [456]. This has been observed, for example, in GWA calculations of the band structure of CdS [456, 457], copper [458], or Cu_2O [459]. The use of self-interaction-free Kohn–Sham functionals, like the EXX, can lead to improvements. This has been shown for example for GaN, ZnO, ZnS, and CdS [457]. However, the most reliable procedure is to treat entire shells as valence electrons, for example $3s$, $3p$, and $3d$ electrons for copper, and not just the $3d$ electrons. This is straightforward in all-electron calculations, but difficult for pseudopotential calculations in a plane-wave basis that are efficient for loosely bound, delocalized electrons.

Core polarization

A way to also include the core polarization is to rewrite Eq. (12.15) as

$$W = W_c + W_c P_v W \qquad \text{with} \qquad W_c = v + v P_c W_c. \qquad (12.18)$$

Now the core-screened interaction W_c replaces the bare Coulomb interaction in the screening equation for the valence electrons. If one expands the right-hand side of Eq. (12.18) in terms of P_c, using the fact that the cores are localized, one obtains W_c as a sum over contributions from one, two, and more atomic cores. This can then be used to represent the core contribution through a core-polarization potential [460–463] in terms of the core polarizability of isolated atoms.

Valence wavefunction in the core region: pseudopotentials and beyond

Even when the core is not explicitly included in a GWA calculation, its presence is manifest in the wavefunction of the valence electrons. In particular, this wavefunction must have wiggles due to the orthogonalization of valence and core states, and it is steep close to the attractive core. This makes the wavefunctions difficult to describe in a plane-wave basis. Therefore often *pseudopotentials* are used, which replace the valence wavefunction in the core region by a smooth function but leave the wavefunction outside the core unchanged. Pseudopotentials are widely used, in particular in DFT ground-state calculations, and they are a topic of [1]. They are frequently used for the calculation of electronic excitations, for example in the GWA. One has to be careful, since the ability to describe excited states is an additional exigency for the transferability of pseudopotentials, which are often designed for the ground state.[8] Moreover, in the calculation of spectra one resolves the oscillator strength of every transition, contrary to the calculation of total energies, which are integrals. The error that is introduced by the use of pseudopotentials in the description of matrix elements is often of the same order of magnitude as the errors due to approximations of core–valence exchange and correlation discussed in the previous subsections. The two kinds of error tend to have opposite sign [455], and in many cases there is partial error canceling. This contributes to the overall success of valence-only GW calculations for materials with a reasonable core–valence separation [454, 464].

Pseudopotential results can be improved by using the projector-augmented-wave (PAW) approach, which avoids the use of pseudo-wavefunctions by reconstructing the full valence wavefunction; see [465] and Sec. 11.11 of [1]. The method starts from a pseudopotential plane-wave approach. Contrary to pseudopotentials called "norm conserving," here one does not constrain the pseudo-wavefunctions in the core region to integrate to the correct charge [466]. The charge and its multipoles in a sphere around the core are recovered by adding local compensation charges. Finally, an accurate description inside the spheres is obtained by replacing pseudo-wavefunction contributions inside the PAW spheres by the corresponding all-electron contributions.

Mathematically, the full wavefunction $\psi_{n\mathbf{k}}$ for a band n and wavevector $\mathbf{k}$ in the BZ is related to the pseudo-wavefunction ψ^{ps} by

$$|\psi_{n\mathbf{k}}\rangle = |\psi_{n\mathbf{k}}^{\mathrm{ps}}\rangle + \sum_i \left(|\phi_i\rangle - |\tilde{\phi}_i\rangle \right) \langle p_i|\psi_{n\mathbf{k}}^{\mathrm{ps}}\rangle. \tag{12.19}$$

The pseudo-wavefunctions are Bloch functions that are conveniently expanded in a plane-wave basis. The index i indicates an atomic site, angular momentum quantum numbers, and a reference energy. The $|\phi_i\rangle$ are solutions of the radial Schrödinger equation for non-spin-polarized reference atoms at the reference energies; they are called "all-electron partial waves." The $|\tilde{\phi}_i\rangle$, called "pseudo-partial waves," are equal to the all-electron partial waves outside the PAW spheres. The two partial waves match continuously inside the spheres. $|p_i\rangle$ are projectors that are localized inside the spheres. They are orthogonal to

---

[8] One could argue that also for ground-state calculations the pseudopotential must be able to describe excited atoms, because atoms in solids are in an excited state with respect to their ground state when they are isolated. Nevertheless, it is recommended to test the quality of a pseudopotential for spectroscopy over a wider energy range than for ground-state calculations.

the pseudo-partial waves, $\langle p_i | \tilde{\phi}_j \rangle = \delta_{ij}$, and depend on the distance from the center of the PAW sphere on which they are localized. Matrix elements of local or semi-local operators $\hat{O}$ read

$$\langle \psi_{n\mathbf{k}} | \hat{O} | \psi_{m\mathbf{k}} \rangle = \langle \psi_{n\mathbf{k}}^{\text{ps}} | \hat{O} | \psi_{m\mathbf{k}}^{\text{ps}} \rangle$$
$$+ \sum_{ij} \langle \psi_{n\mathbf{k}} | p_i \rangle \left[\langle \phi_i | \hat{O} | \phi_j \rangle - \langle | \tilde{\phi}_i \hat{O} | \tilde{\phi}_j \rangle \right] \langle p_j | \psi_{m\mathbf{k}}^{\text{ps}} \rangle. \qquad (12.20)$$

The first term is the one evaluated in a standard pseudopotential scheme, and it can be calculated in a plane-wave basis. The second term involves on-site integration of $\hat{O}$ between all electron and pseudo-partial waves. Expansion in angular momenta or radial meshes constitutes efficient means to evaluate that term. For example, one important expression for GWA calculations is the transition matrix element Eq. (12.51), the Fourier transform of a pair of orbitals, which means that Eq. (12.20) has to be evaluated for $\hat{O} \rightarrow e^{i(\mathbf{q}+\mathbf{G})\mathbf{r}}$. This can be done by expanding the plane wave in spherical Bessel functions. The result is given in [467]. The resulting expressions are more complicated than the Fourier transforms of periodic states that appear when one is working only with plane waves. More about GW calculations using PAW can be found in [468, 469] and references therein, and in [467].

The use of pseudopotentials is completely avoided in all-electron calculations. All-electron GWA calculations can be carried out using localized basis sets such as numeric atom-centered orbitals, or mixed basis sets, for example using the full-potential linearized augmented-plane-wave (LAPW) approach or full-potential linearized muffin-tin orbitals (LMTO), with two contributions: local atom-centered functions confined to muffin-tin spheres and plane waves where the overlap with the atomic functions is projected out. Details about the methods can be found in [1], and Sec. 12.6 contains more specific information about GWA calculations in a basis.

12.4 Different levels of self-consistency

The approximations outlined in Sec. 12.2 suppose that the self-energy is built with a given G^0, resulting from a static mean field. The results of such a G_0W_0 calculation are starting point dependent, which makes the method less predictive. The problem could be attenuated by introducing some degree of self-consistency. Different levels of self-consistency, with their pros and cons, are presented in Sec. 11.3. The present section explains how self-consistent calculations can be performed in practice. For each level of self-consistency, we outline the consequences on the results, which helps to extract some guidelines at the end of the section.

Our starting point is G_0W_0. For the discussion, we suppose that G^0 stems from a KS calculation, with a bandgap that is smaller than the quasi-particle one. This only influences the expected trends of self-energy corrections, not the technical realization, which is the same for all inputs from static mean-field potentials or from static self-energies.[9]

---

[9] An exception is the calculation of matrix elements that involve vanishing momentum transfer, see Eq. (12.56).

$G_0 W_0$ has several bad features:

- it is starting point dependent;
- it violates particle number conservation, see Sec. 11.3;
- it can give a bad description of satellites, see Sec. 11.6;
- it can yield unphysical spectral functions, because Im $\Sigma(\omega)$ is zero in a range of three times the KS gap (see Sec. 11.3) instead of three times the quasi-particle gap.[10]

In the following we will see step by step how some of these shortcomings can be overcome by self-consistency.

Alignment of the chemical potential

Fixing the zero of the energy scale is important in a GWA calculation, contrary to calculations in extended systems using a static potential, because of the energy dependence of the self-energy. The simplest step towards self-consistency is an overall shift $\Delta\varepsilon$ [287] that aligns the chemical potentials μ_0 of G^0 and of the output of a $G_0 W_0$ calculation,[11] $\mu = \mu_0 + \Sigma_{\mathbf{k}_F}(\mu)$.

This means that the input G^0, Eq. (12.6), is replaced by

$$G_\sigma^0(\mathbf{r}_1, \mathbf{r}_2; z) = \sum_n \frac{\psi_{n\sigma}(\mathbf{r}_1)\psi_{n\sigma}^*(\mathbf{r}_2)}{z - \varepsilon_{n\sigma}^0 - \Delta\varepsilon}. \tag{12.21}$$

Equation (11.29) shows that this corresponds to a shift of the frequency argument in the self-energy,

$$\Sigma(z) \to \Sigma(z - \Delta\varepsilon), \tag{12.22}$$

with respect to the case where G^0 is built with $\Delta\varepsilon = 0$. The shift can be determined by demanding self-consistency for the energy of the highest occupied state at μ,

$$\mu = \mu_0 + \Delta\varepsilon = \mu_0 + \Sigma_{\mathbf{k}_F}(\mu). \tag{12.23}$$

If quasi-particle energies are determined to first order and the self-energy is linearized in ω following Eqs. (12.12) and (12.13), the quasi-particle energy of a state ℓ becomes

$$\varepsilon_\ell = \varepsilon_\ell^0 + \Delta\varepsilon + Z_\ell \left(\langle \psi_\ell^0 | (\hat{\Sigma}(\varepsilon_\ell^0) - v_{\mathrm{xc}}) | \psi_\ell^0 \rangle - \Delta\varepsilon \right), \tag{12.24}$$

as shown in Ex. 12.2. In practice, one determines $\Delta\varepsilon$ from a $G_0 W_0$ calculation using Eq. (12.23) in order to evaluate Eq. (12.24). Note that no iteration is needed, because the frequency structure of the GWA self-energy has been used to derive Eq. (12.24) analytically.

---

[10] This estimate supposes that the onset of excitations in W is at the KS or QP gap, respectively.

[11] In a system with a gap, one usually aligns the highest occupied state. This means that if one started with an exact KS calculation, $\Delta\varepsilon$ for the exact self-energy would be zero because of the "DFT Koopmans' theorem" explained in Sec. 4.4.

Because of the frequency dependence of Σ, the self-energy correction is changed in a non-trivial way. In semiconductors often Z_ℓ is only weakly state-dependent; in that case there is only a rigid shift, and in particular no effect on the bandgap.

Self-consistency in the chemical potential (and more generally in the quasi-particle eigenvalues corresponding to a satellite) is *very important for satellite spectra*, since the distance between quasi-particle and satellite is changed, and the screening due to the denominator in the spectral function Eq. (7.22) is modified. The alignment of the chemical potential leads to substantial improvement. Examples are results for a half-filled four-site Hubbard cluster [367] and Hubbard chains [402], and the satellite spectrum of VO_2 [470] discussed in Sec. 13.4. Exercise 12.2 proposes an in-depth exploration of the effects of $\Delta\varepsilon$ on spectral functions.

$Z = 1$ calculations

Suppose that all states have the same shift, or, more realistically, that each matrix element Σ_ℓ is essentially affected by a shift $\Delta\varepsilon_\ell$ of the state ℓ. Then at self-consistency Eq. (12.11) with Eq. (12.22) becomes

$$\varepsilon_\ell = \varepsilon_\ell^0 + \langle \psi_\ell^0 | (\hat{\Sigma}(\varepsilon_\ell - \Delta\varepsilon_\ell) - v_{\mathrm{xc}}) | \psi_\ell^0 \rangle. \tag{12.25}$$

With $\varepsilon_\ell - \Delta\varepsilon_\ell = \varepsilon_\ell^0$ this suggests that one can simulate self-consistency by evaluating the $G_0 W_0$ self-energy at the independent-particle energy, *without* the renormalization factor Z. This simple argument does not take into account a change of the plasmon frequency, nor of wavefunctions. More arguments for using $Z = 1$ in $G_0 W_0$ quasi-particle energy calculations are given in [293, 471–473]. Bandgaps of simple semiconductors are improved with respect to $G_0 W_0$ calculations that use $Z \neq 1$ [471].

Update of quasi-particle energies

A more realistic approach is to update the quasi-particle energies self-consistently in the calculation of the self-energy. Since in general the quasi-particle gap is larger than the KS gap, updating the energies reduces the screening and the final result is an even larger quasi-particle gap, closer to Hartree–Fock. The changes can be important, especially when states are reordered and occupations change; see for example the GWA calculation of surface states in [474]. Note that when one updates the quasi-particle energies to self-consistency, the linearization Eq. (12.12) is not pertinent and no factor Z appears.

Self-consistent COHSEX

In some cases the KS wavefunctions are quite different from the quasi-particle wavefunctions. In d- and f-electron materials, for example, the LDA has a tendency to delocalize electrons and the ground-state density is not well described. In this case one should also update wavefunctions in a self-consistent way. In order to stay in a quasi-particle picture, one needs a static self-energy. In Sec. 11.3 the static COHSEX approximation to GW [43] is introduced. It can be used for self-consistent calculations yielding improved energies and wavefunctions as input for a G_0W_0 calculation [475]. Since COHSEX is an approximation to the GWA, this can be regarded as an approximate self-consistent GW calculation. The COHSEX self-energy is built from occupied states only, and only the static limit of W is needed.[12] Therefore, the calculations are simpler than fully self-consistent ones, or than the QSGW explained in the next subsection.

Usually one solves the quasi-particle equation in the basis of KS orbitals. The underlying hypothesis is that only a limited number of KS states mix to form a quasi-particle state, which leads to relatively small matrices.[13] Results are significantly improved with respect to G_0W_0 using an LDA input. The gap opening in VO_2 [360] described in Sec. 13.4 is a typical example.

Optimized static approximation: the quasi-particle self-consistent GW approach

To do better, one has to optimize the static self-energy that is used in the self-consistency cycle. This is the idea of the *quasi-particle self-consistent GW* (QSGW) approach [291, 292]. It constructs an effective non-local but static hamiltonian H^{QP} [293] that yields a quasi-particle spectrum close to the Fermi level with real eigenvalues ε_ℓ^{QP} and eigenfunctions ψ_ℓ^{QP} as close as possible to the ε_ℓ and $\psi_\ell(\varepsilon_\ell)$ obtained from the quasi-particle Eq. (12.10). Supposing that $\varepsilon_\ell \approx \varepsilon_\ell^{QP}$ and $\psi_\ell(\varepsilon_\ell) \approx \psi_\ell^{QP}$ allows one to replace Re $\hat{\Sigma}(\varepsilon_\ell)|\psi_\ell\rangle$ by Re $\hat{\Sigma}(\varepsilon_\ell^{QP})|\psi_\ell^{QP}\rangle$ in Eq. (12.10). Since the ψ_ℓ^{QP} form an orthonormal set,

$$\hat{R}|\psi_\ell^{QP}\rangle = \text{Re } \hat{\Sigma}(\varepsilon_\ell^{QP})|\psi_\ell^{QP}\rangle \quad \text{with the definition} \quad \hat{R} \equiv \sum_{ij} |\psi_i^{QP}\rangle \text{Re } \Sigma_{ij}(\varepsilon_j^{QP})\langle\psi_j^{QP}|.$$

(12.26)

This means that one can replace Re $\hat{\Sigma}(\varepsilon_\ell^{QP})$ by $\hat{R}$. Symmetrization leads to the effective hermitian hamiltonian

$$H^{QP} = \frac{1}{2} \sum_{ij} |\psi_i^{QP}\rangle \left(\text{Re } \Sigma_{ij}(\varepsilon_i^{QP}) + \text{Re } \Sigma_{ij}(\varepsilon_j^{QP}) \right) \langle\psi_j^{QP}|.$$

(12.27)

The approach is summarized in Fig. 12.2. Results for various materials can be found in Ch. 13; note in particular the bandgaps in Fig. 13.5, and the improved magnetic properties in Sec. 13.5. By definition, the approach is limited to the calculation of quasi-particle wavefunctions and real quasi-particle energies. It can also be used as a starting point for a dynamical G_0W_0 calculation that can yield more properties, such as lifetimes or satellites.

An alternative approach is proposed in [476]. In that work the effective quasi-particle hamiltonian is constructed by orthonormalizing the quasi-particle wavefunctions obtained with the real part of the full frequency-dependent self-energy. The procedure ensures that the obtained orthonormal orbitals are the closest to the original non-orthogonal quasi-particle wavefunctions in the least-square sense. Results are similar to QSGW.

---

[12] However, empty states may still be needed for the calculation of the static W itself.

[13] Note that the self-energy conserves momentum, which means that there is no mixing of **k**-points, contrary to the exciton problem explained in Sec. 14.7.

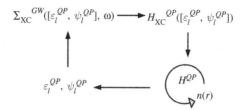

Figure 12.2. Self-consistency loop of the QSGW approach [291, 292]. From a starting dynamical GWA exchange–correlation self-energy an optimized static non-local potential is constructed. This is added to the starting Hartree hamiltonian to yield the full optimized H^{QP}. It is updated with an internal self-consistency loop on the density, which is recalculated with the eigenfunctions of H^{QP}. Finally the new quasi-particle wavefunctions and energies are used as input to update the GWA self-energy. The outer loop is cycled till convergence.

GW_0

In the GW_0 approach the Green's function is calculated fully self-consistently but W_0 is fixed. Contrary to G_0W_0, the GW_0 approach conserves particle number and, contrary to fully self-consistent GW, it does not deteriorate the satellite spectrum of the electron gas [352, 369], see Fig. 11.8. Results are closer to Hartree–Fock than G_0W_0, because of the larger gap, and because G is an interacting Green's function with satellites and a quasi-particle weight smaller than one. This means that spectral weight is on average further away from the Fermi level than in G^0; see Sec. 7.4. This reduces correlation effects. The final Green's function therefore has increased lifetime and quasi-particle weight compared with the G_0W_0 result. GW_0 results for atoms and diatomic molecules [350] and for bandgaps of semiconductors and insulators [477] are promising. Of course, they depend on W_0. Good quasi-particle energies are obtained when the inverse dielectric function in W_0 is similar to the measurable one, defining a "best W" strategy (see Sec. 15.3). This is discussed in [477], and ways to obtain good dielectric functions are presented in Ch. 14.

To do a GW_0 calculation requires the full G, and therefore the solution of the Dyson equation instead of a quasi-particle equation. How to solve a Dyson equation is explained in the second to next subsection.

Fully self-consistent GWA

Fully self-consistent GWA calculations are mostly carried out when one is interested in total energies, since spectral properties in extended systems have a tendency to worsen with respect to GW_0, when W is calculated in the RPA form. Self-consistency introduces some contributions of higher order in W, but others are missing because of the neglect of vertex corrections. The problem can be seen for the HEG in Sec. 11.6, and it is explained in Sec. 11.3. The quality of spectra in finite systems in variable, see, e.g., closed-shell molecules in [478] and results for azabenzenes in [479]. GW results are closer to HF than

GW_0 results, because the screening is reduced; see illustrations for a model in [480]. A simple way to estimate the fully self-consistent quasi-particle renormalization factor Z is given in [352].

Like GW_0, fully self-consistent GW requires the solution of the Dyson equation, which is explained in the next subsection. Moreover, the self-consistent W has a complicated frequency structure, which makes a plasmon pole model inadequate and a careful frequency integration is necessary to obtain the self-energy. Several possibilities are explained in Sec. 12.5. They have been used for self-consistent calculations, for example a convolution of spectral functions (Eq. (12.32)) together with a multipole representation in [351], a sum over Matsubara frequencies (Eq. (12.31)) and an analytic continuation to the real axis using Padé approximants (Eq. (12.34)) in [481], or contour integration (Eq. (12.33)) in [369].

Solution of the Dyson equation

When one asks for the full Green's function, as output of a G_0W_0 calculation or for a self-consistency cycle, one has to solve the Dyson Eq. (11.15) instead of a quasi-particle equation. Since $G(\omega)$ and $\Sigma(\omega)$ enter the equation with the same frequency ω, it is sufficient to write the Dyson equation in a basis, and solve the resulting matrix equation for each frequency.

For a self-consistent solution, the equation has to be iterated. A typical iteration step for the nth iteration is $G^{-1(n+1)} = [G^0]^{-1} - \Sigma[G^{(n)}]$. There are many ways to iterate, and there are several difficulties that one may encounter. One is instabilities; this is a frequent problem in all iterative procedures. It can often be solved[14] by mixing results of previous steps into the iteration, as

$$G^{-1(n+1)} = [G^0]^{-1} - \Sigma[(1 - \alpha)G^{(n)} + \alpha G^{(n-1)}], \qquad (12.28)$$

where α is a parameter that is chosen around $\alpha = 0.5$. It is recommended to align the chemical potential at each step, following the procedure at the beginning of this section. Many useful details can be found in [350].

Different ways to iterate Dyson equations can change the performance of the approach, and sometimes even the results [482, 483]. Some discussion can be found in Sec. G.1. For the solution of the $G - \Sigma$ Dyson equation the scheme in Eq. (12.28) is recommended to avoid bad surprises.

Rules of thumb

The GWA is an approximation, and there can be no single cooking recipe that always leads to the best meal. However, physics suggests some constraints. The GW_0 approach with a "best W" combines spectral functions with the correct analytical properties, particle number conservation, and realistic dynamical screening. Here the *best W* is the one that has a dynamical screening closest to what can be measured. According to the system and

---

[14] It can be solved provided there exists a stable limit and one is close enough to the correct solution.

available resources, it can be calculated in DFT-RPA, TDDFT, or by solving the Bethe–Salpeter equation as discussed in Ch. 14. Screening is underestimated in the RPA-based QSGW or self-consistent COHSEX approaches in systems where the effect of the electron–hole interaction discussed in Ch. 14 is important, which leads to an overestimation of certain transition energies (see Sec. 13.4).

Self-consistency of G is very important, and should be done at least partially. In particular, the alignment of the chemical potential is needed to describe satellites, and self-consistency of the quasi-particle wavefunctions corrects errors in the density. How much one should and can do depends on the system and available resources.

To aim for GW_0 is a guideline for the calculation of spectra. In total energies there are other mechanisms of error canceling, and there are arguments in favor of fully self-consistent RPA GW (see Sec. 15.3).

12.5 Frequency integrations

The frequency integrations are the most delicate part of a GWA calculation, because the functions that are integrated have poles close to the real axis. This section summarizes the most frequently used approaches.

Calculation of P_0

With the frequency Fourier transforms defined in Sec. C.2, omitting real space arguments the Fourier transform of $P_0(1, 2) = -iG(1, 2)G(2, 1^+)$ is[15]

$$P_0(\omega) = -\frac{i}{2\pi} \lim_{\eta \to 0^+} \sum_\sigma \int d\omega' G_\sigma(\omega + \omega') G_\sigma(\omega') e^{i\eta\omega'} \tag{12.29}$$

or, using Matsubara frequencies for $T \neq 0$,

$$P_0(z_n) = \frac{1}{\beta} \sum_m \sum_\sigma G_\sigma(z_m) G_\sigma(z_n + z_m). \tag{12.30}$$

These expressions yield Eq. (12.7) for $G = G^0$, an independent-particle Green's function. One can also calculate the product of the two one-body Green's functions in real space and time or imaginary time, and then perform the numerical Fourier transform. This is done in the space–time approach of [344, 345].

Plasmon pole calculation of Σ

The next step is the frequency integration, for example in Eq. (12.5). The simplest way to calculate Σ is to use a plasmon pole model for W. This is a valid approximation when one is interested in the real part of the self-energy close to the Fermi level, far from the important plasmon energies. With the single-plasmon pole model Eq. (12.8), the matrix

---

[15] See also Eq. (11.42) for the reciprocal space expression in the HEG.

elements of the self-energy are obtained from the general formula Eq. (11.29) by replacing ω_s with one plasmon pole ω^p and the matrix elements $W_{s,kk\sigma}^{p,ii}$ of the polarization part of W by matrix elements of the intensities $\mathcal{A}$.

The self-energy from a full frequency integration

In principle one can calculate the full frequency integral in Eq. (12.5) numerically on the real axis. This requires a very fine frequency mesh, of the order of several hundreds of frequencies, because of the poles of G and W. An example where this has been done is the quasi-particle calculation for copper in [458]. However, there are several alternatives.

Working with Matsubara frequencies. At non-vanishing temperature the standard approach is to make use of the discrete Matsubara frequencies. With the help of App. D, Eq. (12.5) can be written as

$$\Sigma(z_n) = -\frac{1}{\beta} \sum_m G(z_n - v_m) W(v_m) \tag{12.31}$$

for the self-energy evaluated at a complex Matsubara frequency z_n. Note that z_n corresponds to fermionic Matsubara frequencies, and v_m are complex bosonic frequencies; both are defined in Eq. (D.11).

The spectral function method. When working with time-ordered Green's functions at $T = 0$ one can make use of the fact that G and W have the analytic properties of correlation functions, with the Lehmann representations of Eq. (5.36) in terms of the spectral functions. The Green's function is given in terms of its spectral function in Eq. (5.90), and the spectral function in terms of the time-ordered Green's function is $A(\omega) = \frac{1}{\pi} |\text{Im} G(\omega)|$, which is consistent with Eq. (5.93). Similarly, the spectral function of W is $D(\omega) = -\frac{1}{\pi} \text{Im} W(\omega)\text{sgn}(\omega)$. The integration can be performed using Cauchy's residue approach (see App. C.3). The result for the correlation part Σ_c of the self-energy is

$$\Sigma_c(\omega) = \int_{-\infty}^{+\infty} d\omega_1 \int_0^\infty d\omega_2 \frac{A(\omega_1)D(\omega_2)}{\omega + (\omega_2 - i\eta)\,\text{sgn}(\mu - \omega_1) - \omega_1}. \tag{12.32}$$

This is consistent with Eq. (11.29), if one expresses A and D as a sum over poles, as in Eq. (5.101). In general one discretizes the problem, with a discretized **k**-space and a multipole representation of the functions as described in Sec. 12.2. This requires very fine grids, or interpolation techniques like the tetrahedron method [484]. One can smooth the poles by using a non-vanishing imaginary part $i\eta$ in the energy denominators, but this can change the results. The multipole approach has been used, for example, to obtain fully self-consistent results for the homogeneous electron gas [485].

Contour integration. To make the frequency integration at $T = 0$ easier, one can use the pole structure of the time-ordered functions in the complex plane. This is shown in Fig. 12.3 as a function of ω', with the argument of G shifted to $G(\omega'+\omega)$, as it appears in the integral of Eq. (12.5). The poles of the time-ordered $G(\omega)$ lie in the upper half-plane for ω below the chemical potential μ, and in the lower half-plane for $\omega > \mu$. The poles of $W(\omega)$

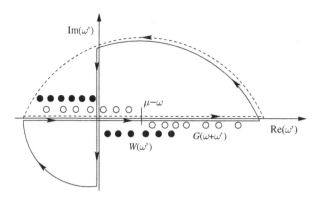

Figure 12.3. Pole structure of G and W, and integration paths in the complex plane to obtain the self-energy. Poles of G are empty circles, poles of W are filled circles. Dashed line: simplest contour corresponding to Eq. (12.5). The path includes poles of both G and W, and the result is a sum over residues. Continuous line: this contour includes only poles of G, and the result is Eq. (12.33).

are positioned in the same way, but with respect to the chemical potential for electron–hole pairs, $\mu = 0$ (see Sec. 5.8). The polarization part W^P vanishes at infinity, $W^P(z) \sim 1/z^2$ for $|z| \to \infty$. The integral of Eq. (12.5) can be calculated by closing a contour in the upper half-plane, and using Cauchy's residue theorem. In this way poles of both G and W are enclosed, and the integral corresponds to the sum of residues. Alternatively, one can choose the contour indicated by the continuous line in Fig. 12.3. It encloses only poles of G. The integrals along the quarter circles vanish. Along the imaginary axis G and W are smooth, and the integration is straightforward. The correlation part of the self-energy becomes [486, 487]

$$\Sigma_c(\omega) = -\frac{i}{2\pi} \int_{+i\infty}^{-i\infty} G(\omega + z) W^P(z) dz + \sum_i f_i(\mathbf{r}) f_i^*(\mathbf{r}') W^P(\mathbf{r}', \mathbf{r}, |\varepsilon_i - \omega| - i\eta)$$
$$\times \left[\Theta(\mu - \varepsilon_i)\Theta(\varepsilon_i - \omega) - \Theta(\varepsilon_i - \mu)\Theta(\omega - \varepsilon_i) \right]. \tag{12.33}$$

In a $G_0 W_0$ calculation f_i and ε_i are independent-particle wavefunctions and energies. One could use the expression also when the Green's function is interacting; in this case, f_i and ε_i have to be interpreted as Dyson amplitudes and electron addition and removal energies. Besides the integral on the imaginary axis, W^P has to be calculated at a series of frequencies close to the real axis. This method avoids the double-frequency integral that is needed in Eq. (12.32), and it is therefore more reliable.

From the imaginary axis to real energies. Having the self-energy Eq. (12.31) on the imaginary frequency axis is very useful for further integration, for example in order to calculate total energies. However, to obtain spectra one has to get back to the real axis in Σ or after the solution of the Dyson equation in G. The same problem arises in the space–time approach of [344, 345], where quantities are evaluated on the imaginary time axis and subsequently transformed to imaginary frequencies.

The move from the imaginary to the real frequency axis can be done by an analytic continuation. One has to fit a set of parameters to the imaginary-axis result, and evaluate the resulting function on the real axis [487, 488]. A very good approach in view of its analytic properties is the use of a Padé approximant

$$R(z) = \frac{\sum_{i=0}^{n} a_i z^i}{\sum_{j=0}^{m} b_j z^j}, \tag{12.34}$$

with parameter sets a, b. The number of terms (n, m) used for the fit depends on the details of the information that should be reproduced; typical numbers are between 5 and 10. The method is reliable in the vicinity of the bandgap, but to get full spectral functions is delicate, and in general the contour integration is more precise. More discussion about analytic continuations can be found in Sec. 25.6.

Alternatively, removal energies can be obtained by using an extended Koopmans' theorem [489, 490]. In this approximate approach, ionization energies are found from an eigenvalue equation that involves the imaginary-time derivative of the Green's function and the density matrix.

GW total energy calculations in practice

Once the Green's function or spectral function is determined, the total energy can be obtained with an additional frequency integration. The Galitskii–Migdal [371] formula of Eq. (5.129) reads

$$E = \frac{1}{2\beta} \sum_{\ell\sigma} \sum_{\nu} (\varepsilon_{\ell\sigma}^0 + z_\nu) G_{\ell\ell\sigma}(z_\nu) \qquad \text{for } T \neq 0 \tag{12.35}$$

$$E = \frac{1}{2} \sum_{\ell\sigma} \int_{-\infty}^{\mu} d\omega \, (\varepsilon_{\ell\sigma}^0 + \omega) A_{\ell\ell\sigma}(\omega) \qquad \text{for } T = 0, \tag{12.36}$$

where ℓ denotes the independent-particle basis of the non-interacting system with energies ε_{ℓ}^0. This method has been used, for example, in total energy calculations [392] for the homogeneous electron gas in Sec. 11.6.

Alternatively, one can evaluate the functionals in Sec. 8.2 to get the grand potential, or the total energy at $T = 0$. Let us concentrate on Eq. (8.14); the other functionals are evaluated accordingly. The first contribution is the Φ-functional. For the GWA it is $\Phi^{GW} = \frac{1}{2} \mathfrak{Tr}(\Sigma^{GW} G)$, and the sum of the first two terms is $-\frac{1}{2} \mathfrak{Tr}(\Sigma^{GW} G)$. To calculate total energies it is convenient to use Matsubara frequencies. In a basis, this reads

$$-\frac{1}{2} \mathfrak{Tr}(\Sigma^{GW} G) = -\frac{1}{2\beta} \sum_{\ell m} \sum_{z_\nu} \Sigma_{\ell m}^{GW}(z_\nu) G_{m\ell}(z_\nu). \tag{12.37}$$

The next contribution to Eq. (8.14) is the logarithmic term $\mathfrak{Tr} \ln(1 - G_0 \Sigma)$. To evaluate the logarithm, one has to work with scalar quantities. This is obtained by diagonalizing $G^0 \Sigma$. If $g_i(z_\nu)$ are the eigenvalues of $G^0 \Sigma$, one gets

$$\mathfrak{Tr} \ln(1 - G^0 \Sigma) = \sum_{z_\nu} e^{z_\nu 0^+} \ln(1 - g_i(z_\nu)). \tag{12.38}$$

Finally according to Eq. (8.4), $\Omega^0 = -\frac{1}{\beta}\sum_\ell(1+e^{-\beta(\varepsilon_\ell^0-\mu)})$, where ε_ℓ^0 are non-interacting single-particle energies as, e.g., in Eq. (12.36). More details and examples can be found in [300, 372–374].

RPA total energy calculations in the framework of DFT are closely related to total energy calculations in the GWA. This is explained in Sec. 11.5. The DFT calculations are simpler, since the KS Green's function has an independent-particle form. Therefore the field is more developed. An overview including computational tricks can be found in [491].

12.6 GW calculations in a basis

The frequency integrations of the previous section are specific to the GWA and other calculations in the framework of many-body perturbation theory, as opposed, e.g., to DFT calculations. In this section we deal with the spatial structure. Although the self-energy is non-local in space, there is not much fundamental difference from mean-field calculations in this respect. We refer to other books like [1] for the details, and concentrate in the following on points that are specific to the topics of this book.

As before, we give the expressions in a form that can be interpreted as imaginary-frequency $z = z_\nu$ Matsubara results at $T \neq 0$, or at $T = 0$ as $z = \omega$ with the appropriate small imaginary part.

Matrix elements

Very often the equations are solved as matrix equations, in a basis.[16] Let us define a set of normalized basis functions $\chi_m(\mathbf{r})$, in which independent-particle or quasi-particle wavefunctions $\psi_i(\mathbf{r})$ are expanded with coefficients b_i^m,

$$\psi_{i\sigma}(\mathbf{r}) = \sum_m b_{i\sigma}^m \chi_m(\mathbf{r}). \tag{12.39}$$

In the following we need matrix elements of two-point functions in this basis,

$$F_{n_1 n_2} = \int d\mathbf{r}_1\, d\mathbf{r}_2\, \chi_{n_1}^*(\mathbf{r}_1)F(\mathbf{r}_1, \mathbf{r}_2,)\chi_{n_2}(\mathbf{r}_2), \tag{12.40}$$

as well as matrix elements in the basis of the independent-particle wavefunctions,

$$F_{ij} = \int d\mathbf{r}_1\, d\mathbf{r}_2\, \psi_i^*(\mathbf{r}_1)F(\mathbf{r}_1, \mathbf{r}_2,)\psi_j(\mathbf{r}_2). \tag{12.41}$$

Four-point functions transform as

$$F_{n_1 n_2 n_3 n_4} = \int d\mathbf{r}_1\, d\mathbf{r}_2\, d\mathbf{r}_3\, d\mathbf{r}_4\, \chi_{n_1}^*(\mathbf{r}_1)\chi_{n_2}^*(\mathbf{r}_2)F(\mathbf{r}_1, \mathbf{r}_2, \mathbf{r}_3, \mathbf{r}_4)\chi_{n_3}(\mathbf{r}_3)\chi_{n_4}(\mathbf{r}_4); \tag{12.42}$$

we also use the equivalent notation

$$F_{n_1 n_2 n_3 n_4} \equiv F_{n_1 n_3}^{n_4 n_2}. \tag{12.43}$$

---

[16] One can also use a discrete grid in real space. This is done in the space–time approach [344, 345].

As worked out in App. F, in general two-body interactions have to be treated like four-point quantities, which corresponds to replacing $v(\mathbf{r}_1, \mathbf{r}_3) \rightarrow \delta(\mathbf{r}_1 - \mathbf{r}_4)\delta(\mathbf{r}_2 - \mathbf{r}_3)v(\mathbf{r}_1, \mathbf{r}_3)$. In particular, we need matrix elements of W in the basis of the independent-particle wavefunctions,

$$W_{ki\ell j} = \int d\mathbf{r}_1 \, d\mathbf{r}_3 \, \psi_k^*(\mathbf{r}_1)\psi_j(\mathbf{r}_1)W(\mathbf{r}_1, \mathbf{r}_3)\psi_\ell(\mathbf{r}_3)\psi_i^*(\mathbf{r}_3). \tag{12.44}$$

These are useful to express matrix elements of the self-energy,

$$\Sigma_{\mathrm{xc},k\ell} = i \sum_{ij} G_{ij} W_{k\ell}^{ij}, \tag{12.45}$$

where we have omitted time and spin for compactness.

In practical calculations it can be convenient to use the basis transformation of Eq. (12.40) for the two-point function $W(\mathbf{r}_1, \mathbf{r}_3)$ that appears in the evaluation of the four-point matrix element $W_{kj}^{\ell i}$ given by Eq. (12.44). For this purpose we define matrix elements of pairs of orbitals,

$$\tilde{\rho}_{nj\sigma}(\mathbf{r}) = \psi_{n\sigma}^*(\mathbf{r})\psi_{j\sigma}(\mathbf{r}) \qquad \text{or}$$
$$\tilde{\rho}_{ij\sigma}^m = \int d\mathbf{r} \, \psi_{i\sigma}^*(\mathbf{r})\psi_{j\sigma}(\mathbf{r})\chi_m(\mathbf{r}). \tag{12.46}$$

This leads to

$$W_{k\ell\sigma}^{ij} = \sum_{mn} \tilde{\rho}_{ik\sigma}^{m*} W_{mn} \tilde{\rho}_{j\ell\sigma}^n, \tag{12.47}$$

which requires the calculation of matrix elements of the independent-particle polarizability

$$P_{mn}^0(z) = \sum_{ij\sigma} (f_{i\sigma} - f_{j\sigma}) \frac{\tilde{\rho}_{ij\sigma}^m \, \tilde{\rho}_{ij\sigma}^{n*}}{z - (\varepsilon_{j\sigma} - \varepsilon_{i\sigma})}. \tag{12.48}$$

Finally, it can be useful to express the independent-particle Green's function in the basis,

$$G_{mn\sigma}^0(z) = \sum_i \frac{b_{i\sigma}^m b_{i\sigma}^{n*}}{z - \varepsilon_{i\sigma}} = (z - \hat{h})_{mn\sigma}^{-1}, \tag{12.49}$$

with

$$\hat{h} = \sum_i \varepsilon_{i\sigma} |\psi_{i\sigma}\rangle\langle\psi_{i\sigma}|. \tag{12.50}$$

According to the basis, it can be convenient to use the second expression in Eq. (12.49) in terms of the hamiltonian, which does not contain a sum over states. Similarly, one can eliminate one of the sums in Eq. (12.48) to write P_0 in terms of $\hat{h}$. This allows one to use iterative inversion techniques for efficient calculations (see, e.g., [492–496]).

Choice of the basis

Plane waves. Plane waves are a basis of choice in *ab initio* spectroscopy calculations for extended systems. One can use fast Fourier transforms, and the results show regular convergence behavior. The bare Coulomb interaction is diagonal and overall, plane waves are easy to program. The first fully *ab initio* GW calculations were performed in a plane-wave basis [418, 421, 422]. In a periodic system and in a plane-wave basis, Eq. (12.46) becomes

$$\tilde{\rho}_{ijk\sigma}(\mathbf{q}+\mathbf{G}) = \frac{1}{\sqrt{\Omega}} \int d\mathbf{r}\, \psi_{ik\sigma}^*(\mathbf{r})\psi_{jk+q\sigma}(\mathbf{r})e^{-i(\mathbf{q}+\mathbf{G})\mathbf{r}}. \tag{12.51}$$

This can be evaluated using fast Fourier transforms. The polarizability is a matrix in reciprocal space,

$$P^0_{\mathbf{G},\mathbf{G}'}(\mathbf{q},\omega) = -\sum_{ijk\sigma}(f_{ik\sigma} - f_{jk+q\sigma})\frac{\tilde{\rho}_{ijk\sigma}(\mathbf{q}+\mathbf{G})\tilde{\rho}_{ijk\sigma}^*(\mathbf{q}+\mathbf{G}')}{z - \varepsilon_{ik\sigma} + \varepsilon_{jk+q\sigma}}, \tag{12.52}$$

and the $T = 0$ expression for self-energy matrix elements becomes (similarly for the Matsubara expression)

$$\Sigma_{ijk\sigma}(\omega) = \frac{i}{2\pi}\sum_s\sum_{\mathbf{GG}'}\int_{BZ}d\mathbf{q}\int_{-\infty}^{+\infty}d\omega'\frac{\tilde{\rho}_{isk\sigma}(\mathbf{q}+\mathbf{G})\tilde{\rho}_{jsk\sigma}^*(\mathbf{q}+\mathbf{G}')W_{\mathbf{GG}'}(\mathbf{q},\omega')e^{i\eta\omega'}}{\omega+\omega'-\varepsilon_s+i\eta\,\mathrm{sgn}(\varepsilon_s-\mu)}, \tag{12.53}$$

where the screened Coulomb interaction is obtained from the inverse dielectric function as

$$W_{\mathbf{GG}'}(\mathbf{q},\omega) = \frac{\epsilon_{\mathbf{GG}'}^{-1}(\mathbf{q},\omega)}{|\mathbf{q}+\mathbf{G}'|^2}. \tag{12.54}$$

The integrals over $\mathbf{q}$ are transformed into discrete sums. As one can see from Eq. (12.51), the set of $\mathbf{q}$ must correspond to differences between $\mathbf{k}$-points for which the wavefunctions are calculated, because the $\tilde{\rho}$ are zero otherwise. This includes the point $\mathbf{q} = 0$, where the Coulomb interaction of Eq. (12.54) diverges when $\mathbf{G}' = 0$. In this case, one can use $\mathbf{k}\cdot\mathbf{p}$-perturbation theory, explained, e.g., in [51], which leads to

$$\tilde{\rho}_{iik\sigma}(\mathbf{q}\to 0) = 1 \tag{12.55}$$

$$\tilde{\rho}_{isk\sigma}(\mathbf{q}\to 0) = -i\lim_{q\to 0}\frac{\mathbf{q}\cdot\langle ik\sigma|\nabla + [\mathbf{r}, v^{nl}]|sk\sigma\rangle}{\varepsilon_{ik\sigma} - \varepsilon_{sk\sigma}} \qquad \text{for } i \neq s, \tag{12.56}$$

where v^{nl} is a non-local potential or self-energy in $\hat{h}$. For example, it can contain the non-local part of a pseudopotential, or a non-local screened Fock term, when the wavefunctions are the result of a COHSEX calculation. There are methods that avoid the explicit calculation of the commutator: for example, one can expand the $\psi(\mathbf{r})$ into a set of wavefunctions stemming from a *local* hamiltonian, before applying $\mathbf{k}\cdot\mathbf{p}$-perturbation theory. In this case, Eq. (12.56) becomes a sum of matrix elements without the commutator, calculated with eigenvalues and eigenfunctions of the local hamiltonian. One can also use two $\mathbf{k}$-point grids that are shifted by a very small fraction of the BZ, to calculate the limit numerically.

It can be important to take the non-locality of the potential into account; corrections to the matrix elements can be of the order of 10% or several times more. This is crucial in the calculations of optical spectra, which are explained in Sec. 14.8. When the $\mathbf{q} \to 0$ contribution appears in an integral that is discretized in real calculations, such as the integral in Eq. (12.53), it tends to have measure zero when the $\mathbf{k}$-point grid is converged. However, a bad treatment of the $\mathbf{q} = 0$ term deteriorates results for small sets of $\mathbf{q}$-points, and therefore it slows down the convergence of the sum. As an indicative value, the neglect of the non-locality of a standard norm-conserving pseudopotential in bulk silicon leads to an error of about 50 meV for the self-energy corrections to valence bands, when 10 $\mathbf{k}$-points in the irreducible Brillouin zone are used in the GW calculation, which corresponds to 19 $\mathbf{q}$-points in the integral. The terms with $\rho_{isk\sigma}(\mathbf{q} \to 0)$, $i \neq s$ are well behaved, because the matrix elements are proportional to q, which cancels the divergence of the Coulomb interaction. However, since the matrix elements can depend on the direction of $\mathbf{q}$, the limit is non-analytic and one has to average over directions.

The terms with $\rho_{iik\sigma}(\mathbf{q} \to 0)$ have to be treated separately. For these terms, one goes back to the original integral over $\mathbf{q}$. One has to integrate over the small volume around $q = 0$, which is represented in the mesh by the point $q = 0$. This integral replaces the divergent $q \to 0$ term.

Localized orbitals. Calculations in a localized-orbital basis are also widely used and have a long history. The first realistic GW calculations were performed based on a linear combination of atomic orbitals (LCAO) [416, 417]. Later the LMTO method [497], which is appropriate for materials with d and f electrons, was adapted from ground-state to spectroscopy calculations by treating valence and conduction states on an equal footing. It gave access to spectroscopic properties of materials like Ni or NiO [498, 499]. GW calculations on a gaussian basis were developed and tested on bulk materials [500] and finite systems [501]. A localized basis allows one to treat in a convenient way core as well as semi-core and valence electrons, finite as well as disordered infinite systems, or transport through molecules between leads. There are many flavors of localized basis sets, which may or may not be orthonormal, including tight-binding, localized Wannier functions [502], wavelets, gaussians, atomic orbitals as in LCAO, or numeric atom-centered orbitals [503, 504].

A difficulty of localized basis sets is the description of the more delocalized empty states. This can sometimes be ameliorated by adding delocalized basis functions, and is important when one looks directly at excitations to delocalized states. Instead, the sums over a large number of empty states that appear in the spectral representations, like Eqs. (12.6) or (12.7), rather stand for a completeness relation representing $(\omega - \hat{h})^{-1}$. In this case, details of the empty states are not important, and a localized basis has no severe problems.

Selected localized basis sets are as follows.

- *The LMTO basis functions,*

$$\chi_n(\mathbf{r}) \quad \to \quad \chi_{\mathbf{R}\ell m}(\mathbf{r}, \mathbf{k}_n), \tag{12.57}$$

where $\mathbf{R}$ labels atom sites and ℓm are indices of spherical harmonics. In a periodic system, this basis is different for each $\mathbf{k}_n$ in the Brillouin zone.

- *Numeric atom-centered orbitals* [504],

$$\chi_n(\mathbf{r}) = \quad \rightarrow \quad \frac{u_n(r)}{r} Y_{\ell m}(\Omega), \tag{12.58}$$

where $u_n(r)$ are numerically tabulated functions, $Y_{\ell m}$ are complex spherical harmonics, with real parts ($m = 0, \dots, \ell$) and imaginary parts ($m = \ell, \dots, 1$), and ℓ, m are implicit functions of the basis function index n.

There also exist implementations (see, e.g., [505]) for the GWA using a *mixed basis*, like the full-potential linearized augmented-plane-wave method (FLAPW) [506], where the space is partitioned into a region around the core that is described by non-overlapping atom-centered muffin-tin spheres, and the interstitial region. The valence electron wavefunctions are expanded in numerical functions inside the spheres, and interstitial plane waves elsewhere.

Optimized basis set for the polarizability. The calculation of the polarizability contributes to a large extent to the computational effort of a GW calculation. Significant speedup can be obtained by introducing an optimized basis for the polarizability. This can be understood by looking at Eq. (12.7): P_0 is built with *pairs* of orbitals that form an overcomplete basis. The aim of a product basis is to minimize the number of basis elements for P_0. In [507] this has been done with a product basis using LMTO, with an application to electron energy loss spectra of Ni and Si. Another example is a product basis [494] based on maximally localized Wannier functions [502] that has been benchmarked for various small and large bulk and finite systems.

Basis for self-consistent calculations. In self-consistent GW calculations on a quasi-particle level it is convenient to express the new quasi-particle wavefunctions in the basis of the eigenfunctions of the independent-particle hamiltonian used to build G^0. Often a relatively small number of eigenfunctions is sufficient, since the self-energy mixes the states over a moderate energy range, of a few eV, and different $\mathbf{k}$-points in the BZ do not mix due to translational symmetry.

12.7 Scaling and convergence

The Coulomb interaction

The long-range Coulomb interaction has to be treated with care. In plane-wave calculations, finite systems are often simulated by a supercell, a large periodically repeated unit cell that contains the object of interest, and much empty space. In ground-state calculations of neutral systems without dipole moment, the use of supercells is relatively straightforward. When excited states are calculated, there are always long-range interactions since electrons are added or removed from the system or dipoles are created. This worsens convergence with supercell size. It is unsatisfactory, since one has to make a huge effort just to fill the empty space with plane waves. One can make the supercells smaller by using a modified Coulomb interaction that is cut off in real space [508, 509].

Since the Coulomb interaction $4\pi/q^2$ is steep in reciprocal space for small $\mathbf{q}$, it also poses problems in extended systems because it requires a fine $\mathbf{q}$-mesh to represent the integral in Eq. (12.53). One can improve the convergence by using two different grids, a coarser one to represent the $\tilde{\rho}$ and the inverse dielectric function, and a denser one to represent the bare Coulomb interaction. In this way, the Coulomb interaction is integrated around each point on the coarse grid, similar to the treatment of the divergence at $\mathbf{q} = 0$ explained after Eq. (12.55). This method is called *improved integration* [510]. It can be used in a similar way to integrate the discontinuity at the Fermi level in metals at low temperatures [458], or to simulate a dense $\mathbf{G}$-space, which means a large supercell when one is interested in finite systems. This is an alternative to the Coulomb cutoff method.

Important convergence parameters

The convergence criteria for spectroscopy calculations are generally quite different from ground-state calculations. In spectroscopy calculations, details of spectra in a certain spectral range are important, whereas ground-state calculations require sum rules to be fulfilled. For example, in a GW_0 quasi-particle or spectral function calculation, it is not necessary to converge the particle number to a very high precision, but to describe well the spectrum of W_0. Important parameters are the following.

- **k-points.** Convergence with **k**-points is less critical than for the optical spectra in Sec. 14.8, but a careful integration of poles still requires rigorous convergence tests.
- **Empty states.** Convergence of quasi-particle energies with the number of empty states in the spectral representations is slow. An exactly solvable model system consisting of two electrons on a sphere shows an error that disappears as $E_{cut}^{-3/2}$ with the cutoff energy E_{cut} [511]. This is confirmed by realistic calculations, e.g., [512, 513].

  The methods based on iterative inversion techniques [492–496] mentioned in the previous section do not sum over bands, and therefore do not have the number of bands as explicit convergence parameter. In these methods, bands and plane waves are implicitly determined by the size of the hamiltonian.

  Another possibility to avoid empty states is the use of the *effective energy technique* (EET) [514]. This consists of replacing the state-dependent energy denominator that appears in expressions like P_0 in Eq. (12.48), by one effective number, the "effective energy," that depends in principle on all arguments of P_0. This allows one to use the closure relation, which makes the calculations often an order of magnitude faster. Approximations for the effective energy in the calculation of the polarizability and the self-energy are given in [514].

- **The number of basis functions** for the expansion of the wavefunctions, and for the summations in the self-energy. These two parameters are in principle the same, but in practice it is convenient to set them separately. For example, for systems where crystal local field effects (LFE, see Sec. 14.3) are small, one can include a smaller number of components $(\mathbf{G}, \mathbf{G}')$ of the inverse[17] dielectric function $\epsilon^{-1}_{\mathbf{G},\mathbf{G}'}(\mathbf{q})$ than the number of plane waves needed to describe the wavefunctions in the matrix elements.

Scaling of GW calculations

A standard approach to calculate quasi-particle energies in the GWA is based on the spectral representation evaluated in a plane-wave basis, Eqs. (12.52) and (12.53). The calculation of each $\tilde{\rho}$ using fast Fourier transforms scales as $N_{at}\ln[N_{at}]$. The calculation of the $N^2_{\mathbf{G}} N_{\mathbf{q}}$ matrix elements of P_0 contributes to an important part of the effort. It scales as $N^4_{at} N^2_{\mathbf{k}}$. For small enough unit cells, where doubling the cell reduces the number of $\mathbf{k}$-points by a factor of two, the scaling with number of atoms is N^2_{at}. Finally, to determine one matrix element of Σ in Eq. (12.53) scales as $N_{\mathbf{q}} N^3_{at}$, or N^2_{at} in the range of small enough unit cells.

One can obtain much better scaling for large unit cells by using alternative approaches based on the nearsightedness principle [515], namely the fact that the relevant quantities are quite localized in real space. For example, one has to consider a range of about 10 a.u. for the extension of P_0 in a semiconductor like silicon. This is used in methods like the space–time approach [344, 345]. The Green's function is calculated in real space and time, which, using the nearsightedness principle, scales as $N_{\mathbf{r}} N_{bands} \rightarrow N^2_{at}$. The calculation of P_0 that is obtained as a product of two Green's functions in real space and time scales linearly with the number of atoms.

12.8 Wrap-up

This chapter has given an overview of the main established computational approaches to put the GW approximation into practice. The expressions can be evaluated from first principles, without using empirical parameters. Still, there are two kinds of choice to be made.

The ingredients. The choice of the ingredients *changes the results*. One has to specify *which* one-body Green's function G and *which* screened Coulomb interaction W is to be computed. More precisely, one has to define the level of self-consistency, and whether W is calculated in the random phase approximation, or beyond. There can be no 100% clear answer from theory, because the GWA is an approximation, and therefore different systems and properties enhance or hide different problems. As a rule of thumb, a safe approach for spectra is to use a static self-consistent approach, like QSGW or COHSEX, as starting point for a $G_0 W_0$ calculation. This is described in Sec. 12.4. W should be calculated as well as possible for spectra. Ideally one would use the Bethe–Salpeter equation described in Ch. 14, but this is expensive. In principle one could also use time-dependent density functional theory, briefly summarized in Sec. 14.11. The RPA evaluated with Kohn–Sham

---

[17] Note that this refers to the number of elements of the *inverse*, not of the matrix before inversion. These numbers can also be different.

ingredients can be regarded as a good approximation to TDDFT in cases where the effect of the exchange–correlation kernel f_{xc} on the inverse dielectric function is weak. Put together, the requirement of a self-consistent Green's function and a "best W" give arguments for using the GW_0 approach, in cases where a good W_0 can be found.

The calculation of total energies is a different problem: these are integrals. One does not look at details of spectra in some window of frequency, but at averages over the whole frequency range. This makes sum rules and conservation laws more important, and there are arguments to perform fully self-consistent RPA calculations to get good total energies.

The technical choices. Once the ingredients of a calculation are established, there are many ways to put the equations into practice. *This should not change the results.* The main technical choice is the basis. Criteria are fast and controllable convergence, easy implementation, and computational efficiency. Plane waves, for example, are easy to handle and fast in periodic systems, but less efficient for finite systems. The expressions can be evaluated using spectral representations. In this case, sums over many empty states have to be calculated, which converge slowly. Tricks like the EET in Sec. 12.7 can often be used to speed up the calculations. One can also calculate bare and interacting Green's functions by solving the Dyson equation, which requires matrix inversions and iterations. There are various methods to do this, and they can be extended for example to calculate elements of the independent-particle polarizability P_0 by iterative matrix inversion.

With all these choices, there is no free lunch, and one has to be careful. For example, it is not safe to check just one physical property, like the bandgap. One number can hide many things, and it is better to look at details, such as a whole band structure or spectral function. The GW approximation is helpful in this respect, because it allows one to calculate a variety of properties including ground-state total energies and spectra, and W on its own is a measurable quantity.

The goal is to enable more and more precise calculations for more and more complex systems. This remains a challenge. Nevertheless, it is exciting to see how much realistic physics one can obtain from first-principles calculations using the, after all quite simple, GW approximation. This is illustrated in the next chapter.

SELECT FURTHER READING

Aryasetiawan, F. and Gunnarsson, O. E., "The GW method," *Rep. Prog. Phys.* **61**, 237, 1998. Discussion of various aspects of and beyond the GWA, illustrated by a number of results.

Aulbur, W. G., Jonsson, L., and Wilkins, J. W., "Quasiparticle calculations in solids," *Solid State Phys.* **54**, 1, 2000. Short introduction to GW and extensive compilation and discussion of results.

Bechstedt, F. *Many-Body Approach to Electronic Excitations; Concepts and Applications.* Springer Series in Solid-State Sciences No. 181 (Springer-Verlag, Berlin, 2015). A recent book on the theory and computation of electronic excitations. It contains many detailed expressions and explanations useful for GW calculations.

The website www.electronicstructure.org contains more information, and links to tutorials and numerical exercises that accompany this book.

Exercises

12.1 Verify that Eq. (12.7) can be obtained directly from the general expression for the retarded electron–hole correlation function of Eq. (5.146) by replacing the interacting matrix elements $f_{\alpha\lambda}$ with pairs of single-particle wavefunctions and the total energy differences in the denominator with the difference between single-particle valence and conduction states, making the appropriate modifications to get the $T = 0$ time-ordered or the $T \neq 0$ Matsubara form.

12.2 Using a linear expansion of the self-energy and a first-order self-energy correction, show that a shift $\Delta\varepsilon$ of the energy zero in the starting G^0 leads to a shift $\Delta\varepsilon$ of the GW quasi-particle energies, plus a state-dependent correction (*hint*: start from Eq. (11.29) for the self-energy). Verify that this correction does not change the bandgap if valence and conduction states have the same Z-factor. Discuss under which conditions a change of the plasmon frequency also leads to a rigid shift of the spectrum. Plot the model spectral function of Eq. (11.51) to see the influence of $\Delta\varepsilon$: for $\Delta\varepsilon = 0$, fix the prefactor of the function β and the plasmon pole position ω_p to reproduce qualitatively the spectral function of silicon in Fig. 13.4. As explained in Sec. 15.7, this spectral function contains a plasmaron, a spurious solution to the quasi-particle Eq. (7.39). Show that the plasmaron is attenuated when $\Delta\varepsilon$ is determined by aligning the chemical potential, as explained in Sec. 12.4. Show that an imaginary part in $\Delta\varepsilon$ also improves the spectrum. Show that the quasi-particle is relatively little affected. Link these findings to a general discussion about self-consistency, and interpret the results for the HEG in Fig. 15.8.

12.3 For the Hubbard dimer at quarter- and half-filling, examine the influence of a shift of the energies in G^0 on the results of a G_0W_0 calculation, as a function of U.

12.4 Starting from the general expression of Eq. (11.12), show that W is spin-independent and given by Eq. (12.2).

12.5 Derive the independent-particle polarizability Eq. (12.7) as a convolution of the two G^0 in frequency space.

12.6 Following App. F, write Hedin's equations in a space–spin basis, with $G(1,2) \rightarrow G_{ij}(t_1, t_2)$. You will find that all interactions and the polarizability become four-point matrices. Work out the expressions for the case where the bare interaction v_c does not depend on spin and therefore G is spin-diagonal.

13

GWA calculations: illustrative results

Summary

The present chapter illustrates what the GWA can do and what it cannot do. It contains results for metals, semiconductors, and insulators, for bulk materials and low-dimensional systems, for materials that reflect features of the homogeneous electron gas and for others that present completely different aspects. It shows the physical soundness and broadness of the approach in a wide range of applications, and in others the need to go beyond.

The GW approximation to the self-energy is a surprisingly simple formula, with a clear physical content as discussed in Ch. 11, and the potential for efficient calculations of electronic properties beyond independent-particle methods as described in Ch. 12. It is now time to look at results for realistic systems and ask: which properties can be described successfully by the GWA, and for which materials? And what can we learn from the results?

The selection presented in this chapter covers only a fraction of what could be cited. The GWA has become a very popular method – for valuable reasons, as the chapter shows: having a good approximation to the self-energy allows one to calculate accurate band structures, electron addition and removal spectra, densities, density matrices, total energies, and many related properties. GWA calculations lead to qualitative and often quantitative agreement with experimental findings.[1] There exist many more applications, but here we have selected examples meant to illustrate content, strong and weak points of the GWA, in its various flavors such as perturbative or non-perturbative calculations. The examples are chosen to stress certain aspects, and they are not necessarily the first or the most recent calculations. A time span of several decades is covered. As a consequence, the computational scheme that has been used, for example the level of self-consistency from G_0W_0 to fully

---

[1] Here we consider the accuracy of addition and removal energies on a scale of 0.1 eV, with total energies on a scale of meV.

self-consistent GW, is not always the same. However, the quality of the results is always sufficient to support the message that the example is supposed to illustrate.

One can find, for instance, a huge number of GW bandgap calculations in the literature: for our purpose, it is sufficient to give just a few representative examples. Far fewer results exist for spectral functions and satellites, although the dynamically screened interaction W contains electron–hole and plasmon excitations. Most GW results have been obtained in the quasi-particle approximation, where satellites are absent *by definition*: this is a choice, one can also perform GWA calculations of the full spectral function. We have chosen and explained a small number of examples where GW can describe satellites, and others where it fails.

For better understanding, we concentrate on a few prototype materials where salient features can be brought out clearly. Bulk silicon is such an example, as a material that has been looked at over and over again. As one can see in Sec. 13.1, it is in many respects close to the homogeneous electron gas for which the GW approximation was first designed. One difference from the HEG is of course the existence of a bandgap in silicon; its calculation constituted one of the early successes of the GWA. The example of silicon demonstrates the variety of properties that can be obtained; an overview for many other materials is given in Sec. 13.2. The performance of the GWA is materials-dependent, and it is interesting to examine cases that are far from the HEG. Therefore, Sec. 13.3 contains results for finite and low-dimensional systems, where screening is weak and states are confined. Another class of very different systems is given by transition metals and their oxides, where the localized d or f-electrons give rise to intriguing properties. It has long been supposed that the GWA could not work well for these "correlated" materials that are thought to require alternative approaches such as dynamical mean-field methods, as brought out by the examples in Ch. 20. However, as Sec. 13.4 shows, one has to distinguish between a mere problem of charge localization that can be overcome by an improved description of the ground-state charge density and a true problem of correlation involving degenerate states, where the GWA indeed meets its limits. Finally, Sec. 13.5 is dedicated to results for GW ground-state calculations. The field is less explored than spectroscopy, partly because a higher precision is required, and because simplifications such as plasmon pole models or perturbative calculations are often not appropriate. Moreover, ground-state calculations compete with DFT ones, since the Kohn–Sham approach does in principle also give direct access to exact ground-state properties. Hence the GWA has to outperform computationally very efficient approximations such as the LDA. However, there are cases that nicely illustrate the physical content of the GWA, such as van der Waals binding: as mentioned in Sec. 2.2, this requires the explicit description of long-range screening, and is hence a natural playground for the GWA. A note on issues linked to temperature in Sec. 13.6, and a wrap-up conclude the chapter.

13.1 From the HEG to a real semiconductor: silicon as a prototype system

Section 11.6 contains results of the GWA for the HEG; indeed, GWA studies on that system are detailed and numerous. However, real systems exhibit properties far from the homogeneous gas – how much of that information is then pertinent? We will first look at silicon, which is maybe *the* prototype application for the GWA. For this material we can expect to rediscover some of the features of the HEG, for example, plasmons. At the same time, silicon is a material with covalent bonds and a bandgap, which is different enough to expect new insights.

Bulk silicon and diamond have been among the earliest examples of GWA calculations on real materials, starting initially from screened exchange and becoming progressively more accurate [192, 346, 418, 421, 422, 516–518]. Important milestones were the evaluation and discussion of the full frequency-dependent GW expression for diamond in 1982 [346] starting from a semi-empirical independent-particle calculation,[2] and the first-principles G_0W_0 calculations for silicon and diamond based on an LDA band structure in 1985 [418]. In bulk silicon the LDA to DFT describes the density quite well. Kohn–Sham LDA is therefore considered to be a reasonable input for G_0W_0 calculations, where an approximate, independent-particle, Green's function is used to construct W and the self-energy, as explained in Sec. 11.3. This simplifies the discussions. The evolution of the calculations, including changes due to self-consistency, can be nicely followed by comparing calculations and results for the bandgap of silicon, for example in [408, 519, 520] and references therein.

Self-energy, spectral function, and quasi-particles

Let us first look at some global features of the frequency-dependent self-energy, and at their consequences on the spectral function, with its quasi-particles and satellites.

Figure 13.1 shows matrix elements $\Sigma_{\ell\ell}(\omega)$ of the self-energy as a function of frequency [521], calculated in G_0W_0 using LDA wavefunctions and energies. The dotted lines represent the function $\omega - \varepsilon_\ell^{KS} - (\text{Re} \, \Sigma_{xc} - v_{xc})_{\ell\ell}$. The quasi-particle condition Eq. (7.23) is fulfilled where this function crosses the real axis.[3]

The figure looks similar to the left and middle panels in Fig. 11.5 for the HEG.[4] In particular, the imaginary part of the self-energy has a well-defined peak. It is a direct consequence of the plasmon peak around 17 eV in the imaginary part of the inverse dielectric function that is shown in the top-right panel of Fig. 5.6. It is useful to look at the expression Eq. (11.28) for the G_0W_0 self-energy to remember that structures in the dynamical screening appear in the imaginary part of the self-energy at the corresponding distance from the quasi-particle energy. The imaginary part of the self-energy vanishes at the Fermi level. As can be seen from Eq. (11.28), the imaginary part of the self-energy must be zero for frequencies closer to the Fermi level than the lowest electron or hole energy plus the lowest

---

[2] Also vertex corrections in the screening were considered in this work.

[3] Note that the independent-particle energy and self-energy correction are taken with respect to a Kohn–Sham band structure. Since in silicon quasi-particle and Kohn–Sham wavefunctions are similar, $(\text{Re} \, \Sigma_{xc} - v_{xc})$ is approximately diagonal in the KS basis, whereas this is not necessarily true for Σ_{xc} alone.

[4] Note that the presentation of the figures is different: Fig. 11.5 shows directly the real part of $\Sigma_k(\omega)$, such that there the quasi-particle condition is fulfilled when the curve crosses the diagonal line.

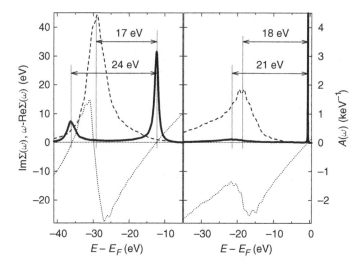

Figure 13.1. Matrix elements $A_{\ell\ell}(\omega)$ of the $G_0 W_0$ spectral function of bulk silicon (continuous lines) at the Γ point for a state ℓ at the bottom (left panel) and top of the valence bands (right panel). The corresponding imaginary part of the matrix elements $\Sigma_{\ell\ell}(\omega)$ of the self-energy is given by the dashed curves. The dotted curve is the function $\omega - \varepsilon_\ell^{KS} - (\mathrm{Re}\,\Sigma_{xc} - v_{xc})_{\ell\ell}$, which crosses zero at the quasi-particle energy. The top of the valence band is set to 0 eV. The figure has been adapted from [521] by M. Guzzo. Note that here selected matrix elements as a function of frequency are displayed. This is different from Fig. 13.2, where each matrix element is taken at the corresponding quasi-particle energy.

neutral excitation energy, in other words, in a region about three times the bandgap (see Sec. 11.3).

Close to the peaks in the imaginary part, the real part shows the strong variation typical for a Kramers–Kronig transform, and it is approximately linear around the Fermi level, as discussed in Ex. 7.6 and Sec. 12.2.

From the self-energy the spectral function is obtained using the relation of Eq. (7.22). Starting from a Kohn–Sham band structure, matrix elements of the spectral function are

$$A_{\ell\ell}(\omega) = \frac{1}{\pi} \frac{|\mathrm{Im}\,\Sigma_{\ell\ell}(\omega)|}{[\omega - \varepsilon_\ell^{KS} - \mathrm{Re}\,(\Sigma(\omega) - v_{xc})_{\ell\ell}]^2 + [\mathrm{Im}\,\Sigma_{\ell\ell}(\omega)]^2}. \qquad (13.1)$$

Figure 13.1 shows that, as discussed in Sec. 7.4, the spectral function has peaks close to the zeros of $\omega - \varepsilon_\ell^{KS} - (\mathrm{Re}\,\Sigma_{xc} - v_{xc})_{\ell\ell}$ and close to the peaks of $\mathrm{Im}\,\Sigma_{\ell\ell}(\omega)$. The former are quasi-particle peaks, the latter are satellite structures. The difference between the maximum of the quasi-particle peak ε_ℓ^{QP} and the corresponding Kohn–Sham eigenvalue ε_ℓ^{KS} is the quasi-particle shift.[5] The left panel of Fig. 13.2 shows the quasi-particle shifts in

---

[5] As discussed in Sec. 7.6, this definition of the quasi-particle energy is slightly different from the quasi-particle condition Eq. (7.23) that searches the zeros of the shifted real part as discussed above, but the results are in general very close.

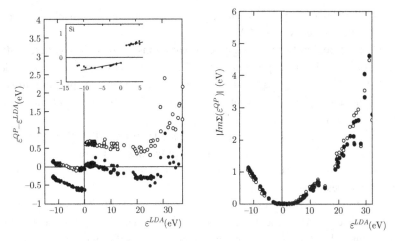

Figure 13.2. Self-energy correction to the Kohn–Sham LDA eigenvalues ε^{LDA} in silicon, as a function of ε^{LDA}, calculated along the L–Γ–X lines. Full circles are the result of a G_0W_0 calculation, empty circles contain an approximate vertex correction discussed in Sec. 15.2. Left: quasi-particle shift $\varepsilon^{QP} - \varepsilon^{LDA}$, determined from the maxima in the respective spectral function [290]. Inset: zoom on the gap region. The zoom is taken from [421], where quasi-particle shifts are calculated in first-order perturbation theory using Eq. (12.11). In this energy range this should yield the same results as taking the maxima of the spectral function. Right: matrix element $\mathrm{Im}\,\Sigma_{\ell\ell}(\varepsilon_{\ell}^{QP})$ of the imaginary part of the self-energy evaluated at the quasi-particle energy, from [290].

silicon obtained with respect to an LDA band structure, as a function of ε^{LDA} [290]. Full circles are the result of a G_0W_0 calculation, empty circles are discussed in Ch. 15 since they go beyond the GWA by including a vertex correction. An energy range of several tens of eV is covered, which is possible because the full frequency integral Eq. (12.5) has been performed.

Within about 5 eV around the Fermi level, the quasi-particle shifts are close to linear with a small slope, as can be seen in the inset in the left panel, taken from [421].[6] In contrast to the HEG, silicon has a bandgap. The minimum indirect gap of Kohn–Sham eigenvalues is much smaller than the experimental value of 1.17 eV, by about 50% in the LDA. This is often called the *bandgap problem*, and it is a typical result for simple semiconductors; see, for example, germanium in Fig. 2.6 or the overview in Fig. 13.5. As discussed in Sec. 4.4, even the exact KS eigenvalues should not be interpreted as electron addition or removal energies. The scope of a GWA calculation is then to correct both this conceptual problem and shortcomings of the approximate density functional itself. G_0W_0 lowers the valence bands by about half an eV, which brings the gap into good agreement with experiment. The

---

[6] These results have been obtained using the perturbative expression Eq. (12.11), which in general does not influence results in the gap region for simple semiconductors.

correction of the "bandgap problem" in many semiconductors and insulators is probably the most prominent success of the GWA.

Beyond the gap region there are significant variations in the quasi-particle shifts, with a minimum followed by a steep rise around 25 eV. The overall shape can be understood from Eq. (11.29) (Ex. 13.1), and the dip and rise related to the bulk plasmon in silicon in the upper-right panel of Fig. 5.6: for high enough energies, there is significant probability of plasmon emission.

The scattering of the results is a main difference from the HEG, Sec. 11.6. There, all wavefunctions are simply plane waves, and all directions of $\mathbf{k}$ are equivalent. In a real crystal such as silicon, matrix elements of Σ_{xc} between states with different spatial and symmetry properties can be very different.

In contrast to v_{xc}, the self-energy has an imaginary part reflecting inelastic losses due to excitations. The matrix elements $\mathrm{Im}\,\Sigma_{\ell\ell}(\varepsilon_{\ell}^{QP})$ are shown in the right panel of Fig. 13.2 [290]. As discussed in Sec. 7.5, the imaginary part of the self-energy is inversely proportional to the lifetime of the quasi-particles. In the GWA, the lifetime of quasi-particles is limited because of impact ionization, which means that the quasi-particle loses energy to other excitations that are contained in the inverse dielectric function. As expected from Eq. (11.29) and consistent with Fig. 13.1, $\mathrm{Im}\,\Sigma$ vanishes in the gap and changes behavior in the vicinity of the plasmon energy. The finite lifetime of the quasi-particles is seen experimentally as a broadening of the photoemission peaks.[7]

From quasi-particles to the band structure

In a crystal, such as silicon, quasi-particle energies are usually reported in a band structure plot where the top valence of the KS and GWA results are aligned. Figure 13.3 shows the band structure of silicon on a path X–Γ–L–Γ–X for comparison of GWA results, QMC calculations, and experiment, and X–W–Γ for further GWA results. The figure has been adapted from [292].

Experimental band structure information can be obtained from angle-resolved photoemission data [522–525], electron momentum spectroscopy [526, 527], and inverse photoemission experiments [528, 529]. For the determination of empty-state energies, optical experiments are frequently used. Experimental results for selected states used in Fig. 13.3 are collected in the theoretical paper [500]. The experimental position of the bottom of the valence band at -12.5 ± 0.6 eV is indicated by the gray strip. In the electron addition or removal experiments full spectra are measured, and quasi-particle energies are extracted as the maxima of quasi-particle peaks.[8] For sharp peaks close to the Fermi level this procedure is more straightforward than for the broader peaks at the bottom of the bands.

---

[7] Note that impact ionization is not the only contribution to broadening; for example, a quasi-particle may decay by exciting phonons. Moreover, extrinsic losses of the outgoing photoelectron contribute to the measured spectra.

[8] This is based on the hypothesis that the spectra are dominated by the *intrinsic* spectral function, which is the one that is calculated (see footnote 7 and the discussion in Sec. 5.7).

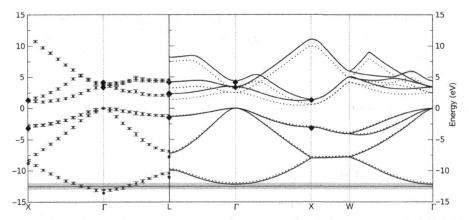

Figure 13.3. Quasi-particle band structure of silicon. Diamonds are experimental results. They correspond to those collected in the theoretical paper [500]. The horizontal gray strip indicates the measured value of the bottom valence energy at Γ and its error bar, -12.5 ± 0.6 eV. Left: bands in direction X–Γ–L, calculated in QMC. Symbols with error bar are AFQMC results [530]. Filled circles at Γ, X, and L are DMC results [531]; the size of the circles corresponds to the statistical error bars. Around the Fermi level DMC results are hidden by AFQMC ones. Right: bands in direction L–Γ–X–W–Γ, data from [292] provided by Mark van Schilfgaarde. Dotted lines are the result of an LDA calculation, the continuous line has been determined with QSGW explained in Sec. 12.4.

Good agreement, with deviations less than 0.2 eV, between QMC, the GWA evaluated in the quasi-particle self-consistent GW approach (QSGW, see Sec. 12.4),[9] and experiment is found in the upper valence and lower conduction bands. The most obvious effect of the GWA with respect to a KS-LDA calculation is a rigid increase of all gaps between valence and conduction states, by about half an eV. This rigid shift is often called a *scissor shift*, or caused by a *scissor operator*, because it corresponds to cutting the band structure plot horizontally inside the gap, and displacing the two pieces vertically by hand.

A closer look shows that the valence bandwidth in the GWA has increased slightly with respect to the LDA. This is the result of several competing effects, which are not shown in the figure: in silicon, self-consistency increases the bandwidth, like in the HEG in Sec. 11.6. Instead, the frequency-dependence of the self-energy leads to band narrowing. Effects that are due to this frequency-dependence are called *dynamical effects*. The fact that dynamical effects lead to band narrowing can be seen by comparing results of the static COHSEX approximation, explained in Sec. 11.3, to fully dynamical GWA results. For silicon this comparison has been done in [421]. It turns out that dynamical effects reduce the absolute magnitude of the self-energy operator by about 20%. Band narrowing due to dynamical effects is also seen in the framework of DMFT, when a frequency-dependent interaction is introduced, as explained in Sec. 21.6. The final GWA bandwidth lies within the experimental error bar. In QMC the bandwidth of silicon is overestimated.

---

[9] Deviations from older results using pseudopotentials [418] and the G_0W_0 scheme are small, partly due to error canceling.

Band structure calculations using QMC are discussed in Sec. 24.8. Here it is important to note that in the case of QMC total energy differences, not spectral functions, are calculated. A quasi-particle peak does not however correspond to a single state. Close to the Fermi energy the peaks are narrow, and there is no ambiguity. At the bottom of the valence band, however, there is significant broadening, as one can see in Fig. 13.1. It is therefore difficult to say *which* total energy difference is representative for the quasi-particle peak.

Beyond quasi-particles: the photoemission spectrum of silicon

Figure 13.4 shows the full valence photoemission spectrum of silicon [521], integrated over the Brillouin zone. Experiment shows quasi-particle structure between the Fermi level at zero and the bottom valence at -12.5 eV. Most of today's GWA calculations focus on that region, which is directly related to the band structure. In the experiment, two prominent satellite features can be seen at higher binding energy, i.e., further away from the Fermi level. Their energetic distance from the quasi-particles is respectively once and twice the plasmon energy of about 17 eV. A very weak triple-plasmon feature is also present. Theoretical results are obtained by summing calculated matrix elements of the spectral function over bands and the Brillouin zone. The G_0W_0 result from an LDA starting point, supplemented with cross-sections and a calculated secondary electrons background to simulate

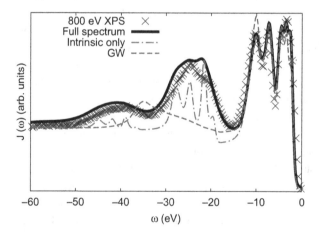

Figure 13.4. Photoemission spectrum of bulk silicon, from [521]. Experimental results are given by crosses, G_0W_0 results by the dashed line. The quasi-particle structure is well described, but only one satellite is found. Thick continuous and dot-dashed lines: cumulant expansion (see Sec. 15.7), intrinsic spectral function only, and including extrinsic losses of the outgoing photoelectron and interference effects, respectively. The quasi-particle structure and the series of plasmon satellites are well reproduced. The intrinsic spectral function has a structure that follows the structure of the valence band quasi-particles; this is smeared out by extrinsic and interference effects in the final spectrum. All calculated spectral functions are supplemented with cross-sections and a secondary electrons background.

the experiment,[10] agrees well with experiment in the quasi-particle region, both concerning peak positions and relative intensities. However, the GWA calculation produces only one satellite, about 1.5 times the plasmon frequency away from the quasi-particles. This failure, which occurs in spite of the fact that W contains in principle a good description of plasmons, is analyzed in Sec. 15.7.

13.2 Materials properties in the GWA: an overview

Each material is different, but some general conclusions can often be drawn by looking at a number of selected cases. This section gives a summary of the results, a bird's eye view on the physics, and the potential of the GWA.

Bandgaps in solids: success of the GWA

The comparison of calculated and experimental bandgaps has been *the* representative demonstration of the impact of the GWA on band structure calculations. The two panels of Fig. 13.5 (from [291]) show calculated gaps for various materials as a function of experimental gaps. For perfect agreement the symbols should lie on the diagonal.

LDA Kohn–Sham eigenvalue gaps lie below the diagonal: Kohn–Sham severely underestimates the gaps. As in the case of silicon, state-of-the-art calculated GWA gaps fall, in general, close to experimental values, for a wide range of semiconductors and insulators, and correct the famous *bandgap problem* of the Kohn–Sham eigenvalues; besides silicon, early examples in the literature are Ge, diamond, GaAs, AlAs, or LiCl [192, 416–424]. In all cases, as discussed in Sec. 11.3, the bandgap is *opened* by the non-local *exchange* in the Fock operator, for which Koopmans' theorem holds. However, the effect is too large. Correlation beyond Hartree–Fock appears in the GWA in the form of dynamical screening, which leads to a gap *reduction*. Still, GWA gaps shown here are always larger than their LDA counterparts.

A closer look shows that in most cases, the $G_0 W_0$ gaps on top of the LDA as shown in the left panel are still too small, sometimes to a significant extent. As one wants increasingly better precision and a broader range of applications, this shortcoming becomes evident, especially for materials with localized electrons. That said, why should one start from the LDA? Some other possibilities are given in Sec. 12.2. In particular, a better treatment of exchange helps to localize the electrons correctly. This is naturally included in a self-consistent calculation, such as the partially self-consistent QSGW described in Sec. 12.4.

---

[10] Cross-sections reflect the probability that an electron is excited by the photon. They are obtained by integrating the spatially non-local spectral function with the final-state photoelectron wavefunction and the dipole operator. Cross-sections can also be found tabulated as a function of photon energy and character of the initial single-particle states [532]; this approach has been used to create the spectra of Fig. 13.4. Secondary electrons are created by scattering of the photoelectron on its way to the detector. Their number is proportional to the number of primary excited electrons at equal or higher kinetic energy. The secondary electron background as a function of binding energy can therefore be obtained by integration of the spectrum from the Fermi level up to the binding energy.

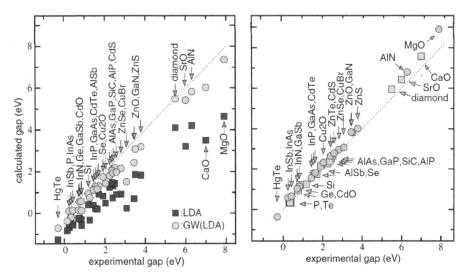

Figure 13.5. Fundamental gaps of *sp* compounds in LDA (squares) and in G_0W_0 starting from LDA (circles) in the left panel and QSGW in the right panel. All results are taken from [291]. The effects of spin–orbit coupling were subtracted from the experimental values. The G_0W_0 gaps improve on the LDA, but are still systematically underestimated and can even remain negative for small-gap semiconductors. For QSGW data, zinc-blende compounds with direct $\Gamma - \Gamma$ transitions are shown as circles; all other gaps are shown as squares. Errors are small and systematic, and would be even smaller if the electron–phonon renormalization were included; see Sec. 13.6. G_0W_0 and QSGW are defined in Sec. 12.4.

This leads in general to significant improvement, as can be seen by the squares and circles in the right panel of Fig. 13.5. Note, however, that there are cases where the gaps come out too large in QSGW, which can be explained by a too weak screening of the self-consistent RPA, as discussed in Secs. 13.4 and 15.3. Moreover, there are materials that are well described in one phase but the GWA completely fails in another phase, even within QSGW. Typically, this happens when correlation is important, but some aspects of it can be captured by symmetry breaking in an ordered phase. A prototype example is NiO, discussed in Sec. 13.3.

Metal or insulator?

In many cases, one may not care about a too small bandgap, when a simple shift in spectra does not spoil our understanding. However, the LDA band structure of some semiconductors comes out metallic, for example germanium in Fig. 2.6, or InN [533]. Also in that case, the action of the non-local exchange operator in the GWA is often sufficient to cure the problem. This holds for some technologically important applications, such as InN or the so-called *switchable mirror* compounds: upon increase of hydrogen pressure, yttrium hydrides switch from metallic to transparent, with an optical gap of 2.8 eV. In this process the material passes from YH_2 to YH_3. LDA yields a metal in all cases. A G_0W_0 calculation on top of LDA for YH_3 removes the band overlap responsible for the erroneous metallic LDA behavior [534–537]. Correlation effects were invoked to explain the effect, but the GWA results show that exchange plays a predominant role [538, 539].

Band structures: some more aspects

Bandwidth. The GWA has been developed in the homogeneous electron gas. There the quantity of interest is of course not a bandgap, but a *bandwidth*. The bandwidth of the HEG and simple metals is discussed in detail in Sec. 11.6. In short, G_0W_0 leads to band narrowing with respect to the LDA [351], which seems to be an improvement compared with experiment. The band is further narrowed by self-consistency using QSGW [291] (see Sec. 12.4), where an optimized *static* effective hamiltonian is used. Instead, when the full frequency-dependent expression of the GWA self-energy is used, self-consistency has the opposite effect. This is not a contradiction: updating *quasi-particle* energies following a G_0W_0 calculation drives the system further in the direction that is taken by G_0W_0. A *dynamical* self-energy leads to a Green's function with satellites far from the Fermi level, which corresponds to a larger effective bandwidth and therefore brings the system closer to Hartree–Fock. Note that this does not mean that QSGW *always* leads to band narrowing. A counter-example is silicon in Fig. 13.3: beyond the HEG and simple metals, wavefunctions and density modulations also play an important role, which leads to several competing effects that make predictions difficult.

Effective mass. The GWA has also been used to calculate effective masses from the dispersion of the bands. Good agreement with experiment has been obtained, for example, in III-V nitrides [540] and lead chalcogenides [541]. Agreement is worse in some small-bandgap compounds [542], where the gap itself has about 20% error. The relation between the gap and the effective mass is explored in Ex. 13.4. A detailed analysis of quasi-particle corrections to the effective mass can be found in [543]. The GWA has also been used to correct the slope of the Dirac cone in graphene with respect to LDA values [544, 545]. The Fermi velocity is increased significantly by a G_0W_0 calculation, which improves agreement with experiment [546].

Beyond the scissor. Many interesting materials contain groups of electrons of different character, in particular, different degrees of localization. Screened exchange has a very important contribution to GW corrections with respect to a local Kohn–Sham potential, as discussed in Sec. 11.3. Since exchange depends on the overlap of wavefunctions and on their degree of localization, the self-energy correction is very sensitive to this property, and states with different localization shift differently. Examples where the simple "scissor operator" picture is no longer adequate are localized states at semiconductor surfaces, such as GaAs(110) [510], or d-electron metals that contain localized d states and more delocalized s and p states, like copper [458] or silver [547].

Band offsets

In the calculation of bandgaps or band structures, all that matters are the relative energies. For example, the bands in a solid are usually defined relative to the highest occupied state.

In the calculation of energy differences between states of similar character, error canceling
helps to converge the calculations, in particular with respect to the number of empty bands
taken into account in screening and self-energy. Moreover, the starting point dependence
of a G_0W_0 calculation cancels partially when differences are taken, see, e.g., [366]. The
same holds for the influence of the plasmon pole model,[11] as worked out in Ex. 12.2. This
is different when one is interested in band offsets at interfaces. Shifts of the band energies
of the two different materials that form the interface are then required, and calculations
are more demanding. When there is less error canceling, one also has to re-examine the
neglect of vertex corrections beyond the GWA (see Sec. 15.2).

An increasing number of GWA calculations for interfaces can be found in the literature. Comparison with experiment may be difficult from a quantitative point of view due
to uncertainties in the structure, or presence of defects, in real materials. However, some
interesting points and trends can be found. For example, in [548] calculations have been
performed for a Si/SiO_2 interface. With well-converged calculations and full frequency
integration, G_0W_0 yields values in close agreement with experiment, only weakly influenced by vertex corrections. In [549] G_0W_0 has been used to calculate the quasi-particle
electronic structure of an Al/GaAs(110) Schottky barrier as a function of distance from
the interface. Near the metal there is narrowing of the semiconductor bandgap, with large
quantum corrections with respect to a classical image-potential approximation.

Lifetimes

In Secs. 7.5 and 7.6 the relation between the imaginary part of the self-energy and the
probability of decay per unit time, or inverse lifetime, of an electron are discussed. Equations (11.28) and (11.29) show that in the GWA, Im Σ stems from the imaginary part of the
screened Coulomb interaction W.[12] Hence, coupling of the quasi-particle to excitations of
the system that are contained in W leads to a finite quasi-particle lifetime [550].

Figure 13.6 shows electron lifetimes in aluminum. Experimental results have been
obtained by a laser pump-and-probe experiment, theoretical results by a G_0W_0 calculation
for two different bands and parts of the Brillouin zone [551]. The result for the homogeneous electron gas, based on a low-frequency expansion of the dielectric function, shows
the quadratic behavior given by Eq. (7.24). Although aluminum is a simple metal, neither
experiment nor GWA calculations coincide with this result. Instead, lifetimes are direction-
dependent and can be larger or smaller than the electron gas result. This is due to band
structure effects.

Semiconductors behave similarly overall; in particular, the direction dependence of the
imaginary part of the self-energy leads to the scattered results for the imaginary part of
the self-energy of silicon in Fig. 13.2, and has already been seen in early calculations on
diamond [416, 417]. The lowest energy at which an additional particle can decay into an

---

[11] The plasmon pole model of [441] shows the best results; see Sec. 12.2.

[12] The contribution of the Green's function to the imaginary part is a shift of the energies, as one can see from
Eq. (11.28).

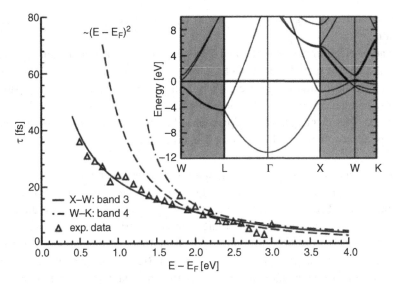

Figure 13.6. Lifetimes of excited electrons in aluminum, as a function of energy. From [551]. Measured lifetimes are triangles. The inset shows the LDA band structure. The conduction bands between X and W and between W and K (thick lines) are the ones that mainly contribute to the measured lifetime. The continuous and dot-dashed lines in the main panel are the lifetimes calculated in G_0W_0 along X–W and W–K, respectively. The dashed line denoted by $(E - E_F)^2$ is the lifetime obtained approximately for the HEG from Fermi-liquid theory with $r_s = 2.07$.

excitation across the bandgap is called the Auger threshold. Below this threshold, energy conservation forbids such a decay, and the imaginary part of the self-energy vanishes. As pointed out in the discussion of silicon in the previous section, only when the absolute value of the energy exceeds the lowest electron or hole energy plus the lowest neutral excitation energy is the imaginary part non-zero, and broadening of the quasi-particle peak in the spectral function is observed.[13]

When moving to noble metals, results depart even more strongly from the homogeneous electron gas. A detailed theoretical investigation of inelastic lifetimes of holes in Cu and Au can be found in [552]. It was shown that the contribution from occupied d states to the screening leads to electron and hole lifetimes that are much longer than those of electrons and holes in the electron gas. In agreement with experiment, d holes close to the Fermi level show longer lifetimes than sp electrons with the same excitation energy, because of the small overlap between d and sp states below the Fermi level (see Ex. 13.5). The d-hole lifetime decreases then rapidly for larger binding energies. Lead also shows a strong departure from the electron gas [553]. Between 5.5 and 8 eV binding energy a strong local-ization effect is found: the lifetime of d states is almost twice as long as the lifetime of p

---

[13] Note that the non-self-consistent G_0W_0 or GW_0 introduces then an inconsistency, in the sense that the thresh-old is determined by the band structure that is used as input, e.g., an LDA one, instead of the quasi-particle energies.

states. Spin–orbit coupling has significant impact with respect to a purely scalar relativistic calculation.

These are d-electron metals with filled shells. For more results on transition metals, see [554]. In the case of partially filled shells, other excitations not contained in W may have an important contribution to the decay possibilities that enter the imaginary part of the self-energy and hence limit the lifetime. In particular, spin fluctuations can be crucial in materials like iron or nickel. There, approaches like the T-matrix discussed in Sec. 15.6 can be applied; see, for example, [555] for T-matrix corrections to the GWA self-energy of nickel.

Excited electrons can also couple to excitations of the lattice, phonons; this adds another piece to the imaginary part of the self-energy. Phonons are a huge topic, and therefore beyond the scope of this book. However, electron–phonon coupling can be understood in a similar way as electron–plasmon coupling. Phonon contributions to measured lifetimes can be important. For example, in n-doped graphene [556] it becomes comparable to the purely electronic contribution at a binding energy of 0.2 eV. See also the remarks about coupling to phonons in conjunction with the discussion about temperature in Sec. 13.6.

Concerning bulk and surface states of metals, more details and results can be found in [551], and an exhaustive review, including calculations beyond the GWA (see Sec. 15.6), is given in [557]. More recent results for semiconductors and comparison with previous results, as well as a discussion of the formalism, is contained in [558].

Can the GWA describe satellites?

Most GWA calculations are for quasi-particle properties. However, in principle the full spectral function beyond the quasi-particles, including some satellite features, can be calculated. Figure 13.4 shows that the GWA has a big problem in describing satellites in silicon. This is also noted in the HEG (Sec. 11.6), where far from the Fermi level sharp structures appear that are not seen in measurements on simple metals. The experimentally observed series of plasmon satellites is instead absent in G_0W_0; there is only one satellite, in the wrong position. In nickel, discussed in Sec. 13.4, the GWA does not give the satellite seen in experiments at 6 eV binding energy. In contrast, that same section shows that the GWA gets satellites in VO_2 reasonably well. Materials-specific details are discussed there. However, one can make a few general points.

- The GWA can create a satellite only when W contains the dominant excitations that contribute to this structure.
- If this is the case, the satellites must be weak enough to be described with the GW self-energy that is first order in W.
- The first-order expression does not lead to a satellite *series*, at least not in G_0W_0.
- As a reminder of Sec. 12.4, a GW calculation must be self-consistent in the quasi-particle energies or at least the chemical potential, in order to yield reliable results for the satellite spectrum (see also Ex. 13.6).

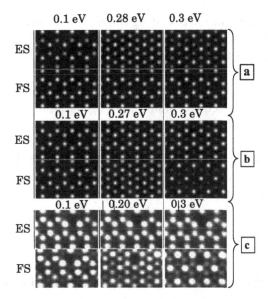

Figure 13.7. STM image of α-Sn/Ge(111). GWA calculations for two different structures in panels (a) and (b) are compared with experiment in panel (c), as a function of bias voltage. ES and FS indicate empty and filled states, respectively. The comparison of calculated and measured images shows that the structure leading to the result in panel (b) is more realistic. (Adapted from [559].)

The first point can be understood from Eq. (11.28) and is discussed for example in Sec. 15.6, an analysis concerning the second and third points can be found in Sec. 15.7.

GW band structure in the calculation of other properties

Many properties of materials depend on the band structure. Therefore, GW corrections are also used as ingredients in other calculations. Important examples are as follows.

- **Scanning tunneling microscopy (STM) images.** Bias-dependent STM images can be obtained as energy-integrated local density of states. They are influenced by the energetic positions of the states. An example is shown in Fig. 13.7, adapted from [559]. It shows an experimental image in panel (c). Images calculated from the GWA band structure for two different surface configurations that were proposed for the α-Sn/Ge(111) surface are shown in panels (a) and (b). Here it is important that the "scissor operator" (see the discussion on p. 316) is not valid: GW shifts are different for surface states related to tin atoms and states related to germanium, and the final result allows one to distinguish the STM pictures of the two configurations. Panel (b) agrees better with experiment than panel (a), confirming also predictions from the comparison of total energies.
- **Optical and electron energy loss properties.** State-of-the-art calculations of optical properties are based on the Bethe–Salpeter equation, described in Ch. 14. The

quasi-particle band structure is a crucial ingredient in these calculations. It is most often evaluated in the GWA. When the GW correction is a scissor shift, GW absorption spectra are simply shifted to higher energies with respect to Kohn–Sham ones. In addition, the GWA can influence the spectral shape. As discussed in Ch. 14, in particular in Sec. 14.11 and Ex. 14.1, the effect of GW corrections is in part canceled by the electron–hole interaction. However, GW corrections can still have a significant effect on the final spectrum; an example is silver [547].

- **Transport.** The one-body Green's function enters the description of transport through a molecule between leads. Equilibrium GWA can be used for the description of a steady state, and the extension to non-equilibrium Green's function explained in App. E makes it possible to understand the dynamics of the problem, for example, when a bias is switched on. A detailed description can be found in the book [301] on non-equilibrium Green's functions.

13.3 Energy levels in finite and low-dimensional systems

The GWA has been developed for extended systems, such as the homogeneous electron gas. What happens when one applies it in lower dimensions, or in finite systems? When a system is confined in one or more dimensions, several characteristic features appear:

- States are on average more localized. This increases the electron–electron repulsion, and in particular the self-interaction contribution to exchange and hence GW corrections, as discussed in Sec. 11.3.
- Screening is lowered. Since the system must overall remain charge neutral, in a confined system positive and negative induced charges are close to each other. Therefore, the effective screening is reduced and its macroscopic average is zero. This can even lead to regions of *antiscreening*, which has been seen in systems such as clusters, one-dimensional surface states, or nanotubes [508, 560–562]. The effect is illustrated in Fig. 13.8 for the case of a carbon nanotube. Note, however, that this does not mean that screening can be neglected in finite systems: screening on a local, microscopic, scale is an important ingredient, even in systems as small as light atoms [563].

GWA corrections tend to be larger in finite and low-dimensional systems than in bulk, due to the localization and reduced screening. Figure 13.9 gives an illustration [564]. It shows results for hexagonal boron nitride starting from its bulk phase, and progressively increasing the interlayer distance: the GWA correction increases steadily. Also, excitonic effects are shown; these are discussed in Sec. 14.9.

- LDA wavefunctions may be quite poor in systems that are finite in one or more directions, so one should go beyond a first-order G_0W_0 correction starting from LDA. The exchange contained in the GWA can lead to charge *localization*; in contrast, the GWA calculation may move a bound LDA state above the vacuum level, with a consequent *delocalization* of charge [565].

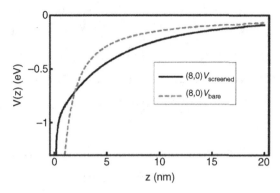

Figure 13.8. Screened interaction (continuous line) as a function of distance along a (8,0) carbon nanotube, in comparison with the bare interaction (dashed). From [562]. There exists a region of *antiscreening*, where the screened interaction is stronger than the bare one.

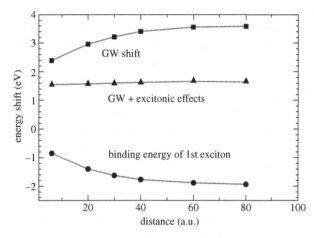

Figure 13.9. Dependence of many-body corrections on interlayer distance in h-BN, from [564]. GW corrections and exciton binding energies increase when going from the bulk to the single sheet.

- There exists an *outside* of the system. There, new effects can be studied, like the image potential, image states, or ionization potentials. The image potential is an effect of long-range screening that is well described by the GWA (see Sec. 11.4). Figure 13.10 shows image states outside sodium clusters (calculated as homogeneous electron gas spheres) that are found in G_0W_0, but absent in the LDA [394]. Since there is so little overlap with the cluster or substrate, there is little probability of decay, and the lifetime of these states is particularly long compared with the states inside.
- In principle, ionization energies could be calculated as the quasi-particle eigenvalue linked to the highest occupied molecular orbital of the N-electron system, or from the lowest unoccupied molecular orbital of the $(N-1)$-electron system, and should be equal to the total energy difference between the system with N and $N-1$-electrons. This can easily be checked in finite systems. It turns out that the GWA [353, 396] yields different

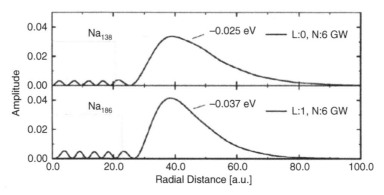

Figure 13.10. Image states outside the clusters Na$_{138}$ and Na$_{186}$, from [394]. These are loosely bound states found in the GWA, but absent in LDA.

results for the three possibilities. As mentioned in Sec. 4.4, the total energy as a function of fractional number of electrons should be a straight line between integer numbers [190], but it is concave in approximations like Hartree–Fock and convex in most DFT functionals. In the GWA it is slightly concave [563], a remnant of Hartree–Fock. This can be linked to an over-localization, as opposed to the delocalization error stemming from most DFT functionals [566, 567].

- In a crystal, an electron with its periodic Bloch wavefunction has equal probability of being in any of the unit cells. Localized orbitals in finite systems indicate a region of high probability for an electron to be. ΔSCF approaches, which are explained in Sec. 11.4, profit in finite systems from the fact that when an electron is added or removed, its Hartree potential induces the essential relaxation effects. Instead, the relaxation induced by the Hartree potential of an electron added in a delocalized state in a bulk solid is insignificant [568]. Therefore, in finite systems ΔSCF is competitive with the GWA. Examples are atoms in the iron series [431, 569] or small sodium clusters [508, 570, 571].

- Finite systems can be located on a substrate; the effect of the substrate must be taken into account. The substrate leads to screening, for example of adsorbed thin films (see, e.g., [572]). Quasi-particle calculations for molecules adsorbed on insulator films show the importance of screening [573]. The GWA accounts for screening in an explicit, flexible, and detailed way, unlike simple approximations to Kohn–Sham DFT, or hybrid functionals [574].

- Finite systems can highlight the problem of multi-determinental ground states and degeneracy. Take the example of the Hubbard molecule with one electron, discussed in Sec. 11.7. In the dissociation limit, bonding and antibonding states become degenerate. At the same time, the probability of the electron hopping from one site to the other decreases. One should no longer consider the square of its wavefunction as a symmetric classical charge density – the electron *is* on one side, but on *which* one? When one adds an electron to this system it can meet an empty, or an occupied, site. The energy difference is the Coulomb interaction U. Therefore, the exact spectral function for electron addition shown in Fig. 11.10 has two peaks, at a distance of U. The GWA simply places half an electron on each site; it contains no explicit correlation between the electron in the system and the additional electron.[14] Therefore, it finds a single peak in the middle. As an example for a real molecule, one can look at C_2 [575].

13.4 Transition metals and their oxides

Most interesting things happen when several aspects of a problem compete. For example, what about materials that are extended, but where some states are localized? This section presents a prototype class of bulk materials where some states are localized in space close to the atoms, and others are not: these are transition metals and their oxides, where one finds localized d or f-electrons. They constitute one of the main applications of dynamical mean-field theory, as discussed in Chs. 19 and 20, and are interesting materials for comparison between methods.

Nickel: a d-electron metal

Let us first look at a metallic system, for example nickel. It differs from simple $s - p$ electron metals in several important points:

- It is ferromagnetic at low temperatures.
- It contains localized d states resulting in a narrow d band close to the Fermi level, with a width of about 3.3 eV.
- The two lowest atomic configurations $3d^8 4s^2$ and $3d^9 4s^1$ are almost degenerate; the difference is only 0.025 eV [576].
- There are local magnetic moments that survive above $T > T_c$, as can be seen in Fig. 2.3 and Sec. 20.4.

For the band dispersion, the GWA does well. Figure 13.11 shows the band structure of nickel measured in angle-resolved photoemission [577, 578] and calculated in G_0W_0 starting from LSDA [505]. Left and right panels are majority and minority spin, respectively. As also found in [293, 484], there is significant improvement with respect to LSDA calculations. The self-energy correction is strongly state-dependent and dispersing over the Brillouin zone. It leads to a reduction of the occupied d-bandwidth by almost 1 eV with respect to LSDA, which brings calculations closer to experiments. This can be seen most clearly at the X point around -3 eV.

In experiment the distance between majority and minority bands, called *exchange splitting*, is about 0.3 eV. The LSDA overestimates this value by about a factor of two. The problem is not cured by the GWA, which yields values close to the LSDA [293, 484]. The

---

[14] In the wrap-up of Ch. 11, the description of an additional particle in the GWA is compared with a boat that creates waves in water. The GW particle (the boat) interacts with the system in an average way. To introduce the missing correlation in this picture, one should take into account the interaction of the boat with the individual water molecules, which becomes important at low density.

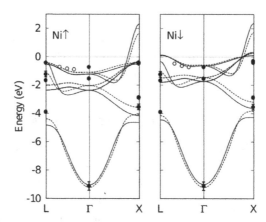

Figure 13.11. Quasi-particle band structure of nickel. Shown are experimental ARPES results (circles) [578] (full circles are spin-averaged, open circles are spin-resolved), LDA eigenvalues (continuous), and results of a G_0W_0 calculation (dashed line). The Fermi level is set to zero. Calculations are from [505].

reason for this is that the self-energy correction of spin-up and spin-down states is too similar: the GWA is determined by W, which is not spin-dependent, since it contains only the excitations of the density–density correlation function. Similar or worse errors in the exchange splitting are also found in f-electron materials like gadolinium [579].

In the experiments for nickel there is also a satellite, 6 eV below the quasi-particle peak, which is absent in the GWA. Since the loss function and hence W of bulk nickel does not show significant structure at that energy, GWA cannot create a peak. One has to go beyond GWA, for example by using the T-matrix approximation (Sec. 15.4) or DMFT (Sec. 20.4), in order to get this feature; see the spectral functions in Figs. 15.6 and 20.6.

NiO: a prototype "strongly correlated" material

NiO is one of the materials often termed *strongly correlated*, or Mott insulator, and its particular behavior was already realized in 1937 [580] (see also Chs. 19 and 20). Its electronic structure and orbital characteristics are analyzed in detail in Sec. 20.7. We have to distinguish two insulating phases in NiO: the antiferromagnetic phase at low temperature, and the paramagnetic phase above 523 K. It is not difficult to understand that the antiferromagnetic phase can be insulating. In the paramagnetic phase the unit cell is only half that of the antiferromagnet. The number of electrons in the unit cell is still even, so that the material could be metallic or insulating in a mean-field picture. However, as explained in Sec. 20.7 and Ex. 20.3, the level splitting due to the symmetry breaking of the crystal, shown in Fig. 19.5, indicates that there are partially filled levels in an independent-particle picture, with the conclusion that all reasonable static mean-field approaches that do not break symmetry must yield a metal for the paramagnetic phase.

Here we concentrate on a few aspects that are interesting in the context of the GWA. In particular, what are the consequences of the presence of localized states for a GWA calculation? And how does the GWA perform for the antiferromagnetic and the paramagnetic phase, respectively?

Let us first look at the ordered, i.e., the antiferromagnetic phase. The calculated density of states and the band structure, calculated in various approaches, are shown in Fig. 20.8, and in Figs. 13.12 and 13.13, respectively. Kohn–Sham calculations using GGA or LSDA yield a small gap[15] of less than 1 eV [581], much smaller than the experimental value of 4.3 eV. In the LSDA and GGA the upper valence bands have d character, whereas the lower valence states are O $2p$. The separation between occupied oxygen and nickel states is clearly seen in the GGA results in Fig. 13.12. The right panel of Fig. 13.12 contains results obtained using the HSE03 hybrid functional. Here the direct gap is 4.5 eV. The dispersion of the upper valence bands is increased with respect to the GGA results, because the almost dispersionless Ni t_{2g} levels are pushed down, so that the top valence acquires more O $2p$ character, in agreement with experiment. Except for a small region around the Γ point, where the bottom conduction state has oxygen s character,[16] the gap is

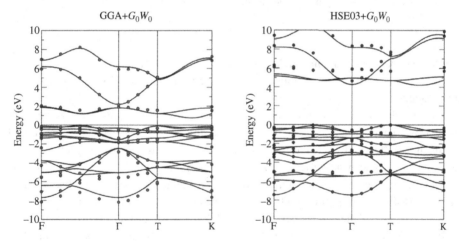

Figure 13.12. Band structure of NiO, adapted from the results of [581] by Claudia Rödl. Left panel: the continuous lines are the result of a GGA calculation. Dots denote the QP energies obtained within G_0W_0 starting from the GGA. Right panel: the continuous lines are the result of a HSE03 hybrid functional calculation. Dots as before. The valence band maximum is set to zero in both calculations. Note the starting-point dependence of the G_0W_0 results. The labels of the high-symmetry points refer to the rhombohedric cell. In the f.c.c. cell this corresponds to L (rhombohedric F), L' (rhombohedric T), and X (rhombohedric K), respectively.

---

[15] The gap is difficult to see in the bottom panel of Fig. 20.8 because of broadening. Anyway, the presence of this small gap has no important consequences; GWA results for materials like CoO or FeO, where the LSDA is metallic, behave in a similar way.

[16] This feature is observed also in MnO, FeO, and CoO.

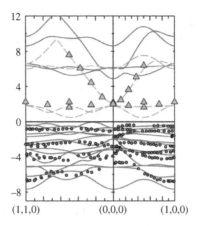

Figure 13.13. Band structure of NiO from [292]. The dashed lines are obtained by LSDA, and triangles denote the QP energies obtained within G_0W_0 starting from the LSDA. Only the conduction states are shown. (Note that Fig. 20.8 contains both the occupied and unoccupied DFT density of states). Continuous lines are valence and conduction states obtained from self-consistent GW in the QSGW approximation, as explained in Sec. 12.4. Circles in the valence bands are experimental results from angular-resolved photoemission. The valence band maximum is set to zero. The labels of the high-symmetry points refer to the f.c.c. cell, where (1,0,0) is the X-point.

between the oxygen $2p$ and nickel $3d$ states, and is therefore called a *charge-transfer* gap. Figure 20.8 also illustrates this point. Note that the valence band maximum is set to zero in all calculations.

These results were used as a starting point for subsequent G_0W_0 calculations. The final G_0W_0 band structures can be compared in the two panels. There is a striking memory effect: in each case, G_0W_0 is relatively close to the respective starting independent-particle band structure, both concerning the gap and the hybridization, and also concerning the order of the bottom conduction states. The HSE03 results, and therefore also HSE03+G_0W_0, are much closer to experimental findings than the GGA and GGA+G_0W_0 ones. The direct gap, for example, is 1.4 eV in GGA+G_0W_0 and 5.2 eV in HSE03+G_0W_0. The reason is that HSE03 contains already some physics of the GWA, although in an approximate way (see Sec. 11.4), whereas the GGA makes NiO too close to a metal: the gap is strongly underestimated, and the d states are too delocalized. Therefore screening is overestimated, with a GGA dielectric constant of more than 25 compared with the experimental value of 5.7. The GWA self-energy correction is roughly inversely proportional to the dielectric constant, and it is therefore unrealistic when calculated with a GGA input. HSE03 yield a slightly underestimated macroscopic dielectric constant of 4.5, which explains the small overestimate of the gap by HSE03+G_0W_0. The risk of starting from GGA or LSDA, especially for materials with localized states, has been pointed out since 1995 [499]. HSE03 hybrid functionals [365], DFT+U with an appropriate U, or self-interaction corrected calculations contain screened non-local exchange in an approximate way. They are closely linked to the GWA, as explained in Sec. 11.4, and can therefore reproduce to some extent the bandgap and trends in hybridization. In NiO and similar materials they are better starting points for G_0W_0 calculations than local or semi-local Kohn–Sham functionals [581].

These findings explain why the results of G_0W_0 calculations in the literature can be very different: the starting point can act like a tunable parameter. *It is important to deal carefully with this aspect*, in order to make G_0W_0 meet the expectations of a first-principles method.

A way to overcome the starting-point dependence is self-consistency. Results of the quasi-particle self-consistent QSGW [292], explained in Sec. 12.4, are shown in Fig. 13.13. The direct gap is 4.8 eV, close to the experimental value of 4.3 eV. The QSGW valence bands are in good agreement with angle-resolved photoemission data. QSGW also improves the magnetic moment from 1.28 in LSDA to 1.78, closer to the experimental value of 1.9. In the self-consistent calculation, exchange localizes the d states with respect to the LSDA or GGA density, which explains the lowering of the t_{2g} level and associated changes in hybridization, as well as improved magnetic moments. Overall the QSGW results for NiO are in good agreement with experiment. Since QSGW is based on an effective static hamiltonian, this indicates that the quasi-particle picture is sufficient to understand the main features of antiferromagnetic NiO near the Fermi energy. Of course, by definition the quasi-particle calculations do not describe the satellites that are observed around 9 eV binding energy in the measured photoemission spectrum.

The remaining discrepancy is that QSGW places the empty Ni d-level too high in energy, by about 0.8 eV compared with experiment. This is a general finding for transition metal oxides, even more pronounced when moving to 4f electrons [579]. One explanation is the fact that the RPA W in QSGW underestimates screening, similar to HSE03, and therefore brings the method too close to Hartree–Fock [292]. This is discussed further in Sec. 15.3.

Concerning G_0W_0, the QSGW results confirm the choice of a HSE03 starting point, as opposed to a GGA or LSDA one. Since the starting-point dependence of G_0W_0 is understood, one can in principle know when it is important to use a quasi-particle self-consistent method like QSGW, or at least restrict the range of possible starting points. For a review of this point, see [582].

The paramagnetic state presents an additional difficulty. In the paramagnet, fluctuating local magnetic moments average to zero, and the average magnetic moment vanishes. Symmetry-restricted mean-field calculations like LDA or Hartree–Fock, and also the GWA, simply neglect spin in the paramagnet. As explained at the beginning of this subsection, in view of the energy level diagram it is not possible to explain the insulating gap in a spinless independent-particle picture.

So, why *is* there a gap in paramagnetic NiO? The order or disorder of local moments is not the aspect that governs the gap: the GWA gap of NiO in a *ferromagnetic* configuration is only about 6% smaller than in the antiferromagnetic one (see also [499, 583]). This, however, does not mean that spin can simply be neglected. To understand electron addition or removal energies, it is useful to look at Fig. 19.4 representing atomic-like d states in a cubic lattice. Two electrons in the same orbital must have opposite spin, and their interaction energy is the Hartree interaction U_0. The interaction energy between electrons of the same spin is reduced by the exchange J. However, electrons of the same spin must occupy different orbitals that have only weak spatial overlap. Therefore, J is not large enough to cancel the effect of the repulsion, and the inter-orbital Hartree interaction U_1 is still significant. In both cases one finds a significant repulsive interaction energy, which leads to the gap opening. Instead, in a picture without spin one can fill one orbital with two electrons, and the interaction energy is not taken into account.

This argument is very local: the important fact is only the existence of a spin on an atom. *How* the spins are ordered is less relevant. However, in order to yield a gap, a theory must "know" whether spins are parallel or antiparallel. In a spin-ordered phase the position of up and down spins is fixed, and such a symmetry breaking captures part of the correlation, as pointed out in several places in this book, starting from the definition of correlation in Sec. 2.1. Otherwise, one needs a spin–spin correlation function somewhere in the formulation. This is not the case for Hartree–Fock, and it is not the case for the GWA either. Therefore, also the GWA gap vanishes in the paramagnetic phase. A way out is to simulate spin disorder directly. This idea is used in a mean-field method called the *disordered local moment* approach (see Sec. 19.6). Calculations in the disordered local moment approach based on self-interaction-corrected Kohn–Sham [584] yield insulating paramagnetic NiO. Descriptions of the paramagnet in terms of disordered spins are in the spirit of the Hubbard III alloy approximation described in App. K.2, and can be seen as an approximation to DMFT (see Secs. 16.4 and 19.6).

VO$_2$: a metal–insulator transition

Vanadium dioxide exhibits a metal–insulator transition at 340 K [585]. Below the transition temperature, it is monoclinic and insulating, but the high-temperature phase is metallic with rutile structure. The role of electronic correlations in this phase transition has been subject to many debates, for example in [586, 587].

In a band picture, the electronic states can be understood as schematized in Fig. 13.14 [588]. Around the Fermi level the states have strong vanadium $3d$ character. Each vanadium atom is surrounded by an oxygen octahedron, and the crystal field splits the d states into t_{2g} and e_g. Since the structure is not cubic, the t_{2g} states are split further into e_g^π and a_{1g}, like in NiO. Each vanadium atom has three $3d$ and two $4s$ valence electrons. The oxygen $2p$ orbitals in VO$_2$ capture four of these electrons, hence VO$_2$ is in d^1 configuration. With one d-electron partially occupying the a_{1g} band, the system is metallic in the rutile phase. In the monoclinic phase there is a Peierls distortion, and two vanadium atoms form a dimer. The unit cell doubles. The a_{1g} orbitals strongly hybridize between the two vanadium atoms, as shown in Fig. 13.14. Now in an independent-particle picture the two electrons fill the bonding a_{1g} state. It is separated by a gap from the unoccupied e_g^π if the splitting is strong enough such that the bands do not overlap.

The Kohn–Sham band structure calculated within LDA predicts a metal for both phases [586, 589], because valence and conduction bands overlap. G_0W_0 starting from LDA, or with self-consistency through an update of energies only, is not sufficient to open a gap. However, this does not mean that the band picture breaks down, like in paramagnetic NiO. The problem is simply due to the fact that the LDA overlap of a_{1g} and e_g^π bands

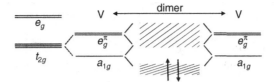

Figure 13.14. Schematic picture of the energy levels in VO_2 corresponding to the crystal field diagram in Fig. 19.5 where the t_{2g} levels for each atom are split into a_{1g} and e_g^π states. The a_{1g} level is half-filled in the rutile phase. In the monoclinic phase, two vanadium atoms form a dimer and the a_{1g} orbital splits into bonding and anti-bonding components. The bonding state is fully occupied and a gap is formed.

is not removed. Indeed, LDA has a tendency to delocalize charge and make the system too isotropic. Only if the charge density is calculated with a method that describes exchange better, and therefore correctly localizes the charge, is the problem overcome. This has been shown by calculations on the GWA level, first in a model approach [359] and subsequently in parameter-free calculations [360–362]. In [360], quasi-particle wavefunctions have been calculated self-consistently in the static COHSEX approximation to GW (explained in Sec. 11.3), followed by one fully dynamical GW step. The self-consistent COHSEX quasi-particle wavefunctions of the d states are superpositions of LDA d conduction and valence states, which leads to an increase of anisotropy in the monoclinic phase. The effect of this anisotropy on the band structure is to remove the overlap of the d bands. Note that VO_2 is also paramagnetic, but in contrast to the case of NiO, the local moments are not the crucial factor for the formation of a bandgap.

There are, however, effects of dynamical correlation beyond the quasi-particle picture in VO_2. This can be seen in photoemission spectra. Looking at the band structure, one would expect to see a peak due to the d-bands close to, or at, the Fermi level, and an O $2p$ structure at higher binding energy, with a gap between the two. The two structures can be seen in the measured spectra shown in Fig. 13.15. Moreover, one observes lifetime broadening and most importantly, in the metallic phase there is an additional structure about 1.5 eV below the d-quasi-particle peak, before the onset of the O $2p$ peak. There is also a shoulder on the O $2p$ peak, at about -9 eV [590, 591]. These are satellites. Experimental electron energy loss spectra [592] as well as RPA calculations of the loss function $-\mathrm{Im}\,\epsilon^{-1}$ [360] display a peak around 1.5 eV in the metallic, but not in the insulating phase. The calculations show that this peak is due to localized electron–hole excitations.[17] The loss function is directly reflected in the screened Coulomb interaction W. As explained in Sec. 11.3, within the GWA peaks in W lead to satellite peaks in the spectral function. The GWA is therefore able to reproduce and explain the observed deviation from the quasi-particle picture [360, 470].

---

[17] Note that the loss function is in general dominated by plasmon excitations, coherent oscillations of all electrons. However, W contains in principle all excitations that do not change particle number, including for example excitons that are described in Ch. 14, or local electron–hole excitations that can be important at low energies.

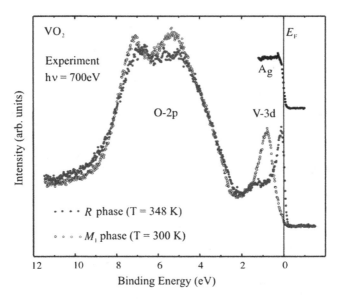

Figure 13.15. Experimental photoemission spectra of metallic (black dots) and insulating (open circles) VO_2, from [590]. The metallic phase shows a satellite around -1.5 eV binding energy, and an increase of intensity with respect to the insulating phase around -9 eV. Both are due to local excitations of d-electrons as response to the removal of an electron.

In order to describe the satellites, *it is important that the calculation is self-consistent in the quasi-particle energies*. If this is not so, satellites come out much too far from the quasi-particle peak, and much too weak. This is a general finding. It has also been seen on model systems that can be calculated numerically exactly, for example Hubbard clusters [402].

The GWA spectral functions at Γ, calculated using the results of [360] in the expression for the spectral function Eq. (13.1), are compared with results of DMFT in a two-site approximation in Fig. 13.16. The c-DMFA calculations are described in Sec. 20.6. The dimer plays a central role here; the single-site DMFA approximation cannot describe VO_2. The left panel shows results for the metallic rutile phase, the right panel for the semiconducting monoclinic phase. The c-DMFA results [593] are shown as an intensity plot along high-symmetry directions of the Brillouin zone. Dark regions correspond to peaks, brighter regions to lower intensity. The scale is logarithmic. One can see quasi-particle bands and additional structure. The Γ-point is at the left of each panel. There, the GWA spectral function at Γ is given as a line plot, with logarithmic intensities on the horizontal axis and energies along the vertical axis, in correspondence to the energies of the c-DMFA results. There is good agreement between GWA and c-DMFA for both the rutile and the monoclinic phase; note, for example, the different spectral shape of the top valence band region in the two phases. In the rutile phase there is weight about 1.5 eV below the quasi-particle d-band in both GWA and c-DMFA; this corresponds to the weak satellite seen in photoemission. In the GWA this satellite is due to a structure in the loss function. The c-DMFA

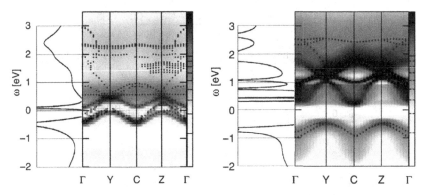

Figure 13.16. Spectral function of VO_2, comparison between the GWA and c-DMFA. Left panel: the metallic rutile phase. Right panel: the semiconducting monoclinic phase. The intensity plots are the results of c-DMFA along high-symmetry directions of the Brillouin zone. At the left of each panel is the Γ-point. There, the GWA spectral function at Γ is given as a line plot, with intensities on the horizontal axis and energies along the vertical energy axis of the c-DMFA results. Intensities are on a logarithmic scale for both GWA and c-DMFA. Both methods produce well-pronounced bands as well as lifetime broadening and satellites. GWA from the calculations published in [360], c-DMFA picture from [593].

contains all local excitations in the subspace of d states, including for example spin fluctuations. The two methods agree when a satellite is dominated by local electron–hole or plasmon excitations, as is the case here. In both approaches, in the monoclinic phase the top valence quasi-particle is pushed down to about -1 eV, with a large broadening. Similar agreement is found in the conduction region. Altogether, both methods agree quite well with each other and with experiment.

The satellite in GWA comes out a little too weak and too far from the quasi-particle,[18] whereas in c-DMFA its strength and position depend on the parameter U; those uncertainties are, however, small on the scale of variations between various experiments [590, 591]. The d states in VO_2 are a case where the GWA and approximations to DMFT, which is designed for strong local correlations, meet and confirm each other. The GWA treats all electrons on the same footing and can therefore detect interaction effects on a broader spectral range. As a consequence in VO_2, it can also elucidate the origin of the shoulder at -9 eV on the oxygen $2p$ peak in the metal [470]: like the satellite at -1.5 eV, it is due to d-band excitations in the loss function.

13.5 GW results for the ground state

There are far fewer ground-state results produced by GWA calculations than excited-state properties. The main reason is probably that in this domain DFT is appropriate, and often successful. However, standard approximate DFT functionals have well-known problems. These are, for example, the spurious self-interaction due to an approximate treatment

---

[18] It may seem strange that the GWA can describe at all satellites in a transition metal oxide, while it fails for silicon. The reason is the fact that the local plasmon in VO_2 is weak. Therefore, contrary to the case of silicon, it does not induce a spurious plasmaron peak. The plasmaron problem is discussed in Sec. 15.7.

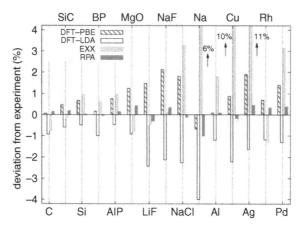

Figure 13.17. Relative error in lattice constants with respect to experimental values obtained from various DFT functionals. The RPA performs best. (From [594].)

of exchange, or the absence of long-range dispersion in local or semi-local functionals. The GWA can ameliorate these problems, since it contains Fock exchange and non-local polarization effects. In the spirit of Ch. 8 one can use the fact that there exist variational functionals that are stationary around the solution. This means that even evaluated with an approximate Green's function, they may give reasonable results. The simplest approximation to a GWA total energy calculation is obtained when the functional of the Green's function Eq. (8.15) on the GW level is evaluated approximately using a Kohn–Sham independent-particle Green's function. As explained in Sec. 11.5, this yields the DFT-RPA total energy. We will therefore start this section by considering some results obtained in that framework.

Total energy in DFT-RPA

The DFT-RPA total energy is given by Eq. (11.38). It has been used to calculate total energies in finite and infinite systems. Lattice constants in solids are systematically improved with respect to local or semi-local functionals. This is illustrated in Fig. 13.17, taken from [594]. There is, however, an overall tendency towards underbinding, i.e., larger lattice constants, which is also reflected in atomization energies [376, 594], or heat of formation [594].

The RPA or related functionals account for van der Waals dispersion interactions, because of the presence of long-range screening. For example, the RPA significantly improves the description of ground-state properties of noble gases [595]; their cohesion is a correlation effect as invoked in Sec. 2.2. Figure 13.18 shows the total energy versus interplane distance of hexagonal boron nitride [596]. The calculation has been done using exact Kohn–Sham exchange (EXX) and RPA with a local correction, which is called RPA+.[19]

---

[19] This correction is chosen such that the result is exact in the homogeneous gas, namely, $E_c^{RPA+} = E_c^{RPA} - (E_c^{LDA-RPA} - E_c^{LDA})$, where $E_c^{LDA-RPA}$ is the local density approximation to the RPA [597].

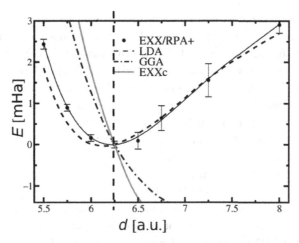

Figure 13.18. Total energy versus interplane distance d of h-BN, adapted from [596]. Shown is the difference of total energy E with respect to the value calculated at the experimental equilibrium distance, which is given by the vertical dashed bar. Different methods are compared: LDA (dashed), GGA ([601], dot-dashed), EXX with a local correlation correction (continuous gray), and RPA using EXX ingredients and an LDA-like correction (dots with error bars due to an extrapolation technique used to determine the values, fitted by the continuous line). The RPA+ result yields an equilibrium distance close to the experimental one.

In-plane and out-of-plane bonding are well described over a wide range of interlayer distances. Also, the equilibrium lattice constant of graphite as well as its C_{33} elastic coefficient and interlayer binding energy have been calculated in the RPA [598]. Results compare well with experiment. Quantum Monte Carlo [599] yields a lattice constant that is only 2% larger, and an about 15% larger interlayer binding energy; this might be due to uncertainties in the QMC calculations, in particular the finite size of the cell. As predicted previously by analytic calculations [600], the interlayer dispersion interaction follows a d^{-3} law at large distances d.[20]

For an overview on RPA and attempts to go beyond, see the review [378].

Total energy from the GWA

Section 11.6 contains self-consistent GWA results for total energies of model systems, namely Hubbard clusters and the homogeneous electron gas. In the 3D gas self-consistent GW calculations show excellent agreement with QMC data, whereas stronger deviations are detected in the 2D case, where the effective interaction is stronger. GW calculations

---

[20] Note that usually the DFT-RPA calculations are not self-consistent, in the sense that eigenvalues and eigenfunctions used to construct the ingredients of the calculation stem from other functionals, like LDA or PBE, and *not* from the local exchange–correlation potential that one would obtain from the functional derivative of the RPA total energy with respect to the density. Therefore, whereas the qualitative improvement due to the physical content of the RPA is evident, its final quantitative reliability still has to be assessed.

for jellium slabs show the presence of the van der Waals interaction also in the absence of self-consistency [395], consistently with the RPA findings cited in the previous subsection.

When one uses the self-consistent GWA Green's function, there are several equivalent possibilities to calculate total energies: the Galitskii–Migdal formula, or one of the functionals in Sec. 8.2. They yield the same result when G and Σ are consistently linked through the Dyson equation. If one wishes to use an approximate G, the Luttinger–Ward functional [36] seems to be more stable than the Klein functional [294] with respect to the choice of an approximate Green's function. It yields results close to self-consistent calculations when a Hartree–Fock Green's function is used for atoms and small molecules (see [373, 375, 602] and Ch. 12).[21] Instead, the Galitskii–Migdal formula depends very much on the approximate input Green's function [602].

Using the functionals, GWA correlation energies for atoms [375] are improved with respect to the RPA, though results are not as good as for the homogeneous electron gas. For molecules, DFT-RPA and the GWA have a problem in the dissociation limit, because they are not size-consistent. This point is addressed in Sec. 11.7.

Overall it must be stated that GWA total energy calculations are less explored than the calculation of spectra. Therefore, the results presented here are snapshots of the situation at the date of writing this book, and it is premature to draw a general conclusion concerning GWA total energies.

Phonons and electron–phonon coupling with GW corrections

In an adiabatic picture, phonon modes are governed by changes of the total energy with respect to changes in the positions of ions. This can be evaluated in DFT-RPA or in the GWA. One can also use an expression that contains the second derivative of the eigenvalues, see, e.g., [604]. This shows in a direct way how GWA corrections to the band structure can influence phonons. This also happens for the electron–phonon coupling. The GWA corrections can be important; examples are phonons in graphite [605], or the electron–phonon coupling in diamond and GaAs [604].

Magnetism

Magnetic moments are particularly sensitive to the localization of electrons. Therefore, in general, methods containing Fock or the full Kohn–Sham exchange do better than methods that approximate the exchange term. This explains why the GWA is successful in this respect. In MnO, for example, the spin magnetic moment has been predicted within the experimental range by model GWA calculations [449]. Also, other methods that do well on the exchange can yield this, for example self-interaction-corrected Kohn–Sham [606], hybrid functional calculations and PBE+U, whereas, e.g., PBE alone underestimates the moment [607]. An increase in magnetic moments with respect to local density functionals

---

[21] The stability of the Luttinger–Ward functional is not linked to the GWA but has also been verified, e.g., in second-order Møller–Plesset theory [603], which is introduced in Sec. 9.1.

has been observed using QSGW [291] for several $3d$ compounds and gadolinium. Results are close to measured ones within about 10%. This also holds true for nickel.[22]

13.6 A comment on temperature

Often GWA calculations are performed at zero temperature, $T = 0$. There are several aspects that enter in the extension of the method to non-zero temperatures, which is important for understanding properties like heat capacity and phase behavior.

Electronic temperature

All correlation functions and Green's functions can be straightforwardly formulated as a function of the electronic temperature, as shown in Ch. 5. Hence, the GWA self-energy can also be calculated at $T \neq 0$. Indeed, to calculate at a non-vanishing temperature is technically easier, because it introduces broadening that can reduce numerical problems linked to sharp poles. Moreover, it allows one to use the Matsubara formalism, where integrations over poles on the real axis are replaced by discrete sums on the imaginary frequency axis, where the functions are smooth. This is sometimes used as a trick in GWA calculations (see, e.g., [481]), although calculations on the imaginary axis are more frequently met in the context of QMC and DMFT.

In other cases the temperature is not a computational trick, but a real aspect of the material. For example, the band structure of silicon and other semiconductors has been calculated [608] as a function of temperature, up to a temperature corresponding to a few eV. With increasing electronic temperature the KS-LDA gaps have a tendency to increase. GWA corrections to the bandgap calculated in QSGW[23] decrease with increasing temperature.

Temperature of the lattice

Electrons are also coupled to the lattice. The temperature can influence the geometry of the system and lead, for example, to thermal expansion. This can be determined in DFT, and the result can indirectly change spectra. Moreover, phonons become important with increasing temperature. The quasi-particles couple to phonons, similar to their coupling to plasmons. Both can be understood qualitatively by the electron–boson coupling model used in Sec. 11.7. Therefore, similar to the electron–plasmon coupling contained in the GWA, electron–phonon coupling leads to a reduction of the bandgap and broadening, and it can cause satellites and strongly coupled electron–phonon states [610]. To

---

[22] Note that, as pointed out in Sec. 13.4, even though the local moment is correct, the exchange splitting is instead overestimated by almost a factor of two.

[23] A high-temperature calculation for the homogeneous gas and metallic aluminum [609] has shown that the quasi-particle peak is broadened because of the increasing number of final states available for quasi-particle decay, but it remains well-defined, which means that a quasi-particle calculation is still meaningful up to temperatures of several tens of thousands of Kelvins.

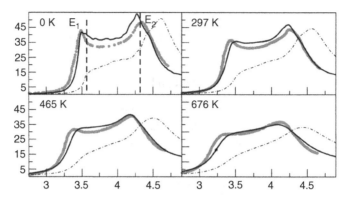

Figure 13.19. Optical absorption of bulk silicon for several temperatures. The experimental spectra [245] (dots) are compared with calculations in the independent-particle approximation to absorption (Eq. (12.7), dot-dashed line) and by solving the Bethe–Salpeter equation (solid line; see Sec. 14.10). Both use quasi-particles calculated in G_0W_0. The width of the absorption peaks reflects the damping of the excitons due to the scattering with phonons. No additional numerical damping is included. Moreover, the main absorption peaks are red-shifted with respect to their position when phonons are neglected (vertical bars; these are the positions of the main absorption peaks in the $T = 0$ calculation in the upper-left panel of Fig. 5.6). (From [611].)

first approximation, the total self-energy is then a sum of two terms: one containing coupling to electron–hole excitations and plasmons, which is the usual GWA self-energy, and a contribution due to phonons. Figure 13.19 shows the result of calculations of the optical spectrum of silicon as a function of temperature [611], starting from a GWA quasi-particle calculation. The independent-particle spectrum Eq. (12.7) is obtained by summing all possible transitions between quasi-particle states. The electron–hole interaction is included by solving the Bethe–Salpeter equation; this is explained in Ch. 14 and Sec. 14.10. Two main absorption peaks are visible. Without the inclusion of phonons, their position is given by the vertical bars. One can observe a red shift of the peaks due to the phonons, and a broadening that increases with temperature, in good agreement with experimental spectra [245]. Here, electron–phonon coupling is included as an additional contribution to the self-energy only, whereas its effects on the electron–hole interaction are neglected.

Transitions to a disordered phase

With increasing temperature a system may undergo a phase transition. When the effect is primarily a modification of the lattice, this does not pose problems. The phase transition in VO_2 in Sec. 13.4 is an example. When, instead, the main effect is disorder in the sense of fluctuating local moments or charges, the GWA has a problem. These are cases like the paramagnetic insulators NiO in Sec. 13.4, or V_2O_3 discussed in Sec. 20.6. As explained in Sec. 13.4, correlation functions other than the electron–hole one, which is contained in W, would be needed to describe such a phase transition. Therefore, *this* kind of temperature effect is not well described by the GWA. The results can be qualitatively wrong, for example the insulating paramagnetic phase of NiO comes out metallic in the GWA.

13.7 Wrap-up

As with any other method, the GW approximation is designed to solve certain problems and not others, and while it works in many cases, it fails in others. The first and most prominent success of the GWA is its ability to yield realistic bandgaps. Compared with the Kohn–Sham gaps, there is a fundamental difference: Kohn–Sham eigenvalues are not meant to reproduce the spectrum of electron addition or removal energies, but GW is an approximation to an exact approach. The inclusion in GW of non-local Fock exchange together with a detailed description of dynamical screening turns out to be sufficient for getting good gaps in a wide range of materials. The same is true for the band structure throughout the Brillouin zone: the literature contains parameter-free predictive GWA band-structure calculations with unit cells containing up to several hundred atoms.

The first-principles inclusion of screening in the electron–electron interaction is a key to the success of the GWA. It allows the description of long-range phenomena such as van der Waals binding, image states, and substrate effects that are difficult to obtain with other methods. *Dynamical* screening allows one to describe effects that clearly go beyond a single-particle picture, such as lifetime broadening and certain photoemission satellites. To get satellites one has to calculate the full spectral function, not just solve the quasi-particle equation. When satellites are due to electron–hole or plasmon excitations, and when they are relatively weak, they may be well described by the GWA. Energy self-consistency, at least of the chemical potential, is particularly important for the quality of these results. In many other cases, for example for strong plasmon satellite series, or for structures due to hole–hole scattering, the GWA meets its limits.

Crucially needed improvements to the calculation of the one-body Green's function beyond the GWA are treated in Ch. 15 and, from a completely different point of view, in the chapters related to dynamical mean-field theory (Chs. 16– 21). Moreover, independently of a preference for this or that approximation, there is physics beyond the one-body Green's function, e.g., in response functions: this is the topic of the next chapter.

SELECT FURTHER READING

Aryasetiawan, F. and Gunnarsson, O. E., "The GW method," *Rep. Prog. Phys.* **61**, 237, 1998. Discussion of various aspects of and beyond the GWA, illustrated by a number of results.

Aulbur, W. G., Jonsson, L., and Wilkins, J. W., "Quasiparticle calculations in solids," *Solid State Phys.* **54**, 1, 2000. Short introduction to GW and extensive compilation and discussion of results.

Hedin, L., "On correlation effects in electron spectroscopies and the GW approximation," *J. Phys. C* **11**, R489–R528, 1999. Nice compact discussion of the GWA with an accent on photoemission, that is helpful for the analysis of results.

The website www.electronicstructure.org contains more information, and links to tutorials and numerical exercises that accompany this book.

Exercises

13.1 Discuss the pictures in Sec. 13.1 using Eq. (11.29).

13.2 Using a linear expansion of the self-energy and a first-order self-energy correction, show that a shift $\Delta\varepsilon$ of the energy zero in the starting G_0 leads to a shift $\Delta\varepsilon$ of the GW quasi-particle energies plus a state-dependent correction. Verify that this correction doesn't affect the bandgap if valence and conduction states have the same Z-factor. Now, use a single-plasmon pole model as introduced in Sec. 12.2, and imagine choosing a bad plasmon pole position. Show that this is approximately equivalent to calculating Σ at a wrong energy. Use the linearity of Σ and the fact that Z is quite state-independent to explain why a bad plasmon pole model gives rise to an overall shift.

13.3 For the two-site Hubbard model at half-filling, examine the influence of a shift of the energies in G_0 on the results of a $G_0 W_0$ calculation, as a function of U.

13.4 Using $\mathbf{k} \cdot \mathbf{p}$ perturbation theory [51], show that in an independent-particle picture with a local potential the inverse effective mass is a constant plus a term proportional to the inverse of the gap, if one assumes that matrix elements of the velocity operator are constant for all states. Show that there is a correction when one uses the non-local GWA self-energy instead of a local potential. This explains why, moving from the LDA to the GWA, one does not find the naively expected drastic change in effective mass.

13.5 Using Eq. (11.29), explain how the overlap of wavefunctions influences the lifetime of excited electrons or holes.

13.6 Calculate the $G_0 W_0$ self-energy and Green's function for the electron–boson coupling hamiltonian Eq. (11.46), to derive the spectral function Eq. (11.51). Play with the parameter $\Delta\varepsilon$ that sets the zero of energy, and compare the resulting spectral function to the exact solution Eq. (11.48). What do you observe concerning the quality of the approximation for the satellite spectrum?

13.7 Carry out the numerical tutorial GWA calculations suggested on www.electronicstructure.org.

14

RPA and beyond: the Bethe–Salpeter equation

Summary

In this chapter we examine the two-particle correlation function. Many of its properties are experimentally accessible, such as the macroscopic dielectric function and optical spectra, or the dynamic structure factor and loss function. Formally, it is determined by the Bethe–Salpeter equation and a first useful approximation is the random phase approximation. Comparison with experiment displays the strong and the weak points of the RPA. Its shortcomings motivate the search for corrections, commonly called *vertex corrections*. We show how better approximations to the BSE can be obtained and how the equation can be solved in practice, and we give some illustrations. A comparison with time-dependent density functional theory completes the chapter.

This chapter is dedicated to the calculation of the two-particle correlation function L, which is one of the central quantities in this book. It contains a wealth of experimentally accessible information: optical spectra, electron energy loss spectra, and the dynamic structure factor that is measured for example by inelastic X-ray scattering, the energy of doubly charged defects, and much more. Moreover, many-body perturbation theory can be formulated in terms of the dynamically screened Coulomb interaction W, which is derived directly from the electron–hole correlation function. The GW approximation is the most prominent example of an approximation based on W. Finally, the total energy can be expressed exactly in terms of L, because the Coulomb interaction v_c is a two-body interaction.

The two-particle correlation function is formally given by the Dyson-like Bethe–Salpeter equation derived in Sec. 10.3. The present chapter starts in Sec. 14.1 with a brief reminder of the links between L and measurable quantities, since in this chapter we are interested in the spectra that are derived from L. Some important formal relations are recalled in Sec. 14.2. The simplest approximation for L is the random phase approximation introduced in Sec. 11.2. Section 14.3 is dedicated to the study of spectra calculated in the RPA: which physics is contained in this approximation? What are its strengths and limitations? We will see that optical spectra, for example, often require a description beyond the RPA.

The shortcomings of the RPA motivate the effort to go beyond. The two-particle correlation function contains information about the propagation of two particles, which can be electron–hole pairs, or two electrons or two holes, as explained in Sec. 5.8. The propagation of the two particles is coupled in the BSE. Beyond the RPA, this leads to a complicated spin and frequency structure of the equation. It is outlined in Sec. 14.4, and constitutes one of the major difficulties. Like the self-energy for the propagation of one particle, the kernel of the BSE, which expresses the effective interaction between the two particles, has to be approximated; the most widely used approximation for the propagation of electron–hole pairs is derived from the GWA. This is the topic of Sec. 14.5. A static approximation for the effective interaction allows one to formulate the problem in terms of a two-body Schrödinger equation. Section 14.6 gives the derivation, leading up to a compact expression for the frequency-dependent dielectric matrix in terms of eigenvalues and eigenfunctions of the two-particle hamiltonian. Section 14.7 provides tools to analyze the physics of the coupled propagation of electrons and holes, including links to models such as Frenkel and Wannier excitons. Section 14.8 explains how the equations can be solved in practice. Illustrations of results can be found in Sec. 14.9. These nine sections are the main part of the chapter. They are meant to give insight into the physics contained in the BSE, and its limitations caused by the most widely used approximations. It mostly focuses on the main target of BSE calculations: the density–density response, or propagation of electron–hole pairs.

The last part of the chapter contains extensions. We discuss attempts to go beyond current approximations and aspects of parts other than electron–hole excitations, namely spin waves and hole–hole interaction, in Sec. 14.10. Section 14.11 contains a different view of the electron–hole BSE, placing it in a time-dependent framework, which naturally also includes contributions beyond linear response. This, in turn, suggests a link to time-dependent density functional theory. A wrap-up concludes the chapter.

14.1 The two-particle correlation function and measurable quantities

Much of this chapter is dedicated to the calculation of quantities that can be derived from the two-particle correlation function, and that can be compared directly with experiment. These are in general spectroscopic measurements, such as optical absorption, electron energy loss experiments, or inelastic X-ray scattering. The experimental results can often be understood in terms of linear response theory. It is therefore useful to recall some important relations that were introduced in Sec. 5.5, where more details can be found.

Many spectroscopic measurements are related to the frequency-dependent dielectric function $\epsilon(\omega)$ or equivalently, to the linear density–density response function $\chi(\omega)$. The linear response is the first-order change of density δn in a system due to an external perturbation v_{ext}, given by Eq. (5.56) as $\delta n = \chi v_{\text{ext}}$. The response function and the inverse dielectric function are related by Eq. (5.61). More specifically in a periodic system, one can use the expression in Eq. (5.62),

$$\epsilon_{GG'}^{-1}(\mathbf{q}; \omega) = \delta_{GG'} + v_c(\mathbf{q} + \mathbf{G})\chi_{GG'}(\mathbf{q}; \omega), \tag{14.1}$$

where $\mathbf{q}$ is a vector in the first Brillouin zone and $\mathbf{G}$ is a reciprocal lattice vector. As explained in Sec. 5.5, spectra are obtained from ϵ or χ by evaluating the following.

- **The loss function** $-\mathrm{Im}\,\epsilon_{GG}^{-1}(\mathbf{q}; \omega) = -v_c(\mathbf{q} + \mathbf{G})\mathrm{Im}\,\chi_{GG}(\mathbf{q}; \omega)$ that is measured, e.g., in momentum-resolved electron energy loss experiments, with a momentum transfer $\mathbf{Q} = \mathbf{q} + \mathbf{G}$.
- **The dynamic structure factor** $S(\mathbf{Q}, \omega) = -\frac{1}{\pi(1 - e^{-\beta\omega})}\mathrm{Im}\,\chi_{GG}(\mathbf{q}, \omega)$ for a momentum transfer $\mathbf{Q} = \mathbf{q} + \mathbf{G}$ that can be measured by inelastic X-ray scattering.
- **The optical absorbtion spectrum** $\mathrm{Im}\,\epsilon_M(\mathbf{q}, \omega)$, given by the macroscopic dielectric function $\epsilon_M(\mathbf{q}, \omega) = [\epsilon_{00}^{-1}(\mathbf{q}; \omega)]^{-1}$ in the limit of long wavelength, $\mathbf{q} \to 0$.

It is also important to recall that ϵ^{-1} yields the screened Coulomb interaction $W = \epsilon^{-1}v_c$, as given by Eq. (11.5). This is the effective interaction between classical charges in a medium, and a key quantity in Hedin's equations in Sec. 11.1.

The Dyson Eq. (11.12) for W is equivalent to a Dyson equation for χ,

$$\chi(1, 2) = P(1, 2) + P(1, \bar{3})v_c(\bar{3}, \bar{4})\chi(\bar{4}, 2), \tag{14.2}$$

with the irreducible polarizability P that expresses the response of the system to the total classical perturbation according to Eq. (11.8).

Optical spectra measure the response to the total *macroscopic* classical perturbation (see footnote 17 on p. 104). It is therefore useful to define the bare Coulomb interaction without its long-wavelength component,

$$\bar{v}_c(\mathbf{G}) = 0 \quad \text{for} \quad \mathbf{G} = 0 \quad \text{and} \quad \bar{v}_c(\mathbf{G}) = v_c(\mathbf{G}) \quad \text{for} \quad \mathbf{G} \neq 0, \tag{14.3}$$

and the response function $\bar{\chi}$ that is determined by the Dyson equation

$$\bar{\chi}(1, 2) = P(1, 2) + P(1, \bar{3})\bar{v}_c(\bar{3}, \bar{4})\bar{\chi}(\bar{4}, 2). \tag{14.4}$$

This is similar to Eq. (14.2), but the long-wavelength component is included in the definition of the perturbation, and only the short-wavelength components $\bar{v}_c$ have to be considered explicitly in the self-consistent response in Eq. (14.4). With the general definition $\epsilon_M(\mathbf{Q}, \omega) = 1/\epsilon_{GG}^{-1}(\mathbf{q}; \omega)$ for arbitrarily large wavevector $\mathbf{Q}$, one can express the relation between ϵ_M and $\bar{\chi}$ as in Eq. (5.69),

$$\epsilon_M(\mathbf{Q}, \omega) = 1 - v_c(\mathbf{Q})\bar{\chi}_{GG'}(\mathbf{q}, \omega). \tag{14.5}$$

This relation with the definition of Eq. (14.4) can be derived as an exercise (Ex. 14.9). More about response functions for optical spectra can be found, for example, in [248, 612].

When P is given, for example in the random phase approximation Eq. (11.16), the solution of Eqs. (14.2) or (14.4) is not a major problem. However, beyond the RPA the irreducible polarizability P contains vertex corrections[1] as expressed by Eq. (11.13). Since the vertex function $\tilde{\Gamma}$ depends on three space, spin, and time arguments, it is determined

---

[1] Alternatively, P or χ can be calculated in TDDFT, as explained in Sec. 14.11.

by an equation (Eq. (11.14)) that is of higher dimension than the Dyson equations for χ or $\bar{\chi}$. This is the main difficulty and difference from the RPA.

Usually, one does not solve Eq. (11.14) for the vertex $\tilde{\Gamma}$ to build P and χ, but the Bethe–Salpeter Eq. (10.23) for the two-particle correlation function L. Then one can use Eq. (10.39), $\chi(1,2) = -iL(1,2;1^+,2^+)$, to obtain χ, or the irreducible polarizability $P(1,2) = -i\tilde{L}(1,2;1^+,2^+)$ from the irreducible two-particle correlation function $\tilde{L}$, as given by Eq. (11.9). Both cases correspond to a situation where the two correlated particles are an electron and a hole, as worked out in Sec. 14.4. Other parts of L are used for example to calculate the correlation between two holes in Sec. 14.10, and still other elements of L appear in the equation of motion for the one-body Green's function Eq. (10.6). However, the main focus of this chapter is on the electron–hole part, the density–density response, for which many comparisons with experiment are possible.

14.2 The two-particle correlation function: basic relations

In principle, Ch. 10 tells us how to get the two-particle correlation function L: it is the exact solution of the Bethe–Salpeter Eq. (10.23). How to solve this equation approximately in practice, what it means, and what it yields are the main topics of this chapter. Therefore, we start by summarizing in this section some basic features, including a brief reminder of points derived in earlier chapters.

Bethe–Salpeter equations

The Bethe–Salpeter[2] equation [615] expresses the linear response[3] of the one-body Green's function to a general non-local source. It takes into account the self-consistent variation of the system internal potentials, namely the Hartree potential and the exchange–correlation self-energy. It describes the propagation of two particles. The BSE can be written as

$$L(1,2,1',2') = L_0(1,2,1',2') + L_0(1,\bar{4},1',\bar{3})\Xi(\bar{3},\bar{5},\bar{4},\bar{6})L(\bar{6},2,\bar{5},2'), \qquad (14.6)$$

with

$$\Xi(3,2,3',2') \equiv -i\delta(3,3')\delta(2',2)v_c(3,2) + \frac{\delta\Sigma_{xc}(3,3')}{\delta G(2',2)}, \qquad (14.7)$$

which is the same as Eq. (10.22), and $L_0 = GG$ (Eq. (10.7)). Equation (14.6) is depicted diagrammatically in Fig. 14.1 for the case of the coupled motion of an electron and a hole.

As explained in Sec. 7.7, a Dyson equation with a kernel that consists of two pieces can be written as two coupled Dyson equations. Using this idea, we introduce the *irreducible* two-particle correlation function $\tilde{L}$,

---

[2] The BSE was first announced at a meeting; see the abstract G10 on p. 309 of [613]. It is a general (relativistic) equation that has been derived to study bound states of two interacting Fermi–Dirac particles. For a history of the Bethe–Salpeter equation, see the notes by E. E. Salpeter [614].

[3] Here the term "response" is not intended in a causal sense, but refers to the variation that is taken on some contour.

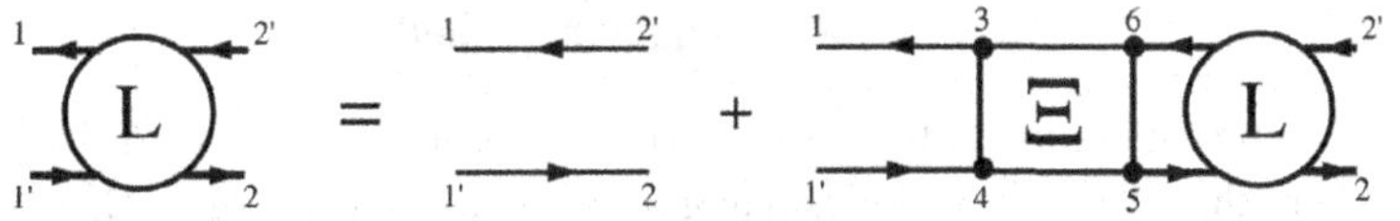

Figure 14.1. The Bethe–Salpeter equation for the coupled motion of an electron–hole pair: the Hartree and exchange–correlation kernel correlates the propagation of the electron and the hole. An approximation for Ξ based on the GWA is drawn in Fig. 14.3.

$$\tilde{L}(1,2,1',2') = L_0(1,2,1',2') + L_0(1,\bar{3}',1',\bar{3})\frac{\delta\Sigma_{xc}(\bar{3},\bar{3}')}{\delta G(\bar{2}',\bar{2})}\tilde{L}(\bar{2}',2,\bar{2},2') \qquad (14.8)$$

and the subsequent equation for the *reducible* correlation function L,

$$L(1,2,1',2') = \tilde{L}(1,2,1',2') + \tilde{L}(1,\bar{3},1',\bar{3})v_c(\bar{3},\bar{2})L(\bar{2},2,\bar{2},2'). \qquad (14.9)$$

By taking diagonal elements directly on Eq. (14.9), one obtains Eq. (14.2) for the reducible χ, as worked out in Ex. 14.4. However, it is not possible to take diagonal elements directly in the full BSE Eq. (14.6), nor in Eq. (14.8) for the irreducible function, unless one uses a local approximation for the exchange–correlation contribution. This means that in principle, one has to solve an equation with four space, spin, and time arguments; only once the result is obtained can one take the diagonal elements. This is the key difficulty in calculations based on the Bethe–Salpeter equation.

There are several ways to express the BSE. None of them avoids the four-point problem. It is nevertheless useful to know them and to realize that they are equivalent, since they are used in different contexts. In Sec. 8.3 a functional is formulated in terms of the four-point vertex $^4\tilde{\Gamma}$ defined in Eq. (8.25), which reads

$$L(1,2,1',2') = G(1,2')G(2,1') + G(1,\bar{3})G(\bar{4},1')^4\tilde{\Gamma}(\bar{3},\bar{5};\bar{4},\bar{6})G(\bar{6},2')G(2,\bar{5}). \qquad (14.10)$$

When $^4\tilde{\Gamma}$ is set to zero, L and the two-particle Green's function $G_2(1,2,1',2') = -L(1,2,1',2') + G(1,1')G(2,2')$ (Eq. (5.134)) contain only products of two one-body Green's functions. One says that the vertex represents the *bound part* of G_2 [248, 616]. The Bethe–Salpeter equation for $^4\tilde{\Gamma}$ is derived in Ex. 14.5. It reads

$$^4\tilde{\Gamma}(1,2;1',2') = \Xi(1,2,1',2') + \Xi(1,\bar{4},1',\bar{3})G(\bar{3},\bar{6})G(\bar{5},\bar{4})^4\tilde{\Gamma}(\bar{6},2;\bar{5},2'). \qquad (14.11)$$

The equation is depicted diagrammatically in the first line of Fig. 15.4.

In $P = -iGG\Gamma$ (Eq. (11.13)) for the irreducible polarizability $P(1,2)$, the coupling of two particles is given by the irreducible three-point vertex $\tilde{\Gamma}$ (Eq. (11.14)). As in Eq. (14.8), the variation of the exchange–correlation self-energy is responsible for this coupling.[4]

---

[4] Remember that when the *reducible* polarizability and vertex Γ are calculated, the Hartree contribution leads to an additional coupling between the two particles, as in Eq. (14.9).

14.3 The RPA: what can it yield?

The simplest approximation to Eq. (14.6) with Eq. (14.7) is the RPA introduced in Sec. 11.2, where the variation of the exchange–correlation self-energy $\delta \Sigma_{xc}/\delta G$ is completely neglected. In spite of this apparently drastic approximation, the RPA is adequate for many purposes, and it has interesting features that are worthwhile to elucidate before moving on. In the present section we concentrate on electron–hole excitations, for which many experimental results are available; examples are optical or electron energy loss spectra. This allows us to highlight strong and weak points of the random phase approximation.

The RPA corresponds to approximating Eq. (14.8) by $\tilde{L}^{RPA} \approx L_0$, where $L_0 \equiv GG$ is defined as in Eq. (10.7). To discuss this approximation, one has to specify *which* one-body Green's function G is used: a non-interacting or mean-field G^0, a quasi-particle approximation or the fully dressed G. This depends on the context. It is important for the quality of the approximation. In particular, the RPA optical gap depends on the gap of G. We will be clear on what is meant in every case.

Experimentally one measures parts of the reducible two-particle correlation function L, for example as stated in the previous section, loss spectra are directly proportional to Im χ and hence to diagonal elements of L. The reducible and the irreducible functions are linked by Eq. (14.9). The Coulomb kernel v_c of this equation is due to the self-consistent variation of the Hartree potential in the response to an external potential. This variation is included in the RPA, which only makes an approximation to $\tilde{L}$. As we will see, it gives important contributions to spectroscopy.

To get optical spectra, one could use χ to calculate $\bar{\chi}$ from Eq. (14.67) in Ex. 14.9. One can also formulate an equation directly, similar to Eq. (14.9),

$$\bar{L}(1,2,1',2') = \tilde{L}(1,2,1',2') + \tilde{L}(1,3,1',\bar{3})\bar{v}_c(\bar{3},\bar{4})\bar{L}(\bar{4},2,\bar{4},2'), \tag{14.12}$$

for a modified two-particle correlation function $\bar{L}$ that gives direct access to optical spectra through the relation

$$\bar{\chi}(1,2) = -i\bar{L}(1,2,1^+,2^+), \tag{14.13}$$

in the same way as χ is obtained from L in Eq. (5.147) and P from the irreducible $\tilde{L}$ in Eq. (11.9). Combined with Eq. (14.8), the Dyson-like Eq. (14.12) is the most frequently used equation in the context of Bethe–Salpeter calculations for optical spectra. The steps that lead to Eq. (14.12) in more detail are the topic of Ex. 14.9.

The independent-particle approximation

The simplest approximation for L or $\bar{L}$ is to neglect the microscopic variations of the Hartee potential completely, by setting $\bar{v}_c$ to zero in Eq. (14.12), together with the RPA where $\bar{L} = L_0$. This *independent-particle approximation* yields $\bar{L} = L_0$. It is frequently used to describe absorption spectra. The top left panel in Fig. 5.6 shows a typical result for silicon. This result has been obtained using LDA Green's functions in L_0. Therefore, the bandgap and onset of absorption are underestimated with respect to experiment. Also, the lineshape differs from the measured one, since there is not enough oscillator strength on the low-energy side. However, one can still recognize a correspondence between calculated and measured spectra.

Long-range component of the induced Hartree potential: plasmons

The right panels of Fig. 5.6 show the dynamic structure factor. With Eq. (5.66), it is proportional to the imaginary part of the density–density response function χ. It has the main structures at much higher energies than the absorption spectrum, and it is completely different from the spectrum of L_0, because χ is a diagonal of the *reducible* correlation function L. The BSE Eq. (14.9) for L contains the long-range part $v_c(\mathbf{G} = 0)$ of the Coulomb kernel, contrary to the case of $\bar{L}$ in Eq. (14.12). This long-range part causes long-range charge oscillations, the plasmons, that give rise to the strong peak in the dynamic structure factor in Fig. 5.6. With $v_c(q) = 4\pi/q^2$, it cannot be neglected for small momentum transfer q in extended systems. At larger momentum transfer it is less important, and the spectra of $\bar{\chi}$ and χ are more similar. Moreover, the energy of the plasmon disperses as a function of momentum transfer, and the plasmon moves into the continuum of electron–hole transitions. As one can see from the lower right panel in Fig. 5.6, it becomes much broader and is no longer a well-defined peak. The physics of plasmons is a topic on its own; it is extensively discussed, for example in [617]. Here it is important to note that the RPA can describe plasmons, because it contains the long-range variation of the Hartree potential.

Short-range components: crystal local field effects

The effects of the microscopic components of v_c are called *crystal local field effects*[5] (LFE) [618–620]. Without these components, the Dyson Eq. (14.2) for χ_{00} in reciprocal and frequency space is scalar, and $\bar{\chi} = P$. When LFE are included, one has to solve a matrix equation in reciprocal space, since all components mix. One can understand this by looking at the relation $\epsilon_M = 1/\epsilon_{\mathbf{GG}}^{-1}$, Eq. (5.67), between the macroscopic dielectric function $\epsilon_M(\mathbf{q}, \omega)$ and the microscopic $\epsilon_{\mathbf{G}=\mathbf{G}'}(\mathbf{q}, \omega)$: neglecting LFE corresponds to neglecting off-diagonal elements of the matrix $\epsilon_{\mathbf{GG}'}$. In that case a diagonal element of the inverse equals the inverse of a diagonal element, and $\epsilon_M = \epsilon_{\mathbf{GG}}$. The physical reason is the self-consistency of the response: an applied potential induces charge fluctuations on all length scales of the system, and these fluctuations, in turn, create induced potentials that are macroscopic *and* microscopic perturbations. Microscopic components of the induced Hartree potential gain in importance when the system is inhomogeneous, and when one probes shorter distances, with increased momentum transfer.

Figure 14.2 shows the electron energy loss spectrum of graphite as a function of the angle of the momentum transfer $\mathbf{Q}$ with respect to the graphite planes, for fixed absolute value of Q. Within the graphene planes the system is rather homogeneous. Therefore,

---

[5] Sometimes they are simply called local field effects, but one has to distinguish them from variations of exchange–correlation contributions, which are also sometimes called local field effects in the literature.

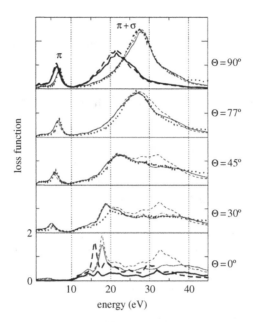

Figure 14.2. Calculated and measured loss function of graphite for a momentum transfer $Q = 0.25A^{-1}$, from [621]. Different panels show results for different angles Θ between **Q** and the direction perpendicular to the planes of hexagonal carbon. Dots in all but the lowest panel are experimental results. Dashed lines are calculated in the RPA without crystal local field effects, continuous lines include LFE. The thick dashed and continuous lines are calculated with double interlayer spacing.

for an in-plane **Q** the LFE are not important. The more **Q** is turned around and has a component perpendicular to the planes, the more the inhomogeneity of the system is detected. Therefore, results of calculations neglecting LFE show increasing discrepancies with respect to experiment, whereas RPA including LFE stays close to experiment.[6] When the distance between the planes is increased, as can easily be done in the calculations', the LFE are enhanced. This corresponds to the fact that at larger spacing the system is more inhomogeneous.

The example of graphite shows a trend that is quite general: the dominant features of the loss spectra are due to variations of the Hartree potential. With $\epsilon^{-1} = 1 + v_c \chi$, Eq. (5.62), they are also the dominant features of the inverse dielectric function and hence of W. The fact that these variations are included in the RPA is one of the reasons for the good performance of the GWA in many cases.

Failure of the RPA

The limitations of the RPA are most obvious when one looks at optical absorption in the first few eV spectral range. Figure 2.10 shows a typical example, the optical absorption

---

[6] Calculations are based on LDA Green's functions.

spectrum of LiF. The experimental spectrum [622] is dominated by a sharp peak at 12.6 eV. The figure caption tells us that the dashed line is obtained in a calculation [71] neglecting the electron–hole interaction: in the more formal language of the present chapter, this corresponds to setting $\bar{L} \approx L_0$, the RPA without LFE. Here the one-body Green's functions are independent-particle Green's functions constructed with quasi-particle energies obtained in the GWA. This leads to an absorption onset that is too high in energy. Moreover, the peak at 12.6 eV is completely absent in the RPA calculation. This qualitative discrepancy is not resolved when LFE are included. It is typical for RPA absorption spectra.[7] The loss spectra contain the same low-energy structures as optical spectra, but they are screened according to Eq. (5.68). Therefore they have the same problems, but the discrepancies are much harder to see.

The screened Coulomb interaction is also used to calculate the GW self-energy. In fully self-consistent GWA, using the RPA leads to problems. As discussed in Sec. 12.4, an RPA W constructed from interacting Green's functions does not fulfill the f-sum rule [623]. This is linked to the fact that $P(1, 2) = -iG(1, 2)G(2, 1)$ with dressed Green's functions is not a response function, in the sense that it cannot be expressed as the functional derivative of the density with respect to some potential [352]. Its spectrum is broad and featureless, which leads in self-consistent GWA to the washed-out spectral functions that can be seen in the HEG, Fig. 11.8. Moreover, it does not have the proper asymptotic behavior for vanishing momentum transfer [351, 624].

These fundamental questions, and the qualitative and quantitative problems in the description of spectra, are strong motivations to go beyond the RPA. The following sections explain how this can be done.

14.4 Beyond the RPA: spin and frequency structure of the BSE

The problem of the BSE is its four-point structure in space, spin, and time. In this section we examine the structure of the equation in spin and frequency. The real-space coordinates will be treated by a basis transformation in Sec. 14.6. We will mostly write the equations for L. The equations for $\bar{L}$ are strictly analogous, with $v_c \rightarrow \bar{v}_c$.

Spin structure

Suppose we have a system with collinear spins and a spin-independent hamiltonian without contributions such as spin–orbit interaction, the same as in Sec. 12.1. Then $G_{\sigma\sigma'} = \delta_{\sigma\sigma'}G_\sigma$ and the self-energies are spin-diagonal. Since v_c is spin-independent, spin can be summed over in several places in the equation. In Sec. F.2 the spin structure of Hedin's equations can be made explicit. The kernel of the BSE consists of two pieces [248], both with a particular spin structure,

$$\Xi_{\sigma_1\sigma_2\sigma_3\sigma_4} = \delta_{\sigma_1\sigma_3}\delta_{\sigma_2\sigma_4} \Xi^a_{\sigma_1\sigma_3} + \delta_{\sigma_1\sigma_4}\delta_{\sigma_2\sigma_3} \Xi^b_{\sigma_1\sigma_2}. \tag{14.14}$$

---

[7] Also in the example of bulk silicon discussed above, the inclusion of LFE does not improve the RPA absorption spectrum significantly.

In the Hartree–Fock approximation for the self-energy, for example, the Hartree part yields $\Xi^a = -iv_c$ and the Fock contribution leads to $\Xi^b = iv_c$. The more general case is examined in Ex. 14.7. One can decouple the equations resulting from the two contributions by a linear combination, which yields singlet and triplet excitations, with the definition

$$\Xi^s \equiv 2\Xi^a + \Xi^b \quad \text{and} \quad \Xi^t \equiv \Xi^b. \tag{14.15}$$

This is worked out in Sec. 14.5 for the case of the GWA to the self-energy.

The BSE in frequency space

The BSE depends on four times, or, in the absence of a time-dependent external potential, on three time differences. In the frequency Fourier transform of the equation one therefore has three frequencies. Omitting space and spin arguments for compactness, Eq. (14.6) becomes [625]

$$L(\omega_1, \omega_2, \omega_3) = L_0(\omega_1, \omega_2, \omega_3) + \frac{L_0(\omega_1, \omega_2, \bar{\omega}_4)}{(2\pi)^2} \Xi(\omega_1, \bar{\omega}_4, \bar{\omega}_5)L(\omega_1, \bar{\omega}_5, \omega_3), \tag{14.16}$$

where the Fourier transforms of all four-point functions C are defined as

$$C(t_1, t_2, t_{1'}, t_{2'}) = \frac{1}{(2\pi)^3} \int d\bar{\omega}_1 d\bar{\omega}_2 d\bar{\omega}_3 \, C(\bar{\omega}_1, \bar{\omega}_2, \bar{\omega}_3)e^{-i\bar{\omega}_1 \tau_1}e^{-i\bar{\omega}_2 \tau_2}e^{-i\bar{\omega}_3 \tau_3}, \tag{14.17}$$

and the time differences in $L(1,2;1',2')$ are defined as [248]

$$\tau_2 = t_1 - t_{1'}, \quad \tau_3 = t_2 - t_{2'}, \quad \tau_1 = \frac{1}{2}[(t_1 + t_{1'}) - (t_2 + t_{2'})]. \tag{14.18}$$

These definitions are intuitive when one wishes to describe the propagation of an electron–hole pair: τ_2 and τ_3 are differences in the time where the electron and the hole are considered, and τ_1 is the average time of propagation. If one is interested in the propagation of two electrons or two holes, another definition of time differences is more suitable. In this chapter we are mainly interested in electron–hole pairs. Appendix G of [248] contains a summary of the information needed to access the electron–electron and hole–hole sectors of L.

Simultaneous propagation of an electron and a hole

In order to obtain the density–density response one has to set $\tau_2 = \tau_3 = 0^-$, $\tau_1 = t_1 - t_2$. This corresponds to the simultaneous propagation of an electron and a hole. In frequency space this equal-time limit reads[8]

$$L(\omega) = \frac{1}{(2\pi)} \int d\omega_2 \, L(\omega, \omega_2) = \frac{1}{(2\pi)^2} \int d\omega_2 \, d\omega_3 \, L(\omega, \omega_2, \omega_3). \tag{14.19}$$

---

[8] For simplicity, in the following the same symbol L represents the different functions with one, two, or three frequency arguments.

Its imaginary part corresponds to the electron–hole spectral function in Eq. (5.144). The integral $d\omega_3$ can be performed immediately in Eq. (14.16). This yields

$$L(\omega_1, \omega_2) = L_0(\omega_1, \omega_2) + \int d\omega_4 \, d\omega_5 \, \frac{L_0(\omega_1, \omega_2, \omega_4)}{(2\pi)^2} \Xi(\omega_1, \omega_4, \omega_5) L(\omega_1, \omega_5), \quad (14.20)$$

where the two-frequency $L_0(\omega_1, \omega_2)$ is defined like $L(\omega_1, \omega_2)$ in Eq. (14.19). It reads

$$L_0(\omega_1, \omega_2) = -iG\left(\omega_2 + \frac{\omega_1}{2}\right) G\left(\omega_2 - \frac{\omega_1}{2}\right). \quad (14.21)$$

Equation (14.20) cannot be simplified further: in principle, one has to solve a two-frequency equation, and only at the end can one perform the last frequency integration that corresponds to the equal-time limit.

14.5 The Bethe–Salpeter equation in the GW approximation

To specify the kernel Ξ of the BSE defined in Eq. (14.7), one needs the exchange–correlation self-energy Σ_{xc} as a functional of the one-body Green's function G. The simplest result is obtained when one approximates $\Sigma_{xc}(1, 2)$ by a local and instantaneous potential that depends only on the density, like the Kohn–Sham potential $\delta(1^+, 2)v_{xc}(1)$. As shown in Ex. 14.3, together with the use of Kohn–Sham Green's functions this leads to the linear response screening equation of TDDFT, Eq. (14.58), which depends only on two space and spin arguments, and on one frequency.

Another mean-field approximation to the self-energy is the Fock exchange Σ_x given in Eq. (10.17). It is non-local in space, but instantaneous in time like v_{xc}. It gives an exchange contribution to the kernel that reads $\Xi_x(1, 2, 3, 4) \equiv i\delta(1, 4)\delta(2, 3)v_c(1^+, 3)$. Since v_c is instantaneous, the equation that one has to solve to determine χ or $\tilde{\chi}$ depends on one frequency only, as worked out in Ex. 14.2. However, it has four space and two spin arguments.[9] When L_0 is built with Hartree–Fock Green's functions, the resulting BSE is equivalent to linear response time-dependent Hartree–Fock [12, 221, 626]. The link to a time-dependent formulation is deepened in Sec. 14.11. As explained in Sec. 11.3, Hartree–Fock in extended systems suffers from the absence of screening. This problem is overcome by moving to the GWA.

Effective interaction from the GWA

Using the GWA to calculate Ξ corresponds to circulating Hedin's pentagon Fig. 11.2 beyond the GWA and recalculating $\tilde{\Gamma}$. The resulting exchange–correlation contribution to the kernel reads

$$\Xi_{xc}^{GWA}(1, 2, 3, 4) = i\delta(1, 4)\delta(2, 3)W(1, 2) + iG(1, 3)\frac{\delta W(1, 3)}{\delta G(4, 2)}. \quad (14.22)$$

---

[9] Note that Ξ_x does not cancel the Hartree contribution $-iv_c$, because the two terms have different space and spin indices, see, e.g., Eq. (14.14).

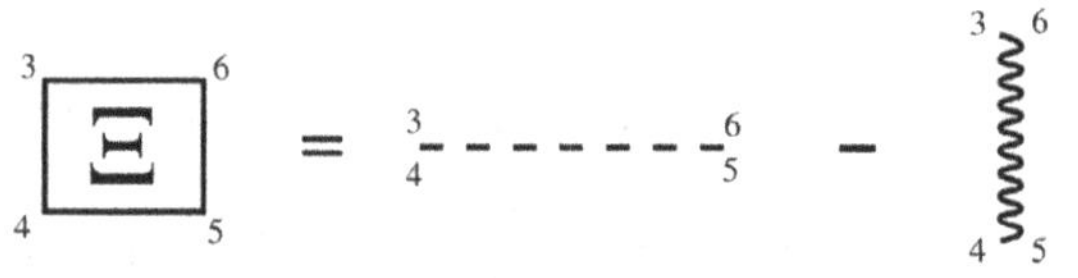

Figure 14.3. Approximate kernel of the Bethe–Salpeter equation derived from the GWA self-energy and neglecting the variation of W.

The first term is of first order in W. It looks like the kernel that one would find in time-dependent Hartree–Fock, but it is screened. The second term contains the information that screening changes when the system is perturbed. It contains higher orders of W [627], and it is most often neglected. The kernel that corresponds to this approximation is depicted diagrammatically in Fig. 14.3.

With the neglect of $\delta W/\delta G$ in Eq. (14.22) the frequency Fourier transform of the kernel reads

$$\Xi_{xc}^{GWA}(\omega_1, \omega_2, \omega_3) \approx iW(\omega_2 - \omega_3). \tag{14.23}$$

Contrary to the kernels derived from Kohn–Sham or Hartree–Fock, here the coupling between the particles is frequency-dependent. The reason is the frequency dependence of screening, caused by the fact that it needs time to build the screening cloud. From Eq. (14.20) we obtain

$$L(\omega_1, \omega_2) = L_0(\omega_1, \omega_2) + \frac{1}{2\pi} L_0(\omega_1, \omega_2) \int d\bar{\omega}_3 \, [v_c - W(\omega_2 - \bar{\omega}_3)] \, L(\omega_1, \bar{\omega}_3), \tag{14.24}$$

with L_0 given by Eq. (14.21), using GWA Green's functions. This equation cannot be simplified further: a kernel that is consistent with the GWA requires the solution of a two-frequency equation.

Static approximation for the interaction

In order to simplify the frequency dependence, a static approximation for W is adopted in most *ab initio* calculations. This is consistent with the screened exchange or the COH-SEX approximations to GW, which are discussed in Sec. 11.3. This approximation to the BSE could be called "linear response time-dependent screened Hartree–Fock" or "linear response time-dependent COHSEX."[10] The resulting equation that one has to solve in order to obtain χ, or the equivalent for $\bar{\chi}$, is

$$L(x_1, x_2, x_{1'}, x_{2'}; \omega) = L_0(x_1, x_2, x_{1'}, x_{2'}; \omega)$$
$$- i \, L_0(x_1, \bar{x}_3, x_{1'}, \bar{x}_3; \omega) v_c(\bar{x}_3, \bar{x}_4) L(\bar{x}_4, x_2, \bar{x}_4, x_{2'}; \omega)$$
$$+ i \, L_0(x_1, \bar{x}_4, x_{1'}, \bar{x}_3; \omega) W(\bar{x}_3, \bar{x}_4) L(\bar{x}_3, x_2, \bar{x}_4, x_{2'}; \omega), \tag{14.25}$$

---

[10] As specified in Sec. 14.8, in practice L_0 is most often obtained from a quasi-particle approximation to the full dynamic GWA, and the static W is only used for the kernel of the BSE. However, this does not change the equations here.

where the frequency-dependent functions are defined as in Eq. (14.19), and W is the screened Coulomb interaction at vanishing frequency.

Equation (14.25) or its equivalent for $\bar{L}$ is the starting point for most *ab initio* Bethe–Salpeter calculations in solid-state physics. We will call this approximation GW-BSE in the following.

Spin structure

The approximate GWA kernel in spin space reads

$$\Xi_{\sigma_1\sigma_2\sigma_3\sigma_4} = -i\delta_{\sigma_1\sigma_3}\delta_{\sigma_2\sigma_4}v_c + i\delta_{\sigma_1\sigma_4}\delta_{\sigma_2\sigma_3}W. \tag{14.26}$$

Similar to Hartree–Fock, the Hartree contribution constitutes Ξ^a in Eq. (14.14), and the exchange–correlation self-energy creates Ξ^b. Since $L_{0\sigma_1\sigma_2\sigma_3\sigma_4} = \delta_{\sigma_1\sigma_4}\delta_{\sigma_3\sigma_2}L_{0\sigma_1\sigma_2}$, the irreducible $\tilde{L}$ reads

$$\tilde{L}_{\sigma_1\sigma_2\sigma_3\sigma_4} = \delta_{\sigma_1\sigma_4}\delta_{\sigma_3\sigma_2}L_{0\sigma_1\sigma_2}/(1 - WL_{0\sigma_1\sigma_2}) \equiv \delta_{\sigma_1\sigma_4}\delta_{\sigma_3\sigma_2}\tilde{L}_{\sigma_1\sigma_2}. \tag{14.27}$$

Equation (14.9) becomes

$$L_{\sigma_1\sigma_2\sigma_3\sigma_4} = \delta_{\sigma_1\sigma_4}\delta_{\sigma_3\sigma_2}\tilde{L}_{\sigma_1,\sigma_2} + \delta_{\sigma_1\sigma_2}\tilde{L}_{\sigma_1,\sigma_2}v_c\sum_{\sigma}L_{\sigma\sigma\sigma_3\sigma_4}. \tag{14.28}$$

From this, one can construct the singlet and triplet contributions of $L_{\sigma_1\sigma_2\sigma_2\sigma_4}$. In the triplet, the electron and hole have a contribution with opposite spin. This is the case where $\sigma_1 \neq \sigma_2$. This cannot happen in an optical experiment with linearly polarized light, because of selection rules. Equation (14.28) shows that there is no contribution from the bare Coulomb kernel v_c when $\sigma_1 \neq \sigma_2$. In the case of the singlet $\sigma_1 = \sigma_2$, and $\sigma_3 = \sigma_4$. By adding and subtracting the equations for $\sigma_1 = \sigma_3$ and $\sigma_1 \neq \sigma_3$, one finds

$$L_\sigma = \tilde{L}_\sigma[1 - 2v_c\tilde{L}_\sigma]^{-1}, \tag{14.29}$$

where $L_\sigma \equiv L_{\sigma\sigma\sigma\sigma}$ and similar for $\tilde{L}_\sigma$: contrary to the case of the triplet, the bare Coulomb term enters the equation for the singlet, with a factor of two. This term is responsible for the so-called *singlet–triplet splitting*. The examples of optical spectra shown in Sec. 14.9 are calculations of singlet excitations.

14.6 A two-body Schrödinger equation

The aim of this section is to transform the approximate GW-BSE Eq. (14.25) into a form that highlights the underlying physics: we will derive an effective two-particle equation. This allows us to make the link to models for excitons in Sec. 14.7, and it is the starting point for real calculations as outlined in Sec. 14.8.

Complete neglect of dynamical effects

The frequency dependence of W in the kernel of the BSE has consequences that are called *dynamical effects*. They are discussed in Sec. 14.10. However, it is important to note that the fully dressed Green's functions in L_0 also stem from a frequency-dependent self-energy, which leads to additional dynamical effects. It has been shown [628] that dynamical effects from L_0 and from W cancel partially, which suggests that a good approximation is to neglect both. In most calculations for real materials one replaces the Green's functions G in L_0 by independent-particle ones, built with a set of orthonormal orbitals $\psi_{n\sigma}(\mathbf{r})$ and real quasi-particle eigenvalues $\varepsilon_{n\sigma}$. The eigenvalues are most often obtained from a GWA calculation. It is important to note that the quasi-particles are normalized to one. If one used just the quasi-particle part of the spectral function, that has weight Z, the overall weight of the spectrum would be wrong by a factor Z^2 [629]. More discussions about cancellations of dynamical effects can be found in Sec. 14.10.

The electron–hole interaction

In the absence of interaction, particles propagate independently. The effect of the interaction is brought out by writing the problem in the basis of independent particles. This is the set of orthonormal orbitals $\psi_{n\sigma}(\mathbf{r})$ with which the independent-particle Green's functions that enter L_0 are built. In the following we concentrate on the spin singlet case, and put a factor 2 for spin sums. The frequency Fourier transform of Eq. (10.7) yields

$$L_0(\mathbf{r}_1, \mathbf{r}_2, \mathbf{r}_3, \mathbf{r}_4; z) = 2i \sum_{ij} (f_j - f_i) \frac{\psi_i(\mathbf{r}_1)\psi_j(\mathbf{r}_2)\psi_j^*(\mathbf{r}_3)\psi_i^*(\mathbf{r}_4)}{z - (\varepsilon_i - \varepsilon_j)}. \tag{14.30}$$

As in Ch. 12, at $T = 0$ this represents the time-ordered correlation function on the real frequency axis $z = \omega$, Eq. (5.105), and one has to replace $z \to \omega + i\eta\,\mathrm{sgn}(f_i - f_j)$ in the denominator. In the grand-canonical ensemble at $T \neq 0$, Eq. (12.6) is the correlation function at the bosonic Matusbara frequencies $z = z_m$ defined in Eq. (D.11). Equation (14.30) is consistent with the generalization to four points in space and spin of Eq. (12.7).

In the basis of the ψ's, and with the definition of the basis transformation Eqs. (12.42) and (12.43),

$$L_{0 n_1 n_2 n_3 n_4}(z) = L_{0 n_1 n_3}^{n_4 n_2}(z) = 2i \frac{\left(f_{n_1} - f_{n_2}\right)\delta_{n_1 n_4}\delta_{n_2 n_3}}{z - (\varepsilon_1 - \varepsilon_2)} \qquad \text{is diagonal.} \tag{14.31}$$

The BSE becomes

$$L_{n_1 n_3}^{n_4 n_2}(z) = \left[L_0^{-1} + \frac{i}{2}\Xi \right]_{n_1 n_3}^{-1\, n_4 n_2} = 2i[H^{2p} - \mathbb{I}z]_{n_1 n_3}^{-1\, n_4 n_2}(f_{n_2} - f_{n_4}), \tag{14.32}$$

where $\mathbb{I}$ is the identity matrix, and H^{2p} the two-particle hamiltonian

$$H^{2p\, n_4 n_2}_{n_1 n_3} = \left(\varepsilon_{n_2} - \varepsilon_{n_1}\right)\delta_{n_1 n_4}\delta_{n_2 n_3} + \left(f_{n_1} - f_{n_3}\right)\Xi_{n_1 n_3}^{n_4 n_2}. \tag{14.33}$$

The matrix element of the Coulomb interaction is defined consistently with Eq. (12.44),

$$v_{n_1 n_2 n_3 n_4} = v_{n_1 n_3}^{n_4 n_2} = \int d\mathbf{r}_1\, d\mathbf{r}_2\, \psi_{n_1}^*(\mathbf{r}_1)\psi_{n_4}(\mathbf{r}_1) v_c(\mathbf{r}_1, \mathbf{r}_2)\psi_{n_3}(\mathbf{r}_2)\psi_{n_2}^*(\mathbf{r}_2), \tag{14.34}$$

and the same for W. The kernel $\Xi = v_c - W$ is defined as

$$\Xi_{n_1 n_3}^{n_4 n_2} = 2v_{n_1 n_4}^{n_3 n_2} - W_{n_1 n_3}^{n_4 n_2}. \tag{14.35}$$

In a non-metal at $T = 0$, only pairs of an occupied and an empty state contribute to an optical spectrum. This is guaranteed by the occupation-number prefactor of Ξ in Eq. (14.33). Therefore the terms that contribute are of the form

$$\Xi_{vc}^{v'c'} = 2v_{vv'}^{cc'} - W_{vc}^{v'c'}. \tag{14.36}$$

The matrix elements of the bare Coulomb interaction couple dipoles; therefore, this contribution is often called *electron–hole exchange*.[11] The second, screened, part is closer to the interaction between charge densities, as one can see by inspecting the diagonal matrix elements $v' = v$, $c' = c$. It is called *direct electron–hole interaction*, although it stems from the variation of an exchange–correlation potential. It is important to keep this in mind, since the terminology easily leads to confusion.

The two-particle hamiltonian

Now we have all the elements in hand to solve the BSE as a linear matrix equation. For further discussion we consider the causal response of semiconductors and insulators at $T = 0$, where the occupation numbers imply that only pairs of occupied states v and empty states c contribute to the spectra. Moreover, we drop the spin index for simplicity, unless needed. H^{2p} consists of four blocks. In the optical ($\mathbf{q} \to 0$) limit,

$$H^{2p} = \begin{pmatrix} H^{res} & H^{coupl} \\ -[H^{coupl}]^* & -[H^{res}]^* \end{pmatrix}, \tag{14.37}$$

where the *resonant part* is

$$H^{res} \equiv H^{2p}{}_{vc}^{v'c'} = (\varepsilon_c - \varepsilon_v)\,\delta_{vv'}\delta_{cc'} + \Xi_{vc}^{v'c'}. \tag{14.38}$$

It corresponds to transitions at positive frequencies. It is hermitian. The *coupling part* alone,

$$H^{coupl} \equiv H^{2p}{}_{vc}^{c'v'} = \Xi_{vc}^{c'v'} = [\Xi_{cv}^{v'c'}]^*, \tag{14.39}$$

is symmetric, as one can see from Eqs. (14.34) and (14.35) and the symmetry of the bare and screened Coulomb interaction under exchange of its two arguments. The part in the lower right of Eq. (14.37),

$$H^{ares} \equiv H^{2p}{}_{cv}^{c'v'} (\varepsilon_v - \varepsilon_c)\,\delta_{vv'}\delta_{cc'} - \Xi_{cv}^{c'v'} = -[H^{res}]^*, \tag{14.40}$$

is called *anti-resonant*. It is linked to the resonant part because of the symmetry $\Xi_{cv}^{c'v'} = [\Xi_{vc}^{v'c'}]^*$. The whole hamiltonian Eq. (14.37) is not hermitian, but pseudo-hermitian. Its eigenvalues are real. More explanations concerning these properties can be found in [630] and references therein and the $\mathbf{q} \neq 0$ case is described in [639].

---

[11] This explains also why it does not contribute to triplet excitations, as discussed in Sec. 14.5.

A two-particle equation

Equation (14.32) can be solved by matrix inversion. Alternatively, one can use the spectral representation of the hamiltonian, obtained by solving the two-particle Schrödinger equation

$$\sum_{n_3 n_4} H^{2p\,n_3 n_4}_{n_1 n_2} A^{n_3 n_4}_\lambda = E_\lambda A^{n_1 n_2}_\lambda. \tag{14.41}$$

This leads for the causal L at real frequencies to

$$L^{n_3 n_4}_{n_1 n_2}(\omega) = 2i \sum_{\lambda \lambda'} \frac{A^{n_1 n_2}_\lambda N^{-1}_{\lambda \lambda'} A^{*\,n_3 n_4}_{\lambda'}}{\omega - E_\lambda + i\eta} (f_{n_4} - f_{n_3}), \qquad \text{with} \tag{14.42}$$

$$N_{\lambda \lambda'} \equiv \sum_{n_1 n_2} A^{*\,n_1 n_2}_\lambda A^{n_1 n_2}_{\lambda'}. \tag{14.43}$$

The overlap matrix N is needed because the full two-particle hamiltonian Eq. (14.37) is not hermitian.

Each couple (nn') corresponds to a pair (vc) or (cv) of an occupied and an empty state. Without the electron–hole interaction, each eigenstate $A^{*\,n_1 n_2}_\lambda$ would correspond either to the resonant or to the anti-resonant sector, and each would pick just one electron–hole pair, $A^{vc}_\lambda = \delta_{vv\lambda} \delta_{cc\lambda}$, with transition energy $\varepsilon_c - \varepsilon_v$ (or $(c \leftrightarrow v)$ for the anti-resonant). The electron–hole interaction, even in the RPA where Ξ consists only of the bare Coulomb contribution, leads to a mixing of transitions. E_λ are the new transition energies that replace $\varepsilon_c - \varepsilon_v$.

The Tamm–Dancoff approximation

The coupling part of Eq. (14.39) mixes transitions of positive and negative energy. It can be neglected when the interaction matrix elements are small with respect to the energy difference between those transitions. Neglecting the coupling is called the Tamm–Dancoff approximation [47, 631, 632]. It allows one to work with hermitian matrices, of half the size. It is successful in the calculation of the optical spectra of bulk semiconductors [633], but less justified for metals and for finite systems [630], as well as for plasmons [243], where the long-range part of v_c gives strong contributions, as explained in Secs. 14.3 and 14.9.

Spectra of solids

To calculate spectra, one has to transform the expressions back to real space and take matrix elements. For optical spectra one needs $\bar{\chi}_{G=G'=0}(\mathbf{q} \to 0, \omega)$, so one calculates $\bar{L}$ as discussed in Sec. 14.3. This is obtained by diagonalizing H^{2p} (Eq. (14.37)), where Ξ contains $\bar{v}_c$ instead of v_c. In the Tamm–Dancoff approximation $\bar{\chi}(\mathbf{q} \to 0, \omega)$ reads

$$\bar{\chi}_{00}(\mathbf{q} \to 0, \omega) = 2 \lim_{\mathbf{q} \to 0} \sum_\lambda \frac{\left| \sum_{vck} A^{vck\mathbf{k}+\mathbf{q}}_\lambda \tilde{\rho}_{vck\mathbf{k}+\mathbf{q}} \right|^2}{\omega - E_\lambda + i\eta}, \tag{14.44}$$

where the transition matrix elements $\tilde{\rho}$ are defined in Eq. (12.46). Momentum conservation in the matrix elements results in the fact that only transitions with momentum transfer $\mathbf{q}$ between the hole in state $v\mathbf{k}$ and the electron in state $c\mathbf{k} + \mathbf{q}$ contribute. Equation (14.44) can be compared with the independent-particle response function χ^0, which is the retarded version of $P_0 = -iG_0G_0$ (Eq. (11.16)) with G_0 an independent-particle Green's function. It reads

$$\chi^0_{00}(\mathbf{q} \to 0, \omega) = 2 \lim_{\mathbf{q} \to 0} \sum_{v c \mathbf{k}} \frac{|\tilde{\rho}_{v\mathbf{k}c\mathbf{k}+\mathbf{q}}|^2}{\omega - (\varepsilon_{c\mathbf{k}+\mathbf{q}} - \varepsilon_{v\mathbf{k}}) + i\eta}. \tag{14.45}$$

With respect to this expression, the coefficients A_λ in Eq. (14.44) mix the formerly independent transitions, and the transition energies are changed.

The same expression as Eq. (14.44) is obtained for the density–density response function χ, but the two-particle hamiltonian is constructed with v_c instead of $\bar{v}_c$. Beyond the Tamm–Dancoff approximation, Eq. (14.44) is generalized following Eq. (14.42).

To calculate spectra of triplet excitations, one has to drop v_c in the kernel of the two-particle hamiltonian following the discussion around Eq. (14.28), and take the appropriate spin combinations.

14.7 Importance and analysis of electron–hole interaction effects

In Sec. 14.9 selected applications of the BSE are discussed. The present section prepares the analysis of the results. We want to know, and be able to tell experimentalists, *why* spectra are what they are, and what is their relation to the geometry and chemistry of a material. We want to be able to judge whether new results are reasonable, and we want to understand potential problems of calculations and of the approximate theory that is used. For this, we need guidelines that tell us in which way the electron–hole interaction is expected to change spectra, and we need analysis tools. These guidelines and tools are the topic of the present section. Some illustrations are given; how these are computed in practice, and more applications, are the topic of the next and subsequent sections.

Transition energies

Our discussion is based on the two-particle Schrödinger Eq. (14.41). For simplicity, let us take a two-level system with an occupied state v and an empty state c. In the Tamm–Dancoff approximation there is no mixing of transitions. The electron–hole interaction simply changes the transition energy from $\varepsilon_c - \varepsilon_v$ to

$$E = \varepsilon_c - \varepsilon_v + 2v^{cc}_{vv} - W^{vc}_{vc}, \tag{14.46}$$

with
$$v^{cc}_{vv} = \int d\mathbf{r}\, d\mathbf{r}'\, \tilde{\rho}^*_{vc}(\mathbf{r})v_c(\mathbf{r} - \mathbf{r}')\tilde{\rho}_{vc}(\mathbf{r}') \tag{14.47}$$

and
$$W^{vc}_{vc} = \int d\mathbf{r}\, d\mathbf{r}'\, \tilde{\rho}^*_{vv}(\mathbf{r})W(\mathbf{r}, \mathbf{r}')\tilde{\rho}_{cc}(\mathbf{r}'). \tag{14.48}$$

The bare Coulomb term of Eq. (14.47) is the interaction energy between dipoles, the $\tilde{\rho}$ defined in Eq. (12.46). It is positive and leads to an increase of the transition energy. One can put this in parallel[12] with the observations in Sec. 14.3: both the macroscopic component of v_c, which creates plasmons, and the microscopic components, which lead to crystal local field effects, shift spectral weight to higher energies.

The screened Coulomb interaction enters the equations with opposite sign: it contains the attractive interaction between the charge densities of an electron and a hole. Since it is an interaction between monopoles, it can have a stronger effect than the dipole–dipole contribution from the bare Coulomb interaction, although it is screened. In this case transition energies appear in the quasi-particle gap $\varepsilon_c - \varepsilon_v$. The difference between the quasi-particle gap and the transition energy is called *exciton binding energy*. Often, bound electron–hole pairs are called *excitons*, or one speaks more generally about *excitonic effects* whenever the direct electron–hole interaction has a significant effect. The screening of the direct electron–hole interaction is important: its neglect, in time-dependent Hartree–Fock, leads to strong overbinding of excitons in solids [634].

More transitions: intensities

Within the Tamm–Dancoff approximation and in the two-level model, there is no mixing of transitions. One can use the two-level model without the TDA to study effects on energies, coefficients, and spectra. This is proposed in Ex. 14.8.

When the independent-particle system has several possible transitions, in general one finds that the coefficients A_λ mix the transitions. Since the A_λ and the matrix elements in Eq. (14.44) can be positive or negative, or even complex, this leads to *interference effects*. The A_λ do not change the transition energies, but in this way they change intensities, and therefore spectral weight is shifted. By just looking at the spectra in the continuum, it is difficult to say whether energies or wavefunctions are changed. When the spectra are calculated by diagonalizing the hamiltonian, one can make this distinction by analyzing energies and wavefunctions separately. The set of E_λ that is obtained from the calculation contains all possible transition energies, independently of matrix elements, even when the intensity is strictly zero. In an optical spectrum, for example, one finds both dipole-allowed, or *bright*, and forbidden transitions, also called *dark*.

It is often interesting to analyze the origin of the peaks in spectra. Very often, one would like to link these structures to independent-particle transitions from a band structure. The coefficients A_λ^{vc} tell us which electron–hole pairs (vc) yield the main contribution to the wavefunction of an electron–hole state λ. The plot of $|A_\lambda^{vck}|^2$ as a function of $(\varepsilon_{ck} - \varepsilon_{vk})$ in the inset to the right panel of Fig. 14.4 shows the quite narrow distribution of independent-particle transitions that contribute to the peak labeled A in the spectrum of Cu_2O.

For a complete analysis it is not enough to look at the absolute values $|A_\lambda^{vck}|^2$, since they do not reflect possible interference effects. A complementary view is obtained by plotting $S_\lambda(\omega) = \sum_{vck} A_\lambda^{vck} \tilde{\rho}_{vck} \delta(\omega - \varepsilon_{ck} + \varepsilon_{vk})$, and $I_\lambda(\omega) = \int_0^\omega d\omega' \, S_\lambda(\omega')$. This is done [635] in

---

[12] This should be seen only as a trend. As explained in the subsection "When are excitonic effects inportant" on p. 363, in a solid the effects rather show up through modifications of the wavefunctions.

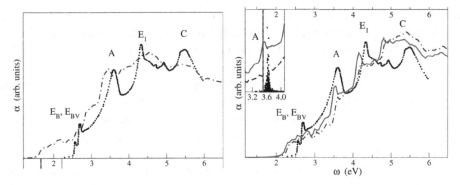

Figure 14.4. Optical absorption of Cu_2O, from [459]. Dots in the main panels are the experimental results. The absorption coefficient α is proportional to $\mathrm{Im}\,\epsilon_M$. Calculations have been performed using GW-BSE. The dot-dashed curve in the left panel uses G_0W_0 results with LDA ingredients. The dot-dashed curve in the right panel is obtained with a QSGW Green's function and W_0, whereas the continuous curves use W from QSGW (see Sec. 12.4 for explanations concerning QSGW). The dashed curve in the inset to the right panel neglects the electron–hole interaction. Dots in the inset are the square of the weight A^{vck} of independent transitions contributing to peak A, as a function of their energy $(\varepsilon_{ck} - \varepsilon_{vk})$.

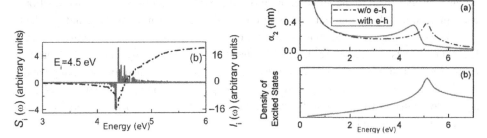

Figure 14.5. GW-BSE calculation for the optical spectrum of graphene, from [635]. The left panel shows the contribution $S_\lambda(\omega)$ of independent transitions with transition energy ω to the intensity of an optically bright state at 4.5 eV, labeled $\lambda = i$ (left axis). $I(\omega)$ (right axis) is the integral; for definitions, see the text. Right panel: optical absorption spectra (a) and density of excited states (b) of graphene with (continuous) and without (dot-dashed) excitonic effects. α_2 is the absorption coefficient, proportional to $\mathrm{Im}\,\epsilon_M$.

the left panel of Fig. 14.5 for the optical spectrum of graphene. The spectrum is shown on the right. In graphene, S is a real function because of inversion symmetry. Still, $S_\lambda(\omega)$ can be positive or negative. As one can see, there is significant canceling among the dominant contributions, and $I(\omega)$ converges more slowly to its final value than one might expect from the strongly peaked $|S|$.

When are excitonic effects important?

In order to find guidelines that indicate whether or not excitonic effects will be important, one can use perturbation theory. To first order in the screened interaction, the change of the transition energy E_λ is given by Eq. (14.46). Obviously, the change of energy is large when the matrix element of W is large. Equation (14.48) shows that this is the case when screening is weak, when charges are localized, and when the electron and hole are close to each other. This holds in finite or low-dimensional systems. When systems are extended in one or more directions, one has to be careful with perturbation theory: each individual matrix element of W is vanishingly small, because the wavefunctions are normalized over the volume Ω, which is infinite. Therefore, *there is no first-order correction to the transition energies in a solid.*[13] This explains the bottom panel on the right in Fig. 14.5, where the density of transition energies $\sum_\lambda \delta(\omega - E_\lambda)$ is compared with the joint density of independent quasi-particle states $\sum_{v c \mathbf{k}} \delta(\omega - (\varepsilon_{c\mathbf{k}} - \varepsilon_{v\mathbf{k}}))$. The two results are virtually indistinguishable. Transition energies start to change with second order, which means that only in materials with a strong electron–hole interaction can one find states well within the quasi-particle gap.

The first-order correction to the coefficients A_λ reads

$$A_\lambda^{(1)t} = \sum_{t'} \frac{W_t^{t'}}{\varepsilon_t - \varepsilon_{t'}}. \tag{14.49}$$

This expression sums over all pairs $t = (v c \mathbf{k})$ in the Brillouin zone, with the transition energy $\varepsilon_t \equiv \varepsilon_{c\mathbf{k}} - \varepsilon_{v\mathbf{k}}$, and similar for finite momentum transfer. Since the number of $\mathbf{k}$-points is proportional to the volume, there *is* a first-order correction to the wavefunctions, even in a solid. The important contributions to the correction stem from independent-particle transitions where the matrix element of W is large, and *the denominator is small*. Excitonic effects are therefore enhanced when one can mix many independent-particle transitions with similar energy. This is the case for flat bands, but also for dispersing bands, provided they are parallel. Equation (14.49) also explains the shape of $S(\omega)$ in Fig. 14.5: its sign is determined by the sign of the denominator, which changes at $t = t'$, if one supposes all other ingredients to be constant.

Link to models for excitons

Bound excitons fall into two broad classes: weakly bound *Wannier–Mott excitons* [636, 637], where the electron and hole are very mobile, and the in general more tightly bound *Frenkel excitons*, [638], which consist of electrons and holes that are localized on atomic sites. The Wannier exciton can be considered as a modified hydrogen atom, consisting of the hole (positive charge) and the electron, which can move freely through the crystal. The Frenkel exciton instead is bound to the lattice, and its center of mass has much less dispersion.

To make a link with the widely used Wannier–Mott model for excitons, we take a homogeneous system with one occupied and one empty dispersing band. The bands are parabolic and centered in the Γ-point, $\varepsilon_{v k} = -k^2/2m_h$ and $\varepsilon_{ck} = E_g + k^2/2m_e$, where the effective

---

[13] This is a good example of a case where solving the Dyson equation (here the BSE $\tilde{L} = L_0 - L_0 W \tilde{L}$) with a first-order kernel (here W) leads to good results, whereas a direct first-order correction corresponding to $\tilde{L} = L_0 - L_0 W L_0$ would not be sufficient.

masses m_h and m_e give the curvature of the quasi-particle bands and E_g is their direct gap. For delocalized states that are well described by a plane wave, $\psi_{n\mathbf{k}}(\mathbf{r}) = \frac{1}{\sqrt{\Omega}}e^{i(\mathbf{G}_n+\mathbf{k})\mathbf{r}}$, the matrix elements of $\bar{v}_c$ (Eq. (14.35)) do not depend on $\mathbf{k}$ and can be neglected. The matrix elements of W following Eq. (14.34) are the Fourier transform of the screened Coulomb interaction. In the Tamm–Dancoff approximation the two-particle Schrödinger equation becomes

$$\left[E_g + \frac{k^2}{2}\left(\frac{1}{m_e} - \frac{1}{m_h}\right)\right]A_\lambda^k - \sum_{k'} W(k-k')A_\lambda^{k'} = E_\lambda A_\lambda^k. \tag{14.50}$$

This is the effective mass approximation for Wannier–Mott excitons [636, 637]. If one neglects the fact that the equation has been derived for $\mathbf{k}$ limited to the first Brillouin zone, transformation to real space yields the Schrödinger equation for two particles of opposite charge interacting through a screened Coulomb interaction: this is the Schrödinger equation for a hydrogen atom, with modified masses and interaction. In this picture the exciton is a bound state of the electron–hole pair. The excitonic transition of lowest energy is the most strongly bound, with a wavefunction that is localized in real space. It therefore extends in $\mathbf{k}$-space, mixing independent transitions over a wider range of the Brillouin zone than the excitons of higher excitation energy that are more localized in $\mathbf{k}$-space. As we will see in Sec. 14.9, bound excitons in solids often show hydrogen-like features. Similar arguments hold for non-vanishing momentum transfer, as discussed in [616].[14]

In the opposite limit one finds the Frenkel exciton, [638]. It appears when bands are flat, electrons localized. The Frenkel exciton is best described using Wannier functions (see [1]) with localized Wannier orbitals on each site. Keeping only the on-site term in the interactions, the kernel matrix elements of the screened interaction become $2v_k^{k'} - W_k^{k'} \approx 1/N_k[2v_0 - W_0]$, which yields

$$(\varepsilon_c - \varepsilon_v)A_\lambda^k + \frac{1}{N_k}(2v_0 - W_0)\sum_{k'}A_\lambda^{k'} = E_\lambda A_\lambda^k, \tag{14.51}$$

or $E_\lambda = \varepsilon_c - \varepsilon_v + 2v_0 - W_0$. Screening at short distances is incomplete, therefore the direct on-site interaction W_0 is larger than the bare dipole–dipole interaction v_0, and one finds a strongly bound exciton.

Looking at excitons: a problem of strong correlation

With the coefficients A_λ one can construct the electron–hole wavefunction of a state λ in terms of the wavefunctions of the single electrons and holes. It reads for a given $\mathbf{q}$:

$$\Psi_\lambda(\mathbf{r}_e, \mathbf{r}_h) = \sum_{vkc} A_\lambda^{vkck+q}\psi_{v\mathbf{k}}^*(\mathbf{r}_e)\psi_{c\mathbf{k}+\mathbf{q}}(\mathbf{r}_h). \tag{14.52}$$

---

[14] In the fundamental work [616], the derivation of the effective mass approximation starts from the full Bethe–Salpeter equation instead of the GW approximation.

This wavefunction is not a simple product: it is a linear combination of electron–hole pairs. This is similar to a configuration interaction approach of quantum chemistry mixing excitations of single electron–hole pairs, which is called "CI singles" [3].[15] The state given by Eq. (14.52) is a *correlated* state: one cannot tell where the electron is if one doesn't know the position of the hole. It is instructive to look at a picture of Eq. (14.52) for a strongly bound exciton. We take as example the excitation corresponding to the strong peak at 12.6 eV in the optical spectrum of LiF, which is shown in Fig. 2.10. The charge density of the electron $n_e^\lambda(\mathbf{r}_e) \equiv |\Psi_\lambda(\mathbf{r}_e, \mathbf{r}_h)|^2$ for the position $\mathbf{r}_h$ of the hole fixed on a fluorine atom in the center of the picture is shown on the cover of this book, which has been adapted from [639]. Shown is a cut along the [111] direction of the charge density of the electron in the first excitonic state in LiF. The picture covers several unit cells. If there were no correlation between hole and electron, the charge density of the electron would be periodic. Instead, the picture shows that charge is attracted close to the hole and on neighboring fluorine atoms. This is a direct view of a strong correlation effect, made possible by theory and computation. Similarly, one can visualize correlation between spins by looking at the spin components of the wavefunctions. The example of antiferromagnetic MnO can be found in [640].

A note on translational symmetry

The absence of translational symmetry in the cover figure is puzzling at first sight, since there is nothing in the optical experiment that should break this symmetry. It happens in the picture, simply because the position of the hole has been fixed. The probability of finding the hole in a given place, however, is perfectly periodic. Had we put the hole in another unit cell, we would have obtained the same picture, simply translated. Hence, there is no symmetry breaking.

Mathematically this is expressed by the sum over **k**-points in Eq. (14.44). All **k**-points are mixed to create an exciton,[16] but the transitions that contribute to this mixing conserve momentum: valence and conduction states in each pair have the same **k** in an optical transition, or more generally, their momentum always differs by the exciton wavevector **q** that corresponds to the momentum transferred by the experiment with which the excitation is probed. Since a hole corresponds to a missing electron, overall momentum is conserved. As explained in Sec. 14.8, the mixing of many **k**-points is often the major computational bottleneck of BSE calculations. It is the price to pay to describe the strong electron–hole correlation. As discussed in Sec. 4.7, sometimes one can capture a part of the correlation by breaking symmetry. In the example of the exciton, one could put a positive charge in some place, and calculate the effect on an excited electron using a mean-field method like ΔSCF

---

[15] There is, however, an important difference: the hamiltonian that describes the electron–hole pair contains screening. In this way, other possible excitations of the system are included in an average way. When one goes beyond the approximation of static screening, as mentioned in Sec. 14.10, the additional excitations appear explicitly and can be detected in the spectra.

[16] Note that this prohibits inserting excitonic levels into an electron addition and removal band structure plot, since one can no longer relate energies to **k**-points.

in Sec. 11.4. This is sometimes done for excitations from core levels, where it is obvious where the hole should be placed [641]. However, since this approach breaks symmetry, one has to use a supercell that is large enough to contain the charge distribution of the electron. The symmetry-conserving BSE calculations, instead, are performed in the unit cell of the crystal.

Cancellations and interpretations

The concept of electron–hole pairs is relative to the ground state of the N-electron system. The electron and the hole in the pairs are described by fully dressed Green's functions in the BSE, Eq. (14.6): they contain all many-body effects that go with the addition or removal of a charge to the N-electron ground state. However, in experiments like optical absorption there is no addition or removal of charge. Therefore, $L_0 = GG$ can be a quite bad starting point to describe electron–hole excitations, and the kernel of the BSE has the paradoxical task to remove many-body effects that have been hard to add in the first place.

Take the excitation of a single electron. Of course there are no interaction effects in this case. However, if one tries to calculate this excitation using the BSE, one first has to calculate electron removal and addition. As shown at the example on the quarter-filled Hubbard dimer in Sec. 11.7, for large distance (small hopping) electron addition can be very hard to describe. The subsequent solution of the BSE has only one task: to remove all the many-body effects that have been added in L_0. This is, of course, an extreme example, but it shows a real trend: self-energy and the electron–hole interaction effect have a tendency to cancel partially. How this comes about in the BSE is illustrated in a simple model in Ex. 14.1.

Solids are far from the extreme case of a single electron, but transitions between strongly localized states show some analogy. One example is NiO in Sec. 14.9. Besides providing ideas for potentially more efficient approximations,[17] this shows that interpretations are not unique: nobody would call the excitation of the single electron a "strongly bound exciton," but what about the Frenkel excitons in the examples in Sec. 14.9? One may look at them as strongly bound excitons, since their energy lies in the gap of a quasi-particle band structure. This is the point of view of the BSE. One may also consider that these are atomic-like excitations embedded in a crystal. This is closer to the point of view of DMFT in Chs. 16–20. Both are correct. This shows that it is not only important to join theoretical and computational efforts, but also to realize that one and the same effect can be expressed in a completely different language, without contradiction.[18]

---

[17] Note that time-dependent density functional theory in Sec. 14.11 is closer to the physical picture of excitations, without going through electron addition and removal. It is an in-principle-exact alternative to the BSE for the calculation of optical spectra.

[18] Of course, for most atoms more than one electron has to be considered, and even for strongly localized electrons the problem does not really reduce to the non-interacting case. The intra-atomic interactions give rise to multiplet effects. They can be seen, for example, in the satellite excitations of cerium in Fig. 2.9.

14.8 Bethe–Salpeter calculations in practice

This section outlines how the BSE can be solved in first-principles calculations, based on the most commonly used approximations introduced in previous sections:

- the kernel Ξ (Eq. (14.7)) is derived from the GWA self-energy;
- the contribution $\delta W/\delta G$ in Eq. (14.22) is neglected;
- $L_0 = GG$ is built with independent-particle Green's functions G that contain quasi-particle energies from the GWA, and most often Kohn–Sham wavefunctions;
- a static approximation to the direct electron–hole interaction W is made.[19]

With these approximations, a typical BSE calculation follows the flowchart in Fig. 14.6. The subsections below give more details about the various steps.

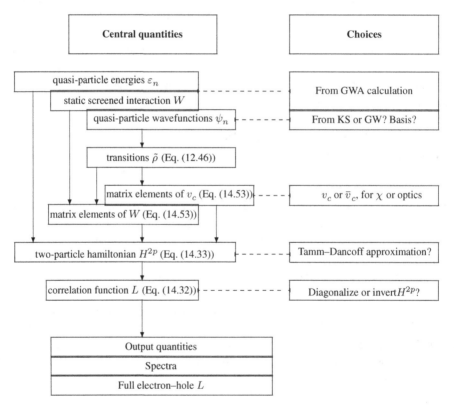

Figure 14.6. Flowchart for a typical BSE calculation starting from the GWA and using a static screened Coulomb interaction W.

---

[19] The frequency dependence of the self-energy is usually taken into account to calculate the quasi-particle energies.

Building the hamiltonian

The BSE calculation comes after a GWA calculation, where one obtains the quasi-particle energies and, in the case of a self-consistent calculation, quasi-particle wavefunctions. With this aim one chooses a basis, as in Sec. 12.6. One also obtains the static screened Coulomb interaction $W(\mathbf{r}, \mathbf{r}')$, the same as the $\omega = 0$ value of the dynamically screened interaction used in the GWA calculation.[20]

The first new step of the BSE is to build the two-particle hamiltonian H^{2p} given in Eq. (14.33). The energy difference on the diagonal is the result of a GWA calculation. It can often be approximated by a rigid shift, called "scissor," as explained in Sec. 13.1. Usually only the real part of the quasi-particle energies is considered.

The matrix elements of v_c and W given in Eq. (14.34) are calculated using the transition elements $\tilde{\rho}$ in a basis, Eq. (12.46), like in the GWA calculation in Sec. 12.6. The only difference is that in the BSE, for the construction of the matrix W, the $\tilde{\rho}_{nn'}$ for pairs of occupied or pairs of empty states are also needed. The expressions to be evaluated are

$$v_{n_1 n_4}^{n_3 n_2} = \sum_{\ell m} \tilde{\rho}_{n_1 n_3}^{\ell} v_c^{\ell m} \tilde{\rho}_{n_4 n_2}^{m*} \quad \text{and} \quad W_{n_1 n_3}^{n_4 n_2} = \sum_{\ell m} \tilde{\rho}_{n_1 n_4}^{\ell} W^{\ell m} \tilde{\rho}_{n_3 n_2}^{m*}, \tag{14.53}$$

where $v_c^{\ell m}$ and $W^{\ell m}$ are matrix elements of v_c and W in the chosen basis, as defined for the two-point functions in Eq. (12.40). More specifically, in a plane-wave basis and for a momentum transfer $\mathbf{q}$, the matrix elements of the resonant part are

$$v_{v k v' k'}^{c k + q c' k' + q} = \sum_{\mathbf{G}} \tilde{\rho}_{v k c k + q}(\mathbf{G}) v_c(\mathbf{q} + \mathbf{G}) \tilde{\rho}_{v' k' c' k' + q}^{*}(\mathbf{G}) \tag{14.54}$$

and

$$W_{v k c k + q}^{v' k' c' k' + q} = \sum_{\mathbf{G} \mathbf{G}'} \tilde{\rho}_{v k v' k'}(\mathbf{G}) W_{\mathbf{G} \mathbf{G}'}(\mathbf{k} - \mathbf{k}') \tilde{\rho}_{c k + q c' k' + q}^{*}(\mathbf{G}'). \tag{14.55}$$

For an optical spectrum, in the limit of vanishing wavevector $\mathbf{q}$, the matrix element of v_c in Eq. (14.54) diverges for $\mathbf{G} = 0$. At the same time the transition elements $\tilde{\rho}$ go to zero proportional to $\mathbf{q}$. The limit $\mathbf{q} \to 0$ can be taken numerically or using $\mathbf{k} \cdot \mathbf{p}$ perturbation theory[21] as in Eq. (12.56). The limit is non-analytical. For a given dipole $(v k c k + \mathbf{q})$, the matrix element $\tilde{\rho}$ is maximum when $\mathbf{q}$ is parallel to the dipole, and it is zero when $\mathbf{q}$ is perpendicular to the dipole. As explained in Sec. 14.1, the $\mathbf{q} \to 0$ contribution distinguishes loss and absorption spectra. Loss spectra correspond to longitudinal excitations, where the direction of the polarization is parallel to the propagation vector. Absorption spectra are called transverse, since the field and the propagation are perpendicular. Therefore, the non-analytic contribution causes the so-called *longitudinal–transverse splitting*.

The screened Coulomb interaction in Eq. (14.55) diverges for $\mathbf{k} = \mathbf{k}'$. Since the space of $\mathbf{k}$-points is in principle continuous, this problem only occurs because of the discretization

---

[20] Sometimes models are used for the screened Coulomb interaction in the BSE kernel, but the results are less transferable.

[21] See the explanations around Eq. (12.55), in particular concerning non-local potentials.

of **k**-space. It is solved by integrating W over a small volume in reciprocal space around the divergence, corresponding to the weight of the points.

   W should be the same as that used in the GWA calculation. This implies that for materials where self-consistency is needed in the GWA, it is also important in the BSE. This is illustrated in Fig. 14.4, which shows BSE results [459] for Cu_2O. The effect of static self-consistency[22] on the quasi-particle energies is an essentially rigid shift of the spectrum; the effect of a self-consistent W is an enhancement of the intensities and sharpening of the structures, leading to better agreement with experiment.

Bands and k-points

The main mixing of transitions occurs over a limited range of energy. This is discussed in the previous section, and illustrated in Figs. 14.4 and 14.5. Because of the finite range of mixing, for a given part of the spectrum only a finite range of bands has to be considered. The mixing of **k**-points is a more difficult problem. In general, the calculation of an optical spectrum requires a sampling that is much denser than for the corresponding ground-state calculations. The reason is that in a spectrum one is interested in details of the electronic structure, not just in an integral. The difficulty is illustrated for the case of an RPA calculation for silicon [642] in Fig. 14.7. The regular grid used in this calculation consists of 2048 **k**-points in the BZ that are centered on high-symmetry points. This enhances the importance of van Hove singularities and makes the spectrum too structured with respect to a fully converged calculation using 400,000 random **k**-points. The effect of the electron–hole interaction is to enhance low-energy structures. This makes the problem even more severe.

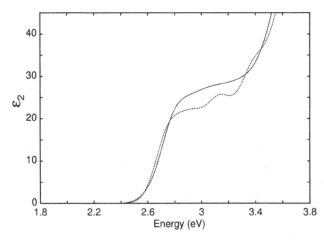

Figure 14.7. Effect of **k**-point sampling on the calculation of an optical spectrum, at the example of an RPA calculation for silicon, from [642]. The dotted line has been obtained using 2048 special points in the Brillouin zone. The continuous curve is the result of 400,000 random **k**-points.

---

[22] Here, self-consistency is performed in the QSGW scheme described in Sec. 12.4.

It is not possible to perform a BSE calculation with the same number of **k**-points as an RPA calculation, since this would lead to untreatably large matrices. Moreover, one cannot limit the calculations to the irreducible part of the Brillouin zone because **k**-points mix all over the BZ. A good strategy is to use a regular **k**-point grid that is shifted in an off-symmetry direction by a fraction of the distance between **k**-points. In this case, all points are non-equivalent, which corresponds to a better sampling. This has been done [651] in order to obtain the spectra in the right panel of Fig. 14.8 later. One can also use two different grids, one to calculate L_0 and one to solve the BSE, as proposed in [644], or variable grids that are denser in the regions of the BZ that are the most important for the excitons, as in [71] or [645]. The **k**-point problem is the reason for the fact that the first fully *ab initio* BSE calculations were done for a cluster [508], and in a solid to determine the binding energy of a strongly bound exciton [646], before *spectra* in solids could be calculated.

Solution of the BSE

Once the hamiltonian is constructed, it has to be diagonalized or inverted. The choice depends on the size of the matrix and the scope. Diagonalization has the advantage that it yields coefficients and transition energies that help to interpret the results, as outlined in the previous section. When many transitions have to be taken into account, the hamiltonian matrix can be too large and cause memory problems, because at least half of the matrix has to be kept in memory for the diagonalization. An alternative is iterative inversion [647, 648]. This is efficient because, as can be understood, e.g., from Eq. (14.44), in the calculation of spectra one does not ask for the full matrix L, but only for scalar products with transition matrix elements. In the iterative approach at each iteration step a matrix–vector product has to be performed, and one never has to keep the full matrix in memory. The original iterative inversion scheme is limited to hermitian matrices, and therefore to the Tamm–Dancoff approximation. Beyond the TDA, one can use the more general approach of [630]. Alternatively to these approaches, one can use the fact that calculating spectra from the BSE corresponds to evaluating the linearized time evolution of the Green's function or density matrix, as explained in Sec. 14.11. In [649], BSE spectra are calculated solving an initial value problem in time. This method requires the calculation of the H^{2p} matrix, and only the method of solution is changed. In [650] an iterative solution of the quantum Liouville equation $id\hat{\rho}/dt = [\hat{h}(t), \hat{\rho}(t)]$ was proposed, where $\hat{\rho}$ is the density matrix and $\hat{h}$ the one-body hamiltonian including the GWA self-energy in a static approximation. This approach avoids the calculation of empty states. Iterative approaches are in general very suitable for massive parallelization, much better than diagonalizations.

Scaling of the calculations

Since the BSE treats a two-body problem, scaling is worse than for the GWA. The most time-consuming part is the calculation of the matrix elements of W in Eq. (14.53) or (14.55). To calculate the $N_k^2 N_v^2 N_c^2$ matrix elements scales with the number of k-points and atoms as $N_k^2 N_{at}^5$, or as N_{at}^3 for unit cells small enough that $N_k \propto 1/N_{at}$. The dimension of H^{2p} grows as $N_k N_{at}^2$, or N_{at} for small unit cells. This determines the scaling of the matrix diagonalization (cube of the dimension, for a brute-force approach) or inital value problem (square of the dimension [649]), and of the iterative inversion approaches, where the calculation of a spectrum scales as N_{at}^4 [648].

14.9 Applications

This section contains applications meant to illustrate the main points. It is particularly useful to read Sec. 14.7 first, in order to acquire the analysis tools. Unless otherwise stated, all BSE calculations have been performed using the same RPA W_0 as in the $G_0 W_0$ calculation that is the starting point of the BSE.

Excitonic effects in the continuum. The BSE has first been used to calculate the optical spectrum of simple semiconductors. The left panel in Fig. 14.8 shows the historical result for silicon of Hanke and Sham [643], based on a semi-empirical band structure. This result shows the main features of the BSE: the crystal local field effects contained in $\bar{v}_c$ shift oscillator strength to higher energies. Up to here, this is the RPA. The direct electron–hole attraction W has the opposite effect. It leads to a strong enhancement of the first significant structure in the spectrum, called E_1. This brings the calculated result in much better agreement with experiment than the RPA. A similar trend has been found in the subsequent *ab initio* calculations [633, 651]. The right panel in Fig. 14.8 shows the result of [651]. The final agreement with experiment is very good.[23]

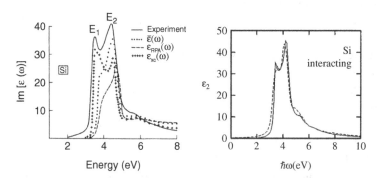

Figure 14.8. Effects of the electron–hole interaction on the optical spectrum of silicon. Left: calculation using a semi-empirical band structure, from [643]. The curves labeled $\bar{\varepsilon}$ and ε_{RPA} are RPA results without and with crystal local fields, respectively. ε_{xc} includes the direct electron–hole attraction. Right: first-principles calculation including excitonic effects (dashed curve), from [651]. In both calculations, the BSE results compare well with experiment (continuous curves in both panels).

---

[23] Like most BSE calculations in the literature, these calculations are carried out at $T = 0$. The temperature-dependent result from [611] is shown in Fig. 13.19.

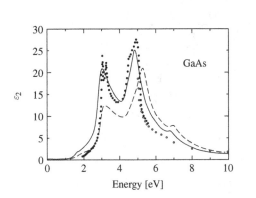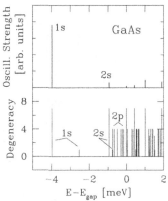

Figure 14.9. Effects of the electron–hole interaction on the optical spectrum of GaAs, from [71]. Left: absorption spectra. Dots are experiment, the dashed line is the independent-particle spectrum based on $G_0 W_0$, and the continuous line is the solution of the BSE. Right: zoom on the weakly bound excitons inside the quasi-particle gap.

One may wonder why W has such a strong effect in silicon, with its large dielectric constant of $\epsilon_M(\omega = 0) \approx 12$ and strongly dispersing bands. Indeed, one cannot see a bound exciton in the figures, since the exciton binding energy is in the meV range, and the density of transition energies $\sum_\lambda \delta(\omega - E_\lambda)$ in the continuum is practically unchanged from the RPA. However, the top valence and bottom conduction bands in silicon are almost parallel between Γ and L and in the Γ–X direction, as one can see in Fig. 13.3. As explained in Sec. 14.7, this leads to strong changes in the electron–hole wavefunction, with consequent interference effects. This happens in many materials. The right panel of Fig. 14.5 shows calculations for graphene [635], a very different system. The density of transition energies in the lower-right panel is the same with and without the electron–hole interaction. Nevertheless, including the interaction there is a prominent peak in the absorption spectrum near 4.5 eV, at lower energy than the absorption peak from independent quasi-particle transitions. Also, the line shape differs significantly.

Weakly bound excitons. Many simple semiconductors show a picture similar to silicon; another example is GaAs, shown in Fig. 14.9. The right panel contains a zoom on the absorption onset. There is a series of very weakly bound peaks, also seen experimentally. These are Wannier–Mott excitons, which are discussed in Sec. 14.7. They are difficult to calculate, because the smaller the binding energy, the smaller is the range of **k**-points over which transitions are mixed. This requires a very fine sampling of the BZ. In [71] this has been achieved by using a fine grid in the specific region of the BZ that contributes to the excitons, together with an interpolation of the hamiltonian matrix elements.

Rydberg series and Frenkel excitons. Rare gas solids are examples where screening is weak, and by consequence W is large. This causes strongly bound excitons in the quasi-particle gap. Figure 14.10 shows the absorption spectrum of solid argon. Several

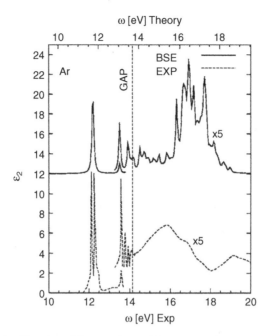

Figure 14.10. Excitonic effects in solid argon, from [652]. Top panel: GW-BSE calculation. Bottom: experiment.

peaks can be seen inside the gap, both in experiment and in the solution of the BSE [652].[24] They are part of a hydrogen-like Rydberg series predicted by the Wannier–Mott model of Eq. (14.50) in Sec. 14.7. Such a series is an obvious correlation effect: none of these peaks can be captured by an independent quasi-particle calculation. The Wannier–Mott model is derived in the limit of weakly bound excitons. Nevertheless, it works surprisingly well even in the case of exciton binding energies in the eV range. A more quantitative analysis of solid argon shows that the first peak corresponds to a very strongly bound pair, which has Frenkel rather than Wannier character. The rest of the series, instead, closely follows the distribution of energies and mixing coefficients predicted by the Wannier–Mott model on the basis of the dielectric constant and effective masses.

The first peaks in the Rydberg series of solid argon are similar to those measured for atoms [653]. This is related to the very localized character of the excitons. The BSE calculation in the solid starts from Bloch states, and has to localize the electron close to the hole. As discussed in Sec. 14.7, one could also think to approach the problem by starting from atomic excitations, which are embedded in a solid. The two pictures are not in contradiction.

Charge transfer excitons. In ionic systems like NaCl the electron and hole are localized, but situated on different atoms. One speaks about a *charge-transfer exciton*. The

---

[24] The interpretation of the results is complicated by spin–orbit splitting and surface effects.

contributing valence and conduction wave functions have little overlap, so the dipole–dipole matrix elements of v_c are small. These are materials that are particularly difficult to describe in time-dependent density functional theory as outlined in Sec. 14.11, because in TDDFT all matrix elements of the kernel $v_c + f_{xc}$ have dipole–dipole character, and f_{xc} must be very strong to compensate this fact. In the BSE, instead, W is a direct interaction between charge densities that can be significant even when there is no overlap. Therefore, the description of charge-transfer excitations is in principle easier in the BSE. This is exploited also for the description of molecules, which shows interesting charge-transfer excitations [654].

Effects of the electron–hole interaction on plasmons. To see plasmons, one has to calculate the loss function $-\operatorname{Im} \epsilon^{-1}(\omega) \propto -\operatorname{Im} \chi(\omega)$, as stated in Sec. 14.1. The density–density correlation function χ is obtained from L. On the low-energy side of the spectrum of χ one finds the screened absorption spectrum, called *interband transitions*, since $\operatorname{Im} \chi$ and $\operatorname{Im} \epsilon$ are linked by Eq. (5.68). It contains excitonic effects, but since it is screened, they are less visible than in the absorption spectrum. Instead, the spectrum is dominated by plasmons, the collective excitations that correspond to zeros of the real part of ϵ. To calculate the real part requires a large number of bands. This makes BSE calculations much more cumbersome. Moreover, because of the strong long-range part of the Coulomb interaction the Tamm–Dancoff approximation is problematic [243], as one can see in Fig. 14.11 for the example of silicon. Note that the optical absorption spectra of silicon with and without the Tamm–Dancoff approximation are instead virtually indistinguishable.

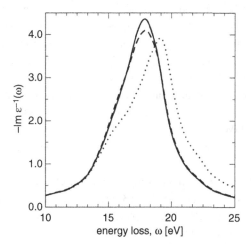

Figure 14.11. Effects of the coupling between resonant and anti-resonant parts of the electron–hole interaction on the electron energy loss spectrum of silicon, from [243]. Dotted line: no coupling (TDA). Continuous line: full calculation. The dashed line includes coupling to first order in the interaction.

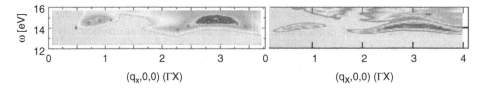

$(q_x,0,0)$ (ΓX) $(q_x,0,0)$ (ΓX)

Figure 14.12. Exciton dispersion in LiF. Shown is the dynamic structure factor as a function of energy (vertical axis) and momentum transfer **q** (horizontal axis) along the Γ–X direction, in units of ΓX. Left panel: experimental data of [656] in arbitrary units. Right panel: GW-BSE, from [639].

Exciton dispersion, disappearance, and reappearance. All previous examples in this section were calculated at vanishing momentum transfer **q**. Interesting effects are accessible when one studies spectra as a function of **q**. This has been done, for example, in the case of LiF [639, 655]. Figure 14.12 shows the dynamic structure factor of LiF as a function of energy and **q** along the Γ–X direction. Shown is the energy region around the exciton; the plasmon is at higher energy, around 25 eV. Like the experimental result in the left panel, the BSE calculation in the right panel shows the screened, but strongly bound, exciton above 13 eV, which corresponds to the sharp peak in the absorption spectrum in Fig. 2.10. The exciton disperses in a periodic way, related to the band structure. Its intensity varies strongly; it is stronger in the second Brillouin zone than in the first, and becomes weaker for even higher momentum transfer.

It is important to be able to calculate excitons at non-vanishing momentum transfer, since many interesting excitations, like the transitions between states stemming from d-orbitals, are dipole forbidden. The sharp peak in the inelastic X-ray scattering spectrum of NiO in Fig. 2.12 is a good example. It is an exciton, as has been shown by an approach based on LDA+U and Wannier functions [657]. The peak cannot be seen in optical absorption, so the BSE at $q = 0$ yields a dark exciton [658]. This peak is an example for an excitation that can alternatively be regarded as a strongly bound exciton, or as a quite atomic-like transition embedded in a solid, similar to the case of argon. A closer look shows that the peak consists of several structures. The BSE result cannot fully explain this multiplet structure. Some excitations are missing in the static approximation based on the GWA, as explained in this chapter, which might be accounted for if dynamical effects were included. How this could be done in principle is the topic of Sec. 14.10.

Confined systems. In lower dimensions, interaction effects are stronger. This is discussed in Sec. 13.3, and illustrated in Fig. 13.9 with the example of GW corrections and exciton binding energies in h-BN. Since screening is weaker and charges more confined, excitons in low-dimensional systems are often more strongly bound.

Let us first consider the surface of a semiconductor like silicon or germanium. These materials have no strongly bound excitons in the bulk. However, bound excitons are found at surfaces. When a germanium crystal is cut in the (111) direction, the surface atoms rearrange and form chains of dimers [659–661]. This doubles the unit cell,

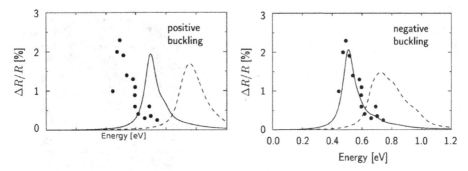

Figure 14.13. Effects of the electron–hole interaction on the differential reflectivity spectrum of Ge(111)2×1, from [662]. Dashed curve is without, continuous curve including the electron–hole interaction in G_0W_0-BSE. Dots are the experimental result. Left: surface reconstruction with positive buckling. Right: surface reconstruction with negative buckling.

which is why the reconstruction is called Ge(111)2×1. The dimers are slightly tilted with respect to the bulk. There are two energetically favorable possibilities for the tilt, called positive and negative buckling, respectively. Optical experiments can help to detect which of the two structures is formed. Figure 14.13 shows the experimental and calculated differential reflectivity spectrum of Ge(111)2×1, from [662]. The effect of the electron–hole interaction is very strong, with an exciton binding energy of $E_B = 0.2$ eV and 0.15 eV for the two possible structures, respectively, in spite of the small quasi-particle gaps of 0.88 and 0.66 eV between surface states. This is due to the weak screening; as illustrated in Sec. 13.3, low-dimensional structures can even show anti-screening, which means regions of space where the interaction is stronger than the bare one. This has been shown to occur along the chains on semiconductor surfaces [663], which explains the strong excitonic effect. The RPA spectra of Ge(111)2×1 are therefore very different from the experimental results. Solving the BSE for the two possible surface reconstructions yields good agreement in the case of negative buckling. This agreement allows one to infer the structure from the optical experiment, using calculations.

Molecules and molecular solids are examples of the difference between isolated objects and solids. Figure 14.14 shows the absorption spectrum of PPV, the conjugated polymer poly-para-phenylenevinylene [664]. In the two left panels the spectra of bulk and isolated molecular chains are compared. There is a strong decrease of exciton binding energy when one moves from the isolated chain to the crystal. The right panel compares the exciton wavefunction Ψ_{eh} of a bright and a dark transition in the solid. Here, $\int d\mathbf{R} \, |\Psi_{eh}(\mathbf{r}_e - \mathbf{r}_h, \mathbf{R})|^2$ is plotted, with the center of mass coordinate $\mathbf{R} = (\mathbf{r}_e + \mathbf{r}_h)/2$. The bright exciton on the left is essentially confined to a single chain, whereas the electron and hole have a tendency to be on different chains in the case of the dark exciton on the right. This explains its small oscillator strength. Dark excitons are hardly seen in the

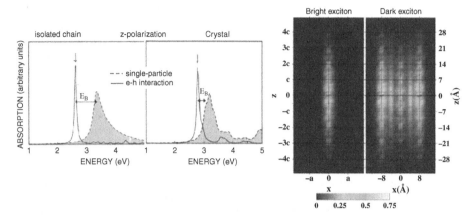

Figure 14.14. Effects of the electron–hole interaction on the optical spectrum of PPV, from [664]. Left: solid-state effect on the spectra. Right: comparison of the electron–hole wavefunction of a bright and a dark state in the crystal. The wavefunction is plotted as a function of the electron–hole distance and averaged over the center of mass coordinate, as explained in the text.

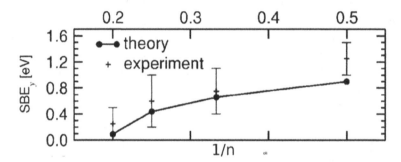

Figure 14.15. Singlet exciton binding energy (SBE) of oligoacenes as a function of chain length given by the number n of phenyl rings in the molecule, from [665].

spectra, but can be detected and analyzed thanks to the diagonalization of the two-particle hamiltonian.

The confinement of the electron–hole pair and its binding energy are also influenced by the chain length. This is illustrated in Fig. 14.15. It shows exciton binding energies in organic molecular crystals made of linear oligoacenes from naphthalene ($n = 2$ phenyl rings) to pentacene ($n = 5$ rings) [665]. In agreement with experiment, the binding energy is larger for shorter chains.

By reducing the size of the objects even more, the effect of the electron–hole interaction cancels to a large extent the self-energy shift of spectra to higher frequencies. This has been seen in small clusters such as Na_4 [508] or SiH_4 [565]. The first dipole-allowed transition energies in the BSE are found to be in good agreement with experiment. Table 14.1 shows a comparison of BSE results for singlet and triplet excitation energies of SiH_4 with results obtained with fixed-node diffusion Monte Carlo calculations, and with the accurate

Table 14.1. Excitation energies (eV) for the first excited singlet and triplet states of SiH_4, from [666]. The experimental result corresponds to the vertical excitation (no effect of geometry relaxation), inferred from the energy of maximum absorption. QMC and CASSCF results are obtained by calculating total energy differences

Method	Singlet S	Triplet T	$\Delta = S - T$
CASSCF	9.1	8.7	0.4
GW-BSE	9.2	8.5	0.7
QMC	9.1(1)	8.7(1)	0.4(1)
EXP	8.8	...	...

quantum chemistry method CASSCF (complete active space self-consistent field) [666]. The methods yield results that are close to each other.[25]

Defects

In complex systems the ability to analyze experiment in terms of theory and calculations is particularly important. A good example is the study of oxygen vacancies (F centers) in MgO: experimentally, the positively charged and the neutral vacancy (F^+ and F^0, respectively) absorb light in the optical range at practically the same energy. Bethe–Salpeter calculations on top of GW quasi-particle energies [667] have reproduced this result, and found that also the two emission spectra are very close. The excitation energies in GW-BSE are similar to those calculated in projector Monte Carlo [668], as described in Ch. 24. The GW+BSE approach allows one to access the spectrum on a wider range, of several eV. The calculations in [667] predict a second absorption feature for the F^+ center, which is situated at 3.6 eV and could therefore be easily distinguished from the main line around 5 eV. It is due to the fact that the empty spin state on the defect can be filled by an electron excited from the valence band. This feature is therefore absent in the F^0 absorption, which could be a fingerprint to distinguish the two defects.

14.10 Extensions

The rich variety of results in the previous section has been obtained by looking only at the electron–hole part of L, and only in the strongly simplified static GW-BSE approach. The present section describes briefly ways to overcome the static approximation, and some

---

[25] As pointed out in Sec. 13.3, one has to be careful concerning the independent-particle wavefunctions in $G_0 W_0$ calculations, since LDA wavefunctions may not be sufficiently accurate in small systems. In order to obtain the good agreement in Tab. 14.1, it was necessary to go beyond the solution of the perturbative quasi-particle Eq. (12.11), and diagonalize the quasi-particle hamiltonian. However, the Dyson equation was not iterated to self-consistency.

results other than electron–hole excitations that can be obtained from the full BSE. The section is not vital for the main concerns of the book, but it is meant to broaden the horizon.

Dynamical effects of the electron–hole interaction

To go beyond the static approximation for the electron–hole interaction requires in principle the solution of Eq. (14.20). In practice, approximations are proposed. The main idea is to obtain an approximate one-frequency equation. In [670], dynamical effects in W have been treated as a perturbation. In [671], dynamical contributions stemming from the dressed Green's functions in $L_0 = GG$ were added to the approach of [670]. In [672, 673] an approximate two-particle equation was derived, with an interaction term that depends on the excitation energy. The equation is based on a quasi-particle approximation for the one- and two-particle Green's functions of discrete levels. It has been used, for example, to determine binding energy shifts and narrowing of Auger widths for core levels [672, 673], and to correct energy levels in molecules [565].

In [669] the Bethe–Salpeter equation is expanded in orders of the electron–hole interaction. With Eq. (14.19) the first-order contribution[26] is

$$\tilde{L}^{(1)}(\omega) = \int d\omega_2 \tilde{L}^{(1)}(\omega, \omega_2) = -\frac{1}{2\pi} \int d\omega_2 \, d\omega_3 L_0(\omega, \omega_2) W(\omega_2 - \omega_3) L_0(\omega, \omega_3). \quad (14.56)$$

One can define an effective one-frequency kernel $K_W(\omega)$ by

$$\tilde{L}^{(1)}(\omega) = -L_0(\omega) K_W(\omega) L_0(\omega). \quad (14.57)$$

Combining Eqs. (14.57) and (14.56), one can solve for K_W. This kernel K_W is then used in the one-frequency BSE Eq. (14.25) instead of the static W. In $L_0 = GG$ a quasi-particle approximation is made, including a factor Z for each of the two Green's functions as prescribed by Eq. (7.28). This leads to a reduction of the spectral weight by a factor Z^2. The results that are obtained when dynamical effects are included consistently in L_0 and in the electron–hole interaction confirm that, as suggested in [628], the factor Z^2 of L_0 and the dynamical effects in W cancel to a large extent, especially in semiconductors and insulators with moderate exciton binding energy. This supports the widely used static approximation that has been the topic of the previous sections. The cancellation is illustrated for the optical spectrum of silicon in Fig. 14.16. The approach was also applied to calculate the electron–hole interaction in the absorption spectra of metals, in good agreement with experiment. In a static approach, electron–hole effects are absent in metals, because of the strong screening of the interaction. When dynamical effects are included, the finite time that is necessary to build the screening cloud allows the electron and hole to interact.

The cancellation of intensities discussed above is not a valid argument when one is interested in satellite structure. As in the case of the one-particle Green's function, a dynamical W in the Bethe–Salpeter equation contains excitations that can show up as extra peaks. These are seen, e.g., in the spectra of molecules. They are named *double excitations*. In

---

[26] The same can be done for higher orders.

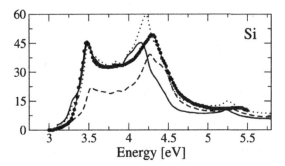

Figure 14.16. Dynamical correlation effects on the optical spectrum of silicon, from [669]. Dotted line: static BSE as in previous sections. Dashed line: static BSE including the quasi-particle renormalization Z^2 in $L_0 = GG$. Continuous line: BSE with dynamical effects included in L_0 and in the electron–hole interaction, using the dynamical kernel defined in Eq. (14.57). Full dots: experiment.

[625, 674] the analogue of [672, 673] has been applied to model molecular systems, and such extra structures were detected. Multiple plasmon peaks can be seen in the dynamic structure factor of simple metals; they have been measured with inelastic X-ray scattering and explained with calculations for the homogeneous electron gas on an approximate dynamical Bethe–Salpeter level [675, 676].

The extra structures clearly go beyond the picture of the static BSE, where one has one electron–hole pair in a medium that screens the pair adiabatically. The static picture has some similarity to the CI singles approach of quantum chemistry [3], which is mentioned in the discussion of the electron–hole wavefunction in Sec. 14.7. The double excitations, and other features that appear when screening is dynamical, correspond to the excitations beyond the singles in a quantum chemistry language.

Bethe–Salpeter equations for other properties

Most first-principles BSE calculations deal with electron–hole excitations. However, by choosing the appropriate time ordering and time differences in analogy to the last subsection of Sec. 14.5, one can also describe the correlation of two holes or two electrons. This means that one has to change Eq. (14.21). A BSE corresponding to the correlation of two holes has been used, for example, to calculate the spectrum of doubly ionized final states in the Auger spectra of hydrocarbon systems [677], or to determine the two-electron distribution function in atoms and molecules [678]. The two-electron distribution function is the square of the electron–electron wavefunction, which is defined analogous to the electron–hole wavefunction in Eq. (14.52). Figure 14.17 shows the result for a CO molecule [678]. The position of one of the two electrons is fixed, and the distribution of the second electron is examined. It depends strongly on the position of the first electron, and one can see the Coulomb hole that is due to the electron–electron repulsion. This is similar to Fig. 6.3 for H_2, and it is opposite to the attractive effect observed in the electron–hole case, shown for the excitonic wavefunction on the cover of this book.

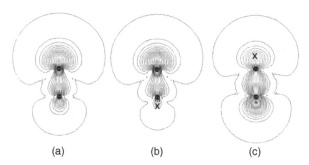

(a)   (b)   (c)

Figure 14.17. Two-electron distribution function in a CO molecule, from [678]. The fat dots are the carbon atom (top) and oxygen atom (bottom). The cross indicates the position of one of the two electrons. The result on the left is from a calculation neglecting the electron–electron correlation. It therefore does not depend on the position of one of the two electrons.

Another quantity that can be obtained from the full L is the spectrum of spin waves. In contrast to plasmons or optical excitations, in this case one has to describe the coupling of electrons and holes with different spins. The formalism has been derived in [679], and used to calculate spin excitations in $3d$ transition metals in [680, 681].

14.11 Linear response using Green's functions or density functionals

In this section we consider the electron–hole BSE from a different point of view, which allows us to make a close link to time-dependent density functional theory. TDDFT is not a central topic of this book, but making the link points out parallels, and it is in line with the idea of combining approaches. More about the relations of the BSE and TDDFT can be found in [682].

Propagating Green's functions in time

The expression for the two-particle correlation function L in Eq. (8.35) shows that it is a linear variation of the Green's function as a functional of an external source. Most often one is interested in the density response to a physical, local potential. This is equivalent to saying that one is interested in χ, the diagonal of L. In the BSE one calculates L, and from this χ. Alternatively, one can calculate the response of the Green's function or the density in time, by applying a potential and letting the system evolve. From the result, one can deduce response functions. This describes the physical, causal, response, which has to be kept in mind when comparing equations. However, since time-ordered and retarded quantities are readily connected as outlined in Ch. 5, it is not an obstacle for the derivations.

The general structure of the relations can be understood by starting from the Dyson equation in the presence of an extra external potential v_{ext}. It can be written as $G_0^{-1}G = 1 + v_{\text{ext}}G + \Sigma G$. With $G_0^{-1} = i\frac{\partial}{\partial t} - \hat{h}$ (Eq. (5.85)) this is an equation for the propagation of the Green's function in time, provided an approximation for Σ is given. It can be evaluated by applying explicitly a time-dependent external potential $v_{\text{ext}}(t)$ and letting the Green's function evolve in time. This must be done using non-equilibrium Green's functions, because of the presence of the time-dependent potential. Detailed explanations can be found in [683]. An example of applications to atoms and molecules, with the second Born approximation for the self-energy, is shown in [684].

Given the self-energy, the time-propagation approach makes in principle no further approximation. Therefore this method can naturally include the dynamical effects, as well as, e.g., the term $\delta W/\delta G$, when the GWA for Σ is used. Moreover, the propagation includes the response to all orders, which means that one has access to linear as well as non-linear response functions.[27] The linear response is equivalent to the BSE. The time-propagation approach in the non-linear regime has been used, for example, to describe excitations of h-BN by a strong pulse [686].

Comparison to linear response TDDFT

The response of the density to an external potential can in principle also be obtained exactly using time-dependent density functional theory, which is briefly introduced in Sec. 4.6. The Runge–Gross theorem [217], the analog to the Hohenberg–Kohn theorem in Sec. 4.3, states that there is a one-to-one correspondence between the time-dependent density and the time-dependent external potential, provided the initial state of the system is given. The initial state is usually the ground state. In this case, there is a universal functional for the exchange–correlation potential that depends on the density at all points in space and past time. The time-dependent Green's function and density functional framework are then strictly analogous, which is an extension of the parallel of the static theories, discussed in Sec. 8.1. In particular, the equivalent to the BSE Eq. (14.6) reads[28] [687, 688]

$$\chi(1,2) = \chi_0(1,2) + \chi_0(1,\bar{3})\left[v_c(\bar{3},\bar{4}) + f_{\text{xc}}(\bar{3},\bar{4})\right]\chi(\bar{4},2), \qquad (14.58)$$

with

$$\chi_0(1,2) = \frac{\delta n(1)}{\delta v_{KS}(2)} \quad \text{and} \quad f_{\text{xc}}(3,4) = \frac{\delta v_{\text{xc}}(3)}{\delta n(4)}, \qquad (14.59)$$

where the derivative is taken in absence of the time-dependent external potential and $v_{KS} = v_{\text{ext}} + v_H + v_{\text{xc}}$ is the total Kohn–Sham potential. An analogous equation holds for $\bar{\chi}$ defined in Eq. (14.4). The Kohn–Sham response function, which is the response to the total Kohn–Sham potential, is $\chi_0(1,2) = -iG_{KS}(1,2)G_{KS}(2,1^+)$. Besides a prefactor $-i$, it is the analog to L_0 in the BSE, but it is made of Kohn–Sham Green's functions G_{KS} instead of the dressed G in L_0. The functional derivative of the exchange–correlation potential is the *exchange–correlation kernel* f_{xc}. It plays the role of the derivative of the self-energy, Ξ_{xc}, in the BSE. The contribution v_c stems from the variation of the Hartree potential: it is the same as in the BSE.

---

[27] Note that one can also formulate Dyson-like equations for each higher-order correlation function, similar to the BSE; see, for example, [685] for the second-order response.

[28] This is spin-resolved. A sum over spin yields the response of the total density.

In frequency space, Eq. (14.58) reads

$$\chi(\omega) = \chi_0(\omega) + \chi_0(\omega)\left[v_c + f_{xc}(\omega)\right]\chi(\omega), \tag{14.60}$$

which is much simpler than the GW-BSE of Eq. (14.24). The complexity of the problem is contained in the unknown exchange–correlation potential and kernel.

One can reformulate Eq. (14.58) as

$$L_{TDDFT}(1, 2, 1', 2') = L_{KS}(1, 2, 1', 2') + L_{KS}(1, \bar{3}, 1', \bar{3}')\delta(\bar{3}, \bar{3}')\delta(\bar{4}, \bar{4}')$$
$$\times(-i)\left[v_c(\bar{3}, \bar{4}) + f_{xc}(\bar{3}, \bar{4})\right]L_{TDDFT}(\bar{4}, 2, \bar{4}', 2'), \tag{14.61}$$

where $L_{KS}(1, 2, 1', 2') \equiv G_{KS}(1, 2')G_{KS}(2, 1')$ analogous to Eq. (10.7), and where the density–density response function is obtained from $\chi(1, 2) = -iL_{TDDFT}(1, 2, 1^+, 2^+)$. For a static f_{xc}, this allows one to transform the TDDFT linear response equation into a two-particle equation [689], similar to the two-particle equation for the BSE in Sec. 14.6. In a solid, this is not convenient, because one would move from an equation with two space arguments, Eq. (14.58), to a four-point equation. However, it can be helpful in small systems, where only a few transitions mix into a given part of the spectrum. In that case, the basis of transitions is smaller than a basis given by points in space or by reciprocal lattice vectors of a supercell.

The main difference from the BSE is the nature of the interaction: in the case of TDDFT there is no direct interaction with a structure like $W_{vv'}^{cc'}$ in Eq. (14.36). All kernel matrix elements are of dipole–dipole form, like $v_{vc}^{v'c'}$, since f_{xc} appears in Eq. (14.61) in the same way as v_c. Therefore, it is difficult to find TDDFT functionals that describe charge-transfer excitations [690], where the occupied and empty states have very little overlap, which means that the dipole is very weak.

One has to be careful if one wishes to interpret the results of the two-particle TDDFT equation, based on Eq. (14.61), along the lines of Sec. 14.7: the response function χ calculated from Eq. (14.61) is in principle the same as χ calculated from the full BSE, but there is no reason for L_{TDDFT} to equal L. Only the electron–hole part of L, and only its diagonal elements $L_{TDDFT}(1, 2, 1^+, 2^+)$ are given in principle correctly by TDDFT.

Combining methods: TDDFT from the BSE

To solve the two-argument TDDFT linear response Eq. (14.58) is easier than to solve the four-argument BSE, but less is known about the exchange–correlation potential v_{xc} that is needed to build χ_0 and about the kernel f_{xc}. In particular, it is difficult to find approximations for f_{xc} that describe excitonic effects in solids. The development of density functionals is a topic on its own. Here we only discuss one aspect, namely the possibility of deriving approximations for TDDFT from many-body perturbation theory, which is in the spirit of this book. Analogously, one can also use the dynamical mean-field approach of Chs. 16–20 to obtain elements of TDDFT, see, e.g., [691].

One can derive f_{xc} by using Eqs. (11.13) and (11.3) to write

$$P = -iGG\tilde{\Gamma} = -iGG + iGG\frac{\delta\Sigma_{xc}}{\delta v_{cl}} = P_0 + iGG\frac{\delta\Sigma_{xc}}{\delta n}P, \qquad (14.62)$$

where we have used the chain rule for the functional derivative, and omitted arguments of the functions for compactness. With Eqs. (14.60) and (14.2), one obtains

$$f_{xc}(1,2) = \chi_0^{-1}(1,2) - P_0^{-1}(1,2) - iP_0^{-1}(1,\bar{6})G(\bar{6},\bar{3})G(\bar{4},\bar{6})\frac{\delta\Sigma_{xc}(\bar{3},\bar{4})}{\delta n(2)}, \qquad (14.63)$$

where $P_0 = -iGG$. The first contribution to f_{xc} is a difference of inverse response functions. It is responsible for the gap opening from the Kohn–Sham to the quasi-particle band structure. The last term $f_{xc}^{(2)} \equiv -iP_0^{-1}GG(\delta\Sigma_{xc}/\delta n)$ contains the electron–hole interaction in the derivative of the self-energy. This separation helps to make the link with the picture of the BSE, but one should keep in mind that in TDDFT only the sum is meaningful, and there can be strong cancellations, as discussed in Sec. 14.7. It is nevertheless interesting to note that by using $f_{xc}^{(2)}$ the irreducible polarizability Eq. (11.13) can be written as

$$P(1,2) = -iG(1,\bar{3})G(\bar{3},1)^2\tilde{\Gamma}(\bar{3},2) \quad \text{with} \quad {}^2\tilde{\Gamma}(3,2) = (1 - f_{xc}^{(2)}P_0)^{-1}(3,2), \quad (14.64)$$

where the effective two-point kernel ${}^2\tilde{\Gamma}$ is built with the electron–hole part $f_{xc}^{(2)}$ only. This shows that the full three-point function $\tilde{\Gamma}$ is not needed in order to describe the screened interaction, even starting from *interacting* Green's functions. This information is helpful for the discussion of cancellations between vertex corrections in Sec. 15.3.

14.12 Wrap-up

The equilibrium one-body Green's function that is discussed in earlier chapters describes electron addition and removal with respect to the N-body ground state or thermal equilibrium. Higher-order Green's functions give access to the addition and removal of more particles: two electrons, two holes, an electron–hole pair, and so on. The creation of electron–hole pairs is a particularly important process, since it describes density fluctuations, for example plasmons. These are the excitations that are measured in spectroscopies such as optical absorption, electron energy loss, or inelastic X-ray scattering. It contains the physics that yields the density–density response function χ, needed to build the screened Coulomb interaction W. Therefore, it is a key ingredient in many-body perturbation theory.

This chapter was mainly, though not exclusively, devoted to electron–hole excitations, also called neutral excitations. The kind of excitation that is given by L depends on the time ordering, and the other sectors of L can be accessed by choosing a time ordering corresponding to the creation of two electrons, or of two holes.

In principle, electron–hole excitations and χ are also accessible in time-dependent density functional theory. However, there are several reasons to investigate the problem in the framework of many-body perturbation theory.

- It may be difficult to obtain good approximations to TDDFT, especially for absorption spectra of solids. A diagrammatic approach suggests approximations with a clear physical picture, suitable for extended systems.
- Diagrammatic approaches ease the combination of methods, in the spirit of Ch. 21.
- MBPT gives access to the full two-particle correlation function L, including the electron–electron and the hole–hole sector.
- With an improved description of the full L, one obtains an improved self-energy, since Σ is given in terms of L in Eq. (10.6). The electron–hole part of L alone only gives an improved description of W, but it is not sufficient to correct the self-energy beyond $\Sigma = iGW$.

In the diagrammatic framework, L is given by the Bethe–Salpeter equation. The BSE is an in-principle-exact Dyson-like equation for the two-particle correlation function. It is a linear response equation, because L is the linear variation of G with respect to a non-local source, a generalized non-local potential. It depends on four space and spin arguments, and on three frequencies.

The kernel of the electron–hole BSE expresses the electron–hole interaction. It contains the bare Coulomb interaction and the functional derivative of Σ_{xc} with respect to the Green's function G. The bare Coulomb contribution is the variation of the Hartree potential with respect to the density. When the exchange–correlation contribution is neglected, one has the random phase approximation.

The variation of Σ_{xc} yields the corrections to the RPA. Since the exact Σ_{xc} is not known, and since this term is responsible for the complexity of the BSE, in practice approximations are needed. A widely used approximation is obtained by deriving the kernel from the GWA, and by neglecting its frequency dependence and the variation $\delta W/\delta G$. This allows one to transform the BSE into an effective two-particle Schrödinger equation. A physically intuitive picture emerges, with an effective electron–hole interaction composed of two parts: a bare electron–hole exchange interaction and a screened attractive direct electron–hole interaction. The attractive electron–hole interaction is responsible for excitonic effects, like the formation of bound excitons that appear in optical spectra as sharp peaks in the quasi-particle gap. Models such as the Frenkel or Wannier–Mott models are contained in this more general first-principles description.

Solving the Bethe–Salpeter two-particle equation allows one to calculate spectra, and to analyze features of the, often strongly correlated, electron–hole pairs, like the spatial distribution of electrons and holes, or the mixing of transitions between quasi-particle states. Some phenomena, such as double excitations in molecules, or multiple plasmon satellites that are observed in inelastic X-ray scattering spectra, cannot be described by the BSE using the static approximation for W. Some ways to go beyond the standard approximations are outlined in this chapter. However, very often even in the simplest approximation, optical, energy loss or inelastic X-ray scattering spectra are found in good agreement with experiment over a wide range of energy, and for a wide range of materials.

Solving the BSE has become a standard instrument in the toolbox of many-body calculations. Insight from the BSE can be used to improve other approaches, in particular

TDDFT. To calculate the full L beyond the RPA is equivalent to calculating vertex corrections; therefore, some of the points made in this chapter lead over to the next chapter that deals with vertex corrections beyond the GW approximation.

SELECT FURTHER READING

Knox, R. S. *Theory of Excitons* (Academic Press, New York, 1963). An old book on optical properties and excitons, that makes it easy to enter the vast field of excitonic effects.

Yu, P. P. and Cardona, M. *Fundamentals of Semiconductors: Physics and Materials Properties*, 4th edn. Graduate Texts in Physics (Springer-Verlag, Berlin, 2010). Great introduction to the physics of semiconductors, with many more modern facts. A textbook for beginners and those who need detailed information.

Hanke, W., "Dielectric theory of elementary excitations in crystals," *Adv. Phys.* **27**, 287 (1978). Rigorous view on microscopic theory of elementary excitations in solids combining Maxwell's equations and self-consistent field approximation. Includes a chapter on electron–phonon interaction.

Strinati, G., "Application of the Green's function method to the study of the optical-properties of semiconductors," *Rivista del Nuovo Cimento* **11**, 1, 1988. Pedagogical review of the theoretical framework underlying today's Bethe–Salpeter calculations. Derivation of the main equations and link to spectroscopy.

Onida, G., Reining, L., and Rubio, A., "Electronic excitations: density-functional versus many-body Greens-function approaches," *Rev. Mod. Phys.* **74**, 601, 2002. Review of *ab initio* calculations of electronic excitations with accent on optical properties and a comparison between Bethe–Salpeter and TDDFT.

Rohlfing, M. and Louie, S. G., "Electron–hole excitations and optical spectra from first principles," *Phys. Rev. B* **62**, 4927, 2000. Useful overview for BSE calculations in practice, and comparison of some results.

The website www.electronicstructure.org contains more information, and links to tutorials and numerical exercises that accompany this book.

Exercises

14.1 Study the following self-energy:

$$\Sigma = v_{\text{xc}} + \Delta \sum_c |c\rangle\langle c|. \tag{14.65}$$

It represents a "scissor" correction (see Eq. (13.1)) to the KS potential v_{xc}. With the hypothesis that $|c>$ are eigenstates of both the Kohn–Sham and the quasi-particle hamiltonian, transform this expression into

$$\Sigma(\mathbf{r}, \mathbf{r}', t, t') = \delta(t' - t^+)(\delta(\mathbf{r} - \mathbf{r}')v_{\text{xc}}(\mathbf{r}) + \Delta(\delta(\mathbf{r} - \mathbf{r}') + iG(\mathbf{r}, \mathbf{r}', t, t'))). \tag{14.66}$$

Calculate the Bethe–Salpeter equation with a kernel derived from this self-energy, and show that quasi-particle corrections and excitonic effects stemming from this non-local part cancel exactly.

14.2 Calculate the frequency Fourier transform of the full kernel Ξ of the Bethe–Salpeter equation using the definition of Eq. (14.17) and Hartree, Hartree–Fock, and GW approximations for the self-energy. Derive Eq. (14.24).

14.3 Derive the TDDFT linear density–density response Eq. (14.58) from the BSE by using the approximation $\Sigma_{xc} \approx v_{xc}$ in Eq. (14.7) and $G \approx G_{KS}$ in L_0. Note that the result is in principle exact, because the errors introduced by the two approximations cancel. With the help of Sec. 14.11, discuss why this is not fortuitous.

14.4 Derive from Eq. (14.12) the Dyson-like linear response Eq. (14.2) that links the irreducible polarizability P and the reducible polarizability χ for the simultaneous propagation of an electron and a hole. Show that this simplification cannot be done to link directly χ and P_0 starting from Eq. (14.6). Try to carry out the analogous job for the simultaneous propagation of two holes or two electrons.

14.5 Derive Eq. (14.11) for the four-point vertex $^4\tilde{\Gamma}$ from the BSE of Eq. (14.6) and the definition Eq. (8.25).

14.6 Calculate the resonant matrix elements of W in Eq. (14.36) in the HEG. Compare the results with and without screening.

14.7 Discuss why the kernel Ξ of the Bethe–Salpeter equation has only two possible spin contributions, given in Eq. (14.14). You can use a graphical approach: since it is a derivative, Ξ is obtained by removing a Green's-function line from the self-energy diagrams. Note that each Green's-function line conserves spin since G is spin-diagonal, and that the interaction lines conserve spin at each vertex, but spins at the two opposite vertices of an interaction line are independent.

14.8 Calculate and discuss transition energies, coefficients, and the spectrum of a two-level model without the Tamm–Dancoff approximation. Compare with the result obtained in the TDA, Eq. (14.46).

14.9 With the definition Eq. (14.3) of the microscopic part $\bar{v}_c$ of the Coulomb interaction, derive the expression Eq. (14.5) for the macroscopic dielectric function from Eq. (14.1) and the Dyson Eq. (14.4). Also show that

$$\bar{\chi}_{GG}(\mathbf{q}, \omega) = \chi_{GG}(\mathbf{q}, \omega) - \chi_{GG}(\mathbf{q}, \omega) v_c(\mathbf{Q}) \bar{\chi}_{GG}(\mathbf{q}, \omega). \qquad (14.67)$$

Finally, derive Eq. (14.12).

15

Beyond the GW approximation

Summary

In this chapter, shortcomings of the GW approximation are analyzed, and different ways are proposed to go beyond. The approaches fall into three broad classes: first, one can iterate Hedin's equations that link the self-energy, the polarizability, the screened Coulomb interaction, and the vertex function with the Dyson equation for the one-body Green's function. Second, one can use other Dyson equations like in the T-matrix approach, and third, one can make an ansatz for the one-body Green's function inspired by physical considerations or insight from models, in particular the idea of electron–boson coupling that leads to the cumulant expansion. Applications are given for illustration. Some more ideas and links to subsequent chapters provide an outlook.

The GW approximation is a widely used approach in computational electronic structure. However, the previous chapters show some severe limitations. One example is systems with low-energy excitations, as illustrated by the Hubbard dimer in the dissociation limit, where bonding and antibonding orbitals become degenerate. This can be seen in Sec. 11.7. Another example is satellite series, for example in the photoemission spectrum of silicon in Sec. 13.1. Moreover, even in a situation where the approximation is well justified, it induces some error, though maybe small, and one eventually wants to do better.

The full self-energy functional is in principle known: it is the Luttinger–Ward sum of skeleton diagrams derived from Eq. (11.35). However, this is an infinite sum, and no closed form of the result is known. Hence, the task is not to *invent* new diagrams: it is to make a *choice* of diagrams or combinations of diagrams, which are evaluated fully or in an approximate way. This is the framework of the present chapter. In some cases we will add diagrams to the GWA ones, which gives rise to *vertex corrections*; in other cases some of the GWA diagrams are neglected while different ones are taken into account. An example is the T-matrix approximation, which sums ladder diagrams instead of bubbles. In general one tries to combine this with the GWA, especially in extended systems where screening is important. Therefore the GWA plays a prominent role, and the whole chapter is called *beyond GW*.

The chapter contains advanced material, that is in part exploratory and refers to current research. It is intended to make links to communities outside the GW world, and to other chapters. It should also serve as a pool of thoughts that may inspire new ideas. We will indicate for each section what is established and what is a snapshot of the moment the book is written, and we will point out what can be read as an introduction, and what can be skipped if one doesn't want to go deeper into a given direction.

Section 15.1 exposes some fundamental ideas, and it can be read without going into details of results of this or that vertex correction. It analyzes *why* one should go beyond GW, in a simple physical picture without complex mathematics. Then we move on to the *how*: the first, most straightforward idea outlined in Sec. 15.2 is to iterate Hedin's equations, Eqs. (11.11)–(11.15). Some interesting things can be learned, although a vertex correction that would lead to a generally valid improvement is not established; this is discussed in Sec. 15.3.

In Sec. 15.4 we put together some insight gained previously. The *T-matrix approximation* is introduced by generalizing the concept of an effective interaction, and by generalizing the idea of a Dyson equation. It is designed to work in the low-density limit where the GWA, which is based on the importance of screening, fails. It is useful to read this section in order to gain some more insight about what is missing in the GWA, and how one might remedy the problems. The T-matrix and GWA views can be combined, as outlined in Sec. 15.5. There are many possibilities, and the section sketches some ideas of how one may play with the formula in order to design new approximations. It is meant to touch upon some aspects without being complete or conclusive. The section can be skipped if one is not particularly interested in the topic. Applications of T-matrix and related approaches can be found in the literature, and some examples are given in Sec. 15.6. The prototype example is bulk nickel and its 6 eV photoemission satellite. It should, however, be stressed that many fewer results for real materials are available than in the case of the GWA, and the situation is not as settled.

In all those sections, Dyson equations are the tool of choice to include an infinite number of diagrams in the Green's function. Alternatively, one can use *cumulants*: these appear as the exponent in an exponential expression for the Green's function. Like the self-energy, a low-order approximation to the cumulant creates infinitely many diagrams in the Green's function. This is explained in Sec. 15.7. The section on cumulants is recommended to all readers, since it shines new light on the GWA, it opens a different perspective on how to approach the many-body problem, and it makes a link to many places in the book, including dynamical mean-field theory in Chs. 16–21.

The chapter also contains a section featuring a point of view that is quite prominent in the design of *density* functionals: this is the idea to use exact constraints for improving functionals. Section 15.8 explores this idea in the context of functionals of Green's functions, suggests some of the constraints that might be used, and gives examples. It is a section that points to forefront research, and it should be understood as a place where ideas can be found, rather than certainties.

A wrap-up is given in the last section. There we also make a point on the ensemble of chapters on many-body perturbation theory that ends here, and we look forward to the following parts.

15.1 The need to go beyond GW: analysis and observations

In many places in this book it is pointed out that screening is the essential ingredient of the GWA, see, e.g., Sec. 11.3. This statement implies the picture of a quasi-particle, which can be an electron or hole, surrounded by electron–hole pairs. Although the GWA is not *per se* a quasi-particle approximation,[1] it is still based on the idea of a quasi-particle in some medium.

Figure 15.1 illustrates the case of photoemission; electron addition can be examined in an analogous way (Ex. 15.1). In photoemission an electron is removed from the system. This induces excitations, given by the creation of electron–hole pairs.[2] The full many-body problem is symbolized by the three-body picture in panel (a) of Fig. 15.1: *all* particles in the system are equivalent, and all are correlated, as indicated by the ellipse.

The energy difference of the system for electron removal is

$$\Delta E_s = E_N - E_{N-1,s}, \tag{15.1}$$

since the system is left with $(N-1)$ electrons in a many-body state labeled s, which can be the ground state or some excited state. A photoemission spectrum displays all accessible ΔE_s.

Let us start by adding and subtracting from Eq. (15.1) the energy $E_{N-1,t}$ of a state t of the $(N-1)$-electron system:

$$\Delta E_s = E_N - E_{N-1,t} + E_{N-1,t} - E_{N-1,s} \approx \varepsilon_t^{QP} - \omega_{st}. \tag{15.2}$$

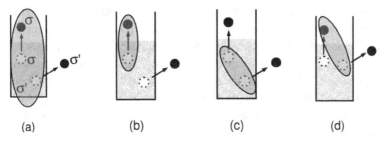

(a)	(b)	(c)	(d)

Figure 15.1. Schematic picture of electron removal, presented as a three-body problem: the removal of an electron from the system leaves a hole, which excites the system and leads to the creation of electron–hole pairs. The ellipse in panel (a) indicates that all particles are correlated, and the spin indices are a reminder of the fact that the excitation conserves the spin. Many current approximations are based on a picture where a quasi-particle is dressed by pairs of particles. This splits the many-body problem into one effective particle and a correlated two-body part. The methods considered in this chapter differ in the choice of how to split the problem. Panel (b) gives the picture that corresponds to the GWA: it treats a hole screened by electron–hole pairs. Note that the electron and hole in this pair must have the same spin. Panel (c) shows the particle–particle T-matrix approximation. It correlates the two holes. Panel (d) is the electron–hole T-matrix. The main difference from the GWA is the fact that here electron and hole can have opposite spin.

---

[1] Remember, for example, the presence of satellites in the spectral functions of Fig. 13.15.
[2] Here we concentrate on the intrinsic one-body spectral function and neglect correlation between the hole and the outgoing electron; see subsection "Link to experiment" in Sec. 5.7 for a short discussion.

If there is a relatively sharp quasi-particle in the system that can be associated[3] with t, one can approximate this relation as a quasi-particle excitation with energy ε_t^{QP} plus an excitation with energy ω_{st} of the $(N-1)$-electron system that does not change particle number; we call such an electron–hole pair excitation a *neutral excitation*. This corresponds to the last expression of Eq. (15.2), and to the picture in panel (b) of Fig. 15.1: the system is described by a hole in the presence of electron–hole pairs. The GWA is based on this approximate picture of a quasi-particle screened by neutral excitations. Moreover, it involves several approximations that can be understood by looking at the equation and the picture.

(i) The neutral excitations in the GWA, contained in W, are those of the N-electron system in its ground state. Instead, Eq. (15.2) shows that it should be the excitations of the $(N-1)$-electron system in state t. This is the origin of the self-screening problem discussed in Sec. 11.7.

(ii) The GWA uses the RPA to calculate the neutral excitations. It therefore misses features such as double excitations.

(iii) The approximate excitation energy $\varepsilon^{QP} - \omega_{st}$ appears in the GWA as a peak in the imaginary part of the self-energy (see Eq. (11.29)), instead of the spectral function of the one-body Green's function. This leads to shifts of peaks and a spurious "plasmaron," as discussed in Sec. 15.7.

Even if these approximations could be avoided, eventually the picture of a screened quasi-particle breaks down. One can see this breakdown, for example, in the Hubbard dimer in Sec. 11.7, where the quasi-particle peak loses weight toward satellites with increasing interaction strength U/t and vanishing bonding–antibonding gap. However, the GWA picture in panel (b) of Fig. 15.1 is not the only possible choice: in panel (c) the excited electron is dressed by a correlated hole–hole pair. This corresponds to inserting $-E_{N+1,t} + E_{N+1,t}$ in Eq. (15.1), as discussed in Ex. 15.8. This picture is appropriate in situations dominated by hole–hole interaction, like the 6 eV satellite in nickel discussed in Secs. 13.4, 15.4, and 20.4. In panel (d) the role of the two holes is exchanged with respect to the GWA. This allows the electron and the hole in the pair to have opposite spin, because the removal of a spin-down electron can excite both spin-up and spin-down pairs. Such a spin-flip contribution is not contained in the GWA, because W does not include excitations with spin-flip. It is important, for example, in determining magnetic phase diagrams [692] or the electron lifetimes in ferromagnets [693], like the experimental data shown in Fig. 2.7. The T-matrix approximations in Sec. 15.4 correspond to the two right-side panels in Fig. 15.1.

The following sections are dedicated to ways to overcome the shortcomings of the GWA. Everything is cast into a formal framework, but one may recognize the different pieces of physics that emerge from the picture discussed here.

---

[3] Note that all total energy differences are real. Quasi-particle broadening comes about because the energy spectrum is dense. Therefore, as explained in Ch. 7, a *group* of many-body states contributes. We do not consider this fact here, in order to keep the analysis simple.

15.2 Iterating Hedin's equations

Formally, the GW approximation to the self-energy $\Sigma = iGW\tilde{\Gamma}$ in Eq. (11.11) is obtained by setting the irreducible vertex $\tilde{\Gamma}$ to unity. One can understand the GWA as a first step in iterating the pentagon Fig. 11.2, starting with $\Sigma = 0$. This leads to $\tilde{\Gamma} = 1$ in Σ and in the polarizability P, which corresponds to the RPA. Iterating the pentagon further seems the obvious thing to do.

Vertex corrections as functionals of a Green's function

Let us iterate the Dyson-like Eq. (11.14) for the vertex $\tilde{\Gamma}$, which is directly related to the Bethe–Salpeter Eq. (14.11). Suppose that Σ_{xc} and the Green's function G are obtained in the iteration step n. Then the $(n+1)$-order expression for the vertex is given by[4]

$$\tilde{\Gamma}^{(n+1)}(1,2;3) = \delta(1,2)\delta(1,3) + \frac{\delta\Sigma_{xc}^{(n)}(1,2)}{\delta G^{(n)}(\bar{4},\bar{5})}G^{(n)}\left(\bar{4},\bar{6}\right)G^{(n)}\left(\bar{7},\bar{5}\right)\tilde{\Gamma}^{(n+1)}\left(\bar{6},\bar{7};3\right). \quad (15.3)$$

Starting from $\Sigma_{xc}^{(1)}(12) = iG^{(1)}W^{(1)}$, and neglecting the functional derivative of $W^{(1)}$ with respect to $G^{(1)}$, the result is $\tilde{\Gamma}^{(2)} = 1 + iW^{(1)}G^{(1)}G^{(1)}\tilde{\Gamma}^{(2)}$. This is a key expression in Ch. 14: it corresponds to the approximate Bethe–Salpeter equation for the two-particle correlation function discussed in Sec. 14.5 with the direct electron–hole interaction kernel given by the screened Coulomb interaction W.

The iteration of a Dyson-like equation can be performed in different ways. An alternative expression is derived in Ex. 15.2. Suppose one can calculate the Green's function G_{vl} of some local potential v_{loc} that is a functional of v_{cl}, as for example the Kohn–Sham potential. This allows one to use the chain rule, which yields

$$\tilde{\Gamma}^{(n+1)}(1,2;3) = \delta(1,2)\delta(1,3) + \frac{\delta\Sigma_{xc}^{(n)}(1,2)}{\delta G_{vl}(\bar{4},\bar{5})}G_{vl}(\bar{4},\bar{6})G_{vl}(\bar{6},\bar{5})\frac{\delta v_{loc}(\bar{6})}{\delta v_{cl}(3)}, \quad (15.4)$$

where v_{cl} is the total classical potential introduced in Sec. 11.1. Note that this is no longer a Dyson-like equation. If one chooses $v_{loc} = v_{cl}$, the Green's function G_{vl} is the Hartree Green's function G_H. In that case, with $\Sigma_{xc}^{(1)} = iG_H W^{(1)}$ and neglecting again the functional derivative of W, Eq. (15.4) yields [627]

$$\tilde{\Gamma}^{(2)}(1,2;3) = \delta(1,2)\delta(1,3) + iW^{(1)}(1^+,2)G_H(1,3)G_H(3,2). \quad (15.5)$$

This corresponds to a first-order expansion in W of Eq. (15.3). As discussed in Sec. 14.7, such a first-order correction cannot yield a polarizability with bound excitons in a solid, contrary to the Bethe–Salpeter Eq. (14.6). However, used in the self-energy $\Sigma_{xc} = iGW\tilde{\Gamma}$, it still adds an important piece of physics to the GWA: it leads to an effective interaction $W\tilde{\Gamma}$ that takes the fermionic nature of the charges into account, for example it contains a

---

[4] A similar equation holds for the *reducible* vertex Γ given by Eq. (11.1). In this case the Hartree potential must be added to Σ_{xc}.

correction to the self-screening problem explained in Sec. 11.7. This is more easily seen in the following subsection.

Vertex corrections from density functionals

It is in principle possible to continue iterating the equations, but the results become clumsy very rapidly, without guarantee of systematic improvement. Another idea is to optimize the starting point. One can, for example, start with a local potential, $\Sigma_{xc}^{(0)}(1,2) = \delta(1,2)v_{loc}(1)$, typically a Kohn–Sham potential $v_{loc} = v_{xc}$ [413, 414, 421, 694, 695]. As shown in Ex. 15.9, in this case Eq. (15.3) leads to

$$\tilde{\Gamma}^{(1)}(1,2;3) = \delta(1,2)(1 - f_{xc}\chi_{KS})^{-1}(1,3), \tag{15.6}$$

where $f_{xc} = \delta v_{xc}/\delta n$ is the exchange–correlation kernel of TDDFT, Eq. (14.59), and χ_{KS} is the KS independent-particle polarizability.[5] Note that this approximation to the vertex $\tilde{\Gamma}$ is local in the first couple of arguments, because v_{xc} is a local potential. When this approximate vertex is used in the polarizability and in the self-energy, the result is

$$\Sigma_{xc}^{(1)}(1,2) = iG_{KS}(1,2)\tilde{W}(1^+,2), \tag{15.7}$$

with the KS Green's function G_{KS}, and the effective test-charge–test-electron interaction

$$\tilde{W}(1,2) = \left[1 - (v_c + f_{xc})\chi_{KS}\right]^{-1}(1,\bar{3})v_c(\bar{3},2) = v_c(1,2) + [(v_c + f_{xc})\chi v_c](1,2). \tag{15.8}$$

The nature of this interaction is clear in the expression on the right side, in terms of the density–density correlation function χ: reading the formula from the right to the left, the Coulomb interaction v_c creates a potential as if it were due to an external charge. This external potential gives rise to an induced charge $n_{ind} = \chi v_c$, which, in turn, creates an induced potential. Its Hartree component is given by $v_c n_{ind} = v_c \chi v_c$, whereas $f_{xc}\chi v_c$ adds an induced exchange–correlation potential [696]. This is physically meaningful in a system of fermions; see the related discussion of Hedin's equations in Sec. 11.1. Exercise 15.10 allows you to understand how this approximation cures the self-screening problem.

15.3 Effects of vertex corrections

At present it would be premature to draw a general conclusion about vertex corrections. This section should be taken as a snapshot of current knowledge. Let us focus on a few important aspects.

Vertex corrections in W

It is reasonable to think that a better W also yields better GW spectral functions. This is suggested by the fact that measured photoemission spectra directly exhibit loss structures,

---

[5] It is understood that the time- or contour-ordered version is used.

for example plasmon satellites. Since W is dynamically screened by the inverse dielectric function, its imaginary part is proportional to the loss function. This suggests a "best W" strategy, which has been supported for example by results for the HEG [386] or for 15 bulk solids ranging from small-gap semiconductors to large-gap insulators [697]. Here "best" means that $-\text{Im}\,\epsilon^{-1}$ is closest to the measurable loss spectrum in the energy range of the plasmon energy, and for small values of the momentum transfer.

The screened interaction W beyond the RPA is the topic of Ch. 14. It shows that excitonic effects, which are created by vertex corrections derived from a GW self-energy as discussed in the previous section, enhance the screening, and lead in general to a dielectric function in better agreement with experiment than a dielectric function calculated from independent quasi-particle transitions. This does not necessarily mean that one always has to go beyond the RPA. There are cancellations between self-consistency and vertex contributions; this is discussed in the last of the following subsections. The cancellations are often not sufficient to yield good optical spectra in the RPA, but W is determined by the *inverse* dielectric function and the spectrum of plasmons. As explained in Sec. 14.11, these can in general be approximated more easily than optical spectra.

Vertex corrections in the self-energy

Moving on to the effect of $\tilde{\Gamma}$ in $\Sigma_{xc} = iGW\tilde{\Gamma}$, let us concentrate on three aspects: effects of the vertex on the shift of single states, the non-locality of the vertex, and vertex corrections to the satellite spectrum.

Ionization potentials. Often one is interested in energy differences, where errors due to the neglect of vertex corrections partially cancel, as discussed in the next subsection. Instead, for the self-energy shift of single states, or for the comparison of shifts of very different states, vertex corrections can be more important. A collection of results for ionization potentials and the position of d states in semiconductors and insulators can be found in [698], where a vertex of the form of Eq. (15.5) was evaluated approximately with a hybrid functional Green's function and screening from PBE. The effect of the vertex is a downshift of localized states and an upshift of delocalized states, leading to results in better agreement with experiment than the GWA, in particular concerning ionization potentials.

Non-locality of the vertex. As discussed after Eq. (15.5), the vertex corrections in Σ create an effective interaction $W\tilde{\Gamma}$ that is more than the measurable interaction W between classical charges. This effective interaction is in principle a function of three arguments, because it contains contributions of non-local exchange.

Only a part of $\tilde{\Gamma}$, that is a function of two arguments, contributes to W. This is an exact result shown in Sec. 14.11.[6] The same is not true for the vertex in the self-energy, which has to create the full non-local $W(1, \bar{3})\tilde{\Gamma}(4, 2; \bar{3})$. Therefore, one has to be careful when discussing results for spectra that have been obtained with an approximate vertex in Σ. Only when the self-energy is integrated, for example in the total energy, do some parts not

---

[6] This is linked to the fact that one can calculate W in principle exactly within TDDFT.

contribute and it can be sufficient to consider a vertex function that depends only on two arguments. This is worked out in Ex. 15.11.

As an example of the fact that, when one calculates spectra, approximations to $\tilde{\Gamma}$ are more critical in the self-energy than in W, let us consider the simplest TDDFT exchange–correlation kernel, f_{xc}^{LDA} of adiabatic LDA (TDLDA). It is static and local, $f_{xc}^{LDA}(\mathbf{r}, \mathbf{r}') \propto \delta(\mathbf{r} - \mathbf{r}')$, and therefore constant in reciprocal space. Often TDLDA yields a test charge–test charge W with a good plasmon spectrum [682]. Its use to calculate the effective test charge–test electron interaction $\tilde{W}$ in Eq. (15.8), instead, is dangerous: since f_{xc}^{LDA} is negative and constant, $(v_c(q) + f_{xc}^{LDA})$ becomes negative for large q, leading to a negative effective interaction and, by consequence, to negative spectral functions[7] (see Ex. 15.5), unphysical quasi-particle dispersion and work function [386, 699].

Effects of vertex corrections on the spectral function. One of the clear shortcomings of the GWA is its failure to describe plasmon satellite series; see the example of silicon in Fig. 13.4. This is a good observable to test vertex corrections in Σ, since, as shown, e.g., in Fig. 5.6, W alone yields a good description of plasmons, often already on an RPA level. Figure 15.2 shows calculations [426] of the spectral function for the electron–boson coupling model, Eq. (11.46). The parameters of the model correspond to the strong coupling regime. The exact solution in the upper panel shows a strong transfer of weight from the quasi-particle to the satellite series. As discussed in Sec. 11.7, G_0W_0 replaces this series with only one average satellite. Self-consistency in the Green's function on the GW level leads to improvement. In the case of the model studied here, the appearance of multiple satellites is clearly seen.[8] This can be understood by looking at the diagrams in the third line of Fig. 15.7. Still, the position of the peaks and the distribution of weight disagree with the exact result. First-order vertex corrections calculated according to Eq. (15.5) lead to moderate improvement. The cumulant approach in Sec. 15.7 is a more efficient way to solve the plasmon satellite problem.

Cancellations of vertex corrections in W and Σ

The GWA is justified when vertex corrections are small, but also when vertex corrections in Σ and W approximately cancel. Let us look at some trends. We start with a vertex derived from a local potential, e.g., v_{xc}, and, as suggested in Ex. 15.6, expand $\tilde{W}$ in Eq. (15.8) around W^{RPA} to first order in $f_{xc}\chi_{KS}/\epsilon_{RPA}$, with ϵ_{RPA} the RPA dielectric function. The result is

$$\tilde{W} \approx W_{RPA} + f_{xc}\chi_{KS}\epsilon_{RPA}^{-1}\left\{1 + v_c\chi_{KS}\epsilon_{RPA}^{-1}\right\}v_c = W_{RPA}\left\{1 + f_{xc}\chi_{KS}\epsilon_{RPA}^{-1}\right\}. \qquad (15.9)$$

All quantities are in principle frequency-dependent matrices in reciprocal space $(\mathbf{G}, \mathbf{G}')$, but for the discussion we limit ourselves to vanishing frequency and the macroscopic element $\mathbf{G} = \mathbf{G}' = 0$.

---

[7] See Sec. 15.8 for more discussions about the problem of negative spectral functions.

[8] The effect is instead not strong enough in the calculation of the HEG in Fig. 11.8.

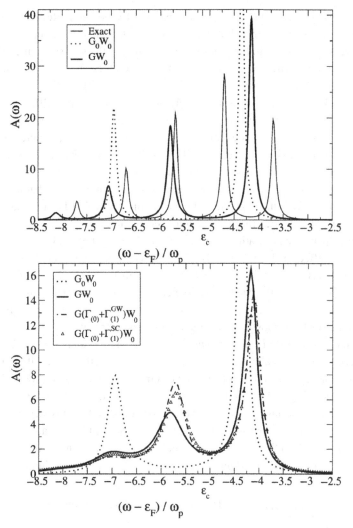

Figure 15.2. Spectral function of the electron–boson coupling model, Eq. (11.46). The parameters correspond to a strong coupling regime, with a density of about $r_s = 4$. (From [426].) Upper panel: comparison of the exact and GWA results. Thin continuous line, exact solution. Dotted line, $G_0 W_0$ with G_0 the zero-coupling solution. Thick continuous line, GW_0. Note that W_0 is fixed by the parameters of the model. Lower panel: dotted and thick continuous lines are the same as in the upper panel. The dot-dashed curve contains effects of vertex corrections calculated from Eq. (15.5). Triangles, the same but evaluated with a self-consistent Green's function. Note that the broadening in the lower panel is larger than that used in the upper panel. Compared with the effect of self-consistency, the vertex corrections only lead to moderate improvement.

Looking at the expression in the middle, the first term of the correction to the RPA stems from the vertex in the self-energy. Since the response functions and f_{xc} are negative, this term is positive, which means that the screened interaction is increased. The second term results from the vertex in the response function; it is negative. Roughly speaking, the vertex in W contains an electron–hole attraction, which increases screening, whereas the vertex in Σ reduces screening, for example by eliminating the self-screening. To convince yourself, look at the electron removal spectrum of a one-electron system, like the Hubbard dimer in Sec. 11.7: Ex. 15.10 shows that in this case using an RPA W in GW is better than using the exact W. For strong screening, with ϵ_{RPA} much larger than 10, $-v_c\chi_{KS} \approx \epsilon_{RPA}$ and $v_c\chi_{KS}\epsilon_{RPA}^{-1} \approx -1$, which means that the cancellation should be almost perfect. This is confirmed by various results. For example, cancellations have been predicted in [413], and found, e.g., in the bandwidth of metals [414]. Results for the HEG in Fig. 11.5 calculated with and without vertex corrections derived from a local potential show only marginal differences. Calculations of quasi-particle energies in silicon show [695] that the top valence state is moved away from the Fermi level by calculating W in TDLDA instead of RPA, and towards the Fermi level when the corresponding correction is made also in the self-energy. There is hence partial cancellation, but the effect of the vertex in W is about four times smaller than the effect of $\tilde{\Gamma}$ in $\Sigma = iGW\tilde{\Gamma}$, which is the trend expected from our discussion. The net effect of the vertex correction, shown in Fig. 13.2, is therefore an upshift of the whole spectrum [290]. Another simple example for incomplete cancellations is the one-electron case in Ex. 15.10. Cancellations are more effective for energy *differences*, like the bandgap, as has been shown, for example, in silicon [290, 695].

In the case of a vertex derived from a *non-local* self-energy, such as Eq. (15.3) or (15.5), the simple arguments used above do not hold anymore. The effect of a fully non-local vertex in the self-energy is studied in [700] for electron lifetimes. A non-local $\tilde{\Gamma}$ is derived from Eq. (15.4) starting with a Kohn–Sham v_{xc}. Results for LiF, using Kohn–Sham Green's functions, show an important effect of the vertex correction in W, consistent with the observations in the absorption spectrum of LiF discussed in Sec. 14.9. Instead, the effect on the electron lifetimes of $\tilde{\Gamma}$ in the self-energy is small, and the vertex in W dominates. This is a case where the error canceling argument in favor of the RPA does not hold.[9]

Cancellation of vertex and self-consistency corrections

In Sec. 12.4 different levels of self-consistency are introduced in the framework of the GWA. Results such as those for the homogeneous electron gas in Sec. 11.6 indicate that a fully self-consistent GWA calculation is not necessarily the best choice to calculate spectra. This implies that vertex and self-consistency contributions beyond G_0W_0 cancel to some extent. Let us first examine W. If the starting point of the calculation is an independent-particle one-body Green's function, calculated in the Kohn–Sham framework in an approximation such as the LDA, the bandgap is too small. The previous chapters show that self-energy corrections tend to open the gap between occupied and empty states, and to localize electrons. This leads to a weaker screening, and to a shift of the plasmon energies to higher frequency. As shown in Sec. 14.10, the vertex correction in the polarizability, or equivalently, in W, brings the plasmon back to lower frequencies, which leads to enhanced

---

[9] Note that hole lifetimes have not been studied in [700]. In that case cancellations could be more efficient, as one may understand from Ex. 15.10.

screening and a result closer to the RPA.[10] This was recognized early on [410], suggesting that self-consistency should be combined with some vertex corrections. Another indication is that the fully self-consistent RPA polarizability does not obey the f-sum rule, as pointed out in Sec. 11.3. Evidence is also given by model calculations of the bandgap for a quasi-one-dimensional semiconducting wire [480], as well as in calculations for bandgaps of several materials ranging from small-gap semiconductors to large-gap insulators [697]. The need for vertex corrections in W in the case of self-consistent GWA calculations can moreover be seen from the gap overestimate in the case of d- or f-electron materials, discussed in Sec. 13.4.

When the effect of $\tilde{\Gamma}$ in the self-energy $iGW\tilde{\Gamma}$ is considered, there are no such simple arguments, and results are diverse. For example, in [701] the first-order vertex Eq. (15.5) was used together with an approximate self-consistency to calculate spectral functions of the homogeneous electron gas. The vertex correction reduces the bandwidth, counterbalancing the effect of self-consistency in the GWA and leading to improved results (see Sec. 11.6). In [702] first-order effects of self-consistency and vertex corrections were examined for silicon and diamond. As expected, they cancel to a large extent in the polarizability. Nevertheless the self-energy changes, and bandgaps increase by as much as 0.7 eV and 0.4 eV for diamond and silicon, respectively. It should be noted, however, that these are only trends: there is no absolute value for a self-energy correction, since the amount of a correction depends by definition on the reference point.

15.4 The T-matrix and related approximations

Perturbation theory is a useful tool when the low orders are suffiently accurate. When this is not sufficient, it may be better to change route. Instead of pushing corrections beyond the GWA to an unrealistic level of complexity, let us therefore examine such an alternative route. Stated in the language of diagrams, instead of summing bubbles and adding some corrections, we are going to sum another class of diagrams. We do, however, keep two successful ideas: effective interactions and Dyson equations. This leads to the T-matrix approximation [47, 133, 299, 703], which is derived in Sec. 10.5. Here we give a short summary of that derivation, before discussing the resulting equations in detail.

The TMA is obtained by starting from the functional differential equation for the self-energy, Eq. (10.36), which can be written

$$\Sigma(1,2) = [v_H + \Sigma_x](12) + iv_c(1^+,\bar{4})G(1,\bar{3})\frac{\delta\Sigma(\bar{3},2)}{\delta G(\bar{6}\bar{5})}L(\bar{6}\bar{4};\bar{5}\bar{4}^+), \qquad (15.10)$$

where L is the two-particle correlation function, Eq. (5.134). The idea is now to work with a generalized effective interaction T, which depends on four arguments in space, spin, and time, instead of the screened interaction W between classical charges that appears in the

---

[10] This is linked to the fact that TDDFT and the Bethe–Salpeter approach give in principle the same result. See the related discussion in Sec. 14.11.

GWA. This effective interaction T is used in the general ansatz Eq. (10.43) for the self-energy, which is expressed as $\Sigma = iGT$, where T is integrated with G. From this, an approximate Dyson equation for T can be obtained. We will examine this in the following.

Dyson equations for the effective interaction

We now insert the ansatz $\Sigma = iGT$ into Eq. (15.10). The T-matrix approximation consists of two hypothesis:

- $\delta T/\delta G \simeq 0$. This truncates the chain of higher-order correlation, similar to the neglect of $\delta W/\delta G$ in the Bethe–Salpeter context; see Sec. 14.5.
- Screening of the new effective interaction is neglected in Eq. (15.10) by setting $L \simeq L_0$, defined in Eq. (10.7). Compared with the GWA, one supposes $W - v_c$, rather than W, to be small.

This leads to Eq. (10.45), or, with $T = T_1 + T_2$, to the two equations

$$iG(\bar{4},\bar{3})T_1(1,\bar{3};2,\bar{4}) = v_H(1)\delta(1,2)$$
$$- G(1,\bar{3})T_1(\bar{3},\bar{5};2,\bar{6})G(\bar{6},\bar{7}^+)G(\bar{7},\bar{5})v_c(1^+,\bar{7}) \quad (15.11)$$
$$iG(\bar{4},\bar{3})T_2(1,\bar{3};2,\bar{4}) = \Sigma_x(1,2)$$
$$- G(1,\bar{3})T_2(\bar{3},\bar{5};2,\bar{6})G(\bar{6},\bar{7}^+)G(\bar{7},\bar{5})v_c(1^+,\bar{7}). \quad (15.12)$$

The TMA for the self-energy corresponds to using the Bethe–Goldstone approximation to the two-particle Green's function in Eq. (10.5). This approximation was originally introduced in the nuclear many-body problem [704]. It is justified in the limit of low density of electrons or holes, i.e., close to completely filled or completely empty bands. It is intuitively clear that this corresponds to a limit of weak screening, and to a situation where one should consider individual particles rather than a mean field.

Schematically, Eqs. (15.11) and (15.12) read $iG(T_i \pm iv_c - iv_c T_i GG) = 0$. By setting the term in parentheses to zero, this yields an equation for T_i. This procedure is not unique: it depends on *which* of the Green's functions is extracted from the last term. The nature of the different solutions will become clearer when their time and spin structure is examined.

Time structure

Let us first examine the version introduced in Sec. 10.5: the Green's function that is extracted is the next to last one, $G(\bar{6},\bar{7}^+)$ in Eqs. (15.11) and (15.12). This yields the **particle–particle** T-matrix approximation given by Eq. (10.48). To understand this name, look at the time structure of T_1 defined in Eq. (10.49); the considerations for T_2 are analogous. In the following, only time arguments are displayed.

Since the bare Coulomb interaction is instantaneous, one can make the ansatz

$$T_1(t_1,t_3;t_2,t_4) = -v_c\delta(t_1^+,t_3)\delta(t_4^+,t_2)T(t_1,t_2). \quad (15.13)$$

Then Eq. (10.49) yields

$$T(t_1, t_2) = \delta(t_1, t_2) + iv_c G(t_1, \bar{t}_3) T(\bar{t}_3, t_2) G(t_1, \bar{t}_3) = \delta(t_1, t_2) - v_c L_0^{pp}(t_1, \bar{t}_3) T(\bar{t}_3, t_2),$$
(15.14)

where we have defined $L_0^{pp}(t_1, t_3) = -iG(t_1, t_3)G(t_1, t_3)$. Note that the two Green's functions have the same time arguments: L_0^{pp} is a *particle–particle* (electron–electron or hole–hole) correlation function. The contribution to the self-energy is

$$\Sigma_1(t_1, t_2) = -iG(t_2, t_1)v_c T(t_1, t_2).$$
(15.15)

Equivalent results are obtained for the contribution of T_2. Equation (15.15) is similar to the GWA expression $\Sigma^{GW}(t_1, t_2) = iG(t_1, t_2)W(t_1, t_2)$. However, the order of times in the Green's function is reversed. This corresponds to the picture in panel (c) of Fig. 15.1: the role of the quasi-particle is played by the excited electron.[11] Moreover, $W = \varepsilon^{-1}v_c$ is replaced by $-Tv_c$, which is constructed from particle–particle excitations. The particle–particle T-matrix approximation to electron removal treats explicitly the correlation of the two holes; similarly, in the case of electron addition, a two-electron correlation function appears.

Let us now come to the second possible solution, the *electron–hole* T-matrix. We extract the last Green's function $G(\bar{7}, \bar{5})$ on the right side of Eqs. (15.11) and (15.12) and obtain

$$T(t_1, t_3; t_2, t_4) = [T_H + T_x](t_1, t_3; t_2, t_4) + iv_c(t_1^+, t_4)G(t_1\bar{t}_6)T(\bar{t}_6, t_3; t_2, \bar{t}_5)G(\bar{t}_5, t_4). \quad (15.16)$$

Again one can split T into T_1 and T_2 defined by the Hartree and exchange interaction, respectively. With the ansatz

$$T_1(t_1, t_3; t_2 t_4) = -v_c \delta(t_1, t_4)\delta(t_3, t_2)T^{eh}(t_1, t_2),$$
(15.17)

one finds that $T^{eh} = 1 + L_0^{eh}T^{eh}$ with the *electron–hole* correlation function $L_0^{eh}(t_1, t_3) = -iG(t_1, t_3)G(t_3, t_1)$. The contribution to the self-energy is[12] $\Sigma_1(t_1, t_2) = -iG(t_1, t_2)v_c T^{eh}(t_1, t_2)$.

The time structure of the electron–hole T-matrix is equal to that of the GWA. However, the two approximations are not the same: the electron–hole TMA corresponds to panel (d) in Fig. 15.1. This can be seen by looking at the spin structure.

Spin structure

The particle–particle and electron–hole TMA have the same spin structure. In both cases

$$\Sigma_\sigma = \Sigma_{\bar{\sigma}_2}iG_{\bar{\sigma}_2}T_{\sigma\bar{\sigma}_2\sigma\bar{\sigma}_2},$$
(15.18)

with

$$T_{1,\sigma\sigma_2\sigma\sigma_2} = -v_c + iv_c G_\sigma T_{1,\sigma\sigma_2\sigma\sigma_2}G_{\sigma_2}$$
(15.19)

---

[11] Contrary to the schematic view expressed through Eq. (15.2), here one would formulate $\Delta E_s = E_N - E_{N+1,t} + E_{N+1,t} - E_{N-1,s} = -\varepsilon_t^{QP} - \omega_{st}$, where ε_t^{QP} is the energy of an additional electron and ω_{st} is a two-hole excitation from a state containing $N+1$ electrons.

[12] Note that the simple relation between T_1 and T_2 given in Sec. 10.5 for the particle–particle T-matrix is due to the fact that two electrons or two holes are indistinguishable. Therefore, it does not hold in the electron–hole case.

and

$$T_{2,\sigma\sigma_2\sigma\sigma_2} = v_c\delta_{\sigma,\sigma_2} + iv_c G_\sigma T_{2,\sigma\sigma_2\sigma\sigma_2}G_{\sigma_2}. \tag{15.20}$$

Unlike the screened interaction in the GWA, the T-matrix is spin-dependent, and in the calculation of the self-energy it mixes Green's functions of opposite spin. This corresponds to panels (c) and (d) of Fig. 15.1: a hole with given spin is created. It couples with a second, excited, particle (electron or hole) that can have spin up or down. In W of the GWA, electron and hole in the correlated pair have necessarily the same spin. Instead, the T-matrix approximation also contains spin-flip. The spin structure of the T-matrix allows important processes like the emission of spin waves in ferromagnetics, or of paramagnons [705].

Diagrams in the TMA

Equations (10.48) and (15.16) are Dyson equations for the particle–particle and electron–hole T-matrix. Therefore, they yield an infinite number of diagrams. They express the scattering of particles, as one can see in Fig. 10.7. If the Dyson equation is iterated once starting from $T^{(0)} = T_H + T_x$, the second Born approximation is obtained.

This is the exact self-energy to second order in the Coulomb interaction shown in Fig. 10.3. The first of the two second-order diagrams also appears in the GWA. To second order, one cannot distinguish scattering diagrams from bubble diagrams. They differ starting from third order. The exchange diagram on the right side of Fig. 10.3 is missing in the GWA. This can be traced back to the fact that the GWA does not treat exchange contributions on the same footing as Hartree contributions: it neglects the exchange part in the calculation of the effective interaction Ξ in Eq. (10.37). Hence, the TMA gives a correction to the GWA already to second order. However, it neglects all but the first bubble diagram, which is a problem whenever screening is important.

15.5 Beyond the T-matrix approximation: combining channels

In general, none of the three "channels" (GWA, particle–particle or electron–hole TMA) gives an exhaustive description. This reflects the dilemma of Fig. 15.1: how to decide which two-particle correlation is the most important one in the description of a (at least) three-particle problem. It suggests working with combinations.

Screening the T-matrix: a combination with the GWA

The improvement of the GWA with respect to Hartree–Fock is due to the screening of the Coulomb interaction. In the TMA the rough assumption $L \simeq L_0$ is made. This can be improved by taking L in the RPA, which is the solution of Eq. (14.9) with $\tilde{L} = L_0$. Equation (10.48) for the particle–particle T-matrix becomes

$$T(1,4^+;2,7) = [T_H + T_x](1,4^+;2,7) + iG(1\bar{3})T(\bar{3},\bar{5};2,7)G(4\bar{5})W^{RPA}(1^+,\bar{4}); \tag{15.21}$$

an equivalent form is found for the screened electron–hole T-matrix. Note that the Hartree and Fock terms remain unscreened. Like in the unscreened case, the resulting self-energy is exact to second order in the Coulomb interaction. Because of the appearance of both v_c and W, the approximation lacks however the necessary symmetry to be conserving. Fully screened versions can also be found in the literature [555, 705, 706]. These require appropriate double-counting corrections; see the example of nickel in Secs. 15.6 and 21.3.

When Eq. (15.21) is iterated once, starting from $T^{(0)} = T_H + T_x$, the self-energy becomes

$$\Sigma(1, 1') = \delta(1, 1')v_H(1) + \Sigma_x(1, 1') + iG(1, 1')W(1, \bar{2})P_0(\bar{2}, \bar{3})v_c(\bar{3}, 1')$$
$$- G(1, \bar{3})W(1, \bar{2})v_c(\bar{3}, 1')G(\bar{2}, 1')G(\bar{3}, \bar{2}), \tag{15.22}$$

where $P_0 = -iGG$ is defined in Eq. (11.16). Since $WP_0v_c = W - v_c$, the first three terms sum to the GWA. The last term is called *second-order screened exchange* (SOSEX) [707]. This may be compared with the self-energy $\Sigma = \Sigma^{GW} - GWWGG$ that is obtained when the first-order vertex Eq. (15.5) is used. In that case, both Coulomb interactions are screened in the last term. As we can see, both results can be justified by a derivation. To find out which one is better demands more work, and the conclusion depends on many requirements, for example positiveness of spectral functions, as is often the case when one has to choose subsets of diagrams.

Combining particle–particle and electron–hole fluctuations

The simplest idea to avoid the choice between the different pictures in Fig. 15.1 is to sum the diagrams that appear in the three channels, the GWA, the particle–particle, and the electron–hole TMA. One only has to avoid the double counting of the first- and second-order contributions. The sum of the three channels is done in the fluctuation exchange (FLEX) approximation [708, 709]. In FLEX one finds the GWA and its exchange counterpart to all orders. Moreover, starting from third order, all unscreened T-matrix particle–particle diagrams, and all unscreened T-matrix electron–hole diagrams [709], are added. The T-matrix and GW approximations can be regarded as an approximation to FLEX, where only one channel is taken into account.

Although FLEX sums all channels, it still does not take into account screening properly: contrary to the screened T-matrix, there are no mixed diagrams. For example, there are no bubbles screening the legs of the ladders in Fig. 10.6. Such a missing diagram is shown in Fig. 15.3. Of course, when FLEX is used to solve the Hubbard model with an effective U, there is some screening built in, because U contains screening effects due to states outside the subspace that is described by the model. One may consider in this case the effective U to be a simple approximation for the insertion of bubbles in the ladder legs. The screening of a subspace is discussed more in Ch. 21. Beyond the summation of independent channels and beyond the use of effective screened parameters, two-step FLEX approaches have been proposed, where one channel enters the calculation of a second channel through an effective screening in the spirit of Eq. (15.21) (see Ex. 15.4) [710–712].

The coupling of channels is formalized in parquet theory [713–715]. It is based on Eq. (10.50) for the self-energy $\Sigma_c = -iv_cGG\,^4\tilde{\Gamma}G$ in terms of the four-point interaction vertex $^4\tilde{\Gamma}$. This interaction vertex is determined by a Bethe–Salpeter equation (14.11) that reads

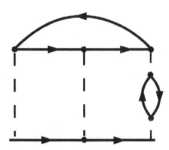

Figure 15.3. One of the diagrams that is missing in FLEX: a third-order scattering diagram where one of the ladder legs is screened with a bubble.

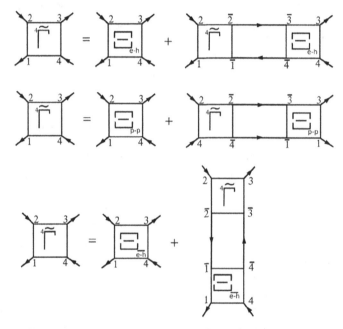

Figure 15.4. The three Bethe–Salpeter equations for the four-point interaction vertex that are used in parquet theory.

$$^4\tilde{\Gamma}(1,2,3,4) = \Xi_{e-h}(1,2,3,4) +^4 \tilde{\Gamma}(1,2;\bar{2},\bar{1})G(\bar{3},\bar{2})G(\bar{1},\bar{4})\Xi_{e-h}(\bar{4},\bar{3},3,4), \quad (15.23)$$

which is the same as Eq. (14.11) since for any Dyson-like equation of the form $X = X_0 + X_0 SX$ one can equivalently write $X = X_0 + XSX_0$. Calculating $^4\tilde{\Gamma}$ is equivalent to calculating the two-particle correlation function using Eq. (14.10). The diagrams are shown in the first row of Fig. 15.4. Note that there is a forward and a backward one-particle Green's function: this is the *electron–hole channel*, which is used, e.g., to calculate W, as discussed in Ch. 14. Everything is exact, but of course the difficulty is to construct $\Xi_{e-h} = \delta\Sigma/\delta G$. In principle, one could iterate the equations in the spirit of Sec. 15.2. Parquet theory goes an alternative way, using the fact that Eq. (15.23) is not the only possible Bethe–Salpeter equation for $^4\tilde{\Gamma}$. Several versions can be formulated, consistent with the fact that there are several ways to derive the two-particle correlation function from a generating functional, as outlined in Sec. 8.4. For every channel, labeled ℓ, the equation schematically reads $^4\tilde{\Gamma} = \Xi_\ell + \Xi_\ell L_{0\ell} \, ^4\tilde{\Gamma}$. Note that $^4\tilde{\Gamma}$ does not depend on ℓ: the different equations yield the same result, provided Ξ_ℓ and $L_{0\ell}$ are consistent. Figure 15.4 shows the diagrams for the three Bethe–Salpeter equations, where the first one is the same electron–hole channel as the BSE, corresponding to panel (b) of Fig. 15.1. The second one is the electron–electron channel. It is equivalent to the hole–hole one (not shown) that represents panel (c) of Fig. 15.1, and the third one is the electron–hole channel that corresponds to panel (d) of Fig. 15.1.

The coupling of the channels is given by the equation

$$\Xi_\ell = \Lambda_{irr} + \, ^4\tilde{\Gamma} \sum_{\ell' \neq \ell} \Xi_{\ell'} L_{0\ell'}, \tag{15.24}$$

where the *fully irreducible vertex* Λ_{irr} is introduced. All functions have four arguments as above, but they will not be shown in the following.

The fully irreducible vertex Λ_{irr} is independent of ℓ, as one can see by combining the equations. It contains only contributions that cannot be separated into two objects by cutting two G-lines,[13] and that are known diagrammatically. Combining Eqs. (15.23) and (15.24), and introducing $\Phi_\ell = \, ^4\tilde{\Gamma} L_{0\ell} \Xi_\ell$, yields the *parquet equations*

$$^4\tilde{\Gamma} = \Lambda_{irr} + \sum_\ell \Phi_\ell \quad \text{and} \quad \Phi_\ell = \, ^4\tilde{\Gamma} L_{0\ell} \Lambda_{irr} + \sum_{\ell' \neq \ell} \, ^4\tilde{\Gamma} L_{0\ell'} \Phi_{\ell'}. \tag{15.25}$$

If Λ_{irr} is given, these equations, together with the Dyson equation for the one-particle Green's function and Eq. (10.50) for the self-energy, are sufficient to determine the one- and two-particle Green's functions. The parquet equations are usually approximated by fixing Λ_{irr} and iterating the set of equations to self-consistency, which creates diagrams that pave the space like a parquet floor. The second set of equations has the structure of a coupled set of Bethe–Salpeter equations. As Ch. 14 shows, with some additional approximations such equations are within reach of first-principles calculations. At the date of this book, they are most often solved for Hubbard or Anderson impurity models using the simplest possible Λ_{irr}.

The lowest-order contribution to Λ_{irr} is the bare Coulomb interaction (direct and exchange terms, as in the TMA). Using this, Λ_{irr} is usually called the *parquet approximation*. The next contribution to Λ_{irr} is of fourth order in v_c. The parquet equations are also used in the DΓA that is introduced in Sec. 21.7: in this approach, a local approximation to Λ_{irr} is made that contains all orders in the bare Coulomb interaction v_c. This has important consequences, in particular, it leads to the formation of Hubbard bands that are absent in the lowest-order parquet approach.

More about the physics, the mathematical structure, and possible solutions of the parquet equations can be understood from the discussion of the electron–hole Bethe–Salpeter

---

[13] This is the reason why it is called "fully irreducible."

equation in Ch. 14 and the T-matrix approximation in the present chapter. More about the parquet equations, references, and the relation to FLEX can be found in [715], and a concise overview is given in Ch. 10 of [716].

Three-body scattering

Instead of choosing a pair of correlated particles as in Fig. 15.1, one may treat the three particles approximately. An example is the three-body scattering approach [717–721]. For electron removal, the many-body wavefunction of the $(N - 1)$-particle state is expanded in a basis that includes configurations with one hole, and with one hole plus an electron–hole pair, starting from the ground state of a single-particle hamiltonian that is optimized by a Kohn–Sham DFT calculation. This leads to a three-body scattering problem, where hole–hole and electron–hole interaction is taken into account. It has been solved for a multi-orbital Hubbard hamiltonian with on-site interaction. This three-body scattering approach has been applied to transition metals [721–723] and cuprates [724, 725]. It overcomes some typical problems of the GWA (see Sec. 13.3), for example the absence of the 6 eV satellite in nickel, and the bad description of the exchange splitting [721, 722].

15.6 T-matrix and related approaches in practice

Although most first-principles calculations focus on the electron–hole channel contained in the GWA, ladder diagrams such as those included in the TMA hold promise for applications where the GWA fails. It is interesting to make some remarks about the approaches in practice and give a few selected examples of applications. Here we mainly focus on the TMA, but the arguments can be generalized.

The computational challenge

The equations of the previous section all have the structure of a Bethe–Salpeter equation, with four-argument functions. Therefore the quantities can in principle be evaluated using the same techniques and similar approximations as in the case of the electron–hole BSE in Ch. 14. In solids one must use a screened version of the TMA. Since screening is dynamical, this introduces an additional frequency dependence, which calls for additional approximations.

The most widely used approximations in TMA calculations are the following.

- **Neglect of self-consistency.** The TMA is conserving when Green's functions are calculated self-consistently, as pointed out in Sec. 10.5. However, similar to the case of the GWA, self-consistency was found to deteriorate spectra [393, 726]. In practice, most calculations of spectra are not self-consistent and, as in the GWA, Sec. 12.4, one uses an optimized mean-field G.

- **Static approximations.** When a screened version of the T-matrix is used in *ab initio* calculations for bulk systems, the screened Coulomb interaction W is often taken in the static limit [555, 693, 706], like in the Bethe–Salpeter approach for the calculation of optical spectra, Sec. 14.5.

- **Local approximations.** If the Coulomb interaction is replaced with a local interaction, the kernel $L_0 = GG$ of the TMA Bethe–Salpeter Eq. (10.48) or (15.16) reduces to a function of two space arguments only. This leads to a self-energy with an effective interaction that acts in a way similar to W, but that is spin-dependent and contains particle–particle or spin-flip excitations, instead of spin-conserving charge excitations [727]. Such an effective interaction has equally been derived from a variational approach [706, 728] in the same spirit as the derivation of a TDDFT kernel from the BSE, Sec. 14.11. A reduction of spatial degrees of freedom has also been obtained by deriving a k-dependent TMA self-energy for a homogeneous system and taking a local limit [729], to be used in the real, inhomogeneous, material [730]. To suppose that the dominant part of the interaction is local is similar to what is done in DMFA.

Tests for model systems

One can get an idea of the performance of the TMA by looking at the Hubbard dimer; results are detailed in [731]. As mentioned in Sec. 11.7, the particle–particle TMA (calculated with G^0) is exact in the case of quarter-filling. This is quite logical: there are at most two electrons (in the case of electron addition), so correlating explicitly two particles solves the problem. The same is not true in the case of half-filling, where there are three electrons in the addition spectral function. The particle–particle TMA performs less well in that case, though better than the GWA, and the electron–hole TMA diverges for $U/t = 2$. Screening the particle–particle TMA following Eq. (15.21) approximately takes into account the third particle and leads to improvement, as has been shown in [731].

Since, as discussed in the previous subsection, a local interaction simplifies the TMA calculations significantly, T-matrix approaches have been used extensively in the context of Hubbard models. Comparison with exact results for clusters has confirmed the quality of the approximation in low-density situations [732]. For example, for a 3×3 square Hubbard array with one orbital per site and two holes (16 electrons) in the ground state, the particle–particle T-matrix yields very good spectral functions over a large range of interactions (examined up to $U/t = 8$), whereas the GWA results deteriorate already at $U/t = 4$ [401]. Calculations on Hubbard clusters also showed that, while the TMA generally performs well at low filling, it is not superior to the GWA at half-filling [733], similar to the result for the Hubbard dimer. Indeed, half-filling does not correspond to the low-density limit for which the TMA is designed.

A prototype application: ferromagnetic nickel and other transition metals

Several first-principles TMA calculations combined with GW have been carried out for lifetimes [705, 734]. The TMA corrections to the GWA are important in transition metals. For example, below 0.7 eV spin-flip contributions to the decay of excited electrons in spin-minority states in ferromagnetic iron turn out to be significant [693].

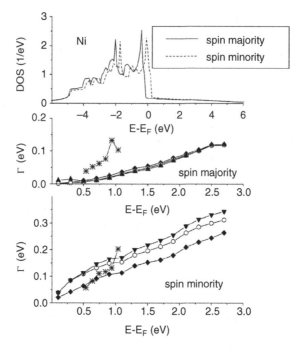

Figure 15.5. Inverse lifetimes Γ of excited electrons in Ni. (From [693].) Stars are the experimental results. The solid diamonds show the GWA contribution, for the open circles non-spin-flip electron–hole TMA contributions have been added. The black triangles are the total result of GWA plus non-spin-flip and spin-flip electron–hole TMA contributions.

Figure 15.5 shows calculations for nickel. There is little effect of the TMA contribution for the majority spin, but the effect is significant for the minority spin. Contrary to iron, the biggest contribution stems from non-spin-flip TMA contributions. This difference with respect to iron is explained by the smaller exchange splitting in nickel. To explain the remaining discrepancy with experiment, especially visible in the majority channel, effects of transport, electron impurity, and electron–phonon scattering are invoked [693].

One of the most prominent success stories of the TMA is the explanation of the existence of a satellite in nickel, which can be seen at 6 eV in the experimental photoemission spectrum in Fig. 15.6. Nickel shows no structure around 6 eV in the loss function, which means that this satellite cannot be explained with excitations contained in W. Therefore, as pointed out in Sec. 13.3, the GWA is not able to produce such a structure. Instead, it is interpreted as due to hole–hole excitations. This interpretation has been widely confirmed, from early calculations based on model hamiltonians [710, 735–737] to more recent results using the GWA+T-matrix with a static screened U [555] or two-step FLEX approach [711]. A bare particle–particle T-matrix calculation does not yield the correct satellite position, since screening is crucial. Several methods have been used to take screening into account. In [710], electron–hole ladders for states

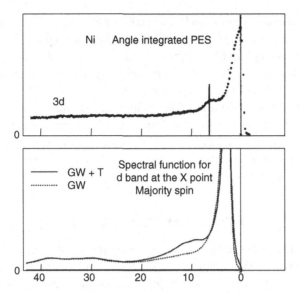

Figure 15.6. Nickel spectral function for the d-band. Upper panel: experiment from [738] showing the satellite at ≈ 6 eV. Lower panel: theoretical spectral functions for the majority spin channel at the X-point of the Brillouin zone, from [555]; the dashed curves are for GWA only and the solid curves include the TMA diagrams for the d states as described in the text.

of opposite spin are introduced into the particle–particle T-matrix diagrams. In [711], a two-step approach has been proposed where the particle–particle channel is used to obtain an effective interaction that is then used to calculate electron–hole contributions. With a Hubbard interaction parameter $U = 2$ eV, the satellite appears in the correct position.

Instead of treating U as a parameter, it can be calculated, e.g., using the constrained DFT method (Sec. 19.5). This has been done in the GWA+T-matrix calculation of [555] mentioned above, where TMA particle–particle diagrams for the d-orbitals only were added to the GWA, using a static effective $U = 5.5$ eV calculated in the LDA. The GWA result in Fig. 15.6 shows a quasi-particle and some weak satellite structure around 30 eV. This corresponds to the broad plasmon that is contained in W, consistent with the screened interaction shown in Fig. 21.6. However, there is no satellite around 6 eV in the GWA result. When screened particle–particle TMA diagrams are added, a structure at the low-energy side of the quasi-particle appears in the spin majority channel. It is due to the hole–hole excitations that are introduced in Eq. (15.14). It can be related to the experimentally observed satellite in the upper panel of Fig. 15.6, although its position is too far from the quasi-particle peak.[14] The exchange splitting and the bandwidth are also improved with respect to the GWA (see Sec. 13.3).

---

[14] The intensity of the calculated satellite with respect to the intensity of the quasi-particle appears to be weaker than in experiment, but we refer the reader to the difference between the calculated intrinsic spectral function and the measured total spectrum, as discussed in Sec. 5.7.

This calculation and its results can be compared with the results of a DMFA calculation with a QMC solver in Fig. 20.6 that also includes only interactions in the d states and a static U calculated using the LDA. However, the value of $U = 3.0$ eV is different and $J = 0.9$ is included. There is a satellite in the spin-majority channel around 6 eV, in general agreement with experiment. Altogether, the GWA+TMA and DMFA results support one another. The difference in the satellite position suggests that screening of the effective interaction is an important issue. Another issue is double counting: this is difficult to determine in the DFT+DMFT approach. For the GW+T-matrix approach, Sec. 21.3 gives more details on how diagrammatic methods can in principle be applied to selected subspaces without double counting.[15]

15.7 Cumulants in electron spectroscopy

The approaches in the previous sections add diagrams to the self-energy in order to improve the results. However, many examples have shown the limits of such a procedure. As discussed in Sec. 10.3, the role of a self-energy is to create contributions to the one-body Green's function to all orders in the interaction through the solution of the Dyson equation, even when Σ is approximated by a low-order expression. The self-energy can often be simplified when one is interested in quasi-particles only. In the following we concentrate on satellites, and in particular on series of multiple satellites that occur, e.g., in the photoemission spectrum of sodium (Fig. 2.8), or silicon (Fig. 13.4). Then the argument in favor of the use of a self-energy given in Sec. 10.3 does not hold, because it is based on the presence of only one pole in the Green's function. Indeed, the GWA cannot describe the satellite series of sodium and silicon. Still, it is desirable to find a simple closed expression that creates contributions to all orders in the interaction.

The approach in the present section is to use an exponential form for the one-body Green's function: $G = G_0 e^C$, where G_0 is an independent-particle or quasi-particle Green's function, and C is called the *cumulant*. The idea of cumulants, i.e., the representation of a quantity as an exponential function or functional, is widespread in physics and mathematics [739]. For example, it corresponds to a procedure discussed in Sec. 6.5 to construct the many-body wavefunction with the generalized Feynman–Kac formula, which leads to the Slater–Jastrow wavefunction, Sec. 6.6. Here we concentrate on its use for our particular problem, electron spectroscopy. To go deeper, we refer the reader to the insightful discussions in [287].

Electron–boson coupling

The cumulant approach to the one-body Green's function is based on the picture of a fermion coupled to bosons. This is an intuitive and widely used picture to describe

---

[15] However, the example of nickel in Sec. 21.3 also illustrates the fact that other problems can arise when one goes beyond the GWA, such as negative spectral functions. These problems are overcome by a different choice of diagrams that are included in the calculations, which makes the discussion more complicated.

electron–phonon coupling [740]. In the present case, where we only consider the system of electrons, it describes a quasi-particle coupled to the excitations of a surrounding medium. This picture makes links to many places in this book: to the general idea of embedding (Ch. 7), to the transformation of the many-body problem into an effective Anderson impurity model in the DMFT (Chs. 16–20), and in particular to the GWA, where a quasi-particle is screened by plasmons and other electron–hole excitations, which play the role of the bosons. The picture is formalized in the electron–boson hamiltonian of Eq. (11.46) discussed in Sec. 11.7. This hamiltonian has led to important insight and early applications in core-level spectroscopy, where a picture of a single fermion line followed by a series of bosonic satellites is particularly appropriate [406, 741–744]. There are reasonable arguments to extend it to the valence region [287, 745]; for example, when the plasmon frequency is much higher than the energy spread of the levels that are coupled through the interaction with the plasmons.

To use this picture for the calculation of real materials, one could think to use the hamiltonian of Eq. (11.46) and determine the model parameters from first-principles calculations. However, projecting a real system onto a model, for example the Hubbard model, always introduces some ambiguity. Alternatively, one can try to find the first-principles expressions that correspond to the model. A realistic cumulant approximation has been derived, for example, by iterating the equation of motion for the one-body Green's function [247]. In the following we sketch a derivation [482, 521] along the lines of Ch. 10, which allows us to put various approximations on the same footing.

Electron–boson coupling from first principles

We start from Eq. (10.12), which is repeated here:

$$G_u(1, 1') = G^0(1, 1') + G^0(1, \bar{2}) \left\{ [u(\bar{2}) + v_H(\bar{2})]G_u(\bar{2}, 1') + iv_c(\bar{2}, \bar{3})\frac{\delta G_u(\bar{2}, 1')}{\delta u(\bar{3}^+)} \right\}.$$

(15.26)

The equation is non-linear, because the Hartree potential depends itself on the Green's function. We linearize the equation by expanding the Hartree potential in the external potential u, which is reasonable since we are only interested in the equilibrium solution where $u = 0$. This yields [482]

$$G_u(1, 1') = G_H(1, 1') + G_H(1, \bar{2})\bar{u}(\bar{2})G(\bar{2}, 1') + iG_H(1, \bar{2})W(\bar{2}, \bar{3})\frac{\delta G(\bar{2}, 1')}{\delta \bar{u}(\bar{3}^+)},$$

(15.27)

where $\bar{u}$ is screened by the inverse dielectric function, and G_H is the Hartree Green's function calculated with the exact density at vanishing $\bar{u}$. This equation contains the screened Coulomb interaction W taken at $\bar{u} = 0$. Now the electrons propagate in a dynamically polarizable medium. The possible excitations of this medium are approximately independent bosons, as discussed in [30]. This is a first-principles equivalent to an electron–boson coupling picture. Even if one limits the equation to a single orbital, the surrounding medium influences the propagation of an electron with respect to the non-interacting case.

If W were instantaneous, Eq. (15.27) for a single orbital would instead describe the propagation of one particle in a static mean field. It is the reaction of the system, the building up of the screening cloud in time, that makes the propagation of a single particle in the screening medium non-trivial: dynamical screening is essential to simulate how a particle is affected by the rest of the system. This is another illustration of the general principle of dynamical embedding introduced in Ch. 7, and found in many places in this book, in particular, it is used in DMFT in Chs. 16–20.

The GWA is an approximate solution to the electron–boson coupling model, as shown in Sec. 11.7. To obtain the GWA Dyson equation from Eq. (15.27) one has to make the RPA-like approximation $\delta G/\delta \bar{u} \approx GG$. In order to remedy the shortcomings of the GWA, one has to overcome this approximation. The simplest way is to decouple the electron and hole branches of the Green's function, and to make an ansatz for the hole branch $G^<$, Eq. (5.73), that is the pertinent quantity for photoemission. The cumulant approximation is obtained using the ansatz

$$G^<_{u,\ell\ell}(t_1, t_2) = G^{GW,QP<}_{u,\ell\ell}(t_1, t_2)e^{C_{\ell\ell}(t_1,t_2)}, \qquad (15.28)$$

where $G^{GW,QP}_\mu$ is the GW QP Green's function in the presence of the external potential $\bar{u}$, the index ℓ denotes the basis of orbitals where $G^{GW,QP}$ at $\bar{u} = 0$ is diagonal, and the cumulant C has to be determined. The decoupling of the electron and hole branches, the fact that we consider diagonal elements only, as well as the assumption that C does not depend on $\bar{u}$, are approximations: the ansatz does not solve the full differential Eq. (15.27). However, one can determine C approximately by inserting Eq. (15.28) into Eq. (15.27) and solving the equation for C at $\bar{u} = 0$. An approximate solution for the cumulant is

$$C_{\ell\ell}(t_1, t_2) = \left[\frac{\partial \Sigma^{GWh}(\omega)}{\partial \omega}\bigg|_{\omega=\varepsilon_i} + \left[\frac{1}{\pi} \int_{\varepsilon_i-\mu}^{\infty} d\omega \frac{\mathrm{Im}\Sigma^{GW}(\varepsilon_i - \omega)}{(\omega - i\eta)^2} e^{(i\omega+\eta)(t_1-t_2)} \right] \right] \qquad (15.29)$$

where the assumption was made that the hole part Σ^{GWh} of the GW self-energy is diagonal in the basis. This is the standard cumulant approximation for a decoupled hole branch of the one-body Green's function. Similar expressions have been used successfully, e.g., in [746] to calculate satellites in sodium. The cumulant is determined by W, which contains the excitations leading to the satellites. Since W appears in the exponent, expansion of the exponential leads to a series. The frequency integral in Eq. (15.29) creates all combinations of times. This yields for each $G_{\ell\ell}$ the series of diagrams shown in Fig. 15.7. W is an input to the cumulant calculations, therefore we do not distinguish W or W_0 here. It is typically determined in an RPA calculation. Note that Eq. (15.29) is only for the diagonal element $G_{\ell\ell}$, and it involves only $\Sigma^{GW}_{\ell\ell}$: this has some similarity to the DMFA, although in the cumulant case the basis functions do not have to be spatially localized, and although of course in the case of DMFA the self-energy contains all local diagrams, well beyond the GWA.

The G_0W_0 approximation is given by the first two rows of Fig. 15.7: at each given time there is only one plasmon. It is contained in W, the wiggly line. Therefore, as observed, G_0W_0 cannot reproduce multiple plasmon satellites. When G is calculated self-consistently, the diagrams in the third row also appear. Now in principle multiple plasmons

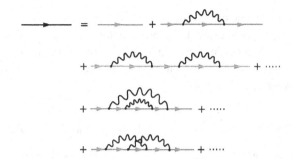

Figure 15.7. Diagrams contained in $G_0 W_0$ (first two lines), GW_0 (including also the third line), and in the cumulant expansion of the one-particle Green's function (all diagrams). W is supposed to be fixed. It is understood that the diagrams are for a diagonal element $G_{\ell\ell}$ and involve only $\Sigma_{\ell\ell}^{GW}$, following Eq. (15.29).

are possible. This is indeed observed when the interaction is strong enough, for example in the model result of Fig. 15.2. In the homogeneous electron gas at the density of simple metals, the effect is not important enough to be visible, as one can see from Fig. 11.8. The cumulant Green's function contains also the diagrams in the last row. To bring out clearly the satellite series, we make a single-plasmon pole approximation for W following Sec. 12.2. Each matrix element reads

$$W(t_1 - t_2) = -i\lambda\omega_p^2 \left[e^{-i\omega_p(t_1-t_2)} \Theta(t_1 - t_2) + e^{i\omega_p(t_1-t_2)} \Theta(t_2 - t_1) \right]. \tag{15.30}$$

The spectral function resulting from Eq. (15.29) is then

$$A(\omega) = \frac{1}{\pi} e^{-\lambda} \left[\frac{\Gamma}{(\omega - \varepsilon^{QP})^2 + \Gamma^2} + \lambda \frac{\Gamma}{(\omega - \epsilon + \omega_p)^2 + \Gamma^2} \right.$$
$$\left. + \frac{1}{2}\lambda^2 \frac{\Gamma}{(\omega - \epsilon + 2\omega_p)^2 + \Gamma^2} + \frac{1}{6}\lambda^3 \frac{\Gamma}{(\omega - \epsilon + 3\omega_p)^2 + \Gamma^2} + \cdots \right]. \tag{15.31}$$

The damping factor $e^{-\lambda}$ contains the weight transfer away from the quasi-particle towards the satellites. The first term in parentheses is the quasi-particle peak at the complex quasi-particle energy $\varepsilon^{QP} + i\Gamma$, followed by the infinite series of satellites at positions $\varepsilon^{QP} - n\omega_p$, with decreasing weight $\frac{1}{n!}\lambda^n$. This is the same result as the solution of the model hamiltonian in Eq. (11.46).

Cumulant calculations in practice

The main ingredient of Eq. (15.29) is the GW self-energy; the rest is simple algebra. Therefore, the procedure and computational cost of a cumulant calculation are the same as for the GWA, and everything that is explained in Ch. 12 also holds here. This includes the choice of approximations. For example, when the plasmon–pole approximation introduced for the GWA in Sec. 12.5 is valid, the imaginary part of the self-energy can be represented by a single pole. This is a good approximation in materials like silicon, where the loss function or dynamic structure factor for small momentum transfer are dominated by one well-defined structure, as one can see in Fig. 5.6. The cumulant calculations for silicon in Fig. 13.4 have been performed using this approximation. In materials with a more complex plasmon spectrum, leading to a more complex Im Σ, one has to perform a full-frequency integration, for example using a multipole representation. An example is graphite, which has two main plasmon structures [747].

The plasmaron problem of the GWA, and cumulant results

The cumulant approach has been applied to the homogeneous electron gas, and compared with the GWA. Results from [748] are shown in Fig. 15.8. G_0W_0 results are given in the left panel. The very sharp satellite predicted by the GWA is called a *plasmaron*, an additional solution to the quasi-particle Eq. (7.39), which is a coupled hole–plasmon mode [383, 412, 425]. It can also be detected in the left panels of Fig. 11.5, by looking at the real part of the self-energy and the resulting spectral function. The plasmaron had been predicted for core levels, based on GWA calculations for the electron–boson model [407]. Later, however, the exact solution of the model was found [406]. It showed that *there should be no such peak*. The model results and general considerations on perturbation theory and Dyson equations indicate that the plasmaron is instead an artefact of G_0W_0 [749, 750]. Self-consistency in GW_0 weakens the plasmaron, as one can see in the right panel of Fig. 15.8.

The cumulant results show two satellites in the energy range of interest, as one would expect from the plasmon energy. The spectral weight of the GW_0 satellite is distributed over the series of satellites. Cumulant results are little affected by self-consistency, because there is no plasmaron. The experimental photoemission or electron momentum spectra of simple metals like lithium, aluminum, and sodium (see Fig. 2.8) can be reproduced and explained with the *ab initio* cumulant expansion calculations [746, 751]. The satellite band that is due to the excitation of the doping charges in doped graphene is also obtained with

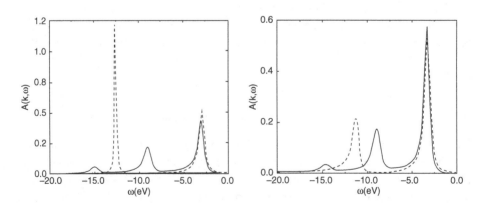

Figure 15.8. GW and cumulant expansion spectral functions of the homogeneous electron gas at the Γ point, $r_s = 4$ (dashed and solid lines, respectively). Left: G_0W_0 calculation. Right: G is calculated self-consistently, in GW_0. (From [748].)

the cumulant approach [752]. The appearance of the plasmaron in the GWA spectrum, instead, is detrimental. It is at the origin of the discrepancies observed between GWA calculations for the homogeneous electron gas and experiments on simple metals, and it explains the poor satellite of the GWA spectrum in silicon, Fig. 13.4.

The cumulant result for silicon [521] is also shown in Fig. 13.4.[16] Now all experimental features, including the satellite series, are well reproduced. Overall, the cumulant expansion performs well for the plasmon satellites. It also improves lineshapes with corrections that can be quite important, e.g., in the case of graphite [747, 753].

Extensions

The separation of the electron and hole branches is justified for core levels, but it is a severe limitation close to the Fermi level. There, the spectral function should be symmetric reflecting particle–hole symmetry, but the standard cumulant approach places all satellites for a given matrix element of the spectral function on one side of the Fermi level, i.e., below the Fermi level for holes, and above the Fermi level for electrons. It is not easy to overcome this limitation, since attempts to go beyond the simplest cumulant often lead to problems such as negative spectral functions [740]. It has been suggested [754] to use the retarded Green's function with an ansatz similar to Eq. (15.28). This ansatz leads indeed to coupling of electron and hole branches, and to symmetric spectral functions. This is a topic of on-going research. Open questions also include extensions to strongly correlated materials, which are not dominated by the plasmon physics contained in W.

Finally, it should be recalled that the electron–boson coupling hamiltonian is very general, and in particular, that it presents a convenient formulation to include coupling to phonons [740], which is used also in *ab initio* calculations [755].

15.8 Use of exact constraints

The choice of diagrams that is at the basis of the methods exposed in the previous sections depends very much on the physics of specific systems and phenomena of interest. However, one can also obtain guidelines from general requirements. In density functional theory, where nothing like the Luttinger–Ward functional is known, exact constraints are often used to design functionals, and contribute to a large extent to the success of simple approximations. A prominent example is the local density approximation, that fulfills the general sum rule on the exchange–correlation hole [175].

In the following we will briefly discuss a series of exact constraints that hold for the Green's functions and self-energies. Only the exact theory can fulfill simultaneously all these constraints, but they may indicate important information about the exact solution,

---

[16] For comparison with experiment, cross-sections, the secondary electron background, and extrinsic and interference effects have been calculated and added to the intrinsic spectral function.

and how to change existing functionals in order to approach it. Much more remains to be explored. The aim of this short section is to motivate new research in this direction.

Conservation laws and sum rules

In Sec. 10.4 the requirements are given for a theory to conserve particle number, total momentum, total angular momentum, and total energy. These conditions are fulfilled when self-energies are limited to expressions that are functional derivatives of a functional of the dressed Green's function, that is in turn constructed as a sum of symmetry-conserving scattering processes as outlined in Ch. 9. Conservation laws motivate the effort to perform calculations self-consistently, since non-self-consistent approaches are usually not conserving.

The conservation of particle number is linked to the conservation of spectral weight, which is the sum rule of Eq. (5.95). One can derive sum rules for all power moments of the spectral function. The zero-order sum rule is Eq. (5.95), and there are sum rules for the first and second momentum like Eqs. (11.30) and (11.31). You can derive these and higher-order sum rules following Ex. 15.12. A simple example for their use is to make an ansatz for the one-body Green's function and determine the parameters from the sum rules. This is worked out in Ex. 15.7.

Crossing symmetry

Condition [B] in Sec. 10.4 is a special case of *crossing symmetry*: the exact expression Eq. (5.133) for the two-particle Green's function tells us that

$$L(1, 2; 1', 2') = -L(2, 1; 1', 2') = -L(1, 2; 2', 1') = L(2, 1; 2', 1'). \qquad (15.32)$$

The exchange of arguments corresponds graphically to a twisting of incoming and outgoing legs in the diagrams, which is the origin of the name "crossing symmetry." Physically, it corresponds to the fact that holes can be described as electrons propagating backwards in time, and it is a consequence of the Pauli principle.

The symmetry of the two-particle correlation function L is reflected in the symmetry of the kernel of the Bethe–Salpeter equation from which it is calculated. Several Bethe–Salpeter equations can be formulated; this is the basis of the parquet equations introduced in Sec. 15.5. The parquet equations are in principle exact, and therefore fulfill conservation laws and crossing symmetry. However, it has been shown [756] that for any approximate fully irreducible vertex the resulting parquet equations cannot be at the same time conserving and show crossing symmetry. This is an example of the fact that there are constraints that cannot be fulfilled at the same time by any approximation. In particular, as stated in Sec. 11.7, the GWA violates crossing symmetry.

Positivity of the spectral function

The analytic properties of the correlation functions in Sec. 5.4 tell us that exact spectral functions are always positive. However, approximations can easily destroy this property.

An example is the TDLDA vertex correction in Sec. 15.3: the vertex correction weakens the effective interaction. If the effect is too strong, the effective interaction changes sign. Therefore, the problem of negative spectral functions is met in attempts to go beyond the GWA.

It is difficult to solve the problem by searching for a better self-energy functional of the dressed Green's function. An idea is instead to formulate approximations in terms of *pieces*[17] of G, for example a functional of $G^>$ and $G^<$. This route is explored in [757].

Koopmans' theorem

Close to the Fermi level, quasi-particles are sharp structures. In particular, the ionization potential of a molecule, which is the difference in total energies of N and $N-1$ particles in the ground state, corresponds to one infinitely sharp quasi-particle peak. The quasi-particle energy for electron removal from the N-particle system and the total energy difference calculated with the same method should therefore coincide. Moreover, the same result should be obtained for the electron addition to the $(N-1)$-particle system. However, problems of approximations, like the self-screening error in the GWA, destroy this property. The question is studied for small sodium clusters in [353]. The GWA performs much better than simple density functionals or Hartree–Fock, but deviations still amount to up to 10% of the ionization energy. Minimizing this discrepancy can be used to optimize the functional.

Ward identities

The Ward identities [758, 759] link the vertex function in the limit of vanishing frequency and momentum transfer to derivatives of the self-energy. They express gauge invariance and local electron-number conservation. Here *local* means that charge is conserved, and moreover, it can only change continuously in space, instead of disappearing in one place and reappearing in another place, as is also stated by the continuity equation. The Ward identities have been used to derive approximations for the vertex function; see, e.g., [760], and the discussion in [346].

These are just examples for the exact properties one may exploit. One can also use limiting cases, like a single electron, or exactly solvable models. The studies of the GWA in the HEG, in the Hubbard dimer, and in the electron–boson coupling model in Secs. 11.6 and 11.7 show the usefulness of simple systems to detect problems, which allows one eventually to go beyond. Much remains to be explored, and much can be learned from other approaches.

15.9 Retrospective and outlook

The group of Chs. 9–15 on many-body perturbation theory has shown both the flexibility and the difficulties of diagrammatic perturbation theory approaches.

---

[17] There are other examples where it is easier to find an approximation in terms of pieces instead of their sum. The most prominent case is the DFT formulation of Kohn and Sham, where the non-interacting kinetic energy is a functional of orbitals, instead of an explicit functional of the density.

On the upside, we have an explicit expression: the Luttinger–Ward formula tells us in principle the self-energy in terms of the dressed Green's function. The GWA is a particular choice of diagrams out of the Luttinger–Ward series. The present chapter has shown how one may go beyond the GWA, for example by iterating Hedin's equations.

On the downside, we have the question of which diagrams to choose. This is a tricky question, because many-body perturbation theory is not a simple perturbation series, where adding a term of next order guarantees improved results. Moreover, when the system comes close to degeneracy, perturbation theory has difficulties. Therefore, the GWA cannot describe, e.g., paramagnetic nickel oxide, and it is not easy to cure the problem. The one-electron Hubbard dimer in the atomic limit in Sec. 11.7, where the ground state becomes degenerate, is another good illustration.

One almost never considers a finite number of diagrams for the Green's functions. Usually one uses *Dyson equations*, where even a low-order approximation to the self-energy yields contributions to all orders in the Green's function. The same is true for the screened interaction that is obtained from a Dyson-like Bethe–Salpeter equation for the two-particle correlation function. The T-matrix approximation is an example of a succesful extension of this strategy, where the classical screened Coulomb interaction W of the GWA is replaced by a generalized interaction T, that is again obtained from a BSE. It includes scattering diagrams, also called ladder diagrams, and it is a useful approach in the limits of almost empty or almost filled bands. Dyson-like equations are also solved in FLEX or in parquet theory, where both bubble and ladder diagrams are included.

As an alternative to the Dyson equations, one can use cumulants in Sec. 15.7: in that case, the Green's function is expressed in terms of an exponential, where the exponent, the cumulant, is approximated to low order. Expansion of the exponential again yields contributions to all orders. The cumulant approach is well suited for the description of electron–boson coupling, for example for plasmon satellites.

Because perturbation theory has problems of convergence when energy differences become small compared with the interaction, it is useful to work with a weaker, screened, interaction, instead of the bare one. *Screening* contains much of the effects that the system of all electrons has on the propagation of an electron or hole. It is a natural bridge to the concept of *embedding*: a small subsystem that experiences the effects of the rest of the system through some effective potential or effective interaction. This concept can be found all over this book, starting from the introductory material in Sec. 7.1. It is the central idea in the following chapters on dynamical mean-field theory.

SELECT FURTHER READING

Csanak, C., Yaris, R., and Taylor, H. S., *Adv. At. Mol. Phys.* **7**, 287, 1971. Overview article that puts several approximation schemes into context.

Fetter, A. L. and Walecka, J. D. *Quantum Theory of Many-Particle Systems* (McGraw-Hill, New York, 1971). Overview and precise discussion of many-body perturbation theory, e.g., the T-matrix approximation.

Kadanoff, L. P. and Baym, G. *Quantum Statistical Mechanics* (W. A. Benjamin, New York, 1964). A short classic with some more discussion about approximations.

The review article by L. Hedin in *J. Phys. C* **11**, R489, 1999, "On correlation effects in electron spectroscopies and the GW approximation" contains a great overview of GW and cumulant approaches to photoemission, with a detailed discussion of electron–boson coupling hamiltonians, solutions, and open problems.

The book of R. D. Mattuck and D. Richard, *A Guide to Feynman Diagrams in the Many-Body Problem* (Dover Books on Physics, 1992) contains a good introduction to various approximations, and additional material on other approaches such as renormalization groups.

Exercises

15.1 Analyze electron addition energies by analogy with the analysis of electron removal energies in Sec. 15.1.

15.2 Derive an iterative expression like Eq. (15.4) for the reducible vertex Γ, starting from its definition Eq. (11.1) and inserting a chain rule with respect to a Green's function G^{vl} that stems from a local potential.

15.3 Expand Eqs. (10.48), (15.16), (10.37) to second order in the Coulomb interaction to show that the unscreened T-matrix approximation in either the particle–particle or the electron–hole channels yields the exact second-order self-energy. Show that this also holds for the screened T-matrix approximation provided the Hartree and Fock terms keep the bare interaction. Examine the situation also to third order.

15.4 Show the effective screening of a channel by another channel in a two-step FLEX approach: split the correlation self-energy into two parts $\Sigma_c = \Sigma_a + \Sigma_b$. Suppose that Σ_a has been determined. Make a T-matrix ansatz for Σ_b and determine T_b following the procedure outlined in Sec. 10.5.

15.5 Show that when one is using a q-independent $f_{xc}(q)$, an unphysical test charge–test electron dielectric function is obtained, whereas the test charge–test charge dielectric function does not show the same pathology.

15.6 Compare the RPA, the test charge–test charge and the test charge–test electron dielectric functions by expanding them to first order in $f_{xc}\chi_{KS}/\epsilon^{RPA}$. Discuss the correction with respect to the RPA, at $\omega = 0$ and for the $\mathbf{G} = \mathbf{G}' = 0$ element, by estimating the sign and magnitude of the correction terms. Show that for materials with large screening the corrections in the test charge–test electron dielectric function cancel and a result close to the RPA is obtained, which may explain partly the success of GW. What happens for materials with weaker screening? These are qualitative discussions, based on static screening and scalar functions.

15.7 Understand the power of exact constraints: make an ansatz for the one-body Green's function as a finite sum over poles, where the positions and weights of the poles are parameters. This ansatz by itself already uses knowledge about the analytic form of the exact Green's function. Determine the parameters of this ansatz by using exact constraints like the sum rules in Sec. 11.3. Which methods in this book might be used to calculate the quantities that appear? Do you have ideas for other exact constraints to be used?

15.8 Add and subtract the energy of an excited $(N + 1)$-electron state in Eq. (15.2). Make the approximation of a quasi-particle in the presence of a pair of particles. What kind of quasi-particle and what kind of pair do you find? Discuss panel (c) of Fig. 15.1. What do you have to do in order to find approximately panel (d)?

15.9 Derive Eqs. (15.6) and (15.7) using $\Sigma_{xc}^{(0)} = v_{xc}$ in Eq. (15.3).

15.10 Discuss the equations of the vertex derived from a Kohn–Sham potential in Sec. 15.2 in the limit of a system with one electron. Show that this vertex correction cures the self-screening problem. Also discuss the difference between using an RPA or the exact W in the GWA. Why is the RPA better for this particular aspect of the problem? What about electron *addition* – do the arguments still hold, or should we use a vertex derived from a non-local potential? To answer this question, imagine a case where the additional electron has no spatial overlap with the occupied state.

15.11 Using the expressions in Sec. 14.11, calculate the correlation contribution $\int dx_1 \int d_2 \Sigma(1,2) G(2,1^+)$ to the total energy. Show that only a part of the vertex $\tilde{\Gamma}$ contributes, and that it is a function of two arguments only, linked to the exact many-body exchange–correlation kernel of Eq. (14.63).

15.12 Derive the sum rules for $A(\omega)$, Eqs. (5.95), as well as $\int d\omega\, \omega A(\omega)$ and $\int d\omega\, \omega^2 A(\omega)$ in the case of the GWA and for the exact many-body system. Compare with Eqs. (11.30) and (11.31). As a hint, for the exact case try to use the closure relation by transforming frequencies into energy differences. You will find expressions containing commutators with the hamiltonian. In the case of the GWA, use the Dyson equation and the behavior of the functions at large frequency that follows from the Lehmann representation Eq. (5.36), following [352].

16

Dynamical mean-field theory

Think globally, act locally.

Summary

Dynamical mean-field theory is designed to treat systems with local effective interactions that are strong compared with the independent-particle terms that lead to
delocalized band-like states. Interactions are taken into account by a many-body
calculation for an auxiliary system, a site embedded in a dynamical mean field,
that is chosen to best represent the coupling to the rest of the crystal. The methods are constructed to be exact in three limits: interacting electrons on isolated sites,
a lattice with no interactions, and the limit of infinite dimensions $d \rightarrow \infty$ where
mean-field theory is exact. This chapter is devoted to the general formulation, the
single-site approximation where the calculation of the self-energy is mapped onto a
self-consistent quantum impurity problem, and instructive examples for the Hubbard
model on a $d \rightarrow \infty$ Bethe lattice. Further developments and applications are the
topics of Chs. 17–21.

One of the most rewarding features of condensed matter theory is the ability to address difficult problems from different points of view. The preceding Chs. 9–15 present an approach
based on perturbation expansions in the Coulomb interaction. In particular, the GW approximation for the self-energy has proven to be extremely successful in describing electronic
spectra of many materials, as described in Ch. 13. The methods can be applied to the
ordered states of materials with d and f states, for example, ferromagnetic Ni and antiferromagnetic NiO, as described in Secs. 13.4 and 20.7. However, the GW and related
approximations have difficulties in treating cases with degenerate or nearly degenerate
states and low-energy excitations. Present-day methods do not describe phenomena like
the fluctuations of local moments in a ferromagnetic material above the Curie temperature
or the insulating character of NiO in the paramagnetic phase; more generally, they have
difficulty describing strong correlation.

The topic of this chapter and Chs. 17–21 is dynamical mean-field theory, which is also a Green's function method in which the key quantity is the self-energy. However, it is designed to treat strong interactions for electrons in localized atomic-like states, such as the d and f states in transition metals, lanthanide and actinide elements and compounds. Instead of a systematic perturbation expansion, the methods are constructed to be correct in three limits: isolated atoms including all effects of interactions, an extended solid with no interactions, and the limit of infinite dimensions ($d \rightarrow \infty$) where mean-field theory is exact. DMFT is designed to treat materials with local moments in a disordered phase, e.g., a magnetic material such as Ni and NiO above the transition temperature; metal–insulator transitions, such as in V_2O_3; and highly renormalized behavior, such as in the heavy fermion material $CeIrIn_5$, where quasi-particles emerge at low temperature ($T \lesssim 10\,\mathrm{K}$) with mass ≈ 100 times that expected in an independent-particle picture. These and other illuminating examples are discussed in Ch. 20. In this chapter and the next, the methods are developed for Hubbard-type models; steps toward quantitative methods for materials are the topics of Chs. 19–21.

The perturbation methods in Chs. 9–15 and the approach in DMFT are by no means exclusive or contradictory; indeed, a theme of this book is the power of combining methods. DMFT is an example of an approach to incorporate auxiliary embedded systems in Green's function methods that can be used in many contexts. Furthermore, application of DMFT to real materials requires that we determine the effective screened interactions, which can be accomplished only if we have a theory of screening such as the RPA in Ch. 11. In addition to the d and f states, the calculations must also treat the rest of the system; these tend to be delocalized bands that are often well described by the GW approximation. If we adhere carefully to fundamental principles, we can use different methods to draw meaningful conclusions in a unified way for broad ranges of materials and phenomena. Methods of the future are likely to combine techniques from both approaches (and/or others). The combination of many-body perturbation theory and DMFT is the topic of Ch. 21.

In this chapter, the first section is devoted to the general approach of using auxiliary systems in Green's function methods and Sec. 16.2 provides an overview of DMFT. The next two sections describe the intuitive approach of expanding about an atomic limit (Sec. 16.3) and pertinent background for mean-field theories (Sec. 16.4). In Sec. 16.5 is the formulation of DMFT in terms of an auxiliary system consisting of a single site or a cell embedded in a dynamic mean field, derivation of the DMFT equations, and the outline of an algorithm for self-consistent calculations. Functionals for the grand potential Ω and a variational derivation of the DMFT equations are presented in Sec. 16.6, and methods to calculate integrated quantities like the total energy are the topic of Sec. 16.7. Instructive examples of calculations for the Hubbard and related models in the single-site approximation are the topics of Secs. 16.8–16.11. The final section is a short wrap-up with a few pointers to further development in later chapters.

16.1 Auxiliary systems and embedding in Green's function methods

Before delving into the specifics of DMFT, it is valuable to point out the overall strategy to utilize auxiliary systems in Green's function methods. The advantages of an auxiliary system are that in principle it can reproduce selected properties of the full interacting many-body system by calculations on a simpler system; in practice it can provide avenues for useful approximations. The power of such an approach can be appreciated from the success of density functional theory. In some ways, DMFT is analogous to the Kohn–Sham method but it involves a dynamic Green's function and self-energy instead of the static density and the exchange–correlation potential. However, this is not merely a generalization to dynamic quantities; as we shall see, an *interacting auxiliary system* provides a way to calculate the local Green's function and to use the entire repertoire of Green's function methods with contributions to the self-energy taken from the auxiliary system rather than from a perturbation expansion as in previous chapters. In fact, the two types of approach (auxiliary systems and many-body perturbation theory) can be combined using the formulation developed in Ch. 7, where the self-energy is the key to combining different contributions to the Green's function.

The first step in constructing an auxiliary system is to identify the part (or parts) of the system to be included. In general, this is a "small" part (S) of the system that is coupled to the rest (R) that is larger. For example, S might contain the states associated with one of the atoms in an infinite crystal. As presented in Eqs. (7.1)–(7.8), the hamiltonian and Green's function matrices can be written as

$$H = \begin{bmatrix} H_S & H_{SR} \\ H_{RS} & H_R \end{bmatrix} \quad \text{and} \quad G^{tot} = \begin{bmatrix} G_S & G_{SR} \\ G_{RS} & G_R \end{bmatrix}, \tag{16.1}$$

where G_S (the part of the full Green's function G^{tot} in the small space) can be expressed in terms of the Green's functions for the decoupled systems $G_S^0(\omega)$ and $G_R^0(\omega)$:

$$G_S(\omega) = \left[\omega - H_S - H_{SR}\, G_R^0(\omega)\, H_{RS}\right]^{-1}. \tag{16.2}$$

This is a very general formulation for "embedding" the smaller system S in the rest, and deriving the properties of S including the fact that it can exchange particles, have interactions, and in general is entangled with the rest of the system. There are no approximations at this point.

An interacting auxiliary system to represent selected properties of S can be defined by keeping the full H_S but replacing the rest, H_R and H_{SR}, by an effective medium (a "bath") chosen so that the resulting many-body problem can be solved. This is depicted in Fig. 16.1, where the circle with lines represents the many-body states of the small system S and the gray region is the bath. There are various choices for the auxiliary system and the properties to be calculated, e.g., the one-particle Green's function for the small system can be determined if we choose the H_{SR} to be hopping terms $t_{\ell,\ell'} c_\ell^\dagger c_{\ell'}$ with states ℓ in S and ℓ' in the bath or vice versa, which add or remove electrons from S. This is the essence

Figure 16.1. Schematic illustration of an auxiliary embedded system. The circle with horizontal lines represents a part of the actual system (labeled S in Eq. (16.1)), which is small enough that the interacting electron problem can be solved. The shaded region represents the bath, an effective medium that is simple enough that the desired properties can be calculated. Coupling with the bath can take various forms, for example, the arrows represent electrons hopping to and from the bath at different times that can be described by a hybridization function $\Delta(\omega)$ (see, e.g., Eq. (16.9)). By construction, the auxiliary embedded system can be solved with no approximation; the approximations enter in the way it is used to represent the actual system of interacting electrons. In DMFT there is an array of auxiliary systems for each site or cell in the crystal, each equivalent to one like the embedded site in Fig. 16.2(c).

of the Anderson impurity model (Sec. 3.5) that is discussed further in Sec. 16.5, where it is pointed out that the bath can be an interacting system and need not be restricted to the independent-particle assumption for the bath in Eq. (3.8).

An essential part of the strategy is that the auxiliary embedded system must be soluble for a range of possible baths and it must be flexible enough to describe the chosen properties of the actual system. In principle, it may be possible to find a bath that reproduces exactly some selected property of the small system S. In general, however, one must make approximations based on physical arguments to find a criterion for the auxiliary system that best represents the actual system. This depends on the problem, and the approach in DMFT is described in general terms in the next section, with references to later sections for details.

There are many precedents for embedded systems, in particular, an enormous body of work in quantum chemistry designed to use high-level many-body methods for a small part of a large system. See, e.g., [761] and references given there. Quantum mechanics/molecular mechanics (QM/MM) methods involve the solution of the interacting electron system in a region that is coupled to a force field representing the rest of the system. Embedding in a quantum system (QM/QM) can be in a spatial sense analogous to DMFT, or it can be a very general approach formulated in terms of matrices like in Eq. (16.1), where a block is chosen to include the most correlated states, which is embedded in the rest of the system that is treated by more approximate methods.

The reader is encouraged to keep an eye toward ways in which the DMFT approach can be used in other problems and ways that experience with other methods can be used to improve current methods in DMFT.

16.2 Overview of DMFT

Dynamical mean-field theory[1] is designed to treat the correlated system of electrons in a crystal with strong local interactions. The primary applications are to materials containing transition metals, lanthanides or actinides with localized atomic-like d or f states, where the effects of interactions are evident from the many phenomena observed experimentally, as illustrated in Ch. 2 and the examples in Chs. 13 and 20. The elements of most interest are indicated in Fig. 19.1, which shows the progression from the most delocalized band-like states at the start of the $5d$ transition series to the most highly localized atomic-like states at the end of the $4f$ lanthanide series. It is most natural to formulate the problem in a basis of localized states (orbitals) centered on the atoms, and we focus on cases where the matrix elements for interactions on a site are large compared with the matrix elements for single-particle hopping or interactions involving orbitals on different sites.

The interacting-electron problem in a crystal is indicated schematically in Fig. 16.2(a), where a circle represents a site (a single atom or a cell with more than one atom) and the lines inside the circles denote many-body states of the interacting system on the site.

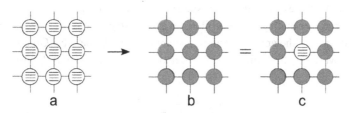

a   b   c

Figure 16.2. Schematic illustration of the DMFT method: the single-site approximation (single-site DMFA) if a site represents a single atom or a cluster approximation (Ch. 17) if a site is replaced by a cell of more than one atom. (a) A crystal with sites shown as circles, independent-particle hopping between sites indicated by connecting lines, and the states of the interacting system on each site indicated by the lines in each circle. (b) Gray circles indicating the one-particle Green's function with the approximation that the self-energy $\Sigma(\omega)$ is restricted to each site (Eq. (16.5)). (c) An interacting site embedded in the rest of the crystal treated as an effective medium with the self-energy $\Sigma(\omega)$ on each site. This is equivalent to an auxiliary system in Fig. 16.1 with a bath that is not yet determined, since $\Sigma(\omega)$ is determined only at the solution. This is a self-consistent problem and the equal sign indicates the condition that the self-energy for a site (the central site in (b)) is represented by the embedded site in (c): $\Sigma(\omega) = \Sigma_{aux}(\omega)$ and $G_{00}(\omega) = \mathcal{G}(\omega)$ (Eq. (16.6)). See Sec. 16.5 for a more detailed description.

---

[1] Two seminal advances have led to the present formulation of DMFT. The first is the work of Metzner and Vollhardt [762, 763] and Müller-Hartmann [764] in 1989, who developed the quantum theory of interacting particles in infinite dimensions. (Related work was done by Brandt and Mielsch [765] and Janis [766].) The second advance is the recognition by Georges and Kotliar [767] and by Jarrell [768] in 1992 that an embedded site can be viewed as an auxiliary system equivalent to an Anderson impurity model. Thus, the analytic and computational methods developed for the AIM are directly applicable as solvers for DMFT calculations, as described in Ch. 18, and many insights developed over the years can be used to understand the phenomena caused by interactions.

For the present discussion we assume interactions are confined to each site[2] and the only coupling between sites in the hamiltonian is independent-particle hopping indicated by the lines connecting the sites in the figure. To keep the notation simple, we describe the theory in terms of sites in the crystal, with the understanding that the same formulas apply whether the site is a single atom or a cell of atoms. There are many details in carrying out the actual calculations, and there are various choices in dealing with cells of more than one atom that are described in Ch. 17, but we can appreciate the general overall approach in DMFT that is the basis for all the methods.

Auxiliary systems and Green's functions

The theory is cast in terms of the single-particle Green's function for the crystal, the same as in previous chapters,

$$G_{\mathbf{k}}(\omega) = [\omega - \varepsilon_{\mathbf{k}} - \Sigma_{\mathbf{k}}(\omega)]^{-1}, \tag{16.3}$$

where each term is a matrix in the basis of local orbitals. The independent-particle bands are given by $\varepsilon_{\mathbf{k}}$ and effects of interactions are included in the self-energy $\Sigma_{\mathbf{k}}(\omega)$. The functions can also be expressed in real space as $G_{ij}(\omega)$ and $\Sigma_{ij}(\omega)$, where i and j are site indices and each is a matrix in the orbitals on a site. Since the local interactions and correlation are central to the approach, an important quantity is the local Green's function for a site $G_{ii}(\omega)$, which is the same at each site and can be expressed as the average over the Brillouin zone

$$G_{00}(\omega) = \frac{1}{N_{\mathbf{k}}} \sum_{\mathbf{k}} G_{\mathbf{k}}(\omega), \tag{16.4}$$

where the sum is over $N_{\mathbf{k}}$ points in the BZ. The theory of Green's functions developed in Chs. 5 and 7–15 applies here as well: the behavior as a function of frequency z in the complex plane, concepts of quasi-particles, functionals for the grand potential Ω in terms of the Green's function or self-energy, and so forth.

The goal is to provide useful methods to calculate the self-energy $\Sigma_{\mathbf{k}}(\omega)$, and the strategy in DMFT is to utilize an array of auxiliary systems on the sites of the crystal. Each is like that depicted in Fig. 16.1, with a site that has the same interacting hamiltonian as a site in the crystal, but the coupling to the rest of the crystal is replaced by coupling to a bath. The only role for the auxiliary system is to provide a way to determine the self-energy and the only results needed from the calculation for the auxiliary system are the Green's function, which is denoted $\mathcal{G}(\omega)$, and the self-energy $\Sigma_{aux}(\omega)$. Methods for the calculations are described in Ch. 18; however, for purposes of developing the method all we need to know is that there are feasible methods to solve the problem.

Before describing methods for DMFT, we can already appreciate that the theory can be constructed to be correct in two limits. Since the auxiliary system is only used to calculate

---

[2] Interactions between electrons in orbitals on different sites can be included in extended DMFT (EDMFT in Sec. 17.5).

the self-energy, there is no approximation of the independent-particle terms $\varepsilon_\mathbf{k}$; thus the results are exact, i.e., the dispersion is given correctly, in the limit of no interactions where $\Sigma_\mathbf{k}(\omega) = 0$. Since the auxiliary system has the same hamiltonian as the crystal for each site, in the limit of no coupling between sites it is just an isolated atom (or a cell of more than one atom) with no bath, which is presumed to be solved exactly.[3]

Practical methods: dynamical mean-field approximations

The approximations enter when we treat both intrasite interactions and dispersion due to intersite hopping. The central tenet of DMFT is that the local correlations are largest and the longer-range correlations are weaker and are approximated in practical methods. This is an example of the nearsightedness principle described by Kohn [515] and in the companion book [1], Ch. 23, where it is emphasized that the concepts apply to the range of correlation in an interacting many-body system. In DMFT there are two different aspects: limitation of the range of the intersite terms considered in the self-energy $\Sigma_\mathbf{k}(\omega)$ and approximation in the way the self-energy is calculated. The first aspect is that the self-energy is restricted to matrix elements only between orbitals on the same site. In the single-site approximation this means that there is no $\mathbf{k}$-dependence:[4]

$$\Sigma_\mathbf{k}(\omega) \rightarrow \Sigma(\omega), \tag{16.5}$$

which is depicted in Fig. 16.2(b) by the shaded circles representing the Green's function with a self-energy that has only on-site components, and coupling between sites that involves only independent-particle hopping as indicated by the lines.

The calculation of the self-energy $\Sigma(\omega)$ is the crux of the method. This is where the effects of interactions are taken into account and it is important to clarify the nature of any approximation. Fortunately, the single-site methods in this chapter and cluster methods in Ch. 17 can all be understood in a unified way. The common features of these approximations are that the lattice self-energy $\Sigma(\omega)$ is set equal to $\Sigma_{aux}(\omega)$ calculated for the auxiliary system and there is a consistency condition that determines the auxiliary system that best represents the crystal. This is illustrated in Fig. 16.2(b, c) and the equal sign between them. Figure 16.2(b) represents the Green's function for the crystal with the approximation that $\Sigma(\omega)$ is restricted to each site; however, $\Sigma(\omega)$ is unknown at this point. Figure 16.2(c) depicts a central site treated as an interacting-electron problem embedded in the rest of the

---

[3] Even if there is an exact solution for an isolated site, this does *not* mean that it is sufficient to determine the ground state of the crystal. As discussed in Sec. 16.3, if the ground state of each site is degenerate, the actual ground state of the crystal is determined by the coupling between sites and one must be careful to take the limit with the correct symmetry of the ground state. See Sec. 16.3.

[4] In the literature "DMFT" is often used to denote only the single-site approximation. Here the term "DMFT" denotes the general theory and various approximations are denoted by an abbreviation ending in "A," for example, the single-site DMFA and the cluster approximations, c-DMFA and DCA, defined in Ch. 17. The approach given here also applies directly to cluster methods where the single site is replaced by a cell with more than one atom, and $\Sigma(\omega)$ is restricted to matrix elements between orbitals on the atoms in the chosen cell. Then Eq. (16.5) becomes a matrix relation expressed in real space (Eq. (17.3)) or expressed as the variation of $\Sigma_\mathbf{k}(\omega)$ in $\mathbf{k}$-space (Eq. (17.4)).

crystal. This is equivalent to an auxiliary system like that depicted in Fig. 16.1 with a bath that is not determined since the Green's function for the crystal is not yet determined.

The equal sign indicates the condition that the Green's function $\mathcal{G}(\omega)$ and self-energy $\Sigma_{aux}(\omega)$ for the embedded site are the same as those on other sites in the crystal where they are given by $G_{00}(\omega)$ and $\Sigma(\omega)$. This is analogous to Weiss mean-field theory in Eq. (4.12) where each atom is in an average field due to the neighbors, with the requirement that its average moment is the same as for the neighbors, which are equivalent. Another way to express the condition is that $G_{00}(\omega)$ and $\Sigma(\omega)$ for a site in the crystal are the same as if it were a site with interactions embedded in the rest of the crystal.[5] In addition, the variational derivation given in Sec. 16.6 shows the sense in which the condition leads to an optimized auxiliary system. Since the calculation of $\mathcal{G}(\omega)$ and $\Sigma_{aux}(\omega)$ depends on the bath (the surrounding medium in Fig. 16.2(c) that is determined by $\Sigma(\omega)$), but $\Sigma(\omega)$ must equal $\Sigma_{aux}(\omega)$, the equations must be solved self-consistently.

Thus the condition that determines the auxiliary system can be written as:

$$G_{00}(\omega) = \mathcal{G}(\omega) \quad \text{with} \quad \Sigma(\omega) = \Sigma_{aux}(\omega), \qquad (16.6)$$

where $G_{00}(\omega)$ is defined in Eq. (16.4). This defines the equations for single-site DMFA and is readily generalized to the expressions that apply to the cluster approximations in Ch. 17. Methods to find the self-consistent solution are described in Sec. 16.5. Methods for calculations for the many-body interacting auxiliary system, called "solvers," are the topic of Ch. 18.

The only role of the auxiliary system is to calculate $\Sigma(\omega)$; there is no approximation for the independent-particle terms in the Green's function $G_{\mathbf{k}}(\omega)$ in Eq. (16.3).

Mean-field approximations for intersite correlation

Figure 16.2(c) also shows the nature of the approximation: the interactions are taken into account explicitly for the central site in the auxiliary system and the correlations on the site are treated exactly to the extent that the bath represents the rest of the crystal. However, correlations between sites are treated only in an average sense. There are no explicit correlations between electrons in orbitals on different sites and the auxiliary system provides no direct information on intersite correlation. The approximation can be described in terms of diagrams, as illustrated in Fig. 21.5. The single-site DMFA corresponds to keeping all diagrams for $\Sigma(\omega)$ on the same site with an effective interaction; this can be contrasted

---

[5] A brief diversion with some history can help bring out the meaning of the equation depicted in Fig. 16.2. A precedent for this approach is the work of Hubbard. As summarized in Sec. K.2, his first approximation (called "Hubbard I") is the assumption that $\Sigma(\omega)$ is the same as for an isolated atom even if there is dispersion. This corresponds to replacing Fig. 16.2(c) by an isolated site. However, Hubbard realized that a much better approximation is an atom in its average environment in the crystal, which is the essence of the "Hubbard III" approximation (see Sec. 16.4 below). Hubbard combined this with the static alloy approximation that is a forerunner of the dynamical interacting many-body embedded site method in DMFT.

with GWA, which includes only certain diagrams but there is no restriction to sites and the screened interaction is calculated as part of the method.

If the sites in Fig. 16.2 are single atoms, correlation is taken into account for electrons in the orbitals on each atom embedded in the average environment provided by its neighbors. This is a mean-field approximation for correlations in space but it is a dynamic quantum theory for correlations in time; comparisons and contrasts with other mean-field methods are described in Sec. 16.4 and listed in Tab. 16.1 later. As discussed in Sec. 16.9, mean-field theory is exact in the limit of a large number of neighbors, and thus we have a third limit in which the solutions are exact, the limit of infinite dimensions $d \to \infty$.

If the sites in Fig. 16.2 are replaced by cells of more than one atom, correlations are taken into account in the auxiliary system electrons in any orbitals within a cell. But the effect of the rest of the crystal outside a cell is treated in a mean-field sense. This amounts to self-consistent mean-field boundary conditions on the many-body problem for the cell. There is more than one way to specify such boundary conditions, and two examples are discussed in Ch. 17: the cellular approximation (c-DMFA) in real space and the dynamic cluster approximation (DCA) where the bath is a function of $\mathbf{k}$ at a discrete set of $\mathbf{k}$-points. Each involves a simulation cell and if the cell is allowed to grow then the results of either method approaches the exact solution. An example using DCA for the 3D Hubbard model is given in Sec. 17.6 (see Fig. 17.8).

Why is it useful to consider this procedure as an auxiliary system instead of just a problem of finding the self-consistent solution for coupled equations in Eq. (16.6)? The auxiliary system clarifies the separation of the problem into two distinct parts: calculation of $\Sigma_{aux}(\omega)$ and its use to find the self-energy for the crystal $\Sigma(\omega)$. The first part is a general approach for embedded systems and can be used in many contexts, as pointed out in Sec. 16.1. The auxiliary system is equivalent to a generalized Anderson impurity model with a bath that must be determined, and we can build upon the extensive body of work on the AIM, which is a prototype system for interacting electrons (see Sec. 3.5). All the examples in this chapter and the single-site DMFA calculations in Ch. 17 have the same auxiliary system that applies for any one-band Hubbard model in any dimension. Moreover, an auxiliary system with one site is not limited to the mean-field single-site approximation or the mean-field boundary conditions on a cluster. In principle, there is a bath that leads to the exact local Green's function $G_{00}(\omega)$, the quantum impurity representation in Sec. 16.5, which provides an underpinning for finding further improved approximations. The same auxiliary system can be used in conjunction with perturbation methods to include intersite correlation, where Eq. (16.6) is replaced by more general conditions, and to include many-body effects in other bands, where $\mathcal{G}_0(\omega)$ is replaced by an interacting Green's function. These are the topics of Sec. 17.5 and Ch. 21.

16.3 Expansion around an atomic limit: low energy scales and strong temperature dependence

Since DMFT is applied to problems where atomic-like behavior occurs in the solid, we start by considering a crystal as built from atoms (or ions) brought together. For isolated atoms the electrons are confined to regions of atomic size and the Coulomb interaction between electrons in the localized states is many electron volts, whether it is helium, silicon, iron, or cerium. As the atoms are brought together, the interactions are screened and the states become delocalized due to hopping between atoms, with very different consequences for different classes of solids.

Closed-shell versus open-shell systems

A closed shell atom has a non-degenerate ground state and an energy gap for all excitations. Binding between atoms is due to van der Waals or dispersion interactions that result from correlated dipole–dipole polarization fluctuations on nearby units. In this case the progression from the atomic limit can be taken literally and the intra-atomic interactions remain much larger than any energy related to interatomic interactions; this is illustrated by the binding energies for rare gases listed in Tab. 2.1. Due to the large gap in all excitations, the form of the weak interactions between atoms is captured by perturbation theory, as discussed in the context of MP2 and GW approximations in Chs. 11 and 13.

An open-shell system has a partially filled valence shell with orbital and/or spin degeneracy. For one electron in a shell of angular momentum L, there are $2(2L + 1)$ states that are degenerate if there is no spin–orbit coupling. When there is more than one electron, Coulomb interactions lead to a set of possible energy levels called a multiplet. An example is shown in the inset to Fig. 19.3 for a V^{3+} ion with two electrons in the $3d$ shell. The ground state of an isolated atom normally obeys Hund's rule of maximum S, maximum L so that it is degenerate and has a magnetic moment. The lines inside each circle in Figs. 16.1 and 16.2 indicate that there are many possible states and non-trivial dynamics of the electrons associated with the atom.

The formation of a solid inherently leads to a competition of local atomic-like interactions versus interatomic hopping matrix elements in the solid. If the latter dominates the states lose their atomic character and form strong bonds and bands of delocalized states. In contrast, atomic-like character can persist in the solid if the local intra-atomic interactions are large compared with the bandwidths due to interatomic hopping. The degenerate atomic states with magnetic moments evolve into "local moments" in the solid that can order at low temperature or be thermally disordered at high temperature. Indeed, local moment behavior is a central feature of all the examples brought out in Ch. 20. This happens in practice in transition metal elements and compounds where there are two different types of atomic states at work: extended states, like s and p, that form bonds to stabilize the solid, and much more localized d or f states that are partially filled, i.e., they act like open-shell atoms weakly coupled to the other states in the solid. The derivation of the hamiltonian for solids is deferred to Chs. 19–21, where there is quantitative consideration of the magnitudes of the interactions in atomic-like states versus the competing interatomic hopping matrix elements.

The atomic limit is not unique for degenerate atomic states

Much confusion can be avoided if we recognize the fact that for open-shell atoms there is not a unique "atomic limit." A lattice of decoupled atoms is massively degenerate and is unstable to perturbations no matter how small. The coupling between the atoms always has a qualitative effect and the interatomic interactions determine whether it will be magnetically ordered, a quantum Fermi liquid, superconductor, or other state. One must start with the appropriate nearly atomic limit (for example, with a small but non-zero interaction that stabilizes the desired state) that is continuously connected to the actual final state of the solid. This becomes a critical issue in the analysis of a Mott metal–insulator transition, especially for low temperature as described in Secs. 16.9 and 16.10.

Correlation, energy scales, and temperature dependence

A signature of systems with strong local interactions in open-shell atoms is the emergence of different energy scales and different types of correlation. The transition elements tend to retain atomic-like character in the solid, with large energies associated with highly correlated electrons in the localized atomic-like states and small energies associated with interactions between the atomic-like states on different sites, for example, the energies for spin waves that depend upon the relative orientation of magnetic moments on neighboring sites. Indeed, the larger the local interactions relative to the bandwidth, the smaller the scale of energies that characterize the correlations between sites. This can be illustrated by the two-site Hubbard model with on-site interaction U and hopping matrix element t; as explained in Exs. 3.5 and 16.4, for large $U \gg t$ the interaction on each site leads to states with high energy $\approx U$, whereas the antiferromagnetic coupling between sites has magnitude $|J| \propto t^2/U$, which is much less than t for strong interactions. In the Anderson impurity model in Sec. 3.5, the characteristic energy scale is $T_K \propto \exp(-\pi U/2\Delta)$ in the regime where U is large compared with the independent-particle impurity resonance width Δ (see Fig. 3.4).

Many of the most compelling reasons for adopting the start-from-atoms point of view are temperature-dependent phenomena due to the low energy scales. Almost every example involving materials with d and f states involves temperature in a crucial way; describing such phenomena requires methods that can predict properties of materials as a function of temperature.

16.4 Background for mean-field theories and auxiliary systems

Mean-field theories and methods (see Ch. 4) are extremely useful and illuminating: witness the impact of the Kohn–Sham approach that uses independent-particle methods or the Weiss mean-field approximation for interacting spins. Similarities and differences of several mean-field methods are compared in Tab. 16.1. Comparison of density functional, Hartree–Fock, and Green's function methods is presented in Tab. 8.1; the difference here is the addition of methods specifically for lattices and the specialization of the Green's function methods to DMFT. Each example illuminates an important feature used in DMFT.

Density functional theory

The Kohn–Sham approach in density functional theory (DFT) is listed first in Tab. 16.1 because it is the quintessential example of an auxiliary system and useful functionals in the

Table 16.1. Comparison of mean-field theories and auxiliary systems based on functionals of a local observable. For DFT, spin systems, and DMFT, in principle the auxiliary system can give the exact values of the local observable, but the static alloy is intrinsically an approximation. In each case there is a canonical local approximation that assumes no correlation between neighboring sites (or at different points in space for the LDA). The LDA+U, LDA+G, and hybrid functional extensions of DFT are listed, even though they are not analogous to the examples in the other columns. Symbols and abbreviations are explained in the text

Theory	Kohn–Sham DFT	Weiss MFT	Hubbard static alloy approximation	DMFT local spectral function
Local observable	Density $n(\mathbf{r})$ at point $\mathbf{r}$	Magnetization $\mathbf{m}_i$ at site i	Disorder averaged $G_{ii}(\omega)$ at site i	Green's function $G_{ii}(\omega)$ at site i
Auxiliary system	Independent particles	Embedded spin	Embedded sites independent particles	Embedded site interacting particles
Effective field	Kohn–Sham local potential $v_{KS}(\mathbf{r})$	Weiss local field $\mathbf{h}_i^{\text{eff}}$	Hybrid function $\Delta(\omega)$ or effective medium $\mathcal{G}_0(\omega)$	Hybrid function $\Delta(\omega)$ or effective medium $\mathcal{G}_0(\omega)$
Exact formulation	Universal $E_{xc}[n]$ $v_{xc}(\mathbf{r}) = \frac{\delta E_{xc}}{\delta n(\mathbf{r})}$	Exact $\Omega[\{\mathbf{m}\}]$ $\mathbf{h}^{\text{eff}} = \frac{\delta \Omega}{\delta \mathbf{m}}$	None	Exact $\Phi_{local}[G_{00}(\omega)]$ $\Sigma(\omega) = \frac{\delta \Phi_{local}}{\delta G_{00}}$
Canonical local approximation	LDA $E_{xc} =$ $\int n(\mathbf{r})\epsilon_{xc}^{hom}(n(\mathbf{r}))$ Sec. 4.3 $T = 0$	Weiss MFA $\mathbf{h}^{\text{eff}} = zJ\langle \mathbf{m}\rangle$ Eq. (4.13) $T \neq 0$	CPA $G_{00}(\omega) = \overline{\mathcal{G}}(\omega)$ Eqs. (16.7) and (K.13) approximation to $T \neq 0$	Single-site DMFA $G_{00}(\omega) = \mathcal{G}(\omega)$ Eq. (16.6) $T = 0$ and $T \neq 0$
Examples: improved approximation and extensions	Gradient expansions LDA+U DFT+G hybrid functionals	Cluster expansions Bethe–Peierls	Cluster expansions Molecular CPA Static DCA	Cluster expansions Cellular DMFA DCA EDMFT

theory of interacting electrons. It is summarized in Ch. 4 and a formulation in terms of the Kohn–Sham Green's function is given in Sec. 8.4. The success of the Kohn–Sham method is largely due to the fact that it defines an auxiliary system designed to describe only part of the properties of the system, the ground-state density and the total energy. Instead of making approximations to the original problem, the auxiliary system is sufficient to describe those properties exactly in principle. Since these are static ground-state properties, it is sufficient to work with a static exchange–correlation potential v_{xc}. A theory such as DMFT that seeks to describe the Green's function $G(\omega)$ must involve a dynamic self-energy $\Sigma(\omega)$ for correlation. The Kohn–Sham auxiliary system consists of non-interacting particles; however, the exchange–correlation functional $E_{xc}[n]$ must be derived for the interacting system. Since DMFT is a method to calculate explicitly the effects of interactions, it must solve an interacting auxiliary system. The correspondences are indicated in Tab. 16.1.

The original proposal of the local density approximation by Kohn and Sham was built upon the idea of the locality of the combined effects of exchange and correlation; the remarkable success of the LDA provides a measure of support for the locality assumed in the single-site DMFA, where the analog of the density $n(\mathbf{r})$ at point $\mathbf{r}$ is local on-site Green's function $G_{ii}(\omega)$ at site i. The cluster methods in Ch. 17 correspond to generalized gradient expansions in DFT; further improvements such as hybrid functionals can be expressed in terms of the density matrix that is also contained in the Green's function; similarly, the DFT+U and DFT+G methods described in Sec. 19.6 can be viewed as static approximations for DMFT; and so forth.

Spectral density functional theory

The generalization of the Kohn–Sham density functional approach to a Green's function functional has been termed spectral density functional theory (SDFT) and a concise summary can be found in [303]. The theory can be constructed formally using the generating functional methods of Sec. 8.4. This is worked out for the density functional in Ex. 8.3 and the arguments are given in the text after Eq. (8.38) for the generalization to the Green's function. The difficulty arises when we try to express the interaction energy as a functional of the local Green's function $G_{00}(\omega)$. The Luttinger–Ward functional $\Phi_{LW}[G(\omega)]$ is defined in terms of the full non-local Green's function $G(\omega)$ by an explicit expression as a series, but there is no such expression for a functional $\Phi_{local}[G_{00}(\omega)]$ of only the local part of the Green's function. This is analogous to the exchange–correlation functional $E_{xc}[n]$ in the Kohn–Sham approach where there is no known expression. However, $E_{xc}[n]$ can be expressed in terms of a coupling constant integration, for example, in Eq. (8.45), and a similar approach can be employed for $\Phi_{local}[G_{00}(\omega)]$, as described in [303]. In the spectral density formulation, the single-site DMFA and improved cluster approximations are approximations to the exact functional. Even though the theory is difficult to employ in practice, it provides the basis for constructing an exact theory for the local spectral function $G_{00}(\omega)$ as listed in Tab. 16.1.

Weiss mean field and the Curie–Weiss approximation

Systems of interacting spins are among the most widely studied models for classical and quantum statistical mechanics, and they are often used to describe magnetic properties of solids. The Heisenberg model is a classic prototype described in Sec. 3.3, and the simplest approximation is the Weiss mean-field theory summarized in Sec. 4.2. In this approximation each spin acts as if it were in a field due to the average spin of its neighbors. Since the spins are really correlated electrons in atomic-like states, they are approximations to the central topic of this book, interacting electrons. Thus the fundamental theory of electrons in materials ultimately should describe the properties of spin systems in appropriate limits and it should go beyond the models in the cases where they are not sufficient or break down completely. Since the spin models are built on the idea of localized degrees of freedom, they are instructive examples. Indeed, $\mathcal{G}_0$ in Eq. (16.8) is called the Weiss function [302] in analogy to $\mathbf{h}^{\text{eff}}$ in the Weiss mean-field theory. Other parallels are brought out in Tab. 16.1.

The Weiss mean field illustrates several properties of mean-field theory useful later. One is that instead of a mean-field approximation, this can be considered to be an auxiliary system of a single spin embedded in a medium that determines an effective field $\mathbf{h}^{\text{eff}}$ in Eq. (4.12). Another is the property that mean-field approximations become exact in the limit of a large number of neighbors N, because the ratio of fluctuations to the average mean field is $\propto 1/\sqrt{N}$ (see Ex. 16.11). This is formally the case for lattices in which the dimensionality d is allowed to become large, i.e., the $d \rightarrow \infty$ limit. Mean-field theory can lead to a phase transition (see discussion after Eq. (4.13)); however, in finite dimensions fluctuations modify the transition temperature and may eliminate the transition at non-zero temperature in one and two dimensions (e.g., the Mermin–Wagner theorem for 2D systems [769]). Improved treatments take into account correlations among neighbors, for example, Bethe and Peierls [172, 173] proposed a finite-size cluster with a central site and its nearest neighbors embedded in a self-consistent mean field, similar to the cluster methods in Ch. 17.

The Hubbard III static alloy approximation

Closely related to dynamical mean-field theory is the work of Hubbard, who developed insightful approaches that treat interacting electrons in ways designed to give the correct spectrum in two limits: the independent-particle limit for vanishing interactions and the limit of isolated atoms including interactions. In particular, the paper often called "Hubbard III" [125] proposed two approximations. In the propagation of an electron, the other electrons are treated as fixed scatterers, i.e., the effect of interactions is like an alloy with a disordered array of potentials assumed to be randomly distributed with no correlation between sites. For the single-band Hubbard model, the energy for an electron on a site is $\varepsilon_0 + U$ if it is occupied by an opposite-spin electron, or ε_0 if it is not. The second approximation is the mean-field coherent potential approximation (CPA), where there is a complex effective potential that represents the average propagation in the disordered alloy. The conceptual structure and the actual equations have much in common with DMFT, and the static disorder can be considered to be an approximation for thermal disorder in DMFT; indeed, an alternative name for the single-site DMFA is the "dynamical CPA" [770, 771].

The equations for the Hubbard III approximation are derived in Sec. K.2 and parallels with DMFT are shown in Tab. 16.1. The correspondence becomes evident by comparing Figs. 16.2 and K.1. Each method is formulated in terms of the on-site Green's function for the effective medium, called $G_{00}^{CPA}(\omega)$ in Eq. (K.11), which is the same as Eqs. (16.3) and (16.4) with Σ replaced by $\Sigma^{CPA}(\omega)$. The CPA amounts to the requirement that $G_{00}^{CPA}(\omega)$ is equal to the weighted average of the Green's functions $\mathcal{G}^+(\omega)$ and $\mathcal{G}^-(\omega)$ for a site with or without an opposite-spin electron. The notation $\mathcal{G}$ denotes a Green's function for a site embedded in an effective medium, in this case the medium with Green's function $G_{00}^{CPA}(\omega)$. This is depicted by the equality in Fig. K.1, and it can be expressed as

$$G_{00}^{CPA}(\omega) = \overline{\mathcal{G}}(\omega) \quad \text{with} \quad \Sigma(\omega) = \Sigma^{CPA}(\omega),$$

where (16.7)

$$\overline{\mathcal{G}}(\omega) = n_{-\sigma}\mathcal{G}^+(\omega) + (1 - n_{-\sigma})\mathcal{G}^-(\omega),$$

for either spin where $n_{-\sigma}$ denotes the occupation by an opposite-spin electron. Since $\mathcal{G}^+(\omega)$ and $\mathcal{G}^-(\omega)$ involve $G_{00}^{CPA}(\omega)$ (the explicit expressions are given in Eq. (K.12)), solution of Eq. (16.7) is a self-consistent problem to find $G_{00}^{CPA}(\omega)$ (or equivalently $\Sigma^{CPA(\omega)}$).

Note the correspondence of Eq. (16.7) with the self-consistent equations for DMFT in Eq. (16.6). The difference from DMFT is the way the self-energy is calculated. In the static alloy approximation, $\Sigma^{CPA}(\omega)$ is determined by the solution of Eq. (16.7), which involves only independent-particle electron methods. The corresponding equation in DMFT is $G_{00}(\omega) = \mathcal{G}(\omega)$ in Eq. (16.6), where $\mathcal{G}(\omega)$ is the solution of the much more difficult dynamic many-body problem for interacting electrons on a site embedded in an effective medium. Of course, Hubbard was aware that the other electrons were not fixed and all electrons should be treated as equivalent. The resonant broadening correction of Eq. (K.14) was an attempt to take into account this dynamical effect, and it is remarkable how close the results are to DMFT results for the test case in Sec. 16.9.

The crucial effect missed in this static, disordered approximation is the low-energy dynamics of the electrons. A symptom of the problem is the finite lifetime for electrons at the Fermi energy in a metal. This is correct in a real alloy with scattering due to disorder and it may be reasonable for a crystal at high temperature, but it is incorrect at $T = 0$ where scattering vanishes at the Fermi energy. In addition it misses the low-energy features such as magnetic excitations, the Kondo effect, etc. The failure is due to the assumption that dynamic correlation can be replaced by static disorder. Two of the most satisfying consequences of DMFT are the proper decrease of the scattering as the energy approaches the Fermi energy and the development of low-energy dynamics in systems with fluctuating local moments.

16.5 Dynamical mean-field equations

Dynamical mean-field theory has much in common with other mean-field methods in Tab. 16.1. The central assumption in the practical development of all these methods is that correlation is strongest at short range and the canonical approximation is that correlation is purely local. In the local density approximation, the exchange–correlation energy is an integral of a quantity (see Eq. (4.22)) that depends only on the density at each point in space with no correlation between different points. In Weiss mean-field theory and the

Hubbard III CPA, fluctuations on different sites are uncorrelated. Like those methods, the single-site approximation ignores correlation between different sites and for each of the methods there are improved approximations such as those listed in Tab. 16.1. However, DMFT is qualitatively different from each of these methods. Unlike the static potential in DFT, the central quantity is the dynamic self-energy. Unlike the classical Weiss mean field, the hybridization function is a quantum field for creation and annihilation of particles. Unlike the static alloy picture of Hubbard, the local interactions are treated by an explicit many-body method that takes into account the local, correlated dynamics of the electrons. This overcomes some of the fundamental deficiencies in the alloy approximation and leads to entirely new features, such as low energy scales akin to the Kondo effect (Sec. 3.5).

The basic ideas of DMFT are laid out in Sec. 16.2. In this section we provide additional information and fill out the details of an algorithm for calculations. The formulation applies also to the cluster methods in Ch. 17 and the combination with many-body perturbation methods in Ch. 21.

The quantum impurity representation

A central part of DMFT is the auxiliary system for each site in the crystal, as depicted in Fig. 16.1: a "small" system with interactions between electrons in the states associated with that site, embedded in an effective medium (a bath). Hopping matrix elements (hybridization) between the states on the site and the medium have the effect of exchange of particles, as indicated by the arrows in Fig. 16.1, which is dynamic, i.e., dependent on the time difference or frequency, since there is a spectrum of energies in the medium. This is analogous to the Anderson impurity model in Sec. 3.5 with a hamiltonian given by Eq. (3.8) in terms of a state on a site with energy ε_0, an interaction U between up and down spins on the site, and coupling to a continuum of states determined by the host crystal. This is the model for the auxiliary system in DMFT, except that the site can represent a single atom or a cell of atoms with many states, the energy ε_0 can denote the matrix for independent-particle energies and hopping between states on the site, and the surrounding medium is a bath that is to be determined. In the AIM hamiltonian the host states are assumed to be independent-particle states, but in the auxiliary system the bath must represent the surrounding crystal where there are interactions on every site.

If the host states are non-interacting, the host Green's function on the site can be written

$$\mathcal{G}_0(\omega) = [\omega - \varepsilon_0 - \Delta(\omega)]^{-1}, \tag{16.8}$$

where the hybridization function $\Delta(\omega)$ can be expressed in the form of Eq. (7.15). However, the only essential requirement is that $\Delta(\omega)$ represents coupling to some system. As shown in Eq. (16.2), the bath can be an interacting system and the effect on the site is given by the generalized self-energy $\Sigma(\omega) = H_{SR} G_R^0(\omega) H_{RS}$ in Eq. (16.2), where $G_R^0(\omega)$ is the decoupled Green's function for the rest of the system. Thus the AIM can be generalized to include interactions in the host crystal by constructing the effective bath function $\Delta(\omega)$ and the Green's function $\mathcal{G}_0(\omega)$, which can be regarded as the effective "bare" Green's function for the AIM in the sense that it does not include the interaction on the site. The solution can be formulated as the problem to calculate the Green's function $\mathcal{G}(\omega)$ for the site including interaction U, given $\mathcal{G}_0(\omega)$; this is the job of the solver and various methods can be found in Ch. 18.

As illustrated in Fig. 16.2, we consider a site embedded in a crystal with one site removed, i.e., the rest of the crystal, which is described by a "cavity Green's function" denoted $G_{ij}^{-}(\omega)$, where i and j denote sites in the crystal.[6] In terms of the Green's function for the lattice $G_{ij}(\omega)$, the cavity function is given by (see Ex. 16.1) $G_{ij}^{-} = G_{ij} - G_{i0}[G_{00}]^{-1}G_{0j}$, and the generalized hybridization function can be expressed as

$$\Delta(\omega) = \sum_{ij} t_{0i} G_{ij}^{-}(\omega) t_{j0}, \qquad (16.9)$$

where the t's are the hopping matrix elements between the central site at 0 and the rest of the crystal. A simplification in infinite dimensions is that G_{ij}^{-} is simply G_{ij} (see Ex. 16.2) and the distinction can be ignored. This is useful in calculations with a model density of states as in Sec. 16.9, where one only needs G. However, the cavity construction adds no complication if the self-consistent algorithm is formulated in terms of $\mathcal{G}_0(\omega)$. As explained later in the text describing the algorithm in Fig. 16.3, at self-consistency the auxiliary system is equivalent to a site embedded in the cavity, i.e., in the rest of the crystal. At that point one can examine $\Delta(\omega)$ and $G^{-}(\omega)$ if desired. Note the correspondence with the expression of Eq. (7.8) for the Green's function in part of a system coupled to the rest.

This defines the auxiliary system: a site with interactions embedded in a bath described by $\mathcal{G}_0(\omega)$. The auxiliary system must be soluble for a range of $\mathcal{G}_0(\omega)$, i.e., it must be possible to find $\mathcal{G}(\omega)$ for any $\mathcal{G}_0(\omega)$ within the range. The relationship of $\mathcal{G}(\omega)$ and $\mathcal{G}_0(\omega)$ is an example of a functional described in Ch. 8, since the result of the calculation $\mathcal{G}(\omega)$ depends on the function $\mathcal{G}_0(\omega)$, i.e., $\mathcal{G}(\omega)$ is a functional of $\mathcal{G}_0(\omega)$.

The self-energy of the auxiliary system is calculated from the relation

$$\Sigma_{aux}(\omega) = \mathcal{G}_0^{-1}(\omega) - \mathcal{G}^{-1}(\omega). \qquad (16.10)$$

Note that in the previous chapters on Green's function methods, the approaches were framed in terms of diagrammatic expansions of the self-energy, which then determines the Green's function. Here the auxiliary system is solved by numerical methods and it is the Green's function that is the direct result, for example, the expression in terms of many-body wavefunctions and energies in Eq. (18.5) or in terms of Monte Carlo sampling in Eqs. (18.14), (18.22), or (18.35). As far as the auxiliary system is concerned, the self-energy plays no essential role; however, it is the essential quantity in order to use the auxiliary system in combination with other Green's function methods, for example, the general theory in Sec. 16.1 and the use in DMFT to construct the lattice Green's function in terms of Σ.

---

[6] The AIM is defined for an added impurity state instead of replacing a host atom by an impurity, but derivations and the use of the techniques for solving the AIM in Ch. 18 apply to both cases.

The auxiliary system is called a "quantum impurity representation" for the local Green's function $G_{00}(\omega)$. Let us rewrite Eq. (16.10) as

$$\mathcal{G}(\omega) = [\mathcal{G}_0^{-1}(\omega) - \Sigma_{aux}(\omega)]^{-1}. \tag{16.11}$$

It is clear that if we regard $\mathcal{G}_0^{-1}(\omega)$ and $\Sigma_{aux}(\omega)$ as variable functions, it is possible to repro-duce any $\mathcal{G}(\omega)$ and thus the exact on-site Green's function $G_{00}(\omega)$ for the crystal. Even if it is not possible to find the exact $G_{00}(\omega)$ in practice, the existence of the functional provides a framework for improved approximations. More aspects of functional relationships are discussed in the following section.

Practical approximations to DMFT: equations and solutions

The DMFT equations follow from the original division of the problem of interacting elec-trons on a lattice into two parts: (1) the periodic Green's function for the lattice, Eq. (16.3), in terms of a self-energy $\Sigma_{\mathbf{k}}(\omega)$ and (2) the calculation of $\Sigma_{\mathbf{k}}(\omega)$. Here we give the expres-sions for the single-site approximation, but the general form carries over to improved approximations. The overall approach is given in the overview in Sec. 16.2, and here we fill in details of the assumptions and the methods. So long as $\Sigma_{\mathbf{k}}(\omega)$ is assumed to be local, i.e., restricted to a single site, it is independent of $\mathbf{k}$ and we can set $\Sigma_{\mathbf{k}}(\omega) \rightarrow \Sigma(\omega)$ (Eq. (16.5)). The strategy is to assume that $\Sigma(\omega)$ is given by some $\Sigma_{aux}(\omega)$ and find a condition to deter-mine the function $\Sigma_{aux}(\omega)$ that best represents the actual crystal. As shown in Sec. 16.2, the single-site DMFA is summarized in the relations in Eq. (16.6): $G_{00}(\omega) = \mathcal{G}(\omega)$ with $\Sigma(\omega) = \Sigma_{aux}(\omega)$. This is the self-consistency condition, depicted in Fig. 16.2, that the Green's function for the central site in Fig. 16.2(c) is the same as for the surrounding sites, which are equivalent sites in the crystal. This is exactly the same idea as in the Weiss mean field, where Eq. (4.13) results from the requirement that the magnetization is the same on a site and its neighbors. An alternative derivation in terms of a functional and a variational equation is given in Sec. 16.6.

Thus the DMFT solution is divided into two parts that are solved separately and related only by the self-consistency condition of Eq. (16.6). A flowchart of the steps in a calculation is given in Fig. 16.3, and the main points can be summarized as follows.

- The embedded impurity problem is the only place where the interactions are treated explicitly and a many-body calculation must be done. This is step 2 in the box "Solve for …"; various solvers are the topic of Ch. 18. This part of a DMFT calculation does not involve the lattice directly. The only input needed is $\mathcal{G}_0(\omega)$ (or equivalently $\Delta(\omega)$) and the only output used in the DMFT equations is the self-energy $\Sigma_{aux}(\omega)$ that is given by Eq. (16.10) in terms of the auxiliary Green's functions.

- Step 3 is the construction of the lattice Green's function Eq. (16.3). This is the only place where the lattice is treated explicitly and it does not involve the interactions directly, only the self-energy with the approximation that $\Sigma_{\mathbf{k}}(\omega)$ is $\mathbf{k}$-independent $\Sigma(\omega)$.

- In the self-consistent loop there must be a way to determine the input to the solver $\mathcal{G}_0(\omega)$ using the functions calculated in previous steps. As indicated in step 1, here the choice

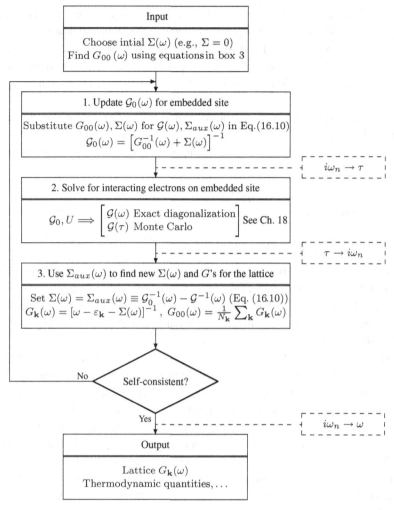

Figure 16.3. Self-consistent loop for iterative solution of DMFT equations. All Green's functions and the self-energy are dynamic functions of frequency: ω, Matsubara ω_n or z defined in the complex plane. For methods that use imaginary time τ, the dashed boxes show the needed transformations $\tau \leftrightarrow \omega_n$ and $\omega_n \to \omega$ (Sec. D.1). The final step requires the difficult transformation from imaginary $i\omega_n$ to real ω. There are various possible algorithms; here the choice is to consider $\Sigma(\omega)$ as the central quantity that is used to construct $G_{00}(\omega)$ (box 3) and update the bath $\mathcal{G}_0(\omega)$ for the auxiliary system (box 1). In each iteration $\Sigma(\omega)$ is set equal to $\Sigma_{aux}(\omega)$ and the cycle continues until $G_{00}(\omega) = \mathcal{G}(\omega)$ with $\Sigma(\omega) = \Sigma_{aux}(\omega)$ (Eq. (16.6)). Note that the interaction appears only in the solver for the embedded site (box 2, called "Solve ...") and the independent-particle terms for the lattice ($\varepsilon_{\mathbf{k}}$) appear only in the calculation of $G_{\mathbf{k}}(\omega)$ in box 3. For multiple bands or a cluster in Ch. 17, each function is a matrix in the orbitals and there are various choices for the cluster. Note the correspondence with the flow diagram for the Kohn–Sham equations in Fig. 9.1 of [1] that can be expressed in terms of the density or potential.

is to substitute $G_{00}(\omega)$ for $\mathcal{G}(\omega)$ and $\Sigma(\omega)$ for $\Sigma_{aux}(\omega)$ and use the Dyson equation in Eq. (16.10). This is the place where we can see that the algorithm leads to a solution at self-consistency, where $\mathcal{G}_0(\omega)$ corresponds to a site embedded in the rest of the crystal (the "cavity construction" described in Eq. (16.9) and the related text and illustrated in Fig. 16.2(c)); it is enforced at each step by the use of $G_{00}(\omega)$ to determine $\mathcal{G}(\omega)$ (Ex. 16.3).

• The single-site approximation can be identified at two points. One is the sum over momenta $\mathbf{k}$ in step 3, which keeps only the on-site part of the lattice Green's function. The other is the assumption that Σ is given by the single-site auxiliary system where there is no correlation between sites.

The final results are the self-energy $\Sigma(\omega)$ and the lattice quantities. The Green's function $G_{\mathbf{k}}(\omega)$ can be written as

$$G_{\mathbf{k}}(\omega) = [\omega - \varepsilon_{\mathbf{k}} - \Sigma(\omega)]^{-1} = [\mathcal{G}(\omega)^{-1} - (\varepsilon_{\mathbf{k}} - \Delta(\omega))]^{-1}, \qquad (16.12)$$

where the second expression brings out the interpretation as the impurity Green's function to which is added the dispersion, i.e., the difference between $\varepsilon_{\mathbf{k}}$ and the on-site hybridization function $\Delta(\omega)$. (See also Eq. (17.13).) In addition, other quantities such as the free energy can be calculated in terms of a functional of $G_{\mathbf{k}}(\omega)$ and the interaction U as discussed in Sec. 16.6.

As pointed out on p. 437 and illustrated in Fig. 16.2, at the solution the auxiliary system is a site embedded in the rest of the crystal treated as an average medium. Because the embedded site is treated in principle exactly, any single-site quantity can be calculated, for example, mean-square fluctuations for spin $\langle \delta S^2 \rangle$ and charge $\langle \delta n^2 \rangle$, time-dependent autocorrelation $\langle S(t)S(0) \rangle$, $\langle n(t)n(0) \rangle$, and other local correlation functions. Examples of fluctuations calculated for several materials are presented in Ch. 20. These are exact for the auxiliary system and can be used to derive approximations for the crystal. In addition, from the correlation functions one can derive susceptibilities that can be used to construct improved approximations, e.g., extended dynamical mean-field theory (EDMFT) in Sec. 17.5.

The methods can be extended to multiple bands, where all quantities become matrices, and to clusters in Ch. 17 that incorporate correlations between different sites within the cluster. Other methods described in Ch. 21 treat intersite correlation in perturbation theory; the auxiliary system is still only for one site, but Eq. (16.6) is replaced by other conditions that take into account intersite correlations.

Iterative algorithm for self-consistent calculations

There are various ways to construct an iterative algorithm to find the self-consistent solution of the equations. At any step in the iteration the functions are not consistent and there are different ways to use the information at one step to determine the input for the next step. The algorithm depicted in Fig. 16.3 is formulated in terms of $\Sigma(\omega)$ and it is often convenient to input an initial $\Sigma(\omega)$, e.g., $\Sigma(\omega) = 0$, which determines the initial lattice $G_{\mathbf{k}}(\omega)$ and

$G_{00}(\omega)$ using the relations in Eqs. (16.3) and (16.4) (also given in box 3 in Fig. 16.3). This is used to find an initial $\mathcal{G}_0(\omega)$ using the relation $\mathcal{G}_0^{-1}(\omega) = \mathcal{G}^{-1}(\omega) + \Sigma(\omega)$ in Eq. (16.10) but with $\mathcal{G}^{-1}(\omega)$ substituted by $[G_{00}(\omega)]^{-1}$. This choice is suggested by the fact that it is the relation that is satisfied at self-consistency where $\mathcal{G}(\omega) = G_{00}(\omega)$.[7] The input to the solver is $\mathcal{G}_0(\omega)$ and the output is $\Sigma_{aux}(\omega)$, again using Eq. (16.10). This leads to a new $\Sigma(\omega) = \Sigma_{aux}(\omega)$ and the procedure is repeated until there is consistency by some measure: the change in $\Sigma(\omega)$ from one step to the next below some tolerance; the difference between $\mathcal{G}(\omega)$ that is output from the solver and $G_{00}(\omega)$ in the last step; or other ways to quantify the approach to self-consistency.

16.6 Self-energy functional and variational equations

In Sec. 16.5 we have only considered the DMFT equations for finding the one-particle Green's function $G_{\mathbf{k}}(\omega)$ and the self-energy $\Sigma(\omega)$ using the physically motivated self-consistency relation $G_{00}(\omega) = \mathcal{G}(\omega)$ in Eq. (16.6). There is more to be gained if we identify a functional for the grand potential Ω and formulate DMFT in terms of variational equations. In Sec. 16.4 it was pointed out that DMFT can be cast in terms of spectral density functional theory (SDFT). Here we consider the grand potential Ω as a functional of the self-energy and derive the equations for the single-site DMFA as an example of a constrained search in a restricted domain.[8]

The essential point is that Ω can be expressed as a sum of two terms, as explained in Secs. 8.2 and I.2. One is a functional that takes into account interactions $F[\Sigma] = \Phi[G] - \mathfrak{Tr}(\Sigma G)$, which is universal, i.e., it has the same functional dependence for all systems with a given interaction.[9] In the present case, this is the on-site interaction that is the same in the actual system and the auxiliary embedded site. (The same approach also applies for a cell of more than one atom.) The second term depends on the system and is specified by the bare Green's function G_0 or $\mathcal{G}_0$. The expressions are most readily derived in terms of the functionals written in the form[10]

$$\Omega[\Sigma] = F[\Sigma] - \mathfrak{Tr}\ln(\Sigma - G_0^{-1}) \tag{16.13}$$

and

$$\Omega_{aux}[\Sigma] = F[\Sigma] - \mathfrak{Tr}\ln(\Sigma - \mathcal{G}_0^{-1}). \tag{16.14}$$

---

[7] After the first iteration, there is a result for $\mathcal{G}(\omega)$ from the solver. The choice of substituting $[G_{00}(\omega)]^{-1}$ for $\mathcal{G}^{-1}(\omega)$ throws away this information. Methods to converge to self-consistency are discussed in [1, Ch. 9], e.g., linear mixing of old and new information, the Broyden method, residual minimization, and other methods.

[8] The discussion here and in Sec. I.2 follows the approach in [772] and [296], which is summarized in [302].

[9] This assumes the functional is well-defined for the range of interactions and systems considered, as discussed in Sec. 8.2.

[10] This is the form given in footnote 7 before Eq. (8.8) and in Eq. (I.5). The $\mathfrak{Tr}\ln$ terms in Eqs. (16.13) and (16.14) are not convergent separately, but the difference in Eq. (16.15) is well-defined (see, e.g., Ex. 8.5). Note that $\mathfrak{Tr}$ denotes a trace over states and an integral or sum over frequencies as defined before Eq. (8.5), and $\mathfrak{Tr}$ has dimensions of energy.

Since $F[\Sigma]$ is the same in both systems, we can subtract Eq. (16.14) from Eq. (16.13) to find

$$\Omega[\Sigma] = \Omega_{aux}[\Sigma] + \mathfrak{Tr} \ln(\Sigma - \mathcal{G}_0^{-1}) - \mathfrak{Tr} \ln(\Sigma - G_0^{-1}). \qquad (16.15)$$

Thus Ω for the actual system is given in terms of Ω_{aux}, which is assumed to be calculated exactly, plus the remaining terms in Eq. (16.15) that are straightforward expressions in terms of Σ, G_0, and $\mathcal{G}_0$. There are no approximations to the functionals; instead, the range of possible functions Σ is restricted to self-energies that can be generated by the auxiliary system for some $\mathcal{G}_0$, i.e., "$\mathcal{G}_0$-representable" functions that can be denoted $\Sigma_{aux}[\mathcal{G}_0]$.

The variational equation[11] for DMFT can be found as the stationary point of $\Omega[\Sigma]$ in Eq. (16.15) in a "constrained search" for Σ restricted to the functions Σ_{aux} generated by some $\mathcal{G}_0$. Following the steps in Sec. I.2 and using the fact that $\Omega[\Sigma_{aux}]$ depends on $\mathcal{G}_0$ only through the variation of $\Sigma_{aux}[\mathcal{G}_0]$, we can use the chain rule and Eq. (16.15) to arrive at the desired result (see Ex. 16.14),

$$\frac{\delta \Omega[\Sigma_{aux}]}{\delta \mathcal{G}_0} = - \left\{ \mathcal{G} - \left[(G_0^{-1} - \Sigma_{aux})^{-1} \right]_{00} \right\} \frac{\delta \Sigma_{aux}}{\delta \mathcal{G}_0} = 0, \qquad (16.16)$$

which is given as explicit sums over matrix elements and Matsubara frequencies in Eq. (I.9). In deriving this equation we have assumed that $\Omega_{aux}[\Sigma]$ is calculated exactly, so that it is at a stationary point where $\delta \Omega_{aux}[\Sigma]/\delta \Sigma = 0$ and $\Sigma = \Sigma_{aux}$. At the solution the term in curly brackets vanishes, which can be recognized as the DMFT condition $\mathcal{G}(\omega) = G_{00}(\omega)$ in Eq. (16.6). In this sense, the single-site approximation for DMFT can be viewed as the best possible local self-energy. The same reasoning applies for the cluster methods in Ch. 17, where a constrained search leads to the best possible self-energy within the space of functions that can be generated by the auxiliary system for a particular cluster approximation.

Note that only on-site elements of Σ (and thus $\delta \Sigma / \delta \mathcal{G}_0$) are non-zero in the single-site approximation; thus the only terms needed in Eq. (16.16) are Σ and $\mathcal{G} = (\mathcal{G}_0^{-1} - \Sigma)^{-1}$, which are calculated for the auxiliary system, and the on-site component of the Green's function $G_{00} = \left[(G_0^{-1} - \Sigma_{aux})^{-1} \right]_{00}$. For a single band this is a scalar equation, but in general each term is a matrix in the basis. For a cluster (Ch. 17), each term is in a matrix in band and site indices. In all cases the term in curly brackets must vanish for every matrix element and frequency in order to satisfy the equations.

16.7 Static properties and density matrix embedding

Static properties, such as the density, total energy, static response functions, etc., can be calculated as integrals over the frequency of spectral functions.[12] General expressions can

---

[11] In the context of DMFT this has been termed the variational cluster approximation (VCA) [296, 772, 773], so called because it has a particular use in cluster methods. See Sec. 17.4.

[12] As explained in Sec. 5.7, there are no such expressions for the entropy; however, it can be determined by a coupling constant integration or integration of specific heat divided by the temperature from a state of known entropy.

be found in Sec. 5.7 (see Eqs. (5.118)–(5.132)) and their use in many-body perturbation theory can be found in Sec. 11.5. In terms of an interacting auxiliary system, expressions for functionals of the Green's function or self-energy are given in Eq. (16.15). The variational character of the functionals makes it possible to find accurate approximations for static quantities with less computational effort than the full DMFT spectral approach [296]. Even the simplest possible bath, a single state, can be used to define a thermodynamically consistent method. Although the spectrum may be rather crude, static quantities are insensitive to the details of the spectrum. In addition, the approach is not limited to DMFT. The functional $\Omega[\Sigma]$ is defined for any self-energy and can be used to find approximations for Ω with various choices for the self-energy. The formulation and examples of the energy for the Hubbard model (and the fact that the stationary solution is a saddle point) can be found in [296].

There are also alternative ways to calculate static quantities. For example, the density and kinetic energy are determined by the one-body density matrix $\rho(x, x')$ (see Sec. 5.2). Since the Coulomb interaction is instantaneous, the interaction energy is a spatial integral over the static pair correlation function given in Eq. (5.15). For a Hubbard model the interaction energy is simply U times the probability of double occupation of a site. There may be ways to determine such quantities directly with much less effort than calculation of the spectrum. Methods such as Lanczos algorithms (see Sec. 18.2) are very efficient at projecting out the ground state and can be used for a huge basis of many-body states, especially if one can take advantage of the sparseness of the matrices. The density matrix renormalization group (DMRG) [774, 775] is designed to find the low-energy states and is well adapted to embedded systems [776] (see also Sec. 18.8). Projection Monte Carlo methods (Ch. 24) such as diffusion Monte Carlo (DMC) start from the best available approximation to the ground state, and hence they can be much more efficient for calculation of ground-state properties than the finite-temperature spectral methods used for DMFT (Ch. 18).

A method to calculate static properties directly using embedding techniques closely related to DMFT is density matrix embedding theory (DMET) [761, 777]. One can already see the basis for such an approach by noting that the density matrix is proportional to the equal-time limit of the Green's function, $\rho(x, x') = -iG(x, t, ; x', t^+)$ in Eq. (5.121) or $\rho(x, x') = G(x, x', \tau^-)$ in Eq. (D.13), which can be found by an integral over frequency of $G(\omega)$. Thus one can also set up an auxiliary system embedded in a static density matrix. The theory and the calculations are much simpler than for DMFT, since the method involves only the density matrix; there is no need to deal with the spectrum, the analytic properties of the functions in the complex plane, etc. DMET is a general approach for a large complex system that is broken into fragments, with calculations done on each fragment embedded in a bath. In principle, an exact solution can be found with only as many bath states as degrees of freedom in the fragment.

The approach in [761, 777] is to construct a mean-field density matrix for the entire system calculated using Harteee–Fock plus a one-body non-local potential $u(\mathbf{r}, \mathbf{r}')$ that is to be determined. The density matrix can be expressed as $\langle \Phi_S[u]|c_i^\dagger c_j|\Phi_S[u]\rangle$, where $\Phi_S[u]$ is a single Slater determinant calculated including the potential u and i, j label the basis functions. For each fragment an embedding function is created using $\Phi_S[u]$ and a high-level calculation is carried out on the embedded fragment to determine a correlated wavefunction $\Psi[u]$. The solution is found by minimizing the mean-square difference

$$\sum |\langle \Phi_S[u]| c_i^\dagger c_j |\Phi_S[u]\rangle - \langle \Psi[u]| c_i^\dagger c_j |\Psi[u]\rangle|^2, \tag{16.17}$$

where the sum is over the fragments and elements of the density matrices confined to the fragments. The theory and the form of the algorithm can be recognized as analogous to DMFT with a static embedding field and efficient methods to calculate static quantities such as DMRG. DMET has been applied to Hubbard-type models [777] where a fragment is a site or cell. For example, the density as a function of chemical potential for the one-dimensional Hubbard model can be compared with the integrated quantities derived from DMFT and known results from exact solutions (see Sec. 17.6). Since DMET is less computationally intensive than DMFT, it should be applicable as a general technique to treat static properties of complex systems including longer-range correlation.

16.8 Single-site DMFA in a two-site model

In this section we consider a test case, the two-site "Hubbard dimer," where the hamiltonian is given in Eq. (3.10) in terms of the on-site interaction U and the hopping matrix element t; the on-site energy $\varepsilon_0 = 0$ for simplicity. For this problem there are exact solutions, which can be classified into singlet and triplet spin, and even and odd parity, as described in Sec. 3.2 and Ex. 3.5. By construction, a two-site cluster DMFT calculation would lead to the exact results for all states; however, the two-site problem is a severe test for the single-site approximation, which is most justified for many neighbors in high dimensions. The final solutions have the proper even and odd symmetry, but the self-energy $\Sigma(\omega)$ calculated for the auxiliary system is the same function for even and odd solutions. This is the analog of the approximation in a crystal, where $\Sigma(\omega)$ is independent of momentum $\mathbf{k}$.

In DMFT there are two steps, the calculation of $\Sigma(\omega)$ in the auxiliary system and the use of $\Sigma(\omega)$ in the actual problem. In the present case there is a nice formulation for the second step using the expressions for coupled systems derived in Ch. 7 and reproduced in Eqs. (16.1) and (16.2). In terms of the self-energy $\Sigma(\omega)$ for a site, if the site were decoupled the Green's function would be $G_S^0(\omega) = [\omega - \Sigma(\omega)]^{-1}$. Then, using Eq. (16.2) expressed in terms of the Green's functions, the Green's function for a site in the coupled system is given by $G_S(\omega) = [G_S^0(\omega)^{-1} - t^2 G_R^0(\omega)]^{-1}$, where $G_R^0(\omega)$ is the Green's function for the "rest." In this case the rest is the other site, which has the same Green's function, leading to the simple relation

$$G_S(\omega) = \left[[G_S^0(\omega)]^{-1} - t^2 G_S^0(\omega) \right]^{-1}, \tag{16.18}$$

where t is the hopping matrix element. Thus the poles of $G_S(\omega)$ occur where the quantity in square brackets is zero, i.e., where $\omega - \Sigma(\omega) = \pm t$.

Consider the Green's function for the dimer in the ground state for one electron with spin ↑ and energy $-t$, the same as in the test of the GWA in Sec. 11.7. The spectrum for adding a ↓ electron to create a singlet is the energies for the difference between the energy of the initial state $-t$ and the four singlet states given in Sec. 3.2 and shown in Fig. 3.2.

The large U/t regime is particularly relevant for tests of DMFT. In this limit the self-energy approaches that for an isolated atom, which is worked out in Sec. K.2 and given explicitly in Eq. (K.9). In the limit $t \to 0$ the single-site DMFA approaches the exact spectrum with poles at $\omega = 0$ and U, unlike the restricted Hartree–Fock or GW approximations that lead to a single average peak at the average energy $U/2$ (see Sec. 11.7).

The limitations of the single-site approximation become apparent if we consider the spectrum and total energy as a function of t. For small t/U we expect the self-energy to be similar to the isolated atom and we can use the "Hubbard I" approximation that $\Sigma(\omega)$ is unchanged as t increases for fixed U. As shown in Ex. 16.4, the result is a spectrum with four poles at energies given by $\frac{1}{2}[U \pm (t \pm \sqrt{(t^2 + U^2)})]$, which are $\approx \pm t/2$ and $\approx U \pm t/2$ for small t/U. Even though the results are correct for $t = 0$, the splitting of the states for small t/U is qualitatively incorrect. The splitting $\pm t/2$ is narrowed from the independent-particle values $\pm t$ by only a factor of 2, whereas the exact splitting is $\pm t^2/U$ for small t/U. The error is because the single-site approximation assumes no correlation between sites: the hopping is reduced by a factor of two since there is 50% probability of finding an opposite-spin electron on the neighboring site. In the actual problem there is near-perfect correlation for large U: if a spin-σ electron is on one site, then the spin-σ electron is on the other site with probability approaching 100%.

The example of $U/t = 4$ is shown in Fig. 11.10, where the singlet addition spectrum shown in the lower-left panel compares the exact and the $G_0 W_0$ spectra. In the Hubbard I approximation the addition energies are given by the above formula minus the energy of the ground state $-t$: $0.94 \pm 1/2$ and $5.06 \pm 1/2$, compared with the exact values 0.58 ± 0.44 and 5.42 ± 0.44 in units of t. Thus the results are qualitatively correct and quantitatively not so far from the true energies even for this difficult case.

In summary, for large U the single-site approximation avoids the severe error in static mean-field methods like Hartree–Fock, where the error is of order U since they average the interaction on a site. However, there are errors of order t due to the mean-field average over configurations on neighboring sites that determines the effective hopping. The two-site problem is a particularly difficult case for a mean-field approximation, which is expected to be better if there are more neighbors, and is exact in the limit of an infinite number of neighbors, as discussed in the next section.

16.9 The Mott transition in infinite dimensions

In this section we illustrate DMFT for a model problem that elucidates features of interacting-electron problems without the details and complications of specific examples. Applications to Hubbard models in one, two and three dimensions are postponed to the following chapter, where the results for the single-site approximation can be compared with cluster calculations that take into account intersite correlation.

A model can be constructed by taking advantage of the fact that for a single-band Hubbard model, the single-site approximation can be formulated solely in terms of the interaction U and the non-interacting density of states (DOS) for the lattice. For a single band, where Σ is a scalar, the Green's function is given by $G_\mathbf{k}(z) = [z - \varepsilon_\mathbf{k} - \Sigma(z)]^{-1} =$

$G_{\mathbf{k}}^0(z - \Sigma(z))$, where z is the frequency in the complex plane and $G_{\mathbf{k}}^0(z - \Sigma(z))$ is the independent-particle Green's function for the lattice evaluated at the frequency z shifted by the self-energy $\Sigma(z)$. Thus the average over the Brillouin zone is the on-site non-interacting Green's function at a shifted frequency

$$G_{00}(z) = G_{00}^0(z - \Sigma(z)), \tag{16.19}$$

which applies for either spin. If one has an analytic expression for a bare Green's function $G^0(z)$, this defines a model problem whether or not $G^0(z)$ corresponds to an on-site $G_{00}^0(z)$ for an actual lattice. (But one must be careful to recognize the limitations of the model in order to draw conclusions about actual physical systems, as emphasized below.)

For a paramagnetic case with the same Green's functions and self-energies for spin up and down, the problem can be formulated as two scalar equations to be solved self-consistently,

$$G^0(z - \Sigma(z)) = \mathcal{G}(z) \quad \text{and} \quad \Sigma(z) = \mathcal{G}_0^{-1}(z) - \mathcal{G}^{-1}(z). \tag{16.20}$$

Of course, one must still carry out the many-body calculation for the auxiliary embedded system to calculate $\mathcal{G}(z)$ and vary the $\mathcal{G}_0(z)$ until both equations are satisfied. This formulation applies if there is only a single band with interactions, even if there is hybridization with additional non-interacting bands, e.g., in Eq. (16.23) below.

Semicircular density of states

In order to compare various methods, we choose the semicircular DOS that is used in the Hubbard and the Gutzwiller–Brinkman–Rice scenarios for the Mott transition in Secs. 3.2 and K.3. In this case there is an analytic expression for the Green's function on a site (see Exs. 16.5–16.7),

$$G^0(z) = \frac{2}{D^2} \left[z - \sqrt{z^2 - D^2} \right], \tag{16.21}$$

which corresponds to a density of states $\rho(\omega)$ for real $z = \omega$,

$$\rho(\omega) = -\frac{1}{\pi} \mathrm{Im} G^0(\omega) = \frac{2}{\pi D^2} \sqrt{D^2 - \omega^2}, \tag{16.22}$$

which is non-zero only for $|\omega| < D$. This is customarily called "semicircular" since $\sqrt{D^2 - \omega^2}$ is the shape of a semicircle and $\rho(\omega)$ is given by scaling the height by the factor $2/\pi D^2$, as shown at the top of Fig. 3.3.

Equation (16.22) is the density of states for a Bethe lattice[13] with an infinite number of neighbors $N \to \infty$ (analogous to dimension $d \to \infty$ where there are $N \propto d$

---

[13] A Bethe lattice is a tree-like structure where each site is connected to N neighbors with no closed rings. Examples of lattices in infinite dimensions are reviewed in [302]. See Exs. 16.8 and 16.9 for a hypercubic lattice.

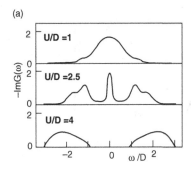
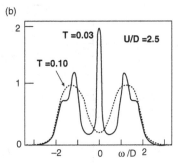

Figure 16.4. Calculated spectral functions for the paramagnetic phase of the half-filled one-band model with a semicircular density of states with frequency ω in units of the half-width D. The left side shows the emergence of a "three-peak structure" (a central peak and upper and lower sidebands) and opening of a gap (taken from [778], an early work that used iterated perturbation theory as a solver). The right side shows the temperature dependence and the disappearance of the peak for T of order the width of the peak, which is above the critical point in Fig. 16.5. Qualitative features, such as the high-energy peaks and the existence of low-energy features that disappear as T is increased, are generic features of systems with strong interactions (see Fig. 20.3 for an example in CeIrIn$_5$), but the low-energy spectra are expected to depend on the specific problem, e.g., the opening of a gap instead of a peak at the Fermi energy in the 2D Hubbard model in Fig. 17.5. (Adapted from figures in [302].)

neighbors) with scaled hopping amplitude $t_0/\sqrt{N}$ and $D = 2t_0$ (see Ex. 16.7); if there is second-neighbor hopping it must be scaled as t_0'/N and the DOS has the same form with $D = 2\sqrt{t_0^2 + t_0'^2}$. The results given below can be interpreted as exact solutions, since mean-field theory is exact in the limit of infinite dimensions. Note, however, that there may be more than one solution with different symmetries, and we consider two examples.

Paramagnetic solution

The solution of the coupled equations in Eq. (16.20) corresponds to a paramagnetic state, i.e., a restricted solution where the Green's function is constrained to be the same for each spin. The one-particle spectra, shown at the left in Fig. 16.4 for representative values of U, show the development of the characteristic "three-peak structure" as U increases. The narrow peak near the Fermi energy corresponds to quasi-particles with total spectral weight Z, and the high-energy excitations above and below the Fermi energy are broadened bands with weight $\propto (1 - Z)$ associated with atomic-like transitions at energies separated by the on-site interaction U. The correlated many-body nature of the low-energy quasi-particle excitations is indicated on the right side of Fig. 16.4, which shows the temperature dependence. The narrow quasi-particle band is well-defined only for temperature $T \ll 2D$ and the peak has turned into a depression with no signatures of quasi-particles for $T = 0.10D$, which is only 5% of the independent-particle bandwidth.

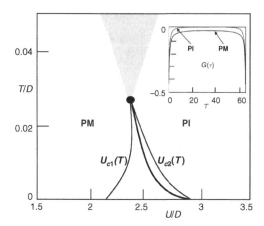

Figure 16.5. Phase diagram for the Hubbard model constrained to be paramagnetic (see Fig. 16.6 for unconstrained solutions) as a function of normalized temperature T/D and interaction U/D for the semicircular density of states Eq. (16.22) (taken from results in [779, 780]). The light lines are spinodals that mark the regions of stability of the two solutions, metallic (PM) for values of U extending up to $U_{c2}(T)$ and insulating (PI) for U larger than $U_{c1}(T)$, as explained in the text. The dark line indicates the first-order transition ending in a critical point indicated by the dot. The shaded region indicates the crossover region where properties vary rapidly, as expected above a critical point. The inset shows the calculated $G(\tau)$ [779] for $\beta = 64$, which shows a clear distinction between metallic and insulating behavior (see text).

It is no accident that the spectra are similar to those for the Anderson impurity model shown in Fig. 3.4, since the single-site DMFA is equivalent to a self-consistent impurity problem. For an impurity in a metallic host, the spectrum always has a peak at the Fermi energy at low temperature $T < T_K$ that goes away at high temperature $T > T_K$, where T_K is the Kondo temperature. The difference in the lattice is that, within the single-site DMFA, the self-consistent problem for the bath can lead to an opening of a gap as U increases and termination of a solution with a peak at the Fermi energy, which occur at different values of U, corresponding to the coexistence region shown in Fig. 16.5. Thus the single-site DMFA calculations show that the Hubbard and Gutzwiller scenarios, illustrated in Fig. 3.3, each capture part of the physics but both miss the most interesting point: the three-peak structure of the spectrum that has both band narrowing (the band at the Fermi energy with width $\approx T_K$ similar to the Kondo effect) and band broadening (the sidebands separated by $\approx U$).

The phase diagram for the system constrained to be paramagnetic is shown in Fig. 16.5 as a function of normalized temperature T/D and interaction strength U/D. Although it is difficult to determine the phase boundaries due to the very small energy differences between competing phases, calculations by different methods agree that there is a first-order transition at $T \neq 0$ with a coexistence region [779, 780]. The metallic solution with a peak at the Fermi energy is stable for small U up to $U_{c2}(T)$, where the quasiparticle weight Z vanishes. This is the explanation for the metal–insulator transition due to

Brinkman and Rice [138], based upon the Gutzwiller approximation (Sec. K.1) and shown by the dashed lines in Fig. 3.3. In contrast, the solution with an insulating gap is found for large U down to U_{c1}, where the gap closes as proposed by Hubbard (Sec. K.2 and illustrated as the solid lines in Fig. 3.3). At low temperature the two boundaries approach $U_{c1}/D \approx 2.35$ and $U_{c2}/D \approx 2.9$ [780]. These are in semi-quantitative agreement with the values $U_{c1}/D = 1.732$, found from the Hubbard alloy approximation, and $U_{c2}/D = 3.395$ from the Gutzwiller approximation (see Fig. 3.3 and related text), which shows the insight contained in those early works.

The inset in Fig. 16.5 shows the Green's function $G(\tau)$ in imaginary time (see App. D for definitions and characteristic forms). The important point for our purposes is the difference between the metallic and insulating phases: in a metal the value at the midpoint indicates the conductivity, whereas decay of $G(\tau)$ to zero indicates the opening of a gap in the spectrum.

The phase transitions for $T \rightarrow 0$ and the nature of the ground state at $T = 0$ require careful consideration. The paramagnetic calculations can be interpreted as an exact solution for the $d \rightarrow \infty$ Bethe lattice, where there is no intersite correlation. Such a state would have massive degeneracy and finite entropy at $T = 0$. One possibility is that the actual ground state is ordered, such as the antiferromagnetic solution in the following paragraphs. The possibility of an insulating state with partially filled bands, a "Mott insulator" with no broken symmetry called a "spin liquid" [114], is among the most fundamental problems in condensed matter physics, as noted in Secs. 3.1 and 3.2. Caution must be exercised in drawing conclusions from a single-site DMFA, which is an approximation in finite dimensions.

Antiferromagnetic solution

There may also be a broken-symmetry antiferromagnetic (AF) phase with lower energy. For a bipartite lattice at half-filling[14] this is expected to be the actual stable phase. The calculated phase diagram for the Bethe lattice with the semicircular density of states at half-filling is shown in Fig. 16.6 (see [302] where there are references to original works). At sufficiently high temperature the solution is a paramagnetic metal with A and B sites equal and no magnetic order. As the temperature is lowered there is transition to an ordered AF insulator at the Néel temperature T_N. As shown in the left panel, if there is only nearest-neighbor hopping ($t' = 0$), T_N is much higher than the paramagnetic critical point and the only stable phases are the PM and AFI. However, if T_N were lower, the metal–insulator transition could occur in the paramagnetic phase for $T > T_N$. This could happen if there is a tendency for frustration of the AF order, for example, by second-neighbor hopping t'.

---

[14] A bipartite lattice, defined on p. 43, can be divided into two sublattices A and B such that the nearest neighbors of an A site are B and vice versa. A Bethe lattice is bipartite since each site has neighbors in shells that can be labeled alternating A and B. Since the two sites are related by time-reversal symmetry (Ex. 16.10) for each site there are two Green's functions $G_{A\sigma}$ and $G_{A-\sigma}$, and $G_{B\sigma} = G_{A-\sigma}$. One must then solve the coupled self-consistent equations.

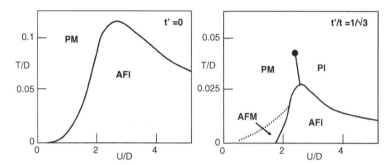

Figure 16.6. Phase diagram for the Hubbard model including the broken-symmetry antiferromagnetic phase as a function of normalized temperature T/D and interaction U/D for the semicircular density of states Eq. (16.22). Left: the Néel temperature T_N for a transition to an ordered AFI, which is the stable phase (note the change in temperature scale from Fig. 16.5). Right: the AF transition can be frustrated by adding a a second-neighbor hopping t', shown here for $t'/t = 1/\sqrt{3}$, where there can be a first-order PM–PI transition for a range of temperatures. Note that the ground state is a PM or has AF order that can be a metal (AFM) or insulator (AFI). (Adapted from figures in [302].)

The lower T_N is illustrated in the right panel of Fig. 16.6 for $t'/t = 1/\sqrt{3}$, where the first-order transition to the paramagnetic insulator extends above the AFI boundary. In this case the phase diagram has a first-order transition that occurs in some range of T, but for all U the ground state is either an ordered insulator (AFI) or a Fermi liquid metal with no order (PM) or AF order (AFM). This has the qualitative features of the phase diagrams for V_2O_3 in Fig. 2.14 and MnO in Fig. 20.10. However, as described in Ch. 20, these materials have many competing effects not captured by a Hubbard model; a combination of experiment and theory is helping to sort out the dominant factors that determine the transition.

16.10 Hybridized bands and consequences for the Mott transition

Actual problems in materials almost always involve more than one band and the many-body problem escalates rapidly as the number of states increases. However, if the added bands are considered as non-interacting (the band-like states called the "rest" in Chs. 19–21), the many-body problem is no more complicated than the one-band case. This can be seen from a two-band example where the hamiltonian can be written as

$$\hat{H} = \sum_{k\sigma} \left[\epsilon_k^1 c_{1k\sigma}^\dagger c_{1k\sigma} + \epsilon_k^2 c_{2k\sigma}^\dagger c_{2k\sigma} + \tilde{t}_k (c_{1k\sigma}^\dagger c_{2k\sigma} + c_{2k\sigma}^\dagger c_{1k\sigma}) \right] + \sum_i U \hat{n}_{1i\uparrow} \hat{n}_{1i\downarrow},$$

(16.23)

where ϵ_k^1 and ϵ_k^2 denote the respective dispersions if there were no coupling, U is the on-site interaction only in band 1, and $\tilde{t}_k$ is the hybridization between the bands.[15] Since there is only one state per site with interactions, the quantum impurity problem is the same

---

[15] If band 1 has no dispersion, it is called the periodic Anderson model or Anderson lattice model, which is composed of a periodic array of localized states coupled to the non-interacting band.

as the one-band problem – a site in a bath where the bath is to be determined. The difference is that the lattice Green's function G^1 for band 1 is changed because the dispersion takes into account coupling to the second band, so that the on-site function is given by (Ex. 16.12)

$$G^1_{00}(\omega) = \sum_{\mathbf{k}} \frac{1}{\omega - \varepsilon^1_{\mathbf{k}} - \Sigma^1(\omega) - |\tilde{t}_{\mathbf{k}}|^2/(\omega - \varepsilon^2_{\mathbf{k}})}, \tag{16.24}$$

which reduces to the one-band expression Eq. (16.4) if $\tilde{t}_{\mathbf{k}} = 0$.

The two-band model has been studied in [781] for cases where a single-site calculation finds a first-order phase transition in the paramagnetic state if the bands are decoupled ($\tilde{t} = 0$). As pointed out in the previous section, such a paramagnetic insulating state is highly degenerate and should be unstable to small perturbations. Indeed, it is found that a non-zero $\tilde{t}$, no matter how weak, leads to a second critical point at low temperature; the first-order transition occurs only in a range of temperatures ending in critical points at both high and low temperature. At lower temperature the coupled bands form a Fermi liquid. This is analogous to the Anderson impurity model, where hybridization, no matter how small, leads to a Fermi liquid state below the Kondo temperature T_K (see Sec. 3.5). In the previous section it was found that a transition to an antiferromagnetic state is a way to avoid the unphysical behavior in the paramagnetic solution for the one-band model. The two-band model shows that another way is to stabilize a metallic Fermi liquid state even for large interactions. This is a model for heavy fermion behavior and a phase transition like that observed in alloys containing Ce (see Sec. 20.2).

If the second band is half occupied (a total of two electrons/cell), the coupling can lead to a bandgap. For large U and small $\tilde{t}$, the gap is small and this is a model for what has been termed a "Kondo insulator" described in Sec. 20.2. This also is a well-defined ground state that eliminates the problem of the degeneracy of isolated spins.

16.11 Interacting bands and spin transitions

The extension of the one-band Hubbard model to multiple interacting bands provides insights into spin transitions that can occur in transition metal compounds. Here we consider the spin on a site that determines the magnitude of the local moment, but there is no information about ordering of the moments, which depends on the particular lattice. An instructive example is two bands with on-site intra- and interband interactions but no hybridization, i.e., no interband hopping matrix elements. The hamiltonian can be written as [782, 783]

$$\hat{H} = \hat{H}^1 + \hat{H}^2 + \sum_i \left[\sum_{\sigma,\sigma'} U' \hat{n}_{1i\sigma} \hat{n}_{2i\sigma'} - J\hat{S}_{1i}\hat{S}_{2i} \right], \tag{16.25}$$

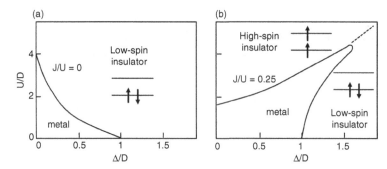

Figure 16.7. Metal–insulator and local spin transitions for two interacting bands as a function of U and Δ, for a semicircular DOS of width $2D$ and splitting 2Δ between the band centers, and two electrons per site [783]. The metal–insulator transition at $J = 0$ (left) evolves to the form shown at the right for $J/U = 0.25$ and an insulating state with local "high-spin" moments. The dashed line is $\Delta = \sqrt{2}J$, suggested by analytic arguments for large Δ (see [783]). Note that these are *local moments* with no long-range order. (Figures adapted from [783].)

where the superscript labels the bands 1 and 2, i denotes sites, and $\hat{H}^m$ is the hamiltonian for band $m = 1, 2$,

$$\hat{H}^m = \sum_{i,\sigma} \varepsilon^m \hat{n}_{mi\sigma} + \frac{1}{2} U \sum_{i,\sigma} \hat{n}_{mi\sigma} \hat{n}_{mi-\sigma} - \sum_{i\neq j,\sigma} t_{ij} c^\dagger_{mi\sigma} c_{mj\sigma}. \qquad (16.26)$$

For simplicity the bands are chosen to be equivalent, except that there can be a relative shift $\varepsilon^1 - \varepsilon^2 \equiv 2\Delta$. The term in brackets in Eq. (16.25) is the most general form for interband interactions with spin-independent interaction U', exchange interaction J, and spin operator $\hat{S}$.[16] For $J > 0$ the energy is lower for parallel spins, favoring the Hund's rule maximum-spin ground state. In order to make this example as close as possible to the problems involving degenerate d states in Sec. 19.3, we can require states 1 and 2 to be related by rotational symmetry, in which case $U' = U - 2J$ as stated after Eq. (19.6).

Consider the case with two electrons per site (half-filling). If there are no interactions the solution is a band overlap metal for small Δ and an insulator with the lower band filled for $\Delta > D$, with $2D$ the bandwidth. For large enough J we expect that the ground state will have parallel spins on each site, and for large U there should be an insulating state for any value of Δ.

The results of single-site DMFA calculations [783] for the semicircular density of states are plotted as a function of Δ and U for two values of J/U in Fig. 16.7. The left side is for $J = 0$, where there are equal interactions between electrons in any two states ($U' = U$), and there is a metal–insulator transition between a paramagnetic metal and an insulator

---

[16] From the definition $\hat{S} = \frac{1}{2} \sum_{\sigma\sigma'} c^\dagger_\sigma \tau_{\sigma\sigma'} c_{\sigma'}$, where τ denotes the Pauli matrices, for each site i and band $m = 1, 2$, one can show (Ex. 16.13) that the term in brackets in Eq. (16.25) can be expressed in the form

$$[\ldots] = U'\hat{n}_{1\sigma}\hat{n}_{2-\sigma} + (U' - J)\hat{n}_{1\sigma}\hat{n}_{2-\sigma} - J(c^\dagger_{1\sigma}c^\dagger_{2-\sigma}c_{2\sigma}c_{1-\sigma} + c^\dagger_{2\sigma}c^\dagger_{2-\sigma}c_{1\sigma}c_{1-\sigma} + c.c.).$$

that smoothly evolves as U increases from the band insulator for $\Delta/D > 1$ and $U = 0$. We can note that for $\Delta = 0$ the bands are degenerate and the metal–insulator transition occurs for larger U than in the one band-case, in agreement with analytic arguments [784]. For $J \neq 0$ the shape of the boundary changes and a new solution emerges with parallel spins. The right side of the figure for $J/U = 0.25$ shows the resulting transition between two insulating states with spin 0 and local spin 1 on each site as a function of U and Δ.

This is a model with the basic features of the spin transitions observed in transition metal oxides, e.g., MnO and FeO in Sec. 20.8. The band splitting Δ is the analog of a crystal field splitting in the oxides, and the fact that it increases under pressure is one of the explanations for the high-spin/low-spin transition that is closely related to metal–insulator transitions. In these systems the spins order at low temperature $T < T_N$; for $T > T_N$ there can be a continuous evolution or a first-order transition.

16.12 Wrap-up

DMFT is an example of a general approach to combine the advantages of auxiliary systems and Green's function methods. The goal is to determine the Green's function for the crystal $G_{\mathbf{k}}(\omega) = [\omega - \varepsilon_{\mathbf{k}} - \Sigma_{\mathbf{k}}(\omega)]^{-1}$ and the approach is to break the problem into two parts: the calculation of the self-energy $\Sigma_{\mathbf{k}}(\omega)$ using an auxiliary system and the use of $\Sigma_{\mathbf{k}}(\omega)$ to find $G_{\mathbf{k}}(\omega)$. The methods are constructed to be exact in three limits: no interactions, isolated atoms including interactions, and infinite dimensions. The simplest implementation is the single-site approximation where the self-energy $\Sigma(\omega)$ is independent of momentum $\mathbf{k}$ and the auxiliary system is an atom with interactions embedded in a bath of independent particles, which is equivalent to an Anderson impurity model. This is analogous to static mean-field approximations such as the Weiss mean-field theory for spin systems, but it includes dynamical local correlation. If the site is replaced by a cell of atoms, the same general formulation carries over to cluster methods that are dealt with in more detail in Ch. 17.

The examples in this chapter are models chosen to bring out important physical results relevant for applications to materials later. In particular, it is illuminating to consider the half-filled Hubbard model for the $d \to \infty$ Bethe lattice (Sec. 16.9), where mean-field theory is exact and we can examine the issues: a metal–insulator transition occurs if the system is constrained to be in a paramagnetic phase with no broken symmetry. But in fact, the stable phase at low temperature is either an ordered antiferromagnet or a paramagnetic metal. Multiband models provide a basis for understanding spin transitions and other phenomena that occur in transition metal systems.

The following chapters are devoted to methods to go beyond the single-site approximation and applications to model systems (Ch. 17), solvers for the interacting auxiliary embedded system (Ch. 18), applications to realistic problems (Chs. 19 and 20), and steps toward first-principles methods combining DMFT and Green's function perturbation methods (Ch. 21).

SELECT FURTHER READING

Georges, A., Kotliar, G., Krauth, W. and Rozenberg, M. J., "Dynamical mean-field theory of strongly correlated fermion systems and the limit of infinite dimensions," *Rev. Mod. Phys.* **68**, 13–125, 1996. A review that brings out the essence of the methods and many of the physical consequences.

Held, K., "Electronic structure calculations using dynamical mean field theory," *Adv. Phys.* **56**, 829–926, 2007. An extensive review including methods for realistic calculations.

Kotliar, G. and Vollhardt, D., "Strongly correlated materials: insights from dynamical mean-field theory," *Phys. Today*, March, 53–59 (2004). A non-technical overview.

Kotliar, G., Savrasov, S. Y., Haule, K., Oudovenko, V. S. and Parcollet, O., and Marianetti, C. A., "Electronic structure calculations with dynamical mean-field theory," *Rev. Mod. Phys.* **78**, 865–952, 2006. A review that emphasizes the formulation in terms of functionals and applications to materials.

Extensive sets of lecture notes with many specific details for DMFT can be found in the lecture notes for Autumn schools at Forschungszentrum in Jülich [785, 786] that are available online at www.cond-mat.de/events.

Two books that include descriptions of DMFT and many examples are in the Springer Series in Solid-State Sciences: Vol. 163, *Electronic Structure of Strongly Correlated Material* by V. I. Anisimov and Y. Izyumov (2010) and Vol. 171, *Strongly Correlated Systems: Theoretical Methods*, edited by A. Avella and F. Mancini (2012).

A book that treats inhomogeneous systems is Fredericks, J. K. *Transport in Multilayered Nanostructures: The Dynamical Mean-Field Theory Approach* (World Scientific, Singapore, 2006).

Exercises

16.1 The cavity Green's function $G_{i,j}^{-}$ with one site (a single atom or a cell, as considered in Ch. 17) removed can be defined by removing the coupling in the hamiltonian between the site and the rest of the lattice. See discussion before Eq. (16.9).

   (a) Show that this leads to the expression

$$G_{ij}^{-} = G_{ij} - G_{i0} \left[G_{00} \right]^{-1} G_{0j},$$

   where $i, j \neq 0$ denote sites in the lattice and each G is a matrix in the state labels m, m'. *Hint*: use the Dyson equation and consider the intersite hopping matrix elements as the perturbation. See also [302].

   (b) The cavity Green's function can be derived in an alternative, simpler way by noting that adding an infinite repulsive potential on a site is equivalent to excluding that site. Show that the expression for G_{ij}^{-} in part (a) follows from the t-matrix expression Ex. K.7 for multiple scattering from a single-site form with the potential on the central site allowed to approach infinity.

   (c) Show that this leads to the expression for the hybridization function given in Eq. (16.9).

16.2 Use the form of G_{ij}^{-} in Ex. 16.1 part (a) to justify the argument that the correction (the second term) vanishes for $d \to \infty$. *Hint*: use the definition that as the coordination increases the hopping matrix element must decrease, as described in Ex. 16.6. See also [302].

16.3 Fill in the steps to show that the algorithm in the self-consistent loop in Fig. 16.3 is equivalent to the cavity construction in Eq. (16.9) without the need to explicitly construct the cavity Green's function $G_{ij}^-(\omega)$ or the hybridization function $\Delta(\omega)$. The essential point is that in each iteration the effective medium Green's function is constructed in box 1 of Fig. 16.3 using the relation $\mathcal{G}_0(\omega) = \left[G_{00}^{-1}(\omega) + \Sigma(\omega)\right]^{-1}$, where $G_{00}(\omega)$ is calculated for the actual lattice. Show that this equation corresponds to removing the self-energy only on one site, which corresponds to Fig. 16.2(c), where the central site is in a cavity with all the other sites having self-energy Σ. This applies at every step in the iteration. Complete the argument that at the self-consistent solution the DMFT equations are satisfied and this corresponds to the equality $b = c$ depicted in Fig. 16.2.

16.4 Find the one-particle Green's function for $\downarrow$ electrons in a Hubbard dimer that has one $\uparrow$ electron in the ground state. Consider only singlet states in the extract spectrum for the system with the added electron.

   (a) Derive the spectrum from differences in the exact energies found in Ex. 3.5.
   (b) Show that in the $U \gg t$ regime, there are pairs of states in the exact spectrum at energies ≈ 0 and $\approx U$ split by an amount $\approx \pm t^2/U$.
   (c) Assuming that the self-energy is the same as in the atom (the Hubbard I approximation), which is given in Eq. (K.9), show that the corresponding pairs of states in the single-site approximation are split by $\pm t/2$. Do this by using the frequency dependence of the self-energy in Eq. (K.9) and the condition for the solution $\omega = \Sigma(\omega) \pm t$ given after Eq. (16.18). *Hint*: consider $\Sigma(\omega)$ near the actual energies and use reasoning similar to Sec. 7.5.
   (d) Derive the full solution for the four energies of the singlet states in the Hubbard I approximation, $\frac{1}{2}[U \pm (t \pm \sqrt{(t^2 + U^2)})]$, by combining Eq. (16.18) with Eq. (K.9) for the self-energy.

16.5 Show that the semicircular form in Eq. (16.21) has the correct analytic form for a retarded function, i.e., it satisfies the Kramers–Kronig relations in Eq. (5.53) for a system with the density of states in Eq. (16.22). (The choice of causal or time-ordered is not crucial since we only need relations between Green's functions in this chapter.) *Hint*: the sign of the square root is fixed by the choice in Eq. (16.22). Show that there are no poles or zeros in the upper half-plane for complex frequency z.

16.6 Show that the density of states for non-interacting particles in the limit of infinite number of neighbors $N \to \infty$ can be taken in a consistent way by scaling the nearest-neighbor hopping by $t = t_0/\sqrt{N}$ with t_0 fixed.

16.7 Consider a Bethe lattice where each site has N neighbors and there is nearest-neighbor hopping $t = t_0/\sqrt{N}$.

   (a) Show that the on-site Green's function is $G_{00}^{-1}(\omega) = \omega - t^2 \sum_i G_{ii}^-(\omega)$, where the sum is over the neighbors i of the central site and G_{ij}^- is the cavity Green's function in Ex. 16.1.

   Hint: in the Bethe lattice each neighbor is the start of an independent tree, so $G_{ij}^- = 0$ if i and j are different neighbors.
   (b) Using part (a) and Ex. 16.2, show that for $N \to \infty$ the Green's function is given by the semicircular form in Eq. (16.21) with width $D = 2t_0$.

(c) If there is second-neighbor hopping on the Bethe lattice, show that the limit $N \to \infty$ is well-defined if the hopping is t_0'/N, where t_0' is a constant. Show that the DOS is the same semicircular form with $D = 2\sqrt{t_0^2 + t_0'^2}$.

16.8 Give an expression for the bandwidth W for a band with nearest-neighbor hopping $-t$ for a cubic lattice in any dimension (line, plane, cube, hypercube, ...). If $t > 0$, find the points in the Brillouin zone at the bottom and top of the band.

16.9 A hypercubic lattice is the generalization of a cubic lattice to any dimension d where each atom has $N = 2d$ neighbors. In the limit $d \to \infty$ with $t = t_0/\sqrt{N}$, show that the DOS approaches a gaussian $D(\varepsilon) = \exp(-\varepsilon^2/2t_0^2)/\sqrt{2\pi t_0^2}$.

16.10 Show that a Bethe lattice is bipartite and that an antiferromagnetic solution can be cast in terms of single-site equations using the fact that the two sites are related by the symmetry under time reversal.

16.11 Derive the result that mean-field theory becomes exact in the limit of infinite dimensions $d \to \infty$, i.e., even if there is interaction between neighboring sites the variation from the mean vanishes for any observable. Show the relation to the central limit theorem in Sec. 22.4.

16.12 Give the argument why the quantum impurity problem for many bands is exactly the same as the one-band problem so long as there are interactions only in one band, and derive Eq. (16.24).

16.13 Show that the term in brackets in Eq. (16.25) is equal to the expression in footnote 16 on the same page.

16.14 Show that the expression in Eq. (16.16) in fact yields the DMFT condition of Eq. (16.6). This requires showing that the 00 component of $(G_0^{-1} - \Sigma)^{-1}$ is the on-site G_{00} in Eq. (16.4). The derivation is not hard but it is instructive for handling such equations. See also Ex. I.1.

17

Beyond the single-site approximation in DMFT

Summary

The previous chapter is devoted to the formulation of DMFT and applications that are exact in the limit of infinite dimensions; however, in finite dimensions this is only a single-site approximation. The present chapter is devoted to clusters and other methods to treat correlation between sites. The equations are derived in a unified way, applicable to a single site or a cluster. For a cluster with more than one site there are various ways to choose the boundary conditions and the embedding procedure, and special care must be taken to satisfy causality and translation invariance. A few selected applications illustrate different techniques and results for Hubbard-type models.

The essential features of DMFT are brought out in the previous chapter in the simplest form: the single-site approximation in which correlation between sites can be neglected. In this case the self-energy due to interactions can be calculated using an auxiliary system of a single site embedded in an effective medium that represents the surrounding crystal. The resulting many-body problem is equivalent to an Anderson impurity model that must be solved self-consistently. The results are exact for infinite dimensions $d \to \infty$; however, in any finite dimension, i.e., real systems, this is just a mean-field approximation analogous to the Weiss mean field in magnetic systems or the CPA for disordered alloys.

There are, however, many important properties and phenomena that require us to go beyond mean-field approximations. For example, basic questions concerning the nature of a metal–insulator transition can be established only by quantitative assessment of the correlations between different sites. Is a single site sufficient (the central idea in the arguments by Mott discussed in Sec. 3.2) or is correlation between sites (for example, an ordered antiferromagnetic state) essential for interactions to lead to an insulating state? A complete theory must establish the range of correlation needed to understand the mechanisms and to make quantitative calculations. A striking example that requires a **k**-dependent self-energy is the opening of a "pseudogap" in only part of the Brillouin zone in a high-temperature superconductor, as illustrated in Fig. 2.16.

This chapter is devoted to methods that go beyond the single-site approximation to take into account interactions and correlations between electrons on sites. There is a long history of cluster approaches in statistical mechanics and the theory of disordered alloys, such as the Bethe–Peierls extensions [172, 173] of the Weiss effective field and cluster generalizations of the coherent potential approximation for alloys [140, 792–794]. A description of cluster methods in DMFT with many earlier references can be found in the review [771]. The theoretical methods are not merely rewriting the formulas for a single site in matrix notation and application of greater computational power. There are serious issues already known from previous work that only become more difficult for the dynamical interacting-electron problem. Development of the methods presents challenges, while it also provides illuminating insights into the interesting physical phenomena.

Sections 17.1–17.4 are devoted to methods that go beyond a single site by applying the DMFT approach to clusters, i.e., auxiliary systems composed of cells of more than one atom embedded in an effective medium. Section 17.1 sets up the problem with the definition of "supercells" and clusters in real space and the corresponding reduced Brillouin zone in momentum space. The cellular approximation (c-DMFA in Sec. 17.2) is the generalization of the single-site approximation in real space expressed directly in terms of matrix elements of the self-energy that couple different sites. The dynamic cluster approximation (DCA in Sec. 17.3) is the generalization in momentum space expressed in terms of a $\mathbf{k}$-dependent self-energy within the Brillouin zone. Alternative approaches cast in terms of funtionals are discussed briefly in Sec. 17.4.

Section 17.5 is devoted to extended DMFT, where the guiding principle is that interaction between sites is smaller than on-site and can be treated by a more approximate approach, while keeping the single-site auxiliary system for the local intra-site interaction. An advantage of such methods is that they are computationally less demanding than cluster calculations, but they lead to frequency-dependent interactions. We return to this in Ch. 21 where such effects arise naturally when the effective interactions are derived from many-body perturbation theory.

Finally, Sec. 17.6 presents examples of Hubbard models that illustrate the methods in cases where it is important to go beyond the single-site approximation, especially at or near half-filling where the system is close to an antiferromagnetic transition.

17.1 Supercells and clusters

The previous chapter described DMFT in terms of a site embedded in a medium that is determined self-consistently, as illustrated schematically in Fig. 16.2. The overall approach also applies to a cell of more than one atom, for example, the self-consistent loop for a DMFT calculation in Fig. 16.3 is the same whether the site refers to a single atom or a cell. It might appear that nothing else is needed to generalize the methods to

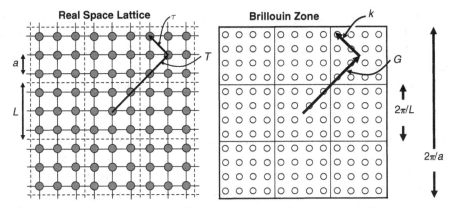

Figure 17.1. Schematic illustration of a crystal with primitive cell dimension a considered as a periodic array of 3×3 supercells with size $L = 3a$, translation vectors $\mathbf{T}_i$ that are multiples of primitive translation vectors and sites in the supercell denoted by τ. On the right is shown the Brillouin zone where $\mathbf{k}$ is defined to be within the RBZ and $\mathbf{G}$ denotes reciprocal lattice vectors for the lattice of supercells. It is often convenient to use a regular grid of $\mathbf{k}$-vectors exemplified here by a 4×4 grid of $\mathbf{k}$-points for the RBZ.

clusters.[1] However, there is more than one way to generalize the equations and algorithms of the previous chapter, each with difficulties that must be resolved. Furthermore, the larger cell is an artificial construction and the final results must respect the actual periodicity of the crystal. In this section we set up the problem that is addressed by different cluster methods in the following sections.

A crystal can be viewed as a periodic array of cells labeled i, displaced by translation vectors $\mathbf{T}_i$. The smallest possible cell is a primitive cell; following the notation in the previous chapter we also refer to it as a site that can denote an atom or primitive cell in general. A "supercell" is a larger unit cell that contains an integral number of primitive cells $j = 1, \ldots, N_c$ at positions τ_j. An example is shown in Fig. 17.1 for a crystal lattice with constant a and a 3×3 supercell with cell edge $L = 3a$, which is composed of nine primitive cells. Viewed as an array of supercells, the translation vectors $\mathbf{T}_i$ are multiples of L in each direction.[2] The reciprocal lattice vectors $\mathbf{G}$ are multiples of $2\pi/L$ in each direction. At the right side of the figure are shown the corresponding Brillouin zones: the BZ for the primitive cell of size $2\pi/a \times 2\pi/a$ and the reduced BZ (RBZ) for the supercell

---

[1] In Chs. 16–21 a cell denotes a periodic unit in a crystal: an atom, a primitive cell, or a supercell containing more than one primitive cell. A cluster denotes a cell embedded in a medium. A cluster does not necessarily have the periodicity of the crystal; however, its only role in the theory is to determine the self-energy that is used to construct a Green's function for the crystal, which is constructed to obey all the crystal symmetries. This last requirement is not automatically satisfied and steps to ensure the results obey the proper symmetries are included in the analysis of various cluster techniques.

[2] In general the supercell can be different multiples of the primitive cells in different directions, e.g., $L_x = N_x a$ and $L_y = N_y a$ with $N_c = N_x \times N_y$ in two dimensions. A sequence of cells argued to be efficient for extrapolation to the infinite size limit has been proposed by Betts *et al.* [795] and described in [771].

of size $2\pi/L \times 2\pi/L$. Integrals over the Brillouin zone can be approximated by a sum over a grid of $\mathbf{k}$-vectors, which can be expressed in any dimension as (see [1, Sec. 4.3]) $\Omega_{BZ}^{-1} \int_{BZ} d\mathbf{k} \rightarrow (1/N_{\mathbf{k}}) \sum_{\mathbf{k}}$, where Ω_{BZ} is the volume of the BZ or RBZ and $N_{\mathbf{k}}$ is the number of $\mathbf{k}$-points.

The hamiltonian can be expressed in the orthogonal tight-binding form (see Sec. 19.2):

$$\hat{H} = \sum h_{ijm,i'j'm'} c_{ijm\sigma}^{\dagger} c_{i'j'm'\sigma} + \hat{U}, \tag{17.1}$$

where $j, j' = 1, \ldots, N_c$ denote primitive cells and $m, m' = 1, \ldots, N_p$ label the states within a primitive cell. Thus $c_{ijm\sigma}^{\dagger}$ creates an electron with spin σ in orbital m in primitive cell j located in supercell i, and similarly for $c_{i'j'm'\sigma}$, and we have assumed the hamiltonian is spin-independent to simplify the notation. Since the independent-particle matrix elements $h_{ijm,i'j'm'}$ depend only on the relative positions of the cells $\mathbf{T}_i - \mathbf{T}_{i'}$, the Fourier transform is

$$h(\mathbf{k}) = \sum_i e^{i\mathbf{k} \cdot \mathbf{T}_i} h_{i0}, \tag{17.2}$$

where $\mathbf{k}$ is defined in the RBZ for a supercell and each term is understood to be a matrix in the orbital indices $jm, j'm'$. The interactions denoted by $\hat{U}$ need not be specified at this point, since the structure of the equations is independent of the detailed form of the interactions.

The expressions for the independent-particle bands in a supercell can be illustrated by the Hubbard model with nearest-neighbor hopping $-t$ on the square lattice. Then $h(\mathbf{k}) = \varepsilon(\mathbf{k}) = -2t(\cos(k_x a) + \cos(k_y a))$ is a scalar with $\mathbf{k}$ in the BZ of size $(2\pi/a)^2$. Exercise 17.1 is to show that this leads to a square Fermi surface for a half-filled band. For a cell with length $L = N_L a$ containing $N_c = N_L^2$ sites, this can be expressed as N_c bands with energy $\varepsilon_{n,n'}(\mathbf{k}) = -2t(\cos(G_x + k_x a) + \cos(G_y + k_y a))$, where $\mathbf{k}$ is in the RBZ of size $(2\pi/L)^2$, and $G_x = 2n\pi/L$, $G_y = 2n'\pi/L$, with $n, n' = 1, \ldots, N_L$. Of course, all formulas reduce to those in the previous chapter for a primitive cell with $N_c = N_L = 1$.

There are various ways to use the supercell construction. Quantum Monte Carlo methods in Chs. 22–25 treat cells with periodic or "twisted" boundary conditions (Sec. 6.2), i.e., allowing $\mathbf{k}$ to vary over the RBZ, and employing techniques to extrapolate to large size to represent condensed matter. In DMFT there are two steps: calculation of the self-energy Σ for a cluster, i.e., a supercell with boundary conditions corresponding to embedding in a medium, and use of Σ to construct the lattice Green's function for any $\mathbf{k}$ in the original BZ. The rest of this chapter is devoted to the use of cluster methods and other approaches to calculate spectra and other properties of interacting electrons in crystals.

17.2 Cellular DMFA

One approach is the cellular dynamical mean-field approximation (c-DMFA) [796–798], which is the logical extension of the single-site approximation in real space:[3] the site in

---

[3] This is the dynamical version of the molecular CPA [792, 799] and is the same as a molecule embedded in an effective medium. An advantage of this direct approach is that it is manifestly causal as long as the hybridization function Δ is causal, since it is the solution of a physical problem.

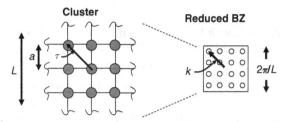

Figure 17.2. Illustration of cellular dynamical mean-field approximation for the 3×3 supercells in Fig. 17.1. At the left is the auxiliary cluster composed of the cell embedded in a surrounding medium with the coupling indicated by the terminating lines, shown here for coupling only to the atoms on the boundaries. In general, the hybridization function $\Delta(\omega)$ is a matrix that can couple to all the orbitals. The RBZ is shown on the right and all sums in the equations are over $\mathbf{k}$ in the RBZ. The corresponding decomposition in the DCA is shown in Fig. 17.3. If the coupling to a medium is omitted, this corresponds to an isolated cell and cluster perturbation theory, as explained in the text.

Fig. 16.2 is replaced by a cell with N_C sites and the cluster is defined by embedding the cell in an effective medium, as illustrated in Fig. 17.2. This is naturally formulated as a division of the hamiltonian expressed in real space,

$$h_{ii'} = h_c\,\delta_{ii'} + \Delta h_{ii'} \quad\longrightarrow\quad \text{no approximation}$$

$$\Sigma_{ii'}(\omega) = \Sigma(\omega)\delta_{ii'} + \Delta\Sigma_{ii'}(\omega) \xrightarrow{c\text{-}DMFA} \Sigma(\omega)\delta_{ii'}, \qquad (17.3)$$

where each term is a matrix in the orbitals $jm, j'm'$ of size $N_s \times N_s$, where $N_s = N_C N_p$ is the total number of states in a cell, and h_c and $\Sigma(\omega)$ are restricted to the central cell. The division of h into h_c and $\Delta h_{ii'}$ is not unique, but the form of the equations is independent of the choice so long as intracell terms, h_c and Σ, act only on the orbitals in the cell, and all intercell matrix elements are included in Δh and $\Delta\Sigma$. The second line in Eq. (17.3) is the generalization of the single-site approximation in Eq. (16.5) to cells, i.e., to assume $\Delta\Sigma_{ii'} = 0$ for any matrix elements between orbitals in different cells $i \neq i'$. Also it is assumed that the interactions are restricted to the cell, i.e., intersite interactions are not included if the sites are in different cells.

In the cellular method, the cluster is composed of the cell embedded in a medium with hybridization function $\Delta(\omega)$. This is indicated by the coupling to each atom at the boundary in Fig. 17.2, but in general $\Delta_{jm,j'm'}(\omega)$ is a matrix that couples different states $jm, j'm'$ including states in the interior of the cell. The calculations are done as if every site were inequivalent, i.e., the same as would be done for a crystal formed by molecules with N_C inequivalent atoms.[4] For example, in the 3×3 cell in Fig. 17.2 the central site is not

---

[4] It may be possible to use some symmetries of the crystal, e.g., the cell in Fig. 17.2 has inversion and four-fold rotation symmetry. In Ch. 16 the formulation of the iterative algorithm in terms of $\mathcal{G}_0$ instead of Δ is only a minor simplification, since they are related by the simple expression $\mathcal{G}_0(\omega) = [\omega - \varepsilon_0 - \Delta(\omega)]^{-1}$ in Eq. (16.8). This is sufficient for infinite dimensions (see Exs. 16.1 and 16.2), but in finite dimensions one must take into account that the surrounding medium should be described in terms of the *cavity Green's function* G^- for the

equivalent to those on the boundary. Thus, the self-consistent equations have exactly the same form as in Sec. 16.5, with $\mathbf{k}$ restricted to the reduced Brillouin zone. The solution can be found by the self-consistency loop in Fig. 16.3. The only difference is that all quantities are matrices in the N_s orbitals j, m and the independent-particle energies $\varepsilon_\mathbf{k}$ are here denoted $h(\mathbf{k})$. The calculation for the cluster corresponds to open-boundary conditions for Σ since it is truncated at the boundary. However, the cluster calculation is used only to determine Σ, and the independent-particle terms are treated with no approximation for any size cell.

If the hybridization function Δ is set to zero, this is termed "cluster perturbation theory" (CPT) (see, e.g., [771, 800]), where Σ is calculated for a cell with the entire hamiltonian truncated at the boundary. This is the generalization of the "Hubbard I approximation" for the self-energy of an isolated atom to an isolated cell of atoms. Since the effect of Δ decreases as the cell becomes larger, CPT can be considered as one way to approach the infinite size limit.

Restoring the translation symmetry

Up to this point all quantities are defined only in the RBZ and the Green's function and self-energy reflect the fact that the cluster does not have the translation symmetry of the crystal. In order to construct a "periodized" translation invariant Σ with the periodicity of the crystal, i.e., a function only of the relative positions of the sites $\tau_j - \tau_{j'}$, one can perform some average over sites j in the supercell. One choice is a uniform average, which introduces an error proportional to the surface-to-volume ratio $\propto 1/L$. This can be improved by a weighted average [797] that depends on the system. Another approach is to use only information for j and j' near the center of the cell. This has a long history in quantum chemistry, where terminated clusters are often used to approximate properties of large systems. For an insulator with an energy gap, the effects of the boundary on the central region decay exponentially for sufficiently large cells. In contrast, for a metal the effects are oscillatory with magnitude decaying as a power law, and the DCA method in the next section may be more appropriate. There is no best method for all cases: the most advantageous approach depends on the problem and the size permitted by feasible calculations.

Once a translation-invariant self-energy $\Sigma_{jm,j'm'}(\omega)$ has been produced, then it is straightforward to construct $\Sigma_{m,m'}(\mathbf{k}, \omega) = \sum_j \exp(i\mathbf{k} \cdot \tau_j)\Sigma_{jm,0m'}(\omega)$ where $\mathbf{k}$ extends over the original Brillouin zone of the crystal. Finally, the lattice Green's function can be written as $G(\mathbf{k}, \omega) = [\omega - h(\mathbf{k}) - \Sigma(\mathbf{k}, \omega)]^{-1}$, where each term is a matrix in the $m = 1, \ldots, N_p$ orbitals in the primitive cell.

---

lattice with one cell removed. In Ex. 16.1 the expressions are derived for G^- and Δ in terms of G^-. The cavity construction is automatically taken into account in $\mathcal{G}_0$, avoiding the needless complications that arise if it is expressed in terms of Δ.

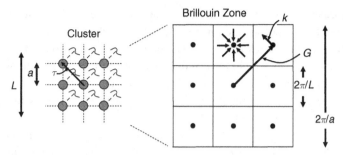

Figure 17.3. Schematic illustration of the DCA approach, which is the extension of the single-site approximation in reciprocal space. The calculation is done for a cluster with periodic boundary conditions, as shown on the left, to determine the auxiliary $\mathcal{G}_{jm,j'm'}(\omega)$, where j, j' are sites in the cluster and $m, m' = 1, \ldots, N_p$, with N_p the number of states per site. The result can be Fourier transformed to find $\mathcal{G}_{mm'}(\mathbf{G}, \omega)$ and $\Sigma_{mm'}(\mathbf{G}, \omega)$ for each $\mathbf{G}$; this is the result from the cluster calculation needed to construct the lattice Green's function in Eq. (17.5). On the right is shown the original BZ divided into reduced zones (RBZ). The average over each RBZ used in the DCA (Eq. (17.7)) is indicated by the arrows (shown in only one zone for clarity).

17.3 Dynamic cluster approximation

The DCA [771, 801] works with a cluster obeying periodic boundary conditions. It can be viewed as the logical extension of the single-site approximation to provide finer resolution in reciprocal space.[5] This can be accomplished by constructing a cluster in which each site is equivalent to a bath that couples equally to each site, as illustrated by the 3×3 cluster shown on the left in Fig. 17.3. This is the auxiliary system used only to calculate the self-energy. For any size cluster the periodicity of the crystal is preserved and the Brillouin zone is the original one for the primitive cell, as indicated on the right side of the figure. However, instead of the single-site approximation where the self-energy $\Sigma(\omega)$ is constant independent of $\mathbf{k}$ in the BZ, the increased size of the cluster makes it possible to resolve the self-energy into functions $\Sigma(\mathbf{G}, \omega)$ that are constant in the reduced zones centered on the reciprocal lattice vectors $\mathbf{G}$ contained in the original BZ, e.g., the 3×3 set of reduced zones shown in Fig. 17.3.

The DCA is most naturally expressed in reciprocal space, where the division analogous to Eq. (17.3) can be expressed as

$$h(\mathbf{G} + \mathbf{k}) = h_c(\mathbf{G}) \quad + \quad \Delta h(\mathbf{G} + \mathbf{k}) \quad \longrightarrow \quad \text{no approximation}$$

$$\Sigma(\mathbf{G} + \mathbf{k}, \omega) = \Sigma(\mathbf{G}, \omega) \quad + \quad \Delta \Sigma(\mathbf{G} + \mathbf{k}, \omega) \xrightarrow{DCA} \Sigma(\mathbf{G}, \omega), \quad (17.4)$$

where each term is on $N_p \times N_p$ matrix with N_p the number of orbitals in a primitive cell. The first terms on the right-hand side of Eq. (17.4), $h_c(\mathbf{G})$, $\Delta(\mathbf{G}, \omega)$, and $\Sigma(\mathbf{G}, \omega)$, are defined

---

[5] This approach does not have a precedent in previous work on disordered systems; instead, it is a new development in DMFT that is now used also in calculations for static disordered alloys [802–804]. A real-space interpretation can be found in [798].

at the reciprocal lattice vectors $\mathbf{G}$ that are in the BZ of the original lattice. The second terms on the right-hand side are the variation with $\mathbf{k}$ defined within each reduced zone, i.e., the difference from the values at $\mathbf{k} = 0$. The division between $h_c(\mathbf{G})$ and $\Delta h(\mathbf{G} + \mathbf{k})$ is not unique as long as they sum to the original independent-particle hamiltonian for lattice $h(\mathbf{G} + \mathbf{k})$. The approximation in the DCA is to ignore the variation of $\Sigma(\mathbf{G} + \mathbf{k}, \omega)$ with $\mathbf{k}$. This is a type of "coarse graining" of $\mathbf{k}$-space, where a continuous function of $\mathbf{k}$ is approximated by the values at a set of discrete points.

Since the DCA requires only $\Sigma(\mathbf{G}, \omega)$ at the discrete $\mathbf{G}$ vectors, it can be determined by a many-body calculation on a cluster with periodic boundary conditions and Fourier components at reciprocal lattice vectors $\mathbf{G}$, e.g., the 3×3 cell in Fig. 17.3. Thus a DCA calculation is done for a cluster in real space, like is done in c-DMFA; the difference is the periodic boundary conditions whereby all quantities are functions of relative position $\tau_j - \tau'_j$. The independent-particle hamiltonian for the cluster in real space is defined by the Fourier transform of $h_c(\mathbf{G})$, $h_{c\,jj'} = \sum_{\mathbf{G}} h_c(\mathbf{G}) \exp(i\mathbf{G} \cdot (\tau_j - \tau'_j))$, which is different from the original hopping matrix elements in the lattice and fixed by the conditions explained below. The independent-particle cluster Green's function is related to the hybridization function by $\mathcal{G}_{0\,jj'}(\omega) = [\omega - h_{c\,jj'} - \Delta_{jj'}(\omega)]^{-1}$. However, it is most convenient to formulate the method directly in terms of $\mathcal{G}_0(\omega)$; if desired, the hybridization function can be calculated as shown in Ex. 17.5. The calculation leads to the cluster Green's function $\mathcal{G}_{jj'}(\omega)$ which can be transformed to find the Fourier components $\mathcal{G}(\mathbf{G}, \omega)$. Finally, the self-energy for the cluster is given by $\Sigma(\mathbf{G}, \omega) = \mathcal{G}_0(\mathbf{G}, \omega)^{-1} - \mathcal{G}(\mathbf{G}, \omega)^{-1}$. Note that each quantity is a matrix in the orbitals m and m' in a primitive cell of the crystal and it is diagonal in $\mathbf{G}$, since $\mathbf{G}$ corresponds to a point in the BZ of the original lattice.

The DCA equations are the generalization of the single-site DMFA equations in Sec. 16.5. From the self-energy $\Sigma(\mathbf{G}, \omega)$, the lattice Green's function can be constructed straightforwardly for all $\mathbf{G} + \mathbf{k}$ in the original BZ for the primitive cell (the generalization of Eq. (16.3)),

$$G(\mathbf{G} + \mathbf{k}, \omega) = [\omega - h(\mathbf{G} + \mathbf{k}) - \Sigma(\mathbf{G}, \omega)]^{-1}, \qquad (17.5)$$

where $\mathbf{k}$ is in the RBZ and $h(\mathbf{G}+\mathbf{k})$ is the original lattice independent-particle hamiltonian. The Green's function is the same at each site, with Fourier components given by

$$G_{00}(\mathbf{G}, \omega) = \frac{1}{N_{\mathbf{k}}} \sum_{\mathbf{k}} G(\mathbf{G} + \mathbf{k}, \omega) = \mathcal{G}(\mathbf{G}, \omega), \qquad (17.6)$$

where the sum is over $\mathbf{k}$ in the RBZ. The second equality is the DMFT condition that the Green's function on the site $G_{00}(\mathbf{G}, \omega)$ and the auxiliary cluster are the same. This is exactly the same as for a single site in Sec. 16.5, except that the equations must be satisfied for each reciprocal lattice vector $\mathbf{G}$.

Note, however, that the lattice Green's function $G(\mathbf{G}+\mathbf{k}, \omega)$ in Eq. (17.5) has unphysical discontinuities in the Brillouin zone since the self-energy is $\Sigma(\mathbf{G})$ independent of $\mathbf{k}$ within each RBZ and changes abruptly at the boundaries of the reduced zones. This does not affect the self-consistent DCA calculation, since $G_{00}(\mathbf{G}, \omega)$ in Eq. (17.6) is an integral over each region. However, it does affect the use for other purposes, since there should be no discontinuities in physical quantities as a function of $\mathbf{k}$ inside the BZ. We will return to this issue and a procedure to reconcile the problem at the end of this section.

Up to this point the analysis applies for any independent-particle terms $h_c(\mathbf{G})$ so long as $h(\mathbf{G} + \mathbf{k}) = h_c(\mathbf{G}) + \Delta h(\mathbf{G} + \mathbf{k})$, where h is the original independent-particle hamiltonian. The choice made in DCA is to define $h_c^{DCA}(\mathbf{G})$ to be the *average* over $\mathbf{k}$ in the RBZ for each $\mathbf{G}$. This is the logical extension of the single-site approximation, where the average is over the full BZ. Thus the two terms in the division of h are the average and the variation with $\mathbf{k}$,

$$h_c^{DCA}(\mathbf{G}) = \frac{1}{N_{\mathbf{k}}} \sum_{\mathbf{k}}^{RBZ} h(\mathbf{G} + \mathbf{k}), \tag{17.7}$$

$$\Delta h^{DCA}(\mathbf{G}, \mathbf{k}) = h(\mathbf{G} + \mathbf{k}) - h_c^{DCA}(\mathbf{G}). \tag{17.8}$$

The procedure of averaging over a region is termed coarse graining of reciprocal space [771]. This means that momentum is not strictly conserved in the interacting system; in the calculation of $\Sigma(\mathbf{G})$, momentum is conserved only on the coarse grid of discrete $\mathbf{G}$. There is no approximation for a non-interacting problem, and the only effect of the coarse graining is in $\Sigma(\mathbf{G})$.

The hopping matrix elements $h_{c,jj'}^{DCA}$ needed in the cluster calculation are modified from the values in the original hamiltonian, and they must be calculated from the Fourier transform of the expressions in Eq. (17.7). This is straightforward to calculate numerically for a general hamiltonian h, and it can be calculated analytically for simple tight-binding cases. The essence can be understood from the example of the one-dimensional Hubbard model defined in Eq. (3.2) or Eq. (K.1). In that case $h(k) \equiv -t(k)$, with $t(k) = 2t\cos(ka)$, and the Fourier transform of $h_c^{DCA}(\mathbf{G})$ in Eq. (17.7) leads to the DCA cluster hamiltonian with cyclic hopping matrix elements (see Ex. 17.3)

$$t_c^{DCA} = -t\frac{L}{\pi}\sin(\pi/L) \tag{17.9}$$

for nearest neighbors $j' = j \pm 1$, which is equivalent to $j' = j + \pm(L-1)$. Thus the matrix elements inside the cluster are always reduced in magnitude and approach $-t$ as $1/L^2$ for large L. The reduction is compensated by the intercluster terms that are the Fourier transform of $t(G+k) - t_c^{DCA}(G)$ for $-\pi/L < k < \pi/L$, which leads to (see Ex. 17.3, part (c))

$$\Delta t^{DCA}(nL) = -t\frac{L}{\pi}\left[\frac{\sin((n \pm 1)\pi/L)}{(n \pm 1)} - \sin(\pi/L)\delta_{n,0}\right] \tag{17.10}$$

between sites j and $j' = j + nL \pm 1$. Thus $\Delta t^{DCA}(nL) = 0$ for $n = 0$ within the cluster, and it is long range, with intercluster terms decreasing as $1/n$. This long-range character applies in any dimension [771] and the form is a consequence of the discontinuities in $\Delta t(\mathbf{G} + \mathbf{k})$ at the boundaries of each RBZ. Although the long-range behavior might appear to be unphysical, it is not a problem since it does not affect the cluster calculation and DCA equations use the expressions in momentum space.

Examples of the DCA calculations for Hubbard models in two and three dimensions are given in Sec. 17.6.

Causality

One of the achievements of the DCA is that it obeys the requirement of causality, which has been a serious obstacle in cluster methods for alloys for many years [140]. The analytic property required is that the spectral weight is positive, i.e., $\mathrm{Im}G(\omega) < 0$ for the retarded Green's function at all real frequencies ω (see Secs. 5.4 and 5.5), which also means that $\mathrm{Im}\Sigma(\omega) < 0$. As pointed out in Sec. 17.2, cellular DMFA is casual since it is equivalent to a calculation for a molecule; the difficulty arises for methods that do not break the translation symmetry. For the DCA, presuming the solver is properly designed, the only place that a problem could occur is in the input to the cluster solver, the non-interacting Green's function $\mathcal{G}^0$. For example, self-consistent iterations in the flow diagram in Fig. 16.3 proceed with $\mathcal{G}_0$ determined by G_{00} and Σ from the previous step, $\mathcal{G}_0(\mathbf{G}, \omega) \leftarrow [G_{00}^{-1}(\mathbf{G}, \omega) + \Sigma(\mathbf{G}, \omega)]^{-1}$. The proof [801] that $\mathcal{G}^0(\mathbf{G}, \omega)$ is causal is an extension of the argument for the single-site approximation, where the average is over the original Brillouin zone, as explained in Ex. 17.4.

Restoring the continuity of $\Sigma(\mathbf{k})$

Finally, we return to the issue of constructing a self-energy that varies smoothly throughout the Brillouin zone. There is no problem for properties that depend only on the values $\Sigma(\mathbf{G}, \omega)$, such as the self-consistent cluster calculation itself. Also there is no problem for integrated quantities that are approximated by a discrete sum over quantities evaluated at the $\mathbf{G}$ vectors. However, an acceptable form for the self-energy on the lattice must vary smoothly throughout the BZ, which is essential in order to obtain smooth non-local quantities such as the Fermi surface or the quasi-particle dispersion. A smoothly varying $\Sigma_{\mathbf{k}}(\omega)$ can be constructed by interpolation. In [771] it is pointed out that an Akima spline [805] is smooth and has the property that it does not overshoot, which is required in the proof (see Ex. 17.4) that the resulting Green's function is causal.

17.4 Variational cluster and nested cluster approximations

The VCA [296, 772, 773] is a realization of the self-energy functional approach formulated in Secs. 16.6 and I.2, in which the grand potential Ω is expressed in terms of an auxiliary system where the self-energy Σ_{aux} and Ω_{aux} can be calculated. The essence is that all problems with the same interaction can be cast in terms of a universal interaction functional $F[\Sigma]$, and $\Omega[\Sigma]$ can be calculated with no approximation to the functional; instead, Σ is restricted to the range of self-energies Σ_{aux} that can be generated by the auxiliary system. The application to DMFT is explained in Sec. 16.6 where the key results are that it provides a useful functional given in Eq. (16.15), a variational derivation of the DMFT equations in Eq. (16.16), and alternative methods that may be simpler than full self-consistent DMFT.

Even though it is very general and is applied to the single-site problem in Sec. 16.6, the appellation "variational cluster" came about because it was derived for clusters where it has a particular use. Because there are many possible clusters, there are many ways to derive useful relations between solutions for different clusters. Examples of VCA calculations can be found in [773, 806, 807] and a variational derivation of the cellular DMFA is given in [806].

In nested cluster approaches (see, e.g., [808] and the reviews [302, 771]) the Luttinger–Ward functional for the interaction $\Phi[G]$ is expressed as a series for the single-site G_{ii}, pair functions G_{ij} with i, j nearest neighbors, a larger cluster with next-nearest neighbors, and so forth. At each level it is translation-invariant and only involves the relative positions of sites. This builds upon methods [140, 794] developed for alloys and the Bethe–Peierls self-consistent cluster approach [172, 173] in statistical mechanics, and in many ways it is the most natural approach to describe non-local correlations. However, it is difficult to ensure causality because each larger cluster contains terms already counted in the smaller clusters [771, 808]. These must be subtracted, which may lead to an unphysical Green's function with $\text{Im}G(\omega) > 0$. Similar issues may arise in the combination of different many-body methods in Ch. 21. Methods to solve or circumvent the causality issues could be an important advance.

17.5 Extended DMFT and auxiliary bosons

An alternative approach is to keep a single-site auxiliary system and treat interactions between electrons on different sites by an approximate method. This is very much in the spirit of DMFT, where the guiding principle is that the intersite interactions are small compared with on-site interactions. Whereas the computational effort in cluster calculations escalates dramatically with the size of the cluster, a criterion for an alternative method is that it involves only a small increase in complexity and computational effort over the single-site DMFA.

The method called "extended DMFT" (EDMFT) is designed to treat intersite interactions by generalizing the auxiliary system.[6] Up to this point the auxiliary system has been used only to calculate the self-energy $\Sigma(\omega)$ for an embedded site; for this purpose it is sufficient for the site to be coupled to the bath by single-body hopping terms that are incorporated in the hybridization function $\Delta(\omega)$, for example, as shown explicitly in Eq. (16.9). However, auxiliary systems can be much more general, and the essence of EDMFT is a site coupled to a bath through both single-body hopping and two-body interactions. EDMFT has much in common with the approach in Ch. 21, where effects of intersite interactions are calculated using many-body perturbation theory. The parallels between DMFT and EDMFT given below mirror the relation of the Luttinger–Ward functional $\Phi[G]$ to the generalized functional $\Psi[G, W]$ in Sec. 8.3.

---

[6] See, for example, [303, 809, 810] that also refer to earlier papers and [811] that places EDMFT in a more general context.

The principle features can be illustrated by the extended Hubbard model,

$$\hat{H} = \varepsilon_0 \sum_{i,\sigma} c_{i\sigma}^{\dagger} c_{i\sigma} - \sum_{i<j,\sigma} t_{ij} c_{i\sigma}^{\dagger} c_{j\sigma}$$

$$+ U \sum_i \hat{n}_{i\uparrow} \hat{n}_{i\downarrow} + \sum_{i<j} \left\{ V_{ij} \delta \hat{n}_i \delta \hat{n}_j + J_{ij} \hat{S}_i \cdot \hat{S}_j \right\}, \qquad (17.11)$$

which is the same as Eq. (3.2) with added interactions between sites $i \neq j$: V_{ij} for charge fluctuations with $\delta \hat{n}_i = \hat{n}_i - \langle \hat{n}_i \rangle$ and J_{ij} for interactions between spins $\hat{S}_i$ and $\hat{S}_j$. Here we give explicit expressions for the charge; spin fluctuations can be treated in an analogous manner [809].

We can follow the same logic as for the one-electron Green's function $G_{\mathbf{k}}(\omega)$ in Ch. 16. The hopping between sites t_{ij} leads to dispersion $t_{\mathbf{k}} = \sum_i t_{0i} \exp(i\mathbf{k} \cdot \mathbf{R}_i)$ and a $\mathbf{k}$-dependent $G_{\mathbf{k}}(\omega)$ where the local on-site $G_{00}(\omega)$ is required to be the same as the Green's function $\mathcal{G}(\omega)$ for the auxiliary system. So also the interaction between sites V_{ij} leads to a momentum-dependent interaction $U_{\mathbf{q}} = U + V_{\mathbf{q}}$ with $V_{\mathbf{q}} = \sum_i V_{0i} \exp(i\mathbf{q} \cdot \mathbf{R}_i)$, where $\mathbf{q}$ is used to denote momentum in the interactions. In analogy with $G_{\mathbf{k}}(\omega)$, the susceptibility $\chi_{\mathbf{q}}(\omega)$ is the Fourier transform of the density–density correlation function $\chi_{ij}(t) = \langle \delta \hat{n}_i(t) \delta \hat{n}_j(0) \rangle$, a type of two-particle Green's function as described in Secs. 5.5 and 5.8 and Exs. 21.1 and 21.2. This suggests that we can define a modified auxiliary system for an embedded site,

$$\hat{H}_{aux} = \varepsilon_0 \sum_{\sigma} c_{0\sigma}^{\dagger} c_{0\sigma} + \sum_{\ell\sigma} t_{0\ell} (c_{\ell\sigma}^{\dagger} c_{0\sigma} + c_{0\sigma}^{\dagger} c_{\ell\sigma}) + \sum_{\ell\sigma} \epsilon_{\ell} c_{\ell\sigma}^{\dagger} c_{\ell\sigma}$$

$$+ U \hat{n}_{0\uparrow} \hat{n}_{0\downarrow} + \sum_{\ell'} \lambda_{\ell'} (b_{\ell'}^{\dagger} + b_{\ell'}) \hat{n}_0 + \sum_{\ell'} \omega_{\ell'} b_{\ell'}^{\dagger} b_{\ell'}, \qquad (17.12)$$

where the operators $b^{\dagger}$ and b create and annihilate auxiliary bosons that represent the coupling to a polarizable medium. This is the same as the Anderson impurity model in Eq. (3.8) with the host having the additional feature of dynamic fields that couple to the occupation of the site. The added terms have the same form as for dynamical interactions in Eq. (21.14) and can be treated by the Monte Carlo solvers in Sec. 18.7. As in Ch. 16, one does not need to specify the actual states labeled ℓ and ℓ' in the hamiltonian for the embedding medium Eq. (17.12); it is sufficient to develop the theory in terms of the hybridization function $\Delta(\omega)$ and an analogous polarizability function $\Lambda(\omega)$ for the embedding medium, where both functions are to be determined simultaneously by self-consistency conditions given below. A difference from the DMFT auxiliary system is that the effective interaction between electrons on the site is modified and we will express it as $\mathcal{U}(\omega) = U - \Lambda(\omega)$ to denote the fact that it is a dynamic function that is not fixed but is varied until the conditions are satisfied. This is an example of the general feature that in order to incorporate the effects of non-local interactions in local equations, the local effective interactions must be frequency-dependent.

Since the role of the polarizable auxiliary system is to mimic the effects of screening of the local on-site interaction U by the response of surrounding sites in the crystal, we need to write the effective on-site interaction in terms of the polarizability of each site. We can follow the approach developed in Chs. 7 and 8 and used throughout Chs. 9–15, where analogous relations apply for the Green's function and the screened interaction W. If we assume that the polarizability $P(\omega)$ is local to each site, in analogy with the self-energy $\Sigma(\omega)$, then $G_{\mathbf{k}}(\omega)$ and $W_{\mathbf{q}}(\omega)$ for the crystal can be expressed in the form

$$G_{\mathbf{k}}(\omega) = [G_{\mathbf{k}}^0(\omega)^{-1} - \Sigma(\omega)]^{-1} \quad \text{and} \quad W_{\mathbf{q}}(\omega) = [U_{\mathbf{q}}^{-1} - P(\omega)]^{-1}, \tag{17.13}$$

where $U_{\mathbf{q}}^{-1}$ is analogous to $G_{\mathbf{k}}^0(\omega)^{-1} = \omega - \varepsilon_0 - t_{\mathbf{k}}$.

In Ch. 16 the condition that determines the auxiliary system is that the local part of the Green's function $G_{00}(\omega)$ is equal to the auxiliary $\mathcal{G}(\omega)$ with the same local self-energy $\Sigma(\omega) = \Sigma_{aux}(\omega)$ (Eq. (16.6)). This is accomplished by varying $\mathcal{G}_0$ until the consistency condition is satisfied. The self-energy is not a direct output of the auxiliary system calculation; instead, it is calculated from the Green's function using the relation $\Sigma_{aux}(\omega) = \mathcal{G}_0^{-1}(\omega) - \mathcal{G}^{-1}(\omega)$ (Eq. (16.10)). The analogous development for the screened interaction is that the local interaction $W_{00}(\omega)$ is equal to the interaction $\mathcal{U}(\omega)$ used in the auxiliary system for the same polarizability $P(\omega) = P_{aux}(\omega)$. Like the self-energy, $P_{aux}(\omega)$ is not given directly in the auxiliary calculation, but can be found from the relation $P_{aux}(\omega)^{-1} = \chi_{aux}(\omega)^{-1} + \mathcal{U}(\omega)$ (see Ex. 17.7). The consistency conditions can be written explicitly,

$$\frac{1}{N_{\mathbf{k}}} \sum_{\mathbf{k}} G_{\mathbf{k}}(\omega) = \mathcal{G}(\omega) \quad \text{and} \quad \frac{1}{N_{\mathbf{q}}} \sum_{\mathbf{q}} W_{\mathbf{q}}(\omega) = \mathcal{U}(\omega). \tag{17.14}$$

Thus the difference from a DMFT calculation is that the many-body calculation must be done in an auxiliary system that has dynamic interactions, the susceptibility $\chi_{aux}(\omega)$ must be calculated to determine the polarizability $P_{aux}(\omega)$, and there is the additional self-consistency condition in Eq. (17.14).

The EDMFT approximation for the density–density correlation function $\chi_{ij}(\tau) = \langle \delta \hat{n}_i(\tau) \delta \hat{n}_j(0) \rangle$ is most conveniently written in terms of the Fourier components,

$$\chi_{\mathbf{q}}(\omega) = [P(\omega)^{-1} - U_{\mathbf{q}}]^{-1}, \tag{17.15}$$

where $P(\omega)$ is the local polarizability. Note that the interaction is the bare $U_{\mathbf{q}} = U + V_{\mathbf{q}}$ and not the effective interaction $\mathcal{U}(\omega)$ in the auxiliary system. The intersite correlation is reflected in the $\mathbf{q}$-dependence of $\chi_{\mathbf{q}}(\omega)$ that is determined by $V_{\mathbf{q}}$. This relation to $V_{\mathbf{q}}$ is brought out if we define an on-site $\chi_{site}(\omega)^{-1} = P(\omega)^{-1} - U$ and express Eq. (17.15) in the form

$$\chi_{\mathbf{q}}(\omega) = [\chi_{site}(\omega)^{-1} - V_{\mathbf{q}}]^{-1}, \tag{17.16}$$

which is the correlation function for a lattice of polarizable units interacting with potential $V_{\mathbf{q}}$.

EDMFT is a forerunner of the methods in Ch. 21. There the combination of DMFT and GW provides a way to calculate all the interactions from first principles using RPA: the dynamic local interaction $\mathcal{U}(\omega)$ and the screened interaction $W_{\mathbf{q}}(\omega)$ that is the generalization of the static $V_{\mathbf{q}}$. The auxiliary system, the consistency conditions in Eq. (17.14), and the RPA calculation for the intersite correlation have the same form as developed here.

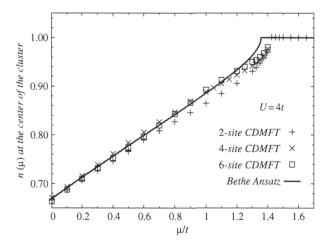

Figure 17.4. Density n as a function of chemical potential μ for the one-dimensional Hubbard model with $U/t = 4$ for clusters of increasing size [812] compared with the exact solution using the Bethe ansatz [134]. The value is measured at central sites of the cluster, which is slightly more accurate than the average over the cluster. Except at half-filling, the solution is a metal with n a smooth increasing function of μ; well away from half-filling the agreement is very good but there are deviations near the onset of the gap, where the correlation length becomes large and there may be instabilities. The range of μ where $n = 1$ indicates the size of the gap, shown here only for the two-site cluster. Figure provided by E. Koch, similar to Fig. 12(b) in [812].

17.6 Results for Hubbard models in one, two, and three dimensions

Hubbard model in one dimension – comparison with exact solutions

The one-dimensional Hubbard model is a challenging case for the DMFT approach, which is most appropriate for high spatial dimensions. For all $U > 0$, there is an exact solution using the Bethe ansatz [134], which shows that the one-dimensional Hubbard model is a metal for any filling except $1/2$ where it is always an insulator with a gap that is exponentially small for small U. Thus the density n increases continuously as a function of the chemical potential μ up to $n = 1$ (counting both spins) and is constant for any μ in the gap. Calculations as a function of cluster size have been carried out in many works, e.g., [773, 812, 813]; results from [812] using the cellular (c-DMFA in Sec. 17.2) method and exact diagonalization are shown in Fig. 17.4, where they are compared with the exact solution. The figure shows that results for such integrated quantities can be quite good for many cases, but there are deviations as the density approaches half-filling, where there are large antiferromagnetic fluctuations with increasing correlation length, and the calculations have instabilities.

For such static properties one can also use the DMET approach (see Sec. 16.7); indeed, results for the one-dimensional Hubbard model are very close to those in Fig. 17.4, except near half-filling, where the DMET calculations also have difficulties and can exhibit non-monotonic behavior [777].

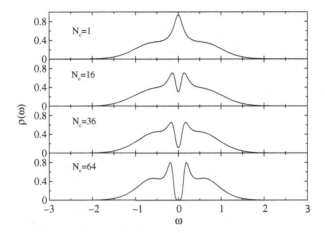

Figure 17.5. Calculated spectra for the half-filled Hubbard model on the 2D square lattice for $U/t = 4$, i.e., $U/W = 0.5$. The energy is in units of the bandwidth $W = 8t$ is the bandwidth [814]. The calculations were done with the DCA method and the Hirsch-Fye QMC solver (Ch. 18) at temperature $T = 0.125t$. The top graph for $N_C = 1$ is the single-site approximation, where the results are similar to Fig. 16.4 for the semicircular density of states. The spectra for larger clusters show that intersite correlations are essential for opening the gap for U/W in this range. Figure from [814].

Hubbard model in two dimensions: spectra and phase transitions at half-filling

There is a vast literature of work on the 2D Hubbard model, especially because of the relevance for high-temperature superconductivity found in materials with square planar structures and a single band per plane, as discussed in Sec. 2.9. The parent compounds are antiferromagnetic insulators corresponding to the half-filled case; superconductivity and other phenomena emerge as the planes are doped to have density less than half-filling. Here we can touch upon only a few aspects of theoretical work, which brings out striking effects of interactions that can only be described by going beyond the single-site approximation.

Figure 17.5 shows the spectra for the half-filled Hubbard model on the 2D square lattice calculated using the DCA (Sec. 17.3). As the size of the cluster is increased, i.e., intersite correlations are treated more accurately, there is a gap in the spectrum even for small values of U [814]. This is the expected result that there is an opening of a gap at all $U > 0$ due to the nesting of the Fermi surface, i.e., parallel parts of the surface that occur for the square Fermi surface in the Hubbard model. However, this is a special case: if the Fermi surface is modified to remove the nesting (for example, due to second-neighbor hopping t' as illustrated in Fig. 17.7 below) it should be a metal for small enough U and a transition to an insulating state would occur at a critical value of U.

The metal–insulator transition for half-filling is shown in Fig. 17.6. In the single-site approximation there is a first-order transition ending in a critical point with a form much like that for a semicircular density of states shown in Fig. 16.5. In particular, the transition temperature decreases with increasing U, whereas it should increase at small U. The cluster

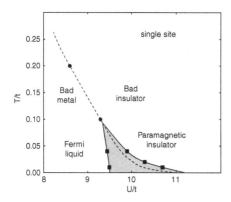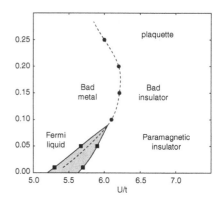

Figure 17.6. Calculated phase diagram for the half-filled 2D Hubbard model comparing the single-site and cluster calculations with a four-site plaquette [815]. Note that the plot for the single site is for $U/t > 9$, but for the plaquette the transition is at $U/t < 6.5$. The shaded areas denote a first-order transition with coexistence regions; the critical point is indicated and at higher temperature the dashed line indicates a crossover behavior. The single-site phase diagram is similar to Fig. 16.5 and the calculation for the cluster shows increased stability of the insulator, with smaller U at low temperature as expected (see text). Figure provided by H. Park, modified from figures in [815].

calculation has qualitatively similar behavior at high temperature, but as the temperature decreases the results are very different. The cluster calculation shows the expected behavior that a small interaction U is sufficient to produce an insulating state, consistent with the gap in Fig. 17.5. This shows the effects of antiferromagnetic correlations even though there is no ordered AF state in 2D at $T \neq 0$. Indeed, at $T = 0$ the ground state is an ordered antiferromagnet (see Secs. 3.3 and 3.2).

The doped 2D Hubbard model

It is especially interesting to consider the 2D Hubbard model with a square lattice away from half-filling. This corresponds to the doped Mott insulator described in Sec. 2.8, and it is the model often proposed to capture the essential features of the electronic structure of the materials that are high-temperature superconductors. Figure 2.16 shows an example of angle-resolved photoemission for $Bi_2Sr_2CaCu_2O_{8+x}$ (Bi-2212) at temperature T above the superconducting T_c, which illustrates remarkable features that occur in the spectra for states near the Fermi energy μ. Along the diagonal direction in the BZ there is a well-defined Fermi surface, but it seems to disappear and a "pseudogap" (a large depression of the spectrum near μ but not a full gap) opens in the part of the Brillouin zone near the boundary at $(0, \pi)$.

Different behavior in different parts of the BZ can be addressed only by calculations that go beyond the single-site approximation. There have been a number of such calculations for clusters of more than one site (see [771, 817] for references to previous work); Fig. 17.7 shows results found using the DCA and a CTQMC solver (Sec. 18.6) [817]. On the left

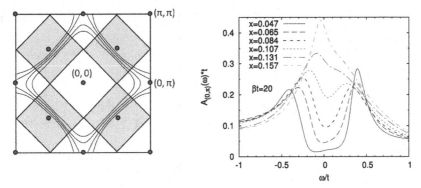

Figure 17.7. Left: patching of the 2D square BZ in a dynamical cluster calculation for a cluster of eight sites. The self-energy is averaged over the patches shown at $(0,0)$, $(0,\pi)$, $(\pi/2,\pi/2)$, and (π,π). The lines indicate the Fermi surface for a non-interacting Hubbard model with $t' = -0.15t$ at half-filling and with dopings of 10, 20, and 30% (adapted from [816]). Right: spectral function from [817] showing $-(1/\pi)\mathrm{Im}G(\omega)$ averaged over the sector containing the $(0,\pi)$ point for the hole dopings indicated, with $U = 7t$ and temperature $\beta^{-1} = t/20 \approx 200$ K. The spectra indicate the opening of a "pseudogap" where the Fermi surface disappears; see discussion in text and compare with the observations shown in Fig. 2.16 for $Bi_2Sr_2CaCu_2O_{8+x}$.

is shown the BZ divided into patches for an eight-site cell. The spectra are calculated as averages over a patch, as described in Sec. 17.3. The figure also shows the Fermi surface for various dopings for the $U = 0$ Hubbard model with nearest-neighbor (nn) and next-nearest-neighbor (nnn) hopping (see Ex. 17.2), $\hat{H}^0 = -\frac{1}{2}\sum_\sigma \left[\sum_{ij}^{nn} tc_{i\sigma}^\dagger c_{j\sigma} + \sum_{ij}^{nnn} t' c_{i\sigma}^\dagger c_{j\sigma}\right]$, with $t' = -0.15t$ chosen to be appropriate for the high-temperature superconductors. When interactions are included ($U = 7t$, comparable with the bandwidth $W = 8t$ (see Ex. 16.8)) there are drastic changes in the spectrum averaged over the patch centered on $(0,\pi)$ as a function of doping x. For large doping (large x far from half-filling) there is a peak near the Fermi energy as expected for a metal, but for low doping near half-filling there is a very large depression of the spectrum averaged over the entire patch, i.e., a "pseudogap."[7] The calculations find no such feature for the patch around $(\pi/2, \pi/2)$. Although the calculations are not directly comparable with the experimental spectra, they lead to the same conclusion that there is an anomalous feature due to interactions.

In order to study a superconducting state with d-symmetry one must go beyond the single-site approximation since the pair function changes sign for different neighbors and cannot be described by a scalar function localized to a site. Numerous calculations, e.g., [819, 820], indicate that the d-symmetry state is stabilized by the repulsive interactions in the Hubbard model with second-neighbor hopping similar to that used to derive the results in Fig. 17.7.

---

[7] The spectra are calculated from imaginary-time data using a maximum entropy method (Sec. 25.6) following [818] where it was found to be preferable to continue the self-energy to real frequencies instead of the Green's function. See also the discussion at the end of Sec. 18.3.

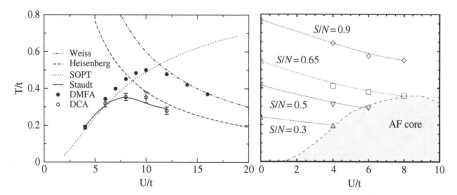

Figure 17.8. Left: antiferromagnetic transition temperature in the 3D Hubbard model using the
DCA and Betts cells with extrapolation (open circles) [822] in good agreement with determinantal
Monte Carlo calculations (see Ch. 25) done with much larger cells (solid line) [823]. The results
agree with second-order perturbation theory (SOPT) for small U and with accurate calculations for
the Heisenberg model for large U. The single-site approximation (denoted DMFA in the figure) is
accurate at small U but has the Weiss mean-field behavior at large U, as explained in the text. Right:
determinantal Monte Carlo calculation [824] of the phase transition for an optical lattice in a
confined trap with an antiferromagnetic core. The lines of constant entropy indicate that a transition
could be observed at accessible values of the entropy. Figures provided by P. Kent and Y. L. Loh,
similar to ones in [822] and [824].

For low doping, spin and charge order is observed experimentally [821]. An example of
calculations using the constrained-path QMC method is shown in Fig. 24.7. Such projector
methods can determine the ground state and treat much larger systems (in this case 512
sites) than the spectral approaches. These works and others indicate the complex phase
diagram and phenomena that may be found in the Hubbard model.

Hubbard model in three dimensions

Figure 17.8 shows the phase diagram for the 3D Hubbard model on a cubic lattice with
half-filling from [822]. The calculations are done with DCA and Betts cells [771, 795] up
to 48 sites. As shown in the figure, the results for the Néel temperature T_N are essentially the same as found using auxiliary field determinantal Monte Carlo (DetMC, see
Sec. 25.5) in [823], where up to 1000 sites were used. For large U the system can be
approximated as a Heisenberg spin model with nearest-neighbor interaction $J = 4t^2/U$,
and indeed the transition temperature in Fig. 17.8 approaches this limit. The figure also
shows the transition found in the single-site approximation DMFA, which is accurate at
small U. However, it overshoots for large interaction and it approaches the Weiss mean-
field result as expected since it is a mean-field approximation for correlation between
sites.

The right side of the figure shows the transition calculated for a Hubbard model in the
central region of a harmonic trap using DetMC [824]. Although there is not a sharp phase

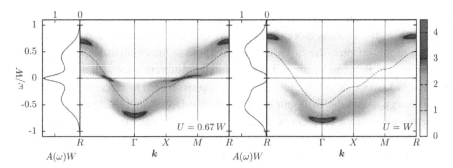

Figure 17.9. The momentum-resolved spectra for the 3D Hubbard model in the paramagnetic phase for two values of U/W, where $W = 12t$ is the bandwidth (see Ex. 16.8). The plot is for **k** along lines between the symmetry points at the centers of the BZ ($\Gamma = (0,0,0)$), face ($X = (0,0,\pi/2)$), edge ($M = (0,\pi/2,\pi/2)$), and corner $R = (\pi/2,\pi/2,\pi/2)$. The integrated spectra $A(\omega)$ shown at the left of each panel can be compared with those in Fig. 17.5 for the 2D Hubbard model. Modified figure from [826] provided by E. Gull.

transition, there is a rapid change at temperatures close to T_N in the homogeneous lattice shown at the left. This is a model for an optical lattice where the relevant comparison with experiments is the change as U is varied at constant entropy S. The calculations find that increasing U can lead to cooling due to redistribution of entropy from the center to the metallic outer regions (see Ex. 17.8 for calculation of entropy). There are also calculations for such problems using DMFT generalized to the inhomogeneous situation such as a trap [825].

Momentum-resolved spectra

The momentum-resolved spectra in Fig. 17.9 show the opening of the gap in a paramagnetic state at temperature $0.24t$, which is close to the Néel temperature T_N as shown in Fig. 17.8. The calculations were done using the DCA with a cell of 18 sites and a solver based on CTQMC expanded in the interaction (see Sec. 18.5). There are satellites at higher energy than the independent-particle bands, as was found by Hubbard (see Fig. 3.3) and in other calculations, e.g., in Figs. 16.4 and 17.5. The momentum resolution shows that they are indeed associated with the dispersion of the independent-particle bands. Perhaps most interesting is the development of the peak at the Fermi energy. The **k**-resolved spectra show that this results from reduced dispersion (heavier mass) around the Fermi surface. Thus the band develops "three-peak" character, like the density of states in the Anderson impurity model in Fig. 3.4; however, here each of the peaks is associated with a different part of the Brillouin zone.

17.7 Wrap-up

Calculations for the interacting electron problem in a cluster, an auxiliary system composed of a cell with more than one atom embedded in an effective medium, provide a direct approach to determine correlations beyond the single-site approximation (single-site DMFA) of Ch. 16. There is more than one way to define the cluster and two methods are described in this chapter: the extension in real space in the cellular method (c-DMFA) to calculate matrix elements of the self-energy linking different sites within the cell, and the extension in **k**-space in the dynamical cluster approximation to calculate the variation of the self-energy within the Brillouin zone. For both c-DMFA and DCA, the calculation for the interacting-electron problem is done for a cluster in real space; the difference is the boundary conditions. In all cases, the role of the many-body calculation is only to calculate the self-energy and there is no approximation for the independent-particle terms in the hamiltonian.

Cluster methods provide a systematic way to approach the exact results for the lattice by calculations for different cells and extrapolation to the infinite size limit. In this chapter we show examples of the cluster methods applied to one-band Hubbard models that illustrate different features. In one dimension, Fig. 17.4 shows results using the c-DMFA that converge toward the known exact solution for the density as a function of chemical potential, but there is slow convergence and instabilities near half-filling. The half-filled model in three dimensions is antiferromagnetic at low temperature; calculations of the transition temperature using the DCA (reported in Fig. 17.8) show the extrapolation to large size that agrees with previous work done with other methods and much larger cells. In two dimensions the opening of the gap and phase transition for the half-filled model are shown in Figs. 17.5 and 17.6. Calculations for a doped Hubbard model in two dimensions (Fig. 17.7) find a remarkable variation of the self-energy as a function of **k** in the Brillouin zone, in qualitative agreement with experimental findings for the copper oxide planes in high-temperature superconductors (see, e.g., Fig. 2.16).

Extended DMFT is an instructive formulation in which intersite single-particle hopping t_{ij} and intersite two-particle interactions V_{ij} are treated on the same footing. This provides a way to include non-local interaction in a perturbation approach while keeping the single-site auxiliary system. This is an intrinsic part of the approach in Ch. 21 to formulate first-principles DMFT methods.

SELECT FURTHER READING

Lichtenstein, A. I. and Hafermann, H., "Non-local correlation effects in solids: beyond DMFT," Ch. 11 in *The LDA+DMFT Approach to Strongly Correlated Materials* (lecture notes of the Autumn School 2011, Forschungszentrum Jülich, eds. E. Pavarini, E. Koch, D. Vollhardt, and A. Lichtenstein). [785] Description of cluster and other methods.

Maier, T., Jarrell, M., Pruschke, T., and Hettler, M. H., "Quantum cluster theories," *Rev. Mod. Phys.* **77**, 1027–1080, 2005. A review of cluster algorithms and methods with applications to Hubbard-type models.

See the list of select further reading at the end of Ch. 16 for additional reading on DMFT and related topics.

Exercises

17.1 Show that for the 2D square lattice, the Hubbard model with nearest-neighbor hopping t has a square Fermi surface at half-filling.

17.2 Show that the dispersion for the Hubbard model with nearest neighbor (nn) and next-nearest-neighbor (nnn) hopping, $\hat{H}^0 = -\frac{1}{2}\sum_\sigma\left[\sum_{ij}^{nn} t c_{i\sigma}^\dagger c_{j\sigma} + \sum_{ij}^{nnn} t' c_{i\sigma}^\dagger c_{j\sigma}\right]$, leads to the dispersion $\varepsilon_{\mathbf{k}} = -2t(\cos(k_x a) + \cos(k_y a)) - 2t'\cos(k_x a)\cos(k_y a)$ for the 2D square lattice. Show that for $t' = -0.15t$ this agrees qualitatively with the surfaces shown in Fig. 17.7, i.e., that the shape is similar to that shown in the figure. What would be the difference (in qualitative terms) in the surfaces for $t' = +0.15t$. Using a computational method find the detailed surfaces and show that they are the same as in the figure. (This requires a calculation to integrate the spectrum and determine the Fermi energy for a given filling.)

17.3 Derive the DCA expressions for matrix elements in Eqs. (17.9) and (17.10). One way uses the following steps.

(a) For $G = n(2\pi/L)$ (with L in units of a), show that $t_{DCA}(G) = t(2\pi/L)[\sin(G + \pi/L) - \sin(G - \pi/L)]$ by averaging $t(G + k)$ over k in the range $\pm\pi/L$.

(b) Fourier transform $t_{DCA}(G)$ to find t_{DCA} for nearest neighbors in the cluster, which leads to the result in Eq. (17.9). *Hint*: use a trigonometric identity for $\sin(G \pm \pi/L)$.

(c) Show that the continuous Fourier transform of $\Delta t(G+k)$ over the range $-\pi/L < k < \pi/L$ leads to the spatial dependence for $\Delta t_{DCA}(n)$ given in Eq. (17.10).

17.4 Show that the DCA is causal, as claimed in Sec. 17.3; that is, show that $\mathrm{Im}G(\omega) < 0$ for all real ω. This can be done using the fact that $\mathrm{Re}\Sigma(\omega)$ is piecewise constant. From the definition of Eq. (17.6), show that the average has the form $r = \sum_k \frac{1}{f(k)+ic}$, where $f(k)$ is real and c is a positive constant. The object is to show that $\mathrm{Im}r < 0$. This can be shown for the average of any range of k and does not depend on the average being a reduced zone. (A simple geometrical argument has been given in [801].)

17.5 Show that the hybridization function Δ in the DCA is most readily expressed as a function of G and can be found from

$$\Delta(\mathbf{G}, \omega) = \frac{\frac{1}{N_{\mathbf{k}}}\sum_{\mathbf{k}}\Delta h(\mathbf{G}+\mathbf{k})G(\mathbf{G}+\mathbf{k})\Delta h(\mathbf{G}+\mathbf{k})}{1 + \frac{1}{N_{\mathbf{k}}}\sum_{\mathbf{k}}\Delta h(\mathbf{G}+\mathbf{k})G(\mathbf{G}+\mathbf{k})}.$$

17.6 In order to compare with Ch. 21, write out a simplified version of the algorithm on p. 563 for the case where there is only an intersite interaction V_{ij}, as in Eq. (17.11), without the other bands and details of the GWA. Recall that the "rest" of the system includes all the other sites that affect a given site; it is essential to verify that the algorithm respects the fact that the screening comes only from the rest with no self-screening. Show that there is sufficient information to calculate $\chi_{\mathbf{q}}(\omega)$ in Eq. (17.15). See also Ex. 17.7.

17.7 For the auxiliary system where $\mathcal{U}(\omega)$ is the effective interaction, derive the relation $P(\omega) = [\chi(\omega)^{-1} + \mathcal{U}(\omega)^{-1}]^{-1}$ from the definition of P and χ in Eq. (8.22) or (21.5). From this derive the expression for $\chi_{\mathbf{q}}(\omega)$ in Eq. (17.15). (See also Ex. 21.1.)

17.8 The thermodynamic expression for the entropy can be expressed as an integral over the specific heat divided by the temperature, $S(T) = \int_0^T C(T')/T' dT'$.

(a) Show that this can be transformed to an integral that involves the internal energy using the relation $C = dE/dT$ and integration by parts. This is more useful in Monte Carlo calculations where statistical noise makes it difficult to find accurate values of the specific heat. Give the expression as an integral from infinite temperature to the actual temperature, and explain why it is more useful to integrate from $T = \infty$ than from $T = 0$.

(b) Give the expression for the entropy for $T \to \infty$ in a system with N electrons in M states per site.

18

Solvers for embedded systems

Summary

The core of a DMFT calculation is the solution of the interacting-electron problem for a site (or a cell with multiple sites) embedded in a medium. Methods to solve this problem include exact diagonalization, quantum Monte Carlo, and other approaches. This chapter is an overview of different approaches and some of the primary aspects of the methods and algorithms.

In this chapter we discuss "solvers," the computational machinery for solving the interacting-electron problem for a system embedded in an effective medium.[1] The solver is concerned only with the embedded system; it does not involve the actual crystal explicitly. It is the only part of a DMFT calculation that involves the electron–electron interaction explicitly, and it is the step in the flow diagram of Fig. 16.3 in the box "Solve interacting electron problem on cluster," expressed rather cryptically as "$\mathcal{G}_0(\omega) \implies \mathcal{G}(\omega)$." The function returned by the solver is the one-particle Green's function $\mathcal{G}(\omega)$ for the cluster as a function of frequency ω, which can be a real frequency ω, a Matsubara frequency $i\omega_n$, or a general point z in the complex frequency plane.[2] The self-energy $\Sigma(\omega)$ is found from the relation $\Sigma(\omega) = \mathcal{G}_0^{-1}(\omega) - \mathcal{G}^{-1}(\omega)$; this is the only information needed to

---

[1] We use the terms "embedded system" or "cluster" to denote a site or a cell with multiple sites embedded in an effective medium (a "bath"). This is often called an "impurity" in the literature, which is brought out in Ch. 16 where the relation to an Anderson impurity model is emphasized. Here we emphasize that the methods apply not only to the AIM but also to general problems of systems in a bath, including the clusters in Ch. 17 and the problems in Chs. 19–21. In realistic problems dealt with in Chs. 19–21, where there are many bands, only the localized states labeled m or m' are included and the other states are considered to be the rest of the system. In general the cluster Green's function $\mathcal{G}$ and self-energy Σ are matrices indexed by m, m', which become scalars for each spin in the one-band Hubbard model.

[2] Note that the Green's function is the central quantity in the calculation for the auxiliary embedded system; the self-energy is needed only for the calculation of the lattice Green's function $G_{\mathbf{k}}(\omega)$. This is in contrast to the perturbation methods of Chs. 9–15, where $\Sigma(\omega)$ is the central quantity that is given in terms of a diagrammatic expansion.

construct the one-particle Green's function for the lattice $G_{\mathbf{k}}(\omega)$ in Eq. (16.3) or multiband generalizations in Secs. 17.2 and 17.3.

Section 18.1 is a succinct statement of the problem that has been developed in Chs. 16 and 17, formulated in a way to lead up to the various methods for solvers. Section 18.2 is focused on methods that work directly with the many-body hamiltonian and wavefunctions of a cluster using exact diagonalization and related techniques. The following sections use Green's function path-integral methods to treat the embedded system in terms of the formulation in Sec. 18.3: auxiliary-field Monte Carlo in Sec. 18.4 and continuous-time quantum Monte Carlo in Secs. 18.5–18.7. Finally, Sec. 18.8 is a very brief summary of a few of the various other approaches.

18.1 The problem(s) to be solved

As described in Chs. 16 and 17, the job of the solver is to solve the interacting many-body problem for a cluster that consists of a cell of the crystal embedded in an effective medium, also called a bath. The cell can denote the localized orbitals on a single site or a cell of more than one site; within the cell the interactions and hopping matrix elements are the same as in the original hamiltonian for the crystal.[3] The bath can be specified in different ways that are determined by the type of solver. This defines the auxiliary system to be solved. The result of the calculation is the self-energy for the states in the cell that will be used in the rest of the DMFT algorithm.

In this chapter the different approaches are illustrated by the example of a single site with one orbital with $\uparrow$ or $\downarrow$ states. This is sufficient for all one-band Hubbard models treated in the single-site DMFA approximation, and the same equations apply for multiple sites and bands where all quantities are matrices, as defined in Eq. (17.1). Of course, the actual calculations become much more difficult for multiple bands and sites, and an important aspect of each method is how it scales to larger problems.

Here we consider two classes of solver that involve two ways to treat the bath. One class treats the problem as a hamiltonian and calculates the eigenstates of the cluster. In this case it is most direct to specify the hamiltonian in the form

$$\hat{H} = [\hat{H}^0 + \hat{H}_{int}]_{cell} + \hat{H}^0_{bath} + \hat{H}^0_{hyb}. \tag{18.1}$$

For one band and a cell of a single site, this is equivalent to the Anderson impurity model in Eq. (3.8), which we repeat here:

$$\hat{H} == \sum_{\sigma} \varepsilon_0 c^\dagger_{0\sigma} c_{0\sigma} + U \hat{n}_{0\uparrow} \hat{n}_{0\downarrow} + \sum_{\ell\sigma} t_{0\ell}(c^\dagger_{\ell\sigma} c_{0\sigma} + c^\dagger_{0\sigma} c_{\ell\sigma}) + \sum_{\ell\sigma} \varepsilon_\ell c^\dagger_{\ell\sigma} c_{\ell\sigma}, \tag{18.2}$$

where 0 denotes the site and ℓ the states of the bath. The only difference from the impurity problem is that the energies and matrix elements ε_ℓ and $t_{0\ell}$ are parameters to be determined. For a cell with more than one state, the problem has the same form with all quantities generalized to matrices. If there is a discrete number set of states labeled by ℓ, this defines

---

[3] In the DCA there are differences that are explained in Sec. 17.3.

a hamiltonian that can be solved by techniques such as exact diagonalization and other methods as described in Sec. 18.2.

A different approach is to consider the problem as a cell embedded in a bath described by a hybridization function $\Delta(\omega)$. The hybridization function is related to the set of ε_ℓ and $t_{0\ell}$ in Eq. (18.2) by the relation in Eq. (7.15); however, it is only the function $\Delta(\omega)$ that is needed, not the specific set of ε_ℓ and $t_{0\ell}$, and $\Delta(\omega)$ can represent a bath with a continuum of states. This is the form most useful in Green's function methods that work with $\Delta(\omega)$, which is a dynamic function instead of a hamiltonian.[4] It is most convenient to work with the non-interacting Green's function $\mathcal{G}_0(\omega)$ given in Eq. (16.8),

$$\mathcal{G}_0(\omega) = [\omega - \varepsilon_0 - \Delta(\omega)]^{-1}, \tag{18.3}$$

which includes both the site ε_0 and $\Delta(\omega)$. The job of the solver is to find the Green's function for the interacting system, i.e., $\mathcal{G}_0(\omega) \Longrightarrow \mathcal{G}(\omega)$. This specifies the task for the methods in Secs. 18.3–18.7.

Development of efficient methods is an on-going effort and it is not appropriate to describe the details of present-day techniques. The following sections are devoted to the formulation of the problem and established approaches, emphasizing the aspects most related to DMFT and pointing out connections to other methods.

18.2 Exact diagonalization and related methods

In terms of the exact many-body eigenstates Ψ_i and energies of the cluster, it is straightforward to write down a formal expression for the Green's function. Using the compact notation $\hat{K} = \hat{H} - \mu \hat{N}$, the partition function in Eq. (5.2) can be written as

$$Z = \mathrm{Tr}\, e^{-\beta \hat{K}} = \mathrm{Tr}\, e^{-\beta(\hat{H}^0 + \hat{H}_{int} - \mu \hat{N})} = \sum_i e^{-\beta K_i}, \tag{18.4}$$

where $K_i = E_i - \mu N_i$ in terms of eigenstates with fixed numbers of particles. Note that this requires all of the eigenstates for all numbers of particles. The Green's function is given in terms of the creation and annihilation operators in the modified Heisenberg picture defined in Eq. (D.6), where $\hat{O}(t) = e^{i\hat{K}t}\hat{O}e^{-i\hat{K}t}$. The retarded, advanced, and time-ordered G's can be constructed as in Eqs. (5.72)–(5.78), e.g., the propagator

$$G^>(x_1, x_2, t_1 - t_2) = \frac{1}{Z}\sum_{ij} e^{-\beta K_i} < i|\psi(x_1)|j > < j|\psi^\dagger(x_2)|i > e^{-i(K_j - K_i)(t_1 - t_2)}, \tag{18.5}$$

where x denotes space and spin variables, $\psi(x_1)$ and $\psi^\dagger(x_2)$ are time-independent operators in the Schrödinger picture, and $K_j - K_i = E_j - E_i - \mu$.

This approach involves treating the entire system, the central site or cell and the bath, so that the sum over i and j includes all states of the system. Evaluation of these expressions

---

[4] In Chs. 16–20 the bath and hybridization are often assumed to be non-interacting, but this is not essential. As developed in Ch. 21, the formulation in terms of the auxiliary Green's function $\mathcal{G}_0$ carries over to an interacting bath, and the effective interactions are dynamic, which can be treated using a solver such as in Sec. 18.7.

leads to significant problems and limitations. The exponential scaling of the computational effort with the number of particles and basis functions renders exact diagonalization feasible only for small systems. To handle the large Hilbert space, a Lanczos or related method can be used to determine the lowest energy states.[5] In general, the number of states needed to determine the desired properties accurately is not known in advance, but it can be established as the iterations proceed. There are very useful relations [830, 831] in terms of a continued fraction with coefficients given by recursion relations, and operators like the Green's function can be determined much more efficiently than straightforward application of Eq. (18.5).

A critical issue is to find an optimal choice of the set of states to represent the bath.[6] Note that this discretization of the bath continuum is *not* due to a cutoff in space that would create a discrete spectrum; instead, it is a selection of states by their energy. This may seem surprising in a method built upon the idea of locality in space, but it is the useful choice to represent phenomena with different energy scales. There must be a careful selection of the states to be kept; the choice will depend on the quantities to be calculated. If it is low-energy phenomena related to the Kondo effect that are of most interest, there must be low-energy bath states; high-energy states can be represented by a coarse set of energies. Integrated quantities are much less sensitive than spectra (e.g., the density vs. chemical potential in Fig. 17.4 where the spectrum is a set of delta functions) and often can be calculated accurately with a few states or even a single state [296, 806].

In addition, there is a problem unique to DMFT. In the iterations to reach self-consistency the bath must evolve: there must be some way to quantify the difference between two sets of bath states and a prescription to specify a new set of bath states at each iteration. Various methods have been proposed. One is to iterate to self-consistency the moments of the spectrum calculated using continued fractions and Lanczos methods [832]. A different approach [302, 833] uses the mean-square difference between the Green's functions in imaginary time to determine the new guess for the updated bath functions. Yet another way is to minimize the total free energy with respect to the set of eigenvalues and hybridization functions for the discretized bath using the variational cluster method [296, 772, 773] in Secs. 17.4 and 16.6.

Alternatives to exact diagonalization

A crushing limitation of exact diagonalization methods is the exponential increase in computational time as the number of basis functions and the number of particles increase. However, there are hosts of methods in quantum chemistry and condensed matter physics

---

[5] See references such as [1, Ch. 23 and App. M], and [827–829]. An example of the methods applied to a Hubbard model is in [812], described in Sec. 17.6.

[6] The simplest approximation is to ignore the bath completely, which corresponds to an isolated atom. As described in Sec. 17.2, this corresponds to the "Hubbard I" approximation and the extension to clusters is called cluster perturbation theory (described in Sec. 17.2). For a cluster the problem can be solved by exact diagonalization, which is still hard unless the cell is very small.

designed expressly to mitigate this problem, e.g., coupled cluster methods. See, for example, texts such as [834, 835]. Examples of applications in DMFT can be found in [836] and [837], which presents many results for Hubbard models. Note that calculation of spectra and other properties in general does not need the extreme precision that is required in many applications to molecules, where it is essential to determine the ground-state structures and excitations in great detail. It may be possible to reduce the number of states drastically by eliminating states with small weight as done, for example, in [838] and [839]. Conversely, methods for embedded systems can be used as alternatives to traditional quantum chemistry methods, e.g., using DMFT in [840] and density matrix embedding in [761]. As mentioned in Sec. 16.7, DMRG methods [774, 775] efficiently project out low-energy states and are well suited for embedded systems (see, e.g., [776]). A different approach is auxiliary-field Monte Carlo methods to sample the space of determinants, as discussed in the following section.

18.3 Path-integral formulation in terms of the action

A different approach is to determine the partition function as a path integral in imaginary time in terms of the action $S = S_0 + S_{int}$, where S_0 includes the independent-particle terms and S_{int} the interactions [841, 842].[7] For a cluster, S_0 can be expressed as a sum of terms derived from the single-body part of the cluster hamiltonian H_c^0 defined in Eq. (17.3) (intra-cluster hopping matrix elements and any mean-field potentials) and bath terms given by $\Delta(\tau)$. Interactions included in S_{int} are assumed to be limited to the cluster. For simplicity we give the expressions for the one-band Hubbard model in the single-site approximation, and indicate the generalizations needed for clusters with multiple sites and/or multiple states per site. The partition function $Z = \text{Tr} \, T_\tau \, e^{-S}$ can be written as

$$Z = \text{Tr} \, T_\tau \, \exp\left[-\sum_\sigma \int_0^\beta d\tau \int_0^\beta d\tau' \psi_\sigma^\dagger(\tau)\mathcal{G}_0^{-1}(\tau - \tau')\psi_\sigma(\tau') - \int_0^\beta d\tau U\hat{n}_\uparrow(\tau)\hat{n}_\downarrow(\tau) \right],$$

(18.6)

which is a variation of Eq. (8.31) with the τ integrals written explicitly. This can also be written in terms of ε_0 and Δ using the relation in Eq. (18.3). Here the operators are expressed in the $T \neq 0$ modified Heisenberg form $\hat{O}(\tau) = e^{\hat{K}\tau}\hat{O}e^{-\hat{K}\tau}$, with $\hat{K} = \hat{H} - \mu\hat{N}$, Tr is the grand-canonical trace over all states of the isolated site or cell, β is the inverse temperature, and T_τ denotes time ordering in τ. In the Hubbard model, the interaction U is instantaneous, i.e., local in time, but dynamical interactions can also be included, as explained in Sec. 18.7.

Note that there is a crucial difference from exact diagonalization: the trace in Eq. (18.6) is only over states in the cell. The coupling to the bath is included in the hybridization

---

[7] The general method is described in Ch. 25 and a good pedagogical description oriented toward the use in DMFT can be found in the review [843]. For embedded systems, the action cannot be represented as a hamiltonian unless the continuum is discretized as in the previous section; however, in some cases S_{int} can be replaced by a hamiltonian, e.g., $S_{int} = H_{int} = U n_\uparrow n_\downarrow$ in the Hubbard model or in a static approximation for the interactions in a material. See Sec. 18.7 and Ch. 21.

function Δ or $\mathcal{G}_0$. In contrast, the trace in Eq. (18.4) is over all states of the cluster, i.e., the cell and the discretized continuum. For example, in the one-band model and the single-site approximation, the trace in Eq. (18.6) involves only four states, whereas many more states may be needed to represent the bath.

The Green's function for the site degrees of freedom can be derived from the general expressions in Sec. 8.4,

$$\mathcal{G}(\tau - \tau') = - < T_\tau \psi(\tau)\psi^\dagger(\tau') >= -\frac{1}{Z}\mathrm{Tr}\left[T_\tau e^{-S}\psi(\tau)\psi^\dagger(\tau')\right] = \frac{\delta\Omega}{\delta\Delta(\tau - \tau')}. \quad (18.7)$$

The final equality follows from Eq. (8.34) since the functional derivative with respect to $\Delta(\tau - \tau')$ is the same as a derivative with respect to a dynamic external field $\mathcal{J}(\tau - \tau')/\beta$ in Eq. (8.34). The functional derivative can also be expressed as $\delta\Omega / \delta\mathcal{G}_0^{-1}(\tau - \tau')$, since $\mathcal{G}_0^{-1}$ and Δ are related as in Eq. (18.3).

In the following sections the time-ordered integrals are written as explicit functions that can be evaluated by Monte Carlo methods:

- The auxiliary field method of Hirsch and Fye for the Anderson impurity model [844, 845]. As summarized in Sec. 18.4, this method applies to potential interactions, such as $Un_\uparrow n_\downarrow$, and is exact except for the time-step error. Determinant Monte Carlo methods in Sec. 25.5 can also be used directly.
- The other approaches are known as continuous-time quantum Monte Carlo (CTQMC) [843], where the partition function is expanded in powers of the interaction or the hybridization, as described in Secs. 18.5 and 18.6. Monte Carlo methods are used to sample both the orders of the expansion and the times τ, which eliminates the time-step error. The expansion in the hybridization is in the spirit of DMFT and can deal with the full set of atomic multiplet states, including double spin-flips, etc. (see Sec. 19.3).

There are fundamental problems introduced by the use of Monte Carlo methods that are brought out in Chs. 22–25. Foremost is the sign problem for systems of fermions. However, there are important examples of embedded systems with no sign problem. As explained after Eq. (18.13), this includes the Anderson impurity model for any filling because products of the factors for two-spin states always form a square. The same reasoning applies for higher spin and angular momentum states so long as the different channels interact only through potential interactions that involve the densities. See Sec. 18.6.

Monte Carlo methods that work with the partition function in imaginary time have other fundamental problems. In order to obtain spectral information in real time or frequency, one must transform to real time. Usually, methods such as maximum entropy (Sec. 25.6) are used. This step can introduce a systematic bias since the transformation from imaginary to real time is numerically highly unstable, particularly with noisy data. In general, the real-time results are systematically broadened and fine details can be lost. An example shown in Fig. 25.7 is the density–density correlation function in ^{4}He where the transformation to real frequency leads to a single broadened peak. Each case must be considered carefully.

For example, often the goal of a simulation is to establish the existence and value of a gap in the spectrum. As noted in App. D, this means that one must show the correlation function goes to zero exponentially for large imaginary time. In this regard a difficulty in CTQMC methods is the reduced probability of the simulation visiting regions where the Green's function is small and the statistics are relatively poor. Such issues are addressed in [818, 843]. Note that static quantities, such as the energy, density, susceptibility, etc., can be obtained with no transformation; other properties, such as the low-frequency conductivity, can be obtained by extrapolation of the calculated values at the Matsubara frequencies ω_n to $\omega = 0$.

In addition, it is hard to reach low temperatures. This varies with the solver and will be discussed in the following. See [843] for comparison of scaling with temperature for various algorithms.

18.4 Auxiliary-field methods and the Hirsch–Fye algorithm

The first definitive, quantitative solution of the Anderson impurity model was carried out by Hirsch and Fye [844, 845], who developed an auxiliary-field Monte Carlo (AFMC) method that is an adaptation of the Blankenbecler, Scalapino, and Sugar (BSS) [846] algorithm described in Sec. 25.5. Before the development of the continuous-time approaches, the Hirsch–Fye algorithm was the primary quantum Monte Carlo method used in DMFT calculations. Many examples are given in reviews such as [302, 303, 771, 785, 788]. In this book the applications include Hubbard models in Sec. 17.6, with results shown in Figs. 17.5, 17.6, and the left panel of 17.8, as well as most of the examples for materials in Ch. 20. In this section we give a brief summary of auxiliary-field methods for embedded sites and clusters; careful derivations can be found in the review [302].

In an auxiliary-field approach (see Sec. 25.5) the integrals over τ in Eq. (18.6) are evaluated as a path of M discrete time steps spaced by $\Delta\tau = \beta/M$. The partition function can be written as in Eq. (25.37),

$$Z = \mathrm{Tr}\, e^{-\beta(\hat{H} - \mu\hat{N})} \approx \mathrm{Tr} \prod_{i=1}^{M} e^{-\Delta\tau(\hat{H}_0 - \mu\hat{N})} e^{-\Delta\tau\hat{H}_{int}}, \tag{18.8}$$

which uses the Trotter breakup of the independent-particle and interaction terms in the hamiltonian. This leads to a time-step error of order $\Delta\tau^2[\hat{H}_0, \hat{H}_{int}]$ (see Ex. 25.7). At each step the particles experience a potential that is drawn from the sample of fields generated by a Hubbard–Stratonovich transformation. As worked out in Sec. 25.5, for a model with interaction $\hat{H}_{int} = U n_\uparrow n_\downarrow$ this can be done with a sum of discrete variables instead of a continuous range of auxiliary fields [847]. If we use the relation $n_\uparrow n_\downarrow = -\frac{1}{2}(n_\uparrow - n_\downarrow)^2 + \frac{1}{2}(n_\uparrow + n_\downarrow)$ in Eq. (25.38), this leads to the form given in Eq. (25.39),

$$e^{-\Delta\tau U n_\uparrow n_\downarrow} = \frac{1}{2} \sum_{s=\pm 1} e^{\lambda s(n_\uparrow - n_\downarrow) - \frac{1}{2}\Delta\tau U(n_\uparrow + n_\downarrow)}, \tag{18.9}$$

where $\cosh(\lambda) = \exp(\Delta\tau U/2)$ (see Ex. 25.9).

In the BSS method for a Hubbard model on a lattice, this transformation is used to generate a set of auxiliary fields defined on each site, and the expression for Z is boiled down to the product of M matrices given in Eq. (25.41). Each matrix has dimensions $N_s \times N_s$, where N_s is the size of the basis for the single-particle states. (For the Hubbard model on a lattice, N_s is the number of sites, but the method is more general, as noted in Sec. 25.5.) Finally, the Monte Carlo algorithm is a sampling over the paths with weights given by the determinants in Eq. (25.42).

Here we are concerned with the Anderson impurity model with the hamiltonian in Eq. (3.8). The problem is simplified since there are interactions only on one site. However, there is a complication due to the continuum of bath states. In order to apply the BSS auxiliary-field method with the Trotter breakup of the partition function in Eq. (18.8), one must specify a basis of N_s discrete states. This is the same problem as in exact diagonalization methods in Sec. 18.2, but it provides an alternative approach (see, e.g., [848]). The auxiliary-field method has the advantage that the computational effort scales as N_s^3 (since it involves operations with matrices of size $N_s \times N_s$), whereas the exact diagonalization methods scale exponentially in N_s. However, there are problems with using the Monte Carlo methods. Even though there is no sign problem in this case, the methods suffer from the difficulty of transforming to real time and the problem in reaching low temperature. An advantage of the BSS approach is that the computational effort scales as $\beta \propto 1/T$ (see Sec. 25.5), whereas the Hirsch–Fye and other methods in this chapter scale as β^3.

The Hirsch–Fye algorithm

The Hirsch–Fye approach is a way to reformulate the auxiliary-field method in terms of the non-interacting Green's function for the embedded site $\mathcal{G}_0(\tau_i - \tau_{i'})$ with the states of the bath integrated out, without the need for a discretization of the bath.[8] Instead of using the Trotter breakup in the form shown in Eq. (18.8), the double integral over time τ in Eq. (18.6) can be discretized as a double sum over time slices $i, i' = 1, \ldots, M$,

$$Z = \text{Tr}\, e^{-\beta(\hat{H}-\mu\hat{N})} \approx \text{Tr} \prod_{i=1}^{M} \left[\prod_{i'=1}^{M} e^{-\Delta\tau^2 \mathcal{G}_0^{-1}(\tau_i - \tau_{i'})} \right] e^{-\Delta\tau \hat{H}_{int}}. \tag{18.10}$$

Using the discrete transformation in Eq. (18.9), this can be written as

$$Z \approx \frac{1}{2^M} \sum_{\mathcal{S}} \text{Tr} \exp \left[\sum_{\sigma ii'} F_{\sigma ii'}(\mathcal{S}) \right], \tag{18.11}$$

where

$$F_{\sigma ii'}(\mathcal{S}) = (\Delta\tau)^2 [\mathcal{G}_0^{-1}]_{\sigma ii'} - \lambda \sigma s_i \delta_{ii'}. \tag{18.12}$$

[8] Here we follow the derivations given by Blümer [785]. A more accurate form, including higher-order corrections, is given by Eq. (18.12), but with $F_{\sigma ii'}(\mathcal{S}) = (\Delta\tau)^2 [\mathcal{G}_0^{-1}]_{\sigma ii'} e^{\lambda \sigma s_{i'}} - \delta_{ii'}(1 - e^{\lambda \sigma s_i})$. This was derived in the original paper [844] and also discussed in detail in the review [302].

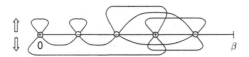

Figure 18.1. Schematic diagram for the Hirsch–Fye auxiliary-field algorithm for an example with time step $\Delta\tau = \beta/5$. The connecting lines in the figure depict propagators $\mathcal{G}^0(\tau_i - \tau_j)$ for one instance of the ways the vertices can be connected, and the circles marked $+$ or $-$ indicate one instance of the stochastically varying $\{\sigma s_i\}$ that determine the auxiliary fields $\lambda\sigma s_i = \pm\lambda$ at times τ_i spaced by $\Delta\tau$ for each spin σ indicated by up and down arrows.

Here S denotes the set of auxiliary variables $\{s_i = \pm 1, i = 1, M\}$ that determine the independent-particle potentials $\lambda\sigma s_i$, where the factor $\sigma = 1$ for $\uparrow$ and $\sigma = -1$ for $\downarrow$ takes into account that the fields have opposite signs for the two spins. Figure 18.1 shows one example of a set of variables $\{s_i\}$ and one way of coupling the different times with the lines representing $\mathcal{G}_0^{-1}(\tau_i - \tau_{i'})$. Since the expressions involve only independent-particle operators, one can use Wick's theorem (see Sec. 9.4 and references such as [842]) to evaluate the trace over all states and all numbers of particles, with the result

$$Z = \frac{1}{2^M} \sum_S detF_\uparrow(S)detF_\downarrow(S). \tag{18.13}$$

The expressions can be evaluated by Monte Carlo sampling with the probability distribution function $detF_\uparrow(S)detF_\downarrow(S)$. There is no sign problem since the product of the two spin factors is a perfect square and hence non-negative [849].

Note also the close relation to expressions in the CTQMC expansion in the hybridization in Sec. 18.6, which involve matrices of the hybridization $\Delta(\tau_i - \tau_j)$. This can be understood since the matrix $F_{\sigma ii'}$ in Eq. (18.12) involves the inverse Green's function. Using the same reasoning as in Eq. (18.35), the Green's function in the Hirsch–Fye algorithm is given by

$$\mathcal{G}_\sigma(\tau_i - \tau_{i'}) = \frac{1}{Z} \sum_S detF_\uparrow(S)detF_\downarrow(S)F_{\sigma ii'}^{-1}(S). \tag{18.14}$$

One of the key issues in auxiliary-field Monte Carlo is reaching low temperature. The algorithm involves operations with matrices of size $M \times M$, where $M = \beta/\delta\tau$, and the computational effort scales as $M^3 \propto \beta^3$. This is the scaling also for the CTQMC expansion in the hybridization (Sec. 18.6). The BSS algorithm has the advantage that it scales as β but it requires a discretized continuum with many more states. Comparison of the scaling for the different algorithms can be found in [843].

18.5 CTQMC: expansion in the interaction

Continuous-time Monte Carlo is an approach in which imaginary times in the interval $0 < \tau < \beta$ are sampled instead of using a fixed time step; this avoids the time-step error.

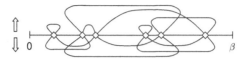

Figure 18.2. Example of CTQMC expansion in the interaction, shown here for sixth order. Vertices (diamonds) denote interactions $Un_\uparrow n_\downarrow$ connected by $\mathcal{G}_\sigma^0(\tau_i - \tau_j)$ for each spin. The times can take any values in the continuum $0 < \tau < \beta$.

Recent developments of techniques for the partition function expanded in powers of the interaction or the hybridization are making possible efficient solvers for DMFT.[9]

We first consider the algorithm that involves the expansion in the interaction. Although the expansion is nominally the same as in many-body perturbation theory, there are qualitative differences. Whereas the methods of Chs. 9–15 involved sums of selected classes of diagrams to all orders, the present methods sample diagrams, including both linked and unlinked, to arbitrary high order. This is possible for a finite system using Monte Carlo methods. There are other differences that are discussed in this and the following sections. To simplify the notation we consider only a single site with interaction $Un_\uparrow n_\downarrow$ in a bath, where the expression for Z can be written in the form

$$Z = \sum_k \frac{(-U)^k}{k!} \int_0^\beta d\tau_1 \cdots \int_0^\beta d\tau_k \, \mathrm{Tr} \left[\mathrm{T}_\tau \, e^{-S_0} n_\uparrow(\tau_1) n_\downarrow(\tau_1) \cdots n_\uparrow(\tau_k) n_\downarrow(\tau_k) \right], \quad (18.15)$$

which is illustrated in Fig. 18.2 for $k = 6$. The electrons propagate as free particles between interactions, which factors into products for $\uparrow$ and $\downarrow$ spins since there are no spin-flip terms.[10] The trace over the time-ordered operators can be evaluated using Wick's theorem (see Sec. 9.4). Since S_0 is the action of a non-interacting system, it factors into a product of independent-particle Green's functions with all permutations, so that Eq. (18.15) becomes (see [843, 857])

$$Z = \sum_k \frac{(-U)^k}{k!} \int_0^\beta d\tau_1 \cdots \int_0^\beta d\tau_k \, D_\uparrow^k D_\downarrow^k, \quad (18.16)$$

[9] There is a history of continuous-time approaches for solving the N-body problem, developed, for example, for lattice problems by Handscomb [850, 851] and Lyklema [852], and "Green's function Monte Carlo" for continuous systems by Kalos [853]. More recently, Prokof'ev and coworkers [854] proposed this approach for exact quantum Monte Carlo simulations for bosons in discrete systems. Techniques for fermions with repulsive interactions have been developed by Rombouts et al. [855] and Rubtsov et al. [856]. Werner and coworkers, in [857] and following papers, have extended the formulation to create efficient solvers for DMFT. The review in [843] provides extensive descriptions and comparison of different methods.

[10] The determinants can be factored in any system where there are different channels $\alpha, \beta, \ldots$ that are distinct by symmetry, and there are only density–density interactions between different channels; for example, $\uparrow$ and $\downarrow$ with interaction U. However, this is not possible in general because the interaction $U_{\alpha,\beta,\gamma,\delta}$ can couple different symmetry independent-particle states.

where $D_\sigma^k(\tau_1, \ldots, \tau_k)$ is the determinant of the $k \times k$ matrix

$$D_\sigma^k(\tau_1, \ldots, \tau_k) = \begin{vmatrix} \mathcal{G}_0(0) & \mathcal{G}_0(\tau_1 - \tau_2) & \mathcal{G}_0(\tau_1 - \tau_3) & \cdots \\ \mathcal{G}_0(\tau_2 - \tau_1) & \mathcal{G}_0(0) & \mathcal{G}_0(\tau_2 - \tau_3) & \cdots \\ \mathcal{G}_0(\tau_3 - \tau_1) & \mathcal{G}_0(\tau_3 - \tau_2) & \mathcal{G}_0(0) & \cdots \\ \cdot & \cdot & \cdot & \cdots \\ \cdot & \cdot & \cdot & \cdots \end{vmatrix}, \quad (18.17)$$

for each spin ↑ and ↓. Here $\mathcal{G}_0$ is the non-interacting Green's function for the embedded system and the determinant takes into account all the possible orderings of the creation and annihilation operators. This is illustrated in Fig. 18.2, where the lines represent $\mathcal{G}_0(\tau_i - \tau_j)$ for ↑ and ↓ spins.

Thus, Z has been written as a sum of k-dimensional integrals over the times $0 < \tau_j \le \beta$. For any term in the integrand in Eq. (18.16), the state of the system can be specified by the order of the expansion k and the time arguments, which can be denoted $S = \{k, \tau_1, \ldots, \tau_k\}$; for example, $D_\sigma^k(\ldots)$ in Eq. (18.17) can be written as

$$D_\sigma(S) = |\mathcal{G}_0(S)|, \quad (18.18)$$

where $\mathcal{G}_0(S)$ is the matrix of bare Green's functions in Eq. (18.17). Any equilibrium property O of the system can be written as an integral of $O(S)$ for each configuration

$$\langle O \rangle = \frac{1}{Z} \int dS \, \Pi(S) \, O(S), \quad \text{with} \quad \Pi(S) = \frac{(-U)^k}{k!} D_\uparrow(S) D_\downarrow(S), \quad (18.19)$$

where the integral denotes the sum over k and integration over time. This can be evaluated by sampling the diagrams using the methods described in Ch. 22. There are special considerations for CTQMC, in particular, sampling of orders k is very different from sampling the times, and efficient algorithms require care in sampling the orders (see, e.g., [857, 858] and the review [843]). In addition, the contributions of different orders reveal much about the physics of the given problem, as discussed in Sec. 18.6.

In Monte Carlo methods the sign of $\Pi(S)$ must be non-negative for all values of S if it is to be considered a probability density.[11] For an unpolarized system, $D_\uparrow(S) = D_\downarrow(S)$, implying a perfect square. However, for a repulsive interaction the factor $(-U)^k$ is negative for odd k, which may lead to a severe loss of efficiency if the sampling must be done for k up to high order.

Since the formulas for the expansion use Wick's theorem to express each term as a determinant of independent-particle Green's functions $\mathcal{G}_0$, the physical cluster Green's function $\mathcal{G}_\sigma(\tau - \tau') = - < T_\tau \psi_\sigma(\tau) \psi_\sigma^\dagger(\tau') >$ for the interacting particles can be computed as a sum of Green's functions for each state $S = \{k, \tau_1, \ldots, \tau_k\}$. This is given by a sum

---

[11] If $\Pi(S)$ is mostly positive then one can sample $|\Pi(S)|$ and use the sign of $\Pi(S)$ as a weight. However, the efficiency of the procedure will be reduced, as described in Secs. 24.3 and 25.3.

over terms like Eq. (18.17), except that for each configuration S the determinant $D_\sigma(S)$ is replaced by the determinant of a matrix with an added row and column,

$$D_\sigma^+(S, \tau - \tau') = \begin{vmatrix} \mathcal{G}_0(S)_{ij} & \mathcal{G}_0(\tau_j - \tau) \\ \mathcal{G}_0(\tau_j - \tau') & \mathcal{G}_0(\tau - \tau') \end{vmatrix}. \tag{18.20}$$

For a determinant of this form, one can show that the ratio of $D_\sigma^+(S, \tau - \tau')$ and $D_\sigma(S)$ is given by (Ex. 18.2)

$$\frac{D_\sigma^+(S, \tau - \tau')}{D_\sigma(S)} = \mathcal{G}_0(\tau - \tau') - \sum_{ij} \mathcal{G}_0(\tau - \tau_i)\mathcal{G}_0^{-1}(S)_{ij}\mathcal{G}_0(\tau_j - \tau'), \tag{18.21}$$

which is independent of spin so that $\mathcal{G}$ can be evaluated as

$$\mathcal{G}(\tau - \tau') = \int dS \,\Pi(S)\,\frac{D_\sigma^+(S, \tau - \tau')}{D_\sigma(S)} \equiv \left\langle \frac{D_\sigma^+(S, \tau - \tau')}{D_\sigma(S)} \right\rangle, \tag{18.22}$$

where $< \cdots >$ denotes the average defined in Eq. (18.19). This gives the one-particle Green's function needed in DMFT; similar expressions can be used to find two-particle Green's functions, correlation functions, response functions, etc.

This approach can be generalized to other cases with density–density interactions by using multiple auxiliary fields. The computational effort has same scaling as for independent-particle methods, since the operations involve independent-particle operators; this is N^3, where N is the number of orbitals per site [843], but in many cases this can be reduced when the matrices are sparse, e.g., for the Hubbard model.

Continuous-time auxiliary-field methods

Auxiliary fields that replace the interaction by sampling independent-particle fields, as in Eq. (18.37), can also be employed in methods that involve an expansion in the potential and continuous-time variables.[12] One way to do this is to add a shift in the interaction with a compensating term in the independent-particle part of the hamiltonian, which can be written as $H = H_0' + H_U'$, where

$$H_U' = U\left(n_\uparrow n_\downarrow - \frac{n_\uparrow + n_\downarrow}{2}\right) - \frac{K}{\beta},$$

$$H_0' = H_0^{AIM} + \frac{K}{\beta}, \tag{18.23}$$

where H_0^{AIM} denotes the independent-particle part of the Anderson impurity model for the embedded site.

---

[12] Here we follow the description in [859]. Methods for lattice models were developed by Rombouts *et al.* [855] and Rubtsov *et al.* [856]. For the Hubbard model these can be shown [860] to be equivalent to one another and, if the time steps are fixed to be equally spaced, they are closely related to the Hirsch–Fye algorithm. (A shift similar to Eq. (18.23) is also proposed as a way to reduce the sign problem [856] in CTQMC, since it leads to an average non-zero sign. However, there is no free lunch; problems can arise in sampling the distribution, as noted in [861].)

The partition function $Z = \text{Tr}\,T_\tau\, e^{-S}$ can be written in the same form as Eq. (18.6), but including the K/β term,

$$Z = \text{Tr}\,T_\tau\,\exp\left[-\sum_\sigma \int_0^\beta d\tau \int_0^\beta d\tau'\,\psi_\sigma^\dagger(\tau)[\mathcal{G}_0^{-1}(\tau - \tau') - K/\beta]\psi_\sigma(\tau') - \int_0^\beta d\tau H_U'\right].$$

(18.24)

An expansion of the exponential in powers of H_U' can be expressed as a power series in K if we define

$$\beta H_U' = -K\left[1 - \frac{\beta}{K}U\left(n_\uparrow n_\downarrow - \frac{n_\uparrow + n_\downarrow}{2}\right)\right].$$

(18.25)

The auxiliary-field decomposition can then be accomplished using the relation [855]

$$1 - \frac{\beta}{K}U\left(n_\uparrow n_\downarrow - \frac{n_\uparrow + n_\downarrow}{2}\right) = \sum_{s=\pm 1} e^{\gamma s(n_\uparrow - n_\downarrow)},$$

(18.26)

where $\cosh(\gamma) = 1 + \frac{\beta}{2K}U$, which can be derived following the same reasoning as for Eq. (25.39) (see Ex. 25.10). Finally, the expression for Z analogous to Eq. (18.15) can be written (see [843, 861] for details)

$$Z = \sum_k \left[\frac{K}{2\beta}\right]^k \sum_{s1\cdots s_k=\pm 1} \int_0^\beta d\tau_1 \cdots \int_{\tau_{k-1}}^\beta d\tau_k\, Z_k(\{s_i, \tau_i\}),$$

(18.27)

with

$$Z_k(\{s_i, \tau_i\}) = \text{Tr} \prod_{i=1,k} e^{-\Delta\tau_i H_0'} e^{\gamma s_i(n_\uparrow - n_\downarrow)},$$

(18.28)

were $\Delta\tau_i = \tau_{i+1} - \tau_i$ for $i < k$ and $\Delta\tau_k = \beta - \tau_k + \tau_1$.

For each k the expressions are similar to those in the auxiliary-field BSS and Hirsch–Fye methods; analogous expressions in terms of determinants can be derived for CTQMC auxiliary-field methods [843, 861]. For large systems, where the hybridization method in the following section becomes very expensive, this method can be the most effective approach since it scales as a polynomial in N instead of an exponential.

18.6 CTQMC: expansion in the hybridization

The formulation of DMFT in Chs. 16 and 17 is in terms of a site or cell embedded in a bath described by a hybridization function Δ. The CTQMC method developed by Werner and coworkers [857] is a direct embodiment of this approach. To this end, Eq. (18.6) can be written as

$$Z = \text{Tr}\,T_\tau\, e^{-S_L}\exp\left[-\sum_\sigma \int_0^\beta d\tau \int_0^\beta d\tau'\,\psi(\tau)_\sigma \Delta_\sigma(\tau - \tau')\psi_\sigma^\dagger(\tau')\right],$$

(18.29)

where S_L is the action for the local region. One of the most important capabilities of this approach is that the local interacting system can be very general and is not limited to

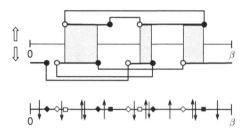

Figure 18.3. Example of CTQMC expansion in the hybridization shown in two representations. Each creation operator (open symbol) is connected to an annihilation operator (filled symbol). Top: explicit illustration of one way the events are coupled by the hybridization function $\Delta_\sigma(\tau_j - \tau_k)$, shown as light lines. Dark line segments represent a state with $n_\sigma = 1$ and the time periods with double occupation, $n_\uparrow = n_\downarrow = 1$ are indicated by the shaded areas. Up and down spins can have different order k and positions of segments. Bottom: a representation as a sequence of events for $\downarrow$ (diamonds) and $\uparrow$ (squares). The occupation of each site is shown on the line between the events. All possible ways the events can be coupled by $\Delta_\sigma(\tau_j - \tau_k)$ are understood but not shown explicitly. The generalization to a site with many states can represent more complicated hamiltonians, i.e., the full multiplet structure of an atom with d or f states, or any other localized states on the site. Figures adapted from [861] and [843].

a potential interaction like $Un_\uparrow n_\downarrow$. The full multiplet structure of a multi-electron atom can be treated by evaluating the expectations values in Eq. (18.34) below, which can be expressed in terms of matrix elements of creation and annihilation operators between many-body states of the atom (denoted X by Hubbard [124]). Of course, Δ is a matrix in general; however, if we ignore spin–orbit interaction, spin is conserved and we indicate the spin index explicitly for Δ_σ. The expansion of the exponential in Eq. (18.29) can be done separately for the two spins, so that the order and the times can be different for $\uparrow$ and $\downarrow$ spins.

The idea proposed in [857] is to combine all terms of the same order with all possible permutations of the times for each spin, so that Z can be written

$$Z = \mathrm{Tr}\, T_\tau \, \exp^{-S_L} \prod_\sigma \sum_{k_\sigma} \frac{1}{k_\sigma!} \int_0^\beta d\tau_1 \int_0^\beta d\tau_1' \cdots \int_0^\beta d\tau_{k_\sigma} \int_0^\beta d\tau_{k_\sigma}'$$

$$\left[\psi_\sigma(\tau_1)\Delta(\tau_1 - \tau_1')\psi_\sigma^\dagger(\tau_1') \times \cdots \times \psi_\sigma(\tau_{k_\sigma})\Delta_\sigma(\tau_{k_\sigma} - \tau_{k_\sigma}')\psi_\sigma^\dagger(\tau_{k_\sigma}') \right]. \quad (18.30)$$

For a given set of operators for spin σ, $\{\psi_\sigma^\dagger(\tau_i'), \psi_\sigma(\tau_i), i = 1, k_\sigma\}$, there are $k_\sigma!$ terms in Eq. (18.30). The combined contribution of these terms is a determinant $D_\sigma(\tau_1 \cdots \tau_{k_\sigma}; \tau_1' \cdots \tau_{k_\sigma}')$ of the $k_\sigma \times k_\sigma$ matrix with element ij given by

$$\Delta_{ij}^\sigma = \Delta_\sigma(\tau_i - \tau_j'), \quad i, j = 1, k_\sigma. \quad (18.31)$$

This is analogous to Eq. (18.18) with the independent-particle $\mathcal{G}_0(\tau_i - \tau_j)$ replaced by the propagator $\Delta_\sigma(\tau_i - \tau_j')$ for non-interacting electrons in the bath. An example of the creation and annihilation operators in the hybridization function $\Delta_\sigma(\tau_j - \tau_k)$ for $k_\uparrow = 2$ and $k_\downarrow = 3$ is shown in two representations in Fig. 18.3.

Thus the partition function can be written in the form

$$Z = \mathrm{Tr}\, \mathrm{T}_\tau \, \exp^{-S_L} \prod_\sigma \sum_{k_\sigma} \int_0^\beta \cdots \left[\psi_\sigma(\tau_1) \psi_\sigma^\dagger(\tau_1') \times \cdots \times \psi_\sigma(\tau_{k_\sigma}) \psi_\sigma^\dagger(\tau_{k_\sigma}') \right] D_\sigma(\cdots),$$

(18.32)

where the integrals are over the times $\tau_1 \cdots \tau_{k_\sigma}; \tau_1' \cdots \tau_{k_\sigma}'$ that are the arguments of D_σ. Detailed expressions are given in [843, 858], including requirements for the limits of the integrals to ensure the proper fermion antisymmetry for $\tau = 0$ and $\tau = \beta$. It is instructive to write the expressions in a condensed form: if we denote the state at any point in the integrand by $\mathcal{S} = (\mathcal{S}_\uparrow, \mathcal{S}_\downarrow)$, with $\mathcal{S}_\sigma = \{k_\sigma; \{\tau_{\sigma i}, \tau_{\sigma i}', i = 1, k_\sigma\}\}$, then Z can be written as

$$Z = Z_L \int d\mathcal{S} \Pi_L(\mathcal{S}) \times D(\mathcal{S}_\uparrow) D(\mathcal{S}_\downarrow),$$

(18.33)

where

$$\Pi_L(\mathcal{S}) = \left\langle \mathrm{T}_\tau \prod_\sigma \psi_\sigma(\tau_{\sigma 1}) \psi_\sigma^\dagger(\tau_{\sigma 1}') \times \cdots \times \psi_\sigma(\tau_{\sigma k_\sigma}) \psi_\sigma^\dagger(\tau_{\sigma k_\sigma}') \right\rangle_L.$$

(18.34)

Here the brackets indicate the local average $< \hat{O} >_L = \mathrm{Tr}\,[e^{-S_L} \hat{O}]/Z_L$ for an operator $\hat{O}$ and $Z_L = \mathrm{Tr}\, e^{-S_L}$ is the partition function for the isolated site or cell. This expression shows succinctly the way Z involves the two types of terms, determinants for each spin and local expectation values that include interactions of both spins. In addition, the multiple integrals and sums indicated in Eq. (18.33) are cast in the generic form appropriate for sampling.

For any case in which the states can be classified into different symmetries with only density–density interactions between the particles in different symmetries, Z can always be written in a form like Eqs. (18.32)–(18.34). For this important set of cases, there is no sign problem and Monte Carlo sampling leads to the exact imaginary-time Green's function within the stochastic error.

The Green's function $\mathcal{G}(\tau)$ can be calculated from the derivative of Z with respect to $\Delta(\tau)$ using the expression in Eq. (18.7).[13] This removes a hybridization function connecting creation and annihilation operators in the expression for Z in Eq. (18.32). Thus the contribution to $\mathcal{G}(\tau)$ of each term with $\tau = \tau_i - \tau_j'$ is given with an added factor $(-1)^{i+j} \frac{\det \Delta^{(ij)}}{\det \Delta}$, where $\Delta^{(ij)}$ is the matrix Δ in Eq. (18.31) with row i and column j removed. Using the relation $(-1)^{i+j} \det \Delta^{(ij)} = \det \Delta \times (\Delta^{-1})_{ji}$, it follows that the one-particle Green's function $\mathcal{G}(\tau)$ is given by the average over configurations $\mathcal{S}$,

$$\mathcal{G}(\tau) = \frac{1}{\beta} \left\langle \sum_{ij} \left[\Delta^{-1} \right]_{ij} \delta(\tau, \tau_i - \tau_j') \right\rangle,$$

(18.35)

---

[13] An alternative is to compute $\mathcal{G}$ for the independent-particle bath. The form is similar to Eqs. (18.20)–(18.22), with an added row and column.

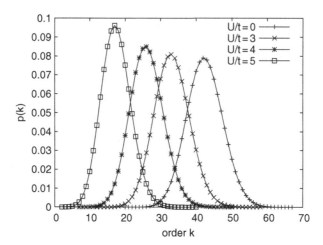

Figure 18.4. The probability distribution $p(k)$ of orders k for the Hubbard model with semicircular density of states at a temperature $0.01t$, where $t = W/4$ and W is the full width of the band. Shown are several values of the interaction strength U/t. (From [857].)

where $< \cdots >$ denotes the Monte Carlo average and the factor $\delta(\tau_1, \tau_2) = \mathrm{sgn}(\tau_2)\delta(\tau_1 - \tau_2 - \Theta(-\tau_2)\beta)$ includes the proper sign and antiperiodicity in β, as explained in [843]. Two-particle correlation functions, response functions, etc. can be calculated using a similar form.

Expansion orders

The diagrammatic expansions are expressed as a sum of orders k that extends to infinity. The remarkable fact is that a finite number of samples can approach arbitrarily close to the exact result, so long as the contribution of higher orders decreases rapidly enough. For CTQMC we have seen that at order k the probability of adding new diagrams of higher order $k \rightarrow k + 1$ is reduced by the factor $\propto [\beta/(k + 1)]^2$, whereas the probability of a decrease in the order $k \rightarrow k - 1$ is increased by $\propto [k/\beta]^2$. Thus the distribution of orders has a probability distribution $p(k)$ that has a maximum at some order k_{max} that scales as the inverse of the temperature $k_{max} \propto \beta = 1/T$. Well above the maximum the probability decreases exponentially, so that sampling of the order k converges rapidly to the exact result. An example of the distribution $p(k)$ of orders k is shown in Fig. 18.4 for the Hubbard model. Note that the larger the interaction, the lower the order of diagrams needed. Conversely, an expansion in U requires an increasing number of diagrams as U increases; examples are given in [843].

It may be surprising that the order required is so large. The reason is that the Monte Carlo sampling is done for all diagrams, linked and unlinked, for the bare interaction. Methods such as the GWA are low order in the *screened* interaction W, which is determined by summing selected sets of diagrams to all orders and is weaker than the bare interaction, a fact emphasized throughout the development of the functionals and perturbation methods in Chs. 7–15.

Generalization to multiple states and interactions

The CMTC method provides a way to treat interactions for many electrons in the atomic-like states (not just U and J, but all matrix elements of the interaction in multi-electron atoms). The effect of the medium can be taken into account by events that involve an electron created or annihilated in the atom. The concept is depicted at the bottom of Fig. 18.3 for the Hubbard model. There each transition is indicated by a symbol (square for ↑ and diamond for ↓ creation or annihilation). This representation has the advantage that it is readily generalized to a problem where there are many possible states of the localized system, e.g., an atom with degenerate $3d$, $4f$, or $5f$ states. It can also be some other system, such as a C_{60} buckyball or a nanostructure with many states that is coupled to its surroundings, etc. Each type of transition that can occur when an electron is created or destroyed can be denoted by a different symbol, and the Monte Carlo method is used to sample the possible transitions.

In this way the CTQMC method can treat much more complicated problems, including double spin-flips due to the Coulomb interaction, and other processes to represent the full multiplet structure of an atom. However, the actual treatment of such a system comes at a very large price, because each step amounts to an operation with a matrix the size of the Hilbert space, which grows exponentially with the number of states, e.g., exact diagonalization for the atomic-like states with all numbers of electrons in the grand-canonical ensemble. This is a small price to pay for problems like the spin-$1/2$ Anderson model, however, it can be forbidding for several sites and/or many states per site, for example, an f shell involves 2^{14} states. Nevertheless, by using symmetry, it is often possible to reduce the problem by breaking it up into factors with much smaller sizes that are manageable [858]. In any case, the scaling for the cluster is much better than the exact diagonalization approach discussed in Sec. 18.2, since that calculation involves the states of the cluster *and* those that represent the bath.

Exact diagonalization may not be essential, however, if there are less expensive methods that are sufficiently accurate. As mentioned at the end of Sec. 18.2, there are hosts of quantum chemistry methods to avoid exact diagonalization and there are applications as solvers in DMFT [836, 837]. It may also be possible to use such methods to solve the local site problem that is a part of the CTQMC algorithm. One choice that may be appropriate for transition-metal problems is the "complete active space" approach, where the states are classified into those considered to be partially occupied and the rest considered to be "virtual" states that are mixed by the Coulomb interaction. For example, the degenerate Hund's rule ground state could capture the essential features of low-energy local moment excitations, the hallmark of open-shell atoms, and the other states would provide corrections and the higher-energy states that are responsible for the full spectrum of excitations.

18.7 Dynamical interactions in CTQMC

Any first-principles method for real materials must face the fact that effective local inter-actions are not constants, but instead they are dynamic functions of time or frequency. This is brought out in Ch. 21, where the screened interactions are calculated by the RPA, or an improved method; the screening is due to the rest of the system and the dynamics is determined by the excitation spectrum. Here we focus on ways to incorporate dynami-cal interactions in the solvers, primarily a method that can be incorporated in a CTQMC algorithm with relatively simple modifications.[14]

As in the previous sections, we consider the one-band Hubbard model. Even though this is a simplified model, a single interaction $U(\omega)$ captures the dominant effect expected in realistic many-band models. As emphasized in Chs. 19 and 21, the primary response of the system is to screen charge fluctuations, i.e., changes in the total number of electrons of a site, which can be described by an effective $U(\omega)$. Various interactions that depend on the different states of an atom (the multiplet splittings) are much less affected and they are much closer to the frequency-independent atomic values.

The dynamic screening of the interaction can be cast in the form of a hamiltonian with the electron states coupled to a set of oscillators. This is related to extended DMFT in Sec. 17.5, where the polarizabilities of neighboring sites lead to a frequency-dependent effective interaction. Here we do the opposite: given a frequency-dependent interaction $U(\omega)$, the goal is to develop methods to treat $U(\omega)$ by mapping it onto a problem of elec-trons coupled to oscillators. One approach is an extension of the exact diagonalization methods in Sec. 18.2: discretize the spectrum in terms of a few oscillators and diagonalize the resulting hamiltonian. An alternative is to modify the action in the methods described in Secs. 18.3–18.6. Then Eq. (18.6) is modified by replacing U with a time-dependent interaction in a form analogous to the time-dependent hybridization,

$$S_{int} = \int_0^\beta d\tau n(\tau) U n(\tau) \quad \longrightarrow \quad \int_0^\beta d\tau \int_0^\beta d\tau' n(\tau) U(\tau - \tau') n(\tau'), \qquad (18.36)$$

where $U(\tau)$ denotes the dynamically screened interaction. In terms of the spectrum of the response, $U(\tau)$ can be expressed as (see Eq. (D.16) and Ex. 18.3)

$$U(\tau) = \int_0^\infty \frac{d\omega'}{\pi} \mathrm{Im} U(\omega') \frac{\cosh(\omega'(\tau - \beta/2))}{\sinh(\omega'\beta/2)}. \qquad (18.37)$$

The essential features of the modified CTQMC method can be brought out by consid-ering the Holstein–Hubbard model described in Sec. 21.6. The hamiltonian in Eq. (21.14) consists of a Hubbard model with bare (unscreened) on-site interaction U_{bare} coupled to a bosonic excitation (such as plasmons or phonons) with frequency ω_0 that act independently at each site i. The algorithm can best be understood by first making the unitary Lang–Firsov transformation to the form given in Eq. (21.15), which is a sum of electron and boson terms with no explicit coupling. This is achieved by writing the expression in terms of

---

[14] Here we follow the description of the methods in the review [843], which cites earlier articles, in particular [862, 863]. An illuminating analysis of a simplified version is described in Sec. 21.6.

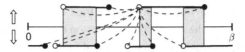

Figure 18.5. Illustration of dynamic interactions included via the boson coupling $K(\tau_i - \tau_j)$ of one time to all the other times for the diagram shown in the upper part of Fig. 18.3 that depicts one state in a CTQMC calculation based on the expansion in the hybridization. Only the coupling to one site is shown for clarity.

modified fermion operators $\tilde{d}_{i\sigma}^{\dagger} = \exp\left[\frac{\lambda}{\omega_0}(b^{\dagger} - b)d_{i\sigma}^{\dagger}\right]$ and $\tilde{d}_{i\sigma} = \exp\left[-\frac{\lambda}{\omega_0}(b^{\dagger} - b)d_{i\sigma}\right]$; also the on-site energy is shifted to $\tilde{\varepsilon}_0 = \varepsilon_0 - \lambda^2/\omega_0$, and there is a reduced static interaction $U_0 = U_{bare} - 2\lambda^2/\omega_0$. After the transformation, the electron and boson factors are decoupled and the expectation value of Eq. (18.34) becomes a product of a term involving electron operators, the same as that computed for the Hubbard model without bosons but with modified parameters, and a term that is the expectation value of a product of exponentials of boson operators. Now the state S includes both boson (b) and fermion (f) configurations and its weight is given by the product of independent thermal expectation values,

$$\Pi_L(S) = \Pi_b(S) \times \Pi_f(S),\tag{18.38}$$

where the boson weight for the nth order is

$$\Pi_b(S) = \left\langle T_\tau \, e^{s_{2n}A(\tau_{2n})} \cdots e^{s_1 A(\tau_1)}\right\rangle_b,\tag{18.39}$$

where $s_i = 1$ if the ith fermion operator is $\tilde{d}_{i\sigma}^{\dagger}$ and $s = -1$ for $\tilde{d}_{i\sigma}$, and $A(\tau) = (\lambda/\omega_0)(e^{i\omega_0\tau}b^{\dagger} - e^{-i\omega_0\tau}b)$. The expectation value for the thermal state of independent bosons can be evaluated (see [862] and Ex. 18.4) to find

$$\Pi_b(S) = \exp\left[\frac{\lambda^2/\omega_0^2}{e^{\beta\omega_0} - 1}\left(n(e^{\beta\omega_0} + 1) + \sum_{i>j=1}^{2n} s_i s_j \left(e^{\omega_0[\beta - (\tau_i - \tau_j)]} + e^{\omega_0(\tau_i - \tau_j)}\right)\right)\right].\tag{18.40}$$

This expression connects a boson factor to each of the points in the simulation, which can be interpreted [863] as an interaction $K(\tau_i - \tau_j)$ between all pairs of operators, where

$$K(\tau) = \frac{\lambda^2}{\omega_0^2} \frac{\cosh(\omega_0(\tau - \beta/2)) - \cosh(\omega_0\beta/2)}{\sinh(\omega_0\beta/2)}.\tag{18.41}$$

This is illustrated in Fig. 18.5, which shows the coupling of one of the operators in Fig. 18.5 to all the others at the same or different times.

The CTQMC simulation with dynamical interactions thus involves the sum over states illustrated by Fig. 18.5. The shaded areas indicate overlap of ↑ and ↓ electrons on the same site, with energy cost the static screened U_0, and the dashed lines depict the interaction $K(\tau_i - \tau_j)$ in Eq. (18.41) that couples all sites (only those involving one site coupled to the others are shown for clarity). An important point is that the computational effort is comparable with a calculation with a static U; the time is predominantly due to the determinants for the fermions and the added sum over boson terms K is small in comparison. A different, more general, approach has been proposed in [864], but it requires significantly more computational effort.

The spectral features of the Green's function can be greatly affected by the dynamical screening with the various energy scales leading to structures that are not simply connected to any single ingredient. Examples are given in [863] and Sec. 21.6 for selected values of the bare interaction, boson frequencies, and coupling constants.

18.8 Other methods

There are difficulties with the solver methods discussed in this chapter. Exact diagonalization is limited to small problems and Monte Carlo methods in general suffer from the sign problem (although there is no sign problem in a number of important cases) and the transformation of the imaginary-time Green's function to real time or frequency. The CTQMC method with an expansion in the hybridization involves diagonalization that scales exponentially with the number of orbitals and becomes very difficult to apply if there are many orbitals per atom or many atoms in a cluster.

Therefore, it is very useful to develop other solvers for embedded systems working at zero or low temperature, which satisfy the following criteria: (i) the solver can capture both the low-energy quasi-particle physics and the high-energy Hubbard bands; (ii) it can give the real-time dynamical properties directly without transformation from imaginary time; (iii) it can easily be generalized to realistic multi-band systems.

There are other techniques that can satisfy some of these criteria, but it is not feasible to describe them here in any detail.

- Perturbative methods at low order are useful for cases where interaction is not so strong. In addition, "iterated perturbation theory" is constructed to agree with a second-order expansion for weak interaction and also to have correct behavior in the large interaction limit, as described in [302, 788]. This has been used for some of the calculations discussed in Ch. 16. Examples of applications to materials are the work on Fe and Ni [865], which finds results in generally good agreement with QMC calculations, as mentioned in Sec. 20.4.
- Diagrammatic methods come in many forms. Extensive discussions of different approximations and examples of applications can be found in the reviews [302, 303, 771, 788] and lecture notes [785]. Any of the methods described in Chs. 9–15 for expansions in terms of the interaction can be used, for example, the T-matrix approximation and FLEX in Ch. 15. There are also expansions in the hybridization that are especially useful for strong interactions, for example, the non-crossing approximation (NCA) that was originally developed for the single-impurity Anderson model [866]; as its name implies, it is the sum of diagrams that do not cross, as explained in detail in [867]. It can be justified as the leading terms in a large-N expansion, where N is the degeneracy of the states on the impurity site. A variational approach is developed in [159]. An

advantage of such methods is that they can treat the full Coulomb interaction matrix and spin–orbit interactions, but they have problems such as violation of Fermi liquid sum rules. Additional diagrams can be included, e.g., in the one-crossing extension [868] and T-matrix approaches [869]. The diagrammatic expansion methods are fast computationally, can be applied at low temperature, and do not suffer from the problems of transformation to real time that is so difficult in Monte Carlo methods, but of course they suffer from other problems and break down in the regimes where they are not applicable.

- The numerical renormalization group method pioneered by Wilson [84] for the Kondo problem has been extended to dynamical methods and can be used as a solver in DMFT [870, 871]. For a review, see [872].
- The DMRG method is applicable to embedded systems, as noted in Sec. 18.2, and is used in the density matrix embedding method (see Sec. 16.7).
- Steps to ameliorate the scaling of exact diagonalization are mentioned at the end of Sec. 18.2 using methods from quantum chemistry and other techniques to reduce the number of states required.
- Methods that build upon the original works of Gutzwiller and Hubbard include the "Hubbard I approximation," where the atomic Green's function is used without a bath, which can be a reasonable approach for wide-band insulators.
- Of course, there are other methods that cannot be mentioned. Good sources for other approaches can be found in the reviews and lecture notes in the select further reading.

18.9 Wrap-up

The essence of dynamical mean-field theory is to divide the problem for interacting particles on a lattice into two parts: a many-body interacting system for each site in the lattice (an atom or a cell of atoms) embedded in a bath that represents the rest of the system, and a separate step to treat the coupling between the sites. The topic of this chapter is solvers that are the computational methods to solve the many-body problem for the embedded system; all the rest of the DMFT algorithms described in Chs. 16 and 17 are algebraic manipulations that provide approximate solutions for the lattice. The solvers fall into two broad categories: exact diagonalization and related approaches that work with eigenstates (Sec. 18.2) and path-integral approaches in imaginary time evaluated by Monte Carlo techniques (Secs. 18.3–18.7). Some of the pluses and minuses of these and other methods are summarized in Sec. 18.8.

Exact diagonalization techniques are very useful for integrated quantities like the density, as illustrated in Fig. 17.4. However, they suffer from the problem that the spectrum for the effective medium must be discretized and the computational effort scales exponentially with the number of states, including both the localized states on the site and the set of states that represent the bath. Several methods to reduce the computational demands are mentioned in Sec. 18.2.

Path-integral methods have much better scaling and there is no need for discretization of the bath, but it is difficult to reach low temperatures and to transform from imaginary time to real time or frequency, which is essential to calculate spectra. These methods include the Hirsch–Fye auxiliary-field method (Sec. 18.4) and two continuous-time CTQMC methods for expansion in the interaction (Sec. 18.5) or in the hybridization (Sec. 18.6). Examples of calculations for model systems are given in Ch. 17 and steps toward realistic calculations for systems with d and f states are the topic of Chs. 19 and 20.

The solver and DMFT algorithms are techniques to deal with a many-body problem; however, if the methods are to become part of a first-principles approach, they must be combined with other methods to determine the effective interactions and other terms that specify the problem. Physically motivated approximations using DFT are described in Chs. 19 and 20, but a much more powerful, satisfying approach is to combine the auxiliary system methods of DMFT with many-body pertubation theory. This is the topic of Ch. 21, where the combined techniques are used to treat the entire system and to derive the effective bath Green's function $\mathcal{G}_0(\omega)$ and dynamical interactions $\mathcal{U}(\omega)$. The CTQMC approach provides a method to treat dynamical interactions (Sec. 18.7) and the ingredients for combined MBPT–DMFT calculations are being developed.

SELECT FURTHER READING

Blümer, N., "Hirsch–Fye quantum Monte Carlo method for dynamical mean field theory," Ch. 9 in the lecture notes. The derivation of the Hirsch–Fye algorithm is followed in this chapter, along with information for actual computations.

Georges, A., Kotliar, G., Krauth, W., and Rozenberg, M. J., "Dynamical mean-field theory of strongly correlated fermion systems and the limit of infinite dimensions," *Rev. Mod. Phys.* **68**, 13–125, 1996. A general exposition of DMFT with detailed dscriptions of exact diagonalization and Hirsch–Fye QMC solvers.

Gull, E., Millis, A. J., Lichtenstein, A. I., Rubtsov, A. N., Troyer, M., and Werner, P., "Continuous-time Monte Carlo methods for quantum impurity models," *Rev. Mod. Phys.* **83**, 349–404, 2011. A review of CTQMC with many references to previous papers.

Gull, E., Werner, P., Millis, A. J., and Troyer, M., "Performance analysis of continuous-time solvers for quantum impurity models," *Phys. Rev. B* **76**, 235123, 2007. An instructive comparison of CTQMC methods with references to previous papers.

Held, K., "Electronic structure calculations using dynamical mean field theory," *Adv. Phys.* **56**, 829–926, 2007. A review that summarizes many methods.

Extensive sets of lecture notes with many specific details for methods of calculations can be found in lecture notes for Autumn schools at Forschungszentrum in Jülich [785, 786] that are available online at www.cond-mat.de/events.

Exercises

18.1 Derive the expressions in terms of a determinant, where the derivation should apply to all the cases: Eqs. (18.13), (18.17) and the determinant $D_\sigma(\tau_1 \cdots \tau_{k_\sigma}; \tau_1' \cdots \tau_{k_\sigma}')$ of the matrix in (18.31). Show that this can be viewed as a consequence of Wick's theorem. It can also be derived as a property of gaussian integrals over Grassmann variables, which is the form used in much of the literature in the field. Work this out or summarize the formulation given in texts such as Negele and Orland [842].

18.2 In order to find the form for the ratio of determinants in Eq. (18.21), consider an $m \times m$ matrix G_{ij} with determinant D and inverse G^{-1}. Make an $(m + 1) \times (m + 1)$ matrix by adding a column U, row V, and a diagonal element g. Find the determinant of this larger matrix. *Hint*: use Cramer's rule for the determinant by expanding along the new row or column.

18.3 Derive the form in Eq. (18.37) using the expressions for the transformation from frequency to time τ in App. C. Show that this reduces to Eq. (18.41) for a single oscillator.

18.4 Fill in the steps to show that the expression in Eq. (18.40) follows from Eq. (18.39) by taking the expectation value for the thermal state of independent bosons.

19

Characteristic hamiltonians for solids with d and f states

Summary

In the previous chapters DMFT has been developed as a methodology for calculation of dynamical and thermodynamic properties of many-body systems, with applications to model systems. In order for DMFT to be a general method for quantitative calculations, there must be systematic procedures to derive all the ingredients from first principles. In this chapter we consider the most prominent examples, materials with partially filled $3d$, $4f$, or $5f$ localized, atomic-like states, and set up the characteristic form of the hamiltonian. This is the basis for DFT+DMFT methods discussed here and used in the examples in Ch. 20, and for the combination of DMFT with many-body perturbation theory in Ch. 21. Simpler techniques, including self-interaction correction, DFT+U, and DFT+Gutzwiller, are also useful and are summarized in the last sections of this chapter.

Up to this point, dynamical mean-field theory has been developed for model systems with Hubbard-type hamiltonians, where the hopping matrix elements and interactions are considered as parameters. For problems with strong interactions, DMFT leads to renormalized bands, satellites in the one-particle spectra, magnetic phase transitions and local moments, metal–insulator transitions, and many other phenomena observed in real materials, as brought out in Ch. 2. The purpose of this chapter is to provide the framework for a quantitative theory of these materials.

These are difficult problems, experimentally, theoretically, and computationally, and the first step is to identify the aspects of the materials that must be treated explicitly as the local strongly interacting system, and the "rest" that are treated by a different method. This division of the problem is useful because a real material involves many degrees of freedom, i.e., the electronic states must be described in a basis sufficiently large to describe all the relevant states. However, the part that can be treated by non-perturbative solvers (Ch. 18) is limited by the computational cost that scales rapidly with the number of states included in the embedded site or cluster. Thus one must choose a small number of localized orbitals that inevitably are non-unique. This may seem like a step backward for quantitative methods; however, it opens a window for methods to treat these interesting problems and it presents a challenge to overcome the limitations.

Dynamical mean-field methods bring the capability to treat electronic properties as a function of temperature and to include strong atomic-like interactions that are essential to understand the properties of materials containing transition atoms with d and f states. However, DMFT must be combined with some other method to determine the bands and effective interactions that are needed in the DMFT calculations. In this chapter we consider a combination with density functional theory to provide the independent-particle hamiltonian and interaction parameters that can be used with the methods described in Chs. 16–18. The topics included in this chapter are not meant to be exhaustive, and they should be regarded as examples of the steps required for constructing first-principles theories that provide both quantitative results and illuminating physical interpretations. These are the methods that are used in the examples in Ch. 20. DMFT can also be combined with other Green's function methods to provide a more systematic set of methods to calculate screened interactions and spectra for the entire system. That is the topic of Ch. 21.

Section 19.1 is an overview of the ingredients needed to describe atomic-like behavior and local moments in solids containing transition elements. The specific form of the hamiltonian for transition metals and oxides with d states is outlined in Secs. 19.2 and 19.3, and the methods for specifying the orbitals are considered in Sec. 19.4. Practical methods for calculation of the effective hamiltonian using density functional theory with selected examples are described in Sec. 19.5. Finally, Sec. 19.6 summarizes static mean-field approximations that are simpler than DMFT and provide insight into the phenomena, and often quantitative information, in the examples in Ch. 20.

19.1 Transition elements: atomic-like behavior and local moments

Figure 19.1 depicts the progression from "delocalized band-like" to "localized atomic-like" observed in the transition elements with d and f states. The elements with highly localized states have atomic-like characteristics that carry over almost unchanged in the solid, with magnetic moments that order at low temperature and have local moment behavior above the transition temperature $T > T_c$. Elements at the lower left form delocalized bands along with s and p states and often are superconductors, the antithesis of magnetism. The competition between the two tendencies leads to remarkable phenomena in the elements on the border, including the "volume collapse" phase transition in Ce (see Fig. 20.1), heavy fermions with mass orders of magnitude larger than band calculations as exemplified in Tab. 2.2, the evolution from band-like to localized moments at Pu in the $5f$ series indicated by the jumps in volume shown in Fig. 2.1, the coexisting band-like and local moment behavior in the ferromagnets Fe and Ni as depicted in Fig. 2.3, and many other examples.

Compounds formed by the transition elements typically exhibit more localized behavior than the elemental solids, since the transition atoms are ionized and further apart with less overlap of the d or f wavefunctions. This is especially true for the oxides, which are the prototypical materials for a variety of phenomena. For example, elemental Cu has filled $3d$

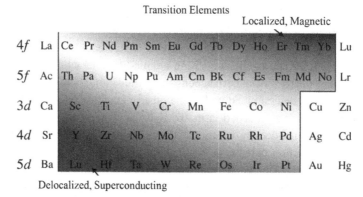

Figure 19.1. Rows of the periodic table for transition elements rearranged so that elements with the most localized states are in the upper right and the most delocalized are in the lower left. Localized states (e.g., the Gd $4f$ states in Fig. 19.2) have magnetic moments that persist in the solid. More delocalized states form well-defined bands and often lead to superconductors instead of magnets. The elements near the border have anomalous properties such as volume collapse in Ce, transition from band-like to localized moments at Pu, and many other examples. In oxides and other compounds the d and f states are more localized than in the elements, leading to the array of phenomena. Provided by K. T. Moore, similar to figure in [874] (originally by J. L. Smith and used with permission).

states and is not at all anomalous; however, in compounds the d shell may not be inert and, indeed, the Cu d states play a crucial role in the high-temperature superconductors, perhaps the most fascinating phenomenon of recent decades. The element vanadium is a relatively benign metal, but vanadium oxides are the classic cases of metal–insulator transitions and are primary examples in Chs. 13 and 20.

The localized nature of the d and f states is illustrated in Fig. 19.2 for the half-filled $3d$ and $4f$ shells in Mn and Gd, respectively. The more extended s and p states (and the $5d$ in Gd) overlap those on the neighboring atoms in a crystal to form bands and bonds that determine the crystal structure and characteristic interatomic distances in the elemental solids. In contrast, the $3d$ and $4f$ states are much more localized so that they retain much of their atomic-like character. In Mn the extent of the $3d$ states is comparable with the filled $3s$ and $3p$ states, which are often considered as core states. The effect is even more dramatic for Gd, where the $4f$ state has very small radial extent compared with the $5d$, $6s$, and $6p$ states.

Interaction versus bandwidth

Our understanding of the consequences of strong interactions in solids has been formed over decades by experimental observations, such as the insulating character of NiO by de Boer and Verwey [580], and perceptive analysis by Mott and Peierls [115, 875] and

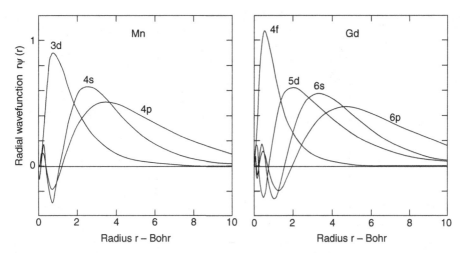

Figure 19.2. Radial wavefunctions $\phi_l(r) = r\psi_l(r)$ for Mn (left, similar to Fig. 10.1 in [1]) and Gd (right) generated by the Octopus website (code originally due to J. L. Martins). For Gd the $4f$ states are much more localized than the other valence states, so that they are almost unchanged in the solid. For Mn the $3d$ states are more localized than the $4s$ and $4p$ states, but not to the same degree as the $4f$ states of Gd or other lanthanides.

many others.[1] The basic ideas are captured by the Hubbard model defined in Sec. 3.2; however, quantitative theories for materials require us to go further, and a good starting point is the series of papers by Hubbard on "Electron correlations in narrow energy bands" [123–128]. Figure 19.3, taken from Hubbard's second paper [124], shows schematically the consequences of interactions for electrons in the $3d$ states of a transition metal atom. The parabola illustrates the primary effect of the dependence of the energy on the occupation N,

$$E(N) = (\varepsilon_0 - \mu)N + \frac{1}{2}UN(N - 1), \tag{19.1}$$

where $U > 0$ denotes the repulsive interaction between electrons. The average energy to add an electron to a state with $N - 1$ electrons,

$$E(N) - E(N - 1) = (\varepsilon_0 + UN) - \mu, \tag{19.2}$$

varies with N and the most stable configuration is the occupation where the addition energy becomes positive, $N = 4$ in this example. The Fermi energy μ represents a reservoir of electrons; in a solid this is fixed by energies relative to the other states, normally the s–p states in an elemental metal or the energy levels of the other atoms in a compound. For example, the ground state has nominal occupation $N = 1$ in VO_2, $N = 2$ in V_2O_3, and $N = 5$ in MnO.

---

[1] Extensive reference to the literature can be found in the book by Herring [121] and the review by Imada *et al.* [91].

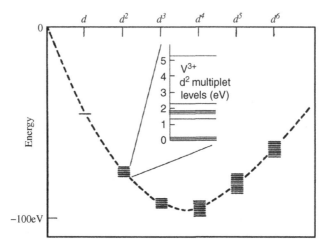

Figure 19.3. Schematic illustration taken from Hubbard's paper [124] showing the energy of an atom as a function of the number N of electrons in $3d$ states with a typical value of the interaction $U = 15$ eV. The levels for each N are atomic multiplets, and the inset shows the actual energy levels for a V^{3+} ion with two electrons in the d shell. As explained in the text, in a solid U is greatly reduced due to screening but the multiplet splitting is not much changed from the atom. (Adapted from [124] with V multiplets added.)

Of course, in a solid the states are mixed and the occupation is not integral. The key point is that if there is large overlap, the states form bands with mixtures of different occupations N; however, if there is small overlap with other states, the energy difference between the states with different occupation leads to stabilization of a particular occupation N with only small admixture of $N - 1$ and $N + 1$. "Mixed valence" [876] denotes the case where two occupations are nearly degenerate, e.g., N and $N - 1$, but U is large enough to reduce the probability of $N - 2$ and $N + 1$ in the ground state. The conclusion is that there is a competition between the interaction U and the bandwidth W, as proposed by Mott and recounted in Sec. 3.2.

The large energy scale in Fig. 19.3 applies for an isolated atom; however, the energy to add or subtract an electron in an atomic-like state is reduced by screening in a solid. This is a major effect and we must devise quantitative methods to determine the relevant interactions in the solid. The approach in Sec. 19.5 using DFT leads to typical effective values $U \approx 2 - 8$ eV. A more fundamental approach is the topic of Ch. 21, where we outline steps toward a truly first-principles approach combining the methods of many-body perturbation theory and dynamical mean-field theory.

Multiplets and local moments

The set of levels for each N denotes the energies of the different states that can be formed by N electrons, called a multiplet. Neglecting for the moment spin–orbit interactions, the atomic state for one electron in a state with angular momentum L has degeneracy $2(2L+1)$.

However, for $N > 1$ there are many possible energies, as shown in the inset to Fig. 19.3 for two electrons in an isolated V^{3+} ion.[2] The multiplet levels for two electrons in the $4f$ state of Ce can be observed by adding an electron in an inverse photoemission experiment, as shown in Fig. 2.9. The bars in the figure comprise the atomic multiplet spectrum that spans a range of several eV and is responsible for the large width of the upper peak in the spectra for Ce and its compounds. Since the states have the same occupation number N, i.e., the same charge, the energy differences between the states in the multiplet are not greatly affected by screening due to the other states. For example, the intra-atomic exchange constant J in Eq. (19.6) is almost the same in the solid as in the atom.[3]

For a free ion, the lowest-energy multiplet state for a given occupation tends to follow Hund's rule with maximum total spin S and angular momentum L, so that it is degenerate and has a magnetic moment. For highly localized states, this leads to "local moments" in the solid[4] corresponding to states that are degenerate in the atom and split to form a spectrum of low-energy excitations in the solid.

From the inset in Fig. 19.3 we see another reason that the atomic-like behavior can persist in the solid. There is a gap of > 1 eV separating the degenerate ground state from the next lowest excited state in the atom. Thus, in addition to the interaction U that tends to stabilize a state with a given occupation N, the intra-atomic Coulomb interactions tend to stabilize the lowest-energy state of the multiplet. Even though this energy is generally smaller than U in the atom, it is not screened and can be comparable in the solid. Thus it is important to include such effects (approximately in terms of J) in a quantitative theory.

19.2 Hamiltonian in a localized basis: crystal fields, bands, Mott–Hubbard vs. charge transfer

Since the goal is to treat atom-like local interactions, the problem is naturally expressed in a localized basis centered on the atoms, e.g., a linear combination of atomic orbitals.[5]

---

[2] Multiplet levels of V^{3+} from A. Kramida, Yu. Ralchenko, J. Reader, and NIST ASD Team (2012). NIST Atomic Spectra Database (ver. 5.0). Available at http://physics.nist.gov/asd National Institute of Standards and Technology, Gaithersburg, MD.

[3] More precisely, the spin-dependent interactions are almost the same as in the atom, but the value of J depends on the orbitals that are not rotationally invariant in the solid. The symbol "J" is also used to denote the inter-atomic coupling between spins on neighboring atoms, e.g., in the Heisenberg model in Eq. (3.4). Of course, this is absent in the atom and is wholly determined by the solid-state environment.

[4] Here the term "local" or "localized" means that the magnitude is large in a region, generally of atomic dimensions, e.g., a local moment or a localized orbital. The spatial extent is a quantitative measure that can be different for different properties. This is different from other uses that have precise meanings: a "local potential" $v(\mathbf{r})$ is a function of position, whereas a non-local potential is a function of more than one position, e.g., $v(\mathbf{r}, \mathbf{r}')$ in Hartree–Fock (see Sec. 4.1). "Localization" has a precise meaning in other contexts, e.g., the distinction between insulators and metals at zero temperature. See Sec. 6.2, [1, Ch. 23], and references such as [260, 262, 877].

[5] See [1, Chs. 14 and 15] for the useful simplifications and notations for the matrix elements given there. The tight-binding hamiltonian is often approximated by the Slater–Koster one- and two-center matrix elements given in detail in [1]: Sec. 14.2 for s, p, and d orbitals and in App. N, which gives formulas valid for arbitrary angular momementa.

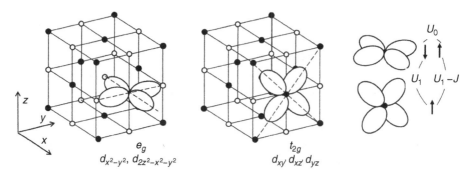

Figure 19.4. Atomic-like d states in a cubic lattice illustrating the symmetry of the t_{2g} and e_g states. The f.c.c. lattice can represent a compound in the NaCl structure, such as NiO (Ni – black, O – white circles); the e_g states hybridize with the O p states, whereas the t_{2g} states have nodes in the directions toward the O atoms and weaker hybridization. The figure also represents an elemental f.c.c. crystal such as Ni if the black symbols denote atoms and the white symbols, positions in the spaces between the atoms. On the right are shown the characteristic effective interactions for electrons in states on the same atom: $U_{mm} \equiv U_0$ for electrons in the same orbital, $U_{mm'} \equiv U_1$ in different orbitals, and $U_1 - J$ for the same spin, as discussed in the text.

In a spherical atom the eigenstates are generally represented as complex functions with definite angular momentum; however, in a solid it is usually more practical to work with real states defined by the symmetry of the site in the crystal. For example, the d states in a cubic crystal depicted in Fig. 19.4 are the appropriate choice for elements such as Ni in the face-centered cubic structure and compounds NiO and MnO in the NaCl structure.[6] Since the d or f states are quite localized, the tight-binding representation is very efficient and provides illuminating interpretations. For example, the primary features of the LDA and GWA bands in Ni shown in Fig. 13.11 can be described by a tight-binding model with one s and five d states and a few parameters, as shown in [1, Fig. 14.7]. An example of the orbitals in SrVO$_3$ is given in Sec. 19.4.

In terms of a set of localized, orthonormal orbitals, the hamiltonian can be written as

$$\hat{H} = \sum h_{im,i'm'} c^{\dagger}_{im} c_{i'm'} + \frac{1}{2} \sum U_{mm'n'n} c^{\dagger}_{im} c^{\dagger}_{im'} c_{in'} c_{in}, \tag{19.3}$$

where the sums are over all the indices, with i and i' denoting the unit cells and $m, m', n, n' = 1, \ldots, N_{states}$ denoting the set of N_{states} basis functions $\chi_m(\mathbf{r})$ centered on an atom at position $\boldsymbol{\tau}_m$ in each cell. The index m includes spin, except spin may be indicated explicitly in cases where it is needed. The basis includes all the states needed to describe the bands: it will be assumed that the interactions are limited to the localized d or f states, and the other states will be referred to as the "rest" whenever it is needed to make

---

[6] A comprehensive overview of the properties of transition-metal oxide structures and independent-particle hamiltonians can be found in the papers by Mattheiss [878, 879]. However, one should realize that the quantitative results in these papers were found using the LDA, where there are significant errors in materials like NiO (see Sec. 20.7).

the distinction. The matrix elements have the same form as the two- and four-point matrix elements in Sec. 12.6. For a crystal with translation vectors $\mathbf{T}_i$, the independent-particle matrix elements $h_{im,i'm'}$ are functions of the relative positions $\mathbf{T}_i - \mathbf{T}_{i'}$ and are given by

$$h_{im,0m'} = \int d\mathbf{r} \chi_m^*(\mathbf{r} - \mathbf{T}_i - \boldsymbol{\tau}_m)[-\frac{1}{2}\nabla^2 + v(\mathbf{r})]\chi_{m'}(\mathbf{r} - \boldsymbol{\tau}_{m'}), \qquad (19.4)$$

where $v(\mathbf{r})$ is a mean-field crystal potential.

The interactions can be expressed as

$$U_{mm'n'n} = \int d\mathbf{r} \, \chi_m^*(\mathbf{r} - \boldsymbol{\tau}_m)\chi_{m'}^*(\mathbf{r}' - \boldsymbol{\tau}_{m'})W_r(\mathbf{r}, \mathbf{r}')\chi_{n'}(\mathbf{r}' - \boldsymbol{\tau}_{n'})\chi_n(\mathbf{r} - \boldsymbol{\tau}_n), \qquad (19.5)$$

where W_r is an effective interaction screened by the rest of the system. This is in the spirit of Chs. 10–15, where the theory is cast in terms of the screened interaction W. However, in general the screening is frequency-dependent, $W_r(\mathbf{r}, \mathbf{r}', \omega)$. This is taken into account in Ch. 21, but here we consider only static interactions such as those found using DFT in Sec. 19.5. Often the four-index interaction matrix $U_{mm'n'n}$ can be simplified using symmetry and there is a very useful approximation for d states on the same atom in terms of two parameters, U and J, defined in Sec. 19.3 and depicted in Fig. 19.4. For simplicity we often give expressions only for a single U, but they can be generalized to treat the full matrix $U_{mm'n'n}$ when needed. Of course, there are also long-range Coulomb interactions that are included in the Hartree potential, which is part of the independent-particle potential $v(\mathbf{r})$ in Eq. (19.4).

Crystal field splitting

In a crystal the degenerate atomic, d or f states are split by the non-spherical potential, hybridization with neighboring atoms, and effects of interactions.[7] It is important to realize that the term "crystal field splitting" is commonly used in different ways. The effects of the potential are included in the on-site matrix elements $h_{im,im}$ in Eq. (19.3), which are often termed crystal fields. However, there is a shift in the energies due to hybridization with states on neighboring atoms that often dominate over the potential effects, especially in the transition metal oxides. The energy difference between the centers of the bands, including all effects, is also called crystal field splitting. We will refer to the splitting as Δ_{cf} and use the term in ways that should be clear from the context, or pointed out explicitly.

In a cubic crystal the d states[8] shown in Fig. 19.4 are split into e_g (two-fold degenerate $2z^2 - x^2 - y^2$, $x^2 - y^2$ with lobes along the cubic axes) and t_{2g} (three-fold

---

[7] The various contributions were discussed by Van Vleck [880] and can be found in references such as [881–883]. The non-spherical terms in the potential are often referred to as "10Dq," which denotes the multipole moments of the Coulomb potential. Ligand field theory includes effects of hybridization with neighbors that lead to effective shifts in the on-site energies.

[8] Here we neglect spin–orbit interactions which are small for the $3d$ states (for example, splitting of states by ≈ 0.07 eV in vanadium) and are ignored in essentially all calculations. The magnitudes are large and important for the $4f$ and $5f$ states of heavy atoms; an example in Sec. 20.2 is Ce, with one electron in the f states and total angular momentum $7/2$ or $5/2$.

Figure 19.5. Schematic illustration of the energy levels of d states in a solid showing the splitting and broadening of the states in a solid. The bands are shown as distinct but they may overlap. For sites with cubic symmetry the bands split into t_{2g} and e_g, with threefold and twofold degeneracy, respectively. For trigonal symmetry the three t_{2g} states are further split (e.g., in V_2O_3 and VO_2 in the low-temperature phase). As discussed in the text, the term "crystal field splitting" and the symbol Δ_{cf} are used in different ways in the literature. Here we adopt the convention that Δ_{cf} denotes the energy difference between the centers of the bands, including all effects.

degenerate xy, xz, yz with lobes along the face diagonals). The energies and broadening into bands are shown schematically in Fig. 19.5. In the oxides the e_g states tend to have higher energy and broader bands due to mixing with the oxygen p states, whereas the t_{2g} states are more localized since they have nodes in the direction toward the oxygen neighbors. Of course, the bands are not purely t_{2g} or e_g, and they are mixed at a general **k**-point in the Brillouin zone; although the precise symmetry designation only applies at high symmetry points, the classifications in terms of the dominant character can be very useful.

In crystals of lower symmetry, the d states are split further, such as the example of $t_{2g} \rightarrow a_{1g} + e_g^{\pi}$ shown in Fig. 19.5. This applies to VO_2 and V_2O_3 and is given in more detail in Fig. 13.14. Whenever the various splittings and bandwidths are comparable in magnitude, there are complications that are difficult to unravel. However, if the splitting is large enough it can lead to a simpler many-body problem. Systems with chain or plane structures may have well-separated bands so that only one is close to the Fermi energy and the other bands are often considered as spectators. For example, the common feature of the high-temperature superconductors is square planes of CuO_2. The Cu $d_{x^2-y^2}$ state is hybridized with oxygen states to form a single mixed Cu–O band. There is a vast amount of work on this problem modeled as a 2D one-band Hubbard model, and some examples are given in Sec. 17.6. Organic compounds that have chains of metal atoms, such as the Bechgaard salt κ-(BEDT-TTF)$_2$X, where X is a halide [884], may be the best examples of a metal–insulator transition in the simplest form envisioned by Mott. See, for example, [885] for a review and [886], which highlights the similarities between organic and cuprate superconductors.

Transition metal oxides: Mott–Hubbard vs. charge transfer

The energy of the d states relative to the rest of the orbitals is also crucial, especially in compounds such as the transition metal oxides. Historically, this has been a point of contention in the development of simplified pictures. Although first-principles calculations should predict the relative energies, the energies of states on different atoms are difficult to

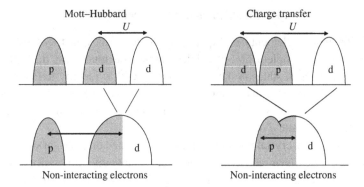

Figure 19.6. Classification of transition metal oxide Mott insulators in two types, Mott–Hubbard and charge transfer [887], in terms of simplified sketches of the density of states. In the former the oxygen p states are tightly bound and are merely spectators to the action happening in the partially filled d states. This is the situation for oxides of the early transition elements, such as vanadium. In oxides of the heavier transition elements, such as NiO, the d states are mixed with the O p states and the $d-d$ interaction may be large enough that the highest filled states have large components of O p. These are the cases that form the fascinating "doped Mott insulators" discussed in Secs. 2.8 and 20.7.

calculate and it is useful to identify the most relevant terms in order to simplify calculations and interpret the results.

Many theoretical models are based on an ionic picture in which the transition metal donates electrons to fill the oxygen bands that are assumed to be inert or can be treated as perturbations. This is illustrated on the left side of Fig. 19.6, where the interaction U leads to a metal–insulator transition within the d bands depicted in the upper diagram. Such a picture is supported by experiments on oxides of transition metals near the start of the series, for example, the spectra for V_2O_3 in Fig. 2.15, which show the small peak near the Fermi energy for four electrons associated with the two vanadium ions in the predominantly d bands separated from the large peak for 18 electrons in the tightly bound oxygen p states. This situation has come to be classified as "Mott–Hubbard," since it fits into the picture of competition between interactions and dispersion in a set of bands of one type ($3d$ states in this case) characteristic of the cases considered by Mott and Hubbard.

However, experiments and theoretical studies have shown that this picture is not sufficient. In the sequence from lighter to heavier $3d$ transition metal elements, the d orbitals become more tightly bound so that the oxygen and d bands are not clearly separated. The difference can be seen in experimental spectra, illustrated in Fig. 20.8 for NiO. The results support a picture like the schematic version on the upper right in Fig. 19.6, where the d states have large weight below the oxygen band. This is called the "charge transfer" regime [887], where the transition from occupied to empty states involves a transfer of charge from the oxygen to the transition metal. The consequence is that the highest occupied states have mixed $p-d$ character, which is a vital part of understanding the doping of holes in a Mott insulator, as brought out in Secs. 2.8 and 20.7.

19.3 Effective interaction hamiltonian

The interaction matrix elements in Eq. (19.5) are four-index matrices in the orbitals $U_{mnn'm'}$. Although a complete first-principles calculation must treat the full matrix of interactions for the actual states in the crystal, it is useful to identify terms that are most important for localized states that retain atomic-like character. For a single band there is only one term U for the on-site effective Coulomb interaction, but the situation is more complicated for degenerate d or f states.[9]

For d states in a cubic crystal the interactions are indicated on the right side of Fig. 19.4. If we assume spherical symmetry, i.e., ignore the quantitative differences in the radial wavefunctions for e_g and t_{2g} states, the intra-atomic interaction in Eq. (19.3) can be written in the form[10]

$$\hat{H}_{int} = U_0 \sum_\ell \hat{n}_\ell^\uparrow \hat{n}_\ell^\downarrow + \sum_{\ell<\ell'\sigma\sigma'} (U_1 - J\delta_{\sigma\sigma'})\hat{n}_\ell^\sigma \hat{n}_{\ell'}^{\sigma'}, \tag{19.6}$$

where orbital index m in Eq. (19.5) is written explicitly as spin σ and orbital index ℓ. The interaction between electrons in the same spatial orbital is U_0, assumed to be the same for all ℓ. For any two different orbitals the interaction is U_1 for opposite spin and $U_1 - J$ for the same spin, where J is the intra-atomic exchange constant. This is simplified to only two parameters using the relation $U_1 = U_0 - 2J$ to recover rotational invariance [785, Ch. 6].

The largest term is the interaction that depends only on the number of electrons. The variation of energy with the number is shown schematically by the parabola in Fig. 19.3, where the curvature is denoted U following the notation for a single orbital. Unfortunately, for multiple orbitals there are different definitions of "U" in the literature. One is the identification of U with the interaction in the same orbital U_0; another is the average interaction over all pairs of orbitals $U_{av} = F_0$, which can be shown [785, Ch. 6] to be $U_{av} = U_0 - (8/7)J$;

---

[9] An extensive discussion of typical cases in crystals can be found in [785, Chs. 3 and 6]. See also [788] and [888, Ch. 11]. There is a vast literature describing atomic spectra, e.g., the book edited by Griffiths [889]. For spherical symmetry the full multiplet spectra can be determined by three parameters: the Racah parameters A, B, C; the Slater integrals F_0, F_2, F_4; or the form denoted U, U', J. The relation of these expressions for the parameters can be found in the review [91, Tab. II], along with effective parameters for a minimum set of effective parameters including crystal fields. For the bare Coulomb interaction in a spherical atom the Slater parameters are given by integrals over the radial wavefunctions $\phi(r)$,

$$F_k = \int dr_1 dr_2 \phi(r_1)^2 \left(\frac{r_<^k}{r_>^{k+1}} \right) \phi(r_2)^2,$$

where $r_<$ and $r_>$ are the lesser and greater of r_1 and r_2, and similarly for the bare exchange integrals.

[10] In addition, there are spin-flip and pair-hopping terms

$$-J \sum_{\ell<\ell'\sigma} c_{\ell\sigma}^\dagger c_{\ell-\sigma} c_{\ell'-\sigma}^\dagger c_{\ell'\sigma} - \tilde{J} \sum_{\ell<\ell'\sigma} c_{\ell\sigma}^\dagger c_{\ell-\sigma}^\dagger c_{\ell'\sigma} c_{\ell'-\sigma},$$

which are essential for rotational invariance. See Sec. 16.11 for the expression in terms of spin operators and the explicit form for two bands. These are examples of terms that are not included in an independent-particle method such as DFT+U and in DMFT methods that involve only density–density interaction. Although they can be treated in the CTQMC methods in Ch. 18, they are not included in any of the examples of DMFT calculations in Ch. 20.

and a third possibility is for the Hund's rule ground state $U_{HR} = U_0 - 3J$, which is a logical choice if the dominant contributions are from the ground states of the ions [91]. Since $J \approx 1$ eV, there are significant differences in the values assigned using different definitions and one must be careful to be consistent. Here we will define U to be the average $U \equiv U_{av} = F_0$, which is the quantity most directly determined in the constrained calculations in the following section and in Sec. 21.5.

19.4 Identification of localized orbitals

The hamiltonian in a localized basis is given in Eq. (19.3), but for a quantitative theory we must have a means to calculate the matrix elements. This requires a specific form of the localized orbitals $\chi_m(\mathbf{r})$ that are centered on an atom at position $\boldsymbol{\tau}_m$ in each cell.

The first step in approaching a complex system is the division into two parts: the subset of states χ_m identified with strong interactions and the rest. *It must be recognized that there is no sharp distinction between the two types of orbitals and there can never be a unique choice.* Nevertheless, theoretical arguments and experimental observations lead to the conclusion that there are strong local, atomic-like interactions in materials containing transition metals, lanthanides, and actinides. The contrast between the localized and extended atomic orbitals is illustrated by the examples shown in Fig. 19.2 for Mn and Gd that are in the middle of the $3d$ and $4f$ transition series: the $4f$ orbitals of Gd are much more localized than the valence $5s$, $5p$, and $4d$ orbitals; the $3d$ orbitals of Mn are more localized than the bonding $4s$ and $4p$ orbitals, although the localization is not as extreme as for the $4f$ states.

One approach is to choose localized functions based on physical reasoning, such as a set of atomic-like orbitals similar to those in Fig. 19.2. This can be viewed as a useful, reasonable form that captures the most important aspects, especially if the states are very localized, e.g., in the lanthanides. The LMTO method (see [890] or [1, Ch. 16]) is appealing since it provides an orthogonal minimal basis of s, p, d, or f functions centered on a site with orthogonal tails on neighboring sites; in essence these are atomic-like functions with tails chosen to represent the state in the crystal.

An alternative is to use a more generally applicable technique such as the construction of Wannier functions[11] that are localized, orthonormal functions derived from a unitary transformation of the extended Bloch eigenstates found by an independent-particle method. Although Wannier functions are not unique, various choices have been employed, such as those found by the NMTO method [891] (see [1, Ch. 16]) and "maximally localized" functions [892], which are appealing since they have the minimum mean-square radius $\langle r^2 \rangle$. However, there are other choices, such as functions chosen to maximize the on-site interaction [893] that are compact and tend to minimize the off-site interactions. The functions also depend on the way they are extracted from "entangled" bands [894] that involve both the localized functions and those for the rest of the system. In an independent-particle method this is not an essential problem because it is merely a unitary transformation, and it

---

[11] See [1] for local orbital bases (Chs. 14 and 15), Wannier functions (Ch. 21), and practical measures for calculations with localized basis functions (Ch. 23).

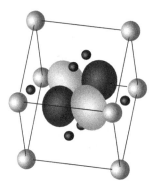 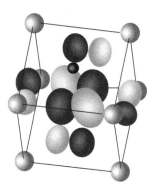

Figure 19.7. Two choices of the d_{xy} Wannier function for the V $3d$ t_{2g} state in SrVO$_3$. On the left is a function calculated from Bloch states that include the $3d$ and oxygen $2p$ bands; a localized $3d$ function is shown, and there are also localized oxygen $2p$ functions (not shown). On the right is a function that describes only the band that is mainly d_{xy} in character which has reduced amplitude on the vanadium site and significant oxygen $2p$ character. Each choice is useful if used appropriately, as discussed in the text. (Adapted from figure provided by J. Kunes.)

is straightforward to test if a given set of functions is sufficiently accurate for a desired result. In an approximate solution of an interacting many-body problem, however, the results depend on the choice of functions included in the cell and those that are relegated to the rest.

An example of two different Wannier functions for a t_{2g} state in SrVO$_3$ is shown in Fig. 19.7. These are both useful functions if used properly. The more localized function shown on the left is atomic-like and is appropriate for a model in which both vanadium d and oxygen p states are taken into account. In a calculation using these states, the $3d$ and $2p$ states are mixed by the p–d hopping terms in the hamiltonian. The Wannier function shown on the right is appropriate for a calculation that involves only the states that have mainly d character; the larger extent is required to take into account the fact that states also have significant oxygen character. This is the type of function used in the calculation of the interaction U in Secs. 19.5 and 21.5.

A shortcoming of the approaches described here is that they are derived from independent-particle methods that may be far from the actual many-body problem. It may be more appropriate to work with "natural orbitals" derived from a many-body density matrix as in Eq. (5.7). A simple example in the $3d$ transition metal oxides is the extent to which the $3d$ and oxygen $2p$ states are hybridized. This depends crucially on the relative energies, which are greatly affected by interactions and may be very different depending on the method. An example of the variations of $3d$ and oxygen $2p$ energies in NiO for several methods is given in Fig. 20.8.

The rest of the system

In real materials there are also the states of the rest of the system, usually more delocalized. There must be some method to determine these states and the appropriate hamiltonian for the entire system. A general procedure is outlined in Ch. 21, where the screened interaction W and the Green's function are calculated by a combination of many-body perturbation and dynamical mean-field methods. A much simpler approach is to combine with an independent-particle method such as the Kohn–Sham equations. Then the only roles of the rest of the system are to screen the interaction and to determine the independent-particle $G_{\mathbf{k}}^0$; the lattice Green's function is given by $G_{\mathbf{k}} = [[G_{\mathbf{k}}^0]^{-1} - \Sigma]^{-1}$, where Σ is provided by the calculation for the embedded auxiliary system as described in Chs. 16–18. However, the use of Kohn–Sham eigenvalues is *ad hoc* in many ways, and there are other problems described in the next section.

19.5 Combining DMFT and DFT

The "DFT+DMFT" method consists of Kohn–Sham DFT with a chosen functional (usually the local density approximation) with additional interactions in the localized states that are also determined by DFT methods.[12] Since DFT is a theory of the ground state and the Kohn–Sham approach has proven to be very successful for total energies, a goal is to formulate DFT+DMFT methods in such a way that they maintain the accuracy for total energies.[13] For excitations the combination with DFT is not as well justified as the combination of DMFT with many-body perturbation methods in Ch. 21, but it is much easier to implement and it suffices to illustrate the capabilities of DMFT. Thus we formulate the method in the next sections and illustrate the consequences in Ch. 20.

In this approach the hamiltonian of Eq. (19.3) can be written in the form

$$\hat{H} = \hat{H}_{KS} - \sum_{im} \Delta\varepsilon_m \hat{n}_{im} + \hat{H}_{int}, \tag{19.7}$$

where i denotes the sites or cells, m labels the localized states in each cell, including the spin σ, and $\hat{H}_{int}$ is the interaction hamiltonian, which is limited to interactions within each site or cell. The independent-particle Kohn–Sham hamiltonian $\hat{H}_{KS}$ is calculated from the full charge density including all orbitals and $\Delta\varepsilon_m$ are shifts in the Kohn–Sham eigenvalues only for the localized orbitals. The $\Delta\varepsilon_m$ are "double-counting" corrections needed to account for the fact that average effects of interactions are included in both the Kohn–Sham potential and $\hat{H}_{int}$. These and related terms in the expressions for the total free energy are explained below after considering the interactions. For simplicity we give explicit expressions only for a single site with one type of orbital, e.g., the $2L + 1$ atomic-like states with angular momentum L, and assuming equal interactions U between electrons in any pair of

---

[12] An extensive set of lecture notes for combined DFT and DMFT methods can be found in the proceedings of a school in 2011 "The LDA+DMFT approach to strongly correlated materials" [785].

[13] There is, however, an important difference in the DMFT and DFT calculations. Thermodynamics involves also the entropy, which is crucial for understanding the properties of systems with thermal disorder, such as fluctuating local moments. This is included in the grand potential Ω, which is defined in Eq. (5.2) and is the quantity calculated in the many-body DMFT method, as formulated in Sec. 16.6 and Ch. 21. It is not at all obvious how to capture such effects as a functional of the density in a Kohn–Sham–Mermin approach, as pointed out in Sec. 8.1.

orbitals. The expressions can be generalized straightforwardly to multiple types of states and sites and a general interaction matrix $U_{mm'n'n}$.

Calculation of effective interactions and energies using constrained DFT

The constrained DFT method is a practical approach to calculate values for the terms in the hamiltonian of Eq. (19.7). Of course, this is an approximation depending on the choice of exchange–correlation functional, and in addition there are caveats and non-uniqueness in the method. DFT is a ground-state theory and the methods described in this section are designed to find static interaction parameters. This is sufficient to illustrate the methods and the types of results, but ultimately one should take into account the effective screened interaction that is frequency-dependent, which is considered in Ch. 21. In addition, the resulting effective hamiltonian in Eq. (19.7) is not unique, since it depends on the choice of localized orbitals, as emphasized in the previous section. In this section we describe general principles for calculation of the interactions and give some typical values found by calculations within the DFT framework. These are only typical numbers; in an actual calculation the orbital energies ε and effective interactions U, J, etc. must be determined in a consistent way and they should be calculated for the same orbitals that are used later in the solution of the many-body problem.

Test of the method: calculations for isolated ions

It is instructive to first consider calculations for isolated ions where constraints are not needed and the results can be compared with experiment. Ignoring for the moment any orbital-dependent terms, the total energy as a function of occupation N is $E(N) = \varepsilon_0 N + \frac{1}{2} U N(N-1)$, as given by Eq. (19.1) (omitting the chemical potential μ, which determines the minimum energy configuration but does not affect the present calculations). Of course, in a real atom U is not a constant for the entire range of N; nevertheless, this expression is meaningful if it is applied only within a limited range close to that in the actual material, for example, $3d^7$, $3d^8$, $3d^9$ in Ni. Other occupations, such as $3d^6$ and $3d^{10}$, are high-energy excitations, with little contribution to ground and low-energy excited states (see Ex. 19.3). If we choose an integer number N of electrons close to the occupation in the solid, the parameters ε_0 and U can be found from

$$\varepsilon_0 + U(N-1) = \frac{1}{2}[E(N+1) - E(N-1)], \quad U = E(N+1) + E(N-1) - 2E(N) \quad (19.8)$$

in terms of energy differences from three separate calculations for $N-1$, N, and $N+1$ electrons, which is shown schematically on the right side of Fig. 19.8. Thus, $\varepsilon_0 + U(N-1)$ is the linear term in the energy relevant for this range of occupation. This is an example of ΔSCF calculations (see Sec. 4.7 and [1, Sec. 10.6]) and density functional theory is appropriate since these are ground-state calculations for each N.

Alternative expressions can be found in terms of the eigenvalues for methods in which the energy is a continuous function of the density $n(\mathbf{r})$, considered as a continuous variable. If we consider DFT with a functional that is continuous, the Kohn–Sham eigenvalue for

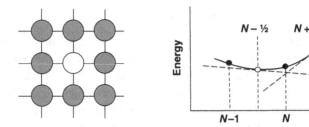

Figure 19.8. Calculation of the effective interaction for localized states at any one of the equivalent sites (central site shown on the left) in the lattice. Right: illustration of constrained occupations used to calculate site energies and screened interaction from the total energies, Eq. (19.8), or eigenvalues at half-integral occupation, Eq. (19.9), while allowing the rest of the electrons to relax and screen the interaction. This approach can be applied for any of the forms of "U" (U_{av}, U_{mm}, etc.) and for J as a difference in energies for $\uparrow\uparrow$ and $\uparrow\downarrow$.

state m is related to the total energy by $\varepsilon_m = dE/dn_m$ (see Eq. (4.24)). Thus the energy differences in Eq. (19.8) can be determined from the Kohn–Sham eigenvalues,

$$\varepsilon_0 + U(N-1) = \frac{1}{2}\left[\varepsilon(N+\frac{1}{2}) + \varepsilon(N-\frac{1}{2})\right], \quad U = \varepsilon(N+\frac{1}{2}) - \varepsilon(N-\frac{1}{2}), \quad (19.9)$$

where the occupation is increased or decreased by $1/2$ in one of the orbitals. The method is readily generalized to multiple orbitals with eigenvalues ε_m and various interactions $U_{mm'}$ (see, e.g., Eq. (19.12) below). This is called the Slater transition rule (see [1, Sec. 10.6]) and it is often much simpler to calculate eigenvalues at half-integral occupations rather than differences in total energies.

The results of calculations for atoms in the lanthanide and $3d$ transition series are remarkably accurate. For example, an LDA calculation [895] for the $3d$ states of a Cu atom in free space yields $U_{av} = E(d^{10}, s^1) + E(d^8, s^1) - 2E(d^9, s^1) = 15.88$ eV compared with the experimental value 16.13 eV for the average of the multiplet energies for the three occupations. This can be considered to be a "bare" interaction for an isolated ion with magnitude similar to that illustrated in Fig. 19.3.

Atomic calculations can also be used to estimate the screened interaction in a solid. The addition of a charge in a d or f state in a solid is screened by the other electrons so that within a very short distance the total charge is reduced to $\approx \pm 1/\epsilon$, where the dielectric constant $\epsilon \gg 1$. This can be mimicked in an atom if the charge in the d or f state is compensated by an electron in an s state [896]. For example, in a Cu atom the energy differences $E(d^{10}, s^0) + E(d^8, s^2) - 2E(d^9, s^1)$ can be interpreted as a screened interaction, with the values 3.96 eV found in an LDA calculation [895] compared with 4.23 eV in experiment. Similar values are found for all $3d$ transition metals. For the lanthanides the result is ≈ 6–7 eV [896, 897]. These are remarkably close to the values for screened effective interactions found in calculations for $3d$ and $4f$ states in solids, as discussed below.

Constrained DFT in a solid

In a solid, constrained DFT calculations [895, 898–903] require a technique to control the occupation of the localized orbitals, as shown schematically in Fig. 19.8. One approach is to artificially remove the hopping matrix elements [895, 899] so that the occupation n_m of a localized orbital can be constrained to be an integer or half-integer. Screening is accomplished by allowing the rest of the system to relax for different occupations n_m. This can be done in practice by a calculation in a supercell in which only the occupation on the central site is constrained. The results approach those for an isolated site in the limit of a large cell, and various methods can be used to reduce the size of the cell required for a converged result (see [1, Sec. F.6]). The effective screened interactions can be found from the eigenvalues at fractional occupation, Eq. (19.9), which is easier to use than Eq. (19.8).

Another way is to enforce the constraint using a Lagrange multiplier [898, 903]

$$E(n_m^c) = \min \left\{ E^{DFT}[n(\mathbf{r})] + V_m(n_m - n_m^c) \right\}, \text{ with } n_m = \int_L d\mathbf{r} n_m(\mathbf{r}), \qquad (19.10)$$

where $n(\mathbf{r})$ is the total density, $n_m(\mathbf{r})$ the density of the localized orbital, n_m its occupation in a region L, n_m^c the desired occupation, and the Lagrange multiplier V_m a variable potential that acts only on the selected orbital. If we identify the interaction U as the quadratic variation of the energy as a function of n_m^c,

$$U = \frac{\partial^2 E(n_m^c)}{\partial n_m^{c\,2}}, \qquad (19.11)$$

the value can be computed by quadratic interpolation [898, 904] or by a response function [903]. As shown in Ex. 19.4, $U = -\chi^{-1}$, where $\chi = \partial^2 E/\partial v_m^2 = \partial n_m/\partial v_m$ is the response to a potential v_m, i.e., a perturbation $v_m c_m^\dagger c_m$ in the hamiltonian. In a solid with many orbitals in addition to the one that is constrained, the energy $E(n_m^c)$ or $E(V_m)$ has various contributions; one should be careful to define U in a way that is consistent with the way it is used.[14]

Examples of calculations of U

It is instructive to survey typical values of the interaction calculated by various methods outlined above and for choices of localized orbitals. All of the results listed here are for the interaction U_{av} averaged over the orbitals. These are typical values, and it is essential to realize that the same orbitals should be used in all parts of the calculation, with careful attention to the difference between the average value U_{av} and the other definitions of the term "U" described in Sec. 19.3.

For elemental metals the values tend to be consistent [905]. However, the transition metal oxides have a larger range of values, which can be anticipated since localized functions can be constructed with different $3d$/oxygen p ratios, and the screening varies for different

---

[14] In [903] a term is subtracted, $U = -(\chi^{-1} - \chi_0^{-1})$, to take into account the response of the system to v_m not related to the interaction; the result is a reduction of the magnitude of U below that given by Eq. (19.11).

methods to treat the rest of the system. Typical values of U_{av} from constrained DFT (and constrained RPA in Sec. 21.5) calculations are as follows.

- Fe and Ni metals: $U \approx 2.2$ [903, 905]–2.8 eV [901] for Fe; ≈ 2.7 [905]–3.0 eV [901] for Ni. Dynamic c-RPA results in Fig. 21.6 indicate a range $\approx 2.5-3.5$ eV for Ni.
- Ce and Yb metals: for Ce, values are in the range ≈ 4.5 [903]–6 eV [901], and for Yb changing from ≈ 6 to 8 eV as n_f varies from ≈ 13.2 to 14.
- MnO, FeO, CoO, and NiO: $\approx 6-8$ eV [902, 906] and lower values $\approx 4-5$ eV in other works [903, 904, 907].
- VO_2, V_2O_3, and $SrVO_3$: $\approx 3.5-4$ eV. $SrVO_3$ is an example where there are calculations of the frequency dependence of the effective interaction; the values at low frequency are similar to those found by constrained DFT but there is strong variation with frequency. (See Fig. 21.6 and the related text.)
- Cu in La_2CuO_4: $\approx 7-8$ eV for Cu d [895, 908, 909]; ≈ 3 eV (the range $\approx 2.4-3.4$ eV is given in [910]) for mixed Cu–O states in a one-band Hubbard model.

Intra-atomic exchange constant J

The spin-dependent interactions between electrons in different orbitals can also be calculated using constrained occupations by a straightforward extension of the approach indicated in Fig. 19.8. The exchange interaction for two orbitals m and m' is the difference in interactions between like and unlike spins, as shown schematically in Fig. 19.4. It can be evaluated using constrained occupation of specific states with $m \to \ell, \sigma$ to indicate space and spin. In analogy with Eq. (19.9), J can be found from the eigenvalues at half-occupation for $\ell \downarrow$ and $\ell \uparrow$,

$$J = \varepsilon_{\ell\downarrow}(n_{\ell\downarrow} = \frac{1}{2}, n_{\ell'\uparrow} = 1) - \varepsilon_{\ell\uparrow}(n_{\ell\uparrow} = \frac{1}{2}, n_{\ell'\uparrow} = 1), \qquad (19.12)$$

when there is a $\uparrow$ electron in state $\ell' \neq \ell$. Similarly, the entire matrix of interactions $U_{mnn'm'}$ in Eq. (19.3) can be calculated. In general, the values depend on the orbitals and an extensive discussion can be found in [785, Ch. 6]. Unlike the energy differences for different charge states in Eq. (19.9) that determine U, these are neutral excitations that are not greatly affected by screening and the magnitudes of the interactions are close to the atomic values, typically $J \approx 0.7 - 1.0$ eV for $3d$ states.

Double-counting corrections

At this point, all the terms in the hamiltonian of Eq. (19.7) are specified except the corrections to the orbital energies $\Delta\varepsilon_m$ needed to take into account that interactions are included twice, in U and in the Kohn–Sham potential. This is called "double counting" and it is a fundamental problem with no simple answer because interactions are treated very differently in DFT and DMFT. The exchange–correlation functionals are non-linear in the density, for example, the leading term in the exchange is $\propto 1/r_s \propto n^{1/3}$. There is no unique mapping to a hamiltonian with operators that are quadratic and quartic in creation and annihilation operators. The combination with many-body perturbation theory in Ch. 21 is much more straightforward and satisfying; however, it is more difficult to implement.

The same issues arise in DFT+U methods in the following section, and there have been various proposals for the correction [303, 356, 788] that can lead to significant consequences for the spectra. For example, the shifts in the d states determine the relative positions of the d and oxygen bands, which is the distinction between Mott–Hubbard and charge-transfer systems brought out in Fig. 19.6. An example is NiO, where there must be large corrections to the LDA bands to have results that agree with experiment. The results for various values of U (and $\Delta\varepsilon$) show the large uncertainty in the results depending on the choices. The effect of different choices of $\Delta\varepsilon$, keeping U fixed, has been investigated in DMFT calculations for NiO [911].

Two proposals for the correction show the reasoning and the fact that there is more than one sensible choice. For simplicity the equations are written with only a single U for all orbitals and with $J = 0$. One approach is to choose a form for which there is no correction to the total energy but the eigenvalues are changed. This is accomplished if the total energy is expressed as

$$E = E^{DFT} + \frac{1}{2} U \left[\sum_{m \neq m'} \langle \hat{n}_m \, \hat{n}_{m'} \rangle - N(N-1) \right]. \tag{19.13}$$

The first term in square brackets is the interaction energy written in terms of density operators for individual orbitals; the second term is the interaction energy independent of the occupation of particular states (the same as in Eq. (19.1)). These two terms cancel and one is left with $E = E^{DFT}$ (see Ex. 19.6). One rationale for this choice is that it is the same as the way U is calculated in Eq. (19.8) or (19.9) where occupations are constrained to be integers. However, the derivatives of these terms do not cancel, as discussed next.

In order to identify the correction to the DFT eigenvalues we must distinguish between the energy due to correlated density *fluctuations* from that due to average occupations of the orbitals n_m. To do this Eq. (19.13) can be written as

$$E = E^{DFT} + \frac{1}{2} U \left[\sum_{m \neq m'} (\langle \delta\hat{n}_m \, \delta\hat{n}_{m'} \rangle + n_m \, n_{m'}) - N(N-1) \right]. \tag{19.14}$$

The eigenvalue is given by the derivative of the energy (see Ex. 19.7) with respect to the average occupation,

$$\varepsilon_m = \left. \frac{\partial E}{\partial n_m} \right|_{n_{m'}, m' \neq m} = \varepsilon_m^{DFT} + U \left(\frac{1}{2} - n_m \right) \equiv \varepsilon_m^{DFT} + \Delta\varepsilon_m. \tag{19.15}$$

Thus the correction in the eigenvalue, $\Delta\varepsilon_m = U \left(\frac{1}{2} - n_m \right)$, raises the energy of empty orbitals ($n_m = 0$) and lowers the energy for filled orbitals ($n_m = 1$). This average term is what is included in the "DFT+U" method described in Secs. 11.4 and 19.6. The correction $\Delta\varepsilon_m$ is also taken into account in a DFT+DMFT calculation; however, DMFT also includes a contribution to the ground-state energy due to the fluctuations $\langle \delta \hat{n}_m \; \delta \hat{n}_{m'} \rangle$ and the DMFT spectra include dynamical effects of correlation that cannot be described by an independent-particle method like DFT+U.

A different proposal for the correction is to identify DFT as a mean-field theory that involves only the average occupations and replace the second term in Eq. (19.13) by $(1/2)U \sum_{m \neq m'} \langle (\hat{n}_m - \bar{n}) \; (\hat{n}_{m'} - \bar{n}) \rangle$, where $\bar{n} = N/M$ is the average for N electrons in M degenerate orbitals. As shown in [912], this leads to the same $\Delta \varepsilon_m$ as above except that there is an average shift of the entire set of M eigenvalues by the energy $-U(\frac{1}{2} - \bar{n})$. If the M localized orbitals are well-separated from other bands, this is just a rigid shift with no consequence; but it can be very important in cases like NiO, where it shifts the Ni d relative to the oxygen p [911].

Self-consistent DFT+DMFT

It is straightforward to formulate a self-consistent algorithm for DFT and DMFT: the DFT calculation determines the orbitals, which, together with the constrained DFT calculation of U, provide the needed information for input to DMFT. In turn, the DMFT calculation modifies the density that is needed in DFT. Self-consistency is important in cases where there are significant changes in the density. For example, in NiO the mixture of $3d$ and oxygen p states changes drastically (see Sec. 20.7) as a function of the relative position of the $3d$ and p bands. A poor starting point such as LDA requires large changes to agree with experiment, whereas a different starting point could be closer to the final solution. In Yb, self-consistency is important [913] because of the large change in density $n(\mathbf{r})$ as the occupation changes from 14 to 13 under pressure (Fig. 20.4).

Conclusions on DFT+DMFT

At this point, we can summarize some of the positives and negatives of the combination of DMFT and DFT.

Positives:

- It is straightforward to include the rest of the states since they are eigenstates in the independent-particle Kohn–Sham hamiltonian.
- Although there are fundamental problems in using DFT for spectra (see negatives below), there is much experience with DFT and DFT+U methods that can be brought to bear when interpreting the results of DFT+DMFT calculations.
- DFT is designed to give accurate total energies. One can make choices to maintain this property, for example, the choice for the energy in Eq. (19.13) is the DFT expression with no correction.
- DMFT can build on the success of DFT for total energies and also include the thermodynamic properties, for example, thermal fluctuations of local moments. In principle, the grand potential Ω_{xc} (see Sec. 8.1) is a functional of the density, but in practice it is very difficult to construct a feasible functional.

Negatives:

- The disadvantages of DFT flow from the fact that the eigenstates and eigenvalues are only auxiliary quantities with no direct physical interpretation as excitations. Thus the on-site energies and hopping matrix elements for all the states (localized states and the rest) are problematic and there is no systematic procedure for correcting them. (This is in addition to the problem that the separation of states into localized and the rest is not unique.)
- A direct consequence is the "double-counting" problem that is intrinsic to the method due to the very different ways interactions are treated in DFT and DMFT. The expression in Eq. (19.15) for the shifts in orbital energies is not unique; we have given one other example and there are other expressions [303, 356, 788] that are qualitatively similar but differ quantitatively.
- There should also be corrections in the hopping matrix elements, but often these are simply taken from a Kohn–Sham calculation with no changes. An example of a systematic way to renormalize the band dispersion is given in Sec. 21.6, where the Green's function for the entire system is calculated by the combination of DMFT and many-body perturbation theory.

19.6 Static mean-field approximations: DFT+U, etc.

This section is devoted to approximations that replace the many-body problem with an effective mean field that incorporates some average effects of the interaction. When coupled with DFT each constitutes a variation of the exchange–correlation functional, while maintaining the simplicity of independent-particle Kohn–Sham -like equations.[15,16] These methods are much less computationally intensive than DMFT, they are very useful in practice, and they give insight into the original interacting electron problem.

Static approximations can lead to unrestricted solutions with lower energy and broken symmetry, as discussed in Sec. 4.7. This may be a phase that can occur in a solid such as a ferro- or antiferromagnet; however, one should be aware that a broken symmetry may be unphysical, the same as in other unrestricted mean-field methods. For example, an unrestricted DFT+U or SIC calculation for the Anderson impurity model (Sec. 3.5) would find an unphysical magnetic solution just as is found in unrestricted Hartree–Fock, which is described in Sec. 4.7 and Ex. 4.7.

DFT+U

The DFT+U method and related techniques can be understood as a static mean-field approximation to DMFT; for that reason the expressions in the previous section were

---

[15] The DFT+U and self-interaction correction methods are described in [1, Ch. 8], as well as the techniques needed for calculations using the Hubbard alloy approximation.

[16] An extensive review of the DFT+U method (often called "LDA+U") can be found in [356] and a summary is given in [1, Sec. 8.6].

written in a way that includes both methods. The static approximation is to ignore the correlated fluctuations $\langle \delta \hat{n}_m \, \delta \hat{n}_{m'} \rangle$ in Eq. (19.14), leaving only interactions due to the average occupations, i.e., a Hartree–Fock approximation for the terms in square brackets in Eqs. (19.14) and (19.15). As stated there, correction to the eigenvalues $\Delta \varepsilon_m = U \left(\frac{1}{2} - n_m \right)$ raises the energy of empty orbitals and lowers the energy for filled orbitals. For a filled shell (such as the $3d$ state in Zn, Ga, As, . . .) the energies are shifted down relative to other states, in better agreement with experiment. A system with partially occupied bands at the Fermi energy may lower its energy by breaking the symmetry with different occupations n_m of otherwise equivalent orbitals, for example, an antiferromagnet with $n_\uparrow \neq n_\downarrow$. So long as the total number of electrons remains the same, the change in the total energy involves only the orbital-dependent energies, Δ_{cf}, J, etc. However, there are large changes in the spectrum involving U that can lead to an insulator, whereas the DFT solution would be metallic.

The DFT+U method can also be viewed as a static approximation to GW, as pointed out in Sec. 11.4. Of course, the GWA includes dynamical many-body effects in all the bands due to the screened interaction; however, the effects are larger for more localized states such as the d and f states. It is this large effect that is approximated by the inclusion of the interaction U in the static DFT+U method. Examples of the comparison of DFT+U and DMFT are given for Ce in Fig. 20.2, and all three approaches (DFT+U, DFT+DMFT, and GW) are compared for NiO in Fig. 20.8.

Self-interaction correction

The self-interaction correction has the straightforward interpretation that it removes the unphysical interaction of an electron with itself [203]. In Hartree–Fock theory it is included in the Hartree energy and canceled exactly by the Fock term. However, the cancellation is not complete in a density functional method with a local or gradient-corrected approximation for exchange. For localized orbitals with integer occupation, one can subtract a self-term for each orbital; this was done by the Hartrees [5] in their original work on atoms in 1928 (see footnote 5 in Sec. 1.3). In a solid the identification of local orbitals is not unique (the same problem as in DFT+U and DMFT) but there is a self-consistent localization method to determine orbitals by energy minimization [914, 915]. The relation between self-interaction corrections and the DFT+U methods can be seen by rewriting Eq. (19.14) using the identity $\sum_{m \neq m'} n_m n_{m'} = \sum_{mm'} n_m n_{m'} - \sum_m n_m^2 = N^2 - \sum_m n_m^2$. The first term $\propto N^2$ includes self-interaction and the last term reduces the energy for each occupied state by subtracting a self-interaction term $\frac{1}{2} U n_m^2$ for each occupied orbital m. Of course, this does not change the results for the total energy and the eigenvalues are still given by Eq. (19.15), but it leads to the interpretation that the shifts in eigenvalues are due to the removal of the self-terms. This relationship can also be turned around: calculation of the self-interaction term is actually what is done in constrained DFT, where U is calculated as the interaction between electrons in the same orbital.

Disordered local moment method

Whereas DFT+U and related methods are restricted to a static mean-field potential with long-range order corresponding to a broken symmetry, Hubbard and Gutzwiller (see App. K) proposed approximate ways to include the effects of interaction without breaking the symmetry. One is the "alloy approximation" (also called Hubbard III) in Sec. K.2 in which the effects of interactions are approximated by an electron moving in a disordered, static array representing the other electrons. This can be viewed as a type of unrestricted approximation with a disordered ground state, even though the hamiltonian has the periodicity of the crystal. The translation invariance is restored in a sense by a second approximation in which the average Green's function for the alloy is determined by the coherent potential approximation, as explained in App. K.

This approach was used by Hubbard to develop a theory of magnetism as a function of temperature with applications to Fe and Ni [916–918]. The combination of this method with density functional theory has been proposed by various authors; a review with references to earlier work can be found in [919]. One approach, termed the "disordered local moment" method [920, 921], was used to develop computational methods to treat ferromagnetic phase transitions. All parameters are calculated using DFT and the partition function is formulated in terms of the energies of an ensemble of configurations of the local moments. Application to Fe and Ni is discussed in Sec. 20.4. This approach is analogous to DFT+DMFT, but the independent-particle CPA calculations for a static alloy are vastly simpler than the many-body calculations in DMFT.[17] This makes possible calculations for much larger systems and inclusion of intersite correlation beyond what is feasible in a DMFT calculation.

This approach has also been used for insulators where there can be an opening of a gap due to interactions, as illustrated for Hubbard's alloy calculation depicted in Fig. 3.3. An example is the band structure of MnO calculated in [584] using a disordered SIC potential (see also Sec. 13.4).

DFT+Gutzwiller approximation (DFT+G)

A generalization of Gutzwiller's work in Sec. K.1 using DFT can be done by identifying localized orbitals and taking into account the reduction in hopping for those states due to interactions. This leads to renormalized bands with reduced width and coupling to the rest of the states. These are not quasi-particles with fractional weight, but rather well-defined bands with no broadening, weight $Z = 1$, and no satellites. The method has been developed for multiband models in [922, 923], and an example of application to Ni can be found in [924].

One can also create a functional that includes the effect of reduced hopping matrix elements [925, 926]. This can be cast in terms of a modified exchange–correlation functional.

---

[17] The CPA can be viewed as a static version of DMFT, and DMFT is sometimes called "dynamical CPA" (see Sec. 16.4).

In analogy to the Kohn–Sham approach, the method is designed to reproduce the total energy; the bands are the states of an auxiliary system and need not have a physical meaning.

19.7 Wrap-up

The primary examples of materials with strong effective interactions are ones containing transition elements with d and f states, and Fig. 19.1 depicts the periodic table of these elements arranged from the most delocalized to the most localized. A compact way to describe these states is a basis of local orbitals, and the first sections of this chapter were devoted to characterizing the hamiltonian for these states in terms of independent-particle matrix elements that determine the splitting of the degenerate atomic states due to the effects of the crystalline environment. The classification of the states into e_g and t_{2g} states for sites with cubic symmetry and further splittings in lower-symmetry cases is useful even if the states are calculated in a plane-wave basis (as in the GWA calculations in Sec. 13.4), because they provide ways to describe and understand the results.

In order to apply DMFT methods to these materials, there must be a way to calculate the effective interactions and the independent-particle matrix elements that determine the bands for the d or f states and the other states in the crystal. The approach considered in this chapter is the combination of DFT and DMFT, where the independent-particle bands are assumed to be given by a Kohn–Sham calculation and the interaction is determined using a constrained DFT method. Different methods are described in Sec. 19.5, along with characteristic examples. This approach has two intrinsic difficulties: the choice of localized states is not unique and the marriage of DFT and DMFT is problematic. Because DFT and DMFT treat interactions differently, there are "double-counting" corrections that are not uniquely determined. Despite these problems, the results are expected to capture the most important effects and they are used in the examples in Ch. 20. However, a truly first-principles method requires that the entire system be treated by a unified approach, and steps in this direction are the topic of Ch. 21.

Static mean-field approximations are often very useful. Section 19.6 is devoted to the methods that are most widely used, with a focus on the ways that they bring out characteristic features on the interacting electron problem.

SELECT FURTHER READING

Anisimov, V. I., Aryasetiawan, F., and Lichtenstein, A. I., "First principles calculations of the electronic structure and spectra of strongly correlated systems: The LDA+U method," *J. Phys.: Condensed Matter* **9**, 767–808, 1997. An extensive review.

Fulde, P. *Electron Correlation in Molecules and Solids*, 3rd edn. (Springer Series in Solid-State Sciences, Book 100, 2013. Presents many theoretical methods and practical formulas.

Imada, M., Fujimori, A., and Tokura, Y., "Metal–insulator transitions," *Rev. Mod. Phys.* **70**, 1039–1263, 1998. Extensive discussion of the form of the hamiltonian and practical information on transition metal compounds.

See the companion book [1, Chs. 8, 14, and 15], for description of atomic-like orbitals and DFT+U, SIC, and other orbital-dependent methods.

See the list of reading at the end of Ch. 16, especially the review by Held [788] and the lecture notes of the Autumn Schools at Forschungszentrum Jülich.

Exercises

19.1 Derive the form of the e_g and t_{2g} symmetry d states in a cubic field given in Sec. 19.2 and draw the orbitals. Which ones point toward oxygen neighbors in a perovskite structure? In a CuO_2 plane that has a square lattice? Which states point to neighbors in an elemental f.c.c. crystal? Based on the answers to these questions, which states does one expect to form wider bands and which states, narrower bands.

19.2 Derive the relation stated in Sec. 19.3 that $U_{mm'} = U_{mm} - 2J$ for spherical geometry. Show also that for M orbitals the average interaction is $U_{av} = U_{mm} - 5(M-1)J/(2M-1)$.

19.3 Show that the maximum probability of finding N electrons on a site is peaked around the average value $\langle N \rangle$ and decreases for N much larger or smaller even if there are no interactions. Argue that the effect is accentuated in the presence of interactions.

19.4 Using the properties of a Legendre transform, show that U in Eq. (19.11) is given by $U = -\chi^{-1}$, where $\chi = \partial^2 E / \partial V_m^2 = \partial n_m / \partial V_m$ is the response to the potential V_m. In addition, give the condition for the system to be stable in terms of χ and show that the above expression leads to a repulsive $U > 0$ if the system is stable.

19.5 Derive the expression corresponding to Eq. (19.13), including the exchange J, and show that the second term in square brackets becomes

$$\frac{1}{2}UN(N-1) - \frac{1}{2}J\left[N^\uparrow(N^\uparrow - 1) + N^\downarrow(N^\downarrow - 1)\right].$$

19.6 Show that the two terms in square brackets in Eq. (19.13) cancel, as stated after that equation.

19.7 Derive the expression in Eq. (19.15) for the eigenvalue.
It may be helpful to first work out part (d), which may provide insight for answering the qualitative questions in parts (a), (b), and (c).
(a) Explain how it is that the total energy is not changed but the eigenvalues are changed.
(b) Give arguments why the eigenvalue is a derivative with respect to occupation, and the derivative of the fluctuation term in Eq. (19.15) is not included.
(c) Using the alternative expression for the correction $(1/2)U \sum_{m \neq m'} \langle (\hat{n}_m - \bar{n})(\hat{n}_{m'} - \bar{n}) \rangle$ given in the paragraph after Eq. (19.15), derive the expression given there for the eigenvalue. Give physical reasons why this form might or might not be preferred.
(d) Work out each case explicitly for the case of one electron in a "Hubbard atom" that has one orbital with $\uparrow$ and $\downarrow$ states.

20

Examples of calculations for solids with d and f states

> In sp solids, correlation effects are not immediately striking, while for the df solids,
> they are glaringly present.
>
> L. Hedin, *J. Phys.: Condens. Matter* **11**, R489–R528 (1999)

Summary

The topics of this chapter are chosen to represent significant classes of materials and phenomena, often called "strongly correlated." Lanthanides and actinides illustrate striking effects such as volume collapse, heavy fermions, and localized-to-delocalized transitions. Transition metals, such as Fe and Ni, are classic problems with both band-like and local moment behavior. Transition metal oxides exhibit a vast array of phenomena including metal–insulator transitions and high–low spin transitions. These are difficult problems involving competing interactions; the examples here and in Ch. 13 are not chosen to demonstrate successes, but rather to illustrate capabilities of different approaches.

This chapter is devoted to representative examples that bring out the range of phenomena observed in materials with strong local atomic-like interactions, and the types of properties that can be calculated with various methods. The foremost examples are elements and compounds of the series of transition elements with d and f states that are localized as depicted in Fig. 19.1. As emphasized in the previous chapter, the theory must deal with the complexities of many competing interactions: direct Coulomb, exchange, spin–orbit, crystal field splitting, hybridization with other orbitals, and other effects, all of which may be essential for understanding any particular material.

Many-body perturbation theory, especially the GW approximation and beyond, is a first-principles method that can be applied directly to ordered states at low temperature. Several examples of applications to transition metal systems are given in Ch. 13 and referred to here to provide a unified picture. Density functional theory and approximate static mean-field approximations also provide useful results and insights. However, many of the most striking phenomena can be understood only by taking into account strong correlation, including large renormalization of electronic states, satellites in excitation spectra, magnetic phase transitions, metal–insulator transitions, and other phenomena. Many phenomena can be understood only if temperature is taken into account. Dynamical mean-field theory brings this capability, but at a price because it is feasible to treat explicitly only a small subset of the degrees of freedom of the electrons. All the results presented here are done in the single-site approximation (single-site DMFA), except the two-site cluster for VO_2 in Sec. 20.6. Each case must be dealt with individually and examined critically to justify the choices made in the calculations.

These are difficult problems and the theoretical methods are under development. The results are not selected to illustrate successes; rather they are chosen to illustrate the capabilities of different approaches. The reader is referred to the wrap-up at the end of this chapter, which summarizes the conclusions with reference to the specific sections.

20.1 Kondo effect in realistic multi-orbital problems

Before we embrace the issues related to d and f states in solids, let us first consider an example that is interesting in itself and is a test bed for methods used later. The Kondo effect is due to electron interactions for impurities in metals, and the solutions for model systems described in Sec. 3.5 are paradigms for qualitative understanding of local moment behavior and other phenomena in solids. However, quantitative understanding requires methods that can treat the actual problem of an impurity in a metal. The theoretical methods described in Ch. 18, especially continuous-time quantum Monte Carlo, make it possible to solve the equations for an embedded site with multiple orbitals, and realistic interactions and embedding functions. In DMFT the embedded site is an auxiliary system that is used to construct an approximate Green's function for the lattice. However, for an actual impurity the results are a test of the other approximations made in applying the methods to a material: the choice of orbitals included in many-body calculation (Secs. 19.2–19.4); the determination of the effective interactions (Sec. 19.5 for methods using DFT); the transformation from imaginary to real frequency (Sec. 25.6); and any other approximations.

An example of a magnetic impurity in a realistic situation is presented in Sec. 2.7: scanning tunneling microscope spectra for a Co atom on a Cu surface. The spectra have the qualitative features expected for an Anderson impurity model and the Kondo effect with a characteristic scale of energy $T_K \approx 50$ K; however, a complete description must deal with the complexities of the actual problem. Calculations [927] have been carried out using a CTQMC solver (Sec. 18.6) for Co atoms on a Cu surface and in a bulk Cu host by treating the full five-orbital impurity problem with interactions and host states taken from density functional calculations following the approach outlined in Secs. 19.3–19.5. The results show that both charge and spin degrees of freedom are relevant and there is not a single T_K, but rather a range for different orbitals. The average T_K is roughly a factor of two to four larger than those extracted from experiment, a level of agreement that is good considering that the independent-particle matrix elements and interactions (of order several eV) enter in an exponent to determine T_K (of order 0.01–0.1 eV), and considering the difficulty of interpreting the experimental data.

This provides a measure of support for application of these methods to other problems involving $3d$ states (and by extension also the more localized $4f$ and $5f$ states in lanthanides and actinides). The methods provide insight into complex realistic problems and they are steps toward accurate first-principles calculations, but there must be judical recognition of the quantitative accuracy that may be expected for various properties of real systems.

20.2 Lanthanides – magnetism, volume collapse, heavy fermions, mixed valence, etc.

The $4f$ states of the lanthanide series,[1] Ce ... Yb, are the most localized states of any partially filled shell in the periodic table, as depicted in Fig. 19.1. The $4f$ wavefunctions are even more compact than the filled core-like $4s$ and $4p$ states, as illustrated for Gd in Fig. 19.2. Thus the division of the valence states into localized and delocalized (in this case $4f$ and $6s-6p-5d$, respectively) is expected to be better justified than in other elements.

Normal elements and magnetism

For most of the series the atomic $4f$ states survive almost intact in the solid and the coupling to the rest of the solid can be treated as a perturbation. For example, the normal elements are those for which the lattice constant is almost independent of the $4f$ occupation, as discussed in Sec. 2.2 and illustrated in Fig. 2.1, where the horizontal axis is the fractional filling of the f shell $n_f/14$. The large interactions within the $4f$ shell lead to nearly integral occupation n_f, e.g., $n_f \approx 7$ for Gd. Any occupation other than n_f is a high-energy excited state: electron removal energies ($n_f \to n_f - 1$) are well below the Fermi energy E_F, and addition energies ($n_f \to n_f + 1$) are well above E_F. For most elements the lowest-energy state is degenerate (obeying Hund's rule of maximum S, maximum L), and the $4f$ states act like spins with Curie–Weiss susceptibility. At low temperature the weak coupling to the more delocalized $s-p-d$ bands leads to the plethora of magnetic structures that are observed [928].

The multiplet structure for more than one electron in the $4f$ orbitals spans several eV, as illustrated in Fig. 2.9 for two electrons in the $4f$ state of Ce. For other elements the effects are even larger [929]. The higher multiplet states are usually multiple determinants that cannot be described by a static mean field, but they can be treated in the hybridization expansion in DMFT (Sec. 18.6). For example, DMFT has been used to study the pnictide compound ErAs, where the ground state of Er^{3+} is $4f^{11}$ with 16-fold degeneracy that retains its atomic-like character with small crystal field splitting, and yet has significant effect upon the Fermi surfaces [930].

Ce: volume collapse and three-peak spectra

The anomalous behavior of cerium has been known for decades [928, 933], since the work of Bridgeman in 1927 [934], who observed the $\alpha-\gamma$ "volume collapse" phase transition

---

[1] An extensive review of their properties can be found in [928].

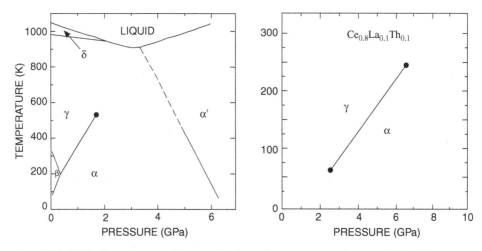

Figure 20.1. Left: phase diagram of Ce as a function of temperature and pressure [928, 931] including the "volume collapse" $\alpha-\gamma$ first-order phase transition that ends in a critical point analogous to a classical liquid–gas transition (like the water–steam transition). The other phases are described in [928]. Right: alloying with Th that expands the lattice (an effective negative pressure) and dilutes the density of Ce atoms, leading to a second critical point [932] at low temperature that can be considered as a quantum liquid–gas transition.

shown in Fig. 20.1, which is analogous to a classical liquid–gas transition with a critical point. The two phases have the same f.c.c. structure, but the volume changes by as much as 15%. The large temperature dependence is due to the entropy of f states that act like decoupled spins (local moments) at large volume with an anomalous decrease of the entropy as the volume decreases. For many years the interpretation[2] was a transition from the γ phase with one $4f$ electron and normal magnetic behavior to the α phase with reduced volume, an empty $4f$ shell and only band-like $s-p-d$ states.

However, later experiments have shown that both α and γ phases have f occupation $n_f \approx$ 1 and have revealed the strongly correlated, quantum nature of α-Ce. As shown in Fig. 2.9 and in the top panels of Fig. 20.2, the spectra for adding ($f^1 \rightarrow f^2$) or removing ($f^1 \rightarrow f^0$) electrons are split by ≈ 6 eV in both phases. In the α phase the narrow feature near the Fermi energy, together with the two high-energy peaks, form the "three-peak" structure that is a signature of correlated-electron behavior and also observed in Ce compounds, e.g., $CeIr_2$ (Fig. 2.9). The effects of correlation are brought to the fore by alloying that eliminates the hexagonal β phase and creates an effective negative pressure along with dilution of the Ce atoms. As shown on the right side of Fig. 20.1, there can be a second critical point at low temperature, where the α and γ phases merge continuously as a correlated Fermi liquid [932]. The transition occurs only at non-zero temperature and cannot be described by a $T = 0$ calculation.

Two mechanisms have been put forth to explain the dramatic volume collapse transition and other anomalous behavior of Ce. One proposal [940, 941] is that the $4f$ states form bands with width W_f and on-site interaction U_f, leading to a Mott transition since the ratio

---

[2]   Pauling [935] and also attributed to W. H. Zachariasen by Lawson and Tang [936].

U_f/W_f is expected to increase with volume. It is not actually a metal–insulator transition due to the presence of the $s-p-d$ bands; otherwise, those bands do not play an essential role. The other proposal is the "Kondo volume collapse" model [942–944] in which a key ingredient is the coupling of the f to the $s-p-d$ states. There is a large value of U_f in both phases and the hybridization with the $s-p-d$ bands leads to greatly renormalized bands, analogous to the Kondo effect.

Dynamical mean-field theory has shown the close connection between these two proposals, e.g., a Kondo-like peak at the Fermi energy that emerges in the analysis of a Mott transition, as discussed in Sec. 16.9. A goal for quantitative calculations is to determine the role of the various terms that distinguish the two models.

Thermodynamic properties

In order to calculate the thermodynamic properties, including the phase boundary terminating in a critical point, one must have a way to determine the energy and entropy. There have been many density functional calculations of the energy at $T = 0$ that find two solutions identified as the two phases, e.g., [945], but the relative stability is sensitive to the choice of functional. Thus, even at low temperature the phase boundary is not well understood from a theoretical perspective and it is not accessible experimentally in pure Ce, as shown on the left side of Fig. 20.1. Thermodynamic properties have been calculated using the same DFT+DMFT methods as for the spectra [937, 938]. As explained in Sec. 16.6, calculation of the entropy requires an integration of the energy (see Ex. 17.8) starting from a state with known entropy. In this case the choice is the high-temperature limit where the entropy is $k_B \ln N$, where N is the number of degrees of freedom, the 14 f states plus the other states included in the calculation. The calculations find that the increased entropy of the spins in the high-volume phase is qualitatively consistent with the temperature dependence of the transition.

Spectra

It is instructive to first consider what can be learned using static mean-field methods. In the lower-left panel of Fig. 20.2 is the $4f$-projected density of states found in an LDA calculation at a volume corresponding to the α phase. The inevitable result of such a calculation with no magnetic order is a partially occupied f band at the Fermi energy, and it turns out that the $4f$ occupation is ≈ 1 and the equilibrium lattice constant is in good agreement with experiment at low temperature. This is an example of the success of DFT for ground-state properties; however, such an independent-particle calculation cannot explain the spectra with dominant satellite peaks well above and below the Fermi energy. The DFT+Gutzwiller approach (Sec. 19.6) can explain the narrowing of the band but still does not describe the high-energy satellites.

On the lower right is the $4f$ density of states calculated with the LDA+U method (the same U as in the DMFA calculations described later) for a magnetically ordered state, which is the stable solution in the calculations at the volume of the γ phase. The high-energy peaks are rather well reproduced; however, such a calculation cannot also describe

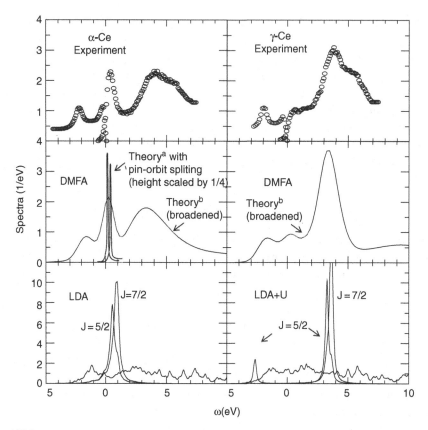

Figure 20.2. Electron removal (photoemission) and addition (inverse photoemission) spectra for α (left) and γ (right) phases of Ce. (Top panels) Experimental spectra at room temperature [68], the same as in Fig. 2.9. (Middle) Single-site DMFA calculations for the $4f$ states with parameters from LDA. Theory "b" [937, 938] is broadened by the experimental resolution and includes the $6s-6p-5d$ states. Theory "a" [939] shows the high-resolution peaks near the Fermi energy including spin–orbit splitting; note that widths are $\approx 10\times$ narrower than the LDA peaks in the bottom panel. (Bottom) The $4f$ projected density of states for static mean-field calculations: LDA for a non-magnetic solution in the α phase and unrestricted LDA+U for a magnetically ordered state in the γ phase. The peaks are $4f$ and the broad spectra are bands formed from $6s-6p-5d$ states. As explained in the text, each of these static mean-field calculations describes an aspect of the spectrum, but cannot explain the three-peak spectrum. (Figure provided by A. K. McMahan, except "theory a" modified from figure by K. Haule.)

a peak near the Fermi energy. In a nutshell, the LDA with no magnetic order and LDA+U magnetically ordered static mean-field calculations each provide useful information on the ground state and excitation spectra, and each misses crucial aspects that can only be captured by a dynamical self-energy.

There have been calculations of the spectra in the GW approximation [946], which find a quasi-particle peak at the Fermi energy in both phases with large weight and width of order 1 eV, comparable with the LDA spectra shown in the lower-left panel of Fig. 20.2. As explained above, such a peak is expected in any calculation at zero temperature with no magnetic order; the test of the method is the quantitative magnitude of the quasi-particle weight, i.e., the renormalization factor Z defined in Eq. (7.26). Since the calculations do not find a large reduction in Z, especially in the γ phase, the authors conclude that the GWA does not capture essential aspects of the spectra for systems with very strong short-range correlations (such as cerium).

Calculations using the single-site DMFA with parameters from DFT have been carried out by several groups [937–939, 947, 948]. All have an on-site $f-f$ interaction $U \approx 5$–$6\,\mathrm{eV}$ from constrained DFT calculations (Sec. 19.5), and all find satellite structures at higher energies in both phases and a "three-peak spectrum" with a narrow Kondo-like feature at the Fermi energy in the α phase. Results from [938] compared with experiment are shown in the middle panels of Fig. 20.2. The calculations were done with a QMC solver using the Hirsch–Fye algorithm at temperatures from 632 K to 1580 K; spectra were obtained by maximum entropy methods (Sec. 25.6) and broadened to compare with experiment. Also shown are sharp lines near the Fermi energy from a later work [939] that includes the spin–orbit interaction. These were done at a temperature of 150 K using a CTQMC solver with self-consistency of the density in the DFT and DMFA calculations.[3] The ground state is primarily $J = 5/2$ in both phases and the peak at the Fermi energy in the calculated α spectrum is for addition and removal of electrons with $J = 5/2$, while a higher energy is required to add an electron in the $J = 7/2$ state. Note that the LDA bands have a width ≈ 1 eV, but the width of the peaks in DMFT are greatly renormalized, a factor ≈ 10 narrower. Calculations for the volume appropriate to the γ phase find disordered spins at room temperature, in agreement with experiment; the f states are essentially decoupled from the other states and have little effect on the Fermi surface. This is the same effect as in the Kondo problem (Sec. 3.5), and leads to the disappearance of the peak at the Fermi energy in agreement with the experimental spectra shown on the right in Fig. 20.2. (See Fig. 16.4 for an example of the temperature dependence in a model calculation.) The difference in spectra for the α and γ phases is due to changes in the hybridization of the f with the $s-p-d$ states, also found in the calculations reported in the right figure of Fig. 20.4 later, supporting this aspect of the "Kondo volume collapse" model.

There is much left to do for quantitative calculations, such as the frequency dependence of the effective interaction U (see Sec. 21.5), and other effects. Nevertheless, the approach provides a unified picture of the spectra that can explain the trends in the whole class of Ce compounds, including intermetallics like $CePd_2$ and CeAl in Fig. 2.9 and heavy fermion materials with very narrow bands like those in Fig. 20.3.

Heavy fermions, Kondo insulators

Examples of materials with very narrow bands and large enhancement of the effective mass (of order 100–1000 m_e, hence the name "heavy fermions") are listed in Tab. 2.2. These are

---

[3] The local orbitals in [938] were taken from an LMTO calculation. See [939] for the orbitals used there and an extensive discussion of various methods to choose the local orbitals and optimize the locality and completeness.

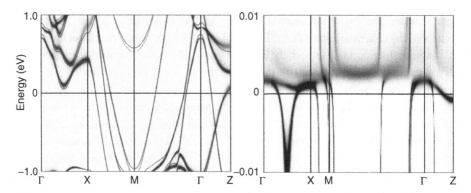

Figure 20.3. Momentum-resolved spectra for CeIrIn$_5$ calculated in the single-site approximation as described in the text [949]. Left: at 300 K where the spectra are almost the same as the light solid lines which are Kohn–Sham bands calculated without the 4f states. Right: at 10 K where the 4f states form narrow bands; note that the energy scale is 100 times smaller than in the left figure. This shows that the 4f states have negligible effect near the Fermi energy at 300 K, but at much lower temperature the spectra are greatly modified to form narrow 4f quasi-particle bands. The small bandwidth ≈ 0.005 eV ≈ 40 K signifies "heavy fermions" with mass ≈ 100 times larger than the DFT mass and of order 1000 times larger than typical s, p, and d bands. (Figures provided by H. C. Choi, adapted from [949].)

compounds where the lanthanides and actinide atoms are widely spaced so that they form narrow f bands with large interactions. Calculations for such extreme cases bring out most clearly effects that also occur in other systems where they are not so pronounced. The effects are illustrated in Fig. 16.4 for a Hubbard model: a narrow quasi-particle band at low temperature that loses intensity as T is increased until there is no longer a peak.

An example is the **k**-resolved spectra at different temperatures for CeIrIn$_5$ shown in Fig. 20.3, which were calculated using the single-site DMFA with parameters from DFT [949]. In this case the auxiliary system is solved approximately by a diagrammatic technique, the "one-crossing approximation" that can be used at low temperature and provides spectra for real frequencies (see Sec. 18.8). At 300 K the spectra (the slightly broadened bands in the left figure) have dispersion that is almost identical to the lines in the figure, which are DFT bands calculated with the f states excluded. This shows that the f states act like local moments and are essentially decoupled and have almost no effect on the Fermi surface at this temperature. The result is a "small Fermi surface" with volume that does not count the f electrons. On the right is shown the spectra at 10 K, somewhat below the characteristic energy that is set by the width of the bands of order 0.005 eV ≈ 40 K. Note that the vertical axis is the energy on a scale 100 times smaller than the left figure. The figure shows 4f quasi-particle bands but even at this temperature they are not very sharp, since 10 K is still a large fraction of the bandwidth. The effect of correlation is to reduce the bandwidth (increase the mass) by a factor ≈ 100 compared with the (already narrow) f bands found in a DFT calculation, and a mass of order 1000 times the free electron mass.

The essential feature is that only below some characteristic energy scale are there well-defined quasi-particles with enhanced mass and a Fermi surface that includes the f electrons, i.e., a "large Fermi surface." This is an instructive example of the Luttinger theorem that is satisfied at temperature low enough for the Fermi surface to be well-defined. In that case it encloses a volume that counts all the electrons. However, the theorem need not be satisfied at non-zero temperature; in this case there are strong deviations at the characteristic temperature ≈ 40 K and the f states act only as disordered spins with little effect on the Fermi surface at 300 K.

A Kondo insulator (see, e.g., [950]) is a material with renormalized bands like heavy fermions but with electronic states and electron counting that lead to an insulator with a small gap. An early example is SmB_6 [951], where the $4f$ states are hybridized with the Sm $5d$ bands to mix states with occupations $4f^6$ and $4f^5$. Taking into account the lowest-energy multiplet states, the symmetry is the same as in a band picture [952]. SmB_6 has been studied using DFT+U [953] and DFT+Gutzwiller methods, and it has been identified as a possible topological insulator [954, 955] (see Sec. 6.3).

Yb, Sm, and Eu: mixed valence

Another class of anomalous lanthanides is indicated by the changes in volume per atom in Fig. 2.1 at the middle and end of the series. This is the signature that the number of $4f$ electrons does not follow the trend of the other elements. In these elements and compounds involving Sm, Eu, and Yb there is a near degeneracy of the 2^+ and 3^+ ionic states, an example of mixed valence [876] discussed in Sec. 19.1. In the spectrum this is signified by either the removal or addition peak close to the Fermi energy. In the case of Yb metal, the nominal valence is 2^+, leading to a full $4f$ shell; however, the trivalent 3^+ state (the normal valence of a lanthanide) with one hole in the $4f$ shell is only slightly higher in energy. Under pressure n_f gradually changes from 14 to ≈ 13.5 at 20 GPa, and is estimated to be fully trivalent with $n_f = 13$ at ≈ 100 GPa. In many ways Yb is complementary to Ce: the $4f$ states in Yb are more localized (the most localized in the periodic table, as indicated in Fig. 19.1) and are more decoupled from the other bands. A consequence is that under pressure the main effect is a continuous transfer of electrons from $4f$ to band states, and Yb behaves as a mixed valence material with a smooth variation in n_f.

This example illustrates the power of a method like DMFT that can calculate both average values and fluctuations in a local quantity, e.g., $n_f = < \hat{n}_f >$ and the fluctuations $< (\delta \hat{n}_f)^2 > = < (\hat{n}_f - < \hat{n}_f >)^2 >$. The left side of Fig. 20.4 shows the calculated decrease in n_f under pressure [913], in good agreement with resonant inelastic X-ray scattering (RIXS) and X-ray absorption spectroscopy (XAS). The same calculations find $< (\delta \hat{n}_f)^2 >$ plotted on the right. An integral value would have no fluctuations and an uncorrelated band would have large fluctuations that increase monotonically in this range. (The difference in fluctuations between non-interacting and interacting cases is illustrated by $< \delta \hat{n}^2 >$ for each site in the two-site Hubbard dimer in Sec. 3.2 and Ex. 3.7.) The non-monotonic behavior in Yb indicates a mixed valence state that is a linear combination of states with

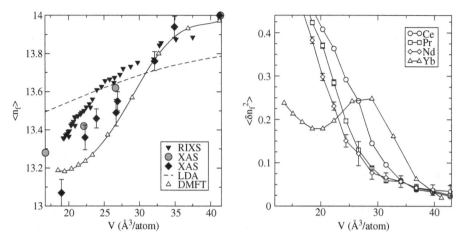

Figure 20.4. Calculated occupation $n_f = <\hat{n}_f>$ and fluctuations $< (\delta\hat{n}_f)^2 >$ for f states in Yb [913]. Left: n_f as a function of volume V, where $n_f \approx 14$ at zero pressure and $n_f \to 13$ at high pressure (reduced volume). The theoretical curves for several temperatures are compared with RIXS and two XAS experiments (see text). Right: calculated $< (\delta\hat{n}_f)^2 >$, which would be very small for a stable valence, where $n_f \approx$ integer. As explained in the text, the form of the increase with pressure is enough to distinguish possibilities: mixed valence with two well-defined states of integral occupation and fluctuations due to hybridization with another band. Both are very different from the fluctuations that would occur if the f bands were uncorrelated (see text). (From [913].)

$n_f = 14$ or 13. In contrast, the early lanthanides Ce, Pr, and Nd have small but increasing fluctuations due to increasing hybridization with the band states, consistent with the calculations for Ce described in the first parts of this section.

20.3 Actinides – transition from band to localized

Actinide elements exhibit a plethora of interesting phenomena due to the $5f$ states.[4] One of the most dramatic transitions in the periodic table is the changeover from band to localized behavior in the actinide series. The juncture occurs in the sequence plutonium (Pu), americium (Am), and curium (Cm), as indicated by the variation in the volume per atom in Fig. 2.1. Plutonium has a very complex phase diagram, with many structures and large changes in volume increasing by $\approx 25\%$ from α to δ phases (even more than Ce). It is particularly appropriate to contrast Pu and Cm, for which there is a simplifying feature that makes possible a direct comparison. Since Am has a non-degenerate $5f$ shell (total angular momentum $J = 0$), the f shell in Cm can be considered to be formed by adding one f electron, and in Pu by adding a hole. In each case the large spin–orbit interactions in these heavy atoms lead to $J = 3 \pm 1/2$ states; however, the consequences are very different.

---

[4] A review of their electronic properties can be found in [874].

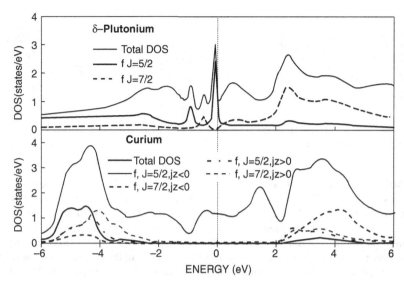

Figure 20.5. Calculated addition and removal spectra showing the change from correlated non-magnetic behavior in plutonium (Pu) to localized magnetic character in curium (Cm) [956]. The former has the three-peak structure analogous to correlated non-magnetic α-Ce in Fig. 20.2, whereas the latter has a half-filled shell with $5f$ electrons strongly bound below the Fermi energy and empty states above. This means a fixed integral number of $5f$ electrons with a magnetic moment that is stable to very high temperature far above the Curie temperature. (Figure provided by J. Z. Shim; similar to figure in [956].)

Figure 20.5 depicts the contrasting one-electron spectra of the two elements calculated using the single-site DMFA [956]. In this case all $14\,f$ states were included in the calculation, with $U = 4.5\,\text{eV}$ and the exchange and other Slater integrals taken from atomic values. For Cm there is essentially no f weight at the Fermi energy in the spectrum; the states are highly localized with a nearly integer number of electrons in the f states because a large energy is required to add or remove f electrons. However, δ-Pu has a spectrum like α-Ce, with a sharp feature at the Fermi energy and broad sidebands showing the highly correlated nature of the electronic states. In the low-volume α phase the $5f$ states are much more delocalized and band-like.

20.4 Transition metals – local moments and ferromagnetism: Fe and Ni

The $3d$ transition metals are quintessential examples of materials that simultaneously have properties that can be explained by itinerant independent-particle bands whereas other properties are characteristic of local moments [54, 121]. As indicated in Fig. 19.1 the series progresses from superconductors at the beginning of the series, itinerant band-like magnetic Cr, to ferromagnetic metals that exhibit local moment behavior (Mn, Fe, Co, and Ni) at the end of the series. The latter elements (Fe, Co, Ni), have ground-state magnetization, specific heat much larger than corresponding metals like Cu, and well-defined Fermi

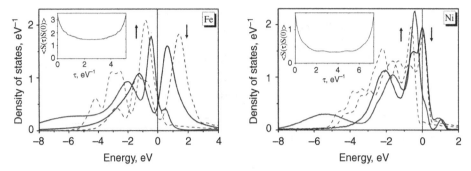

Figure 20.6. The d partial densities of states for $\uparrow$ and $\downarrow$ spins in ferromagnetic Fe (left) and Ni (right) for single-site DMFA (full lines) compared with LSDA results (dashed lines) [55]. Note the narrowing of the bands in both cases and the satellite at $\approx 6\,\text{eV}$ for Ni in the DMFA calculations. The insets show the decay of the spin autocorrelation function in imaginary time, which shows that the moments are more localized in Fe (see Ex. 20.1). The spectra were calculated at temperatures $T = 0.8T_c$ for Fe and $0.9T_c$ for Ni; the correlation functions are for the paramagnetic state at $1.2T_c$ and $1.8T_c$, respectively. (Figures provided by A. Lichtenstein. Similar to Figs. 1 and 2 in [55].)

surfaces at low temperature, which are all explained well by partially filled $3d$ bands mixed with a broad $4s$ band. The $3d$ metals are quite different from the lanthanide and actinide elemental metals, since the $3d$ bands are responsible for the bonding and are much broader than the f bands. The effective interactions are correspondingly weaker, and we can expect that the effects of correlation are smaller.

The bands calculated for ferromagnetic Ni in the GW approximation are shown in Fig. 13.11, where the comparison with the Kohn–Sham bands shows that there is a renormalization of order 30%. As explained in Sec. 13.4, the results are in good agreement with angle-resolved photoemission. There are, however, additional features that require one to go beyond the GWA, for example, a satellite at ≈ 6 eV in the spectrum for Ni as shown in the upper panel of Fig. 15.6. As discussed in Sec. 15.6, the satellite is due to interaction between holes in the filled d bands and it is found in methods that sum diagrams beyond the GWA, e.g., the T-matrix calculations [555] that yield the spectra shown in the lower part of Fig. 15.6. That work calculated the bands within the GWA and included added T-matrix diagrams for the d states using a static effective interaction $U = 5.5$ eV. The calculated satellite is at ≈ 10 eV, which is too high, and may be due to the fact that the value of U is larger than found in other calculations.

As is clear from the magnetic properties described in Sec. 2.3, the local moments survive above the transition temperature in the disordered state. Several DMFT calculations [55, 957–960] have been carried out using the band structure and interactions taken from DFT. There is general agreement on the major points and the results can be illustrated by the results of [55], found in the single-site approximation with a QMC solver and interactions $U = 2.3$ eV for Fe and 3.0 eV for Ni, with $J = 0.9$ eV for both. The resulting spectra in Fig. 20.6 show the key features that the bands are narrowed relative to the Kohn–Sham bands in the local spin density approximation, similar to what is found in the GWA, and for Ni there is a satellite in the majority spin spectrum at ≈ 6 eV, below the energies of the d bands.

The DMFA results for the scaled magnetization as a function of temperature for $T < T_c$ and the susceptibility for $T > T_c$ are given in Fig. 2.3. The local moments above T_c can be determined quantitatively by calculation of the local spin correlation function $<s(\tau)s(0)>$ that is shown in the inserts in Fig. 20.6. The root-mean-square magnitudes found in [55] are $\mu_{\text{eff}} = 3.09(1.50)$ close to experimental values $3.13(1.62)$ for Fe(Ni). The magnetization in the ordered state shown in Fig. 2.3 is given in terms of scaled variables (T/T_c) and magnetization $(M(T)/M(0))$; the form of the temperature dependence agrees with experiment; the calculated T_c is close to experiment for Ni but too large for Fe by almost a factor of 2.

Static mean-field approximations

There are various ways that static mean-field methods can be used to treat magnetism at finite temperature with much less computational effort than DMFT. For example, the disordered local moment method (see Sec. 19.6) replaces the thermal disorder by a static array of disordered orientations of the local spin density. The disorder is treated using the CPA approximation, i.e., Hubbard's "alloy approximation" (Sec. K.2), but formulated in terms of density functional theory with no free parameters. An example is [961], where the calculated T_c is close to the actual transition temperature in Fe but too low for Ni. There are also other approaches such as the equation of motion methods for dynamics of spins (similar to molecular dynamics) [962], where it was concluded that short-range intersite correlations must be included to describe the magnetic properties.

The phase diagram of Fe

The phase diagram of Fe is important for technological applications and for geophysics at temperatures and pressures at the center of the earth. As an illustration of the complexity, the crystal structure of Fe is b.c.c. at ordinary temperature;[5] however, in the paramagnetic phase at 1185 K it transforms to f.c.c., and finally back to b.c.c. This has been studied using the single-site DMFA [959] to find the total energy including a magnetic contribution in the paramagnetic phase that is proportional to the mean-square fluctuation of the local moments (see Ex. 20.2). This is an example of the type of work that can be done using methods like DMFT to determine the local moment fluctuations. The thermodynamics, including lattice vibrations, has been calculated using density functional theory and the DLM method. For example, in [963] it is proposed that a combination of increasing lattice contributions and decreasing effects of correlation (taken from the DMFT calculations) can explain the magnetic transition and the sequence of structure transformations as temperature increases.

---

[5] Density functional calculations using the LDA actually predict that the stable structure of Fe at $T = 0$ is f.c.c. and it is non-magnetic; the correct b.c.c. structure is found using improved GGA functionals (see [1]).

20.5 Transition metal oxides: overview

Transition metal oxides and related compounds are the consummate examples of materials in which strong local correlation in the $3d$ states creates the conditions for many of the most fascinating phenomena in condensed matter physics.[6] Because of the intervening oxygen atoms, the transition metal atoms are more widely spaced and the $3d$ states are more localized than in the elemental transition metals. The interactions between the electrons in the d states is larger due to the decreased screening in the insulating compounds. The consequence is that the d states are intermediate between localized and band-like extended limits, with both types of behavior often coexisting in the same material. Moreover, much of the interest lies in the fact that many of these phenomena can be controlled by small perturbations, temperature, and external influences. A number of instructive examples of compounds in ordered states are included in Sec. 13.4.

These materials are often classified as "Mott insulators" (Sec. 3.2) and we will follow the prescription that this denotes a material with a gap caused by interactions that does not depend on the order, i.e., an insulating state that persists in the disordered phase above the transition temperature even though a static mean-field calculation with no disorder would find a metal.[7] Perhaps the original example is NiO, discussed in Secs. 20.7 and 13.4. The simplest picture is a metal–insulator transition caused by competition between the interaction energy U and the kinetic energy characterized by the bandwidth W, e.g., in the Hubbard model in Sec. 3.2. However, almost none of the actual oxides are like that: the problem involves 10 d states (counting spin) with various interactions, U, J, etc., multiple widths of the different bands, crystal field splittings Δ_{cf},[8] coupling to the oxygen states, structure distortions, etc. Each of the examples in the following sections illustrates ways in which calculations can identify the most relevant factors and, at least potentially, provide satisfying explanations. However, the interesting effects are due to delicate balances of competing factors, and each example also illustrates the fact that it is difficult to make detailed quantitative predictions.

---

[6] There is an enormous literature on the oxides that can be found in sources such as the review by Imada *et al.* [91] and earlier reviews by Goodenough [964] and Brandow [965]. There is a full issue of *Rev. Mod. Phys.* devoted to the 1953 Washington Conference on Magnetism [116] with articles by Slater, Zener, Van Vleck, and many others, and to the 1968 conference on metal–insulator transitions with articles by Mott and many others.

[7] It is important to keep in mind that at non-zero temperature the conductivity is never strictly zero and a gap is not precisely defined. Thus these are qualitative concepts and one should always resort to what is actually measured, e.g., the temperature dependence of the conductivity that is different for metals and insulators, or whether or not there is a significant depression of the spectral weight at the Fermi energy. As pointed out in Sec. 3.2, there are rigorous distinctions at $T = 0$; one must take care in applying $T \neq 0$ concepts to $T = 0$ situations, and vice versa.

[8] As emphasized in Sec. 19.2, the term "crystal field splitting" is commonly used in two ways: the difference in the on-site matrix elements $h_{mm} - h_{m'm'}$ for states m and m', and the splitting of the density of states in the crystal including all contributions. Here Δ_{cf} denotes the latter. Individual references must be checked for the finer details of the individual terms.

20.6 Vanadium compounds and metal–insulator transitions

Vanadium readily forms compounds in two valence states: materials such as $SrVO_3$, $CaVO_3$, and VO_2 with formal ionization state V^{4+} and one electron per V atom in d bands, and others like V_2O_3 with V^{3+} and two electrons in the d bands. The former evoke the picture of one electron per atom in partially filled bands and competition of the intra-atomic interaction U with the bandwidth W and crystal field splitting Δ_{cf}. The latter brings to the fore the additional feature of the intra-atomic exchange J that favors parallel spin for two electrons in d states to form a local moment with spin 1. The competition of J with the independent-particle electron terms W and Δ_{cf} can lead to transitions between states with "high spin" (where intra-atomic interactions dominate) and "low spin" (where independent-particle matrix elements dominate).

In many ways $SrVO_3$, $CaVO_3$, and alloys are the simplest examples of all the transition metal oxides: one electron in a d band well-separated from the filled oxygen bands and no structural complications in perovskite structure (simple cubic with each V surrounded by an octahedron of six oxygen atoms, shown, e.g., in Fig. 4.8 in [1]). Because of the simplicity of the structure and the interpretation when there is only one electron in d states, $SrVO_3$ is used as an example of Wannier functions shown in Fig. 19.7 and dynamic effective interactions in Fig. 21.6.

VO_2 – local versus non-local correlation

As described in Sec. 13.4, there is a first-order transition in VO_2 at $T_c = 340$ K from a metallic rutile phase to an insulating monoclinic structure with V atoms paired to form tilted dimers along the c axis. There is general agreement of GWA and DMFA calculations, as illustrated in Fig. 13.16, which shows the DMFA angle-resolved spectra alongside a selected spectrum from the GWA calculations.

The studies of VO_2 in [587] were carried out using the cellular method (c-DMFA in Ch. 17), with a cluster of two sites. Several values of U were considered and the spectra shown in Fig. 13.16 were found using LDA bands and $U = 4.0$ (a typical value found in constrained DFT calculations) and $J = 0.68$ eV, which is not screened and is similar to the value J in other oxides.[9] It was found that the metal–insulator transition could not be described in a single-site DMFA unless one uses an unphysically large value of U. The results indicate that a V–V dimer is much like the Heitler–London picture of a two-electron bond (Sec. 6.4) and the Hubbard dimer (Sec. 3.2) in the regime where the repulsion significantly reduces the double occupancy. The effect is larger than in the symmetric state, leading to the conclusion that correlation plays an important role in stabilizing the dimer.[10]

---

[9] For our purposes, it is important that essentially the same values are used for V_2O_3 in the following subsection since we want to focus on the difference between VO_2 and V_2O_3.

[10] Note that this means stabilization relative to a restricted Hartree–Fock where there is no correlation. It does not necessarily mean increased stabilization compared with a DFT calculation that includes the correlation energy in an approximate way depending on the choice of the exchange–correlation functional.

Together the GWA and DMFA calculations make a rather convincing case that the explanation of the observed phenomena in VO_2 requires both intersite correlation and crystal field splitting due to the dimerization, and it is not an example of a transition in the simplest form envisioned by Mott.

V_2O_3 – high vs. low spin and the metal–insulator transition

A classic example of a metal–insulator transition is V_2O_3. As described in Sec. 2.8, pure V_2O_3 is an antiferromagnetic insulator with a transition to a paramagnetic metal as temperature or pressure is increased (the right side of Fig. 2.14 for pure V_2O_3 and alloyed with Ti). By alloying with Cr with concentration $\approx 1\%$ there emerges a first-order transition in the paramagnetic phase shown in Fig. 2.14. It is called a metal–insulator transition since there is a large jump in conductivity. The changes in electronic states at the transition can be observed directly by photoemission, as shown in Fig. 2.15 [93]. The figure shows the progression as temperature increases from a gap in the ordered antiferromagnetic insulator, no gap in the metallic phase, and a gap that opens at the metal–insulator transition that is smaller than in the AFI phase. Of course, the difference between the phases decreases with temperature, and above the critical point the conductivity and the spectra vary continuously.

There are similarities with VO_2, but there is a decisive difference. In V_2O_3 there are two d electrons per vanadium atom, which leads to competition between the intra-atomic exchange interactions J that favor a triplet spin-1 state (called "high spin") formed by two electrons on each V atom versus a singlet state ("low spin") that would occur for non-interacting electrons in bonding states. The low-spin picture was the starting point for a model of the metal–insulator transition in which one electron from each V forms a bond in each V pair, leaving one electron in band states. With the addition of an on-site interaction U, this becomes a Hubbard model with a half-filled band proposed to have a Mott transition [967]. The high-spin picture was put forward much later [96], and is supported by the moment inferred from high-temperature susceptibility and by X-ray absorption measurements [95].

Several single-site DMFA [966, 968, 969] and LDA+U [96] calculations have been reported, with agreement on major conclusions: the metal–insulator transition is caused by opening a gap between the crystal field levels with occupied states derived primarily from e_g and smaller occupation of a_{1g} states, as shown in Fig. 20.7. All three phases involve primarily spin-1, which supports the "high-spin" picture and argues against the model of a Mott transition in a half-filled band. The metal–insulator transition is found to occur by the change in overlap of the e_g and a_{1g} bands; however, this is not a simple band overlap that could occur in an independent-particle problem. As shown in the bottom panel of Fig. 20.7, the insulating gap occurs because the interaction splits the e_g band into lower and upper parts with a gap, i.e., a Mott insulator. The small weight of the a_{1g} states is due to hybridization with the e_g bands. In the metallic state shown at the top of Fig. 20.7, the e_g spectrum still has the same basic features and the major change is that there is overlap with a band that has primarily a_{1g} character, i.e., there are two very different types of bands at the Fermi energy.

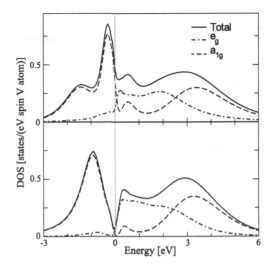

Figure 20.7. Calculated spectra for V_2O_3 for the paramagnetic insulator with a small gap (bottom) and metal with no gap (top). In the insulator the dominant features are described by two electrons in the e_g states that are correlated to form a high-spin local moment with a gap to the higher states. The upper figure is a metal because of overlap with the a_{1g} bands, in agreement with experimental results for the occupation of the a_{1g} states (see text). (Figure provided by Poteryaev; similar to figure in [966].)

The calculations are not at the point where they can pinpoint the transition: variation of the effective interaction U by ± 1 eV can shift the transition drastically [968], and the double-counting terms (see Sec. 19.5) introduce uncertainties. In addition, issues such as the effect of alloying with Cr in the samples have not been addressed. Nevertheless, there are several conclusions that do not depend on a detailed description. The values of the parameters $U = 4.2$ eV and $J = 0.7$ eV used for the results in Fig. 20.7 are essentially the same as U and J used for VO_2. Thus the very different character of the metal–insulator transitions in VO_2 and V_2O_3 emerges naturally without adjustment of parameters.

20.7 NiO – charge-transfer insulator, antiferromagnetism, and doping

NiO has long been recognized as a material with properties dominated by electron interactions; it is often called a Mott insulator and was singled out in the compilation of anomalous materials by de Boer and Verwey [580] in 1937. It is an antiferromagnet with Néel transition temperature $T_N = 525$ K, and a magnetic moment of almost $2\mu_B$, the expected spin moment for d^8 Ni^{++}. The gap is ≈ 4 eV for the d states and the local magnetic moments persist in the disordered state well above T_N. The lattice has the f.c.c. NaCl crystal structure, with only small displacements of the atoms in the antiferromagnetic phase which has a doubled primitive cell.

The electronic states have been studied experimentally using photoemission spectroscopy (PES) and inverse photoemission (IPES, also called BIS, see Sec. 2.4). As shown in Fig. 20.8, the inverse photoemission spectrum for addition of electrons has a single

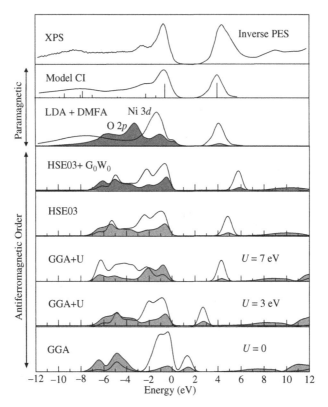

Figure 20.8. Spectra for addition and removal of electrons in NiO. Top: experimental spectra at room temperature in the antiferromagnetic phase from high 1.5 keV X-ray photoemission (from [970], used with permission) and inverse photoemission (inverse PES) [971]. Similar results are found in the paramagnetic phase for $T > T_N$ [972]. The spectra are dominated by the Ni $3d$, showing the gap, the main peak near the Fermi level, and a satellite at ≈ -8 eV. Theoretical spectra as indicated: exact diagonalization (CI) for a parameterized model of a NiO$_6$ cluster [973]; LDA + single-site DMFA [974] in the disordered state for $T \approx 1100$ K $> T_N$, showing the mixed O $2p$ and Ni $3d$ states near the Fermi energy and satellite structure. Calculations of the O $2p$ and Ni $3d$ spectra for $T = 0$ with AF order are shown in the lower panels. The quasi-particle spectra for G_0W_0 [581] starting from the HSE03 hybrid functional results (shown just below) are the angle-integrated spectra corresponding to the **k**-dependent bands in Fig. 13.12. The GGA+U density of states from [581] show the severe disagreement with experiment for GGA ($U = 0$) and improvement in some aspects for $U = 3$ eV and 7 eV, but no value that describes the full spectrum.

dominant peak interpreted as empty d states, whereas photoemission spectra probe the occupied O $2p$ and Ni $3d$ states that are strongly mixed. These are the signatures of a charge transfer compound, illustrated in Fig. 19.6. The PES spectrum shown [970] was taken with high-energy X-rays (XPS), where the cross-section favors the Ni d states. Other work [971, 975] has found similar results and there is little change at the antiferromagnetic transition [972]. Thus, NiO satisfies the criteria for a Mott insulator given in Sec. 20.5: the gap in the spectrum does not depend on the AF order, since it is similar in the ordered and the disordered phase above the transition temperature.

The PES spectrum is interpreted as consisting of two types of excitations: mixed $d–p$ states that form quasi-particle bands near the Fermi energy, and a broad high-energy satellite (also called a sideband or lower Hubbard band) below the Fermi level at ≈ -11 to -8 eV with Ni d character.[11] Such a splitting of the d spectral weight is the signature of correlation that cannot be represented by an independent-particle theory. This interpretation was supported in early work [973] that used exact diagonalization for a small cluster of Ni in an octahedron of six O atoms and a parameterized model fitted to the spectrum [973], with the results shown by the vertical lines in the second-from-the-top panel in Fig. 20.8. The results of various other theoretical methods are presented in Fig. 20.8; all are broadened so that details of the spectra are lost, but the main features are apparent.

There have been many calculations with static mean-field methods for the ordered state, with representative examples shown in Figs. 13.12 and 20.8. Kohn–Sham calculations with traditional functionals find NiO to be antiferromagnetic with magnetization close to experiment (ground-state properties), but the gap is much too small and there are well-separated Ni bands above the O bands, in disagreement with experiment, as shown in the left panel of Fig. 13.12 and the bottom of Fig. 20.8 for a GGA functional. The situation is improved in GGA+U calculations [581], as shown in Fig. 20.8: a small $U = 3$ eV can explain the mixed O p–Ni d bands near the Fermi energy and a large $U = 7$ eV gives a gap comparable to experiment, but no value of U is satisfactory.

Calculations for the antiferromagnetic phase at $T = 0$ have also been done with hybrid functionals and GW methods, with the resulting bands shown in Fig. 13.12. As explained in Sec. 13.4, the G_0W_0 results depend on the starting point and are quite different for the GGA and HSE03 hybrid functionals. The latter is close to the quasi-particle self-consistent GW results, which supports the conclusion that a good choice is G_0W_0 starting from the HSE03 functional. The densities of states for the HSE03 and the corresponding GWA quasi-particle spectra are shown in Fig. 20.8. This is an example that illustrates the applicability of GWA methods to materials that are often called called "Mott insulators." However, such approaches do not capture the satellite peak at ≈ -10 eV and they are limited to an ordered broken-symmetry state. If these methods are applied to the paramagnetic phase ignoring the effect of temperature and disorder, they must lead to a metallic solution. (This is not guaranteed just by electron counting, since there is an even number of electrons per cell, but it can be shown due to band degeneracies as shown in Ex. 20.3. See also Ex. 20.4.)

DMFT provides complementary information and can be used to study the disordered state at high temperature, $T > T_N$, as well as the ordered state. Calculations combining DFT and the single-site DMFA have been carried out by several groups [978–981], with results similar to those from [978, 979] shown in Fig. 20.8. In that work the effective interactions $U = 8$ eV and $J = 1$ eV were determined by constrained DFT [902] (see Sec. 19.5), the calculations were done using a QMC solver (Ch. 18), and the self-energy at real frequencies was determined by maximum entropy methods (Sec. 25.6). The results shown in the figure for the paramagnetic phase at a temperature of $T = 1160$ K reveal

---

[11] There are also angle-resolved photoemission experiments [976, 977] that have found a narrow band at the top of the filled states. Narrow bands are also found in DMFA calculations that are not broadened [978, 979].

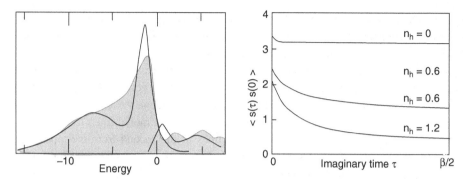

Figure 20.9. Spectra and spin autocorrelation function for doped NiO. Left: experimental spectra [982] for $Li_{0.4}Ni_{0.6}O$ shown by the shaded areas and calculated spectra from [974] (line). Comparison with the large gap for NiO in Fig. 20.8 shows the collapse of the gap with doping, interpreted as the signature of a doped Mott insulator. Right: autocorrelation function of the z-component of the spin [974] in units of μ_B^2, showing the mean-square moment $\langle s(0)^2 \rangle$ and decay as a function of time τ for different doping levels, where $n_h = x/1 - x$ is the number of holes per Ni (see text). The increase in the decay rate with doping indicates delocalization and low-energy excitations, as explained in Ex. 20.1. (Adapted from figures in [974].)

both the quasi-particle bands near the Fermi energy and a satellite at ≈ -8 eV. The gap of ≈ 4 eV is between the mixed $d-p$ occupied states and the empty d states, in agreement with experiment.

The coexistence of band-like and localized character is also revealed by the magnetic behavior. The average ordered moment per site can be found by $T = 0$ calculations. Typical values are $\approx 1 \mu_B$ in the LDA that increases to 1.6 in LDA+U [356] and 1.7 in GW [292] compared with the experimental value of $1.9 \mu_B$. At high $T > T_N$, thermally fluctuating local moments can be determined by correlation functions $\langle s(\tau)s(0) \rangle$; results from the single-site DMFA calculations [978, 979] are shown in the top curve on the right side of Fig. 20.9, where $n_h = 0$ denotes pure, undoped NiO. The local moment $\bar{s} = \sqrt{\langle s(0)^2 \rangle} \approx 1.85 \mu_B$ is essentially the same in the paramagnetic phase as in the ordered phase at $T = 0$. For $\tau \neq 0$, $\langle s(\tau)s(0) \rangle$ provides a measure of the decay time of a fluctuation and slow decay[12] indicates a well-defined localized moment, as explained in Ex. 20.1.

Doping with holes

The consequences of doping a Mott insulator are among the most striking signatures of strong local interactions, as described in Sec. 2.8. In an independent-particle theory, doping an insulator merely shifts the Fermi energy, partially filling the band with only minor effects on the gap and band dispersion. However, strong on-site interactions can lead to transfer of spectral weight [983] that adds to or subtracts from the total weight of a band. Consider doping with holes, the case of interest in oxides such as NiO and CuO_2 planes in

---

[12] The calculations are for $T = 1160$ K ($\beta = 10$ eV^{-1}) so that the decay to zero at very long times cannot be seen.

the high-temperature superconductors. The basic picture is that the gap to the empty d band of NiO shown in Fig. 20.8 is caused by interactions: if an electron is missing, the conduction band can hold one less state and the valence band, one more state. This is not strictly correct since the counting depends on the way the boundaries of the bands are defined; nevertheless, there are large shifts of empty states downward by amounts of order the gap.

Photoemission experiments [982] have been done on $Li_xNi_{1-x}O$, where holes are created by substitution of monovalent Li for divalent Ni. As x increases the gap closes, as shown on the left side of Fig. 20.9 for $n_h = 0.4$, where n_h is the number of holes per Ni atom, $n_h = x/(1-x)$. The proposal that the gap should collapse has been addressed by the DMFA calculation in [979] (see also [974]), keeping the same hamiltonian but reducing the number of electrons per cell. Whereas doping also causes disorder that may complicate the interpretation of the experiments, there is no disorder in the theory and the collapse of the gap occurs as an intrinsic property of fractional occupation. The figure shows that with doping, the local moment per Ni atom decreases and decays more rapidly in the presence of the metallic carriers.

20.8 MnO – metal–insulator and spin transitions

The phase diagrams of transition metal oxides as a function of pressure and temperature are examples of metal–insulator and high-to-low spin ("magnetic–collapse") transitions that are of special interest because of the geophysical implications at high pressures and temperatures in the earth; see, e.g., [986, 987]. An example is MnO, which is an antiferromagnetic insulator with a Néel temperature T_N that is 116 K at zero pressure and increases under pressure. Like NiO, MnO has the B1 (NaCl) structure with a small distortion below T_N. Experiments on MnO under pressure at room temperature [984] have found transitions at the points indicated in Fig. 20.10: at 30 GPa a transition to an AF ordered insulator, at 90 GPa a transition to a paramagnetic insulator with the B8 (NiAs) structure, and at 105 GPa a first-order phase transition where the volume decreases by 6.6% and there is a large increase in conductivity accompanied by a decrease in spin moment. The full phase line has not been measured but it must end in a critical point, as depicted in the figure, if there is no change in symmetry,

In MnO the d states are separated from the oxygen p bands so that (unlike NiO) the problem can be cast in terms of the five bands (not counting spin) that are primarily derived from d states. As described in Sec. 19.2, crystal fields split the d states into threefold degenerate t_{2g} and twofold e_g symmetries with energy difference Δ_{cf}. The diagram at the top of Fig. 20.11 illustrates the competition between crystal field splitting Δ_{cf} that favors occupation of lowest-energy spin-paired states and the exchange J that favors high spin $S = 5/2$. On the left is shown the case if the crystal field splitting Δ_{cf} is small compared with J and the five electrons in the d shell form a high-spin state and an antiferromagnetic insulator as observed in MnO at ordinary pressure. This is borne out by DFT and GW calculations, as described in Sec. 13.4.

There are various possibilities for the spin and metal–insulator transitions. One is the competition between the direct interaction U and the bandwidth as envisioned by Mott,

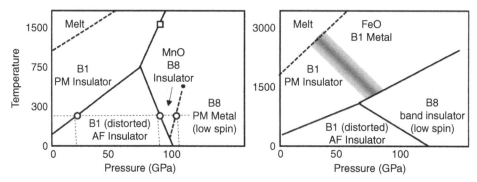

Figure 20.10. Experimental phase diagrams for MnO (left) and FeO (right) for pressures and temperatures relevant to the mantle of the earth. B1 and B8 denote the NaCl and NiAs structures. Only some points on the diagrams are actually measured and the lines indicate the qualitative features of the expected phase boundaries. The light dashed lines for MnO depict room-temperature measurements [984] that find the three transitions indicated by circles in the figure. The square indicates the result of shock measurements. FeO has a metal–insulator transition in the B1 phase that occurs only at high temperature, as discussed in the text. Based on figures in [984] for MnO and [985] for FeO.

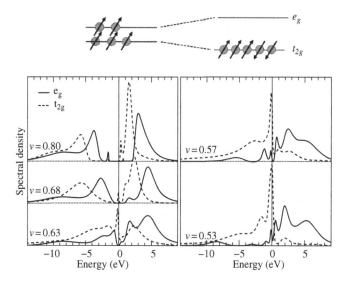

Figure 20.11. Spectra calculated using the single-site DMFA for MnO under pressure, indicated by the reduced volume denoted v, expressed as a fraction of the zero-pressure volume. The sequence shows the evolution of the d states from the high-spin state favored by the atomic exchange J (shown at the top left) to the low-spin state where the electrons are primarily in the lowest crystal field split state, the triply degenerate t_{2g}. (Modified from [988].)

with a relatively minor role played by the complication of various interactions (U and J) and multiple bands. In contrast, the competition of J and Δ_{cf} may be the dominant effect. Under pressure Δ_{cf} is expected to increase, which would lead to the situation shown on the right side of the diagram at the top of Fig. 20.11: partially filled metallic bands in a low-spin $S = 1/2$ configuration. Such a picture is similar to the two-band model for a spin transition in Sec. 16.11. Of course, this is oversimplified and the actual problem involves competition of interactions (exchange and direct Coulomb) and independent-particle terms (crystal field splitting and bandwidths) that must be taken into account for a complete picture.

Calculations have been reported using DFT, DFT+U, and related methods [986, 987, 989], and using the single-site DMFA [988]. All find the fully polarized d^5 state (high spin $S = 5/2$) at low pressure and a transition to a low-spin state with $S = 1/2$ or a metallic state with no identifiable local moments at high pressure. An example is the single-site DMFA calculation[13] of [988], where the independent-particle bands are taken from DFT (which includes effects of the oxygen states, bandwidths, and crystal fields) and interactions $U = 6.9\,\text{eV}$ and $J = 0.86\,\text{eV}$. The results for the density of states with t_{2g} and e_g symmetry are presented in Fig. 20.11 for several volumes. The calculated spectra agree well with experiment at ordinary pressure [990], and the sequence shows a change from a fully polarized d^5 insulator at large volume, with both e_g and t_{2g} states occupied, to a metallic state with mainly t_{2g} occupation at low volume. Note that the calculations find the metallic state emerging as a peak in the gap, as expected from the DMFA results for model systems in Ch. 16. Although there is much left to do for such systems with many competing mechanisms, this is strong evidence that, in addition to the interaction U and bandwidth W, the exchange interaction J and crystal field splitting Δ_{cf} play crucial roles in the metal–insulator and magnetic–collapse transitions in materials such as MnO.

It is instructive to compare FeO and MnO. Each occurs in the B1 (NaCl) structure at low pressure and the crystal field splitting into e_g and t_{2g} states in FeO is the same as that for MnO at the top of Fig. 20.11. However, FeO has one more electron; whereas MnO is an insulator at low pressure and a metal at high pressure, FeO has the opposite tendency in this (over)simplified picture (see Ex. 20.4). In particular, the low-spin state of Fe in FeO is $S = 0$ and the six d electrons are sufficient to fill the t_{2g} states. If the crystal field splitting Δ_{cf} is larger than the bandwidths, there would be filled t_{2g} states and a gap to empty e_g states. It is found experimentally [985] that FeO undergoes a metal–insulator transition at high temperature and pressure, which is either first order or a rapid continuous change, as shown by the shaded lines in Fig. 20.10. DMFA calculations reported in [985] find that at high pressure and low temperature there is a gap with mainly t_{2g} states filled, in qualitative agreement with the crystal field picture. At high temperature the spectrum changes drastically to that of a correlated metal with spectra attributed to fluctuations between high-spin and low-spin states.

20.9 Wrap-up

Elemental solids and compounds containing transition series elements with partially filled d and f shells provide a vast array of phenomena due to interactions. Characteristic examples of experimental observations are brought out in Ch. 2, and the topics of

---

[13] Although the metal–insulator transition occurs in the B8 structure, the local environment of Mn atoms is very similar to B1 and the difference is ignored in the DMFA calculation.

this chapter are examples of calculations and quantitative comparison with experiments. Of course, the calculations involve approximations and the examples are not meant to show successes; instead they illustrate the type of properties that can be calculated with different methods, and they point to the need for further developments in the future.

We have focused on spectra that probe directly effects of interactions that lead to renormalized quasi-particle bands near the Fermi energy and satellites (sidebands) at high energy (typically of order several eVs) above and below the Fermi energy. The effects are intimately associated with temperature dependence: the quasi-particle bands are well-defined only for temperatures below some characteristic scale; at higher temperatures the atomic-like open-shell d and f states act like local moments. The examples illustrate that the characteristic energy scales vary drastically for different materials, which is itself a key factor showing the remarkable effects of interactions.

- Many of the most pronounced effects occur in the lanthanides and actinides, which are used as examples in Secs. 20.2 and 20.3, including volume collapse in Ce and Yb, and band-to-localized transition in the actinides, magnetic phase transitions, crossover from incoherent scattering to coherent bands, etc., which occur with characteristic temperatures varying from > 1000 K to very low temperature (for example, bands that are well-defined only below ≈ 40 K in CeIrIn$_5$) in heavy fermion systems and Kondo insulators. Comparison of DFT, DFT+U, and DMFT methods is illustrated, e.g., in Fig. 20.2 for Ce.
- Ferromagnetic materials like Ni and Fe have bandwidths and temperature scales that are larger than in the lanthanides and actinides. They illustrate coexisting band and local moment character as described in Sec. 20.4, and the band renormalization in the ordered state is described well by the GWA (Sec. 13.4).
- Transition metal oxides exhibit a wide range of phenomena that arise from the competition of single-particle bandwidths and crystal field splitting with direct and exchange interactions. All the examples involve insulating states where correlation plays an essential role, but none are explained by the simplest picture of bandwidth versus interaction.
- The vanadium oxides illustrate metal–insulator transitions with different character: VO$_2$ with one d electron per V has a transition accompanied by a dimerization. This is a case where GWA and two-site cluster DMFA calculations agree (see Sec. 13.4, especially Fig. 13.16). In V$_2$O$_3$ there are two d electrons per V that appear to form "high-spin" states with the metal–insulator transition caused by crystal field splitting of the resulting correlated states (see Sec. 20.6).
- NiO is the classic example of an antiferromagnetic insulator with local moments and insulating character that remains above the transition temperature. Comparison of different methods is given in Secs. 20.7 and 13.4, especially Fig. 20.8, showing that the GWA can describe the ordered state and the single-site DMFA with parameters from DFT explains the high-temperate behavior.

• Transitions at high pressure and temperature important for geophysics are illustrated by MnO and FeO (Sec. 20.8), where there are coincident insulator-to-metal and high-to-low spin transitions and the dominant effects are found to be the competition of exchange and crystal field splitting.

The examples in this chapter show that many-body perturbation methods are very useful for ordered states at low temperature, e.g., the GWA can be quantitatively accurate if the renormalization effects are not too large. Dynamical mean-field theory is designed to deal with local moment and effects of temperature, but it is limited by the size of the correlated embedded system that can be treated. The examples in this chapter use the single-site DMFA approximation with parameters from DFT. The next chapter is devoted to possible avenues for future work to provide truly first-principles methods capable of dealing with these challenging materials and phenomena.

SELECT FURTHER READING

Imada, M., Fujimori, A., and Tokura, Y., "Metal–insulator transitions," *Rev. Mod. Phys.* **70**, 1039–1263, 1998. Extensive references to experimental data and interpretation.

Kuneš, J., Leonov, I., Kollar, M., Byczuk, K., Anisimov, V. I., and Vollhardt, D., "Dynamical mean-field approach to materials with strong electronic correlations," *Eur. Phys. J. Special Topics* **180**, 5–28, 2010. Review with many applications.

See the list of reading at the end of Ch. 16, especially the reviews by Held and Kotliar *et al.* and the lecture notes of the Autumn Schools at Forschungszentrum Jülich.

Exercises

20.1 Give qualitative arguments why a spin autocorrelation function $\langle s(\tau)s(0)\rangle$ (like that shown in Fig. 20.6) should vary less with τ for more localized spins, as stated in the caption to Fig. 20.6. What would be expected for this correlation function in a metal like Na? For an isolated atomic spin? Rapid decay means strong coupling to low-energy excitations. For example, see Fig. 20.9 which shows $\langle s(\tau)s(0)\rangle$ where there is coupling to delocalized states. Check that this is consistent with the expected properties of a thermal Green's function in Sec. D.1 and illustrated in Fig. D.2. See also the inset for the decay of $G(\tau)$ in Fig. 16.5 and the related text.

20.2 Derive the result that the magnetic contribution to the energy can be written as $\frac{1}{4}I < m_z^2 >$ where I is the Stoner parameter (see Ex. 4.4).

20.3 Show that in a static mean-field approximation NiO must be a metal if there is no magnetic order, as stated in Sec. 20.7. There are 14 electrons in the 10 d plus 6 p states per cell. A crystal with an even number of electrons per site can be an insulator unless there is some other condition on the band structure. Show that in this case symmetry considerations lead to the conclusion that NiO cannot be an insulator in a static mean-field approximation. Give arguments that this also applies to the GW calculations.

20.4 Although the transition metal oxides considered here do not have a stable ferromagnetic phase, it is instructive to consider what would be the possibilities for such a state. Show that in a static mean-field approximation MnO and NiO can be insulators whereas CoO or FeO must be metals in the ferromagnetic state. *Hint*: the proofs depend on the assumption that the crystals are cubic, i.e., ignoring the very small distortion in the ferromagnetic state. The demonstration for NiO does not follow from symmetry and it requires also knowledge of the ordering of the t_{2g} and e_g states.

21

Combining Green's functions approaches: an outlook

Have no fear of perfection; you'll never reach it.

Marie Curie

Summary

This chapter is devoted to the combination of many-body perturbation and dynamical mean-field approaches to build more powerful first-principles methods. The GWA and corresponding approximations to the Bethe–Salpeter equation are very successful for one-particle and optical spectra of materials, especially for delocalized states. Dynamical mean-field methods are a set of techniques using an auxiliary embedded system to treat interacting electrons in localized atomic-like states that lead to strong correlation and large temperature dependence due to low-energy excitations. The combination holds promise for feasible approaches to calculate spectra, correlation functions, thermodynamic properties, and many other properties for a wide range of materials.

In this chapter we return to the goal of robust first-principles methods that can treat materials and phenomena, such as the examples of Ch. 2: from covalent semiconductors like silicon to lanthanides with partially filled f shells like cerium; from optical properties in ionic materials like LiF to metal–insulator transitions in transition metal compounds, and much more. A natural approach is Green's function methods; however, it is very difficult to have one method that treats all cases. In fact, this is not the goal. We seek understanding as well as results. We seek combinations of methods that are each firmly rooted in fundamental theory, with the flexibility to take advantage of different capabilities and provide robust, illuminating predictions for the properties of materials.

This is an active area with on-going developments; the topics of this chapter are not meant to be exhaustive, but rather to formulate basic principles and indicators of possible paths for future work. The approach developed here can be used with any method that partitions a system into parts, treats the parts using techniques that may be different for the different parts, and then constructs the solution for the entire system. The formulation in terms of Green's functions, self-energies, and screened interactions is already given in

Ch. 7; here we put this together with the practical methods developed in Chs. 9–20. If the procedures are judiciously chosen, the results have the potential to be more efficient, more accurate, and more instructive than a comparable calculation with a single method.

The body of work presented in the previous chapters suggests that a promising approach is a combination of two types of methods:

- Many-body perturbation theory, developed in Chs. 9–15, is a successful framework for first-principles calculations. In particular, it has been established that the GW and Bethe–Salpeter methods are powerful approaches for describing electrons in solids, especially those with delocalized states. However, as emphasized in Ch. 15, the improvement of the GWA over Hartree–Fock is only a frequency-dependent screening that describes a mean-field response of the system, which is not sufficient for localized electrons with a degenerate or quasi-degenerate ground state. One approach to go beyond the GWA is to include other diagrams only for a subspace of orbitals; an example is the satellite in the photoemission spectrum for Ni described in Sec. 13.4 and the calculations including T-matrix diagrams only for the Ni d states, as described in Sec. 15.6. It is very difficult to construct practical ways to treat thermal fluctuations in phenomena such as local moment behavior and disordered phases above the transition temperature within MBPT.
- Dynamical mean-field theory in Chs. 16–20 can capture essential features of systems with atomic-like states, strong interactions, and local moments, and it can treat disordered states at high temperature. However, in order to become a first-principles, quantitative theory of materials, DMFT must be combined with another theoretical approach. This is essential to determine the effective interactions starting from the long-range Coulomb interaction, and to treat more delocalized states that are present in realistic problems. We must also devise ways to reduce the dependence upon the choice of orbitals in order to make the method more robust and reliable. For accurate methods it is moreover important to go beyond the single-site mean-field approximation and treat intersite correlation.

The combination of MBPT and DMFT methods is an example of a general approach to use auxiliary systems in Green's function methods. This is pointed out at the beginning of Ch. 16; however, Chs. 16–20 were devoted only to the development of DMFT. In this chapter we return to the formulation of unified Green's function methods. As we shall see, the combination of GW and DMFT methods to calculate the one-electron self-energy Σ is a straightforward, transparent generalization of the equations in the previous chapters that dealt with the combination of DMFT and DFT (even though it is not easy to actually do the calculations). However, the way interactions are included is qualitatively different. The combination of DMFT and DFT in the previous two chapters is done in two separate steps: calculation of the parameters in a rather *ad hoc* manner and DMFT calculations using the parameters. In the combination of Green's function methods, DMFT is a part of the overall method in which the self-energy Σ and frequency-dependent screened interaction W are determined consistently for the entire system. Furthermore, there are interactions between electrons on different sites so that the combined method intrinsically goes beyond the single-site approximation with a local Coulomb interaction. Extended DMFT, described in Sec. 17.5, is contained as part of such an approach. Original works that contain many of the ideas for combining GW and DMFT are [810] and [991], and instructive overviews can be found in the lecture notes and review papers listed at the end of this chapter.

Section 21.1 outlines the main ideas, referring to Ch. 7, expressed in a way that is useful in this chapter. In Sec. 21.2 we recall the arguments for partitioning the system into localized atomic-like d or f states and the rest, and we formulate the coupled calculation for a crystal. Section 21.3 describes MBPT methods in which higher-order diagrams are included only for selected states. Section 21.4 outlines an algorithm for "GW+DMFT," and discusses how non-local correlations are taken into account. The next two sections are concerned with analysis of the individual ingredients of the method to understand what is contained in a calculation: the nature of a screened interaction (Sec. 21.5) and effects due to the frequency dependence in the effective interactions (Sec. 21.6). In Sec. 21.7 are short sketches of approaches to extend the DMFT idea from the self-energy and polarizability to the four-point interaction vertex (see Sec. 8.3) and the "dual fermion" approach. Finally, in Sec. 21.8 are conclusions on the overall approach.

21.1 Taking advantage of different Green's function methods

In the following sections we will come to combinations of methods; however, it is enlightening to pose the issues in a general way that brings out the salient points more clearly without the complications that arise in any particular approach. The basic ideas are outlined in Sec. 7.7, where the essential step is to divide the self-energy into two contributions. Here we suppose that they are treated by different methods, $\Sigma(\omega) = \Sigma^1(\omega) + \Sigma^2(\omega)$. For the present purposes it is useful to rewrite the expressions as $\Sigma(\omega) = \Sigma^1(\omega) + \Delta\Sigma(\omega)$, where it is understood that method 1 is applied to the entire system and $\Delta\Sigma(\omega)$ contains the contributions beyond those included in $\Sigma^1(\omega)$. These contributions are only evaluated for a subspace, e.g., $\Delta\Sigma$ restricted to the d orbitals in a transition metal element or compound. Thus the coupled Dyson equations in Eq. (7.41) can be written as

$$G^1(\omega) = G^0(\omega) + G^0(\omega)\Sigma^1(\omega)G^1(\omega) \tag{21.1}$$

$$G(\omega) = G^1(\omega) + G^1(\omega)\Delta\Sigma(\omega)G(\omega), \tag{21.2}$$

where Eq. (21.2) is depicted in Fig. 21.1. Similarly, the screened interaction $W(\omega)$ can be cast in terms of two contributions to the polarizability $P(\omega) = P^1(\omega) + \Delta P(\omega)$, which can be expressed as

$$W^1(\omega) = v_c + v_c P^1(\omega)W^1(\omega) \tag{21.3}$$

$$W(\omega) = W^1(\omega) + W^1(\omega)\Delta P(\omega)W(\omega), \tag{21.4}$$

where v_c is the bare frequency-independent Coulomb interaction. As discussed in Sec. 7.7, these relations can be solved sequentially if Σ^1 is independent of $\Delta\Sigma$, which means that Σ^1 does not depend on the *full* G (and similarly for P). Otherwise, one can devise an iterative method to treat the case where there is feedback and the first equation depends on the solution of the second.

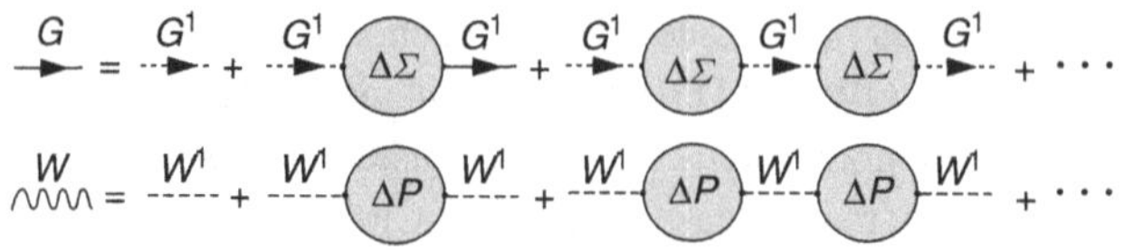

Figure 21.1. Dyson equations for the Green's function $G(\omega)$ and screened interaction $W(\omega)$ analogous to Fig. 7.4 expressed in terms of $\Sigma^1 + \Delta\Sigma$ and $P^1 + \Delta P$, where the entire system is treated with method 1 as discussed in the text. The equations can be used with $\Delta\Sigma$ and ΔP directly or with the steps of subtracting a part of Σ^1 and P^1 and replacing it with the corresponding contributions evaluated by method 2.

In the problems considered here, the system treated by method 2 is intimately coupled to the rest of the system. In this case it is important to distinguish two aspects of the calculations. The first is the division into contributions; this is an issue that must be addressed in each case. The second aspect is the calculation of the self-energies and polarizabilities, which is the difficult part of an interacting many-body problem. In general, calculation of $\Delta\Sigma$ and ΔP requires information from the entire system so that it is not a separate calculation, but instead is a useful way to distinguish parts of the entire calculation. An example is screening of interactions that is critical to all the methods described in Chs. 9–20. In GW, the diagrammatic expansion at any order involves the screened interaction W, and the whole system contributes to the screening of the bare interaction in the self-energy. In DMFT, the bare interaction is replaced by an effective interaction that is screened by the rest of the system. Much of the remainder of this chapter is devoted to identifying approximations that simplify the problem and yet capture the most important ingredients.

Even the definition of $\Delta\Sigma$ and ΔP is not always obvious. In some cases method 2 adds well-defined contributions to method 1, such as added diagrams. In other cases $\Delta\Sigma$ and ΔP may represent a method in which part of Σ^1 is subtracted[1] and replaced by a calculation using a method 2 that cannot easily be compared with method 1. An example is the use of an approximate density functional in method 1 along with a Green's function approach in method 2. This may lead to problems like double counting (Sec. 19.5) that is not uniquely defined.

It is instructive to note that some methods calculate the Green's function and other properties directly without needing a self-energy. An example is the auxiliary system in DMFT where the Green's function $\mathcal{G}(\omega)$ and other correlation functions are calculated directly by exact diagonalization or Monte Carlo methods (see Ch. 18). In order to use such methods in combination with the methods of MBPT, the self-energy can be determined from the Green's function using the relation $\Sigma_{aux}(\omega) = \mathcal{G}_0^{-1}(\omega) - \mathcal{G}^{-1}(\omega)$ as explained following

---

[1] Subtracting components of the self-energy and using different approximations has the possibility to create unphysical results, such as self-energies and Green's functions with negative spectral weights that would violate causality. The issues are discussed in Secs. 17.3 and 17.4, and a similar problem is pointed out for nested clusters where terms in the self-energy are subtracted.

Eq. (16.10). The corresponding analysis for the screened interaction W utilizes the fact that it can be expressed in two ways (see Sec. 7.2),

$$W(\omega) = v_c + v_c P(\omega) W(\omega) = v_c + v_c \chi(\omega) v_c, \qquad (21.5)$$

where $P(\omega)$ is the irreducible polarizability and $\chi(\omega)$ is the full reducible polarizability $\chi = P(1 - v_c P)^{-1}$. As explained following Eq. (21.10), $\chi(\omega)$ is a correlation function that can be calculated directly in the auxiliary system, which then determines the polarizability $P = \chi(1 + v_c \chi)^{-1}$ needed in the combination with MBPT methods.

21.2 Partitioning the system

When is it meaningful to distinguish different parts of a system? Roughly speaking, there must be one or more subspaces that exhibit aspects that are not present, or not important, in the rest of the system. Here we focus on the distinction between localized and delocalized states based on properties described in Chs. 2, 19, and 20. The striking difference in the localization of d and f states compared with typical s and p valence states is shown in Fig. 19.2 for Mn and Gd atoms, and the trends in degree of localization among the transition elements are summarized in Fig. 19.1. Examples of choices of orbitals for calculations in solids are given in Sec. 19.4. Examples of phenomena involving local moments, temperature dependence, and strong renormalization effects are given in Chs. 2 and 20. In this chapter we will consider a crystal with localized atomic-like d or f states on each site and other states that are more delocalized. We will refer to the localized states as the subsystem labeled "d" with the understanding that this can denote states that have atomic-like d or f character. The other states are referred to as the "rest" or "r."

As emphasized in Sec. 19.4, the separation into different orbitals is problematic since the division is not unique. However, the difficulty is different from that encountered in combining DMFT with DFT, where it is an uncontrolled approximation due to the intrinsic difference in the density-based DFT and the Green's function-based DMFT. In a combination of Green's function methods it should always be possible to identify all the terms and account for them, at least in principle.

In this chapter we wish to apply a combination of DMFT and MBPT (or different levels of perturbation theory) to the d states and the rest (r), respectively. There are still choices to be made. If we apply different methods to the d and r states, we have to specify the way that the dr cross terms are treated. This can be done in various ways and it is convenient to adopt the choice made in Eqs. (21.1) and (21.2), where the entire system (rr, dd, and dr) is treated with one method (typically GW) and calculations using method 2 (typically DMFT) are restricted to only the d space.[2] Then $\Sigma(\omega) = \Sigma^1(\omega) + \Delta\Sigma(\omega)$, where "1" denotes GW and $\Delta\Sigma(\omega)$ is the difference between the DMFT calculation and the part of the GW that involves only the d states, $\Sigma_{dd} \equiv \Sigma_d$ and $P_{dddd} \equiv P_d$. Other ways to deal with the dr terms may be useful, but they lead to more cumbersome equations.

---

[2] This is especially useful for the interaction that is a four-index matrix, e.g., $U_{mm'n'n}$ in Eq. (19.5), where $mm'n'n$ can denote any combination of d and r orbitals.

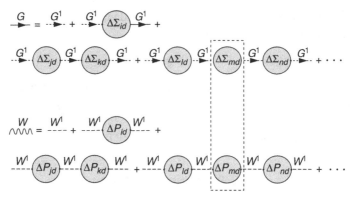

Figure 21.2. Dyson equations for the Green's function $G(\omega)$ and the screened interaction $W(\omega)$, which depicts the generalization of Fig. 21.1 to the case where there is a d system on every site in a crystal. The Green's function G^1 and the screened interaction W^1 couple the sites $i,j,k,\ldots$ but each $\Delta\Sigma_{id}$ and ΔP_{id} is localized on a site. The dashed box indicates a typical site m that is embedded in the rest of the crystal, including the d states on all the other sites. The upper lines for G and Σ are equivalent to the formulation of DMFT in Chs. 16 and 17, with G_0 replaced by G^1. The lower lines for W and P are analogous.

The analysis so far involves two contributions to Σ or P. However, a crystal has localized states on every site and we are led to define a set of systems labeled (id), each of which denotes only those localized d states on site i.[3] Even though $\Delta\Sigma_{id}$ and P_{id} are restricted to site i, they depend on what is happening at other sites. In the language of functionals this can be expressed as $\Delta\Sigma_{id}[G, W]$ and $P_{id}[G, W]$, where G and W are the full non-local Green's function and the interaction W. For example, the Luttinger–Ward form for $\Delta\Sigma_{id}[G]$ is a sum of diagrams that involve elements of G that all couple sites in all possible combinations. In general, one must make approximations, and the approach in this chapter is to restrict the functional dependence of $\Delta\Sigma_{id}[G_{id}, W_{id}]$ and $\Delta P_{id}[G_{id}, W_{id}]$ to only on-site terms, which is an example of the use of the nearsightedness principle [515] discussed in Secs. 16.2, 12.7 and in [1, Ch. 23]. There are still effects of the other sites and of the states outside the d space through the fact that G_{id} is derived from G^1 as in Eq. (21.2).

The expansion of the Green's function G and the screened interaction W are depicted in Fig. 21.2, where G^1 and W^1 connect all the sites in the crystal $i,j,k,l,\ldots$, and there is a $\Delta\Sigma_{id}$ and ΔP_{id} on each site i. The dashed box in the figure indicates a typical site that can be considered to be embedded in the rest of the system, the rest of the states r and the d states on other sites. Finally, in a crystal every site is equivalent so that $\Delta\Sigma_{id} = \Delta\Sigma_{0d} \equiv \Delta\Sigma_d$, and similarly for ΔP_d. Thus the added contribution of the d states is determined by $\Delta\Sigma_d(\omega)$ and $\Delta P_d(\omega)$, which are matrices involving the set of d states for one atom, e.g., the 10 atomic-like d states or 14 f states on a site.

---

[3] The site can also be a cell containing more than one atom with d states. The arguments are readily generalized to multiple sites where there are coupled sets of equations for the different sites.

This analysis is sufficient for the combination of GW (method 1) and a DMFT calculation of $\Delta\Sigma_d$ and ΔP_d using a single-site auxiliary system as described in Sec. 21.4.
However, this does *not* mean that the sites are uncorrelated as in the single-site approximation in previous chapters. The screened interaction W^1 includes interactions between
d states on different sites and leads to a form as in extended DMFT described in
Sec. 17.5. Similarly, the Green's function on one site is affected by the polarizabilities
and self-energies of d states on other sites.

Working in real and k-space

The natural division of the problem is for d to be represented in a localized basis labeled
by a site index i, but for a periodic system it is often most convenient if method 1 is applied
to the entire system in terms of delocalized Bloch states labeled by crystal momentum $\mathbf{k}$.
For an operator $\hat{O}$, the two representations are related by the transformation

$$O_{\mathbf{k}}(\omega) = \sum_{j} e^{i\mathbf{k}\cdot(\mathbf{R}_i - \mathbf{R}_j)} O_{ij}(\omega), \tag{21.6}$$

where O_{ij} and $O_{\mathbf{k}}$ are matrices expressed in a basis of $N_d + N_r$ Wannier functions on a site,
or equivalently in a basis of $N_d + N_r$ Bloch functions that stem from the d and r states, and
O_{ij} depends only on the distance $\mathbf{R}_i - \mathbf{R}_j$. Since part of the calculation is done only in the
space of the d orbitals on a site, the projection of the delocalized states onto the d orbitals
centered on a site i is needed. Using the same conventions as in Eq. (16.4), we can define

$$O_d(\omega) \equiv O_{ii,d}(\omega) = \frac{1}{\Omega_{BZ}} \int d\mathbf{k} O_{\mathbf{k},d}(\omega) \rightarrow \frac{1}{N_{\mathbf{k}}} \sum_{\mathbf{k}} O_{\mathbf{k},d}(\omega), \tag{21.7}$$

where we have used the fact that $O_{ii}(\omega)$ is the same for all sites i in a crystal, Ω_{BZ} is the
volume of the Brillouin zone, and $O_{\mathbf{k},d}$ is the projection of the entire set of states onto the
d orbitals. These relations hold for the hamiltonian, the Green's function G, and the self-
energy Σ, which are two-index matrices. Analogous expressions can also be formulated
for the interaction W and the polarizability P, which are four-index matrices (see, e.g.,
Eqs. (19.5) and (21.10) for U and W expressed in term of localized orbitals $mm'n'n$).

21.3 Combining different levels of diagrammatic approaches

The developments up to this point are valuable relations of the self-energies and Green's
functions expressed as matrices in a basis, with different methods applied to parts of the
system. However, they do not show how to actually implement the equations. Diagrammatic methods provide a practical way to put the ideas into practice. Since the methods are
expressed in terms of selected sets of diagrams for Σ and the polarization P, it is straightforward to devise a scheme in which additional diagrams are added only for the d space
(even though it is often not easy to carry out the calculations). This has the great advantage
that the differences $\Delta\Sigma$ and ΔP are well-defined, and can be calculated directly in terms
of the added diagrams.

To be clear, let us write Eqs. (21.1) and (21.2) in more detail, by introducing a basis with elements labeled $\ell_1, \ell_2, \ldots$ This can be any basis, not necessarily localized orbitals. Within the elements of this basis we choose a subspace; the index denoting elements within this subspace is $d_1, d_2 \ldots$ Hence, for example, $G_{\ell_1 \ell_2}$ denotes any matrix element of the Green's function, whereas $G_{d_1 d_2}$ is a matrix element within the subspace. Now we start from the observation that the full self-energy is a functional of all elements of G, i.e., it can be expressed by a sum of diagrams with lines that represent G's. It is also convenient to consider it to be a functional of the screened interaction W as in Sec. 8.3, so that it can be expressed as $\Sigma([G, W]; \omega)$. Thus, Eqs. (21.1) and (21.2) can be written as

$$G^1_{\ell_1 \ell_2}(\omega) = G^0_{\ell_1 \ell_2}(\omega) + G^0_{\ell_1 \ell_3}(\omega) \Sigma^1_{\ell_3 \ell_4}([G, W]; \omega) G^1_{\ell_4 \ell_2}(\omega) \qquad (21.8)$$

$$G_{\ell_1 \ell_2}(\omega) = G^1_{\ell_1 \ell_2}(\omega) + G^1_{\ell_1 d_3}(\omega) \Delta \Sigma_{d_3 d_4}([G, W]; \omega) G_{d_4 \ell_2}(\omega), \qquad (21.9)$$

where repeated indices are summed over. Even though $\Delta \Sigma_{d_3 d_4}([G, W]; \omega)$ is restricted to matrix elements in the d subspace, in general it is still a functional of the full G and W. The same holds for $\Delta P[G, W]$. This is evident, for example, because the effective interaction between electrons in d states is screened by the rest of the system.

Bubbles and ladders: GW+T-matrix

Let us suppose that a system is well described by the GWA except for some states where particle–particle scattering is important. Nickel with its localized and partially filled d states is an example. The GW+T-matrix calculation of [555], which is discussed in Sec. 15.6, can be used to illustrate the use of Eqs. (21.8) and (21.9).

The formula and diagrams of the GWA and the T-matrix are compared in Sec. 10.5; more details on the GWA are given in Ch. 11, and more details on the T-matrix in Sec. 15.4. We can start by defining Σ^1 to be the GWA self-energy for the whole system. $\Delta \Sigma$ instead should contain T-matrix diagrams built with Green's functions in the subspace of d electrons. This would be the whole story if the two sets of diagrams were clearly distinct, but they are not. As pointed out in Sec. 10.5, to first and second order in the bare interaction v_c, the GWA diagrams are equal to some of the T-matrix diagrams. Therefore, simply taking $\Delta \Sigma$ to be a T-matrix self-energy would lead to double counting. The strong point of diagrams is that one can unambiguously identify the terms that are in common in the two approximations, and exclude them from $\Delta \Sigma$. Such a distinction would not be possible if Σ^1 were, e.g., a density functional Kohn–Sham potential. In the calculation of [555], the double-counting diagrams have been subtracted.[4]

The T-matrix diagrams as given in Sec. 10.5 are unscreened, or, in pictorial language, no bubbles are inserted in the ladder legs. For example, the diagram shown in Fig. 21.3 is missing: all electrons should screen the scattering processes that happen in the d space. The approximation to neglect this screening is not reasonable in a metal. In [555] a model

---

[4] To be precise, only the short-range contributions to the double-counting terms have been subtracted in [555], since otherwise negative spectral functions are obtained. As discussed in Sec. 15.8, there are no approximate functionals that fulfill all exact constraints. In order to obtain positive spectral functions beyond the GWA, compromises have to be made.

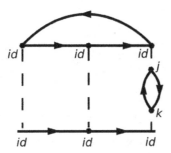

Figure 21.3. A T-matrix diagram for the self-energy that is of fourth order in the bare Coulomb interaction. The diagram is the same as Fig. 15.3, but restricted to the d space. Only the right ladder leg is screened by states that can also be outside this space. Such bubble diagrams have to be included to all orders. This is important in a solid, where s and p electrons give a strong contribution to screening.

screening of the ladder legs has been used. One may view this as an approximation to a screened T-matrix approach such as discussed in Sec. 15.5.

Excitons in a subspace

As pointed out in Sec. 14.2, similar to Eqs. (21.3) and (21.4) also the Bethe–Salpeter equation for the electron–hole correlation function can be split into two contributions, Eqs. (14.8) and (14.9). Equivalently, one can write the two contributions in inverse order, which reads schematically $L^{RPA} = L_0 + L_0 v_c L^{RPA}$ and $L = L^{RPA} + L^{RPA} \Xi_{xc} L$. Here L^{RPA} is the RPA solution, and Ξ_{xc} adds an exchange–correlation contribution to the response, in particular excitonic effects. One can write the equations in a basis of transitions follow- ing Sec. 14.6, and define a subspace of transitions between states where excitonic effects beyond the RPA are expected to be important. As discussed at the end of Sec. 14.9, these are typically transitions between localized states or between parallel bands. In close anal- ogy with Eqs. (21.9) and (21.2), one may then evaluate the second equation in the relevant subspace only. When one is interested in absorption spectra, and when local field effects can be neglected, the RPA in the first equation yields an independent-particle background that is modified by excitonic effects in a limited range of frequency by the solution of the second equation. Otherwise, one has to work harder; this is discussed in Ex. 21.7.

21.4 Combining Green's function methods: GW and DMFT

At this point we return to the goal described in the introduction to this chapter: combining many-body perturbation theory and dynamical mean-field theory to describe systems with localized d and f states. Here we describe the overall approach, referring to previous places where the relevant equations are given, and present the derivation in the cases where they have not already been derived. In an actual calculation one must use a feasible perturbation method; we will consider GW+DMFT as the example for the Green's functions and screened interactions described in the following sections. The explicit formulation is given

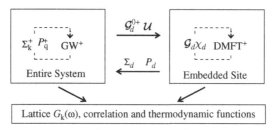

Figure 21.4. Diagram for a combined GW and DMFT calculation that is given in more detail in the box on p. 563. In the notation of Fig. 21.2, method 1 is GW$^+$, which denotes a GW calculation modified only by the replacement of the d polarizability P_d^1 and the self-energy Σ_d^1 by P_d and Σ_d from the DMFT calculation. DMFT$^+$ denotes a calculation as detailed in the flowchart in Fig. 16.3, except that the susceptibility χ is calculated and the inputs are determined by the GW$^+$ calculation: the dynamic interaction function $\mathcal{U}(\omega)$ and the effective $\mathcal{G}_d^{0+}$, which plays the same role as $\mathcal{G}^0$, as discussed in the text and detailed in the box on p. 563. The dashed lines indicate that one can choose to have a self-consistent loop inside the individual parts in addition to the overall self-consistency. The needed relations for $\mathcal{U}$, χ, P_d, etc. are derived in the text.

for GW+DMFT, where it is clear that the two contributions to the self-energy Σ and the polarizability P are distinct and there is no need for double-counting corrections. The approach is more general and one can go beyond the explicit diagrams included in the GWA; but one must be careful how to separate the contributions and check for any ways that the same effects are counted twice.

It is illuminating to first consider the general approach, which can be used in other problems. If we consider Fig. 21.4 without specifying the symbols, the sense of the method is clear: the problem is divided into two parts and the two boxes each represent entirely separate calculations. Each box has no knowledge about the other, except for the information that is passed. Even though the final solution depends on both parts and the result of each calculation is affected by the results of the other calculation, nevertheless, the calculation procedures are completely independent. Each box can be thought of as a functional that has a function (or multiple functions) as its input and a set of results that are functions and scalar values. The final result is determined by a self-consistency condition that the inputs and outputs of each box converge to a stable solution. Of course, this is useful in practice only if the calculations are simplified or if the combined method provides new insights and/or capabilities.

An example of a combined algorithm is shown by the choice of techniques indicated in Fig. 21.4; steps in an actual implementation are given in more detail in the box on p. 563. The notation GW$^+$ indicates a GW calculation, the same as a standard GW calculation, but with the addition of $\Sigma_d(\omega)$ and $P_d(\omega)$ provided by the DMFT calculation. Similarly, DMFT$^+$ indicates a DMFT calculation, the same as in the previous chapters, except that there is a modified effective $\mathcal{G}_d^{0+}$ and a dynamic effective interaction $\mathcal{U}(\omega)$, as explained below. A schematic algorithm for a GW+DMFT calculation is shown in the box and the derivation of equations and steps in the calculations is discussed in the following paragraphs.

GW + DMFT
Combined GW (method 1 for entire system) + DMFT (only for d states)

- Input

  – Form of the interaction v (Coulomb in a first-principles calculation)
  – External potential due to the nuclei and applied fields
  – Choice of orbitals: d orbitals and the rest

- Initial guess: G_k^0 (usually including the Hartree term) and Σ_k for entire system, P_d and Σ_d.

- 1. GW$^+$ calculation for entire system, with $G_k = [(G_k^0)^{-1} - \Sigma_k]^{-1}$.
  After first iteration use P_d and Σ_d from DMFT; G_k^0 (with Hartree term updated) and Σ_k from previous GW calc. (with Σ_d updated).

  – Calculate P_q^1 : $P_q^1 = P_q^{GW} = -iGG$
  – $P_d^1 = \int dq P_{q,d}^1$, $P_q = P_q^1 + [P_d - P_d^1]$, $W_q = [v_q^{-1} - P_q]^{-1}$
  – Calculate the exchange–correlation Σ_k^1 : $\Sigma_k^1 = \Sigma_k^{GW} = iGW$
  – $\Sigma_d^1 = \int dk \Sigma_{k,d}^1$, $\Sigma_k = \Sigma_k^1 + [\Sigma_d - \Sigma_d^1]$, $G_k = [(G_k^0)^{-1} - \Sigma_k]^{-1}$
  – If desired, iterate GW calculation before continuing

- 2. DMFT$^+$ calc. for embedded system of d states with G_k, W_k from GW. After first iteration, use Σ_d and P_d from previous DMFT calc.

  – Specify local embedded system
  – $G_d = \int dk G_{k,d}$, $\mathcal{G}_d^{0+} = [G_d^{-1} + \Sigma_d]^{-1}$ ($\mathcal{G}_d^{0+}$ replaces $\mathcal{G}^0$ in DMFT calc.)
  – $W_d = \int dq W_{q,d}$, $\mathcal{U} = [W_d^{-1} + P_d]^{-1}$ (equiv. to $\mathcal{U} = W_d^{rest}$ in Eq. (21.10))
  – Calculation for embedded system using a solver (Ch. 18) to find:
    Green's function $\mathcal{G}_d$ and $\Sigma_d = [\mathcal{G}_d^{0+}]^{-1} - \mathcal{G}_d^{-1}$
    Reducible polarizability χ_d and
    $W_d = \mathcal{U} + \mathcal{U}\chi_d\mathcal{U}$, $P_d = \mathcal{U}^{-1} - W_d^{-1}$ (see Eq. (21.12))
  – If desired, iterate DMFT calculation before continuing
  – Proceed to GW calculation in step 1

- Iterate until Green's function G_k and screened interaction W_q for entire system are consistent.

- Final results are G_k, W_q and thermodynamic quantities for the entire system.

- Notes:
  The original DMFT results if GW is omitted and $\mathcal{U}$ is a parameter U.
  A multi-method MBPT calculation can be constructed if DMFT is replaced by other diagrams for the d states; see Sec. 21.3.
  A GW calculation can be done with different levels of consistency, as described in Sec. 12.4.

The one-particle Green's function and self-energy

The developments of previous chapters carry over wholly to a combined method. It is most convenient to treat the entire system by the GW method that takes the place of method 1 in Fig. 21.2 with the DMFT calculation used only to find Σ_d, so that $\Delta\Sigma_d = \Sigma_d - \Sigma_d^1$. The GW calculation is expressed in terms of the Green's function given by $G_{\mathbf{k}}(\omega) = [\omega - \varepsilon_{\mathbf{k}} - \Sigma_{\mathbf{k}}(\omega)]^{-1}$ for a crystal. Here the self-energy $\Sigma_{\mathbf{k}}(\omega)$ is given by $\Sigma_{\mathbf{k}}^1(\omega) + \Delta\Sigma_d(\omega)$ and $\Sigma_d^1 = \int d\mathbf{k}\,\Sigma_{\mathbf{k},d}^1$ is the d part of $\Sigma_{\mathbf{k}}^1(\omega)$, as defined in Eq. (21.7).[5] Thus the GW calculation is modified only by the addition of the function $\Delta\Sigma_d$, an algebraic sum that can be included with no other change. In this part of the calculation the self-energy $\Sigma_d(\omega)$ from the DMFT calculation is fixed and one has the choice of whether or not to iterate the GW equations to consistency before entering the DMFT part of the calculation.

No change is needed in the DMFT algorithm given in the flowchart in Fig. 16.3 as far as $\mathcal{G}_d$ and Σ_d are concerned. The actual calculations may be much more difficult, since the DMFT solver must treat the frequency-dependent interaction $\mathcal{U}(\omega)$ and it must also determine the correlation functions $\chi(\omega)$; nevertheless, it is still an interacting system embedded in a bath where $\mathcal{G}^0$ is yet to be determined. The algorithm for the solution of the embedded system does not change if $\mathcal{G}^0$ is replaced by $\mathcal{G}_d^{0+}$. The change is only in the input to the DMFT calculation, $\mathcal{G}_d^{0+} = [G_d^{-1} + \Sigma_d]^{-1}$, where G_d^{-1} is calculated from $G_{\mathbf{k}}$ that is the output of the GW calculation.

It is easy to say that the methods are essentially unchanged and all one must do is iterate to reach consistency, but one should be aware of the pitfalls. Examples can be seen in the model calculations in Ch. 16. Figure 16.5 shows the coexistence of paramagnetic metal and insulator solutions for a range of parameters, and Fig. 16.6 shows the typical situation where the true ground state is an antiferromagnet. The solution depends on the starting point and may converge to the wrong state. There are many more possibilities for error in a calculation with many bands, various interactions that are calculated self-consistently, etc. Great care and attention to fundamental principles is needed to reach robust conclusions.

The screened interaction

The long-range interaction plays an essential role in many applications of MBPT, and it is a major step forward to bring this aspect into play when combining GW with DMFT. From the conceptual side this is straightforward, if one realizes the strict analogy between G and W, and between Σ and P, in Eqs. (21.1)–(21.4), which is also pointed out in Ch. 7.

---

[5] Note that for the method to be consistent, the aim is to remove from Σ^1 the sum of all local contributions, which are then replaced by Σ_d. In the GWA, this sum of all local contributions can be calculated straightforwardly by performing the integral $\Sigma_d^1 \equiv \Sigma_{ii,d}^1 = \int d\mathbf{k}\,\Sigma_{\mathbf{k},d}^1$, because there are no internal diagrams in the GWA self-energy expression that could correspond to an excursion to other sites, when the start and end points of the self-energy are attached to a given site i. This is no longer true in higher-order approximations, which means that it would require a more cumbersome operation than a simple integral $\int d\mathbf{k}$ to identify the on-site contributions. Similarly, for the screened interaction the identification of the purely on-site terms is easy in the RPA, where it is enough to integrate $P_{\mathbf{q},d}$ over momentum, whereas higher-order diagrams would add excursions to other sites between the start and end point that is fixed on a given site.

With this analogy in mind, we can discuss the central relations. As far as the GW calculations are concerned, the only change is that the polarizability is modified by the addition of ΔP_d, so that $P_{\mathbf{q}}(\omega) = P_{\mathbf{q}}^1(\omega) - P_d^1 + P_d$ and the total screened interaction (for the d states and the rest) is given by $W_{\mathbf{q}}(\omega) = [v_c^{-1} + P_{\mathbf{q}}(\omega)]^{-1}$. This is the new interaction to be used in the GW calculation, instead of just $W_{\mathbf{q}}^1$. The screened interaction between electrons in the d states is $W_d = \int d\mathbf{q} W_{\mathbf{q},d}$, where all terms are four-component matrices in the d orbitals as noted following Eq. (21.7). This serves as input to calculate the effective interaction $\mathcal{U}$ for the DMFT part, equivalent to the calculation of the bath $\mathcal{G}^{0+}$ from G_d.

For the DMFT part, this is a generalization of the extended DMFT (EDMFT) approach, which is discussed in Sec. 17.5. The interaction is not simply a parameter, and it must be calculated consistently. Just as $\mathcal{G}_d(\omega)$ is the Green's function for the embedded system, which equals the actual $G_d(\omega)$ only at self-consistency, so also the interaction $\mathcal{U}(\omega)$ is an auxiliary function that is equal to the actual screened interaction within local orbitals at the self-consistent solution. Thus we must derive the equations relating $\mathcal{U}(\omega)$ and the screened interaction $W(\omega)$. As pointed out above, the derivation is facilitated by the parallels with G and Σ that are evident in the corresponding equations for G, Σ and W, P in the algorithm. The result of the GW calculation is the screened interaction $W_d(\omega)$; however, in the DMFT calculation the effective interaction should be screened only by the rest of the system, i.e., the other states and the d states on other sites, and not by the d states on the same site. We call this effective interaction W_d^{rest}. Exercise 21.8 proposes to situate W^{rest} in a framework analogous to Eqs. (21.8) and (21.9). It can be understood in the picture of Fig. 21.2, where a given site labeled md is embedded in the system of all other sites and orbitals. W^{rest} can be found by removing the on-site screening from the fully screened interaction W_d. Using the relation of W and the polarizability P, we find

$$\mathcal{U}(\omega) = W_d^{rest}(\omega) = [W_d(\omega)^{-1} + P_d(\omega)]^{-1}. \tag{21.10}$$

This leaves us with the task of finding $P_d(\omega)$ from the DMFT calculation. What can be calculated directly is the susceptibility, which is given as a function of time by $[\chi_d]_{mm'nn'}(\tau) = <c_m^\dagger(\tau)c_{m'}(\tau)c_n^\dagger(0)c_{n'}(0)>$, and as a function of ω by its Fourier transform $\chi_d(\omega)$. This is the full reducible polarizability and not yet the desired $P_d(\omega)$. The missing relation can be derived using the fact that, as far as the DMFT calculation is concerned, $\mathcal{U}(\omega)$ is the bare interaction between electrons and for this purpose the relations in Eq. (21.5) can be written (see Ex. 21.2) as

$$W_d(\omega) = \mathcal{U}(\omega) + \mathcal{U}(\omega)P_d(\omega)W_d(\omega) = \mathcal{U}(\omega) + \mathcal{U}(\omega)\chi_d(\omega)\mathcal{U}(\omega). \tag{21.11}$$

Finally, these equations can be solved to find the desired relation,

$$P_d(\omega) = \mathcal{U}^{-1}(\omega) - (W_d)^{-1} \tag{21.12}$$

At consistency this equation and Eq. (21.10) are equivalent; however, at a step in an iterative calculation the input $\mathcal{U}$ is given by Eq. (21.10) in terms of P_d and W_d given by the previous calculations. The output of the calculation is χ_d, which determines a new P_d and W_d given by Eq. (21.12). These are the steps indicated in the algorithm given in the box on p. 563.

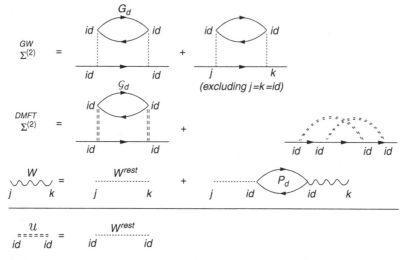

Figure 21.5. Schematic illustration of diagrams for the self-energy Σ, screened interaction W, and local effective screened interaction $\mathcal{U}$, in the combination of GWA and single-site DMFA. For a given site id, the "rest" denotes all other states and d states on other sites $j \neq i$. The third line indicates the screened interaction W for the entire system (step 1 in the box on p. 563), where P_d is the polarizability for site i in Eq. (21.12) and W^{rest} is the screened interaction that contains all polarizability contributions except the d contributions from site i. The top two lines are the self-energies to second order in the interactions involving the d states on site i. The GWA contains only a single id "bubble," but it connects id with all the rest of the system; the DMFA includes all diagrams in the effective interaction $\mathcal{U}$ for the embedded system, such as the exchange diagram not included in the GWA. In addition to the condition $G_d = \mathcal{G}$, the consistency condition for the interactions is that the effective interaction in the DMFA calculation $\mathcal{U}$ is equal to the screened interaction W_d^{rest} (Eq. (21.10)), as shown below the horizontal solid line.

Representation in diagrams

The DMFT methods outlined in the previous chapters use solvers that involve Monte Carlo or exact diagonalization. Thus it is not necessary to ever consider Feynman diagrams; however, the combined algorithm involves the calculation of the interaction function $\mathcal{U}(\omega)$ and it is useful to see what is included that would not be present in the perturbation method alone or the single-site DMFA alone. Figure 21.5 shows examples of diagrams and Dyson equations. In the first line for $\Sigma^{GW}(\omega)$, the dashed lines denote the interaction screened by the rest of the system, $W^{rest}(\omega)$. In the GWA alone, W^{rest} would be calculated in the RPA, i.e., the Coulomb interaction screened by all bubbles except the one connecting id with itself. Including the id bubble to second order in $W^{rest}(\omega)$ leads to the diagrams in the first line of Fig. 21.5. The GWA includes interactions with all other states in the system, but does not include diagrams such as the exchange diagram in the second line (see also Fig. 10.3).

The DMFA calculation for the embedded site includes all on-site diagrams in terms of the screened interaction $\mathcal{U}(\omega)$, e.g., the second line of Fig. 21.5 includes the on-site exchange diagram that is missing in the GW approximation. In Chs. 16–20 the effective interaction U is a parameter; here the interaction function $\mathcal{U}(\omega)$ is an intrinsic part of the problem. At the solution, the effective interaction $\mathcal{U}(\omega)$ is equal to the Coulomb interaction screened by the rest of the system, $\mathcal{U}(\omega) = W_d^{rest}(\omega)$ (Eq. (21.10)), which is indicated at the bottom of Fig. 21.5. In the single-site DMFA, if U is found from a DFT calculation, as in Sec. 19.5, there is no way to identify exactly what diagrams are included or not; this is the source of the problem that the "double-counting" corrections are not unique in the DFT+DMFT methods. In methods such as the constrained RPA, the diagrammatic construction is well-defined and the issues are the topic of Sec. 21.5.

The exchange diagram is indicated in the second line of the figure and higher-order terms include ladders like those in the T-matrix approximation in Figs. 10.6 and 10.7. Thus, for example, the susceptibility χ resulting from the DMFT calculation includes the electron–hole interaction that can lead to bound states, excitons, etc. As pointed out in Secs. 2.6 and 14.7, these effects can be thought of as binding of pairs of particles, or as excitations of the local atomic system. The latter interpretation is most natural in the local DMFT approach, and it includes spin excitations, localized (Frenkel) excitons, etc. These are vertex corrections beyond the RPA and GWA that change the screened interaction and may lead to improvements in the methods. They could in principle also be calculated using the Bethe–Salpeter equation, as discussed in Ch. 14.

Non-local correlation and extended DMFT (EDMFT)

As stated on the first pages of this chapter, in order for DMFT to be a method with general applicability, there must be feasible ways to go beyond the single-site approximation and the assumption that there are only local on-site interactions. The GW+DMFT approach provides a practical method to calculate the screened intersite interactions and the dynamic screened $\mathcal{U}(\omega)$ in a unified way.[6] In fact, the subject of Ex. 17.6 is to show that the GW+DMFT algorithm on p. 563 reduces to the extended DMFT approach outlined in Sec. 17.5 if there are only d states and the hamiltonian has the form in Eq. (17.11). The approach is analogous to the calculation of the self-energy $\Sigma(\omega)$: it can be formulated in terms of an embedding field $\Lambda(\omega)$, as described in Sec. 17.5, but an algorithm for a self-consistent calculation is most directly formulated in terms of the dynamic interaction $\mathcal{U}(\omega)$. The outline of the algorithm on p. 563 is written to show the parallels.

Finally, the calculations provide information on the correlations between electrons in d states on different sites, even though the auxiliary system is a single embedded site. Since the calculation of polarizabilities is an intrinsic part of the GW+DMFT algorithm, the intersite correlation can be calculated from the relation

$$\chi_{\mathbf{q}}(\omega) = \left[P(\omega)^{-1} - W_{\mathbf{q}}^{rest} \right]^{-1}, \tag{21.13}$$

---

[6] In Sec. 17.5 there is also a spin-dependent interaction; this occurs naturally in the MBPT+DMFT approach if spin-dependent processes are included.

which is analogous to Eq. (17.15) in the EDMFT approach. The Fourier transform of $\chi_{\mathbf{q}}(\omega)$ determines the correlation functions in $\chi(i - j, \omega)$ between sites i and j. The result is not merely the correlation that would occur in the perturbation method alone, because the on-site polarizabilities are modified by the DMFT part of the calculation.

21.5 Dynamical interactions and constrained RPA

In this section we illustrate the nature of the local interaction $\mathcal{U}(\omega)$ by an approach called "constrained RPA" (c-RPA) [992]. In this approach the screening is calculated in the random phase approximation, the same as in the GW method and described in Ch. 11. The appellation "constrained" derives from the fact that the d states are constrained not to participate in the screening. In c-RPA, all transitions within the d subspace, including all sites, are excluded from the effective local interaction. This means that the sum over states i, j that yields the irreducible polarizability using Eq. (12.7), has only contributions where i or j, or both of them, do not belong to the d states, without any distinction referring to sites.

At this point we have seen three different ways to include dynamical screening: the RPA $W^{rest}(\omega)$ coming from the GWA, the auxiliary $\mathcal{U}(\omega)$ in extended DMFT, described in Sec. 17.5, and the effective c-RPA screened interaction for which we adopt the conventional notation $U(\omega)$, since it is a fixed interaction and not an auxiliary function that is determined by a self-consistent calculation. In order to avoid confusion, it is useful to make a short comparison. The interaction $U(\omega)$ c-RPA takes into account screening by the other states outside the d space, similar to $W^{rest}(\omega)$ in the GWA+DMFT method, but W^{rest}_d also includes the screening contributions from d states on other sites that are excluded in the c-RPA. In EDMFT, transitions outside the d space are not considered explicitly. However, like W^{rest}, the effective interaction in EDMFT includes screening due to d states on the other sites. One may therefore say that GW+DMFT combines c-RPA and EDMFT in a self-consistent way; moreover, it adds *non-local* polarizabilities $P_{id,jd}$ within the d space that are neither contained in c-RPA nor in EDMFT.

It is useful to consider the c-RPA because it is sufficient to show the general features of the dynamically screened effective interaction and, to the extent that the choice of the subspace is representative of the actual system, c-RPA may be a good approximation to the self-consistent solution. In addition, issues such as the choice of the basis are the same and they can be investigated much more easily than in the full self-consistent calculations.

Illustration of dynamically screened interactions using c-RPA

The c-RPA method can be illustrated by the calculations for vanadium d states in $SrVO_3$ shown in Fig. 21.6, where $U(\omega)$ denotes the diagonal elements $m = n = n' = m'$ of the interaction matrix $U_{mm'n'n}$ defined in Eq. (19.5).[7] The left panel shows the real part

---

[7] There are choices as to how the calculations are done and their interpretation. For example, the polarizability can be separated into different contributions $P = P_{dd} + P_{dr} + P_{rr}$, and it is a choice how the cross terms P_{dr} are treated. This is pointed out in Sec. 21.4, where the choice is made to treat all terms by one method (RPA in

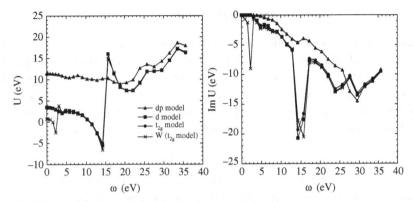

Figure 21.6. Example of frequency-dependent screened interaction calculated for $SrVO_3$ with constrained RPA, showing the real (left) and imaginary (right) parts. The full screened interaction projected on the d orbitals is W and the structure at ≈ 15 eV is due to a sharp plasmon peak. The curves marked "d model" and "t_{2g} model" are almost identical; these are the appropriate choices for a DMFT calculation with all the d states or only the t_{2g} states excluded. The structure in W at ≈ 2 eV is due to excitations in the d states since it is not present when the d's are excluded. The dp model also excludes the oxygen states; there is little screening, which shows the "rest" is mainly the oxygen states. (From [785, Ch. 7], provided by F. Aryasetiawan and T. Miyake.)

of $U(\omega)$ and the dramatic feature at ≈ 15 eV is due to the Kramers–Kronig transform of a sharp plasmon resonance in the imaginary part of U, shown on the right. As noted in Sec. 20.6, the oxygen p states are strongly bound and the vanadium d bands are well separated and partially filled with one electron. This suggests that it is appropriate to work only with the d bands with localized wavefunctions, such as the Wannier function on the right side of Fig. 19.7. The dynamic interactions shown in the curves marked "d model" and "t_{2g} model" are almost identical, because the occupied states are predominantly t_{2g} in character. This is the interaction function $U(\omega)$ that is screened by the oxygen p states and is appropriate to use in a DMFT calculation for the d states. In the low-frequency range it has a smooth variation with ω, with values ≈ 3–4 eV, similar to values found in other methods (see Sec. 19.5).

The other choices are instructive for understanding the nature of the interactions. The curve marked "dp model" in Fig. 21.6 is the result if both the d and the oxygen p states are constrained and do not take part in the screening. Then the screening is due only to the other states and there is very little effect, i.e., the interaction is almost the bare Coulomb interaction. The curve labeled "W" depicts the screened interaction including all states in the RPA calculation with no constraint. The sharp feature at ≈ 3 eV is due to transitions within the d states since it disappears when the d's are constrained and the energy is lower than required for $d-r$ transitions to the oxygen p states.

this case) and only the dd terms by the DMFT calculation. The results shown here are for the corresponding choices in the c-RPA calculations.

21.6 Consequences of dynamical interactions

It is illuminating to examine the effects of dynamical screening without all the intricacies of combined MBPT+DMFT methods. One way to do this is to consider only the change from a static U to a dynamic $U(\omega)$ with no other changes in the methods developed in Chs. 16–20. Screening affects primarily the charge fluctuations that are governed by the interaction $U(\omega)$; intra-atomic excitations, such as spin fluctuations governed by the exchange constant J, are much less affected and are expected to be close to the static atomic values.

An example of dynamic screening is the Holstein–Hubbard model, where electrons interact locally on each site as in the Hubbard model, and they are also locally coupled to bosons as in the Holstein model [993, 994] for electron–phonon interactions. Although the model was originally developed for phonons, the bosons can also represent other excitations such as plasmons that stem from collective excitations of the electrons. In this model the electrons are in a band with their spectrum affected by an excitation with frequency ω_0 that couples to the electron density $n_i = n_{i\uparrow} + n_{i\downarrow}$ at each site i, which can be written

$$H = -\sum_{ij\sigma} t_{ij} d_{i\sigma}^\dagger d_{j\sigma} + + \sum_i \left[U_{bare} n_{i\uparrow} n_{i\downarrow} + (\varepsilon_0 + \lambda[b_i^\dagger + b_i]) n_i + \omega_0 b_i^\dagger b_i \right], \quad (21.14)$$

which is a simplified version of the form that is the basis for the methods to include dynamical interactions in CTQMC solvers in Sec. 18.7.[8] A more useful form can be derived using the Lang–Firsov [995] unitary transformation $\tilde{H} = e^S H e^{-S}$, where $S = -\frac{\lambda}{\omega_0} \sum_i (b_i^\dagger + b_i)(n_{i\uparrow} + n_{i\downarrow})$. This leads to (Ex. 21.4)

$$\tilde{H} = \sum_{ij\sigma} t_{ij} \tilde{d}_{i\sigma}^\dagger \tilde{d}_{j\sigma} + \sum_i \left[\tilde{\varepsilon}_0 \tilde{n}_i + U_0 \tilde{n}_{i\uparrow} \tilde{n}_{i\downarrow} + \omega_0 b_i^\dagger b_i \right], \quad (21.15)$$

where $\tilde{d}_{i\sigma}^\dagger = \exp(\frac{\lambda}{\omega_0}(b^\dagger - b)) d_{i\sigma}^\dagger$, $\tilde{d}_{i\sigma} = \exp(-\frac{\lambda}{\omega_0}(b^\dagger - b)) d_{i\sigma}$, $\tilde{n}_{i\sigma} = \tilde{d}_{i\sigma}^\dagger \tilde{d}_{i\sigma}$, the site energy is shifted, $\tilde{\varepsilon}_0 = \varepsilon_0 - \lambda^2/\omega_0$, and the reduced static interaction is $U_0 = U_{bare} - 2\lambda^2/\omega_0$.

This expression reveals several consequences of dynamic interactions, which are examples of the general approach in Ch. 7. As in the polaron model for core electrons, the operators $\tilde{d}_{i\sigma}^\dagger$ and $\tilde{d}_{i\sigma}$ create and annihilate *quasi-particles* at low energies that have the same quantum numbers as the original particles but reduced weight. For high energies $\approx \omega_0$ there are spectral features like the plasmon satellites that are observed in materials and found in the GW, or improved calculations based on a cumulant representation as explained in Sec. 15.7. So long as the characteristic energy $\omega \ll \omega_0$, we expect that only zero occupation of the bosons is relevant [996], so that the expectation values $\langle 0| \cdots |0\rangle$ of the exponential factors in the modified operators $\tilde{d}^\dagger$ and $\tilde{d}$ can be evaluated. The result is that the bare bandwidth $\propto |t|$ is reduced by the

---

[8] This is closely related to the polaron model in Sec. 11.6, which is used in Secs. 11.7 and 15.7 for a core–hole coupled to the surrounding medium. However, in those cases the bare interaction U_{bare} is not present, since the model is for a single fermion in a bath of bosons.

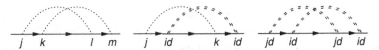

Figure 21.7. Example of diagrams not included in the GW+DMFT approximation shown in Fig. 21.5. The first is the exchange diagram that is omitted in the GW approximation, the same as in Fig. 10.2. The others are crossed diagrams involving d states on different sites or d states and the rest. These are examples of higher-order diagrammatic expansions discussed in Ch. 15, as well as methods like the dynamical vertex approximation and dual fermions described briefly in this chapter.

factor $\exp(-\lambda^2/\omega_0^2)$ (Ex. 21.5). Examples are worked out in [996] showing that the effects can be large, e.g., a bandwidth reduced by a factor of 0.7 in SrVO$_3$, to 0.5 in SrMnO$_3$. Such a bandwidth reduction due to screening is consistent with what is found in GW calculations, see, e.g., the example of the homogeneous electron gas in Sec. 11.6.[9]

21.7 Diagrammatic extensions: dynamical vertex approximation and dual fermions

There are many possible avenues that one might follow to go beyond the GWA and extended DMFT; here we can only point out issues and refer to promising developments. One of the simplest examples is given in Figs. 10.2 and 10.3, where the right-hand diagrams are second-order exchange terms represented by interaction lines that cross. These are omitted in the GWA, even though "bubble" diagrams like those in Fig. 10.4 are summed to all orders. In the single-site EDMFA, interactions are treated exactly including all diagrams for the auxiliary system that represents a single site. Intersite hopping and interactions are included to all orders; however, crossed diagrams for interactions involving two or more sites are not included. Figure 21.7 shows the lowest-order diagrams that are omitted in the GWA for the delocalized r states (left) and in the single-site DMFA for any diagrams that involve the localized d states on more than one site (right) or d and r states (middle).

The approach in EDMFT suggests additional ways that one might proceed. Instead of concentrating only on the one-body Green's function and self-energy, EDMFT considers an additional quantity on the local site: the local polarizability, which is linked to a two-particle Green's function. A systematic way to extend the dynamical mean-field idea is to realize that the analog of the self-energy, also called the *fully irreducible one-particle vertex*, exists on every n-particle level: as a next step, one can move to two-particle vertices such as the four-point (i.e., two-particle) vertex $^4\tilde{\Gamma}$ in Sec. 8.3, or the *fully irreducible two-particle vertex* Λ_{irr}, introduced in Sec. 15.5. Once these quantities

---

[9] As the discussion in Sec. 11.6 shows, this effect also arises when only *static* screening is considered. It is due to the passage from U_{bare} to U_0. Even the static screening is due to the excitations, and the static value of U_0 is meant to apply only in a frequency range much smaller than ω_0.

have been determined in a local approximation, non-local correlations beyond DMFT are obtained diagrammatically, by solving the corresponding Bethe–Salpeter or parquet equations. In the following we briefly summarize the dynamical vertex approximation (DΓA) that is based on Λ_{irr}, and the dual fermion approach, where the key quantity is $^4\tilde{\Gamma}$ [716].

Dynamical vertex approximation (DΓA)

The idea in the dynamical vertex approximation [997, 998] is to determine the fully irreducible two-particle vertex Λ_{irr}, introduced in Sec. 15.5, by a local dynamical calculation. The self-energy, including its non-local part, can then be calculated by an expansion in diagrams that can be constructed from this vertex. The full expansion is obtained by solving the parquet equations as discussed in Sec. 15.5, which yields $^4\tilde{\Gamma}$. This four-point vertex enters the correlation part of the self-energy in the Schwinger–Dyson Eq. (10.50), as depicted in Fig. 10.8. The full parquet equations couple several Bethe–Salpeter equations. Equation (14.11) is called the "particle–hole channel"; it can be solved for $^4\tilde{\Gamma}$ if an approximation for the effective particle–hole interaction Ξ_{e-h} is made, such as, e.g., the diagrams in Fig. 14.3; such an approximation corresponds to a decoupling of the parquet equations. Otherwise, once Λ_{irr} has been determined, the diagrammatic solution of the problem is still quite involved. The approach has been applied to the 2D Hubbard model [997], where there are signatures of a "pseudogap" analogous to that illustrated in Fig. 17.7 and the related text.

Dual fermions

"Dual fermions" denotes an approach in which the problem is divided into the auxiliary system of interacting fermions on an embedded site and propagators between sites described by "dual" fermions. This is an example of a general approach (a variation of the concept of the use of auxiliary systems outlined at the beginning of Sec. 21.4) to replace the problem of coupled systems by independent systems with a constraint that enforces the effects due to the coupling. The general approach has been formulated in works such as [999] and development of methods to go beyond the single-site EDMFA was done in [1000]. A pedagogical presentation can be found in Ch. 11 of [785], and a generalization to dual boson and fermion fields in [811].

In the present application, the only role of the dual fermions is to couple the sites; they are not the same as the actual fermions (electrons) that have strong interactions on each site *and* propagate between sites. The two types of fermions are coupled by the Green's function $\mathcal{G}_d$ that denotes the matrix elements for creation and annihilation of particles on the embedded site. Since the dual fermions only describe hopping between sites, they are only weakly interacting so that perturbation expansions may converge more rapidly than for the electrons. In fact, the single-site DMFA is an example of such an approach, and it is the lowest-order approximation to the equations below. The dual fermion methods go further to include selected intersite diagrams such as those in Fig. 21.7, which go beyond the diagrams for GWA+DMFT.

The first step is the division of the hamiltonian for the lattice into intrasite and intersite parts. For simplicity we give the equations only for the Hubbard model where the action as a function of $\mathbf{k}$ can be written as

$$S_{\mathbf{k}}^0 = [\varepsilon_0 + \Delta(\omega)] + [-t_{\mathbf{k}} - \Delta(\omega)] \equiv S_{loc}^0 + [-t_{\mathbf{k}} - \Delta(\omega)], \qquad (21.16)$$

where $t_{\mathbf{k}} = \frac{1}{2}\sum_i t_{0i} \exp(i\mathbf{k} \cdot \mathbf{R}_i)$. This is equivalent to Eq. (17.3), but it groups all the single-body terms for the embedded site and the remainder $[-t_{\mathbf{k}} - \Delta(\omega)]$, which is the actual hamiltonian for the intersite terms with the on-site term ε_0 replaced by $\Delta(\omega)$. In this approach the full action including interactions can be written as

$$S = S_{loc} + [-t_{\mathbf{k}} - \Delta(\omega)], \qquad (21.17)$$

where S_{loc} and $\Delta(\omega)$ are yet to be determined. Note that the interaction part of S_{loc} is not simply U but rather an effective interaction that may include effects of the other sites.

The dual Green's function can be derived by a Hubbard–Stratonovich transformation of the form in Eq. (24.39), where it is a gaussian integral over auxiliary fermion fields instead of classical fields. However, the structure of the theory can be understood without the detailed derivation. The resulting Green's function for the bare dual fermions (denoted with a tilde) can be written as (see [785, Ch. 11, Eq. (67)]),[10]

$$\tilde{G}_{\mathbf{k}}^0(\omega) = -\mathcal{G}_d(\omega)\left[\mathcal{G}_d(\omega) + (\Delta(\omega) + t_{\mathbf{k}})^{-1}\right]^{-1}\mathcal{G}_d(\omega) = G_{\mathbf{k}}^{DMFA}(\omega) - \mathcal{G}_d(\omega). \qquad (21.18)$$

The middle expression has the form of a particle with local Green's function $\mathcal{G}_d(\omega)$ propagating between sites with the term $\left[\mathcal{G}_d(\omega) + (\Delta(\omega) + t_{\mathbf{k}})^{-1}\right]^{-1}$. This has the form given in Eq. (7.9), with $\mathcal{G}_d$ playing the role of a bare G^0 defined at every site i and $(\Delta(\omega) + t_{\mathbf{k}})$ considered as a $\mathbf{k}$-dependent "self-energy," i.e., the change in G^0 at site i due to t_{ij} that couples i to other sites $j \neq i$. The second equality in Eq. (21.18) can be shown by algebraic manipulations (see Ex. 21.6) and it provides a simple interpretation of $\tilde{G}_{\mathbf{k}}^0(\omega)$ as the *difference* between the single-site DMFA Green's function $G^{DMFA}(\mathbf{k})$ (given by Eq. (17.13)) and the local auxiliary $\mathcal{G}_d(\omega)$.

The single-site DMFA is the first approximation where the average effect of the dual fermions vanishes. Then Eq. (21.18) leads to

$$\frac{1}{N_{\mathbf{k}}}\sum_{\mathbf{k}}\tilde{G}_{\mathbf{k}}^0(\omega) = \frac{1}{N_{\mathbf{k}}}\sum_{\mathbf{k}}G_{\mathbf{k}}^{DMFA}(\omega) - \mathcal{G}_d(\omega) = 0, \qquad (21.19)$$

which can be recognized as the condition for the single-site DMFA in Eq. (16.6) with $G_{00}(\omega)$ defined by Eq. (16.4). The condition that there is no average effect of coupling between sites corresponds to the neglect of correlation between sites, which is the hallmark of the single-site DMFA. Improved approximations take into account coupling between sites, i.e., the $\mathbf{k}$-dependence of the dual fermion Green's function $\tilde{G}_{\mathbf{k}}^0(\omega)$. Expansions in

---

[10] Note that t is defined with opposite sign in [785, Ch. 11].

$\tilde{G}_{\mathbf{k}}^0(\omega)$ lead to the set of parquet diagrams that have the same form as those for the expansion in the usual Green's function $G_{\mathbf{k}}^0(\omega)$.

An advantage of this approach is that many effects of the neighboring sites are included in the auxiliary system. The many-body calculation for the embedded site is unchanged from those in the single-site DMFA, but there are different self-consistency conditions that determine the hybridization function $\Delta(\omega)$. The coupling to the sites leads to interactions between the dual fermions, but they may be smaller than those for bare electrons. This approach has been applied to the 2D Hubbard model [1001]. For a model with next-neighbor hopping and doping away from half-filling, the spectra near the Fermi energy display features in general agreement with cluster calculations (see, e.g., Fig. 17.7) [1002].

21.8 Wrap-up

The combination of Green's function methods holds the potential to provide practical ways to calculate the properties of materials and physical phenomena from first principles. In the most general sense, the approach developed in this chapter is the combination of many-body perturbation theory with the idea of embedding, and the reader is encouraged to look beyond the specific methods described here. Nevertheless, much experience brought out in Chs. 7–20 shows that the combination of GWA and DMFT methods may be a promising path. The GWA is a perturbation method designed to work in cases where interactions are weak or in delocalized band-like states, whereas DMFT methods are designed to work in cases where interactions are strong in localized atomic-like states. The combination builds on the fact that both are Green's function methods in which the essential quantities are the self-energy and the screened interactions. Since this is a developing area of research, this chapter is devoted to the basic aspects of combining the two methods, and the type of effects that one can expect. Diagrammatic analysis gives a clear picture of what is included (see Fig. 21.5) and what is not included (Fig. 21.7), and the last section points to possible ways to incorporate effects that go beyond GW+DMFT approximations.

SELECT FURTHER READING

Biermann, S., "Dynamical screening effects in correlated electron materials – a progress report on combined many-body perturbation and dynamical mean field theory: GW+DMFT," *J. Phys.: Condensed Matter* **26**, 173202 (2014).

Held, K., "Electronic structure calculations using dynamical mean field theory," *Adv. Phys.* **56**, 829–926 (2007). An extensive review including an outline of GW+DMFT.

Lecture notes for the 2011 Autumn school [785] at Forschungszentrum in Jülich available online at www.cond-mat.de/events:

Held, K., *et al.*, "Hedin equations, GW, GW+DMFT, and all that"

Aryasetiawan, F., "The constrained RPA method for calculating the Hubbard U from first principles"

Lichtenstein, A. I. and Hafermann, H., "Non-local correlation effects in solids: Beyond DMFT"

Exercises

21.1 Show that χ can be written as an infinite sum of terms involving P using the relations in Eq. (21.5). Discuss why P is considered to be an irreducible polarizability. (See also Ex. 17.7.)

21.2 Write out expression $U\chi_d U$ in Eq. (21.11) with matrix indices $m, m', \dots$ Together with the definition $[\chi_d]_{mm'nn'}(\tau) = < c_m^\dagger(\tau)c_{m'}(\tau)c_n^\dagger(0)c_{n'}(0) >$, show that χ_d is the correct expression for the response of the local d system to the field $\mathcal{U}$. Explain why the response should involve the screened effective interaction $\mathcal{U}$ and not a bare interaction.

21.3 Verify the statements after Eq. (21.14) that the Holstein model ($U_{bare} = 0$) always leads to an attraction between electrons at low frequency and a reduced (or even negative) effective U for repulsive $U_{bare} > 0$. Without any calculations show that for very large ω_0 the consequence is a reduced $U(\omega)$ that is approximately constant and not frequency-dependent for $\omega << \omega_0$.

21.4 Show that the Lang–Firsov transformation leads to the hamiltonian given in Eq. (21.15).

21.5 Show that the expectation value of the exponential operators in the zero boson occupation state is $\exp(-\lambda^2/\omega^2)$, as stated after Eq. (21.15). A proof should include a careful demonstration that commutation relations of the operators are taken into account.

21.6 Derive the last equality in Eq. (21.18). This only requires the definition of the terms and rearranging the expression.

21.7 Follow the suggestion in Sec. 21.3 and split the Bethe–Salpeter Eq. (10.23) into an RPA equation and the exchange–correlation correction. This is the inverse order of equations with respect to Eqs. (14.8) and (14.9). Suppose you are interested in the optical spectrum, so that the long-range part of the bare Coulomb interaction does not contribute. Write the equations in a basis of transitions between independent-particle states, as done in Sec. 14.6. Suppose that excitonic effects are significant only in a subspace of transitions. Discuss what happens to an absorption spectrum if crystal local field effects are neglected. Which complication do you encounter when one cannot make this approximation? Do you have an idea of how one could deal with this problem?

21.8 Formulate the equivalent of Eqs. (21.8) and (21.9) for the case of the screened interaction. Note that in this case W plays the role of G, and P plays the role of Σ. Can you see a link to the Ψ-functional of Sec. 8.3? Imagine that the set d_i in Eq. (21.9) for G contains only one specific site, and that ℓ_i is the complete set of sites in the d subspace. To which quantity does G^1 correspond in that case in the DMFT approach? What if ℓ_i contains also s and p states? Follow through the same discussion for the W–P equations. Can you find the quantity that plays the role of $\mathcal{U}$? What is the major difference between the bare interaction v_c and the effective $\mathcal{U}$ of the embedded site? How does this link to the discussions in Ch. 7?

PART IV

STOCHASTIC METHODS

22

Introduction to stochastic methods

... a general method, suitable for electronic computing machines, of calculating the properties of any substance which may be considered as composed of interacting individual molecules.

N. Metropolis *et al.*, 1953

Summary

Quantum Monte Carlo methods have been very useful in providing exact results, or, at least, exact constraints on properties of electronic systems, in particular for the homogeneous electron gas. The results are, in many cases, more accurate than those from other quantum many-body methods, and provide unique capabilities and insights. In this chapter we introduce the general properties of stochastic methods and motivate their use on the quantum many-body problem. In particular, we discuss Markov chains and the computation of error estimates.

The methods that we introduce in the next four chapters are quite different from those in Parts II and III: stochastic or quantum Monte Carlo methods. In stochastic methods, instead of solving deterministically for properties of the quantum many-body system, one sets up a random walk that samples for the properties. Historically the most important role of QMC for the electronic structure field has been to provide input into the other methods, most notably the QMC calculation of the HEG [109], used for the exchange–correlation functional in DFT. A second important role has been as benchmarks for other methods such as GW. There are systems for which QMC is uniquely suited, for example the Wigner transition in the low-density electron gas, see Sec. 3.1. In this chapter we introduce general properties of simulations, in particular Markov chains, and error estimates. In the following chapters we will apply this theory to three general classes of quantum Monte Carlo algorithms, namely variational (Ch. 23), projector (Ch. 24), and path-integral Monte

Carlo (Ch. 25); Ch. 18 already introduced the QMC calculation of the impurity Green's function used in the dynamical mean-field method. We note that there are a variety of other QMC methods not covered in this book.

22.1 Simulations

First let us define what we mean by a simulation, since the word has other meanings in applied science. The dimensionality of phase space (i.e., the Hilbert space for a quantum system) is large or infinite. Even a classical system requires the positions and momenta of all particles, and, hence, the phase space for N classical particles has dimensionality $6N$. For more than about five electrons, the dimension of phase space for interacting electrons almost always precludes the explicit representation of wavefunctions or of doing many-body integrations explicitly. By simulation, we will mean that we sample this phase space; we perform matrix elements by averaging over representative points in phase space. This can either be done deterministically with molecular dynamics, or using random numbers: a Monte Carlo procedure. How this is done is the subject of this chapter and the next three chapters.

One might ask why do simulations, as opposed to the other theoretical techniques discussed elsewhere in this book. We answer this question by considering the relation between quantum and classical systems. In general, properties of classical systems can be calculated rigorously only by simulation techniques.[1] Much of the progress in classical statistical mechanics, such as our improved understanding of critical properties, has been crucially aided by numerical experiments involving simulations. By the correspondence principle, quantum systems reduce to classical systems in the high-temperature limit. Thus, unless we find a way to calculate properties of classical many-body systems without using simulations, the general properties of interacting quantum systems at non-zero temperature can also only be calculated with controlled errors by simulation techniques. Although one may develop accurate approximate methods for classes of quantum systems, and nice theoretical models to understand the relationship between various properties, it seems almost certain that simulations are needed for high-accuracy calculations on general quantum systems, especially at finite temperature, and to provide the same sort of support for theory and experiment as simulation does for classical systems. However, it is not clear that such controlled algorithms for quantum systems are even possible for many quantum expectation values.

Simulation is not without serious problems and challenges. Simulations always come with statistical errors, and this may require huge or unobtainable computation resources to achieve the needed accuracy. It is not known which quantum problems can be done on a computer with reasonable effort. For a system involving many electrons, one needs to

---

[1] We do not mean to imply that high-quality classical simulations are easy to do, just that it is possible to get very good results with error bars given enough effort; the errors will shrink in a regular fashion with more computation.

map the quantum expectation value into a probabilistic process equivalent to a problem in classical statistical mechanics. Feynman [1004] established such a mapping using imaginary-time path integrals in his 1953 work on the superfluid transition in liquid helium. No one has determined whether this can be done for many-fermion systems or for quantum dynamical properties. However, it has been found that certain properties of some many-body quantum systems can be computed by stochastic methods: e.g., equilibrium properties of bosons, fermion systems in 1D, and the half-filled Hubbard model. For these systems, all approximations are under control: by using more computer time it is feasible to systematically reduce the error estimates. For other problems there are methods, such as those based on the fixed-node approximation in Sec. 24.4, that scale well with the number of fermions and give sufficiently high accuracy to be used to benchmark other methods.

Our task in this chapter is to learn how to sample a probability distribution:

$$\Pi(s) = \frac{\exp[-S(s)]}{Z}, \tag{22.1}$$

where $S(s)$ is called the *action* function (assumed to be real-valued) and s, the "state," is a point in the space to be sampled and Z, the partition function, normalizes the probability distribution. For classical systems $S(s)$ would be equal to $V(R)/(k_BT)$, with $V(R)$ the potential and T the temperature; $\Pi(s)$ is the classical Boltzmann distribution. The choice of the state space will vary depending on the method and the system, but for now, let s represent the $3N$ coordinates of all the electrons, $s \equiv R \equiv \{\mathbf{r}_1, \mathbf{r}_2, \ldots \mathbf{r}_n\}$. In the following, for simplicity, we treat the variables as discrete. In fact, since computers only do discrete mathematics, the generalization to continuous variables, although it does not pose any difficulty, is not necessary.

22.2 Random walks and Markov chains

Monte Carlo[2] methods for many-body systems are almost without exception examples of Markov processes, also referred to as random walks.[3] The random walk algorithm is one of the most important and pervasive numerical algorithms used on computers: it is included on the "top ten" list [1005] of important numerical algorithms. It is a general method of sampling highly dimensional probability distributions by taking a random walk through phase space. The random walk algorithm was first used in physics by Metropolis *et al.* [1006] in 1953. Here we review the basic concepts.

---

[2] We refer to any numerical procedure that uses random numbers in an essential way as a Monte Carlo algorithm.

[3] The problem with direct Monte Carlo (i.e., independent sampling) methods as contrasted with the random walk method is that their efficiency goes to zero as the dimensionality of the space increases because first of all direct Monte Carlo requires knowing the normalization, i.e., the partition function and second it requires being able to sample the distribution efficiently even if it contains many degrees of freedom.

In a random walk, one changes the state of the system randomly according to a transition rule that depends only on the "present" position, $P(s \to s') \equiv P_{s's} \equiv \hat{P}$. If the random walk is initialized to the state $s^{(0)}$, we sample the next step from the distribution $P(s^{(0)} \to s^{(1)})$ and then iterate, generating a random walk through state space, $\{s^{(0)}, s^{(1)}, s^{(2)} \ldots\}$. The simplicity of the transition rule, since it has no memory of the previous history, makes it easy to analyze mathematically. Let us denote by $f_s^{(n)} \equiv \mathbf{f}^{(n)}$ the probability of the walk being in state s at step n. Both f and P are probability distributions, they are both non-negative and normalized:

$$\sum_s f_s = \sum_{s'} P_{s's} = 1. \tag{22.2}$$

With this notation, we can write the evolution of f as:

$$\mathbf{f}^{(n)} = \hat{P}\mathbf{f}^{(n-1)} = \hat{P}^n \mathbf{f}^{(0)}. \tag{22.3}$$

Here $\hat{P}^n$ is the nth power of the matrix $\hat{P}$. Then what is the equilibrium distribution defined as $\lim_{n \to \infty} f^{(n)}$? Since the evolution is via a linear operator $\hat{P}$, we can analyze the evolution of $\mathbf{f}^{(n)}$ using mathematics similar to that of quantum evolution, the only differences being that $\hat{P}$ is real and non-negative, has the sum rule in Eq. (22.2), and is usually non-symmetric: $P_{s's} \neq P_{ss'}$. Any equilibrium distribution Π will be a right eigenfunction of $\hat{P}$ with eigenvalue one:

$$\hat{P}\Pi = \Pi. \tag{22.4}$$

It can be shown that there is at least one such equilibrium distribution, but can there be more than one? To answer this question we need to define the ergodic property of a random walk: a random walk transition rule is ergodic if, after a certain number of steps, one loses all information about the starting state; the walk could be anywhere in the state space. It can be shown that one only needs to satisfy the following conditions to be ergodic:

(i) One can move from any state to any other state in a finite number of steps with non-zero probability. In other words, there are no barriers that restrict a walk to a subset of the full configuration space. Hence, all matrix elements of $\hat{P}^n$ are greater than zero for some finite n.

(ii) $\hat{P}$ is non-periodic. An example of a periodic rule is nearest-neighbor hopping on a bipartite (checkerboard) lattice; the hopping only connects A sites to B sites, so that after an even number of iterations one is certain to be on the same sublattice as one started from. The rule is guaranteed to be non-periodic if there is some probability of remaining at the same state for two successive steps. That is, $P(s \to s) > 0$.

(iii) The average return time to any state is finite. This will always be true in a finite system, e.g., with periodic or fixed boundary conditions.

For an ergodic matrix the Perron–Frobenius theorem [1007] holds: there is a *unique* stationary distribution, which we denote as Π

$$\lim_{n \to \infty} f_s^{(n)} = \Pi(s). \tag{22.5}$$

Transition rules, P, are often chosen to satisfy the detailed balance property (also called micro-reversibility): the flux of probability from s to s' equals the reverse flux,

$$\Pi(s)P(s \to s') = \Pi(s')P(s' \to s) \quad \text{for all } (s, s'). \tag{22.6}$$

If the pair $\{\Pi, \hat{P}\}$ satisfies detailed balance and if $\hat{P}$ is ergodic, then the random walk has Π and only Π as its equilibrium distribution.[4] Detailed balance is one way of making sure that we sample Π; it is a sufficient condition but it is not necessary; some Monte Carlo algorithms are based directly on Eq. (22.4).

22.3 The Metropolis Monte Carlo method

The Metropolis (rejection) method[5] is a particular way of ensuring that the transition rules satisfy detailed balance. It is appropriate for sampling any known, computable, non-negative function. It does this by splitting the transition probability into an *a priori sampling distribution* $T(s \to s')$ (a probability distribution that we can sample directly because we know its normalization) and an *acceptance probability* $A(s \to s')$, where $0 \le A \le 1$. First, one samples a trial position (state s') from the transition probability $T(s \to s')$, then, this trial move is either accepted or rejected. The overall transition probability[6] is the product of the sampling function and its acceptance:

$$P(s \to s') = T(s \to s')A(s \to s') \quad \text{for } s' \neq s. \tag{22.7}$$

The best and most common acceptance formula[7] is:

$$A(s \to s') = \min\left[1, \frac{\Pi(s')T(s' \to s)}{\Pi(s)T(s \to s')}\right]. \tag{22.8}$$

One can verify detailed balance with this procedure (see Ex. 22.1). Detailed balance then implies that for any ergodic transition probability $T(s \to s')$, the probability will converge to Π after many steps. However, the efficiency in sampling the state space will be determined by the transition probability.

---

[4] To prove, insert Eq. (22.6) into the left-hand side of Eq. (22.4).

[5] Metropolis *et al.* [1006] used ideas of a random walk from statistical mechanics; their contribution was that of enforcing detailed balance with rejection and realizing its application to the many-body problem. Hastings [1008] generalized the algorithm to non-uniform moves.

[6] The probability of not moving (rejection) is determined as $P(s \to s) = 1 - \sum_{s' \neq s} P(s \to s')$.

[7] Exercise 22.1 shows it is optimal. The min (minimum) function guarantees that $A \le 1$ as required of a probability.

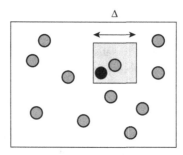

Figure 22.1. Schematic of a random walk algorithm. The trial position of a particle (filled circle) is sampled from a uniform distribution in a square about the old position.

Metropolis algorithm

- Decide what distribution to sample ($\Pi(s)$) and how to move from one state to another, $T(s \to s')$.
- Initialize the state to some value s_0.
- Iterate the following M times.
- To advance the state from $s^{(n)}$ to $s^{(n+1)}$:
  - Sample s' from $T(s^{(n)} \to s')$.
  - Calculate the ratio

$$A = \frac{\Pi(s')T(s' \to s^{(n)})}{\Pi(s^{(n)})T(s^{(n)} \to s')}. \tag{22.9}$$

  - Accept or reject: If $A > u_n$, where u_n is a uniformly distributed random number in $(0, 1)$, set $s^{(n+1)} = s'$. Otherwise, set $s^{(n+1)} = s^{(n)}$.
  - Compute and output properties whether the move is accepted or rejected.

As an example, consider the sampling of a classical Boltzmann distribution, $\exp(-\beta V(s))$, where $V(s)$ is a potential. In the original algorithm [1006], $T(s \to s')$ was chosen to be a constant distribution inside a cube (in three dimensions) of side Δ and zero outside. A single particle was displaced during a move; see the schematic in Fig. 22.1. Because T is a constant, the acceptance formula in Eq. (22.8) simplifies to $A = \min[1, \exp(-\beta(V(s') - V(s)))]$. Moves that lower the potential energy are always accepted, since $A = 1$. Moves that raise the potential energy are often accepted, if the energy cost (relative to $k_B T = 1/\beta$) is small. The random walk does not simply go downhill; thermal fluctuations can drive it uphill.

Some things to note about the Metropolis algorithm:

- If one can move from s to s', then the reverse move must also be possible: $T(s \to s')$ and $T(s' \to s)$ should be zero or non-zero together. To compute the acceptance probability, you need both the forward and reverse probabilities.

- We cannot calculate the normalization Z, the partition function, of Π, nor is it needed. There are a variety of methods that can calculate its logarithm, the free energy; see [1009].
- Moves that are rejected, remain at the same location for at least one more step, but they contribute to averages in the same way as accepted moves.
- The acceptance ratio (number of successful moves/total number of trials) is a key quantity to monitor. If the acceptance ratio is very small, one is doing a lot of computation without moving through phase space. In contrast, if the acceptance ratio is close to 1, one could probably use larger steps and get faster convergence. There is a rule-of-thumb that the acceptance should be 50%, but the correct criterion is to maximize the efficiency, as defined in the next section.
- Often, it is necessary to have several different kinds of moves. This may be required because of the structure of phase space; we will encounter this in path-integral Monte Carlo. Having several types of moves can make the algorithm more robust, since one does not necessarily know which type of move will lead to rapid movement through phase space. We can easily generalize the Metropolis procedure for a menu of possible moves. The simplest implementation is to choose the type of move randomly, according to some fixed probability. For example, one can choose the particle to be updated from a distribution, including in the definition of $T(s \rightarrow s')$ the probability of selecting that move from the menu. A more common procedure is to go through all menu items one-by-one. After one *pass*, all menu items have been attempted exactly once. One can show (Ex. 22.2) that a composition of moves after one pass through the menu will give a random walk with Π as equilibrium distribution as long as each menu item satisfies detailed balance individually.
- Usually, transition rules are local; at a given step only a few coordinates are moved, as in Fig. 22.1. If we try to move too many variables simultaneously, the move will almost certainly be rejected, leading to long correlation times and low efficiency. A general criterion is to maximize the mean-squared displacement per unit of computer time.

22.4 Computing error bars

The statistical error, inherent to Monte Carlo methods, measures the distribution of the sample average about the true mean. It can be reduced by extending or repeating runs, that is by increasing the sample size. Suppose we are trying to calculate a property, $O = \langle \tilde{O}(s) \rangle_\Pi$, for example, the energy, pressure, or a correlation function. The subscript on the expression $\langle \ldots \rangle_\Pi$ indicates the distribution function being sampled. We evaluate the property $\tilde{O}(s)$ at each step of the random walk. Here $\tilde{O}(s)$ is called the estimator of the property O. The curve of $\tilde{O}(s)$, the so-called trace, is very useful for monitoring the run.[8]

---

[8] Some of the questions you should consider in observing a trace are: Are there any trends in the data? What is the correlation between successive values? Are there any spikes or outliers? A positive answer could represent a real physical effect or could be the result of a bug: they should be understood. How long does the trace take

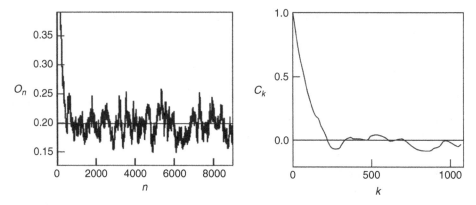

Figure 22.2. A trace of an estimator (left panel) at each step of a random walk with 9000 steps. The first few hundred steps that are strongly influenced by the initial condition of the walk are called the transient. The remainder is the plateau regime, where we take averages. In this example, the first 500 steps were discarded in computing averages: the mean is $\bar{O} = 0.1975$, the fluctuation in the property is estimated to be $\nu^{1/2} = 0.0193$. The autocorrelation function (Eq. (22.14)) shown in the right panel gives an estimate of the autocorrelation time. Its sum is $\kappa = 131$ steps. Thus the run has about $8500/131 = 65$ "effective" uncorrelated samples. The estimated error bar (Eq. (22.12)) on the mean is then $\sigma[\bar{O}] = 0.0024$.

Figure 22.2 shows an example of a trace. Our estimate of O is the mean value of $\tilde{O}(s)$ along the trajectory:

$$O = \langle \tilde{O}(s) \rangle_\Pi = \sum_s \Pi(s)\tilde{O}(s) \;\mapsto\; \bar{O} = \frac{1}{M}\sum_{n=1}^{M} \tilde{O}(s^{(n)}). \tag{22.10}$$

Throughout the remainder of the book, the numerical estimate of a quantity is written with a bar on top: the expression to the left of $\mapsto$ is the precise definition in terms of integrals and averages. We also use the $\langle \tilde{O}(s) \rangle_\Pi$ notation, just as is done with correlation functions in Eq. (5.1). The equation to the right of $\mapsto$ is its numerical evaluation using a random walk where M is the number of sampled states.

The next order of business is to figure out how accurate the estimate $(\bar{O})$ is of the exact value (O). Simulation results without error bars are of little scientific value, since one has no idea of their significance. The fundamental result on which error estimates are based is the central limit theorem of Gauss: the average of any statistical process for large M will have a gaussian (or normal) distribution. The probability distribution of observing $\bar{O}$ is:

$$\mathcal{P}(\bar{O}) = (2\pi\sigma^2)^{-1/2} e^{-(O-\bar{O})^2/(2\sigma^2)}. \tag{22.11}$$

to settle to a plateau, if it does at all? It is very important that the plateau region be much longer than the initial transient and the autocorrelation time to ensure convergence. The beginning of a run is influenced by the initial conditions (the transient, equilibration, or warm-up) and should be discarded in computing the average. See Fig. 22.2 for an illustration of this.

The formula is valid if the variance[9] ν exists (is not infinite), M is large enough that higher-order moments can be neglected,[10] and the run is much longer than the "correlation time" as defined below.[11]

The standard deviation $\sigma[\bar{O}]$ is the error of the estimated value. This is what is quoted as the error bar. One standard deviation means 66% of the time the correct answer is within σ of the sample average, 95% of the time within 2σ and so forth. When we say that Monte Carlo is an exact method for solving many-body problems, it is in this statistical sense: we know with a high degree of certainty a range of probable values of O and this range can be squeezed down by sampling more.

Usually, the error and mean are both determined from the same data. If the values of $\tilde{O}_n$ were uncorrelated, that is $\tilde{O}_n$ is independently sampled from $\tilde{O}_m$ for all $m \neq n$, we would have $\sigma = \sqrt{\nu/M}$. A common misconception is that because simulation data is usually correlated, this invalidates the central limit theorem, and Eq. (22.11). The estimated properties of an ergodic random walk do obey the central limit theorem. Ergodicity means that after a certain correlation time the walk will lose memory of its previous state. After that time, all properties of the system not explicitly conserved are as unpredictable as if they were randomly sampled from Π. To make an error estimate, we must correct for the correlation of values. We can show (see Ex. 22.3) that the standard error of the estimate of the average is:

$$\sigma[\bar{O}] = \sqrt{\frac{\kappa \nu}{M}}, \tag{22.12}$$

where the "correlation time," κ, defined as

$$\kappa = 1 + 2 \sum_{k=1}^{\infty} C_k, \tag{22.13}$$

gives the average number of steps to decorrelate the property $\tilde{O}(s)$ and is defined in terms of the autocorrelation function:[12]

$$C_k = \langle(\tilde{O}(s^{(n)}) - O)(\tilde{O}(s^{(n+k)}) - O)\rangle/\nu \mapsto \bar{C}_k = \frac{\sum_n(\tilde{O}(s^{(n)}) - \bar{O})(\tilde{O}(s^{(n+k)}) - \bar{O})}{\sum_n(\tilde{O}(s^{(n)}) - \bar{O})^2}. \tag{22.14}$$

The number of independent steps that contribute to reducing the error bar from Eq. (22.12) is not M as on the previous page, but $M_{\text{eff}} = M/\kappa$. The correlation time depends crucially on the transition rule and has a value of 1 if successive values are uncorrelated. Note that one must cut off[13] the sum in Eq. (22.13) for large k to avoid fluctuations in the estimate

---

[9] The variance is the mean-squared deviation about the mean: $\nu = \langle(\tilde{O}(s) - O)^2\rangle_\Pi$. The variance depends only on Π, not on the random walk procedure.

[10] The third moment, the skewness of $\mathcal{P}(\bar{O})$, is reduced by a factor of $1/M$, the fourth moment, the kurtosis, by $1/M^2$, etc. M must be large enough that these terms can be neglected.

[11] The correlation is between different values of the estimator at different points on the random walk (not between electrons). The word "time" is shorthand for iterations of the Markov chain, not physical time.

[12] C_k has the properties $C_0 = 1$, $C_k = C_{-k}$, and $\lim_{k\to\infty} C_k = 0$.

[13] For example, one can cut off the sum in Eq. (22.13) to be determined self-consistently at say $5\,\kappa$.

of C_k dominating the estimate of κ. Figure 22.2 shows an example of the computation of the error estimate.

It is important that the total length of the random walk be much greater than κ. Otherwise the estimated result and the estimated error will be unreliable. Runs in which the number of steps $M \gg \kappa$ are called *well converged*. In general, there is no rigorous procedure to determine κ since it is hard to prove that the random walk has converged and that the estimated autocorrelation function, and hence κ, is correct. It is good practice occasionally to do very long runs to test that the results are converged. Also, it is useful to start the random walk from different initial states and see if all properties converge to the same values.

Binning analysis[14] is an alternative procedure for correcting for correlation. Sections of the run are averaged together to make "bins" and errors are computed assuming the bin values are statistically uncorrelated. Then the bin values are themselves averaged together to make new bins and the errors recomputed. The procedure is repeated recursively. If the curve of the computed error versus the bin size shows a plateau, the plateau value is a good estimate of the true error. Bin analysis is more robust than computing the correlation time for well-converged walks, since it does not require cutting off the sum in Eq. (22.13) at large time, however, the analysis using Eqs. (22.12) and (22.13) gives more information about correlation times and more precise error estimates.

The correlation time defined above is an equilibrium average. There is another correlation time relevant to a random walk, namely, how many steps it takes to reach equilibrium from some starting state. Normally this will be at least as long as the equilibrium correlation time, but in some cases it can be much longer. For example, meta-stability, hysteresis, and slow convergence are characteristic of many-body systems near a phase boundary. The simplest way of testing convergence is to start the random walk from several, radically different, starting places and see if a variety of properties converge to the same equilibrium values. Very often one uses for starting states, well-converged states at neighboring densities and temperatures.

The *efficiency* of a random walk is defined as how quickly the error bars decrease as a function of computer time T of a calculation: $\zeta = (T\sigma^2)^{-1}$. For long calculations ζ is independent of the length of the calculation and is the figure-of-merit for a given algorithm, computer, and implementation. For example, this criterion should be used to determine the step size Δ in the Metropolis algorithm illustrated in Fig. 22.1. If we ignore how much computer time a move takes, an optimal transition rule is one that minimizes the autocorrelation time κ.

22.5 The "heat bath" algorithm

The efficiency of a random walk is determined by the transition probability. It is then natural to inquire how we can change the transition rule to improve the efficiency. If one

---

[14] Also called blocking.

could sample any probability distribution in the same amount of computer time one should simply sample $\Pi(s)$ directly, since thereby the correlation time $\kappa = 1$. But such is not usually the case for an interacting many-body system, since in order to directly sample $\Pi(s)$ one needs to know the the free energy of the system. In this section we introduce heat bath sampling: one directly samples $\Pi(s)$ in a local region of phase space.

Given a transition rule, we define the neighborhood,[15] $\mathcal{N}(s)$, for each point in state space as the set of states s' that can be reached in a single move from s. With the *heat bath* transition rule, one samples elements from the neighborhood with a transition probability proportional to their equilibrium distribution,

$$T_{HB}(s \rightarrow s') = \frac{\Pi(s')}{Z_s}. \tag{22.15}$$

The method is so called since conceptually we are putting the system in contact with a "heat bath" but constraining it to remain in the given neighborhood. The transition probability must be normalized to unity; this determines the normalization constant

$$Z_s = \sum_{s'' \in \mathcal{N}(s)} \Pi(s''). \tag{22.16}$$

Note that it depends on the current position of the walk, s. By substitution into Eq. (22.8), the acceptance probability is:

$$A(s \rightarrow s') = \min \left[1, \frac{Z_s}{Z_{s'}} \right]. \tag{22.17}$$

Since the heat bath algorithm comes into equilibrium within the neighborhood in one step, it has optimum stepwise efficiency. If the old and new neighborhoods coincide, $N(s) = N(s')$, then all moves will be accepted.

The heat bath rule is frequently used in discrete models, such as the Ising model, where one can easily compute the normalization constant Z_s needed in Eq. (22.17) and perform the sampling. In continuum systems, the heat bath approach is not often used because Z_s is difficult to compute. Instead, one tries to find transitions that both approximate the heat bath one, so that the correlation time is small, and can be executed quickly; see, for example, smart Monte Carlo in Eq. (23.12).

22.6 Remarks

The Monte Carlo methods are quite different from deterministic algorithms. The topics we have covered in this chapter are the basis for the QMC methods in Ch. 18 and in the following three chapters. We have presented here only the minimum background. There is a great deal of research in developing simulation methods for both quantum and classical systems that will not be covered in this book, in particular algorithms for lattice models.

Because the methods are often quite unfamiliar, we mention a few aspects important in either writing a QMC code or performing a calculation with existing software.

---

[15] It is essential for detailed balance that the neighborhoods be reflexive: $s' \in \mathcal{N}(s) \Leftrightarrow s \in \mathcal{N}(s')$.

- *Random numbers.* Clearly a good source of random numbers is essential for correct QMC calculations. Although, even today, the generation of so-called pseudo-random numbers [276] is not a completely solved problem, high-quality modern packages for the generation of random numbers are almost always satisfactory. Special care is needed for computation in parallel, so that random number streams on different processors or threads are independent.
- *Finite size effects.* Most quantum simulations that we will discuss are done with fewer than a thousand electrons. Thus, correcting for finite size effects to make predictions for bulk systems is very important; see Sec. 23.5.
- *Testing.* Some thought has to be given to how to test QMC calculations. The key quantity needed is the estimated error bar. If possible, one should find a result in the literature to compare with.[16] Metropolis algorithms can be tested to see if the results (but not the efficiency) are independent of the transition probability $T(s \rightarrow s')$.

SELECT FURTHER READING

Feller, W. *An Introduction to Probability Theory and Its Applications, Volume I* (Wiley, New York, 1968).

Hammersley, J. M. and Handscomb, D. C. *Monte Carlo Methods* (Chapman and Hall, London, pp. 113–122, 1964).

Hammond, B. L., Lester, Jr., W. A., and Reynolds, P. J. *Monte Carlo Methods in ab initio Quantum Chemistry* (World Scientific, Singapore, 1994).

Kalos, M. H. and Whitlock, P. A. *Monte Carlo Methods, Volume I: Basics* (Wiley, New York, pp. 73–86, 1986).

Metropolis, N., Rosenbluth, A. W., Rosenbluth, M. N., Teller, A. H., and Teller, E., *J. Chem. Phys.* **21**, 1087, 1953.

Exercises

22.1 (a) Prove detailed balance for the procedure given by Eqs. (22.7) and (22.8). (b) Show that Eq. (22.8) leads to the largest acceptance ratio among formulas satisfying detailed balance. *Hint:* consider maximizing the acceptance probability between states s and s'.

22.2 Suppose that there are two types of moves "a" and "b" in the random walk, with probability given by P_a and P_b, and suppose that each satisfies detailed balance separately, Eq. (22.6). During the random walk we make an "a" move, followed by a "b" move. Let us call the combination a "c" move. What is the transition probability for a "c" move? Does this satisfy detailed balance? Assuming ergodicity, what will the equilibrium distribution be? Prove that this procedure leads to the correct distribution. Now generalize to an arbitrary number of different types of moves.

22.3 Derive Eq. (22.12) in the limit of a long ergodic random walk. *Hint:* see Fig. B.1.

---

[16] Suppose your calculation obtains an energy $\bar{E} \pm \sigma$ and the literature value is $\bar{E}_0 \pm \sigma_0$. The probability of getting the normalized difference (or larger), $x = (\bar{E} - \bar{E}_0)/(\sigma_0^2 + \sigma^2)^{\frac{1}{2}}$, is given by the complementary error function, erfc(x).

22.4 (a) How does the transition probability $T(s \rightarrow s')$ change if we divide the equilibrium distribution by a fixed function $g(s)$? (b) For a random walk with the detailed balance property, find that function $g(s)$ that makes the transition matrix symmetric with respect to the old and new points. (c) Now apply the eigenfunction expansion for symmetric matrices familiar from quantum mechanics to determine how the random walk converges.

23

Variational Monte Carlo

Since the form of P_N is the same as that occurring in the statistical mechanics of the classical gas (replace $f^2(r_{ij})$ by $\exp[-V(r_{ij}/kT)]$), we can use the same integration techniques that are used in the classical problem.

W. L. McMillan, 1965

Summary

Building on the random walk methods developed in the previous chapter, we show how to compute properties of many-body trial wavefunctions using a random walk. This method, called variational Monte Carlo, is the simplest stochastic quantum many-body technique. Whereas mean-field methods are usually limited to single determinants, variational Monte Carlo can treat any correlated trial function, as long as its values are computable. We discuss how to optimize such trial wavefunctions, how to compute their momentum distribution, how to use non-local pseudopotentials, how to compute excited states, and how to correct for the finite size of the simulation cell.

Deterministic quantum methods have difficulties. For example, the Hartree–Fock method assumes the wavefunction is a single Slater determinant, neglecting correlation. If one expands as a sum of determinants, it is very difficult to have the results size-consistent since the number of determinants needed will grow exponentially with the system size. As we have seen, the DMFT method introduced in Ch. 16 assumes locality. In Ch. 6 we discussed general properties of many-body wavefunctions.[1] Using Monte Carlo methods, we can directly incorporate correlation into a wavefunction, without having to make any further approximations other than the form of the correlation factors. In many cases the energy and other properties are very close to the exact results. Some of the usual restrictions on the form of the many-body wavefunction are not an issue in variational Monte Carlo. The most important generalization of the HF wavefunction is to put correlation directly

---

[1] The reader should review the properties of many-body wavefunctions, Ch. 6.

into the wavefunction via the "Jastrow" factor. At next order, one can use the "backflow" wavefunction, in which correlation is also built into the determinant.

The variational Monte Carlo method (VMC) was first used by McMillan [44] to calculate the ground-state properties of superfluid ^{4}He. One of the key problems at that time was whether the observed superfluid properties were a consequence of Bose condensation. For this, Hartree–Fock methods are not reliable[2] since the helium interaction is very different from the electron–electron interaction and diverges much more rapidly at small distances. In the quote at the beginning of this chapter, McMillan observed that the calculation of a quantum boson system using a pair product (Jastrow) wavefunction can be mapped to the evaluation of a classical many-body system of atoms interacting with a pair potential and can be calculated with Monte Carlo techniques. The main differences between the classical and the VMC calculation are the methods for evaluating the kinetic energy and the momentum distribution and the procedure for optimizing the variational parameters. The generalization of the VMC method to many-body fermion systems, which requires antisymmetric trial wavefunctions, was done by Ceperley, Chester, and Kalos [1011].

The VMC method is based on the variational theorem, which asserts that the expectation value of the hamiltonian for an arbitrary[3] trial function, $\Psi(R)$, is greater than the exact ground-state energy

$$E_V = \frac{\int dR \Psi^*(R) \hat{H} \Psi(R)}{\int dR \Psi^*(R) \Psi(R)} \geq E_0, \tag{23.1}$$

where $R \equiv \{\mathbf{r}_1, \mathbf{r}_2, \ldots, \mathbf{r}_N\}$. Note that for some systems, the integrals over electron coordinates need to be augmented with sums over spin degrees of freedom. We do not assume that the trial function is normalized, since that is difficult to accomplish with most correlated many-body functions.[4]

The variational Monte Carlo method is an application of the random walk technique described in the previous chapter to sample the normalized square of the modulus of the trial wavefunction; that is, we use the Metropolis Monte Carlo method to sample the distribution:

$$\Pi(R) = \frac{|\Psi(R)|^2}{\int |\Psi(R)|^2}. \tag{23.2}$$

We rewrite the variational energy bound in Eq. (23.1) as the average value of the "local energy" over the distribution, $\Pi(R)$:

$$E_V = \int dR \Pi(R) E_L(R) = \langle E_L(R) \rangle_\Pi \mapsto \bar{E}_V = \frac{1}{M} \sum_n E_L(R^{(n)}). \tag{23.3}$$

---

[2] The interaction between helium atoms is well approximated by a hard-sphere interaction. HF methods are especially bad since the potential matrix elements are infinite.

[3] There are conditions on the trial function needed to ensure the upper bound property, e.g., the trial function must be antisymmetric under electron exchange. See Sec. 6.5 for other conditions.

[4] We do assume that the integrals in both the numerator and the denominator of Eq. (23.1) are well-defined and finite.

The local energy is defined[5] as

$$E_L(R) = \mathrm{Re}\frac{\hat{H}\Psi(R)}{\Psi(R)}. \tag{23.4}$$

Here $\bar{E}_V$, the average of the local energy over a random walk of M steps, is the VMC estimate of E_V.

Variational Monte Carlo has a very important property: as the trial function approaches an exact eigenfunction, $\Psi \to \Phi_\alpha$, the local energy approaches the eigenvalue everywhere, $E_L(R) \to E_\alpha$, implying that the Monte Carlo estimate of the variational energy converges more rapidly with the number of steps in the random walk. As discussed in the previous chapter, the variance of the estimator[6] is a key ingredient of the efficiency of the calculation:

$$v_T = \frac{\int dR |\Psi|^2 (E_L(R) - E_V)^2}{\int dR |\Psi|^2} \mapsto \bar{v}_T = \frac{1}{M}\sum_{n=1}^{M}(E_L(R^{(n)}) - \bar{E}_V)^2. \tag{23.5}$$

As the trial function gets more accurate, both the statistical error of the variational energy, $\sigma[\bar{E}_V]$ discussed in Sec. 22.4, and the systematic error, $E_V - E_\alpha$, vanish. In fact, both v_T and $E_V - E_\alpha$ vanish quadratically[7] with $\Psi - \Phi_\alpha$, where E_α and Φ_α are an exact energy and wavefunction. It is because of this *zero variance property* that Monte Carlo calculations of energies of quantum systems at zero temperature can be much more precise than those of a classical system at non-zero temperature.

23.1 Details of the variational Monte Carlo method

We now discuss more details of the VMC algorithm. So that we can be definite, we will assume the trial wavefunction has the Slater–Jastrow form as discussed in Sec. 6.6:

$$\Psi(R) = e^{-U(R)}\det(\psi_i(\mathbf{r}_j)) = e^{-U(R)}\det(R). \tag{23.6}$$

Here $U(R)$, the Jastrow factor, is assumed to be real and includes one- and two-body terms as in Eq. (6.23) and possibly three-body terms as in Eq. (6.33). Also the Slater matrix, i.e., $\psi_i(\mathbf{r}_j) = \psi_{ij}$, is assumed to have a spin-up block and a spin-down block, so that the determinant is a product of a spin-up determinant and a spin-down determinant. There is no requirement that the orbitals be orthogonal, only that the matrix not be so ill-conditioned that its determinant cannot be accurately evaluated during the random walk.

The overall random walk procedure has already been given in Ch. 22. We repeat it here, specializing it for VMC.

---

[5] Since the hamiltonian is hermitian, E_V is real and we can neglect the imaginary part of the local energy since it will integrate to zero.

[6] Note that for the central limit theorem to hold, $v_T < \infty$. We are neglecting the autocorrelation of the local energy, see Eq. (22.12).

[7] We assume that the random walk has a finite correlation time that is independent of Ψ.

Variational Monte Carlo algorithm

- Pick the trial wavefunction, i.e., U and the $\psi_k(\mathbf{r})$'s.
- Initialize electron positions $R^{(0)} \equiv \{\mathbf{r}_1^{(0)}, \mathbf{r}_2^{(0)}, \ldots, \mathbf{r}_N^{(0)}\}$ and the inverse Slater matrix, D (see Eq. (23.10)).
- Iterate the loop over n, and electrons, i.
  - Sample a new coordinate $\mathbf{r}'$ for electron i from $T(\mathbf{r}_i^{(n)} \to \mathbf{r}')$.
  - Determine the acceptance probability:

$$A = \min\left[1, \frac{T(\mathbf{r}' \to \mathbf{r}_i^{(n)})}{T(\mathbf{r}_i^{(n)} \to \mathbf{r}')} \left|\frac{\Psi(R_i')}{\Psi(R^{(n)})}\right|^2\right]. \tag{23.7}$$

Here $R_i' = R^{(n)}$ except the coordinate of electron i is replaced: $\mathbf{r}_i^{(n)} \to \mathbf{r}'$.
  - If $A > u$ with $u \in (0, 1)$ a uniform random number, accept the move: update the coordinates $(R^{(n+1)} = R')$ and the inverse matrix, D; see Eq. (23.11). Otherwise, keep the old coordinates: $R^{(n+1)} = R^{(n)}$ and old inverse.
  - Compute and average the local energy $E_L(R^{(n+1)})$ and other properties.
- Adjust the trial wavefunction parameters to minimize the average local energy.

We now elaborate some of these steps:

(i) For more than about 10 electrons, it is more efficient to move the electrons one at a time, by displacing a single electron in a local region about its current position, rather than moving all of the electrons at once.

(ii) Assuming a uniform move within a volume Ω so that $T(\mathbf{r}' \to \mathbf{r}_i^{(n)}) = T(\mathbf{r}_i^{(n)} \to \mathbf{r}') = \Omega^{-1}$, the trial move is accepted with probability

$$A = \min\left[1, |\Psi(R')/\Psi(R)|^2\right] = \min\left[1, e^{-2(U(R')-U(R))} |q|^2\right], \tag{23.8}$$

where q is the ratio of the old and new determinants

$$q \equiv \frac{\det(R')}{\det(R)} = \sum_k D_{ki}\psi_k(\mathbf{r}_i'). \tag{23.9}$$

If only the ith electron is displaced, only one column of the Slater matrix changes, and q can be determined from the transpose of the inverse defined by the linear equations

$$\sum_k D_{ki}\psi_k(\mathbf{r}_j) = \delta_{ij}. \tag{23.10}$$

While the evaluation of a general determinant takes $O(N^3)$ operations, the evaluation of the fermion part of the acceptance ratio, i.e., q, will take $O(N)$ operations if D is available and a number of operations independent of N ($O(N^0)$) if the orbitals $\psi_k(\mathbf{r})$ have a finite range.

(iii) If a move is accepted, we need to update the inverse matrix, D. The Sherman–Morrison–Woodbury formula [1012] gives the update to an inverse when a single column of a Slater matrix is altered:

$$D'_{kj} = \begin{cases} D_{kj} - (D_{ki}/q) \sum_l \psi_l(\mathbf{r}) D_{lj} & j \neq i \\ D_{kj}/q & j = i \end{cases}, \qquad (23.11)$$

where q is defined in Eq. (23.9). The update takes $O(N^2)$ operations if the matrices are full. Hence, to attempt moves for all N electrons (a pass) will take $O(N^3)$ operations, the same power as in a Kohn–Sham independent-particle DFT calculation. The update will be the dominant computational cost for fermion VMC at sufficiently large N. In practice, the inverse update is a fast linear algebra operation, so that calculating the other parts of the trial function, such as the $O(N^2)$ operations to update the orbitals, dominate the computer time. Occasional tests using Eq. (23.10) are useful to see if round-off error has accumulated to such a point that a fresh calculation of D is needed.

(iv) Instead of moving an electron uniformly in a local region, to improve the efficiency of the random walk (see Sec. 22.5) we can sample the new position from a gaussian whose mean is pushed in the direction of increasing probability:

$$T(\mathbf{r}_i \to \mathbf{r}') = (4\pi \Delta\tau)^{-3/2} \exp(-(\mathbf{r}' - \mathbf{r}_i - \frac{1}{2}\Delta\tau \mathbf{F}_i)^2/(2\Delta\tau)). \qquad (23.12)$$

Here the parameter $\Delta\tau$ is the step size. This procedure in classical simulations is known as smart Monte Carlo [1013, 1014]. The new coordinates are computed as:

$$\mathbf{r}' = \mathbf{r}_i + \Delta\tau \mathbf{F}_i + \chi. \qquad (23.13)$$

Here χ is a normally distributed 3-vector, with zero mean and variance equal to $\Delta\tau$. The *quasi-force* $\mathbf{F}_i$, defined and computed as:

$$\mathbf{F}_i = 2\nabla_i \ln \Psi(R) = 2\left(- \nabla_i U + \sum_k D_{ki} \nabla_i \psi_k(\mathbf{r}_i) \right) \qquad (23.14)$$

is proportional to the force needed to sample the quantum many-body distribution using a classical molecular dynamics method. Here $\Delta\tau$ controls how far electrons move in one step; the root-mean-squared distance is $\sqrt{(3\Delta\tau)}$. The step size should be chosen to maximize the efficiency of the sampling, e.g., by maximizing how far electrons move on average per unit of computer time and taking into account rejections. Within VMC, one uses this transition probability to increase the acceptance probability, and, hence, the efficiency. The same transition probability will be used in diffusion Monte Carlo (Ch. 24) to implement the projection method: the identical transition probability can be used in both algorithms.

(v) The local energy, defined in Eq. (23.4) (also Eq. (6.11)), is needed to evaluate the variational energy. For the Slater–Jastrow trial function it is calculated[8] by differentiating the trial function, obtaining:

---

[8] One can conveniently calculate the derivatives by finite differences of the trial wavefunction, however, that will take $O(N)$ times longer and be subject to greater numerical errors. Comparison of the analytical result from

$$E_L(R) = V(R) + \frac{1}{2} \sum_{i=1}^{N} [\nabla_i^2 U - (\nabla_i U)^2 - \sum_k D_{ki} (\nabla_i^2 \psi_k(\mathbf{r}_i) - 2\nabla_i U \cdot \nabla_i \psi_k(\mathbf{r}_i))].$$

(23.15)

There is a further simplification of the computation of the local energy when the orbitals are exact solutions of a non-interacting hamiltonian with potential $V_m(R)$ and total eigenvalue E_m, since then $-\frac{1}{2} \sum_{i,k} D_{ki} \nabla_i^2 \psi_k(\mathbf{r}_i) = E_m - V_m(R)$. One can replace the term $\nabla_i^2 \psi_k$ in the local energy with something quicker to compute: $E_m - V_m(R)$. There exist other formulas [1011] to calculate the variational energy using Green's identities. For example, one can show that:

$$E_V = \left\langle V(R) + \frac{1}{4} \sum_{i=1}^{N} \nabla_i^2 \ln(|\Psi|) \right\rangle$$

(23.16)

(see Ex. 23.3). This estimator is simpler since one no longer has to compute the quasi-force, however, it is less efficient to use. This is because the new estimator no longer has the zero-variance property discussed at the end of the last section, so that its statistical error is much larger than that of Eq. (23.15). However, Eq. (23.16) is very useful to verify that the difference between Eqs. (23.16) and (23.4) is zero within error bars. This is a test of the evaluation of the derivatives, the sampling and continuity of the trial function, and requires little additional coding.

(vi) The Ewald sum method for calculating the long-ranged Coulomb potential in periodic boundary conditions is used extensively for classical systems [1013] and in electronic structure, as discussed in [1, App. F]. For PBC the Coulomb potential cannot simply be truncated at the edge of the supercell but must be made periodic in such a way that the long-wavelength correlations that give rise to plasmons and van der Waals interaction are preserved. In addition, since the optimal pair Jastrow correlation factor $u(r_{ij})$ decays as r_{ij}^{-1} in an extended 3D system (see Eq. (6.22)), the Jastrow factor should also be made periodic consistently with how the potential is made periodic. Using a generalization of the Ewald sum method, we can split the periodic factor $u_P(\mathbf{r})$ into the sum of a short-range spherical function, $u_{sr}(r)$, and the remainder, the long-range potential $u_{lr}(\mathbf{r})$, which we expand in plane waves:

$$u_P(\mathbf{r}) = \sum_n u(\mathbf{r} + \mathbf{L_n}) \mapsto u_{sr}(r_m) + \sum_k u_{lr}(\mathbf{k})e^{i\mathbf{k}\cdot\mathbf{r}}.$$

(23.17)

Here $\mathbf{L_n}$ are the Bravais lattice vectors of the supercell, $\mathbf{k}$ is a vector in the reciprocal lattice of the supercell, and $r_m = \min_n(|\mathbf{r} + \mathbf{L_n}|)$ is the minimum image (the smallest separation of two particles obtained by adding and subtracting lattice vectors). In contrast to the potential, the Jastrow factor is not a pure power law so that standard Ewald formulas cannot be used. Methods for optimizing u_{sr} and u_{lr} to minimize the needed number of $\mathbf{k}$'s are given in [275, 1015]. To ensure the variational bound

---

Eq. (23.15) with the numerical finite difference result is very highly recommended to verify the correctness of the coding of the analytical derivatives.

is upheld, $u_{sr}(r)$ and its derivative must be everywhere continuous; we require that $u_{sr}(r)$ vanish for $r \geq a$, where $a \leq L/2$.

(vii) Although the Slater–Jastrow trial function is the most common form in VMC, generalizations are also used, for example, backflow determinants and pfaffians as discussed in Sec. 6.7. Some generalizations require little modification of the basic algorithm, e.g., multiple determinants are often needed for accurate description of molecular systems; these require only extra computational effort without much change in the underlying code. Fast recursion methods generalizing the update formula of Eq. (23.11) have been developed [1016]. Three-body terms require a similar development, with extra work needed to evaluate the local energy. Replacing the Slater determinant by a backflow determinant significantly increases the computational effort, from $O(N^3)$ to $O(N^4)$. Single-electron updates, Eq. (23.11), and the formulas for the local energy, Eq. (23.15), are no longer applicable. Details are given in [279]. The necessary formulas for pfaffians [283] follow closely those of determinants.

Note that the overwhelming fraction of computational effort in VMC is devoted to computing $\Psi(R)$ (Eqs. (23.6) and (23.11)), the pseudo-force, Eq. (23.14), and the local energy, Eq. (23.15). In contrast to some other applications of Metropolis Monte Carlo, for VMC calculations of electrons, lack of ergodicity or long correlation times have not been observed. One might worry that the many-body wavefunction nodes would present barriers to the random walk. But in fact, the tiling theorem (Sec. 24.4) implies that for the majority of fermion trial functions there is no need for the random walk to cross the node since proper observables are obtainable from walks that never cross the nodes. In any case, random walks are observed to cross the nodes.

23.2 Optimizing trial wavefunctions

Finding an accurate trial function is crucial for the success of the variational method and important for the projector Monte Carlo method described in the next chapter. Although much success has been obtained with understanding the properties of many-body wavefunctions analytically, see Sec. 6.5, the usual approach in quantum chemistry and DFT or independent-particle methods is to express a wavefunction as an expansion in a basis and then to optimize the expansion parameters. We follow that approach in VMC.

First, consider how to write a general Jastrow factor in a form suitable for optimization. We can express a general electron–electron Jastrow function in the form of Eq. (23.17), and represent $u_{sr}(r)$ by a cubic spline on an interval $(0, a)$ with boundary conditions at the ends of the interval. The values of $u_{sr}(r)$ and its derivative(s) on the "knots," as well as the values of $u_{lr}(\mathbf{k})$ for $\mathbf{k}$'s allowed by the PBC, is a general parameterization of the Jastrow–Slater trial function in a periodic box. Orbitals are much more data-intensive since for N electrons, assuming spin symmetry, there are $N/2$ spatial orbitals, each of which is a full 3D function. The individual orbitals are not unique since any non-singular linear transformation leaves the many-body wavefunction the same up to a normalization. For

VMC, the orbitals are typically computed in a DFT or HF calculation, and then modified by multiplying by a one-body term of the form $\exp(-\sum_i u_{el}(\mathbf{r}_i))$. This additional variational freedom is needed because the electron–electron Jastrow pushes the electrons away from each other and makes the electron density unphysically more homogeneous than that of the Slater determinant. The 3D function $u_{el}(\mathbf{r})$ can be expanded in plane waves having the symmetry of the potential, or in some other basis; the expansion coefficients are then variational parameters.

Before we start to optimize the trial function, we need to decide the quantity that is to be minimized: either the variational energy or the variance of the trial function or a linear combination of the two; see the discussion in Sec. 6.5. The variance has a known minimum value, namely zero, which can prove helpful in the optimization. Moreover, the variance is strictly positive; even its fluctuations are positive. This makes the optimization more stable. However, the variance is dominated by high-energy parts of phase space, e.g., electrons in core regions, while the valence region is more important for physical properties. Minimizing the energy, though more difficult, leads to a trial function that has most properties superior to those coming from a minimum-variance trial function.

The simplest way to find the best trial function is to perform independent VMC runs (possibly in parallel) with different variational parameters. Then one fits the energies as a function of parameters to a simple analytic form, does more calculations around the predicted minimum, and continues until convergence in parameter space is attained. McMillan's search done this way for liquid ⁴He is shown in Fig. 23.1. One difficulty is that close to the minimum, the statistical errors mask the variation with respect to the trial function parameters; the change of the variational energy with respect to trial function parameters is poorly calculated. More importantly, it is difficult to optimize a function involving more than a few variational parameters, since the number of needed calculations to bracket the minimum scales as the power of the number of parameters, limiting practical optimizations to a handful of parameters.

The *reweighting* method [1011, 1017], where the output of a single random walk is used to estimate the energy as a function of variational parameters, solves these two problems. One stores a set of configurations $\{R^{(1)}, R^{(2)}, \ldots, R^{(M)}\}$, sampled from a distribution $\Pi_0(R)$. The variational energy[9] for a given set of parameters, denoted as a vector $\mathbf{a}$, is written as:

$$E_V(\mathbf{a}) = \frac{\int \Psi^*(R;\mathbf{a})\hat{H}\Psi(R;\mathbf{a})}{\int |\Psi(R;\mathbf{a})|^2} \mapsto \bar{E}_V(\mathbf{a}) = \frac{\sum_n w_n(\mathbf{a})E_L(R^{(n)};\mathbf{a})}{\sum_n w_n(\mathbf{a})}, \quad (23.18)$$

where the weight of the nth configuration is $w_n(\mathbf{a}) \equiv |\Psi(R^{(n)};\mathbf{a})|^2/\Pi_0(R^{(n)})$. The right equation is the estimate from the finite ensemble of the exact quantity defined in the left equation. The computed estimate of the variational energy, $\bar{E}_V(\mathbf{a})$, is now a smooth function of $\mathbf{a}$, the derivative $\nabla_{\mathbf{a}}E_V(\mathbf{a})$ has finite variance, so efficient minimization methods using gradients [1018] can be employed. A significant problem with this method is that if the overlap between $\Pi_0(R)$ and $|\Psi(R,\mathbf{a})|^2$ gets small during the optimization search,

---

[9] To minimize the variance or a linear combination of the energy and the variance, we substitute for $E_L(R^{(n)},\mathbf{a})$ in Eq. (23.18) a term $E_L(R^{(n)},\mathbf{a}) + \alpha E_L^2(R^{(n)},\mathbf{a})$.

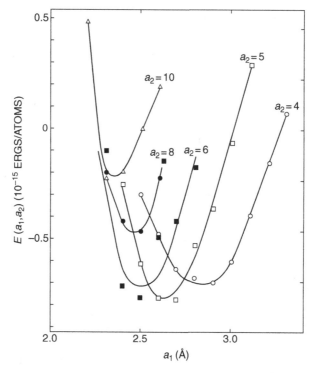

Figure 23.1. The variational Monte Carlo estimate of the energy of liquid ^{4}He as a function of trial function parameters [44] with a Jastrow factor $u(r) = (a_1/r)^{a_2}$. Each symbol is an independent run; the lines are fits through the data for a fixed value of a_2. The optimal Jastrow factor of this form is seen to be $a_1 \sim 2.65$Å and $a_2 \sim 5$.

the error of $\bar{E}_V(\mathbf{a})$ will become large, possibly infinite, and the optimization becomes unstable. By monitoring the number of effective configurations (the overlap), defined as $N_{\text{eff}} = (\sum w_n(\mathbf{a}))^2 / \sum_n w_n(\mathbf{a})^2$, one can stop the optimization process before this happens, reset $\Pi_0(R) = |\Psi(R; \mathbf{a}^*)|^2$, where $\mathbf{a}^*$ is the current best estimate of the optimal parameters, regenerate the set of configurations, and proceed. However, the distance in parameter space ($|\mathbf{a} - \mathbf{a}^*|$), for which the estimate of the gradient is accurate, shrinks as the number of electrons grows.

More advanced methods, see for example [1019–1021], allow one to optimize the variational energy with hundreds or thousands of parameters in a stable fashion. They incorporate a walk in parameter space, along with the walk in configuration space, and use lower-variance estimators for the hessian, the second derivative matrix.

23.3 The momentum distribution and single-particle density matrix

While classical systems in equilibrium always have a Maxwellian momentum distribution, the momentum distribution of a quantum system is different and often intimately related to its order. Delocalized bosons at low temperature will "condense" into the zero-momentum state (BEC) and have long-range order in the single-particle density matrix, a signature of the superfluid transition. Delocalized electrons at low temperature form a Fermi liquid, signalled by a discontinuity in the momentum distribution as shown in Fig. 5.2.

In Ch. 5 we defined the single-particle density matrix, $\rho(\mathbf{r}, \mathbf{r}')$ in Eq. (5.6) and its Fourier transform, the momentum distribution, $\rho(\mathbf{q})$ in Eq. (5.9). Since momentum is conjugate to position, the momentum distribution cannot be calculated as a simple average of a local function of the coordinates of the random walk; we need to evaluate a non-local operator involving an extra integral. McMillan [44] developed an ingenious and efficient algorithm for its calculation: using a many-body coordinate, R sampled from $|\Psi(R)|^2$ as before, we periodically sample a new coordinate "$\mathbf{s}$" (called the ghost particle below) in the simulation cell from a distribution $f(\mathbf{s})$ (usually taken to be uniform: $f(\mathbf{s}) = 1/\Omega$, where Ω is the volume of the simulation cell). Then the single-particle density matrix is estimated as:

$$\rho(\mathbf{r}, \mathbf{r}') = \left\langle \sum_{k=1}^{N} \frac{\delta(\mathbf{r} - \mathbf{r}_k)\delta(\mathbf{r}' - \mathbf{s})}{f(\mathbf{s})} \frac{\Psi(R'_k)}{\Psi(R)} \right\rangle. \tag{23.19}$$

Here $R'_k = R$, except $\mathbf{r}_k$ is replaced by $\mathbf{s}$; we need to do the same computation as for accepting a move in Eq. (23.8). [10]

The momentum distribution is computed either by recording values for $\rho(\mathbf{r}, \mathbf{r}')$ as above and Fourier-transforming or by the "direct" method. If we Fourier-transform Eq. (23.19) with respect to $\mathbf{r}$ and $\mathbf{r}'$ and normalize as in Eq. (5.9), then

$$\rho(\mathbf{q}) = \left\langle \sum_{k=1}^{N} \frac{e^{-i\mathbf{q}\cdot(\mathbf{r}_k-\mathbf{s})}}{(2\pi)^3 f(\mathbf{s})} \frac{\Psi(R'_k)}{\Psi(R)} \right\rangle. \tag{23.20}$$

The direct method is usually preferred to avoid the excessive memory requirements of histogramming $\rho(\mathbf{r}, \mathbf{r}')$ on a fine mesh. An example of such a calculation for the HEG is shown in Fig. 5.2.

In a Bose-condensed system, $\rho(\mathbf{q}) = \rho_0 \delta(\mathbf{q}) + \rho_{nc}(\mathbf{q})$, where ρ_0 is the condensate density. The delta function comes because $\rho(\mathbf{r} - \mathbf{r}') \to \rho_0$ at large $|\mathbf{r} - \mathbf{r}'|$: as we insert the ghost particle at $\mathbf{s}$ far from $\mathbf{r}_k$, the average value of $\Psi(R'_k)/\Psi(R)$ remains non-zero. For fermions such a non-zero value is only possible if at least two fermions are displaced: we can compute a pair condensate by generalizing Eq. (23.19).

23.4 Non-local pseudopotentials

The use of non-local pseudopotentials[11] is often needed to treat atoms with core states to reduce computer time. The computational burden imposed by core electrons is much worse in QMC compared with other methods; in [1022] it is estimated that the computer

---

[10] In the original McMillan algorithm, one gains a factor of M in efficiency for a pair product trial function by doing this operation with M ghost particles at the same time instead of with just one.

[11] The reader is advised to review the material in [1, Ch. 11].

time scales with nuclear charge Z as Z^5 because of the combination of more computer work per step and larger fluctuations of the local energy as Z increases. Hence, a calculation of a system of Ne atoms will take 10^5 times as long as that of the same number of H atoms to reach the same accuracy in the total energy. By using a pseudopotential, both the work per step and the fluctuations in the local energy are reduced but at the cost of introducing an approximation that cannot be easily controlled.

Let us consider how to use a "semi-local form" with a many-body wavefunction in the VMC method; see the discussion in Eqs. (11.15)–(11.17) in [1]. Suppose there is a single atom located at the origin and let $\Psi(R)$ be the many-body trial wavefunction. Then the local pseudopotential matrix element [1023, 1024] is given by the integral:

$$V_{PP} = \sum_{ilm} \int dR d\hat{s} \Psi^*(R_i') Y_{lm}(\hat{s}) v_l(r_i) Y_{lm}(\hat{r}_i) \Psi(R) \mapsto \sum_{i,l} \left\langle P_l(\cos(\theta)) v_l(r_i) \frac{\Psi(R_i')}{\Psi(R)} \right\rangle,$$

(23.21)

where R_i' equals R with the ith electron rotated to point in the direction $\hat{s}$ but keeping the same distance from the atom. Here, $\cos(\theta) = \hat{s} \cdot \hat{r}_i$.

To estimate this matrix element, we carry out a similar operation as we did with the momentum distribution. Suppose we have used the VMC procedure to sample a many-body coordinate R from $|\Psi(R)|^2$. We then estimate V_{PP} using the formula on the right-hand side of Eq. (23.21): rotating each electron i an angle θ about a random axis; the estimator is the quantity inside $\langle \cdots \rangle$. Typically, the non-local pseudopotential $V_l(r)$ is constructed to vanish outside a "core radius" r_c. This implies that only electrons for which $r_i < r_c$ need to be rotated. The local pseudopotential decays as Z/r but does not require a rotation; it is simply a scalar contribution to the local energy. Generalization to systems with many ions is straightforward.

Since V_{PP} depends linearly on the pseudopotential matrix element, there is no systematic error if we sample the angle randomly.[12] To reduce the variance coming from this term, one can perform several rotations of a given electron. The question then arises as to the most efficient procedure. If we choose the rotations to be special points on the surface of a sphere,[13] such as from a regular polyhedron, e.g., the vertices of a cube [1026], we will calculate angular momentum components up to a given order exactly. The orientation of the polyhedron needs to be periodically randomized to eliminate any bias.

We will see other difficulties with non-local pseudopotentials in Sec. 24.6. The reader should be cautioned against using pseudopotentials defined within a single-particle theory such as density functional theory for use with a correlated method such as VMC. It may be that the improvements in accuracy achieved by using a correlated wavefunction are offset by the inaccuracies of the pseudopotential. Similar considerations apply to other correlated methods, such as GW; see Sec. 12.3. Pseudopotentials specifically developed for quantum Monte Carlo calculations are discussed in [1027–1030].

---

[12] However, note that we lose the zero-variance property: there will be noise even with a perfect trial function unless the integral over $\hat{s}$ is performed exactly.

[13] This is an example of a more general Monte Carlo strategy called antithetic variates [1025].

23.5 Finite-size effects

Although systems in nature that we usually want to calculate properties of contain on the order of 10^{23} electrons, typical QMC calculations are limited to fewer than several thousand electrons. To minimize the effects of the small size of the simulation, periodic boundary conditions are often used, as discussed in Section 6.2. However, the remaining dependence of the energy on the size of the simulation cell is often the largest systematic error in QMC calculations.[14] As we discuss below, finite size effects are larger in QMC than in mean-field methods because electrons are explicitly represented. Fortunately, the errors can be minimized by several methods, as we now discuss.

The largest effect for systems with a Fermi surface comes from the kinetic energy because of the filling of the shells in **k**-space for calculations done in periodic boundary conditions. In the original VMC calculations [275] on the HEG, calculations were done with a sequence of N-electron systems, each having a closed shell. Fermi liquid theory was then used to extrapolate to the thermodynamic limit. Although this works well for homogeneous systems with small unit cells, it does not work as well for more complex systems, for example those with a larger unit cell, or even worse for systems such as a surface or a liquid that are not periodic in all three directions.

As described in Sec. 6.2, the most general conditions for the wavefunction with periodic boundary conditions are:

$$\Psi(\mathbf{r}_1 + \mathbf{L}, \mathbf{r}_2, \ldots) = e^{i\theta} \Psi(\mathbf{r}_1, \mathbf{r}_2, \ldots), \qquad (23.22)$$

where **L** is a lattice vector of the supercell. If the twist angle θ is averaged over (twist-averaged boundary conditions, or TABC), most single-particle finite-size effects arising from shell effects in filling the plane-wave orbitals are eliminated [1031], see Fig. 6.1. In fact, if the number of electrons in the supercell is allowed to vary with respect to the twist angle (the grand-canonical ensemble), there will be no single-particle finite-size effects within TABC.

One can understand the kinetic finite-size effects by looking at the momentum distribution defined in Eq. (5.9) since the kinetic energy per electron can be written (Eq. (5.10)) as:

$$t_N = \frac{(2\pi)^3}{2\Omega} \sum_{\mathbf{k}} k^2 \rho_N(\mathbf{k}), \qquad t_\infty = \frac{1}{2} \int d\mathbf{k} k^2 \rho_\infty(\mathbf{k}), \qquad (23.23)$$

where $\rho_N(\mathbf{k})$ and $\rho_\infty(\mathbf{k})$ are the momentum distribution for an N-electron and an infinite-electron system, respectively, and Ω is the volume of the supercell. Twist averaging gets rid of the largest approximation by replacing the summation in the left equation by the integral expression shown in the right equation. We note that the difference between a sum and an integral is dominated by the discontinuities and singularities in the integrand. For a metal the largest effects occur at the Fermi surface because of the discontinuity there. For an insulator, since the momentum distribution is analytic and smooth everywhere, the

---

[14] QMC is not the only computational method with this problem. Any method that computes an accurate energy of a finite correlated system will have similar problems, e.g., the exact diagonalization method.

kinetic energy of a finite system will approach the bulk limit much more rapidly in N. The remaining finite-size error in the kinetic energy comes from using $\rho_N(\mathbf{k})$ instead of $\rho_\infty(\mathbf{k})$.

Now the potential energy per electron can also be written as a sum (for finite N) or integral (for the bulk) in $\mathbf{k}$-space:[15]

$$v_N = \frac{1}{2\Omega} \sum_{\mathbf{k} \neq 0} \frac{4\pi}{k^2} S_N(\mathbf{k}), \qquad v_\infty = \frac{1}{2(2\pi)^3} \int d\mathbf{k} \frac{4\pi}{k^2} S_\infty(\mathbf{k}), \qquad (23.24)$$

where $4\pi k^{-2}$ is the Fourier transform of the Coulomb interaction. The charged structure factor, $S(\mathbf{k})$ (see Eq. (5.17)) is defined as:

$$S(\mathbf{k}) = \langle \frac{1}{N} | \sum_j e_j e^{i\mathbf{r}_j \cdot \mathbf{k}} |^2 \rangle, \qquad (23.25)$$

where the sum is over all particles, electrons, and ions, with charge e_j. See the left panel of Fig. 23.2. It is assumed that the total charge in the supercell is zero. The dominant finite-size error, the difference between the sum and the integral in Eq. (23.24), comes from the contribution near $\mathbf{k} = 0$. For small values of $\mathbf{k}$ the structure factor is given in Eq. (5.19) as $S(\mathbf{k}) = ck^2$, where $c = \left[(1 - \epsilon^{-1})/(16\pi n) \right]^{\frac{1}{2}}$ and ϵ is the bulk dielectric constant ($\epsilon = \infty$ for a metal). To get an estimate[16] of the finite-size error, we can just integrate over the sphere $k < k_0$ and determine k_0 by the condition that its volume equals $(2\pi)^3/\Omega$. One

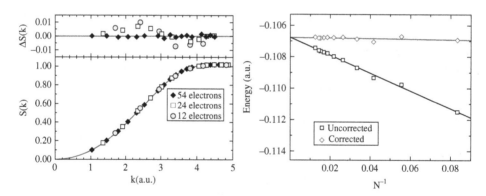

Figure 23.2. Electron gas (HEG) calculations at the density $r_s = 10$ using twist-averaged boundary conditions. The left figure shows comparison of the structure factors for different system sizes; it is seen that the structure factors for different-size cells lie on a common curve with deviations of less than 1%. The right figure is the energy per electron versus the inverse of the number of electrons. The upper symbols and curve have the correction for the potential energy obtained by integrating the structure factor for $k < 2\pi/L$, as discussed in the text. (Both figures from [1032].)

---

[15] Both the sum and the integral diverge for large values of k, but the difference between the two is well-defined. Also note that twist averaging does not convert the sum in Eq. (23.24) to an integral, since $S(\mathbf{k})$ is a two-particle quantity; the values of $\mathbf{k}$ that can be computed are vectors in the reciprocal lattice of the supercell independent of the twist angle.

[16] A more precise estimate of the finite-size error can be found by fitting the computed values of $S_N(\mathbf{k})$ to an appropriate analytic expression and evaluating the difference between the sum and the integral numerically.

gets a correction to the potential energy per electron of $2\pi c/\Omega$. Evaluating for a metal, we get a potential energy correction of $0.866/(Nr_s^{\frac{1}{2}})$ (in Hartrees per electron), where r_s is the electron density parameter. There is an equally important contribution to the kinetic energy because of changes in $\rho(\mathbf{k})$ caused by correlation [1032]. The procedure to do finite-size scaling based on correlation functions was introduced in [1032], and an analysis and comparison with other methods was performed in [1033].

23.6 VMC for lattice models

So far we have discussed QMC methods for electrons in the continuum, but VMC is very commonly used also for lattice models such as the Hubbard model, Sec. 3.2. It would take us too far astray to discuss this work in detail.

At first glance, all we need to do to treat lattice models is to restrict the coordinates of the electrons to lie on lattice sites. This simplifies the description, since the trial function values are only needed on a lattice; rather than dealing with continuous functions, the trial function is parameterized by a discrete set of values. For example, the orbitals or the Jastrow factor are a set of L^3 numbers, where L is the size of the lattice in one direction. Symmetry, if present, will further reduce this size by a large factor. No computation is needed for the orbital or Jastrow, it is simply a memory reference. The local energy is determined by finite differences of the wavefunction values. In addition, there is no need to worry about the continuity of the wavefunction.

The second change is that moves are sampled from a discrete distribution. Typically, a trial move will be to move a single electron to one of its nearest neighbors. Since there are only a finite number of possibilities, the heat bath algorithm discussed in Sec. 22.5 is optimal. However, for fermion systems, the wavefunction nodes occupy a finite fraction of the space; according to some reports, fully a third of all moves for the 2D Hubbard model encounter a zero of the wavefunction [1034]. In contrast, for VMC in the continuum, nodal surfaces are a set of measure zero and never happen. Moves to a node will be rejected, but they raise a concern about the ergodicity of the random walk: is it possible for the random walk to reach all configurations? Procedures for steps that explicitly jump over nodes might be needed.

The VMC method has been applied to numerous lattice models. In [1035] the appropriate long-range Jastrow wavefunctions are described, and calculations for the spin-1/2 Heisenberg model on a square lattice. The VMC results for the Heisenberg model are summarized in [1036, 1037]. For the fermion 2D Hubbard model, VMC calculations for the Mott transition are described in [1038] and results for the $t-t'$ model (see Sec. 17.6) with backflow trial functions are given in [1039]. The simpler Gutzwiller trial function for the Hubbard model is given in Eq. (3.3) and described in App. K.

23.7 Excitations and orthogonality

In this section we generalize VMC to compute excited-state energies. If the excited state is orthogonal by symmetry to lower-lying states then the generalization is not necessary; one just needs a trial wavefunction with the correct symmetry. Otherwise it is necessary to orthogonalize with lower states. One starts with a *basis*, consisting of $(m + 1)$ linearly independent many-body wavefunctions $\{\Psi_i\}$ with $0 \leq i \leq m$. We can generalize the Slater–Jastrow trial function to construct such a basis by making "electron–hole" excitations of the orbitals in the Slater determinant. If the orbitals came from a mean-field calculation, one can systematically substitute valence (occupied) orbitals with conduction (empty) states. Even if the orbitals were orthogonal within HF or DFT, multiplying by a Jastrow factor could cause the orbitals to no longer be orthogonal.

A general linear combination of the basis elements is $\Psi(R) = \sum_{i=0}^{m} d_i \Psi_i(R)$. The Rayleigh quotient for the energy:

$$E_V = \frac{\langle \Psi | \hat{H} | \Psi \rangle}{\langle \Psi | \Psi \rangle} \tag{23.26}$$

has stationary points with respect to variations of d_i. If we set $\partial E_V / \partial d_i = 0$ for all i, we obtain the generalized eigenvalue problem:

$$\sum_{j=1}^{m} \langle \Psi_i | \hat{H} | \Psi_j \rangle d_j^{(k)} = E_V^{(k)} \sum_{j=1}^{m} \langle \Psi_i | \Psi_j \rangle d_j^{(k)}. \tag{23.27}$$

By defining the matrix on the left-hand side as $\hat{H}$ and on the right-hand side as $\hat{N}$, we can write this equation concisely as:

$$\hat{H} d^{(k)} = E_V^{(k)} \hat{N} d^{(k)}. \tag{23.28}$$

There will be $m + 1$ independent solutions $(d^{(k)}, E_V^{(k)})$, unless $\hat{H}$ and $\hat{N}$ happen to be singular.

MacDonald's theorem [1040] asserts that the eigenvalues in the subspace spanned by $\{\psi_i\}$ are upper bounds to the exact excited-state energies of the many-body Schrödinger equation:

$$E_V^{(k)} \geq E_k \quad \text{for all} \quad 0 \leq k \leq m. \tag{23.29}$$

One recovers the usual variational theorem given in Eq. (23.1) for the lowest eigenvalue $E_V^{(0)}$. Note that the bounds are on the excited-state energies, not on energy differences, i.e., the excitation energies. To calculate other observables, an estimate of the kth excited-state wavefunction is $\Phi_k(R) = \sum_j d_j^{(k)} \Psi_j(R)$. If there is a symmetry difference between the states, e.g., they have different momentum, then the matrices $\hat{H}$ and $\hat{N}$ will factor into blocks since matrix elements between basis functions with different symmetries vanish.

To calculate the needed matrix elements of $\hat{H}$ and $\hat{N}$ we use a single Markov random walk. Let us suppose that we sample the function $\Pi(R) = |\Psi_G(R)|^2$, where Ψ_G is known as the guiding function[17] and generate a set of M many-body coordinates $\{R^{(k)}\}$. Using a random walk sampling $|\psi_G|^2$, the two matrices[18] can be estimated as:

$$\bar{N}_{ij} = \frac{1}{M}\sum_{k=1}^{M} f_i^*(R^{(k)})f_j(R^{(k)}), \quad \bar{H}_{ij} = \frac{1}{M}\sum_{k=1}^{M} f_i^*(R^{(k)})f_j(R^{(k)})E_{Lj}(R^{(k)}), \tag{23.30}$$

where $f_j(R) \equiv \Psi_j(R)/\Psi_G(R)$ is the weight of the jth state and $E_{Lj}(R)$ is its local energy. The energy differences will have a much smaller error than the individual energies do, since fluctuations unconnected with the excitation will cancel, an example of what is called in Monte Carlo "correlated sampling." See [1043] for details.

A fundamental problem with excited states is that with higher excitations, the wavefunctions get more complex and the approximations we use for the ground state are less accurate. An accurate variational treatment would require a much larger basis. Other methods, such as path-integral Monte Carlo, described in Ch. 25, which is formulated at finite temperature, are more appropriate when there are many states to sum over. In the correlation function quantum Monte Carlo method (Sec. 25.7) one uses the hamiltonian to project out lower-energy excitations from the wavefunction basis, thus achieving tighter upper bounds. If convergence can be achieved, one attains the exact energy within the statistical error.

As an application of the calculation of excited states, consider the Fermi liquid parameters defined in Sec. 3.4. Using the method described above, we can calculate the ground state and the lowest excitations of the system. For the homogeneous electron gas the ground state at high density consists of filled shells of plane-wave orbitals allowed by periodic boundary conditions; see Fig. 6.1. We consider excited states where a single electron from the last occupied shell is replaced by one in the first unoccupied shell. Because these excitations have different total momenta from each other and from the ground state, the states are orthogonal so calculation of the overlap matrix is not needed. Two different excitations, spin-parallel excitations and spin-antiparallel ones, are possible; the results are reported in [1043]. Because the electron and hole states will interact in a finite system, the excitation energy will have important finite-size corrections. An alternative procedure [1044, 1045] is to add or subtract a single electron from the ground state. This does not have a problem with the electron–hole interaction, but there are other finite-size effects to consider, and one can only calculate parameters having to do with a single-particle excitation such as the effective mass and bandwidth.

The effective mass is defined in Sec. 3.4. Two quite different QMC results for the 2D HEG are shown in Fig. 23.3 and compared with screened RPA and local field method results. The two different QMC calculations were done in a similar way, but the effective

---

[17] Optimizing the variance of the excitation energies [1041] finds that the optimal guiding function will have the form $\Psi_G(R)^2 = \sum_k |\Phi_k(R)|^2$; i.e., it is determined by the states to be calculated. This "guiding function" is non-negative and zero only where all states under consideration have zeros.

[18] Due to fluctuations, H will not be symmetric. One should not symmetrize it, since that will destroy the zero-variance property; see [1042].

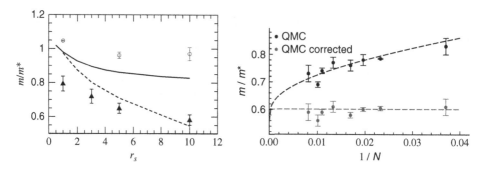

Figure 23.3. The left panel shows the inverse effective mass of the 2D HEG versus r_s as computed with two different ways of extracting the information from QMC; in [1044] (solid triangles), computed using k-values very close to k_F; in [1045] (open circles), averaged over excitations $0 \le k \le 1.3k_F$. Also shown are analytic calculations using a local field factor [1046] (dashed line), or using the screened RPA [1047] (solid line). The right figure shows the finite-size effects in the QMC effective mass calculations at $r_s = 10$, from [1044].

mass differs because of the way it is calculated from the QMC energies. Drummond and Needs [1045] fitted the excitation energy over a wide range of excitations $0 \le k \le 1.3k_F$ and specifically excluded values near the Fermi surface; the effective mass was determined by differentiating the analytic fit. Holzmann *et al.* [1044] only used excitations near the Fermi surface and applied a large finite-size correction to the effective mass, as shown in the right panel of Fig. 23.3. Resolving these different methods of analysis comes down to deciding the range of validity of Fermi liquid theory. For an infinite system, Fermi liquid theory only applies for excitations near the Fermi surface. Since excitations away from k_F acquire a finite lifetime, it is not clear whether the QMC excitation energies of the finite system correspond to quasi-particle peaks in the spectral function and whether using them can introduce a bias in the determination of the effective mass.

23.8 Strengths and weaknesses of VMC

The variational method is very powerful and intuitively pleasing. By only assuming a functional form for the wavefunction, one obtains an upper bound to the energy. In contrast to other many-body correlated methods, no further uncontrolled approximations need to be made. The only restriction on the trial function is that it must be computable. With a Slater–Jastrow trial wavefunction one can do calculations with thousands of electrons. To be sure, the numerical work has to be done very carefully, e.g., the convergence of the random walk has to be tested and dependence on system size needs to be corrected for. To motivate the methods to be described in the next two chapters, we list some of the intrinsic problems with the variational method.

- The variational method is biased to give a relatively lower energy to a simple state over a more complicated state. Consider the estimation of the melting density of the Wigner crystal phase of the homogeneous electron gas. This was estimated by VMC [275] to be at $r_s = 67$, while using more accurate methods [1048], at $r_s = 100$, a difference in density of $(100/67)^3 = 3.3$. This large bias occurs because the Wigner crystal wavefunction is simpler than that of the liquid phase; in the crystal phase the wavefunction is less correlated and dominated by the one-body term, which localizes the electrons around lattice sites. When we compare the liquid and crystal variational energies of the Slater–Jastrow trial function, the crystal energy will be closer to the exact result than that of the liquid, and, hence, the transition will be displaced in density. A second example is given by the ferromagnetic transition in the electron gas. There the wavefunction for the fully polarized system is simpler than for the unpolarized system, which results in an over-estimation of the stability of the ferromagnetic phase within VMC. This implies that the spin susceptibility at zero polarization is too small or even has the wrong sign. As the trial function form becomes closer to the exact wavefunction, this bias becomes smaller but in practice it may be difficult to eliminate it.

- The construction and optimization of trial functions for many-body systems can be time consuming. This allows the possibility of human bias if the optimization is stopped when the expected result is obtained. With improved optimization methods, particularly those that can be run as a "black box," this bias is less of a problem.

- The variational energy is insensitive to long-range order; it is dominated by correlations between nearby electrons and ions. If one is trying to compare the variational energy of a trial function with long-range order to one without, it is important that both functions have the same short-range flexibility and are both well-optimized. Only if this is done, can one have any hope of saying anything about whether long-range order is energetically advantageous. Also note that the error in the variational energy is second order in the accuracy of the trial function, while the error of other properties is first order. Thus, even if variational energies are accurate, correlation functions can be inaccurate.

- The result of a VMC calculation is strongly influenced by the form of the trial wavefunction. For example, suppose the determinant of the trial function describes a metal so that the momentum distribution has a discontinuity at the Fermi surface. Then the momentum distribution after we apply the Jastrow factor will in most cases have a discontinuity at the same place, although its magnitude will change. This does not imply that the true wavefunction has a sharp Fermi surface; it only reflects the properties of the assumed determinant.

In the next two chapters we will describe methods that are less biased by the form of the trial wavefunction.

SELECT FURTHER READING
McMillan, W. L., "Ground state of liquid ^{4}He," *Phys. Rev.* **138**, A442, 1965. The first many-body VMC calculation with a very clear description for a bosonic system.
Ceperley, D., Chester, G. V., and Kalos, M. H., "Monte Carlo simulation of a many-fermion system," *Phys. Rev. B* **16**, 3081, 1977. Discusses the generalization of VMC to fermion systems.

Fahy, S., Wang, X. W., and Louie, S. G., "Variational quantum Monte Carlo nonlocal pseudopotential approach to solids", *Phys. Rev. B* **42**, 3503, 1990. Details of the implementation of non-local pseudopotentials in VMC.

Ceperley, D. M. and Bernu, B., "The calculation of excited state properties with quantum Monte Carlo," *J. Chem. Phys.* **89**, 6316, 1988. Details of the excited-state method.

Exercises

23.1 (a) Determine the variance of the variational energy for the reweighting formula, Eq. (23.18). In particular, how much is the variance increased because $\Pi_0(R) \neq |\Psi(R, \mathbf{a})|^2$? (b) Suppose we wish to calculate the energy difference between two wavefunctions using the reweighting technique, Eq. (23.18). What probability distribution $\Pi_d(R)$ will give the minimum variance for the energy difference? Neglect any correlation between the sampled points.

23.2 Consider the VMC algorithm with the transition probability $(4\pi\delta\tau)^{-3/2}\exp(-(\mathbf{d} - \alpha\mathbf{F})^2/(2\Delta\tau))$, where $\mathbf{d} = \mathbf{r} - \mathbf{r}'$. Optimize the value of the acceptance probability in Eq. (23.7) by doing a Taylor expansion with respect to α.

23.3 Prove Eq. (23.16) using Green's identity. This provides a nice check on the energy evaluation and on the sampling.

23.4 Prove the update formula, Eq. (23.11).

23.5 Go through all the computational steps in the VMC algorithm in Sec. 23.1 for a single "pass" (i.e., moving all of the electrons once). Classify them as $O(1)$, $O(N)$, or $O(N^2)$. Overall, how does the number of computer operations scale with the number of electrons?

23.6 The pressure at zero temperature is the derivative of the energy with respect to volume Ω:

$$P = -\frac{dE_V}{d\Omega} = \frac{1}{3\Omega}(2T_V + V_V). \tag{23.31}$$

Find under what conditions the second equation (the virial form) applies. Here $E_V = T_V + V_V$ is the total energy, and T_V and V_V are the kinetic and potential energies evaluated using a trial wavefunction. Prove these two formulas are equal if all interactions are coulombic and if the energy, E_V, is minimized with respect to all parameters that have units of length.

24

Projector quantum Monte Carlo

... as suggested by Fermi, the time-independent Schrödinger equation ... can be interpreted
as describing the behavior of a system of particles each of which performs a random walk,
i.e., diffuses isotropically and at the same time is subject to multiplication, which is
determined by the value of the point function V.

N. Metropolis and S. Ulam, 1949

Summary

In the projector quantum Monte Carlo method, one uses a function of the hamiltonian
to sample a distribution proportional to the exact ground-state wavefunction, and
thereby computes exact matrix elements of it. An importance sampling transformation makes the algorithm much more efficient. In this chapter we introduce
and develop the diffusion Monte Carlo method, which involves drifting, branching random walks. For any excited state, including any system with more than two
electrons, one encounters the sign problem, limiting the direct application of these
algorithms for most fermion systems. Instead, by using approximate fixed-node or
fixed-phase boundary conditions, one can achieve efficiency similar to variational
Monte Carlo. We also discuss the application of the projector method in a basis of
Slater determinants.

In this chapter, we discuss a different quantum Monte Carlo method, projector Monte Carlo
(PMC). This general method was first suggested by Fermi [1049]; see the quote at the start
of this chapter by two of the inventors of the Monte Carlo method. An implementation
of PMC was tried out in the early days of computing [1050]. Advances in methodology, in particular importance sampling, resulted in a significant large-scale application:
the exact calculation of the ground-state properties of 256 hard-sphere bosons by Kalos,
Levesque, and Verlet [1051] in 1974. Calculations for electronic systems and the fixed-
node approximation were introduced by Anderson [1052, 1053]. One of the most important
projector MC algorithms, the diffusion Monte Carlo algorithm with importance sampling
for fermions, was used to compute the correlation energy of homogeneous electron gas by
Ceperley and Alder [109] in 1980; the resulting HEG correlation energy was crucial in the
development of density functional calculations.

24.1 Types and properties of projectors

In this method, a many-body projector $G(R, R') = \hat{G}$ is repeatedly applied to filter out the exact many-body ground state from an initial state; the operation of the projector is carried out with a random walk, hence the name of this class of methods. Let us denote the initial wavefunction by $\Psi^{(0)}(R)$. Then a sequence of many-body wavefunctions is defined by:

$$\Psi^{(n+1)}(R) = \int dR' G(R, R')\Psi^{(n)}(R'). \qquad (24.1)$$

Three different, but closely related projectors have been used:

$$
\begin{array}{llll}
\hat{G}_D & = e^{-\Delta\tau(\hat{H}-E_T)} & \text{diffusion MC} & \text{DMC} \\
\hat{G}_L & = (1 - \Delta\tau(\hat{H} - E_T)) & \text{lattice projector MC} & \text{LPMC} \\
\hat{G}_R & = (1 + \Delta\tau(\hat{H} - E_T))^{-1} & \text{Green's function MC} & \text{GFMC}
\end{array}
\qquad (24.2)
$$

All three are defined in terms of the many-body hamiltonian, $\hat{H}$, from which a zero of energy, the so-called trial energy E_T, has been subtracted; E_T is used to control the normalization, as discussed below. The time step $\Delta\tau$ has been defined so that the three projectors will have the same convergence rate after many iterations, i.e., at large n in Eq. (24.1). In contrast to the "real-time propagator" in Eq. (5.20), these projectors can be thought of as being propagators in "imaginary time"; they can be sampled as random walks since, as we discuss below, the matrix elements in coordinate space are strictly positive. Real-time propagation is very difficult because of the rapid oscillation of the phase. In Sec. 25.6 we explore the connections between imaginary time, thermal properties, and real time.

To find the effect of applying a projector $\hat{G}(\hat{H})$ (the formulas in this paragraph apply to all three projectors) to the initial function, let E_i and $\Phi_i(R) = |i\rangle$ be the exact many-body energies and normalized many-body eigenfunctions[1] and expand $\Psi^{(0)}$ in this basis:

$$\Psi^{(0)}(R) = \sum_i \Phi_i(R)\langle i|\Psi^{(0)}\rangle. \qquad (24.3)$$

Since the states are eigenfunctions, after n iterations we find:

$$\Psi^{(n)}(R) = \sum_i G(E_i)^n \Phi_i(R)\langle i|\Psi^{(0)}\rangle. \qquad (24.4)$$

As seen in Fig. 24.1 and can be shown by expanding the effect of the projector on a basis of exact eigenfunctions, only the lowest-energy state having a non-zero overlap with the

---

[1] In the following we only need to consider many-body states having non-zero overlap with $\Psi^{(0)}$, thus $|0\rangle$ is the lowest-energy state with $\langle\Psi^{(0)}|0\rangle \neq 0$, $|1\rangle$ the next lowest, etc. For example, if $\Psi^{(0)}$ is a fermion wavefunction with zero momentum and zero spin, we only need to consider other exact fermion states with zero momentum and zero spin.

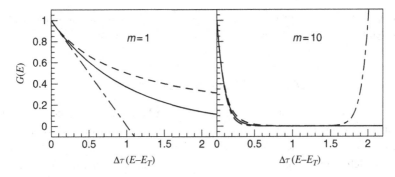

Figure 24.1. The effects of the three projectors after 1 and after 10 iterations. The solid line is $G_D(E)$, the dashed line $G_R(E)$, and the long dash–short dash line $G_L(E)$. The projectors are defined to have the same value and slope at $\Delta\tau = 0$. All three exponentially project out states with energies $E - E_T \gg (n\Delta\tau)^{-1}$ for large n. Note that G_L has a very different behavior at large E.

initial trial function survives after many iterations. The rate of convergence to the ground state is determined by the gap $e_g \equiv E_1 - E_0$ to the next state having an overlap with the initial state and by the (imaginary) "time step" $\Delta\tau$:

$$\lim_{n\to\infty} \Psi^{(n)}(R) = e^{-n\Delta\tau(E_0-E_T)}\left[\Phi_0(R)\langle 0|\Psi^{(0)}\rangle + e^{-n\Delta\tau e_g}\Phi_1(R)\langle 1|\Psi^{(0)}\rangle + \cdots\right]. \quad (24.5)$$

The "trial energy" E_T, and the overlap $\langle 0|\Psi^{(0)}\rangle$, determine the normalization of $\Psi^{(n)}$ since:

$$\lim_{n\to\infty} \langle\Psi^{(n)}|\Psi^{(n)}\rangle = e^{-2n\Delta\tau(E_0-E_T)}|\langle 0|\Psi^{(0)}\rangle|^2. \quad (24.6)$$

To keep the normalization of $\Psi^{(n)}$ asymptotically constant for large n requires that $E_T = E_0$. We can use the dependence of the normalization of $\Psi^{(n)}$ on the iteration count n to estimate both the exact ground-state energy and the overlap of the initial state with the ground state.

Which of the three projectors is the most appropriate to use? For a lattice system, $\hat{G}_L$ is a good choice. Because it is proportional to $\hat{H}$, it can easily be implemented without any time-step errors [1035]. However, for a continuum problem, $\hat{G}_L$ cannot be used since high-energy eigenstates, those for which $\Delta\tau(E_i - E_T) > 2$, will be amplified by the projection. Since a lattice model has finite bandwidth, one can take $\Delta\tau < 2/(E_{max} - E_{min})$ to guarantee convergence, where E_{max} and E_{min} are upper and lower bounds to the many-body eigenvalues.

The GFMC projector $\hat{G}_R$ was the first to be used on a large-scale application to quantum systems [1051]. This and subsequent calculations were done using the "continuous time" Monte Carlo algorithm[2] discussed in Ch. 18, which does not have time-step errors. Since the DMC projector $\hat{G}_D$ is simpler, it is the most common choice for continuum problems,

---

[2] GFMC was not presented as a continuous-time algorithm, and, in fact, the sampling of imaginary time was buried in the procedure to sample positions. See [1054] for a discussion of the general GFMC procedure.

even though it has a time-step error. Although we will confine the discussion to DMC, most of the developments can be carried over directly to the other projectors.

24.2 The diffusion Monte Carlo method

How can we implement the sampling of the diffusion Monte Carlo projector, $\hat{G}_D$? Clearly the expansion in Eq. (24.4) is impractical since, except for a few analytically solvable models, we do not know the many-body eigenfunctions and energies. The idea, attributed to Fermi, sketched in the quote at the start of this chapter, is that the projection can be done by random walks. In the next chapter, Sec. 25.4, we will encounter an alternative but closely related implementation, using path integrals instead of branching. The original applications in the 1950s used the "pure diffusion Monte Carlo" algorithm, which used weighting factors to account for the potential energy. Branching (duplicating or killing random walks) is much more computationally efficient and is commonly used today instead of weighting.

There are two ways to think about this algorithm: either the discrete time (integral formulation) or the continuous time (differential formulation). Please keep in mind that these are two different representations of the same stochastic process. Because an algorithm uses, in practice, discrete time, the integral formulation is used for rigorous analysis, while the differential formulation is more intuitive.

The integral formulation

Since we do not have an explicit expression for $\hat{G}_D$, we need to make a short time expansion. There is a rigorous mathematical result underpinning quantum Monte Carlo: Trotter's formula [1055, 1056]. Consider two operators, $\hat{T}$ and $\hat{V}$. Under general conditions[3] Trotter's formula holds:

$$e^{-\tau(\hat{T}+\hat{V})} = \lim_{n\to\infty} \left(e^{-\Delta\tau\hat{T}}e^{-\Delta\tau\hat{V}}\right)^n,\qquad (24.7)$$

where $\Delta\tau = \tau/n$. An intuitive proof is to note that the corrections to an individual term are proportional to the commutator $[\tau\hat{T}/n, \tau\hat{V}/n]$, which scales as $O(1/n^2)$. The error in the right-hand side of Eq. (24.7) will contain n such corrections, so the total error is $O(1/n)$ and vanishes as $n \to \infty$. Now evaluate the propagators for the potential and kinetic[4] operators $\hat{V}$ and $\hat{T}$ in coordinate space:

$$\langle R|e^{-\Delta\tau\hat{T}}|R''\rangle = (2\pi\,\Delta\tau)^{-3N/2}\exp(-(R-R'')^2/(2\Delta\tau)) \qquad (24.8)$$

$$\langle R''|e^{-\Delta\tau(\hat{V}-E_T)}|R'\rangle = e^{-\Delta\tau V(R')}\delta(R''-R'). \qquad (24.9)$$

---

[3] This will be true for the case that $\hat{T}$ is the non-relativistic kinetic operator and $\hat{V}$ the Coulomb interaction.

[4] The kinetic projector has to be modified in periodic boundary conditions to make it periodic; these effects are negligible when $\Delta\tau \ll L^2$.

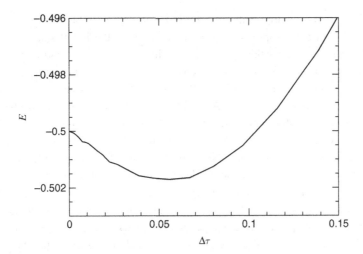

Figure 24.2. The energy of the hydrogen atom as computed using Eqs. (24.10) and (24.20) as a function of time step $\Delta \tau$. By using spherical symmetry, it reduces to a one-dimensional problem. The integral needed in the DMC projection can be done using summation over a uniform radial grid instead of a Monte Carlo method. For stability, the potential was bounded below by $-\Delta \tau^{-1}$.

Putting Eqs. (24.8) and (24.9) together and integrating out the intermediate coordinate, R'', the coordinate-space approximation[5] for the projector is:

$$\hat{G}_D = \langle R | e^{-\Delta \tau (\hat{H} - E_T)} | R' \rangle \sim (2\pi \, \Delta \tau)^{-3N/2} \exp[-\frac{(R - R')^2}{2\Delta \tau} - \Delta \tau (V(R') - E_T)]. \quad (24.10)$$

The first factor is the projector for diffusion and the second is multiplication of the distribution by a positive scalar. Trotter's formula says that if $\Delta \tau$ is sufficiently small, repeated application of Eq. (24.10) to a distribution will result in a distribution arbitrarily close to applying $\hat{G}_D$ to that distribution. This is illustrated for the hydrogen atom in Fig. 24.2. Even though the ionic potential is singular and must be bounded below at distances where the electron closely approaches the proton, the exact energy is obtained for a sufficiently small time step. Later we will make improved algorithms, allowing for a quicker convergence in the time step, $\Delta \tau$.

Note that the diffusion is localized in a radius of $\sim \Delta \tau^{\frac{1}{2}}$, or for n iterations $\sim (n\Delta \tau)^{\frac{1}{2}}$. Hence, the projection corrects the initial wavefunction, $\Psi^{(0)}$, for features on this distance scale. It is important that the initial state be correct on longer-length scales. Typically we would like the number of iterations $n \gg 1/\Delta \tau$; this fixes the physical scale of corrections to the initial distribution to be greater than a bohr, the size of an atom.

We have now represented the projection operator as a high-dimensional integral: for N electrons and n iterations, there are $3Nn$ integrals. Such a high-dimensional integral

[5] This form is not symmetric with respect to R and R'. One can make much better symmetric approximations, that are accurate at larger $\Delta \tau$, but the "primitive" form defined here is sufficient for convergence.

has to be done with a random walk procedure. An initial ensemble of $P^{(0)}$ configurations is constructed with a Metropolis sampling procedure for the initial state $\Psi^{(0)} = \Psi(R)$. For the moment, we discuss the case where the initial state is real and non-negative, for example, for one or two electrons, as discussed in Sec. 6.4. Configurations in the next "generation" are constructed by random diffusion of all the electrons with a mean-squared step size of $\Delta\tau$ for every coordinate. After all $3N$ electron coordinates have been moved, resulting in a new "configuration" R', we make a variable number of copies of R'; on average we make $m = \exp[-\Delta\tau(V(R) - E_T)]$ copies. The simplest and best way of doing this is to add a uniform random number in $(0, 1)$ to m and discard the fractional part. If $V(R) > E_T$ this might result in no copies (death of the walk) or one copy. On the contrary, for $V(R) < E_T$ one might have a single copy or multiple copies. This process of making a variable number of copies of the current configuration is called branching.[6] In future generations, these copies propagate independently of each other. This process of diffusion and branching is repeated until convergence is reached. The average number of random walks, the population, will be asymptotically stable if $E_T = E_0$, as shown in Eq. (24.6).

The above procedure, depicted in Fig. 24.3 for a simple harmonic oscillator, is a Markov process where the state of the walk in the nth generation is $\{P^{(n)}; R_1, R_2, \ldots, R_{P^{(n)}}\}$, where R_n is a "walker," and $P^{(n)}$ is the number of walkers or, in other words, the population. By the theorem discussed in Ch. 22 on Markov processes, it has a unique stationary distribution, constructed to be the exact many-body wavefunction $\Phi_0(R)$. However, the population fluctuates from step to step and needs to be stabilized, as discussed below.

From Eq. (24.10) we see that the projector can be interpreted as a transition probability since $\langle R|\hat{G}_D|R'\rangle \geq 0$ for all (R, R'). Then, if $\Psi^{(0)} \geq 0$, the process described above can be interpreted as a probability distribution and $\Psi^{(n)}(R) \geq 0$ for all n.[7]

The continuous-time picture

Define a continuous imaginary-time variable, $\tau = n\Delta\tau$. Then the projection $\hat{G}_D(\tau)$ applied to the initial wavefunction is:

$$\Psi(R; \tau) \equiv e^{-\tau(\hat{H} - E_T)} \Psi(R; 0). \tag{24.11}$$

Differentiating with respect to τ, we find the Schrödinger equation in imaginary time:

$$-\frac{d\Psi(R; \tau)}{d\tau} = (\hat{H} - E_T)\Psi(R; \tau) = -\frac{1}{2}\sum_{i=1}^{N}\nabla_i^2\Psi(R; \tau) + (V(R) - E_T)\Psi(R; \tau). \tag{24.12}$$

---

[6] In an alternative procedure, pure diffusion Monte Carlo [1057], there is no branching, instead weights are used. This procedure becomes less efficient as the system size increases. In practice, a combined procedure is often used: weights are carried for a certain number of steps, but when the weight either gets too large or too small, branching is triggered.

[7] Because the process can be interpreted as a random walk, we can conclude that the ground state of any hamiltonian of the form $-\frac{1}{2}\nabla^2 + V(R)$ can be assumed to be real and non-negative as long as $V(R)$ is a scalar operator, a "local" potential. If, in addition, the potential is symmetric under particle exchange, then the ground state must also be totally symmetric under particle exchange, i.e., bosonic. The random walk being ergodic is equivalent to the ground state being non-degenerate.

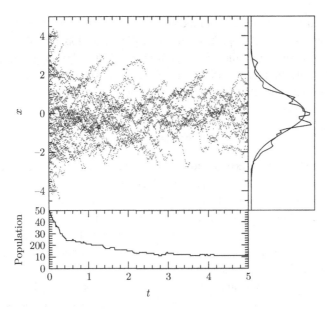

Figure 24.3. The evolution of a projector Monte Carlo calculation for a harmonic oscillator in 1D with a potential $V(x) = \frac{1}{2}x^2$. No importance sampling was used in this illustration. The position of the walkers in imaginary time is shown in the central panel, the population as a function of time in the bottom panel, the estimated final wavefunction in the right panel; the smooth curve represents the exact ground state. Fifty walks were initiated uniformly in $(-4,4)$. Walks with $|x| > 1$ can be killed, walks with $|x| < 1$ can branch to multiple copies. The population drops to 11, there was no control feedback applied as in Eq. (24.6); on average it will equal the initial population times the overlap of the initial with the exact wavefunction; see Eq. (24.5).

Break up this equation at small time into two evolution equations:

$$-\frac{d\Psi(R;\tau)}{d\tau} \sim -\frac{1}{2}\sum_{i=1}^{N}\nabla_i^2\Psi(R;\tau), \quad -\frac{d\Psi(R;\tau)}{d\tau} \sim (V(R) - E_T)\Psi(R;\tau). \quad (24.13)$$

The left equation describes diffusion of a "walker" in $3N$-dimensional space; the right equation, the growth or death (branching) of a walker. Trotter's formula implies that to simulate the evolution of $\Psi(R;\tau)$ in Eq. (24.12), we can divide time into small steps, and alternate between the diffusion process and the branching process. Time-step errors arise because the potential depends on position, so that it varies between the beginning and end of a diffusion step.

A difference between lattice models and continuum models can be understood within the continuous-time picture. In the continuum, the random walk must be continuous in space as a function of time, i.e., $R(\tau)$ must be a continuous function of τ. This implies that if the random walk goes from one region to another, there must be at least one time where the walk is on the surface separating the two regions. This property is implicitly used in the fixed-node method, Sec. 24.4, where we replace antisymmetry with a boundary condition on the nodes separating positive wavefunction values from negative ones. In contrast, if the potential contains a non-local operator, such as from a pseudopotential, the random walk can jump from one point to another in a discontinuous fashion, leading to more complicated boundary conditions. In the Hubbard model, the random walk hops from one lattice site to its neighbors and one does not have the same concept of continuity.

Importance sampling

The projection method introduced up to this point becomes very inefficient as the number of electrons increases, because the branching fluctuations are not controlled. For example, in a system of electrons and ions, the potential can go to both positive and negative infinity, resulting in death or the creation of many new walkers. In the example of the hydrogen atom, Fig. 24.2, we had to limit the magnitude of the potential for small r to get results that converged quickly in the time step, i.e., to "regularize" the singularity at $r = 0$. One can vastly improve the situation by using the technique of importance sampling [1051, 1058]; importance sampling is applicable to any projector method, as well as for many other problems in simulation. *Importance sampling* means to make a judicious choice of the sampling function so as to improve efficiency. Let us multiply Eq. (24.1) by a guiding wavefunction,[8] Ψ_G, and define $f^{(n)}(R) = \Psi_G(R)\Psi_n(R)$. In the integral formulation, the evolution equation for f is:

$$f^{(n+1)}(R) = \int dR' \mathcal{G}(R \leftarrow R') f^{(n)}(R'), \qquad (24.14)$$

where the importance-sampled projector

$$\mathcal{G}(R \leftarrow R') = \Psi_G(R)\langle R|e^{-\Delta\tau(\hat{H}-E_T)}|R'\rangle\Psi_G(R')^{-1}. \qquad (24.15)$$

Since the new importance-sampled projector is not symmetric in its arguments, we use $\mathcal{G}(R \leftarrow R')$ to emphasize that the walk goes from R' to R. The effect of importance sampling is to push the random walk to regions where the guiding function is larger. The steady-state distribution for Eq. (24.14) is now $\lim_{n\to\infty} f^{(n)}(R) = c\Psi_G(R)\Phi_0(R)$; for a guiding function close to the ground state, $\Psi_G(R) \sim \Phi_0(R)$, this distribution is similar to the quantum probability distribution $|\Phi_0|^2$. To make the initial state close to the final state so that it converges with fewer iterations, the initial conditions on the projection are typically taken to be $f^{(0)}(R) \propto \Psi_G(R)\Psi(R)$.

In the continuous-time picture, $f(R, \tau)$ satisfies the evolution equation [109, 1059]:

$$-\frac{\partial f(R; \tau)}{\partial \tau} = -\frac{1}{2}\sum_i \nabla_i(\nabla_i f(R; \tau) - f(R; \tau)\mathbf{F}_i(R)) + [E_L(R) - E_T]f(R; \tau), \quad (24.16)$$

---

[8] We distinguish here between the trial wavefunction Ψ used in VMC and DMC to compute the energy and the guiding function, which must always be real and non-negative. In practice, we often use $\Psi_G = |\Psi|$, but use of a guiding function with nodes will cause problems in some situations.

where $E_L(R) = \Psi_G^{-1} \hat{H} \Psi_G$ is the local energy of Ψ_G and $\mathbf{F}_i(R) = 2\nabla_i \ln \Psi_G(R)$ is the "quantum force"[9] of Ψ_G. The first and last terms on the right-hand side are familiar from Eq. (24.12), i.e., diffusion and branching. The branching is now controlled since it is no longer the bare potential energy but the local energy of the guiding function. As the guiding function approaches the exact eigenfunction, the branching factor $\exp(-\Delta\tau(E_L(R) - E_T))$ approaches unity for all R; thus, a sufficiently good guiding function can control the branching. This means that a larger time step and a smaller population can be used and the estimate of the ground-state energy will have smaller fluctuations.

Let us now understand the effect of the second term in Eq. (24.16) by writing down the evolution equation for just that term:

$$\frac{\partial f(R; \tau)}{\partial \tau} = \frac{1}{2} \sum_i \nabla_i (f(R; \tau) \mathbf{F}_i(R)). \tag{24.17}$$

If $\mathbf{F}_i(R))$ is constant in a region around R_0, then any function of the form $f(R + \tau \mathbf{F}_i(R_0)/2))$ solves this limited evolution equation in that region; it has the form of a uniform drift of the distribution, pushing the random walk in the direction of $\mathbf{F}$. Using Trotter's formula for the effect of the three terms in Eq. (24.16), we obtain

$$\mathcal{G}(R' \leftarrow R) \approx (2\pi\Delta\tau)^{-3N/2} \exp\left[-\frac{(R' - R - \frac{1}{2}\Delta\tau F(R))^2}{2\Delta\tau} - \Delta\tau(E_L(R) - E_T)\right]. \tag{24.18}$$

Note the first term in the exponent is the same as in Eq. (23.12) of variational Monte Carlo; in Ch. 23 it was motivated by getting a higher acceptance probability; here it is the short-time solution of the imaginary-time Schrödinger equation with importance sampling.

Improved propagation

We now discuss several ways to improve the propagation, that is to make the time-step error smaller, allowing larger time steps, and to make the algorithm more robust.

A major improvement is to demand that the propagation satisfy detailed balance [1059, 1060], defined in Eq. (22.6); this sometimes goes by the name *hybrid Monte Carlo*. Using the definition of the importance-sampled projector in Eq. (24.15) and the hermitian property of the hamiltonian, the detailed balance condition follows:

$$\Psi_G^2(R)\mathcal{G}(R' \leftarrow R) = \Psi_G^2(R')\mathcal{G}(R \leftarrow R'). \tag{24.19}$$

An approximate projector such as Eq. (24.18) will not exactly satisfy this equation except in the limit of $\Delta\tau \to 0$. In the hybrid algorithm, we make a trial move using $\mathcal{G}$, but that move is accepted as in the Metropolis algorithm, Eq. (22.8). One might think that accepting using Ψ_G^2 would drive the random walk to the distribution $\Psi_G^2(R)$. It does not because in

---

[9] So named because it is proportional to the force that would be needed for a classical particle to be distributed according to the quantum distribution.

contrast to Metropolis, $\mathcal{G}$ is not a normalized transition probability since the random walks can branch.

With this modification, very bad moves, for example where an electron moves very close to another, are likely to be rejected. If such a move were to be allowed, the succeeding step could also be bad, since $\mathbf{F}_i(R)$ would be large. These bad moves only occur because of time-step error. As the guiding wavefunction approaches an eigenstate, branching becomes very rare; detailed balance is enough to ensure that Ψ_G^2 is sampled exactly, hence, time-step errors will vanish. A second benefit of using rejections is that one has a way of setting the time step internally. The "rule of thumb" is to reduce $\Delta\tau$ until fewer than 1% of the steps are rejected.[10] If precise energies are needed, one needs to perform a careful study of energy versus time steps, as in Fig. 24.2.

Note that there is an unfortunate side-effect of rejection, walkers could get stuck at certain locations; if a configuration is rejected more than a few times in a row, it should be deleted. Other useful procedures are to limit the branching and the quantum force, since if either is too large, it is an indication that the time step is too large in that part of configuration space.

Another improvement to the DMC algorithm is to move electrons one at a time, instead of all simultaneously, just as is done in VMC. The motivation is the same in DMC and in VMC: for a many-electron system, simultaneous movement of all electrons will, with high probability, lead to at least one electron in a very bad position, resulting in either rejection or a bad branching factor. To be successful, the time step $\Delta\tau$ would have to get smaller as the number of electrons increases. On the contrary, if we move electrons one at a time, then we simply reject the trial positions for a few electrons, while most electrons go to new positions, so the necessary time step is independent of the number of electrons. The mathematical justification again comes from Trotter's formula: we can break up the evolution operator on the right-hand side of Eq. (24.16) into drift-diffusion operators for individual electrons but keeping a single local energy term. In effect, this means we apply the VMC algorithm to move the electrons and at the end of one VMC pass, when all of the electrons have had one attempted move, we use the average local energy at the start and end of the "pass" to determine the branching factor of the updated walker.

Further ideas for reducing time-step errors and a detailed description of the DMC algorithm are given in [1061].

Energy estimation

The primary job of projector Monte Carlo is to make an accurate estimate of the ground-state energy. There are two different, but related, ways to estimate it. As shown

---

[10] The idea behind this rule is that most electrons (99%) are in "smooth regions" of configuration space, where an expansion to second order of the propagator is justified. Roughly, we should recover 99% of the correlation energy missing in E_V.

in Eq. (24.6), the normalization of the projected wavefunction is one way. Define the population (i.e., the number of walkers) in the nth iteration (the set of walkers at a given iteration is defined as a *generation*) as $P^{(n)} = \int dRf^{(n)}(R)$. Once convergence is reached, $P^{(n+1)} = \exp(-\Delta\tau(E_0 - E_T))P^{(n)}$. Solving this equation for E_0 defines the *generational estimate* of the ground-state energy:

$$E_0^{(n)} = E_T - \frac{1}{\Delta\tau}\ln(P^{(n+1)}/P^{(n)}). \tag{24.20}$$

Branching and, hence, changes in population arise because of the difference between the local energy and its average, $(E_L(R) - E_T)$.

The *local energy estimate* of ground-state energy is the average local energy[11] of the guiding function in the equilibrated population:

$$E_0^{(n)} = \frac{\int dRf^{(n)}(R)E_L(R)}{\int dRf^{(n)}(R)} \mapsto e_0^{(n)} = \frac{\sum_i^{P^{(n)}} E_L(R_i)}{P^{(n)}}. \tag{24.21}$$

This follows by substituting for $f^{(n)} = \Phi_0\Psi_G$. It is equivalent to the generational estimate in the limit of small $\Delta\tau$.

In DMC there is an approximate relation [1022] between the variance of the ground-state energy estimate and the error of the variational energy:

$$\sigma_{DMC}^2 \simeq [\sigma^2 + 2(E_V - E_0)/\Delta\tau]/P, \tag{24.22}$$

where σ^2 is the variance of the trial function energy defined in Eq. (6.15), E_V is the variational energy (Eq. (6.14)), and P is the total number of walkers used in the evaluation. For small $\Delta\tau$ it is the variational energy, *not* the variance of the trial function, that controls the rate of convergence of the DMC energy. The reason why the variational energy enters instead of the variance is because of the autocorrelation of the local energy in imaginary time arising from both the diffusion step and the branching step. For small time steps, the local energy is correlated over many steps. In order to calculate error estimates for the ground-state energy one needs to take into account this autocorrelation, as discussed in Sec. 22.4. However, often the variational energy and the variance are proportional, e.g., see Fig. 6.6.

Population control

The first time we do a projector MC calculation, we may not have a good estimate of the ground-state energy and thus we do not know what to use for the trial energy. Even if the exact energy were known exactly, it is necessary to actively control the number of walkers

---

[11] It will be useful to distinguish between a trial function Ψ and a guiding function Ψ_G. In case they are different, we need to use a weight factor in taking averages, as in the next section.

since the normalization of the wavefunction is not fixed by the procedure; it is a Markov process, so there is no memory in the algorithm of the initial value of $P^{(0)}$. The population can either shrink in size, possibly even to zero, or it can grow so much that a single iteration will take too long to complete. Active population control is always needed as the number of iterations increases. We discuss two procedures that have been used to control the population.

In the *comb* procedure, branching is delayed until all of the branching factors for the current ensemble of walkers are known. Let w_i for $1 \leq i \leq P^{(0)}$ denote the branching factors for all of the walkers in the current generation. We then make m_i copies of configuration R_i, so that $\sum_i m_i = P^{(0)}$ and $\langle m_i \rangle = P^{(0)} w_i (\sum_i w_i)^{-1}$. The $\langle \cdots \rangle$ denote averages over random numbers used in the procedure. The random numbers can be chosen so that exactly $P^{(0)}$ walkers will be in the next generation (see Ex. 24.6). The comb procedure solves both problems mentioned above, but at the cost of introducing a bias in the distribution, which we analyze next.

Let us divide configuration space into two regions. In region "A," where $E_L(R) < E_0$, excess branching will occur. In the complementary region "B," where $E_L(R) > E_0$, walkers die off on average. Suppose one arrives at a situation where the population has a fluctuation with an excess of A walkers; this will result in excess branching on average but the comb will respond by reducing the branching of all walkers more than is demanded by the projection. Hence, there will be fewer A walkers than is correct, thus raising the estimate of the ground-state energy. If the population B were to be in excess, more B walkers are created than should be, again raising the estimate of the ground-state energy. This positive bias of the energy depends on the population size, since the relative size of these fluctuations is $O(P^{-\frac{1}{2}})$. A VMC calculation can be seen to be the limit of the comb procedure using a single walker.

In the alternative *feedback* procedure, we allow the population to fluctuate, but drive it back to its initial value by continually adjusting E_T. Let the population in the nth generation be $P^{(n)}$ and let the desired steady-state population be $P^{(0)}$. The trial energy in the next step $E_T^{(n+1)}$ is adjusted using Eq. (24.20) to restore the population on average within k generations:

$$E_T^{(n+1)} = E_0 + \frac{\ln(P^{(0)}/P^{(n)})}{k\Delta\tau}. \tag{24.23}$$

This requires an estimate of E_0; typically the average over the last few generations is used. To minimize the bias, the parameter k is chosen to be as large as possible. To ensure a robust evolution of the walkers we need to limit the population fluctuations about $P^{(0)}$, e.g., make sure that relative population fluctuations are less than 20%. The needed value of k depends on the time step, the quality of the trial function, the number of electrons, and $P^{(0)}$. Because the feedback on the population fluctuations is delayed by on the order of k generations, the bias, which is still present and has the same origin as in the comb method, is reduced. This bias has been estimated in [1062] to be inversely proportional to the population size.

The diffusion Monte Carlo algorithm

- Sample $P^{(0)}$ walkers from $|\Psi_G(R)|^2$ using VMC.
- Loop over iterations, n.
- Loop over walkers, j.
  - Walker $R^{(n)}$ advances one step according to the VMC algorithm, Eq. (23.7), generating a new walker $R^{(n+1)}$.
  - Compute new and old local energies to determine branching factor:

$$m = \lfloor e^{-\frac{1}{2}\Delta\tau(E_L(R^{(n)})+E_L(R^{(n+1)})-2E_T^{(n)})} + u \rfloor, \qquad (24.24)$$

  where $u \in (0, 1)$ and $\lfloor x \rfloor$ is the integer part of x.
- Find new population $P^{(n+1)}$ and its average energy $E_0^{(n+1)} = \langle E_L(R^{(n+1)}) \rangle$.
- Adjust trial energy: $E_T^{(n+1)} = E_0^{(n+1)} - (k\Delta\tau)^{-1} \ln(P^{(n+1)}/P^{(0)})$.

24.3 Exact fermion methods: the sign or phase problem

Up to now, we have not discussed how to implement projector Monte Carlo for fermions. A difficulty is immediately apparent for the non-importance-sampled algorithm, since the initial condition is not a probability distribution. Any real-valued fermion wavefunction with more than a single electron of the same spin value will have by symmetry equal hypervolumes of positive and negative regions: it cannot be used as a probability. What we discuss in this section is the simplest exact approach: we use the initial sign (or phase) of the wavefunction as a weight for the random walk. Though exact, it can be very inefficient as we show.

Suppose Ψ_G is a real, non-negative (bosonic) guiding function and Ψ is an antisymmetric trial function. The trial wave function has all of the correct symmetries and boundary conditions of the state we are trying to calculate. It may be real or complex-valued. The *transient estimate* (TE) algorithm [1063] calculates the ratio:

$$E_{TE}(\tau) = \frac{\langle \Psi^*|\hat{H}e^{-\tau\hat{H}}|\Psi\rangle}{\langle \Psi^*|e^{-\tau\hat{H}}|\Psi\rangle} \geq E_0, \qquad (24.25)$$

where τ is the projection time as before and E_0 is the exact ground-state energy. As shown earlier in Eq. (24.6), the energy converges monotonically from above and exponentially fast in τ:

$$\lim_{\tau\to\infty} E_{TE}(\tau) = E_0 + |\langle \Psi|1\rangle|^2 E_1 e^{-\tau e_g} + \cdots, \qquad (24.26)$$

where e_g is the gap to the next excited state, $|1\rangle$, with a non-zero overlap with Ψ. We now write the ratio in Eq. (24.25) in terms of the importance-sampled projector defined in Eq. (24.15):

$$E_{TE}(\tau) = \frac{\int dR dR' w(R, R') E_\Psi \mathcal{G}(R \leftarrow R'; \tau) \Psi_G^2(R')}{\int dR dR' w(R, R') \mathcal{G}(R \leftarrow R'; \tau) \Psi_G^2(R')} \mapsto \frac{\sum w(R, R') E_\Psi}{\sum w(R, R')}, \qquad (24.27)$$

where $w(R, R') = \sigma^*(R)\sigma(R')$, $\sigma(R) = \Psi(R)/\Psi_G(R)$, and $E_\Psi(R) = \Psi^{-1}(R)\hat{H}\Psi(R)$ is the local energy of Ψ. Note that the initial walker ensemble is generated from a VMC calculation of Ψ_G^2.

If we take $\Psi_G = |\Psi|$, then $\sigma(R) = \exp(i\Theta(R))$, where $\Theta(R)$ is the phase of the many-body trial function and $w(R, R') = \exp(i(\Theta(R') - \Theta(R)))$. If the trial function is real, then $w(R, R')$ will be a product of the sign of the trial function at the beginning and end of the walk; it is positive if the walk crosses an even number of nodes (or does not cross at all) and negative if it crosses once or an odd number of times. The transient estimate algorithm for a single electron is depicted in Fig. 24.4.

The transient estimate algorithm

- Sample configuration R' from $|\Psi_G(R')|^2$ with VMC.
- Record the initial amplitude, $\sigma(R') = \Psi(R')/\Psi_G(R')$.
- Propagate the walk forward an amount of time τ with the projector $\mathcal{G}(R \leftarrow R'; \tau)$ using Ψ_G. If a branch occurs, each branch will count separately.
- The weight of the walk arriving at R at time τ is $w = \sigma^*(R)\sigma(R')$. The energy at time τ is computed from Eq. (24.27).

There is a very serious problem with this method, the so-called *fermion sign problem*. To understand this, let us examine how the statistical error of the energy depends on the projection time τ. Note that the values of the numerator and the denominator in Eq. (24.25) are asymptotically proportional to $\exp(-\tau(E_F - E_T))$. To keep the normalization (the denominator) fixed, we must have $E_T = E_F$. But because the guide function allows the walks to cross the nodes, the population will increase as $\exp(\tau(E_T - E_B))$, where E_B is the boson

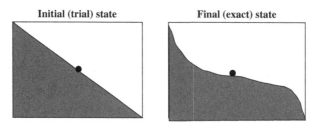

Figure 24.4. Schematic of the TE algorithm in a rectangle with inversion symmetry about the point in the center. We seek to sample the first state with odd parity about the center as analogous to the "fermion" problem. The dark and light regions are separated by a node but the nodal location is not specified by symmetry alone. The trial function nodes are displaced by an unequal diffusion of walkers from the two sides.

energy, the lowest-energy solution of the Schrödinger equation irrespective of symmetry. One can demonstrate quite generally that the signal-to-noise ratio vanishes exponentially fast, see Ex. 24.7. In the TE algorithm, as soon as negative weights are introduced, the statistical efficiency will decrease as $\chi = \exp(-\tau(E_F - E_B))$ because the estimator depends on finding differences between random walks crossing an even or an odd number of times. This is true even if the trial function is exact. At large times the average weight approaches zero: $\langle w \rangle = \langle \sigma^* \rangle \langle \sigma \rangle = 0$. Note that the exponential growth of the efficiency depends on a total energy difference. This implies that the transient estimate algorithm is guaranteed to fail for a sufficiently large system;[12] the statistical errors will be too large. Note that this method does converge to the exact energy; reliable results have been obtained for systems containing 38 electrons [109]. The sign problem has to do with how long it takes to achieve a given error estimate, and, more precisely, how this scales with the number of electrons.

There have been many attempts to "solve" the fermion sign problem. For example, one can try to pair positive and negative random walks in the TE method and thus cancel them off against each other [1064, 1065]. This is difficult in many dimensions simply because configuration space is so large that random walks rarely approach each other. Booth, Thom, and Alavi [1065] introduced an efficient method (full configuration interaction (CI) QMC), where the random walk occurs in the second quantized basis of Slater determinants. One can make the "space" of determinants much more compact by choosing a small basis set in which to expand the orbitals in the determinant. As a consequence, the chances of two walkers ending with the same determinant is much increased, enough so that full CI calculations of small molecules can be converged without being overwhelmed by the sign problem.

24.4 The fixed-node and fixed-phase methods

There is a simple way to avoid the fermion sign problem: forbid moves that change the sign of the trial function. This is the fixed-node (FN) approximation for a real trial function introduced by Anderson [1052].

In a diffusion process, forbidding node crossings puts a zero boundary condition on the evolution equation for the probability. This solves the wave equation with a zero boundary condition at places where the trial function changes sign.[13] As we will demonstrate later in this section, see Eq. (24.29), the *fixed-node* energy is an upper bound to the exact ground-state energy [1048]; the best possible upper bound consistent with the boundary conditions.

---

[12] The projection time τ can be adjusted so that the statistical errors and systematic errors are equal. Then the total error will decrease with the number of steps P as: $\epsilon \propto P^{-\eta}$, where $\eta = \frac{1}{2} e_g / (E_F - E_B + e_g)$ and e_g is the "gap." Only for bosons will $\eta = \frac{1}{2}$. Any excited state will converge at a slower rate. Note that $\eta \propto 1/N$ for a fermion system because E_F and E_B are extensive. Inverting this relation, we find that the computer time needed to achieve a given error will increase exponentially with N.

[13] Another way of saying this: for the time-independent Schrödinger equation, we can replace antisymmetry by boundary conditions and solve inside one nodal pocket. The solution is uniquely determined by the shape of the pocket and the potential inside. This argument is used to generalize the fixed-node algorithm to the density matrix at non-zero temperature [1066].

With the FN method, we do not necessarily have the exact fermion energy, but the results are far superior to the variational energy of the trial function. To get the exact energy in VMC, we need to optimize two-body correlation factors, three-body terms, etc., but since the nodes of the trial function are unchanged by those terms, such optimization is unnecessary in FN-DMC, although in practice the better the trial function the more accurate the results. One solves the wave equation inside the fixed nodal regions, but there is a mismatch of the derivative of the solution across the nodal boundary unless the nodes are exact. Where comparison has been made between the VMC energy, the FN-DMC energy, and the exact answer, one generally finds that the systematic error in the FN calculation is three to ten times smaller than it would be for a well-optimized VMC energy, e.g., see Figs. 6.5 and 6.6.

Let us consider the time-step errors caused by the fixed-node boundary condition. First, consider the situation without importance sampling. Walks that cross the nodes in a given step $\Psi(R)\Psi(R') \leq 0$ need to be killed; their branching factor set to zero. However, it is possible for the walk to have crossed during moving from R to R' but to have recrossed [1053] and end up with $\Psi(R)\Psi(R') > 0$; these walks also need to be killed. What we need is an improved projector in the region near the node, which takes into account the boundary conditions. As we discuss in the next section, the nodes are typically very smooth, so that we can approximate them as planes. As shown in Ex. 24.9, if the distance from the node to point R is d and to point R' is d', we should apply an additional branching factor $1 - \exp(-2dd'/\Delta\tau)$. This vanishes if either d or d' approaches zero, thereby killing that walker, but does nothing if $dd' \gg \Delta\tau$. To apply this formula we need to estimate the nodal distances; this is often done with Newton's method, $d(R) \approx |\nabla \ln(\Psi(R))|^{-1}$.

The situation with importance sampling is more complicated, and may appear paradoxical. On the one hand, importance sampling only multiplies the random walk density by a known function, hence there should still be the same flux of walkers across the nodes. Let us assume that $\Psi_G = |\Psi|$. But then the drift term pushes walkers away from the node of Ψ by a force that diverges inversely with the distance to the node. How then do walkers cross the nodes, so they can be killed? The answer is in the branching term: the local energy of Ψ_G also diverges so that even though most walkers get pushed away, because of the large branching, there is still a finite flux across the node that needs to be eliminated. If we put a maximum on the branching factor, that will kill the node crossing.

Consider now an exact fermion trial function, whose local energy is a constant and hence will not have any branching. What happens to the flux across the node? Because we use the absolute value of the trial function, $\Psi_G = |\Psi|$, the local energy that determines the branching factor has a delta function on the nodal surface so there is a process whereby an infinitesimal flux to the node gets an infinite branching. This has to be eliminated to get the fixed-node distribution. A good approximation is to reject or delete steps that cross the node and to put a cap on branching factors.

The FN algorithm only improves the bosonic correlations of the trial function, not the location of the nodes. There are some straightforward ways of improving on the FN method. For example, if the trial function is parameterized, say assuming that the nodal location depends on a parameter a, one can minimize the FN energy with respect to a.

Examples of such parameters occur in the backflow function (see Sec. 6.7) or in a linear combination of determinants. However, for a many-electron system, it is hard to be certain that a given trial wavefunction form will have nodes close to the exact nodal surfaces for any value of its parameters.

The nodes of the many-body wavefunction

In many situations the ground-state wavefunction can be chosen real (in the absence of magnetic fields and assuming periodic or open boundary conditions). We define its nodes as the set of points where $\Phi(R) = 0$. If the nodes were exactly known, the energy of the many-fermion system could be calcuated by projector Monte Carlo without approximation other than statistical error. The importance of the nodes was pointed out in [1067]. Since the equation defining the nodes is a single equation, the nodes are in general a $(3N - 1)$-dimensional hypersurface. However, we know that when any two electrons with the same spin (σ) are at the same location, the wavefunction vanishes; the "coincident" planes are defined by $\mathbf{r}_i = \mathbf{r}_j$ and for $\{i,j\}$ with $\sigma_i = \sigma_j$. These $(3N - 3)$-dimensional hypersurfaces do not exhaust the nodes; they are a scaffolding on which the nodes are built. The situation is very different in 1D, where the set of nodes is usually equal to the set of coincident hyper-planes.[14] Fermions in 1D are equivalent to 1D bosons with a no-crossing rule between electrons of the same spin. We can solve the many-body problem without approximation in the domain $x_i < x_j$ for $i < j$ and $\sigma_i = \sigma_j$ with a zero boundary condition on the surface of this domain.

In any dimension, the nodal volumes of ground-state wavefunctions possess a tiling property [1066]. To define this property, first pick a many-electron point, R_0, which does not lie on the nodes. Consider the $3N$-dimensional volume $\omega(R_0)$, which is accessible to a fixed-node random walk starting at R_0, i.e., by a continuous path with $\Phi(R) \neq 0$ every-where along the path. We call this volume a "tile." Now look at the set of tiles defined by permuting electrons with $\sigma_i = \sigma_j$. The tiling property says that for an exact many-body ground-state wavefunction,[15] this procedure completely fills phase space, except, of course, for the nodes. If a nodal surface did not have the tiling property we could lower the energy, thus contradicting the assumption that it is the ground state. The tiling property implies that one does not have to worry about where a fixed-node random walk starts; all starting places are equivalent. This property holds for any fermion wavefunction that is the ground state of a local potential. Thus the Slater determinant calculated from a Kohn–Sham DFT calculation will have the tiling property. It is conjectured [1066, 1068] that, except in 1D, there are only two tiles, a positive tile and a negative tile. Excited states that are not the ground state of a given symmetry, such as the $2s$ state of the hydrogen atom, or arbitrary antisymmetric functions, need not have the tiling property. Figure 24.5 shows a cross-section of the many-body nodes for free electrons in 2D. Since the nodes

---

[14] There can be additional nodes, for example caused by degeneracy, boundary conditions, or the interaction potential.

[15] Including ground states with a given quantum number such as momentum.

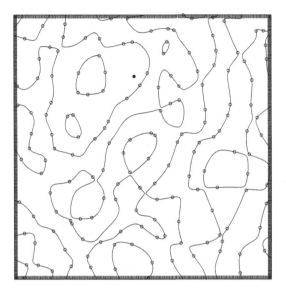

Figure 24.5. A 2D cross-section of the ground-state wavefunction of 161 independent (polarized) electrons in a square box: 160 electrons (open circles) were sampled from the square of the wavefunction. The filled circle indicates the original position of the remaining electron. This last electron is then moved throughout the cell and the curves show where the wavefunction vanishes, delimiting the positive and negative regions. (From [1066].)

pass through the positions of the fixed electrons, the fermion wavefunction connects the open circles with lines. For a non-degenerate ground state the connection is unique. If we displace any electron a distance $n^{-1/3}$ (where n is the electron density), on average we cross a nodal surface. One might guess that long-range properties, such as the existence or non-existence of a Fermi surface, will be determined by the assumed nodes, but such a direct relation has yet to be established.

The fixed-phase (FP) method

To generalize the fixed-node method to complex trial functions [1069], take an arbitrary complex wavefunction and divide its logarithm into real and imaginary parts: $\Psi(R) = \exp(-U(R) + i\Theta(R))$. Now minimize the variational energy with respect to $U(R)$, keeping $\Theta(R)$ fixed; the Euler–Lagrange equation $\delta E/\delta U(R) = 0$ leads to:

$$\left[\frac{1}{2}\sum_i^N(-\nabla_i^2 + [\nabla_i\Theta(R)]^2) + V(R) - E\right]e^{-U(R)} = 0. \qquad (24.28)$$

The quantity in [$\cdots$] defines an effective hamiltonian acting on the wavefunction modulus $e^{-U(R)}$. It is equal to the bare hamiltonian with an additional repulsive potential-like term proportional to the squared gradient of the phase of the trial function. The effective hamiltonian can be solved with the DMC method described earlier in this chapter. However, the resulting solution will not be exact unless the assumed many-body phase $\Theta(R)$ is exact. As long as the conditions on the variational theorem are met, the fixed-phase energy E_{FP} will be an upper bound.[16] Hence

$$E_0 \leq E_{FP} \leq E_V, \tag{24.29}$$

with the right equality holding when the trial modulus U is exact and the left equality holding when the trial phase Θ is exact.

In contrast to the method for real wavefunctions, there is no barrier to the diffusion; for N electrons the nodes of a complex function have dimension $3N - 2$ instead of $3N - 1$ for a real function. However, there are regions where the phase is rapidly varying, leading to a large (but not infinite) effective potential. Consider what happens as a complex-valued trial function continuously approaches a real-valued one (for example as the twist angle tends to the Γ point, see Sec. 6.2). The phase approaches a multiple of π at each point. However, because of antisymmetry, the phase must still change by an odd multiple of π as two electrons with the same spin interchange positions. Hence there is a rapid variation in the phase at some point in the exchange, so that the effective potential becomes very large. Random walks approaching this region will either be pushed out, or killed: the FP effective potential becomes the fixed-node boundary condition.

In practice one uses the many-body phase from the solution of a mean-field equation, such as provided by density functional theory, or theoretical insight, for example the Laughlin wavefunction for the quantum Hall effect [1069]. One can perturb about the mean-field solution using backflow as in Ch. 6, or one can minimize the energy of a parameterized many-body phase.

Release-node/release-phase DMC

The RN algorithm [1048, 1071, 1072] is a variant of the TE method. Instead of beginning the projection from the trial function, one begins the projection from the fixed-node or fixed-phase solution. There are several advantages. First, boson correlations within the fixed nodes are already "solved," thus the projection time is only determined by the time to adjust the position of the nodes. Of course, this will indirectly affect the bosonic correlations. Second, the difference between the exact result and the fixed-node solution can be written in terms of the local energy of walks that cross the nodes. Thus the difference is obtained with more statistical accuracy than either energy alone, allowing the convergence to be carefully monitored. Finally, the release node method can be conveniently integrated into a fixed-node program. The only modifications are to introduce a guiding function, which does not vanish at the nodes, and to keep track of the energy as a function of time since nodal crossing. The Ceperley–Alder results [109] for the HEG were obtained with the RN method. A detailed explanation is given in [1048, 1071].

---

[16] For excited-state calculations, this follows only if Ψ transforms according to a one-dimensional irreducible representation of the symmetry group of the hamiltonian. If the trial function transforms according to a multi-dimensional irreducible representation, corresponding to a degenerate energy level, the DMC energy may lie below the energy of the lowest eigenstate of that symmetry [1070].

The lattice fixed-node method

How do we generalize the fixed-node methods for a lattice model such as the Hubbard model? As we discussed in Sec. 24.1, typically the lattice projector $G_L \equiv \Psi(1 - \Delta\tau(\hat{H} - E_T))\Psi^{-1}$ is used since it is easy to implement without a time-step error. However, because the concept of nodes in a discrete space is quite different from that in the continuum, diffusion in imaginary time resulting from the kinetic energy corresponds to a random jump and not a continuous random walk. But rather surprisingly, the fixed-node method can be generalized [1073]. Let us suppose we have a discrete space whose basis states we label s (equivalent to R in the continuum). One will encounter a sign problem whenever there is a pair of distinct states (s, s') with $s \neq s'$ and $\Psi_s H_{s,s'} \Psi_{s'} > 0$, since this would imply that $G_L(s, s') < 0$ and cannot be used as a probability. [Note, however, that sign changes on the diagonal $s = s'$ can always be avoided by choosing a sufficiently small $\Delta\tau$.] We call this pair of states "sign-flip moves." In the lattice fixed-node procedure, we define an effective hamiltonian $\hat{H}'$ by moving the sign-flip moves to the diagonal:

$$H'_{s,s'} = \begin{cases} H_{s,s'}\Theta(-\Psi_s H_{s,s'} \Psi_{s'}) & : \quad s \neq s' \\ H_{s,s} + \sum_{s'} H_{s,s'} \Psi_{s'} \Psi_s^{-1}\Theta(\Psi_s H_{s,s'} \Psi_{s'}) & : \quad s = s'. \end{cases} \quad (24.30)$$

Here $\Theta(x)$ is the Heaviside function: $\Theta(x) = 1$ for $x > 1$, otherwise $\Theta(x) = 0$. One can show that $E_0 \leq E' \leq E_V$: the exact ground-state energy of the effective hamiltonian lies between the exact ground-state energy and the variational energy. If the trial function is exact, then one gets the exact energy.

Details of the procedure for doing projector Monte Carlo with a lattice model are given in [1035], together with an application to the Heisenberg model. We will discuss a different restriction for a discrete hamiltonian that works in determinant space; see Sec. 24.7.

24.5 Mixed estimators, exact estimators, and the overlap

In the preceding sections of this chapter we have only discussed calculation of the energy. There is a difficulty in calculating other properties, which we now discuss. To obtain ground-state expectations of quantities other than the energy (e.g., the potential energy), one must correct the average over the PMC walk. Although we want to compute the exact estimator, we cannot obtain it directly in projector Monte Carlo.

For any operator $\hat{O}$, we can define three different expectation values as integrals over the trial function Ψ, or the exact wavefunction Φ:

$$\begin{aligned} O_V &= \langle \Psi^* \hat{O} \Psi \rangle \quad \text{variational,} \\ O_M &= \langle \Phi^* \hat{O} \Psi \rangle \quad \text{mixed,} \\ O_X &= \langle \Phi^* \hat{O} \Phi \rangle \quad \text{exact,} \end{aligned} \quad (24.31)$$

known as the "variational estimator," the "mixed estimator," and the "exact estimator," respectively. What is computed naturally in importance-sampled PMC is the mixed estimator. For any operator that commutes with the hamiltonian, we find that $O_M = O_X$, but such is not the case for other operators. Since what is wanted is the exact estimator, we discuss several ways of estimating it.

Perturbative methods for estimators

In the extrapolation method [1017], we assume that Ψ is close to Φ and we expand in the difference.[17] Then we can show (see Ex. 24.8) that the first-order correction can be eliminated by either of the formulas:

$$O_X \approx 2O_M - O_V,$$
$$O_X \approx O_M^2/O_V. \qquad (24.32)$$

Exercise 24.8 shows that to obtain the correct density $n_X(\mathbf{r})$ of a non-interacting system from variational and mixed estimates of the density, one must use the lower Eq. (24.32). This does not follow if there is an interaction, especially if the correlation is large, but the lower equation has the nice property that it gives only positive densities while the upper formula could give regions of negative density. Hence, the lower equation is appropriate for quantities that must be positive, such as the pair correlation function or the structure factor.

It is possible to add terms to the estimators to reduce both the variance and the bias [1074]. This reference shows that there exist estimators for any property with a zero-variance, zero-bias property just like for the energy.

The overlap

We introduced the overlap between a trial function and the ground state in Sec. 6.5. Maximum overlap removes the difference between VMC and DMC [1075]. Suppose we optimize the overlap with respect to the Jastrow function; this leads to the Euler–Lagrange equation: $g_V(r) = g_M(r)$, where $g(r)$ is the pair correlation function defined in Eq. (5.14). Hence, maximizing the overlap of two-body correlation functions implies that the VMC and DMC estimators will agree. For every type of term in the trial function, there is a corresponding operator. For example, optimizing a one-body term implies making the one-body density match.

Forward-walking

Forward-walking is a method for calculating exact estimators [1076]. What is the average population resulting from starting many random walks from the same point R_0, after many generations, assuming importance sampling with a guiding function Ψ_G? It is:

$$P^{(n)}(R_0) = \int dR \mathcal{G}(R \leftarrow R_0; n\Delta\tau) \rightarrow \frac{\Phi(R_0)}{\Psi_G(R_0)} \langle \Phi | \Psi_G \rangle, \qquad (24.33)$$

in the limit $n \rightarrow \infty$.

---

[17] Note that the energy is second order in the difference, while errors of general properties are first order and thus typically much larger.

This method can be used to find the value of the exact ground-state many-body wave-function at a given point and is, in fact, the Feynman–Kac formula as introduced in Eq. (6.12) in Sec. 6.5. Regrouping Eq. (24.33), we find the exact wavefunction at a point is given by:

$$\Phi(R_0) = \Psi_G(R_0)\langle\Phi|\Psi_G\rangle^{-1} \lim_{n\to\infty} \int dR\mathcal{G}(R \leftarrow R_0; n\Delta\tau). \tag{24.34}$$

If we now interpret $\int dR\mathcal{G}(R \leftarrow R_0; n\Delta\tau)$ as a random walk of n steps starting at a point R_0 and drifting under the influence of the guiding function as in Sec. 24.2, and take the integrated local energy as a weight factor, from Eq. (24.18), we end up with Eq. (6.12). In [1077] this method was used to find the probability that a system ended up in a bound state after a nuclear reaction.

Now turning to the problem of finding exact estimators, we use this formula to reweight the mixed average:

$$O_X = \int dR \frac{\Phi\Psi_G}{\langle\Phi\Psi_G\rangle} P(R) \mapsto o_X = \frac{\sum_i P^{(n)}(R_i)O(R_i)}{\sum_i P^{(n)}(R_i)}, \tag{24.35}$$

where $P^{(n)}(R_i)$ is the number of descendants of R_i, n generations later. This can be implemented on the computer by carrying along the past history of a walker. For example, to perform a forward-walking calculation of the forces, one carries along a "stack" of pointers to the forces over the last n generations. Then, when we calculate the force, instead of averaging the force on a walker over all walkers, we delay it by n generations. Walkers that have died do not contribute; those that have branched contribute multiple times. Figure 24.6 shows an example of the convergence of the forces[18] in the molecule LiH with this technique.

The exact average is obtained by taking the generational lag to infinity, however, errors will grow with respect to the lag. We can ask in how many generations most of the population will be a descendant from a single walker. This sets a maximum value for a weight and results in very inefficient sampling.

As we noted earlier, $(n\Delta\tau)^{\frac{1}{2}}$ is a diffusion length and sets the length scale for fixing the trial wavefunction; this highlights the importance of having the correct trial function at long wavelengths. If any expectation values between the variational, mixed, or forward-walking are very different, particularly those involving order parameters or long-range properties, they might not be reliable. Path integrals (or reptation, discussed in the next chapter) leads to a more robust way of computing expectation values.

24.6 Non-local pseudopotentials in PMC

For systems containing atoms with more electrons than neon, it is highly desirable to replace core electrons with pseudopotentials, as discussed previously in Sec. 23.4.

---

[18] Using the Hellman–Feynman theorem, the force is given by the gradient of the charge density. The exact charge density, not the mixed estimator, is needed, thus forward-walking was used. In addition, the contribution of electrons near the nuclei was smoothed by the procedure described in [1078].

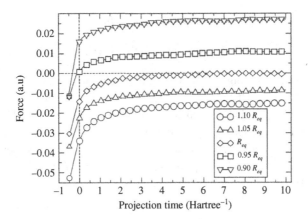

Figure 24.6. The bonding force computed using forward-walking for five different bond lengths in LiH [1078]. Negative value of the projection time represents VMC, zero time, the mixed estimator, and positive time values, with projection.

However, this presents special problems for the FN-DMC method. Let us try to interpret the non-local operator in Eq. (23.21) according to Trotter's formula. This would imply a move that, with probability proportional to the magnitude of the non-local potential, would hop around an ion.[19] We immediately run into a problem with the fixed-node method, since boundary conditions on a surface are no longer sufficient; one also needs to know how the wavefunction changes if any electron inside the "core region" is rotated about the ion. We can use the methods we discussed with lattice models in Sec. 24.4.

In the locality approximation [1026, 1079] one evaluates the non-local potential as in VMC. That is, we construct the local potential:

$$W(R) = \Psi(R)^{-1} \int dR' \langle R|V_{PP}|R'\rangle \Psi(R') \qquad (24.36)$$

and then add $W(R)$ to the local energy of the previous algorithm. This is exact to the limit that the trial function in the core is exact. However, with this approximation, the FN-DMC energy is not necessarily above the exact ground-state energy, and such "upper bound violations" are often encountered. This provides another reason to have well-optimized trial wavefunctions.

Casula [1080] has suggested an improvement based on the lattice fixed-node procedure in Sec. 24.4. Moves are made in the usual DMC algorithm, but prior to branching an additional "heat bath step" is made if an electron is in the range of a non-local pseudopotential. First, one computes an effective non-local potential:

$$\tilde{v}_{PP}(\mathbf{r}';\mathbf{r}) = \Psi(\mathbf{r}')\Psi^{-1}(\mathbf{r})\langle\mathbf{r}'|V_{PP}|\mathbf{r}\rangle. \qquad (24.37)$$

---

[19] If a normal DMC move is like a king's move in chess, the move caused by a non-local pseudopotential is like a knight's move.

From this one constructs the (unnormalized) probability

$$T(\mathbf{r}' \leftarrow \mathbf{r}; \Delta\tau) = \delta(\mathbf{r}' - \mathbf{r}) - \frac{1}{2}\Delta\tau[\tilde{v}_{PP}(\mathbf{r}', \mathbf{r}) - |\tilde{v}_{pp}(\mathbf{r}', \mathbf{r})|] \qquad (24.38)$$

for all values $\mathbf{r}'$, where $\tilde{v}_{PP}$ might be non-zero. In practice this is done on a finite number of points on the sphere around the ion, as in Sec. 23.4. Then we sample $\mathbf{r}'$ from $T(\mathbf{r}' \leftarrow \mathbf{r}; \Delta\tau)$, with the "heat bath" algorithm described in Sec. 22.5. By taking only the negative pieces of $\tilde{v}_{PP}$, we avoid a sign problem. By adding an extra term to the energy, the upper-bound property of the fixed-node energy can be shown to be restored. This method is useful to check the accuracy of the locality approximation. The method has been updated to be size consistent in [1081].

24.7 Projector auxiliary-field quantum Monte Carlo methods

So far we have used the projection operator $\exp(-\tau\hat{H})$ in the coordinate basis. Now we switch to a determinantal basis, in which antisymmetry is explicitly built in. In this basis we can replace the actual electron system by an equivalent one where non-interacting electrons interact with a fluctuating field. This method has been discussed by several authors, starting with [1082, 1083], however, until the constrained path restriction [1084] and importance sampling of the auxiliary fields was made [1085], the method was not practical for general electronic structure problems.

The following section gives a brief overview of the methods. We first discuss the determinantal basis, then explain how the electron–electron interaction can be modeled using random fields. This leads to a branching Monte Carlo algorithm similar to what has been described previously in this chapter. We then explain how the sign problem enters and how the constraint, similar to the fixed-phase condition, enters. For more details, see [1086] and references therein.

Basis of Slater determinants

An arbitrary Slater determinant for N electrons is an antisymmetric product of N single-electron orbitals $\psi(\mathbf{r})$, as in Eq. (4.1). The orbitals can be expressed in terms of a basis with N_{basis} elements as: $\psi_i(\mathbf{r}) = \sum_{m=1}^{N_{basis}} c_{m,i}\chi_m(\mathbf{r})$, where $\chi_m(\mathbf{r})$ are the basis functions. Hence, an arbitrary determinant that we will denote by Ψ_c is represented by the $N \times N_{basis}$ complex matrix $c_{m,i}$. This determinant, which we denote by Ψ_c, is a many-body wavefunction, which is the "walker" in this algorithm.

One can use either a plane-wave basis (for solids) or a gaussian basis set (for molecules). In the plane-wave basis, convergence to the basis set limit is controlled as in DFT calculations by varying the plane-wave cutoff. There will be a basis set error in treating continuum problems. Note that the Slater determinant basis is over-complete, since there is no requirement that the orbitals be orthogonal or even uniquely represented.[20]

---

[20] Multiplication of the Slater matrix by any complex square matrix with non-zero determinant will "rotate" the orbitals but only multiplies the many-body wavefunction by a complex number.

Now consider imaginary-time evolution in this basis: $\Psi_c(\tau) = \exp(-\tau\hat{H}_1)\Psi_c(0)$. For $\hat{H}_1$ the sum of one-body (i.e., independent electron) operators, the time-evolved many-body function $\Psi_c(\tau)$ is still a Slater determinant. The time evolution is performed, as in a Hartree–Fock calculation, by repeatedly applying the kinetic and potential evolution operators on the matrix c. For small enough time steps $\Delta\tau$, Trotter's formula guarantees correctness. Note that it is important to re-orthogonalize the orbitals using a Gram–Schmidt process when needed.

Replacing two-body interaction with random fields

We now add in the electron–electron interaction. Consider the gaussian integral:

$$e^{\frac{1}{2}\hat{A}^2} = \frac{1}{\sqrt{2\pi}} \int_{-\infty}^{\infty} dx e^{-\frac{1}{2}x^2 - x\hat{A}}, \tag{24.39}$$

which holds for a general operator $\hat{A}$. It allows us to decouple the two-electron interaction, $\hat{V}_{ee}$, written in the second-quantized notation (see Eq. (4.8)) into a sum over one-body projectors.[21] This process is known as the Hubbard–Stratonvitch transformation [1087]. Now it is always possible[22] to write a two-electron operator as a sum of squares of one-electron operators:

$$V_{ee} = \frac{1}{2} \sum_{k=1}^{p} \lambda_k \hat{v}_k^2, \tag{24.40}$$

where the $\hat{v}_k$ are one-electron operators. Then, using Trotter's theorem and the gaussian identity, we can represent the imaginary-time evolution of each term in the interaction potential as a separate integral:

$$e^{-\Delta\tau V_{ee}} = \prod_{k=1}^{p} \frac{1}{\sqrt{2\pi}} \int_{-\infty}^{\infty} dx_k \exp(-x_k^2/2 + \sqrt{-\Delta\tau\lambda_k} x_k \hat{v}_k). \tag{24.41}$$

We interpret the first term in the exponent as the probability distribution of the "auxiliary field" x_k: a gaussian distribution with a unit variance; the second term is its interaction with the electrons. Each of the p terms in the sum in Eq. (24.40) will give rise to a different random field. Because the second term is linear in $\hat{v}_k$, it is a one-body operator, and thus can be treated numerically just as any other external potential. However, because λ_k can be positive or negative, the external potential will be real-valued for attractive potentials and imaginary-valued for repulsive interactions.

---

[21] This is a general Monte Carlo strategy: increasing the dimensionality of a problem can make things easier. Adding extra variables does not necessarily cost much more computer time, and in this case, leads to a hamiltonian only involving non-interacting electrons. Note that one can use a sum as well as an integral, though it is not obvious which is advantageous.

[22] The four-index interaction matrix V_{ijkl} in Eq. (4.8) is symmetric under the interchange $(i, l) \leftrightarrow (k, j)$. Such a matrix can always be written in the form $\hat{V} = \hat{R}\hat{\Lambda}\hat{R}^\dagger$, where $\hat{\Lambda}$ is a diagonal matrix.

The auxiliary fields perturb the walker Ψ_c away from the mean-field solution. It is the competition between the auxiliary fields and the one-body hamiltonian that in the long time limit gives rise to the exact many-body ground-state wavefunction

$$\Phi_0 \propto \lim_{\tau \to \infty} \sum_k w_k \Psi_c(\tau), \tag{24.42}$$

where the sum is over all walkers $\Psi_c(\tau)$ after an evolution of time τ with a weight w_k.

Details of how to write the two-electron interaction as a square (Eq. (24.40)) depend on the interaction and the basis set and are discussed in [1086, 1088, 1089]. For the Hubbard model, such a form is given in Eq. (25.38).

Monte Carlo sampling has now entered the algorithm since at each time step τ we need to sample P fields, $\{x_k(\tau)\}$. The normalization of a given walker, $\langle \Psi_c(\tau)|\Psi_c(\tau)\rangle$, is used in computing expectations such as the energy or for branching and population control as discussed previously.

The sign problem in PAF-QMC

The fermion sign/phase problem discussed earlier, Sec. 24.3, does not disappear in this formulation but it is substantially changed. Because the Schrödinger equation is linear, a solution is degenerate with respect to multiplication by an arbitrary complex number. In a random walk, the states can move back and forth between solutions with different values of $e^{i\Theta}$. The continuous stochastic evolution of the orbitals can lead to an exchange without any two orbitals overlapping; this process gives rise to noise and is the manifestation of the sign (or phase) problem.

Realistic applications require[23] a fixed-node-type approximation. One can apply the fixed-node method in the determinant basis since if we can determine when the projection of the walker onto the exact wavefunction vanishes, $\langle \Phi|\Psi_c\rangle = 0$, then the future contributions of that walker to any property would also vanish. Hence, that walker could be eliminated without biasing the results, just as in the fixed-node method. But since we do not know the exact wavefunction, we also cannot be sure when this matrix element will vanish. As before, the procedure is to use a trial wavefunction, Ψ_T, to "kill" walkers. That is, we require that all walkers satisfy $\langle \Psi_T|\Psi_c\rangle > 0$. This is the constrained path condition introduced by Fahy and Hamann [1083] and developed practically in [1084].

As described in Sec. 24.2 for diffusion Monte Carlo, importance sampling is the Monte Carlo technique of changing the sampling distribution in order to increase computational efficiency. Let us introduce an importance-sampled walker:

$$\tilde{\Psi}_c = \max((\langle \Psi_T|\Psi_c\rangle, 0)\Psi_c, \tag{24.43}$$

where Ψ_T is a trial wavefunction. We now perform the evolution of $\tilde{\Psi}_c$. The effect of this will be to push walkers to higher values of the overlap with the trial wavefunction and

---

[23] One can perform exact calculations without "solving" the sign problem, but this will limit the applications to very simple systems.

kill walkers that have a negative overlap with the trial wavefunction. This is called the "constrained path approximation." For efficiency's sake, the trial function is taken as a single determinant, for example the Hartree–Fock solution, thus allowing a quick evaluation of this condition. The constrained phase energy is not necessarily above the exact energy [1090], as it would be in the DMC method.

There are two other sign (or phase) problems that arise in PAF-QMC, which can appear even for interacting bosons. First, the trial function can be complex because of twisted boundary conditions (Secs. 6.2, 23.5) or because one wants to impose a symmetry on the solution. Second, the auxiliary field will be complex whenever $\lambda_k > 0$ in Eq. (24.40) for a two-electron matrix element. In general, repulsive interactions lead to complex weights.

Solutions to these problems for realistic materials were developed in [1085]. It is possible to shift the external field by a constant amount: in Eq. (24.41) we replace x_k by $x_k + \bar{x}_k$, where $\bar{x}_k$ is a walker-dependent complex number. We then rewrite the gaussian integral. This is correct for any value of $\bar{x}_k$ and is similar to the smart Monte Carlo transformation done in VMC, Eq. (23.12). Carrying through the transformation, this introduces an additional contribution of the weight of a walker, $\exp[x_k \bar{x}_k - \bar{x}_k^2 - (x_k - \bar{x}_k)\xi_k \hat{v}_k]$, where $\xi_k = \sqrt{-\Delta\tau\lambda_k}$. The last term shows that the effect of the one-body propagator is shifted. The optimal shift $\bar{x}_k$ is determined by minimizing the fluctuations in the weight factor, giving:

$$\bar{x}_k = \frac{\langle \Psi_T | \xi_k \hat{v}_k | \Psi_c \rangle}{\langle \Psi_T | \Psi_c \rangle}. \tag{24.44}$$

The branching factor now becomes:

$$\exp\left(-\Delta\tau \frac{\langle \Psi_T | \hat{H}_1 | \Psi_c \rangle}{\langle \Psi_T | \Psi_c \rangle}\right). \tag{24.45}$$

The exponent plays the role of local one-body energy.

Note that the algorithm is now "phaseless"; the procedure is independent of the phase of the determinant. However, the phase of the walker can still wander around the complex plane. To control this, an additional step is needed. We multiply the weight of each walker by a factor $\max(\cos(\Delta\Theta), 0)$, where

$$\Delta\Theta = \operatorname{Im} \ln \left(\frac{\langle \Psi_T | \Psi_{c'} \rangle}{\langle \Psi_T | \Psi_c \rangle}\right) \tag{24.46}$$

is the phase picked up during the move from c to c'. See [1086] for details on how to apply these techniques.

In extensive benchmarks of Hubbard models as well as of molecules and solids, the constrained path method has demonstrated accuracy equaling or surpassing that of other QMC approaches. In general, auxiliary field methods will experience more difficulty on very strongly correlated systems such as the Wigner crystal. Basis set error is common to all many-body methods, which treat the hamiltonian written in a finite basis set and convergence in total energy can be slow; see, for example, [1091] for the electron gas at weak coupling, $r_s = 0.5$. Within PAF it is easier to compute imaginary-time dynamics since the constrained-path condition maintains antisymmetry. However, because the trial function typically has no correlation (e.g., a single determinant), the mixed-estimator bias can be severe if the exact wavefunction differs substantially from a determinant. A back-propagation technique [1086] is used instead of the mixed estimator to compute observables and correlation functions.

24.8 Applications of projector MC

We will only discuss a few of the very many applications of PMC that show various capabilities of the methods. More applications are discussed in the readings listed at the end of this chapter.

The homogeneous electron gas

The calculation of the correlation energy of the homogeneous electron gas was one of the first significant calculations using PMC. In fact, the DMC algorithm presented earlier in this chapter was developed to perform this calculation in 1980 [109]. Even today, the HEG calculation is the most-cited QMC application.

The motivation for studying the HEG, its basic properties, and phase diagram in 2D and 3D are detailed in Sec. 3.1. Diffusion Monte Carlo is particularly effective for computing the correlation energy, since one does not have to correct for the trial function bias. Because the system is homogeneous, the Slater–Jastrow trial function using the Gaskell (RPA) Jastrow function (Eq. (6.22)) is very accurate, as shown in [275]. The comparison of the energy using backflow versus HF nodes is shown in Fig. 6.6. To compute accurate correlation energies, it is crucial to make an accurate estimation of finite-size effects as discussed in Sec. 23.5 and illustrated in Fig. 23.2. The computation of the momentum distribution of the 3D HEG is shown in Fig. 5.2. Although other methods, such as GW, are able to treat the HEG for $r_s \leq 5$, at lower density (larger correlation) only DMC is able to determine properties robustly and with high accuracy.

Hubbard model

In the 2D Hubbard model, the energy computed using the PAF method at $U = 4t$ is typically within 0.5% of the exact result [1093]. At half-filling, where the approximation is the largest, the calculated energy per site is $-0.8559(4)$ compared with the exact result[24] of $-0.8603(2)$.

An example of the type of physics that can be extracted is given in Fig. 24.7. The spin wave shown in this figure, when holes are present in the 2D Hubbard model, emerged spontaneously. The spin wave appeared independent of the form of the trial wavefunction.

---

[24] One can obtain exact results at half-filling by removing the constraint, since there is no sign problem there.

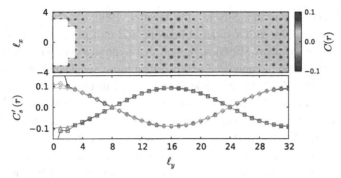

Figure 24.7. The spin–spin correlation function $C(r)$ in the 2D Hubbard model as computed using PAF. Shown is an 8×64 lattice with hole density $1/16$ and $U = 4t$. The upper panel is a 2D plot of the spin–spin correlation function with the peak near the origin cutoff. The lower panel shows the staggered correlation $C'_s(r)$ as a function of l_y, with the two curves plotting even and odd values of l_x. (From [1092].)

The PAF is able to treat much larger lattices than other QMC approaches; the figure shows a system with 512 lattice sites [1092]. The spin wave would not be present in a simulation of a smaller system.

Many-body hydrogen

Hydrogen as the first element in the periodic table is important for a variety of reasons. It is the most common condensed matter constituent of the universe, since it makes up the majority of the giant planets. Wigner and Huntington [1094] proved that it should become a metal at a sufficiently high density and, in one of the first band calculations, they estimated the molecular–atomic transition at 25 GPa. Since then, discovery of this transition has become one of the leading motivations for high-pressure research, but experimental observation of the transition has remained elusive. Dense hydrogen should undergo both an atomic–molecular transition and the metal–insulator transition, see Fig. 25.6 later for a theoretical phase diagram of dense hydrogen. Calculations of dense hydrogen are an important benchmark for many-body methods because they do not have some of the complications of other elements, such as core electrons or spin-orbit effects. On the contrary, they contain very important features of realistic materials, such as the breaking of the molecular bond, and the large contribution of quantum effects of the protons. Calculations on hydrogen are clearly relevant to compounds containing hydrogen, such as water, and to understanding of bond-breaking mechanisms in general.

Ceperley and Alder [1095] used DMC to simulate dense hydrogen, treating the quantum effects of both the electrons and protons. Because the proton/electron mass ratio $M_p/m_e = 1836$ is so large, the protons have a diffusion step much smaller than the electrons. This required very long computation times. They predicted the change from isotropic molecules to oriented molecules at 100 GPa which was subsequently verified in diamond anvil experiments (Lorenzana, Silvera, and Goettel, 1990). Natoli *et al.* [1096, 1097] refined these calculations with a variety of atomic and molecular structures. Subsequent calculations and experiments [1098] suggested that hydrogen will continue to transform to structures with a more complicated unit cell than those considered in the early works. Holzmann *et al.* [284] showed improvements in the trial functions and nodal surfaces using backflow trial functions (see Sec. 6.7), with a final accuracy of better than 0.1 mH/atom. Benchmark calculations [1099] with many-body hydrogen have determined which density functionals perform the best on dense hydrogen systems, in particular, they found that the vdw-DF functional was most accurate. This allows the computationally intensive DMC calculations to have a wider impact by establishing the most accurate functional to use for a given thermodynamic condition.

We will discuss other many-body QMC simulations of hydrogen at elevated temperatures with the RPIMC and the CEIMC algorithms in Secs. 25.3 and 25.4.

Boron nitride as a pressure standard

An important role for electronic structure calculations is to provide benchmarks to guide experiment. An example is a pressure standard. Even though diamond anvil cells can reach high pressures, the pressure is not measured directly. Other properties can be measured, for example, the lattice constants, by scattering experiments. There have been various indirect techniques for determining the pressure, such as the ruby scale [1101], where the variation of the frequency of a given transition with respect to pressure has been calibrated.

The calculation of the pressure versus volume for a known material could take the place of an empirical standard. Cubic boron nitride has been proposed as a standard since it does not have a phase transition up to very high pressure. The Raman active phonon varies with pressure and can be monitored to indicate the pressure. Different density functionals give a variety of results, but without a benchmark it is difficult to decide which is better.

In order to propose a standard, extensive FN-DMC calculations have been carried out with careful control of errors [1100].[25] Figure 24.8 shows the results. The left panel shows the pressure vs. volume compared with experimental data at lower pressures. The inset on a much more expanded scale shows the range of uncertainty from several DMC calculations, of order ± 2 GPa for $P \lesssim 500$ GPa, and ± 5 GPa for P up to 900 GPa. The agreement with experiment within <2 GPa at low pressure indicates the level of confidence. A density functional calculation and a proposed extrapolation of experimental data has differences >15 GPa for $P \lesssim 500$ GPa, with larger errors at higher pressures. In the right panel is

---

[25] Pseudopotential calculations were checked with all-electron calculations. Finite-size effects were extrapolated to infinite size using methods described in Sec. 23.5. Calculations of similar systems have found the fixed-node bias to be small; in addition, excellent agreement with experiment at low pressure was found. The effects of zero-point motion and temperature were calculated, but they are not essential for our present purposes.

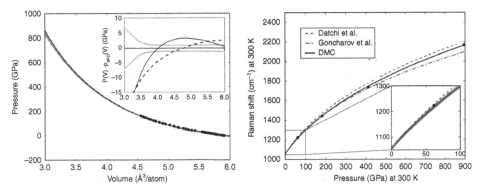

Figure 24.8. Results for DMC calculations for cubic BN from [1100]. Left panel: pressure vs. atomic volume compared with experiment. The inset shows the difference from a DMC calculation (solid line) with the range of other calculations indicated by the broad gray lines. Also shown is a proposed extrapolation of the experimental data and DFT calculation (dashed line); note the large difference at high pressure. Right panel: the calculated frequency of the Raman-active phonon mode (QMC is solid line) compared with similar calculations (dashed lines).

the frequency of the Raman-active optic mode (computed by fitting QMC energies vs. displacements) and compared with other work.

The band structure of silicon

The band structure of silicon is shown in Fig. 13.3, comparing PMC calculations with GW calculations and experiment. The DMC calculations [1102] were done by calculating the energy of an electron–hole excitation in a finite supercell consisting of 16 silicon atoms; LDA calculations with a non-local pseudopotential were used to determine the Slater determinant in the trial wavefunction and hence the nodes of the fixed-node solution for the excited states. Note that relaxation effects have not been included, and it is expected that the nodes of the trial function would be less accurate for higher excitations. The PAF calculations were done in [530]. The lowest band, which corresponds to the highest excitation energies, tends to be too low. The quality of projection methods will decline for higher excitations and a single determinant may not be sufficient within PAF. For both QMC calculations, finite-size effects and excitonic binding energies have been estimated and corrected for in the analysis.

24.9 The pluses and minuses of projector MC

While the projection method is exact for bosonic systems, applied to fermion systems it shares some of the same problems with the variational method discussed at the end of the last chapter. In fact, it is useful to think of the fixed-node projection method as a "super-variational" method: it determines the best upper bound (Eq. (24.29)) consistent with a given sign or phase by allowing all possible variations in the modulus of the wavefunction.

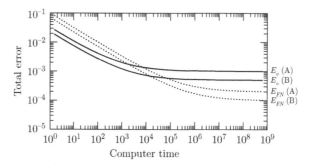

Figure 24.9. The total error (statistical error plus systematic bias) vs. computer time (arbitrary units) for VMC and FN-DMC calculations. At the beginning the statistical error decreases as $t^{-\frac{1}{2}}$ until the bias of a given wavefunction/nodal surface dominates. The solid lines represent the errors of VMC calculations and the dotted lines those of FN-DMC calculations, using either trial function A or B. We assume the time step in the DMC calculation is 10 times smaller than that of VMC, so the statistical error is $\sqrt{10}$ higher for a given trial function. We assume trial function B is more accurate than A by a factor of 2, i.e., it has half the variance and error in the correlation energy.

Figure 24.9 shows how the statistical and systematic errors decrease with computer time in both the variational and fixed-node methods. Both algorithms scale[26] in general as N^3. The bulk of the human time comes from constructing a good trial function and estimating systematic errors coming from boundary conditions, finite-size effects, and pseudopotentials. After a good trial function has been found, the FN-DMC calculations cost very little in extra human effort, but because smaller time steps are used, more computer time is needed. Again, see Fig. 24.9. In both the VMC and DMC methods there is a premium for having good trial wavefunctions; that is the most straightforward way of getting accurate results. Note the lower curves for trial function "B" in Fig. 24.9. In an earlier chapter we discussed how the energy of the homogeneous electron gas [284] was lowered by finding more accurate analytic backflow three-body correlations; see, e.g., Fig. 6.6.

We conclude this chapter with some general features of projection methods:

- The FN energy is guaranteed to be closer to the exact answer than the starting variational trial function. Since the FN algorithm automatically includes bosonic correlation, the results are much less likely to have the human bias that could be present in the VMC method. New things can come from the simulation not present in a VMC calculation. For example, one may observe a particular type of correlation completely absent from the trial function. Hence, it is always good to pay close attention to correlation functions computed by DMC since they can provide a way of learning what is missing in the trial function. One can use the overlap as defined in Eq. (6.17) to improve the trial function by comparing correlation functions computed using PMC and VMC. In general,

---

[26] This is the number of operations needed to move all of the electrons once. For large N it is the update of the inverse matrix that dominates.

PMC is slower with respect to computer time than VMC, because the time step needs to be smaller. But with very large parallel computers available, the necessary computer resources can often be found.

- Although, in principle, the probability distribution does converge to the exact answer, in practice, this does not always occur. Metastable states exist and can have a very long lifetime. The situation is similar to that of a classical simulation near a phase boundary, such as between a solid and a liquid. In addition, importance sampling can bias the mixed-estimator correlation functions. For example, if the trial function describes a solid, even after complete convergence, the density–density correlation functions will show solid-like behavior. Careful observation could reveal liquid-like fluctuations indicating the presence of another phase. The ability to perform simulations in a metastable state is useful, but the results must be interpreted with caution.

- Importance sampling is only a partial cure for the unbounded fluctuations of the branching method. As the number of electrons increases, sooner or later the branching becomes uncontrollable [1103]. To date, most projector Monte Carlo calculations have fewer than a thousand electrons, so this problem has not occurred in practice. Finite-temperature Metropolis methods (discussed in the next chapter) do not suffer from the problem of uncontrolled branching. As a system contains more electrons, it becomes more important to monitor convergence, since the necessary projection times can become larger. It should be kept in mind that the simulation takes place in four-dimensional space-time and as the spatial length L is made larger, the projection time τ must be kept proportional to L.

- Although the fixed-node approximation dramatically improves energies, other properties, such as the momentum distribution, may not be improved. The projector methods can only calculate energies exactly. For all other properties one must extrapolate out the effect of the importance sampling. This is a real problem if one is interested in obtaining the asymptotic behavior of correlation functions. There are ways of getting around some of these problems, such as forward-walking, but none are totally satisfactory. To explore the metal–insulator phase transition with FN-DMC, one must come up with a sequence of nodes spanning the transition and use the upper-bound property of the fixed-node approximation. The path-integral finite-temperature methods discussed in the next chapter are superior to projector Monte Carlo for calculating correlation functions, but are difficult to apply at the temperatures typical of electronic systems. Ground-state path integrals [1104], discussed in Sec. 25.4, allow another way around this problem of mixed estimators, although the scaling versus problem size is still an issue.

- Exact methods, such as transient estimate or release node, only improve the nodes locally. If τ is the release node projection time, then we can move the nodes a distance of roughly $(3N\tau)^{\frac{1}{2}}$ for N electrons. For this reason, it is important to build in exact long-range nodal properties. One can anticipate that with increased computer resources, such exact calculations will be used more frequently.

SELECT FURTHER READING

Foulkes, W. M. C, Mitas, L., Needs, R. J., and Rajagopal, G., "Quantum Monte Carlo simulations of solids," *Rev. Mod. Phys.* **73**, 33, 2001. Introduction of DMC method applied to problems in solids with applications.

Hammond, B. L., Lester Jr., W. A., and Reynolds, P. J. *Monte Carlo Methods in ab initio Quantum Chemistry* (World Scientific, Singapore, 1994).

Kolorenč, J. and Mitas, L. "Applications of quantum Monte Carlo methods in condensed systems," *Rep. Prog. Phys.* **74**, 026502, 2011. Survey of applications on bulk systems.

Needs, R. J., Towler, M. D., Drummond, N. D., and Lopez Rios, P., "Continuum variational and diffusion quantum Monte Carlo calculations," *J. Phys. Cond. Matter* **22**, 023201, 2010. Update of review of DMC method.

Reynolds, P. J., Ceperley, D. M., Alder, B. J., and Lester Jr., W. A., "Fixed-node quantum Monte Carlo for molecules," *J. Chem. Phys.* **77**, 5593, 1982. Detailed description of the FN-DMC algorithm for molecules.

Zhang, S., "Auxiliary-field quantum Monte Carlo for correlated electron systems," in Pavarini, E., Koch, E., and Schollwöck, U., (eds.), *Emergent Phenomena in Correlated Matter*, Autumn School of the Forschungszentrum Jülich 2013.

Exercises

24.1 How can we use DMC with importance sampling to calculate the overlap between an exact wavefunction and a trial wavefunction? *Hint*: see Eq. (24.5). As an alternative, consider the relation derived in part (c) of Ex. 6.6.

24.2 (a) Show that the two methods of energy estimation in Sec. 24.2, Eqs. (24.20) and (24.21), give the same mean value. (b) Show that they give identical fluctuations in the limit $\Delta \tau \to 0$.

24.3 Prove the relation Eq. (24.22) by neglecting correlation between walkers having a common "ancestor." *Hint*: look at [1022].

24.4 Do a DMC calculation of a harmonic oscillator, a hydrogen molecule, or a helium atom.

24.5 Derive Eq. (24.16): the master equation for diffusion Monte Carlo using importance sampling.

24.6 Determine how the comb method in Sec. 24.2 of population control can be done. How can we sample the number of copies while ensuring that the population remains fixed? *Hint*: analyze the case for two walkers, $P^{(0)} = 2$, and then generalize.

24.7 Prove in the transient estimate method that the total error (systematic and statistical) will decrease with the number of steps P as: $\epsilon \propto P^{-\eta}$, where $\eta = \frac{1}{2} e_g / (E_F - E_B + e_g)$, E_F is the energy of the fermion state, E_B is the bosonic energy, and e_g is the "gap," the relevant excitation energy of the fermion system. Make estimates of both the systematic error and the statistical error as a function of projection time and choose the projection time to minimize their sum. See footnote 12 in this chapter.

24.8 (a) Derive the two perturbation formulas in Eq. (24.32). *Hint*: expand either Ψ or $\ln(\Psi)$ about Φ. Remember to keep excited states orthogonal to the ground state. (b) Consider non-interacting particles in an external potential. Justify the formulas for correcting the density, $n(\mathbf{r})$.

24.9 Determine the projector G_D near a node with fixed-node boundary conditions (see Sec. 24.4). *Hint*: first consider a single non-interacting particle in the half-space $z > 0$ with a zero boundary condition on the plane $z = 0$ and determine the exact projector for this boundary condition. How can you generalize this result at small $\Delta \tau$ to include general nodes and interaction?

25

Path-integral Monte Carlo

It is shown from first principles that, in spite of the large interatomic forces, liquid ^{4}He
should exhibit a transition analogous to the transition in an ideal Bose–Einstein gas. The
exact partition function is written as an integral over trajectories, using the space-time
approach to quantum mechanics.

R.P. Feynman, 1953

Summary

In this chapter we discuss imaginary-time path integrals and the path-integral Monte
Carlo method for the calculation of properties of quantum systems at non-zero temperature. We discuss how Fermi and Bose statistics enter, and how to generalize the
fixed-node procedure to non-zero temperature. We then discuss an auxiliary-field
method for the Hubbard model. The path-integral method can be used to perform
ground-state calculations, allowing calculations of properties with less bias than the
projector Monte Carlo method. We also discuss the problem of estimating real-time
response functions using information from imaginary-time correlation functions.

In previous chapters we described two QMC methods, namely variational QMC and projector (diffusion) QMC. Both of these methods are zero temperature, or, more properly,
are formulated for single states. In this chapter we discuss the path-integral Monte Carlo
(PIMC) method, which is explicitly formulated at non-zero temperature. Directly including
temperature is important because many, if not most, measurements and practical applications involve significant thermal effects. One might think that to do calculations at a
non-zero temperature we would have to explicitly sum over excited states. Such a summation would be difficult to accomplish once the temperature is above the energy gap,
because there are so many possible many-body excitations. In addition, the properties for
each excitation are more difficult to calculate than those for the ground state. As we will
see, path-integral methods do not require an explicit sum over excitations. As an added
bonus, they provide an interesting and enlightening window through which to view quantum systems. However, the sign problem, introduced in the previous chapter, is still present
for fermion systems. The fixed-node approximation is used again.

An advantage of PIMC is the absence of a trial wavefunction.[1] As a result, quantum expectation values, including ones not involving the energy, can be computed directly. For the expert, the lack of an importance function may seem a disadvantage; without it one cannot push the simulation in a preferred direction. However, for more complex quantum systems, it becomes increasingly difficult and time-consuming to devise satisfactory trial functions; hence, a procedure that only depends on the hamiltonian can be preferable as it has less human bias. The path-integral method is not limited to non-zero temperatures. A related method can also be used to compute ground-state properties, as we discuss in Sec. 25.4 below; it leads to a different projector Monte Carlo algorithm without the mixed estimator problem.

Applications of PIMC have been to electronic systems at relatively high temperatures [1105], computation of the correlation energy of the homogeneous electron gas [1106, 1107], to bosonic systems [1108], to quantum effects of the nuclei, to few-electron problems [1109], to the Hubbard model, and to the embedded-site models in DMFT. It has been found difficult to converge many-electron PIMC for temperatures much lower than the Fermi energy, $0 < T \ll E_F$, though progress can be expected.

25.1 The path-integral representation

To introduce path integrals, we first review the properties of the thermal N-body density matrix.[2] Its coordinate-space representation is defined in terms of the exact N-body eigenstates $\Phi_i(R)$ and energies E_i:

$$\rho(R, R'; \beta) = \sum_i \Phi_i^*(R) e^{-\beta E_i} \Phi_i(R'). \tag{25.1}$$

In addition to the inverse temperature $\beta = 1/(k_B T)$, the N-body density matrix depends on two sets of N-body coordinates, R and R', where in this and the following sections we neglect the spin dependence. We refer to those matrix elements with $R \neq R'$ as "off-diagonal." The partition function is the trace of $\rho(R, R'; \beta)$, the integral over the diagonal density matrix:[3]

$$Z(\beta) = \int dR \rho(R, R; \beta) = \sum_i e^{-\beta E_i}. \tag{25.2}$$

Thermodynamic properties are obtained as:

$$\langle \hat{O} \rangle = Z(\beta)^{-1} \int dR dR' \langle R|\hat{O}|R' \rangle \rho(R', R; \beta). \tag{25.3}$$

---

[1] However, in the restricted path-integral method, to avoid the fermion sign problem, one introduces trial nodal surfaces, analogous to the trial wavefunction in fixed-node projector Monte Carlo. The bias introduced is only related to errors in the nodal location.

[2] Note that it is not normalized by the partition function and in this chapter we work in the canonical ensemble.

[3] In general this can include tracing over spin or particle number, depending on the ensemble.

The operator identity $\exp(-\beta \hat{H}) = [\exp(-\Delta \tau \hat{H})]^M$, where $\Delta \tau = \beta/M$, relates the density matrix at a temperature $k_B/\Delta \tau$ to the density matrix at a temperature M times *lower* where M, a positive integer, is the number of (imaginary) "time slices." Writing this identity in the coordinate representation gives:

$$\rho(R_0, R_M, \beta) = \int dR_1 \ldots dR_{M-1} \prod_{t=1}^{M} \rho(R_{t-1}, R_t; \Delta \tau). \qquad (25.4)$$

The sequence of intermediate points $\{R_1, R_2, \ldots, R_{M-1}\}$ is the *path*, and $\Delta \tau$ is the *imaginary time step*. Trotter's formula, discussed in Sec. 24.2, allows us to assume that the kinetic $\hat{K}$ and potential $\hat{V}$ operators commute for sufficiently small $\Delta \tau$: $e^{-\Delta \tau \hat{H}} \sim e^{\Delta \tau \hat{T}} e^{-\Delta \tau \hat{V}}$. Define the *action* as $S(R, R'; \Delta \tau) = -\ln[\rho(R, R'; \Delta \tau)]$. The so-called "primitive approximation" to the action is:

$$S_P(R, R'; \Delta \tau) = \frac{3N}{2} \ln(4\pi \lambda \Delta \tau) + \frac{(R - R')^2}{4\lambda \Delta \tau} + \frac{\Delta \tau}{2}(V(R) + V(R')). \qquad (25.5)$$

Here $\lambda \equiv \hbar^2/2m = \frac{1}{2}$ for electrons in atomic units. This is the same form as was derived in Eq. (24.10). Substituting the action Eq. (25.5) into the path-integral expression Eq. (25.4), the partition function is given by:

$$Z_D(\beta) = \lim_{M \to \infty} \int dR_1 \ldots dR_M \exp\left[-\sum_{t=1}^{M} S_P(R_{t-1}, R_t; \beta/M) \right], \qquad (25.6)$$

with the boundary condition $R_0 = R_M$ to find the trace.[4]

If the potential energy is real, the integrand of Eq. (25.6) is non-negative and can thus be interpreted as a classical system with an effective classical potential given by the sum in its exponent. This defines an exact mapping of a quantum system onto a classical equilibrium system: the quantum system of N particles in M time slices becomes an NM-particle classical system. The classical system is composed of N "polymers" each having M "beads" with harmonic springs between neighboring beads (the second term in Eq. (25.5)) and an inter-polymer potential between different polymers (the third term in Eq. (25.5)). To calculate the partition function, and most thermodynamic properties, the polymers must close on themselves. Calculation of the single-particle density matrix and the momentum distribution would require one path to be open. The left panel of Fig. 25.1 shows a typical example of closed paths. A lower temperature means a longer polymer, i.e., with more beads. Note that an individual polymer does not interact with itself except with the springs, and the inter-polymer potential (the third term in Eq. (25.5)) is not like a real polymer: it only interacts with beads with the same path-integral time-slice "index."

To evaluate the properties of the quantum system, we must perform the $3NM$-dimensional integral of Eq. (25.6) over all paths using either Metropolis Monte Carlo

---

[4] In this formula, Boltzmann or distinguishable particle statistics are assumed and its partition function is written as Z_D. We will consider Bose and Fermi statistics in the next section. Also note that we have assumed $\lambda \Delta \tau \ll L^2$ for systems with periodic boundary conditions.

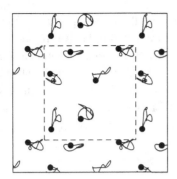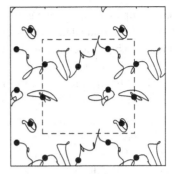

Figure 25.1. Typical paths of six quantum particles in a 2D periodic square. The large black dots represent the positions of the particles at the start of their paths. The paths have been smoothed by zeroing their short-wavelength Fourier components; a picture with all Fourier components would be a space-filling fractal curve, see [1110]. The left panel is the identity permutation, the right panel is a three-particle permutation with a path "winding" in the horizontal direction (see Eq. (25.19)) across the periodic boundaries shown as the inner dotted square.

(Ch. 22) or molecular dynamics. To obtain exact results within the statistical sampling error, one must calculate the results for several values of M and extrapolate[5] to $M \to \infty$. In the following we use the notation for coordinates: $R_t = \{\mathbf{r}_{1,t}, \mathbf{r}_{2,t} \ldots \mathbf{r}_{N,t}\}$. The first index is the particle index, index "t," is the time-slice index.

Single-slice PIMC algorithm for distinguishable particles

- Initialize the state: give $R_1 \ldots R_M$ reasonable values.
- Loop over iterations n.
- Advance the state:
  - Pick time slice t and particle i to update.
  - Sample $\mathbf{r}'_{i,t}$ from the normal distribution:

$$T(\mathbf{r}'_{i,t}) \propto \exp\left[-\Delta\tau^{-1}(\mathbf{r}'_{i,t} - \frac{1}{2}[\mathbf{r}_{i,t-1} + \mathbf{r}_{i,t+1}])^2\right]. \qquad (25.7)$$

  - Calculate acceptance ratio:

$$A = \exp(-\Delta\tau(V(R') - V(R))). \qquad (25.8)$$

  - If $A > u_n$, where u_n is a uniformly distributed random number in $(0, 1)$, set $\mathbf{r}_{i,t} = \mathbf{r}'_{i,t}$. Otherwise, keep path fixed.
  - Accumulate average potential and kinetic energy, acceptance ratio, and other properties.

---

[5] There do exist PIMC methods without a "time-step" error, e.g., continuous-time PIMC in Ch. 18.

For efficient computation, one needs to improve the sampling and use more accurate actions requiring fewer time slices, as described in detail in [1110]. In order to improve the action, it is advisable to use the pair action, i.e., the numerical solution to the two-body problem [1110]. For example, the divergence of the Coulomb potential when two unlike charges approach each other can wreak havoc on the stability of the algorithm, since paths can fall into the region at small r_{ij} and remain for many iterations. A simple approach to solve this problem is to cut off the potential for $r_{ij} < \Delta \tau^{\frac{1}{2}}$. However, it is much better to use the exact two-body density matrix since it does not diverge as the charges approach each other, and has the exact cusp condition (Eq. (6.2)) built in.

To compute the energy there are several approaches [1110]. Differentiating the partition function with respect to temperature gives the *thermodynamic form*. The kinetic energy estimator is:

$$\tilde{K}_T = \frac{3N}{2\Delta\tau} - \sum_{i,t} \left\langle \frac{(\mathbf{r}_{i,t} - \mathbf{r}_{i,t-1})^2}{2\Delta\tau^2} \right\rangle . \tag{25.9}$$

The second term is the negative spring energy of the classical polymer system but has the disadvantage that it diverges at small $\Delta\tau$ as $\Delta\tau^{-2}$, causing loss of efficiency if the time step is extrapolated to zero. A form that has the same average value, the *virial* [1111] estimator

$$\tilde{K}_v = \frac{3Nk_BT}{2} - \frac{1}{2}\left\langle \sum_{i=1}^{N} (\mathbf{r}_{i,t} - \bar{\mathbf{r}}_i) \cdot \mathbf{F}_{i,t} \right\rangle , \tag{25.10}$$

does not have this problem. Here $\bar{\mathbf{r}}_i$ is the centroid position[6] of particle i and $\mathbf{F}_{i,t}$ is the classical force on particle i at time t. Note that the first term is simply the classical kinetic energy. Quantum corrections are given in the second term and will vanish at high temperature, since $|\mathbf{r}_{i,t} - \bar{\mathbf{r}}_i| \to 0$.

To perform an efficient Monte Carlo sampling of Eq. (25.6) one needs to use collective moves because single-bead moves can make the whole procedure slow, particularly in the limit of small $\Delta\tau$. For details on the alternate PIMD approach to sample the paths, see [1112, 1113].

Path integrals for quantum ionic motion

There are many uses of PIMC/PIMD for calculations of many-body quantum systems with significant quantum effects but without effects of Fermi or Bose statistics. The Wigner crystal phase is stable at low density; see Figs. 3.1 and 5.1. Properties within the crystal phase can be computed without assuming Fermi statistics, because the effects of antisymmetry are small and can be ignored if one is not interested in magnetic properties. In [1114]

---

[6] The centroid is the center of mass of a given polymer, $\bar{\mathbf{r}}_i \equiv \beta^{-1} \int_0^\beta d\tau \mathbf{r}_{i,\tau}$. See [1110] for the generalization of Eq. (25.10) to identical particles.

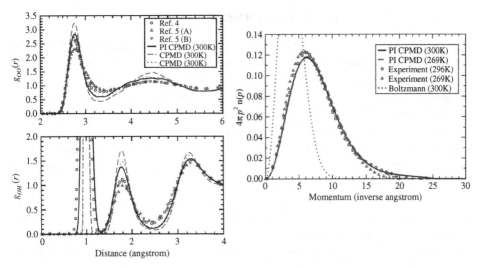

Figure 25.2. Correlation functions in water under ambient conditions using classical and quantum simulations from [1118]. The left figures are the pair correlation functions between the ions, $g_{OO}(r)$ and $g_{OH}(r)$, with measurements shown as symbols, PIMD calculations are shown as solid lines, and classical MD calculations as dashed lines. The right figure shows the classical Maxwellian momentum distribution of the protons in water, while the other curves are experiment and PIMD simulations of the radial proton momentum distribution. CPMD (Carr–Parrinello molecular dynamics) uses DFT forces.

are computed energies of point defects in the 2D Wigner crystal.[7] Details for the transition density and temperature will change if we include Fermi statistics. How to do this is discussed in the following sections of this chapter.

Some examples of computing the zero-point effects of nuclei for liquid and solid hydrogen are given in [1105, 1115–1117]. Many PIMC simulations use an empirical pair potential or a Born–Oppenheimer surface based on an assumed density functional. Water in various conditions of pressure and temperature is being studied intensively; the effects of proton motion in water are significant [1118], as seen in Fig. 25.2. Another example of the use of PIMC is solid lithium [1119]. Although these examples are not concerned with electronic structure, clearly they describe important effects in condensed matter. PIMC or PIMD calculations for ionic zero-point effects are not often done because the computation makes the calculations slower. As computers and software become more capable, the path-integral treatment of ionic motion of light nuclei can become a standard part of the procedures used to simulate materials, since the methodology is a simple generalization of what is done for dynamics of classical nuclei.

---

[7] This calculation is difficult to do using projector Monte Carlo because the trial function is more likely to bias the result.

25.2 Exchange of localized electrons

We now discuss a specialized application of PIMC, namely to compute exchange frequencies between electrons localized on different lattice sites. First we discuss a simple model: a single electron is confined to the interior of the union of two spheres, as shown in Fig. 25.3. Because of mirror symmetry, the quantum states can be classified by parity. The splitting between the lowest even and odd parity states defines the exchange frequency, $J = \frac{1}{2}(E_1 - E_0) > 0$. An electron initially localized in one of the spheres will oscillate back and forth with a frequency given by $J/\hbar$. Let us suppose that the splitting energy is much less than the zero-point energy, so higher excitations can be neglected.

Here we show how to calculate this frequency using path integrals. Figure 25.3 shows diagrams of the imaginary-time paths in the double-sphere model. One sees that the electron spends a long time in a single sphere, but occasionally it "tunnels"[8] across to the other sphere. The tunneling is rapid, since the wavefunction is squeezed as it passes from one sphere to the other.

Let us denote the coordinates of the centers of the two spheres as Z and $\hat{P}Z$; the motivation for this notation will become clear when we discuss the multi-electron generalization. Now define $f_{\mathcal{P}}$ as the ratio of the imaginary-time density matrix connecting Z to $\hat{P}Z$ with that connecting Z to itself:

$$f_{\mathcal{P}}(\beta) = \frac{\rho(Z, \hat{P}Z; \beta)}{\rho(Z, Z; \beta)}.$$ (25.11)

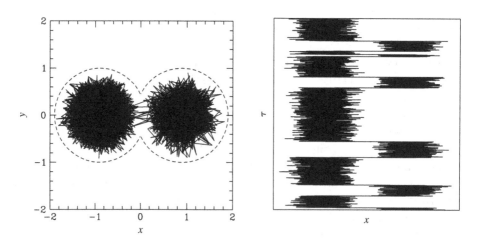

Figure 25.3. Left panel: the imaginary-time path confined to the interior of two spheres (shown by dashed lines). Right panel: the x-coordinate of the same path shown as a function of imaginary time. The electron stays inside one sphere for a long period, until it finds the duct to the other sphere.

---

[8] Tunnels is in quotes because we are in imaginary time not real time. The imaginary-time transversal of the barrier is called an "instanton" because it takes place so quickly.

We now assume that β is large enough so that only the lowest two states contribute to the density matrix. Then, as shown in Ex. 25.2,

$$f_P(\beta) = \tanh[J(\beta - \beta_0)]. \tag{25.12}$$

The rate in imaginary time for the electron to cross from one sphere to the other is J. In the polymer language, J is related to the free energy it takes to pull a single end of a "linear polymer" from one sphere to the other. This can be estimated with special techniques [1120–1122]. The constant β_0 is the average imaginary time to cross the barrier. Note that since $f_P(\beta)$ must be positive for all β, Eq. (25.12) is only valid for $\beta > \beta_0$; higher excited states are needed otherwise.

Now, let us generalize from the two-sphere model to a many-body system. We follow the theory of Thouless [1123], based on the earlier work of Dirac [1124] on electronic exchange. Because exchange is rare for localized electrons, we can label the positions of electrons by localized Wannier orbitals: Z denotes one such assignment of N electrons to N orbitals and $\hat{P}Z \equiv \{\mathbf{z}_{P_1}, \mathbf{z}_{P_2}, \ldots, \mathbf{z}_{P_N}\}$ the effect of applying the permutation $\hat{P}$ to that assignment. If there are N electrons and N Wannier functions, there are $N!$ such assignments, so there is an $N!$ degeneracy of the ground state in the absence of electron exchange. The splitting induced by tunneling between states Z and $\hat{P}Z$ is defined to be $2J_P$, as in the example above. All of the previous discussion concerning how to calculate J_P with path integrals then applies.

Atomic exchange is shown to couple the electron spins on different atoms and results in a Heisenberg spin hamiltonian:

$$\hat{H}_{\text{eff}} = -\sum_P J_P (-1)^P \hat{P}_\sigma \tag{25.13}$$

$$= -J_2 \sum_{(i,j)} \sigma_i \cdot \sigma_j, \tag{25.14}$$

where P ranges over all $N!$ permutations,[9] $(-1)^P$ is the sign of the permutation, and $\hat{P}_\sigma$ permutes spin coordinates. From Eq. (25.13) we see that an exchange of an even number of spins favors antiferromagnetism, while an exchange of an odd number of spins favors ferromagnetism. The second equation, the conventional Heisenberg spin hamiltonian (Eq. (3.4)), applies if the only exchanges allowed involve two- and three-body permutations, see Sec. 3.3. A clear discussion of the origin of the exchange hamiltonian from the "ring exchanges" is given by Roger [1125].

PIMC calculations have been used to determine the Heisenberg exchange coefficients in the Wigner crystals [112, 113] and in solid ^{3}He. The PIMC method to determine the exchange frequencies is much superior to one based on projector Monte Carlo, since one can determine directly the terms in the underlying spin hamiltonian. The results are accurate even if the exchange frequencies are very small.[10] Calculations of the exchange

---

[9] One need only consider cyclic permutations of neighboring electrons, otherwise J_P will be much smaller.

[10] Compare the estimates of the exchange frequencies of the 3D Wigner crystal [112] from PIMC with those using DMC [278].

frequencies of the 2D Wigner crystal suggest a frustrated spin liquid phase may be stable [113]. These methods have not yet been applied to realistic electronic materials.

25.3 Quantum statistics and PIMC

The most interesting consequences of quantum physics (e.g., superfluidity, Bose condensation, superconductivity, and Fermi liquid behavior) come from the Fermi or Bose statistics of the particles. We now consider, in general, how particle statistics are expressed in path integrals. The algorithm for treating quantum statistics is quite simple: we ignore the identities of particles when the imaginary-time paths close. That is, they are allowed to close on a permutation of themselves, i.e., $R_M = \hat{\mathcal{P}}R_0$, where $\hat{\mathcal{P}}R \equiv \{\mathbf{r}_{\mathcal{P}_1}, \mathbf{r}_{\mathcal{P}_2}, \ldots, \mathbf{r}_{\mathcal{P}_N}\}$ is a relabeling of the coordinates. To understand this pictorially, compare the left and right panels of Fig. 25.1.

To prove the correctness of this procedure, we first note that the wavefunctions of fermions (bosons) are antisymmetric (symmetric): their density matrix is defined by summing only over antisymmetric (symmetric) states[11] in Eq. (25.1). In the following, we will denote the statistics of the particles by subscripts: ρ_F will denote the fermion density matrix, ρ_B the boson density matrix, ρ_D the boltzmannon (distinguishable particle) density matrix. The relabeling operator $(N!)^{-1} \sum_{\mathcal{P}}(\pm 1)^{\mathcal{P}}\hat{\mathcal{P}}$ projects out the states of correct symmetry. Here the upper sign $(+1)$ is for bosons and the lower sign (-1) is for fermions, $(-1)^{\mathcal{P}}$ stands for the signature of the permutation.[12] We use the relabeling operator to construct the path-integral expression for bosons or fermions in terms of the Boltzmann density matrix:

$$\rho_{B/F}(R, R'; \beta) = \frac{1}{N!} \sum_{\mathcal{P}}(\pm 1)^{\mathcal{P}} \rho_D(\hat{\mathcal{P}}R, R'; \beta). \qquad (25.15)$$

Note that we could apply this relabeling operator to the first argument, the last argument, or both; since the particles are identical, the resulting density matrix would be the same. The connection between the boltzmannon density matrix and the boson/fermion density matrix is important because it is the boltzmannon density matrix that arises naturally from paths. Including statistics, the path-integral expression of the partition function for Bose or Fermi statistics becomes:

$$Z_{B/F}(\beta) = \frac{1}{N!} \sum_{\mathcal{P}}(\pm 1)^{\mathcal{P}} \int dR_1 \ldots dR_M \exp\left[-\sum_{t=1}^{M} S(R_{t-1}, R_t; \Delta\tau)\right] \qquad (25.16)$$

$$R_0 = \hat{\mathcal{P}}R_M. \qquad (25.17)$$

Bosons

First, we examine the effect of the new boundary conditions for Bose statistics. The sign of the permutation in Eq. (25.16) is always positive, so the combined integration and sign

---

[11] A similar procedure can be used for other symmetries such as momentum or spin.

[12] If a permutation can be written as an odd number of pair exchanges it is negative, otherwise it is positive.

is positive and can be treated as a probability density. For large N, it is very difficult to explicitly perform the permutation sum since it has $N!$ terms. However, we can enlarge the space to be sampled in the Monte Carlo random walk by including how the paths are connected, i.e., $\mathcal{P}$. One such connection is shown in the right panel of Fig. 25.1. With Monte Carlo techniques, this extra sampling does not necessarily slow down the calculation, but one needs to include moves that are ergodic in the combined space of paths and boundary conditions of Eq. (25.16).

Let us examine how this affects the physics of a boson system. One of the fundamental properties of a Bose-condensed system is superfluidity: a superfluid can flow without viscosity similar to how a superconductor can carry current without resistance. The superfluid density is defined experimentally as follows: suppose the walls of a container are moved with a small velocity $\mathbf{V}$ and the momentum acquired by the enclosed system in equilibrium is measured. In a normal liquid or solid, the enclosed system will move with the walls and acquire a momentum $M_0\mathbf{V}$. ($M_0 = Nm_i$ is the total mass of the system, N is the number of bosons of mass m_i.) However, a superfluid can shield itself from the walls. The superfluid fraction n_s/n is defined in terms of the mass **not** contributing to the momentum:

$$\frac{n_s}{n} = 1 - \frac{\mathbf{P}}{M_0\mathbf{V}} \mapsto \frac{m_i\langle\mathbf{W}^2\rangle}{\hbar^2\beta N}. \tag{25.18}$$

The expression on the right is an estimator for the superfluid fraction used with path integrals in periodic boundary conditions [1126]. The winding number of a given path,

$$\mathbf{W} = \sum_{i=1}^{N}\int_0^\beta d\tau\frac{d\mathbf{r}_{i,\tau}}{d\tau}, \tag{25.19}$$

is the number of times a given path wraps around the periodic boundaries. This remarkable formula relates the real-time linear response of moving the boundaries to a topological property of imaginary-time path integrals in periodic boundaries.

Now the size of a path of a single atom is its thermal de Broglie wavelength $\hbar(mk_BT)^{-1/2}$, which is always microscopic even at very low temperatures. Recall that any permutation can be decomposed into permutation cycles, i.e., into two-, three-, ...N-body exchange cycles. Non-zero winding numbers and hence superfluidity are only possible when exchange cycles extending across a macroscopic distance appear, as shown in the right panel of Fig. 25.1. This happens at low temperature. There are interesting connections between the exchange of electrons in an insulator and the exchanges of bosonic paths, as discussed in Sec. 6.3.

Bose condensation is another key property of superfluids that can be interpreted with path integrals. In a 3D superfluid, a certain fraction of the particles will "condense" into the zero-momentum state, or in an inhomogeneous system, into a single "natural orbital" (see Eq. (5.7)). To determine the single-particle density matrix (see Eq. (5.6)), we need to sample paths where one particle does *not* close on itself: the two ends of an open "polymer" need to be free to move around the system independently of each other. For a homogeneous system the momentum distribution is the Fourier transform of $\rho(\mathbf{r}, \mathbf{r}')$, as given in Eq. (5.9).

For a normal (i.e., not Bose condensed) system, the two ends remain within a thermal de Broglie wavelength, implying that the momentum distribution is also localized. However, once macroscopic exchanges in the path can occur, the two ends can separate by a macroscopic distance so that $\lim_{|\mathbf{r}-\mathbf{r}'|\to\infty} \rho(\mathbf{r}, \mathbf{r}') \to n_0 > 0$, implying that $\rho(k) = n_0\delta(k)$, where n_0 is the condensate fraction. The macroscopic exchange of particles is how the phase of the wavefunction is communicated. The macroscopic "percolation" of the polymers (i.e., a network of connected polymers spanning a macroscopic volume) is directly related to superfluidity and Bose condensation.

Using PIMC, one can calculate equilibrium properties of many-body ^{4}He at all temperatures, in the liquid phase above and below the superfluid transition and in the solid phase. For details on the path-integral theory of Bose superfluids and of the PIMC calculations, see [1110]. The worm algorithm [1127] allows the sampling of a superfluid phase to be done more efficiently, particularly for systems with more than a few hundred bosons. It works in the grand-canonical ensemble and can compute unequal-time correlation functions such as the one-particle Green's function in imaginary time defined in Sec. 5.6.

Fermions

Now let us consider how to do a path-integral calculation of a fermion system by summing over permutations just as for bosonic systems. We need to include the factor $(-1)^P$ as a weight in the numerator and denominator of any expectation value, as in Eq. (24.25). In 1965, Feynman and Hibbs [1128] identified the difficulty with this procedure:

> The expression ... for Fermi particles, such as He^3, is also easily written down. However, in the case of liquid He^3, the effect of the potential is very hard to evaluate quantitatively in an accurate manner. The reason for this is that the contribution of a cycle to the sum over permutations is either positive or negative depending on whether the cycle has an odd or even number of atoms in its length L. At very low temperature, the contributions of cycles such as $L = 51$ and $L = 52$ are very nearly equal but opposite in sign, and therefore they very nearly cancel. It is necessary to compute the difference between such terms, and this requires very careful calculation of each term separately. It is very difficult to sum an alternating series of large terms that are decreasing slowly in magnitude when a precise analytic formula for each term is not available. Progress could be made in this problem if it were possible to arrange the mathematics describing a Fermi system in a way that corresponds to a sum of positive terms. Some such schemes have been tried, but the resulting terms appear to be much too hard to evaluate even qualitatively.

If one performs an integration over a function having both positive and negative regions with Monte Carlo, the "signal-to-noise" ratio (i.e., the efficiency) is much reduced. Doing a direct sampling of the boson paths and permutations and using the permutational sign to estimate properties of the fermion system leads [1129] to a computational efficiency of the fermion system (ξ_F) equal to

$$\xi_F = \xi_B \exp[-2N\beta(\mu_F - \mu_B)], \tag{25.20}$$

where μ_F (μ_B) is the free energy per particle of the fermion (boson) system and ξ_B the efficiency of the simulation of the equivalent boson system. For the derivation, see Ex. 25.3. This direct fermion method, while exact, becomes exceedingly inefficient as $N\beta = N/(k_B T)$ increases, precisely when the physics becomes interesting. Note that this bears a strong similarity to what we discussed for projector Monte Carlo in Sec. 24.3.

Restricted path-integral Monte Carlo

The restricted path identity solves Feynman's challenge of keeping only "positive" paths at the cost of making an uncontrolled approximation. It is the generalization of the ground-state fixed-node method discussed in Sec. 24.4. The nodes of the exact fermion density matrix give a rule for deciding which paths can contribute:[13] only use paths that do not cross the nodes of the fermion density matrix. The method is based on the following identity:

$$\rho_F(R_\beta, R_*; \beta) = \int dR_0 \rho_F(R_0, R_*; 0) \oint_{R_0 \to R_\beta \in \Upsilon(R_*)} dR_\tau e^{-S[R_\tau]}, \qquad (25.21)$$

where the subscript on the path integration means that we integrate only over continuous paths starting at R_0, ending at R_β, and which are node-avoiding: those for which $\rho_F(R_\tau, R_*; \tau) \neq 0$ for all $0 \leq \tau \leq \beta$.[14] The "reference point" R_* defines the nodes.

To prove this identity we note that the fermion density matrix satisfies the Bloch equation:

$$\frac{\partial \rho_F(R, R_*; \tau)}{\partial \tau} = \frac{1}{2} \sum_{i=1}^N \nabla_i^2 \rho_F(R, R_*; \tau) - V(R)\rho_F(R, R_*; \tau), \qquad (25.22)$$

with the initial conditions:

$$\rho_F(R, R_*; 0) = \sum_{\mathcal{P}} \frac{(-1)^{\mathcal{P}}}{N!} \delta(R - \hat{P}R_*). \qquad (25.23)$$

Hence, the path starts at a permutation of the reference point, $R_0 = \hat{P}R_*$, and carries a weight $\frac{1}{N!}(-1)^{\mathcal{P}}$. The solution of the Bloch equation is uniquely specified by its boundary conditions, just like the Poisson equation in electrostatics [1130]. Normally, one uses the values at zero imaginary time (i.e., at infinite temperature) for boundary conditions, however, to prove this identity, we take the nodal surfaces, $\rho_F(R, R_*; \tau) = 0$, for boundary conditions as illustrated in Fig. 25.4. We want the solution of the Bloch equation that vanishes on a preselected nodal surface. We enforce this solution by putting an infinite repulsive wall on this surface, or, equivalently, restricting the allowed paths to remain on

---

[13] We need for the path to be continuous. Lattice models or non-local hamiltonians do not have continuous trajectories, see Sec. 24.4.

[14] It can be shown [1129] that one can also use the condition $\rho_F(R_\tau, R_*; \min(\tau, \beta - \tau)) \neq 0$ for all $0 \leq \tau \leq \beta$. This means that smaller time arguments are required (only up to $\beta/2$) and the paths are symmetric under the operation $\tau \leftrightarrow \beta - \tau$.

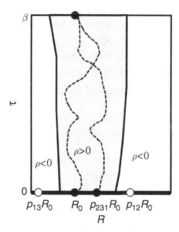

Figure 25.4. Space-time cartoon for proof of the restricted PIMC identity. The horizontal axis represents the spatial coordinates and the vertical axis, imaginary time. The usual boundary conditions are δ-functions at $\tau = 0$, represented on the x-axis; positive (filled dots) for even permutations and negative (open dots) for odd permutations. For thermodynamic properties it is sufficient to work in the strictly positive (shaded) domain. Shown are two allowed paths in this domain, one from the identity permutation and one from a three-body permutation.

the interior of a given nodal domain, $\Upsilon(R_*)$. The solution is exact if the assumed nodes are correct. For the diagonal elements of the density matrix, $R_\beta = R_*$, all contributions must be positive; hence the sum over permutations is restricted to even permutations.[15]

The restricted path picture is a novel way of visualizing fermion systems [1129, 1131]. That a Fermi liquid is equivalent to exchanging paths can be understood by considering the momentum distribution, $\rho(k)$. By definition a Fermi liquid has a discontinuity in $\rho(k)$. This implies that the single-particle density matrix $\rho(\mathbf{r}, \mathbf{r}') \propto |\mathbf{r} - \mathbf{r}'|^{-3}$ at large separation. Such a slow decay can only come from macroscopic exchanges of even permutations. In a superconductor with Cooper pairs of electrons, there will be paired up-spin and down-spin macroscopic exchanges [1132]. Krüger and Zaanen [1133] have interpreted other quantum phase transitions in terms of the restricted path formalism.

The problem we now face is that the unknown fermion density matrix appears on the right-hand side of Eq. (25.21), since it is used to define the criterion of node-avoiding. To apply the formula directly, we would have to self-consistently determine the nodes.

The reference point R_* plays a very special role, since it determines the allowed paths as illustrated in the example below. For boson or distinguishable path integrals, all time slices are equivalent, but restricted paths break this time symmetry. If we write expectation values in terms of derivatives of the partition function (e.g., the kinetic energy is proportional to

---

[15] As shown in the ortho-hydrogen example discussed below, the correct procedure involves more than simply restricting the paths to even permutations. The reason that restriction gives the same result is because negative paths can be paired with positive paths and canceled. The gradient of the density matrix at the node is the flux of the path, and since the gradient is continuous across the node, the positive paths crossing at a given nodal point will precisely cancel against the negative paths.

the mass derivative of the logarithm of the partition function), one recovers an expectation value involving all the time slices, not just one containing the reference point. By a "ground-state" restriction is meant that it does not depend on the reference point. This is achieved by using an antisymmetric trial wavefunction Ψ and requiring that $\Psi(R_\tau) \neq 0$ throughout the path. For sufficiently low temperatures this corresponds to the ground-state fixed-node method described in Sec. 24.4.

Example: restricted paths for molecular hydrogen

One of the simplest examples on how to treat quantum statistics with imaginary-time path integrals is provided by ortho-hydrogen. Consider an isolated hydrogen molecule consisting of two electrons and two hydrogen nuclei that can be either protons or deuterons. Assume the electrons are in their ground state: spatially symmetric and in an electronic spin-singlet. The protonic spin-singlet state, known as para-hydrogen, must have an even value of spatial angular momentum L to keep the molecular wavefunction antisymmetric under proton exchange; the spin-triplet state, ortho-hydrogen, must have odd values of L. In many environments the ions cannot easily change their spin state, so that para- and ortho-hydrogen can be considered as separate chemical species for a time. A third symmetry case arises if the two nuclei are different isotopes, e.g., a proton and a deuteron; they will obey distinguishable particle (Boltzmann) statistics. For simplicity, suppose the nuclei are massive enough that their relative coordinate is fixed at the bond length. This fixes the ion–ion bond vector $\mathbf{r} = \mathbf{r}_1 - \mathbf{r}_2$ to be on the surface of a sphere with radius 0.074 nm.

Now let us consider how to calculate the partition function of the three types of molecular hydrogen with path integrals. Distinguishable particles are the simplest: allow all paths on a sphere periodic in imaginary time, e.g., the left-most sphere of Fig. 25.5. To properly describe ortho- and para-hydrogen, we need to consider wavefunctions $\Phi(\mathbf{r})$ that are symmetric (para) or antisymmetric (ortho) with respect to inversion $\Phi(\mathbf{r}) = \pm\Phi(-\mathbf{r})$. Applying the projection operator as in Eq. (25.16) allows for paths that go from one point on the sphere to its antipode, as shown by the dashed line in the center sphere of Fig. 25.5. For simulations of para-hydrogen, we sum over both types of paths, while for ortho-hydrogen we would *subtract* the contribution of the paths that are antiperiodic in time (dashed) from those that are periodic in time (solid). At low temperature, a given path can range uniformly over the surface of the sphere so these two contributions are nearly equal. As a consequence, the signal-to-noise ratio for ortho-hydrogen will approach zero.

In the RPIMC algorithm for ortho-hydrogen, we only allow paths that are periodic in imaginary time, and only those that lie wholly in the hemisphere within 90° of the reference point,[16] as illustrated in the right sphere of Fig. 25.5. The threefold degeneracy of the ground state of the $L = 1$ state is reflected in the fact that $\mathbf{r}_*$ defines the restriction even at low temperature. Isotropy is restored by averaging $\mathbf{r}_*$ over the surface of the sphere.

---

[16] In relative coordinates, any wavefunction has the angular factor $Y_{lm}(\hat{\mathbf{r}})$; in the sum over the quantum states for different m in Eq. (25.1), using the addition formula for spherical harmonics we obtain a factor $P_l(\hat{\mathbf{r}} \cdot \hat{\mathbf{r}}_*)$. The ortho-density matrix, made from odd Legendre polynomials, will vanish when $\mathbf{r} \cdot \mathbf{r}_* = 0$.

Boltzmann   Bose/Fermi   Restricted Fermi

Figure 25.5. Depiction of path integrals for distinguishable ions (left sphere), ortho- and para-hydrogen (center sphere), and restricted paths for ortho-hydrogen (right sphere). Solid lines are closed paths; the dotted path goes from a point $\mathbf{r}$ to the point $-\mathbf{r}$. The reference point $\mathbf{r}_*$ is the dot on the right sphere. The path in the right sphere is restricted to lie in the northern hemisphere.

The trial density matrix

The accuracy of RPIMC can depend on the assumed restriction; the restriction is an uncontrolled approximation, although even if some of the quantitative details are inaccurate, if we can characterize the nodal restriction sufficiently well, the simulations will be useful in understanding strongly interacting fermion systems.[17] In an actual calculation, one does not use a geometric interpretation of the nodes as we did in the hydrogen example. Instead, one uses a *trial density matrix*, $\rho_T(R, R_*; \beta)$, to decide whether a given path is to be allowed, as we did in the fixed-node algorithm in Sec. 24.4. The trial density matrix can come from theoretical insights or from using the variational principle.

The most common form for the trial density matrix comes from an independent electron function. The distinguishable particle density matrix for independent electrons in an external potential $v_{\text{ext}}(\mathbf{r})$ is a product of solutions of the single-particle Bloch equation:

$$-\frac{dg(\mathbf{r}, \mathbf{r}_*; \tau)}{d\tau} = [-\frac{1}{2}\nabla^2 + v_{\text{ext}}(\mathbf{r})]g(\mathbf{r}, \mathbf{r}_*; \tau), \tag{25.24}$$

with the boundary condition:

$$g(\mathbf{r}, \mathbf{r}_*; 0) = \delta(\mathbf{r} - \mathbf{r}_*). \tag{25.25}$$

In general, the fermion density matrix is related to the distinguishable particle one by:

$$\rho_F(R, R_*; \tau) = \frac{1}{N!} \sum_{\mathcal{P}} (-1)^{\mathcal{P}} \rho_D(\mathcal{P}R, R_*; \tau). \tag{25.26}$$

Then, for independent electrons:

$$\rho_F(R, R_*; \tau) = \frac{1}{N!} \det[g(\mathbf{r}_i, \mathbf{r}_{j*}; \tau)\delta_{\sigma_i, \sigma_j}]. \tag{25.27}$$

---

[17] An analogous situation arises for simulations of classical systems. In principle, classical dynamics is very sensitive to the assumed potential. An approximate potential will give a trajectory that deviates from the true trajectory very rapidly. However, simulations using an approximate potential are still very useful because they provide an understanding of many-body effects and frequently give accurate estimates.

In the absence of an external potential ($v_{ext}(\mathbf{r}) = 0$), the single-particle density matrices are gaussians as in Eq. (24.8):

$$g(\mathbf{r}, \mathbf{r}_*; \tau) = (2\pi\tau)^{-3/2} \exp\left[-\frac{(\mathbf{r} - \mathbf{r}_*)^2}{2\tau}\right]. \qquad (25.28)$$

At high temperatures, $k_B T \geq E_F$, we note that $g(\mathbf{r}, \mathbf{r}_0; \tau)$ vanishes for $(\mathbf{r} - \mathbf{r}_0)^2 \gg \tau$, making the nodal restriction not very important because the individual electron paths do not overlap and hence cannot exchange; see Ex. 25.4.

Militzer and Pollock [1134] went beyond the free-particle form by self-consistently determining the form of the gaussian density matrix in dense hydrogen.

Applications of RPIMC

As discussed in Ch. 3, the homogeneous electron gas is the underlying model needed as input to most density functionals; the energy as a function of density and temperature, needed for the finite-temperature exchange and correlation energy, was computed and tabulated and compared with various perturbation expansions in [1106, 1107].

The second application was to dense liquid hydrogen at relatively high temperatures. The properties of hot dense hydrogen, important in understanding the planetary interiors and inertial confinement fusion, are measured with various types of shock experiments [1098]. Both protons and electrons were represented as path integrals, the proton's path being a factor of $\sqrt{m_p/m_e} = 43$ times smaller than the electron path. Note that hydrogen molecules formed spontaneously in these simulations. The computation of the equation of state for hydrogen at low density, where nodes should not be important, is reported in [1115]. The results at high pressure [1135] were a success story for simulation, since both molecular dynamics based on DFT forces and RPIMC were in disagreement with the initial experimental findings. Later, more accurate experiments gave results in close agreement with RPIMC calculations.

25.4 Ground-state path integrals (GSPI)

We now return to the projector Monte Carlo method discussed in Ch. 24, but this time from the path-integral point of view. As is done in PIMC, a Metropolis random walk will be used instead of a branching random walk. Although the applications of this technique are relatively recent, there are a number of distinct advantages of what we will call the GSPI method [1104, 1110, 1136].[18] It can more easily reach lower temperatures than PIMC and there is less dependence on the trial wavefunction than with DMC, particularly for properties other than the total energy.

As in diffusion Monte Carlo, we introduce a trial wavefunction $\Psi(R)$ and project it for an imaginary time τ with the hamiltonian, defining the projected wavefunction: $\Psi(\tau) \equiv$

---

[18] The method also goes by the names of reptation quantum Monte Carlo (RQMC) and path-integral ground-state (PIGS) methods.

$\exp(-\tau \hat{H})\Psi$. It approaches the ground state at large τ, as in Eq. (24.5). Let us define the following normalization integral:

$$Z_\tau = \langle \Psi(\tau) | \Psi(\tau) \rangle = \langle \Psi | e^{-2\tau \hat{H}} | \Psi \rangle. \tag{25.29}$$

Other expectations can be computed by inserting operators and normalizing. We now divide the projection time τ into M time slices with $\Delta \tau = \tau/M$ and expand the "path" in a coordinate basis. The probability distribution for a path $\{R_0, \ldots, R_{2M}\}$ is:

$$\Pi(R_0, \ldots, R_{2M}; \tau) = Z_\tau^{-1} \exp\left\{ -U(R_0) - U(R_{2M}) - \sum_{t=1}^{2M} S(R_t, R_{t-1}; \Delta \tau) \right\}, \tag{25.30}$$

where $U(R_t) = -\mathrm{Re}[\ln \Psi(R_t)]$ is the contribution of the trial function at the beginning (R_0) and end (R_{2M}) of the path. $S(R, R'; \Delta \tau)$ is the "link action." In contrast to PIMC, the paths are open, i.e., not periodic in imaginary time; this means that R_0 and R_{2M} are independent N-electron coordinates.

As before we use Trotter's formula (Sec. 24.2) to justify an approximation for the action; this can be tested by making M larger while keeping the projection time fixed. The simplest choice of the action is the primitive action defined in Eq. (25.5). But most applications have used an improved action derived from the diffusion Monte Carlo form, as explained in Sec. 24.2. Neglecting unimportant constants in Eq. (24.18), the DMC action is:

$$\left\langle R | e^{-\Delta \tau \hat{H}} | R' \right\rangle = \left| \frac{\Psi(R)}{\Psi(R')} \right| \exp\left[-\Delta \tau E_L(R) - \frac{[R' - R - \frac{1}{2}\Delta \tau F(R)]^2}{2\Delta \tau} \right], \tag{25.31}$$

where the "quantum force" is $F(R) = \nabla U(R)$. If we now symmetrize with respect to R and R', we find:

$$S(R, R'; \Delta \tau) = \frac{\Delta \tau}{2} \left[E_L(R) + E_L(R') + \frac{1}{2}\left(F^2(R) + F^2(R') \right) \right] \tag{25.32}$$

$$+ \frac{(R - R')^2}{2\Delta \tau} + \frac{(R - R') \cdot (F(R) - F(R'))}{2}.$$

For antisymmetric trial functions appropriate to electrons, we again make the fixed-node or fixed-phase approximations. For real trial functions, we set the action to $+\infty$ if $\Psi(R)\Psi(R') \leq 0$. For complex trial functions, we add to the link action an extra term:

$$S_{FP}(R, R'; \Delta \tau) = \frac{1}{2} \int_0^{\Delta \tau} d\tau \, |\nabla \Theta(R_\tau)|^2 \mapsto \frac{\Delta \tau}{2} \left[|\nabla \Theta(R_0))|^2 + |\nabla \Theta(R_{\Delta \tau})|^2 \right], \tag{25.33}$$

where $\Theta(R) = \mathrm{Im} \ln(\Psi(R))$ is the phase of the trial wavefunction. The integral is taken over random walks going from R to R'. In practice, the "end-point" approximation, the right-hand expression, can be used.

Having decided on the distribution, we now use the Metropolis algorithm to sample the paths. Note that this is a change from the diffusion Monte Carlo algorithm in Ch. 24, where we had a population of P independent N-body "walkers" with a dynamics (branching random walks) given by the trial wavefunction and the hamiltonian. In contrast, in

GSPI, the state consists of $(2M + 1)$ time slices, each an N-body system. The dynamics of the random walk must satisfy detailed balance and ergodicity. Within these constraints we consider efficiency and convenience.

In the *reptation*[19] sampling procedure [1104] we designate one end of the path as the head (e.g., R_{2M}) and the other as the tail, R_0. Using the DMC sampling procedure (Eq. (23.7)), we construct a new trial head position R'. The old state of the "reptile" was $\{R_0, \ldots, R_{2M}\}$, while the new trial state is $\{R_1, \ldots, R_{2M}, R'\}$. This trial state is then accepted or rejected according to Eq. (22.8). Only the actions connected with the ends enter into the acceptance probability. For the acceptance formula, see Ex. 25.6. As with DMC, the time step $\Delta\tau$ is chosen to be small enough to get reliable actions, leading to a high acceptance ratio. There is one further refinement that can improve the efficiency by a great deal: it is correct for the reptile to continue to move in one direction until a rejection occurs, then the head and tail switch their roles; this is the bounce algorithm [1137].

The reptation algorithm is quite efficient for small numbers of electrons, however, because it relies on N-electron moves, it becomes slow for $N \gtrsim 100$. For a sufficiently large value of M, the single-electron bisection moves of PIMC will be more advantageous. In contrast to the DMC algorithm, there is no population control problem (see Ch. 24) for large values of N. However, GSPI instead have the necessity to monitor the convergence in projection time.

We now discuss the computation of the energy, which we can calculate as the time derivative of Z_τ:

$$E(\tau) = -\frac{1}{2}\frac{d \ln Z_\tau}{d\tau} = \frac{1}{Z_\tau}\langle\Psi|e^{-2\tau\hat{H}}\hat{H}|\Psi\rangle \mapsto \frac{1}{2}\langle(E_L(R_0) + E_L(R_{2M})\rangle, \qquad (25.34)$$

where the local energy, $E_L(R_\tau) = \mathrm{Re}(\Psi^{-1}(R_\tau)\hat{H}\Psi(R_\tau))$, is computed at the two ends of the path. By the variational theorem (Eq. (23.1)), the energy E_τ is an upper bound to the fixed-node energy for each value of τ, it converges to the fixed-node (fixed-phase) energy at large τ. The variance of the projected trial function is given by the time derivative of the energy:

$$\nu(\tau) = Z_\tau^{-1}\langle\Psi(\tau)(\hat{H} - E(\tau))^2\Psi(\tau)\rangle \qquad (25.35)$$

$$= -\frac{1}{2}\frac{dE(\tau)}{d\tau} \mapsto \langle(E_L(R_0) - E(\tau))(E_L(R_{2M}) - E(\tau))\rangle. \qquad (25.36)$$

Since the variance is strictly positive, the projected energy converges monotonically to the fixed-node energy. The variance tends to zero for large enough τ because the local energies at the two ends of the reptile are decorrelated with each other; it provides a useful signal for the convergence of the energy as a function of projection time.

A big advantage of the GSPI algorithm is an improved ability to estimate observables. It does not suffer from the mixed-estimator problem of diffusion Monte Carlo described in Sec. 24.5. In GSPI, these are computed in the middle of the reptile, i.e., from the distribution $|\Psi(\tau)|^2$. In fact, one can optimize the trial wavefunction by minimizing the difference

---

[19] Reptate means to move like a snake. It is thought that dense solutions of polymers actually move like this.

between the estimator evaluated at the end of the reptile, which gives the mixed estimator (defined in Eq. (24.31)), with that evaluated in the middle of the reptile.

Examples of GSPI calculations

Some of the first applications of GSPI were to bosonic systems where the fermion sign problem does not arise: e.g., calculation of helium droplets [1138]. GSPI has also been used to determine imaginary-time correlations in the Hubbard model [1139]. Several calculations have been performed on the homogeneous electron gas in order to obtain high-accuracy correlation functions without the mixed-estimator bias, e.g., the momentum distribution [390], shown in Fig. 5.2, and the pair distribution function in the 2D HEG [1140].

The most extensive calculations using GSPI of electronic systems have been in the coupled electron ion Monte Carlo (CEIMC) method [1116, 1137, 1142] to perform simulations of dense hydrogen. In the CEIMC method, the movements of classical or quantum ions are determined from a Born–Oppenheimer energy surface. As the ions are moved (in this case protons), one needs to know how the ground-state electronic energy changes. GSPI was used to estimate the electronic energy difference with correlated sampling[20]; one can estimate the energy change with smaller statistical error than that of the total energy [1142]. Correlated sampling is more easily formulated in GSPI because it has an explicit probability distribution, i.e., Eq. (25.30). On the contrary, using correlated sampling in diffusion Monte Carlo is complicated by the effects of branching and rejection [1143]. In addition, it was important to estimate the pressure without the mixed-estimator bias.

The CEIMC method allows simulations of dense fluid hydrogen in a supercell of up to a hundred electrons and protons [1141, 1144]. Although molecular dynamics calculations using DFT forces are able to do larger systems of hydrogen, the results of the hydrogen phase diagram are found to be sensitive to the assumed functional [1099, 1145]. At pressures in the range of 300–500 GPa, hydrogen undergoes a change from a molecular state to an atomic state; the details are difficult for a density functional to get accurately. Figure 25.6 shows the prediction [1141] of a fluid–fluid transition in dense hydrogen, first as calculated with DFT forces and second, with GSPI energies. There is a very significant difference in the transition density between the density functionals that were used. CEIMC simulations provide a useful benchmark.

25.5 Finite-temperature QMC for the Hubbard model

In this section we discuss a finite-temperature quantum Monte Carlo method, determinantal Monte Carlo (DetMC) that is very useful for certain lattice models, especially the Hubbard model, the periodic Anderson model, and the Kondo lattice model.

---

[20] Correlated sampling is a general Monte Carlo technique [1007]. To estimate the change induced by a small change in input parameters, one reuses the same random numbers so that the statistical fluctuations will be positively correlated, leading to a smaller noise in the energy difference. See also Sec. 23.7.

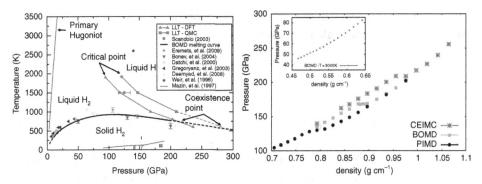

Figure 25.6. Estimates of the liquid–liquid transition in dense hydrogen. The left panel is the phase diagram. The triangles show the prediction for DFT-MD calculations, the squares, the CEIMC calculations. The right panel shows the pressure vs. temperature computed with three different assumptions: circles are path-integral MD with PBE-DFT forces, squares are classical protons with DFT-MD forces, and the "*"s are CEIMC with classical protons. The liquid–liquid transition is indicated by the plateaus. The upper plot, 3000 K, does not show a transition. For further details, see [1141].

The reader may wonder why we do not extend the path-integral method discussed earlier in this chapter to lattice models. A direct implementation of what we have already discussed is called the "world-line method" [1146, 1147]. Because of the sign problem for fermions, the method is not useful at low temperature, except for special cases such as in one dimension. So instead we will describe a method introduced by Blankenbecler, Scalapino, and Sugar (BSS) [846] that uses auxiliary fields and does not have a sign problem for the half-filled Hubbard model. The Hirsch–Fye solver for embedded clusters described in Sec. 18.4 is closely related to the BSS algorithm.

The partition function Z at an inverse temperature β can be written for a general hamiltonian in the grand-canonical ensemble as in Eq. (25.4),

$$Z = \text{Tr} e^{-\beta(\hat{H} - \mu\hat{N})} \approx \text{Tr} \prod_{i=1}^{M} e^{-\Delta\tau(\hat{H}_0 - \mu\hat{N})} e^{-\Delta\tau\hat{H}_{int}}. \tag{25.37}$$

A "Trotter breakup" of imaginary time into M "time slices," spaced by $\Delta\tau = \beta/M$, has been done. The expression will have a time-step error of order $\Delta\tau^2[\hat{H}_0, \hat{H}_{int}]$ (see Ex. 25.7) and is thus controlled.

As in the previous chapter, Sec. 24.7, we introduce auxiliary fields based on the Hubbard–Stratonovich transformation in Eq. (24.39). If the interaction is written as a square of one-particle operators, the factor $\exp(-\Delta\tau\hat{H}_{int})$ can be transformed to a sum or integral over independent-particle systems coupled to random auxiliary fields. The trace in Eq. (25.37) can then be evaluated explicitly at each time step, since it is an independent electron problem; a path of M steps in imaginary time from 0 to β can be considered a sequence of steps in the space of single-particle determinants. We then use Monte Carlo sampling to sum over the auxiliary fields.

For the Hubbard model (see Sec. 3.2), the interaction $H_{int} = U n_\uparrow n_\downarrow$ can be written as a square of operators plus single-body terms:

$$\hat{n}_\uparrow \hat{n}_\downarrow = -\frac{1}{2}(\hat{n}_\uparrow - \hat{n}_\downarrow)^2 + \frac{1}{2}(\hat{n}_\uparrow + \hat{n}_\downarrow)$$

$$= \frac{1}{2}(\hat{n}_\uparrow + \hat{n}_\downarrow)^2 - \frac{1}{2}(\hat{n}_\uparrow + \hat{n}_\downarrow) \tag{25.38}$$

(see Ex. 25.8). The first form leads to real auxiliary fields for repulsive interactions. Instead of the transformation Eq. (24.39) to an integral over a continuous range of auxiliary fields, it was shown by Hirsch [847] that for the Hubbard model it is sufficient to use a discrete field ($s = \pm 1$, the so-called "Ising field"), using the fact that any state has only two allowed occupations $n_\sigma = 0$ and $n_\sigma = 1$. The first expression in Eq. (25.38) leads to:

$$e^{-\Delta \tau U \hat{n}_\uparrow \hat{n}_\downarrow} = \frac{1}{2} \sum_{s=\pm 1} e^{\lambda s(\hat{n}_\uparrow - \hat{n}_\downarrow) - \frac{1}{2}\Delta \tau U(\hat{n}_\uparrow + \hat{n}_\downarrow)}, \tag{25.39}$$

where $\cosh(\lambda) = \exp(\Delta \tau U/2)$. See Ex. 25.9 for a derivation of this and related expressions. The partition function can then be written as a sum of terms with auxiliary fields $s_{k,t} = \pm 1$ at each site k and time step t; the Ising field couples with opposite signs for $\uparrow$ and $\downarrow$ spins.

The combination of Eqs. (25.37) and (25.39) leads to the expression

$$Z \approx \mathrm{Tr} \prod_{t=1}^{M} e^{-\Delta \tau(\hat{H}_0 - \mu \hat{N})} \prod_{k=1}^{N} \left[\frac{1}{2} \sum_{s_{kt}=\pm 1} e^{\lambda s_{kt}(\hat{n}_{k\uparrow} - \hat{n}_{k\downarrow}) - \frac{1}{2}\Delta \tau U(\hat{n}_{k\uparrow} + \hat{n}_{k\downarrow})} \right], \tag{25.40}$$

where $t = 1, \ldots, M$ denote the time slices and $k = 1, \ldots, N$, the lattice sites. By interchanging the order of the sum and products in Eq. (25.40), Z can be written as the sum over all configurations of s_{kt} of the products over time slices. After the Hubbard–Stratonovitch transformation we have independent electrons, and the trace in Eq. (25.40) can be performed analytically. For any sequence of operators of the form $\hat{A}^{(k)} = c_i^\dagger A_{ij}^{(k)} c_j$,

$$\mathrm{Tr}\left[e^{\hat{A}^{(1)}} e^{\hat{A}^{(2)}} e^{\hat{A}^{(3)}} \cdots \right] = \det\left[\hat{I} + e^{\hat{A}^{(1)}} e^{\hat{A}^{(2)}} e^{\hat{A}^{(3)}} \cdots \right], \tag{25.41}$$

with $\hat{I}$ the identity matrix. This is derived in Ex. 25.11 by transforming to eigenstates of the operators and doing the trace over occupations, which are restricted to be 0 or 1 at each site.

Using these relations, the partition function can be written as a sum over all Ising fields $S \equiv \{s_{k,t}, 1 \le k \le N, 1 \le t \le M\}$. For each configuration S of the auxiliary fields there is a product for the two spins, hence, the distribution to be sampled is:

$$\Pi(S) = Z^{-1} \det \hat{O}_\downarrow(S) \det \hat{O}_\uparrow(S), \tag{25.42}$$

$$\hat{O}_\sigma(S) = \hat{I} + \Pi_{t=1}^{M} \hat{B}_{\sigma,t}, \tag{25.43}$$

$$\hat{B}_{\sigma,t} = e^{-\Delta \tau(\hat{H}_0 - \mu \hat{N}) + \sum_k \lambda s_{k,t}(\hat{n}_{k\uparrow} - \hat{n}_{k\downarrow})}. \tag{25.44}$$

The operators $\hat{O}$ and $\hat{B}_{\sigma,t}$ are $N \times N$ matrices.

For problems with one or two sites, the sum over Ising fields can be carried out explicitly. Otherwise one must use Monte Carlo sampling over the Ising fields $s_{k,t}$ with weights given by $\Pi(\mathcal{S})$ in Eq. (25.42), using the Metropolis algorithm as described in Ch. 22. The quantity needed during the random walk is the ratio of determinants in the new and old configurations; the expressions can be simplified if only one s_{ik} is flipped, $s_{k,t} \rightarrow -s_{k,t}$. After an Ising field has changed, the inverse matrices $(\hat{\mathcal{O}}_\sigma^{-1})$ need to be updated and we again use the recursion formulas in Eq. (23.11). More details are given in Sec. 18.4 for the Hirsch–Fye impurity algorithm.

To compute observables we use the inverses of the matrices generated during the sampling. For example, the one-electron Green's function is given as:

$$\hat{G}_\sigma = \hat{\mathcal{O}}_\sigma^{-1}. \tag{25.45}$$

In terms of the Green's function, the density of electrons is given by $\langle \hat{n}_{k\sigma} \rangle = 1 - \langle [G_\sigma]_{kk} \rangle$ while the average double occupancy is $\langle \hat{n}_{k\uparrow} \hat{n}_{k\downarrow} \rangle = \langle (1 - [G_\uparrow]_{kk})(1 - [G_\downarrow]_{kk}) \rangle$.

This formulation is exact in the limit that $\Delta\tau \rightarrow 0$ and within the errors of the Monte Carlo sampling. For the Hubbard model with $U > 0$ at arbitrary filling there is a sign problem since the determinants are not always positive, however, in the case of the half-filled Hubbard model, $\det \hat{\mathcal{O}}_\downarrow = \det \hat{\mathcal{O}}_\uparrow$, so their product is non-negative, because of electron–hole and time-reversal symmetries.

For the Hubbard model, the computational effort scales as βN^3 with N the number of sites. The factor N^3 comes from computing the properties of dense $N \times N$ matrices; the factor β because the number of time slices M must increase $\propto \beta$, since $\Delta\tau$ must be kept fixed; its typical value is $\Delta\tau = (10tU)^{-\frac{1}{2}}$, where t and U are the Hubbard model parameters. This makes low-temperature simulations expensive, however, the scaling is better than the Hirsch–Fye and continuous-time algorithms in Ch. 18, which scale as β^3. A numerical complication is that the products of matrices become very ill-conditioned in the low-temperature limit, so that singular-valued decompositions have to be periodically performed, for example, every tenth time slice. For a detailed discussion of this algorithm, consult [1147].

An example of the use of this algorithm is described in Sec. 17.6 for the Hubbard model. In particular, see the predicted phase diagram of the 3D Hubbard model on a cubic lattice shown in Fig. 17.8. Determinantal Monte Carlo provides a very useful benchmark for other methods.

25.6 Estimating real-time correlation functions

In this section, we discuss the computation of real-time response functions from imaginary-time correlation functions.[21] This is an important topic, since experiments are in real time but computational methods are more easily formulated in imaginary time. Although

---

[21] This is often called analytic continuation, rotation from imaginary to real time, or Euclidean to Minkowski space-time.

practical approaches exist for performing this conversion, it is not, in general, a solved problem.

The stochastic methods, including both projector and path-integral Monte Carlo, perform dynamics in imaginary time.[22] Real-time dynamics is much more difficult using a Monte Carlo approach, since phase oscillations become more severe as time increases. The estimate of real-time response properties is a crucial step in dynamical mean-field theory, see Ch. 18.[23] All of the methods discussed in this book have this issue in one form or another, though the details vary. Within the GW method, it is convenient to perform integrals along the imaginary axis, see, e.g., Sec. 12.6. Note, however, that the source of numerical errors is quite different in a deterministic method such as GW versus stochastic methods such as CT-QMC, PMC, or PIMC.

Since imaginary-time propagation is mathematically related by analytic continuation to real-time dynamics, one would expect that excitations of the quantum system would show up in some form in imaginary-time dynamics. However, propagating modes in real time become diffusive modes in imaginary time, as illustrated in Fig. 25.7 for the density–density response function of superfluid ^{4}He. One can see little in the way of structure in imaginary time, (left panel); to the eye it is a featureless exponential decay up to $\beta/2$, and

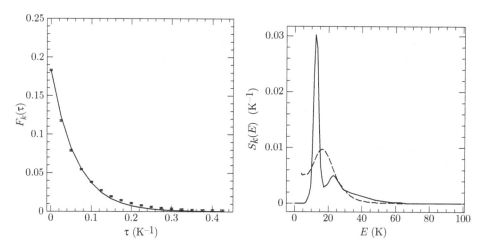

Figure 25.7. Left panel: the density–density response as a function of imaginary time in superfluid ^{4}He at a temperature of 1.2 K and atmospheric pressure as computed with PIMC for $k = 0.76\text{Å}^{-1}$ (error bars). The solid line is the response function for the single-mode approximation with energy 16K. Right panel: the corresponding dynamical structure factor at wavelength $k = 0.76\text{Å}^{-1}$: solid line, as measured by neutron scattering [1148]; dashed curve, as reconstructed from the PIMC imaginary-time response function by the maximum-entropy method [1132].

---

[22] Note that the fixed-node approximation changes the underlying hamiltonian by putting a barrier at the nodal surface, thereby changing the dynamics in imaginary time.

[23] Properties of finite-temperature Green's functions are given in App. D. An example for a non-interacting Green's function is given in Fig. D.2.

then rises again because of the periodicity in imaginary time for bosons (that part is not shown). However, the experimentally measured real-time function (right panel) shows the presence of a very sharp phonon peak, and a secondary peak at higher energy.

The figure shows the dynamic structure factor $S(\mathbf{k}, \omega)$, an important correlation function to calculate and compare with neutron scattering measurements. This is the spatial Fourier transform of the density–density response function and is defined (see also Eq. (5.66)) as:

$$S(\mathbf{k}, \omega) = \frac{1}{2\pi N} \int_{-\infty}^{\infty} dt e^{i\omega t} \langle \hat{\varrho}_\mathbf{k}^*(t) \hat{\varrho}_\mathbf{k}(0) \rangle \tag{25.46}$$

$$= \frac{1}{NZ} \sum_{mn} \delta(\omega - E_m + E_n) e^{-\beta E_n} |\langle m|\hat{\varrho}_\mathbf{k}|n\rangle|^2. \tag{25.47}$$

Here, $\hat{\varrho}_\mathbf{k} = \sum_i e^{i\mathbf{k}\mathbf{r}_i}$ is the Fourier transform of the instantaneous density. The sum in the second equation is over all pairs of many-body states $|n\rangle$ with energy E_n. The imaginary-time density–density response function, defined as

$$F_\mathbf{k}(\tau) = \frac{1}{NZ} \text{Tr} \left\{ \hat{\varrho}_\mathbf{k}^* e^{-\tau \hat{H}} \hat{\varrho}_\mathbf{k} e^{-(\beta-\tau)\hat{H}} \right\} \mapsto \langle \hat{\varrho}_\mathbf{k}^*(0) \hat{\varrho}_\mathbf{k}(\tau) \rangle, \tag{25.48}$$

can be calculated with PIMC or PMC and is shown in the right panel of Fig. 25.7. It is related to $S_k(\omega)$ by a (two-sided) Laplace transform,

$$F_\mathbf{k}(\tau) = \int_{-\infty}^{\infty} d\omega e^{-\tau\omega} S(\mathbf{k}, \omega). \tag{25.49}$$

Mathematically, $F_\mathbf{k}(\tau)$ and $S(\mathbf{k}, \omega)$ are equivalent, since an inverse Laplace transform from imaginary to real time (i.e., equivalent to analytic continuation) is well-defined. But the presence of statistical noise, or indeed any noise coming from approximations or truncations, rules out a direct inversion, since very small features in $F_\mathbf{k}(\tau)$ come from large features in $S(\mathbf{k}, \omega)$. The noise destroys the information needed to do the inversion, so that the errors are not controlled. The numerical inversion of a Laplace transform is a classic ill-conditioned problem. As shown in the figure, one can almost fit the imaginary-time data with a single excitation energy. If more modes are assumed, the PIMC data does not constrain their positions significantly.

Here we only discuss analytic continuation as done with stochastic data. A common approach for the inversion is based on bayesian statistical techniques [1149]. These methods combine QMC-generated imaginary-time response functions with any available theoretical input, and determine the most likely dynamical response function consistent with all the input. A second approach estimates errors by using sampling to determine whether alternative continuations are also consistent with the known information.

In Bayes' theorem, the probability of $S(\mathbf{k}, \omega)$ is proportional to the probability of the QMC data given $S(\mathbf{k}, \omega)$ (the likelihood, $P_L[F|S]$) times the prior knowledge of $S(\mathbf{k}, \omega)$ (the prior function, $P_P[S]$):

$$P[S(\mathbf{k}, \omega)|F_\mathbf{k}(\tau)] \propto P_L[F_\mathbf{k}(\tau)|S(\mathbf{k}, \omega)]P_P[S(\mathbf{k}, \omega)]. \tag{25.50}$$

The central-limit theorem guarantees that the noise in $F_{\mathbf{k}}(\tau)$ from a QMC calculation will be normally distributed after enough iterations:

$$P_L[F_{\mathbf{k}}(\tau)|S(\mathbf{k},\omega)] = \exp\left[-\frac{1}{2}\sum_{\tau,\tau'}\delta F_{\mathbf{k}}(\tau)\sigma_{\mathbf{k}}^{-1}(\tau,\tau')\delta F_{\mathbf{k}}(\tau')\right], \qquad (25.51)$$

where $\delta F_{\mathbf{k}}(\tau) = F_{\mathbf{k}}(\tau) - \langle F_{\mathbf{k}}(\tau)\rangle$.[24] Note that we are not assuming that $\delta F_{\mathbf{k}}(\tau)$ is statistically uncorrelated with $\delta F_{\mathbf{k}}(\tau')$ for $\tau \neq \tau'$. That assumption is not true, even in the limit of good statistics; fluctuations at one point on the path are positively correlated with fluctuations elsewhere.

The form and even the existence of the prior function is controversial. The following thought experiment gives some idea of how it could be defined. Suppose that an experimenter has done a very large number of measurements of dynamical response functions $S(\mathbf{k},\omega)$ of some system at a variety of momentum transfers, temperatures, and pressures, so they have a good feeling for what a real $S(\mathbf{k},\omega)$ should look like. Now consider a new $S^*(\mathbf{k},\omega)$ of unknown provenance. If it matches one of the measured ones within experimental error, or if the experimenter thinks it looks reasonable, they assign the prior a non-zero probability, otherwise, the prior is set to zero. Of course, in practice, we do not have such a data set, and, besides, we want our reconstruction to be independent of experiment.

Theoretical knowledge about response functions can be used to constrain the prior function. For example, we must have that $S(\mathbf{k},\omega) \geq 0$. In addition it must satisfy detailed balance as well as various sum rules mentioned in Ch. 5, and it may have known asymptotic behaviors at large and small energy. The entropic prior is a convenient function with a number of nice properties that has worked well in many applications such as image reconstruction:

$$P_P[S(\mathbf{k},\omega)] \propto \exp\left[\alpha_k\sum_{\omega}S(\mathbf{k},\omega)\ln(S(\mathbf{k},\omega)/m(\mathbf{k},\omega))\right]. \qquad (25.52)$$

It is guaranteed to keep the $S(\mathbf{k},\omega)$ positive and pull it towards a *default model* $m(\mathbf{k},\omega)$. The prior has a maximum when $S(\mathbf{k},\omega) = m(\mathbf{k},\omega)$, but decays very slowly around this maximum. This means that in regions where the PIMC data is good the prior function is unimportant, but in regions where the data is poor, the default model dominates. Here α_k is an adjustable parameter controlling how close $S(\mathbf{k},\omega)$ and $m(\mathbf{k},\omega)$ should be: if it is too small there is a danger of over-fitting the QMC data; if it is too large the QMC data is ignored. In practice, one determines α_k self-consistently so that the probability of the likelihood function (Eq. (25.51)) comes out to have a reasonable value. Together the entropic prior and the likelihood function are convex, so that the posterior probability conveniently has a unique maximum.

---

[24] Note that both the mean value of $F_{\mathbf{k}}(\tau)$ and the error matrix $\sigma_{\mathbf{k}}(\tau,\tau') = \langle\delta F_{\mathbf{k}}(\tau)\delta F_{\mathbf{k}}(\tau')\rangle$ are usually estimated from the same data. If there are M time values, the covariance matrix $\sigma_{\mathbf{k}}(\tau,\tau')$ is an $M\times M$ symmetric matrix; $\sigma_{\mathbf{k}}^{-1}(\tau,\tau')$ is its matrix inverse. Both matrices have on the order of $M^2/8$ independent entries. To get estimates of all those correlations requires at least that many independent estimates of $F_{\mathbf{k}}(\tau)$; in practice, many more are required.

However, the problem with the entropic prior is that every energy appears independently. In quantum systems, energy levels "repel"; if there is an energy level at E there is a reduced probability of energy levels (with the same quantum numbers) in the neighborhood of E. In addition, the entropic prior tends to smooth out spectra since that leads to a larger "entropy."

Having chosen the likelihood and prior function, one can take one of two approaches to using Bayes' theorem. In the *maximum-entropy* approach (MEM) one finds the $S(\mathbf{k}, \omega)$ that is most likely, the one which maximizes $P[S(\mathbf{k}, \omega)|F_{\mathbf{k}}(\tau)]$. This is a good procedure when the probability distribution is narrow. Errors are then estimated by computing the second derivatives at the maximum. The MEM solution gives the smoothest $S(\mathbf{k}, \omega)$ consistent with the QMC data. As a consequence, sharp features are broadened as illustrated in Fig. 25.7.

A second, more intuitive and more rigorous approach, the *average entropy*, is to sample $S(\mathbf{k}, \omega)$ with a probability equal to the posterior probability, $P[S_k(\mathbf{k}, \omega)|F_{\mathbf{k}}(\tau)]$ in Eq. (25.50). This can easily be done with Metropolis Monte Carlo sampling of an arbitrary $S(\mathbf{k}, \omega)$ represented in a basis. This approach is slower, but not nearly as slow as generating the QMC data in the first place and does not rely on assumptions about how narrow the distribution is. To compute errors one simply looks at the fluctuations of $S(\mathbf{k}, \omega)$ coming from this artificial Monte Carlo process. An additional advantage is that the model can be self-consistently defined as $m(\mathbf{k}, \omega) = \langle S(\mathbf{k}, \omega) \rangle$.

There are several ways of improving the result of the bayesian analysis: one can sharpen either the likelihood function or the prior function; one can get more information from the PIMC simulation by running longer or more efficiently; and one can add more theoretical constraints. Also, we can use other estimators as discussed next. But, fundamentally, the bayesian approach is an uncontrolled approximation since one cannot estimate the bias resulting from the assumption of the prior function. Features in the spectrum that are strongly constrained by the imaginary-time data and analytic information will be reliable; other features such as narrowly separated lines are weakly constrained by the data and their extraction is difficult. Other approaches, such as fitting the data to a Padé given in Eq. (12.34), face similar difficulties.

25.7 Correlation-function QMC for excitations

We now discuss a related method to compute excited-state properties: correlation function quantum Monte Carlo (CFQMC) [1041]. It combines using a basis of trial functions as discussed in Sec. 23.7 with the ground-state projection method discussed in Sec. 25.4. We start with a many-body trial basis set consisting of $(m+1)$ linearly independent many-body trial functions: $\Psi_i(R)$ with $0 \le i \le m$. The basis set is now refined by projecting with the hamiltonian to give a "time-dependent" basis: $\Psi_i(\tau) = \exp(-\tau \hat{H}) \Psi_i$. As in Sec. 23.7, the lowest $(m+1)$ exact excited-state energy levels E_k are obtained by solving the generalized eigenvalue problem: $H(\tau)d^{(k)}(\tau) = e(\tau)^{(k)} N(\tau) d^{(k)}(\tau)$, where $e^{(k)}(\tau)$ is the estimate of the kth energy level E_k and $d^{(k)}(\tau)\Psi(\tau)$ converges to the kth wavefunction at large projection time.[25]

---

[25] We have suppressed the basis set indices in the preceding equations for $d(\tau)$, $\Psi(\tau)$ (vectors) and $\hat{H}(\tau)$, $\hat{N}(\tau)$ (matrices).

QMC computes the matrices $\hat{H}(\tau)$ and $\hat{N}(\tau)$ as correlations in imaginary time of the basis sets, thus the name CFQMC. We use one of the QMC methods, either projector Monte Carlo or ground-state path integrals, to compute the overlap integrals of the time-projected basis functions:

$$N_{ij}(\tau) = \langle \Psi_i e^{-2\tau\hat{H}} \Psi_j \rangle \mapsto \langle f_i^*(R_0)f_j(R_{2\tau}) \rangle \qquad (25.53)$$

and the time derivative of this matrix:

$$H_{ij}(\tau) = \langle \Psi_i \hat{H} e^{-2\tau\hat{H}} \Psi_j \rangle \mapsto \langle E_{L,i}(R_0)f_i^*(R_0)f_j(R_{2\tau}) \rangle. \qquad (25.54)$$

Here we assume that a guiding function $\Psi_G(R)$ has been used and define $f_i(R) = \Psi_i(R)/\Psi_G(R)$. The guiding function can only have nodes where all of the desired excited states have nodes. $f(R_0)$ and $f(R_{2\tau})$ mean that the functions are evaluated from coordinates obtained an imaginary time 2τ apart, e.g., the ends of the reptile in GSPI. The matrices $N_{i,j}(\tau)$ and $H_{i,j}(\tau)$ are correlation functions in imaginary time, and contain information about excitations. It is thus a generalization of what was in the previous section and of the transient estimate algorithm for the ground state.[26]

Although the many-body Schrödinger equation is an eigenvalue problem in $3N$-dimensional continuous space, if the basis set has m functions the generalized eigenvalue problem has rank $(m + 1)$, very much smaller and practical even for many electrons. The difficulty with the CFQMC algorithm lies in reducing the noise on the matrices, $\hat{H}$ and $\hat{N}$. Although the projected basis becomes exact in the limit of long projection time, the statistical errors also increase; this increase is how the fermion sign problem shows up in this algorithm. One can show that the statistical error scales like $C_k \exp(\tau(E_k - E_0))$: the exponent is proportional to the excitation energy above the bosonic many-body ground-state energy E_0. The prefactor C_k will vanish if the basis includes the exact excited-state eigenfunction: as the basis set approaches the exact basis set, the systematic and statistical errors go to zero; this is the zero-variance principle. A sequence of CFQMC calculation provides a way to systematically learn about correlated excited states, since improved basis sets will have lower energies and variances. One advantage of this procedure is that it can treat many excitations within a single computation, in principle, one obtains $(m + 1)$ states. As opposed to the MEM method described in the previous section, upper bounds are obtained to the energies of the excited states as given in Eq. (23.29). The original CFQMC approach used the diffusion Monte Carlo algorithm to compute the matrices $\hat{N}$ and $\hat{H}$. However, doing so increases statistical noise as the projection time increases. The GSPI algorithm will eliminate the weights and allow longer projection times. By using all the statistical information in the zeroth, first, and second moments of the hamiltonian versus projection time, one can both reduce the errors and provide an estimation of the statistical and systematic errors [1151]. The CFQMC algorithm is intrinsically highly parallel since the computation of the matrix elements can be done with independent random walks.

---

[26] For a single basis function, we obtain $E_0 = \lim_{\tau\to\infty} H_{00}(\tau)/N_{00}(\tau)$, that is Eq. (24.27).

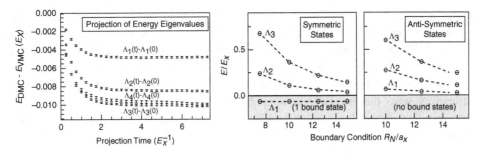

Figure 25.8. The left figure shows the decay of the energy as a function of projection time relative to the variational energy of the basis set for the first four excited states of two excitons inside a sphere. The two figures on the right show how the even and odd parity excited-state energies depend on the sphere size R_N. From these energies one calculates the exciton–exciton cross-sections, using the R-matrix formalism. For details, see [1150].

The expansion in a basis is very similar to the CI method of quantum chemistry, but there are two differences that make CFQMC potentially more powerful. The first is the ability to work with basis sets with the correlation built in. For many-body systems, the natural basis is the set of Slater determinants formed by the various ways of occupying the orbitals from a mean-field calculation multiplied by a Jastrow factor to include correlation. Backflow and other improvements to the wavefunction can also be used, see Ch. 6. The second advantage is that projection in imaginary time automatically removes errors in the basis set for distances less than the diffusion length $\sqrt{N\tau}$.

There have been few applications with this method. Kwon *et al.* [1152] applied the algorithm to compute excitations for the homogeneous electron gas. Shumway and Ceperley [1150] calculated the exciton–exciton scattering cross-sections using CFQMC, see Fig. 25.8. This is a difficult calculation to perform with other methods, since one has four highly quantum particles (two electrons and two holes) and no fixed ions. But the calculation is straightforward using CFQMC.

This section has indicated how to combine analysis of the imaginary-time correlation function with basis-set projection and the ground-state path-integral formulation of QMC. Information about excitations via a correlated basis set can dramatically improve the inversion procedure from imaginary to real time, leading to more accurate benchmark calculations of excited states.

Prospects for PIMC

The path-integral Monte Carlo method is very powerful, capable of doing accurate simulations of both boson and fermion systems. However, applications for electronic structure are much less developed than with other quantum many-body methods and have not been used for general materials. This is due to several roadblocks.

RPIMC simulations are computationally expensive, limiting the number of applications. It is difficult to reach room temperature with PIMC because it starts at the electronic temperature scale. Taking a time step $\Delta \tau = 0.1$ a.u., a time step feasible for hydrogen but not heavier elements, it would take more than 10^4 time slices to reach room temperature. In fact, the lowest temperature that dense many-body hydrogen has been simulated to date with RPIMC is 3000 K. Although projector Monte Carlo is also based on the thermal density matrix and has the same time-step issue, because it uses a trial wavefunction, it effectively starts the projection at a much lower temperature. More effort needs to go into speeding up the dynamics of moving the paths within RPIMC, particularly at low temperature.

There needs to be more investigation of better trial-density matrix nodes, since they determine the ultimate accuracy of RPIMC. The non-interacting density matrix commonly used has certain pathological properties, e.g., the nodes for one spin species do not depend on the location of the electrons of opposite spins. In addition, much needs to be done in understanding fermion path integrals, e.g., how are superfluidity, the Mott transition, and molecular bonding to be interpreted? Relationships discovered for bosonic systems will have counterparts in RPIMC.

One of the big advances in projector quantum Monte Carlo methods has been in the development of algorithms [1026] for non-local pseudopotentials. This causes problems in RPIMC, since paths are not continuous so the fixed-node boundary condition cannot be applied.

Important positive aspects of RPIMC were mentioned in the introduction. First is the ability to perform simulations at non-zero temperatures. Second, there is less of a reliance on the trial wavefunction. This is particularly important in studying phase transitions and complex materials such as liquid hydrogen as it undergoes an atomic–molecular transition. A corollary is the absence of the biased properties with PMC, Sec. 24.5. This has led to the GSPI algorithms described in Sec. 25.4, combining some of the positive points of PMC with PIMC.

SELECT FURTHER READING

Assaad, F. F. and Evertz, H. G., "World line and determinantal quantum Monte Carlo methods for spins, phonons, and electrons," pp. 277–356 in Fehske, H., Schreider, R., and Weisse, A. (eds), *Computational Many-Particle Physics* (Springer-Verlag, Berlin, 2008). Detailed explanation of determinantal quantum Monte Carlo with examples of its use.

Ceperley, D. M., *Rev. Mod. Phys.* **67**, 279, 1995. A review of numerical methods for bosonic path integrals focused on helium.

Ceperley, D. M. and Bernu, B., "The calculation of excited state properties with quantum Monte Carlo," *J. Chem. Phys.* **89**, 6316 (1988). Correlation function QMC is described in detail.

Feynman, R. P. *Statistical Mechanics* (Advanced Book Classics, 1970).

Feynman, R. P. and Hibbs, A. R. *Quantum Mechanics and Path Integrals* (McGraw-Hill, New York, 1965).

Jarrell, M. and Gubernatis, J. E., "Bayseian inference and the analytic continuation of imaginary-time quantum Monte Carlo data," *Phys. Rep.* **269**, 133–195, 1996.

Tuckerman, M. *Statistical Mechanics: Theory and Simulation* (Oxford University Press, Oxford, 2010). Textbook description of path-integral molecular dynamics.

Exercises

25.1 Prove the formulas in Eqs. (25.7) and (25.8). Give an algorithm that satisfies detailed balance as defined in Sec. 22.3.

25.2 Prove Eq. (25.12): $f(\beta) = \tanh[J(\beta - \beta_0)]$ with $\beta_0 = \ln[\phi_1(Z)/\phi_0(Z)]/J$ with ϕ_1 and ϕ_0 the eigenstates corresponding to energies E_1 and E_0. Neglect higher-energy states.

25.3 Prove Eq. (25.20) that gives the efficiency of a direct finite-temperature fermion simulation. *Hint* see [1129].

25.4 Derive the nodal regions for two non-interacting particles with the same spin, e.g., given in Eq. (25.28): what is the condition that the two-particle density matrix vanishes? If we add interaction with the primitive approximation, how does this affect the nodal surface?

25.5 Derive Eq. (25.32).

25.6 What is the acceptance ratio test for GSPI in Sec. 25.4? Assume the distribution to be sampled is given in Eq. (25.30) and assume a transition probability for moving the head $T(R \to R')$.

25.7 Show that the error in the partition function in breaking up the exponent in Eq. (25.37) is of order $\Delta\tau^2[\hat{H}_0, \hat{H}_{int}]$. *Hint:* how can you argue that error terms that have odd powers of τ vanish?

25.8 Derive the operator relations in Eq. (25.38) and find a third form that involves a sum of two squares using two auxiliary fields.

25.9 (a) Derive the identity Eq. (25.39) with $\cosh(\lambda) = \exp(\Delta\tau U/2)$. *Hint:* the essential point is to use the fact that the Hubbard model involves only a discrete set of states. (b) Show that this is equivalent to the relation $\tanh(\lambda) = \sqrt{\tanh(\Delta\tau U/4)}$, from [847].

25.10 Derive the discrete auxiliary-field decomposition in Eq. (18.26). This can be done using reasoning similar to that in (a) of Ex. 25.9.

25.11 Derive the relation of Eq. (25.41) by taking the trace over occupations, which for fermions are restricted to the values 0 or 1.

26

Concluding remarks

These two pages conclude a book of many chapters, that span an arc from fundamental theory to applications, from concepts to computation. This reflects an approach that the book is meant to promote: not the competition between research areas or methods, but the awareness that often exchange and combination leads to the most important advances.

As stated on the first pages of this book, the many-body interacting-electron problem ranks among the most fascinating and fruitful areas of research in science. It combines the intellectual challenges of quantum many-body physics with the opportunities to impact areas of science and technology, from engineering to biology, from archeology to astrophysics. There has been great progress that opens the door for the future, when all these disciplines can be greatly enhanced by quantitative calculations based on the fundamental laws of quantum mechanics.

To describe, understand, and predict the phenomena that are observed in the many-body world requires new concepts and ideas. At the same time, much of the progress in the past has been driven by the advances in computers. Very little described in this book on real materials would have been accomplished on computers from the 1970s, when many of the methods were invented. We expect this trend toward more available processing speed through parallel processing and memory to continue. Of course, advances in algorithms and software are also responsible for progress almost equally with advances in hardware. The field relies on a triangle formed by concepts, techniques, and tools: this triangle should be expanded in all directions in order to make progress.

The methods that translate concepts into feasible approaches and make best use of available hardware are at the heart of this book. We have described different ways to approach the problem. It is important to recognize that the various methods have different capabilities, so that a more complete picture of a given phenomenon or material can be obtained by using a variety of methods. The methods are complementary but not disjunct: there are many touching points that invite us to strive for possible combinations. Such an attitude has a long tradition in the field: Kohn–Sham DFT is used to construct trial wavefunctions for QMC, and QMC results of the homogeneous electron gas are used as input in DFT. The RPA of many-body perturbation theory serves to calculate the screening that is needed to obtain realistic effective interactions for DMFT calculations, and DMFT can be used to improve spectra involving correlated electrons, with respect to a GW calculation. DMFT calculations themselves would often be impossible without the availability of powerful

QMC solvers. This list could be made longer and longer – the take-home message is that many combinations can and should be thought of.

Perhaps most stimulating of all is the coming together of experiment and theory. The conceptual structure and computational methods are continually evolving because the field is relevant to the real world, where experimental discoveries keep the field vital. Even though experiment is the final arbiter for understanding nature, actual experimental observations require interpretation. This interplay of theory and experiment is a motor and a compass for the field, and it will therefore play a crucial role in determining the directions in which we will move and how far we will go.

This is a rapidly progressing field. The website www.electronicstructure.org provides more information about ongoing developments, links to other selected sites, and tutorials and numerical exercises that accompany this book.

PART V

APPENDICES

Appendix A

Second quantization

Summary

"First quantization" leads to the Schrödinger equation for the wavefunction, which replaces the momentum $\mathbf{p}$ in the classical hamiltonian by $-i\hbar\,\nabla$. The wavefunction is a single-particle function ψ or a many-particle function Ψ for a fixed number of particles that evolves in time and carries all the information about the state of the system. In "second quantization," operators are written in terms of creation and annihilation operators or field operators that change the number of particles; commutation or anticommutation rules enforce the proper particle statistics. For many-particle systems this allows us to concentrate on the one- and two-particle matrix elements of interest, avoiding the need for dealing directly with the many-particle wavefunction and the coordinates of all the remaining particles. This is of great value in dealing with the expressions that appear in perturbation theory expansions for interacting-particle systems.

A.1 First quantization

"First quantization" denotes the approach in which the classical hamiltonian $H(\{\mathbf{r}_i\}, \{\mathbf{p}_i\})$ and equations of motion for a system of particles $i = 1, \ldots, N$ becomes the operator in the time-dependent Schrödinger equation $i\hbar\frac{d}{dt}\Psi = \hat{H}\Psi$. The hamiltonian operator $\hat{H}$ is the classical H with the momentum $\mathbf{p}_i$ replaced by $-i\hbar\,\nabla_i$. All information on the quantum N-particle system as it evolves in time is contained in the time-dependent wavefunction $\Psi(\{x_i\}, t)$. Any observable corresponding to an operator $\hat{O}$ can be expressed as the expectation value $\langle\Psi(\{x_i\}, t)|\hat{O}|\Psi(\{x_i\}, t)\rangle$.

First quantization is the most useful form for quantitative solutions of the Schrödinger equation in cases where the particle number is fixed. This is the form used in [1] for independent-particle methods in quantum theory. It is also used in many approaches that treat the full interacting many-body problem. In particular, the real-space continuum quantum Monte Carlo methods described in Chs. 6, 23, and 24 are cast in terms of solving the Schrödinger equation for a fixed number of particles – a differential equation in $3N$ dimensions. The information that the particles are fermions or bosons is encoded in the symmetry of the wavefunction: for bosons the wavefunction is symmetric under interchange of particles, and for fermions the wavefunction must change sign if any two particles are interchanged; as explained in Ch. 6 this is a property of the exact many-body wavefunction and is enforced in trial functions, for example, by the determinant in the Slater–Jastrow form in Sec. 6.6.

A.2 Second quantization

"Second quantization" reformulates the equations of quantum mechanics in a way that facilitates applications to many interacting particles, particularly in perturbation theory that is central to the methods developed in Chs. 7–21. The concepts are essential in relativistic quantum mechanics, where creation and annihilation of particles are intrinsic to the theory. But the formalism is also extremely fruitful in the non-relativistic theory of interacting many-body systems, for example in the description of Green's functions and response functions that allow the addition and removal of electrons. Second quantization can be cast in terms of occupation number representation for indistinguishable particles, with creation and annihilation operators that change the number of particles. Alternatively, it can be formulated in terms of field operators in which the wavefunctions of first-quantized theory become operators that generate the quantum system from the vacuum.

Occupation number representation

States for many indistinguishable particles can be specified in terms of a basis of independent-particle states, which in the first-quantized form are independent-particle wavefunctions ψ_i, $i = 1, \ldots, M$, where M denotes the maximum number of independent-particle wavefunctions considered. Formally, we can let $M \to \infty$ for a complete set of states. A basis for many-particle states can be formed from products of the ψ_i with proper accounting of the fact that the many-body wavefunction must be symmetric for bosons and antisymmetric for fermions. Therefore, as basis states for fermions we can choose determinants built with the independent-particle wavefunctions, called "Slater determinants" (see Eq. (4.1)). The states can be written as $|n_1, n_2, n_3, \ldots, n_k, \ldots, n_M\rangle$, which denotes a state with $\sum_i n_i = N$ particles and n_i is the occupation of the single-particle state i. For example, for $N = 2$ and $M = 3$ the state $|101\rangle$ corresponds to $\Psi(x_1, x_2, t) = 1/\sqrt{2}\,[\psi_1(x_1)\psi_3(x_2) - \psi_3(x_1)\psi_1(x_2)]$ in first quantization. The full space of states for the many-body system can then be expressed as the set of many-particle states with all possible numbers of particles N. Of course, these are not the exact states for many interacting particles, but this defines a basis in which the full many-body problem can be expressed.

Bosons. The wavefunction for indistinguishable bosons must be symmetric under exchange of any two particles. This property follows if one defines time-independent creation and annihilation operators $b^{\dagger}$ and b that satisfy the commutation relations

$$[b_k, b^{\dagger}_{k'}] = \delta_{k,k'}$$
$$[b_k, b_{k'}] = [b^{\dagger}_k, b^{\dagger}_{k'}] = 0, \qquad (A.1)$$

where $[a, b] = ab - ba$. Any given state can be generated from the vacuum state $|0\rangle \equiv |0, \ldots, n_k = 0, \ldots\rangle$ by application of the creation operators

$$|n_1, \ldots, n_k, \ldots\rangle = \ldots (b^{\dagger}_k)^{n_k} \ldots (b^{\dagger}_1)^{n_1} |0\rangle. \qquad (A.2)$$

The commutation relations Eq. (A.1) lead to all the properties of bosons, in particular:

$$b_k |n_1, \ldots, n_k, \ldots\rangle = \sqrt{n_k} \, |n_1, \ldots, n_k - 1, \ldots\rangle$$
$$b^{\dagger}_k |n_1, \ldots, n_k, \ldots\rangle = \sqrt{n_k + 1} |n_1, \ldots, n_k + 1, \ldots\rangle, \qquad (A.3)$$

and the number operator for state k is $\hat{n}_k = b^{\dagger}_k b_k$, since

$$b^{\dagger}_k b_k |n_1, \ldots, n_k, \ldots\rangle = n_k |n_1, \ldots, n_k, \ldots\rangle, \quad n_k = 0, 1, 2, \ldots, \infty. \qquad (A.4)$$

Boson operators are used in this book, for example, in coupled electron–boson problems that have the generic form of polarons (coupled electrons and phonons): electron–plasmon coupling in core photoemission in Sec. 11.7, the cumulant expansion in Sec. 15.7, and dynamic effective interactions in dynamical mean-field theory in Secs. 18.7 and 21.5.

Fermions. The properties of indistinguishable fermions follow from creation and annihilation operators $c^{\dagger}_k$ and c_k that obey the relations (note that the label k includes spin)

$$\{c_k, c^{\dagger}_{k'}\} = \delta_{k,k'}$$
$$\{c_k, c_{k'}\} = \{c^{\dagger}_k, c^{\dagger}_{k'}\} = 0, \qquad (A.5)$$

where $\{a, b\} = ab + ba$ is the anticommutator. In particular, it follows from the second line of Eq. (A.5) that two fermions cannot be in the same state; the occupation of any state k can only be $n_k = 0$ or $n_k = 1$. A state can be generated from the vacuum state $|0\rangle$ by application of the creation operators in the specified order (reading from right to left)

$$|n_1, \ldots, n_k, \ldots\rangle = \ldots (c^{\dagger}_k)^{n_k} \ldots (c^{\dagger}_1)^{n_1} |0\rangle. \qquad (A.6)$$

The application of the operators is analogous to Eq. (A.3); however, the states must be defined with a specified order of the single-particle states with a change of sign whenever any two creation or two annihilation operators are interchanged. From Eqs. (A.5) and (A.6) it follows that

$$c_k |n_1, \ldots, n_k, \ldots\rangle = \begin{cases} (-1)^{S_k} |n_1, \ldots, n_k - 1, \ldots\rangle, & n_k = 1, \\ 0, & n_k = 0, \end{cases}$$

$$c^{\dagger}_k |n_1, \ldots, n_k, \ldots\rangle = \begin{cases} (-1)^{S_k} |n_1, \ldots, n_k + 1, \ldots\rangle, & n_k = 0, \\ 0, & n_k = 1, \end{cases}$$

where the sign factor $(-1)^{S_k}$ with $S_k = \sum_{k'=1}^{k'=k-1} n_{k'}$ takes into account the change of sign each time the operators are exchanged while bringing c_k or $c_k^\dagger$ from the left to the position k. The number operator has the same form as for bosons, $\hat{n}_k = c_k^\dagger c_k$, where

$$c_k^\dagger c_k |n_1, \ldots, n_k, \ldots\rangle = n_k |n_1, \ldots, n_k, \ldots\rangle, \quad n_k = 0, 1. \tag{A.7}$$

One- and two-particle operators

Up to now, these are mainly definitions. The formalism is useful if there is a well-defined and simple way to calculate observables. Expressions for operators in second-quantized form can be derived from the fact that one should obtain the same result as with the first-quantized form in a state with a given particle number. An advantage of the second-quantized representation is that it applies for both bosons and fermions and for any number of particles; the symmetry of the particles is accounted for by the (anti)commutation rules for the operators. Here we give the expressions in terms of c_k and $c_k^\dagger$, with the understanding that they are the same for bosons.

Consider the one-body operator $\hat{O} \equiv \sum_i O(x_i)$, where x_i denotes the position $\mathbf{r}_i$ and spin σ_i. In the general case of N particles, the many-body wavefunction Ψ can be expanded in a basis of Slater determinants Φ_k, and we must require the matrix elements between Slater determinants to be the same in first and second quantization, i.e., $\langle \Phi_k | \hat{O} | \Phi_j \rangle = \int dx_1 dx_2 \ldots \Phi_k^*(x_1, x_2, \ldots) \sum_i O(x_i) \Phi_j(x_1, x_2, \ldots)$. Since for a one-body operator all but one coordinates integrate out, it is enough to look at the case of only one particle to verify this relation. If the wavefunction for the particle is $\phi(x)$, the expectation value in first and second quantization is

$$\langle \phi | \hat{O} | \phi \rangle = \int dx \phi^*(x) O(x) \phi(x). \tag{A.8}$$

We can expand ϕ in the basis of single-particle states $\psi_k(x)$,

$$\phi(x) = \sum_k \alpha_k \psi_k(x) \quad \text{and} \quad |\phi\rangle = \sum_k \alpha_k c_k^\dagger |0\rangle. \tag{A.9}$$

The two expressions are the same if the operator $\hat{O}$ expressed in terms of the second-quantized operators is

$$\hat{O} = \sum_{k_1,k_2} c_{k_1}^\dagger O_{k_1 k_2} c_{k_2}, \quad O_{k_1 k_2} = \int dx \psi_{k_1}^*(x) O(x) \psi_{k_2}(x). \tag{A.10}$$

Similarly, a two-particle interaction term $\frac{1}{2} \sum_{i \neq j} V(x_i, x_j)$ can be expressed as

$$\hat{V} = \frac{1}{2} \sum_{k_1,k_2,k_3,k_4} V_{k_1 k_2 k_3 k_4} c_{k_1}^\dagger c_{k_2}^\dagger c_{k_3} c_{k_4}$$

$$V_{k_1 k_2 k_3 k_4} = \int dx dx' \psi_{k_1}^*(x) \psi_{k_2}^*(x') V(x, x') \psi_{k_3}(x') \psi_{k_4}(x). \tag{A.11}$$

Note the order of the operators, with destruction operators to the right; this is essential for all terms to have zero expectation in the vacuum state $|0\rangle$, which means there is no contribution if there are no particles.

Field operators

Instead of creation and annihilation operators defined for a particular basis of single-particle states, we can define field operators that create or destroy electrons at a given point in space with a given spin. Here we consider only fermions; for bosons the expressions are different. In analogy with Eq. (A.9), define[1]

$$\hat{\psi}(x) = \sum_k \psi_k(x) c_k$$

$$\hat{\psi}^\dagger(x) = \sum_k \psi_k^*(x) c_k^\dagger \qquad (A.12)$$

and assume the sum is over a complete single-particle basis. It follows from the definition of the creation and annihilation operators for fermions that $\hat{\psi}$ and $\hat{\psi}^\dagger$ obey the relations

$$\{\hat{\psi}(x), \hat{\psi}^\dagger(x')\} = \delta(x - x')$$
$$\{\hat{\psi}(x), \hat{\psi}(x')\} = \{\hat{\psi}^\dagger(x), \hat{\psi}^\dagger(x')\} = 0$$
$$n(x) = \hat{\psi}^\dagger(x)\hat{\psi}(x). \qquad (A.13)$$

In terms of the field operators the expressions for one- and two-particle operators can be written in forms analogous to the corresponding expressions above:

$$\hat{O} = \int dx \hat{\psi}^\dagger(x) O(x) \hat{\psi}(x)$$

$$\hat{V} = \frac{1}{2} \int dx dx' \, \hat{\psi}^\dagger(x) \hat{\psi}^\dagger(x') V(x, x') \hat{\psi}(x') \hat{\psi}(x). \qquad (A.14)$$

This suffices to express an interacting-electron hamiltonian that is a sum of one- and two-body terms.

Hamiltonian for electrons in second-quantized form

Here we collect together expressions for electrons with Coulomb interactions, with reference to points in the book where the hamiltonian, or parts of it, are given in different forms. The Coulomb potential $v_c(r) = 1/r$ is independent of spin; it can be written in a basis of plane waves $\exp(i\mathbf{q} \cdot \mathbf{r})$ as $v_c(q) = 4\pi/q^2$.

---

[1] Note that the notation $\hat{\psi}$ and $\hat{\psi}^\dagger$ is also used for field operators in the Heisenberg picture (Sec. B.2). In general, Heisenberg operators are denoted by a subscript H; however, in many places in this book this is not done for the field operators in order to simplify the notation. The field operators can be distinguished since $\hat{\psi}(x, t)$ and $\hat{\psi}^\dagger(x, t)$ with a time argument denote Heisenberg operators whereas the time-independent $\hat{\psi}(x)$ and $\hat{\psi}^\dagger(x)$ are in the Schrödinger picture.

Thus the hamiltonian in Eq. (1.1) (omitting the terms involving only the nuclei) and Eq. (1.2) can be written in terms of field operators:

$$\hat{H} = \int dx_1 \hat{\psi}^\dagger(x_1) h(x_1) \hat{\psi}(x_1) + \frac{1}{2} \int \int dx_1 dx_2 \hat{\psi}^\dagger(x_1) \hat{\psi}^\dagger(x_2) v_c(|\mathbf{r}_1 - \mathbf{r}_2|) \hat{\psi}(x_2) \hat{\psi}(x_1),$$

(A.15)

where $h(x) = -\frac{1}{2}\nabla_{\mathbf{r}}^2 + v_{\text{ext}}(x)$. See, for example, Eq. (10.1).

The hamiltonian with the spin-independent and translation-invariant Coulomb interaction can be written in Fourier space as sums over $\mathbf{k}$ and spin σ,

$$\hat{H} = \sum_{\mathbf{k}_1,\sigma_1} \frac{1}{2}|\mathbf{k}_1|^2 c^\dagger_{\mathbf{k}_1,\sigma_1} c_{\mathbf{k}_1,\sigma_1} + \sum_{\mathbf{k}_2,\sigma_2;\mathbf{k}_1,\sigma_1} v_{\text{ext}}(\mathbf{k}_2,\sigma_2;\mathbf{k}_1,\sigma_1) c^\dagger_{\mathbf{k}_2,\sigma_2} c_{\mathbf{k}_1,\sigma_1}$$

$$+ \frac{1}{2} \sum_{\mathbf{k}_2,\sigma_2;\mathbf{k}_1,\sigma_1} \sum_{\mathbf{q}} v_c(|\mathbf{q}|) c^\dagger_{\mathbf{k}_1+\mathbf{q},\sigma_1} c^\dagger_{\mathbf{k}_2-\mathbf{q},\sigma_2} c_{\mathbf{k}_2,\sigma_2} c_{\mathbf{k}_1,\sigma_1}.$$

(A.16)

Here $\mathbf{k}$ is the wavevector of a general plane wave, and $\mathbf{q}$ is the momentum transfer that corresponds to the difference between two wavevectors. See, for example, Eq. (4.10).

For crystals it is often convenient to write the hamiltonian in terms of the single-particle eigenstates of $\hat{h}$ with energy $\varepsilon_{i\mathbf{k}}$, where $\mathbf{k}$ is restricted to be in the first Brillouin zone and i is a band index that includes the spin:

$$\hat{H} = \sum_{i,\mathbf{k}} \varepsilon_{i\mathbf{k}} c^\dagger_{i\mathbf{k}} c_{i\mathbf{k}} + \hat{V}_{ee}.$$

(A.17)

The interaction $\hat{V}_{ee}$ is cumbersome to express in terms of the band states. The general form is given in Eq. (A.11), (4.8), or (14.34) in terms of a four-center integral. In Fourier space that can be written out in terms of four momenta with the only restriction that the crystal moment $\mathbf{q}$ in the Brillouin zone is conserved. Matrix elements in terms of localized orbitals are defined in Eqs. (19.3)–(19.5).

SELECT FURTHER READING

Feynman, R. P. *Statistical Mechanics: A set of lectures* (Benjamin/Cummings, Reading, MA, 1970).
    Basic principles presented with great insight and with explicit details for simple examples.
See select further reading at end of Ch. 5.

Appendix B

Pictures

Summary

There are different pictures (sometimes called representations) of quantum mechanics that contain the same information expressed in different ways. In the Schrödinger picture the time dependence is carried by the wavefunctions and operators are time independent unless there is an external potential that explicitly depends on time. In the Heisenberg picture the states are fixed and the operators are time dependent. An intermediate description is the interaction (or Dirac) picture, which leads naturally to an expansion in terms of the interaction.

The reference frame that one chooses to describe vectors is arbitrary if one is only interested in scalar products and not in absolute components of a given vector. Similarly in quantum mechanics, the fact that for a physical result one is interested in *matrix elements* $\langle \alpha | \hat{O} | \beta \rangle$ of an operator $\hat{O}$ between states $|\alpha\rangle$ and $|\beta\rangle$ allows one to perform what is called "picture transformations."[1] A picture transformation changes all states $|\alpha\rangle$ and all operators $\hat{O}$ according to

$$|\alpha_R(t)\rangle = R(t)|\alpha\rangle, \quad \hat{O}_R(t) = R(t)\hat{O}R^\dagger(t), \tag{B.1}$$

where $R(t)$ is a unitary operator for each value of the parameter t: $R^\dagger(t)R(t) = R(t)R^\dagger(t) = 1$. In the relevant cases, which are treated here, t represents the time. It is easy to show that $\langle \alpha | \hat{O} | \beta \rangle = \langle \alpha_R(t) | \hat{O}_R(t) | \beta_R(t) \rangle$ for all t. When R and $\hat{O}$ commute, $\hat{O}_R(t) = \hat{O}$.

In the following we distinguish between the "external" time dependence of the problem imposed by a time-dependent potential and the "internal" time dependence that subsists even in absence of the former (remember that there is a rotation of phases in time even when there is no explicit time dependence in the potentials acting on the system). It is only the treatment of this "internal" time dependence that distinguishes the various pictures. We include a factor of $1/\hbar$ in the definition of time so that $\hbar$ never explicitly appears in the formula.

---

[1] Often the term representation is used instead of picture, for example in [149].

B.1 Schrödinger picture

In the Schrödinger picture (denoted by subscript S), states $|\alpha_S(t)\rangle$ are governed by the time-dependent Schrödinger equation

$$i\frac{\partial}{\partial t}|\alpha_S(t)\rangle = \hat{H}_S|\alpha_S(t)\rangle. \tag{B.2}$$

If a time-dependent external potential acts on the system, $\hat{H}_S$ carries that time dependence; otherwise it is static. The remaining time dependence is carried by the states.

Equation (B.2) is a first-order differential equation. Therefore, if one knows $|\alpha_S(t_0)\rangle$ at some arbitrary initial instant t_0, $|\alpha_S(t)\rangle$ at any moment t can be deduced. Since the wavefunction is normalized its evolution can only be a phase rotation, expressed by a time-dependent unitary transformation,

$$|\alpha_S(t)\rangle = U_S(t, t_0)|\alpha_S(t_0)\rangle. \tag{B.3}$$

The differential equation for U_S is obtained by inserting Eq. (B.3) into Eq. (B.2),

$$i\frac{\partial}{\partial t}U_S(t, t_0) = \hat{H}_S U_S(t, t_0), \tag{B.4}$$

with the initial condition $U_S(t_0, t_0) = I$, where I is the identity matrix. It holds that $U_S(t, t_0) = U^\dagger(t_0, t)$ and that U_S is unitary, which is shown in Ex. B.1.

The formal solution of Eq. (B.4) is

$$U_S(t, t_0) = e^{-i\int_{t_0}^{t} dt' \hat{H}_S(t')}, \tag{B.5}$$

which reduces to

$$U_S(t, t_0) = e^{-i(t-t_0)\hat{H}_S} \tag{B.6}$$

when the Hamiltonian is time independent. Note that the exponential of an operator has to be understood as the series

$$e^{\hat{O}} = I + \sum_{n=1}^{\infty} \frac{\hat{O}^n}{n!}. \tag{B.7}$$

B.2 Heisenberg picture

We will now go from the Schrödinger to the Heisenberg picture. This should not change the physical results (matrix elements), but it greatly simplifies the analysis of correlation functions and Green's functions and it is used throughout this book. In addition, it can be convenient for interpretation, in particular it stresses analogies with classical mechanics [1153, 1154].

In the Heisenberg picture, the entire dynamics is contained in the operators, whereas the wavefunctions are static and fixed to be equal to the Schrödinger ones at some time t_a. The Heisenberg picture is obtained by choosing $R = U_S^\dagger$ in Eq. (B.1), referred to the Schrödinger picture, i.e., by the following transformations:

$$|\alpha_H(t)\rangle = U_S^\dagger(t, t_a)|\alpha_S(t)\rangle = |\alpha_S(t_a)\rangle \tag{B.8}$$

$$\hat{O}_H(t) = U_S^\dagger(t, t_a)\hat{O}_S(t)U_S(t, t_a). \tag{B.9}$$

If $\hat{H}_S$ does not depend on time, $\hat{R}$ and $\hat{H}_S$ commute. We can set $t_a = 0$ without loss of generality, and the expressions become

$$|\alpha_H(t)\rangle = e^{i\hat{H}_S t}|\alpha_S(t)\rangle = |\alpha_S\rangle \tag{B.10}$$

$$\hat{O}_H(t) = e^{i\hat{H}_S t}\hat{O}_S(t)\, e^{-i\hat{H}_S t}, \tag{B.11}$$

where $|\alpha_S\rangle \equiv |\alpha_S(t = 0)\rangle$. The equation of motion for $\hat{O}_H(t)$ is obtained by calculating $(d/dt)\hat{O}_H(t)$ from Eq. (B.9) and using Eq. (B.4) as well as its adjoint. The result is the *Heisenberg equation of motion*

$$i\frac{d}{dt}\hat{O}_H(t) = [\hat{O}_H(t), \hat{H}_H(t)] + i\left[\frac{\partial}{\partial t}\hat{O}\right]_H (t), \tag{B.12}$$

with the definition $\left[\frac{\partial}{\partial t}\hat{O}\right]_H \equiv U_S^\dagger\left[\frac{\partial}{\partial t}\hat{O}_S\right]U_S$.

The Heisenberg equations of motion strongly resemble the classical equations of motion [1153, 1154], with the Poisson brackets replaced by a commutator. The classical dynamical variables become operators (although for the equations to be valid a corresponding classical variable does not *have* to exist). More on the analogy with classical mechanics can be found in Ex. B.3.

To conclude this section, it is important to deduce from Eq. (B.9) the expressions for the field operators in the Heisenberg picture. The operators $\hat{\psi}(x)$ and $\hat{\psi}^\dagger(x)$ in the Schrödinger picture are defined in Eq. (A.12), where x denotes space $\mathbf{r}$ and spin σ. They are used frequently and for simplicity we drop the subscript H with the understanding that $\hat{\psi}(x, t)$ and $\hat{\psi}^\dagger(x, t)$ with a time argument denote Heisenberg operators:

$$\hat{\psi}(x, t) = U_S^\dagger(t)\hat{\psi}(x)U_S(t)$$
$$\hat{\psi}^\dagger(x, t) = U_S^\dagger(t)\hat{\psi}^\dagger(x)U_S(t). \tag{B.13}$$

Here we have defined $U_S(t - t') \equiv U_S(t, t')$ in a static external potential. The equation of motion is

$$i\frac{\partial}{\partial t}\hat{\psi}(x, t) = U_S^\dagger(t)\left[\hat{\psi}(x, t), \hat{H}\right]U_S(t), \tag{B.14}$$

and similarly for $\hat{\psi}^\dagger(x, t)$. The form of the one-particle propagator given in Eq. (5.72) can be derived from the above expressions in terms of the evolution operators. For a static external potential it reads

$$G^>(x, t; x', t') = \langle \alpha_S(t)|\hat{\psi}(x)U_S(t - t')\hat{\psi}^\dagger(x')|\alpha_S(t')\rangle$$
$$= \langle \alpha_S|U_S(-t)\hat{\psi}(x)U_S(t)U_S(-t')\hat{\psi}^\dagger(x')U_S(t')|\alpha_S\rangle$$
$$= \langle \alpha_S|\hat{\psi}(x, t)\hat{\psi}^\dagger(x', t')|\alpha_S\rangle, \tag{B.15}$$

as is often represented in the literature. A similar expression holds for $G^<$.

Modified Heisenberg picture

In order to treat the statistical mechanics of many-body systems, it is useful to work in the grand-canonical ensemble and define the operator $\hat{K} = \hat{H} - \mu\hat{N}$, which can be considered to be a modified hamiltonian, with μ the chemical potential. As discussed in App. D, this can be used to define a "modified Heisenberg" picture (see, e.g., [47]). The field operators and other operators are defined as for the usual Heisenberg form but with $\hat{H}$ replaced by $\hat{K}$. This arises naturally in the imaginary-time or frequency expressions, where the operators are in the modified Heisenberg picture (or the corresponding modified interaction picture) unless otherwise specified. We will use the same notation and the distinction between the usual and the modified Heisenberg pictures should be clear from the context. For time-independent hamiltonians, the field operators can be expressed in real time as

$$\hat{\psi}(x,t) = e^{i\hat{K}t}\hat{\psi}(x)e^{-i\hat{K}t}, \quad \hat{\psi}^\dagger(x,t) = e^{i\hat{K}t}\hat{\psi}^\dagger(x)e^{-i\hat{K}t}, \tag{B.16}$$

or in imaginary time, $t \to -i\tau$ where τ is real,

$$\hat{\psi}(x,\tau) = e^{\hat{K}\tau}\hat{\psi}(x)e^{-\hat{K}\tau}, \quad \hat{\psi}^\dagger(x,\tau) = e^{\hat{K}\tau}\hat{\psi}^\dagger(x)e^{-\hat{K}\tau}, \tag{B.17}$$

and for other operators

$$\hat{O}_H(t) \equiv e^{i\hat{K}t}\hat{O}e^{-i\hat{K}t}, \quad \hat{O}_H(\tau) \equiv e^{\hat{K}\tau}\hat{O}e^{-K\tau}. \tag{B.18}$$

Note that the operators are *not* related by simply replacing the real variable t by the real variable τ; they are defined by the above relations and the distinction is determined by the argument t or τ.

B.3 Interaction picture

The Dirac, or interaction, picture is suitable for cases where one cannot solve the problem of some hamiltonian $\hat{H}_S = \hat{H}_0 + \hat{V}$ exactly but wishes to obtain a solution in terms of an expansion in $\hat{V}$. Here $\hat{H}_0$ is supposed to be time independent, and its eigenvalues and eigenstates are supposed to be known. The added term $\hat{V}$ may include a potential that can carry some external time dependence. The name "interaction picture" stems from the fact that this separation is particularly useful when $\hat{H}_0$ is a non-interacting hamiltonian, and $\hat{V}$ contains interactions.

In the interaction picture, both operators *and* wavefunctions are time dependent; only the time dependence of the wavefunction due to $\hat{H}_0$ is removed from the wavefunction and transferred to the operators, as $\hat{V}$ is expressed in the Schrödinger picture (see Eq. (B.3)), which leads to the expressions

$$\begin{aligned} |\alpha_I(t)\rangle &= e^{i\hat{H}_0 t}|\alpha_S(t)\rangle = e^{i\hat{H}_0 t}U_S(t,t_0)|\alpha_S(t_0)\rangle \\ &= e^{i\hat{H}_0 t}U_S(t,t_0)e^{-i\hat{H}_0 t_0}|\alpha_I(t_0)\rangle \\ &\equiv U_I(t,t_0)|\alpha_I(t_0)\rangle \end{aligned} \tag{B.19}$$

and

$$\hat{O}_I(t) = e^{i\hat{H}_0 t}\hat{O}_S(t)e^{-i\hat{H}_0 t}. \tag{B.20}$$

The equations of motion can be derived from Eq. (B.19) (see Ex. B.4):

$$i\frac{\partial}{\partial t}|\alpha_I(t)\rangle = \hat{V}_I(t)|\alpha_I(t)\rangle, \quad \text{with} \quad \hat{V}_I(t) = e^{i\hat{H}_0 t}\hat{V}_S(t)e^{-i\hat{H}_0 t} \tag{B.21}$$

(the Tomonaga–Schwinger equation), and

$$i\frac{d}{dt}\hat{O}_I(t) = [\hat{O}_I(t), \hat{H}_0] + i\left[\frac{\partial}{\partial t}\hat{O}\right]_I(t), \tag{B.22}$$

with $\left[\frac{\partial}{\partial t}\hat{O}\right]_I \equiv e^{i\hat{H}_0 t}\left[\frac{\partial}{\partial t}\hat{O}_S\right]e^{-i\hat{H}_0 t}$. Hence, in the interaction picture wavefunctions obey a differential equation that is similar to the Schrödinger Eq. (B.2), but with the $\hat{H}_0$ part separated out from the beginning.

A complete description of the problem can be obtained from knowledge of the time-evolution operator $U_I(t, t_0) = e^{i\hat{H}_0 t}U_S(t, t_0)e^{-i\hat{H}_0 t_0}$. Together with Eq. (B.4), this relation leads to

$$i\frac{\partial}{\partial t}U_I(t, t_0) = \hat{V}_I(t)U_I(t, t_0). \tag{B.23}$$

The formal solution can be written

$$U_I(t, t_0) = 1 - i\int_{t_0}^{t} dt' \hat{V}_I(t')U_I(t', t_0), \tag{B.24}$$

which obeys the boundary condition $U_I(t_0, t_0) = 1$. Since U_I appears on both sides, one can create an iterative solution, starting with $U_I = 0$ on the right side. This yields

$$U_I(t, t_0) = 1$$
$$- i\int_{t_0}^{t} dt_1 \hat{V}_I(t_1)$$
$$+ (-i)^2 \int_{t_0}^{t} dt_1 \hat{V}_I(t_1) \int_{t_0}^{t_1} dt_2 \hat{V}_I(t_2)$$
$$+ \cdots \tag{B.25}$$

Since $t' < t$ in Eq. (B.24), in the expansion one must have $t_n < t_{n-1} < \cdots < t$.

This expression is still complicated, in particular because the limit of each integral over t_n for $n > 1$ depends on the integration variable t_{n-1}, so that one doesn't simply have a sum of products. Let us study the problem for the example of the term that is quadratic in $\hat{V}_I$ (third line of Eq. (B.25)). The integral is depicted on the left side of Fig. B.1, which shows the two-dimensional region where t_2 runs from t_0 to t_1 for all t_1 between t_0 and t. One obtains the same result by integrating over all $t_0 < t_2 < t$, for t_1 from t_2 to t, as shown on the right. Thus the integral can be expressed as the average

$$\int_{t_0}^{t} dt_1 \int_{t_0}^{t_1} dt_2 \hat{V}_I(t_1)\hat{V}_I(t_2) = \frac{1}{2}\left[\int_{t_0}^{t} dt_1 \int_{t_0}^{t_1} dt_2 \hat{V}_I(t_1)\hat{V}_I(t_2) + \int_{t_0}^{t} dt_2 \int_{t_2}^{t} dt_1 \hat{V}_I(t_1)\hat{V}_I(t_2)\right], \tag{B.26}$$

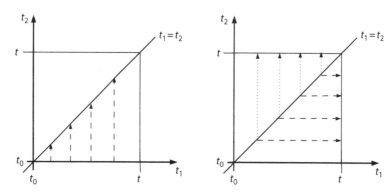

Figure B.1. The second-order term of U_I is an integration over a triangle. The way in which the area is covered by the integration can be changed (dashed arrows in the left or right panel). When the integration variables are changed, $t_1 \leftrightarrow t_2$, the integration runs over the upper triangle (dotted arrows in the left panel). The integration can hence be expressed as half the integral over the whole square, Eq. (B.27).

and a change of integration variables $t_1 \leftrightarrow t_2$ in the second term on the right leads to

$$\int_{t_0}^t dt_1 \int_{t_0}^{t_1} dt_2 \hat{V}_I(t_1)\hat{V}_I(t_2) = \frac{1}{2}\int_{t_0}^t dt_1 \int_{t_0}^{t_1} dt_2 \left[\hat{V}_I(t_1)\hat{V}_I(t_2) + \hat{V}_I(t_2)\hat{V}_I(t_1)\right]. \quad (B.27)$$

If the $\hat{V}_I(t)$ at different times commuted, the result would simply be half the integration over the whole square $\frac{1}{2}\int_{t_0}^t dt_1 \int_{t_0}^t dt_2$. However, in general the $\hat{V}_I(t)$ do *not* commute for different times (recall the definition of $\hat{V}_I(t)$ in Eq. (B.21) that involves $\hat{H}_0$). A simple relation can be found if the order of operators is changed. Formally, this is achieved by introducing the time-ordering operator T:

$$\text{T}\left[A(t_1)B(t_2)\right] \equiv \begin{cases} A(t_1)B(t_2), & t_1 > t_2, \\ B(t_2)A(t_1), & t_1 < t_2. \end{cases} \quad (B.28)$$

Thus, Eq. (B.27) becomes

$$\int_{t_0}^t dt_1 \int_{t_0}^{t_1} dt_2 \hat{V}_I(t_1)\hat{V}_I(t_2) = \frac{1}{2}\int_{t_0}^t dt_1 \int_{t_0}^t dt_2 \text{T}\left[\hat{V}_I(t_1)\hat{V}_I(t_2)\right]. \quad (B.29)$$

The same arguments hold for each order, so that the expansion of U_I can be expressed as

$$U_I(t, t_0) = 1 + \sum_{n=1}^{\infty}\frac{1}{n!}(-i)^n \int_{t_0}^t dt_1 \dots \int_{t_0}^t dt_n \text{T}\left[\hat{V}_I(t_1)\dots\hat{V}_I(t_n)\right], \quad (B.30)$$

where the factor $1/n!$ stems from the fact that an integral over n variables in the range t_0 to t leads to $n!$ times the single term with the limits $t_n < t_{n-1} < \cdots < t$ in the original expression Eq. (B.25).

Formally, this result can be written as

$$U_I(t, t_0) = \text{T}\left[e^{-i\int_{t_0}^t dt'\hat{V}_I(t')}\right]. \quad (B.31)$$

This expansion is the starting point for many-body perturbation theory.

The separation of the hamiltonian into two parts is not limited to a non-interacting hamiltonian plus interactions. For example, in DMFT the hamiltonian is separated into local regions with interactions, which are then coupled by hopping terms. This is manifest in the formulation in Sec. 18.6, where the partition function for the auxiliary system in Eq. (18.29) is expressed in a form analogous to Eq. (B.31) but with the integral over the interaction $\hat{V}_I(t)$ replaced by the double integral over the dynamic hybridization $\psi(\tau)_\sigma \Delta_\sigma(\tau - \tau')\psi_\sigma^\dagger(\tau')$, which is computed in imaginary time τ.

SELECT FURTHER READING
See select further reading at end of Ch. 5.

Exercises

B.1 Show that U_S is a unitary transformation. *Hint*: derive a differential equation for $U_S^\dagger U_S$ from Eq. (B.4) and its adjoint, and use the initial condition $U_S(t_0, t_0) = 1$. Also show that for a static external potential, $U_S(t, t_1)U_S(t_1, t') = U_S(t, t')$.

B.2 Calculate the single-particle propagator $G^>$ of Eq. (B.15) for the case of a free particle. Compare the time dependence of the result in one and three dimensions. Use the classical velocity $\mathbf{v} = \frac{\mathbf{r} - \mathbf{r}'}{t - t_0}$, in order to express the result in three dimensions in terms of the action $S(\mathbf{r}, t; \mathbf{r}', t_0) = \int_{t_0}^t dt' \frac{1}{2}mv^2$ of the free particle.

B.3 Show that for the operators q_j and p_k of two canonically conjugate variables, for which $[q_j, p_k] = i\delta_{jk}$, the formal relations $[q_j, p_k^n] = i\delta_{jk}\frac{\partial}{\partial p_k}p_k^n$ and $[p_k, q_j^n] = -i\delta_{jk}\frac{\partial}{\partial q_j}q_j^n$ hold. Use these findings to show that for a hamiltonian $\hat{H} = \frac{p^2}{2m} + V(\mathbf{r})$, the quantum-mechanical relations $\frac{d\mathbf{p}}{dt} = -\nabla V(\mathbf{r})$ and $\frac{d\mathbf{x}}{dt} = \frac{\mathbf{p}}{m}$ hold, where all operators are understood to be in the Heisenberg picture.

B.4 Derive the equations of motion (Eq. (B.21)) for operators and wavefunctions in the interaction picture. *Hint*: for the operators, proceed as in the case of the Heisenberg picture. For the wavefunctions, calculate $i(\partial/\partial t)|\alpha_I(t)\rangle = i(\partial/\partial t)e^{i\hat{H}_0 t}|\alpha_S(t)\rangle$ and use the Schrödinger Eq. (B.2).

Appendix C

Green's functions: general properties

Summary

This appendix summarizes various aspects of Green's functions that are useful for the main chapters. It links *the* Green's functions, which are the physical objects of interest in this book, to the general mathematical definition of a Green's function. It contains some useful technical details for the transformations between time and frequency. The thermodynamic limit is given particular consideration. The appendix completes the information contained in the main chapters, in particular in Ch. 5.

C.1 Green's functions for differential equations

Many problems in physics can be formulated in terms of an inhomogeneous differential equation, such as

$$\hat{D}_t f(t) = F(t), \tag{C.1}$$

where $f(t)$ is a function of the variable t, and $\hat{D}_t$ is a linear differential operator. One example of such an equation is a driven harmonic oscillator, where $\hat{D}$ describes the intrinsic properties of the oscillator, and F gives the external force. Another example that can be found in this book is the Schrödinger Eq. (B.2) and the corresponding equation of motion for a Heisenberg operator, Eq. (B.12). If one is able to find the solution of

$$\hat{D}_t G(t, t') = \delta(t - t'), \tag{C.2}$$

then the solution of Eq. (C.1) can be cast as an integral equation,

$$f(t) = \int dt' G(t, t') F(t'), \tag{C.3}$$

for any $F(t')$. For $t \neq t'$, the function $G(t, t')$ is a solution to the homogeneous problem

$$\hat{D}_t G(t, t') = 0, \tag{C.4}$$

and it has discontinuous behavior at $t = t'$.

Such a function $G(t, t')$ is a *Green's function*, named after George Green who wrote a first essay in 1828, which had large impact when published between 1850 and 1854 [1155].[1]

Boundary conditions

A differential equation like Eq. (C.1) has more than one solution. The general family of solutions consists of the sum of a particular solution to Eq. (C.1) and any combination of the solutions to the homogeneous equation, Eq. (C.4). Only when a boundary condition is specified, is the solution well-defined. The boundary condition can be included in Eq. (C.2), in other words, in the Green's function. The determination of the solution $f(t)$ corresponding to each $F(t)$ from Eq. (C.3) is then unique.

As a special case, suppose that

$$\hat{D}_t = i\partial/\partial t + \hat{H} \tag{C.5}$$

consists of the first derivative in time plus some part that does not act on time. This is an important case in quantum mechanics, since the Schrödinger equation is a differential equation of first order in time. If one has two solutions $G_{1,2}^{\text{hom}}$ to the homogeneous Eq. (C.4) for which $G_1^{\text{hom}}(t, t) - G_2^{\text{hom}}(t, t) = 1$, one can construct a specific solution to Eq. (C.2) that reads

$$G^R(t, t') = -i\Theta(t - t')\left[G_1^{\text{hom}}(t, t') - G_2^{\text{hom}}(t, t')\right], \tag{C.6}$$

where $\Theta(t)$ is the Heavyside step function $\Theta(t) = 0$ for $t < 0$ and $\Theta(t) = 1$ for $t > 0$. Equation (C.6) corresponds to the retarded Green's function, Eq. (5.76), where t is time and the two homogeneous solutions are the propagators $G^>$ and $G^<$. In the case of retarded Green's functions the boundary condition reflects causality. Similarly, one can construct the advanced and time-ordered solutions of Eqs. (5.77) and (5.78).

C.2 Fourier transforms and spectral representations

The spectral representation is given by the Fourier transform,

$$f(\omega) = \int_{-\infty}^{\infty} dt\, f(t)e^{i\omega t} \quad \text{and} \quad f(t) = \frac{1}{2\pi}\int_{-\infty}^{\infty} d\omega\, f(\omega)e^{-i\omega t}, \tag{C.7}$$

for any function f. The transforms are well-defined, however, only if the integrals are convergent, which poses conditions on the limits of the integrand for $t \to -\infty$ and $t \to \infty$. One can define convergent integrals by separating the time domain into two parts:

$$f(\omega) = \int_{-\infty}^{0} dt\, f(t)e^{i\omega t + \eta t} + \int_{0}^{\infty} dt\, f(t)e^{i\omega t - \eta t}, \tag{C.8}$$

---

[1] For some history of Green's functions, see for example the book *Green's Functions with Applications*, by Dean G. Duffy.

with convergence factors $\exp(\pm\eta t)$, $\eta > 0$. The integrals can be evaluated in the limit $\eta \to 0^+$ with the superscript $^+$ denoting the limit with $\eta > 0$. To be precise, Eq. (C.8) means that one cuts $f(t)$ into two pieces $f(t) = f_+(t) + f_-(t)$, where $f_+(t) = f(t)$ for $t > 0$ and $f_-(t) = f(t)$ for $t < 0$. Then one evaluates $f_+(\omega + i\eta)$ and $f_-(\omega - i\eta)$ at the complex frequencies $z = \omega \pm i\eta$, and one *defines* $f(\omega) \equiv f_+(\omega + i\eta) + f_-(\omega - i\eta)$. This allows one to work with the well-behaved *complex Fourier transforms* of Eq. (C.8) instead of the integrals of Eq. (C.7). One has to keep in mind that this is a new definition of $f(\omega)$. The important point is that one uses the resulting expressions in a consistent way, and that it is clear how to make the link between the definitions and measurable quantities.

Because of the separation into negative and positive times needed to guarantee convergence, the Heavyside Theta function $\Theta(t)$ is an important ingredient in the calculation. Moreover, as explained in the previous section, $\Theta(t)$ appears naturally when one works with the retarded, advanced, or time-ordered functions, which are used in this book. The most useful results can be summarized as (see Ex. C.4 for added explanations)

$$\int_{-\infty}^{\infty} dt \Theta(t) e^{i\omega t - \eta t} = \frac{i}{\omega + i\eta} \quad \text{and} \quad \Theta(t) = \frac{i}{2\pi} \lim_{\eta \to 0^+} \int_{-\infty}^{\infty} d\omega \frac{e^{-i\omega t}}{\omega + i\eta}, \quad (C.9)$$

$$\int_{-\infty}^{\infty} dt \Theta(-t) e^{i\omega t + \eta t} = \frac{-i}{\omega - i\eta} \quad \text{and} \quad \Theta(-t) = \frac{-i}{2\pi} \lim_{\eta \to 0^+} \int_{-\infty}^{\infty} d\omega \frac{e^{-i\omega t}}{\omega - i\eta}, \quad (C.10)$$

$$\int_{-\infty}^{\infty} dt\, e^{-i\omega_0 t} e^{i\omega t} = 2\pi \delta(\omega - \omega_0) \quad \text{and} \quad e^{-i\omega_0 t} = \int_{-\infty}^{\infty} d\omega e^{-i\omega t} \delta(\omega - \omega_0). \quad (C.11)$$

These exact relations support the idea of working with the frequencies $\omega \pm i\eta$ in Eq. (C.8). They imply that the retarded functions should be evaluated at $\omega + i\eta$, the advanced functions at $\omega - i\eta$, with a corresponding combination for the time-ordered functions.

The expressions in frequency space can be separated into a principal part denoted by $\mathcal{P}$ and the singular contribution,

$$\lim_{\eta \to 0^+} \frac{1}{\omega \pm i\eta} = \mathcal{P} \frac{1}{\omega} \mp i\pi \delta(\omega). \quad (C.12)$$

This is especially useful in integrals that contain the Theta function. Moreover, Eq. (C.12) leads to

$$\lim_{\eta \to 0^+} \left[\frac{1}{\omega - i\eta} - \frac{1}{\omega + i\eta} = 2\pi i \,\delta(\omega) \right]. \quad (C.13)$$

Consequences for the Green's functions and spectral functions

In the following, we suppose that $\hat{D}_t$ has the form of Eq. (C.5). The formal solution to the homogeneous equation reads

$$G(t) = e^{iHt}. \quad (C.14)$$

The spectrum of G is therefore given by H.[2] The frequency Fourier transforms of the retarded and advanced Green's functions read

$$G^R = \frac{1}{\omega - H + i\eta} \quad \text{and} \quad G^A = \frac{1}{\omega - H - i\eta}, \tag{C.15}$$

and the corresponding expression for the time-ordered Green's function. The spectrum of H appears as poles of the Green's functions. All retarded functions have poles only in the lower half-pane and are analytic in the upper half-plane where $\operatorname{Im} z > 0$, and conversely for advanced functions. The pole structure is illustrated in Fig. C.1 for the retarded functions in the complex frequency plane.

The *spectral function* is defined as $A(\omega) = \delta(\omega - H)$. According to Eq. (C.13), it can be obtained from the discontinuity of the functions in the limit $\eta \to 0$ that stems from the discontinuity of $\Theta(t)$, Eqs. (C.9) and (C.10). (See the second part of Ex. C.4 for an example where both parts are needed.) Therefore, the spectral function can be obtained from the difference between retarded and advanced Green's functions, Eq. (5.92).

C.3 Frequency integrals

The fundamental equations and derivations, like the Schrödinger equation or many-body perturbation expressions in App. B, are expressed in time but they are often evaluated in frequency space. Frequency space is the natural space when one thinks of spectroscopy, and interpretation of the formula in terms of electronic excitations. It is therefore useful to summarize two main tools: Cauchy's residue approach, and changes of the integration path in the complex plane, called *contour deformation*.

The analytic properties of the Green's function are closely linked to the Theta functions. In the following we examine the case of a retarded function; the case of advanced and time-ordered functions is analogous. For the discussion we suppose that the energy scale is such that there is no pole at $\omega = 0$. Figure C.1 shows the pole structure of the retarded function, analogous to Fig. 5.5.

In order to obtain $G^R(t)$ in the time domain, one has to evaluate Eq. (C.7), an integral along the real axis, as indicated by the horizontal dashed line.

Cauchy's theorem and contour integrals

Integrals over functions that contain poles can be conveniently evaluated using Cauchy's residue theorem. This states that if C is a positively oriented simple closed contour, and f has inside the contour a series of poles a_k labeled k, then

$$\int_C dz f(z) = 2\pi i \sum_k \operatorname{Res}_{z=a_k} f(z), \tag{C.16}$$

---

[2] Note that besides time there can be other variables, such that H and G are matrices in some space. This is not specified here, since one can suppose to work in a basis where H and G are diagonal.

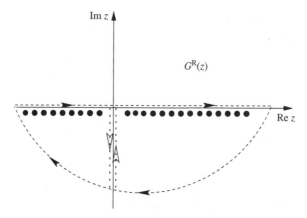

Figure C.1. Schematic illustration of poles for the retarded Green's function, and the integration paths in the complex plane that can be used to obtain the retarded Green's function in the time domain. The poles are isolated singularities for the discrete states of a finite system, or in extended systems, in the case of independent-particle states that are discrete for each momentum **k**. In general, the states of an extended system merge into a continuum, which becomes a branch cut in the complex plane. The retarded function is discontinuous across the branch cut. The retarded Green's function in time is obtained by integrating along the real axis, as indicated by the dashed path. One can use Cauchy's residue theorem by closing the contour. Since for retarded functions $t > 0$, closing the contour at infinity in the lower half-plane does not change the value of the integral. One can use other contours, for example, one can calculate the result as a sum of the two contour integrals that are obtained by following the paths closed by the pieces along the negative imaginary axis indicated by the empty arrows. In this case, the value of the integrals along the vertical part of the contour contributes to each piece, but the final result is the same.

where $\mathrm{Res}_{z=a_k}$ indicates the residue of f with respect to the pole k. For a simple pole,

$$\mathrm{Res}_{z=a_k} f(z) = \lim_{z \to a_k} (z - a_k) f(z), \tag{C.17}$$

and for a pole of order n,

$$\mathrm{Res}_{z=a_k} f(z) = \frac{1}{(n-1)!} \lim_{z \to a_k} \frac{d^{(n-1)}}{dz^{(n-1)}} \left[(z - a_k)^n f(z) \right]. \tag{C.18}$$

In order to use this theorem, we have to close the integration contour. This can in principle be done in an arbitrary way, but one would have to subtract the value of the integral along the extra piece of contour in order to retrieve the desired integral along the real axis. Therefore, one tries to add a piece where the integral vanishes. In the case of the retarded function, this is achieved by closing the contour in the lower half-plane, since $t > 0$. The closed contour is indicated by the dashed lines with full arrows in Fig. C.1.

Similarly, to get advanced functions in time requires closing the contour in the upper half-plane, since $t < 0$.

Contour deformation

Following Cauchy's theorem, the value of the integral is the same for any path that encloses the same poles. This is used in Fig. 12.3 to optimize the integration path for the GWA self-energy. Figure C.1 shows a simple example for an alternative contour, where the final result is obtained as the sum of two pieces.

Some useful expressions

We can now directly write down some useful expressions for quantities of interest or important integrals. In particular, the density and density matrix are equal time limits $t' \to t^+$ of the time-ordered Green's function. For example, the density Eq. (5.118) can also be written in terms of the time-ordered Green's function as

$$n(x) = -iG(x, t, ; x, t^+) = \frac{-i}{2\pi} \int_{-\infty}^{+\infty} d\omega e^{i\omega\eta} G(x, x, \omega). \tag{C.19}$$

Following the above discussion, the contour has to be closed in the upper half-plane. This contour encloses the poles of the time-ordered Green's function in the upper-left quadrant of the complex plane, but not the others, which are situated in the lower-right quadrant (see Fig. 5.5). Using the technique of contour deformation, the same result is obtained by evaluating

$$n(x) = \frac{-i}{2\pi} \int_{-i\infty}^{+i\infty} dz \, e^{\eta z} G(x, x, z) \tag{C.20}$$

along the imaginary axis (see also Eq. (D.14)).

A term containing several correlation functions can contain integrations or multiplications in time. Integrations in time are multiplications in frequency space and do not need special consideration. Multiplications in time occur, for example, in the case of the Hartree–Fock self-energy, which reads $\Sigma^{HF}(t, t') = iG^{HF}(t, t')v_c(t^+, t')$. The infinitesimal in the time argument of v_c guarantees that the contour encloses only occupied states of G^{HF}. Figure 12.3 shows the possibilities that one has to conveniently evaluate the GWA self-energy.

C.4 From many-body to few-body Green's functions

In this short section, we want to clarify a point that can give rise to confusion: up to now, we have not specified $\hat{H}$ in Eq. (C.5). If it is the full many-body hamiltonian, then the resulting Green's function is a huge object, and its calculation is equivalent to the full solution of the many-body problem. Note that the solution of the homogeneous equation corresponds to calculating the time-evolution operator U_S in Eq. (B.4): the full many-body Green's function propagates the wavefunction from a starting time t_0 to the actual time t.

This corresponds to calculating G^{tot} in Ch. 7, and it is not what one wants in general. As discussed in Sec. 9.1, we are only interested in certain matrix elements of G^{tot}, for example, matrix elements between the many-body states $\langle N|c_i$ and $c_j^\dagger|N\rangle$ in the case of a one-body Green's function, or states with more creation and annihilation operators for higher-order Green's functions. Therefore, *the* Green's function in this book rather corresponds to G_S in Ch. 7. Equivalently, the differential operator that determines the one-body Green's function is not built with the full many-body hamiltonian, but with the self-energy, as can be seen from the equation of motion

$$i(\frac{\partial}{\partial t} - h)\,G(t, t') - \int dt''\,\Sigma(t, t'')G(t'', t') = \delta(t - t'),\tag{C.21}$$

where the role of H in D_t is played by the sum of the independent-particle hamiltonian h and the self-energy Σ.

C.5 The thermodynamic limit

We conclude this appendix with a few remarks concerning the thermodynamic limit, which has to be considered with care.

From discrete to continuous spectra

The spectral function Eq. (5.101) of the many-body system displays the possible energy differences between the $(N + 1)$ or $(N - 1)$-particle system and the N-particle one. These excitation energies are much more numerous than the number of excitation energies of one particle alone. For example, in a homogeneous system for a given momentum transfer and a single particle there is only one possible energy difference, but in a many-body system there are many ways to conserve momentum, and the spectrum is in general very rich.

In the thermodynamic limit, the particle number $N \to \infty$ and volume $\Omega \to \infty$ while the concentration N/Ω is fixed. In this case, the many discrete excitations of the many-body system merge into a continuum. Since the excitations appear as poles on the real axis in the Green's functions, going to the thermodynamic limit creates a *branch cut*, a continuum of poles. The Green's function is not defined on this branch cut, but only the limits from above and below the real axis that are introduced in Sec. C.2 and lead to the analytic properties of retarded, advanced, and time-ordered Green's functions.[3] It is important to keep in mind that the limits from above and below are different: the Green's function is discontinuous across the branch cut.

---

[3] To give a different perspective, one can imagine that before going to the thermodynamic limit, one defines a function $G(z)$ for frequencies having an imaginary part $i\eta$, with the property that $\lim_{\eta \to 0+} \tilde{G}(\omega + i\eta) = G(\omega)$ for those frequencies where $G(\omega)$ is defined, i.e., for all real frequencies that do not coincide with the discrete poles. Then one takes the thermodynamic limit, and only *afterwards* lets η become infinitesimally small. This is in line with the fact that the correct procedure to deal with vanishing temperature is to perform the $T \to 0$ limit after the thermodynamic limit. The limiting procedure defines a function $G^+(\omega)$ that coincides with the physical Green's function $G(\omega)$ for those energies where it is defined. Analogously, one can define $G^-(\omega)$ for η approaching zero from below. G^+ and G^- are *not* identical, which reflects the discontinuity across the branch cut.

The spectral function

The situation might seem puzzling: the physical Green's function has poles on the real axis, but they form a continuum and cannot give us useful information about spectral properties of the system. Equations (C.13) or (5.92) show the solution to this dilemma: the information about the spectrum is contained in the *discontinuity* of G across the branch cut. As discussed in Sec. 7.5, the analytic continuation $\mathcal{A}(z)$ has poles in the complex plane, but these are *not* the poles of $G(\omega)$, which must be on the real axis because they are total energy differences. The complex poles of $A(z)$ are picked up by the analytic continuation of the Green's function across the branch cut to higher Riemann sheets.

Interpretation

To interpret a spectral function, one often tries to make a link with an effective non-interacting system. In particular, well-defined, though in general broadened, peaks may appear, which are interpreted as dressed single particles called *quasi-particles*. Rather than looking at all possible total energy differences of the many-body system, which is the information contained in the poles on the real axis, one moves into the complex plane. This corresponds to going over to a description where one focuses on one-particle-like excitations coupled to a bath. These are no longer eigenmodes of the N-particle system, but excitations that can lose energy to the bath given by the rest of the system. Therefore, poles are at complex frequencies, and the imaginary part expresses a damping, a finite lifetime. This is not in contradiction with energy conservation: energy is conserved by being transferred to the bath, which can be described by the self-energy, but the total system conserves energy.

Aren't real samples finite?

One could wonder whether these are just mathematical subtleties without any real interest, since real samples are always of finite extensions, with in principle many discrete excitations. In practice, however, the electrons-only system is coupled to an environment containing other possible excitations of phonons, the vacuum electromagnetic field, etc., which split every discrete electronic excitation into many excitations of the coupled system, corresponding to an effective broadening of the peaks. Realistic samples are finite but still with a huge number of electrons, and the spacing of purely electronic excitation energies is small with respect to this broadening, so that the peaks effectively merge into a continuum.

SELECT FURTHER READING
See select further reading at end of Ch. 5.

Exercises

C.1 Using the residuum theorem of Cauchy, show Eqs. (C.9) and (C.10). Note that, according to the sign of t, the integration path has to be closed in the upper or in the lower complex half-plane.

C.2 Calculate the Fourier transform to the complex frequency plane of the independent-particle Green's functions, Eq. (5.82).

C.3 Show that the retarded Green's function equals the single-particle result $G_0(\omega) = (\omega - \hat{h}_0 - v_{\text{ext}})^{-1}$ when the system consists of only one particle. *Hint*: use the spectral representation of G_0. Consider that G^R describes the propagation of an *additional* particle, with respect to the vacuum state.

C.4 Derive the second equation of Eq. (C.9) in two ways: (1) by contour integration in the complex plane with $\omega \to z$, using the fact that the integrand has a pole at $z = -i\eta$ and (2) using the relation in Eq. (C.12). In the latter case note that one must include both the principal part and the delta function in the integration. (Note that $\int_0^\infty dx \, \sin(mx)/x = \pi/2$ or $-\pi/2$ for m positive or negative.)

C.5 Suppose a system with an interacting Green's function that is diagonal for a part of the spectrum given by the density of states $\Theta(\omega + c)\Theta(c - \omega)$. What does the density of states look like for a non-interacting problem? Calculate the spectral function for the non-interacting and for the interacting case. Show that the analytic continuation of the spectral function $A(z)$ has poles in the complex plane, but not the Green's function. Show that $A(z)$ has many branches. Define the continuation of the Green's function to a higher Riemann sheet by adding the discontinuity $A(\omega)$ when crossing the branch cut. Show that this extended Green's function does have the complex poles of $A(z)$. Make a link to quasi-particles.

Appendix D

Matsubara formulation for Green's functions for $T \neq 0$

Summary

Dynamic correlation functions as a function of temperature are developed in Ch. 5; however, the expressions for $T \neq 0$ are often complicated and difficult to use, as pointed out after Eq. (5.39). In this appendix we present a brief summary of the Matsubara $T \neq 0$ method, which is useful in actual calculations of properties as a function of temperature, as well as the derivation of the Luttinger–Ward functional in Ch. 8, the Luttinger theorem in App. J, and other properties.

The finite-temperature, imaginary-time formulation of Green's functions and other correlation functions is a general approach with advantages and disadvantages compared with the real-time formulation. The central quantities are the values of the functions at equally spaced points $z_n = i\omega_n$ on the imaginary axis in the complex frequency plane, where ω_n are the Matsubara frequencies. Here we derive general properties and show the relation to the correlation functions defined in Ch. 5. The methods are used directly in finite-temperature calculations, e.g., for DMFT in Ch. 18. In addition, the formulation provides a compact, general way to cast the functionals in Ch. 8 and to derive important far-reaching relations such as the Luttinger theorem in App. J. Diagrammatic expansion is described in Sec. 9.6.

D.1 Green's functions at $T \neq 0$: Matsubara frequencies

This section summarizes the finite-temperature Green's function formalism due to Matsubara [234]. Detailed derivations can be found in texts on many-body theory, such as those listed in the select further reading at the end of Ch. 5. Many of the methods developed for real-time Green's functions carry over to imaginary time, for example, a generalized time-evolution operator and a generalized Wick's theorem[1] that makes possible a diagrammatic

---

[1] The $T \neq 0$ generalization of Wick's theorem is a statistical relation valid only in equilibrium, unlike the operator identity derived in Sec. 9.3. There is no notion of normal ordering and it depends on the properties of expectation values in systems in equilibrium in the grand-canonical ensemble.

expansion completely analogous to that for $T = 0$ but with special considerations (see Sec. 9.6). The $T \neq 0$ form is useful for many purposes, for example, it is most convenient to express the functionals in Ch. 8 with the knowledge that the relations carry over to real time. In addition, useful expressions for $T = 0$ calculations can be found in the limit for $T \to 0$.[2] The expressions at $T \neq 0$ are often easier to evaluate and they are used in the many-body solvers for DMFT in Ch. 18; however, it must be realized that the continuation to the real axis is not trivial, and it may be difficult to resolve details of the spectrum in practice.

In order to deal with thermodynamic aspects of a many-particle system, it is convenient to work in the grand-canonical ensemble, and we consider only the case where the hamiltonian $\hat{H}$ has no explicit time dependence. The relations can be expressed in a compact form in terms of $\hat{K} = \hat{H} - \mu \hat{N}$, which can be considered a modified hamiltonian with μ the chemical potential. Thus an expectation value can be expressed as

$$\langle \hat{O} \rangle = \frac{1}{Z} \mathrm{Tr} \left\{ e^{-\beta \hat{K}} \hat{O} \right\}, \quad \text{with} \quad Z = \mathrm{Tr} \left\{ e^{-\beta \hat{K}} \right\} \equiv e^{-\beta \Omega}, \tag{D.1}$$

where $\mathrm{Tr}\{\ldots\}$ denotes the trace over all states, $\beta = 1/[k_B T]$ is the inverse temperature divided by Boltzmann's constant, Z is the partition function, and Ω is the grand potential. One can also define field operators in a modified Heisenberg picture (also called representation, see Sec. B.2, especially Eqs. (B.16)–(B.18)) with the hamiltonian $\hat{H}$ replaced by $\hat{K}$:

$$\hat{\psi}(x, t) = e^{i\hat{K}t} \hat{\psi}(x) e^{-i\hat{K}t}. \tag{D.2}$$

Let us now look at the time-ordered Green's function. For $t > t'$, one has to evaluate

$$\mathrm{Tr} \left[e^{-\beta \hat{K}} \hat{\psi}(x, t) \hat{\psi}^{\dagger}(x', t') \right]$$

$$= \mathrm{Tr} \left[e^{-\beta \hat{K}} \left(e^{i\hat{K}t} \hat{\psi}(x) e^{-i\hat{K}t} \right) \left(e^{i\hat{K}t'} \hat{\psi}^{\dagger}(x') e^{-i\hat{K}t'} \right) \right]. \tag{D.3}$$

Since the trace is cyclic, this expression can be rearranged (see Ex. D.1) to show that for $t > t'$,

$$G(x, t; x', t') = \pm G(x, t + i\beta; x', t'). \tag{D.4}$$

Similarly, for $t < t'$,

$$G(x, t; x', t') = \pm G(x, t - i\beta; x', t'). \tag{D.5}$$

Here the notation $\pm$ is the same as defined in Eq. (5.24), where the sign choice $\pm$ is used to distinguish bosonic operators (upper sign) and fermionic operators (lower sign).

Time-ordered correlation functions are often needed in calculations; however, the expressions in Ch. 5 include thermal factors (see, e.g., Eq. (5.39)) that prevent one from casting it in a form like Eq. (5.35) that would allow an analytic continuation to complex frequencies. The problem can be overcome by taking advantage of the close relation of

---

[2] Care must be taken at this point to take the limits properly and recognize differences, for example, the difference between thermal occupation of degenerate states and a specific choice of a state at $T = 0$.

time and temperature in the Green's function and other expectation values. One may either work with imaginary temperatures and real time or, as we will do in the following, imaginary time and real temperature (expressed in time units by $\beta\hbar$; here we set $\hbar = 1$). (See [40] for an illuminating analysis.) Thus one has a new time parameter, $\tau = it$, where τ is real. Operators can be represented in a modified Heisenberg form (see Eq. (B.18)),

$$\hat{O}_K(\tau) \equiv e^{\hat{K}\tau}\hat{O}e^{-K\tau}, \tag{D.6}$$

and a time ordering in τ can be defined in analogy to real time.

The time-ordered Green's function in imaginary time can be defined by

$$G(\tau - \tau') = -\langle T_\tau \, \hat{\psi}(\tau)\hat{\psi}^\dagger(\tau')\rangle, \tag{D.7}$$

where T_τ is the τ-ordering operator. For a time-independent hamiltonian, the Green's function depends only on the time difference and we can simplify the equations by setting $\tau' = 0$. The Green's function G in Eq. (D.7) can represent a function of position and spin $G(x, x'; \tau) = -\langle T_\tau \, \hat{\psi}(x, \tau)\hat{\psi}^\dagger(x', 0)\rangle$, with $x = (\mathbf{r}, \sigma)$, as in Sec. 9.6, or a matrix expressed in a basis $G_{\ell\ell'}(\tau) = -\langle T_\tau \, c_\ell(\tau)c_{\ell'}^\dagger(0)\rangle$.

Note that the same symbol[3] G is used for the Green's function in real time $G(t)$ and in imaginary time $G(\tau)$; however, we should carefully note the difference. In $G(\tau)$ the operators are defined by $\hat{\psi}(x, \tau) = e^{\hat{K}\tau}\hat{\psi}(x)e^{-\hat{K}\tau}$, etc., as in Eqs. (B.17) and (B.18), and there is no factor of i in Eq. (D.7) corresponding to the fact that the Schrödinger equation is purely real with $i\partial/\partial t \to -\partial/\partial\tau$. The appropriate expression should be apparent from the context and will be distinguished by the argument t or τ.

The imaginary-time formulation has the advantage that the conditions of Eqs. (D.4) and (D.5) translate into

$$G(\tau) = \pm G(\tau - \beta) \tag{D.8}$$

for $0 < \tau \leq \beta$. Similarly, for $-\beta < \tau \leq 0$,

$$G(\tau) = \pm G(\tau + \beta). \tag{D.9}$$

Thus $G(\tau)$ is a periodic function, with a period 2β that can be expressed as a Fourier series (see Ex. D.2)

$$G(\tau) = \beta^{-1} \sum_{n=-\infty}^{+\infty} e^{-i\omega_n \tau} G(i\omega_n). \tag{D.10}$$

Taking into account the additional conditions in Eqs. (D.8) and (D.9), the only non-zero Fourier components are for the Matsubara frequencies

$$\omega_n = \frac{\pi}{\beta} \times \begin{cases} 2n & \text{for bosons,} \\ 2n+1 & \text{for fermions.} \end{cases} \tag{D.11}$$

---

[3] In some sources the Green's function in imaginary time is denoted $\mathcal{G}$ to avoid confusion. Here we use the same symbol G to be compatible with the literature for calculations using the Matsubara method, and we reserve the symbol $\mathcal{G}$ to denote the Green's function for an auxiliary system, in particular, an embedded site or cell in dynamical mean-field theory (Chs. 16–21).

It is the restriction to even or odd integers that encodes the boson or fermion statistics. The inverse transformation can be written

$$G(i\omega_n) = \frac{1}{2} \int_{-\beta}^{\beta} d\tau \, e^{i\omega_n \tau} G(\tau). \tag{D.12}$$

The density and density matrix can be expressed in terms of G in forms analogous to Eqs. (5.118) and (5.120). Expressed as matrices in a basis, the density matrix is given by

$$\rho_{\ell\ell'} = G_{\ell\ell'}(\tau = 0^-) = \beta^{-1} \sum_{n=-\infty}^{+\infty} e^{-i\omega_n 0^-} G_{\ell\ell'}(i\omega_n). \tag{D.13}$$

Here the notation 0^- means the limit $\tau \to 0$ with $\tau < 0$. In the zero-temperature limit the sum becomes an integral that can be written

$$\hat{\rho} \to \frac{1}{2\pi} \int_{-\infty}^{+\infty} d\omega \, e^{-i\omega 0^-} G(i\omega), \tag{D.14}$$

where ω is a real quantity so that the Green's function is integrated over the frequency $z = i\omega$ on the imaginary axis in the complex plane. These are examples of the frequency integrals needed for the density and the energy and grand-potential functionals in Secs. 8.1 and 8.2. See also Sec. C.3 for corresponding expressions derived by transforming from integrals on the real axis to the imaginary axis.

The bottom line is that at non-zero temperature the dynamic Green's function (and other correlation functions) for all frequencies is determined by the values at the Matsubara frequencies on the imaginary axis. For low temperatures the frequencies are closely spaced by $2\pi/\beta = 2\pi k_B T$, which approaches a continuum as $T \to 0$. As the temperature is increased, fewer frequencies are needed to describe the response, i.e., there is less information in the dynamic correlation functions, which become smoother functions of time as the temperature increases. As a function of imaginary time, the function must be determined over an interval $-\beta < \tau < 0$, which decreases as the temperature is increased. The values of the function evaluated at the Matsubara frequencies is a compact way to describe the functions. However, there is no free lunch. Even though it is in principle possible, it is difficult to reconstruct the dynamical functions in real time or the spectrum for real frequencies given only the values at the discrete set of imaginary frequencies. This is discussed further in Secs. D.5 and 25.6.

D.2 Analytic properties in the complex-frequency plane

In this section we show the relation of the expressions in the previous section to the correlation functions and Green's functions for $T \neq 0$ defined in Ch. 5 in terms of spectral function and thermal factors; this also brings out the analytic structure in the complex plane. As in the previous section, we assume the hamiltonian is independent of time and the arguments for space and spin are not shown explicitly.

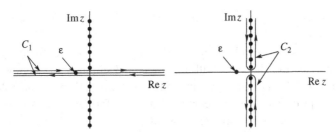

Figure D.1. The analytic structure of $C(z)$ (the analytic continuation of the spectral function $A(\omega)$) and the thermal factor as a function of the complex frequency z. Here we illustrate the structure in $C(z)$ by a simple pole at $z = \varepsilon$, but the conclusions hold for general spectra with continuous structure and branch cuts. The dots on the imaginary axis denote the poles due to the thermal factor for fermions $1/(1 + e^{\beta z})$ that occur at $z = i\omega_n$, where ω_n are the Matsubara frequencies. (As stated in the text, energies are defined relative to the Fermi energy.) The integral Eq. (D.16) on the contour C_1 shown on the left can be carried out by deforming the contour to C_2, which encloses the poles of the thermal factor on the imaginary axis.

In Sec. 5.4 various correlation functions are derived in terms of the spectral function $A(\omega)$ defined for real frequency ω and the function $C(z)$ analytically continued in the complex plane, where z is the complex frequency. The function C is analytic everywhere, except along the real axis for frequencies where $A(\omega)$ is non-zero. For example, a δ-function in $A(\omega)$ at $\omega = \varepsilon$ leads to the pole in $C(z)$ at $z = \varepsilon$ in the complex plane indicated in Fig. D.1. The correlation function $\tilde{C}(t)$ in Eq. (5.24) can be expressed as

$$\tilde{C}(t) = \frac{1}{2\pi} \int d\omega \, e^{-i\omega t} \tilde{C}(\omega) = \pm \frac{1}{2\pi} \int d\omega \, e^{-i\omega t} \frac{A(\omega)}{\mp 1 + e^{\beta\omega}}, \qquad (D.15)$$

where we have used Eq. (5.32). In order to simplify the equations, we have set the chemical potential μ to zero. Since the hamiltonian is taken to be time independent, the correlation functions depend only on the time difference, here denoted by t. Using the relation Eq. (5.31) we can find the analogous expression for $C(t)$ in Eq. (5.22), where there is an additional factor of $\pm e^{-\beta\omega}$ in the integrand. We can now use Eq. (5.38) to reformulate Eq. (D.15) (and an analogous expression for C) as

$$\tilde{C}(t) = \pm \frac{1}{2\pi} \oint_{C_1} dz \, e^{-izt} \frac{C(z)}{\mp 1 + e^{\beta z}}$$

$$\text{(for bosons only)} \ + \frac{i}{\beta} C(0), \qquad (D.16)$$

where z is the frequency defined in the complex plane and the integral is over the contour C_1 shown in the left panel of Fig. D.1, passing above and below the poles and branch cuts on the real axis in the sense indicated. Here we show a single pole enclosed by the contour, but the conclusions apply in general due to the structure in $A(\omega)$. The last term in Eq. (D.16) is due to the pole at $z = 0$ from the thermal factor in the case of bosons.

The only other poles of the integrand stem from the thermal factor $1/(\mp 1 + e^{\beta z})$ for $z_n = i\omega_n$ on the imaginary axis, where the Matsubara frequencies ω_n are multiples of the odd integers $(2n + 1)\pi/\beta$ for fermions and even integers $2n\pi/\beta$ for bosons, the same as found before and given in Eq. (D.11). The positions z_n in the complex plane are illustrated in Fig. D.1 for fermions.

The desired expressions can be found by deforming the contour C_1 to C_2 circling the poles in the upper part of the plane with $\text{Im}\, z > 0$, and similarly for the lower half-plane with $\text{Im}\, z < 0$ as indicated on the right side of Fig. D.1. Since there are no poles anywhere in the plane except on the axes, the contours can be deformed as shown, so long as the integrand vanishes for large $|z|$ sufficiently rapidly.

Using the correspondence of temperature and time $t = -i\tau$,[4] the conditions for convergence can be expressed simply. In the interval

$$-\beta < \tau < 0,$$ (D.17)

the contributions for the complex frequency $|z| \to \infty$ vanish due to the thermal factors in the integrand in Eq. (D.16) (see Ex. D.3). Thus, the integral is the sum of the residues of the poles at $z = i\omega_n$,

$$\tilde{C}(-i\tau) = \frac{i}{\beta} \sum_n e^{-i\omega_n \tau} C(i\omega_n).$$ (D.18)

The sum (and also in Eq. (D.20)) converges because of the oscillating factor $e^{-i\omega_n \tau}$, which is essential to include for the limits $\tau \to 0^{(+)}$ for $\tau > 0$ or $\tau \to 0^{(-)}$ for $\tau < 0$. In order to find the expressions for C, which is related to $\tilde{C}$ by $C(\omega) = \pm e^{+\beta\omega}\tilde{C}(\omega)$ (see Eq. (5.31)), we note that the poles are the same; however, in this case the integrals converge in the interval

$$0 < \tau < \beta$$ (D.19)

and

$$C(-i\tau) = \pm \frac{i}{\beta} \sum_n e^{-i\omega_n(\tau-\beta)} C(i\omega_n).$$ (D.20)

It follows that the imaginary-time-ordered correlation function, defined as

$$C^T(-i\tau) = \begin{cases} \tilde{C}(-i\tau) & \tau < 0, \\ C(-i\tau) & \tau > 0, \end{cases}$$ (D.21)

is periodic for bosons or antiperiodic for fermions for a shift of τ by β,

$$C^T(-i\tau) = \pm C^T(-i\tau - i\beta).$$ (D.22)

The inverse transform is given by

$$C(i\omega_n) = -\frac{i}{2} \int_{-\beta}^{\beta} d\tau\, e^{i\omega_n \tau} C^T(-i\tau).$$ (D.23)

Even though there are thermal factors in the expressions for $\tilde{C}(-i\tau)$ and $C(-i\tau)$, they cancel in the sum of their contributions to the integral in Eq. (D.23) (see Ex. D.4). Thus

---

[4] Here we use the relation $t = -i\tau$ in the argument of a correlation function $C(-i\tau)$, whereas $G(\tau)$ is written with argument τ instead of $-i\tau$ in order to be consistent with the notation for the Green's function in the literature.

we have the simple result that the Fourier transform of the time-ordered function C^T in Eq. (D.21) is $\mathcal{C}(i\omega_n)$, which is the analytic continuation of the spectral function $A(\omega)$ in the complex plane.

Conclusions for the Green's function

The analysis here for a time-ordered correlation function expressed in imaginary time τ corresponds to the Green's function $G(\tau)$ defined in Eqs. (D.7)–(D.12). In the present notation the connection can be seen by comparing Eqs. (D.18) and (D.20) and Eq. (D.23) with Eqs. (D.10) and (D.12), where the Green's function has been defined without the factor of i and we use the convention that the argument of G is τ instead of $-i\tau$. As stated after Eq. (D.11), it is the change of sign of the Green's function for $\beta < \tau < 0$ and $0 < \tau < \beta$ for fermions (no change of sign for bosons) that incorporates the quantum statistics in the imaginary-time formulation of dynamics in the grand-canonical ensemble.

From Eq. (D.23) we see that the Fourier transform $G(i\omega_n)$ of the imaginary-time-ordered Green's function in Eq. (D.7) is simply the analytic continuation of the spectral function $A(\omega)$ to frequencies on the imaginary axis. For static properties such as the density, the proper time order is defined by the definitions above.

D.3 Illustration of the structure of $G^0(i\omega_n)$ and $G^0(\tau)$

Non-interacting fermions

It is instructive to derive the explicit form of the Green's functions for independent particles; here we consider fermions and the expressions for bosons are relegated to Exs. D.5 and D.6. Integrals of products of such functions are needed to expand the partition function in powers of the interaction and many aspects carry over to the general problem of interacting, correlated fermions, as discussed at the end of this section. For simplicity we express G^0 in a basis of eigenstates with energy ε_ℓ. From the definition in Eq. (D.7) and using the fact that at equilibrium the occupation of the state is the Fermi function $f(\varepsilon)$, we find that the Green's function in time can be expressed as

$$G^0_{\ell\ell}(\tau) = e^{-(\varepsilon_\ell - \mu)\tau} \times \begin{cases} f(\varepsilon_\ell) & -\beta < \tau < 0 \\ f(\varepsilon_\ell) - 1 & 0 < \tau < \beta \end{cases} \quad \text{with } f(\varepsilon_\ell) = \frac{1}{1 + e^{\beta(\varepsilon_\ell - \mu)}},$$

$$(D.24)$$

where the chemical potential μ is indicated explicitly. For $T \to 0$ the Green's function $G^0(\tau)$ is also closely related to the real-time $T = 0$ formalism: $G^0_{\ell\ell}(\tau > 0)$ is non-zero only for $\varepsilon_\ell > 0$ above the Fermi energy, and $G^0_{\ell\ell}(\tau < 0)$ is non-zero only for filled states with $\varepsilon_\ell < 0$ below the Fermi energy. Like the $T = 0$ Green's functions in real time, $G^0_{\ell\ell}(\tau)$ has a discontinuity of unity magnitude at $\tau = 0$ due to the T_τ ordering.

The Fourier transform of Eq. (D.12) then leads to the Green's function at the Matsubara frequencies for fermions,

$$G^0_{\ell\ell}(i\omega_n) = \frac{1}{i\omega_n + \mu - \varepsilon_\ell}, \qquad (D.25)$$

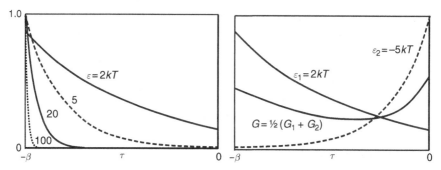

Figure D.2. Examples of the Green's function for non-interacting particles $G^0(\tau)$ as a function of imaginary time τ plotted in the interval $-\beta \leq \tau < 0$; for $\tau > 0$ the Green's function for fermions is determined by $G(\tau) = -G(\tau - \beta)$. Here ε denotes energy relative to the Fermi energy. Left: the exponential decay for positive energies ε above the Fermi energy plotted for various values of ε/kT. High-energy excitations lead to rapid decay whereas low-energy features are characterized by slow variation. Right: $G(\tau)$ for a case with one energy above and one below the Fermi energy. $G(\tau)$ is symmetric if there is electron–hole symmetry, e.g., in the half-filled Hubbard model.

which is the Green's function $G(z) = 1/(z+\mu-\varepsilon_\ell)$ evaluated at $z = i\omega_n$. These expressions also follow from the spectral function $A(\omega)$ (a δ-function in this case) and the formulas as a function of temperature in Ch. 5.

Illustrative examples of $G(\tau)$ are shown in Fig. D.2 for τ in the range $-\beta \leq \tau < 0$. The energy ε is defined relative to the Fermi energy. The contribution from an eigenvalue $\varepsilon > 0$ above the Fermi energy decreases exponentially with increasing τ, whereas there is an exponential increase for $\varepsilon < 0$ below the Fermi energy. On the left side is shown $G(\tau)$ for several values of $\beta\varepsilon = \varepsilon/kT$. The fact that all curves have the same shape but different decay coefficients shows the difficulty of resolving information about the spectrum from information in imaginary time. Energy scales that are very different can be distinguished readily; however, it is still difficult to quantify the results. From the figure we can see immediately that a calculation of $G(\tau)$ with a regular set of points spaced by $\Delta\tau$ for $-\beta < \tau < 0$ requires a very fine grid of points to capture the high-energy feature. But to determine the low-energy behavior requires low temperature and many points $N_\tau = \beta/\Delta\tau$. (Note that $100kT$ is only ≈ 1 eV for $T = 80$ K.) Fortunately, there are many cases where the goal is to determine low-energy phenomena and high-energy structure is not needed in detail. Also, the continuous-time methods (Secs. 18.5 and 18.6) offer an alternative to a regular grid of points. The right-hand part of the figure is an example in which there are two energies, one below and one above the Fermi energy; this is the usual case with $G(\tau)$ that has a minimum in the range $-\beta \leq \tau < 0$. Only if the spectrum is symmetric about the Fermi energy (electron–hole symmetry) is $G(\tau)$ symmetric about $\tau = \pm\beta/2$. For bosons, $G(\tau)$ is always symmetric, e.g., the spin correlation functions in Fig. 20.6 or in Fig. 20.9, where only the range 0 to $\beta/2$ is plotted.

Generalization to interacting fermions

The structure of the Green's function in time and frequency has been derived for independent-particle systems, but the general characteristics depend only on the assumption that ε is the energy to add or remove electrons. There are also discrete energies in an interacting many-body problem in a finite system, and the energies ε_ℓ can be generalized to the energy differences $E_\lambda - E_\alpha$ that are defined in Sec. 5.4 for the interacting system. The difference from the expressions in Eqs. (D.24) and (D.25) is that the numerator is not unity but is a matrix element of the creation or annihilation operators between many-body wavefunctions. Furthermore, the thermal factors in Eq. (D.24) apply not only to the Fermi function in terms of the independent-particle energies ε_ℓ; they occur also for interacting particles where the temperature dependence is incorporated in the sum over Matsubara frequencies that are the poles of the Fermi function. If we let the energies approach a continuum, the arguments carry over to $G(\tau)$ calculated for an interacting many-body system in a solid.

D.4 The grand potential Ω

Calculation of the grand potential Ω provides insight into the expressions in Ch. 8 in terms of a trace $\mathfrak{Tr}$ of the logarithm, and it provides an instructive example of a sum over Matsubara frequencies and convergence issues that should be kept in mind. The expression for the grand potential in terms of the Luttinger–Ward functional given in Eq. (8.8) is

$$\Omega = \Phi[G] - \mathfrak{Tr}(\Sigma G) - \mathfrak{Tr}\ln(1 - G_0\Sigma) + \Omega^0. \tag{D.26}$$

However, as pointed out in footnote 7 before that equation, this is not the actual expression given by LW, who gave an expression with the last two terms combined:

$$\Omega = \Phi[G] - \mathfrak{Tr}(\Sigma G) - \mathfrak{Tr}\ln(\Sigma - G_0^{-1}), \tag{D.27}$$

and Klein or Baym and Kadanoff write the last term simply as $-\mathfrak{Tr}\ln(-G^{-1})$ or $\mathfrak{Tr}\ln(-G)$. However, these expressions are not well-defined because the sums over Matsubara frequencies are only conditionally convergent. They are sufficient for the calculation of functional derivations but they are defined only up to a constant. Here we work out the expressions in a way that brings out the issues in dealing with logarithmic functions, and we show how to rewrite Eq. (D.27) to avoid such problems, which leads directly to Eq. (D.26).

The convergence issues can be addressed by considering only the non-interacting case. This is because at high frequency, large $|z|$, where z is the frequency in the complex plane, $G(z)$ approaches the bare $G_0(z)$ or, equivalently, the self-energy must decrease at high frequency. The sums are conditionally convergent because the absolute value of the argument of the logarithm diverges, since $G(z) \approx z$ for large z, and the sum has a finite value only because of the oscillatory behavior of the logarithm in the complex plane. For this reason LW included a convergence factor. Another approach is to use the expression for the grand potential for the non-interacting system Ω^0 in Eq. (8.4) that is well-defined and does not involve any sums in the complex plane. Then all we need is to find a convergent expression for the difference $\Omega - \Omega^0$.

To this end the original LW expression of Eq. (D.27) can be rewritten by adding and subtracting Ω^0,

$$\Omega = \Phi[G] - \mathfrak{Tr}(\Sigma G) + \left[-\mathfrak{Tr}\ln(\Sigma - G_0^{-1}) - \Omega^0 \right] + \Omega^0, \tag{D.28}$$

and finding the expression for Ω^0 inside the square brackets that renders the trace convergent. The expression for Ω^0 in Eq. (8.4) (the first line of the equations below) can be transformed as follows:

$$
\begin{aligned}
\beta\Omega^0 &= -\sum_i \ln(1 + e^{-\beta(\varepsilon_i^0 - \mu)}) \\
&= -\frac{1}{2\pi i} \sum_i \oint_{C_1} dz \frac{1}{z - \varepsilon_i^0} \ln(1 + e^{-\beta(z-\mu)}) \\
&= -\frac{\beta}{2\pi i} \sum_i \oint_{C_1} dz \ln(\varepsilon_i^0 - z) \frac{1}{e^{\beta(z-\mu)} + 1} \\
&= -\frac{\beta}{2\pi i} \sum_i \oint_{C_2} dz \ln(\varepsilon_i^0 - z) \frac{1}{e^{\beta(z-\mu)} + 1} \\
&= -\sum_i \sum_n \ln(\varepsilon_i^0 - \mu - i\omega_n) = -\beta\mathfrak{Tr}\ln(-G_0^{-1}).
\end{aligned}
\tag{D.29}
$$

The expression in line 2 utilizes the property that the ln function has no singularities enclosed by the contour, since the real part of $e^{-\beta(z-\mu)}$ is always positive.[5] The only singularity is the pole at $z = \varepsilon_i$, so that the integral is simply $\ln(\ldots)$ evaluated at $z = \varepsilon_i$ for each i, which equals the sum in the first line. The third line is established by a partial integration; at this point it is important to invoke convergence factors or rely on the convergence of the integrals for the *difference* $\Omega - \Omega^0$. Since the only non-analytic points in the upper and lower planes are the poles at the Matsubara frequencies for fermions, $z_n = i\omega_n$, the fourth line follows by transforming the integral to contour C_2 that encloses the poles on the imaginary z-axis shown on the right side of Fig. D.1. Finally, the integral can be evaluated as the sum of residues at the poles, which leads to the expression in terms of Green's functions.

Thus the term in square brackets in Eq. (D.28) becomes $-\mathfrak{Tr}[\ln(\Sigma - G_0^{-1}) - \ln(-G_0^{-1})] = -\mathfrak{Tr}\ln(1 - G_0\Sigma)$. Although $\mathfrak{Tr}\ln(\Sigma - G_0^{-1})$ and $\mathfrak{Tr}\ln(-G_0^{-1})$ are each only conditionally convergent, the trace $\mathfrak{Tr}\ln(1 - G_0\Sigma)$ converges, as is clear from the behavior of $G_0(z)$ and $\Sigma(z)$ for large $|z|$. Thus we arrive at the expression for Ω in Eq. (D.26), which is well-defined so long as the last term Ω^0 is calculated by a well-defined method such as the fundamental expression in Eq. (8.4) and the first line in Eq. (D.29).

---

[5] The logarithm is defined with a branch cut on the negative axis, i.e., a discontinuity of $2\pi i$ in the imaginary part. This is not an issue for the integral in line 2, but it must be taken with care in line 3, which is crucial for the derivation. In particular, this determines the sign of the imaginary part of the ln in the final result involving $\ln(-G^{-1}(i\omega_n))$.

D.5 Transformation to real frequencies

The finite-temperature formulation with imaginary time leads to compact expressions with appealing simplicity. However, there are difficulties in actual calculations. If one knows the spectral function $A(\omega)$ for real ω, it is easy to derive $G(i\omega_n)$, e.g., with Eqs. (D.15) and (D.23). But the converse is not true: $G(i\omega_n)$ is known only at a set of points and it is not at all straightforward to invert the relation to find $A(\omega)$. For a general function, the inversion is not unique; however, it has been shown [1156] that there is only one continuation that obeys the sum rule $|G(z)| \approx 1/|z|$ for $|z| \to \infty$.

The temperature Green's functions are very smooth and featureless (see, e.g., Fig. D.2), while the spectra for real ω can have sharp features, especially at low temperatures. This makes it difficult to transform from imaginary time to real frequencies since small errors in $G(i\omega_n)$ will lead to large differences in frequency space. If the imaginary-time correlations are computed with Monte Carlo methods, statistical noise will be present requiring special methods such as maximum entropy to find the spectra for real frequencies. These methods are as described in Sec. 25.6.

SELECT FURTHER READING

See select further reading at end of Ch. 5. Fetter and Walecka give an especially clear description of the imaginary-time formalism.

Exercises

D.1 Derive the relations of Eqs. (D.4) and (D.5) using the cyclic properties of the trace, i.e., the operators can be rearranged by moving the last to be the first.

D.2 Show that the (anti)periodicity of the thermal Green's function for $\tau \to \tau + \beta$ leads to the Matsubara frequencies $\omega_n = (2n+1)\pi/\beta$ for fermions and $\omega_n = (2n)\pi/\beta$ for bosons. *Hint*: recall that each is periodic with period 2β.

D.3 Show that the conditions for convergence in the intervals in Eqs. (D.17) and (D.19) are correct, and give arguments to support the statement after Eq. (D.18) concerning the behavior for $|\tau| \to 0$.

D.4 Show that the combination of $-\beta < \tau < 0$ and $0 < \tau < \beta$ leads to the result stated in Eq. (D.23). This can be derived from the definitions of $\tilde{C}$ in Eq. (D.15) and the relation to C given before Eq. (D.19).

D.5 (a) Derive the expressions for $G(\tau)$ in Eq. (D.24) from the definition of the Green's function in Eq. (D.7).
(b) Show that the zero-temperature limit has the form as discussed after Eq. (D.24).
(c) Derive the corresponding expressions for bosons and sketch the form of $G(\tau)$ corresponding to Fig. D.2.

D.6 Show that the expressions for $G^0(\omega)$ and $G^0(\tau)$ in Eqs. (D.24) and (D.25) follow from the definition of $C(z)$ in Eq. (5.36) and the expression for $\tilde{C}$ as a function of time in Eq. (D.15) along with the corresponding equation for C.

Appendix E

Time ordering, contours, and non-equilibrium

Summary

The perturbation expansion is greatly simplified by the concept of time ordering. This concept can be generalized to the idea of contour ordering, where a contour is a path in a hypothetical time plane. In this way, one can avoid the need for assumptions such as vanishing temperature, adiabatic switching on of the interaction, or equilibrium. In this appendix we summarize the main idea and different realizations.

E.1 The task

Our goal is to calculate averages of operators, defined at zero temperature as an expectation value over the many-body ground state, and at finite temperature as an ensemble average with statistical weights. Even in a system in equilibrium, where there is no time dependence set by an external potential, these averages involve in general the time, because of the phases of wavefunctions or operators. The most important example in this book is the one-body Green's function, which in equilibrium depends on a time difference. Only its equal time limit, the density matrix, is truly time independent in equilibrium. The phases are determined by the many-body hamiltonian, and in App. B it is shown that in order to evaluate the expressions in a perturbative way, it is very useful to work with time-ordered quantities. However, introducing an order implies that one defines a direction. As we will discuss in the following, this requires either some hypothesis, or a new idea.

E.2 The contour interpretation

Let us first take the expression for the propagator $G^>$, Eq. (5.72), at $T = 0$. Using the time evolution of the field operators, Eq. (B.13), it can be written as

$$G^>(x_1, t_1; x_2, t_2) = -i\langle\Psi_0|U_S(t_a, t_1)\hat{\psi}(x_1)U_S(t_1, t_a)U_S(t_a, t_2)\hat{\psi}^\dagger(x_2)U_S(t_2, t_a)|\Psi_0\rangle$$
$$= -i\langle\Psi_0|U_S(t_a, t_1)\hat{\psi}(x_1)U_S(t_1, t_2)\hat{\psi}^\dagger(x_2)U_S(t_2, t_a)|\Psi_0\rangle. \qquad (E.1)$$

One can interpret this expression by saying that the system evolves from the state $|\Psi_0\rangle$ at time t_a with the time-evolution operator to time t_2; this evolution is governed by the many-body hamiltonian. At that point, a particle is inserted, after which the system continues to evolve to t_1, where a particle is annihilated. Then the time-evolution operator takes us back to t_a. As illustrated in the upper panel of Fig. E.1, having the same start and end points necessarily involves two opposite directions, which is a problem if we want to use the idea of time ordering: how can we say what is first and what is next if the direction is not well-defined?

However, the way we have just described Eq. (E.1) actually suggests an interpretation that overcomes this problem: it seems that we are marching on a closed path, but in some chronological ordering. This is a path in a time plane. The lower panel of Fig. E.1 illustrates the idea: instead of going backwards and forwards as indicated by the top panel, we always go forwards on a time path that starts and end at t_a. Of course, the excursion on the horizontal axis must be long enough to include the two times t_1 and t_2 of interest. It can be longer, because if nothing happens (i.e., if there are no operators involving a time $t_3 > t_1$ or t_2), the added pieces cancel. The ordering is now well-defined along the path: it is called *contour ordering*.

Below we will see still other contours. For all of them it holds that one can express the time-evolution operators and the equation of motion using the times on the contour, which means, in particular, that integrals have to be performed on the contour. If this is done, the expressions derived for the Green's functions and the self-energies in Part II of this book remain valid.

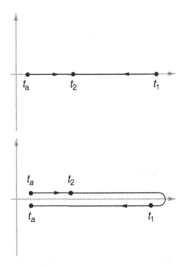

Figure E.1. From going backwards and forwards in time (upper panel) to always going forwards on a contour (lower panel). The contour is plotted slightly off the axis in order to make it visible. This is only a difference of interpretation, but using the contour allows one to transpose the concept of *time ordering* to a concept of *contour ordering*.

E.3 Contours for all purposes

This book contains several kinds of Green's functions, and the concept of a contour is useful in order to put them into a common context, and in order to understand the extension to non-equilibrium, which is a topic of increasing interest.

The first Green's function that is worked out explicitly in Ch. 9 is the time-ordered equilibrium Green's function at zero temperature. Since the aim of that chapter is perturbation expansions, the idea is to start and end with a *non-interacting* instead of an interacting ground state. This requires the hypothesis that one can adiabatically switch on and off the interaction, to bring the system from the non-interacting ground state at a time $-\infty$ to the true interacting ground state at t_a. With respect to the upper panel in Fig. E.1, this is an extension of the path to $-\infty$, indicated by the dashed line in Fig. E.2. Moreover, one has to go *back* to the non-interacting state. In order to be able to always move forwards, the idea in Ch. 9 is to switch off the interaction adiabatically and propagate forwards to $+\infty$. This supposes that one reaches the starting non-interacting ground state by the adiabatic switching off. There must be no degeneracies and level crossing for this to be true. The result, shown in the upper panel of Fig. E.2, is conceptually simple: one has just one time line, with one direction, so that the ordinary real-time ordering is possible. This is what is used in Ch. 9, at the price of several assumptions: zero temperature, equilibrium, adiabatic switching on and off of the interaction starting at, and going back to, the ground state. One could also keep the idea of adiabatic switching on, but avoid the hypothesis that the system goes back to the same state while going forwards in time, by using the idea of the contour. This would correspond to the lower panel in Fig. E.2. Since the path can be as long as we want, provided the times of interest are included, one can suppose that the turning point is at $+\infty$.

As outlined in Sec. 9.6, one can avoid the need to connect to the non-interacting ground state by working at finite temperature. Since there is a trace, it can be performed with non-interacting states; however, the thermal factor $e^{-\beta(\hat{H}-\mu\hat{N})}$ prevents one from making the same perturbation expansion as at zero temperature. The problem is overcome by interpreting β as an imaginary time, such that the exponential can be seen as part of the time evolution. Using the contour concept, this becomes straightforward, as depicted in Fig. E.3. This contour does not require any hypothesis.

As discussed above, one can extend the contour to $+\infty$ with the argument that added pieces cancel, the only requirement for the length of the contour being that the times of interest are included. When one is interested only in equilibrium averages of equal-time quantities such as the density, the "time of interest" can be considered to be zero, and one can shrink the contour such that only the piece on the imaginary axis survives. This shrinking of the real-time axis contour can also be done when one works with the Green's functions defined for imaginary times, which are described in App. D. The Matsubara Green's functions hence lie entirely on the vertical piece of the contour, and one can use time ordering in imaginary time. As discussed in Sec. D.5, going from imaginary time or frequency back to the real axis is difficult, however.

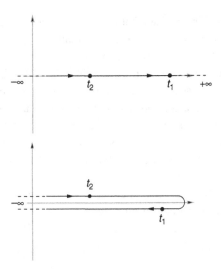

Figure E.2. At $T = 0$ one can connect to the non-interacting ground state by adiabatically switching on the interaction at time $-\infty$. The time period of switching on (off) is given by the dashed lines. The ordinary $T = 0$ time-ordered Green's function also supposes that the system returns to the same state when the interaction is switched off, with real time always going forwards (upper panel). One could also always go forwards on a contour (lower panel), so avoiding the last hypothesis.

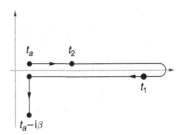

Figure E.3. At non-vanishing temperature the adiabatic hypothesis is not needed, but one has to include the thermal density matrix $e^{-\beta(\hat{H}-\mu\hat{N})}$ in the time evolution. This can be achieved by interpreting β as an imaginary time.

The expressions in Chs. 10–15 are derived using the concept of a generating functional introduced in Sec. 8.4. This also requires ordering of the operators, but it leaves the freedom to choose how this ordering is achieved. The integrals that appear in equations such as Eq. (8.31) are then integrals on the chosen contour, and T_C is the corresponding contour ordering. When explicit equilibrium expressions are given, in this book we suppose either real-time ordering at $T = 0$, or imaginary-time ordering at $T \neq 0$. Note that the functional derivative approach does not seem to assume the adiabatic hypothesis, not even at $T = 0$. However, when the fictitious external potential is sent to zero, one finds analogous problems, in particular the question of which state the system should end up in, in case of degeneracies.

If one wishes to avoid both the adiabatic hypothesis and the problem of getting real spectra from a Green's function on the imaginary frequency axis, one can in principle use the contour in Fig. E.3. Since this involves pieces of a different kind, it requires some "book keeping." The use of this contour, often called the Keldysh contour for the important contributions of L. Keldysh to the development of non-equilibrium Green's functions [1157], is mandatory when one is interested in non-equilibrium problems. In this case the external potential depends on a real time, so that the real-time axis must be part of the contour. Moreover, one cannot suppose that the system will go back to the initial state, which excludes for example the option in the upper panel of Fig. E.2. Non-equilibrium problems are of increasing interest, for example to describe time-resolved spectroscopy or quantum transport beyond the steady state. Also, the real-time propagation of the Green's function mentioned in Sec. 14.11 is a non-equilibrium study. Going out of equilibrium is beyond the scope of this book; however, the general expressions in Chs. 10–15 remain valid, provided one evaluates them on the appropriate contour.

SELECT FURTHER READING

Danielewicz, P., "Quantum-theory of nonequilibrium processes 1," *Ann. Phys.* **152**, 239, 1984. [324]

Stefanucci, G. and van Leeuwen, R. *Non-equilibrium Many-Body Theory of Quantum Systems* (Cambridge University Press, Cambridge, 2013). Extensive introduction to the contour idea and unified presentation of all cases, with critical discussion. Focus on the non-equilibrium case; for example, the "book keeping" rules can be found in this book.

van Leeuwen, R. and Stefanucci, G., "Equilibrium and nonequilibrium many-body perturbation theory: a unified framework based on the Martin–Schwinger hierarchy," *J. Phys.: Conf. Ser.* **427**, 012001, 2013. A paper presenting a unified framework for the various approaches.

van Leeuwen, R., Dahlen, N. E., Stefanucci, G., Almbladh, C.-O., and von Barth, U., "Introduction to the Keldysh formalism," *Lecture Notes in Physics* **706**, 33, 2006. [1158] Detailed and pedagogical introduction to non-equilibrium Green's functions.

Appendix F

Hedin's equations in a basis

Summary

This short appendix presents a generalized form of Hedin's equations. They are derived by adding an exchange–correlation contribution to the total classical potential. Moreover, it is often useful to have equations expressed in a basis. This allows one to appreciate the spin and frequency structure, and it can be used as a starting point for approximations. Therefore, the equations are also written in a basis of spin orbitals. The original Hedin's equations can be retrieved as a special case.

F.1 Generalization of Hedin's equations

In Sec. 11.1 Hedin's equations are derived by starting from Eq. (10.34), where the self-energy is expressed in terms of a variation with respect to an external potential u. However, instead of working with the bare, reducible vertex function $\Gamma = -\delta G^{-1}/\delta u$ (Eq. (11.1)) leading to $\Sigma_{xc} = iGv_c\Gamma$ (Eq. (11.2)), in Hedin's equations a new vertex function $\tilde{\Gamma}$ is defined for which the vertex *corrections* are less important, i.e., where $\tilde{\Gamma}$ is closer to unity. In this way, the vertex can be more efficiently approximated. To achieve this, the total classical potential is introduced. It is defined to be the sum of the external and the Hartree potential. This leads to the definition of the irreducible vertex $\tilde{\Gamma} = -\delta G^{-1}/\delta v_{cl}$ (Eq. (11.3)). It highlights the importance of screening, with the appearance of the inverse dielectric function $\epsilon^{-1} = \delta v_{cl}/\delta u$ (Eq. (11.4)), and the screened Coulomb interaction $W = \epsilon^{-1}v_c$ (Eq. (11.5)). The self-energy can be written as $\Sigma_{xc}(1,2) = iGW\tilde{\Gamma}$ (Eq. (11.6)). The strong point of this formulation is that the simplest approximation, $\tilde{\Gamma} \approx 1$, which leads to the GWA, is much better than the equivalent approximation made on the reducible vertex, $\Gamma \approx 1$, which simply leads to Hartree–Fock.

It is tempting to continue along this road and find an even better starting point for approximations, by including also some exchange–correlation contributions along with the total classical potential. To this end we define the general non-local potential

$$v_{nl}(1,2) \equiv v_{cl}(1)\delta(1,2) + s_{xc}(1,2) \equiv u(1)\delta(1,2) + s(1,2), \tag{F.1}$$

where s_{xc} is some exchange–correlation contribution, that could be for example a part of the exchange–correlation self-energy Σ_{xc}, or the Kohn–Sham v_{xc}, and s is the sum of s_{xc} and the Hartree potential. All steps can now be carried out in close analogy to Sec. 11.1. In particular, using the chain rule, the self-energy becomes

$$\Sigma_{xc}(1,2) = -iG(1,\bar{4})v_c(1,\bar{3})\frac{\delta G^{-1}(\bar{4},2)}{\delta u(\bar{3})} = -iG(1,\bar{4})v_c(1,\bar{3})\frac{\delta G^{-1}(\bar{4},2)}{\delta v_{nl}(\bar{5},\bar{6})}\frac{\delta v_{nl}(\bar{5},\bar{6})}{\delta u(\bar{3})}$$

$$= iG(1,\bar{4})W_{nl}(1,\bar{5};\bar{6})\Gamma_{nl}(\bar{4},2;\bar{6},\bar{5}), \tag{F.2}$$

where we have defined

$$W_{nl}(1,5;6) \equiv v_c(1,\bar{3})\frac{\delta v_{nl}(5,6)}{\delta u(\bar{3})} \tag{F.3}$$

and

$$\Gamma_{nl}(4,2;6,5) \equiv -\frac{\delta G^{-1}(4,2)}{\delta v_{nl}(5,6)}. \tag{F.4}$$

Similarly, with Eq. (F.1) the generalized screened Coulomb interaction of Eq. (F.3) reads

$$W_{nl}(1,5;6) = v_c(1,5)\delta(5,6) + v_c(1,\bar{3})\frac{\delta s(\bar{5},6)}{\delta G(\bar{7},\bar{8})}\frac{\delta G(\bar{7},\bar{8})}{\delta v_{nl}(\bar{4},\bar{9})}\frac{\delta v_{nl}(\bar{4},\bar{9})}{\delta u(\bar{3})}. \tag{F.5}$$

With the definitions

$$\Delta\Sigma_{xc} \equiv \Sigma_{xc} - s_{xc}, \qquad \xi(5,8;6,7) \equiv \frac{\delta s(5,6)}{\delta G(7,8)}, \tag{F.6}$$

and

$$P_{nl}(7,9;8,4) \equiv -i\frac{\delta G(7,8)}{\delta v_{nl}(4,9)}, \tag{F.7}$$

this leads to the generalized Hedin's equations:

$$\Sigma_{xc}(1,2) = iG(1,\bar{4})W_{nl}(1^+,\bar{5};\bar{6})\Gamma_{nl}(\bar{4},2;\bar{6};\bar{5}) \tag{F.8}$$

$$W_{nl}(1,2;3) = v_c(1,2)\delta(2,3) + i\xi(2,\bar{8};3,\bar{7})P_{nl}(\bar{7},\bar{9};\bar{8},\bar{4})W_{nl}(1,\bar{4};\bar{9}) \tag{F.9}$$

$$P_{nl}(1,2;3,4) = -iG(1,\bar{5})G(\bar{6},3)\Gamma_{nl}(\bar{5},\bar{6};2,4) \tag{F.10}$$

$$\Gamma_{nl}(1,2;3,4) = \delta(1,4)\delta(2,3) + \frac{\delta\Delta\Sigma_{xc}(1,2)}{\delta G(\bar{5},\bar{6})}G(\bar{5},\bar{7})G(\bar{8},\bar{6})\Gamma_{nl}(\bar{7},\bar{8};3,4) \tag{F.11}$$

$$G(1,2) = G^0(1,2) + G^0(1,\bar{3})\Sigma(\bar{3},\bar{4})G(\bar{4},2) \tag{F.12}$$

Note that only the Dyson Eq. (F.12) remains unchanged, the same as Eq. (11.15). The five equations have to be completed with the definition of $\Delta\Sigma_{xc}$ and of ξ as a functional derivative of s (Eq. (F.6)). One may interpret ξ as a generalized interaction, in the spirit of the discussions in Sec. 11.1. It reduces to the bare Coulomb interaction when $s_{xc} = 0$, and has an exchange–correlation contribution otherwise. For $s_{xc} = 0$, Hedin's equations, Eqs. (11.11)–(11.15), are recovered. If the variation of $\Delta\Sigma_{xc}$ is smaller than the variation of Σ_{xc} that appears in the original equation, Eq. (11.14), setting Γ_{nl} to unity is a better approximation than the GWA.

F.2 Hedin's equations in a basis

The Green's function $G(1,2)$ is a function of two space, spin and time arguments. Often it is convenient to transform the equations into a complete and orthonormal basis of spin orbitals. Let us first denote these orbitals $\psi_\ell(x_1)$ by a unique subscript ℓ. For the matrix elements, we use the definitions of Eq. (12.40) for two-point functions, Eq. (12.42) for four-point functions, and Eq. (12.43) to make the notation compact.

In the following we omit the time arguments, since they remain unchanged with respect to the above. Moreover, we simplify the notation: Γ stand for Γ_{nl}, P stands for P_{nl}, and W stands for W_{nl}; it is important to keep in mind that these are not the original quantities. The effective interaction Ξ in Eq. (F.16) is the functional derivative of Σ, defined in Eq. (14.7).

With this, the set of Eqs. (F.8)–(F.12) reads (the expressions are summed over repeated indices, independently of subscript or superscript):

$$\Sigma_{ij} = v_{H,ij} + i\, G_{k\ell}\, W_{in}^{km}\, \Gamma_{\ell n}^{mj} \tag{F.13}$$

$$W_{ij}^{k\ell} = v_{c,ij}^{k\ell} + i\xi_{\ell j}^{sm}\, P_{sm}^{tr}\, W_{ir}^{kt} \tag{F.14}$$

$$P_{ik}^{\ell j} = -iG_{in}G_{mk}\, \Gamma_{nj}^{\ell m} \tag{F.15}$$

$$\Gamma_{ij}^{\ell k} = \delta_{i\ell}\delta_{kj} + \Delta\Xi_{ik}^{mn}G_{ms}G_{tn}\, \Gamma_{sj}^{\ell t} \tag{F.16}$$

$$G_{ij} = G_{ij}^0 + G_{ik}^0\, \Sigma_{k\ell}\, G_{\ell j} \tag{F.17}$$

Note that here $\Xi = \frac{\delta\Sigma}{\delta G}$ without prefactors, contrary to the definition of the matrix elements in Ch. 14. Moreover, we define $\Gamma_{n_1 n_3}^{n_4 n_2}$ using $\chi_{n_2}\Gamma\chi_{n_3}^*$ instead of $\chi_{n_2}^*\Gamma\chi_{n_3}$ in Eq. (12.42).

Finally, one can simplify the spin structure of the equations for the case of a spin-independent hamiltonian and collinear spins, and for an exchange–correlation contribution with spin structure $\xi_{ls}^{kt} = \delta_{ls}\delta_{kt}\xi$, where we interpret $i, j, \ldots$ to be just the spin index. This holds for example in the original Hedin's equations, where $\xi = -iv_c$. For such a ξ one can for example find a spin-independent W in Eq. (F.14) by summing over the spins of P.

In order to simplify further, one must take into account the fact that $\Xi_{ij}^{k\ell}$ has only contributions $\Xi_{ij}^{ij}\delta_{ik}\delta_{j\ell}$ and $\Xi_{ii}^{jj}\delta_{ij}\delta_{k\ell}$, as discussed in Sec. 14.4.

SELECT FURTHER READING

Bechstedt, F. *Many-Body Approach to Electronic Excitations; Concepts and Applications*, Springer Series in Solid-State Sciences No. 181 (Springer-Verlag, Berlin, 2015). A recent book on the theory and computation of electronic excitations that contains a careful investigation of the spin structure.

Strinati, G. "Application of the Green's function method to the study of the optical properties of semiconductors," *Rivista del Nuovo Cimento* **11**, 1, 1988. Contains many details concerning the time and spin structure of the main quantities.

Appendix G

Unique solutions in Green's function theory

Summary

The definition of the Green's function as an expectation value in the many-body state with an added particle is unique. However, the Green's function is almost never calculated by evaluating this expectation value. Instead, expressions are derived from differential equations or non-linear integral equations, which have more than one solution. This leads to possible ambiguities that have to be treated with care. First, the equation of motion is a differential equation in time, requiring initial conditions. Second, the Dyson equation with a self-energy that is functional of the dressed Green's function is non-linear, and one has to distinguish the physical from non-physical solutions. This appendix elaborates on these two points. Moreover, it adds some thoughts concerning the convergence of perturbation theory.

G.1 Which G^0? Boundary conditions in time

To understand the main subtlety in getting a well-defined Green's function, it is enough to examine the case of non-interacting particles. The equation of motion Eq. (5.85) for G^0 is a first-order differential equation in time. As stated before Eq. (5.85) and consistent with the discussion in Sec. C.1, the retarded, advanced, and time-ordered G^{0R}, G^{0A}, and G^{0T} obey the *same* equation of motion. This means that the solution is only well-defined when initial conditions in time are specified, for example, causality through $\Theta(t - t')$ in the case of the retarded Green's function.

An additional point with respect to Sec. C.1 is the fact that the Green's function is a function or matrix of two space arguments or orbitals, and two spin arguments. It is built from the spectrum of a hamiltonian. However, the definition of the Green's function Eq. (5.78) shows that the hamiltonian alone does not uniquely define the Green's function, since the Green's function also contains information on how states are occupied, e.g., the ground state at vanishing temperature, or with the thermal equilibrium distribution.

This information does not appear in the equation of motion Eq. (5.85) nor in the corresponding differential equation for the interacting Green's function. In particular, we have the same equation of motion for all temperatures. In the case of an independent-particle Green's function, one often does not realize the problem since one usually calculates all eigenstates of the hamiltonian $\hat{h}$ and explicitly populates the states following Eq. (5.82). In the case of interacting electrons one does not want to do this, since the aim of the Green's functions is to avoid calculating eigenstates of the full many-body hamiltonian $\hat{H}$.

To avoid ambiguities, in terms of mathematics one has to get rid of the solutions of the homogeneous equation, since as discussed in Sec. C.1, they can be added to one particular solution. This is achieved by moving into the complex plane, since $z - \varepsilon_{\mathbf{k}}^0 = 0$ has no solution off the real axis. Once the specific solution, for example $G_{\mathbf{k}}^0(z) = 1/(z - \varepsilon_{\mathbf{k}}^0)$, is determined (similarly for the interacting Green's function), one can go back to the real axis. At $T = 0$, where the analytic continuation of the time-ordered Green's function exists, this leads directly to the desired result. For non-vanishing temperatures one can work with the causal Green's functions, or use the imaginary-time Green's functions of App. D. In imaginary time, the equilibrium distribution is encoded in the initial conditions of Eqs. (D.8) and (D.9). One can see this directly, since for example a change in temperature changes the periodicity of the imaginary-time Green's functions. These initial conditions, together with the equation of motion, uniquely define the Green's function on the imaginary-time or frequency axis. Analytic continuation to the real frequency axis yields the desired result. The last step can be difficult; this is discussed in Sec. 25.6.

When G is obtained from a Dyson equation, the initial condition can be encoded in G^0. The critical step always occurs when one inverts $[G^0]^{-1}$ or G^{-1} to obtain G^0 or G. Therefore, for example, $G = G^0[1 - G^0\Sigma]$ is well-defined if G^0 is well-defined, but $G = [[G^0]^{-1} - \Sigma]^{-1}$ requires the use of initial conditions.

G.2 Which G? Self-consistent Dyson equations

The formulation of Luttinger and Ward, which leads to the self-energy as a functional of the self-consistent interacting Green's function G, is at the basis of much of Part III of this book. However, this approach makes the Dyson equation non-linear, and can therefore lead to multiple solutions. With respect to the previous section, this is an extra problem. Take the simplest example, the Hartree–Fock approximation. Schematically, the Dyson equation reads $G = G^0 - i/2\, G^0 v_c GG$, where the factor $1/2$ expresses the sum of Hartree and Fock terms. Even in the simplest, scalar, case, this is a quadratic equation with two solutions.

In practice, Dyson equations are solved iteratively, by starting from some guess, for example G^0. There are several ways to iterate such an equation, for example, one can begin on the right-hand side and evaluate the left-hand side, or vice versa. The iteration procedure determines the result that is obtained; this is illustrated with a simple example in Ex. G.2. Usually one determines the one-body Green's function by iterating as

$$G^{(n+1)} = G^0 + G^0 \Sigma [G^{(n)}] G^{(n+1)}. \tag{G.1}$$

This procedure seems to work for the most commonly used approximate functionals such as Hartree–Fock and GW, but one should be careful when using other Dyson equations.

The potential introduction of unphysical results should be compared with the unique definition of the Green's function, Eq. (5.78). For a given external potential, there is only one ground-state wavefunction, and therefore only one Green's function. The other way round, the diagonal of a Green's function yields by definition the ground-state density, and therefore uniquely determines the external potential, following the Hohenberg–Kohn theorem. Hence, there is a one-to-one correspondence between the external potential and the Green's function. These arguments suppose that one considers Green's functions that can be expressed as expectation values of a many-body ground-state wavefunction, or similarly for thermal equilibrium. This restricts the domain of definition with respect to all possible functions of two space, spin and time arguments. This restriction is not automatically built into the equations that are used to determine the Green's functions in practice, and that are presented throughout this book. This is closely linked to questions such as the invertibility of the Legendre transform, Eq. (8.10), and it is a good example for the importance of properly defining the domain of a functional.

G.3 Convergence of perturbation expansions and consequences

A perturbative approach is designed for cases where the expansion parameter is small. For larger values, perturbation series converge slowly, or they can be divergent. Let us look at a very simple example, the Dyson equation, Eq. (9.39), and its expansion, Eq. (7.9), with all matrices and functions replaced by simple numbers. The solution of Eq. (9.39) is $G = G^0/(1 - G^0 \Sigma)$, but the expansion converges to this solution only for $G^0 \Sigma < 1$.

The lowest-order approximation to $\Sigma[G^0]$ is $\Sigma_x = i v_c G^0$. The quality of the expansion is therefore determined by the quantity $v_c G^0 G^0$. The product of two independent-particle Green's functions $G^0 G^0$ is closely linked to the independent-particle polarizability P_0 given in Eq. (12.7). Since P^0 is inversely propertional to the bandgap, this implies that the expansion parameter is essentially the ratio of the Coulomb interaction and the bandgap (or U/t in the Hubbard model), which is both intuitive and confirmed by many results: the stronger the interaction, and the smaller the energy to excite the system, the worse we expect perturbation theory in the Coulomb interaction to perform.

There can also be asymptotic convergence (see Ex. G.3). This means that for a given perturbation strength the result of the perturbation series improves up to a certain order, after which it diverges. The larger the perturbation, the lower the order for which this happens. An example for a problem with a structure similar to many-body perturbation theory in terms of the screened interaction is given in Ex. G.3.

When an infinite series is not absolutely convergent, it is dangerous to reorder the terms. A famous example is the logarithm. Its expansion reads

$$\ln(1+x) = -\sum_{n=1}^{\infty} \frac{1}{n}(-x)^n \qquad \text{for} \qquad |x| \le 1. \tag{G.2}$$

In particular,

$$\ln(2) = -\sum_{n=1}^{\infty} \frac{1}{n}(-1)^n = \sum_{n=1}^{\infty} \left[\frac{1}{2n-1} - \frac{1}{2n} \right]. \tag{G.3}$$

This series can be reordered in terms of odd n and even n, multiples of 4 or not:

$$\cdots = \sum_{n=1}^{\infty} \frac{1}{(2n-1)} - \frac{1}{2(2n-1)} - \frac{1}{4n}. \tag{G.4}$$

However, by summing always the first two terms one obtains

$$\cdots = \sum_{n} \frac{1}{2(2n-1)} - \frac{1}{2(2n)} = \frac{1}{2}\ln(2). \tag{G.5}$$

The result has changed by a factor of 2.

This is a warning: the skeleton series for the self-energy $\Sigma[G]$ as a functional of the dressed G is obtained from the original self-energy functional in terms of G^0 by grouping terms. The example shows that this is safe only when the original series is absolutely convergent, which is not guaranteed.

SELECT FURTHER READING

Economu, E. N. *Green's Functions in Quantum Mechanics* (Springer-Verlag, Berlin, 1979). See in particular Chs. 11 and 12 for the question of initial conditions in time.

van Leeuwen, R. and Dahlen, N. E. *Lecture Notes on Non-equilibrium Green's Functions* (https://www.jyu.fi/fysiikka/en/research/material/quantum/NGF.pdf). See in particular the discussion of Hartree–Fock in Ch. 6.

Exercises

G.1 Take the case of vanishing interaction and write the spectral representation of the corresponding independent-particle Green's function, with arbitrary occupation numbers. Check that for any choice of occupation numbers (for example, at $T = 0$ for a Green's function in the ground state, but also in any *excited* state), Eq. (10.4) is fulfilled. Using the $T \neq 0$ imaginary-time Green's function, check that only the equilibrium distribution (or at $T = 0$, ground-state occupation numbers) are compatible with the initial conditions of Eqs. (D.4) and (D.5).

G.2 Consider the quadratic equation $y = y^0 - 1/2y^0y^2$. This corresponds to the scalar version of a Hartree–Fock Dyson equation. Show that the equation has two solutions. Try different ways to iterate this equation. Show that the result does not depend on the starting point, but on the way the iteration is performed. Put your findings in relation to continued fractions. More details can be found in [482, 483].

G.3 Study the scalar differential equation $y(x) = y^0 + y^0 xy - vy^0 \frac{dy}{dx}$ that is a scalar version of the linearized functional differential equation for the Green's function, Eq. (15.27). Find the general family of solutions. Based on the continuity principle, choose the particular solution that connects continuously to $y \rightarrow y^0$ for $v \rightarrow 0$. Show that this solution has an asymptotic expansion in the interaction. Draw the result order by order, as a function of the interaction, and show that for small interaction the first few orders converge towards the exact result, whereas for stronger interaction the result diverges quickly. Details can be found in [482].

Appendix H

Properties of functionals

Summary

This appendix provides a general foundation for functionals and functional equations in Ch. 8. Section H.4 is devoted to statistical mechanics of a spin system that is a model for DMFT (see Ch. 16) and Sec. H.5 provides an example of invertibility that carries over to the derivation of density functional theory in Ch. 8.

H.1 Functionals and functional equations

A functional $F[f]$ is a function that takes functions as its argument; that is, it specifies a mapping of a function $f(x)$ to a result $F[f]$ that can be a value or a function. Three aspects have to be distinguished:

- **The domain** $\mathcal{D}$ of $F[f]$ is a set of functions $f(x)$, which must be specified; for example, the domain might be restricted to continuous, normalized functions, or to functions defined for a specified range of arguments x, or it might be defined to be the set of functions that obey a conservation law such as $\int f(x) = \text{constant}$.
- **The functional dependence** is the mathematical expression of $F[f]$, for example, $F[f] = \int dx' f^2(x')$. F can itself also be a function of variables, for example $F(x, [f]) = \int dx' f^2(x')/(x - x')$. If $F(x, [f])$ depends only on x and $f(x')$ at the point $x' = x$, this is called a local functional, for example $F(x, [f]) = f^2(x)/x$.
- **The value** of the functional is obtained for each specific $f(x)$. Even though the functional form is the same, the value is different if a different function $f'(x)$ is input into the functional.

It is essential to carefully distinguish these aspects of a functional in order to avoid confusion. For example, if one talks about a "universal functional" this means that the functional dependence is universal; it does not mean that the functional takes the same value for every system, because the value depends on the function $f(x)$ that is system-specific.

Familiar functionals include the energy $E = \langle \Psi | \hat{H} | \Psi \rangle$ and the partition function $Z = \operatorname{Tr} e^{-\beta(\hat{H} - \mu \hat{N})}$, defined in terms of the many-body wavefunction $\Psi(\mathbf{r}_1, \mathbf{r}_2, \ldots)$. The calculus of variations is especially relevant for our purposes since the equations describing the systems can often be cast as stationarity conditions on a functional for variations in the specified domain. A review of functionals can be found in [1, App. A] and in [175, App. A], and a more complete exposition of functional derivatives can be found in [1159, 1160].

To illustrate the properties of functionals, we consider a simple example in which F is a weighted integral of an exponential similar to a partition function:

$$I[f] = \int_{x_{min}}^{x_{max}} w(x) \exp(\alpha f(x)) dx. \tag{H.1}$$

For a functional $F[f]$, the functional derivative is defined by a variation of the functional

$$\delta F[f] = F[f + \delta f] - F[f] = \int_{x_{min}}^{x_{max}} \frac{\delta F}{\delta f(x)} \delta f(x) dx, \tag{H.2}$$

where the quantity $\delta F/\delta f(x)$ is the functional derivative of F with respect to variation of $f(x)$ at the point x. In the example of Eq. (H.1), the derivative is given by

$$\frac{\delta I}{\delta f(x)} = w(x)\alpha \exp(\alpha f(x)), \tag{H.3}$$

following the same rules as normal differentiation. In general, the functional derivative at point x depends also on the function $f(x')$ at all other points x'. The expressions are readily extended to many variables and functions $F[f_1, f_2, \ldots]$.

It can be useful to insert derivatives with respect to other functionals. If I is functional of g, and g is functional of f, the chain rule reads

$$\frac{\delta I}{\delta f(x)} = \int dx' \frac{\delta I}{\delta g(x')} \frac{\delta g(x')}{\delta f(x)}.$$

H.2 Legendre transformations and invertibility

The role of a Legendre transformation is to transfer the dependence of a function from one independent variable to another. Here we illustrate the transformation of a function; the formulas are readily generalized to functionals simply by replacing the variables by functions and the derivatives by functional derivatives. If $f(x)$ is a differentiable function with

$$df = u dx \quad \text{where} \quad u = \frac{df}{dx}, \tag{H.4}$$

the Legendre transformation to a new representation $g(u)$ in terms of the slope u is given by

$$g(u) = f(x) - ux \quad \text{where} \quad x = -\frac{dg}{du}. \tag{H.5}$$

In order for the expression of Eq. (H.5) to define $g(u)$ uniquely for all u, the variable x must be given as a function of u. Thus the relation $u = df/dx$ must be inverted to define $x(u)$, which requires that there be a one-to-one correspondence of u and x. Since u is the slope of $f(x)$, it immediately follows that there is a one-to-one correspondence if and only if $f(x)$ is either convex ($d^2f/dx^2 > 0$) for all x or concave ($d^2f/dx^2 < 0$) for all x. (See Sec. H.4 for conditions for convex and concave functionals.)

The condition for invertibility and the properties of $f(x)$ and $g(u)$ can be understood conveniently in terms of the more general function

$$h(x, u) = f(x) - ux - g(u), \tag{H.6}$$

in which x and u are considered to be independent variables. The actual solution with $u = df/dx$ and $x = -dg/du$ is determined by the condition that h be extremal w.r.t. to both x and u,

$$\frac{dh(x, u)}{dx} = \frac{df}{dx} - u = 0 \quad \text{and} \quad \frac{dh(x, u)}{du} = -\frac{dg}{du} - x = 0. \tag{H.7}$$

Since the derivatives of h vanish for the actual solution, h is constant so that the functions obey the relation $g(u) = f(x) - ux$ specified in Eq. (H.5), apart from a constant that is undetermined. Note that since ux is linear in u and x, it follows that $\partial^2 h/\partial x^2 = d^2 f/dx^2$ and $\partial^2 h/\partial u^2 = -d^2 g/du^2$. Since Eq. (H.6) must be satisfied for all x and u, it follows that the stationary point of $h(x, u)$ is a saddle point, i.e., if f is concave then g must be convex, and vice versa. Thus the extremal conditions are a maximum for one function (f or g) and a minimum for the other.

The extremal condition on the function in Eq. (H.7) is a powerful approach for the formulation of physical problems, as illustrated by examples in the following section. However, it is useful to point out that the function defined in Eq. (H.6) is not unique. One can add any function $F(x - u)$, which has the property $F = 0$ and $dF/dx = 0$ at the extremal point. Addition of F does not change the solution, but it can change the curvature at the solution. This can be used to advantage in actual calculations, because it can be used to speed the convergence of an iterative calculation to the solution. Examples of different functionals are given in the following sections and in Secs. 4.3, 8.4, and 8.2.

These properties of the Legendre transform are readily extended to a function of many variables $f(x_1, x_2, x_3, \ldots)$, where the transform can be carried out for any of the variables, and to functionals using the rules of functional derivatives.

H.3 Examples of functionals for the total energy in Kohn–Sham DFT calculations

An expression for the total energy in a Kohn–Sham calculation is given in the second line of Eq. (4.23), written in the same form as Eq. (H.6). This can be used as an example of functionals that have the same minimum energy solution, but they may have different variations away from the minimum. Here we summarize the properties of useful functionals for the energy that illustrate the consequences of various choices:[1]

- The original Kohn–Sham functionals $E_{KS}[n]$ and $E_{KS}[v_{eff}]$ are different functionals[2] that are both variational. The practical functional is $E_{KS}[v_{eff}]$ since the eigenvalues ε_i and the density n are determined by solving the Kohn–Sham equations (Eq. (4.19)) for any given

---

[1] For a more in-depth discussion of the functionals, see [1, Ch. 9].

[2] To avoid complicated notation, the same symbol $E_{KS}[\ldots]$ is used for the two different functionals that are distinguished by the argument, in this case n or v_{eff}.

$v_{eff}(\mathbf{r})$. In principle, the energy can also be written as a functional of density $E_{KS}[n]$; the functional form is not known, however, the proof that such a functional exists is an example of the general derivation in Sec. H.5.

- The Harris–Weinert–Foulkes functional $E_{HWF}[n]$ is defined by interpreting Eq. (4.23) to mean a functional of the density, using Eq. (4.21) to define a potential $v_{eff}[n]$ for any density n even when it is not the equilibrium density. This is a practical functional of the density with the properties derived in [1, Secs. 9.2 and 9.3]. The stationary solution $\delta E_{HWF}/\delta n = 0$ is the same as for the Kohn–Sham functionals; however, it is a saddle point instead of an extremum (Ex. 4.5). The fact that it is not variational makes it difficult to use to iterate the Kohn–Sham equations to consistency. Nevertheless, it can be useful if one has an approximate density, because $E_{HWF}[n]$ is often closer than $E_{KS}[v_{eff}]$ to the final solution.

- The generalized v–n-functional $E[v_{eff}, n]$ is defined by Eq. (4.23) with v_{eff} and n considered as independent variables, which has the classic form of a generalized Legendre function in Eq. (H.6). The solution can be cast in terms of stationarity with respect to each variable v_{eff} and n separately. It is a saddle point that is a maximum w.r.t. v_{eff} and a minimum w.r.t. to n (see Ex. 4.5 and [1, Sec. 9.2]). Since the errors are quadratic in both v_{eff} and n, it is advantageous for evaluation of the energy with approximate forms for both v_{eff} and n.

The functionals of the Green's function G and the self-energy Σ in Sec. 8.2 are analogous; however, the stationary solution is in general a saddle point and not an extremum. The difference can be appreciated from the analysis in the following sections, which depends on inequalities derived for the ground state or thermal equilibrium. The arguments do not apply for dynamic functions.

H.4 Free-energy functionals for spin systems and proof of invertibility

Functionals for statistical mechanics can be illustrated by a system of spins, for example, a classical Heisenberg model with energy $E = -\sum_{i<j} J_{ij}\mathbf{S}_i \cdot \mathbf{S}_j - \sum_i \mathbf{h}_i \cdot \mathbf{S}_i$, which is the classical version of Eq. (3.4) with an added field $\mathbf{h}_i$ acting on the spin at site i.[3] The average moment at each site is given by the thermal average

$$\mathbf{m}_i =< \mathbf{S}_j >= \frac{1}{Z} \text{Tr } e^{-\beta E[\mathbf{h}]}\mathbf{S}_i, \tag{H.8}$$

where $\beta = 1/k_B T$, Tr is the sum over all spin configurations, $\mathbf{h}$ denotes the set of fields $\mathbf{h}_j$, and $Z = \text{Tr } e^{-\beta E[\mathbf{h}]}$ is the partition function. In this expression the moment $\mathbf{m}_i$ at site i is determined by the external fields $\mathbf{h}_j$ at all sites j, which can be denoted as a functional $\mathbf{m}[\mathbf{h}]$. The Gibbs free energy is a functional of the external fields

$$G[\mathbf{h}] = -\beta^{-1} \ln \text{Tr } e^{-\beta E[\mathbf{h}]} \text{ and } \mathbf{m}_i = -\frac{\partial G[\mathbf{h}]}{\partial \mathbf{h}_i}, \tag{H.9}$$

---

[3] The present discussion follows [1161], Sec. 7. The generalization to quantum spins is in the following section.

where the second relation is a convenient way to express Eq. (H.8). The Helmholtz free energy F is the Legendre transform of G that expresses the free energy as a functional of the average internal variables $\mathbf{m}$,

$$F[\mathbf{m}] = G[\mathbf{h}] + \mathbf{m} \cdot \mathbf{h} \quad \text{and} \quad h_i = \frac{\partial F[\mathbf{m}]}{\partial m_i}. \tag{H.10}$$

The second expression defines the external fields $\mathbf{h}$ that correspond to any given set of average moments $\mathbf{m}$.

The key step required for the Legendre transform to be well-defined is invertibility, i.e., that $m_i = \partial G/\partial h_i$ can be inverted to uniquely determine the externally applied fields h_j for a given set of average moments m_i, which can be denoted as a functional $\mathbf{h}[\mathbf{m}]$. Following the previous sections, the functionals are invertible if $G[\mathbf{h}]$ is concave as a functional of $\mathbf{h}$, from which it follows that $F[\mathbf{m}]$ is a convex functional of the average moments $\mathbf{m}$. This is not merely a mathematical relation: if the functionals are invertible, we arrive at a fundamental principle of statistical physics, that the equilibrium state of the system is determined by minimizing the Helmholz free energy F as a functional of the internal variables m_i for a fixed set of externally applied fields h_j.

Proof that $G[\mathbf{h}]$ is a concave functional

A general definition of "concavity" ("convexity" is given by an analogous definition) can be formulated in terms of any two external fields $\mathbf{h}_1$ and $\mathbf{h}_2$, and an intermediate field $\alpha\mathbf{h}_1 + (1-\alpha)\mathbf{h}_2$, $0 < \alpha < 1$. The functional is concave if its value at any intermediate field is always greater than or equal to the straight-line interpolation of G between the end points,

$$G[\alpha\mathbf{h}_1 + (1-\alpha)\mathbf{h}_2] \geq \alpha G[\mathbf{h}_1] + (1-\alpha)G[\mathbf{h}_2]. \tag{H.11}$$

The concavity of $G[\mathbf{h}]$ can be demonstrated using the Hölder inequality[4]

$$\sum_k a_k^\alpha b_k^{(1-\alpha)} \leq \left(\sum_k a_k\right)^\alpha \left(\sum_k b_k\right)^{(1-\alpha)}, \tag{H.12}$$

which holds for any real numbers $0 < a_k < 1$ and $0 < b_k < 1$. For classical spins, the trace over configurations in Eq. (H.9) is a sum over probabilities, which are positive and less than unity when properly normalized. If the energies for external fields h_1 and h_2 are respectively $H_1[\mathbf{S}, \mathbf{h}]$ and $H_2[\mathbf{S}, \mathbf{h}]$, then Eq. (H.12) leads to the inequality

$$\mathrm{Tr}\, e^{-\beta(\alpha H_1 + (1-\alpha)H_2)} = \mathrm{Tr}\left[e^{-\beta\alpha H_1} e^{-\beta(1-\alpha)H_2}\right] \leq \left(\mathrm{Tr}\, e^{-\beta H_1}\right)^\alpha \left(\mathrm{Tr}\, e^{-\beta H_2}\right)^{(1-\alpha)}. \tag{H.13}$$

---

[4] The Hölder inequality [1162] is a generalization of the triangle inequality, which states that the length of any side is less than the sum and greater than the difference of the other sides. The general form is readily checked using examples.

The desired inequality in Eq. (H.11) follows immediately from this relation and the definition of G in Eq. (H.9). Equality holds only if the terms a_k and b_k are 1 or 0. For the classical spin system, this occurs at zero temperature where the spins are fully polarized.[5]

H.5 Extension to quantum spins and density functional theory

It is straightforward to extend the above analysis to quantum spins where $S \to \hat{S}$ and $H \to \hat{H}$ must be treated as operators. The key step is to use the fact that $\hat{H}_1$ and $\hat{H}_2$ are hermitian, in which case we can find a set of eigenstates $\{\Psi_i\}$ of $\alpha \hat{H}_1 + (1 - \alpha)\hat{H}_2$. Then the partition function is given by

$$Z = \text{Tr}\left(e^{-\beta(\alpha \hat{H}_1 + (1-\alpha)\hat{H}_2)}\right) = \sum_i \langle \Psi_i | e^{-\beta(\alpha \hat{H}_1 + (1-\alpha)\hat{H}_2)} | \Psi_i \rangle$$

$$= \sum_i e^{-\beta(\langle \Psi_i | [\alpha \hat{H}_1 + (1-\alpha)\hat{H}_2] | \Psi_i \rangle)} = \sum_i e^{-\beta(\alpha \langle \Psi_i | \hat{H}_1 | \Psi_i \rangle + (1-\alpha)\langle \Psi_i | \hat{H}_2 | \Psi_i \rangle)}. \quad \text{(H.14)}$$

Since the matrix elements $\langle \Psi_i | \hat{H}_1 | \Psi_i \rangle$ and $\langle \Psi_i | \hat{H}_2 | \Psi_i \rangle$ are scalars, we can use the Holder inequality to derive the equation analogous to Eq. (H.13) for the quantum case,

$$\text{Tr}\left(e^{-\beta(\alpha \hat{H}_1 + (1-\alpha)\hat{H}_2)}\right) \leq \left(\sum_i e^{-\beta \langle \Psi_i | \hat{H}_1 | \Psi_i \rangle}\right)^{\alpha} \left(\sum_i e^{-\beta \langle \Psi_i | \hat{H}_2 | \Psi_i \rangle}\right)^{(1-\alpha)}, \quad \text{(H.15)}$$

where equality holds only if $\hat{H}_1$ and $\hat{H}_2$ differ by a constant. In addition, the fact that e^x is a convex function leads to the relation

$$\sum_i e^{\langle \Psi_i | \hat{A} | \Psi_i \rangle} \leq \sum_i \langle \Psi_i | e^{\hat{A}} | \Psi_i \rangle = \text{Tr}\left(e^{\hat{A}}\right), \quad \text{(H.16)}$$

where $\hat{A}$ can represent either $-\beta H_1$ or $-\beta H_2$. Combining this with Eq. (H.15) leads to the penultimate relation

$$\text{Tr}\left(e^{-\beta(\alpha \hat{H}_1 + (1-\alpha)\hat{H}_2)}\right) \leq \left(\text{Tr}\, e^{-\beta \hat{H}_1}\right)^{\alpha} \left(\text{Tr}\, e^{-\beta \hat{H}_2}\right)^{(1-\alpha)}, \quad \text{(H.17)}$$

where equality holds only if $\hat{H}_1 = \hat{H}_2 + constant$, in which case the $\{\Psi_i\}$ are eigenstates of both $\hat{H}_1$ and $\hat{H}_2$. Taking the logarithm of the relation in Eq. (H.17), it is straightforward to verify that G satisfies Eq. (H.11) and therefore is concave. This is sufficient to conclude that the mapping $\mathbf{h} \to \mathbf{m}$ is one-to-one and the relation is invertible in the full quantum theory.

The proof carries over to the relation of the density n and external potential v needed in density functional theory in Sec. 4.3 (see also the discussion after Eq. (8.41) and Ex. 8.3) to demonstrate invertibility in density functional theory. The necessary step is to show that

---

[5] Thus, the classical system is not invertible at $T = 0$, which is easily understood since the system is fully polarized for any non-zero value of the field. Note that the quantum spin system removes this pathology, since the moment is never completely saturated due to quantum fluctuations.

$\Omega[v]$ is a *concave* functional of the external potential v, which means that for any two potentials $v_1 \neq v_2$, the value of Ω at an intermediate point $\alpha v_1 + (1 - \alpha)v_2, 0 < \alpha < 1$, is always greater than the straight-line interpolation for Ω between the end points,

$$\Omega[\alpha v_1 + (1 - \alpha)v_2] > \alpha\Omega[v_1] + (1 - \alpha)\Omega[v_2]. \qquad (H.18)$$

The demonstration in Ex. H.2 follows the steps in Eqs. (H.14)–(H.17).

In the case of DFT, this means that the functional $\Omega[v]$ is *maximum* at the solution $\delta v = 0$, where $\Omega[0] = \Omega$, the grand potential. It follows that $\Gamma[n] = \Omega_{HKM}[n]$ is a *convex functional* since it is the Legendre transform of a concave function. Following Sec. H.2 this is sufficient to show that the expressions are invertible, so there is a one-to-one relation between v and n. Thus, $\Omega_{HKM}[n]$ is the desired functional for which the global minimum is the grand potential Ω (see Ex. 8.3). This derivation using the eigenstates of the hamiltonians does not carry over to functionals of the Green's function and/or Σ, which are in general saddle points at the stationary solution.

SELECT FURTHER READING

Derivation of density funtional theory from the action can be found in the paper by Fukuda *et al.* [1161].

Exercises

H.1 The general expression for the magnetization m as a function of applied field h^{appl} and temperature is given in Eq. (4.13). Look up the definition of the Brillouin function and give the explicit expression for spin $1/2$. Show that there is a monotonic variation of m as a function of h^{appl} so that the expression can be inverted to find h^{appl} as a function of m. Discuss the sense in which this is related to the fact that entropy increases monotonically with temperature.

H.2 This exercise is to carry out the proof of invertibility for density functional theory.
  (a) Derive Eq. (H.18) following the steps in Eqs. (H.14)–(H.17) and show that this guarantees the functional to be concave.
  (b) Discuss why this is the appropriate condition for DFT.
  (c) Show that the resulting Γ is convex and its minimum is Ω, as required for the Hohenberg–Kohn–Mermin functional.

H.3 See also Ex. 8.3.

Appendix I

Auxiliary systems and constrained search

Summary

Here we consider auxiliary systems that are generalizations of the Kohn–Sham construction. In Green's function methods the use of a self-energy from an auxiliary system can be considered as a search in a restricted domain. Dynamical mean-field theory is an example of an interacting auxiliary system.

There are various ways that functionals can be employed to generate auxiliary systems and useful approximations by limiting the range of the functions input to the functional. Each of the functionals of the density, Green's function, or self-energy are defined over a specified *domain* $\mathcal{D}$, e.g., $\Phi[G]$ and $F[\Sigma]$ are defined for all functions that have the required analytic properties for a Green's function or self-energy (see Sec. H.1). Here we consider limiting the domain to a subset of $\mathcal{D}$ denoted by $\mathcal{D}'$. The first examples below derive expressions that provide insights concerning the link of non-locality and frequency dependence, and that lead to useful approximations in the spirit of optimized effective potentials. This is followed by an approach to find an interacting auxiliary system where the full many-body problem can be solved with no approximations, thus providing a self-energy that is exact within the restricted domain of the auxiliary system.

I.1 Auxiliary system to reproduce selected quantities

Let us first consider an auxiliary system that reproduces exactly some quantity of interest. This is similar to the Kohn–Sham approach, but it is more general and it leads to explicit expressions for the functionals in terms of Green's functions [1163].

Suppose we are interested in a quantity $\mathcal{P}$ that is part of the information carried by the Green's function G, symbolically expressed as $\mathcal{P} = p\{G\}$. An example is the density where the "part" to be taken is the diagonal of the one-particle G: $p\{G\} = n(\mathbf{r}) = -iG(\mathbf{r}, \mathbf{r}, t'-t = 0^+)$ or $-G(\mathbf{r}, \mathbf{r}, \tau = 0^-)$. Consider an auxiliary system that has the same bare Green's function G_0 as the original one, but a different effective potential, or a self-energy Σ_P that leads to an auxiliary Green's function G_P. We can use the Dyson equation (see Sec. 7.7) to express the relation to the full Green's function and self-energy as

$$G = G_P + G_P (\Sigma - \Sigma_P) G. \tag{I.1}$$

We now require that $p\{G\} = p\{G_P\}$. So long as the "part" is linear in the Green's function, i.e., $p\{G_1 + G_2\} = p\{G_1\} + p\{G_2\}$ for any G_1 and G_2, it follows that

$$p\{G_P (\Sigma - \Sigma_P) G\} = 0, \tag{I.2}$$

where $G_P = G_0 + G_0 \Sigma_P G_P$. This procedure is meaningful if Eq. (I.2) can be solved for a Σ_P with a simpler form than the full Σ, corresponding to a restricted domain $\mathcal{D}'$ for the self-energy. There is of course a minimum complexity, otherwise one runs into a contradiction when solving $p\{G_P \Sigma_P G\} = p\{G_P \Sigma G\}$ for Σ_P. For example, if one requires G_P to have the full frequency dependence of G, but to be restricted in space, Σ_P cannot be static. Interestingly, this is also true when the original Σ itself is static but non-local. One can verify this statement by taking the example of the Fock self-energy with the requirement that G_P should yield the local spectral function. We can conclude that by restricting the spatial degrees of freedom, non-locality is transformed into frequency dependence, which is an important point for DMFT and consistent with the general ideas in Sec. 7.1.

If $p\{G\}$ is the density, then Σ_P can be local and static. This is a way to construct the Kohn–Sham auxiliary system where Σ_P is the exchange–correlation potential $v_{xc}(\mathbf{r})$, and G_P is the Kohn–Sham Green's function G_{KS}. Then Eq. (I.2) reduces to a relation derived by Sham and Schlüter [1164, 1165], who used the equation to address fundamental issues in density functional theory. Equation (I.2) has been extended to time-dependent external potentials [1166], and employed in many different contexts (see, e.g., [422, 424, 1167]), including the Dyson equation for the two-particle correlation function (see Sec. 14.11).

This construction yields in principle the exact G_P and Σ_P; however, this is not useful in practice since it would require knowledge of the full G and $\Sigma[G]$. Instead, Eq. (I.2) is often linearized, setting $G = G_P$ everywhere, including the construction of Σ:

$$p\{G_P (\Sigma[G_P] - \Sigma_P) G_P\} = 0. \tag{I.3}$$

Here G_P is the Green's function corresponding to Σ_P, i.e., $G_P^{-1} = G_0^{-1} - \Sigma_P$, whereas $\Sigma[G_P]$ is the full interacting self-energy functional evaluated with G_P. Since this is an approximation to Eq. (I.1), it need not reproduce the desired quantity $\mathcal{P}$ exactly. As an example, one obtains a static, local potential $v_{xc}(\mathbf{r})$ by writing Eq. (I.3) for the density,

$$\{G_{OEP}(\Sigma[G_{OEP}] - v_{xc})G_{OEP}\}_{\mathbf{rr}'tt^+} = 0. \tag{I.4}$$

Solving this equation for v_{xc} yields the so-called optimized effective potential [1168, 1169], also mentioned in Sec. 4.5. For a practical method the functional form of $\Sigma[G]$ must be approximated, for example, the "exact exchange" method where $\Sigma[G]$ is replaced by the Hartree–Fock exchange $\Sigma_x[G]$.

I.2 Constrained search with an interacting auxiliary system

Suppose one can find an auxiliary system that has the same interactions as the physical system but is sufficiently simple that it can be solved exactly. The topic of this section is a way to use the resulting self-energy Σ in the functional $\Omega[\Sigma]$ for the actual system. In this approach, there is no approximation in the calculation of Σ in the auxiliary system, and there is no approximation in the functional form of $\Omega[\Sigma]$.[1] Instead, the domain of self-energies is constrained to those functions Σ' that can be computed in the auxiliary system. Here we follow the approach in [772] and [296] (see also [302, p. 24]), and the method is used in Sec. 16.6, where it provides a variational derivation of the equations for DMFT.

The derivation is remarkably simple if one is careful to follow the admonition: "To avoid confusion, it is necessary to be quite explicit about what is assumed and what is to be proved." The final equations are self-consistent requirements on the auxiliary system. The present variational analysis provides insight and suggestions for use in practice.

The constrained search method depends only on the form of the functionals for Ω. The analysis applies to any of the functionals defined in Sec. 8.2; we will use $\Omega_F[\Sigma]$ in Eq. (8.16), which allows the most direct derivation. The key point is that the functional is a sum of two types of terms:

- The first term on the right-hand side of Eq. (8.16) is a *universal functional*. For a given interaction (Coulomb, Hubbard U, etc.), the functional form of $F[\Sigma]$ is independent of the specific problem, i.e., independent of G_0. It can in principle be evaluated for any Σ in its domain.
- A specific system is determined by the bare Green's function G_0. The only way that $\Omega_F[\Sigma]$ depends *explicitly* on G_0 is in the last terms in Eq. (8.16), which can be calculated for any function Σ in the domain.

If the auxiliary system is defined by the independent-particle Green's function G_0', its grand-potential functional can be written[2]

$$\Omega_F'[\Sigma] = F[\Sigma] - \mathfrak{Tr} \ln(\Sigma - (G_0')^{-1}). \tag{I.5}$$

Solving the auxiliary many-body interacting problem leads to Σ' and a Green's function G' that satisfies the condition of Eq. (8.11),

$$G' = \left. \frac{\delta F}{\delta \Sigma} \right|_{\Sigma = \Sigma'}, \tag{I.6}$$

where Σ' and the Green's function G' are functionals of G_0'. The Ω-functional for the original system is given by Eq. (8.16),

$$\Omega_F[\Sigma] = F[\Sigma] - \mathfrak{Tr} \ln(\Sigma - G_0^{-1}). \tag{I.7}$$

In a constrained search, we search for the extremum of this functional only within the domain of those Σ' that we can obtain from the auxiliary system. If we vary G_0' and use the chain rule, the stationary condition can be written

---

[1] The analysis applies also to approximate functionals $F[\Sigma]$, so long as the functional is the same in the actual and auxiliary systems.

[2] Equation (8.16) is derived from Eq. (8.8); see footnote 7 before Eq. (8.8) and Sec. D.4, which explain that it can be written in the simpler form used here, which is sufficient for calculations of derivatives of Ω. See Ex. 8.5.

$$\frac{\delta \Omega_F[\Sigma']}{\delta G_0'} = \left\{ \frac{\delta F[\Sigma']}{\delta \Sigma'} - (\Sigma' - G_0^{-1})^{-1} \right\} \frac{\delta \Sigma'}{\delta G_0'} = 0, \qquad (I.8)$$

where we have used $d \ln(x)/dx = 1/x$. Hence we arrive at the desired condition

$$\sum_{\omega_n} \sum_{km} \left\{ G'(\omega_n) - (G_0^{-1}(\omega_n) - \Sigma'(\omega_n))^{-1} \right\}_{mk} \frac{\delta \Sigma'_{km}(\omega_n)}{\delta G_0' {}_{\ell\ell'}(\omega_{\ell''})} = 0, \qquad (I.9)$$

where the matrix indices and the sum over Matsubara frequencies ω_n are indicated explicitly. For $T = 0$ the sum becomes an integral (see App. D).

Note that this expression involves only quantities determined from the auxiliary system except for the term involving G_0, the independent-particle Green's function for the original system. Thus the solution can be found by varying G_0' and carrying out calculations only on the auxiliary system. This equation must hold for every ℓ and ℓ', and $\omega_{\ell''}$ and the last term $\delta \Sigma'/\delta G_0'$ project the variation of the term in brackets $\{\ldots\}$ onto the space of G_0'-representable self-energies. At the solution, the result is $G = G'$ within the space of restricted functions. This approach is used in Sec. 16.6 to provide a variational derivation of DMFT within the restricted domain of self-energies that can be calculated for the embedded-site auxiliary system. As explained there, the DMFT condition of Eq. (16.16) is equivalent to Eq. (I.9).

Exercises

I.1 Derive the equation for the variational minimum of $\Omega[\Sigma']$ in Eq. (I.9). Show that the derivative with respect to G_0' vanishes. *Hint:* the derivation is straightforward if one chooses the most useful equations. Be sure to carefully derive the sums over matrix indices and frequencies in order to understand what are the independent variables. See also Ex. 16.14.

Appendix J

Derivation of the Luttinger theorem

Summary

This appendix summarizes the arguments of the papers by Luttinger and Ward to derive the Luttinger theorem that is stated in Sec. 3.6. This is *not* a proof that the theorem applies to all possible states of a crystal; it is the derivation of arguments that it applies to all states that can be analytically continued from some non-interacting system, i.e., a "normal state of matter" as defined in Sec. 3.4. The derivation is an example of the use of the $T \neq 0$ Green's functions in App. D and the conclusions for $T = 0$.

The Luttinger theorem is a cornerstone in the theory of condensed matter. As described qualitatively in Sec. 3.6, it requires that the volume enclosed by the Fermi surface is conserved independent of interactions, i.e., it is the same as for a system of non-interacting particles. Similarly, the Friedel sum rule is the requirement that the sum of phase shifts around an impurity is determined by charge neutrality, which was derived by Friedel [163] for non-interacting electrons. This section is devoted to a short summary of the original work of Luttinger and Ward[1] and the extension of the arguments to the Freidel sum rule [166]. Here we explicitly indicate the chemical potential μ, since the variation from μ is essential to the arguments.

There are two key points: in the interacting system the wavevector in the Brillouin zone $\mathbf{k}$ is conserved so that excitations can be labeled $\mathbf{k}$, and the self-energy $\Sigma_{\mathbf{k}}(\omega)$ is purely real at the Fermi energy $\omega = \mu$ at temperature $T = 0$. The latter point is an essential feature of a Fermi liquid or a "normal metal," which is justified by the argument that the phase space for scattering at $T = 0$ vanishes as $\omega \to \mu$ (see Sec. 7.5). Thus, at the Fermi

---

[1] The steps are worked out in a paper by Luttinger and Ward [36] and the interpretation in physical terms is given in a following paper by Luttinger [37]. A good description can be found in the book by Abrikosov, Gorkov, and Dzyaloshinski [38], and detailed proofs of the transformations and convergence of the integrals are given by LW.

energy the Green's function $G_{\mathbf{k}}^{-1}(\mu) = \mu - \varepsilon_{\mathbf{k}}^0 - \Sigma_{\mathbf{k}}(\mu)$ as a function of $\mathbf{k}$ is the same as for an independent-particle problem with eigenvalues $\varepsilon_{\mathbf{k}}^0 + \Sigma_{\mathbf{k}}(\mu)$. (Of course, for an interacting system at any other energy $\omega \neq \mu$, $G_{\mathbf{k}}^{-1}(\omega)$ cannot be described by independent particles.) In an independent-particle system at $T = 0$ the occupation numbers jump from 1 to 0 as a function of $\mathbf{k}$ at the Fermi surface, and in the interacting system there is still a discontinuity in $n_{\mathbf{k}}$ that defines the surface (Sec. 7.5). Another condition that can be used to define the Fermi surface is the locus of points, where the real part of the Green's function evaluated at the Fermi energy changes sign as a function of $\mathbf{k}$; this is readily established for the independent-particle hamiltonian and carries over to the actual $G_{\mathbf{k}}(\omega = \mu)$, which has the same form as in an independent-particle system.

The theorem can be derived using a variation on the arguments in Sec. 8.5 for the conservation of total particle number $N = d\Omega/d\mu$. The functional Ω is stationary with respect to Σ so long as $\Sigma = d\Omega/dG$ and the Dyson equation is satisfied, as shown in Eq. (8.13). Thus, any variation of Ω with Σ vanishes to first order, and Eq. (8.52) can be re-expressed as (see Ex. J.1)

$$N = \frac{d\Omega}{d\mu} = \mathfrak{Tr}\left\{ \Sigma \frac{dG}{d\mu} - \Sigma \frac{dG}{d\mu} - \frac{1}{d\mu} \ln(-G_0^{-1} + \Sigma) \right\}$$

$$= -\beta^{-1} \sum_i \sum_n \frac{1}{d\mu} \ln(-z_n + \varepsilon_i - \mu + \Sigma_i(z_n)), \qquad (\text{J.1})$$

where the first two terms in the middle expression obviously cancel and the derivative of Σ inside the logarithm can be ignored. In the final expression the trace has been written explicitly as a sum over $z_n = i\omega_n$, where the ω_n are the Matsubara frequencies, and a sum over i in a representation where $\varepsilon^0 + \Sigma$ is diagonal; for a crystal this is the wavevector $\mathbf{k}$, but it is useful to emphasize that these steps in the arguments apply also to other cases with conserved quantities – such as point symmetry around an impurity relevant for the Friedel sum rule.

In order to transform Eq. (J.1) to the desired form, we notice that the derivative $\partial/\partial\mu$ can be replaced by $\partial/\partial z_n$ and the sum can be done as an integral over the contour around the poles at $z = z_n + \mu$ shown on the right side of Fig. D.1, which can be deformed to the contour that encircles the real axis as expressed in the following equations:

$$N = -\beta^{-1} \sum_i \sum_n \frac{1}{dz_n} \ln(-z_n + \varepsilon_i^0 - \mu + \Sigma_i(z_n))$$

$$= -\frac{1}{\beta} \frac{1}{2\pi i} \sum_i \oint_{C_2} \frac{1}{e^{\beta z} + 1} \frac{\partial}{\partial z} \ln(-z + \varepsilon_i^0 - \mu + \Sigma_i(z)) \qquad (\text{J.2})$$

$$= -\frac{1}{\beta} \frac{1}{2\pi i} \sum_i \oint_{C_1} \frac{1}{e^{\beta z} + 1} \frac{\partial}{\partial z} \ln(-z + \varepsilon_i^0 - \mu + \Sigma_i(z)), \qquad (\text{J.3})$$

where the contour C_1 encircles the real axis as shown on the left in Fig. D.1. Integrating by parts, the integrand involves the derivative of the Fermi function, which becomes a delta function for $T \to 0$, so that the ln function is evaluated only at the Fermi energy where Σ is real. Finally, one finds the result

$$N = \frac{1}{2\pi i} \sum_i \left[\ln(\varepsilon_i^0 - \mu + \Sigma_i(\mu) + i\eta) - c.c. \right], \tag{J.4}$$

where $\eta = 0^+$ is a positive infinitesimal and the two terms are the contributions from the contour above and below the real axis (see Ex. J.2). Since the expression in square brackets in Eq. (J.4) is the imaginary part of the logarithm, each term in the sum over i is $2\pi i$ if $\varepsilon_i^0 + \Sigma_i(\mu) < \mu$ and zero if $\varepsilon_i^0 + \Sigma_i(\mu) > \mu$. In a crystal, the states are labeled $\mathbf{k}$ and Eq. (J.4) can be written for each spin as

$$N_\sigma = \sum_{\mathbf{k}} \Theta(\mu - \varepsilon_{\mathbf{k},\sigma}^0 - \Sigma_{\mathbf{k},\sigma}(\mu)), \tag{J.5}$$

which is the volume enclosed by the Fermi surface.

As explained in Sec. 3.6, for an impurity the states are labeled by the point symmetry (angular momentum in an isotropic medium) and the corresponding equation is the Friedel sum rule that relates the sum of phase shifts at the Fermi energy to the change in number of electrons due to the impurity. For Coulomb interactions the total number is fixed by charge neutrality to be the change in number of protons due to the replacement of a host atom by the impurity. The sum rule applies to the interacting system, since the phase shifts are real at the Fermi energy based on the same reasoning as for the self-energy in a crystal [166].

It has been argued [161] that an alternative derivation applies to cases like a "Mott insulator" with no order (also called a "spin liquid," see Sec. 3.2), where there is no Fermi surface. This is based on expressions like the first equation in Eq. (J.1); if we use $-G_0^{-1} + \Sigma = -G^{-1}$, the expression depends only on G and not on an analysis in terms of Σ. In this approach the surface is defined by the places where G changes sign, in particular, the zeros of G within a gap. The possible relevance to metal–insulator transitions is mentioned in Sec. 3.6.

SELECT FURTHER READING

Abrikosov, A. A., Gorkov, L. P., and Dzyaloshinski, I. E., *Methods of Quantum Field Theory in Statistical Physics* (Prentice-Hall, Englewood Cliffs, NJ, 1963). See especially Sec. 19.4.

Dzyaloshinskii, I., "Some consequences of the Luttinger theorem: The Luttinger surfaces in non-Fermi liquids and Mott insulators," *Phys. Rev. B* **68**, 085113, 2003. Presents the arguments for the extension of the theorem to systems with a gap at the Fermi energy.

Martin, R. M., "Fermi-surface sum rule and its consequences for periodic Kondo and mixed-valence systems," *Phys. Rev. Lett.* **48**, 362–365, 1982. Shows the relation of the Luttinger theorem and the Friedel sum rule and describes the arguments for analytic continuity with non-interacting systems.

Exercises

J.1 Justify the statement that the first-order variations with Σ vanish in the derivatives needed in Eq. (J.1). This does not require manipulations, but only careful definition of the functionals, since "it is necessary to be quite explicit about what is assumed and what is to be proved," the quote at the beginning of Ch. 8. Then verify the relations in Eq. (J.1).

J.2 Several steps are needed to arrive at Eq. (J.4), which is the central equation for the Luttinger theorem. This exercise is to carefully define integrals to be sure the contribution from part of the contour for large $|z|$ vanishes, and that the result follows from the form of a logarithm for complex arguments. The original paper by LW [36] may be helpful.

Appendix K

Gutzwiller and Hubbard approaches

Summary

The works of Gutzwiller and Hubbard provide prescient insights into the properties of strongly interacting systems. This appendix provides the derivations
needed for the contrasting pictures of the metal–insulator transition in Sec. 3.2; the
"Hubbard III" alloy approximation summarized in Sec. 16.4, which is a precedent
for DMFT; and the disordered local moment and "DFT+Gutzwiller" approximations
in Sec. 19.6. Also there is emphasis on the care needed in interpretation of the results
that is closely related to the issues in DMFT brought out especially in Secs. 16.3
and 16.9.

As background for theoretical methods in dynamical mean-field theory, it is useful to recall
important earlier developments in enough detail to appreciate the insights and the ingenuity
of Gutzwiller and Hubbard in the early 1960s. The key points are revealed using what is
now called the Hubbard hamiltonian (Eq. (3.2)),

$$\hat{H} = \sum_{i,\sigma} \varepsilon_0 \hat{n}_{i,\sigma} + \frac{1}{2} U \sum_{i,\sigma} \hat{n}_{i,\sigma} \hat{n}_{i,-\sigma} - \sum_{i \neq j,\sigma} t_{ij} c^\dagger_{i\sigma} c_{j\sigma}, \qquad (\text{K.1})$$

where i, j label the sites in the lattice and $\hat{n}_{i,\sigma} = c^\dagger_{i\sigma} c_{i\sigma}$. In the simplest version, the sum
over i, j is only over neighboring lattice sites. This appendix summarizes the derivations of
the formulas needed for the description of the metal–insulator transition in Sec. 3.2 and the
precedents for development of dynamical mean-field theory in Chs. 16–21.

It is useful to first review the consequence of the Hartree–Fock approximation in which
the interaction is replaced by the mean-field average (see Sec. 4.1). For the Hubbard model
the only effect is that the energy of a spin ↑ electron is shifted by $U\langle \hat{n}_{i\downarrow} \rangle$ and similarly for
a ↓ electron. The restricted (RHF) solution has no net spin polarization and the solution
is merely a shift in the average energy with no other physical consequence. For example, a fractionally filled band is always a metal with a Fermi surface that is the same as
for $U = 0$. An insulating state can be found only if there is a broken symmetry in an
unrestricted Hartree–Fock approximation with an integer number of electrons of each spin

per cell, e.g., an antiferromagnetic insulator for the half-filled case. This is the essence of one of the classic theories of magnetism often attributed to Slater [122] and Stoner [56], and it is embodied in the "DFT+U" method (see Sec. 19.6), which is a static mean-field approximation with different potentials for ↑ and ↓ electrons.

K.1 Gutzwiller approach in terms of the wavefunction

In a series of papers, Gutzwiller formulated an approach of remarkable simplicity and clarity [129–131], with the goal of describing the correlations beyond Hartree–Fock. This is brought out most clearly in a state without broken symmetry, where the Hartree–Fock approximation yields nothing of interest. Here we derive the equations that are the basis for one of the paradigms for metal–insulator transitions (see Secs. K.3 and 3.2). Extension to more realistic problems is considered in Sec. 19.6.

The Gutzwiller approach consists of two separate approximations. The first is the Gutzwiller wavefunction [129, 130], and the second is known as the Gutzwiller approximation [131] for the evaluation of the kinetic energy for this wavefunction based on neglect of correlations between sites.

Gutzwiller wavefunction

The wavefunction proposed by Gutzwiller [129] is a correlated variational function that penalizes multiple occupation on a site. For a one-band model, it can be written in several ways:[1]

$$|\Psi_G\rangle = g^{\sum_i^N \hat{n}_{i\uparrow}\hat{n}_{i\downarrow}}|\Psi_{HF}\rangle = \prod_i^N g^{\hat{n}_{i,\uparrow}\hat{n}_{i,\downarrow}}|\Psi_{HF}\rangle = \prod_i^N [1 - (1-g)\hat{n}_{i,\uparrow}\hat{n}_{i,\downarrow}]|\Psi_{HF}\rangle, \quad \text{(K.2)}$$

where N is the number of sites, Ψ_{HF} is a single determinant (like the Hartree–Fock wavefunction), and the correlation factor reduces the probability of double occupation. The first expression in Eq. (K.2) involves the total double occupation $\hat{D} = \sum_i^N \hat{n}_{i\uparrow}\hat{n}_{i\downarrow}$, which is equivalent to the middle expression. For a single band, the effect of the correlation factor can also be expressed in the form in the last equality, as shown in Ex. K.1. The strength of the correlation is governed by the parameter g: for $g = 1$ there is no correlation; for $0 \leq g < 1$ the quantity in square brackets is 1 if either $n_{i,\uparrow} = 0$ or $n_{i,\downarrow} = 0$, or it is equal to g if the site is doubly occupied with $n_{i,\uparrow}n_{i,\downarrow} = 1$. The Gutzwiller wave function is a simplified form of the Slater–Jastrow function in Sec. 6.6, appropriate for a lattice model. By reducing the probability of unlike-spin electrons being on the same lattice site, it reduces the interaction energy. At the same time, this increases the kinetic energy since the probability of hopping is reduced.

---

[1] This is the notation in [150]. Some authors use a different convention with $1 - g \to g$, for example in [91]. Generalization to multiple bands requires a matrix of parameters g_{ij} (see references for multiband applications in Sec. 19.6). Other states can be generated by replacing Ψ_{HF} by a different wavefunction, for example, a BCS pair state for a superconductor.

In a variational method, the best wavefunction of this form is found by minimizing the expectation value of the hamiltonian

$$E = \frac{\langle \Psi_G | \hat{H} | \Psi_G \rangle}{\langle \Psi_G | \Psi_G \rangle}, \tag{K.3}$$

with respect to g. This gives the tightest upper bound to the ground-state energy for the Gutzwiller trial function. Numerically exact results for the expression in Eq. (K.3) can be found using the variational Monte Carlo method; see Ch. 23, in particular Sec. 23.6.

However, all results are limited by the form of the wavefunction Eq. (K.2), whether they are found by Monte Carlo methods or by the Gutzwiller approximation given below. Although there can be quantitative improvements in the wavefunction, the symmetry of $|\Psi_G\rangle$ is determined by $|\Psi_{HF}\rangle$, so that crucial aspects of $|\Psi_{HF}\rangle$ carry over to the correlated wavefunction. In particular, for fractional occupation, the independent-particle Slater determinant is a metal with a well-defined Fermi surface. So long as there is a physically sensible solution for the wavefunction at the minimum of Eq. (K.3), the correlated system is also a metal [1170] with a Fermi surface that obeys the Luttinger theorem (see Sec. 3.6).

Gutzwiller approximation

The second step in this approach [131] is the Gutzwiller approximation for the energy and other properties of the Gutzwiller wavefunction. The total energy in Eq. (K.3) can be expressed as

$$E_G/N = q_\uparrow \bar{\varepsilon}_\uparrow + q_\downarrow \bar{\varepsilon}_\downarrow + Ud, \quad \text{where} \quad \bar{\varepsilon}_\sigma = \frac{1}{N_k} \sum_{\mathbf{k}}^{occ} (\varepsilon_{\mathbf{k},\sigma} - \mu), \tag{K.4}$$

where d is the fraction of doubly occupied sites and $\bar{\varepsilon}_\sigma$ is the kinetic energy per cell for uncorrelated particles. Note that $\varepsilon_{\mathbf{k},\sigma} - \mu < 0$ for occupied states, so that $\bar{\varepsilon}_\sigma < 0$ and the factor $q \leq 1$ takes into account the reduced hopping due to repulsive interactions. The approximation is to assume that there is no correlation of spin and charge on neighboring sites, leading to[2] (see Ex. K.2):

$$q_\sigma = \frac{\{[(n_\sigma - d)(1 - n_\sigma - n_{-\sigma} + d)]^{\frac{1}{2}} + [(n_{-\sigma} - d)d]^{\frac{1}{2}}\}^2}{n_\sigma(1 - n_\sigma)}. \tag{K.5}$$

Minimizing the energy in Eq. (K.4) as a function of d leads to remarkably simple conclusions. For any filling of the band other than $\frac{1}{2}$, the Gutzwiller approximation gives a renormalized metal with reduced kinetic energy, which is interpreted as an effective mass increased by the factor $1/q$, and a Fermi surface that is the same as the original unrenormalized Fermi surface. For a half-filled band with $n_\sigma + n_{-\sigma} = 1$, there are two types of solutions as a function of U: a metal with $q \neq 0$ or a solution with $q_\uparrow = q_\downarrow = d = 0$.

---

[2] See the review by Vollhardt [150], who credits Ogawa et al. [1171] with the detailed analysis. The formulation applies also to liquids, and it has been applied to the renormalization of the mass in ^{3}He [150]. The approximation is exact for infinite dimensions [762], a prelude to DMFT in Ch. 16.

Although the latter is an unphysical result for any finite U, the approach to this limit with divergent effective mass is one paradigm for a metal–insulator transition examined in Secs. K.3 and 3.2 and illustrated in Fig. 3.3.

Relation to DFT, DMFT, and other approaches

The Gutzwiller approach in terms of a wavefunction is a lattice version of ground-state variational methods discussed in Ch. 23. Since it is designed to determine properties of the ground state, it has aspects in common with DFT, and the combination is the basis of the "DFT+G" methods in Sec. 19.6. Since the approximation is neglect of intersite correlation, it is closely related to other mean-field methods. In a practical sense, it can be very useful in combination with traditional DFT methods, since it provides a physically motivated method to account for the band narrowing observed in many materials. In a pedagogical sense, the decrease in bandwidth, equivalent to an increase in effective mass, provides a basis for understanding the development of low-energy scales in DMFT, for example, the narrow peak in the DMFT spectrum shown in Fig. 16.4, and for a metal–insulator transition as discussed in Secs. K.3 and 3.2.

K.2 Hubbard approach in terms of the Green's function

In a series of four[3] papers [123–126], Hubbard proposed an approach to interacting electrons in terms of the one-body Green's function. This is a natural way to formulate the problem following the reasoning (see Sec. 3.2) that metal–insulator transitions, magnetism, and other phenomena are governed by the competition between the energies for addition and removal of electrons from the atoms compared with the hopping matrix elements between atoms in a crystal. Hubbard's approach was to first consider decoupled atoms including interactions and treat the coupling between the atoms as a perturbation. This can be formulated in terms of the one-particle Green's function

$$G_{\mathbf{k},\sigma}(\omega) = \frac{1}{[\omega - \varepsilon_0 - \Sigma_\sigma(\omega)] - \Delta\varepsilon_{\mathbf{k}}}, \qquad (\text{K.6})$$

where ε_0 is the independent-particle energy for the localized atomic-like state, $\Sigma_\sigma(\omega)$ includes the effects of intra-atomic interactions, and $\Delta\varepsilon_{\mathbf{k}} = \varepsilon_{\mathbf{k}} - \varepsilon_0$ is the independent-particle dispersion due to interatomic hopping. Instead of developing a systematic perturbation expansion, Hubbard devised methods in which the spectra are exact in two limits: decoupled atoms where $\Delta\varepsilon_{\mathbf{k}} = 0$ and interactions are treated exactly, and the non-interacting limit where $\Sigma = 0$. For intermediate cases he proposed different approximations for the intra-atomic, $\mathbf{k}$-independent self-energy $\Sigma_\sigma(\omega)$, neglecting correlation between sites that would lead to $\mathbf{k}$-dependence. This is also the approach adapted in DMFT: the two approximations due to Hubbard (given below) and the single-site DMFT equations

---

[3] Here we summarize papers I and III [123, 125] that deal with the simplest model, the Hubbard model. Papers II and IV [124, 126] deal with the generalization of the methods to realistic cases with atomic multiplet states.

(Ch. 16) can all be described as different ways to calculate the self-energy Σ. Specific parallels with DMFT are pointed out in Sec. 16.4 and indicated in Tab. 16.1.

For isolated atoms with fractional occupation, Hubbard considered the *average of the Green's functions* instead of the average of the hamiltonian matrix elements as in Hartree–Fock. For the Hubbard model this means the average of atoms that are occupied or not with opposite-spin electrons,

$$\overline{G}_\sigma^{atom}(\omega) = \frac{1 - n_{-\sigma}}{\omega - \varepsilon_0} + \frac{n_{-\sigma}}{\omega - \varepsilon_0 - U}, \tag{K.7}$$

where the notation $\overline{G}$ indicates the average of Green's functions. This has the correct average spectrum with poles at ε_0 and $\varepsilon_0 + U$ with weights $1 - n_{-\sigma}$ and $n_{-\sigma}$. One can also define a self-energy corresponding to this Green's function by

$$\overline{G}_\sigma^{atom}(\omega) = \frac{1}{\omega - \varepsilon_0 - \Sigma_\sigma^{atom}(\omega)}, \tag{K.8}$$

where (see Ex. K.5)

$$\Sigma_\sigma^{atom}(\omega) = U n_{-\sigma} + U^2 \frac{n_{-\sigma}(1 - n_{-\sigma})}{\omega - \varepsilon_0 - U(1 - n_{-\sigma})}. \tag{K.9}$$

It is important to realize that even though an expression like Eq. (K.7) can reproduce the average spectrum, *this does not mean that it is the correct Green's function*. In particular, the average in Eq. (K.7) is independent of the order and is the same for an ordered antiferromagnetic and a disordered array of sites that are empty or occupied. The average Green's function with no broken symmetry can be justified for disordered states at high enough temperature, even though temperature is not included explicitly. However, the interpretation for $T \to 0$ is fraught with dangers for misinterpretation. As pointed out in Sec. 16.3, a system of decoupled atoms has massive degeneracy, and the actual ground state is determined by the coupling between the atoms no matter how small. Here we consider only the paramagnetic case with no broken symmetry, but we must keep in mind the limitations and the fact that in most cases the actual ground state has some broken symmetry, such as an antiferromagnet, for which Eq. (K.7) is *not* the correct atomic limit.

Hubbard's first approximation: bare atomic resonance

The first approximation proposed by Hubbard [123], often called "Hubbard I," assumes that the self-energy is the same as in the atomic limit even in the presence of dispersion, and $\Delta \varepsilon_k$ is unchanged even in the presence of interactions. With these assumptions, the Green's function for the crystal becomes

$$G_{k,\sigma}(\omega) = \frac{1}{\omega - \varepsilon_0 - \Sigma^{atom}(\omega) - \Delta \varepsilon_k}. \tag{K.10}$$

From the form of $\Sigma^{atom}(\omega)$ in Eq. (K.9), which has a singularity unless $n_\sigma = 0$ or 1, it is straightforward to show that the single band always splits into two parts, each with fractional weight (see Ex. K.6), for any interaction $U > 0$ and for any partial filling of the band. As recognized by Hubbard, this is unphysical and totally at variance with experiment and accepted theories for weak interactions.

Nevertheless, the splitting of one band into "upper Hubbard" and "lower Hubbard" bands is intuitively appealing for large U, where the upper band corresponds to adding an electron at a site occupied by an opposite-spin electron, whereas the lower band represents sites with no other electron. In cases where there is broken symmetry, it reduces to an unrestricted Hartree–Fock approximation and is a forerunner of the "DFT+U" approximation in Sec. 19.6. If there is no broken symmetry, the assumption of an average Green's function may be an appropriate starting point for disordered states at high temperature, but the interpretation for $T \to 0$ is far from trivial, as explained above.

The Hubbard III alloy approximation

The "Hubbard III" approach (the third paper in the series [125]) is formulated in terms of the propagation of an electron in the presence of other electrons. Since the interaction U is only between opposite-spin electrons, this is the dynamics of an electron of spin ↑ in the presence of the electrons with spin ↓ and vice versa. There are two separate approximations in this approach. The first is that the opposite-spin electrons can be treated as a random array of fixed scatterers like an alloy. Thus an ↑ electron encounters an array of sites with no ↓ electrons that have an energy ε_0, and other sites occupied by a ↓ electron that have an energy $\varepsilon_0 + U$, and similarly for ↓ electrons.

The second approximation is a simplification of the alloy problem, the coherent potential approximation derived independently by Hubbard in paper III, and by Soven [141] and Velecky [142] for electrons in alloys, and was developed much earlier for waves in disordered media [139]. In this approach one defines a translation-invariant effective medium with a "coherent potential" denoted $\Sigma^{CPA}(\omega)$,[4] so that the band energy $\varepsilon_{\mathbf{k}}$ is replaced by the complex (non-hermitian) function $[\varepsilon_{\mathbf{k}} + \mathrm{Re}\,\Sigma^{CPA}(\omega)] + i\,\mathrm{Im}\,\Sigma^{CPA}(\omega)$ that represents shifts in the energies and scattering due to disorder. The effective medium is depicted on the left side of Fig. K.1 by the shaded circles. Since the system is translation invariant, $\mathbf{k}$ is conserved and we can define the CPA Green's function $G_{\mathbf{k}}^{CPA}(\omega)$ and the on-site $G_{00}^{CPA}(\omega)$ by

$$G_{\mathbf{k}}^{CPA}(\omega) = \frac{1}{\omega - \varepsilon_0 - \Delta\varepsilon_{\mathbf{k}} - \Sigma^{CPA}(\omega)} \quad \text{and} \quad G_{00}^{CPA}(\omega) = \frac{1}{N_k}\sum_{\mathbf{k}} G_{\mathbf{k}}^{CPA}(\omega), \quad \text{(K.11)}$$

which has the same form as Eqs. (16.3) and (16.4) used in DMFT.

The self-energy $\Sigma^{CPA}(\omega)$ is determined by the requirement that it represents the average scattering in the static "alloy." The application to the Hubbard model is illustrated in Fig. K.1, where the Green's function for a spin-σ electron depends on the interactions with

---

[4] Here we omit the spin index with the understanding that the equations apply for each spin interacting with opposite-spin $-\sigma$ electrons. The CPA for alloys is defined for non-interacting electrons as a disordered potential. Here it is an approximation to the self-energy due to interaction and hence is denoted Σ^{CPA}.

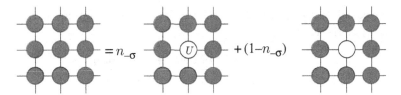

Figure K.1. Schematic illustration of the self-consistency condition for the Green's function in the "Hubbard III" static alloy approximation. For either spin σ the interaction U is, with the opposite-spin electrons, assumed to be distributed randomly with probability $n_{-\sigma}$ on each site whereas there is no interaction with probability $1 - n_{-\sigma}$. The equality represents the CPA where the $G_{00}^{CPA}(\omega)$ is given by the weighted average of the Green's functions $\mathcal{G}^+(\omega)$ and $\mathcal{G}^-(\omega)$ (see Eq. (K.12)) for sites with and without the interaction embedded in the average crystal, which leads to the self-consistent equations in Eq. (16.7) and is explained in more detail in Sec. K.2. Note the similarity to Fig. 16.2, where the static CPA average over sites is replaced by the dynamic interacting electron problem on each site.

the $-\sigma$ electrons. Each site is considered to be embedded in an effective medium with a Green's function

$$\mathcal{G}^{+,-}(\omega) = \frac{1}{\omega - \varepsilon^{+,-} - \Delta(\omega) - \Sigma^{CPA}(\omega)} \tag{K.12}$$

for sites with ($+$) or without ($-$) an opposite-spin electron, with energies $\varepsilon^+ = \varepsilon_0 + U$ and $\varepsilon^- = \varepsilon_0$, respectively. The notation $\mathcal{G}$ is the same as used for an embedded-site Green's function in DMFT and Eq. (K.12) is equivalent to Eq. (16.8), except that here it is convenient to define $\Delta(\omega)$ to be the fixed hybridization for the crystal (due to $\Delta\varepsilon_{\mathbf{k}}$) and consider $\Sigma^{CPA}(\omega)$ to be the function determined by the self-consistency. The CPA condition[5] is that the effective medium is the same as the average over the sites, as depicted in Fig. K.1,

$$G_{00}^{CPA}(\omega) = \overline{\mathcal{G}}(\omega) = n_{-\sigma}\mathcal{G}^+(\omega) + (1 - n_{-\sigma})\mathcal{G}^-(\omega), \tag{K.13}$$

for either spin where $n_{-\sigma}$ denotes the occupation by an opposite-spin electron. This is the same as the average of isolated atomic Green's functions in Eq. (K.7), except that isolated atoms are replaced by renormalized atoms embedded in a bath chosen to best represent the coupling to the rest of the crystal.

The close correspondence with dynamical mean-field theory can be seen by comparing Eq. (K.13) and the corresponding equation for DMFT, $G_{00}(\omega) = \mathcal{G}(\omega)$ in Eq. (16.6) (see also Eq. (16.7)). The difference is the meaning of the self-energy and how it is calculated: here the Green's function $\overline{\mathcal{G}}(\omega)$ is a static average of the simple functions in Eq. (K.12), whereas in DMFT $\mathcal{G}$ is the solution of a difficult many-body problem of a site with interacting electrons embedded in a bath.

---

[5] The condition can also be expressed as the requirement that average scattering vanishes, which can be expressed as the requirement that the average t-matrix is zero. This is worked out in Ex. K.7.

The CPA equations can be expressed in several ways; one form that is equivalent to Eq. (K.13) is (see Ex. K.8)

$$\Sigma(\omega) = \bar{\varepsilon} - \frac{(\varepsilon_0 - \Sigma(\omega))(\varepsilon_0 + U - \Sigma(\omega))}{\omega - \Sigma(\omega) - b\Delta(\omega)}, \tag{K.14}$$

which is solved self-consistently with Eq. (K.11) to find $\Sigma^{CPA}(\omega)$. The factor b in the denominator is unity for the CPA; it is introduced because other values allow the expressions to represent different cases. The limit $b = 0$ corresponds to zero broadening, i.e., the limit where G is the atomic resonance. One can show (see Ex. K.10) that indeed Eq. (K.14) reduces the expression for Σ in Eq. (K.9). Larger values of $b > 1$ correspond to "resonance broadening," considered next.

Resonant broadening correction

Of course, it is not consistent to assume that opposite-spin electrons are in fixed static positions; all electrons should obey the same dynamical equations. In an attempt to take into account the observation that motion of the $-\sigma$ electrons should lead to additional possibilities for hopping for a σ electron, Hubbard devised a "resonance broadening" correction. Hubbard increased the hybridization function in the denominator of Eq. (K.14) by a factor $b > 1$. This is solved self-consistently with Eq. (K.11), which remains the same with no factor of b. For the half-filled case where $n_\uparrow = n_\downarrow = 1/2$, value $b = 3$ was proposed by Hubbard to take into account that the σ electron can hop between two sites without an increase in energy U if a $-\sigma$ electron hops on either site, increasing the total hopping probability by a factor of 3. This is used to estimate the point at which a gap opens, interpreted as a metal–insulator transition in Secs. K.3 and 3.2.

Successes and failures of the Hubbard III approximations

In the Hubbard III approximation the spectrum is correct in two limits, the atomic limit with split bands and the non-interacting limit with independent-particle bands, and there is a gradual development of the upper and lower bands and an opening of a gap, as illustrated in Fig. 3.3. For weak interactions it is a qualitative improvement over the first approximation, where there is a gap for all $U > 0$. As described in Ch. 16, the opening of the gap is in semi-quantitative agreement with DMFT results. If the solution is an insulator with a gap, the CPA has the correct behavior that there is a well-defined gap with no broadening at the band edges [952].

This approach was extended by Hubbard, who developed a theory of ferromagnetism [916–918] as a function of temperature in terms of disordered local moments. This is a forerunner of the disordered local moment method [920, 921] (see Sec. 19.6), which has been applied to magnetic materials as noted in Sec. 20.4 and to oxides such as MnO [584]. It may be a reasonable approximation for DMFT in cases where thermal disorder is the dominant effect.

The failures of the approximation are due to the replacement of the effects of interactions by static disorder. As described in Sec. 16.4, this misses the dynamical properties that are captured, at least approximately, by dynamical mean-field theory.

K.3 Two scenarios for the Mott transition

Hubbard showed that the condition for the opening of a gap can be derived with simple algebra if the density of states is chosen to be the semicircular form in Eq. (16.9). The solution in the alloy approximation is found by solving the self-consistent Eqs. (K.11) and (K.14), (see Ex. K.11). For the half-filled case $n_\uparrow = n_\downarrow = 1/2$, results for different magnitudes of the interaction are shown in Fig. 3.3: there is a characteristic two-peak structure arising from the interaction and the opening of a gap above a critical value U_c^H/D. As shown in Ex. K.12, the opening of the gap occurs for $U_c^H/D = \sqrt{b}$, where b is the broadening factor: $b = 1$ for the CPA and $b = 3$ proposed by Hubbard.

This provides one of the scenarios for the metal–insulator transition in Sec. 3.2, which is best justified as the approach to the transition from the insulating side where the CPA solution has a well-defined gap. For smaller U in the metallic regime the solution may be reasonable for some range of temperature, but it has the drawback that for $T \to 0$ there is unphysical scattering at the Fermi energy due to the assumption of static disorder.

Brinkman–Rice–Gutzwiller approach to the Mott transition

The Gutzwiller approximation is the basis for the other paradigm for the Mott metal–insulator transition in Sec. 3.2. For the case of a half-filled band with no spin polarization, Eq. (K.5) reduces to $q = 8d(1 - d)$. For U less than a critical value U_c^G, minimization of the energy leads to (see Ex. K.3)

$$d = \frac{1}{4}\left[1 - \frac{U}{U_c^G}\right]; \quad q = 1 - \left[\frac{U}{U_c^G}\right]^2, \tag{K.15}$$

with the ground-state energy

$$\frac{E_G}{N} = -|\bar{\varepsilon}|\left[1 - \frac{U}{U_c^G}\right]^2. \tag{K.16}$$

Here $\bar{\varepsilon} = \bar{\varepsilon}_\uparrow + \bar{\varepsilon}_\downarrow$ is the kinetic energy for the uncorrelated system, and the critical value of U is $U_c^G = 8|\bar{\varepsilon}|$. For all $U < U_c^G$, the solution is a metal with a well-defined Fermi surface. As the interaction U increases, the kinetic energy decreases by a factor $q = 1 - \left[\frac{U}{U_c^G}\right]^2$, corresponding to a narrowing of the band and an increase in the effective mass $m^* \sim 1/q$ until the mass diverges at $U = U_c^G$. For $U > U_c^G$ the solution is $d = q = E_g = 0$, i.e., no double occupancy, no hopping, and thus no kinetic energy.

Although there is no physically meaningful solution for $U > U_c^G$ at half-filling, for any other filling there is a solution that is a metal. Thus for $U > U_c^G$ the approach to the transition can be viewed as the divergence of the mass as the filling approaches $1/2$. As described in Sec. 3.2, Brinkman and Rice [138] argued that the approach from the metallic side can be interpreted as a signature for a Mott metal–insulator transition.

For the semicircular density of states defined in Eq. (16.22), the kinetic energy for a half-filled band is $|\bar{\varepsilon}| = (4/3\pi)D$ (see Ex. K.4), so that the critical interaction is given by $U_c^G/D = 32/3\pi \approx 3.40$.

SELECT FURTHER READING

The original papers of Gutzwiller and Hubbard (referred to in the text) are very readable and enlightening. Good descriptions of the Gutzwiller approach can be found in:

Fulde, P. *Electron Correlations in Molecules and Solids* (Springer Series in Solid-State Sciences, 3rd edn, 2003).

Fulde, P., "Wavefunction methods in electronic-structure theory of solids," *Adv. Phys.* **51**, 909–948, 2002.

Vollhardt, D., "Normal ^{3}He: an almost localized Fermi liquid," *Rev. Mod. Phys.* **56**, 99–120, 1984.

Exercises

K.1 Show that the two expressions in Eq. (K.2) are equivalent for fermions where the occupation is 0 or 1.

K.2 The expression for q_σ in Eq. (K.5) is not easy to derive. This exercise is to find the derivation in the book by Fulde [888] or the papers by Vollhardt [150] and Ogawa *et al.* [1171]. Gutzwiller derived the expression by a complicated counting, which was shown by Ogawa to be equivalent to the assumption of no correlation between occupations on neighboring sites.

K.3 Derive the formulas in Eqs. (K.15) and (K.16) and the critical value $U_c^G = 8|\bar{\varepsilon}|$ for the Gutzwiller approximation at half-filling.

K.4 For the semicircular density of states in Eq. (16.22), derive a formula for the kinetic energy $\bar{\varepsilon}$ in Eq. (K.4) (a negative value) as a function of the Fermi energy. For a half-filled band show that $|\bar{\varepsilon}| = (4/3\pi)D$, so that the critical interaction is given by $U_c^G/D = 32/3\pi \sim 3.40$.

K.5 Show that the expression for Σ^{atom} in Eq. (K.9) follows from Eq. (K.7).

K.6 Show that Eq. (K.10) follows from the expression for the atomic Green's function. Is there a definite fractional weight in the upper and lower parts of the band, or are the weights just some fraction that depends on the dispersion $\varepsilon_{\mathbf{k}}$?

K.7 The t-matrix expression is an alternative to the Dyson equation, which can be found from the expansion in Eq. (7.9), $G(\omega) = G^0(\omega) + G^0(\omega)\Sigma(\omega)G^0(\omega) + G^0(\omega)\Sigma(\omega)G^0(\omega)\Sigma(\omega)G^0(\omega) + \cdots \equiv G^0(\omega)t(\omega)G^0(\omega)$. In the example of scattering from a site labeled 0 that differs from other sites in the crystal by a perturbing potential Δv, the Green's function for the system can be written as (we use lowercase notation t since this is an independent-particle problem)

$$G_{ij}^{\Delta v} = G_{ij} + G_{i0}t_{00}^{\Delta v}G_{0j} \;\; \text{with} \;\; t_{00}^{\Delta v} = \frac{\Delta v}{1 - \Delta v G_{00}},$$

where G_{ij}^0 is the Green's function for propagation between sites i and j in the perfect crystal.

(a) Derive these expressions. *Hint:* this sums all multiple scattering events for an electron in an effective medium with the potential difference Δv on site 0, $G_{ij}^{\Delta v} = G_{ij} + G_{i0}\,\Delta v\,G_{0j}$ $+\cdots$

(b) Show that the CPA condition in Eq. (K.13) is equivalent to the requirement that the average of the t-matrices vanishes, i.e., for an alloy with A and B sites, the weighted average of $t^{\Delta v_A}$ and $t^{\Delta v_B}$ scattering vanishes. This can be shown by using only the definition of averaging over $G_{ij}^{\Delta v}$ in the first equation; no algebraic manipulations are needed, and one does not need to use the explicit form for t.

K.8 Show that the form of the CPA equations in Eq. (K.14) with $b = 1$ follows from Eq. (K.13). *Hint:* a useful form may be the alternative to set the average t-matrix to zero (see Ex. K.7). This can also be done by filling in the steps in the derivations in the paper by Hubbard [125] and the CPA papers [141, 142].

K.9 For the half-filled case, derive a simplified form for Eq. (K.14) that is linear in Σ and Δ. This is useful for the special case in Ex. K.11.

K.10 Show that Eq. (K.14) reduces to the atomic resonance limit if $b = 0$, i.e., there is no hybridization. For the one-band Hubbard model at half-filling, show that $\Sigma(\omega) = (U^2/4)(\omega - \bar{\varepsilon})^{-1}$.

K.11 Show that for half-filling the solution for the density of states in the Hubbard III alloy approximation with the semicircular density of states is given by the roots of a cubic equation (Exs. K.9 and K.10 may be useful). The final expression is given in [125].

K.12 Find the critical interaction strength U_c^H/D where the gap opens in the Hubbard III alloy approximation for half-filling with the semicircular density of states. It is not necessary to solve the cubic equation needed in Ex. K.11; we need only consider $\omega = 0$, since the gap must occur at $\omega = 0$ due to the electron–hole symmetry. Furthermore, the transition is signified by the fact that the Green's function changes from real (in the gap) to imaginary (if there is no gap and the density of states is non-zero at $\omega = 0$). Show that this occurs at $U_c^H/D = \sqrt{b}$.

References

[1] R. M. Martin, *Electronic Structure: Basic Theory and Methods,* Cambridge University Press, Cambridge, 2004, reprinted 2005, 2008; Japanese translation, 2010, 2012.

[2] M. Born and J. R. Oppenheimer, "Zur Quantentheorie der Molekeln," *Ann. Physik* 84:457, 1927.

[3] C. J. Cramer, *Essentials of Computational Chemistry,* Wiley, Chichester, 2004.

[4] L. Hoddeson, E. Braun, J. Teichmann, and S. Weart, *Out of the Crystal Maze [Chapters for the History of Solid State Physics],* Oxford University Press, New York, 1992.

[5] D. R. Hartree, "The wave mechanics of an atom with non-Coulombic central field: parts I, II, III," *Proc. Cambridge Phil. Soc.* 24:89,111,426, 1928.

[6] D. R. Hartree, *The Calculation of Atomic Structures,* Wiley, New York, 1957.

[7] E. Fermi, "Sulla quantizzazione del gas perfetto monoatomico (On the quantization of the monoatomic ideal gas)," *Rend. Lincei* 3:145–149, 1926.

[8] A. Zannoni, "Translation of E. Fermi, *Rend. Lincei* 3, 145–9 (1926) On the quantization of the monoatomic ideal gas," arxiv cond-mat/9912229v1, 1999.

[9] E. C. Stoner, "The distribution of electrons among atomic levels," *Phil. Mag.* 48:719, 1924.

[10] W. Pauli, "Uber den Zusammenhang des Abschlusses der Elektronengruppen im Atom mit der Komplex Struktur der Spektren," *Z. Phys.* 31:765, 1925.

[11] P. A. M. Dirac, "On the theory of quantum mechanics," *Proc. Roy. Soc. London Ser. A* 112:661, 1926.

[12] P. A. M. Dirac, "Note on exchange phenomena in the Thomas–Fermi atom," *Proc. Cambridge Phil. Soc.* 26:376–385, 1930.

[13] G. Baym, "Self-consistent approximations in many-body systems," *Phys. Rev.* 127:1391–1401, 1962.

[14] L. D. Landau, "The theory of a Fermi liquid [*Soviet Phys. JETP* 3, 920–925 (1957)]," *Zh. Eksp. i Teor. Fiz.* 30:1058–1064, 1956.

[15] L. D. Landau, "Oscillations in a Fermi liquid [*Soviet Phys. JETP* 5, 101–108 (1957)]," *Zh. Eksp. i Teor. Fiz.* 32:59–66, 1957.

[16] L. D. Landau, "On the theory of the Fermi liquid [*Soviet Phys. JETP* 8, 70–74 (1959)]," *Zh. Eksp. i Teor. Fiz.* 35:97–103, 1958.

[17] C. J. Pethick and H. Smith, *Bose–Einstein Condensation in Dilute Gases,* Cambridge University Press, Cambridge, 2008.

[18] L. D. Landau, "Theory of phase transformations I. Trans: *Zh. Eksp. Teor. Fiz.* 7, 19 (1937)," *Phys. Z. Sowjet* 11:26–47, 1937.

[19] L. D. Landau, "Theory of phase transformations II. Trans: *Zh. Eksp. Teor. Fiz.* 7, 627 (1937)," *Phys. Z. Sowjet* 11:545–555, 1937.

[20] X.-G. Wen, *Quantum Field Theory of Many-Body Systems,* Oxford University Press, Oxford, 2004.

[21] M. Z. Hasan and C. L. Kane, "Colloquium: Topological insulators," *Rev. Mod. Phys.* 82:3045–3067, 2010.

[22] E. P. Wigner, "On the interaction of electrons in metals," *Phys. Rev.* 46:1002–1011, 1934.

[23] J. Bardeen, "Theory of the work function. II. The surface double layer," *Phys. Rev.* 49:653, 1936.

[24] E. A. Hylleraas, "Uber den Grundterm der Zweielektronenprobleme von H^-, He, Li^+, Be^+ usw," *Z. Phys.* 65:209, 1930.

[25] E. P. Wigner and F. Seitz, "On the constitution of metallic sodium," *Phys. Rev.* 43:804, 1933.

[26] E. P. Wigner and F. Seitz, "On the constitution of metallic sodium II," *Phys. Rev.* 46:509, 1934.

[27] R. P. Feynman, "Space-time approach to non-relativistic quantum mechanics," *Rev. Mod. Phys.* 20:367–387, 1948.

[28] R. P. Feynman, "Space-time approach to quantum electrodynamics," *Phys. Rev.* 76:769–789, 1949.

[29] F. J. Dyson, "The S matrix in quantum electrodynamics," *Phys. Rev.* 75:1736–1755, 1949.

[30] D. Bohm and D. Pines, "A collective description of electron interactions. I. Magnetic interactions," *Phys. Rev.* 82:625–634, 1951; II. "Collective vs individual particle aspects of the interactions," *Phys. Rev.* 85:338–353, 1952; III. "Coulomb interactions in a degenerate electron gas," *Phys. Rev.* 92:609–625, 1953.

[31] J. Lindhard, "On the properties of a gas of charged particles," *Kgl. Danske Videnskab. Selskab, Mat.-fys. Medd.* 28:1–57, 1954.

[32] J. Hubbard, "Description of collective motion in terms of many-body perturbation theory. II. The correlation energy of a free electron gas," *Proc. Roy. Soc. London, Ser. A* 243:336–352, 1958.

[33] H. Ehrenreich and M. H. Cohen, "Self-consistent field approach to the many-electron problem," *Phys. Rev.* 115:786–790, 1959.

[34] M. Gell-Mann and K. A. Brueckner, "Correlation energy of an electron gas at high density," *Phys. Rev.* 106:364–368, 1957.

[35] V. M. Galitskii and A. B. Migdal, "Application of quantum field theory methods to the many body problem [translation: *Soviet Phys. JETP* 7, 96–104 (1958)]," *Zh. Eksp. Teor. Fiz.* 34:139–150, 1958.

[36] J. M. Luttinger and J. C. Ward, "Ground-state energy of a many-fermion system. II," *Phys. Rev.* 118:1417–1427, 1960.

[37] J. M. Luttinger, "Fermi surface and some simple equilibrium properties of a system of interacting fermions," *Phys. Rev.* 119:1153–1163, 1960.

[38] A. A. Abrikosov, L. P. Gorkov, and I. E. Dzyaloshinski, *Methods of Quantum Field Theory in Statistical Physics,* Prentice-Hall, Englewood Cliffs, NJ, 1963.

[39] G. Baym and L. P. Kadanoff, "Conservation laws and correlation functions," *Phys. Rev.* 124:287–299, 1961.

[40] P. C. Martin and J. Schwinger, "Theory of many particle systems. I," *Phys. Rev.* 115:1342–1373, 1959.

[41] P. Hohenberg and W. Kohn, "Inhomogeneous electron gas," *Phys. Rev.* 136:B864–871, 1964.

[42] W. Kohn and L. J. Sham, "Self-consistent equations including exchange and correlation effects," *Phys. Rev.* 140:A1133–1138, 1965.

[43] L. Hedin, "New method for calculating the one-particle Green's function with application to the electron-gas problem," *Phys. Rev.* 139:A796–823, 1965.

[44] W. L. McMillan, "Ground state of liquid He^4," *Phys. Rev.* 138:A442–451, 1965.

[45] M. Head-Gordon and E. Artacho, "Chemistry on the computer," *Physics Today* 61:58–63, 2008.

[46] T. Koopmans, "Ueber die zuordnung von wellenfunktionen und eigenwerten zu den einzelnen elektronen eines atoms," *Physica* 1:104–113, 1934.

[47] A. L. Fetter and J. D. Walecka, *Quantum Theory of Many-Particle Systems,* McGraw-Hill, New York, 1971.

[48] V. L. Moruzzi, A. R. Williams, and J. F. Janak, "Local density theory of metallic cohesion," *Phys. Rev. B* 15:2854–2857, 1977.

[49] J. D. van der Waals, *Nobel Lectures in Physics,* Elsevier, Amsterdam, 1964, pp. 254–265.

[50] F. London, "Zur Theorie und Systematik der Molekularkrfte," *Z. Phys. A* 63:245–279, 1930.

[51] N. W. Ashcroft and N. D. Mermin, *Solid State Physics,* W. B. Saunders Company, Philadelphia, PA, 1976.

[52] F. G. Brickwedde, *J. Res. Nat. Bur. Stnd. A* 64:1–4, 1960.

[53] A. Tkatchenko, M. Rossi, V. Blum, J. Ireta, and M. Scheffler, "Unraveling the stability of polypeptide helices: Critical role of van der Waals interactions," *Phys. Rev. Lett.* 106:118102, 2011.

[54] *Ferromagnetic Materials,* edited by E. P. Wolfarth, North Holland, Amsterdam, 1986.

[55] A. I. Lichtenstein, M. I. Katsnelson, and G. Kotliar, "Finite-temperature magnetism of transition metals: An ab initio dynamical mean-field theory," *Phys. Rev. Lett.* 87:067205, 2001.

[56] E. C. Stoner, "Collective electron ferromagnetism. II. Energy and specific heat," *Roy. Soc. London, Proc. Ser. A* 169:339–371, 1939.

[57] S. Huefner, *Photoemission Spectroscopy: Principles and Applications,* Springer-Verlag, Berlin, 2003.

[58] M. Cardona and L. Ley, *Photoemission in Solids I,* Vol. 26 of *Topics in Applied Physics,* Springer-Verlag, Berlin, 1978.

[59] A. Damascelli, Z. Hussain, and Z.-X. Shen, "Angle-resolved photoemission studies of the cuprate superconductors," *Rev. Mod. Phys.* 75:473–541, 2003.

[60] A. L. Wachs, T. Miller, T. C. Hsieh, A. P. Shapiro, and T. C. Chiang, "Angle-resolved photoemission studies of Ge(111)-c(2 x 8), Ge(111)-(1 x 1)H, Si(111)-(7 x 7), and Si(100)-(2 x 1)," *Phys. Rev. B* 32:2326–2333, 1985.

[61] J. E. Ortega and F. J. Himpsel, "Inverse-photoemission study of Ge(100), Si(100), and GaAs(100): Bulk bands and surface states," *Phys. Rev. B* 47:2130–2137, 1993.

[62] M. Rohlfing, P. Krüger, and J. Pollmann, "Quasiparticle band-structure calculations for C, Si, Ge, GaAs, and SiC using gaussian-orbital basis sets," *Phys. Rev. B* 48:17791–17805, 1993.

[63] A. Svane, "Hartree–Fock band-structure calculations with the linear muffin-tin-orbital method: Application to C, Si, Ge, and alpha-Sn," *Phys. Rev. B* 35:5496–5502, 1987.

[64] R. Knorren, K. H. Bennemann, R. Burgermeister, and M. Aeschlimann, "Dynamics of excited electrons in copper and ferromagnetic transition metals: Theory and experiment," *Phys. Rev. B* 61:9427–9440, 2000.

[65] J. Guedde, M. Rohleder, T. Meier, S. W. Koch, and U. Hoefer, "Time-resolved investigation of coherently controlled electric currents at a metal surface," *Science* 318:1287–1291, 2007.

[66] U. Hofer, I. L. Shumay, C. Reuss, U. Thomann, W. Wallauer, and T. Fauster, "Time-resolved coherent photoelectron spectroscopy of quantized electronic states on metal surfaces," *Science* 277:1480–1482, 1997.

[67] P. Steiner, H. Höchst, and S. Hüfner, *Photoemission in Solids II*, Vol. 27 of *Topics in Applied Physics*, Springer-Verlag, Berlin, 1979, pp. 349–372.

[68] M. Grioni, P. Weibel, D. Malterre, Y. Baer, and L. Du'o, "Resonant inverse photoemission in cerium-based materials," *Phys. Rev. B* 55:2056–2067, 1997.

[69] J. W. Allen, S.-J. Oh, M. B. Maple, and M. S. Torikachvili, "Large Fermi-level resonance in the electron-addition spectrum of CeRu$_2$ and CeIr$_2$," *Phys. Rev. B* 28:5347–5349, 1983.

[70] D. M. Roessler and W. C. Walker, "Optical constants of magnesium oxide and lithium fluoride in far ultraviolet," *J. Opt. Soc. Am.* 57:835–836, 1967.

[71] M. Rohlfing and S. G. Louie, "Electron–hole excitations in semiconductors and insulators," *Phys. Rev. Lett.* 81:2312–2315, 1998.

[72] *Epioptics: Linear and Nonlinear Optical Spectroscopy of Surfaces and Interfaces, ESPRIT Basic Research Series*, edited by J. F. McGilp, D. Weaire, and C. Patterson, Springer-Verlag, Berlin, 2011.

[73] W. Schülke, *Electron Dynamics by Inelastic X-Ray Scattering, Oxford Series on Synchrotron Radiation*, Oxford University Press, Oxford, 2007.

[74] W. Schülke, in *Handbook on Synchrotron Radiation*, edited by G. Brown and D. E. Moncton, Elsevier Science, Amsterdam, 1991, pp. 565–637.

[75] T. P. Devereaux and R. Hackl, "Inelastic light scattering from correlated electrons," *Rev. Mod. Phys.* 79:175–233, 2007.

[76] R. F. Egerton, "Electron energy-loss spectroscopy in the TEM," *Rep. Prog. Phys.* 72:016502, 2009.

[77] R. F. Egerton, *Electron Energy Loss Spectroscopy in the Electron Microscope*, Springer-Verlag, New York, 2011.

[78] B. C. Larson, W. Ku, J. Z. Tischler, C.-C. Lee, O. D. Restrepo, A. G. Eguiluz, P. Zschack, and K. D. Finkelstein, "Nonresonant inelastic X-ray scattering and energy-resolved Wannier function investigation of $d-d$ excitations in NiO and CoO," *Phys. Rev. Lett.* 99:026401, 2007.

[79] M. W. Haverkort, A. Tanaka, L. H. Tjeng, and G. A. Sawatzky, "Nonresonant inelastic X-ray scattering involving excitonic excitations: The examples of NiO and CoO," *Phys. Rev. Lett.* 99:257401, 2007.

[80] H. C. Manoharan, C. P. Lutz, and D. M. Eigler, "Quantum mirages formed by coherent projection of electronic structure," *Nature* 403:512–513, 2000.

[81] U. Mizutani, *Introduction to the Electron Theory of Metals*, Cambridge University Press, Cambridge, 2001.

[82] W. J. de Haas, J. H. de Boer, and G. J. van den Berg, "The electrical resistance of gold, copper and lead at low temperatures," *Physica* 1:1115–1124, 1934.

[83] J. Kondo, "Resistance minimum in dilute magnetic alloys," *Prog. Theor. Phys.* 32:37–69, 1964.

[84] K. G. Wilson, "The renormalization group: Critical phenomena and the Kondo problem," *Rev. Mod. Phys.* 47:773–840, 1975.

[85] N. Andrei, K. Furuya, and J. H. Lowenstein, "Solution of the Kondo problem," *Rev. Mod. Phys.* 55:331–402, 1983.

[86] N. Quaas, M. Wenderoth, A. Weismann, R. G. Ulbrich, and K. Schönhammer, "Kondo resonance of single Co atoms embedded in Cu(111)," *Phys. Rev. B* 69:201103, 2004.

[87] Z. Fisk, D. W. Hess, C. J. Pethick, D. Pines, J. L. Smith, J.D. Thompson, and J. O. Willis, "Heavy-electron metals: New highly correlated states of matter," *Science* 239:33–42, 1988.

[88] G. R. Stewart, "Heavy-fermion systems," *Rev. Mod. Phys.* 56:755–787, 1984.

[89] P. Coleman, in *Handbook of Magnetism and Advanced Magnetic Materials,* Wiley, New York, 2007.

[90] D. B. McWhan, A. Menth, J. P. Remeika, W. F. Brinkman, and T. M. Rice, "Metal–insulator transitions in pure and doped V_2O_3," *Phys. Rev. B* 7:1920–1931, 1973.

[91] M. Imada, A. Fujimori, and Y. Tokura, "Metal–insulator transitions," *Rev. Mod. Phys.* 70:1039–1263, 1998.

[92] D. B. McWhan, T. M. Rice, and J. P. Remeika, "Mott transition in Cr-doped V_2O_3," *Phys. Rev. Lett.* 23:1384–1387, 1969.

[93] S.-K. Mo, H.-D. Kim, J. D. Denlinger, J. W. Allen, J.-H. Park, A. Sekiyama, A. Yamasaki, S. Suga, Y. Saitoh, T. Muro, and P. Metcalf, "Photoemission study of $(V_{1-x}M_x)_2O_3$ (M = Cr, Ti)," *Phys. Rev. B* 74:165101, 2006.

[94] E. Papalazarou, M. Gatti, M. Marsi, V. Brouet, F. Iori, L. Reining, E. Annese, I. Vobornik, F. Offi, A. Fondacaro, S. Huotari, P. Lacovig, O. Tjernberg, N. B. Brookes, M. Sacchi, P. Metcalf, and G. Panaccione, "Valence-band electronic structure of V_2O_3: Identification of V and O bands," *Phys. Rev. B* 80:155115, 2009.

[95] J.-H. Park, L. H. Tjeng, A. Tanaka, J. W. Allen, C. T. Chen, P. Metcalf, J. M. Honig, F. M. F. de Groot, and G. A. Sawatzky, "Spin and orbital occupation and phase transitions in V_2O_3," *Phys. Rev. B* 61:11506–11509, 2000.

[96] S. Yu. Ezhov, V. I. Anisimov, D. I. Khomskii, and G. A. Sawatzky, "Orbital occupation, local spin, and exchange interactions in V_2O_3," *Phys. Rev. Lett.* 83:4136–4139, 1999.

[97] M. Hashimoto, I. M. Vishik, T. P. Devereaux, and Z.-X. Shen, "Energy gaps in high-transition-temperature cuprate superconductors," *Nat. Phys.* 10:483–495, 2014.

[98] M. R. Norman, M. Randeria, H. Ding, and J. C. Campuzano, "Phenomenology of the low-energy spectral function in high-T_c superconductors," *Phys. Rev. B* 57:R11093–11096, 1998.

[99] M. R. Norman, A. Kanigel, M. Randeria, U. Chatterjee, and J. C. Campuzano, "Modeling the Fermi arc in underdoped cuprates," *Phys. Rev. B* 76:174501, 2007.

[100] A. Sekiyama, H. Fujiwara, S. Imada, S. Suga, H. Eisaki, S. I. Uchida, K. Takegahara, H. Harima, Y. Saitoh, I. A. Nekrasov, G. Keller, D. E. Kondakov, A. V. Kozhevnikov, Th. Pruschke, K. Held, D. Vollhardt, and V. I. Anisimov, "Mutual experimental and theoretical validation of bulk photoemission spectra of $Sr_{1-x}Ca_xVO_3$," *Phys. Rev. Lett.* 93:156402, 2004.

[101] S.-K. Mo, J. D. Denlinger, H.-D. Kim, J.-H. Park, J. W. Allen, A. Sekiyama, A. Yamasaki, K. Kadono, S. Suga, Y. Saitoh, T. Muro, P. Metcalf, G. Keller, K. Held, V. Eyert, V. I. Anisimov, and D. Vollhardt, "Prominent quasiparticle peak in the photoemission spectrum of the metallic phase of V_2O_3," *Phys. Rev. Lett.* 90:186403, 2003.

[102] I. Lindau and W. E. Spicer, "Probing depth in photoemission and Auger-electron spectroscopy," *J. Electron Spectrosc. Relat. Phenom.* 3:409–413, 1974.

[103] C. J. Powell, "Attenuation lengths of low-energy electrons in solids," *Surface Sci.* 44:29–46, 1974.

[104] G. A. Somorjai, *Chemistry in Two Dimensions: Surfaces, George Fisher Baker non-resident lectureship in chemistry at Cornell University,* Cornell University Press, Ithaca, NY, 1981.

[105] M. Casula, S. Sorella, and G. Senatore, "Ground state properties of the one-dimensional Coulomb gas using the lattice regularized diffusion Monte Carlo method," *Phys. Rev. B* 74:245427, 2006.

[106] G. Giuliani and G. Vignale, *Quantum Theory of the Electron Liquid,* Cambridge University Press, Cambridge, 2005.

[107] D. Pines and P. Nozières, *The Theory of Quantum Liquids,* Vol. I (*Advanced Book Classics,* originally published by W. A. Benjamin, New York, 1966), Westview Press, Boulder, CO, 1999.

[108] B. Tanatar and D. M. Ceperley, "The ground state of the two-dimensional electron gas," *Phys. Rev. B* 39:5005–5016, 1989.

[109] D. M. Ceperley and B. J. Alder, "Ground state of the electron gas by a stochastic method," *Phys. Rev. Lett.* 45:566, 1980.

[110] F. H. Zong, C. Lin, and D. M. Ceperley, "Spin polarization of the low density three-dimensional electron gas," *Phys. Rev. E* 66:036703:1–7, 2002.

[111] B. Spivak and S. A. Kivelson, "Phases intermediate between a two-dimensional electron liquid and Wigner crystal," *Phys. Rev. B* 70:155114, 2004.

[112] L. Candido, B. Bernu, and D. M. Ceperley, "Magnetic ordering of the three dimensional Wigner crystal," *Phys. Rev. B* 70:094413:1–6, 2004.

[113] B. Bernu, L. Candido, and D. M. Ceperley, "Exchange frequencies in the 2D Wigner crystal," *Phys. Rev. Lett.* 86:870–873, 2001.

[114] L. Balents, "Spin liquids in frustrated magnets," *Nature* 464:199–208, 2010.

[115] N. F. Mott, "The basis of the electron theory of metals, with special reference to the transition metals," *Proc. Phys. Soc. A* 62:416–422, 1949.

[116] Proceedings of Washington Conference on Magnetism, *Rev. Mod. Phys.* 25, 1953.

[117] J. C. Thompson, "Introduction and list of participants," *Rev. Mod. Phys.* 40:673–676, 1968.

[118] N. F. Mott, "The transition to the metallic state," *Phil. Mag.* 6:287, 1961.

[119] N. F. Mott, "Electrons in transition metals," *Adv. Phys.* 13:325–422, 1964.

[120] N. F. Mott, *Metal–Insulator Transitions,* Taylor and Francis, London, 1990.

[121] W. C. Herring, *Exchange Interactions among Itinerant Electrons,* Academic Press, New York, 1966.

[122] J. C. Slater, "Magnetic effects and the Hartree–Fock equation," *Phys. Rev.* 82:538–541, 1951.

[123] J. Hubbard, "Electron correlations in narrow energy bands," *Proc. Roy. Soc. London, Ser. A* 276:238–257, 1963.

[124] J. Hubbard, "Electron correlations in narrow energy bands. II. The degenerate band case," *Proc. Roy. Soc. London, Ser. A* 277:237–259, 1964.

[125] J. Hubbard, "Electron correlations in narrow energy bands. III. An improved solution," *Proc. Roy. Soc. London, Ser. A* 281:401–419, 1964.

[126] J. Hubbard, "Electron correlations in narrow energy bands. IV. The atomic representation," *Proc. Roy. Soc. London, Ser. A* 285:542–560, 1965.

[127] J. Hubbard, "Electron correlations in narrow energy bands. V. A perturbation expansion about the atomic limit," *Proc. Roy. Soc. London, Ser. A* 296:82–99, 1967.

[128] J. Hubbard, "Electron correlations in narrow energy bands. VI. The connexion with many-body perturbation theory," *Proc. Roy. Soc. London, Ser. A* 296:100–112, 1967.

[129] M. Gutzwiller, "Effect of correlation on the ferromagnetism of transition metals," *Phys. Rev. Lett.* 10:159–162, 1963.

[130] M. C. Gutzwiller, "Effect of correlation on the ferromagnetism of transition metals," *Phys. Rev.* 134:A923–941, 1964.

[131] M. C. Gutzwiller, "Correlation of electrons in a narrow *s* band," *Phys. Rev.* 137:A1726–1735, 1965.

[132] P. W. Anderson, "New approach to the theory of superexchange interactions," *Phys. Rev.* 115:2–13, 1959.

[133] J. Kanamori, "Electron correlation and ferromagnetism of transition metals," *Prog. Theor. Phys.* 30:275–289, 1963.

[134] E. Lieb and F. Y. Wu, "Absence of Mott transition in an exact solution of the short-range, one-band model in one dimension," *Phys. Rev. Lett.* 20:1445–1448, 1968.

[135] H. A. Bethe, "Zur Theorie der Metalle. Eigenwerte und Eigenfunktionen der linearen Atomkette," *Z. Phys.* 71:205, 1931.

[136] N. Trivedi and D. M. Ceperley, "Ground-state correlations of quantum antiferromagnets: A Green-function Monte Carlo study." *Phys. Rev. B* 41:4552–4569, 1990.

[137] W. Heitler and F. London, "Wechselwirkung neutraler Atome und homopolare Bindung nach der Quantenmechanik," *Z. Phys.* 44:455–472, 1927.

[138] W. F. Brinkman and T. M. Rice, "Application of Gutzwiller's variational method to the metal–insulator transition," *Phys. Rev. B* 2:4302–4304, 1970.

[139] M. Lax, "Multiple scattering of waves," *Rev. Mod. Phys.* 23:287–310, 1951.

[140] R. J. Elliott, J. A. Krumhansl, and P. L. Leath, "The theory and properties of randomly disordered crystals and related physical systems," *Rev. Mod. Phys.* 46:465–543, 1974.

[141] P. Soven, "Coherent-potential model of substitutional disordered alloys," *Phys. Rev.* 156:809–813, 1967.

[142] B. Velicky, S. Kirkpatrick, and H. Ehrenreich, "Single-site approximations in the electronic theory of simple binary alloys," *Phys. Rev.* 175:747–766, 1968.

[143] G. H. Wannier, *Statistical Physics,* Wiley, New York, 1966.

[144] P. M. Chaikin and T. C. Lubensky, *Principles of Condensed Matter Physics,* Cambridge University Press, Cambridge, 1995.

[145] D. Pines, *The Many Body Problem* (*Advanced Book Classics,* originally published in 1961), Addison-Wesley, Reading, MA, 1997.

[146] A. C. Hewson, *The Kondo Problem to Heavy Fermions,* Cambridge University Press, Cambridge, 1993.

[147] G. Baym and C. Pethick, *Theory of Interacting Fermi Systems* (originally published in 1991), Wiley-VCH, Weinheim, 2004.

[148] P. Nozières, *Theory of Interacting Fermi Systems* (*Advanced Book Classics,* originally published in 1964), Addison-Wesley, Reading, MA, 1997.

[149] G. D. Mahan, *Many-Particle Physics, 3rd Ed.,* Kluwer Academic/Plenum Publishers, New York, 2000.

[150] D. Vollhardt, "Normal ^{3}He: An almost localized Fermi liquid," *Rev. Mod. Phys.* 56:99–120, 1984.

[151] C. Kittel, *Introduction to Solid State Physics,* Wiley, New York, 1996.

[152] V. P. Silin, "On the theory of a degenerate electron fluid [*Soviet Phys. JETP* 6, 387 and 985 (1958)]," *Zh. Eksp. i Teor. Fiz.* 33:495 and 1282, 1957.

[153] P. Nozières and J. M. Luttinger, "Derivation of the Landau theory of Fermi liquids. I. Formal preliminaries," *Phys. Rev.* 127:1423–1431, 1962.

[154] J. M. Luttinger and P. Nozières, "Derivation of the Landau theory of Fermi liquids. II. Equilibrium properties and transport equation," *Phys. Rev.* 127:1431–1440, 1962.

[155] P. W. Anderson, "Localized magnetic states in metals," *Phys. Rev.* 124:41–53, 1961.

[156] S. Doniach and E. H. Sondheimer, *Green's Functions for Solid State Physicists* (Reprinted in *Frontiers in Physics Series,* No. 44), W. A. Benjamin, Reading, MA, 1974.

[157] J. R. Schrieffer and P. Wolfe, "Relation between the Anderson and Kondo hamiltonians," *Phys. Rev.* 149:491–492, 1966.

[158] B. Coqblin and J. R. Schrieffer, "Exchange interaction in alloys with cerium impurities," *Phys. Rev.* 185:847–853, 1969.

[159] O. E. Gunnarsson and K. Schönhammer, "Electron spectroscopies for Ce compounds in the impurity model," *Phys. Rev. B* 28:4315–4341, 1983.

[160] K. Yamada, "Perturbation expansion for the Anderson hamiltonian. II," *Prog. Th. Phys.* 53:970–986, 1975.

[161] I. Dzyaloshinskii, "Some consequences of the Luttinger theorem: The Luttinger surfaces in non-Fermi liquids and Mott insulators," *Phys. Rev. B* 68:085113, 2003.

[162] P. Nozières, "A 'Fermi-liquid' description of the Kondo problem at low temperatures," *J. Low Temp. Phys.* 17:31–42, 1974.

[163] J. Friedel, "The distribution of electrons around impurities in monovalent metals," *Phil. Mag.* 43:153–189, 1952.

[164] J. S. Langer and V. Ambegaokar, "Friedel sum rule for a system of interacting electrons," *Phys. Rev.* 121:1090–1092, 1961.

[165] D. C. Langreth, "Friedel sum rule for Anderson's model of localized impurity states," *Phys. Rev.* 150:516–518, 1966.

[166] R. M. Martin, "Fermi-surface sum rule and its consequences for periodic Kondo and mixed-valence systems," *Phys. Rev. Lett.* 48:362–365, 1982.

[167] D. R. Hartree and W. Hartree, "Self-consistent field, with exchange for beryllium," *Proc. Roy. Soc. London, Series A* 150:9–33, 1935.

[168] V. Fock, "Naherungsmethode zur Losung des quanten-mechanischen Mehrkorperprobleme," *Z. Phys.* 61:126–148, 1930.

[169] A. Szabo and N. S. Ostlund, *Modern Quantum Chemistry: Introduction to Advanced Electronic Structure Theory* (Unabridged reprinting of 1989 version), Dover, Mineola, NY, 1996.

[170] R. D. McWeeny and B. T. Sutcliffe, *Methods of Molecular Quantum Mechanics, 2nd Ed.,* Academic Press, New York, 1976.

[171] P. Weiss, "The molecular field hypothesis and ferromagnetism," *J. Phys. Radium* 6:661, 1907.

[172] H. A. Bethe, "Statistical theory of superlattices," *Proc. Roy. Soc. London, Ser. A* 150:552–575, 1935.

[173] R. Peierls, "Statistical theory of superlattices with unequal concentrations of the components," *Proc. Roy. Soc. London, Ser. A* 154:207–222, 1936.

[174] R. Kikuchi, "A theory of cooperative phenomena," *Phys. Rev.* 81:988–1003, 1951.

[175] R. G. Parr and W. Yang, *Density-Functional Theory of Atoms and Molecules,* Oxford University Press, New York, 1989.

[176] *Density Functional Theory: An Approach to the Quantum Many-Body Problem,* edited by R. M. Dreizler and E. K. U. Gross, Springer-Verlag, Berlin, 1990.

[177] *Density Functional Theory,* edited by E. K. U. Gross and R. M. Dreizler, Plenum Press, New York, 1995.

[178] J. D. van der Waals, "Thermodynamische Theorie der Kapillaritat unter Voraussetzung stetiger Dichteanderung," *Z. Phys. Chem.* 13:657, 1894.

[179] J. W. Cahn and J. E. Hilliard, "Free energy of a nonuniform system. I. Interfacial energy," *J. Chem. Phys* 28:258, 1958.

[180] F. Malet and P. Gori-Giorgi, "Strong correlation in Kohn–Sham density functional theory," *Phys. Rev. Lett.* 109:246402, 2012.

[181] P. Hohenberg and W. Kohn, "Inhomogeneous electron gas," *Phys. Rev.* 136:B864–871, 1964.

[182] M. Levy, "Electron densities in search of hamiltonians," *Phys. Rev. A* 26:1200, 1982.

[183] E. Lieb, in *Physics as Natural Philosophy,* edited by A. Shimony and H. Feshbach, MIT Press, Cambridge, MA, 1982, p. 111.

[184] J. T. Chayes, L. Chayes, and M. B. Ruskai, "Density functional-approach to quantum-lattice systems," *J. Stat. Phys.* 38:497–518, 1985.

[185] C. A. Ullrich and W. Kohn, "Degeneracy in density functional theory: Topology in the v and n spaces," *Phys. Rev. Lett.* 89:156401–156404, 2002.

[186] A. D. Becke, "Density-functional exchange-energy approximation with correct asymptotic behavior," *Phys. Rev. A* 38:3098–3100, 1988.

[187] C. Lee, W. Yang, and R. G. Parr, "Development of the Colle–Salvetti correlation-energy formula into a functional of the electron density," *Phys. Rev. B* 37:785–789, 1988.

[188] J. P. Perdew and K. Schmidt, in *Density Functional Theory and Its Applications to Materials,* edited by V. E. van Doren, C. van Alsenoy, and P. Geerlings, *AIP Conference Proceedings,* Vol. 577, American Institute of Physics, College Park, MD, 2001.

[189] N. D. Mermin, "Thermal properties of the inhomogeneous electron gas," *Phys. Rev.* 137:A1441–1443, 1965.

[190] J. P. Perdew, R. G. Parr, M. Levy, and J. L. Balduz, "Density-functional theory for fractional particle number: Derivative discontinuities of the energy," *Phys. Rev. Lett.* 49:1691–1694, 1982.

[191] C.-O. Almbladh and U. von Barth, "Exact results for the charge and spin densities, exchange–correlation potentials, and density-functional eigenvalues," *Phys. Rev. B* 31:3231–3244, 1985.

[192] R. W. Godby, M. Schlüter, and L. J. Sham, "Self-energy operators and exchange–correlation potentials in semiconductors," *Phys. Rev. B* 37:10159–10175, 1988.

[193] L. Hedin, "Electron correlation: Keeping close to an orbital description," *Int. J. Quant. Chemi.* 56:445–452, 1995.

[194] E. J. Baerends, O. V. Gritsenko, and R. van Meer, "The Kohn–Sham gap, the fundamental gap and the optical gap: The physical meaning of occupied and virtual Kohn–Sham orbital energies," *Phys. Chem. Chem. Phys.* 15:16408–16425, 2013.

[195] J. F. Janak, "Proof that $\partial e/\partial n_i = \epsilon_i$ in density-functional theory," *Phys. Rev. B* 18:7165, 1978.

[196] J. P. Perdew, R. G. Parr, M. Levy, and J. L. Balduz Jr., "Density-functional theory for fractional particle number: Derivative discontinuities of the energy," *Phys. Rev. Lett.* 49:1691–1694, 1982.

[197] A. J. Cohen, P. Mori-Sánchez, and W. Yang, "Fractional charge perspective on the band gap in density-functional theory," *Phys. Rev. B* 77:115123, 2008.

[198] A. J. Mori-Sánchez, P. Cohen, and W. Yang, "Discontinuous nature of the exchange–correlation functional in strongly correlated systems," *Phys. Rev. Lett.* 102:066403, 2009.

[199] W. Yang, A. J. Cohen, and P. Mori-Sánchez, "Derivative discontinuity, bandgap and lowest unoccupied molecular orbital in density functional theory," *J. Chem. Phys.* 136, 2012.

[200] E. Teller, "On the stability of molecules in the Thomas–Fermi theory," *Rev. Mod. Phys.* 34:627–631, 1962.

[201] A. Seidl, A. Görling, P. Vogl, J. A. Majewski, and M. Levy, "Generalized Kohn–Sham schemes and the band-gap problem," *Phys. Rev. B* 53:3764–3774, 1996.

[202] P. Rinke, A. Qteish, J. Neugebauer, C. Freysoldt, and M. Scheffler, "Combining GW calculations with exact-exchange density-functional theory: An analysis of valence-band photoemission for compound semiconductors," *New J. Phys.* 7:126, 2005.

[203] J. P. Perdew and A. Zunger, "Self-interaction correction to density-functional approximations for many-electron systems," *Phys Rev. B* 23:5048, 1981.

[204] R. T. Sharp and G. K. Horton, "A variational approach to the unipotential many-electron problem," *Phys. Rev.* 90:317, 1953.

[205] T. Grabo, T. Kreibich, S. Kurth, and E. K. U. Gross, in *Strong Coulomb Correlations in Electronic Structure: Beyond the Local Density Approximation,* edited by V. I. Anisimov, Gordon & Breach, Tokyo, 1998.

[206] K. Burke, "Perspective on density functional theory," *J. Chem. Phys.* 136:150901, 2012.

[207] D. C. Langreth and J. P. Perdew, "Exchange–correlation energy of a metallic surface," *Solid Sate Comm.* 17:1425–1429, 1975.

[208] D. C. Langreth and J. P. Perdew, "Exchange–correlation energy of a metallic surface: Wave-vector analysis," *Phys. Rev. B* 15:2884–2901, 1977.

[209] L. N. Oliveira, E. K. U. Gross, and W. Kohn, "Density-functional theory for superconductors," *Phys. Rev. Lett.* 60:2430–2433, 1988.

[210] M. Lüders, M. A. L. Marques, N. N. Lathiotakis, G. Floris, G. Profeta, L. Fast, A. Continenza, S. Massidda, and E. K. U. Gross, "Ab initio theory of superconductivity. I. Density functional formalism and approximate functionals," *Phys. Rev. B* 72:024545, 2005.

[211] M. A. L. Marques, M. Luders, N. N. Lathiotakis, G. Profeta, A. Floris, L. Fast, A. Continenza, E. K. U. Gross, and S. Massidda, "Ab initio theory of superconductivity. II. Application to elemental metals," *Phys. Rev. B* 72:024546, 2005.

[212] G. M. Eliashberg, "Interactions between electrons and lattice vibrations in a superconductor," *Soviet Phys. JETP* 11:696–702, 1960.

[213] P. Cudazzo, G. Profeta, A. Sanna, A. Floris, A. Continenza, S. Massidda, and E. K. U. Gross, "Ab initio description of high-temperature superconductivity in dense molecular hydrogen," *Phys. Rev. Lett.* 100:257001, 2008.

[214] *Time-Dependent Density Functional Theory, Lecture Notes in Physics,* Vol. 706, edited by F. Nogueira, A. Rubio, K. Burke, E. K. U. Gross, M. A. L. Marques, and C. A. Ullrich, Springer-Verlag, Berlin, 2006.

[215] M. Marques and E. K. U. Gross, "Time-dependent density functional theory," *Ann. Rev. Phys. Chem.* 55:427–455, 2004.

[216] S. Botti, A. Schindlmayr, R. Del Sole, and L. Reining, "Time-dependent density-functional theory for extended systems," *Rep. Prog. Phys.* 70:357–407, 2007.

[217] E. Runge and E. K. U. Gross, "Density-functional theory for time-dependent systems," *Phys. Rev. Lett.* 52:997–1000, 1984.

[218] E. K. U. Gross, C. A. Ullrich, and U. J. Gossmann, in *Density Functional Theory,* edited by E. K. U. Gross and R. M. Dreizler, Plenum Press, New York, 1995, p. 149.

[219] R. van Leeuwen, "Causality and symmetry in time-dependent density-functional theory," *Phys. Rev. Lett.* 80:1280–1283, 1998.

[220] A. Zangwill and P. Soven, "Density-functional approach to local-field effects in finite systems: Photoabsorption in the rare gases," *Phys. Rev. A* 21:1561, 1980.

[221] D. J. Thouless and J. G. Valatin, "Time-dependent Hartree–Fock equations and rotational states of nuclei," *Nucl. Phys.* 31:211, 1962.

[222] G. Vignale and M. Rasolt, "Current- and spin-density-functional theory for inhomogeneous electronic systems in strong magnetic fields," *Phys. Rev. B* 37:10685–10696, 1988.

[223] G. Vignale and W. Kohn, "Current-dependent exchange–correlation potential for dynamical linear response theory," *Phys. Rev. Lett.* 77:2037–2040, 1996.

[224] R. van Leeuwen, "Causality and symmetry in time-dependent density-functional theory," *Phys. Rev. Lett.* 80:1280–1283, 1998.

[225] M. Gatti, "Design of effective kernels for spectroscopy and molecular transport: Time-dependent current-density-functional theory," *J. Chem. Phys.* 134, 2011.

[226] A. W. Overhauser, "New mechanism of antiferromagnetism," *Phys. Rev. Lett.* 3:414–416, 1959.

[227] E. K. U. Gross, E. Runge, and O. Heinonen, *Many-Particle Theory,* Adam-Milger, Bristol, 1991.

[228] P. W. Anderson, "More is different: Broken symmetry and the nature of the heirarchical structure of science," *Science* 177:393–396, 1972.

[229] O. Penrose and L. Onsager, "Bose–Einstein condensation and liquid helium," *Phys. Rev.* 104:576–584, 1956.

[230] S. Huotari, J. A. Soininen, T. Pylkkänen, K. Hämäläinen, A. Issolah, A. Titov, J. McMinis, J. Kim, K. Esler, D. M. Ceperley, M. Holzmann, and V. Olevano, "Momentum distribution and renormalization factor in sodium and the electron gas," *Phys. Rev. Lett.* 105:086403, 2010.

[231] M. Holzmann, B. Bernu, C. Pierleoni, J. McMinis, D. M. Ceperley, V. Olevano, and L. Delle Site, "Momentum distribution of the homogeneous electron gas," *Phys. Rev. Lett.* 107:110402, 2011.

[232] D. Pines and P. Nozières, *The Theory of Quantum Liquids,* Vol. I (*Advanced Book Classics,* originally published in 1966), Addison-Wesley, Redwood City, CA, 1989.

[233] E. N. Economou, *Green's Functions in Quantum Physics, 2nd Ed.,* Springer-Verlag, Berlin, 1992.

[234] T. Matsubara, "A new approach to quantum-statistical mechanics," *Prog. Theor. Phys.* 14:351, 1955.

[235] D. Pines, *Elementary Excitations in Solids,* Wiley, New York, 1964.

[236] M. Marder, *Condensed Matter Physics,* Wiley, New York, 2000.

[237] H. B. Callen and T. A. Welton, "Irreversibility and generalized noise," *Phys. Rev.* 83:34–40, 1951.

[238] R. Kubo, "The fluctuation–dissipation theorem," *Rep. Prog. Phys.* 29:255–284, 1966.

[239] R. Kubo, "Statistical-mechanical theory of irreversible processes. I. General theory and simple applications to magnetic and conduction problems," *Rep. Prog. Phys.* 12:570, 1957.

[240] P. Nozières and D. Pines, "Electron interaction in solids. Characteristic energy loss spectrum," *Phys. Rev.* 113:1254–1267, 1959.

[241] W. Schülke and A. Kaprolat, "Nondiagonal response of Si by inelastic-X-ray-scattering experiments at Bragg position: Evidence for bulk plasmon bands," *Phys. Rev. Lett.* 67:879–882, 1991.

[242] J. Steibling, "Optical-properties of mono-crystalline silicon by electron-energy loss measurements," *Z. Phys. B: Condens. Matter* 31:355–357, 1978.

[243] V. Olevano and L. Reining, "Excitonic effects on the silicon plasmon resonance," *Phys. Rev. Lett.* 86:5962–5965, 2001.

[244] H.-C. Weissker, J. Serrano, S. Huotari, E. Luppi, M. Cazzaniga, F. Bruneval, F. Sottile, G. Monaco, V. Olevano, and L. Reining, "Dynamic structure factor and dielectric function of silicon for finite momentum transfer: Inelastic X-ray scattering experiments and ab initio calculations," *Phys. Rev. B* 81:085104, 2010.

[245] P. Lautenschlager, M. Garriga, L. Vina, and M. Cardona, "Temperature dependence of the dielectric function and interband critical points in silicon," *Phys. Rev. B* 36:4821–4830, 1987.

[246] J. J. Sakurai, *Modern Quantum Mechanics,* Pearson Education, Upper Saddle River, NJ, 1994.

[247] C.-O. Almbladh and L. Hedin, in *Handbook on Synchrotron Radiation,* edited by E. E. Koch, North-Holland, Amsterdam, 1983, Ch. 8.

[248] G. Strinati, "Application of the Green's-functions method to the study of the optical-properties of semiconductors," *Rivista del Nuovo Cimento* 11:1–86, 1988.

[249] S. Boffi, *Da Heisenberg a Landau. Introduzione alla fisica dei sistemi a molte particelle. Quaderni di Fisica Teorica,* Bibliopolis, Napoli, via Arangio Ruiz 83, 2004.

[250] U. von Barth, N. E. Dahlen, R. van Leeuwen, and G. Stefanucci, "Conserving approximations in time-dependent density functional theory," *Phys. Rev. B* 72:235109, 2005.

[251] T. G. Carlson, *Photoelectron and Auger Spectroscopy,* Plenum Press, New York, 1975.

[252] R. R. Rye, T. E. Madey, J. E. Houston, and P. H. Holloway, "Chemical-state effects in Auger-electron spectroscopy," *J. Chem. Phys.* 69:1504–1512, 1978.

[253] P. G. Fournier, J. Fournier, F. Salama, D. Stärk, S. D. Peyerimhoff, and J. H. D. Eland, "Theoretical and experimental studies of the electronic states of the diatomic cation Cl_2^{2+}," *Phys. Rev. A* 34:1657–1666, 1986.

[254] J. Appell, J. Durup, F. C Fehsenfe, and P. Fournier, "Double charge-transfer spectroscopy of diatomic-molecules," *J. Phys. B* 6:197–205, 1973.

[255] J. Appell, J. Durup, F. C. Fehsenfe, and P. Fournier, "Doubly ionized states of some polyatomic-molecules studied by double charge-transfer spectroscopy," *J. Phys. B* 7:406–414, 1974.

[256] G. Csanak, H. S. Taylor, and R. Yaris, "Green's function technique in atomic and molecular physics," *Adv. At. Mol. Phys.* 7:287–361, 1971.

[257] T. Kato, "On the eigenfunctions of many-particle systems in quantum mechanics," *Commun. Pure Appl. Math* 10:151–177, 1957.

[258] R. T. Pack and W. Byers Brown, "Cusp conditions for molecular wavefunctions," *J. Chem. Phys.* 45:556–559, 1966.

[259] R. Resta, "The insulating state of matter: A geometric approach," *Eur. Phys. J. B* 79:121–137, 2011.

[260] W. Kohn, "Theory of the insulating state," *Phys. Rev.* 133:A171–181, 1964.

[261] D. J. Scalapino, S. R. White, and S. Zhang, "Insulator, metal, or superconductor: The criteria," *Phys. Rev. B* 47:7995, 1993.

[262] I. Souza, T. J. Wilkens, and R. M. Martin, "Polarization and localization in insulators: Generating function approach," *Phys. Rev. B* 62:1666–1683, 2000.

[263] R. D. King-Smith and D. Vanderbilt, "Theory of polarization of crystalline solids," *Phys. Rev. B* 47:1651–1654, 1993.

[264] G. Ortiz and R. M. Martin, "Macroscopic polarization as a geometric quantum phase: Many-body formulation," *Phys. Rev. B* 49:14202–14210, 1994.

[265] M. Stone, *Quantum Hall Effect,* World Scientific, Singapore, 1992.

[266] M. Oshikawa and T. Senthil, "Fractionalization, topological order, and quasiparticle statistics," *Phys. Rev. Lett.* 96:060601, 2006.

[267] G. Temple, "The theory of Rayleigh's principle as applied to continuous systems," *Proc. Roy. Soc. London, Ser. A* 119:276–293, 1928.

[268] E. Hylleraas, "Neue Berectnumg der Energie des Heeliums im Grundzustande, sowie tiefsten Terms von Ortho-Helium," *Z. Phys.* 54:347, 1929.

[269] A. Bijl, "The lowest wave function of the symmetrical many particles system," *Physica* VII:869, 1940.

[270] R. B. Dingle, "The zero-point energy of a system of particles," *Phil. Mag.* 40:573, 1949.

[271] R. Jastrow, "Many-body problem with strong forces," *Phys. Rev.* 98:1479, 1955.

[272] R. B. Laughlin, "Quantized Hall conductivity in two dimensions," *Phys. Rev. B* 23:5632–5633, 1981.

[273] S. F. Boys and N. C. Handy, *Proc. Roy. Soc. London, Ser. A* 309:209, 1969.

[274] T. Gaskell, "The collective treatment of a Fermi gas: II," *Proc. Phys. Soc.* 77:1182, 1961.

[275] D. M. Ceperley, "Ground state of the fermion one-component plasma: A Monte Carlo study in two and three dimensions," *Phys. Rev. B* 18:3126–3138, 1978.

[276] W. H. Press and S. A. Teukolsky, *Numerical Recipes,* Cambridge University Press, Cambridge, 1992.

[277] E. Buendía, F. J. Gálvez, P. Maldonado, and A. Sarsa, "Quantum Monte Carlo ground state energies for the atoms Li through Ar," *J. Chem. Phys.* 131:044115, 2009.

[278] N. D. Drummond, Z. Radnai, J. R. Trail, M. D. Towler, and R. J. Needs, "Diffusion quantum Monte Carlo study of three-dimensional Wigner crystals," *Phys. Rev. B* 69:085116, 2004.

[279] Y. Kwon, D. M. Ceperley, and R. M. Martin, "Effects of three-body and backflow correlations in the 2D electron gas," *Phys. Rev. B.* 48:12037, 1993.

[280] M. Casula and S. Sorella, "Geminal wave functions with Jastrow correlation: A first application to atoms," *J. Chem. Phys.* 119:6500–6511, 2003.

[281] M. Casula, C. Attaccalite, and S. Sorella, "Correlated geminal wave function for molecules: An efficient resonating valence bond approach," *J. Chem. Phys.* 121:7110–7126, 2004.

[282] A. J. Leggett, *Quantum Liquids,* Oxford University Press, Oxford, 2006.

[283] M. Bajdich, L. Mitas, L. K. Wagner, and K. E. Schmidt, "Pfaffian pairing and backflow wavefunctions for electronic structure quantum Monte Carlo methods," *Phys. Rev. B* 77:115112, 2008.

[284] M. Holzmann, D. M. Ceperley, C. Pierleoni, and K. Esler, "Backflow correlations for the electron gas and metallic hydrogen," *Phys. Rev. E* 68:046707:1–15, 2003.

[285] Y. Kwon, D. M. Ceperley, and R. M. Martin, "Effects of backflow correlation in the three-dimensional electron gas: Quantum Monte Carlo study," *Phys. Rev. B* 58:6800–6806, 1998.

[286] P. López Ríos, A. Ma, N. D. Drummond, M. D. Towler, and R. J. Needs, "Inhomogeneous backflow transformations in quantum Monte Carlo calculations," *Phys. Rev. E* 74:066701, 2006.

[287] L. Hedin, "On correlation effects in electron spectroscopies and the GW approximation," *J. Phys. C* 11:R489–528, 1999.

[288] B. Farid, in *Electron Correlation in the Solid State*, edited by N. H. March, Imperial College Press, London, 1999, p. 103.

[289] A. J. Layzer, "Properties of the one-particle Green's function for nonuniform many-fermion systems," *Phys. Rev.* 129:897–907, 1963.

[290] A. Fleszar and W. Hanke, "Spectral properties of quasiparticles in a semiconductor," *Phys. Rev. B* 56:10228–10232, 1997.

[291] M. van Schilfgaarde, T. Kotani, and S. Faleev, "Quasiparticle self-consistent GW theory," *Phys. Rev. Lett.* 96:226402–226402, 2006.

[292] S. V. Faleev, M. van Schilfgaarde, and T. Kotani, "All-electron self-consistent GW approximation: Application to Si, MnO, and NiO," *Phys. Rev. Lett.* 93:126406, 2004.

[293] T. Kotani, M. van Schilfgaarde, and S. V. Faleev, "Quasiparticle self-consistent GW method: A basis for the independent-particle approximation," *Phys. Rev. B* 76:165106, 2007.

[294] A. Klein, "Perturbation theory for an infinite medium of fermions. II," *Phys. Rev.* 121:950–956, 1961.

[295] A. Klein and R. E. Prange, "Perturbation theory for an infinite medium of fermions," *Phys. Rev.* 112:994–1007, 1958.

[296] M. Potthoff, "Self-energy-functional approach to systems of correlated electrons," *Euro. Phys. J. B* 32:429–436, 2003.

[297] R. Chitra and G. Kotliar, "Effective-action approach to strongly correlated fermion systems," *Phys. Rev. B* 63:115110, 2001.

[298] C.-O. Almbladh, U. von Barth, and R. van Leeuwen, "Variational total energies from ϕ- and ψ-derivable theories," *Int. J. Mod. Phys.* 13:535–541, 1999.

[299] G. Baym and L. P. Kadanoff, "Conservation laws and correlation functions," *Phys. Rev.* 124:287–299, 1961.

[300] R. van Leeuwen, N. E. Dahlen, and A. Stan, "Total energies from variational functionals of the Green function and the renormalized four-point vertex," *Phys. Rev. B* 74:195105, 2006.

[301] G. Stefanucci and R. van Leeuwen, *Many-Body Theory of Quantum Systems*, Cambridge University Press, Cambridge, 2013.

[302] A. Georges, G. Kotliar, W. Krauth, and M. J. Rozenberg, "Dynamical mean-field theory of strongly correlated fermion systems and the limit of infinite dimensions," *Rev. Mod. Phys.* 68:13–125, 1996.

[303] G. Kotliar, S. Y. Savrasov, K. Haule, V. S. Oudovenko, O. Parcollet, and C. A. Marianetti, "Electronic structure calculations with dynamical mean-field theory," *Rev. Mod. Phys.* 78:865, 2006.

[304] B. A. Lippmann and J. Schwinger, "Variational principles for scattering processes. I," *Phys. Rev.* 79:469–480, 1950.

[305] L. D. Landau and E. M. Lifshitz, *Quantum Mechanics: Non-relativistic theory*, Pergamon Press, Oxford, 1977.

[306] Chr. Møller and M. S. Plesset, "Note on an approximation treatment for many-electron systems," *Phys. Rev.* 46:618–622, 1934.

[307] X. Ren, A. Tkatchenko, P. Rinke, and M. Scheffler, "Beyond the random-phase approximation for the electron correlation energy: The importance of single excitations," *Phys. Rev. Lett.* 106:153003, 2011.

[308] S. Y. Willow, K. S. Kim, and S. Hirata, "Stochastic evaluation of second-order many-body perturbation energies," *J. Chem. Phys.* 137:204122, 2012.

[309] R. J. Bartlett, "Many-body perturbation-theory and coupled cluster theory for electron correlation in molecules," *Ann. Rev. Phys. Chem.* 32:359–401, 1981.

[310] R. J. Bartlett and M. Musial, "Coupled-cluster theory in quantum chemistry," *Rev. Mod. Phys.* 79:291, 2007.

[311] M. Gell-Mann and F. Low, "Bound states in quantum field theory," *Phys. Rev.* 84:350–354, 1951.

[312] R. D. Mattuck, *A Guide to Feynman Diagrams in the Many-Body Problem,* Dover Books on Physics, Mineola, NY, 1992.

[313] G. C. Wick, "The evaluation of the collision matrix," *Phys. Rev.* 80:268–272, 1950.

[314] A. Houriet and A. Kind, "Classification invariante des termes de la matrice-s," *Helv. Phys. Acta* 22:319, 1949.

[315] J. Goldstone, "Derivation of the Brueckner many-body theory," *Proc. Roy. Soc. London, Ser. A* 239:267–279, 1957.

[316] W. Kohn and J. M. Luttinger, "Ground-state energy of a many-fermion system," *Phys. Rev.* 118:41–45, 1960.

[317] L. P. Kadanoff and G. Baym, *Quantum Statistical Mechanics,* W. A. Benjamin, New York, 1964.

[318] F. Aryasetiawan and S. Biermann, "Generalized Hedin's equations for quantum many-body systems with spin-dependent interactions," *Phys. Rev. Lett.* 100:116402, 2008.

[319] P. C. Martin and J. Schwinger, "Theory of many-particle systems. I," *Phys. Rev.* 115:1342–1373, 1959.

[320] J. Schwinger, "On the Green's functions of quantized fields .1." *Proc. Natl. Acad. Sci. USA* 37:452–455, 1951.

[321] G. Baym, "Self-consistent approximations in many-body systems," *Phys. Rev.* 127:1391–1401, 1962.

[322] C.-O. Almbladh, in *Progress in Nonequilibrium Green's Functions III,* Vol. 35 of *Journal of Physics Conference Series,* edited by M. Bonitz, and A. Filinov, IOP Publishing, Bristol, 2006, pp. 127–144.

[323] L. P. Kadanoff and P. C. Martin, "Hydrodynamic equations and correlation functions (Reprinted from *Ann. Phys.* 24:419–469, 1963)," *Ann. Phys.* 281:800–852, 2000.

[324] P. Danielewicz, "Quantum-theory of nonequilibrium processes. 1." *Ann. Phys.* 152:239–304, 1984.

[325] J. Lindhard, "On the properties of a gas of charged particles," *Kgl. Danske Videnskab. Selskab, Mat.-fys. Medd.* 28:1–57, 1954.

[326] J. Hubbard, "The description of collective motions in terms of many-body perturbation theory," *Proc. Roy. Soc. London, Ser. A* 240:539–560, 1957.

[327] J. J. Quinn and R. A. Ferrell, "Electron self-energy approach to correlation in a degenerate electron gas," *Phys. Rev.* 112:812–827, 1958; *Phys. Rev. Lett.* 1:303, 1958.

[328] R. H. Ritchie, "Interaction of charged particles with a degenerate Fermi–Dirac electron gas," *Phys. Rev.* 114:644–654, 1959.

[329] J. J. Quinn, "Range of excited electrons in metals," *Phys. Rev.* 1:1453, 1962.

[330] S. L. Adler, "Theory of range of hot electrons in real metals," *Phys. Rev.* 130:1654, 1963.

[331] J. J. Quinn, "The range of hot electrons and holes in metals," *App. Phys. Lett.* 2:167–169, 1963.

[332] E. P. Wigner, "Effects of the electron interaction on the energy levels of electrons in metals," *Trans. Faraday Soc.* 34:678, 1938.

[333] H. Stoll and A. Savin, *Density Functional Methods in Physics,* Plenum Press, New York, 1985.

[334] L. Hedin and S. Lundqvist, "Effects of electron–electron and electron–phonon interactions on the one-electron states of solids," *Solid State Phys.* 23:1, 1969.

[335] M. E. Casida and D. P. Chong, "Physical interpretation and assessment of the Coulomb-hole and screened-exchange approximation for molecules," *Phys. Rev. A* 40:4837–4848, 1989.

[336] M. Born and W. Heisenberg, "Über den Einfluss der Deformierbarekit der Ionen auf optische und chemische Konstanten. I." *Z. Phys.* 23:388, 1924.

[337] S. Ismail-Beigi and T. A. Arias, "Locality of the density matrix in metals, semiconductors, and insulators," *Phys. Rev. Lett.* 82:2127–2130, 1999.

[338] L. X. He and D. Vanderbilt, "Exponential decay properties of Wannier functions and related quantities," *Phys. Rev. Lett.* 86:5341–5344, 2001.

[339] S. N. Taraskin, D. A. Drabold, and S. R. Elliott, "Spatial decay of the single-particle density matrix in insulators: Analytic results in two and three dimensions," *Phys. Rev. Lett.* 88:196405–196408, 2002.

[340] S. N. Taraskin, P. A. Fry, X. Zhang, D. A. Drabold, and S. R. Elliott, "Spatial decay of the single-particle density matrix in tight-binding metals: Analytic results in two dimensions," *Phys. Rev. B* 66:233101, 2002.

[341] L. J. Sham and W. Kohn, "1-particle properties of an inhomogeneous interacting electron gas," *Phys. Rev.* 145:561–567, 1966.

[342] F. Gygi and A. Baldereschi, "Quasiparticle energies in semiconductors: Self-energy correction to the local-density approximation," *Phys. Rev. Lett.* 62:2160–2163, 1989.

[343] N. E. Zein, S. Y. Savrasov, and G. Kotliar, "Local self-energy approach for electronic structure calculations," *Phys. Rev. Lett.* 96:226403, 2006.

[344] H. N. Rojas, R. W. Godby, and R. J. Needs, "Space-time method for ab-initio calculations of self-energies and dielectric response functions of solids," *Phys. Rev. Lett.* 74:1827–1830, 1995.

[345] M. M. Rieger, L. Steinbeck, I. D. White, H. N. Rojas, and R. W. Godby, "The GW space-time method for the self-energy of large systems," *Comput. Phys. Commun.* 117:211–228, 1999.

[346] G. Strinati, H. J. Mattausch, and W. Hanke, "Dynamical aspects of correlation corrections in a covalent crystal," *Phys. Rev. B* 25:2867–2888, 1982.

[347] F. Aryasetiawan and O. Gunnarsson, "The GW method," *Rep. Prog. Phys.* 61:237–312, 1998.

[348] F. Bechstedt, M. Fiedler, C. Kress, and R. Del Sole, "Dynamical screening and quasi-particle spectral functions for nonmetals," *Phys. Rev. B* 49:7357–7362, 1994.

[349] N. E. Dahlen, A. Stan, and R. van Leeuwen, in *Progress in Nonequilibrium Green's Functions III,* Vol. 35 of *Journal of Physics Conference Series,* edited by M. Bonitz and A. Filinov, IOP Publishing, Bristol, 2006, pp. 324–339.

[350] A. Stan, N. E. Dahlen, and R. van Leeuwen, "Levels of self-consistency in the GW approximation," *J. Chem. Phys.* 130:224101, 2009.

[351] B. Holm and U. von Barth, "Fully self-consistent GW self-energy of the electron gas," *Phys. Rev. B* 57:2108–2117, 1998.

[352] U. von Barth and B. Holm, "Self-consistent GW results for the electron gas: Fixed screened potential W within the random-phase approximation," *Phys. Rev. B* 54:8411–8419, 1996 [Erratum: *Phys. Rev. B* 55:10120–10122, 1997].

[353] F. Bruneval, "GW approximation of the many-body problem and changes in the particle number," *Phys. Rev. Lett.* 103:176403, 2009.

[354] K. Frey, J. C. Idrobo, M. L. Tiago, F. Reboredo, and S. Öğüt, "Quasiparticle gaps and exciton Coulomb energies in Si nanoshells: First-principles calculations," *Phys. Rev. B* 80:153411, 2009.

[355] C. Massobrio, A. Pasquarello, and R. Car, "First principles study of photoelectron spectra of Cu_n^- clusters," *Phys. Rev. Lett.* 75:2104–2107, 1995.

[356] V. I. Anisimov, F. Aryasetiawan, and A. I. Lichtenstein, "First-principles calculations of the electronic structure and spectra of strongly correlated systems: The LDA+U method," *J. Phys. Condens. Matter* 9:767–808, 1997.

[357] A. Liebsch, H. Ishida, and G. Bihlmayer, "Coulomb correlations and orbital polarization in the metal–insulator transition of VO_2," *Phys. Rev. B* 71:085109, 2005.

[358] M. A. Korotin, N. A. Skorikov, and V. I. Anisimov, "Variation of orbital symmetry of the localized 3d(1) electron of the V^{4+} ion upon the metal–insulator transition in VO_2," *Phys. Met. Metall.* 94:17–23, 2002.

[359] A. Continenza, S. Massidda, and M. Posternak, "Self-energy corrections in VO_2 within a model GW scheme," *Phys. Rev. B* 60:15699–15704, 1999.

[360] M. Gatti, F. Bruneval, V. Olevano, and L. Reining, "Understanding correlations in vanadium dioxide from first principles," *Phys. Rev. Lett.* 99:266402, 2007.

[361] R. Sakuma, T. Miyake, and F. Aryasetiawan, "First-principles study of correlation effects in VO_2," *Phys. Rev. B* 78:075106, 2008.

[362] R. Sakuma, T. Miyake, and F. Aryasetiawan, "Quasiparticle band structure of vanadium dioxide," *J. Phys. Condens. Matter* 21:474212, 2009.

[363] J. P. Perdew and A. Zunger, "Self-interaction correction to density-functional approximations for many-electron systems," *Phys. Rev. B* 23:5048–5079, 1981.

[364] D. M. Bylander and L. Kleinman, "Good semiconductor band-gaps with a modified local-density approximation," *Phys. Rev. B* 41:7868–7871, 1990.

[365] J. Heyd, G. E. Scuseria, and M. Ernzerhof, "Hybrid functionals based on a screened Coulomb potential (Erratum: *J. Chem. Phys.* 124:219906, 2006)," *J. Chem. Phys.* 118:8207–8215, 2003.

[366] J. Vidal, S. Botti, P. Olsson, J.-F. Guillemoles, and L. Reining, "Strong interplay between structure and electronic properties in CuIn(S, Se)(2): A first-principles study," *Phys. Rev. Lett.* 104:056401, 2010.

[367] A. Schindlmayr, "Violation of particle number conservation in the GW approximation," *Phys. Rev. B* 56:3528–3531, 1997.

[368] A. Schindlmayr, P. García-González, and R. W. Godby, "Diagrammatic self-energy approximations and the total particle number," *Phys. Rev. B* 64:235106, 2001.

[369] P. García-González and R. W. Godby, "Self-consistent calculation of total energies of the electron gas using many-body perturbation theory," *Phys. Rev. B* 63:075112, 2001.

[370] M. M. Rieger and R. W. Godby, "Charge density of semiconductors in the GW approximation," *Phys. Rev. B* 58:1343–1348, 1998.

[371] V. M. Galitskii and A. B. Migdal, "Application of quantum field theory methods to the many body problem," *Soviet Phys. JETP* 7:96–104, 1958.

[372] C. O. Almbladh, U. von Barth, and R. van Leeuwen, "Variational total energies from Phi- and Psi-derivable theories," *Int. J. Mod. Phys. B* 13:535–541, 1999.

[373] N. E. Dahlen, R. van Leeuwen, and U. von Barth, "Variational energy functionals of the Green function and of the density tested on molecules," *Phys. Rev. A* 73:012511, 2006.

[374] M. P. Agnihotri, W. Apel, and W. Weller, "The Luttinger–Ward method applied to the 2D Coulomb gas," *Phys. Stat. Sol. B* 245:421–427, 2008.

[375] N. E. Dahlen and U. von Barth, "Variational energy functionals tested on atoms," *Phys. Rev. B* 69:195102, 2004.

[376] F. Furche, "Molecular tests of the random phase approximation to the exchange–correlation energy functional," *Phys. Rev. B* 64:195120, 2001.

[377] M. Fuchs and X. Gonze, "Accurate density functionals: Approaches using the adiabatic-connection fluctuation–dissipation theorem," *Phys. Rev. B* 65:235109, 2002.

[378] J. Paier, X. Ren, P. Rinke, G. E. Scuseria, A. Grueneis, G. Kresse, and M. Scheffler, "Assessment of correlation energies based on the random-phase approximation," *New J. Phys.* 14:053020, 2012.

[379] F. Aryasetiawan, T. Miyake, and K. Terakura, "Total energy method from many-body formulation," *Phys. Rev. Lett.* 88:166401, 2002.

[380] T. Miyake, F. Aryasetiawan, T. Kotani, M. van Schilfgaarde, M. Usuda, and K. Terakura, "Total energy of solids: An exchange and random-phase approximation correlation study," *Phys. Rev. B* 66:245103, 2002.

[381] M. Fuchs, K. Burke, Y.-M. Niquet, and X. Gonze, "Comment on 'Total energy method from many-body formulation,'" *Phys. Rev. Lett.* 90:189701, 2003.

[382] F. Aryasetiawan, T. Miyake, and K. Terakura, "Aryasetiawan, Miyake, and Terakura reply," *Phys. Rev. Lett.* 90:189702, 2003.

[383] B. I. Lundqvist, "Single-particle spectrum of the degenerate electron gas," *Z. Phys. B: Condens. Matter* 6:193–205, 1967.

[384] B. I. Lundqvist, "Single-particle spectrum of degenerate electron gas 2. Numerical results for electrons coupled to plasmons," *Phys. Kondens. Mater.* 6:206, 1967.

[385] H. O. Frota and G. D. Mahan, "Band tails and bandwidth in simple metals," *Phys. Rev. B* 45:6243–6246, 1992.

[386] A. J. Morris, M. Stankovski, K. T. Delaney, P. Rinke, P. García-González, and R. W. Godby, "Vertex corrections in localized and extended systems," *Phys. Rev. B* 76:155106, 2007.

[387] R. Maezono, M. D. Towler, Y. Lee, and R. J. Needs, "Quantum Monte Carlo study of sodium," *Phys. Rev. B* 68:165103, 2003.

[388] I.-W. Lyo and E. W. Plummer, "Quasiparticle band structure of Na and simple metals," *Phys. Rev. Lett.* 60:1558–1561, 1988.

[389] E. Jensen and E. W. Plummer, "Experimental band structure of Na," *Phys. Rev. Lett.* 55:1912–1915, 1985.

[390] M. Holzmann, B. Bernu, C. Pierleoni, J. McMinis, D. M. Ceperley, V. Olevano, and L. Delle Site, "Momentum distribution of the homogeneous electron gas," *Phys. Rev. Lett.* 107:110402, 2011.

[391] M. Vogt, R. Zimmermann, and R. J. Needs, "Spectral moments in the homogeneous electron gas," *Phys. Rev. B* 69:045113, 2004.

[392] B. Holm, "Total energies from GW calculations," *Phys. Rev. Lett.* 83:788–791, 1999.

[393] A. Schindlmayr, T. J. Pollehn, and R. W. Godby, "Spectra and total energies from self-consistent many-body perturbation theory," *Phys. Rev. B* 58:12684–12690, 1998.

[394] P. Rinke, K. Delaney, P. Garcia-Gonzalez, and R. W. Godby, "Image states in metal clusters," *Phys. Rev. A* 70:063201, 2004.

[395] P. García-González and R. W. Godby, "Many-body GW calculations of ground-state properties: Quasi-2D electron systems and van der Waals forces," *Phys. Rev. Lett.* 88:056406, 2002.

[396] P. Romaniello, S. Guyot, and L. Reining, "The self-energy beyond GW: Local and nonlocal vertex corrections," *J. Chem. Phys.* 131:154111, 2009.

[397] *The LDA+DMFT Approach to Strongly Correlated Materials: Lecture Notes of the Autumn School 2011, Forschungszentrum Juelich,* edited by E. Pavarini, E. Koch, D. Vollhardt, and A. Lichtenstein, Forschungszentrum Juelich GmbH and Institute for Advanced Simulations, Juelich, Germany, 2011.

[398] W. Nelson, P. Bokes, P. Rinke, and R. W. Godby, "Self-interaction in Green's-function theory of the hydrogen atom," *Phys. Rev. A* 75:032505, 2007.

[399] J. J. Fernandez, "GW calculations in an exactly solvable model system at different dilution regimes: The problem of the self-interaction in the correlation part," *Phys. Rev. A* 79:052513, 2009.

[400] F. Aryasetiawan, R. Sakuma, and K. Karlsson, "GW approximation with self-screening correction," *Phys. Rev. B* 85:035106, 2012.

[401] C. Verdozzi, R. W. Godby, and S. Holloway, "Evaluation of GW approximations for the self-energy of a Hubbard cluster," *Phys. Rev. Lett.* 74:2327–2330, 1995.

[402] T. J. Pollehn, A Schindlmayr, and R. W. Godby, "Assessment of the GW approximation using Hubbard chains," *J. Phys. Condens. Matter* 10:1273–1283, 1998.

[403] X. Wang, C. D. Spataru, M. S. Hybertsen, and A. J. Millis, "Electronic correlation in nanoscale junctions: Comparison of the GW approximation to a numerically exact solution of the single-impurity Anderson model," *Phys. Rev. B* 77:045119, 2008.

[404] F. Caruso, D. R. Rohr, M. Hellgren, X. Ren, P. Rinke, A. Rubio, and M. Scheffler, "Bond breaking and bond formation: How electron correlation is captured in many-body perturbation theory and density-functional theory," *Phys. Rev. Lett.* 110:146403, 2013.

[405] L. Wolniewicz, "Relativistic energies of the ground state of the hydrogen molecule," *J. Chem. Phys.* 99(3):1851–1868, 1993.

[406] D. C. Langreth, "Singularities in X-ray spectra of metals," *Phys. Rev. B* 1:471, 1970.

[407] B. I. Lundqvist, "Characteristic structure in core electron spectra of metals due to electron–plasmon coupling," *Phys. Kondens. Mater.* 9:236, 1969.

[408] W. G. Aulbur, L. Jonsson, and J. W. Wilkins, "Quasiparticle calculations in solids," *Solid State Phys.* 54:1–218, 2000.

[409] J. J. Quinn and R. A. Ferrell, "Electron self-energy approach to correlation in a degenerate electron gas," *Phys. Rev.* 112:812–827, 1958.

[410] D. F. DuBois, "Electron interactions: Part I. Field theory of a degenerate electron gas," *Ann. Phys.* 7:174–237, 1959.

[411] D. F. DuBois, "Electron interactions: Part II. Properties of a dense electron gas," *Ann. Phys.* 8:24–77, 1959.

[412] B. I. Lundqvist, "Single-particle spectrum of degenerate electron gas 3. Numerical results in random phase approximation," *Phys. Kondens. Mater.* 7:117, 1968.

[413] T. M. Rice, "Effects of electron–electron interaction on properties of metals," *Ann. Phys.* 31:100, 1965.

[414] G. D. Mahan and B. E. Sernelius, "Electron–electron interactions and the bandwidth of metals," *Phys. Rev. Lett.* 62:2718–2720, 1989.

[415] H. O. Frota and G. D. Mahan, "Band tails and bandwidth in simple metals," *Phys. Rev. B* 45:6243–6246, 1992.

[416] G. Strinati, H. J. Mattausch, and W. Hanke, "Dynamical correlation effects on the quasiparticle Bloch states of a covalent crystal," *Phys. Rev. Lett.* 45:290–294, 1980.

[417] G. Strinati, H. J. Mattausch, and W. Hanke, "Dynamical aspects of correlation corrections in a covalent crystal," *Phys. Rev. B* 25:2867–2888, 1982.

[418] M. S. Hybertsen and S. G. Louie, "First-principles theory of quasiparticles: Calculation of band gaps in semiconductors and insulators," *Phys. Rev. Lett.* 55:1418–1421, 1985.

[419] M. S. Hybertsen and S. G. Louie, "Electron correlation and the band gap in ionic crystals," *Phys. Rev. B* 32:7005–7008, 1985.

[420] M. S. Hybertsen and S. G. Louie, "Erratum: Electron correlation and the band gap in ionic crystals," *Phys. Rev. B* 35:9308, 1987.

[421] M. S. Hybertsen and S. G. Louie, "Electron correlation in semiconductors and insulators: Band gaps and quasiparticle energies," *Phys. Rev. B* 34:5390–5413, 1986.

[422] R. W. Godby, M. Schlüter, and L. J. Sham, "Accurate exchange–correlation potential for silicon and its discontinuity on addition of an electron," *Phys. Rev. Lett.* 56:2415–2418, 1986.

[423] R. W. Godby, M. Schlüter, and L. J. Sham, "Quasiparticle energies in GaAs and AlAs," *Phys. Rev. B* 35:4170–4171, 1987.

[424] R. W. Godby, M. Schlüter, and L. J. Sham, "Trends in self-energy operators and their corresponding exchange–correlation potentials," *Phys. Rev. B* 36:6497–6500, 1987.

[425] L. Hedin, B. I. Lundqvist, and S. Lundqvist, "New structure in single-particle spectrum of an electron gas," *Solid State Commun.* 5:237–239, 1967.

[426] H. Ness, L. K. Dash, M. Stankovski, and R. W. Godby, "GW approximations and vertex corrections on the Keldysh time-loop contour: Application for model systems at equilibrium," *Phys. Rev. B* 84:195114, 2011.

[427] B. Farid, "Self-consistent density-functional approach to the correlated ground states and an unrestricted many-body perturbation theory," *Phil. Mag. B* 76:145–192, 1996.

[428] L. Hedin, "Electron correlation – keeping close to an orbital description," *Int. J. Quantum Chem.* 56:445–452, 1995.

[429] J.-L. Li, G.-M. Rignanese, E. K. Chang, X. Blase, and S. G. Louie, "GW study of the metal–insulator transition of bcc hydrogen," *Phys. Rev. B* 66:035102, 2002.

[430] P. Rinke, A. Qteish, J. Neugebauer, C. Freysoldt, and M. Scheffler, "Combining GW calculations with exact-exchange density-functional theory: An analysis of valence-band photoemission for compound semiconductors," *New J. Phys.* 7:126, 2005.

[431] E. L. Shirley and R. M. Martin, "GW quasiparticle calculations in atoms," *Phys. Rev. B* 47:15404–15412, 1993.

[432] A. D. Becke, "A new mixing of Hartree–Fock and local density-functional theories," *J. Chem. Phys.* 98:1372–1377, 1993.

[433] J. Heyd, G. E. Scuseria, and M. Ernzerhof, "Hybrid functionals based on a screened Coulomb potential," *J. Chem. Phys.* 118:8207–8215, 2003.

[434] F. Fuchs, J. Furthmueller, F. Bechstedt, M. Shishkin, and G. Kresse, "Quasiparticle band structure based on a generalized Kohn–Sham scheme," *Phys. Rev. B* 76:115109, 2007.

[435] V. I. Anisimov, I. V. Solovyev, M. A. Korotin, M. T. Czyzyk, and G. A. Sawatzky, "Density functional theory and NiO photoemission spectra," *Phys. Rev. B* 48:16929–16934, 1993.

[436] T. Miyake, P. Zhang, M. L. Cohen, and S. G. Louie, "Quasiparticle energy of semicore d electrons in ZnS: Combined LDA+U and GW approach," *Phys. Rev. B* 74:245213, 2006.

[437] E. Kioupakis, P. Zhang, M. L. Cohen, and S. G. Louie, "GW quasiparticle corrections to the LDA+U and GGA+U electronic structure of bcc hydrogen," *Phys. Rev. B* 77:155114, 2008.

[438] S. Kobayashi, Y. Nohara, S. Yamamoto, and T. Fujiwara, "GW approximation with LSDA+U method and applications to NiO, MnO, and V_2O_3," *Phys. Rev. B* 78:155112, 2008.

[439] H. Jiang, R. I. Gomez-Abal, P. Rinke, and M. Scheffler, "Localized and itinerant states in lanthanide oxides united by GW @ LDA+U," *Phys. Rev. Lett.* 102:126403, 2009.

[440] B.I. Lundqvist, *Phys. Kondens. Mater.* 6:206, 1967.

[441] R. W. Godby and R. J. Needs, "Metal–insulator transition in Kohn–Sham theory and quasiparticle theory," *Phys. Rev. Lett.* 62:1169–1172, 1989.

[442] R. Shaltaf, G. M. Rignanese, X. Gonze, F. Giustino, and A. Pasquarello, "Band offsets at the Si/SiO_2 interface from many-body perturbation theory," *Phys. Rev. Lett.* 100:186401, 2008.

[443] J. A. Soininen, J. J. Rehr, and E. L. Shirley, "Electron self-energy calculation using a general multi-pole approximation," *J. Phys. Condens. Matter* 15:2573–2586, 2003.

[444] J. A. Soininen, J. J. Rehr, and E. L. Shirley, "Multi-pole representation of the dielectric matrix," *Physica Scripta* T115:243–245, 2005.

[445] Z. H. Levine and S. G. Louie, "New model dielectric function and exchange–correlation potential for semiconductors and insulators," *Phys. Rev. B* 25:6310–6316, 1982.

[446] M. S. Hybertsen and S. G. Louie, "Model dielectric matrices for quasiparticle self-energy calculations," *Phys. Rev. B* 37:2733–2736, 1988.

[447] F. Bechstedt, R. Delsole, G. Cappellini, and L. Reining, "An efficient method for calculating quasi-particle energies in semiconductors," *Solid State Commun.* 84:765–770, 1992.

[448] W. E. Pickett and C. S. Wang, "Local-density approximation for dynamical correlation corrections to single-particle excitations in insulators," *Phys. Rev. B* 30:4719–4733, 1984.

[449] S. Massidda, A. Continenza, M. Posternak, and A. Baldereschi, "Band-structure picture for MnO reexplored: A model GW calculation," *Phys. Rev. Lett.* 74:2323–2326, 1995.

[450] S. Massidda, A. Continenza, M. Posternak, and A. Baldereschi, "Quasiparticle energy bands of transition-metal oxides within a model GW scheme," *Phys. Rev. B* 55:13494–13502, 1997.

[451] V. Fiorentini and A. Baldereschi, "Dielectric scaling of the self-energy scissor operator in semiconductors and insulators," *Phys. Rev. B* 51:17196–17198, 1995.

[452] J. C. Phillips, "Generalized Koopmans' theorem," *Phys. Rev.* 123:420–424, 1961.

[453] G. Arbman and U. von Barth, "Correlated potentials for simple metals: Aluminium," *J. Phys. F: Met. Phys.* 5:1155, 1975.

[454] M. L. Tiago, S. Ismail-Beigi, and S. G. Louie, "Effect of semicore orbitals on the electronic band gaps of Si, Ge, and GaAs within the GW approximation," *Phys. Rev. B* 69:125212, 2004.

[455] R. Gomez-Abal, X. Li, M. Scheffler, and C. Ambrosch-Draxl, "Influence of the core–valence interaction and of the pseudopotential approximation on the electron self-energy in semiconductors," *Phys. Rev. Lett.* 101:106404, 2008.

[456] M. Rohlfing, P. Krüger, and J. Pollmann, "Quasiparticle band structure of CdS," *Phys. Rev. Lett.* 75:3489–3492, 1995.

[457] P. Rinke, A. Qteish, J. Neugebauer, C. Freysoldt, and M. Scheffler, "Combining GW calculations with exact-exchange density-functional theory: An analysis of valence-band photoemission for compound semiconductors," *New J. Phys.* 7:126, 2005.

[458] A. Marini, G. Onida, and R. Del Sole, "Quasiparticle electronic structure of copper in the GW approximation," *Phys. Rev. Lett.* 88:016403, 2001.

[459] F. Bruneval, N. Vast, L. Reining, M. Izquierdo, F. Sirotti, and N. Barrett, "Exchange and correlation effects in electronic excitations of Cu_2O," *Phys. Rev. Lett.* 97:267601, 2006.

[460] W. Muller and W. Meyer, "Ground-state properties of alkali dimers and their cations (including the elements Li, Na, and K) from ab initio calculations with effective core polarization potentials," *J. Chem. Phys.* 80:3311–3320, 1984.

[461] W. Muller, J. Flesch, and W. Meyer, "Treatment of intershell correlation-effects in ab initio calculations by use of core polarization potentials – method and application to alkali and alkaline-earth atoms," *J. Chem. Phys.* 80:3297–3310, 1984.

[462] E. L. Shirley, X. Zhu, and S. G. Louie, "Core polarization in semiconductors: Effects on quasiparticle energies," *Phys. Rev. Lett.* 69:2955–2958, 1992.

[463] E. L. Shirley and R. M. Martin, "Many-body core-valence partitioning," *Phys. Rev. B* 47:15413–15427, 1993.

[464] E. Luppi, H.-C. Weissker, S. Bottaro, F. Sottile, V. Veniard, L. Reining, and G. Onida, "Accuracy of the pseudopotential approximation in *ab initio* theoretical spectroscopies," *Phys. Rev. B* 78:245124, 2008.

[465] P. E. Blöchl, "Projector augmented-wave method," *Phys. Rev. B* 50:17953–17979, 1994.

[466] G. Kresse and D. Joubert, "From ultrasoft pseudopotentials to the projector augmented-wave method," *Phys. Rev. B* 59:1758–1775, 1999.

[467] B. Arnaud and M. Alouani, "All-electron projector-augmented-wave GW approximation: Application to the electronic properties of semiconductors," *Phys. Rev. B* 62:4464–4476, 2000.

[468] M. Gajdoš, K. Hummer, G. Kresse, J. Furthmüller, and F. Bechstedt, "Linear optical properties in the projector-augmented wave methodology," *Phys. Rev. B* 73:045112, 2006.

[469] M. Shishkin and G. Kresse, "Implementation and performance of the frequency-dependent GW method within the PAW framework," *Phys. Rev. B* 74:035101, 2006.

[470] M. Gatti, G. Panaccione, and L. Reining, "Effects of low-energy excitations on spectral properties at higher binding energy: The metal–insulator transition of VO)$_2$," *Phys. Rev. Lett.* 114:116402, 2015.

[471] M. van Schilfgaarde, T. Kotani, and S. V. Faleev, "Adequacy of approximations in GW theory," *Phys. Rev. B* 74:245125, 2006.

[472] Y. M. Niquet and X. Gonze, "Band-gap energy in the random-phase approximation to density-functional theory," *Phys. Rev. B* 70:245115, 2004.

[473] G. D. Mahan, *Many-Particle Physics,* Plenum Press, New York, 1990.

[474] M. Marsili, O. Pulci, F. Bechstedt, and R. Del Sole, "Electronic structure of the C(111) surface: Solution by self-consistent many-body calculations," *Phys. Rev. B* 72:115415, 2005.

[475] F. Bruneval, N. Vast, and L. Reining, "Effect of self-consistency on quasiparticles in solids," *Phys. Rev. B* 74:045102, 2006.

[476] R. Sakuma, T. Miyake, and F. Aryasetiawan, "Effective quasiparticle Hamiltonian based on Löwdin's orthogonalization," *Phys. Rev. B* 80:235128, 2009.

[477] M. Shishkin and G. Kresse, "Self-consistent GW calculations for semiconductors and insulators," *Phys. Rev. B* 75:235102, 2007.

[478] F. Caruso, P. Rinke, X. Ren, M. Scheffler, and A. Rubio, "Unified description of ground and excited states of finite systems: The self-consistent GW approach," *Phys. Rev. B* 86:081102, 2012.

[479] N. Marom, F. Caruso, X. Ren, O. T. Hofmann, T. Körzdörfer, J. R. Chelikowsky, A. Rubio, M. Scheffler, and P. Rinke, "Benchmark of GW methods for azabenzenes," *Phys. Rev. B* 86:245127, 2012.

[480] H. J. de Groot, P. A. Bobbert, and W. van Haeringen, "Self-consistent GW for a quasi-one-dimensional semiconductor," *Phys. Rev. B* 52:11000–11007, 1995.

[481] W.-D. Schöne and A. G. Eguiluz, "Self-consistent calculations of quasiparticle states in metals and semiconductors," *Phys. Rev. Lett.* 81:1662–1665, 1998.

[482] G. Lani, P. Romaniello, and L. Reining, "Approximations for many-body Green's functions: Insights from the fundamental equations," *New J. Phys.* 14, 2012.

[483] J. A. Berger, P. Romaniello, F. Tandetzky, B. S. Mendoza, C. Brouder, and L. Reining, "Solution to the many-body problem in one point," *New J. Phys.* 16:113025, 2014.

[484] F. Aryasetiawan, "Self-energy of ferromagnetic nickel in the GW approximation," *Phys. Rev. B* 46:13051–13064, 1992.

[485] Y. Dewulf, D. Van Neck, and M. Waroquier, "Discrete approach to self-consistent GW calculations in an electron gas," *Phys. Rev. B* 71:245122, 2005.

[486] B. Farid, R. Daling, D. Lenstra, and W. van Haeringen, "GW approach to the calculation of electron self-energies in semiconductors," *Phys. Rev. B* 38:7530–7534, 1988.

[487] S. Lebègue, B. Arnaud, M. Alouani, and P. E. Bloechl, "Implementation of an all-electron GW approximation based on the projector augmented wave method without plasmon pole approximation: Application to Si, SiC, AlAs, InAs, NaH, and KH," *Phys. Rev. B* 67:155208, 2003.

[488] H. N. Rojas, R. W. Godby, and R. J. Needs, "Space-time method for ab initio calculations of self-energies and dielectric response functions of solids," *Phys. Rev. Lett.* 74:1827–1830, 1995.

[489] O. W. Day, D. W. Smith, and C. Garrod, "Generalization of Hartree–Fock one-particle potential," *Int. J. Quant. Chem.* 8:501–509, 1974.

[490] M. M. Morrell, R.G. Parr, and M. Levy, "Calculation of ionization-potentials from density matrices and natural functions, and long-range behavior of natural orbitals and electron-density," *J. Chem. Phys.* 62(2):549–554, 1975.

[491] H. Eshuis, J. E. Bates, and F. Furche, "Electron correlation methods based on the random phase approximation," *Theor. Chem. Acc.* 131(1):1084, 2012.

[492] H. F. Wilson, F. Gygi, and G. Galli, "Efficient iterative method for calculations of dielectric matrices," *Phys. Rev. B* 78:113303, 2008.

[493] M. Govoni and G. Galli, "Large scale GW calculations," *J. Chem. Theory Comput.* 11:2680–2696, 2015.

[494] P. Umari, G. Stenuit, and S. Baroni, "Optimal representation of the polarization propagator for large-scale GW calculations," *Phys. Rev. B* 79:201104, 2009.

[495] P. Umari, G. Stenuit, and S. Baroni, "GW quasiparticle spectra from occupied states only," *Phys. Rev. B* 81:115104, 2010.

[496] F. Giustino, M. L. Cohen, and S. G. Louie, "GW method with the self-consistent Sternheimer equation," *Phys. Rev. B* 81:115105, 2010.

[497] O. K. Andersen, "Linear methods in band theory," *Phys. Rev. B* 12:3060–3083, 1975.

[498] F. Aryasetiawan and O. Gunnarsson, "Linear-muffin-tin-orbital method with multiple orbitals per L channel," *Phys. Rev. B* 49:7219–7232, 1994.

[499] F. Aryasetiawan and O. Gunnarsson, "Electronic structure of NiO in the GW approximation," *Phys. Rev. Lett.* 74:3221–3224, 1995.

[500] M. Rohlfing, P. Krüger, and J. Pollmann, "Quasiparticle band-structure calculations for C, Si, Ge, GaAs, and SiC using Gaussian-orbital basis sets," *Phys. Rev. B* 48:17791–17805, 1993.

[501] X. Blase, C. Attaccalite, and V. Olevano, "First-principles GW calculations for fullerenes, porphyrins, phtalocyanine, and other molecules of interest for organic photovoltaic applications," *Phys. Rev. B* 83:115103, 2011.

[502] N. Marzari and D. Vanderbilt, "Maximally localized generalized Wannier functions for composite energy bands," *Phys. Rev. B* 56:12847–12865, 1997.

[503] J. M. Soler, E. Artacho, J. D. Gale, A. Garcia, J. Junquera, P. Ordejon, and D. Sanchez-Portal, "The SIESTA method for ab initio order-*N* materials simulation," *J. Phys. Condens. Matter* 14:2745–2779, 2002.

[504] V. Blum, R. Gehrke, F. Hanke, P. Havu, V. Havu, X. Ren, K. Reuter, and M. Scheffler, "Ab initio molecular simulations with numeric atom-centered orbitals," *Comput. Phys. Commun.* 180:2175–2196, 2009.

[505] C. Friedrich, S. Blügel, and A. Schindlmayr, "Efficient implementation of the GW approximation within the all-electron FLAPW method," *Phys. Rev. B* 81:125102, 2010.

[506] M. Weinert, E. Wimmer, and A. J. Freeman, "Total-energy all-electron density functional method for bulk solids and surfaces," *Phys. Rev. B* 26:4571, 1982.

[507] F. Aryasetiawan and O. Gunnarsson, "Product-basis method for calculating dielectric matrices," *Phys. Rev. B* 49:16214–16222, 1994.

[508] G. Onida, L. Reining, R. W. Godby, R. Del Sole, and W. Andreoni, "*Ab initio* calculations of the quasiparticle and absorption spectra of clusters: The sodium tetramer," *Phys. Rev. Lett.* 75:818–821, 1995.

[509] S. Ismail-Beigi, "Truncation of periodic image interactions for confined systems," *Phys. Rev. B* 73:233103, 2006.

[510] O. Pulci, G. Onida, R. Del Sole, and L. Reining, "Ab initio calculation of self-energy effects on optical properties of GaAs(110)," *Phys. Rev. Lett.* 81:5374–5377, 1998.

[511] A. Schindlmayr, "Analytic evaluation of the electronic self-energy in the GW approximation for two electrons on a sphere," *Phys. Rev. B* 87:075104, 2013.

[512] J. Klimes, M. Kaltak, and G. Kresse, "Predictive GW calculations using plane waves and pseudopotentials," *Phys. Rev. B* 90:075125, 2014.

[513] K. Delaney, P. Garcia-Gonzalez, A. Rubio, P. Rinke, and R. W. Godby, "Comment on 'Band-gap problem in semiconductors revisited: Effects of core states and many-body self-consistency,'" *Phys. Rev. Lett.* 93:249701, 2004.

[514] J. A. Berger, L. Reining, and F. Sottile, "Ab initio calculations of electronic excitations: Collapsing spectral sums," *Phys. Rev. B* 82:041103(R), 2010.

[515] E. Prodan and W. Kohn, "Nearsightedness of electronic matter," *Proc. Natl. Acad. Sci. USA* 102:11635–11638, 2005.

[516] J. C. Phillips and L. Kleinman, "Crystal potential and energy bands of semiconductors. IV. Exchange and correlation," *Phys. Rev.* 128:2098–2102, 1962.

[517] W. Brinkman and B. Goodman, "Crystal potential and correlation for energy bands in valence semiconductors," *Phys. Rev.* 149:597–613, 1966.

[518] E. O. Kane, "Comparison of screened exchange with the Slater approximation for silicon," *Phys. Rev. B* 5:1493–1499, 1972.

[519] W. Ku and A. G. Eguiluz, "Band-gap problem in semiconductors revisited: Effects of core states and many-body self-consistency," *Phys. Rev. Lett.* 89:126401, 2002.

[520] R. Gomez-Abal, X. Li, M. Scheffler, and C. Ambrosch-Draxl, "Influence of the core–valence interaction and of the pseudopotential approximation on the electron self-energy in semiconductors," *Phys. Rev. Lett.* 101:106404, 2008.

[521] M. Guzzo, G. Lani, F. Sottile, P. Romaniello, M. Gatti, J. J. Kas, J. J. Rehr, M. G. Silly, F. Sirotti, and L. Reining, "Valence electron photoemission spectrum of semiconductors: *Ab initio* description of multiple satellites," *Phys. Rev. Lett.* 107:166401, 2011.

[522] F. J. Himpsel, P. Heimann, and D. E. Eastman, "Surface-states on Si(111)-(2x1)," *Phys. Rev. B* 24:2003–2008, 1981.

[523] A. L. Wachs, T. Miller, T. C. Hsieh, A. P. Shapiro, and T.-C. Chiang, "Angle-resolved photoemission studies of Ge(111)-c(2X1), Ge(111)-(1X1)H, Si(111)-(7X7), and Si(100)-(2X1)," *Phys. Rev. B* 32:2326–2333, 1985.

[524] D. H. Rich, T. Miller, G. E. Franklin, and T. C. Chiang, "Sb-induced bulk band transitions in Si(111) and Si(001) observed in synchrotron photoemission-studies," *Phys. Rev. B* 39:1438–1441, 1989.

[525] D. H. Rich, G. E. Franklin, F. M. Leibsle, T. Miller, and T.-C. Chiang, "Synchrotron photoemission studies of the Sb-passivated Si surfaces: Degenerate doping and bulk band dispersions," *Phys. Rev. B* 40:11804–11816, 1989.

[526] C. Bowles, A. S. Kheifets, V. A. Sashin, M. Vos, and E. Weigold, "The direct measurement of spectral momentum densities of silicon with high energy (e, 2e) spectroscopy," *J. Electron Spectrosc. Relat. Phenom.* 141:95–104, 2004.

[527] M. Vos, C. Bowles, A. S. Kheifets, and M. R. Went, "Band structure of silicon as measured in extended momentum space," *Phys. Rev. B* 73:085207, 2006.

[528] J. E. Ortega and F. J. Himpsel, "Inverse-photoemission study of Ge(100), Si(100), and GaAs(100) – bulk bands and surface-states," *Phys. Rev. B* 47:2130–2137, 1993.

[529] D. Straub, L. Ley, and F. J. Himpsel, "Inverse-photoemission study of unoccupied electronic states in Ge and Si – bulk energy-bands," *Phys. Rev. B* 33:2607–2614, 1986.

[530] F. Ma, S. Zhang, and H. Krakauer, "Excited state calculations in solids by auxiliary-field quantum Monte Carlo," *New J. Phys.* 15:093017, 2013.

[531] A. J. Williamson, R. Q. Hood, R. J. Needs, and G. Rajagopal, "Diffusion quantum Monte Carlo calculations of the excited states of silicon," *Phys. Rev. B* 57:12140–12144, 1998.

[532] J. J. Yeh and I. Lindau, "Atomic subshell photoionization cross sections and asymmetry parameters: $1 \leq Z \leq 103$," *Atom. Data Nucl. Data Tab.* 32(1):1–155, 1985.

[533] P. Rinke, M. Scheffler, A. Qteish, M. Winkelnkemper, D. Bimberg, and J. Neugebauer, "Band gap and band parameters of InN and GaN from quasiparticle energy calculations based on exact-exchange density-functional theory," *Appl. Phys. Lett.* 89:161919, 2006.

[534] P. van Gelderen, P. A. Bobbert, P. J. Kelly, and G. Brocks, "Parameter-free quasiparticle calculations for YH_3," *Phys. Rev. Lett.* 85:2989–2992, 2000.

[535] T. Miyake, F. Aryasetiawan, H. Kino, and K. Terakura, "GW quasiparticle band structure of YH_3," *Phys. Rev. B* 61:16491–16496, 2000.

[536] P. van Gelderen, P. A. Bobbert, P. J. Kelly, G. Brocks, and R. Tolboom, "Parameter-free calculation of single-particle electronic excitations in YH_3," *Phys. Rev. B* 66:075104, 2002.

[537] J. A. Alford, M. Y. Chou, E. K. Chang, and S. G. Louie, "First-principles studies of quasiparticle band structures of cubic YH_3 and LaH_3," *Phys. Rev. B* 67:125110, 2003.

[538] K. K. Ng, F. C. Zhang, V. I. Anisimov, and T. M. Rice, "Electronic structure of lanthanum hydrides with switchable optical properties," *Phys. Rev. Lett.* 78:1311–1314, 1997.

[539] R. Eder, H. F. Pen, and G. A. Sawatzky, "Kondo-lattice-like effects of hydrogen in transition metals," *Phys. Rev. B* 56:10115–10120, 1997.

[540] A. Svane, N. E. Christensen, I. Gorczyca, M. van Schilfgaarde, A. N. Chantis, and T. Kotani, "Quasiparticle self-consistent GW theory of III–V nitride semiconductors: Bands, gap bowing, and effective masses," *Phys. Rev. B* 82:115102, 2010.

[541] A. Svane, N. E. Christensen, M. Cardona, A. N. Chantis, M. van Schilfgaarde, and T. Kotani, "Quasiparticle self-consistent GW calculations for PbS, PbSe, and PbTe: Band structure and pressure coefficients," *Phys. Rev. B* 81:245120, 2010.

[542] A. N. Chantis, M. van Schilfgaarde, and T. Kotani, "*Ab initio* prediction of conduction band spin splitting in zinc blende semiconductors," *Phys. Rev. Lett.* 96:086405, 2006.

[543] M. Oshikiri, F. Aryasetiawan, Y. Imanaka, and G. Kido, "Quasiparticle effective-mass theory in semiconductors," *Phys. Rev. B* 66:125204, 2002.

[544] P. E. Trevisanutto, C. Giorgetti, L. Reining, Massimo Ladisa, and V. Olevano, "Ab initio GW many-body effects in graphene," *Phys. Rev. Lett.* 101:226405, 2008.

[545] D. A. Siegel, C.-H. Park, C. Hwang, J. Deslippe, A. V. Fedorov, S. G. Louie, and A. Lanzara, "Many-body interactions in quasi-freestanding graphene," *Proc. Natl. Acad. Sci. USA* 108:11365–11369, 2011.

[546] Y. B. Zhang, Y. W. Tan, H. L. Stormer, and P. Kim, "Experimental observation of the quantum Hall effect and Berry's phase in graphene," *Nature* 438:201–204, 2005.

[547] A. Marini, R. Del Sole, and G. Onida, "First-principles calculation of the plasmon resonance and of the reflectance spectrum of silver in the GW approximation," *Phys. Rev. B* 66:115101, 2002.

[548] R. Shaltaf, G.-M. Rignanese, X. Gonze, F. Giustino, and A. Pasquarello, "Band offsets at the Si/SiO_2 interface from many-body perturbation theory," *Phys. Rev. Lett.* 100:186401, 2008.

[549] J. P. A. Charlesworth, R. W. Godby, and R. J. Needs, "First-principles calculations of many-body band-gap narrowing at an Al/GaAs(110) interface," *Phys. Rev. Lett.* 70:1685–1688, 1993.

[550] P. M. Echenique, J. M. Pitarke, E. V. Chulkov, and A. Rubio, "Theory of inelastic lifetimes of low-energy electrons in metals," *Chem. Phys.* 251:1–35, 2000.

[551] W.-D. Schöne, "Theoretical determination of electronic lifetimes in metals," *J. Progr. Surf. Sci.* 82:161–192, 2007.

[552] A. Gerlach, K. Berge, A. Goldmann, I. Campillo, A. Rubio, J. M. Pitarke, and P. M. Echenique, "Lifetime of d holes at Cu surfaces: Theory and experiment," *Phys. Rev. B* 64:085423, 2001.

[553] X. Zubizarreta, V. M. Silkin, and E. V. Chulkov, "First-principles quasiparticle damping rates in bulk lead," *Phys. Rev. B* 84:115144, 2011.

[554] V. P. Zhukov, F. Aryasetiawan, E. V. Chulkov, and P. M. Echenique, "Lifetimes of quasiparticle excitations in 4d transition metals: Scattering theory and LMTO-RPA-GW approaches," *Phys. Rev. B* 65:115116, 2002.

[555] M. Springer, F. Aryasetiawan, and K. Karlsson, "First-principles T-matrix theory with application to the 6 eV satellite in Ni," *Phys. Rev. Lett.* 80:2389–2392, 1998.

[556] C.-H. Park, F. Giustino, C. D. Spataru, M. L. Cohen, and S. G. Louie, "First-principles study of electron linewidths in graphene," *Phys. Rev. Lett.* 102:076803, 2009.

[557] E. V. Chulkov, A. G. Borisov, J. P. Gauyacq, D. Sanchez-Portal, V. M. Silkin, V. P. Zhukov, and P. M. Echenique, "Electronic excitations in metals and at metal surfaces," *Chem. Rev.* 106:4160–4206, 2006.

[558] T. Kotani and M. van Schilfgaarde, "Impact ionization rates for Si, GaAs, InAs, ZnS, and GaN in the GW approximation," *Phys. Rev. B* 81:125201, 2010.

[559] P. Gori, F. Ronci, S. Colonna, A. Cricenti, O. Pulci, and G. Le Lay, "First-principles calculations and bias-dependent STM measurements at the alpha-Sn/Ge(111) surface," *Europhys. Lett.* 85:66001, 2009.

[560] L. Reining and R. Del Sole, "Screening properties of surface states at Si(111) 2x1," *Phys. Rev. B* 38:12768–12771, 1988.

[561] J. van den Brink and G. A. Sawatzky, "Non-conventional screening of the Coulomb interaction in low-dimensional and finite-size systems," *Europhys. Lett.* 50:447–453, 2000.

[562] J. Deslippe, M. Dipoppa, D. Prendergast, M. V. O. Moutinho, R. B. Capaz, and S. G. Louie, "Electron–hole interaction in carbon nanotubes: Novel screening and exciton excitation spectra," *Nano. Lett.* 9:1330–1334, 2009.

[563] F. Bruneval, "Ionization energy of atoms obtained from GW self-energy or from random phase approximation total energies," *J. Chem. Phys.* 136:194107, 2012.

[564] L. Wirtz, A. Marini, and A. Rubio, "Excitons in boron nitride nanotubes: Dimensionality effects," *Phys. Rev. Lett.* 96:12604, 2006.

[565] M. Rohlfing and S. G. Louie, "Electron–hole excitations and optical spectra from first principles," *Phys. Rev. B* 62:4927–4944, 2000.

[566] P. Mori-Sánchez, A. J. Cohen, and W. Yang, "Localization and delocalization errors in density functional theory and implications for band-gap prediction," *Phys. Rev. Lett.* 100:146401, 2008.

[567] A. J. Cohen, P. Mori-Sanchez, and W. Yang, "Insights into current limitations of density functional theory," *Science* 321:792–794, 2008.

[568] R. W. Godby and I. D. White, "Density-relaxation part of the self-energy," *Phys. Rev. Lett.* 80:3161–3161, 1998.

[569] F. R. Vukajlovic, E. L. Shirley, and R. M. Martin, "Single-body methods in 3d transition-metal atoms," *Phys. Rev. B* 43:3994–4001, 1991.

[570] J. L. Martins, J. Buttet, and R. Car, "Electronic and structural properties of sodium clusters," *Phys. Rev. B* 31:1804–1816, 1985.

[571] S. Saito, S. B. Zhang, S. G. Louie, and M. L. Cohen, "Quasiparticle energies in small metal clusters," *Phys. Rev. B* 40:3643–3646, 1989.

[572] R. Hesper, L. H. Tjeng, and G. A. Sawatzky, "Strongly reduced band gap in a correlated insulator in close proximity to a metal," *Europhys. Lett.* 40:177–182, 1997.

[573] C. Freysoldt, P. Rinke, and M. Scheffler, "Controlling polarization at insulating surfaces: Quasiparticle calculations for molecules adsorbed on insulator films," *Phys. Rev. Lett.* 103:056803, 2009.

[574] J. M. Garcia-Lastra, C. Rostgaard, A. Rubio, and K. S. Thygesen, "Polarization-induced renormalization of molecular levels at metallic and semiconducting surfaces," *Phys. Rev. B* 80:245427, 2009.

[575] Y. Pavlyukh and W. Hübner, "Configuration interaction approach for the computation of the electronic self-energy," *Phys. Rev. B* 75:205129, 2007.

[576] U. Litzen, J. W. Brault, and A. P. Thorne, "Spectrum and term system of neutral nickel, Ni-I," *Physica Scripta* 47:628–673, 1993.

[577] H. Mårtensson and P. O. Nilsson, "Investigation of the electronic structure of Ni by angle-resolved UV photoelectron spectroscopy," *Phys. Rev. B* 30:3047–3054, 1984.

[578] J. Bunemann, F. Gebhard, T. Ohm, R. Umstatter, S. Weiser, W. Weber, R. Claessen, D. Ehm, A. Harasawa, A. Kakizaki, A. Kimura, G. Nicolay, S. Shin, and V. N. Strocov, "Atomic

correlations in itinerant ferromagnets: Quasi-particle bands of nickel," *Europhys. Lett.* 61:667–673, 2003.

[579] A. N. Chantis, M. van Schilfgaarde, and T. Kotani, "Quasiparticle self-consistent GW method applied to localized 4f electron systems," *Phys. Rev. B* 76:165126, 2007.

[580] J. H. de Boer and E. J. W Verwey, "Semi-conductors with partially and with completely filled 3d-lattice bands," *Proc. Phys. Soc.* 49:59–71, 1937.

[581] C. Rödl, F. Fuchs, J. Furthmüller, and F. Bechstedt, "Quasiparticle band structures of the antiferromagnetic transition-metal oxides MnO, FeO, CoO, and NiO," *Phys. Rev. B* 79:235114, 2009.

[582] F. Bechstedt, F. Fuchs, and G. Kresse, "Ab-initio theory of semiconductor band structures: New developments and progress," *Phys. Stat. Sol. B* 246:1877–1892, 2009.

[583] D. Ködderitzsch, W. Hergert, W. M. Temmerman, Z. Szotek, A. Ernst, and H. Winter, "Exchange interactions in NiO and at the NiO(100) surface," *Phys. Rev. B* 66:064434, 2002.

[584] I. D. Hughes, M. Dine, A. Ernst, W. Hergert, M. Luders, J. B. Staunton, Z. Szotek, and W. M. Temmerman, "Onset of magnetic order in strongly-correlated systems from ab initio electronic structure calculations: Application to transition metal oxides," *New J. Phys.* 10:063010, 2008.

[585] F. J. Morin, "Oxides which show a metal-to-insulator transition at the Néel temperature," *Phys. Rev. Lett.* 3:34–36, 1959.

[586] R. M. Wentzcovitch, W. W. Schulz, and P. B. Allen, "VO_2: Peierls or Mott–Hubbard? A view from band theory," *Phys. Rev. Lett.* 72:3389–3392, 1994.

[587] S. Biermann, A. Poteryaev, A. I. Lichtenstein, and A. Georges, "Dynamical singlets and correlation-assisted Peierls transition in VO_2," *Phys. Rev. Lett.* 94:026404, 2005.

[588] J. B. Goodenough, "The two components of crystallographic transition in VO_2," *J. Solid State Chem.* 3:490–500, 1971.

[589] V. Eyert, "The metal–insulator transitions of VO_2: A band theoretical approach," *Ann. Phys.* 11:650–702, 2002.

[590] T. C. Koethe, Z. Hu, M. W. Haverkort, C. Schüßler-Langeheine, F. Venturini, N. B. Brookes, O. Tjernberg, W. Reichelt, H. H. Hsieh, H.-J. Lin, C. T. Chen, and L. H. Tjeng, "Transfer of spectral weight and symmetry across the metal–insulator transition in VO_2," *Phys. Rev. Lett.* 97:116402, 2006.

[591] S. Suga, A. Sekiyama, S. Imada, T. Miyamachi, H. Fujiwara, A. Yamasaki, K. Yoshimura, K. Okada, M. Yabashi, K. Tamasaku, A. Higashiya, and T. Ishikawa, "~8 keV photoemission of the metal–insulator transition system VO_2," *New J. Phys.* 11:103015, 2009.

[592] H. Abe, M. Terauchi, M. Tanaka, S. Shin, and Y. Ueda, "Electron energy-loss spectroscopy study of the metal–insulator transition in VO_2," *Japanese J. Appl. Phys.* 36:165–169, 1997.

[593] J. M. Tomczak, F. Aryasetiawan, and S. Biermann, "Effective bandstructure in the insulating phase versus strong dynamical correlations in metallic VO_2," *Phys. Rev. B* 78:115103, 2008.

[594] J. Harl and G. Kresse, "Accurate bulk properties from approximate many-body techniques," *Phys. Rev. Lett.* 103:056401, 2009.

[595] J. Harl and G. Kresse, "Cohesive energy curves for noble gas solids calculated by adiabatic connection fluctuation–dissipation theory," *Phys. Rev. B* 77:045136, 2008.

[596] A. Marini, P. Garcia-Gonzalez, and A. Rubio, "First-principles description of correlation effects in layered materials," *Phys. Rev. Lett.* 96:136404, 2006.

[597] S. Kurth and J. P. Perdew, "Density-functional correction of random-phase-approximation correlation with results for jellium surface energies," *Phys. Rev. B* 59:10461–10468, 1999.

[598] S. Lebègue, J. Harl, T. Gould, J. G. Ángyán, G. Kresse, and J. F. Dobson, "Cohesive properties and asymptotics of the dispersion interaction in graphite by the random phase approximation," *Phys. Rev. Lett.* 105:196401, 2010.

[599] L. Spanu, S. Sorella, and G. Galli, "Nature and strength of interlayer binding in graphite," *Phys. Rev. Lett.* 103:196401, 2009.

[600] J. F. Dobson, A. White, and A. Rubio, "Asymptotics of the dispersion interaction: Analytic benchmarks for van der Waals energy functionals," *Phys. Rev. Lett.* 96:073201, 2006.

[601] J. P. Perdew, K. Burke, and M. Ernzerhof, "Generalized gradient approximation made simple," *Phys. Rev. Lett.* 77:3865–3868, 1996.

[602] A. Stan, N. E. Dahlen, and R. van Leeuwen, "Fully self-consistent GW calculations for atoms and molecules," *Europhys. Lett.* 76:298–304, 2006.

[603] N. E. Dahlen and U. von Barth, "Variational second-order Moller–Plesset theory based on the Luttinger–Ward functional," *J. Chem. Phys.* 120:6826–6831, 2004.

[604] G. Antonius, S. Poncé, P. Boulanger, M. Côté, and X. Gonze, "Many-body effects on the zero-point renormalization of the band structure," *Phys. Rev. Lett.* 112:215501, 2014.

[605] M. Lazzeri, C. Attaccalite, L. Wirtz, and F. Mauri, "Impact of the electron–electron correlation on phonon dispersion: Failure of LDA and GGA DFT functionals in graphene and graphite," *Phys. Rev. B* 78:081406, 2008.

[606] Z. Szotek, W. M. Temmerman, and H. Winter, "Application of the self-interaction correction to transition-metal oxides," *Phys. Rev. B* 47:4029–4032, 1993.

[607] C. Franchini, V. Bayer, R. Podloucky, J. Paier, and G. Kresse, "Density functional theory study of MnO by a hybrid functional approach," *Phys. Rev. B* 72:045132, 2005.

[608] S. V. Faleev, M. van Schilfgaarde, T. Kotani, F. Léonard, and M. P. Desjarlais, "Finite-temperature quasiparticle self-consistent GW approximation," *Phys. Rev. B* 74:033101, 2006.

[609] L. X. Benedict, C. D. Spataru, and S. G. Louie, "Quasiparticle properties of a simple metal at high electron temperatures," *Phys. Rev. B* 66:085116, 2002.

[610] A. Eiguren, C. Ambrosch-Draxl, and P. M. Echenique, "Self-consistently renormalized quasiparticles under the electron–phonon interaction," *Phys. Rev. B* 79:245103, 2009.

[611] A. Marini, "*Ab initio* finite-temperature excitons," *Phys. Rev. Lett.* 101:106405, 2008.

[612] W. Hanke, "Dielectric theory of elementary excitations in crystals," *Adv. Phys.* 27:287–341, 1978.

[613] H. A. Bethe and E. E. Salpeter, "A relativistic equation for bound state problems," *Phys. Rev.* 82:309–310, 1951.

[614] "Bethe–Salpeter equation (origins)," *Scholarpedia* 3(11):7483, 2008.

[615] E. E. Salpeter and H. A. Bethe, "A relativistic equation for bound-state problems," *Phys. Rev.* 84:1232–1242, 1951.

[616] L. J. Sham and T. M. Rice, "Many-particle derivation of the effective-mass equation for the Wannier exciton," *Phys. Rev.* 144:708–714, 1966.

[617] G. Giuliani and G. Vignale, *Quantum Theory of the Electron Liquid*, Cambridge University Press, Cambridge, 2005.

[618] V. Ambegaokar and W. Kohn, "Electromagnetic properties of insulators. I," *Phys. Rev.* 117:423–431, 1960.

[619] S. L. Adler, "Quantum theory of the dielectric constant in real solids," *Phys. Rev.* 126:413–420, 1962.

[620] N. Wiser, "Dielectric constant with local field effects included," *Phys. Rev.* 129:62–69, 1963.

[621] A. G. Marinopoulos, L. Reining, V. Olevano, A. Rubio, T. Pichler, X. Liu, M. Knupfer, and J. Fink, "Anisotropy and interplane interactions in the dielectric response of graphite," *Phys. Rev. Lett.* 89:266406, 2002.

[622] L. X. Benedict, E. L. Shirley, and R. B. Bohn, "Optical absorption of insulators and the electron–hole interaction: An ab initio calculation," *Phys. Rev. Lett.* 80:4514–4517, 1998.

[623] D. Tamme, R. Schepe, and K. Henneberger, "Comment on 'self-consistent calculations of quasiparticle states in metals and semiconductors,'" *Phys. Rev. Lett.* 83:241, 1999.

[624] G. D. Mahan, "GW approximations," *Comments Condens. Matter Phys.* 16:333, 1994.

[625] P. Romaniello, D. Sangalli, J. A. Berger, F. Sottile, L. G. Molinari, L. Reining, and G. Onida, "Double excitations in finite systems," *J. Chem. Phys.* 130:044108, 2009.

[626] T. Ando, A. Fowler, and F. Stern, "Density-functional calculation of sub-band structure in accumulation and inversion layers," *Phys. Rev. B* 13:3468–3477, 1976.

[627] A. Schindlmayr and R. W. Godby, "Systematic vertex corrections through iterative solution of Hedin's equations beyond the GW approximation," *Phys. Rev. Lett.* 80:1702–1705, 1998.

[628] F. Bechstedt, K. Tenelsen, B. Adolph, and R. Del Sole, "Compensation of dynamical quasiparticle and vertex corrections in optical spectra," *Phys. Rev. Lett.* 78:1528–1531, 1997.

[629] R. Del Sole and R. Girlanda, "Optical properties of solids within the independent-quasiparticle approximation: Dynamical effects," *Phys. Rev. B* 54:14376–14380, 1996.

[630] M. Gruning, A. Marini, and X. Gonze, "Exciton-plasmon states in nanoscale materials: Breakdown of the Tamm–Dancoff approximation," *Nano Lett.* 9:2820–2824, 2009.

[631] I. Tamm, "Relativistic interaction of elementary particles," *J. Phys. (USSR)* 9:449, 1945.

[632] S. M. Dancoff, "Non-adiabatic meson theory of nuclear forces," *Phys. Rev.* 78:382–385, 1950.

[633] S. Albrecht, L. Reining, R. Del Sole, and G. Onida, "*Ab initio* calculation of excitonic effects in the optical spectra of semiconductors," *Phys. Rev. Lett.* 80:4510–4513, 1998.

[634] F. Bruneval, F. Sottile, V. Olevano, and L. Reining, "Beyond time-dependent exact exchange: The need for long-range correlation," *J. Chem. Phys.* 124:144113, 2006.

[635] L. Yang, J. Deslippe, C.-H. Park, M. L. Cohen, and S. G. Louie, "Excitonic effects on the optical response of graphene and bilayer graphene," *Phys. Rev. Lett.* 103:186802, 2009.

[636] G. H. Wannier, "The structure of electronic excitation levels in insulating crystals," *Phys. Rev.* 52:191–197, 1937.

[637] N. F. Mott, "Conduction in polar crystals. II. The conduction band and ultra-violet absorption of alkali-halide crystals," *Trans. Faraday Soc.* 34:500, 1938.

[638] J. Frenkel, "On the transformation of light into heat in solids. I," *Phys. Rev.* 37:17–44, 1931.

[639] M. Gatti and F. Sottile, "Exciton dispersion from first principles," *Phys. Rev. B* 88:155113, 2013.

[640] C. Rödl, F. Fuchs, J. Furthmüller, and F. Bechstedt, "Ab initio theory of excitons and optical properties for spin-polarized systems: Application to antiferromagnetic MnO," *Phys. Rev. B* 77:184408, 2008.

[641] J. J. Rehr, J. A. Soininen, and E. L. Shirley, "Final-state rule vs the Bethe–Salpeter equation for deep-core X-ray absorption spectra," *Physica Scripta* T115:207–211, 2005.

[642] S. Albrecht, L. Reining, G. Onida, V. Olevano, and R. Del Sole, "Albrecht *et al.* reply," *Phys. Rev. Lett.* 83:3971–3971, 1999.

[643] W. Hanke and L. J. Sham, "Many-particle effects in the optical excitations of a semiconductor," *Phys. Rev. Lett.* 43:387–390, 1979.

[644] D. Kammerlander, S. Botti, M. A. L. Marques, A. Marini, and C. Attaccalite, "Speeding up the solution of the Bethe–Salpeter equation by a double-grid method and Wannier interpolation," *Phys. Rev. B* 86:125203, 2012.

[645] F. Fuchs, C. Rödl, A. Schleife, and F. Bechstedt, "Efficient $\mathcal{O}(N^2)$ approach to solve the Bethe–Salpeter equation for excitonic bound states," *Phys. Rev. B* 78:085103, 2008.

[646] S. Albrecht, G. Onida, and L. Reining, "Ab initio calculation of the quasiparticle spectrum and excitonic effects in Li_2O," *Phys. Rev. B* 55:10278–10281, 1997.

[647] R. Haydock, "The recursive solution of the Schrodinger equation," *Comput. Phys. Commun.* 20:11–16, 1980.

[648] L. X. Benedict and E. L. Shirley, "Ab initio calculation of $\epsilon 2(\omega)$ including the electron–hole interaction: Application to GaN and CaF_2," *Phys. Rev. B* 59:5441–5451, 1999.

[649] W. G. Schmidt, S. Glutsch, P. H. Hahn, and F. Bechstedt, "Efficient O(N^2) method to solve the Bethe–Salpeter equation," *Phys. Rev. B* 67:085307, 2003.

[650] D. Rocca, D. Lu, and G. Galli, "Ab initio calculations of optical absorption spectra: Solution of the Bethe–Salpeter equation within density matrix perturbation theory," *J. Chem. Phys.* 133:164109, 2010.

[651] L. X. Benedict, E. L. Shirley, and R. B. Bohn, "Theory of optical absorption in diamond, Si, Ge, and GaAs," *Phys. Rev. B* 57:R9385–9387, 1998.

[652] S. Galamić-Mulaomerović and C. H. Patterson, "Ab initio many-body calculation of excitons in solid Ne and Ar," *Phys. Rev. B* 72:035127, 2005.

[653] K. Yoshino, "Absorption spectrum of argon atom in vacuum-ultraviolet region," *J. Opt. Soc. Am.* 60:1220, 1970.

[654] I. Duchemin, T. Deutsch, and X. Blase, "Short-range to long-range charge-transfer excitations in the zincbacteriochlorin–bacteriochlorin complex: A Bethe–Salpeter study," *Phys. Rev. Lett.* 109:167801, 2012.

[655] A. Marini, R. Del Sole, and A. Rubio, "Bound excitons in time-dependent density-functional theory: Optical and energy-loss spectra," *Phys. Rev. Lett.* 91:256402, 2003.

[656] P. Abbamonte, T. Graber, J. P. Reed, S. Smadici, C.-L. Yeh, A. Shukla, J.-P. Rueff, and W. Ku, "Dynamical reconstruction of the exciton in LiF with inelastic X-ray scattering," *Proc. Natl. Acad. Sci. USA* 105:12159–12163, 2008.

[657] B. C. Larson, W. Ku, J. Z. Tischler, C.-C. Lee, O. D. Restrepo, A. G. Eguiluz, P. Zschack, and K. D. Finkelstein, "Nonresonant inelastic X-ray scattering and energy-resolved Wannier function investigation of d–d excitations in NiO and CoO," *Phys. Rev. Lett.* 99:026401, 2007.

[658] C. Rödl and F. Bechstedt, "Optical and energy-loss spectra of the antiferromagnetic transition metal oxides MnO, FeO, CoO, and NiO including quasiparticle and excitonic effects," *Phys. Rev. B* 86:235122, 2012.

[659] N. Takeuchi, A. Selloni, A. I. Shkrebtii, and E. Tosatti, "Structural and electronic properties of the (111)2x1 surface of Ge from first-principles calculations," *Phys. Rev. B* 44:13611–13617, 1991.

[660] K. C. Pandey, "New pi-bonded chain model for Si(111)-(2x1) surface," *Phys. Rev. Lett.* 47:1913–1917, 1981.

[661] K. C. Pandey, "Reconstruction of semiconductor surfaces: Buckling, ionicity, and pi-bonded chains," *Phys. Rev. Lett.* 49:223–226, 1982.

[662] M. Rohlfing, M. Palummo, G. Onida, and R. Del Sole, "Structural and optical properties of the Ge(111)-(2 x 1) surface," *Phys. Rev. Lett.* 85:5440–5443, 2000.

[663] L. Reining and R. Del Sole, "Screened Coulomb interaction at Si(111)2×1," *Phys. Rev. B* 44:12918–12926, 1991.

[664] A. Ruini, M. J. Caldas, G. Bussi, and E. Molinari, "Solid state effects on exciton states and optical properties of PPV," *Phys. Rev. Lett.* 88:206403, 2002.

[665] K. Hummer and C. Ambrosch-Draxl, "Oligoacene exciton binding energies: Their dependence on molecular size," *Phys. Rev. B* 71:081202, 2005.

[666] J. C. Grossman, M. Rohlfing, L. Mitas, S. G. Louie, and M. L. Cohen, "High accuracy many-body calculational approaches for excitations in molecules," *Phys. Rev. Lett.* 86:472–475, 2001.

[667] P. Rinke, A. Schleife, E. Kioupakis, A. Janotti, C. Rödl, F. Bechstedt, M. Scheffler, and C. G. Van de Walle, "First-principles optical spectra for f centers in MgO," *Phys. Rev. Lett.* 108:126404, 2012.

[668] E. Ertekin, L. K. Wagner, and J. C. Grossman, "Point-defect optical transitions and thermal ionization energies from quantum Monte Carlo methods: Application to the F-center defect in MgO," *Phys. Rev. B* 87:155210, 2013.

[669] A. Marini and R. Del Sole, "Dynamical excitonic effects in metals and semiconductors," *Phys. Rev. Lett.* 91:176402, 2003.

[670] K. Shindo, "Effective electron–hole interaction in shallow excitons," *J. Phys. Soc. Japan* 29:287, 1970.

[671] R. Zimmermann, K. Kilimann, W. D. Kraeft, D. Kremp, and G. Ropke, "Dynamical screening and self-energy of excitons in electron–hole plasma," *Phys. Stat. Sol. B* 90:175–187, 1978.

[672] G. Strinati, "Dynamical shift and broadening of core excitons in semiconductors," *Phys. Rev. Lett.* 49:1519–1522, 1982.

[673] G. Strinati, "Effects of dynamical screening on resonances at inner-shell thresholds in semiconductors," *Phys. Rev. B* 29:5718–5726, 1984.

[674] D. Sangalli, P. Romaniello, G. Onida, and A. Marini, "Double excitations in correlated systems: A many-body approach," *J. Chem. Phys.* 134:034115, 2011.

[675] C. Sternemann, S. Huotari, G. Vankó, G. Monaco, A. Gusarov, H. Lustfeld, K. Sturm, and W. Schülke, "Correlation-induced double-plasmon excitation in simple metals studied by inelastic X-ray scattering," *Phys. Rev. Lett.* 95:157401, 2005.

[676] S. Huotari, C. Sternemann, W. Schülke, K. Sturm, H. Lustfeld, H. Sternemann, M. Volmer, A. Gusarov, H. Müller, and G. Monaco, "Electron-density dependence of double-plasmon excitations in simple metals," *Phys. Rev. B* 77:195125, 2008.

[677] Y. Noguchi, S. Ishii, K. Ohno, I. Solovyev, and T. Sasaki, "First principles t-matrix calculations for Auger spectra of hydrocarbon systems," *Phys. Rev. B* 77:035132, 2008.

[678] Y. Noguchi, S. Ishii, and K. Ohno, "Two-electron distribution functions and short-range electron correlations of atoms and molecules by first principles T-matrix calculations," *J. Chem. Phys.* 125:114108, 2006.

[679] F. Aryasetiawan and K. Karlsson, "Green's function formalism for calculating spin-wave spectra," *Phys. Rev. B* 60:7419–7428, 1999.

[680] E. Şaşıoğlu, A. Schindlmayr, C. Friedrich, F. Freimuth, and S. Blügel, "Wannier-function approach to spin excitations in solids," *Phys. Rev. B* 81:054434, 2010.

[681] K. Karlsson and F. Aryasetiawan, "Spin-wave excitation spectra of nickel and iron," *Phys. Rev. B* 62:3006–3009, 2000.

[682] G. Onida, L. Reining, and A. Rubio, "Electronic excitations: Density-functional versus many-body Green's-function approaches," *Rev. Mod. Phys.* 74:601–659, 2002.

[683] N. Sakkinen, M. Manninen, and R. van Leeuwen, "The Kadanoff–Baym approach to double excitations in finite systems," *New J. Phys.* 14:013032, 2012.

[684] N. E. Dahlen, R. van Leeuwen, and A. Stan, in *Progress in Nonequilibrium Green's Functions III*, Vol. 35 of *Journal of Physics Conference Series,* edited by M. Bonitz and A. Filinov, IOP Publishing, Bristol, 2006, pp. 340–348.

[685] H. Hübener, "Second-order response Bethe–Salpeter equation," *Phys. Rev. A* 83:062122, 2011.

[686] C. Attaccalite, M. Grüning, and A. Marini, "Real-time approach to the optical properties of solids and nanostructures: Time-dependent Bethe–Salpeter equation," *Phys. Rev. B* 84:245110, 2011.

[687] M. Petersilka, U. J. Gossmann, and E. K. U. Gross, "Excitation energies from time-dependent density-functional theory," *Phys. Rev. Lett.* 76:1212–1215, 1996.

[688] E. K. U. Gross, J. F. Dobson, and M. Petersilka, *Density Functional Theory II,* Vol. 181 of *Topics in Current Chemistry,* Springer-Verlag, Berlin, 1996, pp. 81–172.

[689] M. Casida, in *Recent Advances in Density Functional Methods I,* edited by P. Chong, World Scientific, Singapore, 1995, p. 155.

[690] A. Dreuw, J. L. Weisman, and M. Head-Gordon, "Long-range charge-transfer excited states in time-dependent density functional theory require non-local exchange," *J. Chem. Phys.* 119:2943–2946, 2003.

[691] D. Karlsson, A. Privitera, and C. Verdozzi, "Time-dependent density-functional theory meets dynamical mean-field theory: Real-time dynamics for the 3d Hubbard model," *Phys. Rev. Lett.* 106:116401, 2011.

[692] M. Fleck, A. M. Oleś, and L. Hedin, "Magnetic phases near the Van Hove singularity in *s*- and *d*-band Hubbard models," *Phys. Rev. B* 56:3159–3166, 1997.

[693] V. P. Zhukov, E. V. Chulkov, and P. M. Echenique, "Lifetimes of excited electrons in Fe and Ni: First-principles GW and the *T*-matrix theory," *Phys. Rev. Lett.* 93:096401, 2004.

[694] M. Hindgren and C. O. Almbladh, "Improved local-field corrections to the G(0)W approximation in jellium: Importance of consistency relations," *Phys. Rev. B* 56:12832–12839, 1997.

[695] R. Delsole, L. Reining, and R. W. Godby, "GWΓ approximation for electron self-energies in semiconductors and insulators," *Phys. Rev. B* 49:8024–8028, 1994.

[696] C. A. Kukkonen and A. W. Overhauser, "Electron–electron interaction in simple metals," *Phys. Rev. B* 20:550–557, 1979.

[697] M. Shishkin, M. Marsman, and G. Kresse, "Accurate quasiparticle spectra from self-consistent GW calculations with vertex corrections," *Phys. Rev. Lett.* 99:246403, 2007.

[698] A. Grueneis, G. Kresse, Y. Hinuma, and F. Oba, "Ionization potentials of solids: The importance of vertex corrections," *Phys. Rev. Lett.* 112:096401, 2014.

[699] M. Hindgren and C.-O. Almbladh, "Improved local-field corrections to the G_0W approximation in jellium: Importance of consistency relations," *Phys. Rev. B* 56:12832–12839, 1997.

[700] A. Marini and A. Rubio, "Electron linewidths of wide-gap insulators: Excitonic effects in LiF," *Phys. Rev. B* 70:081103, 2004.

[701] E. L. Shirley, "Self-consistent GW and higher-order calculations of electron states in metals," *Phys. Rev. B* 54:7758–7764, 1996.

[702] R. T. M. Ummels, P. A. Bobbert, and W. van Haeringen, "First-order corrections to random-phase approximation GW calculations in silicon and diamond," *Phys. Rev. B* 57:11962–11973, 1998.

[703] V. M. Galitskii, "The energy spectrum of a non-ideal Fermi gas," *Soviet Phys. JETP* 7:104–112, 1958.

[704] H. A. Bethe and J. Goldstone, "Effect of a repulsive core in the theory of complex nuclei," *Proc. Roy. Soc. London, Ser. A* 238:551–567, 1957.

[705] V. P. Zhukov, E. V. Chulkov, and P. M. Echenique, "GW+T theory of excited electron lifetimes in metals," *Phys. Rev. B* 72:155109, 2005.

[706] I. A. Nechaev and E. V. Chulkov, "Multiple electron–hole scattering effect on quasiparticle properties in a homogeneous electron gas," *Phys. Rev. B* 73:165112, 2006.

[707] A. Grueneis, M. Marsman, J. Harl, L. Schimka, and G. Kresse, "Making the random phase approximation to electronic correlation accurate," *J. Chem. Phys.* 131:154115, 2009.

[708] N. E. Bickers and D. J. Scalapino, "Conserving approximations for strongly fluctuating electron-systems. 1. Formalism and calculational approach," *Ann. Phys.* 193:206–251, 1989.

[709] N. E. Bickers, D. J. Scalapino, and S. R. White, "Conserving approximations for strongly correlated electron systems: Bethe–Salpeter equation and dynamics for the two-dimensional Hubbard model," *Phys. Rev. Lett.* 62:961–964, 1989.

[710] A. Liebsch, "Ni d-band self-energy beyond the low-density limit," *Phys. Rev. B* 23:5203–5212, 1981.

[711] M. I. Katsnelson and A. I. Lichtenstein, "LDA++ approach to the electronic structure of magnets: Correlation effects in iron," *J. Phys. Condens. Matter* 11:1037–1048, 1999.

[712] M. I. Katsnelson and A. I. Lichtenstein, "Electronic structure and magnetic properties of correlated metals – A local self-consistent perturbation scheme," *Euro. Phys. J. B* 30:9–15, 2002.

[713] A. Lande and R. A. Smith, "Two-body and three-body parquet theory," *Phys. Rev. A* 45:913–921, 1992.

[714] A. D. Jackson, A. Lande, and R. A. Smith, "Variational and perturbation theories made planar," *Phys. Lett. Phys. Rep.* 86:55–111, 1982.

[715] N. E. Bickers and S. R. White, "Conserving approximations for strongly fluctuating electron systems. II. Numerical results and parquet extension," *Phys. Rev. B* 43:8044–8064, 1991.

[716] *DMFT at 25: Infinite Dimensions: Lecture Notes of the Autumn School 2014, Forschungszentrum Juelich,* edited by E. Pavarini, E. Koch, D. Vollhardt, and A. Lichtenstein, Forschungszentrum Juelich GmbH and Institute for Advanced Simulations, Juelich, Germany, 2014.

[717] J. Igarashi, "Three body problem in transition-metals – application to nickel," *J. Phys. Soc. Japan* 52:2827–2837, 1983.

[718] J. Igarashi, "Three-body problem in the one-dimensional Hubbard-model," *J. Phys. Soc. Japan* 54:260–268, 1985.

[719] J. Igarashi, P. Unger, K. Hirai, and P. Fulde, "Local approach to electron correlations in ferromagnetic nickel," *Phys. Rev. B* 49:16181–16190, 1994.

[720] C. Calandra and F. Manghi, "Three-body scattering theory of correlated hole and electron states," *Phys. Rev. B* 50:2061–2074, 1994.

[721] F. Manghi, V. Bellini, and C. Arcangeli, "On-site correlation in valence and core states of ferromagnetic nickel," *Phys. Rev. B* 56:7149–7161, 1997.

[722] F. Manghi, V. Bellini, J. Osterwalder, T. J. Kreutz, P. Aebi, and C. Arcangeli, "Correlation effects in the low-energy region of nickel photoemission spectra," *Phys. Rev. B* 59:R10409–10412, 1999.

[723] S. Monastra, F. Manghi, C. A. Rozzi, C. Arcangeli, E. Wetli, H.-J. Neff, T. Greber, and J. Osterwalder, "Quenching of majority-channel quasiparticle excitations in cobalt," *Phys. Rev. Lett.* 88:236402, 2002.

[724] S. Monastra, F. Manghi, and C. Ambrosch-Draxl, "Role of electron–electron correlation in the valence states of $YBa_2Cu_3O_7$: Low-energy excitations and Fermi surface," *Phys. Rev. B* 64:020507, 2001.

[725] V. Bellini, C. A. Rozzi, and F. Manghi, "Correlation effects on the electronic properties of $Bi_2Sr_2CaCu_2O_8$," *J. Phys. Chem. Solids* 67:286–288, 2006.

[726] M. Cini and C. Verdozzi, "Many-body effects in the electron spectroscopies of incompletely filled bands," *Nuovo Cimento D* 9:1–21, 1987.

[727] T. K. Ng and K. S. Singwi, "Effective interactions for self-energy. I. Theory," *Phys. Rev. B* 34:7738–7742, 1986.

[728] I. A. Nechaev and E. V. Chulkov, "Variational solution of the T-matrix integral equation," *Phys. Rev. B* 71:115104, 2005.

[729] M. Cini and C. Verdozzi, "Photoemission and Auger-spectra of incompletely filled bands – intermediate-coupling theory and application to palladium metal," *J. Phys. Condens. Matter* 1:7457–7470, 1989.

[730] C. Verdozzi, P. J. Durham, R. J. Cole, and P. Weightman, "Correlation and disorder effects in photoelectron and Auger spectra: The late transition metals and their alloys," *Phys. Rev. B* 55:16143–16158, 1997.

[731] P. Romaniello, F. Bechstedt, and L. Reining, "Beyond the GW approximation: Combining correlation channels," *Phys. Rev. B* 85:155131, 2012.

[732] M. Cini and C. Verdozzi, "Photoemission and Auger CVV spectra of partially filled bands – a cluster approach," *Solid State Commun.* 57:657–660, 1986.

[733] M. Puig von Friesen, C. Verdozzi, and C.-O. Almbladh, "Kadanoff–Baym dynamics of Hubbard clusters: Performance of many-body schemes, correlation-induced damping and multiple steady and quasi-steady states," *Phys. Rev. B* 82:155108, 2010.

[734] I. A. Nechaev, I. Yu. Sklyadneva, V. M. Silkin, P. M. Echenique, and E. V. Chulkov, "Theoretical study of quasiparticle inelastic lifetimes as applied to aluminum," *Phys. Rev. B* 78:085113, 2008.

[735] D. R. Penn, "Effect of bound hole pairs on the d-band photoemission spectrum of Ni," *Phys. Rev. Lett.* 42:921–925, 1979.

[736] A. Liebsch, "Effect of self-energy corrections on the valence-band photoemission spectra of Ni," *Phys. Rev. Lett.* 43:1431–1434, 1979.

[737] M. Cini, "Theory of Auger-XVV spectra of solids – many-body effects in incompletely filled bands," *Surface Sci.* 87:483–500, 1979.

[738] F. J. Himpsel, J. A. Knapp, and D. E. Eastman, "Experimental energy-band dispersions and exchange splitting for Ni," *Phys. Rev. B* 19:2919–2927, 1979.

[739] R. Kubo, "Generalized cumulant expansion method," *J. Phys. Soc. Japan* 17:1100–1120, 1962.

[740] O. Gunnarsson, V. Meden, and K. Schönhammer, "Corrections to Migdal's theorem for spectral functions: A cumulant treatment of the time-dependent Green's function," *Phys. Rev. B* 50:10462–10473, 1994.

[741] P. Noziéres and C. T. De Dominicis, "Singularities in the X-ray absorption and emission of metals. III. One-body theory exact solution," *Phys. Rev.* 178:1097–1107, 1969.

[742] F. Bechstedt, "On the theory of plasmon satellite structures in the photoelectron-spectra of non-metallic solids. 1. Soluble model," *Phys. Stat. Sol. B* 101:275–286, 1980.

[743] F. Bechstedt, "Electronic relaxation effects in core level spectra of solids," *Phys. Stat. Sol. B* 112:9–49, 1982.

[744] F. Bechstedt, R. Enderlein, and M. Koch, "Theory of core excitons in semiconductors," *Phys. Stat. Sol. B* 99:61–70, 1980.

[745] L. Hedin, "Effects of recoil on shake-up spectra in metals," *Physica Scripta* 21:477–480, 1980.

[746] F. Aryasetiawan, L. Hedin, and K. Karlsson, "Multiple plasmon satellites in Na and Al spectral functions from ab initio cumulant expansion," *Phys. Rev. Lett.* 77:2268–2271, 1996.

[747] M. Guzzo, J. J. Kas, L. Sponza, C. Giorgetti, F. Sottile, D. Pierucci, M. G. Silly, F. Sirotti, J. J. Rehr, and L. Reining, "Multiple satellites in materials with complex plasmon spectra: From graphite to graphene," *Phys. Rev. B* 89:085425, 2014.

[748] B. Holm and F. Aryasetiawan, "Self-consistent cumulant expansion for the electron gas," *Phys. Rev. B* 56:12825–12831, 1997.

[749] C. Blomberg and B. Bergerse, "Spurious structure from approximations to Dyson equation," *Can. J. Phys.* 50:2286+, 1972.

[750] B. Bergerse, F. W. Kus, and C. Blomberg, "Single-particle Green's function in electron–plasmon approximation," *Can. J. Phys.* 51:102–110, 1973.

[751] M. Vos, A. S. Kheifets, V. A. Sashin, E. Weigold, M. Usuda, and F. Aryasetiawan, "Quantitative measurement of the spectral function of aluminum and lithium by electron momentum spectroscopy," *Phys. Rev. B* 66:155414, 2002.

[752] J. Lischner, D. Vigil-Fowler, and S. G. Louie, "Physical origin of satellites in photoemission of doped graphene: An *ab initio* GW plus cumulant study," *Phys. Rev. Lett.* 110:146801, 2013.

[753] M. Vos, A. S. Kheifets, E. Weigold, and F. Aryasetiawan, "Electron correlation effects in the spectral momentum density of graphite," *Phys. Rev. B* 63:033108, 2001.

[754] J. J. Kas, J. J. Rehr, and L. Reining, "Cumulant expansion of the retarded one-electron Green function," *Phys. Rev. B* 90:085112, 2014.

[755] S. M. Story, J. J. Kas, F. D. Vila, M. J. Verstraete, and J. J. Rehr, "Cumulant expansion for phonon contributions to the electron spectral function," *Phys. Rev. B* 90:195135, 2014.

[756] A. D. Jackson and R. A. Smith, "High cost of consistency in Green's-function expansions," *Phys. Rev. A* 36:2517–2518, 1987.

[757] G. Stefanucci, Y. Pavlyukh, A.-M. Uimonen, and R. van Leeuwen, "Diagrammatic expansion for positive spectral functions beyond GW: Application to vertex corrections in the electron gas," *Phys. Rev. B* 90:115134, 2014.

[758] J. C. Ward, "An identity in quantum electrodynamics," *Phys. Rev.* 78:182–182, 1950.

[759] Y. Takahashi, "On the generalized Ward identity," *Nuovo Cimento* 6:371–375, 1957.

[760] Y. Takada, "Inclusion of vertex corrections in the self-consistent calculation of quasiparticles in metals," *Phys. Rev. Lett.* 87, 2001.

[761] G. Knizia and G. K.-L. Chan, "Density matrix embedding: A strong-coupling quantum embedding theory," *J. Chem. Theory Comput.* 9:1428–1432, 2013.

[762] W. Metzner and D. Vollhardt, "Analytic calculation of ground-state properties of correlated fermions with the Gutzwiller wave function," *Phys. Rev. B* 37:7382–7399, 1988.

[763] W. Metzner and D. Vollhardt, "Correlated lattice fermions in $d = \infty$ dimensions," *Phys. Rev. Lett.* 62:324–327, 1989.

[764] E. Müller-Hartmann, "Correlated fermions on a lattice in high dimensions," *Z. Phys. B* 74:507–512, 1989.

[765] U. Brandt and C. Mielsch, "Thermodynamics and correlation functions of the Falicov–Kimball model in large dimensions," *Z. Phys. B: Condens. Matter* 75:365–370, 1989.

[766] V. Janis, "A new construction of thermodynamic mean-field theories of itinerant fermions: Application to the Falicov–Kimball model," *Z. Phys. B: Condens. Matter* 83:227–235, 1992.

[767] A. Georges and G. Kotliar, "Hubbard model in infinite dimensions," *Phys. Rev. B* 45:6479–6483, 1992.

[768] M. Jarrell, "Hubbard model in infinite dimensions: A quantum Monte Carlo study," *Phys. Rev. Lett.* 69:168–171, 1992.

[769] N. D. Mermin and H. Wagner, "Absence of ferromagnetism or antiferromagnetism in one- or two-dimensional isotropic Heisenberg models," *Phys. Rev. Lett.* 17:1133–1136, 1966.

[770] Y. Kakehashi, "Electron correlations and many-body techniques in magnetism," *Adv. Phys.* 53:497–536, 2004.

[771] T. Maier, M. Jarrell, T. Pruschke, and M. H. Hettler, "Quantum cluster theories," *Rev. Mod. Phys.* 77:1027, 2005.

[772] U. Brandt and C. Mielsch, "Free energy of the Falicov–Kimball model in large dimensions," *Z. Phys. B: Condens. Matter* 82:37–41, 1991.

[773] M. Balzer, W. Hanke, and M. Potthoff, "Mott transition in one dimension: Benchmarking dynamical cluster approaches," *Phys. Rev. B* 77:045133, 2008.

[774] S. R. White, "Density matrix formulation for quantum renormalization groups," *Phys. Rev. Lett.* 69:2863–2866, 1992.

[775] S. R. White and R. L. Martin, "Ab initio quantum chemistry using the density matrix renormalization group," *J. Chem. Phys.* 110:4127–4130, 1999.

[776] R. Olivares-Amaya, W. Hu, N. Nakatani, S. Sharma, J. Yang, and G. K.-L. Chan, "The ab-initio density matrix renormalization group in practice," *J. Chem. Phys.* 142:034102, 2015.

[777] G. Knizia and G. K.-L. Chan, "Density matrix embedding: A simple alternative to dynamical mean-field theory," *Phys. Rev. Lett.* 109:186404, 2012.

[778] X. Y. Zhang, M. J. Rozenberg, and G. Kotliar, "Mott transition in the $d = \infty$ Hubbard model at zero temperature," *Phys. Rev. Lett.* 70:1666–1669, 1993.

[779] J. Joo and V. Oudovenko, "Quantum Monte Carlo calculation of the finite temperature Mott–Hubbard transition," *Phys. Rev. B* 64:193102, 2001.

[780] P. Werner and A. J. Millis, "Doping-driven Mott transition in the one-band Hubbard model," *Phys. Rev. B* 75:085108, 2007.

[781] L. de' Medici, A. Georges, G. Kotliar, and S. Biermann, "Mott transition and Kondo screening in f-electron metals," *Phys. Rev. Lett.* 95:066402, 2005.

[782] A. Koga, Y. Imai, and N. Kawakami, "Stability of a metallic state in the two-orbital Hubbard model," *Phys. Rev. B* 66:165107, 2002.

[783] P. Werner and A. J. Millis, "High-spin to low-spin and orbital polarization transitions in multiorbital Mott systems," *Phys. Rev. Lett.* 99:126405, 2007.

[784] O. E. Gunnarsson, E. Koch, and R. M. Martin, "Mott transition in degenerate Hubbard models: Application to doped fullerenes," *Phys. Rev. B* 54:R11026–11029, 1996.

[785] *The LDA+DMFT Approach to Strongly Correlated Materials: Lecture Notes of the Autumn School 2011, Forschungszentrum Juelich,* edited by E. Pavarini, E. Koch, D. Vollhardt, and A. Lichtenstein, Forschungszentrum Juelich GmbH and Institute for Advanced Simulations, Juelich, Germany, 2011.

[786] *Corrrelated Electrons: Models to Materials: Lecture Notes of the Autumn School 2012, Forschungszentrum Juelich,* edited by E. Pavarini, E. Koch, F. Anders, and M. Jarrell, Forschungszentrum Juelich GmbH and Institute for Advanced Simulations, Juelich, Germany, 2012.

[787] G. Kottliar and D. Vollhardt, "Strongly correlated materials: Insights from dynamical mean-field theory," *Phys. Today* March 53–59, 2004.

[788] K. Held, "Electronic structure calculations using dynamical mean field theory," *Adv. Phys.* 56:829–926, 2007.

[789] V. I. Anisimov and Y. Izyumov, *Springer Series in Solid-State Sciences,* Vol. 163, *Electronic Structure of Strongly Correlated Materials,* Springer-Verlag, Berlin, 2010.

[790] *Springer Series in Solid-State Sciences,* Vol. 171, *Strongly Correlated Systems: Theoretical methods,* edited by A. Avella and F. Mancini, Springer-Verlag, Berlin, 2012.

[791] J. K. Freericks, *Transport in Multilayered Nanostructures: The Dynamical Mean-field Theory Approach,* World Scientific, Singapore, 2006.

[792] F. Ducastelle, "Analytic properties of the coherent potential approximation and of its molecular generalizations," *J. Phys. C* 7:1795–1816, 1974.

[793] A. Gonis, in *Studies in Mathematical Physics,* Vol. 4, edited by E. van Groesen and E. M. DeJager, North Holland, Amsterdam, 1992.

[794] T. Morita, "Formal structure of the cluster variation method," *Prog. Theor. Phys. Suppl.* 15:27–39, 1994.

[795] D. D. Betts, S. Masui, N. Vats, and G. E. Stewart, "Improved finite-lattice method for estimating the zero-temperature properties of two-dimensional lattice models," *Can. J. Phys.* 74:54–64, 1996.

[796] G. Kotliar, S. Y. Savrasov, G. Pálsson, and G. Biroli, "Cellular dynamical mean field approach to strongly correlated systems," *Phys. Rev. Lett.* 87:186401–186404, 2001.

[797] G. Biroli and G. Kotliar, "Cluster methods for strongly correlated electron systems," *Phys. Rev. B* 65:155112–155116, 2002.

[798] G. Biroli, O. Parcollet, and G. Kotliar, "Cluster dynamical mean-field theories: Causality and classical limit," *Phys. Rev. B* 69:205108, 2004.

[799] M. Tsukada, "A new method for the electronic structure of random lattice – the coexistence of the local and the band character," *J. Phys. Soc. Japan* 26:684–696, 1969.

[800] D. Sénéchal, D. Perez, and D. Plouffe, "Cluster perturbation theory for Hubbard models," *Phys. Rev. B* 66:075129, 2002.

[801] M. H. Hettler, A. N. Tahvildar-Zadeh, M. Jarrell, T. Pruschke, and H. R. Krishnamurthy, "Dynamical cluster approximation: Nonlocal dynamics of correlated electron systems," *Phys. Rev. B* 61:12739–12742, 2000.

[802] M. Jarrell and H. R. Krishnamurthy, "Systematic and causal corrections to the coherent potential approximation," *Phys. Rev. B* 63:125102–125131, 2001.

[803] D. A. Rowlands, J. B. Staunton, and B. L. Gyorffy, "Korringa–Kohn–Rostoker nonlocal coherent-potential approximation," *Phys. Rev. B* 67:115109, 2003.

[804] D. A. Biava *et al.*, "Systematic, multisite short-range-order corrections to the electronic structure of disordered alloys from first principles: The KKR nonlocal CPA from the dynamical cluster approximation," *Phys. Rev. B* 72:113105, 2005.

[805] H. Akima, "A new method of interpolation and smooth curve fitting based on local procedures," *J. ACM* 17:589–602, 1970.

[806] M. Potthoff, M. Aichhorn, and C. Dahnken, "Variational cluster approach to correlated electron systems in low dimensions," *Phys. Rev. Lett.* 91:206402, 2003.

[807] R. Eder, "From cluster to solid: Variational cluster approximation applied to NiO," *Phys. Rev. B* 76:241103, 2007.

[808] A. Schiller and K. Ingersent, "Systematic $1/d$ corrections to the infinite-dimensional limit of correlated lattice electron models," *Phys. Rev. Lett.* 75:113–116, 1995.

[809] J. L. Smith and Q. Si, "Spatial correlations in dynamical mean-field theory," *Phys. Rev. B* 61:5184–5193, 2000.

[810] P. Sun and G. Kotliar, "Extended dynamical mean-field theory and GW method," *Phys. Rev. B* 66:085120, 2002.

[811] A. N. Rubtsov, M. I. Katsnelson, and A. I. Lichtenstein, "Dual boson approach to collective excitations in correlated fermionic systems," *Ann. Phys.* 327:1320–1335, 2012.

[812] E. Koch, G. Sangiovanni, and O. E. Gunnarsson, "Sum rules and bath parametrization for quantum cluster theories," *Phys. Rev. B* 78:115102, 2008.

[813] A. Go and G. S. Jeon, "Properties of the one-dimensional Hubbard model: Cellular dynamical mean-field description," *J. Phys.: Condens. Matter* 21:485602, 2009.

[814] S. Moukouri and M. Jarrell, "Absence of a Slater transition in the two-dimensional Hubbard model," *Phys. Rev. Lett.* 87:167010, 2001.

[815] H. Park, K. Haule, and G. Kotliar, "Cluster dynamical mean field theory of the Mott transition," *Phys. Rev. Lett.* 101:186403, 2008.

[816] E. Gull, M. Ferrero, O. Parcollet, A. Georges, and A. J. Millis, "Momentum-space anisotropy and pseudogaps: A comparative cluster dynamical mean-field analysis of the doping-driven metal–insulator transition in the two-dimensional Hubbard model," *Phys. Rev. B* 82:155101, 2010.

[817] N. Lin, E. Gull, and A. J. Millis, "Physics of the pseudogap in eight-site cluster dynamical mean-field theory: Photoemission, Raman scattering, and in-plane and c-axis conductivity," *Phys. Rev. B* 82:045104, 2010.

[818] X. Wang, E. Gull, L. de' Medici, M. Capone, and A. J. Millis, "Antiferromagnetism and the gap of a Mott insulator: Results from analytic continuation of the self-energy," *Phys. Rev. B* 80:045101, 2009.

[819] T. A. Maier, M. Jarrell, T. C. Schulthess, P. R. C. Kent, and J. B. White, "Systematic study of d-wave superconductivity in the 2D repulsive Hubbard model," *Phys. Rev. Lett.* 95:237001, 2005.

[820] E. Gull and A. J. Millis, "Energetics of superconductivity in the two-dimensional Hubbard model," *Phys. Rev. B* 86:241106, 2012.

[821] S. A. Kivelson, I. P. Bindloss, E. Fradkin, V. Oganesyan, J. M. Tranquada, A. Kapitulnik, and C. Howald, "How to detect fluctuating stripes in the high-temperature superconductors," *Rev. Mod. Phys.* 75:1201–1241, 2003.

[822] P. R. C. Kent, M. Jarrell, T. A. Maier, and Th. Pruschke, "Efficient calculation of the antiferromagnetic phase diagram of the three-dimensional Hubbard model," *Phys. Rev. B* 72:060411, 2005.

[823] R. Staudt, M. Dzierzawa, and A. Muramatsu, "Phase diagram of the three-dimensional Hubbard model at half filling," *Euro. J. Phys. B* 17:411–415, 2000.

[824] T. Paiva, Y. L. Loh, M. Randeria, R. T. Scalettar, and N. Trivedi, "Fermions in 3D optical lattices: Cooling protocol to obtain antiferromagnetism," *Phys. Rev. Lett.* 107:086401, 2011.

[825] M. Snoek, I. Titvinidze, C. Toke, K. Byczuk, and W. Hofstetter, "Antiferromagnetic order of strongly interacting fermions in a trap: Real-space dynamical mean-field analysis," *New J. Phys.* 10:093008, 2008.

[826] S. Fuchs, E. Gull, M. Troyer, M. Jarrell, and T. Pruschke, "Spectral properties of the three-dimensional Hubbard model," *Phys. Rev. B* 83:235113, 2011.

[827] C. Lanczos, *Applied Analysis,* Prentice Hall, New York, 1956.

[828] G. H. Golub and C. F. Van Loan, *Matrix Computations,* Baltimore University Press, Baltimore, MD, 1983.

[829] H. Q. Lin and J. E. Gubernatis, "Exact diagonalization methods for quantum systems," *Comput. Phys.* 7:7400–407, 1993.

[830] E. R. Gagliano, E. Dagotto, A. Moreo, and F. C. Alcaraz, "Correlation functions of the antiferromagnetic Heisenberg model using a modified Lanczos method," *Phys. Rev. B* 34:1677–1682, 1986.

[831] E. R. Gagliano and C. A. Balseiro, "Dynamical properties of quantum many-body systems at zero temperature," *Phys. Rev. Lett.* 59:2999–3002, 1987.

[832] Q. Si, M. J. Rozenberg, G. Kotliar, and A. E. Ruckenstein, "Correlation induced insulator to metal transitions," *Phys. Rev. Lett.* 72:2761–2764, 1994.

[833] M. Caffarel and W. Krauth, "Exact diagonalization approach to correlated fermions in infinite dimensions: Mott transition and superconductivity," *Phys. Rev. Lett.* 72:1545–1548, 1994.

[834] T. Helgaker, J. Olsen, and P. Jorgensen, *Molecular Electronic-Structure Theory,* Wiley, Chichester, 1996.

[835] I. R. Levine, *Quantum Chemistry, 6th Ed.,* Prentice Hall, Upper Saddle River, NJ, 1996.

[836] D. Zgid and G. K.-L. Chan, "Dynamical mean-field theory from a quantum chemical perspective," *J. Chem. Phys.* 134:094115, 2011.

[837] D. Zgid, E. Gull, and G. K.-L. Chan, "Truncated configuration interaction expansions as solvers for correlated quantum impurity models and dynamical mean-field theory," *Phys. Rev. B* 86:165128, 2012.

[838] C. Lin and A. A. Demkov, "Efficient variational approach to the impurity problem and its application to the dynamical mean-field theory," *Phys. Rev. B* 88:035123, 2013.

[839] Y. Lu, M. Höppner, O. Gunnarsson, and M. W. Haverkort, "Efficient real-frequency solver for dynamical mean-field theory," *Phys. Rev. B* 90:085102, 2014.

[840] N. Lin, C. A. Marianetti, A. J. Millis, and D. R. Reichman, "Dynamical mean-field theory for quantum chemistry," *Phys. Rev. Lett.* 106:096402, 2011.

[841] R. P. Feynman, *Statistical Mechanics: A Set of Lectures,* Benjamin/Cummings, Reading, MA, 1970.

[842] J. W. Negele and H. Orland, *Quantum Many-Particle Systems (Advanced Book Classics,* originally published in 1988 by Westview Press, Boulder, CO), Addison-Wesley, Reading, MA, 1995.

[843] E. Gull, A. J. Millis, A. I. Lichtenstein, A. N. Rubtsov, M. Troyer, and P. Werner, "Continuous-time Monte Carlo methods for quantum impurity models," *Rev. Mod. Phys.* 83:349–404, 2011.

[844] J. E. Hirsch and R. M. Fye, "Monte Carlo method for magnetic impurities in metals," *Phys. Rev. Lett.* 56:2521–2524, 1986.

[845] R. M. Fye and J. E. Hirsch, "Monte Carlo study of the symmetric Anderson-impurity model," *Phys. Rev. B* 38:433–441, 1988.

[846] R. Blankenbecler, D. J. Scalapino, and R. L. Sugar, "Monte Carlo calculations of coupled boson–fermion systems. I," *Phys. Rev. D* 24:2278–2286, 1981.

[847] J. E. Hirsch, "Discrete Hubbard–Stratonovich transformation for fermion lattice models," *Phys. Rev. B* 28:4059–4061, 1983.

[848] D. Rost, F. Assaad, and N. Blümer, "Quasi-continuous-time impurity solver for the dynamical mean-field theory with linear scaling in the inverse temperature," *Phys. Rev. E* 87:053305, 2013.

[849] J. Yoo, S. Chandrasekharan, R. K. Kaul, D. Ullmo, and H. U. Baranger, "On the sign problem in the Hirsch–Fye algorithm for impurity problems," *J. Phys. A: Math. Gen.* 38:10307–10310, 2005.

[850] D. C. Handscomb, "The Monte Carlo method in quantum statistical mechanics," *Math. Proc. Cambridge Philos. Soc.* 58:594–598, 1962.

[851] D. C. Handscomb, "A rigorous lower bound for the efficiency of Monte Carlo techniques," *Math. Proc. Cambridge Philos. Soc.* 60:357–358, 1964.

[852] J. W. Lyklema, "Quantum-statistical Monte Carlo method for Heisenberg spins," *Phys. Rev. Lett.* 49:88–90, 1982.

[853] M. H. Kalos, "Monte Carlo calculations of the ground state of three- and four-body nuclei," *Phys. Rev.* 128:1791, 1962.

[854] N. V. Prokof'ev, B. V. Svistunov, and I. S. Tupitsyn, "Exact quantum Monte Carlo process for the statistics of discrete systems (translation: *JETP Lett.* 64:911, 1996)," *Pis'ma Zh. Eksp. Teor. Fiz.* 64:853, 1996.

[855] S. M. A. Rombouts, K. Heyde, and N. Jachowicz, "Quantum Monte Carlo method for fermions, free of discretization errors," *Phys. Rev. Lett.* 82:4155–4159, 1999.

[856] A. N. Rubtsov, V. V. Savkin, and A. I. Lichtenstein, "Continuous-time quantum Monte Carlo method for fermions," *Phys. Rev. B* 72:035122, 2005.

[857] P. Werner, A. Comanac, L. de' Medici, M. Troyer, and A. J. Millis, "Continuous-time solver for quantum impurity models," *Phys. Rev. Lett.* 97:076405, 2006.

[858] K. Haule, "Quantum Monte Carlo impurity solver for cluster dynamical mean-field theory and electronic structure calculations with adjustable cluster base," *Phys. Rev. B* 75:155113, 2007.

[859] E. Gull, P. Werner, O. Parcollet, and M. Troyer, "Continuous-time auxiliary-field Monte Carlo for quantum impurity models," *Europhysics Lett.* 82:57003, 2008.

[860] K. Mikelsons, A. Macridin, and M. Jarrell, "Relationship between Hirsch–Fye and weak-coupling diagrammatic quantum Monte Carlo methods," *Phys. Rev. E* 79:057701, 2009.

[861] E. Gull, P. Werner, A. J. Millis, and M. Troyer, "Performance analysis of continuous-time solvers for quantum impurity models," *Phys. Rev. B* 76:235123, 2007.

[862] P. Werner and A. J. Millis, "Efficient dynamical mean field simulation of the Holstein–Hubbard model," *Phys. Rev. Lett.* 99:146404, 2007.

[863] P. Werner and A. J. Millis, "Dynamical screening in correlated electron materials," *Phys. Rev. Lett.* 104:146401, 2010.

[864] F. F. Assaad and T. C. Lang, "Diagrammatic determinantal quantum Monte Carlo methods: Projective schemes and applications to the Hubbard–Holstein model," *Phys. Rev. B* 76:035116, 2007.

[865] O. Miura and T. Fujiwara, "Electronic structure and effects of dynamical electron correlation in ferromagnetic bcc Fe, fcc Ni, and antiferromagnetic NiO," *Phys. Rev. B* 77:195124, 2008.

[866] H. Keiter and J. C. Kimball, "Perturbation technique for the Anderson hamiltonian," *Phys. Rev. Lett.* 25:672–675, 1970.

[867] N. E. Bickers, "Review of techniques in the large-n expansion for dilute magnetic alloys," *Rev. Mod. Phys.* 59:845–939, 1987.

[868] K. Haule, V. Oudovenko, S. Y. Savrasov, and G. Kotliar, "The $\alpha-\gamma$ transition in Ce: A theoretical view from optical spectroscopy," *Phys. Rev. Lett.* 94:036401, 2005.

[869] S. Kirchner, J. Kroha, and P. Wölfle, "Dynamical properties of the Anderson impurity model within a diagrammatic pseudoparticle approach," *Phys. Rev. B* 70:165102, 2004.

[870] D. J. García, K. Hallberg, and M. J. Rozenberg, "Dynamical mean field theory with the density matrix renormalization group," *Phys. Rev. Lett.* 93:246403, 2004.

[871] S. Nishimoto, F. Gebhard, and E. Jeckelmann, "Dynamical mean-field theory calculation with the dynamical density-matrix renormalization group," *Physica B: Condens. Matter* 378–380:283–285, 2006.

[872] R. Bulla, T. A. Costi, and T. Pruschke, "Numerical renormalization group method for quantum impurity systems," *Rev. Mod. Phys.* 80:395–450, 2008.

[873] J. H. Van Vleck, "Models of exchange coupling in ferromagnetic media," *Rev. Mod. Phys.* 25:220–227, 1953.

[874] K. T. Moore and G. van der Laan, "Nature of the $5f$ states in actinide metals," *Rev. Mod. Phys.* 81:235–298, 2009.

[875] N. F. Mott and R. Peierls, "Discussion of the paper by de Boer and Verwey," *Proc.Phys. Soc.* 49:72, 1937.

[876] C. M. Varma, "Mixed-valence compounds," *Rev. Mod. Phys.* 48:219–238, 1976.

[877] R. Resta, "Why are insulators insulating and metals conducting?" *J. Phys.: Condens. Matter* 14:R625–656, 2002.

[878] L. F. Mattheiss, "Electronic structure of the $3d$ transition-metal monoxides. I. Energy-band results," *Phys. Rev. B* 5:290–306, 1972.

[879] L. F. Mattheiss, "Electronic structure of the $3d$ transition-metal monoxides. II. Interpretation," *Phys. Rev. B* 5:306–315, 1972.

[880] J. H. Van Vleck, "Valence strength and the magnetism of complex salts," *J. Chem. Phys.* 3:807–813, 1935.

[881] F. A. Cotton, *The Crystal Field Theory. Chemical Applications of Group Theory, 3rd Ed.*, Wiley, New York, 1990.

[882] R. J. Elliott, *Magnetic Properties of Rare Earth Metals*, Plenum Press, New York, 1972.

[883] J. S. Griffith and L. E. Orgel, "Ligand-field theory," *Q. Rev. Chem. Soc.* 11:381–393, 1957.

[884] S. Lefebvre, P. Wzietek, S. Brown, C. Bourbonnais, D. Jérome, C. Mézière, M. Fourmigué, and P. Batail, "Mott transition, antiferromagnetism, and unconventional superconductivity in layered organic superconductors," *Phys. Rev. Lett.* 85:5420–5423, 2000.

[885] K. Kanoda and R. Kato, "Mott physics in organic conductors with triangular lattices," *Ann. Rev. Condens. Matter Phys.* 2:167–188, 2011.

[886] R. H. McKenzie, "Similarities between organic and cuprate superconductors," *Science* 278:820–821, 1997.

[887] J. Zaanen, G. A. Sawatzky, and J. W. Allen, "Band gaps and electronic structure of transition-metal compounds," *Phys. Rev. Lett.* 55:418–421, 1985.

[888] P. Fulde, *Electron Correlation in Molecules and Solids, 3nd Ed.,* Springer-Verlag, Berlin, 2003.

[889] *The Theory of Transition-Metal Ions,* edited by J. S. Griffith, Cambridge University Press, Cambridge, 1961.

[890] H. Skriver, *The LMTO Method,* Springer-Verlag, New York, 1984.

[891] E. Pavarini, S. Biermann, A. Poteryaev, A. I. Lichtenstein, A. Georges, and O. K. Andersen, "Mott transition and suppression of orbital fluctuations in orthorhombic $3d^1$ perovskites," *Phys. Rev. Lett.* 92:176403, 2004.

[892] N. Marzari and D. Vanderbilt, "Maximally localized generalized Wannier functions for composite energy bands," *Phys. Rev. B* 56:12847–12865, 1997.

[893] C. Edmiston and K. Ruedenberg, "Localized atomic and molecular orbitals," *Rev. Mod. Phys.* 35:457–464, 1963.

[894] I. Souza, N. Marzari, and D. Vanderbilt, "Maximally localized Wannier functions for entangled energy bands," *Phys. Rev. B* 65:035109, 2002.

[895] A. K. McMahan, R. M. Martin, and S. Satpathy, "Calculated effective hamiltonian for La_2CuO_4 and solution in the impurity Anderson approximation," *Phys. Rev. B* 38:6650, 1988.

[896] J. F. Herbst, D. N. Lowy, and R. E. Watson, "Single-electron energies, many-electron effects, and the renormalized-atom scheme as applied to rare-earth metals," *Phys. Rev. B* 6:1913–1924, 1972.

[897] J. F. Herbst, R. E. Watson, and J. W. Wilkins, "Relativistic calculations of 4f excitation energies in the rare-earth metals: Further results," *Phys. Rev. B* 17:3089–3098, 1978.

[898] P. H. Dederichs, S. Blügel, R. Zeller, and H. Akai, "Ground states of constrained systems: Application to cerium impurities," *Phys. Rev. Lett.* 53:2512–2515, 1984.

[899] O. E. Gunnarsson, O. K. Andersen, O. Jepsen, and J. Zaanen, "Density-functional calculation of the parameters in the Anderson model: Application to Mn in CdTe," *Phys. Rev. B* 39:1708–1722, 1989.

[900] O. E. Gunnarsson, "Calculation of parameters in model hamiltonians," *Phys. Rev. B* 41:514–518, 1990.

[901] V. I. Anisimov and O. E. Gunnarsson, "Density-functional calculation of effective Coulomb interactions in metals," *Phys. Rev. B* 43:7570–7574, 1991.

[902] V. I. Anisimov, J. Zaanen, and O. K. Andersen, "Band theory and Mott insulators: Hubbard U instead of Stoner I," *Phys. Rev. B* 44:943–954, 1991.

[903] M. Cococcioni and S. de Gironcoli, "Linear response approach to the calculation of the effective interaction parameters in the LDA+U method," *Phys. Rev. B* 71:035105, 2005.

[904] W. E. Pickett, S. C. Erwin, and E. C. Ethridge, "Reformulation of the LDA+U method for a local-orbital basis," *Phys. Rev. B* 58:1201–1209, 1998.

[905] T. Bandyopadhyay and D. D. Sarma, "Calculation of Coulomb interaction strengths for 3d transition metals and actinides," *Phys. Rev. B* 39:3517–3521, 1989.

[906] K. Karlsson, F. Aryasetiawan, and O. Jepsen, "Method for calculating the electronic structure of correlated materials from a truly first-principles LDA+U scheme," *Phys. Rev. B* 81:245113, 2010.

[907] H. Jiang, R. I. Gomez-Abal, P. Rinke, and M. Scheffler, "First-principles modeling of localized d states with the GW @ LDA+U approach," *Phys. Rev. B* 82:045108, 2010.

[908] M. S. Hybertsen, M. Schlüter, and N. E. Christensen, "Calculation of Coulomb interaction parameters for La_2CuO_4 using a constrained-density-functional approach," *Phys. Rev. B* 39:9028, 1989.

[909] A. K. McMahan, J. F. Annett, and R. M. Martin, "Cuprate parameters from numerical Wannier functions." *Phys. Rev. B* 42:6268, 1990.

[910] S. B. Bacci, E. R. Gagliano, R. M. Martin, and J. F. Annett, "Derivation of a one-band Hubbard model for CuO planar materials." *Phys. Rev. B* 44:7504, 1991.

[911] M. Karolak, G. Ulm, T. Wehling, V. Mazurenko, A. Poteryaev, and A. Lichtenstein, "Double counting in LDA+DMFT – the example of NiO," *J. Electron Spectrosc. Relat. Phenom.* 181:11–15, 2010.

[912] M. T. Czyzyk and G. A. Sawatzky, "Local-density functional and on-site correlations: The electronic structure of La_2CuO_4 and $LaCuO_3$," *Phys. Rev. B* 49:14211–14228, 1994.

[913] E. R. Ylvisaker, J. Kuneš A. K. McMahan, and W. E. Pickett, "Charge fluctuations and the valence transition in Yb under pressure," *Phys. Rev. Lett.* 102:246401, 2009.

[914] A. Svane and O. Gunnarsson, "Localization in the self-interaction-corrected density-functional formalism," *Phys Rev. B* 37:9919, 1988.

[915] A. Svane and O. Gunnarsson, "Transition-metal oxides in the self-interaction-corrected density functional formalism," *Phys Rev. Lett.* 65:1148–1151, 1990.

[916] J. Hubbard, "The magnetism of iron," *Phys. Rev. B* 19:2626–2636, 1979.

[917] J. Hubbard, "Magnetism of iron. II," *Phys. Rev. B* 20:4584–4595, 1979.

[918] J. Hubbard, "Magnetism of nickel," *Phys. Rev. B* 23:5974–5977, 1981.

[919] J. B. Staunton, "The electronic structure of magnetic transition metallic materials," *Rep. Prog. Phys.* 57:1289, 1994.

[920] J. Staunton, B. L. Gyorffy, A. J. Pindor, G. M. Stocks, and H. Winter, "The 'disordered local moment' picture of itinerant magnetism at finite temperatures," *J. Magn. Magn. Mater.* 45:15–22, 1984.

[921] B. L. Gyorffy, A. J. Pindor, J. Staunton, G. M. Stocks, and H. Winter, "A first-principles theory of ferromagnetic phase transitions in metals," *J. Phys. F: Met. Phys.* 15:1337, 1985.

[922] J. Bünemann, W. Weber, and F. Gebhard, "Multiband Gutzwiller wave functions for general on-site interactions," *Phys. Rev. B* 57:6896–6916, 1998.

[923] J. Bünemann, F. Gebhard, and W. Weber, "Gutzwiller-correlated wave functions for degenerate bands: Exact results in infinite dimensions," *J. Phys.: Condens. Matter* 9:7343, 1997.

[924] J. Bünemann, F. Gebhard, T. Ohm, S. Weiser, and W. Weber, "Spin–orbit coupling in ferromagnetic nickel," *Phys. Rev. Lett.* 101:236404, 2008.

[925] K. M. Ho, J. Schmalian, and C. Z. Wang, "Gutzwiller density functional theory for correlated electron systems," *Phys. Rev. B* 77:073101, 2008.

[926] X. Deng, L. Wang, X. Dai, and Z. Fang, "Local density approximation combined with Gutzwiller method for correlated electron systems: Formalism and applications," *Phys. Rev. B* 79:075114, 2009.

[927] B. Surer, M. Troyer, P. Werner, T. O. Wehling, A. M. Läuchli, A. Wilhelm, and A. I. Lichtenstein, "Multiorbital Kondo physics of Co in Cu hosts," *Phys. Rev. B* 85:085114, 2012.

[928] K. A. Gschneidner Jr. and L. R. Eyring, *Handbook on the Physics and Chemistry of Rare Earths,* North-Holland, Amsterdam, 1978.

[929] J. K. Lang, Y. Baer, and P. A. Cox, "Study of the 4f and valence band density of states in rare-earth metals. II. Experiment and results," *J. Phys. F: Met. Phys.* 11:121–138, 1981.

[930] L. V. Pourovskii, K. T. Delaney, C. G. Van de Walle, N. A. Spaldin, and A. Georges, "Role of atomic multiplets in the electronic structure of rare-earth semiconductors and semimetals," *Phys. Rev. Lett.* 102:096401, 2009.

[931] M. J. Lipp, D. Jackson, H. Cynn, C. Aracne, W. J. Evans, and A. K. McMahan, "Thermal signatures of the Kondo volume collapse in cerium," *Phys. Rev. Lett.* 101:165703, 2008.

[932] J. D. Thompson, Z. Fisk, J. M. Lawrence, J. L. Smith, and R. M. Martin, "Two critical points on the gamma–alpha phase boundary of cerium alloys," *Phys. Rev. Lett.* 50:1081–1084, 1983.

[933] J. W. Allen, S. J. Ott, O. E. Gunnarsson, K. Schönhammer, M. B. Maple, M. S. Torikachvili, and I. Lindau, "Electronic structure of Ce and light rare earth intermetallics," *Adv. Phys.* 35:275–316, 1986.

[934] P. W. Bridgeman, "The compressibility and pressure coefficient of resistance of ten elements," *Proc. Am. Acad. Arts Sci.* 62, 1927.

[935] L. Pauling, "Atomic radii and interatomic distances in metals," *J. Am. Chem. Soc.* 69:542, 1947.

[936] A. W. Lawson and T.-Y. Tang, "Concerning the high pressure allotropic modification of cerium," *Phys. Rev.* 76:301–302, 1949.

[937] K. Held, A. K. McMahan, and R. T. Scalettar, "Cerium volume collapse: Results from the merger of dynamical mean-field theory and local density approximation," *Phys. Rev. Lett.* 87:276404, 2001.

[938] A. K. McMahan, K. Held, and R. T. Scalettar, "Thermodynamic and spectral properties of compressed Ce calculated using a combined local-density approximation and dynamical mean-field theory," *Phys. Rev. B* 67:075108, 2003.

[939] K. Haule, C.-H. Yee, and K. Kim, "Dynamical mean-field theory within the full-potential methods: Electronic structure of $CeIrIn_5$, $CeCoIn_5$, and $CeRhIn_5$," *Phys. Rev. B* 81:195107, 2010.

[940] B. Johansson, "The α–γ transition in cerium is a Mott transition," *Phil. Mag.* 30:469–482, 1974.

[941] B. Johansson, I. A. Abrikosov, M. Aldén, A. V. Ruban, and H. L. Skriver, "Calculated phase diagram for the $\gamma \rightleftharpoons \alpha$ transition in Ce," *Phys. Rev. Lett.* 74:2335–2338, 1995.

[942] J. W. Allen and R. M. Martin, "Kondo volume collapse and the gamma to alpha transition in cerium," *Phys. Rev. Lett.* 49:1106–1110, 1982.

[943] C. Lacroix, M. Lavagna, and M. Cyrot, "Volume collapse in the Kondo lattice," *Phys. Lett. A* 90:210–212, 1982.

[944] J. W. Allen and L. Z. Liu, "Alpha–gamma transition in Ce. II. A detailed analysis of the Kondo volume-collapse model," *Phys. Rev. B* 46:5047–5054, 1992.

[945] M. Casadei, X. Ren, P. Rinke, A. Rubio, and M. Scheffler, "Density-functional theory for f-electron systems: The α–γ phase transition in cerium," *Phys. Rev. Lett.* 109:146402, 2012.

[946] R. Sakuma, T. Miyake, and F. Aryasetiawan, "Self-energy and spectral function of Ce within the GW approximation," *Phys. Rev. B* 86:245126, 2012.

[947] M. B. Zölfl, I. A. Nekrasov, Th. Pruschke, V. I. Anisimov, and J. Keller, "Spectral and magnetic properties of α- and γ-Ce from dynamical mean-field theory and local density approximation," *Phys. Rev. Lett.* 87:276403, 2001.

[948] B. Amadon, S. Biermann, A. Georges, and F. Aryasetiawan, "The alpha–gamma transition of cerium is entropy driven," *Phys. Rev. Lett.* 96:066402, 2006.

[949] H. C. Choi, B. I. Min, J. H. Shim, K. Haule, and G. Kotliar, "Temperature-dependent Fermi surface evolution in heavy fermion CeIrIn$_5$," *Phys. Rev. Lett.* 108:016402, 2012.

[950] P. S. Riseborough, "Heavy fermion semiconductors," *Adv. Phys.* 49:257–320, 2000.

[951] A. Menth, E. Buehler, and T. H. Geballe, "Magnetic and semiconducting properties of SmB$_6$," *Phys. Rev. Lett.* 22:295–298, 1969.

[952] R. M. Martin and J. W. Allen, "Theory of mixed valence: Metals or small gap insulators," *J. Appl. Phys.* 50:7561–7566, 1979.

[953] V. N. Antonov, B. N. Harmon, and A. N. Yaresko, "Electronic structure of mixed-valence semiconductors in the LSDA+U approximation. II. SmB$_6$ and YbB$_{12}$," *Phys. Rev. B* 66:165209, 2002.

[954] M. Dzero, K. Sun, V. Galitski, and P. Coleman, "Topological Kondo insulators," *Phys. Rev. Lett.* 104:106408, 2010.

[955] S. Wolgast *et al.*, "Low-temperature surface conduction in the Kondo insulator SmB$_6$," *Phys. Rev. B* 88:180405, 2013.

[956] J. H. Shim, K. Haule, and G. Kotliar, "Fluctuating valence in a correlated solid and the anomalous properties of delta-plutonium," *Nature* 446:513–516, 2007.

[957] J. Minar, L. Chioncel, A. Perlov, H. Ebert, M. I. Katsnelson, and A. I. Lichtenstein, "Multiple-scattering formalism for correlated systems: A KKR-DMFT approach," *Phys. Rev. B* 72:045125, 2005.

[958] J. Sánchez-Barriga, J. Braun, J. Minár, I. Di Marco, A. Varykhalov, O. Rader, V. Boni, V. Bellini, F. Manghi, H. Ebert, M. I. Katsnelson, A. I. Lichtenstein, O. Eriksson, W. Eberhardt, H. A. Dürr, and J. Fink, "Effects of spin-dependent quasiparticle renormalization in Fe, Co, and Ni photoemission spectra: An experimental and theoretical study," *Phys. Rev. B* 85:205109, 2012.

[959] I. Leonov, D. Korotin, N. Binggeli, V. I. Anisimov, and D. Vollhardt, "Computation of correlation-induced atomic displacements and structural transformations in paramagnetic KCuF$_3$ and LaMnO$_3$," *Phys. Rev. B* 81:075109, 2010.

[960] Y. Kakehashi, M. Atiqur, R. Patoary, and T. Tamashiro, "Dynamical coherent-potential approximation approach to excitation spectra in 3d transition metals," *Phys. Rev. B* 81:245133, 2010.

[961] J. B. Staunton and B. L. Gyorffy, "Onsager cavity fields in itinerant-electron paramagnets," *Phys. Rev. Lett.* 69:371–374, 1992.

[962] V. Antropov, "Magnetic short-range order above the Curie temperature of Fe and Ni," *Phys. Rev. B* 72:140406, 2005.

[963] H. Zhang, B. Johansson, and L. Vitos, "Density-functional study of paramagnetic iron," *Phys. Rev. B* 84:140411, 2011.

[964] J. B. Goodenough, "Metallic oxides," *Prog. Solid State Chem.* 5:145–399, 1971.

[965] B. H. Brandow, "Electronic structure of Mott insulators," *Adv. Phys.* 26:651–808, 1977.

[966] A. I. Poteryaev, J. M. Tomczak, S. Biermann, A. Georges, A. I. Lichtenstein, A. N. Rubtsov, T. Saha-Dasgupta, and O. K. Andersen, "Enhanced crystal-field splitting and orbital-selective coherence induced by strong correlations in V$_2$O$_3$," *Phys. Rev. B* 76:085127, 2007.

[967] C. Castellani, C. R. Natoli, and J. Ranninger, "Magnetic structure of V_2O_3 in the insulating phase," *Phys. Rev. B* 18:4945–4966, 1978.

[968] G. Keller, K. Held, V. Eyert, D. Vollhardt, and V. I. Anisimov, "Electronic structure of paramagnetic V_2O_3: Strongly correlated metallic and Mott insulating phase," *Phys. Rev. B* 70:205116, 2004.

[969] A. Toschi, P. Hansmann, G. Sangiovanni, T. Saha-Dasgupta, O. K. Andersen, and K. Held, "Spectral properties of the Mott Hubbard insulator $(Cr_{0.011} V_{0.989})_2 O_3$ calculated by LDA+DMFT," *J. Phys.: Conf. Series* 200:012208, 2010.

[970] J. Weinen, Ph.D. thesis, Universität zu Köln, "Hard X-ray photoelectron spectroscopy: New opportunities for soild state research," 2015.

[971] G. A. Sawatzky and J. W. Allen, "Magnitude and origin of the band gap in NiO," *Phys. Rev. Lett.* 53:2339–2342, 1984.

[972] O. Tjernberg, S. Söderholm, G. Chiaia, R. Girard, U. O. Karlsson, H. Nylén, and I. Lindau, "Influence of magnetic ordering on the NiO valence band," *Phys. Rev. B* 54:10245–10248, 1996.

[973] A. Fujimori, F. Minami, and S. Sugano, "Multielectron satellites and spin polarization in photoemission from Ni compounds," *Phys. Rev. B* 29:5225–5227, 1984.

[974] J. Kuneš I. Leonov, M. Kollar, K. Byczuk, V. I. Anisimov, and D. Vollhardt, "Dynamical mean-field approach to materials with strong electronic correlations," *Eur. Phys. J. Special Topics* 180:5–28, 2009.

[975] S. J. Oh, J. W. Allen, I. Lindau, and J. C. Mikkelsen, "Resonant valence-band satellites and polar fluctuations in nickel and its compounds," *Phys. Rev. B* 26:4845–4856, 1982.

[976] Z.-X. Shen, C. K. Shih, O. Jepsen, W. E. Spicer, I. Lindau, and J. W. Allen, "Aspects of the correlation effects, antiferromagnetic order, and translational symmetry of the electronic structure of NiO and CoO," *Phys. Rev. Lett.* 64:2442–2445, 1990.

[977] Z.-X. Shen, R. S. List, D. S. Dessau, B. O. Wells, O. Jepsen, A. J. Arko, R. Barttlet, C. K. Shih, F. Parmigiani, J. C. Huang, and P. A. P. Lindberg, "Electronic structure of NiO: Correlation and band effects," *Phys. Rev. B* 44:3604–3626, 1991.

[978] J. Kuneš V. I. Anisimov, S. L. Skornyakov, A. V. Lukoyanov, and D. Vollhardt, "NiO: Correlated band structure of a charge-transfer insulator," *Phys. Rev. Lett.* 99:156404, 2007.

[979] J. Kuneš V. I. Anisimov, A. V. Lukoyanov, and D. Vollhardt, "Local correlations and hole doping in NiO: A dynamical mean-field study," *Phys. Rev. B* 75:165115, 2007.

[980] S. Y. Savrasov and G. Kotliar, "Linear response calculations of lattice dynamics in strongly correlated systems," *Phys. Rev. Lett.* 90:056401, 2003.

[981] X. Ren, I. Leonov, G. Keller, M. Kollar, I. Nekrasov, and D. Vollhardt, "LDA+DMFT computation of the electronic spectrum of NiO," *Phys. Rev. B* 74:195114, 2006.

[982] J. van Elp, H. Eskes, P. Kuiper, and G. A. Sawatzky, "Electronic structure of Li-doped NiO," *Phys. Rev. B* 45:1612–1622, 1992.

[983] H. Eskes, M. B. J. Meinders, and G. A. Sawatzky, "Anomalous transfer of spectral weight in doped strongly correlated systems," *Phys. Rev. Lett.* 67:1035–1038, 1991.

[984] C. S. Yoo, B. Maddox, J.-H. P. Klepeis, V. Iota, W. Evans, A. McMahan, M. Y. Hu, P. Chow, M. Somayazulu, D. Hausermann, R. T. Scalettar, and W. E. Pickett, "First-order isostructural Mott transition in highly compressed MnO," *Phys. Rev. Lett.* 94:115502, 2005.

[985] K. Ohta, R. E. Cohen, K. Hirose, K. Haule, K. Shimizu, and Y. Ohishi, "Experimental and theoretical evidence for pressure-induced metallization in FeO with rocksalt-type structure," *Phys. Rev. Lett.* 108:026403, 2012.

[986] R. E. Cohen, I. I. Mazin, and D. G. Isaak, "Magnetic collapse in transition metal oxides at high pressure: Implications for the earth," *Science* 275:654–657, 1997.

[987] K. Umemoto and R. M. Wentzcovitch, "Multi-mbar phase transitions in minerals," *Rev. Mineral. Geochem.* 71:299–314, 2010.

[988] J. Kuneš A. V. Lukoyanov, V. I. Anisimov, R. T. Scalettar, and W. E. Pickett, "Collapse of magnetic moment drives the Mott transition in MnO," *Nat. Mater.* 7:198–202, 2008.

[989] D. Kasinathan, J. Kuneš K. Koepernik, C. V. Diaconu, R. L. Martin, I. D. Prodan, G. E. Scuseria, N. Spaldin, L. Petit, T. C. Schulthess, and W. E. Pickett, "Mott transition of MnO under pressure: A comparison of correlated band theories," *Phys. Rev. B* 74:195110, 2006.

[990] J. van Elp, R. H. Potze, H. Eskes, R. Berger, and G. A. Sawatzky, "Electronic structure of MnO," *Phys. Rev. B* 44:1530–1537, 1991.

[991] S. Biermann, F. Aryasetiawan, and A. Georges, "First-principles approach to the electronic structure of strongly correlated systems: Combining the GW approximation and dynamical mean-field theory," *Phys. Rev. Lett.* 90:086402, 2003.

[992] F. Aryasetiawan, M. Imada, A. Georges, G. Kotliar, S. Biermann, and A. I. Lichtenstein, "Frequency-dependent local interactions and low-energy effective models from electronic structure calculations," *Phys. Rev. B* 70:195104, 2004.

[993] T. Holstein, "Studies of polaron motion: Part I. The molecular-crystal model," *Ann. Phys.* 8:325–342, 1959.

[994] T. Holstein, "Studies of polaron motion: Part II. The 'small' polaron," *Ann. Phys.* 8:343–389, 1959.

[995] I. G. Lang and Y. A. Firsov, "Kinetic theory of semiconductors with low mobility (*Zh. ksp. Teor. Fiz.* 43:1843, 1962)," *Sov. Phys. JETP* 16:1301, 1963.

[996] M. Casula, Ph. Werner, L. Vaugier, F. Aryasetiawan, T. Miyake, A. J. Millis, and S. Biermann, "Low-energy models for correlated materials: Bandwidth renormalization from Coulombic screening," *Phys. Rev. Lett.* 109:126408, 2012.

[997] A. Toschi, A. A. Katanin, and K. Held, "Dynamical vertex approximation: A step beyond dynamical mean-field theory," *Phys. Rev. B* 75:045118, 2007.

[998] H. Kusunose, "Influence of spatial correlations in strongly correlated electron systems: Extension to dynamical mean field approximation," *J. Phys. Soc. Japan* 75:054713, 2006.

[999] S. K. Sarker, "A new functional integral formalism for strongly correlated Fermi systems," *J. Phys. C: Solid State Phys.* 21:L667, 1988.

[1000] A. N. Rubtsov, M. I. Katsnelson, and A. I. Lichtenstein, "Dual fermion approach to nonlocal correlations in the Hubbard model," *Phys. Rev. B* 77:033101, 2008.

[1001] H. Hafermann, G. Li, A. N. Rubtsov, M. I. Katsnelson, A. I. Lichtenstein, and H. Monien, "Efficient perturbation theory for quantum lattice models," *Phys. Rev. Lett.* 102:206401, 2009.

[1002] A. N. Rubtsov, M. I. Katsnelson, A. I. Lichtenstein, and A. Georges, "Dual fermion approach to the two-dimensional Hubbard model: Antiferromagnetic fluctuations and Fermi arcs," *Phys. Rev. B* 79:045133, 2009.

[1003] S. Biermann, "Dynamical screening effects in correlated electron materials – a progress report on combined many-body perturbation and dynamical mean field theory: GW+DMFT," *J. Phys.: Condens. Matter* 26:173202, 2014.

[1004] R. P. Feynman, "Atomic theory of the λ transition in helium," *Phys. Rev.* 91:1291–1301, 1953.

[1005] J. Dongarra and F. Sullivan, "Guest editors' introduction: The top 10 algorithms," *Comput. Sci. Eng.* 2:22–23, 2000.

[1006] N. Metropolis, A. Rosenbluth, M. Rosenbluth, A. Teller, and E. Teller, "Equation of state calculations by fast computing machines," *J. Chem. Phys.* 21:1087, 1953.

[1007] J. M. Hammersley and D. C. Handscomb, *Monte Carlo Methods,* Chapman and Hall, London, 1964.

[1008] W. K. Hastings, "Monte Carlo sampling methods using Markov chains and their applications," *Biometrika* 57:97–109, 1970.

[1009] D. Frenkel and B. Smit, *Understanding Molecular Simulation, From Algorithms to Applications,* Academic Press, San Diego, CA, 2002.

[1010] B. L. Hammond, W. A. Lester Jr., and P. J. Reynolds, *Monte Carlo Methods in ab initio Quantum Chemistry,* World Scientific, Singapore, 1994.

[1011] D. Ceperley, G. V. Chester, and M. H. Kalos, "Monte Carlo simulation of a many-fermion system," *Phys. Rev. B* 16:3081–3099, 1977.

[1012] G. H. Golub and C. F. Van Loan, *Matrix Computations,* Johns Hopkins University Press, Baltimore, MD, 1980.

[1013] M. Allen and D. Tildesley, *Computer Simulation of Liquids,* Oxford University Press, New York, 1989.

[1014] P. Rossky, J. D. Doll, and H. L. Friedman, "Brownian dynamics as smart Monte Carlo simulation," *J. Chem. Phys.* 69:4628–33, 1978.

[1015] V. Natoli and D. M. Ceperley, "An optimized method for treating arbitrary long-ranged potentials," *J. Comput. Phys.* 117:171, 1995.

[1016] B. K. Clark, M. A. Morales, J. McMinis, J. Kim, and G. E. Scuseria, "Computing the energy of a water molecule using multideterminants: A simple, efficient algorithm," *J. Chem. Phys.* 135:244105, 2011.

[1017] D. M. Ceperley and M. H. Kalos, in *Monte Carlo Methods in Statistical Physics, Topics in Condensed Matter Physics,* edited by K. Binder, Springer-Verlag, Berlin, 1986, pp. 145–194.

[1018] W. M. Press, B. P. Flannery, S. A. Teukolsky, and W. T. Vetterling, *Numerical Recipes,* Cambridge University Press, Cambridge, 1986.

[1019] C. J. Umrigar, J. Toulouse, C. Filippi, S. Sorella, and R. G. Hennig, "Alleviation of the fermion-sign problem by optimization of many-body wave functions," *Phys. Rev. Lett.* 98:110201, 2007.

[1020] J. Toulouse and C. J. Umrigar, "Optimization of quantum Monte Carlo wave functions by energy minimization," *J. Chem. Phys.* 126:084102, 2007.

[1021] J. Toulouse and C. J. Umrigar, "Full optimization of Jastrow–Slater wave functions with application to the first-row atoms and homonuclear diatomic molecules," *J. Chem. Phys.* 128:174101, 2008.

[1022] D. M. Ceperley, "The statistical error of Green's function Monte Carlo," *J. Stat. Phys.* 43:815, 1986.

[1023] S. Fahy, X. W. Wang, and S. G. Louie, "Variational quantum Monte Carlo nonlocal pseudopotential approach to solids: Cohesive and structural properties of diamond," *Phys. Rev. Lett* 61:1631–1634, 1988.

[1024] S. Fahy, X. W. Wang, and S. G. Louie, "Variational quantum Monte Carlo nonlocal pseudopotential approach to solids: Formulation and application to diamond, graphite, and silicon," *Phys. Rev. B* 42:3503–3522, 1990.

[1025] M. H. Kalos and P. Whitlock, *Monte Carlo Methods,* Wiley, New York, 1986.

[1026] L. Mitas, E. L. Shirley, and D. M. Ceperley, "Nonlocal pseudopotentials and diffusion Monte Carlo." *J. Chem. Phys.* 95:3467, 1991.

[1027] P. H. Acioli and D. Ceperley, "Generation of pseudopotentials from correlated wave functions," *J. Chem. Phys.* 100:1, 1994.

[1028] J. R. Trail and R. J. Needs, "Smooth relativistic Hartree–Fock pseudopotentials for H to Ba and Lu to Hg," *J. Chem. Phys.* 122:174109, 2005.

[1029] M. Burkatzki, C. Filippi, and M. Dolg, "Energy-consistent pseudopotentials for quantum Monte Carlo calculations," *J. Chem. Phys.* 126:234105, 2007.

[1030] M. Burkatzki, C. Filippi, and M. Dolg, "Energy-consistent small-core pseudopotentials for 3d-transition metals adapted to quantum Monte Carlo calculations," *J. Chem. Phys.* 129:164115, 2008.

[1031] C. Lin, F.-H. Zong, and D. M. Ceperley, "Twist-averaged boundary conditions in continuum Quantum Monte Carlo algorithms," *Phys. Rev. E* 64:016702, 2001.

[1032] S. Chiesa, D. M. Ceperley, R. M. Martin, and M. Holzmann, "Finite-size error in many-body simulations with long-range interactions," *Phys. Rev. Lett.* 97:076404, 2006.

[1033] N. D. Drummond, R. J. Needs, A. Sorouri, and W. M. C. Foulkes, "Finite-size errors in continuum quantum Monte Carlo calculations," *Phys. Rev. B* 78:125106, 2008.

[1034] C. S. Hellberg and E. Manousakis, "Green's-function Monte Carlo for lattice fermions: Application to the t–J model," *Phys. Rev. B* 61:11787–11806, 2000.

[1035] N. Trivedi and D. M. Ceperley, "Ground-state correlations of quantum antiferromagnets: A Green-function Monte Carlo study," *Phys. Rev. B* 40:2737, 1989.

[1036] E. Manousakis, "The spin-1/2 Heisenberg antiferromagnet on a square lattice and its application to the cuprous oxides," *Rev. Mod. Phys.* 63:1–62, 1991.

[1037] H. De Raedt and W. von der Linden, in *The Monte Carlo Method in Condensed Matter Physics,* edited by K. Binder, Springer-Verlag, Berlin, 1995, Vol. 71, pp. 249–284.

[1038] M. Capello, F. Becca, M. Fabrizio, S. Sorella, and E. Tosatti, "Variational description of Mott insulators," *Phys. Rev. Lett.* 94:026406, 2005.

[1039] L. F. Tocchio, F. Becca, A. Parola, and S. Sorella, "Role of backflow correlations for the nonmagnetic phase of the t–t' Hubbard model," *Phys. Rev. B* 78:041101, 2008.

[1040] J. K. L. MacDonald, "Successive approximations by the Rayleigh–Ritz variation method," *Phys. Rev.* 43:830, 1933.

[1041] D. M. Ceperley and B. Bernu, "The calculation of excited state properties with quantum Monte Carlo." *J. Chem. Phys.* 89:6316, 1988.

[1042] M. P. Nightingale and V. Melik-Alaverdian, "Optimization of ground- and excited-state wave functions and van der Waals clusters," *Phys. Rev. Lett.* 87:043401, 2001.

[1043] Y. Kwon, D. M. Ceperley, and R. M. Martin, "Quantum Monte Carlo calculation of the Fermi liquid parameters in the two-dimensional electron gas." *Phys. Rev. B* 50:1684–1694, 1994.

[1044] M. Holzmann, B. Bernu, V. Olevano, R. M. Martin, and D. M. Ceperley, "Renormalization factor and effective mass of the two-dimensional electron gas," *Phys. Rev. B* 79:041308, 2009.

[1045] N. D. Drummond and R. J. Needs, "Diffusion quantum Monte Carlo calculation of the quasi-particle effective mass of the two-dimensional homogeneous electron gas," *Phys. Rev. B* 87:045131, 2013.

[1046] R. Asgari, B. Davoudi, M. Polini, G. F. Giuliani, M. P. Tosi, and G. Vignale, "Quasiparticle self-energy and many-body effective mass enhancement in a two-dimensional electron liquid," *Phys. Rev. B* 71:045323, 2005.

[1047] H.-J. Schulze, P. Schuck, and N. Van Giai, "Two-dimensional electron gas in the random-phase approximation with exchange and self-energy corrections," *Phys. Rev. B* 61:8026–8032, 2000.

[1048] D. Ceperley, in *Recent Progress in Many-Body Theories,* edited by J. G. Zabolitsky, Springer-Verlag, Berlin, 1981, p. 262.

[1049] N. Metropolis and S. Ulam, "The Monte Carlo method," *J. Am. Stat. Assoc.* 247:335, 1949.

[1050] M. D. Donsker and M. Kac, "A sampling method for determining the lowest eigenvalue and the principal eigenfunction of Schrodinger's equation," *J. Res. Nat. Bur. Stand.* 44:551–557, 1950.

[1051] M. H. Kalos, D. Levesque, and L. Verlet, "Helium at zero temperature with hard-sphere and other forces," *Phys. Rev. A* 9:2178–2195, 1974.

[1052] J. B. Anderson, "A random-walk simulation of the Schrödinger equation: H_3^+," *J. Chem. Phys.* 63:1499–1503, 1975.

[1053] J. B. Anderson, "Quantum chemistry by random walk," *J. Chem. Phys.* 65:4121–4127, 1976.

[1054] D. M. Ceperley, "The simulation of quantum systems with random walks: A new algorithm for charged systems," *J. Comput. Phys.* 51:404, 1983.

[1055] H. F. Trotter, "On the product of semi-groups of operators," *Proc. Am. Math. Soc.* 10:545–551, 1959.

[1056] E. Nelson, "Feynman integrals and the Schrödinger equation," *J. Math. Phys.* 5:332–343, 1964.

[1057] M. Caffarel and P. Claverie, "Development of a pure diffusion quantum Monte Carlo method using a full generalized Feynman–Kac formula. I. Formalism," *J. Chem. Phys.* 88:1088–1099, 1988.

[1058] R. C. Grimm and R. G. Storer, "Monte Carlo solution of Schrödinger's equation," *J. Comput. Phys.* 7:134–156, 1971.

[1059] P. J. Reynolds, D. M. Ceperley, B. J. Alder, and W. A. Lester, "Fixed-node quantum Monte Carlo for molecules," *J. Chem. Phys.* 77:5593, 1982.

[1060] D. M. Ceperley, M. H. Kalos, and J. L. Lebowitz, "The computer simulation of the static and dynamic properties of a polymer chain," *Macromolecules* 14:1472, 1981.

[1061] C. J. Umrigar, M. P. Nightingale, and K. J. Runge, "A diffusion Monte Carlo algorithm with very small time-step errors," *J. Chem. Phys.* 99:2865–2890, 1993.

[1062] J. T. Krogel and D. M. Ceperley, in *Advances in Quantum Monte Carlo,* edited by S. Rothstein, S. Tanaka, and W. A. Lester Jr., ACS Symposium Series, Washington, D.C., 2012, Vol. 1094, Ch. 3, pp. 13–26.

[1063] M. A. Lee, K. E. Schmidt, M. H. Kalos, and G. V. Chester, "Green's function Monte Carlo method for liquid ^{3}He," *Phys. Rev. Lett.* 46:728–731, 1981.

[1064] D. M. Arnow, M. H. Kalos, M. A. Lee, and K. E. Schmidt, "Green's function Monte Carlo for few fermion problems," *J. Chem. Phys.* 77:5562–5572, 1982.

[1065] G. H. Booth, A. J. W. Thom, and A. Alavi, "Fermion Monte Carlo without fixed nodes: A game of life, death, and annihilation in Slater determinant space," *J. Chem. Phys.* 131:054106, 2009.

[1066] D. M. Ceperley, "Fermion nodes," *J. Stat. Phys.* 63:1237, 1991.

[1067] D. J. Klein and H. M. Pickett, "Nodal hypersurfaces and Anderson's random-walk simulation of the Schrödinger equation," *J. Chem. Phys.* 64:4811–4812, 1976.

[1068] L. Mitas, "Structure of fermion nodes and nodal cells," *Phys. Rev. Lett.* 96:240402, 2006.

[1069] G. Ortiz, D. M. Ceperley, and R. M. Martin, "New stochastic method for systems with broken time-reversal symmetry – 2d fermions in a magnetic field," *Phys. Rev. Lett.* 71:2777–2780, 1993.

[1070] W. M. C. Foulkes, R. Q. Hood, and R. J. Needs, "Symmetry constraints and variational principles in diffusion quantum Monte Carlo calculations of excited-state energies," *Phys. Rev. B* 60:4558–4570, 1999.

[1071] D. M. Ceperley and B. J. Alder, "Quantum Monte Carlo for molecules: Green's function and nodal release," *J. Chem. Phys.* 81:5833, 1984.

[1072] M. D. Jones, G. Ortiz, and D. M. Ceperley, "Released-phase quantum Monte Carlo." *Phys. Rev. E* 55:6202–6210, 1997.

[1073] D. F. B. ten Haaf, H. J. M. van Bemmel, J. M. J. van Leeuwen, W. van Saarloos, and D. M. Ceperley, "Fixed-node QMC method for lattice fermions," *Phys. Rev. B* 51:13039, 1995.

[1074] R. Assaraf and M. Caffarel, "Zero-variance zero-bias principle for observables in quantum Monte Carlo: Application to forces," *J. Chem. Phys.* 119:10536–10552, 2003.

[1075] L. Reatto, "Structure of the ground-state wave function of quantum fluids and 'exact' numerical methods," *Phys. Rev. B* 26:130–137, 1982.

[1076] K. S. Liu, M. H. Kalos, and G. V. Chester, "Quantum hard spheres in a channel," *Phys. Rev. A* 10:303–308, 1974.

[1077] D. M. Ceperley and B. J. Alder, "Muon alpha sticking probability in muon catalyzed fusion," *Phys. Rev. A* 31:1999, 1985.

[1078] S. Chiesa, D. M. Ceperley, and S.-W. Zhang, "Accurate, efficient and simple forces with quantum Monte Carlo methods," *Phys. Rev. Lett.* 94:036404:1–4, 2005.

[1079] M. M. Hurley and P. A. Christiansen, "Relativistic effective potentials in quantum Monte Carlo calculations," *J. Chem. Phys.* 86:1069–1070, 1987.

[1080] M. Casula, "Beyond the locality approximation in the standard diffusion Monte Carlo method," *Phys. Rev. B* 74:161102, 2006.

[1081] M. Casula, S. Moroni, S. Sorella, and C. Filippi, "Size-consistent variational approaches to nonlocal pseudopotentials: Standard and lattice regularized diffusion Monte Carlo methods revisited," *J. Chem. Phys.* 132:154113, 2010.

[1082] G. Sugiyama and S. E. Koonin, *Ann. Phys. N. Y.* 168:1, 1986.

[1083] S. B. Fahy and D. R. Hamann, "Positive-projection Monte Carlo simulation: A new variational approach to strongly interacting fermion systems," *Phys. Rev. Lett.* 65:3437–3440, 1990.

[1084] S. Zhang, J. Carlson, and J. E. Gubernatis, "Constrained path Monte Carlo method for fermion ground states," *Phys. Rev. B* 55:7464–7477, 1997.

[1085] S. Zhang and H. Krakauer, "Quantum Monte Carlo method using phase-free random walks with Slater determinants," *Phys. Rev. Lett.* 90:136401, 2003.

[1086] S. Zhang, in *Emergent Phenomena in Correlated Matter: Lecture Notes of the Autumn School 2013, Forschungszentrum Julich,* edited by E. Pavarini, E. Koch, and U. Schollwock, Forschungszentrum Juelich GmbH and Institute for Advanced Simulations, Juelich, Germany, 2013.

[1087] R. L. Stratonovich, "On a method of calculating quantum distribution functions [translation: *Soviet phys. doklady* 2:416, 1958]," *Doklady Akad. Nauk S.S.S.R.* 115:1097, 1957.

[1088] M. Suewattana, W. Purwanto, S. Zhang, H. Krakauer, and E. J. Walter, "Phaseless auxiliary-field quantum Monte Carlo calculations with plane waves and pseudopotentials: Applications to atoms and molecules," *Phys. Rev. B* 75:245123, 2007.

[1089] W. Purwanto, H. Krakauer, Y. Virgus, and S. Zhang, "Assessing weak hydrogen binding on Ca+ centers: An accurate many-body study with large basis sets," *J. Chem. Phys.* 135:164105, 2011.

[1090] J. Carlson, J. E. Gubernatis, G. Ortiz, and S. Zhang, "Issues and observations on applications of the constrained-path Monte Carlo method to many-fermion systems," *Phys. Rev. B* 59:12788–12798, 1999.

[1091] J. J. Shepherd, G. H. Booth, and A. Alavi, "Investigation of the full configuration interaction quantum Monte Carlo method using homogeneous electron gas models," *J. Chem. Phys.* 136:244101, 2012.

[1092] C.-C. Chang and S. Zhang, "Spin and charge order in the doped Hubbard model: Long-wavelength collective modes," *Phys. Rev. Lett.* 104:116402, 2010.

[1093] C.-C. Chang and S. Zhang, "Spatially inhomogeneous phase in the two-dimensional repulsive Hubbard model," *Phys. Rev. B* 78:165101, 2008.

[1094] E. Wigner and H. B. Huntington, "On the possibility of a metallic modification of hydrogen," *J. Chem. Phys.* 3:764, 1935.

[1095] D. M. Ceperley and B. J. Alder, "Ground state of solid hydrogen at high pressures," *Phys. Rev. B* 36:2092, 1987.

[1096] V. Natoli, R. M. Martin, and D. M. Ceperley, "Crystal structure of atomic hydrogen," *Phys. Rev. Lett.* 70:1952–1955, 1993.

[1097] V. Natoli, R. M. Martin, and D. M. Ceperley, "Crystal structure of molecular hydrogen at high pressure," *Phys. Rev. Lett.* 74:1601–1604, 1995.

[1098] J. M. McMahon, M. A. Morales, C. Pierleoni, and D. M. Ceperley, "The properties of hydrogen and helium under extreme conditions," *Rev. Mod. Phys.* 84:1607–1653, 2012.

[1099] R. C. Clay, J. McMinis, J. M. McMahon, C. Pierleoni, D. M. Ceperley, and M. A. Morales, "Benchmarking exchange–correlation functionals for hydrogen at high pressures using quantum Monte Carlo," *Phys. Rev. B* 89:184106, 2014.

[1100] K. P. Esler, R. E. Cohen, B. Militzer, J. Kim, R. J. Needs, and M. D. Towler, "Fundamental high-pressure calibration from all-electron quantum Monte Carlo calculations," *Phys. Rev. Lett.* 104:185702, 2010 (see also Supplemental Material).

[1101] H. K. Mao, J. Xu, and P. M. Bell, "Calibration of the ruby pressure gauge to 800 kbar under quasi-hydrostatic conditions," *J. Geophys. Res.: Solid Earth* 91:4673–4676, 1986.

[1102] A. J. Williamson, R. Q. Hood, R. J. Needs, and G. Rajagopal, "Diffusion quantum Monte Carlo calculations of the excited states of silicon," *Phys. Rev. B* 57:12140–12144, 1998.

[1103] N. Nemec, "Diffusion Monte Carlo: Exponential scaling of computational cost for large systems," *Phys. Rev. B* 81:035119, 2010.

[1104] S. Baroni and S. Moroni, "Reptation quantum Monte Carlo: A method for unbiased ground-state averages and imaginary-time correlations," *Phys. Rev. Lett.* 82:4745–4748, 1999.

[1105] C. Pierleoni, B. Bernu, D. M. Ceperley, and W. R. Magro, "Equation of state of the hydrogen plasma by path integral Monte Carlo simulation." *Phys. Rev. Lett.* 73:2145–2148, 1994.

[1106] E. W. Brown, B. K. Clark, J. L. DuBois, and D. M. Ceperley, "Path-integral Monte Carlo simulation of the warm dense homogeneous electron gas," *Phys. Rev. Lett.* 110:146405, 2013.

[1107] E. W. Brown, J. L. DuBois, M. Holzmann, and D. M. Ceperley, "Exchange-correlation energy for the three-dimensional homogeneous electron gas at arbitrary temperature," *Phys. Rev. B* 88:081102, 2013.

[1108] E. L. Pollock and D. M. Ceperley, "Simulation of quantum many-body systems by path integral methods," *Phys. Rev. B* 30:2555, 1984.

[1109] I. Kylänpää and T. T. Rantala, "Finite temperature quantum statistics of H_3^+ molecular ion," *J. Chem. Phys.* 133:044312, 2010.

[1110] D. M. Ceperley, "Path integrals in the theory of condensed helium," *Rev. Mod. Phys.* 67:279, 1995.

[1111] M. F. Herman, E. J. Bruskin, and B. J. Berne, "On path integral Monte Carlo simulations," *J. Chem. Phys.* 76:5150–5155, 1982.

[1112] M. E. Tuckerman, *Statistical Mechanics: Theory and Molecular Simulation,* Oxford University Press, Oxford, 2010.

[1113] M. Ceriotti, D. E. Manolopoulos, and M. Parrinello, "Accelerating the convergence of path integral dynamics with a generalized Langevin equation," *J. Chem. Phys.* 134:084104, 2011.

[1114] L. Candido, P. Phillips, and D. M. Ceperley, "Single and paired point defects in a 2D Wigner crystal," *Phys. Rev. Lett.* 86:492–495, 2000.

[1115] B. Militzer and D. M. Ceperley, "Path integral Monte Carlo simulation of the low-density hydrogen plasma," *Phys. Rev. E* 63:66404, 2001.

[1116] C. Pierleoni, D. M. Ceperley, and M. Holzmann, "Coupled electron ion Monte Carlo calculations of dense metallic hydrogen," *Phys. Rev. Lett.* 93:146402:1–4, 2004.

[1117] K. Delaney, C. Pierleoni, and D. M. Ceperley, "Quantum Monte Carlo simulation of the high pressure molecular–atomic transition in fluid hydrogen," *Phys. Rev. Lett.* 97, 2006.

[1118] J. A. Morrone and R. Car, "Nuclear quantum effects in water," *Phys. Rev. Lett.* 101:017801, 2008.

[1119] C. Filippi and D. M. Ceperley, "Path integral Monte Carlo calculation of the kinetic energy of condensed lithium." *Phys. Rev. B* 57:252–257, 1998.

[1120] C. H Bennett, "Efficient estimation of free energy differences from Monte Carlo data," *J. Comput. Phys.* 22:245–268, 1976.

[1121] D. M. Ceperley and G. Jacucci, "The calculation of exchange frequencies in bcc ^{3}He with path integral Monte Carlo method." *Phys. Rev. Lett.* 58:1648–1651, 1987.

[1122] B. Bernu and D. Ceperley, in *Quantum Monte Carlo Methods in Physics and Chemistry,* edited by M. P. Nightingale and C. J. Umrigar, Kluwer, Dordrecht, 1999, pp. 161–182.

[1123] D. J. Thouless, "Exchange in solid ^{3}He and the Heisenberg hamiltonian," *Proc. Phys. Soc. London* 86:893, 1965.

[1124] P. A. M. Dirac, "Quantum mechanics of many-electron systems," *Proc. Roy. Soc. London, Ser. A* 123:714–733, 1929.

[1125] M. Roger, "Multiple-spin exchange in strongly correlated fermion systems," *J. Low Temp. Phys.* 162:625–637, 2011.

[1126] E. L. Pollock and D. M. Ceperley, "Path-integral computations of superfluid densities," *Phys. Rev. B* 36:8343, 1987.

[1127] M. Boninsegni, N. V. Prokof'ev, and B. V. Svistunov, "Worm algorithm and diagrammatic Monte Carlo: A new approach to continuous-space path integral Monte Carlo simulations," *Phys. Rev. E* 74:036701, 2006.

[1128] R. P. Feynman and A. R. Hibbs, *Quantum Mechanics and Path Integrals,* McGraw-Hill, New York, 1965.

[1129] D. M. Ceperley, in *Monte Carlo and Molecular Dynamics of Condensed Matter Systems,* edited by K. Binder and G. Ciccotti, Editrice Compositori, Bologna, 1996.

[1130] J. D. Jackson, *Classical Electrodynamics,* Wiley, New York, 1962.

[1131] D. M. Ceperley, "Path integral calculations of normal liquid ^{3}He," *Phys. Rev. Lett.* 69:331, 1992.

[1132] M. Boninsegni and D. M. Ceperley, "Density fluctuations in liquid ^{4}He: Path integrals and maximum entropy," *J. Low Temp. Phys.* 104:339, 1996.

[1133] F. Krüger and J. Zaanen, "Fermionic quantum criticality and the fractal nodal surface," *Phys. Rev. B* 78:035104, 2008.

[1134] B. Militzer and E. L. Pollock, "Variational density matrix method for warm, condensed matter: Application to dense hydrogen," *Phys. Rev. E* 61:3470–3482, 2000.

[1135] B. Militzer, D. M. Ceperley, J. D. Kress, J. D. Johnson, L. A. Collins, and S. Mazevet, "Calculation of a deuterium double shock Hugoniot from ab initio simulations," *Phys. Rev. Lett.* 87:275502, 2001.

[1136] A. Sarsa, K. E. Schmidt, and W. R. Magro, "A path integral ground state method," *J. Chem. Phys.* 113:1366–1371, 2000.

[1137] C. Pierleoni and D. M. Ceperley, "Computational methods in coupled electron ion Monte Carlo," *ChemPhysChem* 6:1–8, 2005, (arXiv:physics/0501013).

[1138] S. Moroni, A. Sarsa, S. Fantoni, K. E. Schmidt, and S. Baroni, "Structure, rotational dynamics, and superfluidity of small OCS-doped He clusters," *Phys. Rev. Lett.* 90:143401, 2003.

[1139] G. Carleo, S. Moroni, F. Becca, and S. Baroni, "Itinerant ferromagnetic phase of the Hubbard model," *Phys. Rev. B* 83:060411, 2011.

[1140] P. Gori-Giorgi, S. Moroni, and G. B. Bachelet, "Pair-distribution functions of the two-dimensional electron gas," *Phys. Rev. B* 70:115102, 2004.

[1141] M. A. Morales, C. Pierleoni, E. Schwegler, and D. M. Ceperley, "Evidence for a first-order liquid–liquid transition in high-pressure hydrogen from ab initio simulations," *Proc. Natl. Acad. Sci.* 107:12799–12803, 2010.

[1142] D. Ceperley, M. Dewing, and C. Pierleoni, in *Bridging the Time Scales,* edited by P. Nielaba, M. Mareschal, and G. Ciccotti, Springer-Verlag, Berlin, 2002, pp. 473–499.

[1143] C. Filippi and C. J. Umrigar, "Correlated sampling in quantum Monte Carlo: A route to forces," *Phys. Rev. B* 61:R16291–16294, 2000.

[1144] M. A. Morales, C. Pierleoni, and D. M. Ceperley, "Equation of state of metallic hydrogen from coupled electron–ion Monte Carlo simulations," *Phys. Rev. E* 81:021202, 2010.

[1145] M. A. Morales, J. M. McMahon, C. Pierleoni, and D. M. Ceperley, "Towards a predictive first-principles description of solid molecular hydrogen with density functional theory," *Phys. Rev. B* 87:184107, 2013.

[1146] J. E. Hirsch, R. L. Sugar, D. J. Scalapino, and R. Blankenbecler, "Monte Carlo simulations of one-dimensional fermion systems," *Phys. Rev. B* 26:5033–5055, 1982.

[1147] F. F. Assaad and H. G. Evertz, in *Computational Many-Particle Physics,* edited by H. Fehske, R. Schneider, and A. Weisse, Springer-Verlag, New York, 2008, pp. 277–356.

[1148] E. C. Svensson, P. Martel, V. F. Sears, and A. D. B. Woods, "Neutron scattering studies of the dynamic structure of liquid ^{4}He," *Can. J. Phys.* 54:2178–2192, 1976.

[1149] M. Jarrell and J. E. Gubernatis, "Bayesian inference and the analytic continuation of imaginary-time quantum Monte Carlo data," *Phys. Rep.* 269:133–195, 1996.

[1150] J. Shumway and D. M. Ceperley, "Quantum Monte Carlo treatment of elastic exciton-exciton scattering," *Phys. Rev. E* B63:165209–165215, 2001.

[1151] M. Caffarel and D. M. Ceperley, "A Bayesian analysis of Green's function Monte Carlo correlation functions," *J. Chem. Phys.* 97:8415, 1992.

[1152] Y. Kwon, D. M. Ceperley, and R. M. Martin, "Transient estimate Monte Carlo in the two-dimensional electron gas," *Phys. Rev. B* 53:7376–7382, 1996.

[1153] L. D. Landau and E. M. Lifshitz, *Mechanics,* Pergamon Press, Oxford, 1969.

[1154] J. L. Safko, H. Goldstein, and C. P. Poole, *Classical Mechanics, 3rd Ed.,* Addison-Wesley, Boston, MA, 2001.

[1155] G. Green, *An Essay on the Application of Mathematical Analysis to the Theory of Electricity and Magnetism,* Printed for the author by T. Wheelhouse, London. Sold by Hamilton, Adams & Co., Nottingham, 1828. Reprinted in three parts in *J. reine abgewand. Math.* 39:73, 1850; 44:356, 1852; and 47:161, 1854.

[1156] G. Baym and N. D. Mermin, "Determination of thermodynamic Green's functions," *J. Math. Phys.* 2:232–234, 1961.

[1157] L. V. Keldysh, "Diagram technique for nonequilibrium processes (translation of *Zh. Eksp. Teor. Fiz.,* 47:1515, 1965)," *Soviet Phys. JETP* 20:1018–1026, 1965.

[1158] G. Stefanucci, C.-O. Almbladh, R. van Leeuwen, N. E. Dahlen, and U. von Barth, "Introduction to the Keldysh formalism," *Lecture Notes in Physics* 706:33, 2006.

[1159] G. C. Evans, *Functionals and Their Applications,* Dover, New York, 1964.

[1160] J. Matthews and R. L. Walker, *Mathmatical Methods of Physics,* W. A. Benjamin, New York, 1964.

[1161] R. Fukuda, M. Komachiya, S. Yokojima, Y. Suzuki, K. Okumura, and T. Inagaki, "Novel use of Legendre transformation in field theory and many particle systems," *Prog. Theor. Phys. Suppl.* 121:1, 1995.

[1162] O. Hölder, "Über einen Mittelwerthsatz," *Nachr. Ges. Wiss. Gottingen* 4:38–47, 1889.

[1163] M. Gatti, V. Olevano, L. Reining, and I. V. Tokatly, "Transforming nonlocality into a frequency dependence: A shortcut to spectroscopy," *Phys. Rev. Lett.* 99:057401, 2007.

[1164] L. J. Sham and M. Schlüter, "Density-functional theory of the energy gap," *Phys. Rev. Lett.* 51:1888–1891, 1983.

[1165] L. J. Sham, "Exchange and correlation in density-functional theory," *Phys. Rev. B* 32:3876–3882, 1985.

[1166] R. van Leeuwen, "The Sham–Schlüter equation in time-dependent density-functional theory," *Phys. Rev. Lett.* 76:3610–3613, 1996.

[1167] I. V. Tokatly and O. Pankratov, "Many-body diagrammatic expansion in a Kohn–Sham basis: Implications for time-dependent density functional theory of excited states," *Phys. Rev. Lett.* 86:2078–2081, 2001.

[1168] R. T. Sharp and G. K. Horton, "A variational approach to the unipotential many-electron problem," *Phys. Rev.* 90:317, 1953.

[1169] J. D. Talman and W. F. Shadwick, "Optimized effective atomic central potential," *Phys. Rev. A* 14:36–40, 1976.

[1170] A. J. Millis and S. N. Coppersmith, "Variational wave functions and the Mott transition," *Phys. Rev. B* 43:13770–13773, 1991.

[1171] T. Ogawa, K. Kanda, and T. Matsubara, "Gutzwiller approximation for antiferromagnetism in Hubbard model," *Prog. Theor. Phys.* 53:614–633, 1975.

Index